ELEMENTS OF
ECOLOGY

CANADIAN EDITION

THOMAS M. SMITH

ROBERT LEO SMITH

ISOBEL WATERS

PEARSON

Toronto

Vice-President, Editorial Director: Gary Bennett
Senior Acquisitions Editor: Lisa Rahn
Senior Marketing Manager: Kim Ukrainec
Senior Developmental Editor: John Polanszky
Project Manager: Jessica Hellen
Manufacturing Specialist: Jane Schell
Production Editor: Niraj Bhatt, Aptara®, Inc.
Copy Editor: Julia Cochrane

Proofreader: Dawn Hunter
Compositor: Aptara®, Inc.
Photo and Permissions Researcher: Monika Schurmann
Art Director: Julia Hall
Cover and Interior Designer: Miriam Blier
Cover Image: Jason Carter / Wapati at Ha Ling

Credits and acknowledgments of material borrowed from other sources and reproduced, with permission, in this textbook appear on the appropriate page within the text and on p. C-1.

10 9 8 7 6 5 4 3 2 1 [CKV]

Library and Archives Canada Cataloguing in Publication

Smith, T. M. (Thomas Michael), 1955-
        Elements of ecology / Thomas M. Smith, Robert Leo
Smith and Isobel Waters. - 1st ed.
Includes bibliographical references and index.
        ISBN 978-0-321-51201-7
                1. Ecology.   I. Smith, Robert Leo
II. Waters, Isobel   III. Title.
QH541.S65 2012   577   C2012-903757-5

Environmental Statement

This book is carefully crafted to minimize environmental impact. Pearson Canada is proud to report that the materials used to manufacture this book originated from sources committed to sustainable forestry practices, tree harvesting, and associated land management. The binding, cover, and paper come from facilities that minimize waste, energy usage, and the use of harmful chemicals.

Equally important, Pearson Canada closes the loop by recycling every out-of-date text returned to our warehouse. We pulp the books, and the pulp is used to produce other items such as paper coffee cups or shopping bags.

The future holds great promise for reducing our impact on Earth's environment, and Pearson Canada is proud to be leading the way in this initiative. From production of the book to putting a copy in your hands, we strive to publish the best books with the most up-to-date and accurate content, and to do so in ways that minimize our impact on Earth.

PEARSON

ISBN 978-0-321-51201-7

Brief Contents

Contents

Preface

Writing the first Canadian edition of *Elements of Ecology* represents a responsibility as well as a challenge. It is a responsibility because, as one of the first and most prominent textbooks in the field, *Elements of Ecology* has a long and distinguished history that is paramount to uphold. The challenge resides in incorporating content important to Canadian courses while maintaining the book's outstanding quality. The goal throughout has been to write an ecology textbook that is as much North American and indeed global as Canadian in its perspective. Natural ecosystems know no human boundaries, and ecological principles apply at all scales of space and time. Imposing a strict national boundary on ecological topics does no favour to students. Accordingly, the Canadian edition retains certain American examples as well as incorporating others from around the world.

The first U.S. edition of *Elements of Ecology* appeared in 1976 as a short version of *Ecology and Field Biology.* Since then, it has evolved into a textbook intended for use in a one-semester introductory ecology course. Although the primary readership will be students majoring in the life sciences, it is the belief that ecology should be part of a broadly based education that guided the writing of this Canadian edition. Students who major in such diverse fields as economics, sociology, engineering, political studies, history, law, literature, languages, and the like should have a basic understanding of ecology not only because it impinges on their lives, but also because their lives impinge on the ecology of the biosphere.

NEW IN THE FIRST CANADIAN EDITION

- Examples of ecosystems and/or species that are familiar to Canadians, accompanied by new photographic images.

- Expanded treatment of topics that are of particular importance in a Canadian context, such as adaptations of organisms to snow and extreme cold and predicted changes in Canadian ecoclimate zones with climate change.

- Focus on issues that are of concern to Canadians, using Canadian case studies, such as eutrophication of freshwater lakes and degradation of wetlands in Ontario; the impact of invasive species such as the balsam fir woolly adelgid, emerald ash borer, and Asian carp; the clash between science and politics regarding pesticide use in Maritime forests; the decline of East Coast fisheries; the impact of fragmentation and habitat loss on iconic species such as the polar bear and woodland caribou; the impact of global warming on matter cycling in Arctic sea ice; and the impact of oil sands development on the MacKenzie River system.

- Profiles of fifteen scientists working at Canadian universities.

- Citations of publications of over 250 scientists from institutions across Canada, including many younger as well as established researchers. As many papers have multiple authors from many universities, and as many authors move between universities, standard citation style is used, so this aspect of the Canadian content may be less obvious.

- Dual focus on both ecosystems and individuals. *Elements of Ecology* stresses the individual as the basic unit of ecology. An important objective of the text in its many editions is to use the concept of adaptation through natural selection as a framework for unifying ecological study, linking pattern and process across the hierarchy of organisms, populations, communities, and ecosystems. While embracing this perspective, and maintaining the strong focus on evolutionary ecology, the Canadian edition recognizes that the ecosystem is an equally fundamental ecological unit. The Canadian edition includes a new chapter on ecosystem structure and function, with emphasis on ecosystem regulation via feedback mechanisms, and an updated treatment of key ecosystem-related debates, such as the relationship between ecosystem diversity and stability. The book emphasizes the ecosystem level throughout, particularly in discussing topics such as common mycorrhizal networks, coexistence mechanisms, and diffuse and indirect interactions involving cooperation.

- Expanded treatment of the ecological roles of fungi. Like most introductory textbooks, *Elements of Ecology* stresses the adaptations of plants and animals. In recognition of the importance of fungi to ecosystem function, the Canadian edition goes beyond the role of fungi in decomposition and mutualisms to examine the ecological diversity of this vital group.

- Incorporation of topics previously not covered in U.S. editions, including ecological footprint; neutral model; metacommunity concept; altered sex ratios in social insects; compensatory growth; common mycorrhizal networks; coexistence mechanisms; criteria of old growth; micro-, meso-, and macrocosms; and many others.

- More detailed focus on sustainable solutions to resource use issues, including variable retention harvesting (developed in British Columbia); integrated multi-trophic-level aquaculture; and traditional ecosystem knowledge, which addresses agricultural and fisheries-related issues from the perspective of Aboriginal peoples and other stakeholders.

- Up-to-date data and research in the chapters in Part Eight dealing with population growth, conservation ecology, and global climate change to reflect our current understanding of these important issues.

Expanded Quantitative Features

Ecology is a science rich in concepts. Yet, as with all science, it is quantitative. A major goal of any science course is the development of basic skills relating to the analysis and interpretation of empirical data. The Quantifying Ecology features show students how concepts introduced in the text are quantified. In many chapters, the Quantifying Ecology boxes help students interpret graphs, mathematical models, or methods introduced within the main body of the text. The Interpreting Ecological Data segments further aid in developing quantitative skills, by featuring two or more questions that relate directly to the interpretation of data presented in an associated figure or table.

Annotations embellish figures and tables for particularly complex graphics. Where present, these annotations are not intended to be a ready-made explanation. Rather, we believe it is more important to ask questions that will both encourage and assist in the interpretation and comprehension of data. Only then will students begin to build the basic skills needed to move beyond the specific examples—skills that will allow them to explore the wealth of ecological studies published in the books and journals referenced throughout. Developing these quantitative and interpretative skills is as important as understanding the body of concepts that form the framework of ecology. Students will find answers to questions in these features on EcologyPlace, the text's accompanying website.

Changes to Organization

The overall framework of the U.S. edition has been retained, with some reorganization of key topics. Many instructors teach both exponential and logistic growth together. To facilitate this, both models now appear in Chapter 10, rather than keeping logistic growth with intraspecific regulation in Chapter 11. Similarly, many instructors treat community structure along with the factors that influence it; these topics now appear in a single chapter (Chapter 16). The U.S. edition has a separate chapter on large-scale patterns of biodiversity, whereas the Canadian edition integrates this topic into the discussions of landscape ecology in Chapter 18, systems ecology in Chapter 19, and conservation biology in Chapter 27. While the U.S. 8th edition introduced a new chapter on the evolutionary impact of species interactions, the Canadian edition incorporates this material into the existing chapters. As a result of these changes, the Canadian edition features 28 chapters, as opposed to 30 chapters in the U.S. 8th edition. The number of study questions has been increased in most chapters, and the glossary has been expanded to include cross-references as a study aid. A list of Key Terms has been added at the end of each chapter, to help the student focus their reading and comprehension.

STRUCTURE AND CONTENT

Three principles guide the structure and content of this text: (1) the individual is a fundamental unit of ecology; (2) the concept of adaptation through natural selection provides a framework for unifying ecological study at all levels of the ecological hierarchy; and (3) individuals interact in the context of the ecosystem, which is an equally fundamental unit of ecology. A central recurring theme is the idea of trade-offs—that the set of adaptations (heritable traits) that enable an organism to survive, grow, and reproduce under one set of environmental factors invariably imposes constraints on its ability to function equally well under different environmental factors. These factors include both the non-living (abiotic) environment and the variety of organisms (of both the same and different species) that share the habitat. This simple framework not only provides a basis for understanding population dynamics on both an evolutionary and a demographic timescale, but also underpins the interrelations within and among populations that determine community structure and ultimately ecosystem function.

Following an introduction to the science of ecology in Chapter 1, eight parts divide the remainder of the text. Part One examines the constraints imposed on organisms by climate and other aspects of the abiotic environment, both aquatic and terrestrial. Part Two considers how these environmental constraints function as agents of change through the process of natural selection, the evolutionary mechanism that gives rise to adaptations. The rest of Part Two explores specific adaptations of organisms to their abiotic environment, considering both organisms that derive their energy from the Sun (autotrophs) and those that derive their energy from consumption of organic matter (heterotrophs). Because life history patterns describe adaptations relating to reproduction and longevity, they provide the link between the individual and the population—the biological entity that evolves.

Part Three examines populations, with an emphasis on how traits expressed at the level of the individual determine the collective dynamics of the population. After describing population properties, we consider different patterns of population growth and the intraspecific interactions involved in population regulation. Finally, this part discusses metapopulations, the arrangement of local subpopulations in a landscape. Part Four extends the discussion from intraspecific interactions (between individuals within a species) to interspecific interactions (among individuals of different species). These chapters expand our view of adaptations from traits affecting response to abiotic factors to traits that influence biotic interactions (competition, predation, parasitism, and mutualism), as they affect both natural selection and ecosystem function.

Part Five explores the ecological community, drawing upon topics covered in Parts Two to Four to examine the factors that influence the distribution and abundance of species across environmental gradients in both space and time. Part Five concludes with a consideration of the interactions between adjoining communities in a landscape. Part Six combines the discussions of the ecological community in Part Five and the abiotic environment in Part One to develop the concept of the ecosystem. After discussing feedback regulation, we explore the link between ecosystem properties and function, focusing on the flow of energy and cycling of matter both within biotic ecosystem components and as part of biogeochemical cycles between biotic and abiotic components. Part Seven considers ecosystems in the context of biogeography, examining the major properties and distribution of terrestrial, aquatic, and transitional ecosystems around the world.

Part Eight focuses on the interactions between humans and the environment. Here we examine important environmental issues relating to population growth, sustainability, resource use, declining biological diversity, invasive species, and global climate change. These chapters explore the contributions of ecology to both understanding and addressing critical environmental issues.

Throughout the text we explore these topics by drawing upon research (both current and historically important) into various fields of ecology to provide examples that help the reader develop an understanding of natural history, the ecology of place (specific ecosystems), and the fundamental processes of science.

ASSOCIATED MATERIALS

Instructor resources are password protected and available for download from the Pearson online catalogue at http://catalogue.pearsoned.ca.

Instructor's Resource Manual with Solutions

This useful teaching aid provides commentaries, learning objectives, and additional exercises and discussion topics for each chapter. The manual also includes sample syllabi for courses of different durations, a guide for using the manual, and answer guidelines and solutions for all the study questions found in the textbook.

Microsoft PowerPoint Slides

Microsoft PowerPoint presentations combine graphics and text into pre-made lecture slides.

Image Library

The image library showcases the figures and tables that appear in the text, allowing instructors to easily incorporate this material into their lectures.

Test Item File

The Test Item File provides a broad range of questions to accompany the content of the Canadian edition. This question bank is available in both Microsoft Word and MyTest formats.

MyTest

MyTest is a powerful assessment generation program with which instructors can easily create and print quizzes, tests, and exams online, allowing flexibility and the ability to manage assessments at any time and from anywhere.

EcologyPlace

EcologyPlace, a text-specific website found at www.pearsoncanada.ca/ecologyplace, offers tutorial animations, interactive maps, practice quizzes, an ecological footprint calculator, an electronic version of the text, and more.

Study on the Go

At the end of every chapter, students will find a QR code (also known as a quick response code) that links to Study on the Go mobile content. Students can access text-specific resources, including quizzes and flashcards, through their smartphones, allowing them to study whenever and wherever they wish!

Students can go to one of the sites below to see how to download a free app to their smartphone that facilitates access to these resources. Once the app is installed, the phone will scan the code and link to a website containing *Elements of Ecology*'s Study on the Go content.

ScanLife

http://getscanlife.com

NeoReader

http://get.neoreader.com

QuickMark

http://www.quickmark.com.tw

ACKNOWLEDGMENTS

No textbook is the product of the authors alone. The material that this book covers represents the work of hundreds of ecological researchers who have spent lifetimes in the field and the laboratory. Their published experimental results, observations, and conceptual thinking provide the raw material out of which this textbook is fashioned. We particularly thank the fifteen ecologists featured in the *Ecological Studies* boxes. Their cooperation in providing artwork, photographs, and comments is greatly appreciated.

Textbooks depend heavily on the inputs of peer reviewers. I took suggestions seriously and implemented most recommendations. I am deeply grateful to the following reviewers for their helpful comments and suggestions on how to improve this edition:

Yuguang Bai, *University of Saskatchewan*

Marc Cadotte, *University of Toronto (Scarborough)*

Jennifer A. Chiang, *Redeemer University College*

David Clements, *Trinity Western University*

Gregor Fussmann, *McGill University*

Sharon L. Gillies, *University of the Fraser Valley*

Stephen J. Hecnar, *Lakehead University*

Tafazzal Hoque, *York University (Glendon)*

Penny L. Humby, *Crandall University*

Rod Lastra, *University of Manitoba*

Andrew Laursen, *Ryerson University*

Lisa M. Poirier, *University of Northern British Columbia*

Bernard D. Roitberg, *Simon Fraser University*

Joan Sharp, *Simon Fraser University*

Barry R. Taylor, *St. Francis Xavier University*

Frank Williams, *Langara College*

Connie Zehr, *Centennial College*

Additionally, we wish to thank Lisa M. Poirier, the subject matter expert, who went through the manuscript thoroughly to ensure accuracy.

We also thank the following scientists for their prompt and helpful replies to inquiries into their research, or into general topic areas:

Tom Booth	Charles Krebs
Ryan Brook	Robin Peterson Lewis
Lynda Bunting	John Markham
Jennifer Chesworth	Peter Morin
Thierry Chopin	Shahid Naeem
Darwyn Coxon	Serge Payette
David Currie	James Reid
Allison Krause Danielsen	Gordon Robinson
John Ewel	Sylvie Rondeau
Edward O. Garton	Cynthia Ross
Lane Graham	Spencer Sealy
Brenda Hann	Michael Sumner
Michael Huston	Mark Velland

Publishing a textbook requires the work of many individuals to handle the specialized tasks of development, photography, graphic design, illustration, copy editing, and production, to name only a few. I would like to thank Michelle Sartor, who signed the Canadian edition; Lisa Rahn, Senior Acquisitions Editor, who saw this project through to its completion; John Polanszky, Senior Developmental Editor, for his editorial guidance, calming advice, and expertise; and Miriam Blier, Designer, whose creativity can be seen inside and outside the book and who found the cover image by the Canadian Aboriginal artist Jason Carter. I also thank the rest of the talented team: Jessica Hellen, in-house Project Manager; Maureen de Sousa, Media Developer; Jill Renaud, Associate Editor; and Kim Ukrainec, Senior Marketing Manager. I am deeply indebted to my copy editor, Julia Cochrane, not only for her painstaking attention to detail but also for her insightful comments. They spurred me to improve the book's content as well as its form. I am grateful to Monika Schurmann, who was relentless (and cheerful) in her pursuit of images and permissions. It was an extremely rewarding experience to work with the production team at Aptara, including Niraj Bhatt. They were endlessly patient with my requests to make yet one more change to the figures or the text, and were as desirous as I to make the book as good as it could be.

I thank my husband, Grant, who has done much to help me through the rigours of book production. I appreciate the support of my department head, Judy Anderson, throughout the project. I am particularly grateful to my children (Susan and Matthew); family (especially the dinners provided by Gordon and Linda); friends (Brenda, Darlene, Diane, Emily, Michael, and Rob); and colleagues (Bruce, Mike, and Anne), who each in their own way made me feel it would all work out in the end.

Isobel Waters

THE NATURE OF ECOLOGY

As part of an ongoing research project, wildlife biologists in Alaska fit a Barren-ground caribou (*Rangifer tarandus groenlandicus*) with a radio collar to track the animal and map its patterns of movement and habitat use.

n June 2010 at the Hay Festival of Literature and the Arts in Wales, the following contribution by Marc MacKenzie, a medical physicist at the University of Alberta, was declared the most beautiful Twitter message ever composed: "I believe we can build a better world! Of course, it'll take a whole lot of rock, water & dirt. Also, not sure where to put it."

A "tweet" seems far removed from a textbook, but the serious idea implicit in MacKenzie's ironic message—that we inhabit a planet that we have just barely begun to understand, and that we are more adept at harming than conserving—is at the heart of ecology. Nothing is more urgent than our need to understand the natural world that we both cherish and threaten. David Suzuki, noted Canadian environmentalist and geneticist, says it best: "The human brain now holds the key to our future. We have to recall the image of the planet from outer space: a single entity in which air, water, and continents are interconnected. That is our home."

Western civilization has not had this perspective for long. For most of human history, we have regarded our home planet as a limitless frontier. But by the time the eloquent image of Earthrise was beamed from the *Apollo 8* spaceship in 1968 (Figure 1.1), Earth had shrunk to a fragile and finite sphere.

Figure 1.1 Photograph of Earthrise taken by astronaut William Anders on December 24, 1968. Environmentalist Galen Rowell called it "the most influential environmental photograph ever taken."

"Spaceship Earth," as evolutionary economist Kenneth Boulding dubbed it, was suddenly perceived as limited in resources, crowded by an ever-expanding human population, and threatened by our use of the oceans and the atmosphere (even outer space) as repositories for our wastes. Nor was this awareness restricted to academics and activists. Ordinary citizens were increasingly concerned, and the environmental movement was born.

At the core of this global movement was a recognition of the need to redefine our relationship with nature from one of exploitation to one of stewardship. The field called upon to provide the road map for this new direction was ecology. Until the 1970s familiar only to biologists, *ecology* sprang to the forefront of public consciousness. A half century later, as the world watched oil gushing from the Deepwater Horizon spill in the Gulf of Mexico for three months in 2010, it was obvious that our need for understanding—and paying heed to—the ecology of Earth is as critical now as it has ever been (see Ecological Issues: The Human Factor, p. 13).

1.1 ECOLOGY DEFINED: Ecology Studies Organisms Interacting with Their Environment

Before starting this course, how would you have defined *ecology*? The term is often confused with *environment* and *environmentalism.* Although it has connections to both, ecology is neither. *Environmentalism* is activism focused on protecting the natural *environment*, particularly from the impacts of human activities. This activism often takes the form of public education programs, advocacy, and legislation. **Ecology**, in contrast, is a science: the scientific study of the relationships between organisms and their environment.

This definition is satisfactory only if one considers *environment* and *relationships* in their fullest sense. The **environment** of an organism consists of all the external factors that influence its survival, growth, and/or reproduction, either directly or indirectly. It includes not only the surroundings in which an organism lives (its **habitat**), along with all the **abiotic** (non-living) physical factors at work within it, but also the **biotic** (living) components with which it interacts. *Relationships* include an organism's interactions with its physical surroundings as well as with members of its own and other species. Ecology, then, is the science of life on Earth, with emphasis on how life and Earth interact.

Whether abiotic or biotic, environmental factors are either **resources**, which are consumed by the organism, making them less available for others (e.g., food, water, or mates), or **conditions**, which influence an organism but are not consumed (e.g., temperature, day length, or acidity). The term *environmental conditions* is often used collectively to refer to both, but the distinction is important. Organisms can compete for resources, but not for conditions. Conditions can affect competition indirectly by altering availability of a resource (e.g., freezing temperature affecting water) or by altering an organism's ability to use a resource. A third category, **hazards**, includes factors that can only affect an organism negatively if they are present. Some hazards are substances (e.g., DDT), while others are disturbance events (e.g., earthquakes or floods).

Note that the environment is defined relative to the individual. What constitutes part of the environment of one individual (and by extension its population) may not be part of the environment of another. For example, floods are hazards to many species, causing harm or even death, whereas a species that has evolved with exposure to flooding may not only tolerate but even require periodic flooding for survival.

1.2 HISTORY OF ECOLOGY: Ecology Is Rooted in Many Scientific Disciplines

As a science, ecology is not only crucial to our future but also comparatively young. German zoologist Ernst Haeckel coined the term in 1866, deriving it from the Greek *oikos*, meaning "household." It has the same origin as *economics*, a link to which Haeckel refers explicitly when defining this emerging science: "By ecology we mean the body of knowledge concerning the economy of nature—the investigation of the total relations of the animal both to its inorganic and to its organic environment . . . [E]cology is the study of all those complex interrelationships referred to by Darwin as the conditions of the struggle for existence." Haeckel was a follower of Charles Darwin, who had published *The Origin of Species* just a few years earlier, in 1859. Darwin's revolutionary theory of evolution via natural selection provided a cornerstone for the new science. Natural selection was the mechanism allowing ecology to go beyond descriptive natural history to the processes controlling what would emerge as the major early focus of ecology: the distribution and abundance of organisms.

Of course, the study of life on Earth predated Haeckel and Darwin. Biologists from the time of ancient Greece have pondered the mechanisms whereby organisms survive and grow. Geographers and soil scientists also studied the abiotic components of the ecological equation. Sometimes these biotic and abiotic perspectives came together, with fruitful outcomes. Darwin's work was underpinned not only by geographer Charles Lyell's theories about the age of Earth and by Thomas Malthus's groundbreaking population concepts, but also by the uncovering of the fossil record in the 19th century.

Although less celebrated, the plant geography movement of the 18th and 19th centuries by European scientists, such as Alexander von Humboldt, furthered ecological understanding by noting similarities in the vegetation of widely separated regions with similar climate and soil. Rather than merely naming new species (all the rage during the European voyages of "discovery"), these scientists described correlations between these species and their native habitat. Yet as vital as these studies were, they were strictly observational, and few considered the dynamic interactions involved. It took the theoretical framework of evolution to give impetus to ecology, and the environmental crises of the 20th and 21st centuries to lend it urgency.

Western science has no monopoly on ecology. Our survival as individuals and as societies has always depended on understanding interactions between organisms and their environment. This statement is obviously true for hunter–gatherer cultures, for which securing prey and avoiding being preyed upon is crucial. In Canada, our First Nations peoples have a tradition of revering the natural world. The idea of Earth as a living, interacting entity is fundamental both to their spirituality and to their use of natural resources. Many have likened this view to the philosophy espoused in the film *Avatar* (2010). Whatever its origins and contemporary expressions, such traditional knowledge complements the more conventional ecological science that is the focus of this book (see Research in Ecology: Using Traditional Ecological Knowledge to Manage the Pacific Lamprey, p. 587). Ecological awareness is equally crucial for agricultural societies, which manipulate the interactions between domesticated species and their environment. Indeed, Darwin's theories were inspired as much by what he called "artificial selection" in rural England as by the natural world encountered on his voyages.

Ecology, then, is rooted in several sciences, and interacts with many non-scientific disciplines to further its goal of understanding how nature works. Its youth as a discipline, coupled with its urgency, makes it a dynamic area of study, not just for scientists but for students like you.

1.3 ORGANISMS AND ECOSYSTEMS: Organisms Interact with the Environment within Ecosystems

Organisms interact with their environment in many ways. Prevailing abiotic conditions, such as temperature and acidity, influence physiological processes crucial to survival and growth. An organism must also acquire essential resources from its environment, such as light for a plant or food for an animal, while at the same time protecting itself from hazards, such as falling prey to other organisms. It must recognize friend from foe, differentiating between potential mates and possible competitors or predators. All this effort is geared towards the ultimate "goal" of all organisms: to pass on their genes to future generations. As these activities are not consciously goal directed, this description applies to plants, fungi, and microbes as well as animals.

The environment in which organisms experience what Darwin called the "struggle for existence" is in part a place or *habitat*—a physical location in time and space. It can be as large and stable as an ocean or as small and transient as a puddle after a spring rain. But the environment is more than a habitat; it also includes any coexisting organisms. This complex entity is what ecologists call the **ecosystem**: the biotic community and its abiotic environment. Ecosystems are the units of nature in which organisms interact with each other and with their surroundings. The prefix *eco-* relates to the surroundings or "household," and *-system* to the idea of a collection of parts that function as an integrated whole. A car engine is a system: its components work together to achieve its purpose of making the wheels turn. Your body is also a system, but unlike the engine, it is a living system that is not a product of design. An ecosystem also consists of interacting parts that function together as a unit. Like your body, its structure is not the result of design, but unlike your body, an ecosystem has both biotic and abiotic components.

Consider a forest (Figure 1.2). Its abiotic components are air, soil, and water, as well as climate, which reflects interactions among these components. Its biotic components are the many organisms—plants, animals, fungi, and microbes—inhabiting it. Relationships are complex. Each organism not only responds to its abiotic environment but also modifies it and, in doing so, becomes part of the biotic environment of other organisms. The trees in the canopy intercept sunlight, using its energy to drive photosynthesis. As they grow, they modify the environment of plants below them, reducing light and lowering temperature and wind speed. Herbivores consume some of their leaves and are themselves consumed by predators. These feeding relationships reverberate through the forest. Birds foraging on insects in the fallen leaves reduce insect

Figure 1.2 Forest ecosystem on the Pacific northwest coast of North America. Sitka spruce (*Picea sitchensis*) trees form the canopy, and mosses and lichens cover the dead branches extending from the canopy to the ground. Many shrubs and herbaceous species occupy the understory. Mosses grow on the forest floor, using nutrients released by fungi that decompose organic matter at the soil surface. This forest is home to many animals, including bald eagles, black-tailed deer, and brown bears.

numbers and modify the environment for other organisms that depend on this shared food resource. By reducing the numbers of their insect prey, the birds indirectly influence competition among insect species on the forest floor. When individuals die, others decompose their remains, recycling the nutrients in their tissues back to the soil. Throughout this book, we explore these interactions between the living and non-living components of ecosystems.

1.4 ECOLOGICAL HIERARCHY: Ecology Operates at Many Levels, from the Individual to the Biosphere

Biology is hierarchical. There are many levels below that of the individual: organ systems, organs, tissues, cells, subcellular components, etc. These levels are vital to ecology because they help explain physiological responses, but the starting point of ecology is normally the individual (Figure 1.3), which in turn belongs to a population. Although the term has many meanings, in ecology a **population** is a group of individuals of the same species occupying a given area at a given time. Populations, such as spruce trees or bald eagles in our forest example, rarely function independently of each other. As collectives of individuals, some populations compete with others for limited resources, such as food or space. Some prey on others, or are themselves prey. Nor are interactions always negative—two populations may mutually benefit each other. Collectively, all populations of all species living and interacting within an ecosystem are called a **community**.

Levels above the community include abiotic components. As discussed, the *ecosystem* (the community and its abiotic environment) is next. In turn, ecosystems interact in the broader context of the **landscape**—an area of land (or water; *seascape* if marine) composed of a patchwork of ecosystems. Ecosystems are linked through the dispersal of organisms and exchange of materials and energy. Each ecosystem represents a unique combination of physical conditions (topography and soils) and an associated community, but regions with similar geology and climate support similar ecosystems. For example, warm temperatures, abundant rain, and a lack of seasons typify tropical regions and support tropical rain forests. Broad-scale regions in which the landscape is dominated by similar ecosystems are called **biomes**.

The highest level of ecological organization is the **biosphere**—the narrow interface at Earth's surface that contains and supports life. Within the biosphere, all ecosystems, both terrestrial and aquatic, are linked through their interactions (particularly exchanges of materials and energy) with the global abiotic spheres: (1) **atmosphere** (layers of air surrounding Earth), (2) **hydrosphere** (water at or near Earth's surface, including the soil solution), and (3) **lithosphere** (solid earth and soil). Each sphere supplies ecosystems with resources and determines prevailing conditions and hazards (Table 1.1). Ecology studies the interactions between organisms and their environment at all levels of this hierarchy, from individuals to the biosphere.

Individual

What adaptations allow the polar bear (*Ursus maritimus*) to survive, grow, and reproduce in its Arctic environment?

Population

As affected by its birth and death rates, is the polar bear population increasing, decreasing, or remaining relatively constant from decade to decade?

Community

How does the polar bear interact with other species of animals and plants in the tundra environment?

Ecosystem

How is a warming temperature regime affecting ice cover in the Arctic?

Landscape

By its predatory activity, how does the polar bear influence energy and nutrient exchange between the tundra and the adjacent marine ecosystem?

Biome

What features of geology and regional climate determine the transition from tundra to boreal forest in North America?

Biosphere

What is the role of the Arctic landscape (tundra and ocean) in the global carbon cycle?

Figure 1.3 Ecological hierarchy.

Table 1.1 Abiotic Spheres of the Earth System

Examples of environmental factors associated with each sphere are provided.

Sphere	Resources	Conditions	Hazards
Atmosphere	Gaseous nutrients (CO_2, O_2, N_2); light	Temperature, humidity, wind speed	Air pollutants, tornados
Hydrosphere	H_2O, dissolved nutrients, light, habitat space	Temperature, pH, salinity	Dissolved pollutants, tsunami
Lithosphere	Mineral nutrients, habitat space	Temperature, pH	Toxic heavy metals, earthquakes, volcanoes

1.5 IMPORTANCE OF SCALE: Ecologists Study Pattern and Process at Different Scales

As we traverse the ecological hierarchy, different patterns and processes emerge. Different questions arise, as well as different methods for addressing them. Ecology contains many subdisciplines. Each focuses on a particular level, while remaining attuned to the linkages with, and significance for, other levels. Choice of hierarchical level is an important aspect of a study's **scale**: the level of resolution in time and/or space over which a pattern or process is investigated. Let's consider these subdisciplines by using the polar bear (*Ursus maritimus*) as an example.

Ecologists who focus on the individual examine how its traits influence an organism's ability to survive, grow, and reproduce in its environment. They also consider how these same traits constrain the organism's ability to succeed in other environments. By stressing the traits of individuals, **autecology** (both *structural* and *physiological ecology*) and **behavioural ecology** explore factors affecting species distribution. An autecologist investigates which polar bear traits facilitate its success in its northern habitat. In general terms, a larger body size retains heat more efficiently, but how do specific traits, such as its limb proportions; the dermal papillae on the pads of its paws; and its hollow, transparent guard hairs affect the polar bear's ability to function on ice or in water? Which aspects of its metabolism, including its so-called walking hibernation, are most crucial? All traits entail trade-offs, so how might its adaptive traits prove handicaps in warmer conditions resulting from global climate change? A behavioural ecologist might study the polar bear's hunting and social activities, including those relating to reproduction and rearing of young, which overlap with population ecology.

For individuals, birth and death are discrete events, yet for a collective of individuals, these processes are continuous, as individuals are born and die. In **population ecology**, the focus shifts to population size and how it changes over time. What are the patterns of fertility and mortality for our polar bear population? Are its numbers increasing or decreasing? What are the causal factors behind any population changes, and what will be the likely effect of retreating Arctic ice (see Ecological Issues:

Conserving a Canadian Icon, p. 604). Populations also have patterns in space. How are polar bears distributed in an area, and do their population traits differ among locations? Behavioural ecology, which investigates how social organization influences spatial pattern, again overlaps with population ecology. So does **evolutionary ecology**—the study of changes in population genetics in response to evolutionary processes, particularly natural selection.

As we expand our perspective to encompass the community, new questions arise. **Community ecology** studies community patterns and the processes that contribute to them. What interactions occur among coexisting species, and how do these interactions influence community structure and function? A community ecologist might investigate the impacts of predation by polar bears on ringed seals, and of competition with other predators on species composition.

The community modifies as well as responds to its abiotic environment. **Ecosystem ecology** studies ecosystem structure and function, stressing the flow of energy and nutrients through its abiotic and biotic components. At what rate are energy and nutrients converted into living organisms in the tundra and adjacent Arctic waters? As a top predator, does the polar bear regulate ecosystem function? What processes govern the rate of decomposition whereby energy and nutrients locked in organic matter are released in inorganic forms? What abiotic factors limit energy flow and matter cycling in Arctic systems? Increasingly, ecosystem ecologists are concerned with the consequences of community traits, such as species composition and diversity, on the long-term stability and persistence of ecosystems.

As we expand our perspective still further, the landscape emerges as a patchwork of ecosystems whose boundaries are defined by changes in the physical environment or species composition. **Landscape ecology** investigates the factors influencing the spatial extent and arrangement of interacting ecosystems and the impacts of these patterns on such processes as dispersal of organisms, exchange of energy and nutrients, and spread of disturbances, such as fire or disease. For a landscape of forests interspersed with grasslands, these linkages are easy to visualize, but in our polar bear example, the landscape juxtaposes a terrestrial system (tundra) with an aquatic system (ocean). Does the polar bear affect energy transfer between tundra and ocean? How is the ongoing reduction of Arctic ice affecting these and other landscape processes? At a continental scale, landscape questions focus on the distribution of biomes. How does diversity vary among biomes? Why do tropical rain forests support more species than temperate deciduous forests? What factors determine the distribution of biomes, such as forest, grassland, and desert? In our example, is the tundra boundary shifting with climate change as the boreal forest moves north?

Finally, at the biosphere level, the emphasis is on the links between ecosystems and other components of the Earth system. **Global ecology** studies how the exchange of energy and matter between ecosystems and the atmosphere, hydrosphere, and lithosphere influences global conditions. How will climate change affect, and be affected by, ecological processes in the tundra? Certain key processes, such as carbon transfer between ecosystems and the atmosphere, can be studied only at a global scale, requiring ecologists to collaborate with oceanographers, geologists, and atmospheric scientists. Yet even at this scale, biota cannot be ignored. Scientists are discovering that the activity of microbes occupying sea ice has profound implications for carbon transfer (see Research in Ecology: Altered Matter Cycling in a Warmer Arctic, p. 482).

We use this hierarchical approach—individual, population, community, ecosystem, landscape, biome, and biosphere—as our organizing framework. Each level has its own ecological subdiscipline, and each uses specialized approaches and methods to address its unique questions. However, patterns and processes at one level are linked in an intricate web of cause and effect with those at other levels of the hierarchy. For example, traits of individuals, such as body size, longevity, reproductive age, and degree of parental care, directly influence population rates of birth and death. At the community level, a population is influenced both positively and negatively by interactions with other populations. In turn, the mix of species in a community affects energy and nutrient exchange in the ecosystem. These links have spawned exciting new subdisciplines that span several levels, including **conservation ecology** and **restoration ecology**, which study the factors affecting the conservation and restoration of species and/or ecosystems.

1.6 ECOLOGY AS A SCIENCE: Ecologists Investigate Nature by Using the Scientific Method

Although each subdiscipline has a unique set of questions, all ecological studies have one thing in common: they employ the scientific method (Figure 1.4). This approach sounds intimidating, but taken individually, each step involves commonplace procedures and reasoning. Taken together, these steps provide a powerful tool for understanding nature.

1. *Observation.* All science begins with observation. Indeed, this vital first step defines the domain of science: if something cannot be observed, it cannot be scientifically investigated. Observations need not be direct. Although scientists cannot directly observe the nucleus of an atom, they can explore its structure indirectly by various means. (What is unobservable to one generation may be observable to the next, thanks to technological advances.) Observations must be *replicable*—able to be repeated at different times by different observers. This constraint helps minimize bias, in which someone observes what they "want" or think they "ought" to observe.

In ecology the observation stage is often prolonged and complex. A behavioural ecologist might spend several field seasons recording auditory and visual observations of a bird species in its native habitat. A community ecologist might take multiple measurements of the growth rate and productivity of grassland species in many sites across the North American prairie biome, from Oklahoma to Saskatchewan. Each ecologist makes different kinds of observations, using different devices. Their observations will be multiple, usually quantitative as well as qualitative, and often over an extended period, with the objective of revealing an ecological *pattern* (repeated consistency).

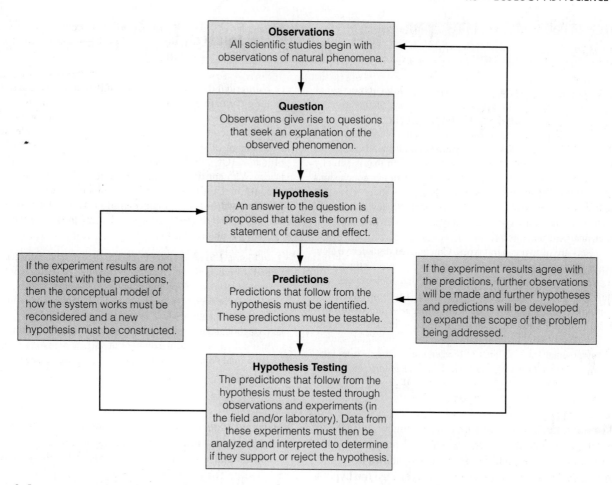

Figure 1.4 Simple representation of the scientific method.

Contrary to what many assume, experimentation is not involved at this stage. Many variables may be measured and sophisticated data analysis used, but no variables are controlled or manipulated. Indeed, researchers take pains not to alter the environment, minimizing the risk that their observations will reflect these alterations.

2. *Hypothesis generation.* The next step is to define a problem, usually posed as a question about the observations. An ecologist may ask, "What factors cause the variations in productivity in grasslands?" The question seeks an explanation for an observed pattern in terms of one or more mechanisms or *processes*. After posing a question, the scientist then generates a **hypothesis**: a proposed explanation of what the answer may be. *Hypothesis generation* is guided by researchers' experience and knowledge of their own and others' research, while trying to avoid bias. Above all, a hypothesis must be a statement about cause and effect that can be tested.

For example, based on the knowledge that nitrogen often limits plant growth and that soils differ in nitrogen content, the ecologist might develop this hypothesis: *Variations in growth rate and productivity of grasses across the prairie biome are due to differences in soil nitrogen.* For statistical reasons, a scientist would state the negative of this hypothesis (a "null" hypothesis, such as H_0: Grassland productivity does not vary with soil nitrogen) and then attempt to support the "alternative"

hypothesis (H_A) by rejecting H_0. As a statement of cause and effect, a hypothesis generates *predictions*. If nitrogen is indeed determining grassland productivity, then productivity should be greater in areas with higher soil nitrogen.

3. *Hypothesis testing.* Next, the ecologist tests the hypothesis to see if its predictions hold true. This step requires gathering more data (see Quantifying Ecology 1.1: Classifying Ecological Data, p. 8), in several possible ways. The first approach might be a non-experimental field study examining how soil nitrogen and productivity co-vary (vary together). If nitrogen is the controlling factor, productivity should increase with higher soil nitrogen. The ecologist would measure both variables at various prairie sites, using suitable sampling procedures. If the ecologist gathered such data in the initial phase, he or she would supplement it with more data.

The relationship between soil nitrogen and productivity is then explored graphically, using a scatter plot (see Quantifying Ecology 1.2: Displaying Ecological Data, p. 11). Nitrogen is placed on the horizontal or *x*-axis, plant productivity on the vertical or *y*-axis (Figure 1.5, p. 8). The arrangement is important; it implies that nitrogen is the cause and productivity the effect. Because nitrogen (*x*) is assumed the cause, we call it the **independent variable**, which in turn affects the **dependent variable**—in this case, productivity (*y*). (Go to QUANTIFYit! at **www.pearsoncanada.ca/ecologyplace** for a tutorial on reading and interpreting graphs.)

QUANTIFYING ECOLOGY 1.1 Classifying Ecological Data

All ecological studies involve collecting *data*—observations for testing hypotheses and drawing conclusions about a **statistical population** (not a biological population but a set of entities—biological or not—about which inferences are drawn). A researcher rarely makes observations on all members, so the **sample** is the portion of a statistical population that is actually observed. From sample data, the investigator draws conclusions about the set of entities, whether it is a population or another object of interest, such as soil nitrogen content. However, not all data are of the same type, and data type influences data presentation, analyses, and interpretation.

Data are of two general types. **Categorical data** derive from observations of qualitative variables that fall into distinct, non-numerical groupings. Examples include hair colour, sex, and developmental stage (pre-reproductive, reproductive, post-reproductive). Categorical data are either *nominal* or *ordinal*. With **nominal data**, the categories are unordered, such as hair colour or sex. With **ordinal data**, the order is important, such as developmental stage. If only two categories exist, such as presence or absence of a trait, categorical data are *binary*.

Numerical data derive from quantitative observations. The resulting data are a set of numbers, such as height, offspring number, or mass. Numerical data are either *discrete* or *continuous*. With **discrete data**, only integer values are possible, such as number of offspring, seeds, or daily visits to a flower by a pollinator. (Of course, means of discrete data have decimal places.) With **continuous data**, any value in an interval is possible, limited only by the measurement device. Examples include height, mass, and concentration. Your ruler may measure only to the nearest millimetre, but much smaller subdivisions are possible.

1. In Figure 1.5, what type of data are (a) available nitrogen and (b) productivity: categorical (nominal or ordinal) or numerical (discrete or continuous)? What other variables might have been measured on the plants that would represent other data types?
2. Categorical data may have a numerical basis. For example, hair colour is due to the amount of various pigments. What numerical variables likely underlie reproductive stage?

From Figure 1.5, it is apparent that productivity in North American grasslands does increase with soil nitrogen. Hence, the data support the hypothesis. Had the data shown no significant relationship, the ecologist would have rejected the hypothesis and sought a new explanation for the observed trends. However, the data do not *prove* nitrogen is controlling productivity. Some other **confounding factor** (a factor that varies with another factor and whose effects can be confused with it) that varies with nitrogen, such as soil moisture or pH, may be responsible. At best, this kind of non-manipulative test can establish *correlation*, not cause.

4. *Experimentation.* To establish if the relationship is cause-and-effect, the ecologist performs an **experiment**: a test

conducted under controlled conditions to determine the validity of a hypothesis. In designing the test, the scientist isolates the presumed cause—in this case, nitrogen level—to see if it is responsible for the effect. There are two options. He or she might conduct a *field experiment* (Figure 1.6), adding nitrogen to some replicated plots and not to others. The ecologist manipulates levels of the independent variable (nitrogen) in a predetermined way to reflect realistic variations in soil nitrogen across the biome, and then monitors the dependent variable (productivity). By observing differences in productivity between fertilized and unfertilized plots, the investigator tests whether nitrogen is the causal agent.

In choosing experimental sites, the ecologist must ensure that other factors affecting productivity are similar. Otherwise, he or she cannot be sure what is causing the effects. Confounding factors can thus still be a problem, but proper site selection can minimize this problem compared with a non-experimental study. Yet, even with careful site selection, the vagaries of the natural world might weaken the test. There might be several seasons in which rainfall is far below normal. During drought, productivity might respond weakly to nitrogen addition compared with normal years.

Alternatively, the ecologist might conduct a *laboratory experiment*, which allows more control over abiotic factors. He or she may grow the grasses in a greenhouse or growth chamber with controlled temperature, soil pH, and moisture (as shown for *Eucalyptus* in Figure 1.7). If productivity increases with nitrogen addition, more evidence supports the hypothesis. Despite their advantages, laboratory tests have an inherent limitation: the results are not directly applicable to the field. The response in a laboratory may differ from that in nature, so some realism is lost. Laboratory conditions are not only controlled

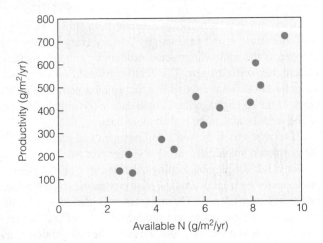

Figure 1.5 Response of grassland productivity to nitrogen availability. Nitrogen, the independent variable, is on the *x*-axis, and productivity, the dependent variable, is on the *y*-axis.

We must remember that plants in nature are part of an ecosystem in which they interact with other organisms as well as with the abiotic environment. A herbivorous insect may affect productivity by selective grazing. Even if known, such biotic variables are hard to incorporate into a lab test. Nevertheless, the ecologist has accumulated more data regarding the response of the plants to nitrogen, data that supplement the evidence pertaining to the hypothesis.

5. *Reiteration.* What happens if the results of the field studies and/or experiments are inconsistent with the predictions? Even if productivity does increase with nitrogen, the variability of the results may engender enough statistical uncertainty that the scientist cannot reject the null hypothesis and cannot establish a causal relationship between soil nitrogen and prairie productivity. The problem may be that the sample size is too small. Repeating the experiment with more samples may allow the researcher to achieve results that are significant at an acceptable probability level (*p*-value). However, if further experimentation fails to validate the hypothesis, then the researcher must reconsider his or her conceptual idea of how the system works, generate a new hypothesis, and repeat the process. The ability to deal with and regroup after such "negative results" is often the true test of a scientist's research potential.

6. *Theory generation.* Having conducted many studies that investigate the link between grassland productivity and nitrogen, the ecologist may explore how the relationship is influenced by other factors, such as soil type, rainfall, and herbivory. Once again, hypotheses are developed, predictions made, and experiments conducted. As the ecologist develops a more detailed understanding of how a host of abiotic and biotic factors interact with nitrogen to control productivity, a **theory**—an integrated set of hypotheses that attempts to explain a broader set of observed phenomena than does any single hypothesis—may emerge. In our example, it might be a general theory of the determinants of North American prairie productivity. Theories are the goal of scientific endeavour. Nowhere have they proved more elusive than in ecology, where the variation inherent in the natural world makes generalization difficult.

Not all ecological studies involve the sequence of steps just outlined. Ecologists often employ **natural experiments**, in which they monitor response to a disturbance or other event, such as the widespread fires in British Columbia in 2010. Despite their name, these are not experiments in the scientific sense, since the researcher does not manipulate treatments, and the only controls are similar areas that were unaffected. Although they have limitations, natural experiments are valuable not only for hypothesis generation but also because they provide an opportunity to observe the impacts of factors that may be impossible to investigate experimentally.

1.7 MODELS IN ECOLOGY: Mathematical and Descriptive Models Allow Predictions

Scientists use the understanding derived from observations and experiments to develop models. Data are limited to the special case of what happened when the measurements were made.

Figure 1.6 Ongoing for 156 years, the Park Grass Experiment at Rothamsted Research in the United Kingdom is testing the impact of different types of fertilizers on grassland productivity and diversity. See Figure 28.17 for a North American example of a field experiment.

but optimal. Non-treatment variables are kept at levels that will not limit response to the treatment variable. In nature, many factors vary simultaneously. Rarely (if ever) will all factors other than nitrogen be optimal.

Figure 1.7 *Eucalyptus* seedlings in a greenhouse experiment. The researcher is using a portable instrument to measure photosynthesis of plants treated with different levels of nitrogen.

Like photographs, data are tied to a given place and time. Models use the understanding gained from data to predict what will happen in some other place and time.

Models are abstract, simplified representations of real systems that allow scientists to predict a response. As with hypotheses, these predictions should be testable by further observations or experiments. Models are either verbally descriptive, like Darwin's theory of natural selection, or mathematical. Hypotheses are models, although some reserve the term *model* for situations in which the hypothesis has at least some support from observations and/or experiments. The hypothesis relating grassland productivity to nitrogen availability is a model. It predicts that productivity will increase with increasing nitrogen. As stated, this prediction is qualitative—it does not predict *how much* productivity will increase. In contrast, mathematical models allow quantitative predictions. For example, from the data in Figure 1.5, we can develop a regression (a type of mathematical model) to predict the productivity per unit of soil nitrogen (Figure 1.8). (See QUANTIFYit! at **www.pearsoncanada.ca/ecologyplace** to review regression analysis.)

Of growing importance in ecology are **null models**, which assume that patterns (in individual traits, population dynamics, community structure, etc.) are generated by random forces in the absence of a particular ecological process (Gotelli and Graves 1996). The specifics of a null model vary with the process under study, but a null model guards against concluding that an observed pattern is due to the mechanism we assume to be at work when it may have arisen purely by chance.

All of the approaches just discussed—observation, hypothesis generation, experimentation, and modelling—

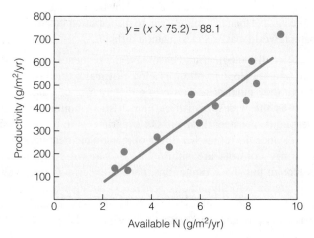

Figure 1.8 Linear regression model predicting plant productivity (*y*-axis) as a function of nitrogen availability (*x*-axis). The general form of the equation is $y = (x \times b) + a$, where *b* is the slope of the line (75.2) and *a* is the *y*-intercept (−88.1), or the value of *y* where the line intersects the *y*-axis (when *x* = 0). The investigator would employ statistical tests to see how well the data "fit" the regression model.

appear in the following chapters to illustrate ecological concepts and relationships. In each chapter, figures and tables present the observations and data used to test specific ecological hypotheses. Being able to analyze and interpret such data is essential. To develop these skills, we have annotated selected figures and tables (labelled "Interpreting Ecological Data"). We pose questions to help you interpret, analyze, and draw conclusions from the data presented. These skills of data interpretation can then be applied to any figure or table in the text.

1.8 UNCERTAINTY AND VARIATION: Ecological Phenomena Exhibit Uncertainty and Variation

Collecting observations, developing and testing hypotheses, and constructing predictive models form the backbone of the scientific method. It is a continuous process of testing and correcting hypotheses and models to arrive at explanations for the variation we observe in the real world, thereby unifying observations that at first seem unconnected. The difference between science and art is that, although both pursuits involve the creation of concepts, science limits its exploration of concepts to the facts. In science, there is no test of concepts other than their empirical truth.

Science has its frustrations. We must simplify to understand, which limits us to inspecting only a part of nature. When designing experiments, we control what we believe to be pertinent factors and eliminate other factors that may confuse the results. Our intent is to focus on a subset of nature about which we hope to establish cause and effect. But there is a trade-off: whatever causal relation we succeed in establishing represents only a partial perspective. So, even when experiments and observations support our hypotheses, and even when the predictions of our models are verified, our job is incomplete. We must loosen the constraints imposed by the need to simplify in order to investigate a broader view of nature. We expand our hypothesis to cover a greater range of conditions and once again test its ability to explain our new observations.

It may sound odd at first, but scientists search for evidence that proves their theories wrong. Rarely is there only one possible explanation for an observation. Many hypotheses may be consistent with an observation, so accumulating data that support a hypothesis does not prove it to be true. The real goal of hypothesis testing is to disprove incorrect ideas. To achieve this end, we must follow a never-ending process of elimination, searching for evidence that proves a hypothesis wrong. Science is thus a self-correcting activity, dependent on ongoing debate. Dissent is the essence of science, fuelled by free inquiry and independence of thought. We explore an example of how debate promotes scientific understanding in Research in Ecology: Prehistoric Snakes as Paleothermometers, p. 143.

To the non-scientist, such debate may seem counterproductive. We depend on science to develop technology and to solve problems. Given the world's growing environmental concerns, scientific uncertainty can be discomforting. However, we must not mistake uncertainty for confusion, nor allow disagreement among scientists—for example, about global climate change—to become an excuse for inaction. Instead, we need to understand uncertainty so that we can balance it against the risks of inaction.

So, why are predictive models and theories so difficult to establish in ecology? Variation contributes to uncertainty in all scientific endeavours. We've discussed how experiments differ from the natural world. As no experiment can duplicate the factors at work in nature, no findings we obtain—however conclusive—can be extrapolated with certainty to the real world. Granted, statistical analysis helps us cope with this

QUANTIFYING ECOLOGY 1.2 Displaying Ecological Data

Whichever type of data an observer collects, data interpretation typically begins with graphically displaying a set of observations. A common method of displaying a single data set involves constructing a *frequency distribution*: a count of the number of observations having a given value. Consider these observations of flower colour in a sample of 100 pea plants:

Flower colour	Purple	Pink	White
Frequency	50	35	15

These data are categorical and nominal, as the colour categories have no inherent order.

Frequency distributions are also used to display continuous numerical data. The following data set represents body lengths (cm) of 20 sunfish sampled from a pond:

8.83, 9.25, 8.77, 10.38, 9.31, 8.92, 10.22, 7.95, 9.74, 9.51, 9.66, 10.42, 10.35, 8.82, 9.45, 7.84, 11.24, 11.06, 9.84, 10.75

With continuous data, the frequency of each value is often a single instance, as multiple measurements are unlikely to be identi-

cal. Therefore, continuous data are often grouped into discrete categories, with each category representing a defined range of values. Each category must be non-overlapping so that each observation belongs to only one category. For example, the body length data could be grouped into discrete categories of length intervals:

Body length (intervals, cm)	Number of individuals
7.00–7.99	2
8.00–8.99	4
9.00–9.99	7
10.00–10.99	5
11.00–11.99	2

Once the observations have been grouped into categories, the resulting frequency distribution is displayed as a *histogram* (type of bar graph; Figure 1a). The *x*-axis represents discrete body length

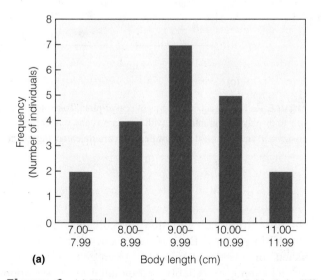

(a)

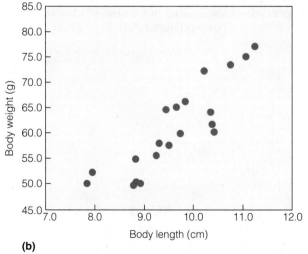

(b)

Figure 1 (a) Histogram relating number of individuals in different body length categories in a sample of a sunfish population. (b) Scatter plot relating body length (*x*-axis) and body weight (*y*-axis) for sample presented in (a).

continued on page 12

intervals, and the *y*-axis represents the number of individuals whose length falls within each interval. In essence, the continuous data are transformed into categorical data for graphical display. Unless there are prior reasons, defining intervals is part of the interpretation process—the search for pattern. For example, how would the pattern depicted by Figure 1a differ if the intervals were in units of 1, starting with 7.50 instead of 7 (7.50–8.49, 8.50–9.49, etc.)?

Researchers often examine the relationship between two variables. When both variables are numerical, the most common method of display is a *scatter plot*, with *x*- and *y*-axes representing the sets of observations of the two variables. Suppose the researcher who measured sunfish body length also measured their weight in grams in order to study the relationship between these two size traits. Body length would be on the *x*-axis (independent variable) and weight on the *y*-axis (dependent variable). Each sunfish individual is plotted as a point on the graph defined by its respective values of length and weight (Figure 1b).

Scatter plots can belong to one of three general patterns. In Figure 2a, *y* increases with increasing values of *x*. The relationship between *x* and *y* is *positive* (as with sunfish length and weight). In Figure 2b the pattern is reversed, and *y* decreases with increasing values of *x*. The relationship between *x* and *y* is *negative*, or inverse. (In Figures 2a and 2b the relationships are also linear, but need not be.) In Figure 2c there is no apparent relationship between *x* and *y*.

Many of the graphs in this text will be histograms or scatter plots. No matter which type is used, ask yourself these questions to help interpret the results. Apply these questions now to Figure 1.

1. What variables do each of the axes represent, what are their units (cm, g, etc.), and what data type are they?
2. How do values of *y* (dependent variable) vary with values of *x* (independent variable)? Is it obvious which of two variables in a non-experimental study should be considered dependent?

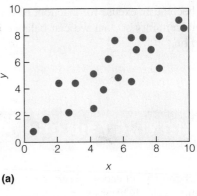

(a)

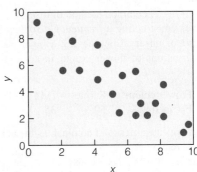

(b)

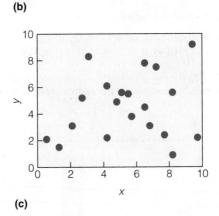

(c)

Figure 2 Three general patterns of scatter plots. To explore further how to display data graphically with histograms and scatter plots, go to QUANTIFYit! and GRAPHit! at **www.pearsoncanada.ca/ecologyplace**.

source of variation, increasing the chances that we will reject false hypotheses by increasing sample size or by improving experimental design.

But in ecology, there is another source of variation that is not a factor in physics or chemistry: *biological variability*. Organisms differ in genetic composition, and these differences affect data collected in both observational and experimental studies. We discuss genetic variation in Chapter 5,

but this biological variation complicates the search for "truth" in ecology. Even if cause and effect can be established for one population of a given genetic makeup, the results cannot be reliably extrapolated to another population. Yet as much as this variation adds to the uncertainty of ecological research, it represents the most valuable and essential attribute of life in its struggle to cope with a changing environment.

ECOLOGICAL ISSUES The Human Factor

Ecologists distinguish between the science of ecology—the study of the interactions of organisms with their environment—and the application of ecology to understanding human interactions with the environment. While ecology studies the natural world, environmental science focuses on the effects of humans on the natural world. However, this distinction is increasingly hard to justify, both in theory and in practice. Ecologists find themselves perplexed by the issue of what constitutes the "natural world" and the extent of human impact on it.

Our species has an ever-growing influence on Earth's environment. The human population now exceeds 7 billion. Like our numbers, our collective environmental impact continues to grow. We use over half of all freshwater resources and have transformed 30 to 40 percent of the terrestrial surface for the production of food, fuel, and fibre. Although air pollution has long been an issue, atmospheric effects resulting from burning fossil fuels are now changing Earth's climate. Accordingly, in the title of his 1989 book, environmentalist Bill McKibben declared *The End of Nature*. His point was that humans have so altered the environment that nature, "the separate and wild province, the world apart from man," no longer exists. In his 2010 work *Eaarth* (the extra "a" is intentional), McKibben goes further, arguing that our activities have created a "tough new planet." Some may disagree, but it has become impossible to study the natural world without considering the impact of human activities, past and present.

For example, by the end of the 19th century many eastern North America forests had been cleared for settlement and farming. In the 1930s and 1940s many of these lands were abandoned as agriculture moved west, leading to extensive forest regrowth. Ecologists cannot study these forests without considering their history. We cannot understand the distribution and abundance of tree species across the region without understanding past land-use patterns. We cannot study nutrient cycling within these forests without understanding how rapidly nitrogen and other nutrients are being deposited from anthropogenic sources. Nor can we understand the causes of population decline in bird species inhabiting eastern forests (an issue that first came to public attention in Rachel Carson's 1962 work *Silent Spring*) without understanding how fragmentation of forests by rural and urban development has altered dispersal, susceptibility to predation and disease, and availability of habitat.

Many questions in ecology today relate directly to the realized and potential effects of human activities on terrestrial and aquatic ecosystems. Throughout the text, we highlight these topics as "Ecological Issues" to illustrate the contribution of ecology to understanding the relationship of humans with the environment in which we play so dominant a role. Working together, ecologists and environmental scientists are quantifying human impacts. Consider the idea of the **ecological footprint**: the demand imposed on Earth's systems, either on a per capita basis or by a country as a whole (Wackernagel and Rees 1996). This concept is widely used, not only for comparing the impacts of countries but for promoting sustainability. As calculated by the International Global Footprint Network, the ecological footprint estimates the land and marine area needed to supply the services consumed by a country, including manufacturing and activities other than providing food and fuel. Expressed as global hectares per capita, footprint values vary widely (Table 1).

The footprint of the average Canadian is typical of industrial countries. The U.S. footprint is even higher, while those of European countries are substantively lower. Although China's per capita footprint is under the global average of 2.7, its total impact is substantial because of its large population. Moreover, China's per capita footprint is growing. Equally important as an index of sustainability is the *ecological balance*—a country's rate of renewal of resources and ecosystem services minus their consumption, also expressed per capita. If the balance is negative, the country uses more resources than its landmass furnishes and is an *ecological debtor*, such as the United Kingdom; if positive, it is an *ecological creditor*, such as Canada. Among developed countries, Sweden is one of the few that has both a relatively low footprint and a creditor status.

Table 1 Global Footprints of Selected Countries

Country	Ecological footprint (gha per capita)	Ecological balance (gha per capita)
Qatar	10.51	−8.00
United States	8.00	−4.13
Canada	7.01	+7.91
Australia	6.84	+7.87
Sweden	5.88	+3.87
France	5.01	−2.01
United Kingdom	4.89	−3.55
South Korea	4.87	−4.54
Mexico	3.00	−1.53
China	2.21	−1.23
Ethiopia	1.10	−0.44

(Adapted from *Ecological Footprint Atlas 2010*, based on 2007 data. Units are global hectares per capita.)

1. How would you define nature? Does your definition include humans? Why?
2. What do you consider the most important current environmental issue? What role might the science of ecology (as you know it) play in helping us understand this issue?
3. Since Canada is an ecological creditor, should we conclude that our per capita resource use is justified?

1.9 AN INTERDISCIPLINARY SCIENCE: Ecology Has Strong Ties to Many Other Fields

The complex web of interactions within ecosystems involves all kinds of physical, chemical, and biological processes. To study these interactions, ecologists must draw on other sciences. This dependency makes ecology a strongly interdisciplinary field. Although in upcoming chapters we explore topics that are typically studied by biochemists, physiologists, and geneticists, we do so only in the context of understanding the interplay of organisms with their environment. How plants absorb CO_2 and transpire water, for example, is a subject for plant physiology. Ecologists consider how such processes respond to variations in light, moisture, and temperature, and to interactions with other species. This information helps them understand the distribution and abundance of plant species and, by extension, the structure and function of terrestrial ecosystems.

Likewise, ecologists draw upon many physical sciences, such as geology, meteorology, and hydrology, to help them chart other ways organisms and environments interact. For instance, as plants take up water, they influence soil moisture and surface water flow. As they lose water to the atmosphere, they increase humidity and affect local precipitation. Similarly, regional geology influences the availability of nutrients and water for plant growth. In each case, other disciplines are crucial to understanding how organisms both respond to and shape their environment.

With the environmental crises of the 20th and 21st centuries, the traditional scope of ecology has expanded to encompass the pervasive role of humans in nature. Among the many environmental problems facing humanity, four broad and interrelated areas are perhaps most crucial: human population growth, conservation of biodiversity, sustainability of natural and human systems, and global climate change. As our numbers increased from about 500 million to more than 7 billion in the past two centuries, dramatic changes in land use altered Earth's surface. Clearing of forests for agriculture has destroyed many natural habitats, causing a rate of species extinction that is unprecedented in Earth's history. Moreover, this expanding human population is exploiting resources at unsustainable levels. Because of the accelerating demand for fossil fuels to drive economic growth, Earth's atmosphere is changing in ways that are altering global climate.

Although Canada suffers fewer immediate effects of human overpopulation, we share the same environmental issues and concerns as the rest of the planet. Issues that seem uniquely Canadian—clear-cutting in the temperate rain forests of British Columbia, mining of the Alberta oilsands, eutrophication of freshwater lakes, such as Lake Winnipeg, and collapse of cod stocks in the Atlantic provinces, to name a few—are local manifestations of global problems. These problems (each of which we consider in later chapters) are inherently ecological because they involve organisms responding to their biotic and/or abiotic environment. Ecological science is thus essential to understanding their causes and mitigating their impacts. Addressing such issues requires an interdisciplinary framework to grasp their historical, social, political, and ethical aspects. That framework is **environmental science**: the study of the impact of humans on the environment. Ecology lays the groundwork for environmental science.

1.10 DUAL FOCUS: Both the Individual and the Ecosystem Are Basic Units of Nature

As we have seen, ecology encompasses a broad area of investigation, from the organism to the biosphere. Historically, the individual—the entity that senses and responds to the environment—is considered the basic unit in ecology. Ecology stresses the processes individuals undergo and the constraints they face in maintaining life in varying environments. After all, the collective birth and death rates of individuals drive population dynamics, and the interactions between individuals of different species influence the community. It is the individual that passes on genetic information to future generations, thereby affecting future populations, communities, and ecosystems. At the individual level we begin to grasp the mechanisms that generate the diversity of life on Earth—mechanisms governed primarily by natural selection operating on individuals. So the individual organism is indeed fundamental.

Complementary rather than antithetical to the centrality of the individual, the ecosystem has assumed increasing importance as another basic unit of nature. Ecologists acknowledge that ecosystem function is vital to biosphere sustainability. This new focus has a long history. The late Stan Rowe made vital early contributions to the "deep ecology" movement. His influential "Manifesto for Earth," originally published in *Biodiversity* in 2004 and reprinted in *Earth Alive: Essays on Ecology* in 2006, set out the principles of *ecocentrism*. We develop these ideas more fully in later chapters, but an example of their growing prominence is *A New Ecology: Systems Perspective* (Jorgenson et al. 2007). By exploring what Jorgenson identifies as the key ecosystem traits of openness, directionality, and connectivity, this approach stresses the dynamic nature of ecosystems in terms of their growth, development, and environmental response, much as traditional ecology does for individuals.

No doubt it is easier for students to relate to individuals than to ecosystems. Ecosystems are also harder to investigate experimentally, as the recent debate about the difficulty of replicating ecosystem experiments sparked by an article in *Science* attests (Carpenter et al. 2011). But the difficulties of the systems approach are matched by its relevance. In light of this new direction, we endeavour throughout to link ecological phenomena at all levels to the ecosystem. But first, in Part One, before stepping onto the first rung of the ecological hierarchy, we examine the characteristics of the abiotic environment that both sustain and constrain life on Earth.

EcologyPlace

Visit EcologyPlace at www.pearsoncanada.ca/ecologyplace to access online resources that complement your textbook, and help you to apply and to review the information in this chapter. EcologyPlace includes

- an eText version of the book
- self-grading quizzes
- glossary flashcards
- and more!

Go to www.pearsoncanada.ca/ecologyplace and follow the registration instructions on the Student Access Code Card included with this text. If your book does not have a Student Access Code Card, you can purchase access to it at www.pearsoncanada.ca/ecologyplace.

SUMMARY

Ecology Defined 1.1

Ecology is the scientific study of the relationships between organisms and their environment. The environment includes abiotic factors (resources, conditions, and hazards) associated with the atmosphere, hydrosphere, and lithosphere, as well as biotic components. Relationships include interactions with the physical world as well as with members of the same and other species.

History of Ecology 1.2

Although ecology is a young science, it is rooted in older disciplines. Darwin's evolutionary theories provided the immediate impetus, but biology and geography made vital contributions. More recently, environmental concerns have raised ecology's public profile. Traditional knowledge can be complementary to Western scientific approaches.

Organisms and Ecosystems 1.3

Organisms interact with their environment in the context of the ecosystem, which consists of biotic and abiotic components interacting as a functioning unit. Biotic components influence abiotic components, which in turn affect biotic components.

Ecological Hierarchy 1.4

Ecology operates on a hierarchy from individuals to the biosphere. Organisms of the same species that inhabit a given location make up a population. Populations of different organisms interact with members of their own and other species in the community. The community and its abiotic environment make up the ecosystem. The landscape consists of the different ecosystems in a region. Regions with similar geology and climate support similar biomes. The highest level is the biosphere—the thin layer around Earth that supports and contains life.

Importance of Scale 1.5

At each hierarchical level, a different set of patterns and processes emerges, requiring a different set of questions and approaches. Each level is associated with its own ecological subdiscipline.

Ecology as a Science 1.6

All ecological studies employ the scientific method. Science begins with observation, from which questions emerge. The next step is developing a hypothesis—a proposed answer to the question. The hypothesis must be testable through observation and experiments, whether in the field or in the laboratory. A theory composed of integrated hypotheses is the ultimate goal.

Models in Ecology 1.7

From research data, ecologists develop models that allow predictions. Models are simplified abstractions of natural phenomena and may be mathematical or non-mathematical. Such simplification is essential to understanding nature. Ecologists use null models to distinguish patterns that result from random forces.

Uncertainty and Variation 1.8

Uncertainty is inherent to science. Ecologists can focus only on a subset of nature, generating an incomplete perspective. Because many hypotheses may be consistent with an observation, the fact that data are consistent with a given hypothesis does not prove it is true. Hypothesis testing tries to eliminate incorrect ideas. Biological variability is an important source of uncertainty.

An Interdisciplinary Science 1.9

Ecology is interdisciplinary because the interactions of organisms with their environment involve an array of responses, the study of which draws upon many non-biological fields.

Dual Focus 1.10

Traditionally, the individual is considered the basic unit in ecology. Individuals respond to the environment and pass on their genes to future generations. Their collective birth and death rates determine population dynamics, and interactions among individuals define communities. Yet the ecosystem is also a vital unit of nature, given its importance in sustaining the biosphere.

KEY TERMS

abiotic	condition	ecosystem ecology	hydrosphere	numerical data
atmosphere	confounding factor	environment	hypothesis	ordinal data
autecology	conservation	environmental	independent variable	population
behavioural ecology	ecology	science	landscape	population ecology
biome	continuous data	evolutionary	landscape ecology	resource
biosphere	dependent variable	ecology	lithosphere	restoration ecology
biotic	discrete data	experiment	model	sample
categorical data	ecological footprint	global ecology	natural experiment	scale
community	ecology	habitat	nominal data	statistical population
community ecology	ecosystem	hazard	null model	theory

STUDY QUESTIONS

1. How do ecology and environmentalism differ? How does environmentalism depend on ecology?
2. Define *population*, *community*, *ecosystem*, *landscape*, *biome*, and *biosphere*.
3. Distinguish between *habitat* and *environment*. Why is this distinction important?
4. How does including the abiotic environment in the ecosystem help ecologists better understand the interaction of organisms with their environment?
5. What is a hypothesis? What is the role of hypotheses in science? What is a model? What is the relationship between hypotheses and models?
6. An ecologist observes that the diet of a bird species consists primarily of large seeds. She hypothesizes that the birds choose larger seeds because they contain more protein. To test the hypothesis, the ecologist measures seed protein content and establishes that the large seeds are indeed richer in protein. Did she prove the hypothesis? Why or why not?
7. You propose a study of the response of forests in the Maritimes to Hurricane Earl in 2010. What type of study would this be, and what are its advantages and disadvantages?
8. What are the causes of uncertainty in science? What is the special significance of biological variability to uncertainty in ecological studies?
9. Given the importance of ecological research in making political and economic decisions regarding current environmental issues, such as global climate change, how do you think scientists should communicate uncertainties in their results to policy makers and the public?
10. Consider an environmental issue of concern in your region of Canada. What sorts of ecological studies would be an essential part of an environmental impact assessment? What other types of disciplines would need to be involved?

FURTHER READINGS

Bates, M. 1956. *The nature of natural history.* New York: Random House.
   A classic for those interested in current environmental issues, written by a lone voice in 1956.

Bronowski, J. 1956. *Science and human values.* New York: Harper & Row.
   Great discussion of science as a human endeavour written by a physicist and poet.

Canadian Society for Ecology and Evolution. www.ecoevo.ca.
   Valuable resource for investigating research by ecologists and evolutionary biologists working in Canada on issues of concern to both academics and the general public.

Carson, R. 1962. *Silent spring.* Boston: Houghton Mifflin.
   One of the first books to bring a major ecological issue (the impact of pesticides, particularly DDT, on North American songbirds) to widespread attention. Reaction was so great that Carson's book is often cited as contributing to passage of the 1973 U.S. Endangered Species Act.

Cronon, W. 1996. The trouble with wilderness; or, getting back to the wrong nature. Pages 69–90 in Cronon, W. (ed.). *Uncommon ground: Rethinking the human place in nature.* New York: Norton.

   This paper, which stirred up considerable debate, proposes that the idea of pristine nature is a human construct and that ecology is not "natural" without considering humans an integral part.

Jorgenson, S. E., B. Fath, S. Bastiononi, J. C. Marques, F. Muller, S. N. Nielsen, B. C. Patten, E. Tiezzi, and R. E. Ulanowicz. 2007. *A new ecology: Systems perspective.* Amsterdam: Elsevier.

Most recent and influential rallying cry for the adoption of a thermodynamic systems approach to ecology, stressing ecosystem properties and feedback mechanisms.

McKibben, W. 1989. *The end of nature.* New York: Random House.

Explores the philosophies and technologies that have brought humans to their current impasse with nature. McKibben's 2010 book, *Eaarth: Making a life on a tough new planet* (Toronto: Alfred A. Knopf) is even more provocative.

Rowe, S. 2006. *Earth alive: Essays on ecology.* Edmonton: Newest Press.

Contains the influential "Manifesto for Earth" (co-written with Ted Mosquin), setting out the principles of ecocentrism.

Worster, D. 1994. *Nature's economy.* Cambridge: Cambridge University Press.

History of ecology written from the perspective of a leading figure in environmental history.

This QR code appears at the end of every chapter, and provides learning resources that you can use with your smartphone to study-on-the-go. Access self-review quizzes, flashcards, and more!

Go to one of the sites below to download an app to your smartphone for free. After installation, your phone will scan the code and link to a website containing Pearson's Study on the Go content.

ScanLife
http://get.scanlife.com

NeoReader
http://get.neoreader.com

QuickMark
http://www.quickmark.com.tw

THE PHYSICAL ENVIRONMENT

In January 2004, two small, robotic vehicles landed on Mars. Their mission was to explore the planet's surface for evidence of whether life ever arose there. The rovers were not looking for living organisms or even for fossils within the rocks littering the terrain. Instead, they were exploring the surface geology to determine the history of water on Mars. Although there is no liquid water on Mars today, a record of past water activity exists in the planet's minerals and landforms, particularly in those that can form only in the presence of water.

Why search for water as evidence of life? The answer is telling: *Life as we know it is impossible without conditions that allow for the existence of liquid water.* The history of water on Mars is thus crucial to determining if the planet's conditions were ever conducive to life. The mission of the Mars rovers was not to search for direct evidence of life but rather to provide information on whether the Martian environment is habitable.

Studying the physical environment is the focus of geology, meteorology, and hydrology, but the idea of **habitability**—the ability of an environment to support life—links these sciences with ecology. To illustrate this link, we move from Mars to a chain of islands off the South American coast—the Galápagos Islands, which so influenced the thinking of the young Charles Darwin.

When we think of penguins, the frozen landscape of Antarctica generally comes to mind. Yet the Galápagos Islands, which lie on the equator, are home to the smallest penguin species: *Spheniscus mendiculus*, the Galápagos penguin, which stands only 40 to 45 cm tall and weighs 2 to 2.5 kg. Found only on the Galápagos, they live the farthest north of all penguins. Galápagos penguins eat mostly small fish, such as mullet and sardines, brought to their feeding grounds by ocean currents flowing from cooler southern waters. Darwin himself noted the importance of these currents: "Considering that these islands are placed directly under the equator, the climate is far from being excessively hot; this seems chiefly caused by the singularly low temperature of the surrounding water, brought here by the great southern Polar current" (*Voyage of the Beagle*).

However, the prevailing winds giving rise to the cool waters that bathe these tropical islands are unpredictable. Periodically, the westward-flowing trade winds stall, and the waters of the Galápagos warm, dramatically reducing fish populations. Such an occurrence (an El Niño event; see Section 2.10) caused a severe food shortage in the early 1980s, when more than 70 percent of the Galápagos penguins died. Since then their numbers have rebounded, and the population was estimated at 1351 in a 2003 census (Vargas et al. 2005). As far back as the 16th century, the region's fishermen have recorded similar warming periods. These times of severe food shortage have undoubtedly affected the penguin population since their ancestors first colonized the islands. Another El Niño event in 1998 caused another setback.

Even from this brief example, it is clear that understanding the ecology of Galápagos penguins depends on an understanding of the abiotic environmental factors that influence habitability for the penguin population. Ecologists use two very different timescales in viewing interactions between organisms and their physical environment. Over many generations, the environment is a shaping force in natural selection, favouring individuals with certain traits over those with other traits. On a shorter timescale, the physical environment influences the physiological performance of individuals as well as the availability of resources, both of which directly influence

A Mars rover. Scientists from the University of Guelph are co-investigators in the project.

The Galápagos penguin (*Spheniscus mendiculus*).

Aerial photo of the Galápagos Islands.

the survival, growth, and reproduction of individuals and the maintenance of the ecosystem. The Galápagos penguin example illustrates both timescales.

Over evolutionary time, the environment of the Galápagos Islands has influenced penguin traits and behaviour through natural selection. First, the penguin's small size lets it easily dissipate heat—an adaptive trait in this tropical habitat. Second, unlike most penguins, Galápagos penguins have no particular breeding season and may have up to three clutches per year. This adaptation allows them to cope with a variable and unreliable food supply, which in turn reflects unpredictable changes in prevailing winds and ocean currents. To understand these traits, we must understand the environment that

has shaped the species over time. Over the shorter term, yearly variations in currents and water temperature influence the penguin's population dynamics by affecting food supply.

In Part One, we examine the features of Earth's physical environments—both terrestrial and aquatic—that directly influence its habitability. We begin with climate (Chapter 2)—the broad-scale patterns of temperature, precipitation, wind, seasonality, and ocean currents. We then turn our attention to the dominant physical characteristics of aquatic (Chapter 3) and terrestrial (Chapter 4) environments. These chapters will set the stage for our discussion in Part Two of the adaptations of organisms, particularly plants and animals, to their physical environment.

As the Sun rises, warming the morning air in this tropical rain forest on the island of Borneo, fog that formed in the cool night air evaporates.

What determines whether a given region will support a tropical forest, a prairie, or a desert? The aspect of the physical environment that most influences ecosystem type by placing the greatest constraints on organisms is climate. *Climate* is a term we use loosely, often confusing it with *weather*. **Weather** is the combination of temperature, precipitation, wind, cloudiness, humidity, and other atmospheric conditions occurring at a specific place and time. **Climate** is the long-term average pattern of weather and may be described on a local, regional, or global scale.

The structure of terrestrial ecosystems is defined by the dominant plants, which in turn reflect the prevailing abiotic environment. Geographic variations in climate, particularly temperature and precipitation, govern plant distributions and hence ecosystem type (Figure 2.1). In this chapter, we examine the major factors that determine these climatic variations. Responses of individuals and ecosystems are covered in later chapters.

2.1 SOLAR RADIATION: Solar Energy Is Subject to Various Fates Before It Is Intercepted by Earth

If asked to describe the climate in our part of Canada, we would likely focus on two things: how warm it is over the year, and how much rain and snow typically falls. These two factors—temperature and precipitation—are the most important climatic determinants of the kind of ecosystem that occupies a particular region. But both factors—along with other, less obvious aspects of climate—derive directly or indirectly from solar energy.

Solar radiation, the electromagnetic energy emanating from the Sun, travels relatively unimpeded through space until it reaches the atmosphere. There, molecular interactions create thermal patterns that, coupled with Earth's rotation and revolution, generate the prevailing winds and ocean currents. These air and water movements influence weather, including the distribution of rainfall. Before considering these complex influences, we first discuss the nature of light.

Scientists describe radiation as a stream of photons, or packets of energy, that—in one of the great scientific paradoxes—behave either as waves or as particles, depending on how they are observed. Radiation varies in *wavelength* (λ, distance between successive crests) and *frequency* (v, number of crests passing a given point per second). All objects radiate energy in a range of wavelengths depending on the object's temperature (Figure 2.2, p. 22): the hotter the object, the more energetic the photons, and the shorter their wavelength. A very hot object, like the Sun, emits primarily **shortwave radiation** (cut-offs vary from 0.1 to 4 μm [micrometres]), whereas cooler objects, such as Earth, emit primarily **longwave (thermal) radiation** (more than 4 μm).

Of the radiation reaching Earth's atmosphere (called the **solar constant**, though it varies up to 15 percent with sunspot activity), only 51 percent makes it to Earth's surface. What happens to the remaining energy, and why is it important ecologically? Radiation may be reflected, scattered, or absorbed in the atmosphere. If 100 units of radiation reach the atmosphere, then clouds and the atmosphere reflect and/or scatter 26 units on average. Earth's **albedo** (reflectivity) is highly variable, but its surface reflects an additional 4 units, for a total of 30 units returned to space (Figure 2.3, p. 22). The atmosphere and clouds absorb another 19, leaving 51 units of **insolation**: direct and indirect solar radiation that reaches Earth's surface without being reflected.

Of the 51 units that reach Earth's surface as insolation, 23 units evaporate water, driving the water cycle (see Section 3.7). Another 7 units heat the air next to Earth's surface, affecting thermal conditions for organisms. The remaining 21 units heat the landmasses and oceans, which in turn emit radiation back to the atmosphere as thermal radiation. Yet the amount of radiation emitted by Earth exceeds the 21 units of solar radiation that are absorbed. In fact, some 117 units

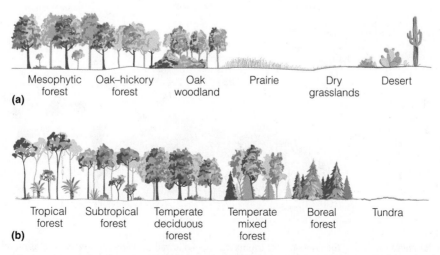

(a)
Mesophytic forest   Oak–hickory forest   Oak woodland   Prairie   Dry grasslands   Desert

(b)
Tropical forest   Subtropical forest   Temperate deciduous forest   Temperate mixed forest   Boreal forest   Tundra

Figure 2.1 Gradients of North American vegetation. **(a)** The east–west gradient reflects declining yearly precipitation, up to the Rockies. **(b)** The south–north gradient reflects declining annual temperature. See Chapter 23 for a detailed discussion of terrestrial ecosystems.

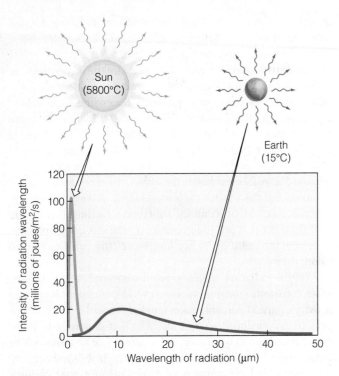

Figure 2.2 Wavelength of radiation emitted by an object varies with its surface temperature. The Sun (~ 5800°C) emits relatively shortwave radiation compared with Earth (~ 15°C).

in total are emitted. How is this possible, and what is its impact?

Although Earth receives shortwave radiation only by day, it emits longwave radiation both day and night. The atmosphere allows only a small fraction of this energy (6 units) to escape into the "thermal sink" of outer space. Most (111 units) is absorbed by H_2O vapour and CO_2 in the atmosphere, and by clouds. This selective absorption of longwave radiation by atmospheric gases, which then radiate the energy back to Earth as heat, is called the **greenhouse effect**. It is crucial to maintaining Earth's surface warmth as well as being the driver of global climate change (see Chapter 28). The amount radiated back in this manner is some 96 units. Thus, Earth receives nearly twice as much longwave radiation from the atmosphere as shortwave radiation from the Sun. In these exchanges, the energy lost at Earth's surface (30 + 117 = 147 units) equals the energy gained (51 + 96 = 147 units). Earth's radiation budget is thus in balance.

One portion of the electromagnetic radiation emitted by the Sun (Figure 2.4) is of particular ecological significance. The wavelength band from 400 to 700 nm (more precisely, 380 to 740 nm; 10^9 nanometre = 1 m) is collectively termed **photosynthetically active radiation (PAR)** because it includes the wavelengths that plants use in photosynthesis. (Radiation of this approximate range of wavelengths is also called *visible light*, because it can be detected by the human eye.) Although PAR makes up only some 25 percent of the solar constant, it makes up almost half of insolation because shorter wavelengths are preferentially absorbed in the atmosphere prior to reaching Earth. Insolation is thus said to be *enriched* in PAR.

The shorter wavelengths flanking visible light are **ultraviolet (UV) radiation**: UV-A (315–380 nm) and UV-B (280–315 nm). UV, especially UV-B, is hazardous to all life but plays a key role in evolution by inducing mutations. **Infrared radiation** is of wavelengths longer than the visible range and is also of two types: *near-infrared* (740–4000 nm), which acts as a signal affecting biological processes such as flowering, and *far-infrared (thermal) radiation* (4000–100 000 nm). Thermal radiation greatly affects habitability by creating conditions that affect life.

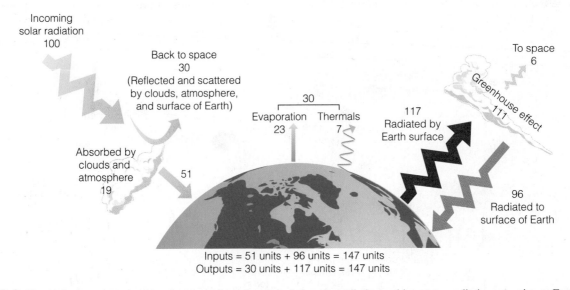

Figure 2.3 Fates of solar energy reaching the atmosphere. Inputs include solar radiation and longwave radiation returning to Earth because of the greenhouse effect. Outputs include heat from evaporation and thermals and longwave energy radiated by Earth's surface.

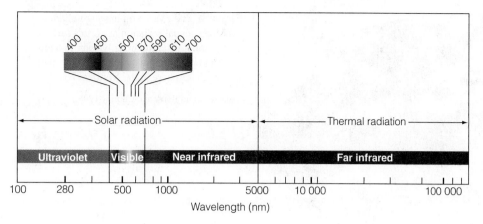

Figure 2.4 Portion of the Sun's electromagnetic spectrum, separated into shortwave and thermal radiation. Ultraviolet, visible, and infrared radiation represent only a small portion. To the left of UV are X-rays and gamma rays (not shown).

(Adapted from Halverson and Smith 1979.)

2.2 RADIATION AND TEMPERATURE: Latitudinal and Seasonal Radiation Patterns Affect Temperature

So far, we have been speaking in mean global percentages. Yet the amount of energy intercepted at any point on Earth's surface varies greatly with latitude (Figure 2.5). Two factors are involved: (1) at higher latitudes, radiation hits the surface at a steeper angle, spreading light over more area, and (2) radiation penetrating the atmosphere at a steeper angle travels through more air and encounters more atmospheric particles, which reflect more of it back into space.

Although variation in intercepted radiation with latitude explains the decreasing temperatures from the equator to the poles, it does not explain systematic seasonal variation. What causes the seasons? Why do the hot days of summer give way to the cool temperatures of fall, or the cold temperatures of

winter to the onset of spring? The answer is simple—because Earth is tilted.

Earth is subject to two distinct motions. While it orbits around the Sun, Earth rotates on an axis that passes through the North and South Poles, giving rise to day and night (the diurnal cycle). Earth (and the other planets) circumnavigates the Sun in a plane called the *ecliptic*. However, Earth's axis is tilted at an angle of 23.5° to the ecliptic. This tilt causes seasonal variations in temperature and day length. Only at the equator are there exactly 12 hours of light and dark every day year-round. At the spring and fall equinoxes (March 21 and September 22, respectively), solar radiation falls directly on the equator (Figure 2.6a). The equatorial region is then heated most intensely, and every place on Earth receives 12 hours of daylight and night.

At the summer solstice (June 20) in the Northern Hemisphere, solar rays fall directly on the Tropic of Cancer (23.5° north; Figure 2.6b). At this time, days are longest and the Sun

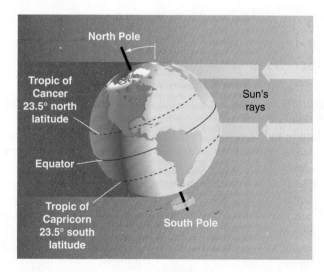

Figure 2.5 At high latitudes, solar radiation strikes Earth at an oblique angle and spreads over a wider area, making it less intense than at the equator.

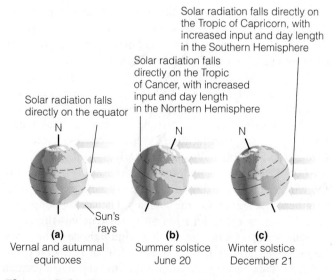

Figure 2.6 Differences in the angle of the Sun and the circle of illumination over the year.

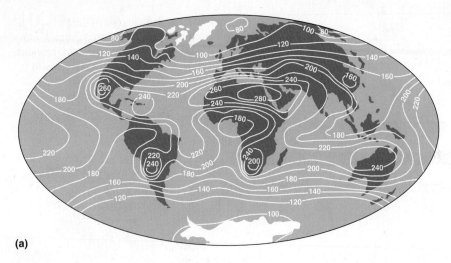

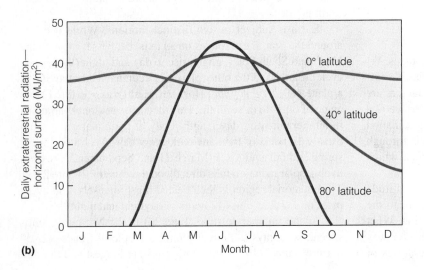

Figure 2.7 Variation in solar radiation. **(a)** Mean annual radiation at Earth's surface. **(b)** Seasonal variation in daily solar radiation incident on a horizontal surface at the top of Earth's atmosphere in the Northern Hemisphere.

(Adapted from Barry and Chorley 1998.)

heats the surface intensely, whereas the Southern Hemisphere experiences winter. At the winter solstice (December 21) in the Northern Hemisphere, solar rays fall directly on the Tropic of Capricorn (23.5° south; Figure 2.6c), bringing shorter days and colder temperatures while the Southern Hemisphere enjoys summer. Summer and winter solstices are thus reversed in the hemispheres.

The seasonality of radiation, temperature, and day length increases with latitude. At the Arctic and Antarctic Circles (66.5° north and south, respectively), day length varies from 0 to 24 hours over the year. The days shorten until the winter solstice, a day of continuous darkness, and then lengthen with spring. On the summer solstice, the Sun never sets. In theory every location on Earth receives the same amount of daylight in a year, but at high latitudes where the Sun is never directly overhead, annual input is lowest (Figure 2.7). This varying exposure explains why mean annual temperature is highest in the tropics and declines towards the poles (Figure 2.8).

2.3 ELEVATION AND TEMPERATURE: Air Temperature in the Troposphere Decreases with Elevation

Whereas varying exposure to radiation explains changes in latitudinal, seasonal, and daily temperatures, it does not explain why air gets cooler at higher elevation. The terms *altitude* and *elevation* are not synonymous; **altitude** refers to distance above Earth's surface, **elevation** to distance above sea level. They can be equal, but only if a landform arises at sea level. You would not be surprised to find snow capping a mountain in British Columbia. Yet Mount Kilimanjaro, which rises from the hot plain of tropical Africa, is also snow-capped (Figure 2.9, p. 26). The explanation of the seeming oddity of snow in the tropics lies in the physical properties of air.

The approximate total mass of air molecules surrounding Earth is a staggering 5×10^{18} kg. The force that this huge mass exerts over a given area of Earth's surface is called

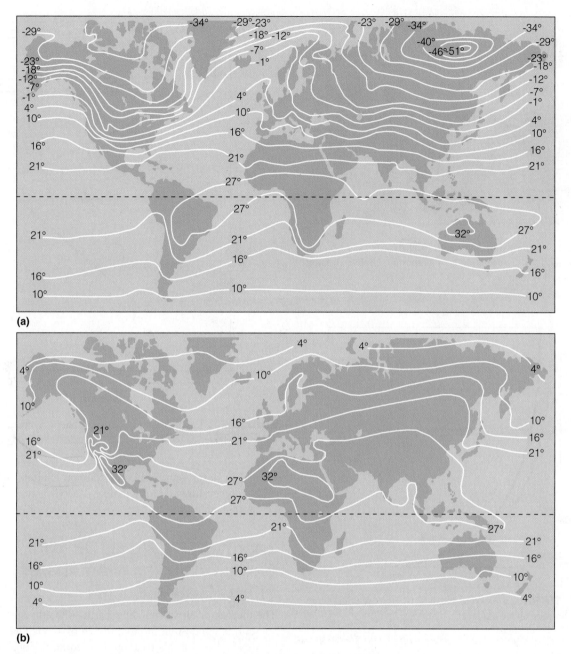

Figure 2.8 Mean sea-level temperatures (°C) in (a) January and (b) July. Note the reversal in temperature pattern between the Northern and Southern Hemispheres. Isotherm lines indicate areas of equal mean temperature.

atmospheric (air) pressure. Envision a vertical column of air. Air pressure at any point is measured as the total air mass above that point. As elevation increases, the air mass decreases, and pressure drops. Although air pressure decreases continuously, the rate of decline slows with elevation (Figure 2.10, p. 26), in parallel with air density (number of molecules by volume). By 50 km, air pressure is only 0.1 percent of that at sea level. The atmosphere extends upwards for hundreds more kilometres, gradually becoming thinner until it merges into outer space.

Air temperature has a more complicated vertical profile. It decreases from Earth's surface to an elevation of about 11 km.

This decrease in temperature with elevation is caused by two factors: (1) Greater air pressure at the surface causes air molecules to move faster. Since air temperature is a measure of the speed of air molecules (their kinetic energy), higher speed means warmer temperatures near the surface. Decreasing pressure with elevation lowers the speed of air molecules, reducing air temperature. (2) The warming effect of Earth's surface declines with elevation. Recall that absorbing solar radiation warms Earth by longwave radiation emitted from its surface. This heat transfer continues upwards as heat flows spontaneously from warmer to cooler areas, but at a continuously declining rate as the energy is dissipated.

Figure 2.9 Mount Kilimanjaro in tropical Africa is snow-capped and supports tundra-like vegetation near its summit. Global warming is causing its snowcap to melt.

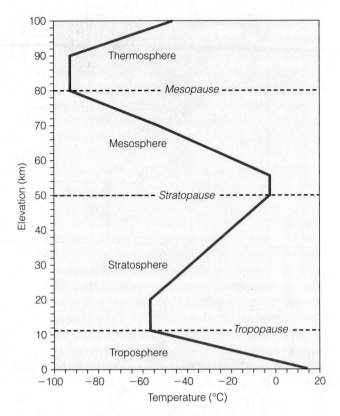

Figure 2.11 Changes in global atmospheric temperature with elevation above sea level.

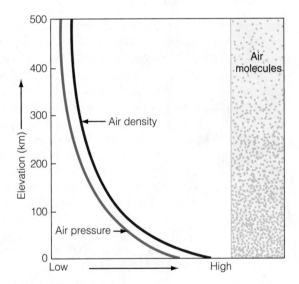

Figure 2.10 Both air pressure and density decrease with increasing elevation above sea level.

Unlike pressure and density, air temperature does not continue to decline indefinitely with elevation. At certain elevations, abrupt temperature shifts occur. These specific elevations distinguish atmospheric regions (Figure 2.11): *troposphere*, *stratosphere*, *mesosphere*, and *thermosphere*, with boundary zones between. The troposphere and stratosphere are of most direct importance for climate and hence for life on Earth.

So far, our discussion of changing temperature with elevation has assumed no vertical movement of air between atmospheric regions. However, when a parcel of surface air warms, it becomes buoyant and rises, like a hot-air balloon. As the parcel of air rises, decreasing pressure causes it to expand and cool. Decrease in air temperature through expansion rather than heat loss is called *adiabatic cooling*. The rate of adiabatic cooling with elevation (**adiabatic lapse rate**) depends on humidity. Dry air cools more rapidly (~ 10°C per 1000 m elevation) than moist air (~ 6°C per 1000 m). This phenomenon is particularly

critical in deserts and especially alpine habitats, where day/night temperatures can fluctuate by 20°C or more, imposing physiological stress.

2.4 ATMOSPHERIC CIRCULATION: Global Circulation of Air Masses Affects Climate

The blanket of air surrounding Earth—the atmosphere—is not static. It is in constant motion, driven by rising and sinking air masses and by Earth's rotation. Equatorial regions receive the most radiation annually. As warm air rises, air heated in the tropics rises to the top of the troposphere, creating a low-pressure zone at Earth's surface (Figure 2.12a). More air rising under it forces the mass to spread north and south towards the poles. As air masses move to the poles, they cool, become heavier, and sink. This sinking air flows towards the low-pressure equatorial zone, replacing the rising warm tropical air and closing the air circulation pattern.

If Earth were stationary and had regular landmasses, the atmosphere would circulate as in Figure 2.12a. But Earth spins on its axis from west to east. Each point on Earth's surface makes a complete rotation every 24 hours, but the speed of rotation varies with circumference and latitude. At the equator (circumference = 40 075 km), the speed of rotation is 1670 km h⁻¹, but 60° north or south, both the Earth's circumference and the rotation speed are roughly half that at the equator. A rotating Earth changes these air circulation patterns (Figure 2.12b). Why?

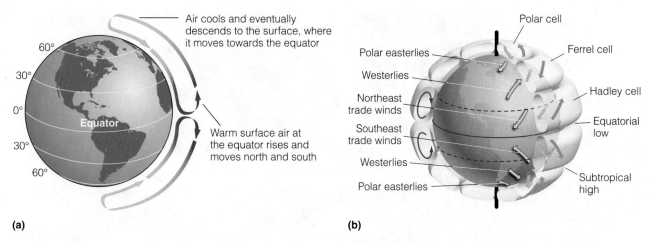

(a)

(b)

Figure 2.12 Circulation of air cells and prevailing winds on **(a)** an imaginary, non-rotating Earth compared with **(b)** a rotating Earth.

According to physical laws, an object moving from a greater to a lesser circumference deflects in the direction of the spin, while one moving to a greater circumference deflects in the direction opposite to that of the spin. Thus, air masses (and all moving objects) deflect clockwise (to the right) in the Northern Hemisphere and counterclockwise in the Southern Hemisphere. This deflection, called the **Coriolis effect** (Figure 2.13), prevents a simple flow of air from the equator to the poles. Instead, it creates belts of prevailing winds, named for their direction of origin. These belts break the flow of surface air into six cells (three per hemisphere), which produce areas of low and high pressure as air masses ascend from and descend to the surface.

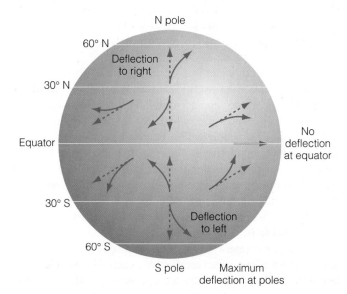

Figure 2.13 Impact of the Coriolis effect on wind direction. There is no effect at the equator, where linear velocity is greatest (465 m/s). The effect increases towards the poles. If an air mass moves north at a constant speed, it speeds up because Earth moves more slowly (403 m/s at 30° latitude, 233 m/s at 60° latitude, 0 m/s at the poles) than the object. An object's path appears to deflect to the right in the Northern Hemisphere and to the left in the Southern Hemisphere.

Let's trace this flow of air as it circulates the globe. Air heated in the tropics rises, creating a low-pressure zone near the surface—the *equatorial low*. This upward flow is balanced by a flow from the north and south towards the equator. As the warm air mass rises, it spreads, diverging towards the North and South Poles, cooling as it goes. In the Northern Hemisphere, the Coriolis effect forces this air to the east, slowing its progress north. At ~ 30° N latitude, the cool air sinks, closing the *Hadley cell*. The descending air mass forms a semi-permanent high-pressure belt encircling Earth—the *subtropical high*.

Upon descending, the cool air warms and splits into two surface currents. One moves north to the pole, diverted by the Coriolis effect to become the *prevailing westerlies*. The other moves south towards the equator, becoming the strong winds called the *trades* (or *easterlies*) by the 17th-century sailors who used them to reach the Americas. In the Northern Hemisphere, they are called the *northeast trades*. The mild westerlies encounter cold air moving down from the pole at about 60° N. These air masses of contrasting temperature do not mix. They are separated by the *polar front*—a zone of low pressure (the *subpolar low*) where surface air converges and rises. Some of the rising air moves south until it reaches about 30° N latitude (the *subtropical high*), where it sinks back to the surface and completes the *Ferrel cell*. This mid-latitude cell is an intermediary between those on either side and is a zone of mixing.

As the air reaches the pole, it sinks to the surface and flows back south towards the polar front, completing the last of the three cells—the *polar cell*. This south-moving air is deflected to the right by the Coriolis effect, giving rise to the *polar easterlies*. Similar flows (and cells of the same name) occur in the Southern Hemisphere (see Figure 2.12b).

2.5 OCEAN CURRENTS: Solar Energy, Wind, and Earth's Rotation Create Ocean Currents

The global pattern of prevailing winds also determines the **currents**: patterns of surface water movements in oceans or other water bodies. Until they encounter a continent, ocean currents

Figure 2.14 Ocean currents. Circulation is influenced by Coriolis forces (clockwise in the Northern Hemisphere, counterclockwise in the Southern Hemisphere) and continental landmasses. Blue arrows represent cool water, and red arrows represent warm water.

mimic movement of the air currents above. Ocean currents give rise to **gyres**: giant circular water motions. Within each gyre, ocean currents move clockwise in the Northern Hemisphere and counterclockwise in the Southern Hemisphere (Figure 2.14).

Along the equator, trade winds push warm surface waters westward. When they encounter the eastern margins of continents, these waters split into north- and south-flowing currents along the coasts, forming north and south gyres. As the currents move farther from the equator, the water cools. Eventually, they encounter the westerlies at higher latitudes (30–60° N and 30–60° S), which produce eastward-moving currents. When these currents encounter the western margins of continents, they form cool currents that follow the coast towards the equator. North of Antarctica, ocean waters circulate unimpeded around the globe.

Ocean currents have substantial ecological impacts on marine environments. They affect productivity by influencing temperature and nutrients (see Section 20.3). Also, changes in currents during irregular climatic phenomena, such as El Niño events (see Section 2.10), contribute to large fluctuations and even collapses in marine populations (see Section 26.10).

2.6 HUMIDITY: Temperature Influences the Moisture Content of Air

Air temperature plays a crucial role in the exchange of water between the atmosphere and Earth. Whenever matter, including water, changes from one state to another, energy is either absorbed or released. **Latent heat** is the energy released or absorbed per gram during a state change. In going from a more ordered (liquid) to a less ordered (gas) state, energy is absorbed to break the bonds between molecules. **Evaporation** (transformation of water from liquid to gas) requires 2260 J (joules) of energy per gram of water. (*Condensation*, the transformation of water vapour to a liquid,

releases an equivalent amount.) When air comes into contact with liquid water, water molecules are freely exchanged between the air and the surface. When the evaporation and condensation rates are equal, the air is said to be *saturated*.

In air, water vapour acts as an independent gas that has mass and exerts pressure. **Vapour pressure** is the amount of pressure that water vapour exerts independent of the pressure of dry air, in pascals (Pa). The water vapour content of air at saturation is the **saturation vapour pressure**. If this pressure is exceeded, condensation occurs and vapour pressure declines. Saturation vapour pressure increases with temperature because of increasing kinetic energy. The warmer the air, the greater its capacity for holding water vapour.

The amount of water in a given volume of air is its *absolute humidity*. A more useful measure is **relative humidity**: the amount of water vapour expressed as a percentage of its saturation vapour pressure, which has a relative humidity of 100 percent. If air cools while its absolute moisture content stays constant, its relative humidity increases because its saturation vapour pressure declines. If the air cools to a point where the actual vapour pressure exceeds the saturation value, moisture condenses and forms clouds. As soon as particles of water or ice become too heavy to remain suspended in air, precipitation falls.

For a given water content of an air parcel, the temperature at which saturation vapour pressure is achieved is the **dew point temperature**. What causes dew to form? As nightfall approaches, temperatures drop and relative humidity rises. If cool night air temperatures reach the dew point, water condenses and dew forms, lowering the water content of the air. As the Sun rises, the air temperature warms and the water vapour capacity (saturation pressure) increases. The dew evaporates, increasing vapour pressure, but before it evaporates, dew can be an important source of moisture for low-growing plants and small animals.

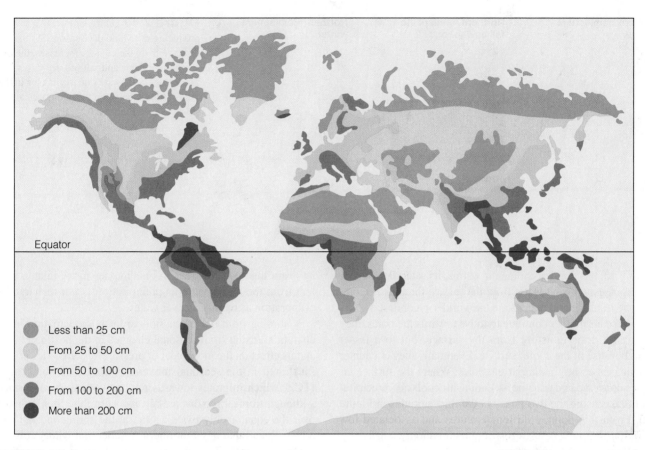

Figure 2.15 Annual global precipitation. Relate the wettest and driest areas to mountain ranges, ocean currents, and winds.

2.7 PRECIPITATION: Movements of Air Masses Create Global Precipitation Patterns

Annual precipitation is among the most influential factors affecting biome distribution, and understanding global precipitation involves integrating patterns of temperature, winds, and ocean currents. **Precipitation** (all forms of water reaching Earth, including not just rain and snow but also fog and mist) is not evenly distributed. At first, annual precipitation trends seem to lack any discernible regularity (Figure 2.15). But if we examine variation with latitude (Figure 2.16), a pattern emerges. Precipitation is highest near the equator, declining as one moves north and south. This decline is not continuous—two peaks occur in the mid-latitudes followed by a further decline towards the poles. This sequence of peaks and troughs corresponds to the rising and falling air masses associated with the belts of prevailing winds in Figure 2.12b.

As the warm trade winds cross the tropical oceans, they gather moisture. Near the equator, the northeast trades meet the southeast trades in a narrow, high-precipitation region called the **intertropical convergence zone (ITCZ)**. Where the two air masses meet, air piles up, and the warm, humid air rises and cools. When the dew point is reached, clouds form, and precipitation falls. This pattern accounts for the high rainfall in tropical regions worldwide (see Figure 2.15).

Having lost much of its moisture, the ascending air mass continues to cool as it splits and moves north and south. In the subtropical high regions (~ 30° N and S), where the cool air descends, two belts of dry climate encircle the globe (the two troughs in

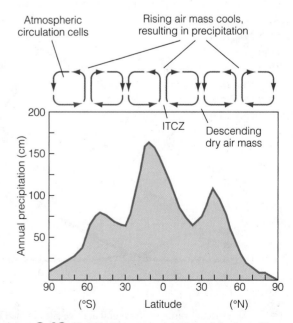

Figure 2.16 Variation in mean annual precipitation with latitude. Peaks correspond to rising air masses, such as in the ITCZ, whereas troughs are associated with descending dry air masses.

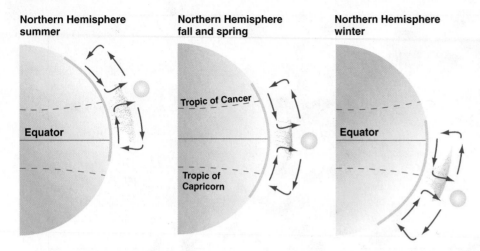

Northern Hemisphere summer

Northern Hemisphere fall and spring

Tropic of Cancer

Equator

Tropic of Capricorn

Northern Hemisphere winter

Equator

Figure 2.17 Shifts of the ITCZ, producing rainy and dry seasons. As the distance from the equator increases, the dry season lengthens and rainfall decreases. These oscillations are due to changes in the Sun's altitude between the equinoxes and the solstices (see Figure 2.6).

Figure 2.16). As the descending air warms, its saturation vapour pressure rises, drawing water from the surface through evaporation. The resulting arid regions are the world's major deserts.

As the air masses continue to move towards the poles, they once again draw moisture from the surface, but to a lesser extent because of the cooler surface. Eventually they encounter cold air masses originating at the poles. Where the surface air masses converge and rise, the ascending air cools and precipitation falls, causing the two peaks in the mid-latitudes in Figure 2.16. From this point, cold temperatures and associated low saturation vapour pressures restrict precipitation, contributing to the "polar desert" of northern Canada.

Another important pattern emerges from Figure 2.16. Rainfall is higher in the Southern Hemisphere (note the southern shift in the peak associated with the ITCZ) because the oceans cover a greater proportion of the hemisphere, and water evaporates more readily from water bodies than from soil or vegetation. This also explains why continental interiors receive less rain than coastal areas. As air masses move inland, water lost from the atmosphere as precipitation is not recharged by evaporation as readily as over water.

Missing from our discussion so far is the variation in precipitation arising from seasonal changes in the heating of Earth and its effect on the movement of pressure systems. This effect is illustrated in the seasonal movement north and south of the ITCZ, which migrates towards regions with warmer surfaces. Although tropical regions are always warm, the Sun is directly over the equator only twice a year, at the spring and fall equinoxes. Recall that at the northern summer and winter solstices, the Sun is over the Tropics of Cancer and Capricorn, respectively. As a result, the ITCZ moves north, invading the subtropical highs in the northern summer, while in winter it moves south, leaving clear, dry weather behind (Figure 2.17) and bringing rain to the southern summer. Thus, as the ITCZ shifts north and south, following the apparent migration of the direct rays of the Sun, it triggers the wet and dry seasons in the tropics (Figure 2.18).

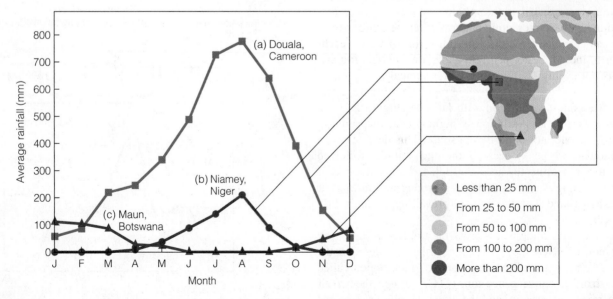

Less than 25 mm
From 25 to 50 mm
From 50 to 100 mm
From 100 to 200 mm
More than 200 mm

Figure 2.18 Seasonal variation in rainfall at three sites in the ITCZ. Although site **(a)** has seasonal variation, precipitation exceeds 50 mm monthly. Sites **(b)** and **(c)** experience distinct rainy and dry seasons, which are six months out of phase because of their location in different hemispheres.

2.8 SNOW AND ICE: Frozen Precipitation Has Many Ecological Effects

So far, our discussion has focused on precipitation in the form of rain. But in high-latitude regions, frozen precipitation (ice and snow; both are solid water, but ice forms from liquid water and snow from crystallization of water vapour) is common and has ecological effects quite apart from its contribution to total precipitation. Percentages vary widely, but in the central prairies, snow can be 25 percent of total precipitation. (Because snow is much less dense than liquid water, snowfall is converted by a factor of 0.1 when included in total precipitation.) Snowfall amounts are much greater in eastern Canada, particularly in regions that receive "lake effect" snow because of their proximity to the Great Lakes and other large water bodies.

Winter precipitation might seem far less important for an ecosystem than rain received in the growing season, but the recharging effect of melting snow on soil moisture is critical for early spring growth, particularly in the prairies, where springs can be warm and dry. Soils with higher clay content (see Section 4.5) retain moisture from snowmelt better than lighter, sandy soils.

Snow has many other ecological effects. Because of its low density, it is an effective insulator for both soils and ground vegetation. Even if two winters are equally cold, frost penetration in soils is much less when snow cover is (1) greater and (2) established earlier in winter. Frost depth in turn affects both winter survival and onset of regrowth in the spring. If frost penetration is shallow, soil warms more quickly once the snow cover is gone.

Because of its high albedo (reflectivity), snow reflects solar radiation and can cause browning of evergreen leaves in late winter in exposed microsites. On a global scale, changing ice cover at the poles both annually and on a longer timescale affects climate. Because ice reflects radiation, it increases the amount of solar energy returned to the atmosphere. In turn, this increased reflection means Earth's surface heats up less, which allows the polar ice caps to increase in size. Even more solar radiation is then reflected, and so Earth cools further. Essentially, a type of positive feedback loop (see Section 19.2) takes over that leads to global cooling—the opposite effect of the positive feedback that promotes global warming when the ice caps are retreating, as they are now.

All forms of precipitation, including rain and snow, can become hazards if they are heavy and/or accompanied by high winds. Although it is typically warmer during a blizzard than under the high-pressure conditions that prevail on clear winter days, high wind chills and snow-load damage can increase mortality. The greatest precipitation-related damage is associated with less frequent events. In summer, hailstorms cause significant damage to living plants, and ice storms in winter not only break branches but topple entire trees. One of the most severe ice storms in recent history hit eastern Canada in 1998, causing widespread mortality (Figure 2.19).

Figure 2.19 A severe ice storm in January 1998 caused great damage to trees in Montréal (shown here) as well as other parts of eastern North America.

2.9 EFFECTS OF TOPOGRAPHY: Landforms Influence Regional and Local Precipitation

Topography, the physical structure of landforms, strongly influences precipitation. At a regional scale, mountainous terrain intercepts air flow. Air ascends, cools, becomes saturated with water vapour (because of lower saturation pressure), and releases much of its moisture at upper altitudes on the windward side. As the cool, dry air descends the leeward side, it warms and gradually picks up moisture. Windward slopes therefore support denser vegetation and different species of plants and other organisms than do leeward slopes, where desert-like conditions, called **rain shadow**, prevail (Figure 2.20, p. 32). Thus, westerly winds blowing over the Coast Mountains in British Columbia support lush forest on west-facing slopes, while eastern slopes are semi-desert (Figure 2.21, p. 32). On the leeward side of the Rocky Mountains, this same topographic effect contributes to the dry prairie conditions in southern Alberta.

2.10 IRREGULAR CLIMATIC VARIATION: Earth Experiences Significant Irregular Variation in Climate

The climatic variation that we have discussed so far occurs at predictable intervals—seasonal changes in temperature with Earth's rotation about the Sun, and in rainfall resulting from migration of the ITCZ. But not all climatic patterns occur so regularly. Earth's climate is highly variable on both spatial (regional and global) and temporal (short- and long-term) scales.

The Little Ice Age, which lasted from the mid-14th (its onset is as controversial as its causes) to the mid-19th century, brought bitterly cold winters to the Northern Hemisphere, affecting health, agriculture, politics, emigration, and even art and literature. Norse colonies in Greenland starved from crop failure in the 15th century, and in the mid-17th century, glaciers in the Alps advanced, engulfing farms and crushing villages. In 1607–1608, ice persisted

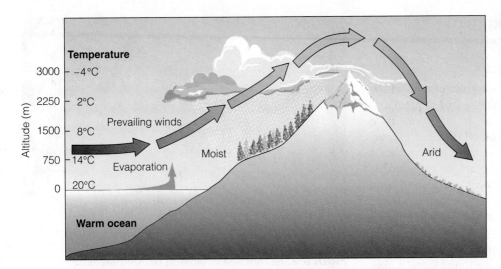

Figure 2.20 Formation of a rain shadow. As air rises, the air mass cools and loses moisture on the windward side. Descending dry air picks up moisture on the leeward side.

(a)

(b)

Figure 2.21 Rain shadow effect in the coastal wet-belt forests in British Columbia. **(a)** Lush forests clothe windward slopes in the upper Fraser Valley. **(b)** Sparser forests grow on the leeward side, as in the Chilcotin valley in central interior British Columbia.

on Lake Superior until June, and the Thames River froze yearly from 1607 to 1814 (Lamb 1995). In fact, the image of a white Christmas evoked by Dickens reflects the cold winters of this era. The climate has since warmed sufficiently that a white Christmas in much of North America and Europe is now an anomaly.

As another example, central North America has undergone periodic droughts over the last 5000 to 8000 years, but the homesteaders of the early 20th century settled the prairies at a time of relatively wet summers. They assumed these conditions were normal and employed farming methods used in the moister Eastern climate. But the cycle of drought returned, and the prairies became a dust bowl in the 1930s, profoundly affecting the economic and social history of North America. As shortgrass prairies were most affected, the greatest damage in Canada occurred in southern Saskatchewan's Palliser triangle.

These examples reflect the variability in Earth's climate systems, which operate on timescales ranging from decades to millennia, driven in part by changes in Earth's orbit. Changes in the tilt of its axis and its path about the Sun affect climate by altering seasonal radiation. Occurring over tens of thousands of years, these changes are associated with glacial advance and retreat.

Variation in solar input to Earth's surface is also due to sunspot activity—huge magnetic storms on the Sun that occur in

cycles, with the number and size peaking every 11 years. Researchers have related sunspot activity to periods of summer drought and winter warming in the Northern Hemisphere. Sunspot activity also complicates predictions of global climate change.

Interactions between the oceans and the atmosphere cause regional climatic variations. Documents reveal that, as far back as 1525, South American fishermen experienced periods of unusually warm water. Peruvians called these El Niño (Spanish for the Christ Child, since they often occur at Christmas). El Niño is the phenomenon in the waters of the Galápagos Islands described in the Part One introduction. The Southern Oscillation, a more recent discovery, is an oscillation in surface air pressure between the tropical Pacific and Australian–Indonesian regions. The **El Niño–Southern Oscillation (ENSO)** refers to these combined events.

The ENSO phenomenon is well known, but weather reports rarely describe its cause. Recall that the trade winds blow west across the tropical Pacific. As a result, surface currents in the tropical oceans flow west, bringing cold, deeper waters to the coastal waters of Peru in an *upwelling* (see Section 3.8). This upwelling, together with the cold current flowing north along the west coast of South America, normally makes this ocean region colder than expected, given its equatorial location (Figure 2.22a). As these surface currents move westward, the water warms, making the

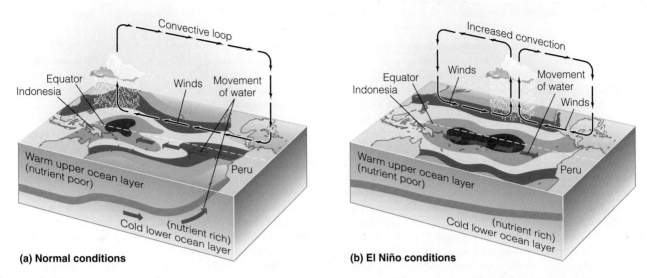

Figure 2.22 Schematic of the El Niño–Southern Oscillation (ENSO) off the west coast of South America. **(a)** Normally, strong trade winds move surface waters westward. The resulting warm waters of the western Pacific cause maritime air to rise and cool, bringing abundant rain. **(b)** Under ENSO conditions, the trade winds slacken, reducing westward flow of surface currents and bringing more precipitation to Peru.

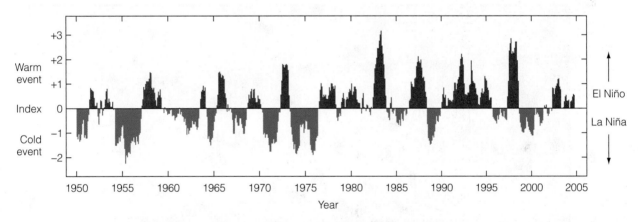

Figure 2.23 El Niño–La Niña events during the second half of the 20th century and the beginning of the 21st. The y-axis represents the ENSO index, which combines six factors: air temperature, surface water temperature, sea-level pressure, cloudiness, wind speed, and wind direction. Warm episodes are in red, cold in blue. Index values greater than +1 represents an El Niño, less than −1 a La Niña.

western Pacific the warmest ocean on Earth. These warm waters cause moist maritime air to rise and cool, bringing abundant rain to the region. In contrast, the cooler waters of the eastern Pacific produce relatively dry conditions along the Peruvian coast.

This describes the normal situation, but during an El Niño event, the trade winds slacken, reducing the westward flow of surface currents (Figure 2.22b). The result is a reduced upwelling, making the eastern Pacific warmer than usual. Rainfall follows the warm water eastward, with associated flooding in Peru and drought in Indonesia and Australia. This eastward displacement also changes global atmospheric circulation, influencing weather in regions as far away from the tropical Pacific as central Canada, where it typically has a warming effect.

Because of the widespread influences of ocean currents, ENSO fluctuations induce large upheavals in both the structure and function of marine food webs. They were a factor in the collapse of the overexploited sardine population along the Pacific

coast of North America in the 1970s (see Section 26.10) and have also played a role in the history of cod off Canada's east coast.

At other times, an event called **La Niña** occurs: an injection of cold water that is more intense than usual, causing a cooling of the eastern Pacific (Figure 2.23). La Niña causes droughts in South America, heavy rain in eastern Australia, and colder temperatures in North America.

2.11 MICROCLIMATES: Local Microclimates Define the Conditions Experienced by Individuals

Many organisms experience local conditions that do not match the general regional climate. Today's weather report may state that the temperature is 28°C under a clear sky. However, the forecaster is painting only a general picture. Actual conditions

ECOLOGICAL ISSUES   Urban Microclimates

Most of us inhabit cities and towns, where environmental conditions vary greatly from the surrounding area. In fact, cities create their own microclimates, with significant differences in temperature, precipitation, and wind flow compared with nearby rural areas. In general, urban microclimates increase energy use and lower air quality, with adverse effects on public health. These differences have sparked a distinct subdiscipline called *urban ecology*.

On warm summer days with little or no wind, urban air temperature can be significantly higher than in the surrounding countryside. These areas, known as **heat islands**, are warmer than surrounding environments because of their energy balance: the difference between the amounts of energy gained and lost. In rural environments, solar energy absorbed by vegetation and by soil is partially dissipated by evaporation from these same surfaces. Urban areas contain less vegetation, so buildings, streets, and sidewalks (which are generally non-reflective) absorb most of the solar energy. In areas with narrow streets and tall buildings, the building walls radiate heat towards one another instead of skyward.

Further, because manufactured surfaces of asphalt, cement, and brick are non-porous, most rain is lost as runoff to storm drains before evaporation can cool the air. Waste heat from cars, buses, city buildings, and industrial activity also increases the energy input. Although this waste heat eventually makes its way into the atmosphere, it can contribute as much as one-third of the heat received from solar energy. Adding to the problem are construction materials that are better heat conductors than is the vegetation that dominates the landscapes of surrounding rural areas. At night, these materials slowly give off heat that was stored during the day.

This heat island effect can raise temperatures 5°C to 8°C above those in the surrounding countryside. The phenomenon is worldwide, and Canadian cities are no exception. Yves Baudouin, a geographer at the University of Québec at Montréal, and his graduate student, Camilo Pérez Arrau, have been studying "micro-urban heat islands" (areas within a city with temperatures more than 5°C above the city's mean temperature—in effect, heat islands within a heat island). These "hot spots" are growing in Toronto and Montréal (Figure 1), but Vancouver experienced the largest increase from the 1980s to the present.

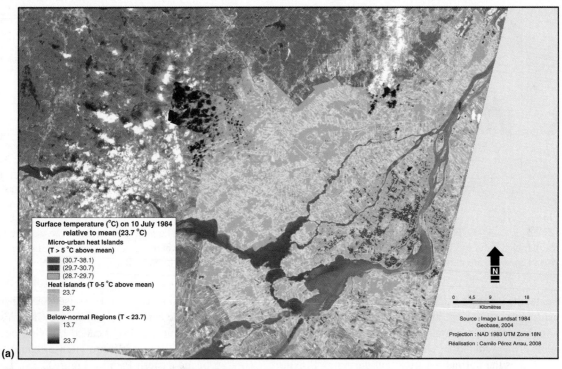

Figure 1   Micro-urban heat islands in Montréal in **(a)** 1984 and **(b)** 2003, relative to the July 10 mean surface temperatures (Figure 1b is on p. 35). Landsat images developed by Camilo Pérez Arrau.

at specific sites may differ considerably depending on whether they are underground or on the surface, beneath vegetation or on exposed soil, or on mountain slopes or at the seashore. As well as such short-term weather variation, light, heat, moisture, and air movement can all vary greatly between locations, influencing heat transfer and creating a wide range of localized climates or **microclimates**. Microclimates define the conditions in which organisms actually live (see Ecological Issues: Urban Microclimates, above).

On a sunny but chilly day in early spring, flies are attracted to sap oozing from the stump of a maple tree. The flies are active despite the near-freezing temperature because during the day, the stump's surface (especially if south facing) absorbs solar radiation, heating a thin layer of air at the surface. On a still day, this warm air remains close to the surface, with the temperature falling sharply above and below. A similar phenomenon occurs when frozen ground absorbs sunlight on a clear, late-winter day: the ground starts to thaw even though the air is cold.

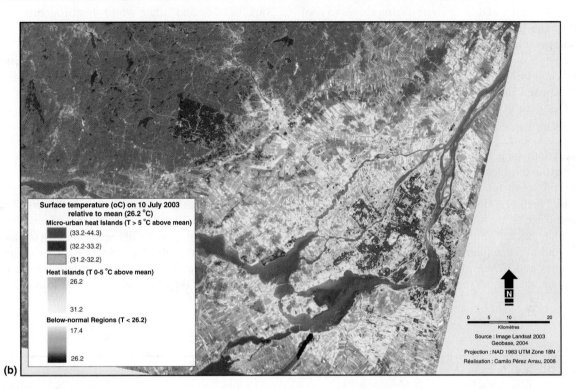

(b)

Figure 1 (*continued*)

Typically, the highest urban temperatures are in core areas of highest population density and activity, with temperatures declining towards the outskirts. However, patterns are far from uniform and vary within a city depending on land use. Ryerson University's Claus Rinner and Mushtaq Hussain (2011) report that the mean temperature of commercial and industrial sites in Toronto is significantly above that of parks and recreational areas. Toronto's wooded ravines—not typical of most cities—were associated with "temperature valleys." There is temporal variation too. The heat island effect is most pronounced during summer and early winter, particularly at night, when heat stored by pavement and buildings re-radiates into the air.

The heat island effect also affects air quality. Year-round, cities are blanketed with particulate matter and pollutants from fossil fuel combustion in cars and buildings and by industrial activity. Smog is created by photochemical reactions of atmospheric pollutants. At higher temperatures, these reactions occur at an increasing rate. In Los Angeles, for example, for every degree the temperature rises above 20°C, smog increases by 3 percent. Cold winters afford Canadians some relief, but according to Environment Canada, Toronto suffered 48 smog-alert days in 2005, the worst year on record.

Smog contains ozone, a pollutant that is harmful at elevated levels in the air we breathe, even though it protects us from UV radiation when naturally present in the stratosphere. This ground-level ozone also damages vegetation. The heat island effect exacerbates air pollution effects, as higher ambient temperatures during the summer months not only promote smog formation but also increase air-conditioning energy use. As power plants burn more fossil fuels, they increase both pollution levels and energy-related impacts. The urban microclimates created by humans thus generate a vicious cycle of energy use and declining air quality.

Bibliography

Rinner, C., and M. Hussain. 2011. Toronto's urban heat island—Exploring the relationship between land use and surface temperature. *Remote Sensing* 3:1251–1265.

1. What simple steps could be taken to decrease the heat island effect in urban areas?
2. How might the urban heat island effect influence surrounding rural environments?

Anyone who has walked into a forest on a summer day has experienced microclimate differences first-hand. By altering temperature, moisture, and wind movement, vegetation moderates and creates microclimates, especially near the ground. Shaded areas are cooler at ground level than are exposed areas. On a summer day in a spot 25 mm above-ground, dense forest cover can lower the daily temperature range by 7°C to 12°C compared with the soil temperature in adjacent open fields. We all know that forest interiors are less windy than open fields, but even under the shelter of heavy grass, air at ground level can be quite calm. This calm—typical of microclimates in densely vegetated sites—influences temperature and humidity, creating favourable conditions for insects and other ground animals.

Topography, particularly *aspect* (direction that a slope faces), also influences microclimate. In the Northern Hemisphere, south-facing slopes receive the most solar energy, north-facing slopes the least. At other aspects, energy varies between these extremes, with western exposures receiving more than eastern. In turn, differing exposure to radiation affects temperature and moisture. Microclimates range from warm, dry, and variable on south-facing slopes to cool, moist, and more uniform on north-facing slopes. Because warmer temperatures draw more moisture from soil and plants, evaporation on south-facing slopes is often 50 percent higher, and soil moisture is much lower. Conditions are driest on exposed tops of south-facing slopes, where air movement is greatest, and wettest at the base of north-facing slopes.

On a finer scale, microclimatic variations occur on north- and south-facing slopes of ant hills, soil mounds, dunes, and ground ridges in otherwise flat terrain, as well as on north- and south-facing sides of buildings, trees, and logs. In the Northern Hemisphere, buildings are always warmer and drier on their south sides—a consideration for landscape planners and gardeners. North sides of tree trunks are cooler and moister than south sides and support more vigorous moss and lichen growth, especially in drier forests. In winter, the temperature on the north side of a tree may be below freezing, while the south side, heated by the Sun, is warm. This temperature difference may cause frost cracks in the bark as sap, thawed by day, freezes at night. Bark beetles and other wood-dwelling insects that seek cool, moist areas for laying their eggs prefer such north-facing locations.

Microclimatic extremes also occur in depressions in the ground and on concave surfaces of valleys, protected from strong wind. Heated by sunlight during the day and cooled by vegetation at night, this air often becomes stagnant. These sites experience cooler nights in winter, warmer days in summer, and higher relative humidity. If the temperature drops low enough, frost pockets form. These microsites support different plant species than does the surrounding higher ground.

So, although global and regional patterns of climate are critical in the distribution and abundance of species and ecosystems, local microclimates define the actual conditions experienced by individuals. By affecting survival, growth, and reproduction, microclimates thus determine the distribution and activities of organisms in a particular locality.

EcologyPlace

Visit EcologyPlace at www.pearsoncanada.ca/ecologyplace to access online resources that complement your textbook, and help you to apply and to review the information in this chapter. EcologyPlace includes

- an eText version of the book
- self-grading quizzes
- glossary flashcards
- and more!

Go to www.pearsoncanada.ca/ecologyplace and follow the registration instructions on the Student Access Code Card included with this text. If your book does not have a Student Access Code Card, you can purchase access to it at www.pearsoncanada.ca/ecologyplace.

SUMMARY

Solar Radiation 2.1

Solar radiation is subject to various fates in the atmosphere, including reflection, absorption, and scattering. Insolation reaching Earth is enriched in photosynthetically active radiation (PAR). Earth emits energy back as longwave radiation, which cannot readily pass through the atmosphere and so returns to Earth, producing the greenhouse effect.

Radiation and Temperature 2.2

The amount of solar radiation intercepted varies with latitude. Tropical regions receive the most, high latitudes the least. The tilt of Earth's axis causes seasonal differences in radiation, which cause seasonal variation in temperature and rainfall. The mean temperature declines from the tropics to the poles.

Elevation and Temperature 2.3

Temperature declines with elevation in the lower atmosphere because of decreasing pressure and increasing distance from Earth's surface. Adiabatic cooling involves expansion caused by declining pressure rather than heat transfer.

Atmospheric Circulation 2.4

Vertical movements of air masses generate global air circulation patterns. Earth's spin causes the Coriolis effect, which deflects air currents to the right in the Northern Hemisphere and to the left in the Southern Hemisphere. Three cells of global air flow occur in each hemisphere.

Ocean Currents 2.5

Global wind patterns and the Coriolis effect cause ocean currents. Currents give rise to large circular patterns (gyres), which move clockwise in the Northern Hemisphere and counterclockwise in the Southern Hemisphere.

Humidity 2.6

The maximum moisture that air can hold at a given temperature is the saturation vapour pressure, which increases with temperature. Relative humidity is the amount of water in the air expressed as a percentage of the maximum amount the air could hold at a given temperature.

Precipitation 2.7

Wind, temperature, and currents produce global precipitation patterns, including the high rainfall of the tropics associated with the intertropical convergence zone, and dry belts at ~ 30° N and S.

Snow and Ice 2.8

As well as being part of total precipitation, snow and ice have many ecological effects. Snow recharges soil moisture and affects frost penetration by its insulating properties. Severe events, such as hailstorms and ice storms, can cause great damage. On a global scale, changes in polar ice caps are associated with global cooling or global warming.

Effects of Topography 2.9

Mountainous topography influences precipitation patterns. As an air mass ascends a mountain, it cools, becomes saturated, and releases much of its moisture at upper altitudes on the windward side. A dry rain shadow forms on the leeward side.

Irregular Climatic Variation 2.10

Earth has experienced large irregular climatic fluctuations at different timescales, such as the Ice Ages. On a shorter timescale, irregular trade winds give rise to periods of unusually warm waters off the coast of western South America. ENSO is a global phenomenon arising from interactions between oceans and the atmosphere.

Microclimates 2.11

The actual climatic conditions that organisms experience vary within a region. Local microclimates reflect the impact of topography, vegetative cover, exposure, and other factors at all spatial scales. Angles of solar radiation cause marked differences between north- and south-facing slopes, whether on mountains, sand dunes, or ant mounds.

KEY TERMS

adiabatic lapse rate	El Niño–Southern	insolation	photosynthetically	solar radiation
albedo	Oscillation	intertropical	active radiation	topography
altitude	elevation	convergence	precipitation	ultraviolet radiation
atmospheric (air)	evaporation	zone	rain shadow	vapour pressure
pressure	greenhouse effect	La Niña	relative humidity	weather
climate	gyre	latent heat	saturation vapour	
Coriolis effect	habitability	longwave (thermal)	pressure	
current	heat island	radiation	shortwave radiation	
dew point temperature	infrared radiation	microclimate	solar constant	

STUDY QUESTIONS

1. Why does the equator receive more solar radiation than the polar regions? What is the consequence of latitudinal radiation patterns?
2. What are the different fates of solar radiation as it approaches Earth? Why is insolation "enriched" in photosynthetically active radiation (PAR)?
3. What is the greenhouse effect, and how does it influence the energy balance (and hence the temperature) of Earth?
4. The 23.5° tilt of Earth gives rise to the seasons (review Figure 2.6). How would the pattern of seasons differ if the tilt were 90°? How would this affect the diurnal (night/day) cycle?
5. At noon at the same location, the air temperature was 5°C on January 20 and 25°C on July 20. The relative humidity on both days was 75 percent. On which day was there more water vapour in the air, and why?
6. How and why might the relative humidity of an air parcel change as it moves up a mountain? What are the consequences of these changes for precipitation?

7. What is the intertropical convergence zone (ITCZ), and how does it explain seasonal precipitation patterns in the tropics?

8. Regina is at a very similar latitude (50° N) as is London, England (51° N). Why is Regina's climate so much (i) colder and (ii) drier?

9. How does snowfall (amount and timing) affect an ecosystem, even if vegetation is dormant when it falls?

10. What feature of global atmospheric circulation causes the deserts of the mid-latitudes?

11. The Sifton Bog near London, Ontario, is dominated by *Sphagnum* mosses and black spruce (*Picea mariana*), which typically grow in colder sites farther north. What does the spruce presence suggest about this site? What non-climatic factors might be involved?

FURTHER READINGS

Ahrens, C. D. 2003. *Meteorology today: An introduction to weather, climate, and the environment.* 6th ed. Belmont, CA: Brooks/Cole.

Excellent, clearly written, and well-illustrated introductory text on climate.

Fagen, B. 2001. *The Little Ice Age: How climate made history, 1300–1850.* New York: Basic Books.

Enjoyable account of the effects of the Little Ice Age on human history.

Graedel, T. E., and P. J. Crutzen. 1997. *Atmosphere, climate and change.* New York: Scientific American Library.

Introduction to climate written for the general public, stressing air pollution and climate change.

Jones, H. G., J. W. Pomeroy, D. A. Walker, and R. W. Hoham (eds.). 2001. *Snow ecology: an interdisciplinary examination of snow-covered ecosystems.* Cambridge: Cambridge University Press.

Comprehensive treatment of all aspects of snow ecology.

Philander, G. 1989. El Niño and La Niña. *American Scientist* 77:451–459.

Good introduction to the El Niño–La Niña phenomenon.

A rainstorm over the ocean—part of the water cycle.

Water is the essential substance of life, the major component of all living organisms. Some 75 to 95 percent of the mass of living cells is water, and there is hardly a physiological process in which it is not fundamentally important. Covering some 75 percent of the planet's surface, water is also the dominant habitat on Earth. Because *salinity* (salt concentration) is a major influence on the adaptations of aquatic organisms, aquatic systems are divided into **marine** (saltwater) and **freshwater**. These categories are further divided into various ecosystems based on water depth and flow, substrate, and type of dominant organisms (typically plants). In Chapter 24, we explore the diversity of aquatic environments. Here, we consider how the unique properties of water interact to define different aquatic environments and constrain the evolution of the organisms occupying them. Chapters 6 and 7 consider specific adaptations to life in aquatic habitats.

3.1 PROPERTIES OF WATER: The Unique Properties of Water Determine Its Ecological Importance

Water is so familiar to us that we take it for granted, but the physical arrangement of its atoms makes water a unique substance. A molecule of water (H_2O) consists of two hydrogen (H) atoms joined to one oxygen (O) atom by asymmetric covalent bonds (Figure 3.1a). The two H atoms occupy one end of the molecule, which is positively charged, and the O atom occupies the other, negatively charged, end (Figure 3.1b). Though uncharged overall, water is thus a *polar* molecule. As a result of water's polarity, each molecule binds weakly with its neighbours such that the positive (H) end of one molecule attracts the negative (O) end of another (Figure 3.1c). The angle between H atoms results in an open, tetrahedral arrangement (Figure 3.1d). This situation, where H atoms act as connecting links between molecules, is called *hydrogen bonding*. Compared with the *polar covalent bonds* between H and O, hydrogen bonds are weak and easily broken and reformed.

An important property of water that is related to its hydrogen bonds is its high **specific heat**—the number of calories needed to raise 1 g of water by 1°C. The specific heat of water is assigned a value of 1, with all other substances given a value relative to water. Water's high specific heat means that large amounts of energy must be absorbed before the temperature of ponds and lakes rises just 1°C. These waters warm up slowly in spring and cool off just as slowly in fall. This property has a vital moderating effect, preventing the wide seasonal temperature flux so typical of continental land habitats. Water's high specific heat is also vital for thermal regulation of organisms. Because water makes up so much of the mass of living cells, any change in body temperature is moderated relative to the change in ambient temperature.

As a result of water's high specific heat, large amounts of energy are absorbed or released when it changes state (*latent heat*). Removing 1 cal of energy lowers the temperature of 1 g of water from 2°C to 1°C, but about 80 times as much energy is needed to convert that same 1 g of water at 1°C to its freezing point (0°C). Likewise, it takes 536 cal to overcome the attraction between molecules and convert 1 g of water at its boiling point (100°C) to vapour.

The lattice arrangement of its molecules gives water an unusual density–temperature relationship. Most liquids become denser as they cool. At their freezing temperature they enter the solid phase, which is denser than the liquid. This description is not true for water. Pure water does become denser as it is cooled to 4°C, but if cooled further, its density decreases (Figure 3.2). At 0°C, freezing occurs and the lattice structure is complete—each O atom is connected to four other O atoms by H atoms. The result is a large, open lattice of greatly decreased density (Figure 3.1e). Because water molecules occupy more space when frozen, ice floats on liquid water.

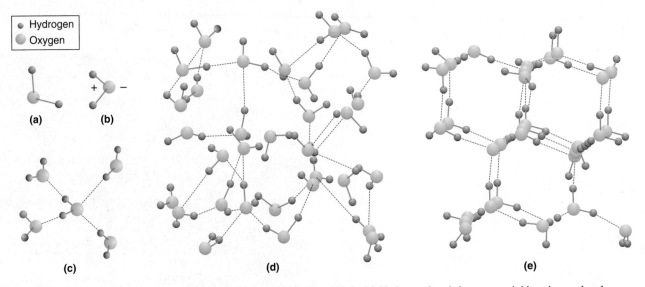

Figure 3.1 Water structure. **(a)** A single H_2O molecule. **(b)** Polarity of H_2O. **(c)** Hydrogen bonds between neighbouring molecules. **(d)** Structure of liquid H_2O. **(e)** Open lattice structure of ice.

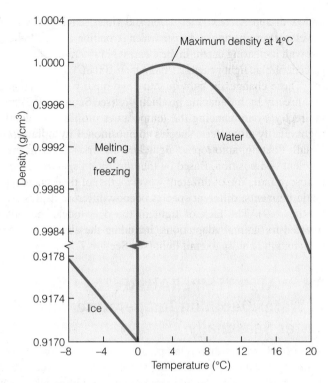

Figure 3.2 Density of pure water as a function of temperature. The maximum density is at 4°C, declining sharply as water changes from liquid to solid (ice).

This property is crucial to life in aquatic habitats. Surface ice insulates, helping to keep larger water bodies from freezing solid. Very large lakes, such as the Great Lakes, may not completely freeze over in winter. In a normal year, the surface of Lake Erie, which is smaller and shallower, freezes to a greater extent (80–90 percent) than Lake Ontario (20–25 percent).

As a result of hydrogen bonding, water molecules exhibit **cohesion**: the tendency to stick together, resisting external forces that would pull them apart. Water's cohesiveness (along with its ability to adhere to polar surfaces) allows a column of water to rise to the top of a tall tree, even when subject to con-

siderable suction (tension) from transpiration. In a water body, these cohesive forces are similar on all sides, but at the surface conditions differ. Water exhibits **surface tension**—a taut surface resulting from the stronger attraction of water molecules to each other than to the air above. This property is vital to aquatic organisms. It allows a pond to support small objects and animals, such as water striders (*Gerridae* spp.) and water spiders (*Dolomedes* spp.), that run across its surface (Figure 3.3). To other small organisms, surface tension is a barrier, whether they want to penetrate the water below or escape into the air above. For some, the surface tension is too great to break; for others, it is a trap to avoid while skimming the surface to feed or lay eggs. If caught by surface tension, a small insect may flounder. Nymphs of mayflies (*Ephemeroptera* spp.) and caddisflies (*Trichoptera* spp.) that live in water and transform into winged adults are hampered by surface tension when emerging. While slowed down at the surface, they become easy prey for fish.

Cohesion is also responsible for water's **viscosity**—the ability of a fluid to resist a force necessary to separate its molecules and cause it to flow or allow an object to pass through it. The high viscosity of water is largely a consequence of its density, which is some 860 times that of air. It greatly increases the frictional resistance for objects moving through water instead of air. The streamlined body shape of most fish and marine mammals reduces their resistance. Replacement of water in the space left behind by a moving animal increases drag on the body. An animal streamlined in reverse, with a short, rounded front and a rapidly tapering body, such as the sperm whale (*Physeter macrocephalus*; Figure 3.4), meets the least resistance.

Although water's viscosity limits the mobility of aquatic organisms, it also benefits them. If a submerged body weighs less than the water it displaces, it is subject to **buoyancy**—the ability of a liquid to exert an upward force on a body placed in it. Because most aquatic organisms are similar in density to water, they are close to neutral buoyancy and do not require structural material, such as skeletons or cell walls, to resist gravity. In addition, swimming requires much less energy than terrestrial animals need to raise their mass against gravity on land.

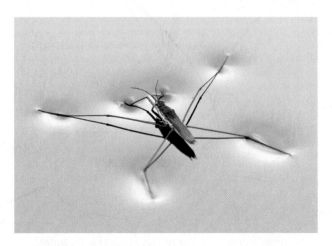

Figure 3.3 Surface tension allows the water strider (*Gerris remigis*) to glide across a pond.

Figure 3.4 The sperm whale (*Physeter macrocephalus*) is streamlined in reverse. Its short, rounded front and tapering body lower frictional resistance.

Water's high density exerts profound constraints on deep-sea organisms. Because of its greater density, water undergoes greater changes in pressure with depth than does air. At sea level, the weight of the column of air from the top of the atmosphere to the sea surface is 1 kg/cm^2, or 1 atmosphere (atm; 1 atm = 0.1 megapascals). Pressure increases 1 atm for each 10 m in depth. Ocean depth varies from a few hundred metres to more than 10 000 m in deep ocean trenches, so pressure at the seafloor varies from 20 atm to more than 1000 atm. Proteins and membranes are pressure sensitive and must be modified in organisms occupying the deep ocean. We explore the unique challenges of deep-water habitats in Research in Ecology: The Extreme Environment of Hydrothermal Vents (pp. 46–47).

3.2 LIGHT IN WATER: Light Quantity and Quality Vary with Water Depth

In Section 2.1, we discussed the fate of light in the atmosphere. Similar losses occur when light passes through water or through vegetation (see Section 4.2). In fact, light is much more strongly *attenuated* (reduced) in water than in air, with important ecological consequences.

When light strikes a water body, a portion is reflected back to the atmosphere, depending on the angle of incidence—the lower the angle, the more light is reflected. Thus, the amount of light reflected from water varies daily and seasonally from the equator to the poles. Light entering water is reduced further by two processes: (1) Suspended particles, both alive and dead, intercept light and either absorb or scatter it. Scattering of light increases its path through water, causing more attenuation. (2) Water itself absorbs light (Figure 3.5a), with some wavelengths absorbed more than others (Figure 3.5b). Visible red and near-infrared radiation are absorbed first, reducing the total energy by about half.

Yellow disappears next, then green and violet, leaving only blue wavelengths to penetrate deeper water. A portion of blue light is lost with increasing depth. In the clearest seawater, only about 10 percent of blue light penetrates more than 100 m.

These changes in light quantity and quality affect aquatic life directly by influencing productivity (see Section 20.3) and indirectly by influencing the temperature profile with depth. Light quality also affects species composition. **Phytoplankton** (small, floating, autotrophic algae and bacteria) differ in their pigment composition. Based on the *absorption spectra* (ability to absorb radiation of different wavelengths) of their photosynthetic pigments, different species occupy different depths (see Section 6.9). The lack of light in the deep ocean has also selected for animal adaptations, including the absence of pigment and the ability to emit light (see Section 7.15).

3.3 WATER TEMPERATURE: The Effect of Water Depth on Temperature Varies Seasonally

As on land, water surface temperature reflects the balance of incoming and outgoing radiation. As solar radiation is absorbed in the water column, the temperature profile with depth might be expected to resemble that of light, as shown in Figure 3.5, that is, to decrease exponentially. However, the high density of water modifies this pattern.

As light is absorbed, surface waters heat up. Wind and waves mix surface waters, distributing heat vertically. The decline in temperature with depth thus lags behind the decline in solar radiation. Below this top layer, temperatures drop rapidly in the transition zone called the **thermocline** (Figure 3.6a). The depth of the thermocline depends on radiation input and on mixing by wind and waves. Below the thermocline, temperature declines

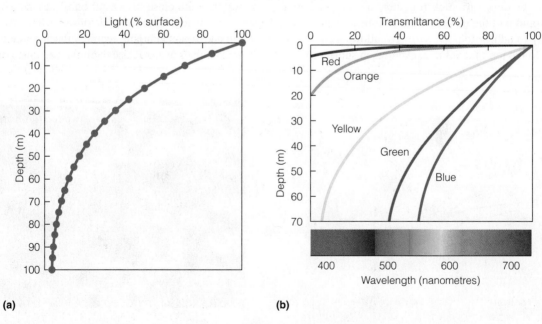

(a)   **(b)**

Figure 3.5 Behaviour of light in water. **(a)** Light attenuates with depth in pure water, expressed as a percentage of light at the surface. Estimates assume a light extinction coefficient of $k_w = 0.035$ (see Quantifying Ecology 4.1: Measuring Light Attenuation, p. 61). **(b)** Passage through water modifies the spectral distribution. Red light is attenuated more rapidly than green and blue.

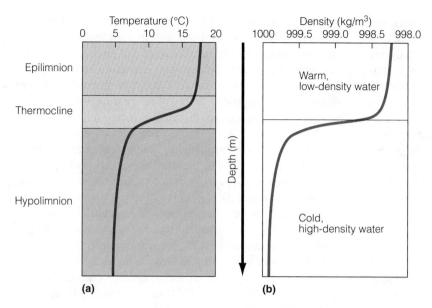

Figure 3.6 Temperature and density profiles of an open water body. **(a)** Vertical mixing of surface waters transports heat to lower waters. Below the epilimnion, temperatures decline rapidly in the thermocline. In the hypolimnion, temperatures drop more slowly. **(b)** The rapid temperature decline in the thermocline causes a sharp difference in density between the warmer, low-density epilimnion and the cooler, high-density hypolimnion.

further with depth but at a slower rate, creating a distinct thermal zonation. The warm, well-mixed surface layer above the thermocline is the **epilimnion**. Its density is much lower than that of the **hypolimnion**, the deeper layer of cold, denser water (Figure 3.6b). This density change at the thermocline is a barrier that prevents mixing of the epilimnion and the hypolimnion.

Seasonal variation in radiation generates seasonal changes not only in temperature on land but also in the vertical temperature profile of water. In tropical waters, the relatively constant radiation year-round makes the thermocline a permanent feature, whereas a distinct thermocline exists only during summer in temperate waters. In fall, as radiation and air temperature decrease, surface waters cool, become denser, and sink, displacing warmer water below to the surface, where it cools in turn (Figure 3.7). As

the density difference between the epilimnion and the hypolimnion decreases, winds mix the profile to greater depths. This mixing continues until the temperature is uniform, allowing water to circulate throughout the pond or lake. This vertical circulation, called **fall turnover**, is important for nutrient dynamics in open-water systems. Stirred by wind, vertical mixing continues until ice forms at the surface.

Then comes winter. If the surface water cools below 4°C, it becomes lighter again and remains on the surface. If the winter is cold enough, the surface water freezes; otherwise, it remains close to 0°C. Now the warmest place in the water body is on the bottom. In spring, ice breakup can cause significant mechanical damage to shoreline vegetation (Figure 3.8). Following the thaw, heating of surface water with increasing solar radiation again causes the water to stratify.

Seasonal turnover is not typical of oceans. The thermocline simply descends or rises seasonally, and bottom waters never mix with the top layer. In deep lakes, such as Lake Huron, turnover occurs once annually, in fall. In shallow lakes and ponds, multiple temporary stratifications of short duration may occur if the depth is sufficient. In windy areas, surface mixing can prevent a thermocline from developing even in lakes as deep as 25 m. However, some form of thermal stratification occurs in all open water bodies.

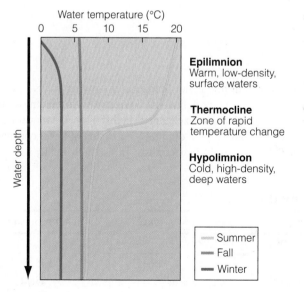

Figure 3.7 Seasonal temperature changes with depth for an open water body. In fall, surface water cools and sinks until temperature is uniform with depth. With the onset of winter, surface water cools further and ice may form. In spring, the process reverses and a thermocline re-forms.

Figure 3.8 Spring ice breakup damages vegetation on riverbanks and lake shorelines.

The temperature of flowing water systems is more variable. Temperatures of small, shallow streams tend to track, but lag behind, air temperatures. They warm and cool with the seasons but fall below freezing only in very cold regions, and not even then if flow rates are high. In summer, streams with large areas exposed to sunlight are warmer than those shaded by vegetation and high banks. Temperature affects the stream community, influencing the presence of cool-water and warm-water organisms. Dominant predatory fish shift from species such as trout and smallmouth bass, which require cooler water with more O_2, to species such as suckers and catfish, which require warmer water and tolerate lower O_2 levels.

3.4 WATER AS A SOLVENT: Many Substances Dissolve Readily in Water

A spoonful of sugar dissolves in a glass of water, forming a **solution**: a homogeneous liquid mixture of two or more substances. The dissolving agent is the *solvent*, and the substance dissolved is the *solute*. A solution in which water is the solvent is an *aqueous solution*. Water is an excellent solvent, capable of dissolving more substances than any other liquid. This ability makes water biologically crucial. Besides its role in thermal regulation, water dissolves and transports nutrients and wastes, both inside and outside organisms, and maintains chemical equilibrium in living cells.

Like many properties of water, its solvent ability reflects its polarity. Because the H atoms are bonded to the O atoms asymmetrically (see Figure 3.1a), the opposite sides of water molecules have positive and negative charges. Opposite charges attract, so water molecules are attracted not only to each other but also to other polar molecules and to **ions**: electrically charged atoms or groups of atoms. Sodium chloride (table salt), for example, is composed of a crystal lattice of sodium ions (Na^+) and chloride ions (Cl^-). The attractions between the negative and positive charges of water molecules and those of the Na^+ and Cl^- ions are greater than the ionic bonds holding the salt crystals together. Thus, salt crystals readily dissociate in water: they dissolve.

The solvent properties of water explain the presence of most minerals in aquatic habitats. When water condenses into clouds, it is nearly pure except for dissolved gases. In falling as precipitation, water acquires substances from particulates suspended in air. Water falling on land flows over the surface and percolates into the soil, picking up more solutes. Streams and rivers accumulate solutes from the materials they flow through or over. Most rivers and lakes contain 0.01 to 0.02 percent dissolved minerals, with relative concentrations reflecting their substrates. For example, waters that flow through limestone formations (composed primarily of calcium carbonate, $CaCO_3$) have high concentrations of calcium (Ca^{2+}) and bicarbonate (HCO_3^-).

Oceans have a much higher solute concentration, given evaporation of pure water from the surface. In effect, the ocean acts as a giant still. Although inflow of freshwater dilutes oceanic solute concentration, it also brings in more minerals. When the concentration of specific elements reaches the limit set by the solubility of the compounds they form, excess amounts precipitate out in sediments. Calcium, for example, readily forms $CaCO_3$ in ocean waters. But the maximum solubility of $CaCO_3$ is $0.014 \text{ g} \cdot L^{-1}$ of water, a concentration reached early in the history of the oceans. Thus, Ca^+ ions are deposited in sediments on the ocean floor.

In contrast, the solubility of sodium chloride (NaCl) is very high ($360 \text{ g} \cdot L^{-1}$). In fact, these two elements, Na^+ and Cl^-, make up some 86 percent of sea salt, with the remainder being small amounts of other elements, such as sulphur, magnesium, potassium, and calcium (Table 3.1). Determination of the most abundant element, chlorine, is used as an index of **salinity** (salt concentration). Salinity is expressed in *practical salinity units (psu)*, represented as parts per thousand (‰) and measured as grams of Cl per kilogram of H_2O. Oceanic salinity is fairly constant (~ 35‰), but over millions of years it has continued to increase. The salinity of freshwater is more variable, ranging from 0.065‰ to 0.30‰. Regulating water and solute content of their tissues is a major challenge for organisms in both marine and freshwater habitats.

3.5 OXYGEN IN WATER: Abiotic and Biotic Factors Affect the O_2 Content of Water

Water's solvent role is not limited to dissolving solids. The surface of a water body defines its boundary with the atmosphere, across which gases are exchanged via **diffusion**: the movement of molecules from a region of higher to a region of lower concentration. Diffusion results in net transfer of two biologically important

Table 3.1 Composition of Seawater of 35 Practical Salinity Units (psu)

Element	g/kg	Millimoles/kg	Element	g/kg	Millimoles/kg
Cations			*Anions*		
Sodium	10.75	467.56	Chlorine	19.35	545.59
Magnesium	1.30	53.25	Sulphate	2.70	28.12
Calcium	0.42	10.38	Bicarbonate	0.15	2.38
Potassium	0.40	10.10	Bromine	0.07	0.83
Strontium	0.008	0.09	Boric acid	0.03	0.44
			Fluorine	0.001	0.07

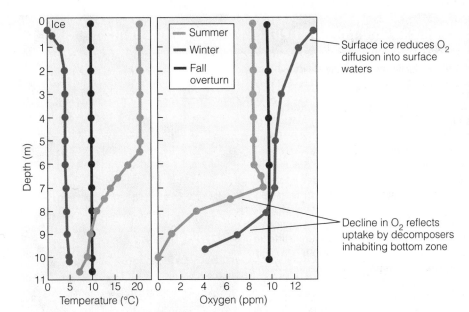

Figure 3.9 Seasonal O_2 changes in a northern lake. Fall turnover causes uniform distribution of O_2. In summer, stratification of both temperature and O_2 occurs. O_2 declines sharply in the thermocline and is non-existent in sediments because of decomposer uptake. In winter, O_2 is also stratified, but its concentration is low in deep water.

(Adapted from Likens 1985.)

gases, oxygen (O_2) and carbon dioxide (CO_2), from the atmosphere, where they are present in higher concentration, into surface waters. The diffusion rate of O_2 is controlled by its solubility in water and the steepness of the gradient in O_2 content between the air and surface waters. The solubility of gases in water is a function of temperature, pressure, and salinity. The saturation value of O_2 is greater in cold water because the solubility of a gas decreases as water temperature rises. Solubility also increases with atmospheric pressure and decreases with salinity, which is significant in marine but not in freshwater systems.

Once O_2 enters surface waters, it continues diffusing into the waters below, where its concentration is lower. Water, with its greater density and viscosity relative to air, limits how quickly gases diffuse through it. In fact, gases diffuse some 10 000 times as fast in air as in water. In addition to diffusion, turbulence and currents mix the O_2 absorbed into surface waters with deeper water. In shallow, rapidly flowing water and in wind-driven sprays, O_2 may even reach supersaturated levels because of the increase in absorptive surfaces at the air–water interface. O_2 is lost from water as the temperature rises (lowering solubility), as well as by biotic uptake.

Seasonal changes in O_2 reflect both abiotic and biotic factors. In summer, O_2, like temperature, may become stratified in water bodies. O_2 content is greatest near the surface, where interchange between water and air, stimulated by wind, occurs (Figure 3.9). Besides entering water by diffusion, O_2 is also released in photosynthesis, which is largely restricted to shallow waters by light limitations. Unlike temperature, O_2 levels also decrease with depth because of the O_2 demand of decomposers inhabiting bottom sediments. During seasonal turnovers, when water circulates throughout, O_2 levels are replenished in deep water. In winter, the reduction of O_2 in unfrozen water is slight, for two reasons: (1) solubility increases, and (2) O_2 demand is reduced in the cold. Under ice, O_2 depletion may be severe because diffusion from the air has ceased while demands continue.

As with lakes, O_2 is not distributed uniformly in oceans (Figure 3.10). A typical profile shows maximum amounts in

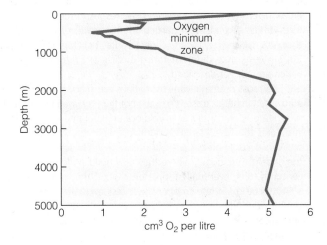

Figure 3.10 Vertical O_2 profile in the tropical Atlantic. O_2 declines to the minimum zone, below which it increases because of the influx of cold O_2-rich waters that sank in polar waters.

the upper 10 to 20 m, where photosynthetic activity and diffusion from the atmosphere may cause saturation. At greater depth, O_2 declines. In open waters, its concentration reaches a minimum at 500 to 1000 m, called the **oxygen minimum zone**. Unlike in lakes, where seasonal breakdown of the thermocline leads to a dynamic gradient of temperature and O_2, the limited depth of surface mixing in the oceans maintains an O_2 gradient year-round.

The availability of O_2 in flowing water is quite different. The constant churning of stream water over riffles and falls increases contact with air, often raising O_2 to saturation levels for the prevailing temperature. Only in deep holes or polluted waters does dissolved O_2 decline significantly. However, the supply of O_2 even at saturation is often limiting to organisms, whereas it is rarely so on land. Compared with its 21 percent content in the atmosphere, O_2 reaches a maximum solubility of 1 percent in freshwater at 0°C, and much less in warmer water. As a result, O_2 often limits metabolic activity in aquatic habitats.

RESEARCH IN ECOLOGY — The Extreme Environment of Hydrothermal Vents

Verena Tunnicliffe, University of Victoria.

We think of oceans as stable habitats, where the unique properties of water shield organisms from environmental extremes. Yet there are some deep-sea habitats, called *hydrothermal vents,* where conditions are as extreme as anywhere on Earth. They support a diverse community of species, with an energy base unlike that of other ecosystems. For two decades, the University of Victoria's Verena Tunnicliffe has been one of the pioneers in the study of deep-sea vents.

What are conditions like in these unique habitats? How do these conditions influence the structure and function of the ecosystems they support? Unknown until the discovery in 1977 of the East Pacific Rise, deep-sea **hydrothermal vents** form where geothermal heat is released from beneath the seafloor through cracks associated with volcanic activity or tectonic plates. (Land examples include geysers and hot springs.) In oceans, they occur along tectonic ridges, where plates are either diverging (called *spreading centres,* as in the East Pacific Rise, which spreads at a particularly fast rate) or converging (*back-arc spreading,* as in the western Pacific) (Tunnicliffe et al. 2003). Vents take different forms in different ridges, many of which are not yet described.

The most dramatic vent formations are *black smokers,* created when minerals dissolved in the hot fluid released by the vent precipitate upon contact with cold ocean water into chimney-like structures that spew large amounts of sulphide. Although black smokers occur in areas with less intense tectonic activity (such as the mid-Atlantic ridge), where they may build up for decades or centuries, they eventually topple over when they reach heights of 40 to 60 m. *White smokers* have less extreme temperature profiles and release lighter-coloured minerals, such as calcium and silicon. *Cold seeps* occur along continental margins, where thermal gradients are minimal.

Just how hot does it get in a black smoker? The maximum temperature of high-velocity jets emerging directly through sealed conduits (Figure 1) is 400°C. Above this temperature, phase separation occurs, causing boiling in shallow sites. In deeper water, these jets precipitate out, adding to the growing chimney. The water cools rapidly away from the jet, and response of organisms to this thermal gradient (which can differ by an order of magnitude within a few centimetres) is critical. Although early reports suggested vent microorganisms could grow at temperatures as high as 250°C, current evidence indicates an upper lethal limit for **hyperthermophilic** ("high-temperature loving") bacteria of 90°C to 115°C (Tunnicliffe et al. 2003), with most activity in cooler regions. Less extreme temperatures from 2°C (ambient ocean temperature) to 100°C are typical of flows that have been thermally diluted by moving through cracks in the seafloor.

How do vent organisms cope with extreme heat? In general, vent animals are more responsive to thermal gradients than are other aquatic invertebrates. They do have higher upper limits, but by actively seeking cooler regions, they limit the risks of rapid temperature flux of hydrothermal fluids, as well as the associated toxicity (Bates et al. 2010).

Conditions other than temperature are harsh, too. Substrates are extremely hard; water pressure is great; flow rates are high; and acidity is severe, with pH as low as 3. Even moderate pH values of 5.4 to 7.3 damage metabolism and shell deposition of molluscs on volcanic arcs (Tunnicliffe et al. 2009). On land, such values would not be considered acidic, but oceanic pH is generally around 8, so lower values can be quite harmful. Like temperature, most of these and other abiotic factors fluctuate spatially (with distance from the vent source) and temporally (with time since release), creating a complex and changeable dynamic of microhabitats in space and time.

An abiotic factor of these habitats that correlates with temperature but is arguably even more critical is their reducing nature. Conditions vary from completely **anoxic** (O_2-absent) waters in which reducing substances accumulate and that only anaerobic prokaryotes can tolerate, to **hypoxic** (O_2-deficient) waters permitting limited aerobic activity. Although anoxic conditions can prevail in hydrothermal basins, most vent habitats have a gradient of reduc-

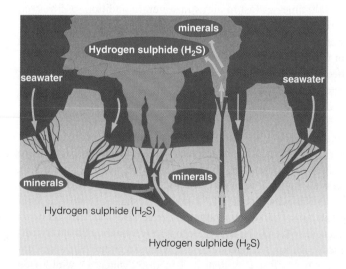

Figure 1 Simplified representation of a hydrothermal vent. Heat escapes both directly, through sealed conduits, and after moving through cracks in the seafloor. Sulphides in the water jets precipitate out to form the black smoker.

(Adapted from Tunnicliffe 1992.)

ing conditions. O_2 is always limited and continually scavenged, but mixing can sustain some aerobic life even below the seafloor. Types of reducing compounds vary, but black smokers are dominated by hydrogen sulphide (H_2S) in micromole (μM) to millimole (mM) concentrations. Methane (CH_4) dominates at cold seeps, where it is of inorganic origin. It also occurs at black smokers, but at lower concentrations and often of organic origin, as a result of the presence of *methanogenic* (CH_4-producing) bacteria. Vent fluids contain H_2, Fe^{2+}, and Mn^{2+}, all of which support chemosynthesis.

Herein lies the groundbreaking ecological importance of hydrothermal vents. For decades, long before these habitats were explored, ecologists presumed that any organisms occupying the deep-sea floor, far below the depth of light penetration, must subsist on organic matter settling down from consumers of photosynthesis-based food webs. And some do; the dead bodies of large consumers, such as whales, form the basis of complex food webs that also involve chemosynthetic bacteria oxidizing reduced organic compounds released during decomposition. But not all deep-sea habitats are light based. In hydrothermal vents, chemosynthetic bacteria (both anaerobic and aerobic) generate ATP by oxidizing compounds of inorganic origin. Because most of the sulphides present result from the reaction of seawater with core materials under heat and pressure, deep-sea vents are powered not by the Sun through photosynthesis, but by the heat of Earth's core through **chemosynthesis**, or more specifically *chemolithoautotrophy* because of the integral role of Earth. Thus, despite being far removed from light, vent habitats provide abundant energy—but only for those organisms that can tolerate the harsh conditions.

Deep-sea vent communities do not consist only of chemosynthetic bacteria. Copepods and amphipods feed on them directly, and in turn support larger invertebrates, including tubeworms, clams, mussels, and shrimp. Vertebrates are rare, though eels and flatfish can be present. Many vent invertebrates live nowhere else—several hundred new species have been identified to date, of which Tunnicliffe's lab has been involved in the discovery of over 75—and many have adaptations that are habitat specific. Among the most remarkable are giant tubeworms, such as *Riftia pachyptila* (Figure 2), which have mutualistic symbioses with chemosynthetic bacteria. Lacking a mouth or digestive tract, these tubeworms have red plumes of hemoglobin, which transfers both H_2S and O_2 to aerobic chemosynthetic bacteria harboured in a specialized organ called a *trophosome*.

Mutualistic interactions are common in other vent invertebrates, but their nature varies. Vent clams also have reduced digestive tracts, but they house their symbionts in large modified gills. Vent mussels are less specialized; they have less reduced digestive tracts and lack blood proteins for binding sulphides and O_2. Some vent mussels house both sulphide- and methane-oxidizing bacteria in the same cell (Fisher 1997). As scientists learn more about vent systems, the focus is shifting from identifying their species composition to understanding their physiological adaptations and community dynamics, including succession and dispersal.

Although their food webs are not light based, vent communities are not wholly independent of photosynthetic communities. Some chemosynthetic bacteria are anaerobic, but others are aerobic and need O_2 to oxidize H_2S and other reduced compounds. Vent consumers are also aerobic, so they too depend on O_2 generated by photosynthesis. Moreover, although the energy produced by hydrothermal vents is a minute fraction of that generated by light-driven aquatic systems, the ecological significance of vents may far exceed their present contribution to Earth's energy budget. Indeed, life may have begun at hydrothermal vents (Corliss et al. 1981).

The more scientists such as Tunnicliffe discover about the environment of deep-sea vents, as well as biotic responses to it, the more we comprehend the crucial impact of abiotic factors not only on existing communities but also on the emergence of life—and perhaps on its future on this planet.

Bibliography

Bates, A. E., R. W. Lee, V. Tunnicliffe, and M. D. Lamare. 2010. Deep-sea hydrothermal vent animals seek cool fluids in a highly variable thermal environment. *Nature Communications* 1:14.

Corliss, J. B., J. A. Baross, and E. E. Hoffman. 1981. An hypothesis concerning the relationship between submarine hot springs and the origin of life on Earth. *Oceanological Acta* 1:59–69.

Fisher, C. R. 1997. Ecophysiology of primary production at deep-sea vents and seeps. Pages 313–336 in Uiblein, F. J., and M. Stachowitisch (eds.). *Deep-sea and extreme shallow-waters habitats: Affinities and adaptations*. Biosystematics and Ecology, Series 11.

Tunnicliffe, V. 1992. Hydrothermal vent biota: Who are they and how did they get there? *American Scientist* 80:336–349.

Tunnicliffe, V., S. K. Juniper, and M. Sibuet. 2003. Reducing environments on the deepsea floor. Pages 81–110 in Tyler, P. A. (ed.). *Ecosystems of the world: The deep sea*. Amsterdam: Elsevier.

Tunnicliffe, V., K. T. Davies, D. A. Butterfield, R. W. Embley, J. M. Rose, and W. W. Chadwick, Jr. 2009. Survival of mussels in extremely acidic waters on a submarine volcano. *Nature Geosciences* 2:344–348.

Figure 2 Colony of giant tubeworms with vent fish and crabs, all specialized for the extreme environment of hydrothermal vents.

3.6 WATER ACIDITY: A Complex Sequence of Carbon Reactions Affects Aquatic pH

Carbon dioxide (CO_2) undergoes complex chemical reactions in water, with important consequences for acidity and carbon cycling. Water has a large capacity to absorb CO_2, which is plentiful in both freshwater and saltwater. Upon diffusing into surface waters, CO_2 not only is present as a dissolved gas (as is O_2) but also reacts with water to produce carbonic acid (H_2CO_3):

$$CO_2 + H_2O \leftrightarrow H_2CO_3$$

Carbonic acid further dissociates into an H^+ ion and a bicarbonate (HCO_3^-) ion:

$$H_2CO_3 \leftrightarrow HCO_3^- + H^+$$

Bicarbonate may further dissociate into another H^+ ion and a carbonate (CO_3^{2-}) ion:

$$HCO_3^- \leftrightarrow H^+ + CO_3^{2-}$$

The CO_2–H_2CO_3–HCO_3^- system is a complex chemical sequence that tends towards equilibrium. (Note that the arrows in the equations go both ways.) If CO_2 is removed from water, the equilibrium is disturbed and the equations shift to the left, with H_2CO_3 and HCO_3^- producing more CO_2 until a new equilibrium is achieved.

These chemical reactions either release or absorb hydrogen ions (H^+). The amount of H^+ in a solution is a measure of its **acidity**: the more H^+, the more acidic the solution. Alkaline solutions, in contrast, have more OH^- (hydroxyl ions) and fewer H^+ ions. The measure of acidity is pH, the negative logarithm of the H^+ concentration. In pure water, a small fraction dissociates—$H_2O \leftrightarrow H^+ + OH^-$—and the ratio of H^+ to OH^- is 1:1. As both have a concentration of 10^{-7} moles L^{-1}, the pH of a neutral solution is 7 (because $-\log(10^{-7}) = 7$).

A solution departs from neutrality when one ion increases and the other decreases. A gain of H^+ ions to 10^{-6} moles L^{-1} means a decrease of OH^- ions to 10^{-8} moles L^{-1}, and the pH of the solution is 6. The pH scale goes from 1 to 14; pH > 7 denotes an *alkaline* solution (greater OH^- concentration) and pH < 7 an *acidic* solution (greater H^+ concentration). However, precipitation is not considered acidic unless its pH is below 5.6.

Although pure water is neutral because it dissociates into equal numbers of H^+ and OH^- ions, the presence of CO_2 alters water acidity by either releasing or absorbing H^+. Thus, the dynamics of the CO_2–H_2CO_3–HCO_3^- system directly affect the pH of aquatic habitats. The system acts as a **buffer** (as discussed for soils in Section 4.8), keeping pH within a narrow range. It absorbs H^+ when in excess (moving the sequence to the left and producing H_2CO_3 and HCO_3^-) and releases H^+ when it is in short supply (moving the sequence to the right and producing HCO_3^- and CO_3^{2-}. At neutral pH, most CO_2 is present as HCO_3^- (Figure 3.11). At high pH, more CO_2 is present as CO_3^{2-} than at low pH, where more CO_2 occurs in the free form. CO_2 addition or removal thus affects pH, and a change in pH in turn affects CO_2.

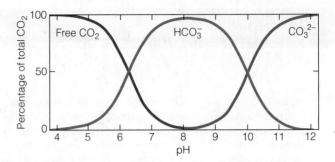

Figure 3.11 Theoretical percentages of CO_2 in water of differing pH. In acidic waters, most CO_2 is in its free form. HCO_3^- dominates in neutral waters and CO_3^{2-} in alkaline waters.

The pH of natural waters ranges between 2 and 12. Waters draining from regions dominated by limestone have higher pH and are better buffered than watersheds dominated by acid sandstone and granite, as in the Precambrian Shield. The abundance of ions such as sodium, potassium, and calcium in oceans causes seawater to be somewhat alkaline, ranging in pH from 7.5 to 8.4.

Acidity influences the distribution and abundance of aquatic species. Increased acidity affects organisms directly by influencing physiological processes and indirectly by influencing concentrations of toxic heavy metals. Tolerance limits vary, but many organisms cannot survive and reproduce below pH 4.5. Aquatic organisms may be intolerant of lower pH largely because acidic waters contain considerable aluminum, which is insoluble at neutral or basic pH. Although insoluble aluminum content is high in rocks and soils, concentrations in lakes are low at normal pH. As waters acidify, aluminum dissolves, raising its concentration. Aluminum is toxic to many species, causing sharp population declines. Much of the damage caused by acid precipitation reflects this indirect effect of pH on heavy metals.

3.7 WATER CYCLE: Water Circulates between Earth and the Atmosphere

Until this point, we have discussed the properties of water and their ecological consequences. Now we consider the global and regional movements of water—movements that affect the availability not only of water itself but also of the resources that water transports.

All marine and freshwater aquatic environments are linked, directly or indirectly, by the **water (hydrological) cycle** (Figure 3.12): the totality of processes by which water travels between Earth and the atmosphere. Solar radiation, which heats the atmosphere and provides energy for evaporating water, is the driving force. *Precipitation* sets it in motion. Water vapour circulating in the atmosphere eventually falls in some form of precipitation. Some lands directly on soil and bodies of water. **Interception** is the amount of precipitation that lands on vegetation, dead organic matter on the ground, and other structures. Such interception can be considerable. Portions of it never enter the ground, instead evaporating directly back to the atmosphere.

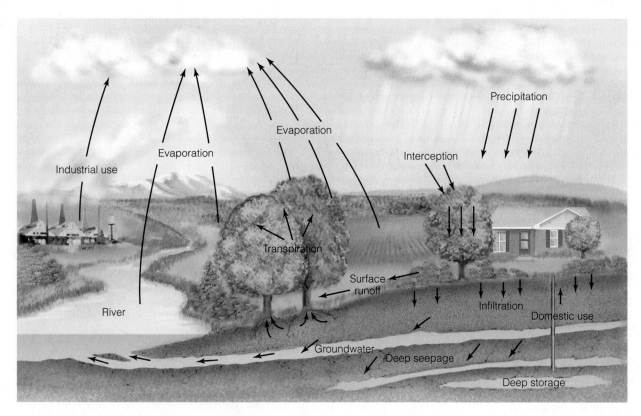

Figure 3.12 Local water cycle, showing major pathways of water movement.

Some of the precipitation that reaches the soil—either directly or after dripping off trunks and branches or leaves—enters the ground by **infiltration**. The infiltration rate depends on the soil type, slope, vegetation, and intensity of precipitation. During heavy rains when the soil is saturated, excess water flows across the ground as **surface runoff** (overland flow). In places, it collects in depressions and gullies, and the flow changes from sheet to channelized flow—a process common on city streets as water moves across the pavement into gutters. As a result of low infiltration, runoff from urban areas can be as much as 85 percent of precipitation.

Some water entering the soil travels by **percolation** to impervious layers of clay or rock and collects as **groundwater**, a portion of which seeps down into deep-storage areas, including water-containing rock formations called **aquifers** (see Ecological Issues: Our Disappearing Groundwater Resources, pp. 51–52). Groundwater eventually finds its way into springs and streams, which coalesce into rivers as they follow the landscape contours. Lakes and wetlands form in basins and floodplains. Rivers eventually flow to the sea, making the transition to marine systems.

Water remaining on the surface of the ground and in the upper soil, and collected on vegetation surfaces, as well as water in surface layers of streams, lakes, and oceans, returns to the atmosphere by *evaporation*. The evaporation rate is governed by how much vapour is in the air relative to the saturation vapour pressure (see Section 2.6). Plants generate substantial additional water losses. Through their roots, they absorb water from soil and release it via **transpiration**: evaporation of water from internal surfaces of leaves and stems through openings called

stomata (singular: *stoma*). The total water lost from the ground and vegetation (surface evaporation plus transpiration) is called **evapotranspiration**. Thus, although the local water cycle is an abiotic process, organisms affect its rates of exchange through interception and evapotranspiration. Indeed, destruction of tropical rain forests will likely permanently alter precipitation patterns, as well as increase the amount of surface runoff.

The totality of local water processes determines the global water cycle (Figure 3.13, p. 50). It involves various reservoirs (given as 10^3 km³) and fluxes (exchanges between reservoirs; given as km³ y⁻¹). The total water volume on Earth is some 1.4×10^9 km³. More than 97 percent is in the oceans, another 2 percent in polar ice caps and glaciers, and 0.3 percent in groundwater. Over the oceans, evaporation exceeds precipitation by 40 000 km³. Much of the water evaporated from oceans moves over land as water vapour and falls as precipitation. Of the 111 000 km³ falling on land, some 71 000 km³ returns to the atmosphere by evapotranspiration. The remaining 40 000 km³ is carried as runoff by rivers and eventually reaches the oceans, balancing the net evaporative loss.

The relatively small size of the atmospheric reservoir (13×10^3 km³) belies its importance. Flux between the atmosphere and the oceans and land surface is very large relative to the amount in the atmosphere at any given time. The importance of the atmosphere is better reflected by its turnover time, calculated by dividing reservoir size by rate of output (flux out): about 0.024 years. In other words, the entire water content of the atmosphere is replaced every nine days. In contrast, the turnover time for the ocean is the size of the reservoir (1.37×10^9 km³) divided by the evaporation rate (425 000 km³ y⁻¹)—more than 3000 years.

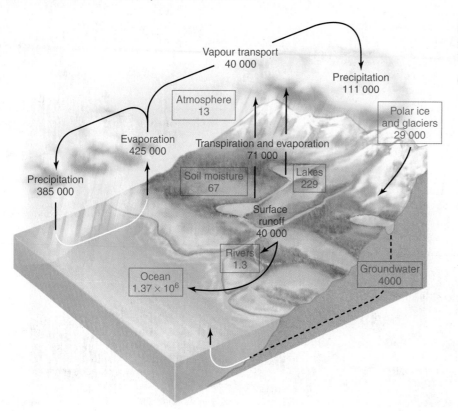

Figure 3.13 Global water cycle. Values for reservoirs (blue boxes) are in 10^3 km^3. Values for fluxes (in red) are in km^3 y^{-1}.

(Adapted from Reeburgh 1997.)

3.8 WATER MOVEMENTS: Currents and Waves Shape Freshwater and Marine Habitats

The movement of water—currents in streams and waves in an open water body or breaking on a shore—determines the nature of many aquatic habitats. Current velocity moulds stream structure. In turn, velocity is affected by the shape and steepness of the channel, its width and depth and the roughness of its bot-

tom, and the intensity of rainfall and rapidity of snowmelt. The velocity of fast streams is 50 cm · s^{-1} or higher. At this velocity, the current removes all particles under 5 mm in diameter, leaving a stony bottom. High water volume increases velocity; it moves stones and rubble, scours the streambed, and cuts new banks and channels. As stream width, depth, and volume increase, silt and decaying organic matter accumulate on the bottom. Thus, stream character changes from fast water to slow, with important effects on aquatic life (Figure 3.14).

(a)

(b)

Figure 3.14 Flowing water ecosystems. **(a)** A fast mountain stream with a steep elevation profile. Fast-flowing water scours the bottom, leaving bedrock. **(b)** A slow stream meandering through willow growth. The flat topography reduces flow, allowing fine sediments to accumulate on the bottom. As a result of their different habitats, these streams support different communities.

ECOLOGICAL ISSUES Our Disappearing Groundwater Resources

One resource with which Canada is well endowed is freshwater. Canada contains an estimated 7 percent of Earth's freshwater resources. When one looks at a map of Canada, or travels across it on a long-distance trip, the huge number of lakes and rivers is very noticeable. But the majority of our freshwater lies below the surface, as groundwater. Given the strategic importance of water, conserving the amount and protecting the quality of this vital resource is a priority.

Groundwater includes all water located under Earth's surface. Some of it is in the soil solution, between soil particles. But the largest amounts are in subterranean formations called *aquifers*: layers of water-holding, permeable rock, sand, or gravel, of which the top layer is called the *water table*. The depth to the water table varies with many factors, including precipitation, but at and below the water table the rock is saturated with water. Much of the water in aquifers is in cracks in fractured bedrock, which acts like a giant sponge.

In natural ecosystems, aquifers play key roles. Aquifer water connects to surface water in lakes and streams, and also emerges as springs. Through participation in the water cycle, aquifers are important reservoirs that help sustain wetlands. Because rock formations vary from place to place, the depth, extent, and permeability of aquifers vary across the country. So do patterns of human use. According to Natural Resources Canada, about 26 percent of Canadians depend on groundwater, with two-thirds of users in rural areas. In global terms, this percentage is quite small; in comparison, 50 percent of Americans rely on groundwater, including almost all of the rural population. In Canada, only the Maritimes have a similarly high reliance on groundwater.

Humans tap into aquifers for more than drinking water. In much of the world, they are used for irrigation. In dry regions, adding water to fields is essential. However, even in regions where mean precipitation is adequate, irrigation provides a steady water supply, increasing yields and making crop growth less dependent on weather. Although only 15 percent of Earth's cultivated land is irrigated, irrigated lands account for 35 to 40 percent of the global food harvest. In Canada, irrigation is most common in the drier western prairies and in the British Columbia interior. Alberta alone contains 60 percent of Canada's irrigated lands, with most of its irrigation water coming from the Paskapoo aquifer (Figure 1). Situated between Edmonton and Calgary, the Paskapoo formation supports more groundwater wells than any other aquifer in the prairies and is under increasing pressure (Grasby et al. 2008). Drilling for both conventional and unconventional natural gas sources poses a contamination risk for this already-stressed aquifer.

In the United States, 30 percent of groundwater used for irrigation comes from one source: the High Plains–Ogallala aquifer, extending 450 000 km^2 across eight Midwestern states. (The Paskapoo, in contrast, covers only 10 000 km^2.) The buried sands and gravel that hold the aquifer water originated from rivers flowing east from the Rockies over the past several million years. The water it contains is so-called fossil water, dating to the last ice age. Many Canadian aquifers also originated from material deposited by glacial rivers, including those supplying British Columbia's lower Fraser Valley, Ontario's Kitchener–Waterloo

Figure 1 The Paskapoo aquifer in southern Alberta covers 10 percent of the province and supports 103 000 bedrock wells (28 percent of the provincial total).

region, and the Fredericton region in New Brunswick. The Carberry aquifer in western Manitoba is an old delta originating from glacial Lake Agassiz. Prince Edward Island taps a sandstone aquifer for its entire water supply. In contrast, the granitic Precambrian Shield that dominates northern Canada is poor at storing water.

Heavy use of aquifers—whether for drinking water, irrigation, or industry—threatens their long-term sustainability by withdrawing water faster than it is recharged. The more permeable the aquifer, the greater the risk. The unusually high permeability of the Ogallala allows large amounts to be pumped rapidly. In fact, water is being pumped from more than 200 000 wells at a rate some 50 times as fast as the aquifer is being recharged. For decades, the supply was assumed to be endless, but intensive pumping since the 1940s has steadily lowered the water level. Withdrawals in 1990 for irrigation alone exceeded 50 billion litres daily. During the drought of the mid-1990s, the decline in the water level of the aquifer averaged 40 cm per year.

Pumping has also decreased the volume of water discharged into streams and springs. Besides the impacts on agriculture and streamflow, the decline of the Ogallala aquifer affects the quantity and quality of drinking water. Although 95 percent of the water is used for irrigation, 80 percent of people in the region rely on the aquifer for drinking water. Water in some areas fails U.S. federal standards for human consumption, and declining levels reduce the quality of remaining waters as levels of salts and other solutes increase. The recent development of the Bakken shale oil formation in North Dakota and Montana (extending into Saskatchewan) poses yet another threat to Ogallala water quality and quantity. Shale-oil processing not only releases large amounts of hydrocarbon pollutants but also uses large volumes of water. In

continued on page 52

ECOLOGICAL ISSUES continued

Estonia, for example, more than 90 percent of water use is associated with shale-oil processing.

Because Canadians are on the whole less dependent on groundwater, it is easy to dismiss these problems as someone else's concerns. Certainly, higher population levels in the United States, coupled with smaller supplies of surface water and a greater need for irrigation, intensify its water problems. But the water issues south of the border are useful warnings. Groundwater use is increasing in Canada, and water quality is declining. According to Environment Canada, some of our groundwater sources have already been contaminated by proximity to point sources, including landfills and industrial waste disposal sites such as in Ville Mercier, Québec, and Elmira, Ontario. More general contamination from agricultural pesticides and fertilizers is common.

Alberta's oilsands deposits pose a further contamination risk for North American groundwater. Many of the deposits lie under a vast groundwater channel in northeast Alberta, flowing into the Athabasca River. Called the Empress Formation, it is the largest freshwater aquifer in Canada. Blowouts have the potential to pollute this relatively pristine aquifer, as well as the overlying boreal forest. There is also growing concern that the pipelines proposed to carry oil to U.S. markets will contaminate the already-

stressed Ogallala aquifer (see Ecological Issues: Oilsands: Collateral Damage of Our Fossil Fuel Dependence, pp. 626–627). National boundaries have no relevance for aquifers and the ecosystems that they influence. North Americans need to address the growing cross-border threats to our collective groundwater resources.

Bibliography

Grasby, S. E., Z. Chen, A. P. Hamblin, P. R. Wozniak, and A. R. Sweet. 2008. Regional characterization of the Paskapoo bedrock aquifer system, southern Alberta. *Canadian Journal of Earth Sciences* 45:1501–1516.

1. Knowing that water use in one region can influence the long-term availability of water in another region, how would you propose that water resources be managed?
2. Many regard this issue as an example of what Garrett Hardin (1968) calls the **tragedy of the commons**: a situation in which it makes economic sense for one user to exploit a shared resource to the utmost, even though the combined effect of all users is resource degradation and depletion. Discuss. What other environmental issues involve a similar effect?

Wind generates waves on large lakes and on the open sea (Figure 3.15). Frictional drag on smooth water causes ripples. As the wind continues to blow, it applies more pressure to the steep side of the ripple, and wave size grows. As the wind strengthens, choppy waves of all sizes develop and grow as they absorb more energy. When the wind energy equals the energy lost by the breaking waves, whitecaps develop.

Waves breaking on a beach do not contain water from distant seas. Each particle of water remains largely in the same place, following an elliptical orbit with the passage of the wave.

Figure 3.15 Waves breaking on a rocky shore.

As a wave moves forward, it loses energy to the waves behind and disappears, its place taken by another. The swells breaking on a beach are thus distant descendants of waves generated far out at sea. As the waves approach land, they advance into increasingly shallow water. The height of each wave rises until the wave front grows too steep and topples over. As the waves break on shore, they dissipate their energy, pounding rocky shores or tearing away sandy beaches in one location and building up new beaches elsewhere.

In Section 2.5, we discussed surface ocean currents, as influenced by prevailing winds interacting with the Coriolis effect. Deep ocean waters move differently. Because they are isolated from wind, their motion does not depend on it. Yet the movement of deep waters does reflect surface changes. Seawater increases in density with lower temperature and higher salinity. As the warm, saline surface currents of tropical waters move north and south (see Figure 2.14), they cool, become denser, and sink. These cold, dense waters originated at the surface and so are O_2 rich. After sinking, these waters return to the tropics as deep-water currents. When these currents meet in equatorial waters, they form regions of **upwelling** where deep, O_2-rich waters rise to the surface (Figure 3.16a). Upwellings also occur in non-tropical coastal areas, where winds blowing parallel to the coast (and also subject to the Coriolis effect) move surface waters offshore. Water moving upwards replaces this water, causing a coastal upwelling (Figure 3.16b).

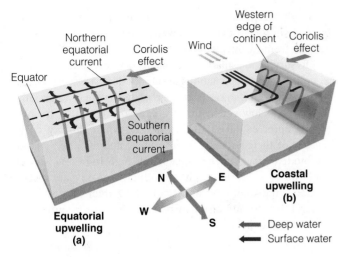

Figure 3.16 Impact of the Coriolis effect on oceans. **(a)** At the equator, the Coriolis effect pulls westward-flowing currents to the north and south (purple arrows), resulting in an upwelling of cold waters to the surface. **(b)** Along the western margins of continents, the Coriolis effect causes surface waters to move offshore (purple arrows), also causing an upwelling of deeper, colder waters. The example shown is for the Northern Hemisphere.

3.9 TIDES: Tide Action Dominates Marine Coastal Habitats

The rhythm of life on ocean shores is determined by **tides**—water movements resulting from the gravitational pulls of the Sun and the Moon. Each causes two bulges (tides) in oceanic waters. The lunar tides occur at the same time on opposite sides of Earth. As Earth rotates eastward, the tides advance westward. The Sun also generates two daily tides on opposite sides of Earth, but because the Sun has a weaker gravitational pull, solar tides are partially masked by lunar tides—except for twice monthly, when the Moon is full and when it is new. Then, Earth, the Moon, and the Sun are nearly in line, and the gravitational pulls of the Sun and Moon are additive, producing the exceptionally large *spring tides* (referring not to the season but to the brimming fullness of the water). When the Moon's pull is at right angles to the pull of the Sun, the two forces interfere with each other, producing the weak *neap tides*.

Tides are neither entirely regular nor the same worldwide. They vary from day to day in the same place, following the waxing and waning of the Moon. The Atlantic experiences semi-daily tides, whereas in the Gulf of Mexico and off the Alaskan coast there is a single daily tide. Mixed tides in which low tides differ in height through the cycle are common in the Pacific and Indian Oceans. Other factors contributing to local tidal differences include variations in the gravitational pull of the Moon and the Sun with Earth's elliptical orbit, the angle of the Moon in relation to Earth's axis, water depth, shore contour, winds, and wave action.

The area lying between the lines of high and low tide, called the **intertidal zone**, is a habitat of extremes that under-goes dramatic daily shifts in conditions. As the tide recedes, the uppermost layers of this zone are exposed to air, temperature fluctuations, intense radiation, and desiccation, whereas the lowest portions are exposed only briefly before high tide submerges them again. Temperatures on tidal flats may reach 38°C when exposed to direct sunlight and drop to 10°C within a few hours when the flats are again submerged. Organisms living in sand and mud do not experience the same extreme temperature flux as those on rocky shores. Although the sand surface at noon may be 10°C warmer than the returning seawater, the temperature a few centimetres below the surface remains almost constant year-round.

3.10 ESTUARIES: Transition Areas between Freshwater and Saltwater Are Unique Habitats

Water from streams and rivers eventually drains into the sea. The place where freshwater joins and mixes with saltwater is an **estuary** (see Section 24.7). Temperatures in estuaries fluctuate considerably, both daily and seasonally. Sunlight and inflowing currents heat the water. High tide on the mudflats may heat or cool the water, depending on the season. The upper layer of estuarine water may be cooler in winter and warmer in summer than the bottom—a condition that, as in lakes, generates spring and fall turnovers.

The interaction of inflowing freshwater and tidal saltwater influences estuary salinity, which varies vertically and horizontally, often within one tidal cycle (Figure 3.17). Salinity may be similar from top to bottom, or it may be stratified, with a layer of freshwater on top and a layer of dense, salty water on the bottom. Salinity is homogeneous when currents are strong enough to mix the water column. In some estuaries, salinity is homogeneous at low tide, but at high tide a surface wedge of seawater moves upstream more rapidly than the bottom water. Salinity is

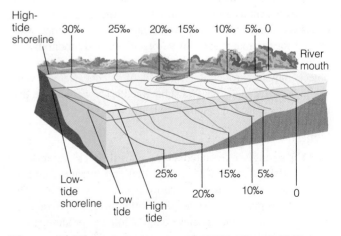

Figure 3.17 Vertical and horizontal stratification of salinity (psu) from river mouth to estuary at high (brown lines) and low (blue lines) tide. At high tide, incoming seawater increases salinity towards the river mouth. At low tide, salinity drops. Salinity increases with depth because lighter freshwater flows over denser saltwater.

then unstable, and density is inverted: the seawater on the surface tends to sink as lighter freshwater rises, generating a phenomenon called **tidal overmixing**. Strong winds also mix salt with freshwater in estuaries, but when the winds are still, river water flows seaward on a shallow surface over an upstream flow of seawater, more gradually mixing with saltwater.

Horizontally, the least saline waters are at the river mouth, and the most saline at the sea. Incoming and outgoing currents deflect this pattern. In all estuaries in the Northern Hemisphere, outward-flowing freshwater and inward-flowing seawater are deflected to the right (relative to the axis of flow from river to ocean) because of Earth's rotation. In addition, the concentration of metallic ions carried by rivers varies among drainage areas, and salinity and chemistry differ among estuaries.

Although the portion of dissolved salts in estuarine waters remains similar to that of seawater, concentration varies in a gradient from freshwater to marine.

To survive in estuaries, organisms have evolved physiological or behavioural adaptations to salinity flux (see Section 7.14). Marine fish often enter estuaries when the freshwater influx from rivers is low and salinity is highest. Conversely, freshwater fish enter estuaries during floods, when salinity drops. Because of their stressful conditions, estuaries often support low biodiversity. On the other hand, high light and nutrient levels make estuaries highly productive. They are critical interfaces between land and sea (see Chapter 25), and their vulnerability to human disturbance, such as oil spills, is of grave global concern.

EcologyPlace

Visit EcologyPlace at www.pearsoncanada.ca/ecologyplace to access online resources that complement your textbook, and help you to apply and to review the information in this chapter. EcologyPlace includes

- an eText version of the book
- self-grading quizzes
- glossary flashcards
- and more!

Go to www.pearsoncanada.ca/ecologyplace and follow the registration instructions on the Student Access Code Card included with this text. If your book does not have a Student Access Code Card, you can purchase access to it at www.pearsoncanada.ca/ecologyplace.

SUMMARY

Properties of Water 3.1

Water has a unique structure. The side of the molecule where the H atoms are located is positively charged. The side where O is located is negatively charged, polarizing the molecule. This polarity causes water molecules to be linked by hydrogen bonds in a lattice arrangement. Depending on temperature and pressure, water is present as a liquid, solid, or gas. Because it absorbs or releases considerable energy with changes in temperature, water is vital for thermal regulation of organisms and moderation of habitat temperature. Water's high viscosity creates resistance for objects moving through it. As a result of its cohesiveness, water has high surface tension. If a submerged body weighs less than the water displaced, it is subject to buoyancy.

Light in Water 3.2

Light is strongly attenuated as it passes through water. Light quality also changes with depth. Red and infrared light are absorbed first, followed by yellow, green, and violet. Blue wavelengths penetrate the deepest. Differences in light quantity and quality affect the species composition and productivity of the phytoplankton community.

Water Temperature 3.3

Many lakes and ponds experience seasonal temperature shifts. In summer a vertical gradient develops, separating warm surface waters (epilimnion) from colder waters (hypolimnion) below the thermocline. When surface waters cool in fall, the temperature becomes uniform and water mixes throughout the basin. A similar turnover occurs in spring in some lakes. In oceans, the thermocline simply descends or rises seasonally. The temperature of flowing water varies seasonally, and with current velocity, depth, and exposure to sunlight.

Water as a Solvent 3.4

Water can dissolve more substances than any other liquid and transports solutes both within organisms and in the abiotic environment. Most rivers and lakes contain low concentrations of dissolved minerals, determined largely by underlying bedrock. The oceans have much higher concentrations, as a result of evaporation of water into the atmosphere. Excess amounts of minerals such as Ca^+ precipitate out in ocean sediments, whereas NaCl is highly soluble.

Oxygen in Water 3.5

Oxygen enters surface waters from the atmosphere via diffusion. The solubility of O_2 in water depends on temperature, pressure, and salinity. In lakes, turbulence mixes O_2 absorbed in surface water with deeper water. In summer, O_2 may become stratified, decreasing with depth as a result of decomposer demand. During spring and fall turnover, O_2 is replenished in deep water. Swirling of stream water increases its contact with air, raising the O_2 content.

Water Acidity 3.6

Acidity is measured as pH, the negative logarithm of the concentration of H^+ ions in solution. In aquatic environments, a close relationship exists between diffusion of CO_2 into surface waters and pH. A complex sequence of reactions starting with carbonic acid formation buffers water against changes in acidity. Because acidity influences the availability of nutrients and toxic heavy metals, it restricts the success of many organisms.

Water Cycle 3.7

Water cycles between Earth and the atmosphere by cloud formation, precipitation, interception, infiltration, and percolation into soils. Water eventually reaches groundwater, springs, streams, and lakes, from which evaporation occurs, returning water to the atmosphere. Vegetation affects water cycling by intercepting precipitation and by transpiration. All of Earth's environments are linked directly or indirectly by the water cycle. The largest reservoir in the global water cycle is the oceans. Although the atmosphere is a small reservoir, it turns over rapidly and is a vital link.

Water Movements 3.8

Currents in streams and rivers, as well as waves in open seas and breaking on shores, determine the nature of many aquatic environments. Current velocity shapes the flowing water environment. Waves pound rocky shores and tear away and build up sandy beaches. Movement of water in surface currents of the ocean affects deep-water circulation. As the equatorial currents move north- and southward, deep waters travel to the surface, forming upwellings. In coastal regions, winds blowing parallel to the coast create coastal upwellings.

Tides 3.9

Rising and falling tides, caused by the gravitational pull of the Moon and the Sun, shape the environment and influence the rhythm of life in coastal intertidal zones.

Estuaries 3.10

Water from streams and rivers eventually drains into the sea. Freshwater joins and mixes with saltwater in an estuary. Temperatures in estuaries fluctuate considerably, daily and seasonally. The interaction of inflowing freshwater and tidal saltwater influences estuarine salinity, which varies vertically and horizontally, often within one tidal cycle.

KEY TERMS

acidity	estuary	infiltration	solution	transpiration
anoxic	evapotranspiration	interception	specific heat	upwelling
aquifer	fall turnover	intertidal zone	stomata	viscosity
buffer	freshwater	ion	surface runoff	water (hydrological)
buoyancy	groundwater	marine	surface tension	cycle
chemosynthesis	hydrothermal vent	oxygen minimum zone	thermocline	
cohesion	hyperthermophilic	percolation	tidal overmixing	
diffusion	hypolimnion	phytoplankton	tide	
epilimnion	hypoxic	salinity	tragedy of the commons	

STUDY QUESTIONS

1. Explain the structural reasons for the polarity of water, and discuss the consequences for the (i) density–temperature relationship, (ii) cohesiveness, and (iii) viscosity of water. Consider the ecological implications of each of these properties.
2. Explain the significance of water's high specific heat for (i) thermal regulation of organisms and (ii) temperature moderation in abiotic environments. Which kind of environment (aquatic or terrestrial) is more affected, and why?
3. Which property of water allows aquatic organisms to function with far fewer supportive structures (tissues) than terrestrial organisms have?
4. What is the fate of visible light (quantity and quality) in water? What are the ecological consequences of these changes in light through the water column?
5. What is the thermocline, and what causes its development?
6. Discuss how seasonal stratification in lakes differs for temperature and O_2.
7. Discuss the impact of biological organisms on O_2 availability in aquatic ecosystems.
8. How will increasing the CO_2 concentration of water affect its pH? Explain.
9. Explain why the most damaging effects of acidity on organisms are often indirect.

10. The concentration of which element is used to define the salinity of water?
11. Draw a simple diagram of the local water cycle and describe the processes involved.
12. Explain the ways in which vegetation affects the local water cycle.
13. What causes upwellings in the equatorial oceans? Coastal areas? Why are upwellings important ecologically?
14. What causes the tides? What is their ecological significance?
15. Estuaries are unique environments. Explain why in terms of their salinity.

FURTHER READINGS

Garrison, T. 2001. *Oceanography: An invitation to marine science*. Belmont, CA: Brooks/Cole.
Clearly written, well-illustrated introduction for those interested in more detail on the subject.

Hutchinson, G. E. 1957–1967. *A treatise on limnology. Vol. 1: Geography, physics, and chemistry*. New York: Wiley.
Classic reference by an ecologist noted for his work on niche theory as well as limnology.

Hynes, H. B. N. 2001. *The ecology of running waters*. Caldwell, NJ: Blackburn Press.
A reprint of a classic work, this major reference continues to be influential.

Nybakken, J. W. 2005. *Marine biology: An ecological approach*. 6th ed. San Francisco: Benjamin Cummings.
Chapter 1 and Chapter 6 provide an excellent introduction to the physical environment of the oceans.

THE TERRESTRIAL ENVIRONMENT

The input and decomposition of dead organic matter is a key factor in the development of forest soils.

What first comes to mind when we think of terrestrial environments? The physical and chemical properties of water dominated our discussion of aquatic habitats, but with land habitats we do not usually think first of abiotic factors. What we visualize is vegetation: the tall, dense forests of the tropics; the changing colours of fall in a temperate deciduous forest; the broad expanses of grass in the prairies. Because animals depend on plants for food and cover, the type of plant life present limits animal distribution and abundance. Yet, as in water habitats, the abiotic factors prevailing in land habitats are critical. Plant life reflects climate and soil, but even if suitable vegetation is present, abiotic factors impose the ultimate constraints on other life forms.

These are not one-way influences. Although the prevailing conditions of light, moisture, and soil determine vegetation type and distribution, vegetation influences the abiotic habitat by absorbing and filtering light, intercepting and transpiring water, and modifying soil composition and properties. We explore these and other reciprocal loops of influence throughout the text, but first we consider key features of terrestrial environments that directly influence life on land.

4.1 LIFE ON LAND: Terrestrial Habitats Present Challenges and Opportunities

Life first colonized land more than a billion years ago. The transition from life in water opened up new opportunities of light and space but posed significant constraints for organisms adapted to life in water. To understand these constraints, we must consider the differences between terrestrial and aquatic habitats and the problems these differences create. Many adaptations of land organisms reflect the selective influence of (1) desiccation (drying out) and (2) gravity.

Desiccation is perhaps the greatest constraint for life on land. Living cells are 75 to 95 percent water. Unless the air is saturated with moisture, water readily evaporates from cell surfaces. The cell must replace this water to remain hydrated and functional. Maintaining this **water balance** with the environment has been a strong selective force in the evolution of life on land. For example, plants have evolved a waxy *cuticle* on the epidermal tissues of their stems and leaves to reduce water loss (and to minimize damage from UV radiation, an unavoidable downside of the increased light available on land). But the cuticle blocks gas exchange, so terrestrial plants have evolved *stomata* that allow CO_2 and O_2 to diffuse into the leaf interior.

To stay hydrated, an organism must replace water lost through evaporation or through bodily processes such as urination. Terrestrial animals acquire water by drinking and eating and by metabolic water generation. Some animals have evolved structural adaptations, such as *kidneys*, to regulate water reabsorption. For plants, acquiring water is passive. Early in their evolutionary history, land plants evolved vascular tissues (*xylem*) consisting of cells joined into tubes that transport water and minerals absorbed by roots to other plant parts. Water balance and the array of adaptations terrestrial plants and animals have

(a)

(b)

Figure 4.1 Structural constraints on organisms in aquatic versus land habitats. **(a)** The giant kelp grows 30 m or more without support tissues, thanks to gas-filled bladders. **(b)** A Douglas-fir of similar height must invest most of its biomass in woody support tissues.

evolved to overcome their shared problem of water loss are discussed in more detail in Chapters 6 and 7.

Desiccation is not the only problem for life on land. Because air is less dense than water, it imposes less frictional resistance on body movement but greatly increases the impact of gravity. Structural devices that enhance buoyancy help organisms in aquatic habitats, but the need to remain erect on land requires a significant investment in structures such as *skeletons* (internal or exoskeletons) and *cell walls* rich in cellulose and lignin (plants) or chitin (fungi).

Consider the giant kelp (*Macrocystis pyrifera*), which grows in dense stands in Pacific coastal waters from British Columbia to California (Figure 4.1a). Anchored to the substrate, kelp can grow up to 30 m tall. Kept afloat by gas-filled bladders attached to its blades, these macroalgae collapse if they are out of the water. Lacking tissues strengthened by cellulose and lignin, kelp cannot support its weight against gravity. In contrast, a Douglas-fir tree (*Pseudotsuga menziesii*) of similar height inhabiting a nearby coastal forest (Figure 4.1b) allocates more than 80 percent of its mass to support tissues.

Another feature of land habitats is their variability. Temperature fluctuates less in water than in air because of water's high specific heat (see Section 3.1), reducing daily and seasonal variation. In contrast, temperature flux typifies land habitats, especially in non-tropical sites. Likewise, the amount and timing of precipitation limit water availability for land organisms, as well as their ability to maintain a functional water balance.

These fluxes in temperature and moisture affect metabolism in the short term and species distribution and evolution in the long term. As we saw in Chapter 2, some of this variation is seasonal and predictable, but some is irregular. Ultimately, climatic variation governs the distribution of plants and thereby the terrestrial ecosystems for which plants provide the energy base.

4.2 LIGHT AND VEGETATION: Vegetative Cover Influences the Vertical Light Profile

Entering a forest in summer, one immediately notices a decline in light (Figure 4.2a). A similar if less pronounced effect occurs in grasslands (Figure 4.2b). In aquatic habitats, absorption of radiation by water itself creates a distinct light profile, but the main factor affecting the light gradient on land is absorption and reflection of light by plants. The uppermost layer of vegetation is the *canopy* (see Section 16.9; a forest canopy consists of the upper portions or "crowns" of trees, but the term applies to any ecosystem in which the upper layer forms a distinct habitat). The amount and quality (wavelength; see Section 2.1) of radiation that penetrate the canopy vary with the quantity and orientation of leaves. Moving down through the canopy and lower layers of vegetation, the number of leaves above increases, so the amount of light decreases.

However, because leaves vary in size and shape, leaf number is not the best measure of leaf quantity. Leaf quantity is often expressed as leaf area. Most leaves are flat, so leaf area is measured as the upper surface, if held horizontally. (If the

leaves are vertical, as with grasses, both upper and lower surfaces may be included.) For three-dimensional leaves such as pine needles, the entire surface area is measured. As a predictor of the impact of leaf quantity on available light, the **leaf area index (LAI)** is calculated as the total leaf area (m^2) per unit ground area (m^2) (Figure 4.3, p. 60). An LAI of 3 indicates that there are 3 m^2 of leaf area over each 1 m^2 of ground. The higher the LAI, the less light reaches the surface. Moving through a canopy, total leaf area and hence LAI increase, and light declines. This relationship between light and LAI is described by Beer's law (see Quantifying Ecology 4.1: Measuring Light Attenuation, p. 61).

Leaf orientation as well as area affects attenuation. If a leaf held perpendicular to the Sun's rays absorbs 1.0 unit of light energy per unit leaf area per unit time, the same leaf held at a 60° angle absorbs only 0.5 units. Why? The same leaf area represents only half the projected surface area and intercepts only half as much light (Figure 4.4, p. 60). Leaf angle thus affects the vertical distribution of light, as well as the total amount of light absorbed and reflected.

The angle of the Sun varies daily and seasonally at any given location. Thus, different leaf angles are more effective at intercepting light in different locations and/or times. At high latitudes, where the solar angle is low, leaves held at an angle absorb light more effectively. Leaves held at an angle are also typical of arid habitats, where they reduce light interception at midday when temperature and water demand peak. Nor is leaf angle static; many plants adjust their leaf angle either to increase interception of light or to decrease it when light levels are sufficiently high to cause light damage or overheating.

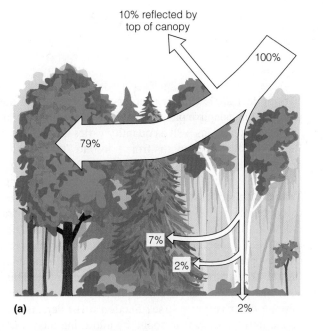
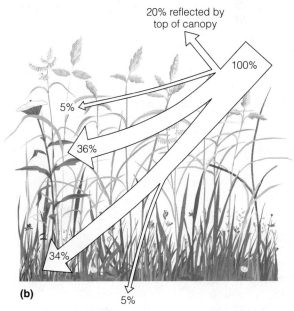

Figure 4.2 Absorption and reflection of light by vegetation. **(a)** A mixed coniferous–deciduous forest reflects ~ 10 percent of incident photosynthetically active radiation (PAR) from the upper portion of the canopy and absorbs most of the remaining PAR within the canopy. **(b)** A meadow reflects ~ 20 percent of incident PAR reaching the upper surface. The densely foliated middle and lower regions absorb most of the rest, with only 2 to 5 percent reaching the ground.

(Adapted from Larcher 1996.)

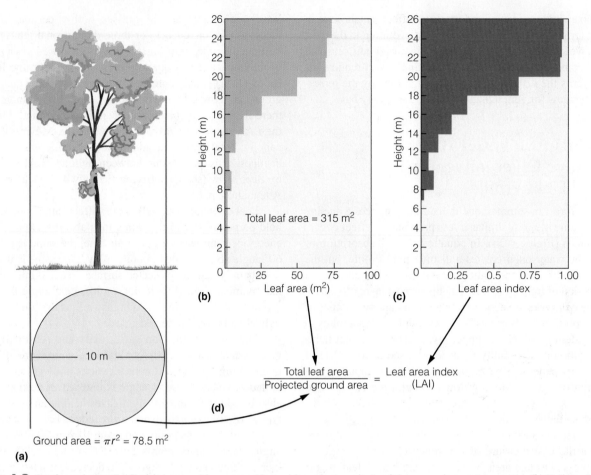

Figure 4.3 Leaf area index (LAI). **(a)** A tree with a crown 10 m wide projects a circle of the same size on the ground. **(b)** Foliage density at various heights above the ground. **(c)** Contributions of crown layers to LAI. **(d)** Calculation of LAI. Total leaf area is 315 m². Projected ground area is 78.5 m². LAI is 4.

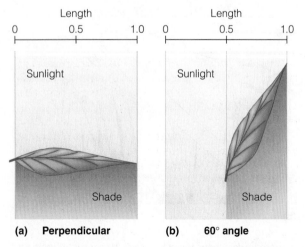

Figure 4.4 Influence of leaf orientation on light interception. If a leaf held perpendicular to a light source **(a)** intercepts 1.0 unit of light energy, the same leaf held at a 60° angle **(b)** intercepts 0.5 units. The angled leaf projects less surface area relative to the light source.

Varying leaf angle (through turgor-driven changes in the leaf petiole) is an adaptive plant "behaviour" that is highly plastic.

Light quality as well as quantity varies through a canopy. Recall that the wavelengths from ~ 400 to 700 nm are called *photosynthetically active radiation* (PAR; see Figure 2.4) because they include the wavelengths used in photosynthesis. **Transmittance** (passage without absorption) of PAR is typically less than 10 percent, whereas transmittance of far-red radiation (730 nm) is much greater. As a result, the ratio (not the absolute amount) of red (660 nm) to far-red radiation (R:FR) decreases through the canopy, and the radiation reaching the forest floor is said to be *enriched* in FR. This shift activates **phytochrome**, a non-photosynthetic pigment that (among other responses; see Section 6.14) detects shading. Shade-intolerant plants respond by allocating more carbon to leaves instead of roots and by elongating their shoots to access more light.

Although part of PAR, green light (~ 550 nm) is not strongly absorbed by chlorophyll *a* and *b*, the two main photosynthetic pigments of plants. Some green light is absorbed

QUANTIFYING ECOLOGY 4.1   Measuring Light Attenuation

As a result of light absorption and reflection by leaves, there is a distinct light profile from the top of the canopy to the ground. The greater the leaf area, the less light that penetrates. Beer's law quantifies light attenuation through a homogeneous medium (in this case, a leaf canopy):

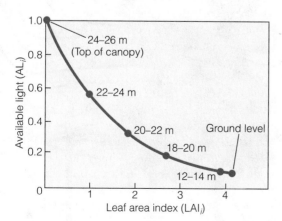

where i = canopy height (in this case, metres);

e = natural logarithm (2.718); and

k = light extinction coefficient (amount of light attenuated per unit of LAI).

The **light extinction coefficient** is a measure of the degree to which leaves absorb and/or reflect light; it varies with leaf angle (see Figure 4.4) and optical properties. Available light (AL_i) is expressed as a proportion of the light reaching the top of the canopy. Light quantity at any level is calculated by multiplying the AL_i value by the actual measured quantity of light quanta (or PAR) reaching the top of the canopy as expressed by photon flux density (μmol/m^2/s).

We can now construct a curve describing available light at any height in the Figure 4.3 example. The light extinction coefficient, k, is 0.6, a typical value for a temperate deciduous forest. We graph vertical positions from the top of the canopy to the ground. Knowing the LAI above any height i in the canopy, we use the equation to calculate available light (Figure 1).

Figure 1 Relationship between leaf area index at various canopy heights and available light, expressed as a proportion of PAR at the top of the canopy (1500 μmol/m^2/s).

Why is available light important? It is as simple as this: at lower levels in the canopy, less light reaches the leaves, and net photosynthesis declines (Figure 2).

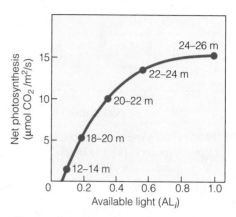

Figure 2 Relationship between net photosynthetic rate at various canopy heights and available light, expressed as a proportion of PAR at the top of the canopy (1500 μmol/m^2/s).

As Beer's law also describes light attenuation in water, we include it here. However, the extinction coefficient (k) is far more complex. Declining light reflects attenuation by (1) water itself, (2) phytoplankton (microscopic algae, as quantified by the chlorophyll a concentration of the water), (3) dissolved substances, and (4) suspended particulates. The extinction coefficients of all three factors are summed to give the overall extinction coefficient (k_T):

$$k_T = k_w + k_c + k_d + k_p$$

with labels: Chlorophyll, Particulates, Total, Water, Dissolved substances.

Whereas the light extinction coefficient for leaf area expresses light attenuation per unit of LAI, these values of k express light attenuation per unit of water depth. As before, Beer's law then predicts the amount of light reaching any depth (z): $AL_z = e^{-k_T z}$.

For aquatic habitats dominated by algae, no further calculations are needed, but if the ecosystem supports rooted submerged vegetation such as kelp, this equation merely quantifies the light available at the top of the canopy. From then on, ecologists use the previous equation describing attenuation as a function of LAI to calculate the light decline from the canopy to the substrate.

1. If we assume that the value of $k = 0.6$ is for a canopy in which leaves are held horizontally, would k be higher or lower if leaves were oriented at a 60° angle (see Figure 4.4)?

2. In shallow water, storms can suspend sediment particulates for some time before they once again settle to the bottom. How would this situation affect the value of k_T and hence light attenuation in the water profile?

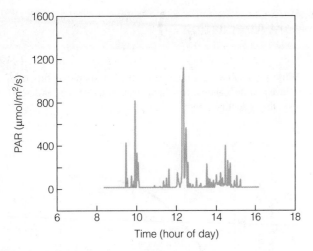

Figure 4.5 PAR at ground level in a redwood forest over a single day. The spikes represent sunflecks (median length 2 s in an otherwise low-light environment (mean PAR = 30 µmol/m²/s)).

(Adapted from Pfitsch and Pearcy 1989.)

by *carotenoids*, accessory pigments that pass some energy on at a longer wavelength for use by chlorophyll. But much of the green light hitting the canopy is either reflected (giving vegetation its green hue) or transmitted. Some direct sunlight does penetrate openings in the canopy, especially in windy conditions, as **sunflecks**. Unaltered in wavelength because they have not been transmitted through leaves, sunflecks account for 70 to 80 percent of solar energy reaching a forest floor (Figure 4.5). They are a vital (if variable) resource for low-growing species.

Figure 4.6 PAR in a yellow poplar (*Liriodendron tulipifera*) stand. The isopleth lines define the PAR gradient. Radiation is most intense in summer, but little reaches the floor. The most PAR reaches the floor in spring, when trees are leafless, and the least in winter, with its low solar angle and short day length.

(Adapted from Hutchinson and Matt 1977.)

Seasonal changes strongly influence leaf area and light penetration, especially in temperate regions where many trees are deciduous and shed their leaves in autumn (Figure 4.6). In spring, as leaves are still expanding, 20 to 50 percent of incoming light may reach the forest floor, and much less in mid-summer. In regions with wet-dry seasons, there is more light at ground level during the dry season when many species shed their leaves. These changes provide opportunities for **ephemerals**—species (typically low-growing) that complete their annual life cycle in a short period when light and/or moisture are available.

Daily changes in light are also substantial. The low angle of the Sun in the morning and evening reduces light penetration through vegetation and alters its spectral quality. *Phototropism*—movements of plant parts either towards or away from light, as discussed for leaf angle—is a particularly effective strategy for dealing with such short-term changes.

So far we have implied that light limitations are always the issue for terrestrial ecosystems. Certainly, light often does limit plant growth, particularly for species occupying the lower layers of vegetation, but too much light can also cause damage. Plants have evolved many adaptations to protect themselves from too much as well as too little light (see Section 6.9).

4.3 SOIL DEFINED: Soil Is the Foundation upon Which All Terrestrial Life Depends

Soil is many things—a medium for plant growth; the main factor controlling the fate of water in terrestrial environments; nature's recycling system, which breaks down the waste

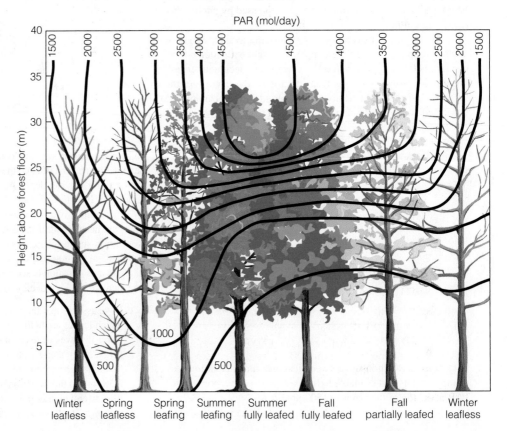

products of organisms and transforms them into their basic elements; and the habitat for a diversity of animals and fungi, from small mammals to innumerable microorganisms. Yet, as familiar as it is and as ecologically vital as its many functions are, soil is hard to define. Some define **soil** as a natural product formed by the weathering of rocks and the actions of living organisms. Others define it as a collection of natural bodies of earth, composed of mineral and organic matter and capable of supporting plant growth. Of one thing scientists are sure. Soil is not just an abiotic medium for plants. It is teeming with life—billions of minute and not-so-minute animals, fungi, and bacteria. This interaction between the biotic and the abiotic makes the soil a living system.

However defined, soil is three-dimensional, with length, width, and depth. In most places, exposed rock has broken down to produce a **regolith**—a layer of unconsolidated debris overlying unweathered rock. The depth of the regolith varies from very thin to tens of metres. It is at this interface between rock, air, water, and living organisms that soil formation begins.

4.4 SOIL FORMATION: Weathering Initiates Soil Formation, Which Is Affected by Six Factors

Weathering of rocks and minerals initiates soil development. **Mechanical weathering** is the breakdown of rock into smaller particles from the combined action of water, wind, and plants. Water seeps into crevices, freezes, expands, and cracks the rock. Frost action is particularly influential at high latitudes. Wind-borne particles, such as dust and sand, wear away at the rock surface, causing it to flake. Growing roots of trees split rock apart. Even a germinating seed can crack pavement, and the impact of mechanical forces acting on rock over millennia is great. However much it fragments rocks, mechanical weathering does not alter their composition.

Simultaneously, these particles are chemically altered and broken down further by **chemical weathering**. The presence of water, oxygen, and acids from the activities of soil microbes and plant roots, plus continual inputs of organic matter, enhances chemical weathering. Rain falling on and filtering through this mix of organic and mineral substances initiates chemical reactions that transform the composition of the original rocks and minerals.

Six interdependent factors affect soil formation, both during weathering and subsequently: *parent material*, *site history*, *climate*, *biota*, *topography*, and *time*.

1. **Parent material** is the matter from which soil develops—either underlying bedrock, as reflected in the *regolith*, or from material deposited by (i) glacial ice (*till*), (ii) wind (*eolian*), (iii) gravity (*colluvium*), and (iv) flowing water (*fluvium*). The physical and chemical nature of parent material determines many soil traits, especially during the initial stages of soil development.

2. *Site history* greatly influences parent material. Soils in much of Canada bear the imprint of glaciation. In the Precambrian Shield, granitic parent material appears as rock outcrops exposed by retreating glaciers. Surrounding boreal soils vary in type and depth depending on the deposition of glacial till in various configurations (*moraines*). This till overlies bedrock, often differing greatly from it. Ice retreat is not the only glacial influence. Many areas of the Shield are covered in deep deposits originating from melting glaciers. Non-boreal regions are also affected. Much of central Manitoba, for example, was once the bed of glacial Lake Agassiz.

3. *Climate* influences soil development in many ways. Temperature, wind, and precipitation directly affect physical and chemical weathering and subsequent **leaching** (movement of solutes) and **eluviation** (movement of suspended materials) by water. Water affects chemical weathering as well as transport of substances, so the deeper the water percolates, the greater the depth of soil development. Temperature affects chemical weathering as well as biochemical reaction rates, affecting the balance between accumulation and breakdown of organic materials. In warm, moist conditions, the combined impacts of weathering, leaching, and plant growth on soil development are maximized, whereas in cold, dry conditions, their influence is much weaker. Soils develop much more slowly in Newfoundland than in southern Ontario, where warm, humid conditions speed up both abiotic and biotic processes. Within an area, microclimatic differences alter soil development. Climate also influences soil development indirectly, by affecting soil biota.

4. *Biota*—plants, animals, bacteria, and fungi—affect soil formation. Roots break up parent material, enhancing mechanical weathering. They also stabilize soil and reduce erosion. Roots move nutrients up from deeper soil, recirculating leached minerals. To facilitate mineral uptake, roots release H^+ and organic exudates, which alter soil pH and participate in chemical weathering. Via photosynthesis, plants capture solar energy and add some of it to the soil in organic compounds. Microorganisms break down the remains of dead organisms, eventually incorporating their organic residues into soil (see Chapter 21).

5. *Topography*, the contour of the land, interacts with the influence of climate. More water runs off and less water enters soil on steep slopes than on level land, whereas water draining from slopes enters the soil on low, flat surfaces. Steep slopes are more subject to **soil erosion**: movement of soil material through wind, water, frost action, or gravity (the latter either in rapid movements, such as mudslides, or slow movements called *soil creep*).

6. *Time* is a crucial factor in soil formation: all of the other factors listed above assert their influence over time. Weathering of rock; accumulation, decomposition, and mineralization of organic material; and downward movement of materials all require considerable time. A well-developed soil may take 2000 to 20 000 years to form—which is why ecologists are concerned when the rate of soil loss in

agricultural regions far exceeds the rate of soil formation (see Ecological Issues: Have We Passed Peak Soil?, pp. 67–68).

4.5 SOIL PROPERTIES: Soils Differ in Physical Traits

Soils differ from each other in their physical properties, including colour, texture, and depth. They also vary in their chemical properties, as discussed in Section 4.8.

1. *Colour* is one of the most easily defined soil traits. Although of little direct influence on function, it is a useful indicator of more important soil properties. Organic matter (particularly humus) makes soil dark or black. Prairie soils, which are naturally rich in humus from grass residues, are typically black. Other colours indicate the chemical content of the parent material from which the soil was formed. Iron oxides render a soil yellowish-brown to red, as in the soils of Prince Edward Island or the southeastern United States, whereas manganese oxides turn soil purplish. Quartz, kaolin, gypsum, and carbonates of calcium and magnesium give soils a white or greyish cast. Yellow-brown or grey blotches indicate poorly drained soils or soils saturated by water. Soils are classified by colour using standardized Munsell colour charts.

2. **Soil texture** is the proportion (by dry mass) of gravel, sand, silt, and clay. Texture reflects both parent material and soil-forming processes. Gravel consists of particles larger than 2.0 mm and is not part of the fine soil fraction. Sand particles range from 0.05 to 2.0 mm and feel gritty. Silt particles range from 0.002 to 0.05 mm, can scarcely be seen by the naked eye, and resemble flour. Clay particles are less than 0.002 mm—too small to be seen under a light microscope. When moist, they can be rubbed between the fingers to form a "ribbon." Clay determines the most critical functional traits of soils, including water-holding capacity and fertility (see Sections 4.7 and Section 4.8). Soils rarely contain just one particle size. They are divided into texture classes based on their proportions of each (Figure 4.7).

   Why does soil texture matter? As well as affecting soil fertility (see Section 4.8), texture determines pore space, which in turn affects air and water movement and root penetration. In an "ideal" soil, pore space makes up half of total soil volume. It includes spaces within and between soil particles, as well as old root channels and animal burrows. Coarse-textured soils (with high sand content) have large pore spaces that favour rapid water infiltration, percolation, and drainage. In general, the finer the texture (higher proportion of silt and clay), the smaller the pores, and the greater the surface for water adhesion and chemical activity. But small pores, even if they have higher total pore volume, have lower infiltration rates and drain poorly. Water tends to puddle on the surface of fine-textured soils after a hard rain. Fine-textured soils are also easily compacted if ploughed or walked on. They are then poorly aerated and resist root penetration.

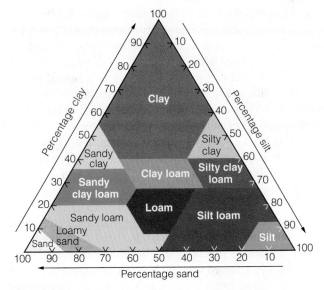

Figure 4.7 Percentage of clay, silt, and sand in soil texture classes. A soil with 60 percent sand, 30 percent silt, and 10 percent clay is a sandy loam.

INTERPRETING ECOLOGICAL DATA

Q1. What is the texture class of a soil with (a) 60 percent silt, 35 percent clay, and 5 percent sand? (b) 60 percent clay and 40 percent silt?

3. *Soil depth* varies across a landscape, depending on slope, weathering, parent materials, and vegetation. In prairies, much of the soil organic matter originates from the deep, fibrous roots of grasses. By contrast, leaves provide most of the organic debris in forests. As a result, prairie soils tend to be several metres deep, whereas forest soils are typically shallow. On flat ground at the base of slopes and on alluvial plains, soils tend to be deep, whereas soils on ridges and steep slopes are shallow, with bedrock close to the surface. Soil depth influences how deeply roots penetrate, affecting the stability of vegetation, especially trees. A pine tree growing on a rocky lakeshore is highly susceptible to wind damage, because its roots provide little anchorage. Deeper soils also have more capacity for holding water and minerals.

4.6 SOIL HORIZONS: Soil Has Distinct Horizontal Layers

Initially, soil develops from undifferentiated parent material. Over time, changes occur from the surface down, through accumulation of organic matter and downward movement of material. These changes give rise to horizontal layers differentiated by physical, chemical, and biological properties. A sequence of layers or **soil horizons** constitutes a **soil profile**. Soil horizons are easily visible along a roadside cut (Figure 4.8). A typical profile has four horizons: O, A, B, and C (Figure 4.9). Each horizon can have many subdivisions, but we will describe a simple general model based on the Canadian soil classification system.

Figure 4.8 Soil horizons in a cut along a road bank. The parent material of this shallow soil is close to the surface.

The surface layer is the **O horizon** (organic layer), consisting of plant litter at various stages of decomposition. It is subdivided into a surface layer, L, composed of leaves and twigs that can still be identified; a middle layer, F, composed of fibrous material that is partially decomposed; and a bottom layer, H, consisting of dark brown to black homogeneous organic material (**humus**) that is resistant to further breakdown.

The O horizon is often absent in agricultural soils but is visible on a forest floor. In temperate regions, the O horizon is thickest in fall, when new leaf litter accumulates, and thinnest in summer after decomposition has taken place. In coniferous forests, it is highly acidic from the presence of conifer needles, and contains much material derived from mosses and lichens. In all soils, it is a site of intense microbial and animal activity.

Below the organic layer is the first of the mineral layers—the **A horizon**, or *topsoil*. Humus accumulates from above, giving it a darker colour than lower layers. Downward movement of water sometimes creates a lighter-coloured **Ae horizon**, or *zone of eluviation*, with maximum removal of suspended particles, as well as leaching of solutes. Ae horizons are common in forests but rarely occur in grasslands as a result of lower rainfall.

Below the A (or Ae if present) horizon is the **B horizon**, or *subsoil*. Containing less organic matter than the A horizon, the B horizon accumulates clays, iron oxides, and salts by **illuviation** (deposition of materials moved by either eluviation or leaching). The B horizon is usually denser (less porous) than the A horizon, making it more difficult for roots to penetrate. B horizons differ in their colour and structure, and in the kinds of material accumulated from other horizons.

Figure 4.9 Generalized soil profile. Changes occur from the surface down, forming horizons.

Organic layer: dominated by organic material, consisting of undecomposed or partially decomposed plant materials, such as dead leaves

Topsoil: largely mineral soil developed from parent material; organic matter leached from above gives this horizon a distinctive dark colour

Subsoil: accumulation of mineral particles, such as clay and salts accumulated from topsoil; distinguished based on colour, structure, and kind of material accumulated from leaching and eluviation

Unconsolidated material derived from the original parent material from which the soil developed

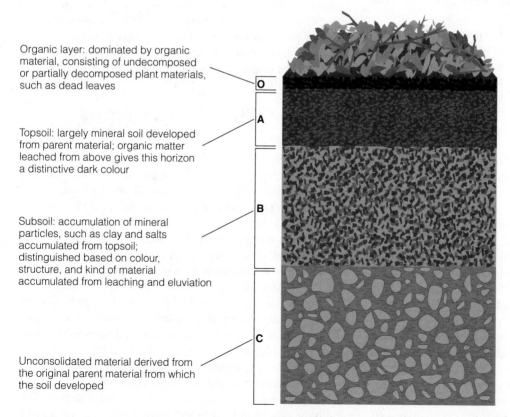

The **C horizon** is the unconsolidated material (*regolith*) underlying the subsoil. Because it is below the zones of greatest biotic activity and weathering and has been little altered by soil-forming processes, it typically retains many of the properties of the parent materials from which it was formed. Below the C horizon lies the bedrock. However, in some situations—including the large portion of the Canadian landscape affected by glacial activity—the soil profile does not derive from underlying bedrock but has been transported from elsewhere by ice, wind, or water.

4.7 SOIL MOISTURE: Water-Holding Capacity Is a Vital Property of Soils

Given the importance of water to the success of plants on land, the capacity of a soil to absorb and retain water is arguably its most ecologically critical feature. Prior to entry, topography affects water movement on, and later in, soil (see Section 3.7). Water drains downslope, leaving soils on ridge-tops relatively dry and creating a moisture gradient from top to bottom. Even after a soaking rain, there is a sharp transition between the wet surface soil and the drier soil below. As rain falls on the soil surface, it enters by infiltration. Gravity transports water into pore spaces, with the size and spacing of soil particles determining how much water enters. Wide pore spacing facilitates water entry, so water infiltrates sandy soils faster than clay soils.

If there is more water than the pore space can hold, the soil is *saturated*, and excess water drains freely under the influence of gravity (assuming there is no underlying hardpan or permafrost that impedes drainage). Heavy rains in regions with saturated soils cause extensive runoff. On the prairies or other flat regions, spring floods are more severe when the ground is saturated from snowmelt.

A key property of soil is its **field capacity (FC)**: the percentage of the soil mass occupied by water after the soil has been saturated and gravitational drainage is complete, compared with the oven-dried mass of the soil at a standard temperature. Field capacity varies with texture. Water drains through sandy soil as quickly as it infiltrates it. Fine clay soil has smaller pores and holds much more water, even though water infiltrates it much more slowly.

Capillary water is water held between soil particles by capillary forces. As evaporation and roots extract capillary water from pores, soil water content declines. When soil moisture falls to the point where roots can no longer extract water, the soil has reached its **wilting point (WP)**. The amount of water retained by a soil between its field capacity and its wilting point is the **available water capacity (AWC)** (Figure 4.10). The AWC is an estimate of the water available for plant uptake. Although water still remains in the soil—occupying up to 25 percent of pore spaces—soil particles hold it tightly, making it difficult to extract.

As Figure 4.10 illustrates, soil texture affects wilting point as well as field capacity. Particle size directly influences pore space and the surface area to which water adheres. Sand has 30 to 40 percent of its volume in pore space, whereas clays and loams range from 40 to 60 percent. As a result, fine-textured soils have a higher field capacity than sandy soils, but the increased surface area results in a higher wilting point as well.

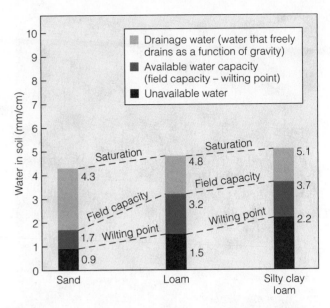

Figure 4.10 Water content of three soils of differing texture at saturation, wilting point, and field capacity. Both FC and WP increase in fine soils, so AWC is highest in intermediate soils.

INTERPRETING ECOLOGICAL DATA

Q1. Although fine-textured soils have a greater overall water-holding capacity, their high wilting point means they must be near or at field capacity for plants to extract the water they hold. In arid regions, low and infrequent rainfall may keep soil water content well below field capacity for most of the season. If the soil water content at a site is 1.6 mm per cm of soil, which of the three soil types shown would have the most available water? What if the water content were 2.4 mm per cm of soil?

Conversely, coarse-textured sandy soils have both a low field capacity and a low wilting point. Thus, AWC is highest in intermediate (loam) soils. Humus content also increases a soil's AWC. Organic matter is removed yearly from agricultural soils, reducing their humus content and their ability to retain moisture.

4.8 SOIL FERTILITY: Ion Exchange Capacity Affects Soil Fertility

Plant don't rely on soils just for water; the soil is also their major source of minerals. Chemical nutrients dissolved in the **soil solution** (water held in the soil matrix) are the most readily available for uptake—and also the most susceptible to leaching. Minerals in solution are also constantly interchanging with exchangeable minerals adsorbed (attached) to the outer surfaces of negatively charged soil particles, particularly clay and humus. Here is a critical instance when a physical property of soil—its texture—greatly affects a chemical property—its fertility.

Recall that *ions* are charged particles. Positively charged ions are *cations*, and negatively charged ions are *anions*. Chemical substances in the soil solution include cations such as calcium (Ca^{2+}), magnesium (Mg^{2+}), and ammonium (NH_4^+), and anions such as nitrate (NO_3^-), phosphate (PO_4^{4-}), and sulphate (SO_4^{2-}). Because clay and humus are negatively charged, only cations participate in ion exchange in most temperate soils. The ability of cations to bind to soil particles depends on a soil's

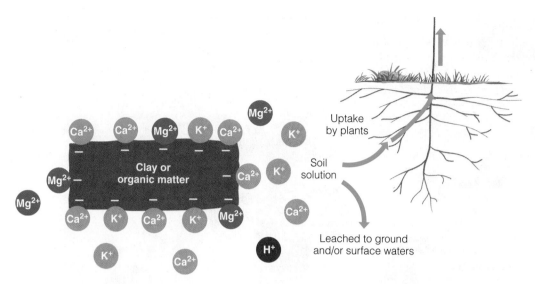

Figure 4.11 Cation exchange in soils. Cations occupying negatively charged sites are in a state of dynamic equilibrium with similar cations in the soil solution. Cations in the soil solution are taken up by plants and/or leached to ground and surface waters.

cation exchange capacity (CEC): the total number of negatively charged sites within a volume of soil. The higher the CEC, the fewer cations are leached into lower soil horizons. The CEC is a rough measure of soil fertility and increases with clay and organic matter content. Anions, in contrast, are not retained on soil exchange sites. They are present in the soil solution and are easily leached if not taken up by plants.

Cations occupying negatively charged sites on soil particles are continuously replaced by or exchanged with cations in the soil solution (Figure 4.11). The relative abundance of ions on exchange sites reflects their (1) concentration in the soil solution and (2) relative affinity for exchange sites. The smaller the ion and the greater its positive charge, the more tightly it is held. The **lyotropic series** lists major soil cations in order of their bonding strength to exchange sites:

$$Al^{3+} > H^+ > Ca^{2+} > Mg^{2+} > K^+ = NH_4^+ > Na^+$$

Higher concentrations in the soil solution can overcome these differences in affinity. H^+ ions in rainfall or released by roots and microorganisms increase the H^+ concentration in the soil solution, displacing other cations on soil exchange sites. As more and more H^+ ions replace other cations, the soil becomes increasingly acidic. Acidity is one of the most critical chemical properties of soil. Typically, soils range from pH 3 to 4 (extremely acidic, typical of bogs, such as those dominated by black spruce and *Sphagnum* moss) to pH 8 to 8.5 (strongly alkaline, such as Ca-rich prairie soils). An acidic soil is considered one with pH < 5.6. Vegetation type influences soil pH. As a result of its acidic needle-fall, a coniferous forest soil has a lower pH than that of a deciduous forest, even if other soil properties are similar.

As acidity increases, the proportion of exchangeable aluminum (Al^{3+}) rises, while exchangeable Ca^{2+}, Na^+, and other cations decline. Al^{3+} is toxic to plants. Affected roots are short and stubby and have dead tips. The impacts of acid rain on nutrient imbalances in individuals and ecosystems are discussed in Section 22.10, but differences in a soil's *buffering capacity* (ability to resist acidification) explain why some regions are more affected than others. Prairie soils of high clay content, with their many exchange sites, are better buffered than sandy soils, which have fewer exchange sites and are already more acidic naturally, especially under conifers.

ECOLOGICAL ISSUES Have We Passed Peak Soil?

Anyone with environmental concerns will know that "peak oil" is the time of maximum global production of petroleum. Some claim we have already reached peak oil, some that it will happen in the next decade, and others that we will push back peak oil by new discoveries and/or technologies. In *Peak Everything: Waking up to the Century of Declines*, Richard Heinberg (2007) argues that oil is just one of many crucial entities for which a peak is looming or has passed. Indeed, we may be ignoring the most critical of these junctures in the troubled relationship between humans and their environment: "peak soil."

Soil is a dynamic entity, continually generated by abiotic and biotic forces acting on a mineral substrate. But the very forces that create new soil—water, weathering, and climate—also contribute to its degradation. Old, highly weathered soils like those in the tropics are less fertile and more dependent on internal cycling to support vegetation. Ironically, the most lush vegetation in the world—tropical rain forests—grows on an impoverished mineral base. In temperate regions, soils are generally more fertile, but long-term cropping, particularly in grasslands, has exposed soils to higher rates of degradation and erosion than would occur naturally. Agricultural ecologist David Pimental contends that since humans began practising agriculture some 10 000 years ago, we have been degrading soils at a rate of 55 000 ha per year—one to two orders of magnitude faster than new soil is being created. The pace has accelerated; by the end of the 20th century, arable land was being lost at a rate of 10 million

continued on page 68

ECOLOGICAL ISSUES continued

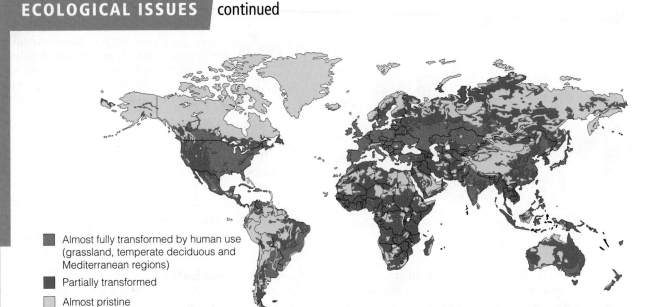

Almost fully transformed by human use (grassland, temperate deciduous and Mediterranean regions)

Partially transformed

Almost pristine

Figure 1 Current state of the world's soils. Go to GRAPHIT! at www.pearsoncanada.ca/ecologyplace to graph global soil degradation.

(Adapted from AAAS Atlas of Population and Environment, ESRI 2000.)

ha yearly (Pimental et al. 1995). Peak soil has thus come and gone, and, as Ward Chesworth (2010) warns, our continued attempts to "mine" the soil to feed an ever-growing population will only worsen the already deteriorating state of soils worldwide (Figure 1).

How can this be? We assume that modern agriculture practices improve soil, and hence the productivity of crops. In the short term, we do enhance soil fertility and increase its moisture content by irrigating and adding fertilizer. In 1970, Norman Borlaug was awarded the Nobel Peace Prize for his contribution to the so-called *Green Revolution*: the dramatic rise in agricultural production in less developed countries by the introduction of hybrid crops bred to respond to fertilizer and water application with large increases in productivity. But as we will see in Section 26.6, industrialized agriculture's reliance on external inputs of water, fertilizers, and pesticides comes at a high long-term ecological cost. Much of that cost is borne by soil.

Stimulating crops by adding mineral fertilizer alters mineral cycling. When added as highly mobile nitrate (NO_3^-), nitrogen is rapidly leached, and the processes by which soil nitrogen is maintained by microorganisms are compromised. Cycling of other minerals is affected too, and by large-scale removal of biomass in crops, organic matter is continuously depleted. Not only does this loss increase soil density, making it a less favourable habitat, but less organic matter also lowers mineral- and water-holding capacity. Similarly, long-term pesticide use doesn't just reduce populations of "pests" such as weeds or crop-eating insects. It also kills beneficial soil-dwelling organisms that improve soil quality for both vegetation and soil fauna. The damaging effects of agriculture on soil can only intensify as human population growth imposes increasing demands on food production worldwide.

The peak soil debate is by no means independent of the more publicized oil debate. After all, use of fossil fuels both to power machinery and to manufacture fertilizers and pesticides is

a large part of the global oil demand. Ironically, some environmentalists propose biofuels as a solution. Yet expecting agriculture to supply not only food but also fuel for an expanding population would only exacerbate our already substantial negative impact on the world's soils. Arguments in the "food or fuel" debate usually focus on the consequences for world hunger. If land currently used to grow food is increasingly converted (as in Brazil) to biofuel production, then our capacity to feed even our current population is doubtful. But this aspect of the debate, while critical from a social and ethical perspective, is less fundamental than the fact that intensified use of soils for *any* purpose—feeding our growing numbers and/or furnishing biofuels to replace conventional fuels—will speed up the global decline in soil quality.

None of these issues admit of easy solutions. We must start by acknowledging that the soil is the foundation of any ecosystem, be it a forest or a farmer's field. The health of a soil, which relates to its flora and fauna as well as to its abiotic properties, is crucial to the sustainability of the ecosystem it supports. The future of our civilization depends on protecting our soils as much as it does on ensuring our future energy supply.

Bibliography

Chesworth, W. 2010. Peak soil: Does civilization have a future? *Earthmagazine.* www.earthmagazine.org.

Chesworth, W. (ed.). 2008. *Encyclopedia of soil science.* New York: Springer.

Heinberg, R. 2007. *Peak everything: Waking up to the century of declines.* Gabriola Island, B.C.: New Society Publishers.

Pimental, D., C. Harvey, P. Resosudarmo, K. Sinclair, D. Kurz, M. McNair, S. Crist, L. Shpritz, L. Fitton, R. Saffouri, and R. Blair. 1995. Environmental and economic costs of soil erosion and conservation benefits. *Science* 267: 1117–1123.

4.9 SOIL TYPES: Different Processes Produce Different Soil Types

Regional differences in geology, climate, and vegetation generate different soil types. The broadest level of classification is the **soil order**. Each order has distinctive features (Figure 4.12) and its own distribution both globally and in Canada (Figure 4.13, p. 70). Soil scientists recognize five major soil-forming processes that contribute to different soil types: *laterization, calcification, salinization, podzolization,* and *gleization.*

Laterization is common in the humid tropics and subtropics. Warm, rainy conditions accelerate weathering and facilitate leaching. Many of the compounds released by weathering (with the exception of iron and aluminum oxides) are transported out of the profile if not taken up by plants. Iron oxides give tropical soils their unique reddish colour. See the *ultisol* profile in Figure 4.12. Heavy leaching makes these soils acidic through the loss of cations other than H^+.

Calcification occurs in temperate grasslands, such as southern portions of the prairie provinces. Evaporation and water uptake by plants exceed rainfall, leading to upward movement of dissolved alkaline salts, typically calcium carbonate from groundwater. At the same time, infiltration of surface water causes downward movement of salts. The net result is a buildup of deposits in the B horizon, sometimes forming a hard white layer or *caliche* (Figure 4.14a, p. 70). In moister areas, a humus-rich topsoil forms. See the *chernozem* profile in Figure 4.12.

Salinization functions similarly to calcification, but in drier climates. Salt deposition occurs at or near the soil surface, forming a salt crust (Figure 4.14b). Saline soils are common in deserts, but also in coastal regions exposed to sea spray. Salinization is a growing problem in dry agricultural areas where irrigation is practised.

 **Regosol (Entisol)** Immature soils that lack vertical development of horizons; associated with recently deposited sediments

 **Chernozem (Mollisol)** Surface horizons dark brown to black with soft consistency; rich in bases; soils of semi-humid regions; prone to the process of calcification

 **Luvisol (Alfisol)** Shallow penetration of humus; translocation of clay; well-developed horizons

 **Andisol** Developed from volcanic parent material; not highly weathered; upper layers dark coloured; low bulk density

 **Solonetzic (Aridisol)** Develop in very dry environments; low in organic matter; high in base content; prone to the process of salinization

 **Brunisol (Inceptisol)** Young soils that are more developed than regosols; often shallow; moderate development of horizons

 **Organic (Histosol)** High content of organic matter; formed in areas with poor drainage; bog and muck soils

 **Oxisol** Highly weathered soils with nearly featureless profile; red, yellow, or grey; rich in kaolinate, iron oxides, and often humus; in tropics and subtropics

 **Gleysol (Vertisol)** Dark clay soils that show significant expansion and contraction due to wetting and drying

 **Podzol (Spodosol)** Light grey, whitish surface horizon on top of black or reddish B horizon; high in extractable iron and aluminum; formed through process of podzolization

 **Ultisol** Intensely leached; strong clay translocation; low base content; humid warm climate; formed by process of laterization

 **Cryosol (Gelisol)** Presence of permafrost or soil temperature of 0°C or less within 2 m of the surface; formed through the process of gleization

Figure 4.12 Canadian soil orders. Several soil types absent from the Canadian system but important globally (ultisols, andisols, and oxisols) are also shown. Where equivalent soil orders exist in the American system, their names are given in parentheses.

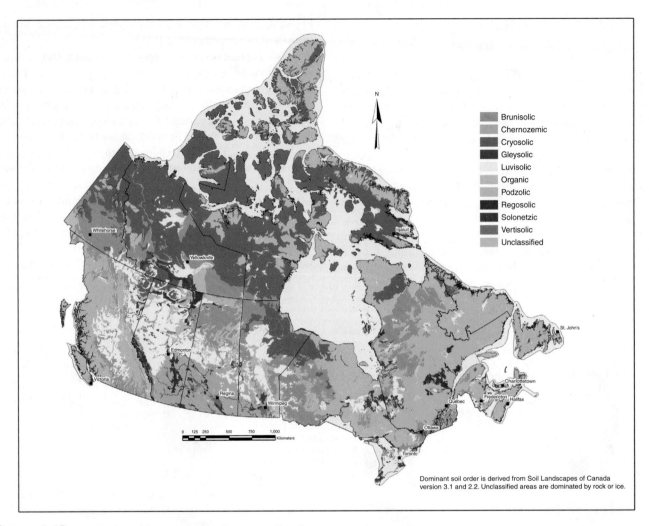

Figure 4.13 Distribution of the major soil orders across Canada.

(Source: Agriculture Canada.)

(a)

(b)

Figure 4.14 Soil-forming processes. **(a)** Calcification occurs when calcium carbonates precipitate out from water moving down through the soil or from capillary water moving up from below. The result is an accumulation of calcium in a white layer in the B horizon. **(b)** In arid regions, salinization occurs when salts accumulate near the soil surface as a result of evaporation.

Podzolization occurs in cool, moist climates of mid-latitudes where coniferous vegetation dominates. Coniferous organic matter, such as needles, creates acidic conditions, which enhance leaching and eluviation of cations and iron and aluminum compounds from the A horizon. This process, common in boreal forests as well as in some regions of southern British Columbia, creates a white- to grey-coloured Ae horizon. See the *podzol* profile in Figure 4.12.

Gleization occurs in regions with high rainfall or in low-lying areas that are poorly drained. Waterlogged conditions slow decomposition, allowing organic matter to accumulate in upper soil layers. This organic matter releases acids that react with soil iron, giving the soil a black to bluish-grey colour. See the *gleysol* profile in Figure 4.12.

Other soil types are not identified with a single process. Some, such as *regosols* and *brunisols*, have had very little development at all. They occur in coastal British Columbia and parts of the Atlantic provinces. *Organic* soils are common in northern Canada. They lack mineral content and form in areas, such as peat bogs, with poor drainage and slow decomposition. Under deciduous forests in eastern Canada, *luvisols* develop, with movement of clays but little penetration of humus. The tundra is dominated by *cryosols*, poorly drained soils underlain by permafrost.

Soil development involves the integration of biotic, climatic, and *edaphic* (soil or substrate) factors, ultimately giving rise to the diversity of soils that so profoundly influences the distribution, abundance, and productivity of terrestrial ecosystems.

EcologyPlace

Visit EcologyPlace at www.pearsoncanada.ca/ecologyplace to access online resources that complement your textbook, and help you to apply and to review the information in this chapter. EcologyPlace includes

- an eText version of the book
- self-grading quizzes
- glossary flashcards
- and more!

Go to www.pearsoncanada.ca/ecologyplace and follow the registration instructions on the Student Access Code Card included with this text. If your book does not have a Student Access Code Card, you can purchase access to it at www.pearsoncanada.ca/ecologyplace.

SUMMARY

Life on Land 4.1

Maintaining a water balance between organisms and their environment has been a major influence on life on land. The need to remain erect against gravity requires a significant investment in supportive tissues. Variation in temperature and precipitation influences metabolism in the short term and distribution and evolution of species in the long term. Patterns of terrestrial ecosystems reflect geographic gradients of temperature and precipitation.

Light and Vegetation 4.2

Light is attenuated as it passes through vegetation. Leaf quantity (expressed as the leaf area index) and orientation influence the amount of light reaching the ground. Sunflecks increase the light that reaches the forest floor. Seasonal and daily changes in solar angle also affect light penetration. As light passes through the canopy, the decrease in the ratio of red to far-red radiation is detected by phytochrome. In open sites, too much light can cause damage.

Soil Defined 4.3

Soil is the unconsolidated mineral substrate at Earth's surface. It is the medium for plant growth, the principal factor controlling the fate of water in terrestrial environments, the site of nature's recycling system, and the habitat for a diversity of life.

Soil Formation 4.4

Soil formation begins with weathering. In mechanical weathering, water, wind, temperature, and plants break down rock and minerals. In chemical weathering, the activity of soil organisms, the acids they produce, and rainwater modify soil minerals. Soil formation is affected by six factors. Parent material provides the substrate. Site history (particularly glaciation) affects the type of substrate and its deposition pattern. Climate shapes soil development through temperature, precipitation, and its influence on vegetation. Biota add organic matter and influence soil through their physiology and behaviour. Topography affects the amount of water entering the soil and the rate of erosion. Time is required to develop distinctive soils.

Soil Properties 4.5

Soils differ in colour, texture, and depth. Colour is indicative of chemical and physical properties. Soil texture (proportion of sand, silt, and clay) affects the amount of pore space and water- and mineral-holding capacity. Soil depth varies with slope, weathering, parent material, and vegetation.

Soil Horizons 4.6

Four horizons may be present in soils: O horizon (organic layer); A horizon (topsoil), characterized by accumulation of organic matter; B horizon (subsoil), in which minerals accumulate; and C horizon, the unconsolidated material extending down to bedrock.

Soil Moisture 4.7

The amount of water a soil can hold is critical. When water fills all pore spaces, the soil is saturated. When a soil holds the maximum amount of water possible against the force of gravity, it is at field capacity. Capillary water is held between soil particles by capillary forces. When plants cannot extract water, the soil has reached its wilting point. The available water capacity (the difference between field capacity and wilting point) is largely determined by texture.

Soil Fertility 4.8

Soil particles, particularly clay and humus, affect nutrient availability. The cation exchange capacity of a soil is the num-ber of negatively charged sites. Cations occupying exchange sites are in a state of dynamic equilibrium with cations in the soil solution. Anions are only present in the soil solution. Soil acidification occurs naturally as a result of the release of H^+ by roots, but it is exacerbated by acid precipitation. Soils differ in buffering capacity.

Soil Types 4.9

Regional differences in geology, climate, and vegetation give rise to different soils. The broadest level of classification is the order, which reflects the influence of five soil-forming pro-cesses: laterization, calcification, salinization, podzolization, and gleization. The Canadian soil system has 10 orders, which occur in different regions of the country.

KEY TERMS

A horizon
Ae horizon
available water
   capacity
B horizon
C horizon
calcification
capillary
   water

cation exchange
   capacity
chemical weathering
eluviation
ephemeral
field capacity
gleization
humus
illuviation

laterization
leaching
leaf area index
light extinction
   coefficient
lyotropic series
mechanical
   weathering
O horizon

parent material
phytochrome
podzolization
regolith
salinization
soil
soil erosion
soil horizon
soil order

soil profile
soil solution
soil texture
sunflecks
transmittance
water balance
wilting point

STUDY QUESTIONS

1. Explain two major constraints imposed on organisms by the transition to land habitats.
2. Assume two forests have the same LAI. In one, the leaves are held horizontally, whereas in the other the leaves are held at a 60° angle. How would light availability at the forest floor differ for these forests at noon? Which forest floor would receive more light at mid-morning?
3. Explain why varying leaf angle can be an adaptive strat-egy in different abiotic conditions, including low light, high light, high temperatures, and low moisture.
4. Contrast the impact of the two types of weathering on soil development.
5. What six major factors affect soil formation, and how? In general terms, describe the influence of each factor in your region of Canada.
6. Many invertebrates (e.g., earthworms and nematodes) live in soil, as do some small mammals (ground squirrels). How would their activities affect specific soil properties?
7. In Figure 4.10, which soil holds more moisture at field capacity? At wilting point? Which soil has more available water when the soil water content is 2.5 mm per cm (y-axis)?
8. What distinguishes the O and A horizons? Why is the A horizon so much deeper and more fertile in a prairie soil compared with a typical forest soil?
9. Why do clay soils typically have more cation exchange capacity than do sandy soils?
10. According to the lyotropic series, H^+ ions are held more tightly than most other cations. How would release of H^+ by roots facilitate uptake of Mg^{2+} and K^+? Over time, how might this H^+ release reduce soil fertility through leaching?
11. Why is salinization more prevalent in dry areas? How does irrigation accelerate salinization? (Remember that solutes diffuse from areas of high to areas of low concen-tration.) What practices might reduce salinization?
12. What soil-forming process dominates in the wet tropics, and how does it affect nutrient levels in the A horizon? How do these soils support lush vegetation, despite their low fertility?

FURTHER READINGS

Agriculture Canada, Soil Classification Working Group. 1998. *The Canadian system of soil classification.* 3rd ed. Agriculture and Agri-Food Canada Publication 1646. Excellent resource for soils in Canada, available on the web.

Brady, N. C., and R. W. Weil. 1999. *The nature and properties of soils.* 12th ed. Upper Saddle River, NJ: Prentice Hall. Classic introductory textbook on soils.

Kohnke, H., and D. P. Franzmeier. 1994. *Soil science simplified.* Prospect Heights, IL: Wavelength Press. Well-written overview of concepts and principles of soil science for the general reader.

Patton, T. R. 1996. *Soils: A new global view.* New Haven, CT: Yale University Press. Outlines a new approach to studying soil formation at a global scale.

THE ORGANISM
AND ITS ENVIRONMENT

The Namib Desert, stretching for 1700 km along Africa's southwest coast, is home to the highest sand dunes in the world. Rainfall is a rare event here. But each morning as the Sun rises, the cool, moist air of this coastal desert begins to warm, and the Namib becomes shrouded in fog. And each morning, black thumbnail-sized beetles perform one of nature's more bizarre behaviours (Figure 1). These tenebrionid beetles (*Stenocara* spp.) upend their bodies into what looks like a handstand. The beetle stays in this position as fog droplets collect on its back and roll down the wing case (the elytra) into its mouth. If you viewed the bumps on its back (Figure 2) through an electron microscope, you would see a wax-coated carpet of tiny nodules covering the bumps and the valleys between them that channel water into the beetle's mouth.

Tenebrionid beetles illustrate two fundamental ecological concepts: (1) the relationship between structure and function and (2) how that relationship reflects adaptations of organisms to their environment. The structure of the beetle's back and the beetle's behaviour of standing on its head in the morning fog together help the beetle acquire water, a scarce resource in this arid environment. These same traits, however, are unlikely to be useful for acquiring water in deserts of the continental interior, where morning fog does not form, or in wet environments, such as a tropical rain forest, where standing pools of water are readily available.

Key questions arise from such observations: *What controls the distribution and abundance of species? What enables a species to succeed in one environment but not another?* The link between structure and function provides a vital clue. The traits of an organism—its physiology, morphology, behaviour, and life history (lifetime pattern of development and reproduction)—reflect adaptations to its environment. Each environment

Figure 1 Tenebrionid beetle perched on a sand dune in the Namib Desert.

Figure 2 Fog droplets on the wing case (elytra) of a tenebrionid beetle.

presents a different set of constraints on survival, growth, and reproduction. Traits that enable an organism to succeed in one environment typically preclude it from doing equally well in another. Different evolutionary solutions to life in various environments are largely the result of trade-offs. In nature, one size does not fit all. Earth's diverse environments are inhabited by some 1.9 million known species—in effect, 1.9 million different ways that life exists on this planet.

Despite the diversity of species, all organisms—from single-celled bacteria to the largest of all animals, the blue whale—represent solutions to the same three basic needs: (1) assimilation, (2) reproduction, and (3) response to external stimuli. Organisms must assimilate energy and matter in order to produce new tissue. To maintain the continuity of life, some of this assimilated energy and matter must be allocated to reproduction. Finally, organisms must respond to stimuli relating to environmental factors, both abiotic (such as temperature and humidity) and biotic (such as recognition of potential mates or predators).

Perhaps the most fundamental constraint on life is the acquisition of energy. It takes energy to assimilate essen-

tial nutrients and to perform the fundamental processes of life—growth, maintenance, and reproduction. Chemical energy is generated by the breakdown of carbon compounds in all living cells, via *aerobic* or **anaerobic respiration**. But the ultimate source of energy for life on Earth is the Sun (with the exception of ecosystems based on *chemosynthesis*, such as hydrothermal vents; see Research in Ecology: The Extreme Environment of Hydrothermal Vents, pp. 46–47). Solar energy fuels photosynthesis, the process whereby plants, algae, and some bacteria convert inorganic carbon into carbohydrates (see Section 6.1). By consuming plant and animal tissues, other organisms use energy derived from photosynthesis. The energy source of an organism is a fundamental ecological distinction. Those that derive their energy directly from sunlight are **photoautotrophs (primary producers)**. Those that derive their energy from consuming other organisms are **heterotrophs (secondary producers)**.

These two modes of energy acquisition impose fundamentally different evolutionary constraints. We discuss the ecological adaptations of autotrophs and heterotrophs in Chapters 6 and 7, respectively. Of the innumerable adaptations possessed by organisms, these chapters focus on those facilitating the exchange of energy, carbon, nutrients, and water between an organism and its environment, thereby affecting survival and growth. Chapter 8 treats adaptations relating to reproduction—how organisms use resources to try to ensure the continuity of life. The idea of trade-offs is a common theme, linking the "problems" imposed by different environments and the "solutions" reflected in the diverse traits of species.

But first, in Chapter 5, we turn our attention to ecological genetics and in particular to natural selection, the unifying concept that is fundamental not only to understanding the evolution of adaptations but also to linking pattern and process at all levels, from organisms to ecosystems.

(a)

(b)

(c)

Structural variations in the hind legs of beetles. **(a)** In diving beetles (Dytiscidae), the last pair of legs bear long hairs that aid in swimming. **(b)** Dung beetles (Scarabaeidae) have wide legs with spines, used to fashion dung into circular structures in which the female lays a single egg. **(c)** The enlarged hind legs of flea beetles (Chrysomelidae) aid in jumping.

One of every four known animal species is a beetle. There are some 400 000 beetle species in the order Coleoptera (meaning "sheathed wing," referring to the hardened forewings covering the delicate hind wings). Scientists estimate the actual number of beetle species to be in the millions. Beetles live almost everywhere, from the Arctic tundra to deserts, and exploit almost every type of habitat and food. This diversity of habitats and diets is paralleled by an equally impressive diversity of morphology and behaviour. While most beetles use their legs for walking, legs have been modified for other uses. In diving beetles, the last pair of legs bears rows of long hairs that aid in swimming. Others have wide legs with spines for digging. Dung beetles use their legs to fashion dung into circular structures in which the female lays a single egg. Flea beetles use their enlarged hind legs for jumping. These structural variations represent *adaptations*—traits that enable an organism to exploit a particular resource or thrive in a given environment.

As evolutionary ecologist Ernst Mayr notes, examples like these signified to the early 19th-century mindset the "wise laws that brought about the perfect adaptation of all organisms one to another and to their environment" (2000). Adaptation seemingly implied design, and design a designer. Natural history was the task of cataloguing the creations of the divine architect. But by the mid-1800s, a revolutionary idea emerged that would forever change our view of nature:

> [I]t is quite conceivable that a naturalist . . . might come to the conclusion that species had not been independently created, but had descended . . . from other species . . . [S]uch a conclusion . . . would be unsatisfactory, until it could be shown how the innumerable species, inhabiting this world, have been modified, so as to acquire that perfection of structure and coadaptation which justly excites our admiration.

This passage is from *The Origin of Species*, first published in 1859. With this seminal work, Charles Darwin altered the history of biology by questioning a worldview that had been held for millennia. He proposed a mechanism—natural selection—to explain how organisms have acquired traits that facilitate their survival and reproduction. Its beauty lay in its simplicity: natural selection works not by design but by eliminating "inferior" individuals.

5.1 ADAPTATIONS: Natural Selection Promotes Adaptive Traits

Stated precisely, **natural selection** is the differential success (in survival and reproduction) of individuals that arises from their interactions with their environment. Natural selection has two necessary conditions. (1) Variation exists among individuals in a population in some heritable trait. (2) This variation causes individuals to differ in survival and reproduction. In essence, natural selection is a numbers game. Individuals with more surviving offspring than others are considered more fit. In evolutionary terms, **fitness** is the proportionate contribution an

individual makes to future generations. In any given environment, individuals with traits that enable them to survive and reproduce, eventually passing those traits on to future generations, are selected for. Individuals without these traits are selected against. Natural selection thus leads to **evolution**: changes in the gene frequency of a population over generations.

Changes in gene frequencies from parental to offspring generations reflect differences in the fitness (survival and reproduction) of individuals in the parental generation. But if fitness is an individual's relative contribution to future generations, why mention survival? An individual has to survive to reproduce, so survival is necessary for fitness but does not determine it. Consider two individuals, one of which lives much longer than the other, and may even be more vigorous. If the shorter-lived individual produces more viable offspring, then it has greater fitness. This doesn't mean that the longer-lived individual is ecologically unimportant. Over its lifetime, it may interact with more individuals of its own or other species as a predator, competitor, or mutualist. In doing so, it exerts natural selection pressure on traits that affect their fitness.

To recap—an **adaptation** is a heritable behavioural, morphological, or physiological trait that has evolved over time by natural selection such that it maintains or increases the fitness of an organism under a given set of environmental factors. The concept of adaptation by natural selection is central to ecology because adaptations affect how successfully an organism interacts with its environment. Not only do adaptive traits enable an organism to survive and reproduce under prevailing conditions; they also influence the interactions of the organism with others, both of the same and different species. How adaptations enable an organism to function in its prevailing environment—and conversely, how those same adaptations limit its success in other environments—is key to understanding species distribution and abundance.

5.2 GENES: Adaptations Are Coded by Genes, the Basic Units of Inheritance

Adaptations are heritable—they pass from parent to offspring. So, to understand adaptations, we must understand inheritance: how traits are passed between generations, and what forces bring about changes in these traits. At the root of all similarities and differences among organisms is the information contained within DNA (deoxyribonucleic acid) molecules. Recall from basic biology that DNA is organized into discrete informational subunits called **genes**: stretches of DNA coding for a polypeptide (sequence of amino acids), where one or more polypeptides make up a protein. **Alleles** are alternative forms of a gene. The **genome** is the totality of the DNA in a cell. In gene expression, proteins that determine traits are produced from the DNA blueprint.

In eukaryotic organisms, genes occur on microscopic, thread-like **chromosomes** in the nucleus (Figure 5.1, p. 78). The **locus** is the position occupied by a gene on a chromosome. In most multicellular organisms, each cell is **diploid** ($2n$) and has two copies of each chromosome, one inherited from its female parent through

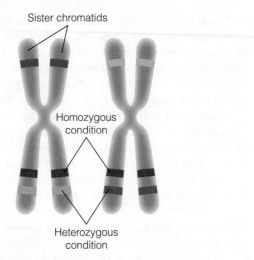

Figure 5.1 A pair of homologous chromosomes. For the gene shown in red, this individual possesses two identical alleles and is homozygous at that locus; for the genes shown in green, it possesses differing alleles and is heterozygous. In eukaryotic organisms, genes are separated on chromosomes by regions called *introns*.

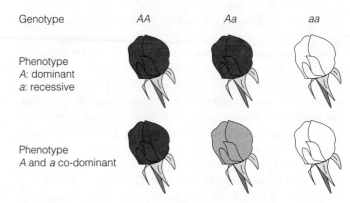

Figure 5.2 Modes of gene expression at a single locus. Assume flower colour is controlled by a single gene with two alleles, *A* and *a*. The *A* allele codes for production of red pigment, and the *a* allele codes for no pigment. In the first case, heterozygous (*Aa*) individuals exhibit the same phenotype as homozygous-dominant (*AA*) individuals. The recessive allele (*a*) is expressed only in homozygous-recessive (*aa*) individuals. In the second case, heterozygous individuals are intermediate to the homozygotes. The alleles are co-dominant, with each allele having a proportionate effect on the phenotype.

the ovum (egg) and one from its male parent through the sperm. Each diploid individual thus contains two alleles of each gene, one at each corresponding locus of its two **homologous chromosomes**. If the two alleles are identical, the individual is **homozygous** at that locus. If they differ, the individual is **heterozygous** at that locus. The two alleles present at a locus define the **genotype** of an individual for the trait coded by that gene.

5.3 PHENOTYPES: Genotype Expression Is Affected by the Environment

What we see when we observe an organism is not its genotype but its **phenotype**: the observable expression of a trait as determined by the genotype interacting with the environment. Although the phenotype for structural traits such as hair colour is easily visible, not all traits are so obvious. The phenotype of a structural trait may involve a feature that is internal or microscopic. The phenotype of a physiological trait is typically not manifest in the organism's appearance.

In straightforward Mendelian genetics, different genotypes may or may not produce different phenotypes. For a heterozygous individual, one allele (**dominant allele**) may mask expression of the other (**recessive allele**) (Figure 5.2). In this case, the phenotype of the heterozygote is identical to that of an individual that is homozygous dominant for that trait. Only a homozygous recessive individual expresses the recessive allele. In other cases, a heterozygote produces a phenotype that is intermediate between those of the homozygotes. The alleles are then said to be *co-dominant*, with each allele having a proportionate effect on the phenotype (Figure 5.2).

Phenotypic traits that fall into a limited number of distinct categories, such as the flower colour example in Figure 5.2, are **qualitative traits**. Although all genetic variation is discrete in that it is coded by separate genes, many phenotypic traits have

a continuous distribution (e.g., height), and are **quantitative traits**. There are two reasons for a continuous distribution. First, many traits are affected by more than one locus. If flower colour is controlled by genes at two loci, each with two alleles (*A, a* and *B, b*), there are nine possible genotypes (Figure 5.3). Instead of two distinct colour phenotypes arising from a single

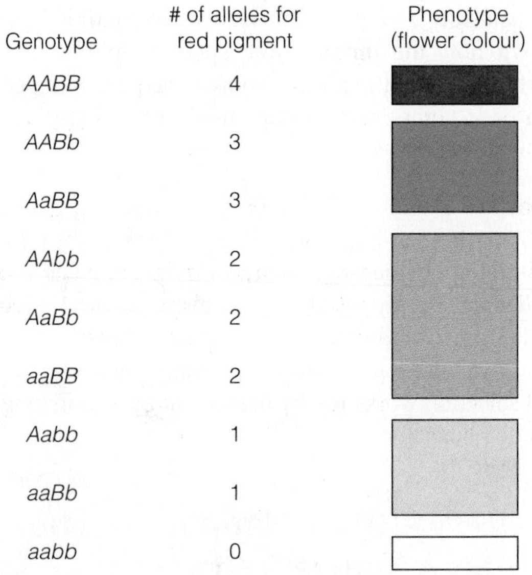

Genotype	# of alleles for red pigment	Phenotype (flower colour)
AABB	4	
AABb	3	
AaBB	3	
AAbb	2	
AaBb	2	
aaBB	2	
Aabb	1	
aaBb	1	
aabb	0	

Figure 5.3 A phenotypic trait controlled by two loci. Assume flower colour is controlled by two genes, each having two alleles (*A:a* and *B:b*). Both *A* and *B* alleles code for production of red pigment, while *a* and *b* alleles do not. There are nine possible genotypes, with the number of pigment-coding alleles ranging from four (*AABB*) to zero (*aabb*). Resulting phenotypes fall into five categories ranging from dark red to white depending on the number of pigment-coding alleles. The intermediate colour (two alleles) is the most abundant. The number of possible phenotypes increases with the number of loci controlling phenotypic expression.

locus with complete dominance, there is now a colour range from dark red to white, depending on the number of loci.

The second reason why many traits (even those with a single locus) show continuous phenotypic variation is because gene expression is often affected by the environment. Since temperature, rainfall, light, and predation usually vary continuously, the environment can cause the phenotype of a given genotype to vary continuously. Using the example of flower colour controlled by two loci, assume pigment production is affected by temperature. If cool temperatures reduce the expression of *A* and *B* alleles, temperature flux during development further increases the colour range produced by the nine genotypes. We examine gene–environment interactions (*phenotypic plasticity*) in more detail in Section 5.12.

5.4 GENETIC VARIATION: Allele and Genotype Frequencies Differ within and among Populations

Adaptations are traits of individuals that result from natural selection acting on the interaction of their genes with the environment. Yet even though selection reflects the success or failure of individuals, the genetic makeup of a population changes through time, as individuals either succeed or fail to pass on their genes. So, to understand adaptation through natural selection, we must consider how genetic variation is structured within populations.

A species rarely consists of a single continuous interbreeding population. Instead, a population is often organized as a **metapopulation**: a group of subpopulations (local populations) of interbreeding individuals, linked to each other by movements of individuals (see Chapter 12). Genetic variation can thus occur at two hierarchical levels: *within* and *among* local populations. Genetic variation *among* local populations of a species is called **genetic differentiation**.

The **gene pool** is the sum of genetic information present in all individuals in a population. It represents the total genetic variation of the population, as quantified by (1) **allele (gene) frequency** and (2) **genotype frequency**, where *frequency* refers to the proportion of a given allele or genotype among all the alleles or genotypes at that locus in a population. Quantifying the gene pool is critical for understanding not only the present status of a population but also its potential to respond to selective pressures exerted by the environment in the future.

5.5 TYPES OF NATURAL SELECTION: Natural Selection Can Be Directional, Stabilizing, or Disruptive

Earlier we defined evolution as changes in the gene frequency of populations over successive generations. More specifically, phenotypic evolution is a change in the mean and/or variance of a phenotypic trait across generations as a result of changes in gene frequencies. (See QUANTIFY it! at **www.pearsoncanada.ca/ecologyplace** for a discussion of statistical methods for estimating central tendency and variation in populations.) In favouring one phenotype over another, natural selection acts on the individual, but in doing so, it changes the gene frequency of the population.

Given the long timeline of evolution, how do scientists detect natural selection at work? Research often focuses on "model" organisms, one of which is the ground finch of the Galápagos Islands, the same islands whose diverse biota so influenced Darwin when he was the naturalist aboard the HMS *Beagle*. This research has revealed a dramatic shift in a physical trait of finches inhabiting these islands during a period of extreme climate change.

Recall from Section 5.1 that natural selection requires (1) variation among individuals in a population for some heritable trait and (2) differential effects of that trait on fitness. Beak size varies in Darwin's medium ground finch (*Geospiza fortis*) on the 40-ha islet of Daphne Major (Figure 5.4) (Boag and Grant 1984). A comparison of beak size between parents and offspring (Figure 5.5, p. 80) established that this trait is heritable (Boag 1983), thus meeting the first condition.

Demonstrating the second condition of natural selection—that variation in a trait affects fitness—is more difficult. As a structural trait, beak size has the potential to affect fitness by influencing feeding behaviour. Ground finches with large beaks feed on a wide range of seed sizes, whereas birds with smaller beaks consume only smaller, softer seeds (Boag and Grant

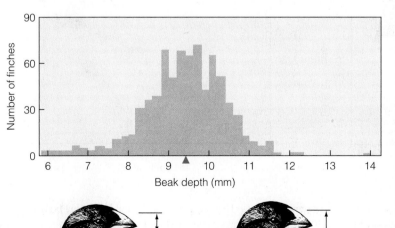

Figure 5.4 Variation in beak size in Galápagos medium ground finches on Daphne Major. The *y*-axis shows the number of individuals sampled in 1976 in each bill depth category (*x*-axis). The blue triangle is the estimated population mean.

(Adapted from Grant 1999, after Boag and Grant 1984.)

█ INTERPRETING ECOLOGICAL DATA

Q1. What type of data do the original measures of bill depth represent? (See Quantifying Ecology 1.1: Classifying Ecological Data, p. 8.)

Q2. How have the original measurements of bill depth been transformed in Figure 5.4? What type of graph is Figure 5.4?

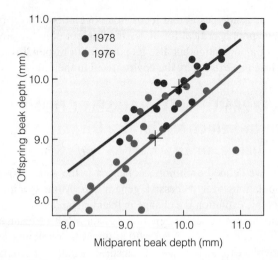

Figure 5.5 Relationship between beak size of offspring and parents in medium ground finches on Daphne Major. The slope estimates heritability. Blue is 1976 data, red 1978. The mean is indicated by +. Beak size of offspring increased in 1978, but beak size of parents and offspring was strongly correlated in both years.

(Adapted from Grant 1999, after Boag 1983.)

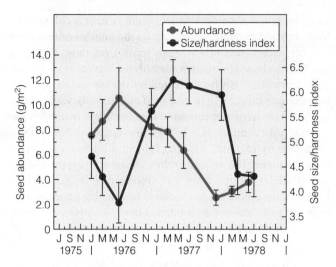

Figure 5.6 Changes in seed abundance and seed traits on Daphne Major from 1975 to 1978. Bars represent 95% confidence intervals. Seed size and hardness index is the square root of seed depth × hardness.

(Adapted from Grant 1999, after Boag and Grant 1981.)

1984). But it took a departure from normal conditions to show that this trait affects fitness.

During the early 1970s, the island received normal annual rainfall (127–137 mm), supporting abundant seed production and a large finch population (~ 1500). In 1977, a shift in the climate of the eastern Pacific (La Niña; see Section 2.10) caused a severe drought. Only 24 mm of rain fell that season, and seed production plummeted. Small seeds declined more than large seeds, increasing the mean size and hardness of seeds available (Figure 5.6). This decline in seed abundance precipitated an 85 percent drop in finch numbers through mortality and emigration. However, mortality effects were unequally distributed, and the body size of sur-

viving finches increased (Figure 5.7). Small birds had difficulty finding food, while large birds, especially large-beaked males, survived best because they could crack large, hard seeds.

The second condition of natural selection is satisfied: differences in fitness are associated with differences in a phenotypic trait (beak size). (The researchers based fitness on survival, not reproductive success, given that only surviving birds can pass on their genes.) The phenotypic trait that selection acts upon is the **target of selection**—in this case, beak size. The environmental cause of fitness differences among organisms with different phenotypes is the **selective agent**—in this case, a change in food abundance and size.

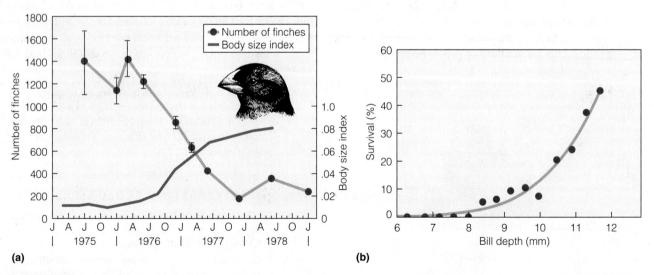

Figure 5.7 Natural selection in the medium ground finch population on Daphne Major. **(a)** Declining population and increasing body size after the 1977 drought. Points are means plus 95 percent confidence intervals. **(b)** Birds with larger beaks survived better because they could feed on the more available larger seeds.

(Adapted from Grant 1999, after Boag and Grant 1981.)

Birds with larger beaks survived better, causing a shift in beak size distribution (Figure 5.8) (Boag and Grant 1984). This type of selection, where the mean value of a trait shifts towards one extreme over another, is **directional selection** (Figure 5.9a). (In this case, larger size was more advantageous, but bigger is not always better; for some traits, small size increases fitness.) In **stabilizing selection**, natural selection favours individuals near the mean at the expense of those at the extremes (Figure 5.9b). **Divergent (disruptive) selection** favours both extremes of a trait, even if not to the same degree, generating a bimodal trait distribution (Figure 5.9c). Divergent selection occurs when a population is subjected to different selection pressures.

All three types of selection can alter allele frequency, but whereas directional and stabilizing selection are generally associated with evolution *within* a species, divergent selection can cause evolution of new species (see Section 5.9 and Research in Ecology: Dynamic Evolution in Sticklebacks, pp. 87–89). A precondition of **sympatric speciation** (evolution of new species in the same location) by divergent selection is **reproductive isolation**—separation of (sub)populations by an inability to produce viable offspring, either by *pre-mating* or *post-mating isolating mechanisms*.

It is tempting to assume that once selection is set in motion, its effect will continue. But the direction of selection is not inevitable: environmental change can halt or even reverse its effects. By feeding finches and introducing non-native plants, the human presence on Santa Cruz has altered selection pressures. Compared with a less-populated site on Santa Cruz where there is divergent selection for beak size, selection for finches with intermediate beak size is occurring at Academy Bay (Hendry et al. 2006). In this case, humans may be altering selection from divergent to stabilizing, thereby halting incipient speciation by weakening the bimodality.

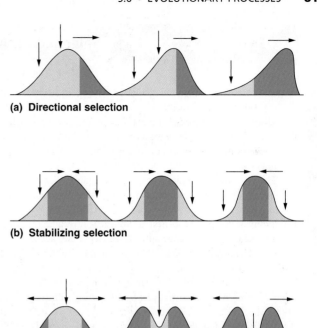

(a) **Directional selection**

(b) **Stabilizing selection**

(c) **Divergent selection**

Figure 5.9 Three types of selection. The curves represent the changing distribution of phenotypes in a population over time. The *x*-axis shows the range of a phenotype in a population, the *y*-axis the number of individuals with that phenotype. (**a**) Directional selection moves the mean towards one extreme. (**b**) Stabilizing selection favours organisms with values near the mean. (**c**) Divergent selection increases the frequency of both extremes. Vertical arrows represent selection pressures, horizontal the direction of change.

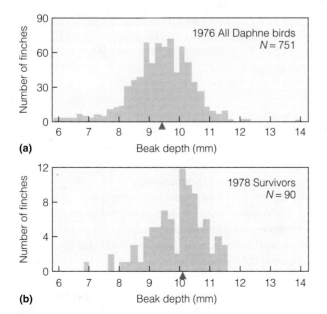

(a) Beak depth (mm)

(b) Beak depth (mm)

Figure 5.8 Beak size of medium ground finches on Daphne Major (**a**) before and (**b**) after drought. Blue triangles indicate mean beak depth.

(Adapted from Grant 1999, after Boag and Grant 1984.)

Throughout this text, adaptation arising from natural selection is a unifying concept for understanding species distribution and abundance. We explore the selective pressures giving rise to the adaptations that define the diversity of life, as well as the trade-offs associated with these adaptations under different conditions. (In Ecological Issues: Selecting for Antibiotic Resistance, p. 82, we explore an instance where humans have unwittingly caused natural selection to work against our interests as a species.) In Chapters 6 and 7 we consider adaptations of organisms to key features of the abiotic environment that directly influence their survival and growth. In Chapter 8 we consider the trade-offs involved in the evolution of life history traits relating to reproduction. Part Four examines the role of species interactions as agents of natural selection. Ultimately, we consider how trade-offs inherent in adaptations affect the patterns and processes of communities and ecosystems, as environmental factors change in space and time.

5.6 EVOLUTIONARY PROCESSES: Various Processes Disrupt the Hardy–Weinberg Equilibrium

Natural selection is the only evolutionary process that causes adaptations because it is the only one in which changes in allele frequency between generations reflect differences in the relative

ECOLOGICAL ISSUES   Selecting for Antibiotic Resistance

As we have seen from the change in bill size in finches, natural selection can act quickly when populations are exposed to sudden shifts in the environment. The more negative the shift, the more intense selection pressure can be. In bacteria, humans have accelerated the evolution of antibiotic resistance by our attempts to eradicate disease-causing bacteria.

The discovery of antibiotics is one of medicine's greatest triumphs. Since their introduction in the 1940s, antibiotics have saved many lives. But soon after companies began mass-producing penicillin (the first antibiotic), resistant microbes began appearing. Resistance spread fast. By 1987, according to the U.S. Centers for Disease Control and Prevention (CDC), 0.02 percent of pneumonia-causing bacterial strains were penicillin resistant. By 1998, that percentage had risen to 24, with many strains resistant to several antibiotics (Whitney et al. 2000). In 2002, the CDC estimated that over 44 000 North Americans die annually from antibiotic-resistant bacterial infections. In the summer of 2011, 17 deaths were attributed to an outbreak of drug-resistant *Clostridium difficile* in the Niagara, Ontario, area. Although hospitals pose the greatest risk of exposure, Lowe and Romney (2011) report the presence of antibiotic-resistant bacteria in bed bugs in Vancouver.

This alarming increase is a direct outcome of evolution. Widespread antibiotic use for both humans and livestock has accelerated selection in bacterial populations. All populations contain genetic variations that influence traits—in this case, the ability of a bacterium to withstand an antibiotic. When a person takes an antibiotic, the drug kills the vast majority of bacteria. But resistant bacteria survive. These resistant strains then multiply, increasing their numbers—up to a million-fold per day. Thanks to positive feedback (see Section 19.2), the process accelerates, with more and more bacteria developing resistance to more and more antibiotics. Selection favours resistant individuals, whose increased fitness leads to a greater proportion of resistant individuals in each generation. The rapidity of bacterial growth speeds up evolution even more.

An individual can develop an antibiotic-resistant infection not only by becoming infected with a resistant strain but also by having a resistant strain emerge in the body once antibiotic use begins. Bacteria acquire resistance genes in three ways: (1) bacterial DNA may mutate spontaneously (likely the most common way); (2) one bacterium may gain DNA from another by a process called *transformation*; or (3) resistance may be acquired from a small circle of DNA called a *plasmid*, which can be transmitted between bacteria. A single plasmid can provide multiple resistance. In 1968, over 12 000 people died in Guatemala in a diarrhea epidemic. The bacterial strain responsible harboured a plasmid that carried resistance to four different antibiotics.

Although antibiotic resistance is a natural phenomenon, inappropriate antibiotic use exacerbates the problem. Doctors are pressured to prescribe antibiotics for viral infections. Patients may not take the complete course of an antibiotic, increasing resistance. Giving antibiotics to healthy livestock to prevent disease or to increase weight gain is another concern. Although government agencies limit the amount of drug residue allowed in meats, these drugs accelerate the development of antibiotic resistance, making human illnesses harder to treat.

Evolution of antibiotic resistance is not a deliberate case of **artificial selection** (manipulation by humans of the gene pool, as with selective breeding of crops and livestock), but it is a dramatic illustration of the power of selection to increase the prevalence of a trait that is as adaptive for bacteria as it is deleterious for humans.

Bibliography

Lowe, C. E., and M. G. Romney. 2011. Bedbugs as vectors for drug-resistant bacteria. *Emerging Infectious Diseases* 17 [letter].

Whitney, C. G., M. M. Farley, J. Hadler, I. H. Harrison, C. Lexau, and A. Reingold. 2000. Increasing prevalence of multidrug-resistant *Streptococcus pneumoniae* in the United States. *New England Journal of Medicine* 343:1917–1924.

---

fitness of individuals in a population. But not all phenotypic traits represent adaptations, and processes other than natural selection alter both allele and genotype frequencies. The following evolutionary processes may act on a population concurrently with natural selection, or in its absence: (1) *mutation*, (2) *genetic drift*, (3) *gene flow*, and (4) *non-random mating*.

Before we explore these processes, we must consider a puzzling question: *How are harmful recessive alleles maintained in a population?* The answer lies in a fundamental tenet of population genetics, named for the German and British scientists who published it independently in 1908. The **Hardy–Weinberg principle** provides mathematical proof that both allele and genotype frequencies remain unchanged in successive generations of a sexually reproducing population if certain criteria are met: (1) natural selection does not occur, (2) the mutation rate is very low, (3) the population is sufficiently large that genetic drift is not a factor, (4) there is

no net movement of individuals, and (5) mating is random. In sum: No evolutionary change occurs strictly as a result of sexual reproduction. Go to QUANTIFYit! at **www.pearsoncanada.ca/ecologyplace** to work through an example showing how the Hardy–Weinberg equilibrium works.

In natural populations, these assumptions are rarely met. Natural selection *does* occur, mutations *do* happen, populations *may* be small, individuals *do* move between populations, and mating is *not* always random. Any or all of these circumstances can change the frequencies of either alleles or genotypes between generations. This reality might seem to make the Hardy–Weinberg principle irrelevant to a discussion of evolution. But the utility of the principle is that it provides a default position or null model, deviations from which elucidate the evolutionary processes at work within a population. We have already considered natural selection, which violates the first assumption, but when any one of the other four evolutionary processes discussed below occurs, it violates another assumption and thereby disrupts the Hardy–Weinberg equilibrium.

1. *Mutations.* The ultimate source of the genetic variation upon which natural selection acts is **mutation**: heritable change in the structure of genes (*micromutation*) or chromosomes (*macromutation*). The term can refer to the process of altering the material as well as to the altered state. Although mutation is a random process, the mutation rate is affected by many factors, including UV radiation. An altered phenotypic trait resulting from a mutation may be beneficial, neutral, or harmful. Moreover, a mutation that enhances fitness in one environment may reduce it in another. Most mutations that have large effects are harmful, but such mutations often do not survive long. Natural selection eliminates most deleterious alleles from the gene pool, leaving behind alleles that enhance (or at least do not harm) an organism's ability to survive, grow, and reproduce. Whatever their impact, the mere fact of mutations alters allele frequencies.

2. *Genetic drift.* Changes in allele frequencies that occur in small populations as a result of chance are called **genetic drift**. Because allele recombination in sexual reproduction is random, such recombination is subject to the laws of probability. The offspring represent only a subset of the parents' alleles. If the parents have only a few offspring, not all the parental alleles will be passed on to their progeny through the assortment of chromosomes at meiosis and recombination during gamete fusion. Genetic drift is the evolutionary equivalent of sampling error, with each successive generation representing only a subset of the gene pool from the previous generation.

In a large population, genetic drift has little effect, but in a small population its impact can be rapid and significant. Consider tossing a coin. With a single toss, the probability of each of the two possible outcomes, heads or tails, is equal (50 percent). With a series of four tosses, the probability of the outcome being two heads and two tails is 50 percent. But each outcome is independent, and so, in a series of four tosses, there is also a probability of 0.0625 (6.25 percent) that the outcome will be four heads. The probability of the outcome being all heads drops to 8.88×10^{-16} for 50 tosses. Likewise, the probability of heterozygous (*Aa*) individuals producing only homozygous (either *aa* or *AA*) offspring, and hence "losing" one of the alleles to genetic drift, decreases with increasing population size.

There are two ecological situations in which genetic drift is particularly important. When a few individuals colonize an area, their genetic composition can have a long-lasting **founder effect** on the resident population. Some assume the founder effect is an example of natural selection, because the colonizing individuals must have traits that have been selected for in the new habitat. While this can be true, colonizing ability doesn't necessarily correlate with traits that improve fitness in the new environment, if there is a large element of chance involved in which individuals colonize. Non-colonizing individuals might actually be better adapted to the new site. Some interesting examples of the founder effect in humans include high rates of deafness in Martha's Vineyard, an island off the coast of Massachusetts, and the incidence of myotonic dystrophy in the Lac

Saint Jean region of Québec, where the condition affects 1 in 550 compared with 1 in 5000 to 50 000 in the general population (Yotova et al. 2005). This medical condition has not been selected for in this environment; it just happened that the initial colonizers had a high incidence, and the founder effect has persisted in future generations.

Genetic drift is also involved in an **evolutionary bottleneck**, which results when a disturbance wipes out most of a population, leaving a few individuals from which the population re-establishes (Figure 5.10). Although these individuals did not survive as a result of any adaptive traits (if they had, it would have been an instance of natural selection) but merely by chance, their genetic composition is disproportionately represented in future generations. If a population remains small after a bottleneck, as on an island, it is vulnerable to further genetic drift.

Even if the population recovers, its genetic diversity is often reduced, increasing its vulnerability to environmental change. The northern elephant seal (*Mirounga angustirostris*) occupying the Pacific coast of North America was hunted almost to extinction, with only 20 remaining by 1890. Now a protected species, its numbers recovered to some 30 000 by the early 1970s. However, its genetic variation is highly compromised in contrast to that of the southern elephant seal (*M. leonina*), which was never severely reduced and retained high genetic variability (Bonnell and Selander 1974). So far, this reduced genetic variation has not prevented its ongoing recovery—numbers now exceed 100 000—but it could prove critical in the long term if the alleles needed to cope with future environmental stress are absent from the gene pool.

3. *Gene flow.* Genetic variation within a population is also influenced by **gene flow**: movement of genes between populations as a result of **migration**: (movement of individuals into (**immigration**) or out of (**emigration**) local populations. Many

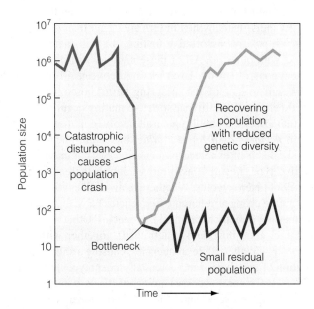

Figure 5.10 An evolutionary bottleneck caused by a catastrophe. Only a small portion of the gene pool survives. If the population recovers, it lacks genetic diversity. If it remains small, it is subject to genetic drift.

populations are structured as metapopulations—groups of local populations of interbreeding individuals linked by movements of individuals. Because individuals carry genes, the terms *gene flow* and *migration* are often considered synonymous. However, if an individual moves into a population but does not successfully reproduce, its genes are not introduced into the gene pool. In circumstances where individuals do reproduce, migration is a potent force in reducing genetic differences among local populations (assuming their genetic composition differs). Barriers to movement, in contrast, facilitate (but do not in themselves cause) *allopatric speciation* (see Section 5.9).

4. *Non-random mating.* The processes discussed so far—natural selection, mutation, genetic drift, and gene flow—violate the first four assumptions of the Hardy–Weinberg principle. The final process violates the final assumption, that mating is random—that is, the chance that an individual mates with an individual of a given genotype is equal to the frequency of that genotype in the population. In reality, individuals often practise **assortative (non-random) mating**, whereby they choose mates non-randomly based on a phenotypic trait or traits. The most recognized example, *female mate choice*, involves a female bias towards mates based on specific phenotypic traits such as body size or markings (see Section 8.4). Assortative mating is not restricted to animals; plants or fungi do not choose mates in a behavioural sense, but their mating is often non-random and correlated with phenotypic traits.

In *positive assortative mating*, mates are phenotypically more similar than expected by chance. Consider the timing of plant reproduction. In populations with an extended flowering time, early-flowering plants may no longer be flowering when late-flowering plants are in bloom. Plants are more likely to mate with individuals with similar timing. Positive assortative mating increases the frequency of homozygotes over heterozygotes. Imagine a locus where *AA* individuals tend to be larger than *Aa* individuals, which in turn are larger than *aa* individuals. With positive assortative mating, *AA* will mate with *AA*, and *aa* with other *aa*. All of these matings will produce only homozygous offspring. Even mating between *Aa* individuals will result in half of the offspring being homozygous. The genetic effects of positive assortative mating occur only at the loci that affect the phenotypic trait involved in mate choice.

In *negative assortative mating*, mates are phenotypically less similar than expected by chance. Less common than the positive type, negative assortative mating increases the frequency of heterozygotes. Whether positive or negative, assortative mating changes genotypic frequencies between generations but does not directly alter allele frequencies. Other processes—mutation, gene flow, and genetic drift, together with natural selection—*do* alter allele frequencies, causing a shift in the distribution of genotypes (and potentially phenotypes) in the population. All four processes are evolutionary mechanisms. However, natural selection is the only evolutionary process that generates adaptations; the others can only speed up or slow down the evolution of adaptations.

A special case of non-random mating is **inbreeding**—mating of individuals that are more closely related than

expected by chance. (They are also more likely to be phenotypically similar, although this is not necessarily the case.) Unlike positive assortative mating, inbreeding increases homozygosity at all loci, because related individuals are genetically similar by common ancestry and are thus more likely than unrelated individuals to share alleles throughout the genome. Inbreeding tends to be detrimental. Offspring are more likely to inherit rare, recessive, deleterious alleles that may decrease fertility or cause loss of vigour and even death. These negative consequences, which all reduce fitness, are called *inbreeding depression*.

Although we have treated them as distinct, natural selection often occurs concurrently with non-random mating, if individuals select mates on the basis of traits that will enhance their probability of survival in a particular environment. Obviously, if an individual is chosen as a mate, its fitness is potentially higher than that of another individual that is not chosen, even if that potential is not realized. Sexual selection (see Section 8.4) is thus a type of natural selection. Other instances of natural selection, however, do not directly involve mate choice.

5.7 GENETIC DIFFERENTIATION: Natural Selection Alters Genetic Variation among Populations of Widely Distributed Species

As described in Section 5.5, natural selection at work in medium ground finches inhabiting Daphne Major caused a shift in the distribution of phenotypes. This shift in mean beak size reflects a change in allele and genotype frequencies *within* a population, involving disruption of the Hardy–Weinberg equilibrium. Natural selection can also alter genetic variation *among* populations, through *genetic differentiation*. Species with a wide geographic range often encounter a broader range of abiotic conditions than do more restricted species. This variation often gives rise to a corresponding variation in morphological, physiological, and behavioural traits. As a result, significant differences often exist among local populations of a species inhabiting different regions. The greater the distance between the populations, the more pronounced the differences often become, as each population adapts to the locality it inhabits.

Such geographic variation among populations within a species can result in *clines*, *ecotypes*, and *geographic isolates*. A **cline** is a measurable, gradual change over a geographic region in the mean of some phenotypic trait, such as size (it can also refer to a gradient in genotype frequency). Clines are usually associated with gradients of abiotic factors such as temperature, moisture, light, or altitude. Continuous variation results from gene flow from one population to another along the gradient. Because environmental constraints influencing natural selection vary along the gradient, any one population will differ genetically to some degree from another, with their genetic difference increasing with the distance between populations.

Clines occur in traits such as body size, colour, and physiology. For example, the white-tailed deer (*Odocoileus virginianus*) exhibits clinal variation in body weight (Figure 5.11; see

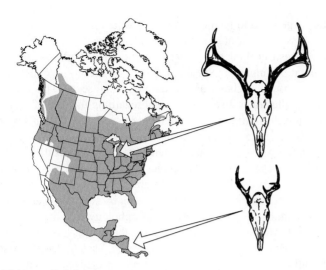

Figure 5.11 Skull sizes of white-tailed deer across North America.

(Adapted from Baker 1984.)

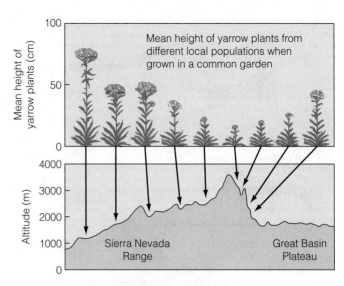

Figure 5.12 Local yarrow populations in the Sierra Nevadas exhibit an inverse relationship between altitude and plant height. When seeds from local populations were grown together, height differences persisted, indicating genetic differentiation.

(From Campbell and Reece 2005, after Clausen et al. 1948.)

Bergmann's rule, Section 7.9). Deer in Canada and the northern United States are heaviest, weighing on average over 136 kg. Individuals in more southerly populations are considerably smaller: 93 kg in Kansas and 60 kg in Louisiana. The smallest individuals, the Key deer in Florida, weigh less than 23 kg.

Clinal variation can show marked discontinuities. Such an abrupt change, or *step cline*, often reflects abrupt changes in local environments and may be called an **ecotype**: a population that is genetically distinct and adapted to its unique local environment. For example, a population inhabiting a mountaintop may differ from a population of the same species in a nearby valley. When several habitats to which the species is adapted recur throughout the species' range, ecotypes may be scattered like a mosaic across the landscape.

Yarrow (*Achillea millefolium*) is a plant native to temperate and sub-Arctic regions of the Northern Hemisphere with a large number of ecotypes differing in height. Populations living at lower altitudes are much taller than *montane* (mountain-dwelling) populations. In a classic study, seeds were collected from local populations in the Sierra Nevada mountains. To remove the confounding effects of environmental differences with elevation, the researchers planted seeds in a common garden (Figure 5.12). Local populations maintained the differences in height and other traits observed in their native habitats, illustrating the role of genetic as well as environmental factors in phenotypic variation with altitude (Clausen et al. 1948). Later experiments showed that differences in height and leaf morphology reflected underlying physiological adaptations to the climatic constraints imposed by altitude.

Ecotypes often represent distinct but related points on a continuum. In some species, variation is far more disjoint. The southern Appalachians are noted for their diversity of salamanders (Figure 5.13, p. 86), as fostered by the rugged terrain, varying environment, and limited dispersal ability of these amphibians. Populations become isolated, preventing the free flow of genes. *Plethodon jordani* exists as several iso-

lated populations, each occupying a particular region (Hairston and Pope 1948). Such populations are **geographic isolates**, among which gene flow is prevented by some extrinsic barrier—in this case, rivers and mountain ridges. The degree of isolation depends on the barrier, but rarely is isolation complete. These isolates are often classed as **subspecies**, a more general taxonomic term for populations that are distinguishable by one or more traits.

Unlike clines, geographic isolates can theoretically be separated by lines on a map, although practically it may be difficult. Unlike ecotypes, geographic isolates may not possess adaptations specific to their local environment. Both ecotypes and isolates can become new species if they become (and remain) reproductively isolated (see *allopatric speciation*, Section 5.9).

5.8 TRADE-OFFS AND CONSTRAINTS: Adaptations Entail Ecological Compromises

If Earth were one large homogeneous environment, perhaps a single phenotype with a single set of traits might allow all organisms to survive, grow, and reproduce. But such is hardly the case. Abiotic factors that directly influence life (such as temperature and precipitation) vary in space and time, producing a diversity of land habitats. Likewise, variations in depth, salinity, pH, and dissolved O_2 define an array of aquatic habitats. Each combination of factors represents a unique set of constraints on the ability of organisms to maintain processes essential to survival, growth, and reproduction. Thus, as environments change, so do the traits that enhance success. Natural selection favours different phenotypes under different conditions. Traits that maximize fitness in one environment often limit fitness in another.

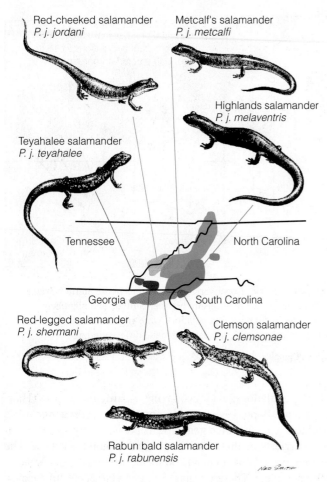

Red-cheeked salamander
P. j. jordani

Metcalf's salamander
P. j. metcalfi

Highlands salamander
P. j. melaventris

Teyahalee salamander
P. j. teyahalee

Tennessee

North Carolina

Georgia

South Carolina

Red-legged salamander
P. j. shermani

Clemson salamander
P. j. clemsonae

Rabun bald salamander
P. j. rabunensis

Figure 5.13 Geographical isolates of *Plethodon jordani* in the Appalachians. When the eastern population of *P. yonahlossee* was separated by the French Broad River valley, it developed into *P. j. metcalfi*, which spread north, the only direction with suitable conditions (in other directions, the mountains end abruptly). Another subspecies, *P. j. jordani*, became isolated by the deepening of the Little Tennessee River. Other subspecies are somewhat connected.

(Adapted from Hairston and Pope 1948.)

The concept of ecological trade-offs is illustrated in selection for beak size in Darwin's medium ground finch (see Section 5.5). Recall that the ability to eat seeds of different size and hardness is related to beak size. These differences in diet as a function of beak size reflect a trade-off in morphological traits that facilitate exploitation of different food items. Such trade-offs are even more apparent if we compare differences in beak morphology and use of seed resources for other species of Darwin's ground finch that inhabit Santa Cruz, another island in the Galápagos.

As their common names suggest, mean beak size increases from the small (*Geospiza fuliginosa*) to the medium (*G. fortis*) and large (*G. magnirostris*) ground finch (Figure 5.14a). The proportion of seeds of different sizes in their diets (Figure 5.14b) reflects these differences, with mean seed size increasing with beak size. Small beaks restrict the ability of the small finch to consume larger seeds, whereas large beaks allow

the large finch to feed on a range of seed sizes. However, they are less efficient at exploiting smaller seeds, so large-beaked birds restrict their diet to large seeds. **Profitability**—energy gained per unit time spent handling a diet item—declines for large ground finches when feeding on small seeds. This inefficiency illustrates a second concept regarding the role of trade-offs in selection: individual traits are usually components of a larger adaptive complex involving multiple traits (and gene loci). Beak size is just one of many interrelated traits that affect foraging behaviour and diet choice. Larger beak size correlates with larger body size and with differences in skull structure and head musculature that influence feeding by affecting bite force.

In sum, the relation between morphology and diet in seed-eating birds exemplifies the trade-off that constrains evolution of adaptations relating to food acquisition. Such trade-offs exist for all adaptations. Typically, the more specialized the adaptation, the more severe the trade-offs.

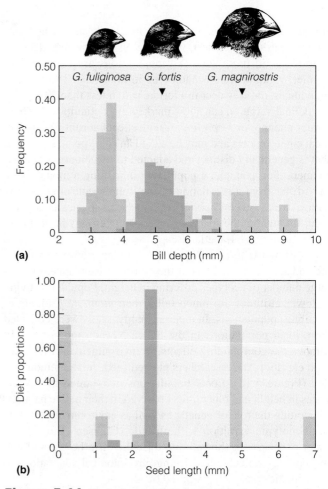

Figure 5.14 Beak size and diet of ground finch species on Santa Cruz. **(a)** Frequency distribution of beak depths of adult males of the small (*Geospiza fuliginosa*), medium (*G. fortis*), and large (*G. magnirostris*) ground finch. Mean values are indicated by solid triangles. **(b)** Differences in beak size correlate with differences in the proportion of seed sizes in the diet.

(Adapted from Grant 1999, after Schluter 1982.)

RESEARCH IN ECOLOGY Dynamic Evolution in Sticklebacks

Andrew Hendry, McGill University.

Beren Robinson, University of Guelph.

Most of us think of evolution as incredibly slow, requiring millions of years. And in many cases, evolution—particularly speciation—is, as Darwin suggested, very gradual. Recently, however, researchers are discovering that evolution can be surprisingly rapid in circumstances in which selection pressures are intense. Some of the evidence for non-gradual evolution comes from fossils. The concept of punctuated equilibria, for example, posits long periods of relatively stable genetic equilibria interrupted by periods of dramatic change associated with, and in response to, environmental disruption (Eldredge and Gould 1972). But rapid evolution is not just a phenomenon of the past—it is happening all around us.

One of the most valuable model systems for studying this ongoing evolution is a group of fish inhabiting freshwater lakes and streams worldwide: three-spine sticklebacks (*Gasterosteus* spp.). Pioneered by J. D. McPhail and Dolph Schluter, research into stickleback evolution has been ongoing at Canadian universities and around the world for over three decades. We focus on the contributions of Beren Robinson and Andrew Hendry. Although they differ in approach and perspective, both stress the dynamic interactions between ecology and evolutionary biology.

Many lakes and streams scattered over northern regions of North America were formed after the immense ice sheets that covered the region retreated about 15 000 years ago. Sticklebacks inhabiting lakes in coastal British Columbia are among the youngest species on Earth, making them particularly useful for speciation studies because they carry less of the genetic "baggage" (residual genetic material) of older species. No more than two species occur in any one lake, and pairs of species in several different lakes have evolved independently from marine ancestors. In every lake where a pair of stickleback species occurs, the two species differ in habitat use and diet. One species feeds on zooplankton in open-water habitats (*limnetic form*), while the other exploits larger invertebrate prey from the sediments and submerged vegeta-

tion of shallow inshore waters (*benthic form*). Morphological differences between species correlate with these differences in habitat and diet. Whether selective pressure imposed by these distinct environments is responsible for the evolution of species pairs is a focus of Robinson's research.

Only a small number of lakes in a limited geographic area are occupied by these species pairs. Most lakes contain only a single species, which tends to be intermediate in morphology and habitat use to limnetic and benthic members of the species pairs. In sampling the single-species Cranby Lake, Robinson found that open-water individuals differed morphologically from those in inshore habitats. These morphological differences paralleled the differences between the species that occupy these two habitats in the species-pairs lakes (Figure 1). Robinson hypothesized that these distinct phenotypes reflect natural selection promoting divergence within the population. Disruptive selection can occur when individuals face trade-offs, whereby performance of one task (feeding on zooplankton in open water) entails a cost to performance of another task (feeding in inshore sediments). Such trade-offs can cause disruptive selection that favours resource specialization, setting the stage for speciation.

To test his hypothesis, Robinson needed to establish that two conditions were met. (1) The morphological differences between

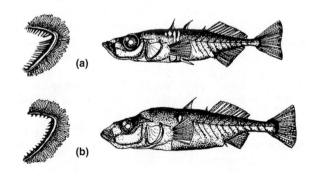

Figure 1 Morphological differences in the gill rakers and body depth of sticklebacks between **(a)** limnetic and **(b)** benthic forms. Illustration by Laura Nagel.

(From Schluter 1993.)

continued on pages 88–89

RESEARCH IN ECOLOGY   continued

the two forms are heritable, rather than an expression of phenotypic plasticity in response to different habitats or diets. (2) These morphological differences represent a trade-off, by influencing foraging efficiency.

To test the first condition, Robinson reared offspring of the two forms under identical laboratory conditions and diet. Although there was some phenotypic plasticity, the benthic form (BF) had shorter and deeper bodies, wider mouths, more dorsal spines, and fewer gill rakers (bony structures on a gill that divert solid substances away from the gill) than the limnetic form (LF). Thus, the observed morphological differences are heritable.

To test the second condition, Robinson conducted feeding trials of foraging efficiency in artificial limnetic and benthic habitats, using two food types. Brine shrimp larvae (*Artemia*), a proxy of zooplankton in these lakes, were placed in the limnetic habitats, and larger amphipods that forage on dead organic matter in the benthic habitats. A trial consisted of releasing a single fish into the aquarium and observing it for a period of time, after which the total number of items eaten was determined. There were two measures of foraging success: (1) intake rate (number of prey consumed per minute) and (2) capture effort (mean number of bites per prey item).

There were significant differences in foraging success. LF individuals were more successful at foraging on shrimp larvae. They had a higher consumption rate and used half the number of bites per item as compared with BF individuals, which had a higher intake rate for amphipods and consumed larger amphipods than did LF individuals. Robinson determined that the higher intake of shrimp larvae by LF individuals was related to their greater number of gill rakers, whereas greater mouth width was related to the higher intake of amphipods by BF individuals. Thus, foraging efficiency was associated with morphological differences, suggesting trade-offs in traits related to successful exploitation of the two habitats and their associated food resources.

Robinson's findings suggest that disruptive selection is occurring in this population and was therefore likely present in the early stages of speciation in the species-pairs lakes. Previous studies of stickleback species pairs had indicated that opposing selective pressures in open-water and inshore habitats were a major evolutionary factor. Yet Robinson's work was unique in illustrating natural selection at work in a single population, resulting in distinct phenotypes inhabiting distinct habitats. It is also one of few studies that have quantified trade-offs in resource use within a population, and it provides critical insights into the mechanisms at work in the evolution of diversity in this taxonomic group. (See QUANTIFYit! at **www. pearsoncanada.ca/ecologyplace** to calculate summary statistics on stickleback data.)

Working with the same system, but in a different context, Andrew Hendry has pursued a contrasting if complementary course. Given that speciation into species pairs has occurred only in a small number of lakes despite the fact that morphological divergence is common within single species, his focus was on what other factors may prevent speciation from occurring in

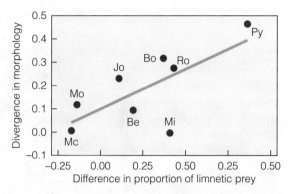

Figure 2 Relationship between the divergence in stickleback morphology and diet. The *y*-axis is a composite measure based on foraging and feeding traits; the *x*-axis is the proportion of limnetic species in the diet. Each point represents a watershed. Mi = Misty Lake watershed.

(Adapted from Berner et al. 2008.)

many locations. He posed the following questions: *Are the similar selective pressures of lake (open-water, zooplankton-feeding) habitats versus stream (near-shore, benthic-feeding) habitats in different watersheds sufficient to explain observed differences in stickleback morphology and foraging behaviour? Or are other evolutionary processes playing a role?*

Hendry's work involved some of the same experimental methods as Robinson's, but the portion discussed here was observational. He and his colleague David Berner sampled a large number of sticklebacks in eight watersheds on Vancouver Island. As well as measuring morphological traits, they analyzed stomach contents to estimate diet composition. Given that the aim was in part to test the consistency of the lake/stream dichotomy (analogous to Robinson's open-water/inshore dichotomy), this approach was appropriate to compare the spectrum of morphology and diet among watersheds, rather than studying response to one diet item in artificial habitats.

Like Robinson, Berner and Hendry found that habitat-related disruptive selection has influenced morphology, with available prey as the major selective force. However, by investigating a range of watersheds, their study could draw broader conclusions. Morphological divergence varied among watersheds and was not always fully explained by the open water/inshore dichotomy. Overall, there was a strong correlation between differences in diet (proportion of limnetic species in stomach contents) and morphology (Figure 2). However, the Misty Lake watershed had more intermediate forms than expected based on diet differences—a finding attributed to the greater role of gene flow in this watershed (Berner et al. 2008). This study not only supports the idea of ecologically driven morphological divergence in coexisting related groups such as sticklebacks, but also stresses the importance of multiple evolutionary processes. In sticklebacks, gene flow may explain why ongoing disruptive selection has not always led to the evolution of new species.

RESEARCH IN ECOLOGY continued

Bibliography

Berner, D., D. C. Adams, A. C. Grandchamp, and A. P. Hendry. 2008. Natural selection drives patterns of lake-stream divergence in stickleback foraging morphology. *Journal of Evolutionary Biology* 21:1653–1665.

Eldredge, N., and S. J. Gould. 1972. Punctuated equilibria: An alternative to phyletic gradualism. Pages 82–115 in Schopf, T. J. M. (ed.). *Models in paleobiology*. San Francisco: Freeman Cooper.

Robinson, B. W. 2000. Trade offs in habitat-specific foraging efficiency and the nascent adaptive divergence of sticklebacks in lakes. *Behaviour* 137:865–888.

Robinson, B. W., and S. Wardrop. 2002. Experimentally manipulated growth rate in threespine sticklebacks: Assessing trade-offs with developmental stability. *Environmental Biology of Fishes* 63:67–78.

Schluter, D. 1993. Adaptive radiation in sticklebacks: Size, shape, and habitat use efficiency. *Ecology* 74:699–709.

1. In the Robinson study, how might the observed differences in morphology, diet, and habitat of the two phenotypic forms lead to reproductive isolation and eventual speciation?
2. Suppose that an intermediate phenotype was present, and that this intermediate was equally capable and efficient at feeding in both open-water and inshore environments compared with the other two phenotypes. How would this alter the interpretation of the results?
3. How does the Robinson study differ from that of Berner and Hendry, and why are these differing approaches appropriate to their differing objectives? In relation to our discussion of the scientific method in Chapter 1, how do these two studies complement each other?

5.9 SPECIATION: Evolution Can Lead to the Emergence of New Species

The ground finch example used throughout this chapter illustrates how natural selection operates on genetic variation on three levels (see Section 5.4): (1) *within* a population, (2) *among* populations of the *same* species, and (3) *among* populations of *different* species.

First, natural selection operating *within* the medium ground finch population on Daphne Major during a drought increased beak size in response to a change in seed abundance and size. Second, natural selection led to differences in beak size *among* populations of the medium ground finch inhabiting Daphne Major and Santa Cruz. The larger beak size of the Santa Cruz population may reflect competition with the small ground finch, which is absent on Daphne Major. The presence of the smaller species made smaller, softer seeds less available, which increased the fitness of medium ground finches with larger beaks that can feed on larger, harder seeds.

Both of these levels involve evolution, but neither involves **speciation** (evolution of new species), although they may set the stage for it in the future. In contrast, speciation is involved in the third level—selection *among* populations of *different* species. Studies suggest that natural selection is responsible for genetic differentiation among the many species of Darwin's finches inhabiting the Galápagos Islands (Figure 5.15, p. 90). At some point, these species had a common ancestor, but over time natural selection has given rise to distinct species. **Adaptive radiation** is the process by which one species gives rise to multiple species that exploit different features of the environment, such as food resources or habitats. Different factors exert selection pressures that push populations in divergent directions with respect to their traits. Reproductive isolation (a necessary condition of speciation) is often a by-product of the changes in morphology, behaviour, or habitat preference that are the immediate targets of selection.

How does adaptive radiation relate to the more general concept of speciation? Some reserve the term for instances where many new species proliferate from a common ancestor in a sudden burst, in response to an opening up of new niche opportunities. This type of adaptive radiation has been observed in animals (as with Darwin's finches) and is also well documented in plants. The silverswords are a group over 30 closely related species in three genera (most are *Dubautia* spp.) native to the Hawaiian Islands. The diversity of habitats and the isolation of the Hawaiian archipelago facilitated the proliferation of species varying in form from trees to woody vines to small mat plants, with a corresponding variation in leaf morphology and physiology (Figure 5.16, p. 90). This diversification correlates with the ability to tolerate different habitats, from dry, open volcanic sites to moist, shady rain forests (Raven et al. 2005). Many of these **endemics** (species with a highly localized range) are at risk of extinction through human impacts or loss of specialized pollinators (see Section 27.3).

Not all speciation involves the sudden proliferation of new species. But even if a single species gives rise to only two new species, the same question arises: *How does the gene pool diverge, and how do the new species, which are likely very similar initially, remain reproductively isolated?*

Most speciation is **allopatric**; that is, it occurs as a result of geographic separation of a population into two or more subpopulations with no movement between them. *Allopatric speciation* is easily envisioned (Figure 5.17a, p. 91). Subpopulations are isolated by a geographic barrier, and so any existing differences resulting from genetic drift will not be eliminated by subsequent interbreeding. In their different locations, they will likely experience different environmental conditions, which will exert different types and intensities of selective pressure. Over time, the two groups will diverge more and more, until interbreeding is no longer possible. At that point, even if they are brought back into contact, they no longer rely upon geographic isolation to ensure reproductive isolation. Adaptive radiation in ground finches (see Figure 5.15) involved allopatric speciation, where isolated island habitats provided opportunities for new species.

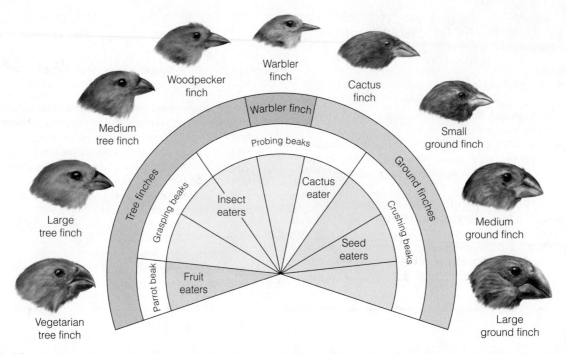

Figure 5.15 Adaptive radiation of Darwin's ground finches in the Galápagos Islands. Genetic studies show that they arose from a common ancestral species.

(From Patel 2006.)

Sympatric speciation, in contrast, occurs without geographic barriers (Figure 5.17b). Although there are relatively few documented examples, sympatric speciation can follow the emergence of **polymorphism**: the existence of more than one distinct form of individuals in a population, possibly reflecting disruptive selection operating in the same location. However, since the incipient species coexist, disruptive selection can progress to speciation only if intrinsic reproductive barriers develop, involving differences in reproductive morphology, timing, or behaviour.

Figure 5.16 Hawaiian silverswords represent an example of adaptive radiation in plants.

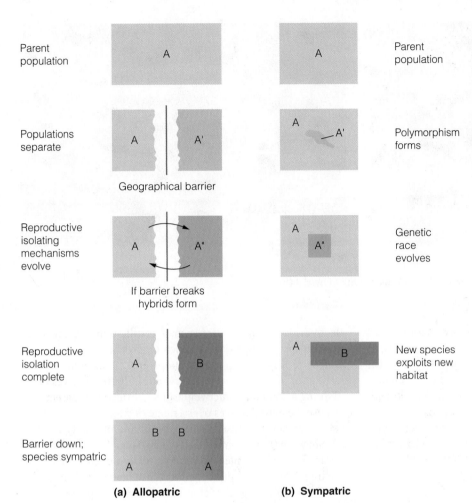

Figure 5.17 Types of speciation. **(a)** Allopatric speciation occurs when two populations are isolated over a long time period by a geographic barrier. Different selective pressures cause the populations to diverge into subspecies, which may become new species. **(b)** Sympatric speciation occurs in the same location when disruptive selection (often reflecting patchy habitat) causes polymorphism, followed by the development of intrinsic barriers to reproduction.

If the traits that prevent interbreeding result from selection for ecological adaptations, the process is called **ecological speciation** (Schluter 2000). However, ecological speciation is difficult to establish. The findings of Andrew Hendry, whose research we featured in Research in Ecology: Dynamic Evolution in Sticklebacks (pp. 87–89), are often cited in support of the phenomenon, but Hendry himself sounds a cautionary note, citing a tendency to overinterpret results and to assume ecological speciation is ubiquitous. Despite growing evidence that ecological speciation is a powerful force promoting diversity, Hendry (2009) sets out strict criteria to establish its necessary condition: reproductive isolation resulting from adaptation to differing habitats and/or resources. Ecological speciation is thus a continuum involving different degrees (and reversibility) of reproductive isolation. In contrast, when sympatric speciation arises from hybridization (see Section 5.10), the resulting reproductive isolation is virtually instantaneous.

5.10 HYBRIDIZATION: Hybridization Can Prevent or Promote Speciation

The successful mating of two species, **interspecific hybridization**, has a complex relationship to speciation. In some cases, it reverses speciation. If two species produce fertile offspring that can then mate with one or other of the parent species, the line between the original species is blurred. After all, reproductive isolation is the cornerstone of the **biological species** concept, and the existence of viable, fertile hybrids can collapse species into so-called *hybrid swarms*—groups of individuals of hybrid origin with varying degrees of backcrossing to either parent species. Particularly in circumstances when humans have disturbed or, as Edgar Anderson described it, "hybridized the habitats" (1948), these genetic amalgams of previously distinct species may exploit new ecological opportunities in which they may surpass the parents.

How often does this happen? Hybrids are well-known in domestic species, such as the mule (donkey × horse). However, animal hybrids are often sterile, and given that most animals have no alternative to sexual reproduction, there is little evolutionary consequence for the parent species. Some animal hybrids do have limited fertility. High-profile instances include a cross between a grizzly bear (*Ursus arctos horribilis*) and a polar bear (*U. maritimus*). In 2010, a bear shot by an Inuit hunter was identified by DNA analysis by the Northwest Territories Environment and Natural Resources Department as a second-generation offspring of a hybrid mother and grizzly father.

Such instances raise the possibility of a greater role for hybridization in animal evolution than once thought. Even in

(a)

(b)

Figure 5.18 Male American black duck (**a**) and male black duck–mallard hybrid (**b**).

mammals, as much as 6 percent of European species form natural hybrids (Mallet 2005), and in some bird and insect groups, it is almost as common as in vascular plants. An example is the American black duck (*Anas rubripes*) (Figure 5.18). A century ago, biologists noted a decline in black ducks in Ontario and Québec, coincident with range extension of mallards (*A. platyrhynchos*). Atlantic Canada is a stronghold for the species, but as mallards extend eastward, their ability to hybridize with other ducks raised concerns about the future of the black duck. Studies in Cape Breton suggest that, although hybridization is occurring, black duck populations in urban parks are relatively stable (McCorquodale and Knapton 2003).

The black duck study relied on visual detection of hybridization. DNA analysis of another well-known instance of animal hybridization, between wolves and coyotes (*Canis* spp., see Figure 12.12), indicates an extremely complex genetic situation (Wilson et al. 2009), humorously dubbed "Canis soup." Species in this group diverged relatively recently in evolutionary terms and share the same ploidy level ($2n = 78$). They hybridize freely, particularly in central Ontario, where the coyote is expanding its range eastward. Other interspecific hybrids may become more common as climate change causes ranges to expand for some species and contract for others. When they occur, these hybrids blur the lines between existing species, rather than promote new species.

Zoologists often use *hybridization* to refer to within-species matings between genetically divergent individuals, such as between limnetic and benthic forms of a stickleback species (see Research in Ecology: Dynamic Evolution in Sticklebacks, pp. 87–89). Such intraspecific hybridization can only forestall speciation. In contrast, hybridization has dramatically promoted speciation in plants. Estimates vary, but as many as 70 percent of flowering plants are thought to be of hybrid origin, and potentially an even higher percentage of some non-flowering groups, notably ferns (Raven et al. 2005). What explains the promoting effect of hybridization on plant speciation?

First, interspecific hybridization is extremely common in plants. It varies widely, but many groups, such as saskatoons (*Amelanchier* spp.) and willows (*Salix* spp.), readily form fer-

tile hybrids. Hybrid swarms may result, blurring the lines between species. In North America the hybrid cattail (*Typha* × *glauca*) readily backcrosses to its parents (*T. latifolia* and the introduced *T. angustifolia*) (Snow et al. 2010). The resulting offspring are not reproductively isolated from either parent, and because the hybrid thrives over a range of water depths in freshwater marshes as a result of its phenotypic plasticity (Waters and Shay 1990), its dominance is increasing.

In other cases, interspecific hybridization leads to speciation that is not only rapid but almost instantaneous. How is this possible? The second factor that promotes speciation via hybridization is **polyploidy**: multiplication of chromosome number that allows hybrid offspring to be fertile but isolates them from their parents. Evolutionary biologists now believe that polyploidy has occurred in virtually all eukaryotic groups, but it is particularly common in plants, as reflected in the high chromosome number of many non-flowering species. The fern *Ophioglossum reticulatum* has the most chromosomes (1260) reported for any species (Raven et al. 2005).

Polyploidy can result from spontaneous chromosome doubling after fertilization but most often arises from the union of unreduced gametes (Ramsey and Schemske 2002). If the unreduced gametes of two diploid ($2n$) species unite, the offspring will be tetraploid ($4n$) and reproductively isolated from either parent. As long as the offspring can reproduce with other $4n$ individuals of similar origin (or reproduce asexually), the taxon has achieved species status. If this $4n$ species then hybridizes with another $4n$ species, an octoploid ($8n$) species may result, and so on. High rates of fertile hybrids in Hawaiian silverswords (see Figure 5.16) suggest that adaptive radiation in plants is facilitated by polyploidy. New species formed in this way are called *allopolyploids*. Wheat (*Triticum aestivum*) is a hexaploid (six sets of chromosomes) that arose naturally as an allopolyploid (Figure 5.19). Many plant species arise from polyploidy occurring within a species and are called *autopolyploids*. Bananas, for example, are autotriploids.

A third factor involves the fate of sterile interspecific hybrids. Unlike sterile animal hybrids, plant hybrids can repro-

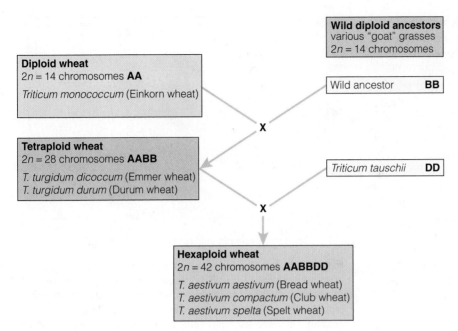

Figure 5.19 Suggested polyploid origin of wheat.

duce asexually, by vegetative (clonal) reproduction (see Section 9.1). So a sterile plant hybrid is less likely to be an evolutionary dead end. Most kinds of asexual reproduction are vegetative, such as a shoot arising from the base of its parent plant. Genetic studies of the groundnut (*Apios americana*, a leguminous vine) indicate that autotriploidy has occurred multiple times in its evolutionary history (Joly and Bruneau 2004). As is common with triploids, the offspring are sterile. However, they spread asexually through underground tubers (a food used by native North Americans) and are common in the northern portion of the species' geographic range, where they have a selective ecological advantage over diploid individuals.

Other sterile plant hybrids mimic sexual reproduction, thus partaking of some of its ecological benefits. Consider the grass that dominates most front lawns: Kentucky bluegrass (*Poa pratensis*). As a result of hybridization with related species, the genetic composition of this species (likely introduced from Europe in the 1700s) is highly complex (Porceddu et al. 2002). It consists of many distinct races, most of which reproduce by **apomixis**: asexual reproduction in which seeds are formed without meiosis or sexual recombination. Like most seeds, these can be dispersed by various vectors, but unlike seeds formed by sexual reproduction, the embryo contained within is genetically identical to its parent. Should we call these individuals, originally of hybrid origin and now reproductively isolated, distinct species? Surely not, but the point remains that even if a plant hybrid is sterile, asexual reproduction allows it a future denied to sterile animal hybrids.

Speciation resulting strictly from hybridization is not ecological speciation (see Section 5.9) because the reproductive isolation has a genetic cause (polyploidy), not a divergence in ecological function. But even if a species arises from hybridization, it can persist only if its individuals can survive, grow, and reproduce. So natural selection (and other evolutionary processes) will act upon it, even if its speciation was not ecologi-

cally driven. Indeed, the evolutionary role of natural hybridization in plants—and, in an ecological context, its ability to transfer and give rise to adaptations—is receiving renewed attention (Arnold 2004).

5.11 ABSENCE OF PERFECTION: No Organism Is Perfectly Adapted to Its Environment

Earlier in this chapter, we quoted Mayr's observation that prior to Darwin, the prevailing view was that organisms were perfectly adapted to their environment. Despite Darwin's impact, many continue to promote this idea, which is mistaken. Despite the undeniable power of natural selection to shape populations, we should never expect a "perfect match" between any organism and its environment. The reasons are many.

1. *Trade-offs.* Even if the environment were to remain the same, a trait that is optimal for one environmental factor may prove a handicap for another.

2. *Environmental change.* Given that the environment even in the same place is always changing, being "perfectly adapted" for one state of the environment entails that an organism is imperfectly adapted for some other state. Consider the analogy of a moving target: how can an organism be perfectly adapted to an environment that is always changing? There is always a time lag. A particular phenotype may be selected under one set of factors, but by the time individuals with that trait have successfully reproduced, their offspring (which may or may not retain the trait) may experience altered environmental conditions for which this trait is no longer optimal.

3. *Genetic variation.* Natural selection can "choose" only from available phenotypes, which in turn reflect the genetic diversity of a population. If a frigid winter exposes a plant population to lows of −45°C, only individuals with the genetically determined potential for extreme cold hardiness will be selected, no matter how vigorous or well adapted they are in other respects. If the genes that code for a particular adaptive trait are not present, they cannot be selected. Natural selection acts as a filter for existing phenotypes but is not in itself a creative force.

4. *Chance.* Individuals with more adaptive traits may not survive simply because they are exposed to a hazard that other individuals have evaded through no morphological, physiological, or behavioural trait of their own. For example, a tornado may kill animals in a local subpopulation that are otherwise well adapted to their local environment. Another subpopulation that is less well adapted overall may escape harm through sheer chance—being in the right place at the right time—rather than because they possess traits that enhanced their survival. (Recall our discussion of genetic drift in Section 5.6.) If these surviving individuals reproduce, they will have higher relative fitness, but not because they are better suited to their environment.

5. *Genetic factors.* An allele that codes for an adaptive trait may occur on the same chromosome as an allele coding for a trait that is neutral, or even negative in adaptive value. Such genetic linkages are inevitable. Similarly, a trait that is deleterious in a recessive condition may confer some selective advantage in the heterozygous state. Consider sickle cell anemia. The homozygous recessive state codes for susceptibility to the disease, but the heterozygous state confers resistance to malaria. In the African tropics, where malaria is a strong selective force, a higher proportion of humans carry this deleterious allele.

So what is the bottom line? Evolution is a process of perpetual change taking place against a backdrop of environmental variation and random events, both extrinsic and intrinsic to the organism. Perfection, in contrast, implies a static end-state that has never existed, nor ever will exist. Natural selection increases fitness but does not generate perfection.

5.12 PHENOTYPIC PLASTICITY: Organisms Respond to Environmental Variation at the Individual and Population Levels

The interaction between genes and the environment is fundamental to ecology. Organisms respond to changes in the environment in space and time, at the level of both the individual and the population. Over generations, and in response to changing environmental conditions, natural selection favours certain phenotypes over others, causing a shift in the distribution of phenotypes (and in allele and genotype frequencies) within the population. This evolutionary process involves adaptation to environmental change through natural selection operating on individuals, but having collective impacts on the population.

This process also occurs spatially. As we saw in Section 5.7, local populations of the same species inhabiting different environments and experiencing different selective pressures may undergo genetic differentiation. This differenti-

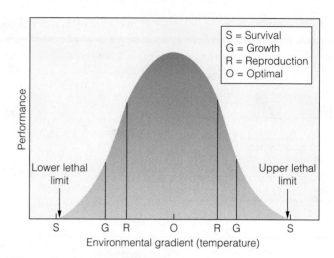

Figure 5.20 Generalized response of an individual to a gradient of an environmental factor. The end points represent the zone of intolerance, on either side of the lethal limits. Maximal growth and reproduction occur in the optimum range. On either side of the optimum, in the zones of physiological stress, survival is possible but growth and reproduction are impaired.

ation also causes a divergence in the distribution of genotypes and associated phenotypes. The result can be distinct ecotypes—populations with genetically determined adaptations to their unique local environments.

But individuals as well as populations respond to environmental change. For mobile organisms, the simplest response to a disadvantageous change is to move to a more suitable location. This behaviour, which is common in animals, is not an option for sessile organisms such as plants and fungi. Even for mobile organisms, distance or other obstacles may limit an individual's ability to reach more suitable habitat, assuming it is available. Is there another way to cope?

We start by considering a concept that is fundamental to the ecology of the individual: the **tolerance range**, which depicts an individual's response to an environmental factor (Figure 5.20). Think of any example—a plant responding to light, an animal to temperature, a fungus to humidity. For every factor, there is an **optimum range**—the level at which the individual does best, which (in most cases) means it experiences its maximal growth and reproduction. The optimum is usually a range rather than a specific value. At levels of the factor above and below the optimum, in the **zone of physiological stress**, the organism can survive, but growth and reproduction are impaired. At even lower or higher levels, we reach the **zone of intolerance**: levels at which the individual cannot survive for any extended time. These intolerance zones are bounded by *lower* and *upper lethal limits*.

For some resources or conditions, the tolerance range is a bell-shaped curve, as in Figure 5.20. But this is not always the case; an organism might have a skewed response, with a long plateau followed by a sharp decline at one of its lethal limits but not the other. For a hazard, the optimum level will be at the origin of the *x*-axis, since the most favourable level of a hazard is its absence. Higher levels of the hazard will have increasingly severe effects until the upper lethal limit is reached.

Nor will response be the same for an individual at all stages of its life cycle; the larval stage of an insect might be more sensitive to temperature than the adult. Response differs among individuals, reflecting differences in genotypes. A generalized population response curve could be constructed from the aggregate of the individuals' responses. In this case, the variable on the y-axis would be a population parameter such as abundance and/or overall reproduction.

An individual (or a population) will have a response curve for each environmental factor that influences it. Is the response to some factors more important than the response to others? In general, the factor that is farthest from the optimum will be the **limiting factor** for the individual in that environment: the factor that limits its success. A factor can be limiting because it is too high (above the upper lethal limit) or too low (below the lower lethal limit). Making one factor more favourable cannot compensate for the limiting effect of another. For example, if a plant is below its lower lethal limit for light, increasing the temperature to be closer to its optimum will have little impact on its growth or reproduction because it is already limited by light.

The limiting factor differs not only for individuals of different species but also for the same individual over time. Imagine a deciduous forest anywhere from western Canada to the Maritimes. For a plant growing on the forest floor, low temperature might limit its success in early spring, then moisture if rainfall is less than usual that spring, then nitrogen if the moisture shortage is alleviated by late spring rains, then light as the leaves in the canopy thicken by early summer, then high temperature during an August heat spell, and then low temperature in late fall.

The generalized response curve in Figure 5.20 illustrates the concept of **phenotypic plasticity**: the ability of a genotype to alter its phenotypic expression under different environmental conditions. Of course, not all instances of phenotypic plasticity are adaptive, such as stunted growth in response to inadequate nutrition. However, under many ecological circumstances, phenotypic plasticity is another mechanism (apart from leaving) whereby individuals cope with environmental change, particularly if they can shift their optima.

The set of phenotypes expressed by a single genotype across a range of environments (as opposed to different genotypes adapted to different environments) is called the **reaction norm** (Figure 5.21). If the shift in expression is adaptive, it increases an individual's fitness by improving its ability to survive, grow, and reproduce under the prevailing conditions. Reaction norms are not always parallel to each other, as in Figure 5.21. If two different genotypes express the same phenotype in the same environment, their lines will intersect. We cannot assume that two individuals share the same genotype simply because they look or function the same (i.e., share the same phenotype) in the same environment.

Plants provide many examples of phenotypic plasticity, largely because they are sessile and cannot avoid the effects of environmental change by moving. Plant size, the ratio of reproductive (flowering) to vegetative tissue, and leaf shape often vary widely with nutrition, moisture, temperature, and light. In a study of the phenotypic response of the annual spotted lady's

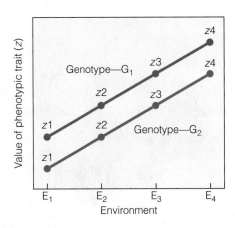

FIGURE 5.21 Reaction norm for two genotypes in four environments. Lines represent the phenotypic traits ($z1$–$z4$) exhibited by genotypes G1 and G2 in environments E_1–E_4.

INTERPRETING ECOLOGICAL DATA

Q1. Is there any environment in which the two genotypes express the same phenotype? Is it possible for the two genotypes to exhibit the same phenotype in different environments?

Q2. Suppose rather than the two lines representing the reaction norms being parallel, they intersected to form an X. How would this change your answers to question 1?

thumb (*Polygonum persicaria*) to soil moisture, eight genotypes were grown under four water treatments, with other factors held constant and non-limiting. All eight genotypes varied in phenotypic response, reflecting differences in biomass allocation to different organs (Sultan and Bazzaz 1993). This type of plasticity, which occurs during growth and is irreversible, is called **developmental plasticity**. Understanding how organisms develop in real-life environments is an emerging field called *ecological development* or "eco-devo" (Sultan 2010).

Large plastic differences in growth are less common in animals, but phenotypic plasticity during development can be dramatic. As with other social insects, ants can develop into different castes depending on their exposure to intrinsic (hormonal) and extrinsic factors. In particular, the diverse ant genus *Pheidole* has two wingless worker subcastes: minor workers that forage and work in the nest, and soldiers that defend the nest and process food. Some *Pheidole* species native to deserts also produce "supersoldiers" with extremely large heads that block the nest entry and guard against raids by co-occurring army ants. Expression of this supersoldier phenotype (Figure 5.22, p. 96) can be induced both in the wild and with environmental manipulation in *Pheidole* species in which it had not previously been observed but that are derived from species containing supersoldiers (Rajakumar et al. 2012). Species may thus retain "hidden" ancestral genetic potential, the phenotypic expression of which can be highly adaptive when exposed to novel environmental stresses.

Non-developmental forms of phenotypic plasticity are often reversible. Fish have upper and lower tolerance limits for temperature (Figure 5.23, p. 96), which change seasonally. During summer, the upper lethal temperature of the bullhead catfish *Ictalurus nebulosus* is 36°C, but it drops to 28°C in winter

Figure 5.22 Phenotypic expression of ancestral genetic potential in *Pheidole morrisi*. A normal soldier phenotype is shown with a supersoldier, as induced by methoprene, a hormone analogue.

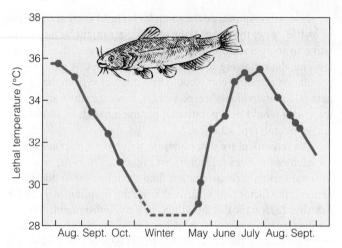

Figure 5.23 Thermal acclimation in the bullhead catfish. Tolerance limits shift as temperatures change gradually. Exposure to higher temperatures is lethal when the fish is acclimated to lower temperatures. The dashed line is an estimate. Each point indicates the temperature that is fatal to 50 percent of the population after exposure for 15 hours.

(Adapted from Fry 1947.)

(Fry 1947). As water temperatures change, shifts in enzyme and membrane structure allow the individual's physiology to adjust slowly, influencing heart rate, metabolic rate, neural activity, and enzymatic rates. Such reversible changes in response to changing conditions are called **acclimation** (sometimes *acclimatization* if in response to multiple factors) and can involve adjustments in physiology, morphology, and/or behaviour.

Acclimation occurs to some degree in all organisms. Plants typically adjust their physiology and/or morphology, whereas animals typically adjust their physiology and/or behaviour. Primates are noted for behavioural acclimation, but athletes preparing for the 2008 Olympics in Beijing trained at high temperature and humidity to adjust their physiological response.

Acclimation can be extremely rapid. Changes in enzyme and hormonal systems allow organisms to adjust their physiology and behaviour very quickly. For example, plants adapted to the shaded forest floor open their stomata in response to sunflecks in seconds. Seasonal adjustments of these same systems allow individuals to acclimate to periodic and predictable environmental changes. Finally, a combination of acclimation and developmental plasticity allows an individual maximum flexibility to alter the expression of its phenotypic traits during development to suit the conditions encountered. These phenotypic responses are stimulated by external signals that must be received and processed by cells, and are ultimately under genetic control.

Factors such as light, temperature, and moisture vary on both a daily and a seasonal basis in a predictable way, facilitating acclimation. But the environment varies on many timescales that directly affect organism function, and year-to-year variations in these same factors are often unpredictable, giving rise to periods of drought or flooding or unusual cold or heat that are more difficult for organisms to acclimate to. Over a longer timescale, geological processes and climatic variations change conditions at a regional to global scale. At these broader spatial and/or temporal scales, response to changing conditions is increasingly likely to involve evolutionary change as well as acclimation by existing genotypes.

Some assume that because changes in phenotypic expression (developmental or acclimation) are not heritable, phenotypic plasticity is not evolutionarily important. It has sometimes been misinterpreted as an alternative to natural selection: if individuals cope with changing environments by altering phenotypic expression, then selection leading to a change in allele or genotypic frequencies might seem unnecessary. But the *capacity* for phenotypic plasticity—whether a subtle difference in physiology or a dramatic difference in the size of a soldier ant's head—is itself an adaptation subject to natural selection. If a genotype has little capacity for altering its expression in adaptive ways, individuals of that genotype will likely be less fit (unless the environment is uniform). Ultimately all adaptations—phenotypic traits that facilitate an organism's success in prevailing conditions, including the capacity to alter expression of those traits—reflect natural selection operating on genetic variation within and among populations.

At this point, one might ask: *Does phenotypic plasticity represent a way in which an organism can be "perfectly adapted" to its environment, even if that environment is always changing?* While plastic responses help organisms cope with change, they are not a recipe for perfection. Any genotype is limited in how much it can alter its phenotypic expression, and how quickly such alteration can occur. The catfish may be able to lower its lower lethal limit, but it is highly unlikely that it could acclimate sufficiently to survive extreme cold, no matter how gradually this lower temperature came about.

With this background in ecological genetics, we are ready to explore the adaptive traits of organisms that allow them to survive, grow, and reproduce in the diversity of habitats on Earth.

EcologyPlace

Visit EcologyPlace at www.pearsoncanada.ca/ecologyplace to access online resources that complement your textbook, and help you to apply and to review the information in this chapter. EcologyPlace includes

- an eText version of the book
- self-grading quizzes
- glossary flashcards
- and more!

Go to www.pearsoncanada.ca/ecologyplace and follow the registration instructions on the Student Access Code Card included with this text. If your book does not have a Student Access Code Card, you can purchase access to it at www.pearsoncanada.ca/ecologyplace.

SUMMARY

Adaptations 5.1

Adaptations, heritable traits that enable an organism to thrive in a given environment, arise from natural selection. Individuals whose traits are better adapted to their environment have higher relative fitness, as measured by their proportionate contribution to future generations. Evolution involves changes in gene frequencies of populations over successive generations.

Genes 5.2

The units of heredity are genes, which are linearly arranged on chromosomes. Alternative forms of a gene are alleles. The pair of alleles present at a given locus defines the genotype for that trait. If both alleles are the same, the individual is homozygous. If the alleles differ, the individual is heterozygous. The genome is the sum of the individual's heritable information.

Phenotypes 5.3

The phenotype is the physical expression of the genotype. The manner in which the genotype affects the phenotype is the mode of gene action. When heterozygous individuals exhibit the same phenotype as one of the homozygotes, the expressed allele is dominant and the masked allele is recessive. If the phenotype of the heterozygote is intermediate, the alleles are co-dominant. Most phenotypic traits have a continuous distribution because most traits are affected by more than one locus and because expression is usually affected by the environment.

Genetic Variation 5.4

Genetic variation occurs at three levels: within populations, among populations, and among species. The sum of genetic information across all individuals in the population is the gene pool. Genetic variation in populations is measured by allele frequency and genotype frequency.

Types of Natural Selection 5.5

Natural selection acts on the phenotype, altering both the allele and the genotype frequencies of the population. Selection can be directional, stabilizing, or divergent (disruptive). Divergent selection is most likely to cause speciation. The target of selection is the trait that selection acts upon, and the selective agent is the external cause of differences in fitness among individuals.

Evolutionary Processes 5.6

Although natural selection is the only evolutionary process that generates adaptations, other processes alter gene frequency. These are (1) mutation, heritable changes in a gene or chromosome; (2) genetic drift, change in allele frequency through chance; (3) gene flow, movement of genes between populations as a result of movement of individuals; and (4) non-random mating. Non-random mating on the basis of phenotypic traits is called assortative mating and can be either positive (mates more similar than expected by chance) or negative (less similar). Inbreeding involves mating of individuals that are more closely related than expected by chance.

Genetic Differentiation 5.7

Natural selection alters genetic variation among populations. Widely distributed species often encounter a broader range of environmental factors than do more restricted species. This variation in selective pressures may generate a corresponding variation in morphological, physiological, and behavioural traits. Such variation can be associated with clines (if continuous) or with geographic isolates or ecotypes (if discontinuous). Geographic isolates and ecotypes can become new species if they are reproductively isolated.

Trade-offs and Constraints 5.8

Environmental factors vary in both space and time. Traits enabling a species to succeed under one set of conditions may limit its success under different environmental conditions. Trade-offs can affect the profitability of a trait, causing different species to adopt different strategies.

Speciation 5.9

Natural selection (along with other evolutionary processes) can lead to speciation. Adaptive radiation involves the proliferation of species from a common ancestor when there are opportunities for adapting to diverse habitats. Allopatric speciation occurs in different geographic locations. Selective pressures usually differ among locations, and reproductive isolation is assured by geographic barriers. Sympatric speciation occurs in the same location, through disruptive selection. If reproductive isolation occurs as a result of selection for traits that represent adaptation to environmental differences (either allopatric or sympatric), it is called ecological speciation.

Hybridization 5.10

Interspecific animal hybrids may not produce fertile offspring. Where hybridization occurs between populations of the same animal species, it may prevent speciation. Plants often form fertile interspecific hybrids. If accompanied by polyploidy, hybridization promotes sympatric speciation. Some sterile plant hybrids persist through asexual reproduction.

Absence of Perfection 5.11

Despite the power of natural selection as an evolutionary mechanism, no species is perfectly adapted to its environment. The reasons are many, including ecological trade-offs, speed and unpredictability of environmental change, absence of particular alleles in a population, random events, and genetic factors such as gene linkage.

Phenotypic Plasticity 5.12

Phenotypic plasticity is the ability of a genotype to express different phenotypes in different environments. For every factor, the individual has a tolerance range, with an optimum and zones of stress and intolerance. The limiting factor limits the success of an individual by being farthest from the optimum. Not all phenotypic plasticity is adaptive, but individuals that can shift their lethal limits and/or optima may be at a selective advantage. The range of phenotypes expressed by an individual is the norm of reaction. Developmental plasticity is irreversible. Acclimation is reversible phenotypic change in response to changing conditions, particularly if the changes are gradual and predictable. The capacity for plasticity is itself subject to natural selection.

KEY TERMS

acclimation, adaptation, adaptive radiation, allele, allele (gene) frequency, allopatric speciation, anaerobic respiration, apomixis, artificial selection, assortative (non-random) mating, biological species, chromosome, cline, developmental plasticity, diploid, directional selection, divergent (disruptive) selection, dominant allele, ecological speciation, ecotype, emigration, endemic, evolution, evolutionary bottleneck, fitness, founder effect, gene, gene flow, gene pool, genetic differentiation, genetic drift, genome, genotype, genotype frequency, geographic isolate, Hardy–Weinberg principle, heterotroph (secondary producer), heterozygous, homologous chromosomes, homozygous, immigration, inbreeding, interspecific hybridization, limiting factor, locus, metapopulation, migration, mutation, natural selection, optimum range, phenotype, phenotypic plasticity, photoautotroph (primary producer), polymorphism, polyploidy, profitability, qualitative trait, quantitative trait, reaction norm, recessive allele, reproductive isolation, selective agent, speciation, stabilizing selection, subspecies, sympatric speciation, target of selection, tolerance range, zone of intolerance, zone of physiological stress

STUDY QUESTIONS

1. What is natural selection? What conditions are necessary for natural selection to occur?
2. Distinguish between the terms *gene* and *allele*.
3. What is the relationship between an individual's genotype and its phenotype?
4. With regard to mode of gene action, distinguish between *dominance* and *co-dominance*.
5. What is the difference between the allele frequency and the genotype frequency of a population? Why might it be important to distinguish between these two measures? Hint: Consider the genetic impact of positive assortative mating.
6. What is the rationale for using survival as an indicator of fitness in Figure 5.7? Under what circumstances might this rationale prove inadequate?
7. What is the Hardy–Weinberg principle, and why is it important to evolution, even though its assumptions are rarely if ever met?
8. David Reznick studied natural selection in guppies (small freshwater fish) in Trinidad. Populations at lower elevations face the hazard of predatory fish, whereas those at higher elevations avoid predation because few predators can move upstream past the waterfalls. The mean size of guppies is

higher in high-elevation sites. Reznick hypothesized that smaller fish in lower-elevation populations reflected increased rates of predation on larger individuals; predation was selecting for smaller individuals. To test this hypothesis, Reznick moved individuals from the lower elevations to unoccupied pools upstream, where predation was absent. Eleven years later, the size of individuals increased in upstream populations (Reznick et al. 1997). Does this study illustrate natural selection (does it meet the necessary conditions)? If so, what type of selection does it represent? Can you think of alternative hypotheses to explain why the mean size of individuals may have shifted as a result of moving the population to the upstream environment?

9. Why does the development of antibiotic-resistant bacterial strains represent evolution by natural selection rather than by genetic drift (bottleneck type)? What type of selection is involved (directional, stabilizing, or divergent)?

10. Why are small populations more prone to genetic drift? How might genetic drift and inbreeding affect conservation of endangered species?

11. Why is natural selection the only process that can lead to adaptations? Why is natural selection not considered creative in and of itself?

12. Distinguish among *clines*, *ecotypes*, and *geographic isolates* (subspecies). Under what circumstances could a subspecies also be classed as an ecotype?

13. Does variation in the body size of guppies from upstream and downstream populations presented in Question 8 represent a trade-off similar to that involved in beak size of medium ground finches? What is the presumed selective agent in the guppy example?

14. Why is reproductive isolation critical for sympatric evolution?

15. Does plant speciation by hybridization involve sympatric or allopatric speciation? Why is speciation by hybridization not necessarily ecological speciation? What role does natural selection play in speciation by hybridization?

16. Andrew Hendry argues for caution in concluding that ecological speciation is occurring, even though he is a proponent of the importance of the process. How does this comment exemplify good science?

17. What is phenotypic plasticity? Contrast the two types of phenotypic plasticity.

18. Plastic differences in phenotypic expression are not always adaptive. Explain why not with reference to the tolerance range of an individual for an environmental factor. Why is the ability to shift the optimum and/or the lethal limits so important?

19. If two individuals have the same phenotype for a given trait, we cannot assume that they possess the same genotype for that trait. Explain why, with reference to the norm of reaction.

FURTHER READINGS

Conner, J. K., and D. L. Hartl. 2004. *A primer of ecological genetics*. Sunderland, MA: Sinauer.
Excellent introduction to population and quantitative genetics for the ecologist.

Desmond, A., and J. Moore. 1991. *Darwin: The life of a tormented evolutionist*. New York: Norton.
Written by two historians, this book provides an introduction to the man and his works.

Gould, S. J. 1992. *Ever since Darwin: Reflections in natural history*. New York: Norton.
A collection of Gould's humorous but insightful popular science essays. Other collections include *The Panda's Thumb*, *The Flamingo's Smile*, and *Dinosaur in a Haystack*.

Hendry, A. P. 2009. Ecological speciation! Or the lack thereof? *Canadian Journal of Fisheries and Aquatic Sciences* 66:1383–1398.
Thought-provoking article on the difficulty of establishing ecological speciation.

Mayr, E. 2001. *What evolution is*. New York: Basic Books.
Primer on the topics of natural selection and evolution by a leading figure in evolution. Well written and accessible to the general reader.

Ramsey, J., and D. W. Schemske. 2002. Neopolyploidy in flowering plants. *Annual Review of Ecology and Systematics* 33:589–639.
Review paper outlining the role of polyploidy in speciation of flowering plants.

Reznick, D. N., F. H. Shaw, F. H. Rodd, and R. G. Shaw. 1997. Evaluation of the rate of evolution in natural populations of guppies (*Poecilia reticulata*). *Science* 275:1934–1937.
Beautifully designed experiment for evaluating the role of natural selection in evolution.

Sultan, S. 2010. Plant developmental responses to the environment: Eco-devo insights. *Current Opinion in Plant Biology* 13:96–101.
This review article highlights the importance of developmental plasticity.

CHAPTER

6

PLANT ADAPTATIONS TO THE ENVIRONMENT

These Kokerboom trees in the desert region of Bloedkoppie, Namibia (southwestern Africa), use the CAM photosynthetic pathway to conserve water in this harsh environment.

Life on Earth is carbon based. Organisms are constructed of complex molecules built on a framework of carbon atoms, which bond with other carbon atoms to form long-chain molecules. In general terms, **assimilation** is the process of converting carbon into living tissues. The carbon building blocks needed for assimilation derive from various sources, and the means by which organisms acquire carbon represent the most fundamental of all adaptations. Animals and fungi are *heterotrophs*—organisms that acquire carbon by consuming other organisms. But the ultimate source of carbon is carbon dioxide (CO_2). Not all organisms can use this abundant form of carbon directly; only **autotrophs** can assimilate CO_2 into organic molecules and living tissue.

There are two types of autotrophs, based on their energy source. **Chemoautotrophs** convert CO_2 into organic matter by using energy derived from oxidizing inorganic molecules (e.g., hydrogen sulphide) or methane. Chemoautotrophic bacteria are the dominant primary producers in hydrothermal oceanic vents (see Research in Ecology: The Extreme Environment of Hydrothermal Vents, pp. 46–47). Far more abundant are *photoautotrophs*, which use solar energy to convert CO_2 into organic compounds using photosynthesis. This process, as performed by green plants, algae, and some bacteria, is essential for maintaining life on Earth.

Although all green plants and algae derive carbon from photosynthesis, the way in which they allocate the products of photosynthesis varies greatly, from the most minute of flowering plants (duckweed family, Lemnaceae) to the largest trees. These differences represent solutions to the problem of being a plant—acquiring the carbon, light, water, and minerals needed to support life. In this chapter, we examine the many adaptations that allow plants to survive, grow, and reproduce across the range of environments on Earth. But first, we review photosynthesis, the process so essential to life on Earth.

6.1 CARBON METABOLISM: Photosynthesis Converts CO_2 into Simple Sugars

Photosynthesis is the process by which plants, algae, and some bacteria use photosynthetically active radiation (PAR; see Section 2.1) to drive a complex sequence of reactions that reduce CO_2 to carbohydrates, with oxygen (O_2) as a by-product. Photosynthesis is summarized as follows:

$$6CO_2 + 12H_2O \xrightarrow[\text{chlorophyll}]{\text{light}} C_6H_{12}O_6 + 6O_2 + 6H_2O$$

The net effect of photosynthesis is the utilization of six molecules of water (H_2O) and the release of six molecules of O_2 for every six molecules of CO_2 that are reduced to one molecule of glucose ($C_6H_{12}O_6$). Other carbon-based compounds, including complex carbohydrates, fatty acids, and proteins, are synthesized from these initial products in leaves and stems.

Photosynthesis consists of two distinct phases. (1) The **light-dependent reactions** (often called the "light reactions") begin when **chlorophyll**, a pigment embedded in the internal membranes (*thylakoids*) of chloroplasts, absorbs light energy. Absorbing a photon raises the electrons of the chlorophyll molecule to an unstable excited state. In antennae chlorophyll molecules, these electrons rapidly return to the ground state, releasing the absorbed energy. This energy is transferred to other chlorophylls in a process called *resonance transfer* until it reaches the reaction centre chlorophylls. When these specialized chlorophyll molecules are excited, their electrons do not return to the ground state but are instead transferred to an electron acceptor, initiating electron transport. This process synthesizes two key compounds: the high-energy molecule adenosine triphosphate (ATP) from ADP, and the reducing agent nicotinamide adenine dinucleotide phosphate (NADPH) from NADP$^+$. To replace the "holes" in the reaction centre chlorophylls, H_2O is oxidized to provide electrons, releasing O_2 as a by-product.

ATP and NADPH produced in the light-dependent reactions are used in the second phase of photosynthesis—(2) the **light-independent reactions**, which incorporate CO_2 into sugars. (This phase was formerly called the "dark reactions" as it does not use light directly. But it does occur in light and requires the products of the light-dependent reactions.) In most plants, assimilating CO_2 begins when the five-carbon compound ribulose biphosphate (RuBP) combines with CO_2 to form two molecules of the three-carbon compound phosphoglycerate (3-PGA):

$$\underset{\substack{\text{1-carbon} \\ \text{molecule}}}{CO_2} + \underset{\substack{\text{5-carbon} \\ \text{molecule}}}{RuBP} \rightarrow \underset{\substack{\text{3-carbon} \\ \text{molecule}}}{2\ 3\text{-PGA}}$$

This reaction is catalyzed by the most abundant protein on Earth—the enzyme ribulose biphosphate carboxylase-oxygenase (**Rubisco**). The resulting 3-PGA is converted into glyceraldehyde 3-phosphate (G3P) using ATP and NADPH from the light-dependent reactions. Some of this G3P is converted to glucose ($C_6H_{12}O_6$), starch, and other carbohydrates, with the remainder used to regenerate RuBP to continue the process. Regenerating RuBP requires more ATP. Thus, light supplies the ATP and NADPH needed for the light-independent reactions. This pathway, involving initial fixation of CO_2 into the three-carbon compound PGA, is also called the *Calvin–Benson cycle*, and plants employing it are called C_3 **plants** (Figure 6.1, p. 102).

The C_3 pathway has a major drawback. Rubisco, which catalyzes the carboxylation step, also acts as an oxygenase, fixing O_2 to RuBP. The oxygenation of RuBP competes with its carboxylation, leading to loss of CO_2 in **photorespiration**, which lowers photosynthetic efficiency (see Section 6.10 for a discussion of how some plants counter this problem).

Some of the carbohydrates produced in photosynthesis are used by the plant in **aerobic (cellular) respiration**, which harvests energy as ATP from the oxidation of sugars in the presence of O_2. It occurs in the mitochondria of eukaryotic cells, and is summarized as

$$C_6H_{12}O_6 + 6O_2 \rightarrow 6CO_2 + 6H_2O + \text{energy (as ATP)}$$

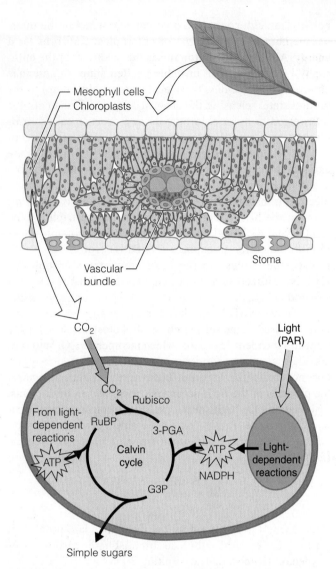

Figure 6.1 Simplified representation of the C_3 pathway. Photosynthesis occurs in chloroplasts inside mesophyll cells. ATP and NADPH produced in the light-dependent reactions are used to synthesize the energy-rich sugar G3P and to regenerate RuBP in the light-independent (Calvin cycle) reactions. (PAR = photosynthetically active radiation.)

Because leaves use CO_2 in photosynthesis and produce CO_2 in respiration, the difference in the rates of these two processes is the net carbon gain, or **net photosynthesis**:

Net photosynthesis = Gross photosynthesis − Respiration

In the absence of O_2, **fermentation** (a type of **anaerobic metabolism**) enables some cells to convert glucose into lactic acid and ATP:

$$C_6H_{12}O_6 \rightarrow 2C_3H_6O_3 + 2ATP$$

Rates of photosynthesis and respiration are typically measured in moles of CO_2 per unit leaf area per unit time (e.g., μmol $CO_2/m^2/s$). The ecosystem equivalent, resulting from the photosynthesis of all the autotrophs in a community, is *net primary production* (see Section 20.2).

6.2 IMPACT OF LIGHT: Photosynthetic Rate Varies with Light Intensity

Because solar radiation provides the energy to convert CO_2 into simple sugars, availability of light of an appropriate wavelength directly influences photosynthetic rate. We first consider a general plant response curve to light (Figure 6.2). Later, in Section 6.9, we consider how the response varies for species adapted to different light environments.

As light levels decline, photosynthetic carbon gain drops until it equals the rate of respiratory carbon loss. Net photosynthetic rate is then zero. The light level (value of PAR) at which this occurs is the **light compensation point (LCP)**. The LCP is a breakeven point and varies greatly among species. At light levels below the LCP, losses from respiration exceed uptake in photosynthesis, causing a net loss of CO_2. The rate of carbon loss in darkness (PAR = 0) provides an estimate of mitochondrial respiration rate.

At light levels above LCP, photosynthetic rate increases with increasing PAR, and the light-dependent reactions limit photosynthesis. Eventually, photosynthesis becomes light saturated, and photosynthetic rate no longer increases with increasing PAR. Now the light-independent reactions are limiting. The level of PAR above which no further increase in photosynthesis occurs is the **light saturation point (LSP)**. In species adapted to deep shade, photosynthesis declines rapidly as light exceeds saturation. This damaging effect of high light, called **photoinhibition**, reflects "overloading" of processes in the light-dependent reactions.

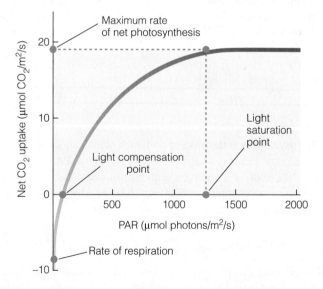

Figure 6.2 Response of net photosynthesis to available light (PAR = photosynthetically active radiation, expressed as photon flux density). Net photosynthetic rate increases to a maximum as PAR approaches the light saturation point. The light compensation point is equal to the PAR value at which photosynthetic CO_2 uptake equals respiratory CO_2 loss. The net CO_2 exchange at PAR = 0 (darkness) estimates the respiratory rate.

6.3 GAS EXCHANGE: Plants Exchange CO_2 and H_2O Vapour with the Atmosphere

Photosynthesis occurs in specialized leaf cells called **mesophyll** cells ("middle of the leaf"; see Figure 6.1). For photosynthesis to occur, CO_2 must move from the atmosphere into the leaf interior. In terrestrial plants, CO_2 diffuses into leaves through regulated epidermal pores called *stomata* (singular: *stoma*) (Figure 6.3). *Diffusion* is movement of a substance from areas of higher to areas of lower concentration; the greater the concentration difference, the faster the diffusion. CO_2 concentration is often measured in parts per million (ppm) of air. So, 355 ppm CO_2 corresponds to 355 CO_2 molecules for every million air molecules. Substances diffuse between two areas until the concentrations are equal. As long as the concentration of CO_2 outside the leaf is greater than that inside, CO_2 will continue diffusing in through the stomata.

Why, then, don't CO_2 concentrations inside and outside the leaf reach equilibrium, halting further net diffusion? As CO_2 is consumed in photosynthesis, the concentration inside the leaf declines, maintaining the gradient. If photosynthesis stops and the stomata stay open, CO_2 will diffuse into the leaf only until the internal concentration equals that of the atmosphere. In fact, when photosynthetic CO_2 demand declines for any reason, stomata close, reducing diffusive entry. Why do stomata close in this circumstance? Stomatal closure is a vital protective measure with respect to another critical resource: water. As CO_2 diffuses into a leaf, water vapour diffuses out through the same stomatal pores in *transpiration* (see Section 3.7). Transpiration can pose a serious risk to plant survival. Although the plant is forced to lose water in order to access CO_2, it is advantageous to limit gas exchange when water supply is low even if stomatal closure limits photosynthesis by reducing supplies of CO_2.

How rapidly water moves from inside the leaf, through the stomata, and into the air depends on the gradient of water vapour. Like CO_2, water vapour diffuses from areas of higher to areas of lower concentration. The air inside the leaf is saturated with water, so outflow is determined by the water vapour in the air—the *relative humidity* (see Section 2.6). The drier the air (lower relative humidity), the more rapidly water diffuses through the stomata into the surrounding air. Therein lies the risk: the plant must replace water lost in transpiration, or it will wilt and die.

6.4 WATER MOVEMENT: Water Follows a Free-Energy Gradient from the Soil through the Plant to the Atmosphere

All organisms require water, but the sensitivity of plants to water loss relates partly to their rigid cell walls. **Turgor pressure** is the force exerted on the cell wall by the water contained in the cell. Growth and physiological efficiency of plant cells are optimal when the cells are at maximum turgor, that is, fully hydrated. When cell water content declines, turgor pressure drops and water stress occurs, ranging from wilting to dehydration and death. For leaves to remain turgid, they must replace water lost to the atmosphere in transpiration with water absorbed through the roots from the soil and transported to the leaves.

All water movement obeys the basic laws of physics. Recall that "work"—displacement of matter, including water transport—requires energy. The measure of energy available ("free") to do work is called Gibbs free energy (G). In *active transport*, such as the pumping of blood by the heart or (to give an abiotic example) the transport of water from the ground to an elevated tank using an electric pump, energy must be added to make the process happen because it is going against a free-energy gradient. In contrast, movement of water through the soil–plant–atmosphere continuum involves *passive transport*, a spontaneous process that follows a free-energy gradient.

All reactions proceed spontaneously from a state of high free energy to one of lower free energy. If we define the difference in free energy (ΔG) between two states, A and B, as

$$\Delta G = G_B - G_A$$

where G_A and G_B are the free energy of states A and B, respectively, we can predict the direction that a reaction between these states will take. If ΔG is negative (i.e., $G_A > G_B$), the reaction will proceed spontaneously from state A to state B. To reverse the reaction, that is, to go from B to A, we must add free energy. If the free energies of A and B are equal ($\Delta G = 0$), the reaction is at equilibrium.

The transition between states A and B can represent either a chemical or a mechanical reaction. Photosynthesis is a chemical reaction requiring input of free energy as solar radiation, because the total free energy of the products (G_B) is greater than that of the reactants (G_A). In contrast, movement of water from the soil to the atmosphere is a spontaneous mechanical reaction. But if water movement from the soil into the roots, from the roots into the leaf, and from the leaf into the atmo-

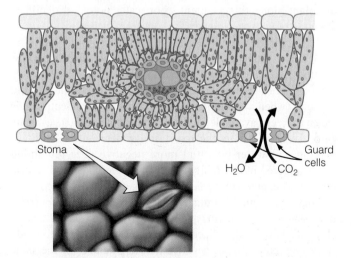

Figure 6.3 Cross section of a C_3 leaf. Stomata (pores on the epidermis flanked by guard cells) facilitate diffusive exchange of CO_2 and H_2O between the atmosphere and the leaf interior.

Stoma

Guard cells

H_2O CO_2

sphere is spontaneous, what provides the free-energy gradient that makes it happen?

To answer this question, we begin by considering *transpiration*—movement of water from the leaf to the atmosphere through the stomata. Recall that transpiration involves diffusion of water down a concentration gradient from regions of higher concentration (inside the leaf) to regions of lower concentration (in the surrounding air). The free energy that allows this work to happen is the kinetic energy associated with the random movement and collision of water molecules. Diffusion of water follows a concentration gradient, which is a type of free-energy gradient. A region with a higher concentration of water (greater density of water molecules) has more free energy from random motion and collision of water molecules than does a region of lower concentration. There will be net movement of water molecules from the region of higher to the region of lower free energy until the free energy is the same in both (i.e., $\Delta G = 0$). Then, the concentration of water molecules is the same in both, and net exchange of energy (and molecules) ceases. Equilibrium is reached.

The measure of the free energy of water at any point along the soil–plant–air continuum is **water potential** (ψ). Pure water (no solutes), which has maximal free energy because it has the highest possible density of water molecules, is assigned a water potential of zero ($\psi = 0$) units of pressure (megapascals or MPa). (It may seem puzzling that a substance with high free energy is assigned a value of zero, but lower water potentials are assigned negative values because movement is always away from pure water.)

Using this terminology, we reconsider transpiration. When atmospheric relative humidity drops below 100 percent, the free energy of water declines and the water potential of the atmosphere (ψ_{atm}) becomes a progressively larger negative value (Figure 6.4a). Under most conditions, the air inside the leaf is at or near saturation (relative humidity ~ 100 percent). As long as the relative humidity of the air is below 100 percent, a steep gradient of water potential between the leaf (ψ_{leaf}) and the atmosphere (ψ_{atm}) will drive transpiration. Water vapour will move from the region of higher water potential (leaf interior) to that of lower water potential (atmosphere), along this free-energy gradient. The lower the humidity, the steeper the water potential gradient, and the faster the transpiration. As water is evaporating from a liquid to a gas, which lowers its water potential further, the water potential difference is the greatest at this point in the overall path of water movement.

Meanwhile, what happens to the water potential of the plant? Unlike atmospheric water potential, which is determined by relative humidity, several factors determine water potential within the plant. As water is lost in transpiration, the leaf cell's water content decreases, thereby increasing its solute concentration. More solutes lower the cell's water potential because the concentration of water molecules declines. The component of plant water potential associated with solutes is called **osmotic potential** (ψ_π), because the difference in solute content inside and outside a cell drives water movement by **osmosis**: movement of a liquid across a semi-permeable membrane. As we have seen, plant osmotic potential declines with transpiration, but osmotic potential can be manipulated. If the cell breaks down starch into

sucrose, its osmotic potential declines (becomes a larger negative number), reducing the cell's overall water potential.

A second factor affecting water potential of a plant is **pressure potential**: positive pressure exerted by the rigid cell wall (ψ_p). It increases the plant's water potential. The decrease in turgor resulting from transpiration lowers the water potential of leaf cells, making it easier to take up water from other, more hydrated cells. Finally, the surfaces of some macromolecules, such as cellulose in plant cell walls, attract water. **Matric potential** (ψ_m) is the component of water potential associated with the tendency of water to adhere to *hydrophilic* (water-loving) surfaces. Like osmotic potential, it lowers the free energy of water, reducing overall water potential.

Overall water potential, ψ, at any point, from leaf to stem to root, is the sum of these components:

$$\psi = \psi_\pi + \psi_p + \psi_m$$

Osmotic and matric potentials will always be negative or zero, while turgor pressure can be either positive (in a turgid cell with a rigid wall) or negative (in xylem tissue when subject to suction from transpiration). Theoretically, a cell's overall water potential (ψ) can be either positive or negative, depending on the values of its components. However, water potential values at any point along the continuum (soil, root, leaf, and atmosphere) are typically negative, and movement of water will be from regions of higher (zero or less negative) to lower (more negative) potential. Water movement from the soil to the root, to the leaf, and to the air depends on maintaining a gradient of increasingly negative water potential at each point (Figure 6.4b):

$$\psi_{atm} < \psi_{leaf} < \psi_{root} < \psi_{soil}$$

Drawn by the low water potential of air, water evaporates from the surface of and between mesophyll cells, escaping through the stomata. This water potential gradient is transmitted to the mesophyll cells and on to the water-conducting xylem tissue in the leaf veins. A gradient of increasingly negative water potential extends down through the stem to the fine roots, which are in contact with soil particles. As water moves through the plant along this gradient, root water potential declines, allowing more water to enter the root from the soil.

Transpiration continues as long as (1) the energy striking the leaf is enough to supply the *latent heat of evaporation* (see Section 2.6), (2) soil moisture is available, and (3) the roots have a more negative water potential than the soil. At field capacity (when a soil holds the most moisture; see Section 4.7) water is freely available, and soil water potential (ψ_{soil}) nears zero. As water enters the roots, soil water content declines, making soil water potential more negative. The remaining water adheres more tightly to soil particles, and soil matric potential becomes more negative. For a given water content, soil matric potential is influenced strongly by *soil texture* (relative content of sand, silt, and clay; see Section 4.5). Because clay soils have more surface area per unit volume than do sandy soils, they have a more negative matric potential for the same water content, making it harder for the roots to extract water.

As soil water potential becomes more negative, root and leaf water potentials must decline (become even more negative

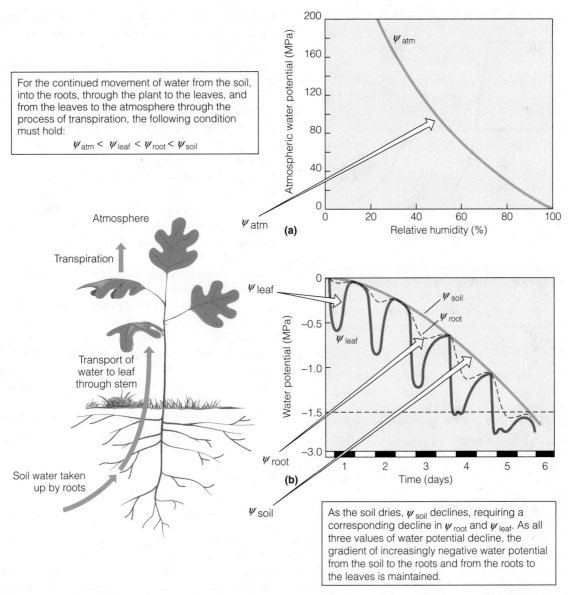

For the continued movement of water from the soil, into the roots, through the plant to the leaves, and from the leaves to the atmosphere through the process of transpiration, the following condition must hold:

$$\psi_{atm} < \psi_{leaf} < \psi_{root} < \psi_{soil}$$

As the soil dries, ψ_{soil} declines, requiring a corresponding decline in ψ_{root} and ψ_{leaf}. As all three values of water potential decline, the gradient of increasingly negative water potential from the soil to the roots and from the roots to the leaves is maintained.

Figure 6.4 Transport of water along a water potential gradient from soil to plant to air. Water moves from a region of high to a region of low (more negative) water potential. **(a)** As the relative humidity drops below 100 percent, a steep decline in atmospheric water potential maintains the gradient from leaf to air. **(b)** As the soil dries, the soil water potential becomes increasingly negative. This decline requires a corresponding decline in the water potential of the roots and leaves to maintain the gradient sustaining water movement. The graph depicts changes in the plant's leaf and root water potential in response to declining soil moisture over six days. Note the diurnal flux in leaf water potential, which peaks at night when the stomata are closed.

(Adapted from Pessarakli 2002.)

than that of the soil) to maintain the gradient for water movement. If rain does not recharge soil water and soil water potential continues to fall, the plant will eventually not be able to maintain the gradient. The stomata will then close to stop further transpirational loss. This closure also halts CO_2 uptake. The soil water potential at which the stomata close is determined by the plant's ability to reduce leaf water potential without disrupting leaf physiology. The value of leaf water potential at which the stomata close and net photosynthesis ceases varies among species based on their **drought tolerance** (Figure 6.5, p. 106).

We can now better appreciate the dilemma faced by land plants. The inevitable trade-off between CO_2 uptake and water loss establishes a crucial link between soil water and photosynthesis. To photosynthesize, the plant must open its stomata, but when its stomata are open, it loses water, which must be replaced by the roots. If soil water is scarce, the plant must compromise, partially closing its stomata to minimize water loss while still allowing some uptake of CO_2. Regulating stomatal opening through a complex mechanism involving the guard cells (see Figure 6.3) is the plant's strategy for regulating water loss, but at the expense of photosynthetic capacity.

This balancing act between photosynthesis and transpiration is the single most important constraint affecting evolution of land plants. It also directly influences the productivity of

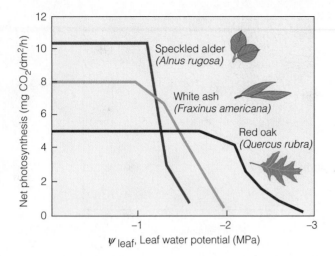

Figure 6.5 Changes in net photosynthesis as a function of leaf water potential for three North American woody species of increasing drought tolerance. Lower net photosynthesis with declining leaf water potential results from partial stomatal closure. Drought-tolerant species can continue photosynthesis at lower leaf water potentials, until the stomata close entirely.

ecosystems. For the organism, its importance is reflected in **water-use efficiency (WUE)**: the ratio of net carbon fixed in photosynthesis per unit of water lost in transpiration. Drought-tolerant plants have a higher WUE as a result of natural selection. Yet, despite being a necessary evil, transpiration is not without its benefits. It cools the plant (see Section 6.6), and movement of water in the xylem transports minerals absorbed by the roots.

The theory of water movement just described is called **transpirational pull** because it postulates that water loss through the stomata is the first step in establishing the water potential gradient that drives upward movement of water. The cohesive properties of water help maintain the water column in the xylem, which is under strong negative pressure (tension). Physiologists have supplemented this theory with the idea of **hydraulic redistribution**. Soil regions often differ in water content, and when water travels from a deeper, wetter soil to an upper, drier region, some water exits the roots into the soil, especially at night when transpiration ceases (Richards and Caldwell 1987). This movement also follows a water potential gradient but diverts water away from the soil–plant–air continuum, increasing water availability for plants in the upper reaches of soil under large trees. The movement is normally upward, but it can be downward in desert soils after a heavy rain (Schulze et al. 1998).

6.5 AQUATIC PHOTOSYNTHESIS: Some Aquatic Photoautotrophs Utilize Bicarbonate

A major difference in CO_2 uptake between submerged aquatic and terrestrial plants is their lack of stomata (although some species have non-functional vestigial stomata). CO_2 diffuses from the atmosphere into surface waters, where it mixes into the water column and diffuses across cell membranes. Photosynthesis then proceeds in much the same way as outlined in Section 6.1.

However, some aquatic plants and algae can also use bicarbonate (HCO_3^-), which forms when dissolved CO_2 reacts with H_2O. This reaction is reversible, and CO_2 and HCO_3^- concentrations tend towards a dynamic equilibrium as part of a complex reaction sequence (see Section 3.6). Interestingly, plants that use HCO_3^- must convert it back to CO_2 using the enzyme carbonic anhydrase, either (1) by active uptake of HCO_3^- into the leaf, followed by enzymatic conversion, or (2) by excreting the enzyme into the water, followed by uptake of converted CO_2. Any use of carbon by aquatic plants, either as CO_2 or HCO_3^-, lowers the CO_2 concentration of the adjacent water. Because CO_2 diffuses in air at 10 000 times the rate in water, it can easily become depleted, reducing uptake and photosynthesis. This constraint is particularly important in still waters such as dense sea-grass beds or rocky intertidal pools.

Marine *phytoplankton* (small, floating photosynthetic organisms, either algae or photosynthetic bacteria and archaea; see Section 24.9) reduce CO_2 levels in the ocean (and ultimately the atmosphere) not only by photosynthesis. Some also deposit large amounts of carbon as $CaCO_3$ in tiny scales covering their cells. This calcification leads to accumulations on the ocean floor, affecting the carbon cycle (see Section 22.4) and potentially ameliorating global warming (see Chapter 28), as might the sulphur oxides released by some phytoplankton.

6.6 PLANT TEMPERATURE: Plants Exchange Heat Energy with Their Environment

Photosynthesis and mitochondrial respiration respond differently to temperature (Figure 6.6). As temperatures rise above freezing, gross photosynthetic rate initially increases faster than does respiration. As temperatures rise further, respiration increases more rapidly than gross photosynthesis, which

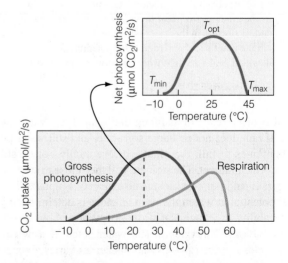

Figure 6.6 Temperature response of photosynthesis and respiration. Here, the temperature optimum for net photosynthesis (gross photosynthesis—respiration) is between 20°C and 30°C.

reaches a maximum, reflecting the temperature optimum of Rubisco. At even higher temperatures, gross photosynthesis declines while respiration continues to increase up to an upper critical limit beyond which respiration ceases as a result of breakdown of enzymes and other macromolecules. The temperature response of net photosynthesis is the difference between the responses of gross photosynthesis and respiration.

Three *cardinal values* describe this temperature response curve. T_{min} and T_{max} are the minimum and maximum temperatures, respectively, at which net photosynthesis approaches zero (no net carbon uptake). T_{opt} is the temperature (typically a range) over which net carbon uptake is at its maximum (optimal). This curve resembles the general response curve in Figure 5.20, except that T_{min} and T_{max} are not lethal limits for the plant but rather the limits between which photosynthesis is possible.

The temperature of the leaf, not the air, controls the rates of photosynthesis and respiration, and leaf temperature depends on the exchange of energy between the leaf and its environment. Of the radiation they intercept, plants reflect some shortwave radiation and (after absorption) re-radiate some longwave radiation back to the atmosphere. The difference between the amount of radiation a plant (or other object) receives and the amount it reflects and/or emits is the **net energy balance** (R_n). If it is positive, the plant heats up; if negative, the plant cools down.

Some absorbed radiation is used in plant metabolism and stored in chemical bonds through the totality of plant biochemical processes, including photosynthesis and respiration. This quantity is quite small, typically less than 5 percent of R_n. The remaining energy heats the plant and the surrounding air. On a clear sunny day, the energy absorbed can raise internal leaf temperatures well above ambient air or water temperature. Internal leaf temperatures may exceed the optimum for photosynthesis and possibly reach damaging levels.

Like fungi and many animals, plants are *poikilotherms* (organisms whose internal temperature varies with that of their surroundings; see Section 7.7). To keep internal temperatures within their range of tolerance, plants exchange heat with their environment. Aquatic plants exchange heat mainly by *convection* (transfer of heat energy between a solid and a moving fluid; see Quantifying Ecology 7.1: Heat Exchange and Temperature Regulation, pp. 136–137). Convective heat loss depends on the difference between the temperature of the leaf and the surrounding fluid (air or water). If the leaf is warmer than the surrounding water, there is a net heat loss from the leaf.

Transpiring plants also exchange heat by *evaporation*. As plants lose water through their stomata, their temperature declines through evaporative cooling. The ability to dissipate heat in this way depends on the transpiration rate, which is influenced by atmospheric relative humidity and by water availability (see Section 6.3).

Heat transfer is influenced by the **boundary layer**, a layer of still air (or water) adjacent to the leaf surface. The environment of the boundary layer is modified from that of its surroundings by diffusion of heat, water, and CO_2 from plant surfaces. As water exits the stomata, the humidity of the boundary layer rises, reducing transpiration. Likewise, as heat radiates from the leaf, the boundary layer temperature increases,

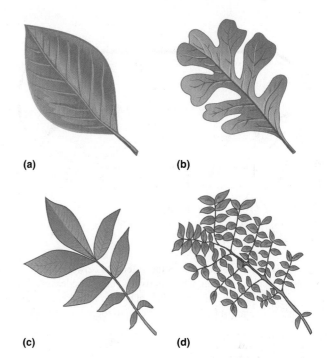

(a)   **(b)**

(c)   **(d)**

Figure 6.7 Leaf morphology. Leaves are either simple, as in **(a)** and **(b)**; or compound, as in **(c)** and **(d)**. Shape is highly variable, including smooth (entire) margins, as in **(a)**, or lobed margins, as in **(b)**. Small, lobed leaves reduce the boundary layer, increasing convective heat exchange.

reducing heat transfer. Under still conditions with little air or water flow, the boundary layer thickens, reducing transfer of heat, water, and CO_2 between the leaf and its surroundings. Wind or water flow reduces the thickness of the boundary layer, facilitating mixing with air or water and re-establishing the diffusion gradient between the leaf surface and its surroundings.

Factors intrinsic to the plant, such as leaf size and shape, influence boundary layer dynamics and hence the ability of plants to exchange heat. Air tends to move more smoothly over a larger surface than a smaller one, so the boundary layer tends to be thicker and more intact in larger leaves, especially those with entire (smooth) margins (Figure 6.7a). Deeply lobed leaves, like those of most oaks (Figure 6.7b), and small, compound leaves (Figure 6.7d) disrupt air flow, causing turbulence that reduces the boundary layer and increases exchange of heat and water vapour.

6.7 CARBON ALLOCATION: Plants Allocate Net Photosynthetic Carbon Gain to New Tissues

Because leaves both absorb and release CO_2, a simple economic approach called the **carbon balance model** describes the balance between CO_2 uptake in photosynthesis and its loss in respiration. So far, our discussion has focused on net photosynthesis of leaves. But plants are not just composed of leaves; they also have roots, stems, and (in flowering species) flowers and fruits. In keeping with the carbon balance model, net carbon uptake of the whole plant is the total photosynthetic uptake minus total respiratory loss from all organs (Figure 6.8).

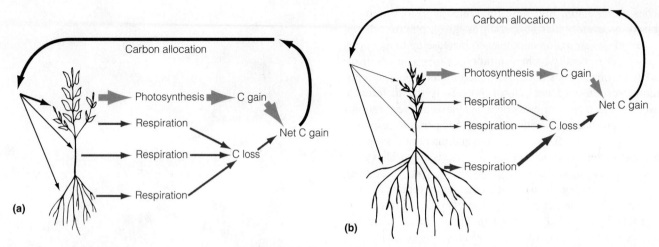

Figure 6.8 Net carbon gain is the difference between photosynthetic gain and respiratory loss. Because leaves (photosynthetic tissues) are responsible for carbon gain, while all tissues (including leaves) respire, net gain is affected by allocation to different tissues. Relative thickness of the arrows represents relative values of carbon gain, loss, or translocation. **(a)** Increased allocation to leaves increases gain relative to loss, increasing net gain. **(b)** Increased allocation to roots has the opposite effect, decreasing gain relative to loss and lowering net gain.

Total carbon uptake (gain per unit time) is the average rate of carbon uptake per unit leaf area (photosynthetic surface) multiplied by total leaf surface area. Because all living cells respire (not just photosynthetic cells), total carbon loss per unit time reflects the total mass of living tissue—the sum of leaf, stem, root, and reproductive tissues. (Each of these organs also contains some dead and hence non-respiring cells, such as water-conducting cells in the xylem, but we will ignore these to keep the model simple.) Net carbon gain for the whole plant is apportioned by **carbon allocation** to various processes, including maintenance and synthesis of new tissues.

Carbon allocation greatly influences plant survival, growth, and reproduction. Acquiring the resources needed for photosynthesis and growth involves different tissues. Leaves provide the photosynthetic surface, gaining access to light and CO_2. As well as anchoring the plant in the soil, roots provide access to soil water and minerals. Stems provide vertical support, elevating leaves (and flowers) and increasing access to light by reducing shading by taller plants. Stems also conduct water and nutrients from the roots to other plant parts. Availability of essential resources influences carbon allocation to production of these various plant tissues.

Under ideal conditions with no resource limitations, allocating carbon to leaves promotes the fastest growth (Figure 6.8a). Producing more leaves increases the photosynthetic surface, which increases carbon uptake as well as respiratory loss. Allocation to all other tissues, such as stems and roots, increases carbon loss without directly increasing the capacity for uptake. The result is reduced net carbon gain (Figure 6.8b). Given this reality, it might seem non-adaptive to allocate carbon to any tissues other than leaves. However, producing stems and roots is essential for acquiring resources other than light. As water and minerals become scarce, it is increasingly necessary to allocate carbon to non-photosynthetic tissues at the expense of leaves. We consider the implications of these shifts in allocation in the following sections.

6.8 INTERDEPENDENCE OF ADAPTATIONS: Plant Adaptations Reflect Responses to Multiple Factors

Physical environments vary widely in their geology and soils; spatial and temporal patterns of precipitation and temperature; and salinity, depth, and flow of water on their surfaces. In all but the most extreme of Earth's environments, autotrophs harness solar energy in photosynthesis. To survive, grow, and reproduce, plants must maintain a positive carbon balance, converting enough CO_2 into carbohydrates to offset respiratory losses. To achieve this end, a plant must acquire the essential resources of light, CO_2, water, and minerals, while tolerating abiotic conditions such as temperature, salinity, and pH that directly affect plant processes. Although discussed as if independent of each other, most plant adaptations are interdependent, for reasons relating both to the physical environment and to the plants themselves.

Many abiotic factors, such as light, temperature, and moisture, are linked. The amount of solar radiation not only affects the availability of light for photosynthesis but also directly influences the temperature of the leaf and its surroundings. In turn, temperature affects humidity, a key factor influencing transpiration. As a result, there are correlations in the adaptations of plants to variations in these interrelated factors. Plants adapted to dry, sunny habitats must cope with the high water demand associated with warm temperatures and low humidity. They have traits, such as small, thick leaves and more allocation to roots, that represent adaptations to multiple factors.

There are critical trade-offs in the ability of plants to adapt to limitations imposed by multiple abiotic factors, particularly acquisition of above- versus belowground resources. Allocating carbon to leaves increases access to light and CO_2, but at the expense of allocating carbon to roots. Likewise, allocating carbon to roots increases access to water and minerals but limits allocation to leaves. Adaptations that allow a plant to succeed under one set of environmental conditions inevitably limit its ability to do equally well under other conditions. We will now explore the consequences of this simple premise of ecological trade-offs.

6.9 ADAPTATIONS TO LIGHT: Shade Tolerance Reflects Genotype and Acclimation

The solar radiation reaching Earth's surface varies daily, seasonally, and geographically. However, a major microhabitat factor influencing the amount of PAR a plant receives is shading by nearby plants. Although the amount of light reaching an individual plant varies continuously as a function of the area of leaves above it (LAI; see Section 4.2), plants occupy qualitatively different light environments—sun or shade—depending on whether they are overtopped by other plants. Plants have evolved an array of physiological and morphological adaptations that allow them to survive, grow, and reproduce in these contrasting light habitats.

The relationship between light and photosynthesis varies among plants (Figure 6.9). Assuming they have the capacity for shade tolerance, shade-grown plants have lower light compensation points (LCP), light saturation points (LSP), and maximum photosynthetic rates than plants grown in high light (Valladares and Niinemets 2008). These differences reflect lower respiration rates in shade-grown plants. Plants that can lower their respiratory costs can survive with less light. Plants expend considerable energy and nutrients to produce Rubisco and other parts of the photosynthetic apparatus. Low light, not availability of Rubisco to catalyze CO_2 fixation, limits photosynthesis in shade, so shade-grown plants produce less Rubisco. In turn, lower Rubisco content is correlated with lower respiration rates (Björkman 1981). Because the LCP is the value of PAR where gross photosynthetic gain equals respiratory loss, shade-grown plants have a lower LCP. However, reduced Rubisco limits the ability to increase photosynthesis in high light, lowering both the LSP and the maximum photosynthetic rate of shade-grown plants.

This trade-off is evident in an experimental study of seedlings of nine *Macaranga* species (rain forest trees in Borneo) grown in a greenhouse under high and low light (created by using shade cloths). After six months, the leaf respiratory rate, light compensation point, and maximum net photosynthetic rate of seedlings grown in low light were all significantly lower than for seedlings of the same species grown in high light (Figure 6.10a-c, p. 110) (Davies 1998).

These shifts in photosynthetic response were accompanied by altered leaf morphology. The ratio of surface area to mass for a leaf is its **specific leaf area** (SLA; cm^2/g). SLA increased for all species when grown in low light (Figure 6.10d) (Davies 1998). In general, shade-grown leaves are larger in area and thinner than leaves grown in high light (Figure 6.11, p. 110). This shift increases the surface for capturing light (a limiting factor in shade).

As well as producing broader, thinner leaves, shade-grown plants allocate more carbon to leaves than to roots (Figure 6.12,

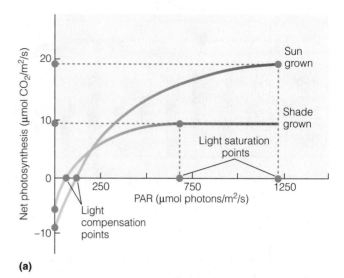

(a)

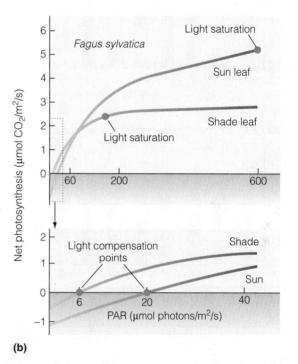

(b)

Figure 6.9 Photosynthetic response to light availability (PAR) for shade- and sun-grown plants. **(a)** Shade-grown plants typically have lower light compensation and light saturation points than do sun-grown plants of the same species. Similar differences occur in shade-tolerant versus shade-intolerant species. **(b)** Shift in photosynthetic response of sun and shade leaves of copper beech (*Fagus sylvatica*). The sun leaf is from the periphery of the canopy, exposed to full sun. The shade leaf is from the shaded canopy interior. The insert shows the regions of the response curves below a photon flux density of 50 µmol photons/m^2/s, showing the differences in LCP.

(Adapted from Larcher 1996.)

p. 110). This shift further increases the surface area for light capture, partially offsetting the decrease in photosynthetic rate per leaf area in low light. (Investing less in roots might be risky when soil water is limited, but it is safe in most shady habitats, where water is less likely limiting.) As well as these morphological

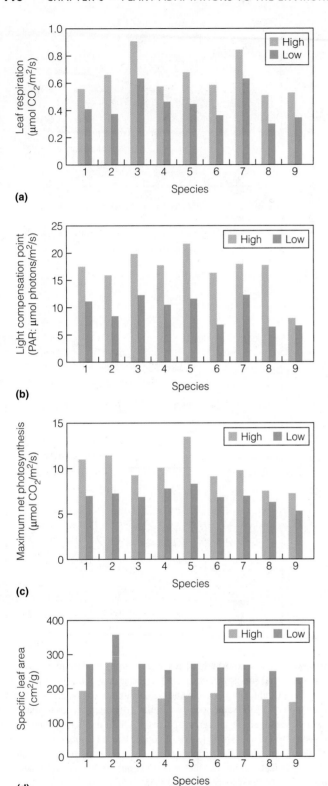

(a)

(b)

(c)

(d)

Figure 6.10 Variation in **(a)** leaf respiration, **(b)** light compensation point, **(c)** maximum net photosynthesis, and **(d)** specific leaf area for nine *Macaranga* species grown in high and low light. (1) *M. hosei*, (2) *M. winkleri*, (3) *M. gigantea*, (4) *M. hypoleuca*, (5) *M. beccariana*, (6) *M. triloba*, (7) *M. trachyphylla*, (8) *M. hullettii*, and (9) *M. lamellata*.

(Adapted from Davies 1998.)

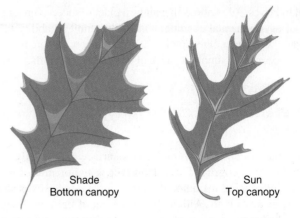

Shade
Bottom canopy

Sun
Top canopy

Figure 6.11 Response of leaf morphology to light. Red oak (*Quercus rubra*) leaves in the upper canopy experience higher light and warmer temperatures than lower leaves. Upper (sun) leaves are smaller and thicker, which reduces water loss, and more lobed, which facilitates cooling. Lower (shade) leaves are larger and thinner, increasing the surface area for light capture.

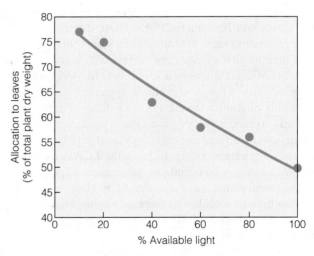

Figure 6.12 Changes in allocation to leaves for broadleaf peppermint (*Eucalyptus dives*) grown in different light levels. Allocation to leaves is expressed as leaf mass as a percentage of the total dry mass of five seedlings. Light level is the percentage of full sun. Shading was controlled by a shade cloth. Increased leaf allocation, along with thinner leaves, increases surface area for light capture.

INTERPRETING ECOLOGICAL DATA

Q1. What is the mean allocation to leaves for seedlings grown in full sun? How does allocation to leaves vary with light availability?

Q2. Assume that allocation to stems does not vary with light level. Using the carbon allocation model in Figure 6.8, how would the graph of allocation to roots under the different light environments differ from the graph of allocation to leaves in Figure 6.12?

QUANTIFYING ECOLOGY 6.1   Calculating Plant Growth Rate

We think of growth rate as change in size over some time period, such as change in mass over a week. Yet this conventional measure is misleading when comparing individuals of different sizes or when tracking the growth of an individual through time. Larger individuals may have a greater absolute mass gain than smaller individuals, but a different pattern may emerge when gain is expressed as a proportion of total body mass. An alternative growth measure is the mass-specific or **relative growth rate (RGR)**, which expresses growth as a function of the size of the individual. RGR is calculated by dividing the mass increase during a time interval by the size of the individual at the beginning of that time period (mass gain in grams/total mass in grams), and then dividing this value by the time period. This calculation expresses the mass change as a rate (g/g/time).

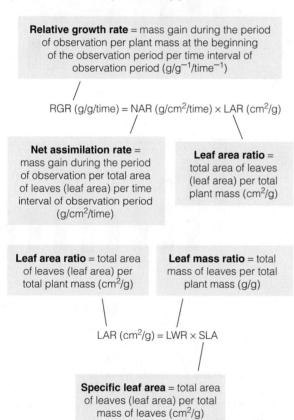

Using RGR to quantify plant growth has another benefit; it can be partitioned into components reflecting the influence of allocation. **Net assimilation rate (NAR)** quantifies new tissues produced per unit leaf area (g/cm^2/time) and reflects net photosynthesis. Total leaf area as a proportion of total plant mass (cm^2/g) is the **leaf area ratio (LAR)** and reflects overall carbon allocation to leaf area. LAR is further partitioned into two components that describe leaf allocation: (1) *leaf weight ratio (LWR)*, total leaf mass (g) as a proportion of total plant mass (g), and (2) *specific leaf area (SLA)*, a measure of leaf thickness, expressed as leaf area per leaf mass (g). For the same mass, a thinner leaf has a higher SLA.

Partitioning RGR allows comparisons among individuals of the same or different species under different conditions. The data in

Table 1 are from a greenhouse experiment where *Acacia tortilis* seedlings (a tree native to African savannahs) were grown under two light levels: full sun and shaded (50 percent full sun). Seedlings were harvested at four and six weeks, and total plant mass, leaf mass, and leaf area were measured. From the means, LAR, SLA, LWR, RGR, and NAR were calculated. RGR calculations used total plant mass at four and six weeks. NAR was calculated by dividing RGR by LAR. Because LAR varies between weeks 4 and 6, the mean of LAR at four and six weeks was used in estimating RGR.

Table 1 Growth Measures of *Acacia tortilis* Seedlings Grown in Sun and Shade

	Week 4		Week 6	
	Sun	Shade	Sun	Shade
Leaf area (cm^2)	18.65	12.45	42.00	24.00
Leaf mass (g)	0.06	0.03	0.13	0.06
Stem mass (g)	0.09	0.06	0.28	0.14
Root mass (g)	0.10	0.04	0.24	0.09
Total mass (g)	0.25	0.13	0.65	0.29
LAR (cm^2/g)	76.00	93.75	64.85	83.30
SLA (cm^2/g)	334	392		
LWR (g/g)	0.23	0.24	0.19	0.21
RGR (g/g/week)			0.47	0.38
NAR (g/cm^2/week)			0.007	0.004

The mean size (mass and leaf area) of sun-grown seedlings is about twice that of those grown in shade. Despite this difference, and despite the lower light available for growth, the difference in RGR between sun- and shade-grown seedlings is only about 20 percent. By examining the RGR components, we can see how shade-grown seedlings achieve this feat. Low light reduces photosynthesis, lowering NAR for shade-grown seedlings. But these plants compensate by increasing allocation to leaves (higher LWR) and producing thinner leaves (higher SLA) than do sun-grown seedlings. Shade-grown seedlings have a greater LAR (photosynthetic area relative to mass), offsetting their lower NAR and maintaining comparable (albeit lower) RGR.

These results illustrate the value of the RGR approach. By partitioning growth into components related to morphology, allocation, and photosynthesis, we can begin to understand how plants of the same or different species can both acclimate and adapt to differing environmental conditions.

1. When plants are grown under dry conditions, carbon allocation shifts to root production at the expense of leaves. How would this shift influence the plant's leaf area ratio (LAR)?
2. Nitrogen availability can directly influence net photosynthetic rate. Assuming no change in carbon allocation or leaf morphology, how would higher net photosynthesis with increased nitrogen influence RGR? Which component of RGR would be influenced by this increase?
3. Dry mass is used in RGR studies but requires destructive harvesting. Why is dry mass preferable to fresh mass? How could you make repeated measures of dry mass over time if destructive harvesting is used?

adjustments, shade plants typically produce more chlorophyll (Valladares and Niinemets 2008), allowing them to "mop up" more of the limited light. These and other subcellular adjustments in the photosynthetic apparatus (including orientation of the chloroplasts themselves) enhance a shade-grown plant's ability to maintain a positive carbon balance and continue growth—or to survive without growing in extreme shade.

How does a plant "know" how to respond to low light? For many species, exposure to differing light levels is itself responsible. But changes in light quality—in particular, the decrease in the red:far-red ratio as light is transmitted through the canopy (see Section 4.2)—may be involved. When detected by the non-photosynthetic pigment *phytochrome*, altered light quality may trigger acclimation in capable species (Walters 2005). Acclimating to high light may also involve complex signalling. In a study in which some but not all leaves on a shade-adapted *Aradopsis* plant were exposed to high light, the unexposed (shaded) leaves acquired photoprotection (increased levels of anti-oxidant defences) (Karpinski et al. 1999). The authors call this phenomenon *systemic acquired acclimation* (analogous to systemic acquired resistance to herbivore attack; see Section 14.12) and suggest that hydrogen peroxide may be the signal molecule.

Similar photosynthetic or morphological responses to variations in light occur not only among members of the same species grown in different light levels (see Figure 6.12) but also among leaves growing on the same plant with different exposure to light (see Figures 6.9b and 6.11). When expressed by a single genotype, such changes involve phenotypic plasticity (see Section 5.12). Differences are even more dramatic for species adapted to different light habitats and represent among-species genetic variation. Plants adapted to high light are called **shade-intolerant** (*sun-adapted*) species; those adapted to low light are **shade-tolerant** (*shade*) species.

Shade-tolerant and shade-intolerant species differ greatly in light response. In a study of nine North American boreal tree species, shade-tolerant species such as white pine (*Pinus strobus*) have lower rates of maximum net photosynthesis, leaf respiration, and RGR than shade-intolerant species such as trembling aspen (*Populus tremuloides*) (Figure 6.13) (Reich et al. 1998). These differences affect the role of a species in succession (see Chapter 17) of boreal forests.

Shade tolerance influences not only carbon gain and growth but also survival. Shade-tolerant species benefit minimally if at all when grown in full sun. Most (particularly obligate shade species) are not highly plastic and hence cannot take advantage of increased light (Murchie and Horton 1997). Many are damaged by light levels below full sun. Similarly, survival of shade-intolerant species (especially obligate sun species) declines sharply in shade. These differences are a direct result of photosynthetic traits and carbon allocation. The higher photosynthetic maxima of shade-intolerant species support faster growth in full sun, but the associated high leaf respiratory rates and light compensation points reduce their survival in shade. In contrast, the lower rates of respiration and maximum photosynthesis that allow survival of shade-tolerant species in low light limit their growth in high light, where they are at a competitive disadvantage.

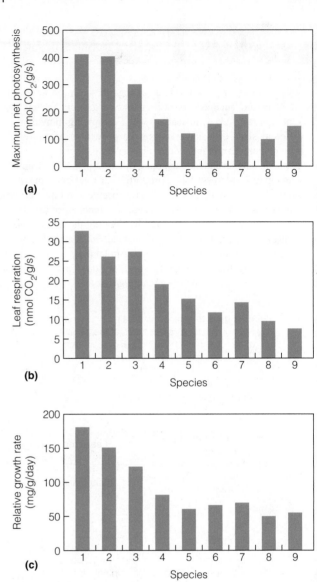

Figure 6.13 Differences in rates of **(a)** maximum net photosynthesis, **(b)** leaf respiration, and **(c)** relative growth for nine North American boreal trees. Species are ranked from lowest to highest shade tolerance. (1) *Populus tremuloides*, (2) *Betula papyrifera*, (3) *Betula allegheniensis*, (4) *Larix laricina*, (5) *Pinus banksiana*, (6) *Picea glauca*, (7) *Picea mariana*, (8) *Pinus strobus*, and (9) *Thuja occidentalis*.

(Adapted from Reich et al. 1998.)

INTERPRETING ECOLOGICAL DATA

Q1. How do net photosynthesis and leaf respiration vary with increasing shade tolerance for the nine species? What does this imply about the corresponding pattern of gross photosynthesis with increasing shade tolerance for these species?

Q2. Based on the data in Figure 6.13, how would you expect the light compensation point to differ between *Populus tremuloides* and *Picea glauca*?

To recap, the dichotomy between shade-tolerant and shade-intolerant species reflects a fundamental trade-off. Changes in physiology, morphology, and carbon allocation of shade-tolerant species enable them to survive and grow in low light. These same traits limit their ability to increase photosynthesis and growth in high light. Similarly, sun species sustain high rates of photosynthesis and growth in high light, but at the expense of their success in shade.

Phytoplankton also acclimate, with lower saturation points in low light. Not surprisingly, given their diversity, there are also genetic differences, with green algae generally more adapted to high light than diatoms or dinoflagellates. In high light, accessory carotenoid pigments, which help protect land plants, play a similar role in green algae. Diatoms also produce a carotenoid, fucoxanthin, but its function seems to be more related to its role as an accessory pigment, increasing the ability to absorb more of the blue-green light that penetrates deeper into the water column (Neori et al. 1984). Because phytoplankton can move in the water column, many species (along with the herbivorous zooplankton that feed on them) show a plastic behavioural response, moving up and down on a daily cycle to areas where light levels are favourable.

Studies of photoacclimation in a unicellular green alga, *Euglena gracilis* (Figure 6.14) (a "sun-alga" that has been proposed as a biofuel), suggest that different phototrophs have different ways of coping with light. Although shade-grown *Euglena* exhibits responses typical of shade-grown plants—decreased growth, lower light compensation and saturation points, increased chlorophyll, and lower respiration—it increases its chlorophyll *a:b* ratio (Beneragama and Goto 2010). This trend is the reverse of that reported for most shade-grown plants (Givnish 1988), including other green algae (Yamazaki et al. 2005). The more that physiological ecologists delve into plant adaptations to light quantity and quality, the more apparent it is that multiple strategies have evolved to cope with this most critical of all plant resources—their energy source.

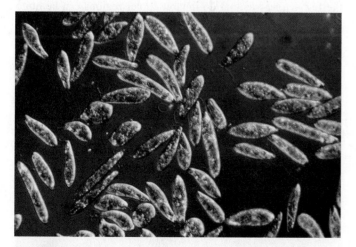

Figure 6.14 *Euglena gracilis*, a common flagellated freshwater green alga that can convert from a photosynthetic to a heterotrophic lifestyle in the absence of light, has a high capacity for photoacclimation.

6.10 ADAPTATIONS TO DROUGHT: Many Adaptations Help Plants Cope with Limited Water

It should be no surprise that terrestrial plants have evolved a diverse array of adaptations in response to variations in precipitation and soil moisture. Recall that water demand is linked to temperature. Saturation vapour pressure rises with air temperature, lowering relative humidity and increasing the water potential gradient between the leaf interior and the outside air. Transpiration increases, as does the amount of water required to offset these losses. Water stress has been, and continues to be, a powerful selective force in the evolution of terrestrial plants.

When the air or soil is dry, plants respond by partially closing their stomata. In the early stages of water stress, a plant closes its stomata during the hottest part of the day when relative humidity is lowest, reopening them in the later afternoon. As water becomes scarcer, the plant opens its stomata only in the cooler, more humid conditions of morning. Closing stomata reduces transpiration, but it also reduces CO_2 diffusion into the leaf and dissipation of heat through evaporative cooling. As a result, photosynthesis declines and leaf temperature rises.

Some species, such as rhododendrons and other members of the Ericaceae family, respond to water stress by curling their evergreen leaves. Others wilt as a result of reduced turgor. Although curling and wilting allow leaves to reduce water loss and heat gain by reducing surface area, they are only short-term strategies. Prolonged moisture stress inhibits chlorophyll production, causing leaves to turn yellow. As conditions worsen, deciduous species may shed their leaves early, with the oldest leaves dying first. Such premature shedding can cause dieback of twigs and branches. This response reflects damage rather than an adaptive response, but in tropical regions with distinct and predictable wet and dry seasons, some woody species drop their leaves at the onset of the dry season. For these **drought deciduous** species, new leaves emerge just before the rainy season begins.

Some plant species, known as C4 plants and CAM plants, have a modified photosynthetic pathway that greatly increases their water-use efficiency in hot, dry habitats. This genetically controlled modification involves an additional step in fixation of CO_2 into sugars. In C3 plants, capture of light energy and conversion of CO_2 into sugars both occur in mesophyll cells. The sugar products move into the vascular bundles in leaf veins, to be transported to other plant parts (see Figure 6.1). In contrast, **C4 plants** have a different leaf anatomy, with two distinct types of photosynthetic cells—mesophyll cells and **bundle sheath cells**, which surround the vascular bundles (Figure 6.15a, p. 114). In C3 plants, bundle sheath cells exist but have no photosynthetic function. C4 plants partition photosynthesis between their mesophyll and bundle sheath cells.

In the mesophyll cells of C4 plants, CO_2 is fixed not to RuBP but to phosphoenolpyruvate (PEP), a three-carbon compound. This reaction is catalyzed by the enzyme PEP carboxylase and produces oxaloacetate (OAA), which is then transformed into the four-carbon compounds malic and aspartic acid (hence the name *C4 photosynthesis*). These organic acids are then transported to the bundle sheath cells (Figure 6.15b),

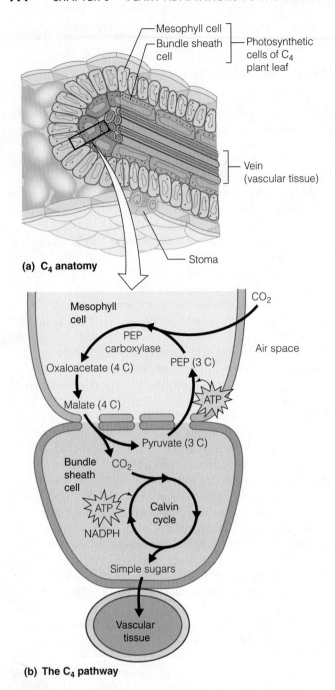

(a) C₄ anatomy

(b) The C₄ pathway

Figure 6.15 Simplified representation of C₄ photosynthesis. (a) C₄ anatomy. (b) C₄ pathway. Compare with the C₃ pathway in Figure 6.1 (PEP = phosphoenolpyruvate; OAA = oxaloacetate).

where enzymes reverse the process, breaking down the organic acids to release CO_2. In the bundle sheath cells, the CO_2 is then transformed into sugars using the Calvin cycle of the C_3 pathway, again involving RuBP and Rubisco.

This additional step, involving initial fixation of CO_2 to PEP, gives C_4 plants certain ecological advantages. First, PEP does not interact with O_2 as does RuBP, eliminating the inefficiency that occurs in C_3 plants when oxygenation catalyzed by Rubisco stimulates photorespiration (see Section 6.1). Second, release of CO_2 from malic and aspartic acids in the bundle sheath concentrates CO_2 around Rubisco, which increases

Calvin cycle efficiency and supports a higher maximum photosynthetic rate. (Note that natural selection for improved photosynthetic efficiency of C_4 plants has manipulated the CO_2 environment in which Rubisco operates, rather than altering Rubisco itself.) So it is clear how C_4 photosynthesis is an adaptation that concentrates the carbon resource for plants, but how does it represent an adaptation to drought?

Recall the trade-off between CO_2 uptake and water loss through the stomata. C_4 plants have greater water-use efficiency (WUE); for a given degree of stomatal opening, C_4 plants typically fix more carbon relative to water lost. Higher WUE is a great asset in hot, dry climates where water often limits growth. But it comes at a price. C_4 photosynthesis uses more ATP in order to regenerate PEP. This extra cost is not a serious drawback in hot, sunny conditions, which are also the conditions where photorespiration losses are highest for C_3 species. Yet it can be a severe drawback for C_4 species in shade. The C_4 grass big bluestem (*Andropogon gerardii*, the dominant species in tallgrass prairie) (Figure 6.16) acclimates less well to tree canopy cover than does the C_3 grass *Bromus inermis* (Awada et al. 2003). In addition, C_4 plants perform poorly in chilling temperatures (0°C–8°C), which limits their success in cool climates and in early spring. So C_4 photosynthesis is not better than C_3 photosynthesis—it is simply better adapted to some environmental conditions and less so to others.

C_4 photosynthesis does not occur in algae, bryophytes (mosses and related plants), ferns, gymnosperms (including conifers), or the more primitive flowering plants (angiosperms). Although they occur in several angiosperm families, most C_4 plants are grasses native to tropical and subtropical regions as well as some shrubs characteristic of arid and saline habitats, such as *Larrea* (creosote) and *Atriplex* (saltbush), which dominate the deserts of southwestern North America.

The distribution of North American C_4 grasses (*warm season grasses*, versus C_3 or *cool season grasses*) reflects the advantage of the C_4 pathway in hot, dry climates. The proportion of C_4 grasses increases from north to south, peaking in the dry American southwest (Teeri and Stowe 1976). However, the few C_4 grasses that do inhabit cooler regions can be quite

Figure 6.16 Big bluestem, the dominant grass in tallgrass prairies from Manitoba to Kansas, is a C_4 species.

successful. In both the tall- and shortgrass prairies of the American Midwest and southern Canada, the dominant grasses, big bluestem (*Andropogon gerardii*), buffalograss (*Bouteloua dactyloides*), and blue grama (*B. gracilis*), are all C_4. Nor are C_4 species restricted to prairies. Spiked muhly (*Muhlenbergia glomerata*) is a C_4 grass occupying boreal fens across Canada. It prefers raised moss hummocks, where its high water-use efficiency is beneficial (Kubien and Sage 2003).

In hot deserts, environmental conditions are particularly severe. Solar radiation and temperatures are high, and water is scarce. To cope with these conditions, a small group of desert plants, mostly succulents in the Cactaceae (cacti), Euphorbiaceae, and Crassulaceae families, use a third type of photosynthetic pathway—crassulacean acid metabolism (CAM). There are even a few CAM species that grow in Canada, of which the most common is the prickly pear cactus (*Opuntia fragilis*), found as far north as the Peace River district of Alberta and British Columbia.

CAM plants have a photosynthetic pathway similar to C_4 plants in that CO_2 is first transformed into four-carbon compounds using PEP carboxylase and later converted back into CO_2 for use in the Calvin cycle. However, unlike C_4 plants, in which these two steps are spatially separated in mesophyll and bundle sheath cells, both steps occur in the mesophyll cells of CAM species, but at separate times (Figure 6.17). CAM plants open their stomata at night, taking up CO_2 and converting it to malic acid, which accumulates in large quantities in the mesophyll. During the day, the plant closes its stomata and reconverts malic acid back into CO_2, which then enters the Calvin cycle. Relative to both C_3 and C_4 plants, CAM plants are slow and inefficient in fixing CO_2, largely because they are limited by how much malic acid they can accumulate during the night. But by opening their stomata at night when temperatures are lowest and humidity is highest, they dramatically reduce transpirational loss, increasing their water-use efficiency well above that of C_4 species.

Plants respond to declining soil water availability in other ways. Many increase their allocation to roots (Figure 6.18a; another example of phenotypic plasticity), which lets them forage for water in a larger soil volume. The associated reduction in leaf area decreases the solar radiation the plant intercepts. This effect might seem a disadvantage, but light is unlikely to limit growth in deserts. The combined effect of increased root allocation and decreased leaf surface area is to increase water uptake per unit leaf area while minimizing transpirational water loss. This shift in allocation, however beneficial for alleviating water stress, has consequences for plant growth by decreasing photosynthetic carbon gain relative to respiratory loss (see Figure 6.8). The inevitable result of this trade-off is a reduced growth rate.

The decline in leaf area with decreasing water availability is a combined effect of reduced allocation to leaves and changes in leaf size and shape. Leaves grown in **xeric** (dry) conditions tend to be smaller and thicker with a lower specific leaf area (see Quantifying Ecology 6.1: Calculating Plant Growth Rate, p. 111) than leaves grown in more **mesic** (moist) environments. Xeric leaves also tend to have hardened cell walls, tiny stomata, and a dense vascular system for transporting water. Some xeric leaves are covered with hairs that scatter incoming radiation and increase the thickness of the boundary layer (see Section 6.6). Others are coated with waxes that reflect light and reduce evaporation. Some (such as pine needles) locate their stomata in sunken pits or crypts. These and other structural traits help species cope with low water either directly, by reducing water lost through evapotranspiration, or indirectly, by reducing the heat load.

As with shifts in response to light (see Section 6.9), these changes in leaf shape and carbon allocation in response to water availability are instances of phenotypic plasticity. Even more pronounced are genetically controlled differences reflecting selection both within and among species adapted to mesic and xeric habitats. CAM photosynthesis is an example of a fixed response, but some responses combine plastic and genetically fixed aspects. Although the two species depicted in

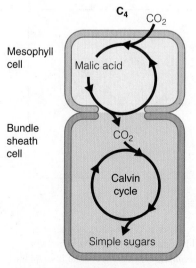

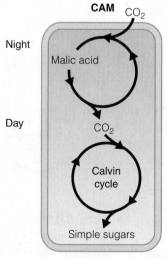

(a) Spatial separation of steps.
In C_4 plants, carbon fixation and the Calvin cycle occur in different types of cells.

(b) Temporal separation of steps.
In CAM plants, carbon fixation and the Calvin cycle occur in the same cells at different times.

Figure 6.17 Simplified comparison of C_4 and CAM photosynthesis. **(a)** In C_4 plants, CO_2 is incorporated into four-carbon organic acids in mesophyll cells. These organic acids release CO_2 to the Calvin cycle, which takes place in bundle sheath cells. **(b)** At night in CAM plants, the stomata open, the plant transpires, and CO_2 diffuses into the leaf. CO_2 is stored as malate in the mesophyll, to be used in photosynthesis by day. During the day, when the stomata are closed, the stored CO_2 is re-fixed in the mesophyll cells using the Calvin cycle.

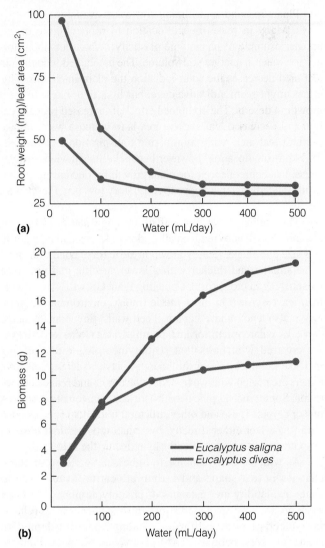

(a)

(b)

Figure 6.18 Carbon allocation and growth of two *Eucalyptus* species along a moisture gradient. **(a)** With declining water, both species increase their root mass to leaf area ratio. This response, which increases the surface area for water uptake and decreases the leaf area for transpirational loss, is much more pronounced for the xeric *E. dives* than for the mesic *E. saligna*. **(b)** *E. saligna*'s biomass gain continues to increase with greater water availability, while that of *E. dives* plateaus at intermediate water treatments.

Figure 6.18a exhibit a similar (plastic) trend of increased allocation to roots with declining water, the xeric *Eucalyptus dives* has consistently greater root investment than the mesic *E. saligna*. Yet *E. dives* is less able to increase its growth when more water is available (Figure 16.18b), illustrating the trade-off in its drought-tolerance strategy.

In areas where water stress is especially severe, plants have evolved multi-faceted strategies involving morphology, physiology, and stomatal behaviour. The most extreme example is **succulence**, whereby either leaves (e.g., Euphorbiaceae) or stems (e.g., Cactaceae) are specialized for storing water, which is available only intermittently. Succulents further enhance their water-use efficiency by their three-dimensional shape and by using CAM photosynthesis.

6.11 ADAPTATIONS TO TEMPERATURE: Plants Have Various Adaptations for Surviving Extreme Temperatures

Species vary in the temperature range over which net photosynthesis is maximal, depending on the thermal regime of their native habitat (Figure 6.19). Cool-climate species have lower T_{min}, T_{opt}, and T_{max} values than species from warmer climates. These differences reflect adaptations that shift the temperature responses of both gross photosynthesis and respiration towards prevailing temperatures, and are most pronounced between C_3 and C_4 plants. C_4 plants inhabit warmer, drier habitats and have higher optimal temperatures for photosynthesis (30°C–40°C) than do C_3 plants (Figure 6.20).

Although species that have evolved in different thermal environments respond differently to temperature, these responses are not entirely fixed. Despite the strong genetic component, exposing individuals of the same species to different temperatures alters their response curves (Figure 6.21). Typically, T_{opt} shifts in the direction of the growth temperature. This shift also occurs seasonally (Figure 6.22) through *acclimation* (reversible response; see Section 5.12).

As well as its influence on metabolism, temperature can directly damage plant tissues. Plants that inhabit seasonally cold habitats where temperatures drop below freezing have evolved adaptations for surviving frost damage. **Cold (frost) hardiness**, the ability to tolerate extreme cold, is a genetically controlled trait that varies among species as well as among populations of the same species. It is also subject to acclimation, in that its expression varies seasonally. Cold hardiness develops through the fall, achieving its maximum in early winter.

Plants acquire some cold hardiness by accumulating protective compounds that prevent formation of lethal ice crystals inside living cells. Species requiring only moderate cold hardiness synthesize substances such as sugars, amino acids (notably

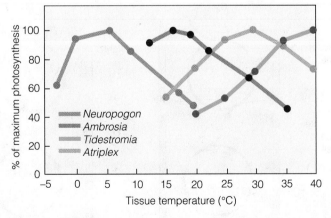

Figure 6.19 Temperature response of net photosynthesis of plants from dissimilar thermal habitats: *Neuropogon acromelanus* (an Arctic lichen), *Ambrosia chamissonis* (a cool, coastal dune plant), *Atriplex hymenelytra* (an evergreen desert shrub), and *Tidestromia oblongifolia* (a summer-active desert perennial).

(Adapted from Mooney et al. 1976.)

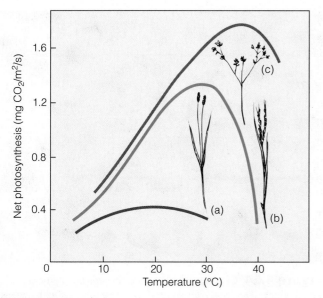

Figure 6.20 Effect of leaf temperature on net photosynthesis. **(a)** Photosynthetic rate of a C₃ plant, the temperate grass *Sesleria caerulea*, declines with increasing leaf temperature. **(b)** A C₄ temperate grass, *Spartina anglica*. **(c)** A C₄ shrub of the North American hot desert, *Tidestromia oblongifolia* (Arizona honeysweet). C₄ species have higher optimal temperatures for photosynthesis than do C₃ species.

(Adapted from Björkman 1981.)

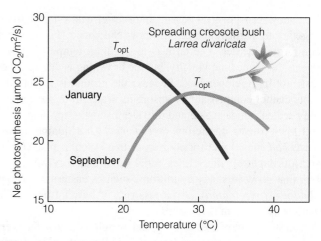

Figure 6.22 Seasonal shift in temperature response of net photosynthesis in field-grown creosote bush (*Larrea divaricata*).

(Adapted from Mooney et al. 1976.)

proline), and other compounds that lower the temperature at which freezing occurs. Even if temperatures drop below this lowered freezing point, another process called **supercooling** (not to be confused with freezing-point depression) allows

plants (and other ectotherms) to avoid frost damage at temperatures as low as $-30°C$ to $-35°C$. Supercooling involves specialized antifreeze proteins that bind to the surface of ice crystals, inhibiting further ice crystal growth (Hopkins and Hüner 2009). Typical of any overwintering plant tissue, supercooling is particularly critical for floral buds of fruit trees. In protected spots, it may be sufficient to avoid damage, but where plants are exposed to severe winter winds, shaking of the branches can induce lethal ice to form inside the cells.

Freezing-point depression and supercooling are both *avoidance* strategies, but for species inhabiting colder Canadian climates where temperatures drop as low as $-45°C$ to $-50°C$, avoidance is not enough. These species also undergo a prolonged multi-step process called **deep hardening** that alters the properties of cell membranes and walls, allowing the cells to export water to sites between cells where ice crystals can safely form. This strategy involves tolerating freezing, not avoiding it. For species capable of it, deep hardening allows survival to temperatures well below $-50°C$. In fact, the biggest risk is the dehydration that accompanies deep hardening.

Whatever their degree of hardiness, plants lose it quickly once growth resumes in spring, making them particularly vulnerable. Even a mild winter can result in severe frost damage if occasional severe lows occur in the late fall or early spring. Cold-temperature damage (either mid-winter or during an unseasonably cold spring) is a critical factor affecting tree survival in cold temperate and boreal forests and influences the location of the treeline (see Section 23.10).

Acquiring frost hardiness demands a large expenditure of energy and nutrients. In cold climates, **winter-deciduous** species avoid these costs by shedding their leaves in the fall and replacing them during the spring, when conditions are once again favourable. (There are, however, still living cells in their woody parts that must acquire frost hardiness.) In contrast, **needle-leaf evergreen** species, such as pine (*Pinus* spp.) and spruce (*Picea* spp.), invest in protective compounds and structural changes that allow their needles to survive the winter. As we will see in Section 6.12, this investment also conserves mineral resources in infertile sites.

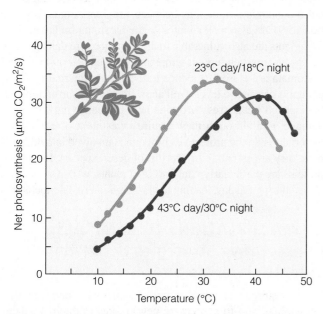

Figure 6.21 Effect of temperature on net photosynthesis of cloned big saltbush (*Atriplex lentiformis*) plants grown in different day/night temperature regimes. The shift in T_{opt} corresponds to the temperature conditions in which the plants were grown.

(Adapted from Pearcy 1977.)

Snow can have damaging indirect effects on evergreen vegetation. Light reflected off snow can cause "burning" of coniferous needles. The term is misleading; temperatures do rise in the immediate vicinity, but the browning is due to water loss, not high temperature. Because the ground is frozen, the roots cannot compensate by absorbing water from the soil, so the needles can desiccate and die. Deep snow cover helps protect low-growing coniferous evergreens such as juniper (*Juniperus communis*), but for coniferous trees such as pine (*Pinus* spp.), drought tolerance is essential for winter survival. Conifers that grow in shady conditions, such as eastern red cedar (*Juniperus virginiana*), are at less risk of burning unless the canopy above them is removed.

In heavy snow years, the sheer weight of snow on evergreens can cause their branches to break. Particularly in eastern North America, where snow tends to be denser, the physical stress is considerable. Some argue that the conical shape of evergreens such as spruce (*Picea* spp). is an adaptation that minimizes snow damage by shedding snow (Figure 6.23). However, many coniferous species evolved in habitats where snow load was not a selective factor, and some of our most common northern conifers—for example, jack pine (*Pinus banksiana*)—lack a strongly conical habit.

Tundra plants obviously require extreme frost hardiness to endure their harsh habitat. Most are broad-leaved evergreens, often members of the heath family (Ericaceae) (Figure 6.24). By retaining their leaves, they adopt the conservative strategy of a *stress-tolerator* (see Section 8.10). Growth may be slow, but they retain nutrients. Tundra plants have temperature-related stresses beyond winter lows. Temperature is not only cool but variable, so they have low, broad temperature optima. The tundra is a windy habitat, and the ground-hugging form of tundra plants minimizes the cooling effect of winds while retaining heat emitted by the dark soil. Growing in clumps also retains heat, and some have dish-like flowers that trap heat in the early spring. Leaves of many tundra plants are covered with hairs that not only trap warm air closer to the plant surface but

Figure 6.24 Common tundra species include bearberry or kinnikinnick (*Arctostaphylos* spp.), shown here growing with lichens.

also reduce water loss in a habitat that alternates between being waterlogged from impermeable permafrost, and being as dry as a desert.

Many traits of tundra plants can be linked to several abiotic factors. They are often dark green or purplish in colour, which increases heat gain. But this colouring (which also occurs in many boreal species, especially in spring) is due to accumulation of anthocyanin, a non-photosynthetic pigment that limits photoinhibition at low temperature (Pietrini et al. 2002). The tundra is dark for most of the winter, but there is no tree cover, and light levels can be high in the short growing season. Perhaps no other plant group better illustrates the interdependence of adaptations, as well as their trade-offs. Lichens (not plants but mutualistic interactions between fungi and algae; see Section 15.9) are also well adapted to the harsh tundra habitat.

Plants are also vulnerable to high-temperature damage. Desert species must endure air temperatures in excess of 45°C, and, if stomata are closed to reduce transpiration, internal leaf temperatures can rise to levels well above that of the surrounding air. Specialized **heat-stress proteins** help protect living cells from damage, and although they require investment of energy and materials as do the protective compounds involved in cold hardiness, they are essential for survival of desert species. Heat-stress proteins are particularly vital for CAM plants, whose stomata are closed during the day, leading to very high internal temperatures.

6.12 ADAPTATIONS TO MINERALS: Conservative Strategies Help Plants Tolerate Infertile Sites

Plants require a variety of chemical elements to carry out their metabolism and to synthesize new tissues (Table 6.1). Thus, nutrient availability has direct effects on plant survival, growth, and reproduction. Some of these elements, called **macronutrients**, are needed in large amounts, often for structural purposes. Others, called **micronutrients** (trace elements), are needed in lesser, often minute quantities, for specialized physiological

Figure 6.23 The conical shape of many conifers, such as these spruces in Castle Mountain, Alberta, may reduce snow damage.

Table 6.1 Essential Elements in Plants

Element	Major functions
Macronutrients	
Carbon (C)	
Hydrogen (H)	Basic constituents of all organic matter.
Oxygen (O)	
Nitrogen (N)	Used only in a fixed form: nitrates, nitrites, ammonium. Component of chlorophyll, enzymes (such as Rubisco), and amino acids, the building blocks of protein.
Calcium (Ca)	In plants, combines with pectin to give rigidity to cell walls; activates many enzymes; regulates many responses of cells to stimuli; essential to root growth.
Phosphorus (P)	Component of nucleic acids, phospholipids, ATP, and several enzymes.
Magnesium (Mg)	Essential for maximum rates of enzymatic reactions; integral part of chlorophyll; involved in protein synthesis.
Sulphur (S)	Constituent of some amino acids.
Potassium (K)	Involved in osmosis and ionic balance; activates many enzymes; involved in stomatal control.
Micronutrients	
Chlorine (Cl)	Enhances electron transfer from water to chlorophyll in plants.
Iron (Fe)	Involved in the production of chlorophyll; part of the complex proteins that activate and carry O_2 and transport electrons in mitochondria and chloroplasts.
Manganese (Mn)	Enhances electron transfer from water to chlorophyll and activates enzymes in fatty-acid synthesis.
Boron (B)	Many known functions in plants, including cell division, pollen tube germination, carbohydrate metabolism, maintenance of conductive tissue, and translocation of sugar.
Copper (Cu)	Concentrates in chloroplasts; influences photosynthetic rates; activates enzymes.
Molybdenum (Mo)	Essential for symbiotic relationship with N-fixing bacteria.
Zinc (Zn)	Helps form growth hormones (auxins); associated with water relations; affects chlorophyll formation; component of several enzyme systems.
Nickel (Ni)	Necessary for enzyme function in nitrogen metabolism.

roles. The prefixes *micro-* and *macro-* refer to the quantity needed, not their importance. If a micronutrient such as boron is lacking, plants fail as completely as if they lacked nitrogen, calcium, or any other macronutrient.

Of the macronutrients, carbon (C), hydrogen (H), and oxygen (O) form the majority of plant tissues. These elements are derived from CO_2 and H_2O and are made available through photosynthesis and respiration. The remaining six macronutrients—nitrogen (N), phosphorus (P), potassium (K), calcium (Ca), magnesium (Mg), and sulphur (S)—exist in varying states in soil and water. Their availability is affected by various processes depending on location. In terrestrial environments, plants take up nutrients from the soil. In aquatic environments, plants take up nutrients from the substrate or from water. Nitrogen is a special case. Required in large amounts, it is available to plants through the activities of microorganisms (see Section 22.5). The selective pressure of low nitrogen has promoted mutualistic associations of plants with bacteria and fungi (see Section 15.9).

An instructive example of the direct link between nutrient availability and plant performance involves nitrogen. In Section 6.1, we discussed two key compounds in photosynthesis: Rubisco, which catalyzes the reduction of CO_2 to sugars, and chlorophyll, which absorbs light energy. Nitrogen is a major component of both. In fact, over 50 percent of leaf nitrogen content is in some way directly involved with photosynthesis, with much of it tied up in these two compounds. Thus the maximum (light-saturated) rate of photosynthesis for a species is correlated with the nitrogen content of its leaves (Figure 6.25, p. 120).

Nutrient uptake depends on supply and demand. Uptake of a nutrient is positively correlated with its concentration in the soil, until some maximum rate is achieved (Figure 6.26a, p. 120). No further increase in uptake occurs above this level, because the uptake system is saturated for that nutrient and the plant is likely meeting its demand. Low nitrogen concentration in soil or water restricts the plant's uptake, which in turn decreases the nitrogen content of the leaf (Figure 6.26b). Low leaf nitrogen content results in suboptimal levels of Rubisco and chlorophyll, limiting photosynthesis and growth. A similar pattern holds for other nutrients.

Geology, climate, and biotic activity alter nutrient availability in the soil. Some environments are relatively nutrient rich, and others are nutrient poor. How do plants succeed in low-nutrient sites? Because plants require nutrients for synthesizing tissue, growth rate influences a plant's nutrient demand. In turn, the plant's rate of nutrient uptake influences growth. This relationship may seem circular, but the key point is that not all plants have the same inherent maximum potential growth rate. In Section 6.9, we saw how shade-tolerant plants have inherently lower rates of photosynthesis and growth than

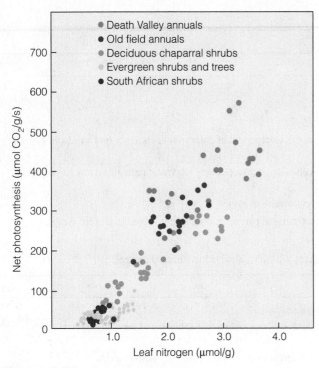

Figure 6.25 Effect of leaf nitrogen content on maximum net photosynthetic rates of species from differing habitats.

(Adapted from Field and Mooney 1986.)

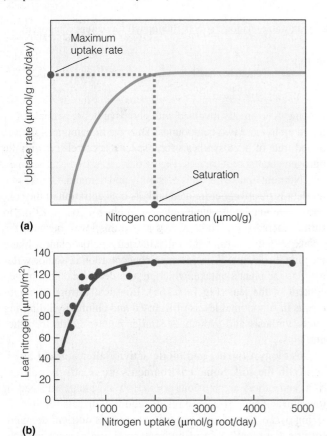

Figure 6.26 Effects of soil nitrogen content. **(a)** Uptake by roots increases with soil nitrogen content until the plant achieves maximum uptake. **(b)** Leaf nitrogen content increases with root uptake.

(Adapted from Woodward and Smith 1993.)

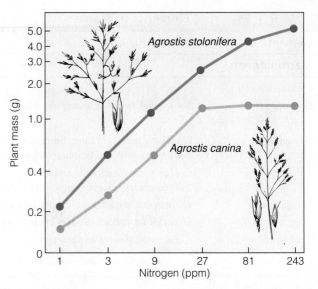

Figure 6.27 Growth responses of grasses to nitrogen fertilizer. Creeping bentgrass (*Agrostis stolonifera*) is from high-nutrient sites, and velvet bentgrass (*A. canina*) is from low-nutrient sites. *A. canina* responds up to a certain level only.

(Adapted from Bradshaw et al. 1964.)

shade-intolerant plants, even in high light. These lower rates mean less demand for resources, including minerals.

The same pattern of reduced photosynthesis occurs among plants in low-nutrient sites. A species that normally grows in high-nitrogen sites keeps increasing its growth rate with increasing soil nitrogen, whereas a species native to low-nitrogen sites reaches its maximum growth at moderate nitrogen levels (Figure 6.27). It does not respond to additional nitrogen, just as xeric species respond weakly to additional water (see Figure 6.18). Some suggest that a low growth rate is in itself an adaptation to low-nutrient sites. By employing a conservative strategy, slower-growing plants minimize nutrient stress while maintaining favourable rates of photosynthesis and other processes crucial for growth when nutrients are scarce. A plant with an inherently high growth rate will more likely suffer from nutrient deficiency.

A second conservative adaptation to low nutrient levels is leaf longevity (Figure 6.28). Producing leaves is expensive in terms of the energy, carbon, and other nutrients required. At low photosynthetic rates, a leaf needs a longer time to "pay back" its production costs. Hence, plants inhabiting low-nutrient sites tend to have longer-lived leaves. A good example is the dominance of pines on nutrient-poor, sandy soils in the southern boreal forest. In contrast to deciduous species, which shed their leaves annually, needle-leaf evergreens retain their needles for several years. This conservative strategy uses limited nutrients efficiently, even though such long-lived leaves must be capable of surviving winter frosts.

Like water, minerals are a belowground resource for land plants, and their ability to exploit them is related to their root mass. One way plants compensate in infertile soils is by producing more roots, which—as with water limitation—is

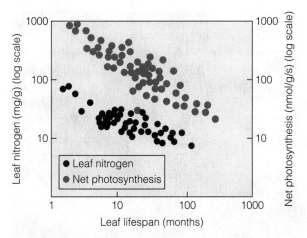

Figure 6.28 Relationship among leaf longevity, leaf nitrogen, and maximum rate of net photosynthesis for species from different environments. Points represent different species. Longer-lived leaves have lower N and photosynthetic rates.

(Adapted from Reich et al. 1992.)

at the expense of leaf production. Less leaf area decreases photosynthetic carbon gain relative to respiratory carbon loss, resulting in slower growth. So, all things considered, slow growth can be a critical adaptation, allowing plants to occupy environments that are limited in many resources, including light, water, and nutrients.

6.13 ADAPTATIONS TO FLOODING: Submerged Substrates Impose O₂ Stress on Plants

In wetlands where soils are saturated with water for most or all of the year, or in areas along waterways that are prone to flooding, too much water induces as much stress as too little. Plants differ in their ability to deal with the stresses of flooding and **hydric** (waterlogged) soils. Although the causes differ, the symptoms of flood intolerance are similar to those of drought: stomatal closure; leaf yellowing, wilting, and premature drop; and rapid photosynthetic decline.

Plants need adequate water and gas exchange with their environment. Much of this exchange occurs between roots and soil air spaces. When water fills soil pores, gas exchange slows. Deprived of O₂, plants asphyxiate. Roots need O₂ for aerobic respiration. Flooded roots shift to less efficient *anaerobic metabolism* (see Section 6.1), which, if prolonged, causes a buildup of fermentation by-products such as ethanol that are toxic to plants (Kozlowski 1982).

In flooded sites, some plants accumulate a gaseous hormone called *ethylene* in their roots. Ethylene is highly insoluble in water and is normally produced in small amounts. In water, ethylene diffusion slows, allowing it to accumulate. High ethylene levels stimulate root cells to self-destruct, forming interconnected, gas-filled chambers in a tissue called **aerenchyma** (Figure 6.29a). In species that emerge above the water, such as cattails (*Typha* spp.), aerenchyma permits O₂ to diffuse into submerged tissues from leaves. In plants with floating leaves (e.g., water lilies (*Nympheae* spp., Figure 6.29b)), aerenchyma ventilates the entire plant, with interconnected gas spaces occupying nearly 50 percent of the plant and also aiding buoyancy.

In other plants, especially woody species, the original roots die. In their place, **adventitious roots** (roots arising from non-root structures) develop from the submerged lower stem. Replacing the functions of the original roots, adventitious roots spread horizontally along the soil surface where O₂ is more available. When red maple (*Acer rubrum*, Figure 6.30a, p. 122) grows in poorly drained soils, it develops shallow, horizontal roots with adventitious roots to cope with flooding, making it highly susceptible to **windthrow** (toppling of trees or branches by wind). Some woody species, such as bald cypress (*Taxodium distichum*), mangroves, and willows (*Salix* spp.) grow on sites that are flooded, either permanently or in tidal cycles. They develop **pneumatophores**, root growths (Figure 6.30b) that facilitate gas exchange.

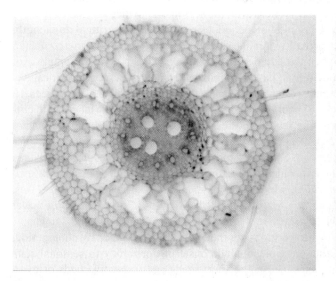

Figure 6.29 Plant adaptations to waterlogging. **(a)** Cross section of a tall fescue (*Festuca arundinacea*) root in flooded soil, showing spongy aerenchyma. A common feature of aquatic plants, as well as land plants in waterlogged soils, aerenchyma facilitates O₂ transport to submerged tissue. **(b)** Water lily (*Nymphaea odorata*) growing in a lake near Sudbury, Ontario.

(a)   **(b)**

(a)

(b)

Figure 6.30 Adaptations of woody species to flooding. **(a)** Species such as red maple that inhabit flooded sites often develop adventitious roots as well as shallow roots. **(b)** Pneumatophores, or "knees," are typical features of a bald cypress swamp.

Plants of salt marshes and other saline habitats grow in a physiologically dry environment where salinity severely limits the amount of water that roots can absorb. Called **halophytes**, saline-tolerant plants accumulate salts in their cells, especially in the leaves. These high salt levels, which may equal or exceed that of seawater, allow halophytes to maintain a tolerable cell water content despite the low external osmotic potential (and hence low water potential) of saline soils.

Such high salt content can damage leaf function. After taking up saline water, some halophytes dilute it with water stored in their tissues. Some have salt-secreting glands that deposit salt on the leaves to be washed away by rain. Others filter salts at the root cell membranes. Depending on their adaptations, halophytes vary in tolerance. Some species, such as salt marsh hay grass (*Spartina patens*), grow best at low salinities. Others, such as salt marsh cordgrass (*S. alternifolia*), thrive at moderate salinities. A few, such as glasswort (*Salicornia* spp.), tolerate high salinities and are common in saline sites across North America (see Figure 25.6).

6.14 PERIODICITY: Many Plants Have Predictable Daily and Seasonal Activity Patterns

Anyone who has ever experienced jet lag, or noticed that many animals give birth only at a certain time of the year, realizes that animals have daily and seasonal activity patterns. What is less obvious is that all living organisms (with the exception of bacteria) possess **biological clocks**—internal mechanisms that use timing devices to control periodicity. In fact, given their sessile habit, plants are particularly reliant on periodicity. In Section 6.9, we considered plant adaptations to light as a resource. But, as predictable changes in light are the means by which organisms "tell time," so to speak, light is also a condition for plants.

Biological processes fluctuate in cycles ranging from minutes to months or even years. Processes that cycle on a 24-hour basis are called *daily rhythms*. When a daily rhythm results from a physiological response to an external environmental cycle, it is called a **circadian rhythm**. In plants, processes that exhibit circadian rhythms include stomatal opening and onset of photosynthesis and other metabolic activities. Some species also time flower movements on a 24-hour cycle. Tulips open their flowers during the day, while tobacco (*Nicotiana* spp.) flowers open at night. These behaviours are timed to coordinate with the activities of their pollinators, which also exhibit circadian rhythms. If a flower is pollinated by moths (as is tobacco), then night opening is optimal.

Circadian rhythms are controlled by the biological clock. But where is the clock located? Although it varies among organisms, its location must expose it to its time-setter: light. We consider the general principles of biological clocks in Section 7.16, but in plants the receptor is a pigment we have mentioned before—*phytochrome* (see Sections 4.2 and 6.9). The importance of phytochrome as a device for processing information about the environment is clear.

Periodicity is not confined to daily rhythms. In mid- and upper latitudes, **photoperiod** (length of the light period) lengthens and shortens with the seasons. Activities are geared to these changes in photoperiod, using the signal of **critical day length**. In actuality, it is the dark period that is the determining factor. When the duration of darkness reaches a certain proportion of the 24-hour day, it inhibits or promotes a response. Critical day length varies among species, but it typically falls between 10 and 14 hours. Organisms "compare" actual day length with critical day length and respond accordingly. *Day-neutral species* respond to factors other than day length, such as rainfall or temperature. *Short-day species* flower when nights are longer than the critical length. Examples include cocklebur (*Xanthium strumarium*), which colonizes the shores of many Canadian lakes. *Long-day species* such as bellflowers (*Campanula* spp.) flower when nights are shorter than a critical length, in mid-summer.

Flowering is not the only activity affected by photoperiod. Other examples include onset of **dormancy** (a seasonal state of suspended activity) in seeds and vegetative buds of overwintering plants, leaf fall, and initiation of germination and growth. Dormancy allows species in higher latitudes to avoid an unfavourable season when resources are limiting and conditions are harsh. It is no surprise that organisms are so attuned to photoperiod, as it is a predictable cue for changes in other, less predictable factors.

We have only scratched the surface of the many ways in which plants and algae have adapted to their environment, both by evolving specialized traits through natural selection and by adjusting phenotypic trait expression. Each species has its own adaptations, some of which resemble those of other species, and others of which are unique and as yet unknown to science. Yet, no matter how successful these adaptations are, they all have their trade-offs as well as their limitations.

EcologyPlace

Visit EcologyPlace at www.pearsoncanada.ca/ecologyplace to access online resources that complement your textbook, and help you to apply and to review the information in this chapter. EcologyPlace includes

- an eText version of the book
- self-grading quizzes
- glossary flashcards
- and more!

Go to www.pearsoncanada.ca/ecologyplace and follow the registration instructions on the Student Access Code Card included with this text. If your book does not have a Student Access Code Card, you can purchase access to it at www.pearsoncanada.ca/ecologyplace.

SUMMARY

Carbon Metabolism 6.1

Photosynthetic autotrophs harness solar energy to convert CO_2 and H_2O into sugars, using ATP and NADPH produced in the light-dependent reactions. Rubisco catalyzes fixation of CO_2 in the Calvin cycle. Cellular respiration, which occurs in the living cells of all organisms, oxidizes carbohydrates to yield energy (as ATP), H_2O, and CO_2.

Impact of Light 6.2

Photosynthetic rate varies with the amount of light. The light level at which the rate of CO_2 uptake in photosynthesis equals the rate of CO_2 loss in respiration is the light compensation point. The light level at which a further increase in light no longer increases photosynthetic rate is the light saturation point. Light levels above saturation may cause photoinhibition.

Gas Exchange 6.3

Gas exchange is vital to plants. CO_2 diffuses into the leaf through stomata. Transpiration is water loss by diffusion through stomata. As photosynthesis slows and demand for CO_2 lessens, stomata close to reduce transpirational loss, which is also affected by relative humidity. Water lost in transpiration must be replaced by water taken up from the soil.

Water Movement 6.4

Water moves from the soil into roots, up through stems and leaves, and into the atmosphere along a water potential (free-energy) gradient. Plants absorb water from the soil, where water potential is highest, and release it to the air, where it is lowest. Transpiration reduces plant water potential, causing more water to enter the roots, provided soil water is available. Plants face a dilemma; they must open their stomata to absorb CO_2, but at the risk of dehydration.

Aquatic Photosynthesis 6.5

In contrast with land plants, submerged aquatic plants lack functional stomata. CO_2 diffuses directly from the waters adjacent to the leaf across the cell membrane. Some aquatic species (including algae) utilize bicarbonate as a carbon source, but convert it to CO_2 for use in photosynthesis. Phytoplankton also remove carbon from water by calcification.

Plant Temperature 6.6

Plants have an optimal temperature range for photosynthesis. Respiration also increases with temperature, but at a faster rate. Internal plant temperature is influenced by heat exchange. Plants absorb radiation and reflect or re-emit some of it back to the environment. The difference between energy gain and loss is the plant's net energy balance. The plant uses some absorbed radiation in photosynthesis, with the remainder stored as heat or dissipated through evapotranspiration and/or convection.

Carbon Allocation 6.7

Net carbon gain is the difference between carbon uptake in photosynthesis and carbon loss in respiration. Carbon gain is allocated to various processes, including production of new tissues. Because leaves perform the majority of photosynthesis while all living tissues respire, net carbon gain (and subsequently growth) is strongly influenced by carbon allocation.

Interdependence of Adaptations 6.8

Plants have evolved a variety of adaptations to differing environments. These adaptations are not independent, for reasons relating to the physical environment and to the plants themselves. Trade-offs are involved wherein an adaptation to one factor constrains response to another.

Adaptations to Light 6.9

Plants exhibit a variety of genetic adaptations and phenotypic responses to light. Shade-tolerant plants have lower rates of photosynthesis, respiration, and growth than shade-intolerant plants, but survive better in shade. Leaves of sun plants tend to be small, lobed, and thick, whereas leaves of shade plants tend to be large and thin.

Adaptations to Drought 6.10

In the short term, plants respond to water stress by closing their stomata. Longer-term responses include the deciduous habit. C_4 photosynthesis is an adaptation to carbon uptake that increases water-use efficiency. Using PEP carboxylase, C_4 plants fix CO_2 into malate in mesophyll cells. They transfer these acids to bundle sheath cells, where they release CO_2 for use in the Calvin cycle. Many desert succulents employ CAM photosynthesis. They open their stomata at night to absorb CO_2, forming malate. By day, they close their stomata and complete the Calvin cycle. Other xeric traits include modified leaves and increased root allocation.

Adaptations to Temperature 6.11

As well as genetic differences correlated with their natural environment, plants shift their temperature optimum for photosynthesis when grown in different temperatures. Plants have evolved adaptations to extreme temperatures. Although subject to acclimation, cold hardiness is genetically controlled and varies among species. Plants acquire moderate hardiness by forming protective compounds that lower the freezing point or prevent growth of ice crystals. Tolerance of extreme cold requires deep hardening, which involves structural changes to cell walls and membranes that allow ice to form between cells. Tundra species cope with low temperatures in the growing season by trapping heat, growing low to the ground, and having dark-coloured, hairy leaves. Plants tolerate high temperature by producing heat-stress proteins.

Adaptations to Minerals 6.12

Plants require both macro- and micronutrients. Their roots absorb minerals from soil. As mineral supplies are depleted, diffusion of water and nutrients through the soil replaces them. Nutrient availability affects plant survival and growth. Nitrogen is particularly important, as Rubisco and chlorophyll are N-based compounds needed for photosynthesis. Mineral uptake is subject to supply and demand. Plants with high demands grow poorly in low-nutrient sites. Plants with low demands survive and grow slowly in low-nutrient sites and have longer-lived leaves.

Adaptations to Flooding 6.13

Many plants are intolerant of flooding, which reduces O_2 supply, disrupting metabolism and altering root growth. In water-logged conditions, ethylene levels increase, stimulating formation of gas-filled aerenchyma, which facilitates O_2 exchange to submerged tissues. Woody wetland species often have pneumatophores that facilitate gas exchange.

Periodicity 6.14

Plants time their daily and seasonal activities using a biological clock. The receptor for light timing is phytochrome. Activities that exhibit circadian rhythms include stomatal and flowering opening, and onset of metabolism. Plants respond to seasonal photoperiodic changes by comparing night length to a critical value. Activities affected include flowering and onset and release from dormancy of seeds and vegetative buds. Predictable changes in photoperiod allow sessile organisms to anticipate less predictable changes in other abiotic factors.

KEY TERMS

adventitious roots	carbon balance model	hydric	needle-leaf evergreen	shade-intolerant
aerenchyma	chemoautotroph	leaf area ratio	net assimilation rate	shade-tolerant
aerobic (cellular) respiration	chlorophyll	light compensation point	net energy balance	specific leaf area
	circadian rhythm		net photosynthesis	succulence
anaerobic metabolism	cold (frost) hardiness	light-dependent reactions	osmosis	supercooling
assimilation	critical day length		osmotic potential	transpirational pull
autotroph	deep hardening	light-independent reactions	photoinhibition	turgor pressure
biological clock	dormancy		photoperiod	water potential
boundary layer	drought-deciduous	light saturation point	photorespiration	water-use efficiency
bundle sheath cell	drought tolerance	macronutrient	photosynthesis	windthrow
C_3 plant	fermentation	matric potential	pneumatophore	winter-deciduous
C_4 plant	halophyte	mesic	pressure potential	xeric
CAM plant	heat-stress proteins	mesophyll	relative growth rate	
carbon allocation	hydraulic redistribution	micronutrient	Rubisco	

STUDY QUESTIONS

1. What does it mean to say that life is carbon based?
2. Distinguish between *photosynthesis* and *assimilation*. How are they related?
3. What is the function of respiration?
4. What is the specific role of light (PAR) in photosynthesis?
5. In the relationship between net photosynthesis and light in Figure 6.2, there is a net loss of CO_2 at light levels below the light compensation point. Why? Based on this relationship, how will net photosynthesis vary over the day?
6. How does diffusion control both uptake of CO_2 and loss of water from the leaf?
7. How does water availability constrain photosynthesis? Is transpiration a "necessary evil"; that is, what are its benefits as well as its risks?
8. Given their high solute content, saline soils have a lower water potential than typical soils. Using water potential terminology, explain how this fact affects water uptake by plants.
9. What are the advantages of C_4 photosynthesis compared with the C_3 pathway? How do these advantages influence where these species occur? What ecological drawbacks make it unlikely that all plant species will evolve C_4 photosynthesis, despite its advantages?
10. What is the advantage of a lower light compensation point (LCP) for plant species adapted to low-light environments? What is the trade-off of maintaining a low LCP?
11. How do plants growing in shade increase their photosynthetic surface area?
12. How does a decrease in water availability influence carbon allocation during growth?
13. Explain how differences in leaf morphology (size and shape) affect response to temperature, light, and moisture.
14. Why are plants so plastic in their temperature response, especially in temperate regions?
15. Given that winter-deciduous trees avoid severe winter conditions, why do deciduous forests give way to coniferous-dominated forests in higher latitudes?
16. What different strategies or adaptations are involved in the response of tundra species to temperature in winter versus in summer? How do their growing season adaptations reflect the interdependence of other factors, particularly wind and moisture, with temperature?
17. What explains the relationship between leaf nitrogen and photosynthetic rate in Figure 6.25? What consequence does this relationship have for plants growing in sites that have abundant light but low nitrogen availability?
18. Explain how a conservative strategy can be adaptive in low-nutrient sites.
19. In Canada, unusually warm weather can occur even in October. Why do native species generally not grow despite these favourable conditions, whereas some non-native species do?
20. Water lilies (see Figure 6.29) have several adaptations to life in water. (i) Describe two distinct functions of aerenchyma in these species. (ii) Unlike most leaves, which concentrate their stomata on the lower surface, floating water lily leaves have stomata on the upper surface only. Why? (iii) Water lilies have retained the xylem tissue of their terrestrial ancestors, but have much less than is typical for land plants. Why is their xylem reduced, and what residual function might xylem play in these rooted plants? Hint: Recall that xylem supplies more than just water.

FURTHER READINGS

Dale, J. E. 1992. How do leaves grow? *Bioscience* 42:423–432.
   Provides a basic understanding of how environmental conditions affect leaf growth.

Grime, J. 1971. *Plant strategies and vegetative processes*. New York: Wiley.
   Integrated overview of plant adaptations. Describes the constraints imposed by the environment on survival, growth, and reproduction throughout a plant's life history.

Schulze, E. D., R. H. Robichaux, J. Grace, P. W. Randel, and J. R. Ehleringer. 1987. Plant water balance. *BioScience* 37:30–37.
   Well-written and illustrated introduction to plant water balance.

Valladares, F., and U. Niinemets. 2008. Shade tolerance, a key plant feature of complex nature and consequences. *Annual Review of Ecology, Evolution and Systematics* 39: 237–257.
   Excellent review of the complexities and consequences of shade tolerance.

Walker, D. 1992. *Energy, plants and man*. East Sussex, UK: Packard.
   Humorous overview of plant biology. The photosynthesis sections are easy to read and well illustrated.

A male white-fronted brown lemur (*Eulemur fulvus albifrons*) in a rain forest in Madagascar (Africa). This omnivore feeds on fruits, leaves, and insects in the forest canopy, where it lives in small, cohesive groups of 3 to 12 individuals.

All photoautotrophs, whether tiny phytoplankton or giant Douglas-fir, derive their energy from one process—photosynthesis. The story is very different for animals. Because heterotrophs derive their energy and most of their nutrients from organic compounds, they encounter thousands of potential food items, packaged as the diversity of species on Earth. Yet, despite the many complex adaptations that help animals acquire and digest different foods, certain challenges are common to all aerobic organisms: absorbing O_2, maintaining a water and temperature balance, and adapting to a changing photoperiod. Solving these common problems in aquatic versus terrestrial habitats imposes fundamentally different evolutionary constraints.

In this chapter, we examine the many adaptations whereby animals survive, grow, and reproduce in the diversity of Earth's environments. As always, we focus on the benefits and limits of specific adaptations, and how these trade-offs affect success in different environments.

7.1 BODY SIZE: Size Imposes a Fundamental Constraint on Evolution

Animals differ greatly in size (Figure 7.1). The smallest weigh some 2 to 10 µg, while the largest are mammals: in marine habitats, the blue whale at over 100 000 kg, and on land, the African elephant at 5000 kg. Each taxonomic group has its own size range. Some, such as Bryozoa, contain species within one or two orders of magnitude, while mammal size is much more variable. The smallest mammal is a shrew that weighs about 2 g fully grown, whereas the blue whale is about 10^8 times this size. Why and how does size matter? To understand how size influences which adaptations are ecologically feasible, we must explore how size affects the relationship between structure and function in different environments.

We begin with the physics of size. Most morphological traits change predictably with body size—a phenomenon called **scaling**. Consider a mouse and an elephant. In absolute terms, the elephant is much larger in surface area (*SA*) and volume (*V*). But surface area increases as the square of body length (l^2), while volume increases as a cubic function (l^3), so the elephant's surface area is less than that of the mouse in relative terms (**surface area to volume ratio, *SA/V***) (Figure 7.2, p. 128).

This relationship between surface area and volume imposes a constraint on increasing body size. The metabolic processes of assimilation and respiration require exchange of energy and materials between an organism's interior and its environment. All aerobic organisms (including plants and fungi as well as animals) need O_2 for respiration. Every living cell thus requires a steady, ongoing supply. O_2 is a relatively small molecule that readily diffuses across cell surfaces; in seconds, it can penetrate a millimetre of living tissue. So, if we imagine a spherical organism with a 1-mm radius, the centre of this tiny creature is close enough to the surface that as O_2 is consumed, it is rapidly replenished by diffusion across its surface from air or water.

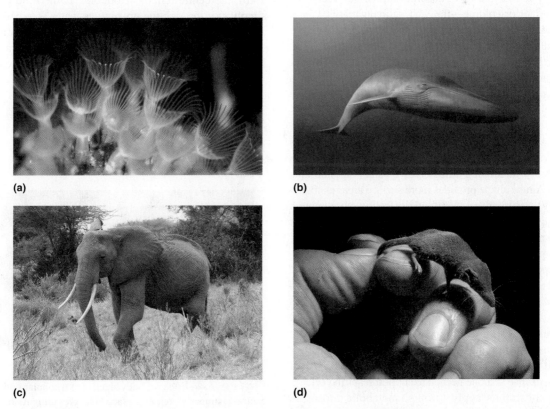

(a)

(b)

(c)

(d)

Figure 7.1 Range of body size in animals. **(a)** Bryozoa are aquatic colonial animals. Individuals weigh as little as 5 µg. Mammals have a huge size range. The blue whale **(b)** is the largest marine mammal. Land mammals range from **(c)** the African elephant to **(d)** the pygmy shrew.

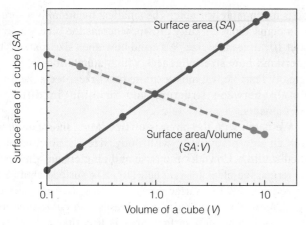

Figure 7.2 Surface area (*SA*) versus volume (*V*) of a cube (solid line; both variables on a $\log_{10}$ scale). The surface area to volume ratio (*SA/V*; dashed line) decreases with increasing size. This relationship applies to any geometrically similar (isometric) objects and constrains an organism's ability to exchange energy and matter with the environment as its body size increases.

Now imagine a spherical organism with the radius of a golf ball—about 21 cm. It would take over an hour for O_2 to diffuse into its centre. Cells near the surface would receive enough, but depletion of O_2 as it diffuses towards the core, and the greater distance over which it must travel, would cause death of interior cells (and eventually the organism) from O_2 starvation. Simply put, the problem is this: as an organism gets bigger, the surface area across which O_2 (and other substances) diffuses into its body mass increases in absolute terms, but decreases relative to its interior volume—the *SA/V* ratio decreases. Given that organisms are subject to this scaling effect, how do large animals maintain an adequate flow of O_2 to their interior?

One solution is to alter surface contour. The more convoluted an organism's surface, the greater its surface area for a similar volume (Figure 7.3). As a result, there is more surface area over which O_2 can diffuse, and no point in its interior is more than a few millimetres from the surface.

Another solution is to alter body shape. Many small animals are tubular, with a central chamber (Figure 7.4). Water is drawn into this chamber, facilitating diffusion of O_2 and other nutrients into the core. A tube shape provides more surface area relative to volume (*SA/V*), assuring that every interior cell is close enough to the surface to receive adequate O_2. This solution works for small organisms, but larger animals need a different strategy.

Large organisms have evolved complex structural adaptations to transport substances into their interior. *Lungs* are internal chambers that bring O_2 close to blood vessels, where it is transferred to *hemoglobin* for transport in the blood. A circulatory system with a heart acting as a pump ensures that oxygenated blood is actively transported into the minute capillaries that permeate the body. This system increases the surface area for exchange, ensuring that all cells are within the maximum distance over which O_2 can diffuse at a rate needed to support respiring cells.

Similar constraints apply to other metabolic processes requiring exchange with the environment. Nutrients are also absorbed across surfaces. In most animals, the food canal is a

Figure 7.3 Effect of body contour. Objects **(a)** and **(b)** have the same volume, but given its more convoluted surface, (b) has more surface area (and higher *SA/V*) than (a). Note the similarity between (b) and the sea anemone **(c)**.

tube in which digestion occurs and through which substances are absorbed into the circulatory system. In the smallest animals, such as Bryozoa and tube worms (see Figures 7.1a and 7.4b), the central chamber doubles as the food canal. Wastes simply exit through the opening as water is expelled. In larger animals, the food canal is a tube extending from the mouth to the anus (see Section 7.2). Food is digested as it travels through, and nutrients are absorbed. The greater the surface of the canal,

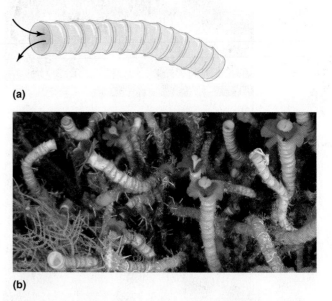

Figure 7.4 Effect of body shape. **(a)** A tubular body with a central chamber increases surface area, providing more exchange surface and reducing the distance from the surface to the interior. Note the similarity between (a) and the shape of tube worms **(b)**.

the greater its ability to absorb nutrients. As a result of scaling, larger animals must increase their food canal length proportionately in order to maintain a workable *SA/V* ratio.

Clearly, increasing body size requires compensatory structural adaptations that manipulate the relationship between the cell volume that must be supplied with resources, and the surface area across which this supply occurs. We will examine specific adaptations that allow animals to exchange nutrients, O_2, water, and heat with their environment, noting how these adaptations are constrained not only by body size but also by the habitat in which the animal lives.

7.2 FOOD ACQUISITION: Animals Acquire Energy and Nutrients by Different Strategies

Animal species have many potential diet items. This diversity of diet requires an equally diverse array of physiological, morphological, and behavioural adaptations that enable animals to acquire (Figure 7.5) and assimilate (see Figure 7.6, p. 130) food. The simplest ecological classification of animals is based on their diet. **Herbivores** eat plants. **Carnivores** eat animals. **Omnivores** eat both plants and animals. **Detritus feeders** consume **detritus** (dead organic matter). Each group has specialized traits related to its particular diet.

Herbivory: Herbivores Must Cope with the Low Digestibility of Plant Tissue

Given that plants are both abundant and sessile, herbivory seems the most sensible strategy for a heterotroph. However, the chemical composition of plant and animal tissue differs so much that the problem facing herbivores is how to convert plant tissue into animal tissue. Animal tissue is high in protein, whereas plant tissue is low in protein and high in complex carbohydrates—notably cellulose and lignin in cell walls—that are hard to digest. Nitrogen content is particularly relevant, as it is a major constituent of protein. In plant tissue, the ratio of carbon to nitrogen (C:N) is 30 to 50:1 (as high as 300:1 for wood) versus 10 to 15:1 for animals. This discrepancy creates a major difficulty for the digestive efficiency of herbivores.

Herbivores are categorized based on the type of tissue consumed. **Grazers** eat leaves, often of grasses. **Browsers** feed on woody tissue, especially young twigs and bark. **Granivores** eat seeds and **frugivores** eat fruit. **Sap-feeders** such as sapsucker birds and sucking insects (e.g., aphids) feed on phloem sap, and **nectivores** such as hummingbirds, butterflies, and some moths eat nectar.

Typically, the diet of grazers and browsers has the least protein and the most cellulose. They have evolved symbiotic interactions with bacteria and protozoa inhabiting their digestive tracts, not only to digest cellulose but also to synthesize fatty acids, amino acids, proteins, and vitamins. In vertebrate grazers, microorganisms occupy either the *foregut* (from mouth to intestines; Figure 7.6a) or *hindgut* (intestines; Figure 7.6b). In these O_2-free environments, aerobic respiration is replaced by *fermentation*, an anaerobic process that converts sugars to lactic acids or alcohol.

Ruminants such as caribou and deer are anatomically specialized for digesting cellulose. They have a four-chambered stomach (Figure 7.6a) and a long intestine. As they graze, they quickly chew their food, which enters the *rumen* and *reticulum* to be moistened, kneaded by muscular action, and fermented by anaerobic bacteria and protozoans. These gut microflora degrade cellulose into sugars (which are then converted into fatty acids), synthesize B-vitamins and amino acids, and expel methane. Later, ruminants regurgitate and chew their cud, further reducing particle size (increasing the surface area for microbial action), and then re-swallow. The mass re-enters the rumen, with finer material going to the reticulum. Contractions force the material into the *omasum* for further digestion, and then to the *abomasum*

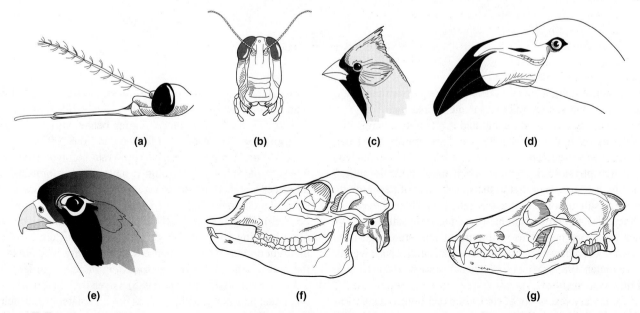

(a)   (b)   (c)   (d)

(e)   (f)   (g)

Figure 7.5 Mouthparts reflect how animals obtain food. **(a)** Piercing mouthparts (mosquito). **(b)** Chewing mouthparts (grasshopper). **(c)** Conical bill (seed-eating bird). **(d)** Straining bill (flamingo). **(e)** Tearing beak (hawk). **(f)** Grinding molars (deer). **(g)** Shearing teeth (coyote).

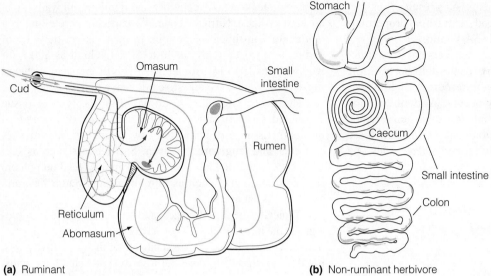

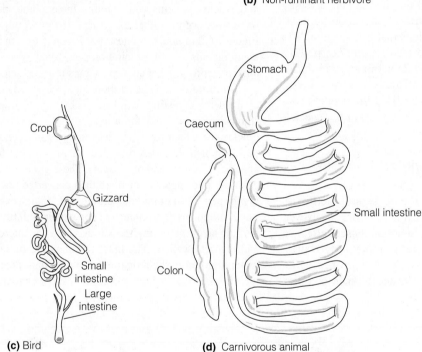

Figure 7.6 Types of digestive tracts. **(a)** Ruminants have a four-compartment stomach: rumen, reticulum, omasum, and abomasum. **(b)** Non-ruminant herbivores have a long colon and a large caecum. **(c)** Birds have a crop, a glandular stomach, and a gizzard. **(d)** Carnivorous mammals have simple tracts, with a small caecum and colon.

(glandular stomach). As well as deriving energy from fermentation, the ruminant digests microbes in the abomasum.

In cold climates, food quality and availability change with the seasons, as do the needs and responses of ruminants. Deer, bighorn sheep (*Ovis canadensis*), and elk (*Cervus elaphus*) experience physiological stress when they suddenly switch from a high-fibre winter diet to the spring flush of new growth. Ruminant saliva contains bicarbonates, which buffer rumen pH. On a low-fibre diet, ruminants produce less saliva, and the rumen pH drops below the optimum for microbial digestion. This pH decline causes intestinal difficulties that reduce weight gain (Kreulen 1985). Mineral deficiencies are also common at this time (see Section 7.3).

The dietary responses of ruminants can be behavioural as well as physiological. White-tailed deer fawns that were fed poor-quality winter forage in experimental enclosures on Anti-costi Island in the Gulf of St. Lawrence were less aggressive than control fawns, suggesting that deer exhibit plasticity in their social as well as in their foraging behaviour in response to deteriorating food resources (Taillon and Côté 2007).

Digestion in non-ruminant vertebrate herbivores occurs in the hindgut. Horses, for example, have simple stomachs, long intestines that increase surface area and slow food passage, and a fermentation pouch called a *caecum* attached to the colon (Figure 7.6b). They digest less efficiently than ruminants but can continue foraging without regurgitating. Some non-ruminants alter their gut dimensions in response to food quality. The hindgut of the prairie vole (*Microtus ochrogaster*) increases in length on low-quality food (Gross et al. 1985). Given that this small rodent (abundant across the prairies) does not hibernate, its plasticity is an adaptive strategy for increasing dietary efficiency in winter when energy demands are high and food availability is low.

Lagomorphs (rabbits, hares, and pikas) employ **coprophagy**: ingestion of fecal matter for further digestion. Some food enters the caecum, with the rest entering the intestine to form dry pellets. In the caecum, microorganisms ferment the material, which enters the large intestine, from which it is expelled as moist pellets. Lagomorphs re-ingest these soft pellets, which contain more protein and less fibre than the dry pellets. Re-ingestion recycles 50 to 80 percent of fecal matter and also provides bacterial-origin B vitamins.

Coprophagy is also widespread among invertebrate detritus-feeders such as wood-eating beetles and millipedes. Many herbivorous insects rely on bacteria and protozoa in the hindgut, but some wood-consuming beetles and wasps utilize fungi, which possess the enzymes to break down plant tissue (see Section 21.3). They carry fungal spores with them when invading new wood.

Seed-eating birds—gallinaceous (chicken-like) birds, pigeons, doves, and many songbirds—have digestive tracts with three separate chambers. The first is the *crop*, a pouch in the esophagus for food, which then moves to the glandular stomach (Figure 7.6c) and finally the *gizzard* (muscular stomach). Birds assist the gizzard's grinding action by swallowing small pebbles and gravel or grit. Many reptiles, amphibians, and invertebrates also have gizzards.

Most herbivorous marine fish are small and inhabit coral reefs, where they make up 25 to 40 percent of total fish biomass. They consume algae, which lack the lignin and other structural compounds that make land plants hard to digest. Marine fish access algal nutrients by various means, including (1) low stomach pH that weakens algal cell walls, allowing digestive enzymes access to cell contents; (2) gizzard-like stomachs containing inorganic matter that mechanically shreds algal cells; (3) specialized jaws; and (4) microbial fermentation in the hindgut. These types are not exclusive. Some fish combine low stomach pH or grinding with fermentation.

Carnivory: Food Quantity, Not Quality, Is the Major Problem Facing Carnivores

Unlike herbivores, carnivores avoid the problems of cellulose digestion. The composition of their tissues resembles that of their prey, so carnivores assimilate nutrients easily. Thus, their main problem is not food quality, but quantity. Carnivores often have difficulty getting enough food. Not needing to digest cellulose, carnivores have short digestive tracts through which food passes rapidly, and simple stomachs (Figure 7.6d). In mammalian carnivores, the stomach is just an expanded tube with muscular walls. It mixes foods, adding mucus, enzymes, and acids to aid digestion. In raptors such as hawks and owls, the gizzard is an extendable pocket with reduced muscles in which digestion, started in the glandular stomach, continues. The gizzard is also a barrier for hair, bones, and feathers, which these birds regurgitate and expel orally in pellets.

How do carnivores obtain the vitamins that humans get from eating fruits and vegetables? With some exceptions (e.g., primates and guinea pigs), most animals synthesize their own vitamin C.

Omnivory: Given Their Mixed Diet, Omnivores Have Less Specialized Digestive Adaptations

Omnivores feed on both plants and animals, though a minimum amount of each must be part of their normal diet to qualify. Accordingly, their digestive adaptations are less specialized. Anatomically, omnivorous mammals resemble carnivores more than herbivores. They tend to eat plant items such as seeds or fruits that are relatively low in cellulose, with high-cellulose items important mainly for roughage. However, microbial digestion (in the hindgut) utilizing a diverse flora of bacteria, protozoans, and even fungi is more important in vertebrates than previously realized (Stevens and Hume 1998).

An omnivore's feeding habits vary with developmental stage and seasonal availability and are often opportunistic. For example, the red fox (*Vulpes vulpes*) feeds on berries, apples, acorns, cherries, grasses, grasshoppers, crickets, beetles, and small rodents. The black bear (*Ursus americanus*) feeds more heavily on plants—buds, leaves, nuts, berries, bark—than many omnivores, but also eats bees, beetles, ants, fish, and even small mammals. The number of species known to be omnivores is growing, and the idea of a "typical" omnivore diet is changing. Some species previously thought to be herbivores or carnivores are opportunistic omnivores. The red squirrel (*Tamiasciurus hudsonicus*), a granivore specializing in conifer seeds, preys heavily on bird nests in Yukon Territory and northern British Columbia (Willson et al. 2003). They also eat young snowshoe hares (Krebs et al. 2001) and spruce bark beetles (Pretzlaw et al. 2006) in northern Canadian forests.

7.3 NUTRITIONAL REQUIREMENTS: Animals Meet Most Nutritional Needs from Their Diet

The needs of animals for various minerals (Table 7.1) and amino acids are very similar to those of plants (see Table 6.1). But whereas plants can synthesize all 20 required amino acids, most animals produce only half, assuming their diet includes organic nitrogen. They obtain the remaining acids that they cannot produce (*essential amino acids*) in food. Mineral needs are similar for vertebrates and invertebrates, although insects require more K, P, and Mg and less Ca, Na, and Cl. For animals, plants are the ultimate source of most minerals. Quantity and quality of plant tissue thus affect herbivore nutrition directly and carnivore nutrition indirectly. When food is scarce, consumers suffer malnutrition, leave the area, or starve. Even when there is enough food to relieve hunger, mineral shortages affect animal health, reproduction, and longevity.

High-quality plant food is rich in protein. As dietary N increases, herbivore assimilation improves, increasing growth, reproduction, and survival. Because N is highest in growing tips and new leaves (which also have less lignin), herbivores preferentially exploit new growth. Insect larvae are most abundant in early spring, completing their growth before leaves mature. Similarly, vertebrate herbivores often give birth in spring, when the most protein-rich diet is available.

How do animals know what to eat? Herbivores detect N-rich plants by taste and odour. Beavers (*Castor canadensis*)

Table 7.1 Essential Minerals in Animals

Element	Role
Carbon (C)	Constituent of all organic matter; particularly important in energy compounds.
Hydrogen (H)	Constituent of all organic matter; involved in operation of ion pumps.
Oxygen (O)	Constituent of most organic compounds; oxidizing agent in mitochondria.
Nitrogen (N)	Building block of proteins and nucleic acids.
Calcium (Ca)	Needed for acid–base relationships, blood clotting, contraction and relaxation of heart muscles; controls movement of fluid through cells; gives rigidity to vertebrate skeletons; forms shells of molluscs, arthropods, and one-celled foraminiferans.
Phosphorus (P)	Needed for energy transfer, acid–base balance, and bone and tooth formation; component of cell nuclear material.
Magnesium (Mg)	Metallic co-factor of many enzymes.
Sulphur (S)	Basic constituent of protein.
Sodium (Na)	Needed for maintaining acid–base balance, osmotic homeostasis, formation and flow of gastric and intestinal secretions, nerve transmission, lactation, and growth.
Potassium (K)	Involved in protein synthesis, growth, and carbohydrate metabolism.
Chlorine (Cl)	Similar roles as sodium.
Fluorine (F)	Maintenance of tooth and bone structure.
Iron (Fe)	Component of hemoglobin in blood of vertebrates and hemolymph of insects; electron carrier in energy metabolism.
Manganese (Mn)	Metallic co-factor of many enzymes.
Selenium (Se)	As does vitamin E, functions as an antioxidant.
Cobalt (Co)	Required by ruminants for synthesis of vitamin B_{12} by rumen bacteria.
Copper (Cu)	Involved in iron metabolism, melanin synthesis, and electron transport.
Molybdenum (Mo)	Metallic co-factor of many enzymes.
Zinc (Zn)	Metallic co-factor of the enzyme carbonic anhydrase in red blood cells as well as of some digestive enzymes.
Iodine (I)	Involved in thyroid metabolism.
Chromium (Cr)	Involved in glucose and energy metabolism.

prefer willow (*Salix* spp.) and aspen (*Populus* spp.), both of which are N-rich. Chemical receptors in a deer's nose and mouth encourage or discourage consumption of certain foods. During drought, nitrogenous compounds accumulate in some plants, making them more attractive. But preference for certain plants means little if they are unavailable. Food selection thus reflects availability as well as food quality and dietary preferences.

The need for high-quality foods differs among herbivores. Ruminants can subsist on lower-quality food because bacteria synthesize vitamin B_{12} and certain amino acids. Thus, the caloric content of a food might not reflect its true nutritive value. Non-ruminant herbivores must consume a larger amount of complex proteins. Seed-eaters exploit the high nutrient content of seeds. Among carnivores, quantity is more important than quality because their prey have already re-synthesized protein and other nutrients into animal tissue.

Soil mineral availability indirectly influences animal abundance and fitness. As one of very few minerals required by animals that are not required by all plants (only in trace amounts by some species), sodium (Na) is one of the least available nutrients in land habitats. In areas with low-Na soils, the distribution, behaviour, and physiology of many species, especially herbivores, are affected. Spatial distribution of elephants (*Loxodonta africana*) in Wankie National Park in Africa is positively correlated with the Na content of waterholes. Moose (*Alces alces*) and white-tailed deer (*Odocoileus virginianus*)

suffer Na deficiency in parts of their range. Moose preferentially consume aquatic plants, which contain more Na than land plants (Figure 7.7).

The seasonal switch to a spring diet can cause severe mineral deficiencies for herbivores. New leaves and buds are high in K relative to Ca and Mg. High K stimulates secretion of *aldosterone*, a hormone that causes the kidneys to retain Na and excrete K and Mg. Because they have low body stores of Mg, herbivores then experience Mg deficiency, which causes diarrhea

Figure 7.7 A moose feeding on Na-rich aquatic plants.

and muscle spasms. This deficiency comes at a time of high mineral demand—late in gestation for females, and at the onset of antler growth for male deer and elk. Deer on mineral-deficient diets have stunted growth, with males developing thin antlers.

To counteract this problem, ruminants seek out **mineral licks**—sites where animals ingest or lick soil. Although NaCl is typically high at licks, access to Mg and other minerals is as critical as access to Na. Wet and dry licks in north-central British Columbia are also concentrated in carbonates, which alleviate the rumen pH problem for elk, moose, and other ruminants (Ayotte et al. 2006). Nor is this behavioural adaptation confined to spring. Caribou lick ice on frozen lakes in the Northwest Territories. Areas licked are significantly higher in minerals than adjacent unused areas (Heard and Williams 1990). In this case, use of licks compensates for the low mineral content of lichens, caribou's winter staple.

7.4 OXYGEN UPTAKE: Structural Adaptations Facilitate O₂ Uptake

Animals obtain energy through aerobic respiration, which requires O_2. For air-breathing animals, O_2 is readily available, but in water O_2 may be limiting and its acquisition difficult. Differences between terrestrial and aquatic animals in O_2 acquisition reflect differences in O_2 supply. Small terrestrial invertebrates simply absorb O_2 by diffusion across their body surfaces. More specialized terrestrial adaptations involve *invaginations*—ingrowths that allow O_2 to be absorbed across a protected, moist internal surface. The **tracheae** of insects, for example, are tubular structures that carry O_2 directly from the body surface to their cells. Tracheae open out to the air through openings called **spiracles** on the body wall (Figure 7.8a).

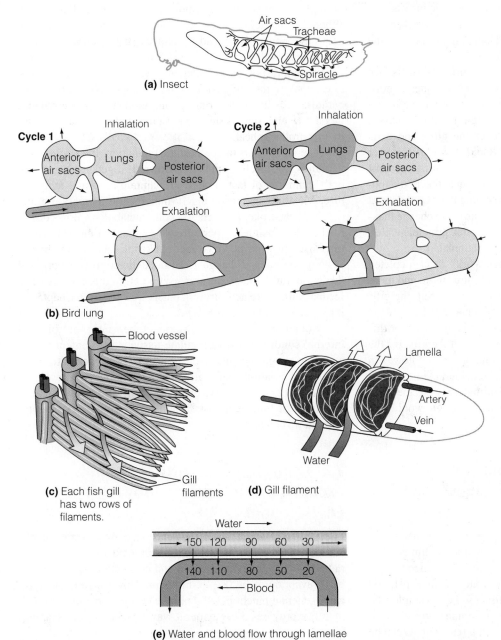

Figure 7.8 Respiratory systems. **(a)** Air enters the tracheal tubes of a grasshopper through spiracles on the body wall. **(b)** A bird lung uses two inhalation cycles. The first inhalation flows past the lungs into a posterior air sac. The next inhalation enters the anterior air sac while the posterior sac draws in more air, maximizing O_2 uptake. **(c)** Fish gills have flattened plates called lamellae. **(d)** Blood flowing through lamellae capillaries absorbs O_2 using countercurrent exchange. Water flows through the gills in a direction opposite to blood flow, facilitating efficient uptake. **(e)** Depleted blood entering the gills picks up O_2 as it flows through the lamellae. Numbers refer to O_2 content (percent saturation; values can exceed 100 percent as a result of aquatic photosynthesis).

Air sacs
Tracheae
Spiracle
(a) Insect

Inhalation
Cycle 1
Anterior air sacs
Lungs
Posterior air sacs
Exhalation

Cycle 2
Inhalation
Anterior air sacs
Lungs
Posterior air sacs
Exhalation

(b) Bird lung

Blood vessel
Gill filaments
(c) Each fish gill has two rows of filaments.

Lamella
Artery
Vein
Water
(d) Gill filament

Water ⟶
| ⟶ | 150 | 120 | 90 | 60 | 30 | ⟶ |
| ⟶ | 140 | 110 | 80 | 50 | 20 | |
⟵ Blood
(e) Water and blood flow through lamellae

Unable to get enough O_2 by diffusion across their surfaces, larger land animals (mammals, birds, and reptiles) have a more complex type of invagination—**lungs**. (Lungless salamanders are an exception; they absorb O_2 only through the skin.) Unlike tracheae, which branch throughout the insect body, lungs are internal chambers found in one location. Lungs have many small sacs that even in the driest habitats provide a moist surface across which O_2 diffuses into the blood. Amphibians absorb O_2 both into simple lungs and by diffusion across their vascularized skin. To meet their high O_2 demand during flight, birds have accessory air sacs that act as bellows to keep air flowing through the lungs in a one-way circuit, whether inhaling or exhaling (Figure 7.8b).

In aquatic habitats, organisms take in O_2 from water or air. Zooplankton are small enough for diffusive uptake. Having evolved from terrestrial ancestors, marine mammals such as whales and dolphins retain their lungs and must come to the surface to expel CO_2 and breathe air. Some aquatic insects also fill their tracheae at the surface. Others, like diving beetles, carry an air bubble when submerged. Held under the wings, the bubble contacts spiracles on the abdomen.

Unlike the invaginations of land animals, **gills** are *evaginations*—outgrowths of internal surfaces. In water, a moist environment is ensured, and outgrowths allow for more efficient uptake. Fish pump in water through their mouth. It passes through slits in the pharynx, flows over the gills, and exits through the gill covers (Figure 7.8c). Rapid flow of water over the gills allows efficient uptake (Figure 7.8d) via **countercurrent exchange**: water passing over the gills flows in a direction opposite to blood flow, creating a favourable gradient for O_2 uptake at all points of contact. (In this case, countercurrent exchange maximizes the efficiency of O_2 uptake, but the same principle underlies heat retention and kidney function in mammals.)

How does countercurrent exchange work? As O_2-depleted blood from the body enters the gills, it acquires more and more O_2. This blood encounters water that is just entering the gills and hence contains more and more O_2 because it has not yet given up any O_2. So there is still a favourable gradient for diffusive uptake even as the blood is about to exit the gills for the rest of the body (Figure 7.8e). If both water and blood flowed in the same direction, the gradient would be favourable for uptake only when both first enter the gills. Ultimately, were the principle of countercurrent exchange not involved, O_2 would exit the blood into the water.

fatal. The maintenance of a relatively constant internal environment in a varying external environment is called **homeostasis**.

Whatever the specific processes, homeostasis depends on **negative feedback**—the means by which a system returns to a set point or equilibrium after deviating from it. Consider a furnace thermostat set at 20°C. When the temperature falls below 20°C, the thermostat activates the furnace. When the temperature reaches the set point, the thermostat shuts off the furnace. If the negative feedback device (the thermostat) fails, the furnace continues heating, the temperature continues rising, and the furnace overheats, causing a fire or a mechanical breakdown.

In living systems, the set point is rarely so firmly fixed. Instead, organisms have a set-point range, a **homeostatic plateau**. Homeostatic systems work within a minimum and maximum, using negative feedback to regulate activity. Control mechanisms can be both physiological and behavioural. An example is temperature regulation of a homeotherm (Figure 7.9). When external temperature rises, sensory receptors in the skin detect the change. They send a message to the brain, which relays it to receptors that increase blood flow to the skin, stimulating behavioural responses and inducing sweating. The sweat evaporates, cooling the body back to the set point. When body temperature falls below a certain point, another response reduces blood flow and triggers shivering, an involuntary muscular activity that produces heat.

There are limits to homeostatic regulation. In extreme temperatures, the mechanism breaks down. If it gets too warm, the body cannot cool fast enough to maintain a normal temperature. Metabolism speeds up, further raising temperature and causing heatstroke. If it gets too cold, metabolism slows down, further decreasing temperature. Hypothermia ensues. These responses involve **positive feedback**: the means by which a system continues to change in the same direction as the original deviation, rather than return to a set point. In this case, positive feedback may threaten survival, but as we will see in Chapter 19 it plays a key role in ecosystems.

Not all organisms employ strict homeostatic control of all internal conditions. Most regulate tissue water content within a certain range, but many do not practise homeostatic temperature regulation, especially in aquatic habitats, where temperature flux is less extreme than on land. We consider examples of both homeostatic and non-homeostatic adaptations in this chapter.

7.5 HOMEOSTASIS: Regulation of Internal Conditions Involves Negative Feedback

In a changing environment, many organisms must maintain a relatively constant internal environment within the limits tolerated by their organs, cells, and enzymes. They need some means of regulating body temperature, water balance, pH, and the salt content of their fluids and tissues. For example, the human body must maintain its internal temperature close to 37°C. A sustained deviation of more than a few degrees can be

7.6 THERMAL EXCHANGE: Animals Exchange Heat Energy with Their Environment

In theory, an animal's energy balance resembles that of plants. Animals gain heat from their environment by (i) absorbing radiation and by (ii) conduction and convection, and lose heat by (i) evaporation and (ii) re-radiation. However, animals typically generate much more metabolic heat than plants, and their mobility provides a behavioural way of seeking (or escaping) heat or cold.

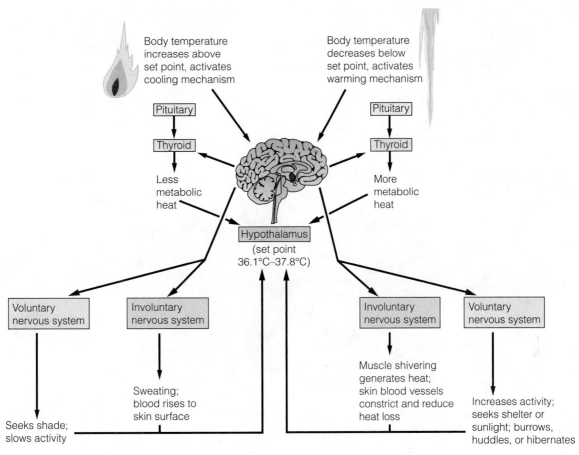

Figure 7.9 Homeostatic thermoregulation. The hypothalamus receives information regarding the temperature of blood arriving from the core. If core temperature rises, the hypothalamus activates the autonomic (involuntary) and voluntary nervous systems and the endocrine system.

Body structure influences heat exchange between animals and their environment, as illustrated with a simple model (Figure 7.10a). The core temperature of many animals must be regulated within a set range, whereas the temperature of its surroundings may vary far beyond this range. Moreover, its surface temperature varies from the temperature of the air or water in which the animal lives. It equals the temperature of the *boundary layer*, the thin layer of air or water at the body surface just above and within the animal's covering materials. Thus, surface temperature (T_s) differs from that of the air or water (T_a) and the body core (T_b). Between the core and the surface are insulating layers of muscle and fat, across which the temperature gradually changes. This layer affects the organism's **thermal conductivity**—its ability to conduct heat.

Let's apply this model to an actual animal—a frog sitting on a rock (Figure 7.10b) (Tracy 1976). The frog's heat budget is determined by heat exchange and/or changes in metabolism. Its core exchanges metabolically generated heat with the surface by *conduction*. The thickness and conductivity of its tissues and the movement of blood to its surface influence this exchange. The frog also exchanges heat from its surface by *convection* (of either water or air currents, depending on its location), *conduction* (with the rock or solid ground), *radiation* (either by absorbing light or re-radiating heat if it is warmer than its surroundings), and *evaporation* (by losing water through its skin). Each of these

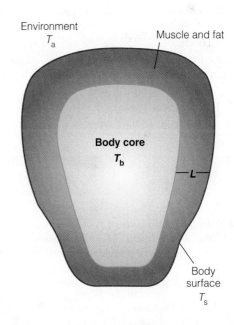

(a)

Figure 7.10 Heat budgets in animals. **(a)** A thermal model of an animal body. The body core temperature is T_b, the environmental temperature is T_a, the surface temperature is T_s, and L is the thickness of the body coverings.

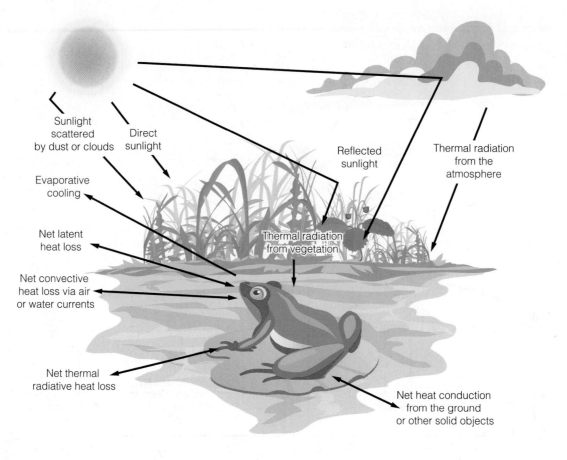

Sunlight scattered by dust or clouds

Direct sunlight

Reflected sunlight

Thermal radiation from the atmosphere

Evaporative cooling

Net latent heat loss

Net convective heat loss via air or water currents

Thermal radiation from vegetation

Net thermal radiative heat loss

Net heat conduction from the ground or other solid objects

(b)

Figure 7.10   *continued* **(b)** Thermal exchange of a frog.

(Adapted from Tracy 1976).

processes is influenced by the properties of its body coverings. (See Quantifying Ecology 7.1: Heat Exchange and Temperature Regulation.)

External conditions influence the thermal stress that an animal experiences. Air has less specific heat and absorbs less radiation than water does, so land animals face more radical temperature changes. Incoming solar radiation can produce lethal heat. Radiant heat loss, especially at night, can cause deadly cold. In contrast, aquatic animals live in a more stable thermal environment, but they are typically less tolerant of rapid temperature change.

With this background, we can now consider the categories of thermal regulation in animals—a classification as important ecologically as that based on diet. To regulate internal temperature, some animals rely on generating heat metabolically, a strategy called **endothermy**. Endothermy is usually associated with **homeothermy**: maintenance of a fairly constant internal temperature independent of external temperature. Other

QUANTIFYING ECOLOGY 7.1   Heat Exchange and Temperature Regulation

nternal temperature regulation is crucial for animals, be they endotherms or ectotherms. Thermal regulation involves balancing heat inputs from the environment and from metabolism with losses to the environment. The total heat stored by the body (H_{stored}) is the sum of these gains and losses:

$$H_{stored} = H_{metabolism} + H_{conduction} + H_{convection} + H_{radiation} + H_{evaporation}$$

Heat energy from metabolism ($H_{metabolism}$) will always be positive. In contrast, heat exchange with the environment through conduction, convection, radiation, and evaporation can

be positive (gain) or negative (loss). Physical laws determine these transfers, in which energy always moves from hot to cold—from higher to lower energy. Heat transfer by evaporation (latent heat exchange) is discussed in Section 3.1, and by radiation in Section 2.1. The other two processes, conduction and convection, are vital to thermal regulation of animals in both terrestrial and aquatic habitats.

Conduction is heat transfer through or between solids that are in direct contact. Heat transfer between the surface and core of a body involves conduction (see Figure 7.10). As ever, energy flows from a region of high to a region of low

QUANTIFYING ECOLOGY 7.1   continued

temperature. The rate of conductive transfer, $H_{conduction}$, is described as

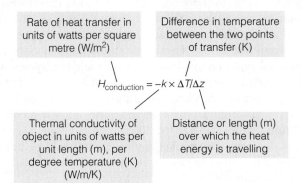

The symbol Δ refers to a difference. Thus, ΔT is the difference in temperature between two regions (e.g., body core and surface), and Δz is the change in position (z)—the distance between the two regions. The thermal conductivity of an object (k) quantifies its ability to conduct heat, as influenced by various factors. Insulation—fat, fur, or feathers—lowers thermal conductivity.

Consider conductive transfer through an animal. In an endotherm, the core temperature is maintained by metabolism, and heat is transferred from the core to the surface, which is usually cooler. If we assume that the thermal conductivity (k) of the body is 1.25 W/m/K and the distance Δz from the core to the surface is 10 cm, we can calculate the heat transfer (Figure 1). If $H_{conduction}$ is positive, heat flows outward. If negative, heat flows inwards, from the surface to the core, as with a reptile basking on a hot rock. (We can also use this approach to calculate heat conduction between two different objects in direct contact, such as the frog and the rock in Figure 7.10b.)

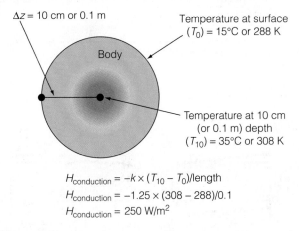

$$H_{conduction} = -k \times (T_{10} - T_0)/\text{length}$$
$$H_{conduction} = -1.25 \times (308 - 288)/0.1$$
$$H_{conduction} = 250 \text{ W/m}^2$$

Figure 1 Heat transfer between the surface and the core of an organism.

Convection is heat transfer between a solid and a moving fluid (air or water). The rate of convective transfer ($H_{convection}$) reflects the temperature gradient between an object and the surrounding fluid:

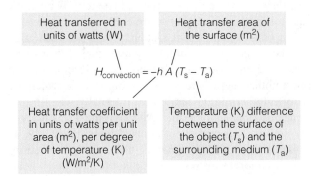

The **heat transfer coefficient** (h) quantifies heat movement through a fluid. Its value depends on the fluid (gas or liquid) and its flow rate, viscosity, and thermal properties. An object's SA/V ratio is critical. Heat energy is transferred across an organism's surface, so the greater its SA/V ratio, the faster the convective transfer. Smaller bodies thus exchange heat faster than larger bodies. Shape also influences heat exchange. Consider two objects of the same size (Figure 2). Object (b) has much more surface area than object (a), so (b) has a greater capacity for convective exchange. The effects of shape and size on heat transfer are vital for poikilotherms, which lack thermoregulation, but they constrain the evolution of all organisms, including plants and fungi as well as animals.

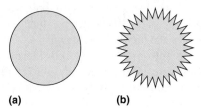

Figure 2 The shape of similarly sized organisms affects heat transfer.

1. Figure 6.11 illustrates the differences in size and shape of leaves on the same plant that are exposed to sunlight versus shade. How might these morphological differences influence the ability of sun versus shade leaves to dissipate heat through convection?

2. Given the importance of conductive and convective exchange for body temperature of poikilotherms, how might body shape differ between homeotherms and poikilotherms of similar size?

animals acquire heat from their environment, a strategy called **ectothermy**. Ectothermy is associated with **poikilothermy**: variation of body temperature with environmental temperature. Although poikilotherms do not practise homeostatic regulation, they often regulate their temperature through behaviour.

Birds and mammals are homeotherms and are often called *warm blooded*. Fish, amphibians, reptiles, and invertebrates are poikilotherms and are often called *cold blooded* because they can be cool to the touch. (The term is misleading, as poikilotherms can be warmer than homeotherms, depending on the environment.) A third group, **heterotherms**, practise both ecto- and endothermy over their life history. Hibernating homeotherms relax homeostatic control and become ectothermic, whereas some flying insects, which are normally ectothermic, generate heat internally when flying and hence become endothermic. Although the terms *homeothermy* and *endothermy* are often used interchangeably, as are *poikilothermy* and *ectothermy*, there is a distinction. *Ecto-* and *endothermy* refer to the source of body heat, whereas *homeo-* and *poikilothermy* refer to the type of body temperature variation.

7.7 POIKILOTHERMY: Poikilotherms Rely on External Heat Sources

Poikilotherms gain heat easily from their environment but lose it just as quickly. They typically have high thermal conductivity, as signalled by their lack of insulating coverings. External heat controls their metabolism and activity. Rising temperatures increase body temperature, which in turn increases enzyme activity and metabolism (Figure 7.11). Poikilotherm metabolism has a Q_{10} (increase in metabolism for a 10°C rise in temperature) of 2—that is, rates roughly double. They are active only when the temperature is sufficiently warm. At cool temperatures, poikilotherm metabolism declines and they become sluggish.

There are exceptions. Arctic poikilotherms typically have lower temperate minima and optima (Danks 2004), with significant locomotor activity below 5°C (Hodkinson

2003). Arctic springtails (*Onychiurus arcticus*) do not achieve these activity levels with higher basal metabolism. Instead, they have unusually high and non-linear Q_{10} values ($Q_{10} = 7$ from 0°C–10°C and $Q_{10} < 2$ from 10°C–20°C), allowing them to become active quickly when temperatures rise even slightly (Block et al. 1994). Nor are they unusual; mean Q_{10} of invertebrates on Devon Island (Northwest Territories) was 4.0 over the range from 2°C to 12°C (MacLean 1981)—far higher than the value of 2 assumed to be typical for poikilotherms. These metabolic adjustments allow adapted poikilotherms to complete their life cycles in the short Arctic season. But metabolism of Arctic poikilotherms responds to more than temperature. It is affected by food supply and cold acclimation and varies greatly among and within species (Hodkinson 2003). The more we learn about poikilotherms in extreme environments (cold or hot), the clearer it is that we must view generalizations about "typical" animals with caution.

Although rates vary, poikilotherms have low metabolic rates. Normally, they use aerobic respiration. However, while pursuing prey, they often cannot obtain enough O_2, so their metabolism becomes anaerobic, depleting their energy and accumulating lactic acid in their muscles. Rapid physical exhaustion follows, limiting poikilotherms to short bursts of activity.

Despite lacking homeostatic thermoregulation, some poikilotherms maintain a relatively constant daytime temperature by behaviour, such as seeking sun or shade. Lizards and snakes may vary in temperature by only 4°C to 5°C (Figure 7.12), whereas amphibians may vary by 10°C. The **operative temperature range** is the range of body temperatures over which a poikilotherm is active.

In contrast, there is typically little difference between the temperature of aquatic poikilotherms and that of the surrounding water when immersed. Like all poikilotherms, fish are poorly insulated. This trait is adaptive for organisms that rely on external heat. Any heat produced in the muscles

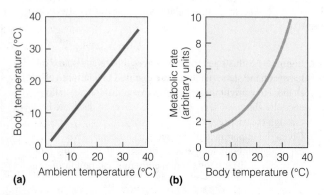

(a) **(b)**

Figure 7.11 In poikilotherms, **(a)** body temperature is a function of ambient temperature and **(b)** resting metabolism is a function of body temperature.

(Adapted from Hill and Wyse 1989.)

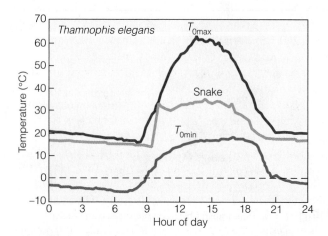

Figure 7.12 Daily temperature variation in the western garter snake (*Thamnophis elegans*) within its operative temperature range (T_{0min} and T_{0max}). Its temperature stays fairly constant during the day.

(Adapted from Peterson et al. 1993.)

(a)   **(b)**

Figure 7.13 The South American frog *Bokermannohyla alvarengai* changes in colour from **(a)** dark to **(b)** light when moved from a dark to a light environment.

moves to the blood and on to the gills and skin, followed by rapid convective transfer to the water. Exceptions are sharks and tunas. They have a **rete**, a network of veins that allows them to keep internal temperatures higher than their surroundings (see Section 7.11). Despite their high conductivity, fish and aquatic invertebrates maintain a fairly constant temperature in any given season because the water temperature is quite stable. They adjust to temperature change by *acclimation*—physiological adjustment to changing conditions (see Figure 5.21).

Upper and lower tolerance limits for temperature vary among poikilotherms. At the upper end, poikilotherms adjust their physiology at the expense of being able to tolerate lower temperatures. Similarly, during colder periods, poikilotherms shift physiological function to a lower thermal range. Aquatic poikilotherms adjust slowly. Fish are very sensitive to rapid temperature change and may die of thermal shock if the temperature changes faster than they are able to acclimate.

To maintain a fairly constant body temperature when active, terrestrial and amphibious poikilotherms use behavioural thermoregulation by seeking out appropriate microclimates. Butterflies, moths, and dragonflies bask in the sunlight to raise their temperature to the level necessary for activity, and seek shade if they are too warm. Semi-terrestrial frogs, such as bullfrogs (*Lithobates catesbeianus*, formerly *Rana catesbeiana*) and green frogs (*L. clamitans*), raise their body temperature as much as 10°C above the ambient temperature by basking (Lillywhite 1970). By changing position or by seeking warmer or cooler substrates, they maintain body temperatures within a narrow range. The trade-offs of these behavioural adjustments include increased predation risk as well as higher evaporative

losses, so amphibians remain either near or partially submerged in water.

In the South American frog, *Bokermannohyla alvarengai*, plastic structural changes supplement behavioural thermoregulation. In both field and laboratory studies, skin colour lightens significantly in response to incident light (Tattersall et al. 2006). Darker frogs heat up quickly when they begin basking, but any further rise in temperature is slowed by the increased reflection from their progressively lighter skin colour (Figure 7.13). The risk of overheating and excessive evaporation is presumably reduced, although not eliminated.

Most reptiles are terrestrial and experience changing temperatures. Like frogs, they expose themselves to solar rays (**heliothermy**) to warm up. Snakes heat up rapidly in the morning (Figure 7.14, p. 140). Once at its preferred temperature, the snake resumes normal activity, retreating to shade if needed. In the evening, it cools slowly, with its night temperature varying with location. Snakes alter their behaviour to achieve a preferred temperature. Pregnant female cottonmouth snakes (*Agkristodon piscivorus*) seek warmer microhabitats than non-pregnant females, illustrating a critical link between habitat use and thermoregulation (Crane and Greene 2008).

Lizards also bask, seek shade, or burrow to adjust their temperature. They raise and lower their bodies and even change shape to alter conductive exchange with the rocks or soil they rest upon. Desert beetles, locusts, and scorpions raise their legs to reduce contact with the ground, reducing conduction and increasing convection by exposing their body surfaces. Body temperatures of poikilotherms are thus rarely equal to ambient temperature, despite a lack of physiological homeostatic regulation.

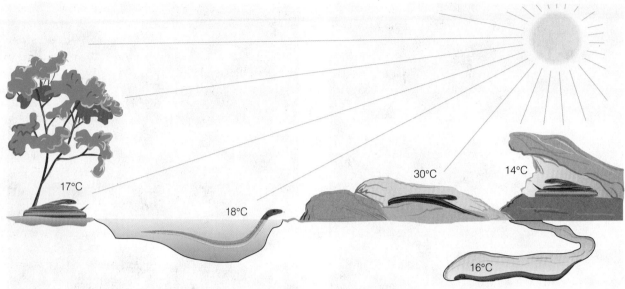

Figure 7.14 Snakes use microclimates to regulate body temperature.

(Adapted from Peterson et al. 1993).

7.8 HOMEOTHERMY: Homeotherms Reduce Thermal Constraints by Regulating Body Temperature

Through endothermy, birds and mammals minimize the thermal constraints of their environment. With homeostatic regulation, they maintain a constant body temperature using heat generated by metabolic oxidation. Oxidation is not entirely efficient; as well as producing chemical energy as ATP, some energy is converted to heat. Homeotherms rely on this "waste" heat. Their thermostat is set fairly high because their enzymes operate with a high temperature optimum (~ 39°C).

The **thermoneutral zone** is the range of external temperatures within which the metabolic rates of a homeotherm are minimal and constant (Figure 7.15; "minimal" is relative; the metabolic rates are much higher than that of a poikilotherm). Outside this zone, marked by upper and lower **critical temperatures**, metabolism increases, in contrast with the response of poikilotherms (see Figure 7.11). With efficient cardiovascular and respiratory systems to transport O_2, homeotherms support high aerobic respiratory rates. They can sustain prolonged activity independently of external temperature, allowing them to exploit more thermal habitats than poikilotherms—but see Section 7.7 regarding Arctic poikilotherms. Homeotherms can generate energy rapidly when needed, whether escaping from predators or pursuing prey, without resorting to anaerobic metabolism.

Like poikilotherms, homeotherms exchange heat with their environment, but because they do not rely on external heat, it is in a homeotherm's interests to retain expensive, metabolically

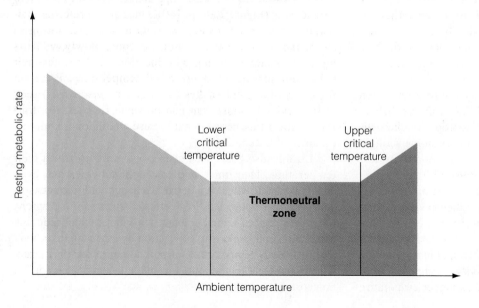

Figure 7.15 Temperature response of the resting metabolic rate of a homeotherm. Within the thermoneutral zone, resting metabolic rate remains constant. Beyond these limits, metabolic rate increases sharply as a result of feedback (see Figure 7.9).

(Adapted from Schmidt-Nielsen 1997.)

generated heat by minimizing heat exchange. So homeotherms invest in insulation—fur, feathers, hair, fat, or blubber. For mammals, fur slows heat loss, but its insulation value varies with its thickness. Small mammals can only carry so much fur, because a heavy coat restricts their mobility. Many acclimate by changing the thickness of their fur with the season. Aquatic mammals and some Arctic and Antarctic birds, such as auks (Alcidae) and penguins, have a thick layer of fat beneath the skin. Birds reduce heat loss by fluffing their feathers and drawing in their feet, making the body a feathered ball. Some Arctic birds, such as ptarmigan (*Lagopus* spp.), have feathered feet, unlike most birds, which have scaled feet that radiate heat easily.

Insulation also functions to keep heat out. In hot climates, an animal has to either rid itself of heat or prevent heat from being absorbed in the first place. One way is to reflect radiation from light-coloured fur or feathers. Another way is to grow a heavy coat that heat cannot easily penetrate. Some large desert mammals, such as the camel, employ this method.

When insulation fails, many animals resort to **shivering**— involuntary muscular activity that generates heat. But shivering has its trade-offs; it reduces control of activities such as foraging that are necessary for survival. Many mammals generate heat without shivering by oxidizing highly vascularized brown **adipose fat**. Found about the head, upper body, and major blood vessels, adipose fat is prominent in hibernators, such as bats and groundhogs (*Marmota monax*).

If the heat load is excessive, homeotherms resort to evaporative cooling. When their body temperature nears the upper critical limit, birds and mammals accelerate evaporative cooling by sweating or panting. Relatively few mammals (examples are horses and humans) have sweat glands. Although sweating is an effective cooling strategy, it increases the risk of dehydration and is risky in dry environments. Panting in mammals and gular fluttering in birds increase air flow over moist surfaces in the mouth and pharynx, but with less dehydration risk. Many mammals use behaviours such as wallowing in water or wet mud to cool down.

7.9 TRADE-OFFS IN THERMAL REGULATION: Endothermy and Ectothermy Have Benefits and Costs

Thermoregulation is a prime example of the trade-offs inherent in all adaptations. Endothermy and ectothermy (and by extension homeothermy and poikilothermy) have advantages and disadvantages that enable organisms to excel under different conditions—and even under the same conditions. Endothermy allows animals to remain active regardless of external temperature, which may limit the activity of ectotherms. Yet this freedom comes at a cost. To generate enough heat, endotherms must have a high food intake, most of which they must allocate to respiration, not growth or reproduction. Because much of the heat generated is lost to the environment, metabolic costs weigh heavily on endotherms, especially in regions with food shortages.

In contrast, ectotherms can allocate more energy to growth and reproduction. Not relying on metabolism as a heat source,

ectotherms require far fewer calories per unit body weight and can curtail metabolism in times of food and water shortages or temperature extremes. Although poikilotherms occupy virtually all habitats, their low energy demands enable some to occupy areas that have insufficient food and water to support endotherms.

Endotherms and ectotherms have different size constraints. Recall from Section 7.6 that a body exchanges heat with its environment in proportion to the surface area exposed. In contrast, it is the entire body volume that is being heated. The *surface area to volume ratio (SA/V)* is thus a key factor in controlling uptake of heat and maintenance of body temperature. As organism size increases, *SA/V* decreases. Because the organism has to absorb sufficient energy across its surface to warm its entire body volume, the amount of energy and/or length of time required to raise body temperature also increases. For this reason, ectothermy restricts the maximum body size of poikilotherms and limits the distribution of larger poikilotherms to warm, aseasonal tropical and subtropical regions. It is no accident that large reptiles such as alligators, crocodiles, iguanas, and pythons occupy tropical habitats. Large aquatic poikilotherms such as sharks escape this maximum size constraint because water facilitates convective exchange, and because temperature changes so much slower in water than in air. Small size, in contrast, is an ecological asset for ectotherms because it facilitates heat exchange by increasing *SA/V*.

The constraint that size imposes on endotherms is the reverse. For endotherms, the body volume generates heat, while heat is lost across the surface. *SA/V* decreases with size for endotherms just as it does for ectotherms—it is the ecological consequences that differ. Smaller endotherms have greater relative heat loss, which must be offset by increased metabolism. Thus, small endotherms have a higher mass-specific metabolic rate (rate per unit body mass; Figure 7.16, p. 142) and must consume more food per unit body weight than do large endotherms. Small shrews (*Sorex* spp.), ranging from 2 to 29 g, must consume their own weight in food daily. Small endotherms are forced to spend most of their time seeking and eating food, which limits their success in many habitats and increases their predation risk.

The mass-specific metabolic rate of small endotherms rises so rapidly that below a certain size, they simply cannot meet their energy needs. On average, 2 g is about as small as an endotherm can be and still maintain its heat balance. Some shrews and hummingbirds undergo daily torpor (see Section 7.10) to reduce their costs. Given the conflicting demands of body temperature and growth, the young of most birds and small mammals (which may weigh less than 2 g) begin life as ectotherms. Born in an *altricial* state (blind, naked, and helpless; see Section 8.6), they rely on the bodies of their endothermic parents (their biotic environment) as a heat source. This strategy allows them to allocate most of their energy to growth instead of heat generation.

Although they have strict minimum size constraints, endotherms have weaker constraints on maximum size than do ectotherms. Large size makes it easier to retain metabolically generated heat, and so selection for large size in homeotherms is common, especially in cold climates. This trend is stated as **Bergmann's rule**: among related homeotherms, those inhabiting colder climates tend to have larger-sized individuals. Consider bears: the

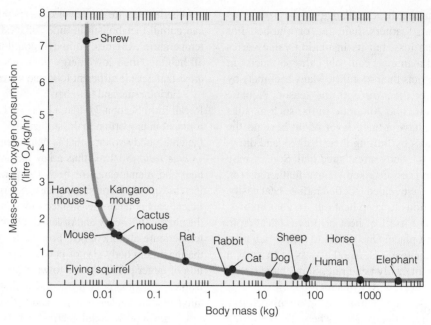

Figure 7.16 Negative correlation between mass-specific metabolic rate (O_2 kg^{-1} h^{-1}) and body mass (log$_{10}$) for various mammals. (Adapted from Schmidt-Nielson 1997.)

INTERPRETING ECOLOGICAL DATA

Q1. How would this relationship differ if metabolic rate (O_2 consumed per hour) were graphed on the *y*-axis instead of mass-specific metabolic rate (O_2 consumed per kg body mass per hour)? How would this affect the positioning of the elephant versus the mouse?

Q2. How would this graph differ if the *y*-axis were plotted on a log$_{10}$ scale, as is the *x*-axis?

polar bear (*Ursus maritimus*) is larger than the black bear (*U. americanus*), and across the black bear's range, mean size of individuals in northern Canada is greater than in southerly populations. This "rule," first formulated by biogeographers in the 19th century, has recently assumed new significance given its relevance for current and future global climate change (Millien et al. 2006).

Although its generality is disputed (Huston and Wolverton 2011), Bergmann's rule is supported reasonably well by variation both within and among species in the mid-latitudes. The factors responsible, however, are controversial. The original explanation focused on thermoregulation and heat exchange, but interactions between temperature and moisture (a common confounding factor) may be as or more important, as indicated by geographic variation in North American bobcats (Wigginton and Dobson 1999). The relationship between latitude and size is stronger for some species than others. White-tailed deer (*Odocoileus virginianus*) conform well to Bergmann's rule, whereas the relationship is weak for moose (*Alces alces*) and reverses for caribou (*Rangifer rangifer*), with mean body size declining north of 60° latitude (Geist 1998) (Figure 7.17). Huston and Wolverton (2011) argue that body size is independent of latitude and depends instead on food availability (particularly plant tissue) during the growing season. Whatever the cause(s), how much of body size variation is genotypic versus plastic is also critical. Moreover, gravity reinforces maximum size constraints in land habitats, but in a buoyant aquatic environment, homeotherms, such as whales, can attain huge size.

Size considerations can have far-reaching and controversial implications; see Research in Ecology: Prehistoric Snakes as Paleothermometers. Nor is the debate confined to snakes. For decades scientists argued about the physiology of dinosaurs.

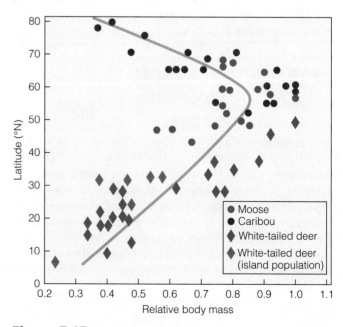

Figure 7.17 Distribution of body size in North American Cervidae. (Adapted from Geist 1998).

RESEARCH IN ECOLOGY | Prehistoric Snakes as Paleothermometers

Just how far-reaching are the ecological implications of body size? The concept of scaling (see Section 7.1) is a fundamental aspect of biomechanics that profoundly influences an organism's energy budget. In essence, it is the factor that distinguishes the way in which homeotherms and poikilotherms relate to their environment, contributing to the minimum size constraints of the former and the maximum size constraints of the latter. But how fluid are these constraints, and what do they reveal about the changing environment in which organisms live, and in response to which taxa evolve? Let's explore these fascinating questions about the significance of body size in the context of a recent paleontological advance: identification from fossil remains of a snake species, *Titanoboa cerrejonensis*. Like many discoveries, this one has a story attached—one that involves serendipity, collaborative endeavour, and controversy.

In 2009, Carlos Jaramillo, a paleobotanist at the Smithsonian Tropical Research Institute, uncovered a fossilized snake in a remote area of northeastern Columbia. Just how big was this extinct species? Comparing its vertebrae with those of its nearest living relatives, the green anaconda (*Eunectes murinus*) (Figure 1) and the African python (*Python natalensis*), Head et al. (2009a) estimated that *Titanoboa* averaged 12.8 m in length, with a mass of 1135 kg. (The mean length of its living relatives rarely exceeds 6 m.) Head had studied other prehistoric snakes, including *Sanajeh indicus*, but *Titanoboa* was much larger. Its feeding niche differed too; *Sanajeh* lived in the Cretaceous and fed on dinosaur eggs, whereas *Titanoboa* lived 58 million to 60 million years ago, in the early Paleocene, after non-avian dinosaurs went extinct. With dinosaur eggs no longer available, *Titanoboa* preyed on ancient turtles and crocodiles.

If the story stopped here, it would have been remarkable enough. But the University of Toronto's Jason Head was intrigued by what *Titanoboa's* size might reveal about the climate in which it lived. Russian ecologist Anastassia Makarieva had argued for a universal scaling factor that allowed extrapolation of prevailing mean annual temperature from dimensions of poikilotherms across a range of taxa (Makarieva et al. 2005). The argument goes as follows. Organisms must maintain a minimum mass-specific metabolic rate to support activity. Mass-specific metabolic rate declines with increasing body mass for all organisms, but for poikilotherms it increases with environmental temperature (see Figure 7.11). Thus, warmer temperatures compensate for the drop in the mass-specific metabolic rate of larger poikilotherms, relaxing maximum size constraints and allowing evolution of larger forms. In essence, Makarieva's formula uses the ratio of maximum body size of similar animals from different localities as a "paleothermometer" to reconstruct past temperature regimes.

Reconstructing past climates on evidence provided by organisms (living or fossil) is not new. By evaluating tree rings, dendrochronologists infer climatic conditions of the past 5000 to 10 000 years. Fossil evidence allows inferences on a much longer timeframe. Paleobotanists base predictions of past climates on analysis of leaf margins (Greenwood et al. 2004). Makarieva's formula (an example of a "nearest living relative" method) expands

Figure 1 The green anaconda, shown here basking on a termite mound near the Orinoco River in Venezuela, is one of the largest living relatives of *Titanoboa*.

the evidence base for climate reconstruction to a wider spectrum of organisms.

Using Makarieva's formula, Head and his colleagues estimated that a snake of the size of *Titanoboa* would have required a minimum mean annual temperature of 30°C to 34°C to survive—about 6°C to 8°C warmer than most current estimates for the early Paleocene. If accurate, the consequences for climate theory would be profound, particularly as the Paleocene (55 million–65 million years ago) was the first phase of the Paleogene, a greenhouse era of rising CO_2. As well as altering our understanding of Earth's past climate, *Titanoboa* could furnish important clues about the selective pressures animals would be subject to as a result of future global warming.

As we should expect with good science, Head's paper met with controversy. We consider each of three responses that appeared in *Nature*, along with Head's reply, as indicative of the constructive role of debate. Sniderman (2009) questioned the validity of Head's temperature estimates by disputing Makarieva's assumption that extant (existing) species are at their maximum body size for prevailing ambient temperatures. If this assumption is incorrect, then the thermometer would be biased and a snake as large as *Titanoboa* might have lived in a climate little warmer than today's. Applying the methodology to a comparison of lizards, Sniderman argues that Head ignored the possible role of competition with mammalian carnivores in restricting the maximum size of present-day poikilotherms. In reply, Head et al. (2009b) argued that whereas Makarieva's assumption might not apply to *Vanarus* lizards, for which body sizes are known to be restricted by biotic interactions, there are no ecological factors apart from ambient temperature that are known to restrict body size in boid snakes.

Denny et al. (2009) contended that Head did not factor in the potential for a large snake like *Titanoboa* to control body temperature through behaviour. *Titanoboa's* large size would allow it to raise its temperature as much as 4.3°C above ambient temperature by coiling on an insulating substrate. At the temperatures predicted by Head, a coiled *Titanoboa* would

continued on page 144

RESEARCH IN ECOLOGY continued

risk overheating (Denny et al. 2009). In response, Head et al. (2009b) agreed that behavioural thermoregulation combined with thermal inertia does maintain higher temperatures in large poikilotherms, but that such thermal regulation is not unique to snakes in general or *Titanoboa* in particular and hence does not falsify paleothermic predictions based on body size comparisons.

Makarieva, the ecologist whose model Head employs, also weighed in on the debate. She argued that her own use of a scaling factor of "possible universality" is justified in comparative analysis of many diverse taxa, but that a study based on a single taxonomic group requires a taxon-specific coefficient. For snakes, this coefficient is at the lower end of the range. (The paper doesn't suggest a reason, but it seems likely that the elongated snake body would weaken maximum size constraints at any environmental temperature because they have greater *SA/V* ratios than do organisms of similar overall size.) Making this adjustment, Makarieva et al. (2009) re-calibrated the snake paleothermometer (Figure 2), estimating a mean annual temperature in the Paleocene of 28°C to 31°C. In reply, Head et al. (2009b) contend that even if this snake-specific coefficient is accepted, the resulting 2°C to 3°C decrease from the earlier estimate still suggests a significantly warmer equatorial climate 60 million years ago than previously thought.

Does this controversy diminish the importance of Head's work? Recall from Section 1.8 that dissent is essential to science. That these scientists criticize Head's paper using detailed arguments, to which he can respond in like manner, illustrates the progress such debate not only makes possible but also promotes. Rather than weakening his findings, the debate on the pages of *Nature* strengthens them by building on the collective brain-power of scientists from around the world. The point here is not just that body size—and its consequences for an animal's heat budget as a result of scaling—is an important ecological trait for both extant and extinct taxa, be they poikilotherms or homeotherms. The broader significance is that science progresses through discovery and debate, which are essential to acquiring and testing knowledge.

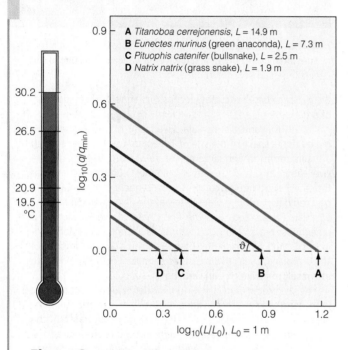

Figure 2   Re-calibrated snake thermometer based on the relationship between mass-specific metabolic rate (q/q_{min}) (*y*-axis) and body length (L/L_0 (*x*-axis; both axes on a log scale and expressed relative to a minimum value). The points at which the lines cross the horizontal dashed line $q = q_{min}$ correspond to body lengths of the largest snakes in the Paleocene neotropics (A), South America (B), Colorado (C), and the United Kingdom (D). Temperatures reconstructed from the metabolic rate-size relationship (assuming that the anaconda lives at 26.5°C) are marked on the thermometer, with the difference between the modern and Paleocene neotropics shown in blue. Reconstructed temperature differences pertain to typical "lifestyle" temperatures of the species considered and can differ significantly from mean annual temperature in seasonal climates.

(Adapted from Makarieva et al. 2009.)

Bibliography

Denny, M. W., B. L. Lockwood, and G. N. Somero. 2009. Can the giant snake predict palaeoclimate? *Nature* 460:E3–E4.

Greenwood, D. R., P. Wolf, S. L. Wing, and D. C. Christophel. 2004. Paleotemperature estimates using leaf margin analysis: Is Australia different? *Palaios* 19:129–142.

Head, J. J., J. I. Bloch, A. K. Hastings, J. R. Bourque, E. A. Cadena, F. A. Herrera, P. D. Polly, and C. A. Jaramillo. 2009a. Giant boid snake from the Palaeocene neotropics reveals hotter past equatorial temperatures. *Nature* 457:715–717.

Head, J. J., J. L. Bloch, A. K. Hastings, J. R. Bourque, E. A. Cadena, F. A. Herrera, P. D. Polly, and C. A. Jaramillo. 2009b. Reply. *Nature* 460:E4–E5.

Makarieva, A. M., V. G. Gorshkov, and B.-L. Li. 2005. Gigantism, temperature and metabolic rate in terrestrial poikilotherms. *Proceedings of the Royal Society of London* 272:2325–2328.

Makarieva, A. M., V. G. Gorshkov, and B.-L. Li. 2009. Re-calibrating the snake palaeothermometer. *Nature* 460:E2–E3.

Sniderman, J. M. K. 2009. Biased reptilian palaeothermometer? *Nature* 460:E1–E2.

Many assumed that dinosaurs were lizard-like poikilotherms, with large, sluggish bodies that were slow to warm (and cool). Others contended that their large body size indicated that they could not have been efficient poikilotherms, as they violated the maximum body size constraints of terrestrial poikilotherms in even the warmest of climates. Moreover, many large dinosaurs lived in cold climates, where large body size is only a clear advantage for homeotherms, because it retains expensive, metabolically generated heat.

Robert Bakker (1972), reviving views first proposed in the 19th century, argued that dinosaurs were homeotherms capable of sustained aerobic metabolic activity. Over time, and in response to innumerable studies of dinosaur physiology, anatomy, and behaviour, a consensus emerged that dinosaurs were indeed endotherms. But the debate has recently shifted to what kinds and degree of thermoregulation they practised: thermal inertia, behavioural regulation, or homeostatic control of metabolic rate. Controversy still rages about whether birds are "feathered dinosaurs" or merely evolved from a common theropod ancestor (Feduccia et al. 2005), but it is likely that dinosaurs as a group employed various thermoregulation strategies.

On balance, neither homeothermy nor poikilothermy is a "better" strategy. Homeothermy might seem superior because it is associated with sophisticated homeostatic regulation, but worldwide far more organisms (species and individuals) are poikilotherms. Even though one strategy might seem to suit some habitats more than others (e.g., homeothermy in cold climates), most if not all habitats support organisms of both types. The key is that each species, whatever its thermal strategy, must find an available *niche*, or means of functioning in its habitat (see Section 13.5).

7.10 HETEROTHERMY: Some Species Practise Both Ectothermy and Endothermy

Heterotherms are species that regulate their body temperature at some times but not others. At different stages of their daily or seasonal cycle, heterotherms practise either endothermy or ectothermy. At such times, they undergo rapid, drastic, repeated temperature changes. Insects, for example, are classic ectothermic poikilotherms, yet as adults, most flying insects are heterothermic. With wings beating up to 200 times per second, they have high metabolic rates, generating as much heat on a mass basis as do many homeotherms. They support their high metabolism by taking in O_2 through openings in the body wall and transporting it via tracheae (see Section 7.4).

Why do flying insects employ endothermy? Most cannot fly if the temperature of their flight muscles is below 30°C or above 44°C. This constraint means that an insect has to warm up before taking off and rid itself of excess heat in flight. Butterflies and dragonflies warm up by orienting their bodies and spreading their wings to the rays of the Sun. Others shiver their flight muscles. Moths and butterflies vibrate their wings. Bumblebees pump their abdomens with no external wing movements, allowing them to fly in the cold days of early spring. To

(a)

(b)

Figure 7.18 Body coverings retain internally generated heat in flying insects such as bumblebees **(a)** and moths **(b)**.

retain heat generated in flight, some insects, notably moths, bees, and bumblebees, have a dense, fur-like coat on their thorax (Figure 7.18). Although these insects practise endothermy, they are not homeothermic. They do not maintain a physiological set point, and they cool to near ambient temperature when not in flight.

The most familiar heterotherms are homeotherms that temporarily relax homeostatic control. To reduce metabolic costs during inactive periods, some small homeotherms enter daily **torpor**, in which their body temperature drops to near ambient temperature. Examples are small birds such as poorwills (*Phalaenoptilus nuttallii*) and small mammals such as bats and many mice. Daily torpor reduces energy demands during inactivity. Nocturnal mammals (e.g., bats) enter torpor by day, and diurnal animals (e.g., hummingbirds, Trochilidae) by night. The animal relaxes homeostatic control, allowing its temperature to drop to within a few degrees of ambient temperature. Body temperature quickly returns to normal when the animal is roused and resumes generating metabolic heat. Only small animals practise torpor, because the lower *SA/V* ratio of a larger animal means it would take too long to cool down and heat up on a daily basis.

To escape the rigours of long, cold winters, many terrestrial poikilotherms and some mammals enter a prolonged seasonal torpor or **hibernation**, characterized by the near cessation of cellular and bodily activity. Hibernating poikilotherms undergo complex physiological changes such as decreased blood sugar; increased liver glycogen; altered blood levels of hemoglobin, CO_2, and O_2; altered muscle tone; and darkened skin. Hibernating homeotherms invoke controlled **hypothermia** (reduction of body temperature). As with daily torpor, homeostatic regulation is relaxed and body temperature approaches ambient. Entering hibernation is a tightly controlled process that differs among species. Some, such as the groundhog (*Marmota monax*), feed heavily in late summer to build up fat reserves, from which they draw during hibernation. Others, such as the chipmunk (*Tamias striatus*), lay up a store of food.

Energy costs drop sharply during hibernation. Heart rate, respiration, and metabolism plummet, and body temperature sinks below 10°C. Blood CO_2 levels rise, which lowers blood pH. This state, called *acidosis*, lowers the threshold for shivering, further reducing metabolism. However, hibernators must rouse periodically to rid themselves of bodily wastes, rehydrate their body tissues, or warm up if their body temperature approaches freezing. At these times, hibernating homeotherms warm spontaneously using internally generated heat, after which they re-enter torpor. In general, species with food stores, such as chipmunks and ground squirrels, spend less time in torpor than do species with bodily fat reserves.

Despite its obvious advantages, there are ecological trade-offs associated with the energy savings of remaining in deep torpor for long periods versus the physiological costs of waste accumulation and body dehydration. Humphries et al. (2003) hypothesized that torpor expression should reflect a hibernator's energy reserves. Food supplementation studies support his *torpor optimization hypothesis* by indicating that hibernators alter the depth and duration of torpor in response to energy supply.

With its complex mix of behaviour and physiology, hibernation entails ecological consequences beyond its importance in energy savings. In many species, it is closely integrated with mating and reproduction. Hence, hibernation may have population consequences apart from winter survival of individuals. For example, the small brown bat (*Myotis lucifugus*) (Figure 7.19) travels considerable distances to reach its *hibernacula* (singular *hibernaculum*: location chosen for hibernation) in caves or abandoned mines. In fall, males and females mate promiscuously in a swarm near the cave entrance. Females store the sperm over winter, to be used in fertilization upon emergence in spring. The female's reproductive success is thus strongly tied to spring energy reserves, whereas the male's is not. Unlike most other small mammals, bats cannot alter their litter size when energy reserves are low because they produce only one offspring yearly.

Figure 7.19 The small brown bat practises both daily and seasonal torpor.

The torpor optimization hypothesis suggests that the fitness of female bats is maximized by conservative use of body reserves during hibernation. By increasing the depth and duration of torpor, female bats are more likely to retain enough reserves for successful reproduction. In a study of over 400 bats in Manitoba, females entered hibernation with greater fat reserves and depleted them more slowly than males or juveniles (Jonasson and Willis 2011). A further complication involves a link between torpor duration and white-nose syndrome, a fungal pathogen (*Geomyces destructans*) that is devastating bat populations across North America. The disease is thought to increase the length of time bats spend out of the torpor state, thus compromising winter survival. The more conservative energy use of the "thrifty female" may lower mortality from white-nose syndrome, but their spring reproductive success will likely suffer (Jonasson and Willis 2011).

Contrary to popular belief, black bears, grizzlies, and female polar bears are not true hibernators. Instead, they enter a unique winter sleep during which their body temperature only falls a few degrees below normal. During this period, bears do not eat, drink, urinate, or defecate, and females give birth to and nurse young. Yet they maintain a metabolism that, while lower than normal, is nowhere near as low as that of a true hibernator. They rouse easily from external stimuli but lack a true hibernator's need for spontaneous awakening, because they have not relaxed homeostatic temperature control and are in no danger of freezing. To remain sleeping, bears recycle urea (normally excreted in urine) through the blood, where it is degraded into amino acids that are reincorporated into plasma proteins.

Why don't bears hibernate? Their winter sleep conserves energy, but far less than hibernation would. The bear's large size makes true hibernation both less efficient and less feasible than a winter sleep, given the time needed to heat up and cool down. Hibernation is most beneficial to small homeotherms, for whom maintaining body temperature during periods of cold and limited food is costly. It is far more efficient to reduce metabolism by relaxing homeothermy, thus eliminating the need to obtain scarce food resources.

Hibernation is far less prevalent in polar than in temperate regions with shorter winters. There are relatively few true hibernators in the Arctic because winters are so long that hibernation severely restricts the time available for reproduction and rearing of young. An exception is the Arctic ground squirrel (*Urocitellus parryii*). During hibernation, which lasts from September to late April, they spontaneously and periodically arouse from and re-cool to core temperatures as low as −2.9°C without freezing (Barnes 1989). Core temperatures of mammalian hibernators in more temperate habitats typically fall to 1°C to 2°C before they awaken. The Arctic ground squirrel's low minimum temperature is likely due to supercooling (see Section 7.11).

7.11 TEMPERATURE EXTREMES: Animals Have Adaptations for Extreme Heat or Cold

Ectothermy and endothermy represent general strategies for heat gain, but animals have also evolved many specialized physiological and structural adaptations for maintaining thermal balance, particularly when faced with temperature extremes.

Given an animal's limited heat tolerance, retaining body heat does not seem like a sound option. But some mammals, such as the camel and oryx, do just that. The camel stores body heat by day and dissipates it by night. Its temperature can rise from 34°C to 41°C from morning to late afternoon. By storing heat, these animals reduce the need for evaporative cooling and conserve both water and food. Such controlled **hyperthermia** (rising of body temperature) illustrates how different abiotic factors—in this case, temperature as a condition and water as a resource—can interact in selecting for adaptive traits.

On the other end of the thermal scale, many cold-climate ectotherms withstand long periods of freezing temperatures through adaptations that either (i) avoid or (ii) tolerate ice formation. *Supercooling* of body fluids occurs when the body temperature falls below the freezing point without actually freezing, reflecting the presence of substances that reduce sites where ice crystals can form. (As in plants, *supercooling* is not to be confused with **freezing point reduction** as a result of higher solute content, another avoidance strategy of overwintering ectotherms, but one that offers less protection from extreme cold.) Some Arctic marine fish, as well as some temperate and cold-climate insects and reptiles, employ both supercooling and freezing point reduction by increasing the glycerol content of their body fluids. Many amphibians, including wood frogs (*Lithobates sylvaticus*), spring peepers (*Pseudacris crucifer*), and grey tree frogs (*Hyla versicolor*), overwinter beneath the leaf litter by accumulating glycerol.

The polar desert that dominates the Arctic landscape is an especially severe thermal environment, given its combination of extreme cold, short summers with variable temperatures, and limited water. Poikilotherms, which cannot compensate for cold by increasing metabolic heat, might seem poorly adapted to its rigours. Yet over 2200 insect and related species exist north of the treeline, with many more undiscovered (Danks

Figure 7.20 Larvae of the goldenrod gall fly (*Eurosta solidaginis*) tolerate much of the water in their tissues in the form of ice.

2004). This number is small relative to temperate fauna, but their range of adaptations is impressive (see also Section 7.7).

To cope with freezing temperatures, some Arctic poikilotherms use supercooling, which can provide protection up to −60°C. However, most survive cold not by avoiding but by tolerating freezing (Figure 7.20). These insects provide nucleating sites—the opposite of a supercooling strategy—that initiate freezing at a relatively high temperature (Duman 2001). Over 90 percent of their body fluids may freeze, with the remaining fluids highly concentrated in solutes. Ice forms outside their shrunken cells, distorting their muscles and organs. They thaw as temperatures rise, quickly regaining their normal shape. Arctic springtails survive winter by dehydrating their tissues, while the Antarctic nematode *Panagrolaimus davidi* even tolerates ice formation within its cells (Smith et al. 2008). These strategies of avoiding or tolerating freezing are comparable to the strategies of overwintering fungi and plants (see Section 6.11), both of which are also ectothermic.

Arctic insects also exhibit a range of behavioural responses relating to microhabitat selection. Some seek out protected sites, but some moth caterpillars seek out exposed ridges in order to take advantage of the earlier thaw (Danks 2004). In this case, the selective advantage of the longer period of sufficient warmth to complete their life cycle outweighs the drawback of the increased exposure to extreme cold.

For homeotherms, large size is itself an adaptation to cold. Altered body proportions also help retain heat. A concentrated body mass with small appendages such as ears retains heat, as in the Arctic fox (*Vulpes lagopus*) (Figure 7.21a, p. 148). The reverse holds in hot climates. The large ears of the Saharan desert fox (*V. zerda*) act as **thermal windows** to dissipate heat (Figure 7.21b). Insulating coverings also help, particularly if they vary in colour and thickness with the seasons.

To conserve heat and/or to cool vital body parts, many animals use *countercurrent heat exchange* (Figure 7.22, p. 148)—the same principle underlying O$_2$ retention in Section 7.4. The porpoise (*Phocaena* spp.) is insulated with blubber

(a)

(b)

Figure 7.21 The ears of the Arctic fox **(a)** are much smaller relative to its body size than those of the desert fox **(b)**.

for swimming in Arctic waters, but could lose much heat through its uninsulated flukes and flippers. It maintains its core temperature by exchanging heat between arterial and venous blood in these structures (Figure 7.23). Veins surround the arteries, allowing warm arterial blood from the lungs to lose heat to cool venous blood returning to the core. Little heat escapes to the environment. Blood entering the flippers cools, while blood returning to the core warms. In warm waters, where the porpoise must lose excess heat, blood bypasses these heat exchangers and returns through veins close to the skin, cooling the core.

Countercurrent heat exchange devices also keep heat out. The oryx (*Oryx beisa*), an African desert antelope, tolerates elevated body temperatures (another instance of hyperthermia), yet keeps its heat-sensitive brain cool with a rete. The external carotid artery passes through a cavernous sinus filled with venous blood that is cooled by evaporation from the moist mucous membranes of the nasal passages (Figure 7.24). Arterial blood passing through the sinus cools en route to the

brain, reducing its temperature to 2°C to 3°C below that of the body core.

Countercurrent heat exchangers are not restricted to homeotherms. Poikilotherms that practise limited endothermy employ the same strategy. Tuna (*Thunnus* spp.) and mackerel sharks (Lamnidae) possess a rete in a band of dark muscle tissue used for sustained swimming. Metabolic heat produced in the muscle warms the venous blood, which gives up heat to the adjoining oxygenated blood returning from the gills. Such countercurrent heat exchange increases the power of the muscles because warm muscles contract and relax more rapidly.

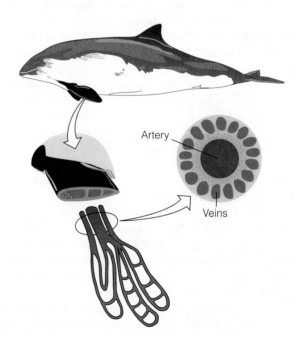

Artery

Veins

Figure 7.23 Porpoises and whales use flippers and flukes as temperature-regulating devices. Veins in these appendages surround the arteries. Venous blood returning to the body core is warmed through heat transfer, retaining body heat.

(Adapted from Schmidt-Nielson 1997.)

(a)

37°C   32°C   28°C

16°C   18°C   21°C   24°C

(b)

37°C   29°C   22°C

                    15°C

36°C   28°C   21°C

Figure 7.22 Countercurrent flow in a mammal's limb, showing hypothetical blood temperature in the **(a)** absence and **(b)** presence of countercurrent heat exchange.

(a)

Figure 7.25 The large hind feet of the snowshoe hare allow it to move effectively on snow.

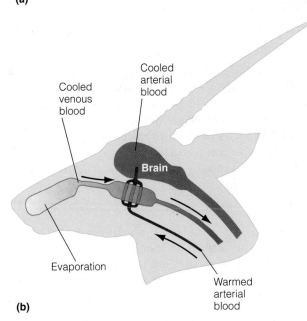

(b)

Figure 7.24 Countercurrent exchange as a cooling device. (a) The African oryx (*Oryx beisa*) keeps its brain cool despite a high core temperature by use of a rete. (b) Arterial blood passes through a pool of venous blood that is cooled by evaporation as it drains from the nasal region.

7.12 SNOW: Adaptations Allow Animals to Benefit from Snow Cover and Reduce Its Negative Impacts

With freezing temperatures comes snow accumulation. The amount varies, and can have various effects, as we saw for plants in Section 6.11. Species that have evolved in habitats with significant snow cover have evolved to take advantage of snow's beneficial effects while reducing its negative impacts by various structural and behavioural means.

Small mammals depend on snow for winter cover and are affected by its density as well as its depth. Non-hibernating species are especially vulnerable in years with little snow. Their small size makes heat retention difficult, but heavy dense snow-fall reduces O_2 supply. Some winter-active species (including the snowshoe hare, *Lepus americanus*, and predators such as the Arctic fox, *Vulpes lagopus*; see Figure 7.21) change their fur colour with the seasons—an example of a phenotypically plastic structural adaptation.

Snow depth can impede feeding by herbivores, but as long as the animal can access food, the benefits of snow in preserving plant tissue often outweigh the disadvantages. Rabbits benefit from deep snow if it lets them access bark at a greater height than they could normally reach. Many large herbivores, such as white-tailed deer, switch their diet in winter to browse young branches of trees and shrubs that they can access in deep snow. Snow also provides shelter, especially in forests, but a particularly heavy snow load not only reduces food supply but also makes escaping predators more difficult. The snowshoe hare has large hind feet that distribute its weight, letting it move quickly on top of snow rather than crashing through it (Figure 7.25). This adaptation reflects selection pressure from predation, but predators like the Arctic fox have evolved similar adaptations—an instance of coevolution (see Section 14.9).

Herbivores have evolved behavioural responses to the combined effects of snow and food supply. In southeastern Québec (at the northern limit of their range), white-tailed deer compensate for higher locomotion costs in deep snow by adjusting three variables: travel distance, forage intake, and cropping rate (Dumont et al. 2005). Deer are less selective in their feeding as winter progresses and snow depth increases. Energy, not nutrient content, determines their intake. They prefer deciduous twigs, consuming coniferous browse such as balsam fir (*Abies balsamea*) only when snow is especially deep. Cropping rate increases with the sinking depth in snow.

7.13 WATER BALANCE: Terrestrial Animals Are Constrained by Water Uptake and Loss

Living cells, be they plant, animal, or fungal, are 75 to 95 percent water. Water is involved in virtually all biochemical reactions and acts as a medium for excreting wastes and dissipating excess heat through evaporation. To stay hydrated, an organism must offset these losses by water uptake. The relationship between water

(a)

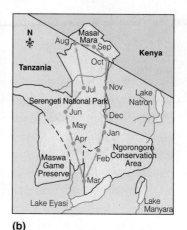

(b)

Figure 7.26 Migration as a drought-avoidance strategy. **(a)** Many large ungulates in semi-arid regions of Africa, such as the wildebeest (*Connochaetes* spp.), migrate over the year, following seasonal shifts in rainfall. **(b)** Seasonal migration of wildebeest populations in East Africa ensures access to food and water.

uptake and loss is an organism's *water balance*. There are numerous ways to maintain a water balance, but all involve a combination of increasing uptake and/or decreasing losses through structural, physiological, and behavioural adaptations.

Terrestrial animals gain water (and solutes) directly by drinking and eating, and indirectly by producing water during metabolism. They lose water and solutes through wastes (urine and feces), evaporation from the skin, and exhalation of moist air. Animals have many adaptations that manipulate these gains and losses. Birds and reptiles have a *cloaca*—a common receptacle for the digestive, urinary, and reproductive tracts, from which they reabsorb water into the body. Mammals have *kidneys* that conserve water by concentrating urinary wastes. This adaptation—which uses the familiar principle of countercurrent exchange, in this case to retain water—is strongly selected in dry climates. In relative terms, kidneys of desert animals have much longer loops of Henle than do kidneys of animals from moister environments.

In arid habitats, animals face severe water balance problems, which they solve either by evading drought or by avoiding or tolerating its effects. Some animals, including African *ungulates* (hooved mammals; Figure 7.26) and many birds, evade drought by leaving during a dry season. Some ectotherms practice **estivation**, a dormant stage induced by drought. The spadefoot toad (*Scaphiopus couchii*) remains dormant belowground, emerging only when the rains return. Some pond invertebrates, such as the flatworm *Phagocata vernalis*, retreat into hardened casings if the pond dries up. Other semi-aquatic animals retreat deep into the soil until they reach groundwater. Many insects undergo **diapause**, a seasonal dormant stage from which they emerge when conditions improve (see Section 7.16).

Others remain active in the dry season but avoid drought effects by reducing water loss by slowing their breathing. Small desert rodents cool their exhaled air by passing moist air from the lungs over cooled nasal membranes, leaving condensed water on the walls. This water humidifies and cools the warm, dry air that the rodent inhales. Controlled hyperthermia, which sets the body's thermostat to a higher temperature, further reduces evaporative losses.

There are still other adaptive solutions to the water balance problem. Small desert mammals, such as the gerbil (*Dipodillus*

campestris), native to Middle Eastern deserts, reduce water loss by being nocturnal. By remaining in burrows by day, they avoid activity during the hottest period, thus conserving water by reducing evaporative cooling. Upon returning to their burrows, they cool their bodies rapidly by direct conductive transfer with the ground, further conserving water.

Many desert mammals extract water from the food they eat, either directly from consumed tissue or indirectly from metabolic water produced during respiration. The camel's hump stores not water but fat, which yields copious metabolic water when oxidized. Desert mammals also produce highly concentrated urine and dry feces. In addition to avoiding drought effects by conserving water, some desert mammals tolerate dehydration. Desert rabbits withstand water losses of up to 50 percent and camels of up to 27 percent of their body weight.

7.14 OSMOTIC BALANCE: Aquatic Animals Face Problems of Water and Salt Balance

Aquatic animals face continual exchange of water with their surroundings through *osmosis*. As with passive water transport in plants (see Section 6.4; water potential terminology is not normally used with animals because their lack of a cell wall means there is no need to factor in pressure potential), osmotic gradients move water through cell membranes from the side of greater water concentration (fewer solutes) to that of lesser water concentration (more solutes).

Freshwater animals are **hyperosmotic**—the salt concentration in their bodies is higher than that of their surroundings. Water enters while salts exit. Because animals lack a cell wall that prevents bursting, they must either limit water uptake or rid themselves of excess water while replacing lost salts. Freshwater fish maintain osmotic balance by absorbing and retaining salts in special cells in the gills and by producing large amounts of watery urine. Amphibians balance salt loss through their skin by absorbing ions directly from water and transporting them across the skin and gill membranes. In their terrestrial stage, amphibians store water from the kidneys in the bladder and, if needed, reabsorb it across the bladder wall.

Marine fish face the opposite problem. These organisms are **hypoosmotic**, with a lower body salt content than the surrounding water. Osmosis draws water out of the body, increasing their risk of dehydration. In marine and brackish (mixed fresh- and saltwater) habitats, organisms must inhibit water loss and prevent excessive salt accumulation. Marine invertebrates solve this problem by being **isoosmotic**, allowing the osmotic pressure of their cells to fluctuate with their surroundings. In contrast, marine bony (teleost) fish absorb saltwater into the gut and secrete magnesium and calcium through the kidneys as a semi-crystalline paste.

Typically, fish excrete sodium and chloride by pumping them across special membranes in the gills. This process involves active transport to move salts against their gradient and has a high energy cost. Sharks and rays solve the problem differently. They retain enough urea to maintain a higher concentration of solutes in the body than in the surrounding seawater. Seabirds and sea turtles can consume seawater because they possess salt-secreting nasal glands. Seabirds of the order Procellariiformes (albatrosses, shearwaters, and petrels) excrete fluids of over 5 percent salt from these glands, either by forcibly ejecting fluids through the nostrils or by dripping fluids through the nares. In marine mammals, the kidney is the main route for eliminating salts.

7.15 WATER DEPTH: Aquatic Animals Have Specialized Adaptations to Deep Water

Deep-water portions of aquatic habitats present special problems relating to light and pressure. Organisms inhabiting depths of 200 to 1000 m are typically silvery grey or deep black, whereas those occupying deeper waters (below 1000 m) often lack pigment entirely. Another adaptation in deep water is large eyes, which maximize light-gathering ability. Flatfishes (Heterosomata) have protruding eyes and an interesting type of asymmetry. One eye migrates to the other side of the head during development, so that both eyes face upward in the mature organism. Some organisms have evolved organs that produce their own light through chemical reactions known as **bioluminescence**.

Aquatic animals have evolved various strategies to stay at a desired depth. Most marine animals maintain neutral buoyancy by having overall body densities similar to that of seawater. Because living tissues are denser than water, animals have lower-density areas to counter their higher-density tissues. Most fish have a **swim bladder** (Figure 7.27a) occupying 5 to 10 percent of body volume. These fish regulate the amount of gas in the bladder, allowing them to adjust their depth. Lungs in air-breathing aquatic animals are another way to maintain neutral buoyancy.

Some marine animals maintain neutral buoyancy by replacing heavy chemical ions in body fluids with lighter ions. Squid (*Loligo* spp.) have body cavities in which lighter ammonium ions replace heavier sodium ions. Their body fluid is less dense than the same volume of seawater. Another mechanism involves storing lipids, which are less dense than seawater.

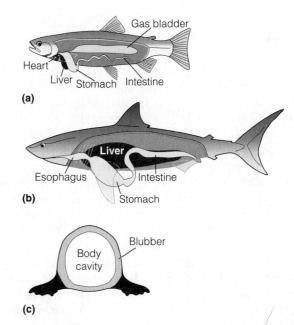

Figure 7.27 Buoyancy adaptations of animals. **(a)** Gas bladder in a fish. **(b)** Large, fat-filled liver of a shark. **(c)** Blubber surrounding the body cavity of a seal.

Fishes that lack swim bladders (e.g., sharks, mackerels, and bluefish) store large lipid deposits in their muscles, internal organs, and body cavity (Figure 7.27b).

Many marine mammals have **blubber**, a fat layer just below the skin (Figure 7.27c). Besides aiding in buoyancy, blubber provides insulation and energy storage. For seals that dive to great depths, blubber is beneficial because it retains its insulating properties with depth. In contrast, hair is an ineffective insulator at the increased pressures experienced in deep water. Blubber is a trait that is adaptive for a specific set of abiotic factors. Although it might seem the ideal insulator in water, it too has trade-offs; it must be shed periodically on land, increasing the animal's predation risk.

Maintaining neutral buoyancy in water has the added benefit of relieving the constraints imposed by gravity on land. Jellyfish, squid, and octopus quickly lose their form when removed from water. Larger animals benefit the most. Beached whales suffocate when they can no longer support their body weight. It is no coincidence that the largest animals inhabit the seas.

7.16 PERIODICITY: Daily and Seasonal Light Cycles Influence Animal Activity

Although light is an important indirect resource for animals, the major direct influence of light on animals is as a condition, in its role in vision but also in the timing of daily and seasonal activities such as feeding, food storage, reproduction, and migration. In Section 6.14, we discussed *biological clocks* as internal devices that control periodicity. In animals, these devices influence hormones that regulate processes as diverse as the sleep cycle and body temperature control.

As do plants, animals exhibit many daily rhythms that reflect physiological responses to external environmental cycles. Many involuntary processes in animals exhibit these circadian rhythms, including body temperature control, cardiovascular function, secretion of *melatonin* (a hormone regulating the sleep–wake cycle) and *cortisol* (primary stress hormone), and metabolism. Circadian rhythms are controlled by the biological clock. But where is the clock located? Although it varies among organisms, its location must expose it to its time-setter: light.

Circadian rhythms include three elements: (1) an *input pathway* that transmits the light signal, (2) a *pacemaker* (clock), and (3) an *output pathway* by which the clock regulates the rhythm. In mammals, the input pathway begins with photoreceptors in the retina of the eye. These cells, which contain a photosensitive pigment, transmit light information to the clock: the *suprachiasmic nuclei* (SCN), a group of cells in the hypothalamus region of the brain. The SCN interprets and transmits the information to the *pineal gland*, located in the brain's epithalamus region. The pineal gland provides the output pathway by secreting melatonin, which communicates the signal to various parts of the body, affecting functions from sleep to reproduction. Secretion of melatonin peaks at night and declines during the day, entraining the circadian rhythm.

The eye is thus the photosensitive organ in mammals, but the pineal gland of other vertebrates, such as birds and some lizards, is located on the brain surface, directly under the skull. In these species, the pineal gland contains photoreceptors that act as the input pathway. Lizards, frogs, lampreys, and some fish have a **parietal eye**, a photosensory organ connected to the pineal gland (Figure 7.28). In some fish, such as tuna and pelagic sharks, the parietal eye is a light-sensitive spot on top of the head. A less-developed version, the *parapineal gland*, occurs in salamanders.

As intriguing as the physiological mechanism is, ecologists focus on the adaptive value of biological clocks, which provide a time-dependent mechanism enabling the organism to prepare for predictable environmental changes. Circadian rhythms help organisms cope with more than light or dark. For example, a rise in humidity and a drop in temperature often accompany the transition from light to dark. Anyone who turns over a stone or rotting log discovers wood lice, centipedes, and millipedes. They lose water rapidly in dry air and so spend the day in these dark, damp microhabitats. At dusk they emerge, when the air is more humid. Typically, the strength of their response to low humidity decreases with darkness. Thus, they safely emerge at night in places too dry for them by day, and as quickly retreat as light returns.

Circadian rhythms may relate to an organism's biotic as well as its abiotic environment. Insectivorous bats and birds must match their feeding activity to the activities of their prey. Moths and bees must seek nectar when flowers are open. Flowers must open when insects that pollinate them are flying. The circadian clock lets these and other organisms coordinate many aspects of their daily activities, as well as make most efficient use of available resources.

Periodicity is not confined to diurnal responses. In middle and upper latitudes, the *photoperiod* (daily period of light and dark) lengthens and shortens over the year. Activities of animals (as well as of plants and fungi) are geared to these seasonal changes. The flying squirrel (*Glaucomys volans*), a more common species in forests across eastern Canada than most realize as a result of its nocturnal habit, becomes active at dusk. As spring days lengthen, it becomes active a little later each day (DeCoursey 1960) (Figure 7.29). Experimental studies indicate that photoperiod also synchronizes flying squirrel reproduction and growth (Lee and Zucker 1990).

Most animals of temperate regions have reproductive periods that track seasonal changes in photoperiod. For most birds, the breeding season corresponds to the lengthening days of spring, whereas for deer, the mating season is the shortening days of fall. As with plants, the signal for these responses is *critical day length* (see Section 6.14). *Diapause*, the hibernating stage of arrested growth over winter in insects of temperate regions, is a complex example of photoperiodism

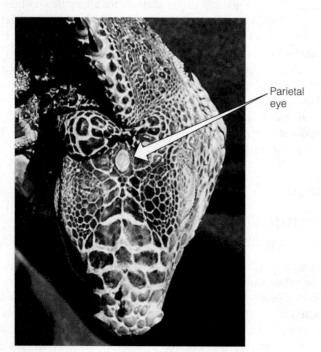

Parietal eye

Figure 7.28 The iguana has a parietal eye—a photosensory organ connected to the pineal gland.

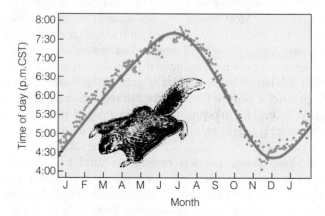

Figure 7.29 Onset of activity in flying squirrels.
(Adapted from DeCoursey 1960.)

in animals. Time measurement is precise, usually between 12 and 13 hours of light. A 15-minute difference determines if an insect enters diapause. The shortening days of late summer induce diapause, whereas lengthening spring days signal the insect to resume development, pupate, emerge as an adult, and reproduce. Diapause and hibernation allow organisms to avoid an unfavourable period when resources are limiting and conditions are harsh.

Photoperiodic responses are multi-faceted. In birds, longer days induce migration, stimulate gonad development, and initiate reproduction. After the breeding season (which may involve multiple clutches), the gonads regress spontaneously and no longer respond to photoperiod. In mammals such as white-tailed deer, photoperiod influences a range of activities from food storage to reproduction (Figure 7.30). Melatonin initiates the reproductive cycle. More melatonin is produced in darkness, so deer receive a higher dose in the shortening days of fall. This increase reduces the sensitivity of the pituitary gland to negative feedback effects of hormones from the ovaries and testes. Lacking this feedback, the pituitary releases pulses of *luteinizing hormone*, which stimulates sperm production and the growth of ova.

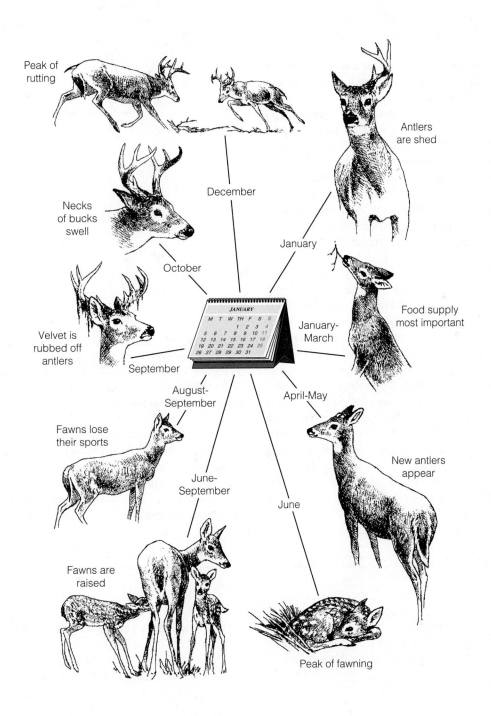

Figure 7.30 Activities affected by photoperiod in white-tailed deer.

Animal activities reflect these and other seasonal responses to changing day length, known as **circa-annual rhythms**. Such rhythms often relate to food supply. White-tailed deer mate in fall, and the young are born in spring when the highest-quality food for lactating mothers and young is available. In tropical rain forests of Central America, home to many species of frugivorous bats, bat reproductive activities track seasonal availability of food. They give birth during the peak fruiting period. In contrast, insectivorous bats occupying Costa Rican forests give birth earlier in the rainy season, when insects and other arthropods reach their greatest biomass. However, animals may be responding to factors that are confounded with photoperiod. In Mexico's Yucatan peninsula, rainfall seasonality appears to be the most important factor affecting reproductive cycles of fruit-eating bats (Monteil et al. 2011).

Periodicity applies to conditions other than photoperiod. Tidal cycles are predictable features of coastal habitats, and many intertidal species—from diatoms, green algae, crustaceans, and periwinkles, to fish such as blennies (*Malacoctenus,* spp.) and cottids (Cottidae)—respond to both tidal and circadian cycles. Fiddler crabs (*Uca* spp.) swarm across the exposed mud of salt marshes and mangrove swamps at low tide, retreating to burrows at high tide. In a laboratory with constant temperature and light and without tidal cues, they exhibit the same rhythm (Figure 7.31) (Palmer 1990), mimicking tidal ebb and flow every 12.4 hours. In constant conditions, the crabs also exhibit a circadian rhythm of changing colour, turning dark by day and light by night.

But is the fiddler crab's clock (1) unimodal, with a 12.4-hour cycle, or (2) bimodal, with a 24.8-hour cycle, close to the period of the circadian clock? If bimodal, does one clock keep a solar-day rhythm of about 24 hours and another a lunar-day rhythm of 24.8 hours? Palmer's results suggest that a single solar-day clock synchronizes daily activities, while two tightly coupled lunar-day clocks synchronize tidal activity. Each lunar-day clock controls its own tidal peak. If one clock quits running in the absence of external cues, the other clock still runs. This feature allows tidal organisms to synchronize their activities in a variable tidal environment. Day–night cycles reset solar-day rhythms, and tidal changes reset tidal rhythms. Even at the cellular level, organisms do not depend on one clock any more than we keep a single clock at home. Organisms have built-in redundancies that enable the clocks to run at different speeds, governing different processes with slightly differing periodicities.

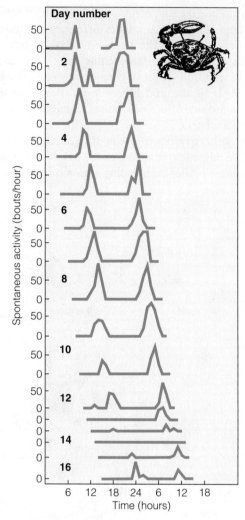

Figure 7.31 Tidal rhythm of a fiddler crab held in constant light and temperature (22°C) for 16 days. Because the lunar day is 51 minutes longer than the solar day, the tides occur 51 minutes later each solar day, causing peaks of activity to move to the right.

(Adapted from Palmer 1990.)

Why have we elaborated on this example? The point is not to memorize the details of tidal and circadian rhythms in fiddler crabs, but to illustrate the astounding complexity and fine-tuning that has evolved in the responses of many organisms—even one so "primitive" as a fiddler crab—to environmental cues, particularly those that correlate with changes in critical abiotic factors.

EcologyPlace

SUMMARY

Body Size 7.1

Size has critical consequences for animal structure and function and exerts a fundamental constraint on their evolution. For objects of similar shape, the ratio of surface area to volume decreases with increasing body size, limiting exchange of materials and energy. Animals have evolved many adaptations to facilitate exchange between interior cells and the environment.

Food Acquisition 7.2

Animals are classified based on the type of food they eat. Herbivores consume plants; they rarely face supply problems, but low digestibility and nutrient content can depress their growth, development, and reproduction. Herbivores have evolved various strategies for digesting plant tissues, including mutualistic associations with bacteria. Carnivores consume animals. They have few problems with food quality but often face shortages. Omnivores, which eat both plants and animals, have less specialized feeding adaptations. Detritus feeders consume dead organic matter.

Nutritional Requirements 7.3

Animals have similar mineral needs as plants. Nitrogen is critical, especially for herbivores. Sodium, calcium, and magnesium often influence distribution, growth, and reproduction of grazing animals. Grazers use mineral licks when vegetation contains insufficient levels.

Oxygen Uptake 7.4

Animals generate energy by oxidizing organic compounds through aerobic respiration. Terrestrial and aquatic animals differ in their means of acquiring O_2. Insects use tracheae, whereas most terrestrial vertebrates have some form of lungs. Fish use gills to absorb O_2 from water using countercurrent exchange.

Homeostasis 7.5

To cope with daily and seasonal environmental change, many animals practise homeostasis: maintaining a relatively constant internal environment in a variable environment by means of negative feedback. Using sensory mechanisms, an organism responds physiologically and/or behaviourally to maintain an optimal internal environment.

Thermal Exchange 7.6

All animals gain and lose heat energy via conduction, convection, radiation, and evaporation. They use behavioural and physiological means to maintain their heat balance in a variable environment. Muscle, fat, and body coverings insulate the body core. Land animals must cope with a more variable thermal environment than aquatic animals. There are three main strategies. Poikilotherms are ectothermic and have variable body temperatures. Homeotherms maintain a constant body temperature by relying on internally produced metabolic heat (endothermy). Heterotherms function as endotherms or ectotherms, depending on circumstances.

Poikilothermy 7.7

Poikilotherms have low metabolic rates and high conductance. External temperature controls their metabolism. Except for species adapted to polar habitats, they are active only at moderate temperatures and are sluggish at cool temperatures. Fish maintain little difference between body and water temperatures and can acclimate to gradual temperature change. Many poikilotherms use behaviour to regulate body temperature. They exploit microclimates by basking or by seeking shade, or by moving in and out of water. Insects and desert reptiles alter heat exchange by raising or lowering their bodies. Small desert animals often spend hot periods in burrows.

Homeothermy 7.8

Homeotherms maintain body temperature by oxidizing energy-rich compounds. They have high metabolic rates and low conductance. Insulating body coverings reduce heat loss. Some desert mammals use fur to keep out both heat and cold. Many homeotherms shiver or burn fat reserves. Evaporative cooling by sweating, panting, and/or wallowing dissipates body heat.

Trade-offs in Thermal Regulation 7.9

Both poikilothermy and homeothermy involve trade-offs. Although homeotherms remain active in a wider range of external temperatures, they have high metabolic costs, which impose lower limits on body size. Larger body size may confer an advantage in colder climates. Given the low energy costs of ectothermy, poikilotherms can inhabit areas with limited food and water, and can curtail metabolic activity when food is scarce or temperatures extreme. Because they rely on heat exchange, terrestrial poikilotherms have strict maximum size constraints.

Heterothermy 7.10

Heterotherms practise both endothermy and ectothermy. Many flying insects increase their metabolism to generate heat prior to flight. Some homeotherms become ectothermic either daily or seasonally to reduce energy costs. In seasonal torpor (hibernation), metabolism, heartbeat, and breathing decline greatly, and body temperature drops below 10°C. The benefits to hibernators in energy savings are offset by physiological costs. Hibernators alter their torpor expression in response to energy supply. Bears are not true hibernators. Their metabolism drops greatly during their winter sleep, but body temperature stays close to normal.

Temperature Extremes 7.11

Many homeotherms employ countercurrent circulation to exchange body heat between arterial and venous blood reaching the extremities. This exchange works both ways, reducing heat loss or cooling blood flowing to vital organs. Some desert mammals use hyperthermia to reduce their need for evaporative cooling. Some invertebrate poikilotherms avoid freezing by supercooling by accumulating glycerol in their body fluids.

Others tolerate ice formation outside their cells. Arctic poikilotherms employ a variety of adaptations to extreme cold.

Snow 7.12

Animals have evolved to take advantage of snow's beneficial effects while reducing its negative impacts by structural and behavioural adaptations. Snow provides winter cover for small mammals. Deep snow can impede feeding by herbivores and affects predation. Some animals change colour in winter. Others have large hind feet for moving on snow.

Water Balance 7.13

Terrestrial animals offset water loss from evaporation, respiration, and waste excretion by consuming and/or conserving water. Terrestrial animals gain water by drinking, eating, and producing metabolic water. Animals of arid regions reduce water loss by various strategies, including nocturnal activity, producing highly concentrated urine and dry feces, generating metabolic water, practising hyperthermy, and tolerating dehydration of body tissues.

Osmotic Balance 7.14

Aquatic animals must maintain optimal tissue levels of both water and salts. Freshwater fish maintain osmotic balance by absorbing and retaining salts in specialized cells and rid themselves of excess water by producing copious watery urine. Many

marine invertebrates maintain their cells at the same osmotic pressure as seawater. Marine fish drink water and secrete excess salt and other ions through kidneys or across gill membranes.

Water Depth 7.15

Some animals living in deep water have large eyes to capture more light. Others generate light by bioluminescence. Aquatic animals have little need for structural adaptations to support themselves against gravity, but they have evolved adaptations to maintain neutral buoyancy at their preferred depth, including swim bladders and lipid deposits. Blubber is a more effective insulator in deep water than hair or fur.

Periodicity 7.16

Most organisms have innate circadian rhythms that are synchronized to a 24-hour day by light cues. Onset and cessation of activity depend on whether the organism is day or night active. Circadian rhythms in animals are regulated by biological clocks in the brain. In the dark, animals produce more melatonin, which measures day length. Seasonal activities such as reproduction and migration are controlled by photoperiod. These circa-annual rhythms initiate reproduction at the most favourable season for offspring survival. Activities of intertidal animals respond to daily rhythms, controlled by a solar-day clock, and tidal cycles, controlled by two lunar-day clocks.

KEY TERMS

adipose fat	detritus	heliothermy	lung	scaling
Bergmann's rule	detritus feeder	herbivore	mineral lick	shivering
bioluminescence	diapause	heterotherm	nectivore	spiracle
blubber	ectothermy	hibernation	negative feedback	surface area
browser	endothermy	homeostasis	omnivore	to volume ratio
carnivore	estivation	homeostatic	operative temperature	swim bladder
circa-annual rhythm	freezing point	plateau	range	thermal
conduction	reduction	homeothermy	parietal eye	conductivity
convection	frugivore	hyperosmotic	poikilothermy	thermal window
coprophagy	gill	hyperthermia	positive feedback	thermoneutral zone
countercurrent	granivore	hypoosmotic	rete	torpor
exchange	grazer	hypothermia	ruminant	trachea
critical temperature	heat transfer coefficient	isoosmotic	sap-feeder	

STUDY QUESTIONS

1. Why does the surface area to volume ratio of an organism decrease with increasing size?
2. What are the major dietary constraints facing (i) herbivores and (ii) carnivores?
3. Digestion of non-ruminant herbivores is less efficient than that of ruminants. Why might they be at a selective advantage over ruminants in some ecological situations? Hint: Consider not only digestive efficiency but also feeding behaviour and predation risks.
4. Describe two differences between the ways ruminants and lagomorphs digest food.

5. In suburban areas, people often put out high-quality plant food for deer in winter. Why might this be a bad idea from (i) a nutritional and (ii) a behavioural standpoint?

6. Humans and other primates cannot synthesize vitamin C. What does this fact suggest about our diet in the natural state?

7. What is homeostasis? What is the general role of negative feedback in homeostasis?

8. In Figure 7.16, why does the mass-specific metabolic rate increase with decreasing body mass? How would this relationship change if metabolic rate were not mass-specific?

9. Why are the largest poikilotherms found in tropical and subtropical regions? How and why does mean size of related homeotherms vary from the tropics to the poles?

10. Why might it be easier to capture a snake in the early morning than in midday?

11. Why do homeotherms have more insulation than poikilotherms even in warm climates?

12. How does body size and shape influence an animal's ability to exchange heat with its environment? How does this factor influence size constraints of ectotherms versus endotherms?

13. How and why do poikilotherms and homeotherms differ in their metabolic response to temperature? Hint: Compare the *operative temperature range* with the *thermoneutral zone*.

14. Plants practise aerobic respiration and are rich in carbohydrates, yet it is more ecologically advantageous for them to be poikilothermic than homeothermic. Why?

15. What behaviours help poikilotherms maintain a fairly constant temperature while active?

16. Contrast the problem of maintaining water balance for freshwater and marine fishes.

17. How does supercooling enable some insects, amphibians, and fish to survive freezing?

18. Explain the principle of countercurrent exchange, and several different animal adaptations that employ it.

19. What are the ecological trade-offs involved in seasonal torpor (hibernation)?

20. Consider a fish population living below a power plant that discharges heated water. The plant shuts down for three days in winter. How and why would this affect the fish?

21. Both the oryx and the gerbil are desert-dwelling mammals. Compare their strategies for coping with drought. Consider how they manipulate their water budget, and the structural, physiological, and/or behavioural adaptations that they have evolved.

22. Compare *circadian* and *circa-annual rhythms*. Why do some animals have both? What is the adaptive advantage of relying on photoperiod as a cue for seasonal activity?

FURTHER READINGS

Bonner, J. T. 2006. *Why size matters*. Princeton: Princeton University Press.
Excellent overview of the constraints imposed by body size in animal evolution and ecology.

Danks, H. 2004. Seasonal adaptations in Arctic insects. *Integrative and Comparative Biology* 44:85–94.
Up-to-date review of adaptations of Arctic insects to cold temperature and other seasonal stresses, written by a scientist at the Canadian Museum of Nature in Ottawa.

French, A. R. 1988. The patterns of mammalian hibernation. *American Scientist* 76:569–575.
Excellent, easy-to-read, and well-illustrated overview of hibernation.

Heinrich, B. 1996. *The thermal warriors: Strategies of insect survival*. Cambridge: Harvard University Press.
This enjoyable book describes the variety of thermal strategies used by insects. It is full of strange and wonderful examples of insect evolution.

Schmidt-Neilsen, K. 1997. *Animal physiology: Adaptation and environment*. 5th ed. New York: Cambridge University Press.
This comprehensive text gives more information about animal physiology and adaptations in various environments.

Storey, K. B., and J. M. Storey. 1996. Natural freezing survival in animals. *Annual Review of Ecology and Systematics* 27:365–386.
Excellent review of the adaptations that allow animals to cope with freezing temperatures.

Takahashi, J. S., and M. Hoffman. 1995. Molecular biological clocks. *American Scientist* 83:158–165.
Highly readable overview of the mechanisms involved in biological clocks.

A female brown bear (*Ursus arctos*) with her cubs. Litter size ranges from one to four cubs, born between January and March. Cubs remain with their mothers for over two years, so females can only breed at most every three years. Brown bears inhabit Eurasia as well as North America.

An organism's **life history** is its lifetime pattern of growth, development, and reproduction. Life histories combine adaptations involving behavioural, physiological, and morphological responses to both the abiotic and biotic environment—an organism's interactions with others. Perhaps the most critical life history traits concern reproduction, because they link the individual with the next rung on the ecological hierarchy—the population.

Evolution results from differential reproduction of individuals, as influenced by natural selection and other processes (see Section 5.6). The ultimate measure of reproductive success is *fitness*: the relative number of offspring that survive to reproduce. Imagine an organism designed to maximize fitness. It would reproduce as soon as possible and continuously thereafter, producing many large offspring that it would nurture and protect. Such an "ideal" organism is not possible. Each individual has limited resources to allocate to specific tasks, and allocation to reproduction reduces allocation to growth. Given this reality, should an individual reproduce early, or delay reproduction? For a given reproductive allocation, should it produce many small, or fewer but larger, offspring? How much energy should it allocate to care of its young?

This dilemma shows that organisms face trade-offs in life history traits relating to reproduction, just as they do for traits relating to carbon, water, and energy. These trade-offs reflect constraints imposed by physiology, energetics, and the environment. In this chapter, we explore these trade-offs and the strategies that have evolved to maximize success at the one task essential for continuing life: reproduction.

8.1 TYPES OF REPRODUCTION: Reproduction May Be Sexual or Asexual

In Chapter 5, we explored how genetic variation within a population arises from assortment and recombination of chromosomes in sexual reproduction. Given the enormous number of possible recombinations, sexual reproduction is the main source of genetic variability upon which selection acts. Yet not all reproduction is sexual. Many organisms (including many that reproduce sexually) reproduce asexually, producing offspring with no involvement of gametes.

Asexual reproduction takes many forms, but the offspring are always genetically identical clones of the parent. The protozoan *Paramecium* simply divides in two (**binary fission**), as do bacteria. Freshwater hydras reproduce by **budding** (Figure 8.1), in which a bud pinches off as a new individual. In spring, wingless female aphids emerge from eggs produced by other wingless females through **parthenogenesis**, in which the ovum develops into an organism without fertilization. Some form of parthenogenesis is common in insects and other invertebrates, including crayfish (Scholtz et al. 2003). It has even been reported in a vertebrate, the Komodo dragon lizard (*Varanus komodoensis*) (Watts et al. 2006).

Figure 8.1 Asexual budding of freshwater hydra.

Most fungi produce both sexual and asexual spores. Virtually all plants use asexual reproduction, but usually involving vegetative structures, not unfertilized eggs. Strawberry (*Fragaria* spp.), both the native species inhabiting forests across North America and its cultivated forms, spreads by *stolons*, aboveground runners from which new shoots sprout. Weeds such as quackgrass (*Elytrigia repens*) often spread by underground runners called *rhizomes*. Some plants, including dandelion (*Taraxacum officinale*) and the bluegrass (*Poa* spp.) that grows alongside it in most city lawns, produce unfertilized seeds via *apomixis*, a type of parthenogenesis (see Section 5.10).

Many plants maximize their ecological flexibility by reproducing both sexually and asexually simultaneously. Trembling aspen flowers and sets seed while sending up suckers from its roots (see Figure 9.1). Pussytoes (*Antennaria parlinii*) is a common herbaceous perennial in eastern North America. Of polyploid origin (see Section 5.10), it contains both sexual and apomictic races. Apomictic plants produce more but smaller seeds (Figure 8.2a, p. 160) that are less likely to survive as seedlings (Bazzaz 1996) but are better adapted for dispersal. In high light, the population invests more in apomictic reproduction (Figure 8.2b), increasing its colonization potential.

Species that usually reproduce asexually occasionally revert to sexual reproduction, often induced by environmental change. During warm periods, hydras reproduce sexually to produce overwintering eggs. The young hydra that emerge in spring then reproduce asexually. Aphid reproduction is particularly complex, involving a mix of sexual and asexual reproduction that is tied to the seasons. Details vary among species, but most offspring are asexual. Females undergo a modified meiosis to produce many generations of cloned wingless females by *vivipary* (live birth). With declining food quantity or quality, females may produce a generation of winged females that migrate to other plants, establish, and again reproduce asexually.

In fall, aphids typically switch to sexual reproduction. Females again give birth asexually, but to sexual forms—

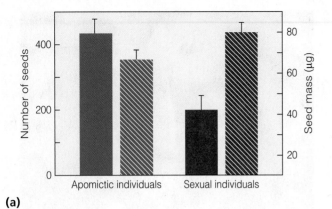

(a)

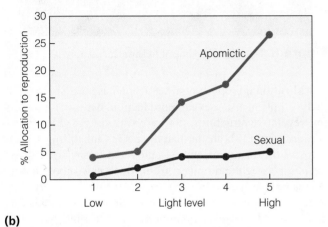

(b)

Figure 8.2 Reproduction in *Antennaria parlinii.* **(a)** Apomictic individuals produce more seeds, but total seed mass is greater in sexual individuals. Solid bars indicate seed number; hatched bars indicate seed mass. **(b)** Apomictic investment in reproduction increases with light.

(Adapted from Bazzaz 1996, as modified from Michaels and Bazzaz 1989.)

males that are identical to their female parent except for having a single X chromosome, and females that practise *ovipary* (egg laying), laying overwintering eggs. Because males can only produce sperm with an X chromosome, the fertilized eggs that hatch the next spring develop into wingless females that produce more females asexually. In this case, sexual reproduction is more vital for generating new genetic combinations than for maintaining population size, which is achieved asexually.

Both types of reproduction have advantages and drawbacks. The ability to survive, grow, and reproduce indicates that an organism is well adapted to prevailing conditions. Its asexual offspring will share its combination of locally adapted genes. Asexual reproduction also allows rapid population growth. Its trade-off is low genetic variability. The population may respond more uniformly to a changing environment than would a sexually reproducing, variable population. If an environmental change is detrimental, the effect can be catastrophic, as with the susceptibility of genetically uniform crops such as corn to pest outbreaks.

In contrast, each individual in a sexually reproducing population is unique. This variability allows a wider range of responses, making it more likely that some individuals will survive environmental change. In evolutionary terms, it guarantees more material upon which natural selection can act. But this variability comes at a cost. Only half the genetic material of each offspring comes from each parent. By chance, the particular gene combination present in one or both parents might be better suited to local conditions than are the combinations in their offspring.

Sexual reproduction also requires specialized reproductive organs that may be neutral to, or even impair, survival. Gamete production, courtship, and mating are energetically expensive and may increase the risk of predation or weaken competitive ability. Nor are reproductive expenses shared equally. Eggs are larger and more costly than sperm, whereas males often face heavy courtship costs. As we will soon discover, such sex-related differences in investment and risk have important implications for the evolution of life history traits.

In harsh climates, asexual reproduction can be a mainstay if growing seasons are too short to complete sexual processes. Tundra plants often do not flower every year, spreading instead by clonal reproduction. Danks (2004) notes that parthenogenesis is very common in Arctic insects, including mayflies, caddisflies, and midges. He argues that asexual reproduction is particularly important in favourable years, when sexual reproduction in rapid-cycling species might foster short-term natural selection for genetic recombinations that might be ill adapted when harsh conditions return. Interestingly, the Arctic aphid *Acyrthosiphon svalbardicum* is less reliant on parthenogenesis than are more temperate aphid species. The females that hatch from overwintering, sexually generated eggs immediately produce sexual forms parthenogenetically, instead of bearing several generations of wingless females viviparously (Strathdee et al. 1993).

8.2 MODES OF SEXUAL REPRODUCTION: Sexual Organisms Are Unisexual or Hermaphroditic

There are several modes of sexual reproduction. **Unisexual** species have separate male and female individuals, as do most animals. Unisexual plants are called **dioecious** (Figure 8.3a; e.g., ash trees [*Fraxinus* spp.]).

Individuals of **hermaphroditic** species possess both male and female organs. Plants can be hermaphroditic in two ways: (1) Some have **bisexual** (*perfect*) flowers, with both male organs (*stamens*) and female organs (*carpels*) (Figure 8.3b; e.g., lilies [*Lilium* spp.]). Carpels contain *ovaries*, which develop into *fruits*. Self-fertilization can occur, but if sperm and eggs mature at different times, crossing is more likely. (2) Others are **monoecious**, with separate male and female flowers on the same plant (Figure 8.3c; e.g., birch [*Betula* spp.]). Such flowers are *imperfect*, as are flowers of dioecious species. Hermaphroditism can be ecologically adaptive. A single self-fertilized hermaphrodite can quickly establish a population in a new habitat.

Hermaphroditic animals are bisexual, with both male and female sex organs (*testes* and *ovaries*). This condition is

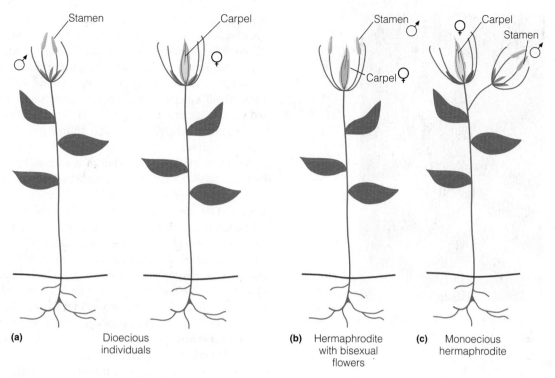

(a) Dioecious individuals

(b) Hermaphrodite with bisexual flowers

(c) Monoecious hermaphrodite

Figure 8.3 Modes of sexual reproduction in flowering plants. **(a)** Dioecious plant with separate male and female flowers on different individuals. **(b)** Hermaphroditic plant with bisexual flowers. **(c)** Monoecious plant with separate male and female flowers on one individual.

common in invertebrates such as earthworms (Figure 8.4). In *simultaneous hermaphrodites*, the male organ of one individual mates with the female organ of the other, and vice versa. The population can thus produce twice as many offspring as a unisexual population.

Sequential hermaphrodites, such as some molluscs in the Gastropoda (snails and slugs) and Bivalvia (clams and mussels) groups, undergo a sex change as individuals grow larger, typically changing from male to female. An altered population sex ratio (see Section 9.7) may stimulate sex change. Removal of female parrotfish (Scaridae) stimulates males to become female. Less often, removal of males stimulates a female to become a dominant male (Figure 8.5).

Some plants also undergo sex change. Jack-in-the-pulpit (*Arisaema triphyllum*), a herbaceous perennial native to eastern North America (Figure 8.6, p. 162), may produce male flowers one year, an asexual shoot the next, and female flowers the next, all on the same plant. An asexual stage usually follows a sex change and is triggered by the high energy cost of female flowers (and their subsequent fruits). Few individuals have sufficient resources to produce female flowers in successive years—an example of the interaction of life history traits with energy availability.

Figure 8.5 Sex change in parrotfishes in coral reefs. When a predator or researcher removes a dominant male, the largest female becomes a dominant male in days and takes over the harem.

Figure 8.4 Hermaphroditic earthworms mating.

Figure 8.6 Female stage of jack-in-the-pulpit, which changes sex depending on its energy reserves. The species, which grows across eastern Canada, gets its name from the hood-like sheath enclosing its flower stem.

8.3 MATING SYSTEMS: Resource Supply Affects the Adaptive Value of Mating Systems

On a brushy rise of ground at the edge of a forest, a pair of red foxes (*Vulpes vulpes*) have their burrow. Inside are the female and her pups. At the entrance are scattered bits of fur and bones, leftovers of meals brought by the male for his family. Back in the woods, a white-tailed deer doe has hidden her fawn in a patch of ferns. The father is gone and has no knowledge of the fawn's existence. His only interaction with the mother occurred the previous fall, when she was one of several females he mated with over several days.

The fox and the deer represent extremes in **mating system**—the pattern of mating between males and females (Table 8.1). They range from **monogamy**, involving a pair bond between one male and one female, to **polygamy**, in which one individual has two or more mates, with each of which it forms a pair bond, to **promiscuity**, in which both sexes mate with many individuals, forming no pair bond. In plants, whose mating

systems are strongly influenced by floral structure (see Section 8.2), the primary mating system is **outcrossing** (in which pollen from one individual fertilizes the egg of another) and **autogamy** (self-fertilization, the extreme of *inbreeding*). However, mixed mating systems are common. As sessile organisms, plants cannot always rely on dispersing pollen to other individuals, particularly in protected environments where wind may not be a reliable vector and animal pollinators may not always be present.

Why is the mating system of a species important? Each mating system has ecological correlates and consequences. The mating system of a given species influences the allocation to reproduction, particularly in males. Competition among males for mates, courtship behaviour, territorial defence, and parental care can represent a significant investment of time and energy. Moreover, the degree of parental care, which differs among mating systems, directly affects offspring survival. Thus, the mating system is both influenced by and influences population properties such as fecundity and mortality (see Chapter 10). But a mating system does not develop independently of the environment; different patterns are more likely not only in some taxonomic groups but also in some environments.

What combination of factors influences the likelihood of different mating systems in animals? Monogamy exists mostly in species in which the cooperation of both parents is needed to raise the young. Most birds are seasonally monogamous. Newly hatched birds are helpless and need food, warmth, and protection. The mother is no better suited than the father to provide these needs. Instead of seeking other mates, the male increases his fitness more by continuing his investment. Without him, the young carrying his genes are less likely to survive. In a few species, such as swans, pair bonds last a lifetime. Monogamy is rarer in mammals, occurring in only 5 percent (van Schaik and Dunbar 1990). North American examples include carnivores such as foxes and weasels (*Mustela* spp.), and herbivores such as beavers (*Castor* spp.), muskrats (*Ondatra zibethica*), and prairie voles (*Microtus ochrogaster*). The percentage of monogamous primates is higher (15 percent) (van Schaik and Dunbar 1990). Among large primates, only gibbons (*Hylobates* spp.) are monogamous. In humans, monogamy is not universal.

Why is monogamy rarer in mammals than in birds? Females lactate, providing milk for the young. Males contribute little or nothing to offspring survival, so it is in their best interest to mate with as many females as possible. Exceptions are foxes, wolves, and other canids, in which the male provides

TABLE 8.1 Types of Mating Systems in Animals

Type	Bond type	Care of young	Habitat correlates
Monogamy	Single pair bond	Both parents	Homogeneous; similar territories.
Polygamy			
(i) Polygyny	1 male and > 1 female	Female only	Heterogeneous; males defend territories of differing quality.
(ii) Polyandry	1 female and > 1 male	Male only	Heterogeneous; females defend territory of differing quality.
Promiscuity	No pair bond	Female only	Males defend no resources for young.

for the family and defends a *territory* (area defended for exclusive use and access to resources; see Section 11.9). Both parents regurgitate food for the weaning young.

Monogamy does have another side. Both sexes may "cheat" by engaging in *extra-pair copulation*, while maintaining the bond with the primary mate for the purposes of caring for the young. This behaviour is more common than first thought. Based on recent molecular evidence, ecologists now estimate that approximately 90 percent of monogamous bird species engage in extra-pair copulation (Griffith et al. 2002). Typically, about 11 percent of the offspring of monogamous species result from extra-pair paternity, and in some species, over 50 percent.

Timing of sexual receptivity may contribute to interspecific differences in extra-pair copulation. For populations in which females are all sexually receptive simultaneously, the female may be better able to compare males and choose additional partners (Stutchbury 1998). Although this and other hypotheses have engendered much debate (Griffith et al. 2002), the high frequency of extra-pair copulation suggests that it is advantageous for monogamous species. By copulating with other males, the female may increase her fitness by rearing young sired by two or more males, while the male increases his fitness by producing offspring with several females.

Polygamy is the acquisition by an individual of two or more mates—either one male and several females (**polygyny**) or one female and several males (**polyandry**). Unlike promiscuity, a pair bond exists between the individual and each mate. Unlike monogamy, only one member of the pair cares for the young. Freed from parental duty, the individual with multiple mates can devote more time and energy to competing for mates and resources. In polygyny, which is common in mammals, harem size depends on the synchronicity of sexual receptivity. If females are sexually active for only a brief period, as with white-tailed deer, the number a male can monopolize is limited. But if females are receptive over a long period, as with elk (*Cervus elaphus*), harem size depends on the availability of females and the number the male can defend.

Polyandry is best known in three bird groups, the jacanas (Jacanidae; Figure 8.7), phalaropes (*Phalaropus* spp.), and sandpipers (Scolopacidae). As with polygyny, its adaptive value is affected by the distribution and defensibility of resources. As well as competing for mates, the female competes for and defends resources essential for the male. She produces multiple clutches of eggs, each with a different male. After the female lays a clutch, the male begins incubation and becomes sexually inactive. A subtype called *convenience polyandry* is common in some insects. One female mates with many males, but as a result of sexual conflict, not active female choice (see Research in Ecology: Sexual Conflict and the Coevolutionary Arms Race, pp. 165–166).

The evolution of mating systems is influenced by the environment, especially the availability and distribution of resources; and by behaviour, especially the ability to control access to resources. If the habitat is sufficiently uniform that territories differ little in quality, selection favours monogamy because female fitness is similar in all habitats. But if the habitat is diverse, with some parts more productive than others, competition may be intense, and some males will have poorer territories. It may be more advantageous for a female to join another female in the territory of a male defending a rich resource than to settle alone with a male in a poorer territory. Selection then favours polygamy, even though the male does not help feed the young.

Promiscuity is selected if the male neither helps care for the young nor defends any resources available to them. The female gains no advantage by staying with him. Likewise, the male does not increase his fitness by staying with the female. The ruby-throated hummingbird (*Archilochus colubrus*) is the most widely distributed hummingbird in Eastern Canada, occurring also on the prairies. It is considered promiscuous, although polygamy may also occur (Sandilands 2005). Male ruby-throats vigorously defend feeding territories—squabbles over hummingbird feeders are a common sight—but do not provide any food resources to their offspring.

8.4 SEXUAL SELECTION: Acquiring Mates Involves Intrasexual and/or Intersexual Selection

The flamboyant plumage of the peacock (*Pavo cristatus*; Figure 8.8) was a troubling problem for Darwin. Its tail feathers are big and clumsy and require considerable energy to support. They are also very conspicuous and a hindrance when trying to escape predators. In terms of natural selection, what accounts for the peacock's tail? Of what possible benefit could

Figure 8.7 Male African jacana defending its young. Males also incubate eggs.

Figure 8.8 Male peacock in a courtship display.

Figure 8.9 Large bull elk bugles a challenge to other males in a contest to control a harem.

it be? Similar questions arise regarding the horns and antlers of some polygamous species. To explain why males and females of a given species often differ in size, ornamentation, and colour (termed **sexual dimorphism**), Darwin (1871) proposed (1) intrasexual and (2) intersexual selection.

Intrasexual selection involves male-to-male (or in polyandry, female-to-female) competition for mates. It promotes exaggerated secondary sex traits, such as aggressiveness, large size, and threat organs such as antlers and horns (Figure 8.9). **Intersexual selection** typically involves differential attractiveness of one sex to the other, although in species where females resist matings, intersexual selection can be antagonistic (see Research in Ecology: Sexual Conflict and the Coevolutionary Arms Race, pp. 165–166). The targets of intersexual selection are such traits as bright or elaborate plumage for sexual displays, as well as traits that also figure in intrasexual selection (such as horns and antlers). Intersexual selection leads to *assortative mating* (see Section 5.6), wherein the female selects a mate based on specific phenotypic traits. Rivalry is intense, with the female ultimately selecting a mate. Fitness increases for the chosen, shifting the distribution of male phenotypes in favour of the traits on which females base their choice.

But do traits such as bright colours and elaborate plumage (collectively called **ornamentation**) really affect mate selection, and what are the trade-offs? Investment in such traits may limit energy for activities that affect fitness, such as foraging, growth, defence, and reproduction. Consider sexual dimorphism in swordtails (*Xiphophorus* spp.). Males have a sword—a colourful, elongated appendage of the caudal fin (Figure 8.10a) that is a visual cue during courtship. In tests that allowed females to choose between a pair of green swordtail (*X. helleri*) males with differing sword length, females preferred males with longer swords (Basolo 1990). Intersexual selection for sword length increases male fitness by increasing the chance of mating.

But what are the costs? Locomotion uses much of a fish's energy budget, and a longer sword may increase mating success while decreasing swimming efficiency. In a test on the

Montezuma swordtail (*X. montezumae*), swords were removed from males, and respiratory costs compared for males with and without swords for routine and courtship swimming. These costs were 30 percent higher for all males when females were present, indicating something we all know—courtship is energy demanding. However, males with swords had higher costs than males without swords for both routine and courtship swimming (Figure 8.10b) (Basolo and Alcarez 2003). So, although sexual selection via female choice favours long swords, males with longer swords have higher swimming costs. They are also more vulnerable to predation. Swords are significantly shorter in green swordtails that coexist with predatory fish than in populations where predators are absent. In this case, intersexual and natural selection have opposing effects (Basolo and Alcarez 2003), but the fitness advantage of long swords must offset the combined energy and survival costs.

Similarly, removing tail feathers from male peacocks (*Pavo cristatus*) reduced their mating success (Petrie 1994).

(a)

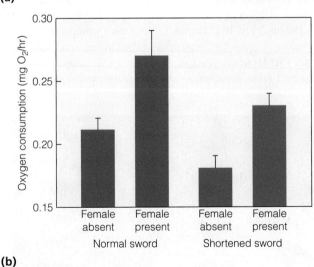

(b)

Figure 8.10 The colourful, elongated sword on the caudal fin of males of the green swordtail **(a)** is a visual cue to attract females. **(b)** Mean O_2 consumption of males with normal and shortened swords in routine and courtship swimming.

(Adapted from Basolo and Alcarez 2003.)

RESEARCH IN ECOLOGY   Sexual Conflict and the Coevolutionary Arms Race

Locke Rowe, University of Toronto.

For a population, reproduction is essential. Sexual selection increases the frequency of traits of males and females that enhance their fitness as individuals, but it is in the interest of both sexes that mating produce viable offspring. Yet the ecological interests of the sexes can be at odds, and intersexual selection, usually discussed in terms of female choice, operates differently for species in which female resistance to mating must be overcome by male persistence. Intersexual conflicts then occur, as opposed to (or as well as) intrasexual conflicts for mating opportunities.

Antagonistic intersexual selection favours (1) traits of the male that increase his ability to force mating on an unwilling female and (2) traits of the female that increase her ability to resist mating. The result is a coevolutionary arms race in which changes in the traits of one sex are met with changes in the traits of the other (Arnqvist and Rowe 2002). Female choice still applies, but it is a negative choice reflecting the ability of the female to *not* mate with a particular male. Sexual conflict and its role in the evolution of mating systems is a research focus of the University of Toronto's Locke Rowe and Swedish ecologist Göran Arnqvist, using semi-aquatic insects called water striders (Gerridae) as a model system (Figure 1).

Mating behaviour in water striders is prolonged, dramatic, and fraught with conflict. Males chase, harass, and leap upon females in frantic mating efforts. Females evade or flee, but if a male does manage to land, the female struggles vigorously, using somersaults and other gymnastic actions to dislodge him. Rarely, a male endures female resistance, and copulation occurs. After copulation, a guarding phase ensues, in which the male remains attached to the female until dislodged. Questions arise: *What are the factors, intrinsic and extrinsic, that underlie this struggle? What are its evolutionary outcomes, both within and among species?*

Because a female water strider stores viable sperm for at least 10 days, she needs few matings over her reproductive life to acquire enough sperm. Moreover, the male provides no nutritional benefit, either as a nuptial gift or by contributing his body for an act of sexual cannibalism, as does the male black widow spider (*Latrodectus mactans*). Thus, female water striders increase their fitness by increasing their food supply or avoiding predation, not by engaging in more mating events. Indeed, additional matings are superfluous to her interests and may pose substantive risks. The female's ability to forage is compromised by time spent resisting, copulating, and guarding post-coitus. The struggle uses considerable energy and makes her more vulnerable to predation. No wonder females are reluctant to mate, given the minimal benefits and considerable costs. Optimal female behaviour balances these costs by varying resistance, as indicated by the duration and intensity of pre-mating struggles. Females submit to superfluous matings only to reduce the costs of further harassment and/or struggles. This mating system, called *convenience polyandry*, is common in insects.

What about the interests of the male? As in many species, males are limited by the availability of fertilizable eggs. Food and predation are also factors, but male fitness is served by increasing the number of matings. There is also a last-minute fertilization advantage (Arnqvist 1988), which increases the utility of additional matings and of guarding the female after mating to reduce the risk of other males pursuing a similar advantage. The interests of the sexes are thus at odds, setting the stage for antagonistic intersexual selection.

Several population and environmental parameters influence the outcome of sexual conflict in water striders, especially in the short term. When the *sex ratio* (proportion of males to females; see Section 9.7) is experimentally increased, female resistance decreases (Rowe and Arnqvist 2002). The energy expended becomes counterproductive, and both the duration and the intensity of the struggle decline. A similar response occurs with increased *density* (number of individuals per unit area; see Section 9.3). The reverse occurs in female-biased or low-density populations.

For species that continue feeding while mating, the net energy cost of female resistance is lower (and the advantage to the female of prolonging the struggle is higher) than in species where feeding ceases. Also, when predatory pressure drops, selection for female resistance increases. As both food supply and predation risk vary greatly, selection for resistance waxes and wanes. Environmental variability thus not only affects conflict outcomes but also influences plasticity and diversification of mating behaviour within and among species (Rowe et al. 1994).

Figure 1 Male and female water striders mating.

continued on page 166

RESEARCH IN ECOLOGY   continued

But what about the impact of sexual conflict on heritable traits? A suite of traits arises from antagonistic coevolution. In males, these include a flattened abdomen and enlarged clasping genitalia that allow the male to grasp the female more firmly. Females have evolved **counteradaptations** (traits that counter the effect of the adaptations of other species, or in this case of the opposite sex), such as elongated abdominal spines and a downward tilting of the abdomen that make it harder for males to maintain a hold. *But if the advantages gained by one sex are matched by counteradaptations in the other, will the outcome be an evolutionary stand-off, with no net differences in mating rate or activity?* To address this question, Rowe and Arnqvist analyzed the morphology of 15 water strider species collected in North America and Europe, focusing on the "armament" traits associated with mating. They also quantified the following behaviours of the different species in experimental ponds: struggle duration, male struggle success, time spent by females in mating, and number of matings per female.

The authors constructed a "coevolutionary trajectory" (Figure 2), where the axes are composites based on morphological traits related to male persistence (*x*-axis) and female resistance (*y*-axis). Each point represents a species. The line describes a coevolutionary path along which the level of arms in males and females is balanced, with no relative advantage. The intensity of the arms race differs among species, as indicated by how far along the trajectory they have travelled over evolutionary time. For example, one species may have a low absolute armament level, whereas another species has more developed armament (and counterarmament) traits. Such differences reflect environmental variation that affects the strength of intersexual selection.

Species also differ in how far they deviate from the trajectory. Whereas some species are relatively balanced, despite their differences in absolute armament level, other species have evolved either a strong female or male advantage. The resulting deviations from the line, where species have evolved to a point where one sex has gained a relative advantage over the other, provide insight into the evolutionary consequences of an arms race between the sexes. Why is this so? The behavioural experiments indicate that only the relative armament level affects the outcome of mating conflicts. When females have the relative advantage (i.e., are above the line), struggles last longer, and male struggle success and female mating rate decline (Arnqvist and Rowe 2002). Females dislodge males more successfully and so mate less often and spend less time mating. The reverse applies when males have the relative advantage.

These experiments show that antagonistic intersexual selection leading to coevolution does occur in water striders. Small deviations in relative advantage between the sexes can propel significant changes in traits over generations, against the backdrop of fluctuating environments. Rowe and Arnqvist combine experimental and observational approaches to reveal the complex interactions between behaviour and morphology that propel the evolution of mating systems.

Bibliography

Arnqvist, G. 1988. Mate guarding and sperm displacement in the water strider *Gerris lateralis* Schumm. (Heteroptera: Gerridae). *Freshwater Biology* 19:269–274.

Arnqvist, G., and L. Rowe. 2002. Antagonistic coevolution between the sexes in a group of insects. *Letters to Nature* 415:787–789.

Rowe, L., and G. Arnqvist. 2002. Sexually antagonistic coevolution in a mating system: Combining experimental and comparative approaches to address evolutionary processes. *Evolution* 56:754–767.

Rowe, L., G. Arnqvist, A. Sih, and J. J. Krupa. 1994. Sexual conflict and the evolutionary ecology of mating systems: Water striders as a model system. *Trends in Ecology and Evolution* 9:289–293.

1. Scientists often stress mean response. Yet in Figure 2, variation around the mean, including the outliers, is no less important than the trajectory itself. Explain.
2. Compare convenience polyandry with polyandry in terms of (i) female choice and (ii) impact on female traits. Does sexual dimorphism apply, and how does it differ from sexual dimorphism in typical polyandry? How does convenience polyandry compare with promiscuity?

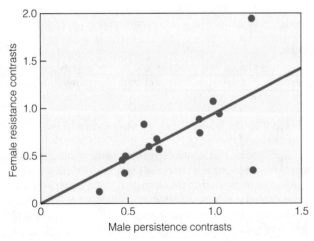

Figure 2 Relationship between male persistence and female resistance traits in 15 water strider species. The axes are based on multivariate trait analysis.

(Adapted from Arnqvist and Rowe 2002.)

So, in the case of the peacock's tail and the swordtail's caudal fin, size really does matter. Such traits may be important because of what they imply about the individual. A large, conspicuous tail makes the male more vulnerable to predation, or in other ways reduces his survival. A male that survives despite such handicaps shows his health and genetic vigour. Females who select males with large tail feathers will produce offspring that are more likely to inherit these "good genes." Offspring of female peacocks that mated with large-tailed males had increased survival and growth over offspring of short-tailed males (Petrie 1994). Thus, the evolution of exaggerated secondary sex traits in males is driven by female mate selection.

A similar mechanism may underlie plumage of male birds. Hamilton and Zuk (1982) proposed that males with low parasitic infection have the brightest plumage. Females that select males based on differences in plumage are thus selecting males that are the most disease resistant. Other hypotheses abound, and some argue that females are to some extent "fooled" or manipulated by sensory stimulation, without any real benefit accruing. Sexual dimorphism is also promoted by negative assortative mating (see Section 5.6), whereby females preferentially mate with males with less genetic similarity to themselves: opposites attract. Most likely different (and multiple) factors are at work in different species and for different traits.

The examples of inter- and intrasexual selection considered so far involve male traits as the target of selection. Why is sexual selection often more intense for males? For males, sexual selection typically involves competition to mate with as many females as possible, in order to fertilize more eggs. At any one time, few females may be available, intensifying the competition. In contrast, engaging in multiple matings can reduce female fitness. Copulation has energy costs and may increase exposure to disease and predation. Females typically invest more time and energy in offspring, so female fitness may be better served by limiting the number of matings. These differences in what constitutes optimal behaviour for males and females select for morphological and/or behavioural traits relating to display for males and to choosiness and/or resistance for females (see Research in Ecology: Sexual Conflict and the Coevolutionary Arms Race, pp. 165–166).

In polyandrous species, the situation reverses and sexual selection becomes more intense for females. Male reproductive investment increases in terms of time and energy spent caring for the young. Now males are less available, and females compete for both territories and males. Sexual dimorphism develops, but with females becoming larger and more brightly coloured. In species where parental investment is similar, sexual dimorphism tends to be less pronounced.

8.5 IMPACT OF RESOURCES: Females May Choose Mates Based on Resource Availability

In Section 8.4, we saw that a female may choose a mate by selecting for traits that are indirect indicators of the male's health and/or genetic quality. Alternatively, she may evaluate a

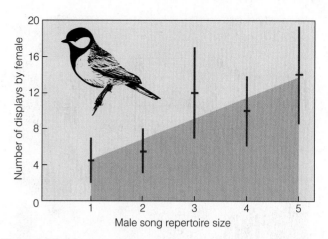

Figure 8.11 Mean number of copulation–solicitation displays given by 11 female great tits increases with the complexity of the male song, as measured by repertoire size (ranging from 1 to 5 song types).

(Adapted from Baker et al. 1986.)

potential mate's ability to provide access to resources, such as habitat or food, that will likely improve her fitness. For monogamous species, the criterion is more often resource acquisition, based on the quality of territory defended by the male. But does the female select the male and accept the territory that goes with him, or select the territory and accept the male that goes with it?

Female songbirds may base their choice in part on the male's song repertoire. In an aviary study, female great tits (*Parus major*) were more receptive to males with more varied or elaborate songs (Figure 8.11) (Baker et al. 1986). In a field study in England's Thames Valley, female sedge warblers (*Acrocephalus schoenobaenus*) based their selection on multiple cues: repertoire complexity, territory size, and song flight behaviour (Buchanan and Catchpole 1997). Females based their selection not on a male's morphological traits—males are not brightly coloured—but on his behavioural traits, as an indirect indicator of territory quality.

For polygamous species, the question is more complex. If the female acquires a resource along with the male, the situation resembles monogamy. In polygamous birds, females strongly prefer males with high-quality territories. On territories with good nesting cover and abundant food, females can successfully reproduce even if they share the territory with other females. But when polygamous males offer no resources, the female has little information upon which to act. She might employ sexual selection, choosing a winner among males that engage in combat; or based on the intensity of courtship display or on some other trait that reflects a male's genetic superiority or vitality. In some polygamous species, it is hard to see female choice at work. In elk and seals, a dominant male commands a harem. However, females still express some choice. Protests by female elephant seals over the attention of a male may attract other males nearby, who try to replace him. Such behaviour increases the chance that females will mate with the highest-ranking male.

Figure 8.12 Courtship display of the male sage grouse on a communal courting ground or lek.

In other polygamous species, females are in control, with males putting on a display for their benefit. Such advertising can be costly; given their conspicuous behaviour, males may be more vulnerable to predation. Female choice is clearly involved in species in which males aggregate in groups on communal courtship arenas or **leks** (Figure 8.12). Leks are common in ground birds such as the sage grouse (*Centrocercus urophasianus*), native to North America, including western Canada, and the prairie chicken (*Tympanuchus cupido*), now extirpated in Canada. Leks provide an opportunity for females to choose a mate. Advertising their presence by colourful visual and in some cases vocal displays, males defend small temporary territories that hold no resources. Females visit the leks, select a male, mate, and move on. A similar behaviour occurs in the "sand-castle" building displays of African cichlid fish (Cichlidae) (Tweddle et al. 1998).

Although few species use the lek system, it is widespread among animal types, from insects to frogs to birds and mammals. Males defend small, clustered mating territories, whereas females have large overlapping ranges that males cannot economically defend. Dominant males occupy central portions of the lek. Congregating about dominant males with the most effective displays, subdominant males steal occasional mating chances. Most matings involve a small percentage of males in a dominance hierarchy formed in the absence of females.

8.6 REPRODUCTIVE EFFORT: Species Vary in the Time and Energy Allocated to Reproduction

Just as plants allocate carbon to various tissues (see Section 6.7), animals must allocate their limited energy to meet many demands, including growth, maintenance, food acquisition, defence, escape from predators, and reproduction. The time and energy allocated to reproduction make up an organism's **reproductive effort**. As always, trade-offs are inevitable. The more energy expended on reproduction, the less is available for maintenance, growth, and defence.

How does this theoretical concept play out in actual organisms? Reproducing female woodlice (*Armadillidium vulgare*) grow more slowly than do non-reproducing females, which devote as much energy to growth as reproducing females do to both growth and reproduction. Likewise, there is a negative relationship between annual growth and cone production of Douglas-fir (*Pseudotsuga menziesii*; Figure 8.13a) (Eis et al. 1965). The more cones a tree produces, the smaller its annual trunk growth. Moreover, the demands of female and male

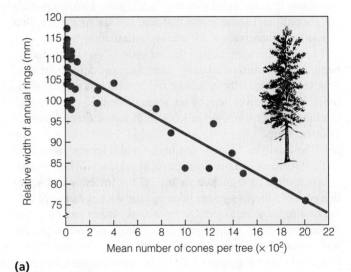

(a)

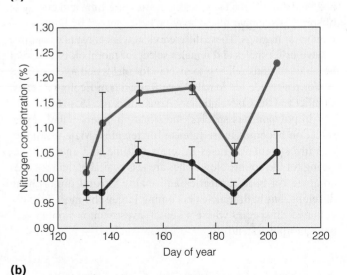

(b)

Figure 8.13 Reproductive allocation in Douglas-fir. **(a)** Relationship between number of cones and annual radial growth. **(b)** Mean N concentration (% ± 1 standard error) of needles on shoots with female cones (red) and without (blue) female cones.

(Adapted from (a) Eis et al. 1965 and (b) McDowell et al. 2000.)

cones differ in this monoecious species. Female Douglas-fir cones use 92 and 78 percent, respectively, of the carbon and nitrogen allocated to reproduction, depleting the nitrogen levels of nearby foliage (Figure 8.13b) (McDowell et al. 2000). Life histories inevitably involve compromise. By allocating more energy to reproduction, the organism—be it woodlouse or fir tree—reduces its own growth, which may make it less able to survive, compete for resources, or reproduce in the future.

Energy invested in reproduction varies greatly (Reekie and Bazzaz 2005). Herbaceous perennials invest 15 to 20 percent of annual net production in reproduction (sexual and asexual), whereas undomesticated annuals that reproduce only once expend slightly more (15–30 percent). Even for similar organisms, investment varies widely; common lizard (*Lacerta vivipara*) females invest 7 to 9 percent of their assimilated energy in reproduction (Avery 1975) versus 48 percent for the Allegheny Mountain salamander (*Desmognathus ochrophaeus*) (Fitzpatrick 1973).

Reproductive investment includes the costs of care and nourishment as well as the costs of producing offspring. This investment affects an individual's fitness, which is ultimately determined by the number of offspring that survive to reproduce themselves. All aspects of reproductive allocation—timing, number and size of offspring, and care provided after birth—interact within the context of the environment to determine the return in terms of fitness (see Quantifying Ecology 8.1: Interpreting Trade-offs, p. 170).

A given allocation can produce many small offspring or a few large ones. There is a fundamental trade-off between the number and size of offspring: the more offspring, the smaller their size (Figure 8.14). In plants, this *seed size/number trade-off* influences community composition in different habitats (Leishman 2001). Species that opt to produce many small offspring often inhabit unpredictable or disturbed environments, or (in animals) habitats such as the open ocean where parental care is difficult at best. By dividing their allocation among as many young as possible, parents increase the chances that some will survive, establish, and reproduce. This choice increases parental fitness at the expense of the fitness of their young. Parents that produce few young invest more in each, increasing the fitness of their offspring at the expense of their own.

Parental investment also varies with development at birth. Some organisms expend little energy during incubation. These animals, such as young mice or birds, are **altricial**: their young are born or hatched in a helpless condition and require much parental care. Others have longer gestations, so the young are born in a more advanced stage, known as **precocial**. They can move about and forage for themselves soon after birth. Examples are gallinaceous birds, such as turkeys, and ungulate (hooved) mammals, such as deer.

Parental care in mammals is well known, but it varies greatly in other animal groups. Some fish, such as cod (*Gadus morhua*), lay millions of eggs that drift freely in the ocean. Other species, such as bass, lay eggs in the hundreds

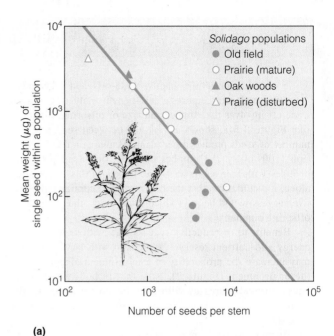

(a)

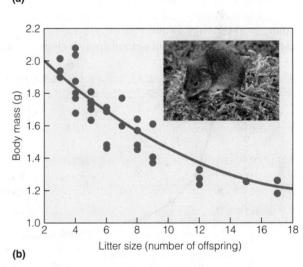

(b)

Figure 8.14 Trade-off between number and size of offspring. **(a)** Mean seed weight and number of seeds per stem of goldenrod (*Solidago* spp.) populations in different habitats. **(b)** Litter size and mean body mass of offspring at birth for litters of bank vole (*Clethrionomys glareolus*).

(Adapted from (a) Werner and Platt 1976 and (b) Oksanen et al. 2002.)

and provide some parental care. Among amphibians, parental care is common in tropical toads and frogs and some salamanders. Among reptiles, which rarely exhibit parental care, crocodiles are an exception. They actively defend the nest and young for some time. Some invertebrates also exhibit parental care. Octupi; crustaceans such as lobsters, crayfish, and shrimp; and some amphipods, such as millipedes, brood and defend eggs. Parental care is well developed in social insects (bees, wasps, ants, and termites), which perform all functions of parental care, including feeding, defence, heating, and sanitation.

QUANTIFYING ECOLOGY 8.1 Interpreting Trade-offs

Most life history traits involve trade-offs, and understanding trade-offs requires a cost–benefit analysis. One such trade-off involves the number and size of offspring. Figure 1, like Figure 8.14a, shows a trade-off between seed size and number of seeds produced per plant. Assuming a fixed allocation (100 units), the number of seeds produced per plant declines with increasing seed size. Based on this information alone, it would seem that the best way to maximize reproductive success would be to produce small seeds, thus increasing offspring number.

Benefits to reproductive success vary with seed size. The energy and nutrient reserves associated with large seed size may increase the probability of establishment, depending on the environment (Figure 2). Note that in both wet and dry sites, survival increases with size, but in dry habitats, the probability of survival declines dramatically with decreasing seed size.

By multiplying the number of seeds produced by the probability of survival, we can calculate the reproductive success (number of surviving offspring per plant) for plants producing seeds of a given size in both wet and dry sites (Figure 3).

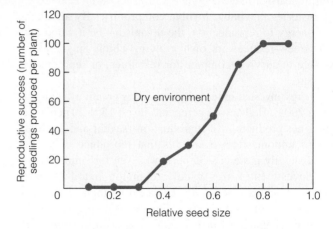

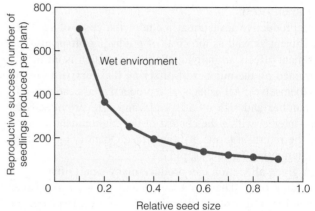

Figure 3 Effect of seed size on reproductive success in contrasting environments.

In wet sites, where all seed sizes have comparable survival chances, producing many small seeds increases reproductive success, thus maximizing fitness. But the strategy of producing large seeds is more successful in dry habitats, where larger seeds have a greater chance of survival. Nor is this pattern invariable. In some species, smaller seeds can be more successful in dry sites, if they are more tolerant of desiccation. Thus, interpreting the trade-offs involved in life history traits requires understanding how those trade-offs vary with the environment. The diversity of life history traits attests to the fact that there is no single best solution—one size does not fit all.

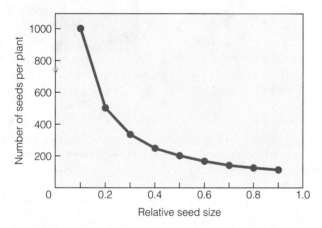

Figure 1 Trade-off between seed size and number of seeds per plant.

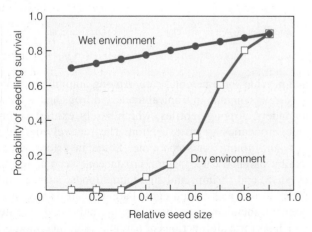

Figure 2 Effect of seed size on the probability of seedling survival in contrasting environments.

1. In this example, natural selection should favour small-seeded plants in wet habitats and large-seeded plants in dry habitats, causing a difference in mean seed size. What might happen in a habitat where rainfall is high most years but in which drought may persist for several years?
2. Seeds of shade-tolerant species are typically larger than those of shade-intolerant species. How might this difference reflect a trade-off relating to establishment in sun and shade?
3. Despite the advantage of large seeds in dry habitats, species native to moist habitats may also produce large seeds. An example is the large nuts of the beech (*Fagus grandifolia*), common in temperate forests of eastern North America. What factors might be responsible? Hint: Consider question 2.

8.7 REPRODUCTIVE TIMING: Differences in Timing of Reproduction Entail Trade-offs

How should an organism invest in reproduction over time? One tactic is to invest all of its energy initially into growth and storage, followed by one massive reproductive effort before death. In this strategy, called **semelparity** or "big bang" reproduction, an organism sacrifices future prospects by concentrating its energy on one suicidal reproductive act. Semelparity is common among insects and other invertebrates; some fish (notably salmon); and many plants, particularly annuals, biennials, and even some perennials. Semelparous plants, such as ragweed (*Ambrosia* spp.), are often small and short-lived, occupying ephemeral (temporary) or disturbed sites. It does not pay such species to hold out for future reproduction, as survival chances are slim. They attain maximum fitness by expending all their energy in a single bout of reproduction.

Other semelparous organisms are long-lived and delay reproduction. Mayflies (Ephemeroptera) may spend several years as aquatic larvae before emerging for an adult life of several days devoted to reproduction. Periodical cicadas (*Magicicada* spp.) spend 13 to 17 years belowground before they emerge as adults for a single massive reproductive event. Some bamboo delay flowering for over 100 years, produce one massive seed crop, and die. Semelparous species must increase offspring number enough to compensate for the loss of repeated reproductive opportunities.

Iteroparity is the strategy whereby an organism has repeated reproduction events over its lifetime. Examples include most vertebrates, herbaceous perennials, and woody shrubs and trees. Iteroparous invertebrates include molluscs and some insects, including some species that live for less than a year. Mosquitoes, for example, typically reproduce several times in their short life. With iteroparity, the problem is the trade-off involved in reproductive timing. Early reproduction means less growth, earlier maturity, and lower survivorship. Later reproduction means more growth, later maturity, and higher survivorship, but less overall time for reproduction. To maximize fitness, an organism must balance immediate reproductive gains against future prospects, including the cost to its total offspring production as well as to its own survival.

8.8 REPRODUCTIVE OUTPUT: Parental Traits and Food Supply Affect Fecundity

For a given species, the range of offspring number and size is genetically determined, but within that range, the actual number varies with the age and size of the parent. These differences reflect phenotypic plasticity (see Section 5.12). Many plants and poikilothermic animals exhibit **indeterminate growth**: the pattern whereby individuals grow throughout their lives, usually at a declining relative rate, with no typical final size. Among plants, whose indeterminate growth reflects a modular body plan (see Section 9.1), perennials delay flowering until they reach a sufficiently large size and leaf area to support seed production. Many biennials delay flowering beyond the usual two-year lifespan if conditions are unfavourable. In annuals, which show little relationship between size and percentage of energy devoted to reproduction, size differences affect only the number of seeds produced. Small plants produce fewer seeds, even though the plants may be allocating the same proportion of energy to reproduction as larger plants do.

Similar patterns exist in poikilothermic animals. In fish, reproduction increases with size, which increases with age. Because early fecundity reduces both growth and reproductive success, fish may gain a selective advantage by delaying maturation. A study of 216 North American fish species (freshwater and marine) revealed a consistent pattern of later-maturing fish producing more and smaller eggs in fewer reproductive events than early-maturing fish (Winemiller and Rose 1992). An individual might not survive until a later age, in which case its fitness would be compromised, but despite this trade-off it may pay to delay sexual maturity. This age effect relates to body size. Likewise, body size of female big-handed crabs (*Heterozius rotundifrons*) in New Zealand is correlated with number of young (Figure 8.15a, p. 172) (Jones 1978). Larger female mammals reproduce more successfully, and more of their young survive. Body mass of female European red squirrels (*Sciurus vulgaris*) correlates strongly with lifetime reproductive output (Figure 8.15b) (Wauters and Dhondf 1989). Few squirrels that weigh less than 300 g reproduce.

Food supply often affects fecundity. In species with indeterminate growth, food supply directly influences body size and hence reproductive output. In habitats where food supply is variable, the number of offspring that can be produced may exceed the number that can be provided for in lean years. Reducing the number of young may then be advisable.

Many birds reduce the number of offspring competing for limited resources by *asynchronous hatching* (in which the young hatch at different times) or by allowing siblicide. If the young hatch at different times, older siblings beg more vigorously for food. Parents may ignore younger siblings, which perish. The grackle (*Quiscalus quiscula*), a large bluish-black bird common in backyards across North America, begins incubating before its entire clutch of five eggs is laid. The eggs laid last are heavier, and these hatchlings grow fast. But if food is scarce, the parents ignore these late offspring in favour of the vigorous begging by older siblings. The last-hatched young starve. Thus, asynchronous hatching favours early hatchlings, increasing the chance that some offspring survive under adverse conditions.

Some birds, including raptors, herons, egrets, boobies, and skuas, practise **siblicide**, in which older or more vigorous young kill weaker siblings. The parents normally lay two eggs, possibly to insure against infertility of a single egg. The larger of the two hatchlings kills the smaller sibling, or *runt*, and all

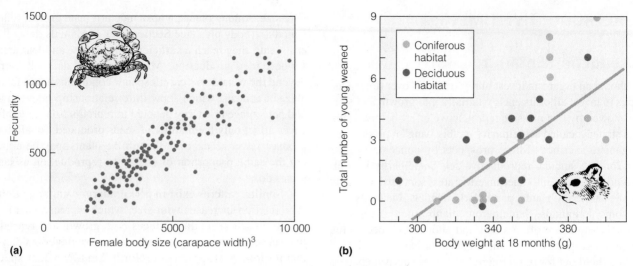

(a)

(b)

Figure 8.15 Effect of body size on reproductive output. **(a)** Annual production of young by female big-handed crabs increases with body size. **(b)** Lifetime reproductive success of the European red squirrel in Belgian forests is correlated with body weight in the first winter as an adult.

(Adapted from (a) Jones 1978 and (b) Wauters and Dhondf 1989.)

resources are redirected to the survivor (Figure 8.16). Similarly, females of parasitic wasps lay two or more eggs in a host, and the larvae fight until only one survives.

Reproductive effort often exhibits geographic trends, which may reflect food supply. Birds at higher latitudes have larger clutches (number of eggs) than do tropical birds (Figure 8.17), and mammals have larger litters. Similarly, lizard species at lower latitudes reproduce at an earlier age and have smaller clutches than do high-latitude species. Insects have a different pattern. Milkweed beetles (*Oncopeltus* spp.) have similar clutch sizes, length of egg stage, egg survivorship, developmental rate, and age at maturity, but tropical species lay fewer clutches and produce 40 percent less output than do temperate species (Landahl and Root 1969).

Some plant species also follow latitudinal trends. Three species of cattail grow in North and Central America. The wide-leaved cattail (*Typha latifolia*) grows from the Arctic to the equator. The narrow-leaved cattail (*T. angustifolia*) inhabits higher latitudes, whereas *T. domingensis* grows only in the south. These species show a latitudinal gradient in allocation of energy to asexual reproduction. *T. angustifolia* and northern populations

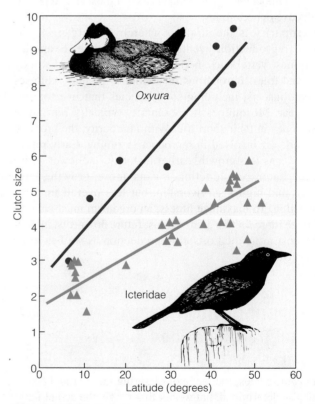

Figure 8.17 Relationship between clutch size and latitude in birds. Represented are the family Icteridae (blackbirds, orioles, and meadowlarks) in North and South America and the worldwide genus *Oxyura* (ruddy and masked ducks) of the family Anatidae.

(Adapted from Cody 1966.)

Figure 8.16 Siblicide in the masked booby (*Sula dactylatra*), in which the larger of two offspring kills the smaller sibling. Parents collaborate by ignoring the battle.

of *T. latifolia* grow earlier in spring and produce more but smaller clonal shoots than does *T. domingensis*. McNaughton (1975) used this example to illustrate the *r versus K* model discussed in Section 8.10.

What explains these latitudinal patterns? Lack (1954) proposed that clutch size in birds is an adaptation to food supply and has evolved in relation to the most young that parents can feed. Temperate species have larger clutches because longer spring days allow more foraging time to support large broods. In the tropics, where day length is roughly 12 hours, foraging time is more limited. Cody (1966) modified Lack's ideas by proposing that clutch size reflects differential allocation to egg production, predator avoidance, and competition. In temperate regions, periodic climatic disturbances (such as a harsh winter or a summer drought) hold a population below the level that the resources could support—that is, below its carrying capacity (see Section 10.7). Organisms respond with larger clutches and faster population growth. In tropical regions, with predictable, favourable climates and increased survival chances, there is less need to allocate more energy to producing more young because populations are at or near carrying capacity.

A third hypothesis contends that clutch size varies in proportion to seasonal variation in food supply (Ashmole 1963). Population density is regulated by mortality in winter, when resources are scarce. Greater winter mortality means more food for survivors in the breeding season, which results in larger clutches. Geographical variation in both clutch size and breeding populations is thus inversely related to winter food supply. This factor is also associated with greater fluctuations in populations of high-latitude species (see Section 10.9).

Although reproductive output does tend to increase at higher latitudes, the number of comparable species for which data exist is too small to pinpoint the cause. Different factors are likely at work in different taxonomic groups and/or under different conditions. As clutch size is correlated with body size, Huston and Wolverton's 2011 re-evaluation of Bergmann's rule suggests that available productivity in the growing season may be critical, as Lack first suggested.

8.9 HABITAT SELECTION: Habitat Choice Influences Reproductive Success

Reproductive success depends on choice of *habitat*—the place where an organism lives. Habitat provides access to resources, nesting sites, and cover from predators. If habitat also influences the ability to attract a mate, then settling on an unfavourable habitat can reduce fitness. **Habitat selection** is the process by which organisms actively choose a specific location to live. *Given the importance for fitness, how do organisms assess habitat quality? What do they seek in a location?* Such questions have long intrigued ecologists. Understanding the relationship among habitat selection, reproductive success, and population dynamics is more important than ever, given the loss of habitat for many species from human impacts (see Chapter 27).

Habitat selection has been most studied in birds, particularly species that defend breeding territories (see Section 11.9). Ecologists delineate territories and contrast their habitat features with adjacent areas that have not been chosen. Many studies show a strong correlation between habitat selection and structural features of vegetation, suggesting that habitat selection involves a hierarchical approach. Birds first assess general landscape features—type of terrain; presence of water bodies; vegetation type, such as grassland, shrubland, and forest type; and extent. Once in an area, birds respond to more specific features, particularly the presence or absence of vertical layers (shrubs, small trees, tall canopy) and patchiness (Figure 8.18). The term *niche gestalt* (James 1971) describes the vegetation profile associated with the breeding territory of a species. In some birds, such as the hooded warbler (*Wilsonia citrina*), males and females select different habitats when not breeding (Morton 1990).

Structural features of the vegetation that define its suitability for a species relate to its particular needs, including food, cover, and nesting. A lack of song perches prevents some birds from colonizing otherwise suitable habitat. An adequate nesting site is another key requirement. Animals need shelter to protect themselves and their young from enemies and adverse weather. Cavity-nesting

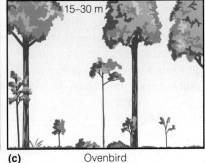

(a) Yellowthroat (b) Hooded warbler (c) Ovenbird

Figure 8.18 Vegetation structure of the habitat of three warbler species. **(a)** The common yellowthroat (*Geothlypis trichas*) inhabits shrubby margins of woodlands and wetlands from southern Canada to Mexico. **(b)** The hooded warbler (*Wilsonia citrina*), a species at risk in Canada, occupies small forest openings. It occurs only in mature forests in southwestern Ontario. **(c)** The ovenbird (*Seiurus aurocapilla*) inhabits deciduous or mixed conifer–deciduous forests with open forest floor. It is widely distributed in North America, occurring from Alberta to the Maritimes. Labels refer to the height of vegetation.

(Adapted from James 1971.)

animals, such as woodpeckers, require suitable dead trees or other structures. In areas lacking such sites, populations of birds and squirrels increase dramatically with provision of nest and den boxes, which has become part of many conservation programs. Besides the physical structure, the actual plant species present is important, either as food items or because they influence the type and quantity of insects available as prey for insectivores.

First studied in birds, habitat selection is a common behaviour of other vertebrates. Fish, reptiles, amphibians, and mammals furnish numerous examples. Garter snakes (*Thamnophis elegans*) living along lakeshores in sagebrush–ponderosa pine habitat in northern California select rocks of intermediate thickness (20–30 cm) as their retreat sites. Shelter under thinner rocks becomes lethally hot; shelter under thicker rocks prevents the snakes from warming to their preferred body temperature. Under rocks of intermediate thickness, snakes can best achieve and maintain their preferred body temperature (Huey et al. 1989). Habitat selection in snakes may vary across a species' geographic range. Ratsnakes (*Elaphe obsoleta*) use forest edges more in Ontario than in Illinois, where they prefer forest interiors (Carfagno and Weatherhead 2006). Such differences may also reflect thermoregulation, with northern ecotypes preferring more exposed sites.

Insects also cue in on habitat features. The gall-forming aphid (*Pemphigus betae*), which parasitizes the narrow-leaf cottonwood (*Populus angustifolia*), selects the largest leaves and the best positions on the leaf for tapping into the phloem (Whitham 1980). The ability to assess habitat depends upon the ability of a species to respond to visual, olfactory, and other cues. Habitat selection thus involves physiology and morphology as well as behaviour. Insectivorous bats, which use *echolocation* to detect prey, select habitat in part based on the degree of interference or "clutter" arising from vegetation of differing density. The northern long-eared bat (*Myotis septentrionalis*), which selects denser forests, can detect echolocation signals over a longer distance than the little brown bat (*M. lucifugus*), which selects less dense forests near the open water of lakes (Broders et al. 2004).

But even when a species can detect a suitable habitat, it may not choose it. The presence or absence of others of the same species may cause individuals to choose or avoid a site. In social species such as the herring gull (*Larus argentatus*), an animal selects a site only if others of the same species are nearby. In a study of herring gulls in Newfoundland, high densities in the preferred but crowded rocky habitats can increase socially induced mortality (Pierotti 1982).

The presence of predators or humans may discourage selection of otherwise suitable habitat. In boreal lakes, habitat selection by the common loon (*Gavia immer*) in areas of human activity, particularly watercraft use, may reduce fitness. In Ontario's Muskoka Lakes, hatchling success of loons declined with the number of cottages within 150 m of the nest (Heimberger et al. 1983). However, habituation to humans was occurring, indicating that habitat choice is highly plastic. Indeed, most species exhibit some flexibility in habitat selection. Otherwise, individuals would never settle in what seem like less suitable habitats, nor would they ever colonize new habitats. Individuals are often forced to make this choice. Available habitats range from optimal to marginal. Like good seats at a concert, optimal habitats fill up fast. Marginal habitats go next, and latecomers and subdominant

individuals are left with poor habitats, where they may have little chance of reproducing successfully. Chance events can alter habitat quality. If a hurricane destroys vegetation in what was otherwise a high-quality habitat, the individual in the initially poorer habitat may be better off if it escapes hurricane damage.

Do plants actively select habitats, and if so how? Plants, like animals, fare better in certain habitats, depending on factors such as light, moisture, nutrients, and herbivore abundance. But if their present habitat is less than optimal, plants cannot get up and move to a better one. Their main recourse is to evolve dispersal strategies that increase the probability that their offspring will arrive at a place suitable for germination and establishment. Habitat selection for plants involves—besides an element of chance—their ability to disperse via wind, water, or animal vectors to preferred habitat patches. Many plants occupying the forest floor, such as *Trillium cernuum*, have oil-rich bodies called *elaiosomes* attached to their seeds (Figure 8.19). Their only function is as a bribe to ants, which drag their seeds to underground nests, consume the elaiosomes, and discard the seed unharmed. Although the parent plant incurs an energy cost, it has increased the chances that its seeds will find themselves in a site suitable for germinating.

(a)

(b)

Figure 8.19 Directed dispersal of seeds. **(a)** Nodding wake-robin, *Trillium cernuum*, occupies deciduous forests from Manitoba to the Maritimes. Its seeds have oil-rich elaiosomes **(b)** that attract ants, which disperse the seeds to suitable sites.

8.10 LIFE HISTORY TRAITS: Extrinsic Conditions Influence Life History Strategies

Thus far, we have considered some of the many ways in which species differ in life history traits. They may be large or small; mature early or late; produce only a few offspring over their lifetimes or thousands in a single reproductive event. Given the interdependence and trade-offs involved in these and other traits—most obviously in the size and number of offspring—it seems clear that the set of life history traits exhibited by any particular species is not a random assemblage. These traits are the product of evolution by natural selection, with the outcomes moulded by the abiotic and biotic environment and constrained by trade-offs reflecting fundamental physiological and developmental processes. So, if life history traits are adaptations to the prevailing environment, do similar habitats support species with similar suites of life history traits—in effect, with similar strategies regarding allocation of time and energy to reproduction versus the potentially competing demands of survival and growth?

One way of classifying life history traits is in relation to temporal habitat variability. Consider habitats that are (1) variable in time or ephemeral versus habitats that are (2) relatively stable (long lasting and constant), with little random fluctuation. This dichotomy is the basis of the concept of r- and K-selection (Pianka 1972). The terms r and K relate to the logistic population growth model (see Section 10.7: r is the intrinsic rate of growth; K the carrying capacity). This theory predicts that species adapted to these different habitats will differ in their size, fecundity, age at first reproduction, number of reproductive events over a lifetime, and lifespan.

In this model, **r-strategists** are short-lived, reproduce at an early age, and have high reproductive output (often semelparous) with minimal parental care. They are typically small in size, and their many offspring have low survivorship. Such species inhabit temporary habitats, with unstable or unpredictable environments that can cause catastrophic mortality that is independent of population density. In such habitats, resources are rarely limiting, and r-strategists exploit these non-competitive situations while they last. Weeds and short-lived insects are typical r-species. They are well adapted for dispersal and colonization and respond rapidly to disturbance.

K-strategists, in contrast, have traits that support stable populations of long-lived individuals. They have a slower growth rate at low densities but can better maintain growth at high densities. K-strategists delay reproduction and are typically iteroparous, with relatively few offspring produced each time. Their offspring develop more slowly, but are larger at maturity. Among animal K-strategists, such as large mammals, parents care for the young; among plants, seeds possess abundant stored food reserves that are the plant equivalent of parental care. They are good competitors in the sense that they are efficient users of resources, but their populations are at or near *carrying capacity* (maximum sustainable population size) and are resource limited. Their mortality relates more to density than to unpredictable abiotic conditions. These traits, combined with restricted dispersal, make K-strategists poor colonizers of disturbed habitats.

The r versus K model works best in comparing organisms that are taxonomically or functionally similar. The comparison of *Typha* spp. in Section 8.8 is an example (McNaughton 1975). Similarly, the spotted (*Ambystoma maculatum*) and northern redback (*Plethodon cinereus*) salamanders (Figure 8.20) have contrasting life histories, although (somewhat at odds with the theory) they occupy similar habitats in mixed-wood forests in eastern North America, occurring in Canada from the Great Lakes to the Maritimes. The spotted salamander, an r-strategist, lives under logs and damp leaves. In late winter, individuals migrate to ponds to reproduce. After mating, females lay up to 250 eggs in large masses attached to twigs just below the water surface. After mating, adults provide no parental care of eggs or young. In contrast, females of the redback salamander, a K-strategist, deposit just 4 to 10 eggs in a cluster in a crevice of a rotting log or stump. The female curls about the cluster, guarding the eggs until they hatch. It is the first salamander reported to practise social monogamy, wherein a male and female form a lasting social bond even if it does not always entail mating monogamy (Gillette et al. 2000).

Using the r versus K model to compare species across a wide range of body sizes is of less value. The correlation of body size with metabolic rate and longevity in homeotherms

(a)

(b)

Figure 8.20 Contrasting life history strategies in salamanders. The spotted salamander **(a)** abandons its huge egg mass, whereas the redback **(b)** guards its few eggs until hatching.

results in species with small bodies generally being considered *r*-strategists and those with large bodies *K*-strategists, even though species with similarly sized individuals often differ in life history strategies. Comparisons among unrelated taxonomic groups are even less valuable. The brown lemming (*Lemmus trimucronatus*), a common Arctic species, has three litters a year and up to nine pups per litter. It is an *r*-strategist, but only compared with other mammals. An invertebrate species with a similar reproductive output would be a *K*-strategist. Although useful, the *r* versus *K* dichotomy is oversimplified, with no gradations between types, and no allowance for the fact that a species might alter its strategy with age, developmental stage, or prevailing conditions.

More recently, a model called the *fast–slow continuum hypothesis* (Oli 2004) attempts to explain co-variation in life history traits in a way that is independent of body size. Typically applied to mammals and birds rather than plants, this hypothesis stresses the selective force exerted by ecological factors (extrinsic and intrinsic) on the pace of life. In contrast with the simplified dichotomy of the *r* versus *K* model, species are assessed along a continuum from fast to slow. For a given body size, "fast" species mature earlier and have higher resting metabolic rates and reproductive outputs than species at the "slow" end of the spectrum. Because of its size-independence, the model can be applied both among and within species, with age at first reproduction often used as a ranking criterion.

The coevolution of life history traits, metabolism, and behaviour are an intriguing aspect of this theory. Wolf et al. (2007) suggest there is a trade-off between an organism's age at first reproduction and the thoroughness with which it explores its environment. Fast-species individuals explore their habitat more superficially and exhibit personality traits of boldness and aggressiveness. Slow-species individuals are more thorough in their exploratory behaviour and exhibit more timidity. From an

evolutionary standpoint, neither lifestyle is better or worse, but may be more or less adaptive in different habitats depending on food resources and predation risk.

Based on a comparative analysis of behavioural studies of 19 muroid rodents (including various vole and field mouse species), Careau et al. (2009) provided the first empirical evidence for these proposed links among personality, life history, metabolic expenditure, and behaviour. Exploratory thoroughness was positively correlated with age at first reproduction, with more thorough explorers reproducing later and having lower basal metabolic rates. Far from being a disadvantage, this "slow" strategy may prove beneficial in unproductive and/or unpredictable environments because more thorough exploration may increase the probability of finding scarce resources, and lower metabolic and reproductive costs may increase survival during resource shortages (Careau et al. 2009). Although not normally applied to plants, this fast–slow continuum fits the contrast between the herbaceous habit of fast-growing grasses in a fertile prairie versus the conservative habit of a coniferous evergreen in a nutrient-poor boreal site (see Chapter 23).

A life history classification developed specifically for plants is based not only on habitat but also on ecosystem properties: (1) frequency and severity of disturbance; (2) intensity of competition, as resources become more limited; and (3) environmental stress or harshness (Figure 8.21a; Grime 1977). There are three primary strategies. Highly disturbed sites are colonized by **ruderals** (*R*). Typically small and short-lived, ruderals allocate most of their resources to early and prolific reproduction and have traits that facilitate dispersal to disturbed sites. Ruderals, of which annuals are good examples, are essentially *r*-strategists. Predictable, favourable habitats with abundant resources select for species that allocate resources to growth over reproduction, favouring species with efficient resource acquisition and competitive ability. These **competitors** (*C*) resemble

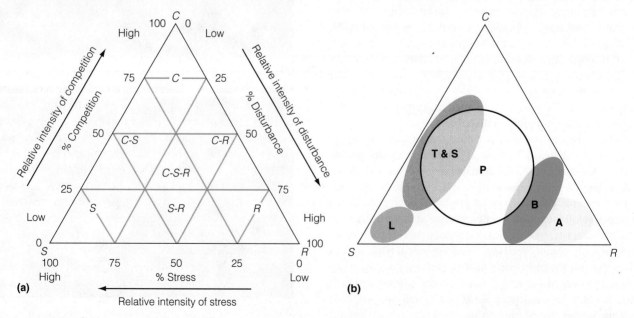

Figure 8.21 Grime's model of life history variation in plants. **(a)** Ruderals (*R*), competitors (*C*), and stress-tolerators (*S*) define the three points of a triangle. Intermediate strategies involve combinations (e.g., *CS*, *CR*, *CSR*, and *SR*). **(b)** Life history strategies of trees and shrubs (T & S), lichens (L), biennials (B), perennial herbs (P), and annuals (A).

(Adapted from Grime 1977.)

K-strategists and tend to be large trees or other long-lived perennials with delayed reproduction that can cope with crowded habitats subject to resource limitations.

The third category, **stress-tolerators** (*S*) has no equivalent in the *r* versus *K* model. They occupy habitats where resources are limited not because population sizes are high (as with competitors) but because conditions are harsh, as on bare rock or in deserts. These conditions favour slow-growing, conservative species that allocate resources to maintenance over growth and reproduce rarely or to a limited extent. Typical examples, such as lichens, tend to be small, but unlike ruderals, they are often long-lived. Moreover, some tree species, despite their large size, are stress-tolerators. Black spruce (*Picea mariana*), the most common conifer in Canada's northern boreal forests, is a classic stress-tolerator, coping with pH as low as 3.5 in bogs where resources, particularly nitrogen, are extremely limiting.

The Grime system also has shortcomings, of which perhaps the most serious is its tendency to view competitive ability as an absolute trait, rather than something that varies under different environmental conditions (Grace 1991; see Section 13.8). However, compared with the *r* versus *K* model, the Grime system is more flexible and less tied to size differences. The three strategies form the end points of a triangular model that allows for intermediate strategies such as *CR* (competitive ruderals), depending on resource availability and disturbance frequency. Its flexibility is apparent from Grime's assessment of perennials, which occupy a continuum of intermediate strategies (Figure 8.21b). Both life history models are important in community development (*succession*; see Section 17.6).

However they are interpreted, life history traits represent a vital link between the individual and the next level on the ecological hierarchy—the population, the subject of Part 3. No matter how well adapted an individual might be to its abiotic and biotic environment, it is only by allocating energy to reproduction that it can help maintain or increase its population. But how it allocates its reproductive energy is affected by the ecosystem in which it has evolved, particularly its resource availability, harshness, and disturbance regime.

EcologyPlace

Visit EcologyPlace at www.pearsoncanada.ca/ecologyplace to access online resources that complement your textbook, and help you to apply and to review the information in this chapter. EcologyPlace includes

- an eText version of the book
- self-grading quizzes
- glossary flashcards
- and more!

Go to www.pearsoncanada.ca/ecologyplace and follow the registration instructions on the Student Access Code Card included with this text. If your book does not have a Student Access Code Card, you can purchase access to it at www.pearsoncanada.ca/ecologyplace.

SUMMARY

Types of Reproduction 8.1

Reproductive success is measured by fitness: the relative number of an individual's offspring that survive to reproduce. Reproduction can be asexual or sexual. Asexual reproduction (cloning) produces offspring that are genetically identical to the parent. Sexual reproduction, involving genetic recombination through gamete fusion, produces genetic variability among offspring.

Modes of Sexual Reproduction 8.2

Sexual reproduction takes many forms. Most animals are unisexual. Plants with separate male and female individuals are called dioecious. An organism with both male and female sex organs is hermaphroditic. Animals may be simultaneous or sequential hermaphrodites. Plant hermaphrodites have bisexual flowers or, if monoecious, separate male and female flowers on the same individual. Some plants and animals change sex.

Mating Systems 8.3

Mating systems in plants include cross-fertilization and autogamy. Mating systems in animals include monogamy, polygamy, and promiscuity. Different mating systems are favoured in different environments, influencing life history traits and population properties. Monogamy, which involves a lasting pair bond, is selected for when the care of both parents is beneficial for rearing young. There are two forms of polygamy. In polygyny, the male acquires more than one female; in polyandry, the female acquires more than one male. Competitive mating and sexual selection are generally more intense in polygamy than in monogamy, but the high frequency of extra-pair copulation in monogamous species suggests it is more important than once thought.

Sexual Selection 8.4

In sexual selection, males generally compete with males for mating opportunities, but females typically choose mates. Sexual selection favours traits that enhance mating success, even if they handicap the male by making him more vulnerable to predation. Intrasexual selection involves competition among individuals of the same sex, whereas intersexual selection involves differential attractiveness of individuals of one sex to the other. By choosing males with "good genes," females may increase their own fitness. Intersexual selection can be antagonistic in species where females resist additional matings, leading to a coevolutionary arms race.

Impact of Resources 8.5

Females may choose mates based not on their phenotypic traits but on the resources they offer, usually a defended territory. By choosing a male with a high-quality territory, the female may increase her fitness. Male traits may be indirect indicators of territory quality.

Reproductive Effort 8.6

Reproductive effort is the time and energy allotted to reproduction and the care of offspring. It varies greatly among species and involves trade-offs: the more the reproductive effort, the less energy is available for growth and maintenance. Organisms that produce many offspring invest little in each, thereby increasing parental fitness but decreasing the fitness of the young. Organisms that produce few young invest more in each, thereby increasing the fitness of the young at the expense of the parents. Altricial offspring require more post-birth care than precocial offspring.

Reproductive Timing 8.7

To maximize fitness, an organism must balance its immediate reproductive effort against future prospects. One strategy, semelparity, invests maximum energy in a single reproductive effort. The alternative, iteroparity, allocates less energy each time to repeated reproductive efforts.

Reproductive Output 8.8

Both parental traits and food supply affect reproductive output. Within a related group, there is a direct relationship between size and fecundity—larger organisms produce more young.

Offspring production may reflect food supply. When food is scarce, parents may fail to feed some offspring, especially if hatching is asynchronous. In some cases, more vigorous young kill weaker siblings. Clutch and litter sizes typically increase from the tropics to the poles. This pattern may reflect (1) length of daylight, which influences foraging time; (2) climatic instability; or (3) greater winter mortality in higher latitudes, keeping populations below carrying capacity.

Habitat Selection 8.9

Reproductive success depends heavily on habitat choice. Most studies have focused on birds that defend breeding territories. Habitat selection involves a hierarchical approach, with initial selection based on general landscape features. Within this general area, individuals respond to specific structural features, which relate to species needs and their response capabilities.

Life History Traits 8.10

Various models attempt to explain correlated patterns of life history traits. According to the *r* versus *K* model, *r*-strategists occupy variable or ephemeral environments or face heavy predation. They produce many young, increasing the chance that some will survive. *K*-species are large, long-lived species that live in more stable environments. They produce few young and are iteroparous, adjusting the number of young in response to environmental conditions and availability of resources. The fast–slow continuum hypothesis proposes that, for organisms of a given size, age at first reproduction is positively correlated with exploratory thoroughness. Fast species explore their environment more superficially, are more aggressive, reproduce earlier, and have higher metabolic rates. Grime's model classifies plants as ruderals, stress-tolerators, competitors, or an intermediate type based on their exposure and response to disturbance, resources, and stress.

KEY TERMS

altricial	dioecious	lek	polyandry	semelparity
asexual reproduction	habitat selection	life history	polygamy	sexual dimorphism
autogamy	hermaphroditic	mating system	polygyny	siblicide
binary fission	indeterminate growth	monoecious	precocial	stress-tolerator
bisexual	intersexual selection	monogamy	promiscuity	unisexual
budding	intrasexual selection	ornamentation	*r*-strategist	
competitor	iteroparity	outcrossing	reproductive effort	
counteradaptation	*K*-strategist	parthenogenesis	ruderal	

STUDY QUESTIONS

1. Why might sexual reproduction be advantageous in a variable environment? Despite this possible advantage, many northern species rely heavily on asexual reproduction. Why?
2. Contrast *dioecious* and *monoecious* plants.
3. What are the advantages of hermaphroditism? Why is bisexuality more common in plants than in animals?

4. Distinguish among *monogamy*, *polygyny*, and *polyandry*. What type or types of mating system are selected for in a patchy habitat, and why?
5. Why is monogamy more common in birds than in mammals? Why do monogamous species engage in extra-pair copulation, and how does this differ from promiscuity?

6. How might female preference for a male trait, such as colour or body size, drive selection in a direction counter to that of natural selection? What hypothesis tries to reconcile this seeming contradiction?

7. Contrast *intrasexual selection* and *intersexual selection*, giving examples of traits selected for by each. Can the same traits be the target of both types of selection?

8. Why are the impacts of sexual selection (both intersexual and intrasexual) stronger on male traits than female traits? Under what circumstances is the reverse true, and why?

9. In a number of insects and arachnids (e.g., spiders), sexual cannibalism occurs in which the female eats the male during mating. How does sexual cannibalism relate to ecological efficiency? What kind of sexual dimorphism is likely to be associated with it?

10. Discuss the trade-off in the number of offspring produced and the amount of parental care, including biotic and abiotic factors that influence this trade-off.

11. What conditions favour semelparity over iteroparity?

12. For a given allocation to reproduction, there is a trade-off between the number and size of offspring (see Figure 8.14a). What type(s) of environment would favour plant species with a strategy of producing many small seeds rather than few large ones?

13. Can you think of examples where life history traits of (i) plant or (ii) animal species have been actively manipulated for human purposes? Which traits have been manipulated in each case, and what have been the likely ecological trade-offs for the species affected?

14. What life history traits of humans have been altered by technology or other aspects of modern culture? Think of life history in the broadest sense, including cultural factors that influence lifetime patterns of growth, development, and reproduction.

15. Describe three competing hypotheses regarding the trend to increased reproductive output in higher latitudes.

16. Why is habitat selection important for fitness? How do plants select habitat?

17. Contrast *r*-selected and *K*-selected organisms. Which strategy is more prevalent in unpredictable environments? Why is it preferable to apply this comparison to related species?

18. What types of environments might favour a "fast" species, and why?

19. Explain the three factors that are involved in Grime's model. What are its advantages over the *r* versus *K* strategy model?

FURTHER READINGS

Alcock, J. 2001. *Animal behavior: An evolutionary approach.* 7th ed. Sunderland, MA: Sinauer.
Excellent treatment of topics covered in this chapter, and a good reference for students who want to pursue specific topics in behavioural ecology.

Andersson, M., and Y. Iwasa. 1996. Sexual selection. *Trends in Ecology and Evolution* 11:53–58.
Excellent but technical review of sexual selection.

Buss, D. M. 1994. The strategies of human mating. *American Scientist* 82:238–249.
This article, an application of sexual selection theory to humans, is a fun read for students. It explores the question of whether mate selection by females has influenced male traits in humans.

Krebs, J. R., and N. D. Davies. 1993. *An introduction to behavioral ecology.* 3rd ed. Oxford: Blackwell Scientific.
Comprehensive discussion of behavioural topics covered in this chapter.

Policansky, D. 1982. Sex change in plants and animals. *Annual Review of Ecology and Systematics* 13:471–495.
Review article that explores many examples of individuals that change sex over their lifetime. It considers the cues as well as the mechanisms involved.

Reekie, E. G., and F. A. Bazzaz (eds.). 2005. *Reproductive allocation in plants.* New York: Academic Press.
Written by a team of international experts, this volume examines the evolutionary and ecological processes underlying differences in reproductive allocation by plants.

Stearns, S. C. 1992. *The evolution of life histories.* Oxford: Oxford University Press.
Explores the link between natural selection and life history, illustrating how both biotic and abiotic factors interact to influence evolution of specific life history traits.

The Copper River in southeast Alaska originates at the Copper Glacier on Mount Wrangell. From its headwaters, the river flows 480 km westward, dropping 1100 m before reaching the Gulf of Alaska. En route, it flows past 12 major glaciers, each adding to the volume of water and sediments that give the Copper River its distinctive grey tone. The colour suggests lifelessness, but each spring the Copper River comes alive as millions of Chinook salmon (*Oncorhynchus tshawytscha*) enter its mouth from the Gulf of Alaska and make their way upriver to spawn.

Upon entering freshwater, the salmon stop eating. Their bodies darken, losing their silvery sheen. In battling the current, their physical condition deteriorates. Their abundance and weakened state make them easy prey for brown bears and bald eagles. Most die before their task is completed. The survivors continue their trek to the headwaters and on to the streams that feed the river.

Upon arrival, the female chooses a site and digs a nest or *redd* with her caudal fin (tail). She deposits 3000 to 5000 eggs that are then fertilized by one or more males. Within a week, the adults die. The hatchlings that emerge live in freshwater for up to a year before heading downriver. Once in the sea, the salmon move thousands of kilometres as they mature. In two to five years, the adults return to the Copper River, and the cycle begins again.

This cycle has repeated itself for millennia, and the Copper River population of Chinook salmon has persisted. If on average each individual that spawns produces one offspring that also reproduces successfully, the population will persist. The Copper River population depends on various factors that affect survival, such as food abundance and abiotic conditions in the river and streams that are home to the hatchlings and in the open ocean that is home to the adults. In unusually warm and dry years, streams and tributaries can become so shallow that access to spawning areas is severely restricted, reducing reproduction.

This story is repeated along the British Columbia coast and as far south as northern California. (The species also occupies waters off Russia and Japan.) Because mature salmon return to their parent streams to spawn, the larger population of the north Pacific is a collection of local, genetically distinct subpopulations. The dynamics of the larger population, both demographic and genetic, reflect the collective dynamics of these local breeding populations.

The salmon's life history illustrates the broader concept of population dynamics. Birth and death are discrete events for the individual, but for the population, they are collective properties that affect population size over time. If the number born exceeds the number dying, the population increases. If deaths exceed births, the population declines. These changes can be dramatic. After decades of steady decline, culminating in a run of only one million in British Columbia's Fraser River in 2009, the sockeye salmon (*O. nerka*) was in serious trouble. In 2010, scientists such as John Reynolds, a salmon conservation biologist at Simon Fraser University, were astonished when the sockeye had its biggest run in 97 years: 30 million. Sockeye runs also reached records elsewhere along the west coast.

Various factors, both biotic and abiotic, influence individual survival and reproductive success. The net effect of these factors determines population growth. In the case of the turnaround in the Fraser River sockeye salmon, an abiotic factor may be responsible. Ash from the eruption of the Kasatochi volcano in Alaska in 2008 was dispersed over great distances and mixed in coastal waters by heavy storms. Iron contained in the ash fertilized north Pacific waters, causing a large algal bloom (Hamme et al. 2010) that some scientists believe may have boosted the sockeye's food supply. However, this possible causal link has yet to be established. Similarly, whether the surge in sockeye numbers is the harbinger of a long-term recovery remains to be seen.

Chinook salmon swim against the current to the upper reaches of the river, where they spawn.

Whatever their short-term fluctuations, the collective of sockeye salmon (or of Chinook salmon) in the Pacific Northwest does not form one homogeneous, inter-breeding population. Rather, the regional population consists of local populations connected by occasional exchange of individuals. Thus, the regional population is governed by processes operating at two spatial scales: the dynamics of local populations and of the collective. In Part Three, we explore these basic population concepts. In Chapter 9, we examine the fundamental properties of populations. In Chapter 10, we consider how the collective properties of birth and death govern the growth of local populations. In Chapter 11, we examine factors, both environmental and behavioural, that regulate population growth. Finally, in Chapter 12, we explore the processes that influence interactions among local populations and thus affect the dynamics of the regional metapopulation.

PROPERTIES OF POPULATIONS

This field of daisies represents a population—a group of individuals of the same species inhabiting a given area.

How do we perceive the organisms we encounter? Typically, we regard a friend, a squirrel, or an oak tree as an individual. Rarely do we consider it as part of a larger unit—the **population**: a group of individuals of the same species that inhabit a given area at a given time. This definition has two key points. First, because its members belong to the same species, the population is a genetic unit with the potential for interbreeding. So, whereas natural selection acts upon the individual, it is the population that evolves. Second, the population is a spatial and temporal concept, with a defined boundary in space and time—the population of sockeye salmon in the Fraser River, or the population of medium ground finches on Daphne Major in the Galápagos.

As aggregates of individuals, populations have unique properties such as *density* (number of individuals per unit area), *age structure* (proportion of individuals in various age classes), and *dispersion* (spacing of individuals relative to each other). Populations also exhibit *dynamics*—changes in number over time as a result of birth, death, and movement of individuals. In this chapter, we explore population structure in preparation for examining population growth in Chapter 10.

9.1 TYPES OF INDIVIDUALS: Individuals Are Either Unitary or Modular

A population is a group of individuals, but what is an individual? In most cases, defining an individual seems easy enough. We are individuals, and so are cats, insects, and so on. What defines us as individuals is the fact that we are **unitary organisms** whose form, development, growth, and longevity are predictable and determinate from conception on. The zygote, which forms when gametes fuse, grows into an embryo, which gives rise to a genetically unique organism that is easy to distinguish from others. Admittedly, this view is oversimplified; some unitary organisms (e.g., fish) are indeterminate and continue growing over their lifetime. However, the concept of an individual as a discrete unit cannot be applied to modular organisms.

In **modular organisms**, the zygote (which again arises from the fusion of gametes during sexual reproduction) develops into a unit of construction (a *module)* that then produces more, similar modules. Most plants and fungi are modular. A tree, shrub, or herbaceous plant grown from a seed (which contains an embryo that arose from a zygote) is a genetically distinct individual. But instead of growing as a single unit, it grows as a series of repeated modules, adding on leaves and buds throughout its life from regions of active growth called **meristems** (Figure 9.1a, p. 184). Unlike unitary organisms, modular organisms often have a wide range of final size. In addition, many modular organisms, including trees such as trembling aspen (*Populus tremuloides*), shrubs, and many herbaceous perennials, also send up asexual shoots or suckers that either remain attached or live independently (Figure 9.1b).

These new clonal modules may cover a considerable area and resemble discrete individuals.

What constitutes an individual in a modular species? A **genet** is a genetically distinct, free-living organism that arises from a fertilized egg. (Genets can also arise from parthenogenesis; see Section 8.1.) A **ramet** is a module that is produced asexually by a genet and is capable of independent existence. Ramets—which may or may not remain physically linked to their parent genet—can flower, produce seeds, and give rise to their own ramets. Whether living independently or attached, all ramets are genetically identical to their parent genet. (Indeed, the term *genet* is also used for the totality of ramets that are produced by a genet.) Thus, by reproducing asexually, a genet can cover a large area, horizontally and vertically, and greatly extend its life. Some modules die, others live, and new ones are added.

Plants and fungi are obviously modular, but there are also modular animals, such as corals, sponges, and bryozoans (Figure 9.2, p. 184). Whatever the taxonomic group, modularity complicates population study by generating two levels of population structure: (1) the genetic individual (genet) and (2) the module (ramet). Consider an aspen population anywhere in Canada. There may be only two or three genets in a stand of hundreds of trunks, most of which arose as suckers. Counting genets is worthwhile if genetic variability is the focus of a study, but for most purposes ramets are treated as individual members of a population.

9.2 DISTRIBUTION AND RANGE: Many Factors Influence the Spatial Extent of a Population

The **distribution** of a population describes its spatial location, the area in which all its members live. Distribution is based on the presence or absence of organisms. If each red dot is the position of an individual on the landscape (Figure 9.3, p. 185), we can draw a line defining the distribution of the population. If the area encompasses all individuals of a species (total area occupied by all its populations), it describes the distribution of the species (also called its **geographic range**).

A population's distribution reflects the occurrence of suitable abiotic conditions. Red maple (*Acer rubrum*) is one of the most widespread deciduous tree species in eastern North America (Figure 9.4, p. 185). Its northern limit occurs in parts of southern Canada where winter lows drop to −40°C. Its southern limit is the Gulf Coast and southern Florida, and dry conditions halt its westward extent (Burns and Honkala 1990). Within this range, red maple is a generalist, growing on soils of differing texture, moisture, and pH on a variety of elevations from wooded swamps to dry ridges. Thus, red maple's broad environmental tolerance facilitates its extensive geographic range—a range that will likely move farther north with global climate change, assuming sufficient moisture (see Section 28.6).

Red maple illustrates another critical factor limiting species range: geographic barriers. Although it inhabits islands

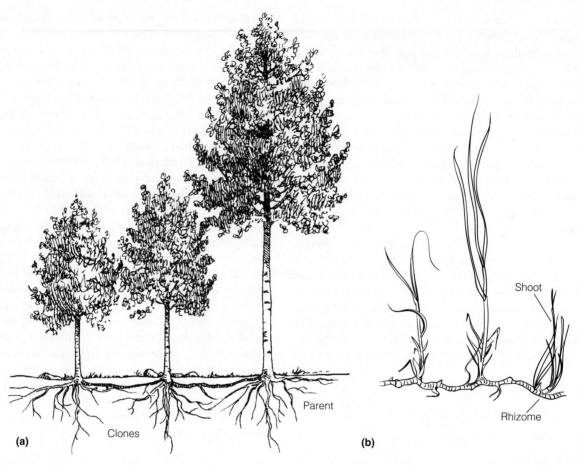

Figure 9.1 Modular growth in plants. **(a)** Aspen produces root modules, which give rise to clones of varying ages, with the youngest suckers forming the leading edge of growth away from the parent genet. **(b)** Rhizomes of the sea grass *Halodule beaudettei* grow laterally below bottom sediments, with new shoots produced at regular intervals.

south of mainland Florida, its southern and eastern limits correspond to the Gulf and Atlantic coasts. There may be suitable sites for red maple in Europe and Asia, but geographic barriers block it from colonizing them. Other barriers to dispersal, such as mountains and extensive areas of unsuitable habitat, and

biotic interactions, such as competition and predation, may also restrict the geographic range of a species.

Within a population, individuals do not occur everywhere. They occupy only those locations that meet their resource needs and where all conditions fall within their tolerance range

Figure 9.2 Modular animal species: **(a)** corals and **(b)** sponges.

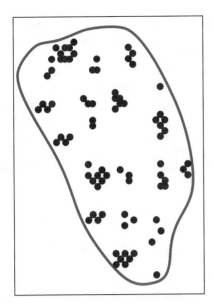

Figure 9.3 A hypothetical population. Each red dot represents an individual. The blue line defines the population's distribution.

(see Section 5.12). We can describe the distribution of the moss *Tetraphis pellucida* at many scales, ranging from its global extent to a clump on a dead conifer stump (Figure 9.5) (Forman 1964). *Tetraphis* grows only where temperature, humidity, and pH are suitable. Different factors limit it at different scales.

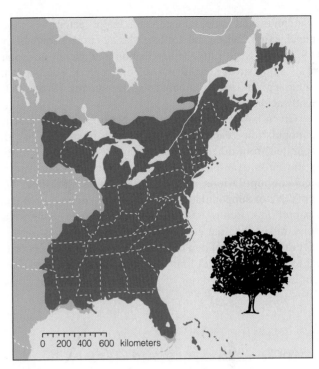

Figure 9.4 Red maple's widespread distribution reflects its ability to thrive on a wider range of soils and elevations than any other North American tree species.

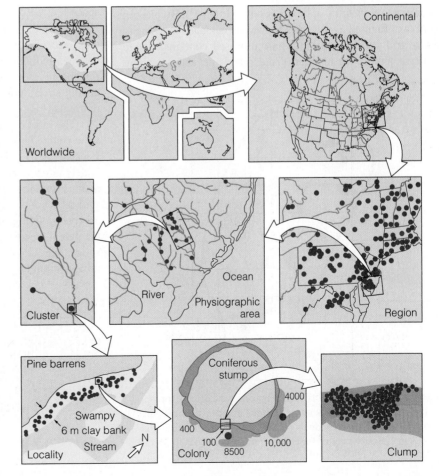

Figure 9.5 Distribution of the moss *Tetraphis pellucida* at various spatial scales, from its global range to colonies on a tree stump.

(Adapted from Forman 1964; as illustrated in Krebs 2001.)

At a continental scale, climate is the major factor. It is widespread across Canada and north to Alaska, but rare south of the U.S. border. Regionally, its distribution is limited to microclimates along stream banks, where conifer trees are abundant. Within a given locality, it only colonizes stumps that are sufficiently acidic.

Given habitat heterogeneity, most populations exist as subpopulations, each occupying a patch of suitable habitat separated from other subpopulations by unsuitable habitat. In the moss example, the regional population consists of **local subpopulations**, each in an associated watershed. The collective of subpopulations is called a *metapopulation* (see Chapter 12). Ecologists typically study local subpopulations, rather than the entire population of a species over its geographic range. When discussing a population, it is critical to define its spatial limits, such as the population of red maples in Nova Scotia or of sockeye salmon in the Fraser River.

9.3 POPULATION SIZE: Populations Differ in Abundance and Density

Whereas distribution defines a population's spatial extent, **abundance** defines its size (total number of individuals). In Figure 9.3, the abundance is the total number of red dots (individuals) within the blue line outlining the distribution. Abundance reflects two factors: (1) the area over which a population occurs and (2) **population density**: the number of individuals per unit area (m^2, km^2, or ha) or volume (L or m^3). By placing a grid over Figure 9.3 (Figure 9.6), we can calculate the density of any grid cell by counting the red dots it contains. *Crude density* is simply the number of individuals per unit area. But

individuals are usually not equally numerous everywhere, because not all areas are suitable or accessible. Crude density then varies widely.

To account for such variation, ecologists may estimate **ecological density**: the number of individuals per unit of suitable habitat. For a bird species that occupies hedgerows, biologists might express density as the number of birds per unit length of hedgerow, rather than as birds per hectare. However, ecologists rarely estimate ecological density because determining what portion of an area constitutes suitable habitat is often difficult or subject to bias.

9.4 SPATIAL DISPERSION: Individuals May Be Spaced Randomly, Uniformly, or in Clumps

Whatever their abundance, populations can also differ in their pattern of **spatial dispersion**: the position of individuals relative to each other. Individuals may be spaced randomly, uniformly, or in clumps (Figure 9.7). The dispersion pattern may reveal abiotic and biotic forces at work in the population. A *random dispersion* pattern, in which each individual's position is independent of other individuals, indicates either an absence of biotic interactions, or random availability of resources. As these two possibilities rarely apply, random dispersion is rare.

A *uniform dispersion*, in which individuals are evenly spaced, often reflects negative interactions among individuals. For example, *intraspecific competition* (within the same species; see Chapter 11) tends to create a minimum distance between individuals. Uniform dispersion is common in territorial species (see Section 11.9) and in plant populations experiencing severe competition for soil resources such as water or nutrients (Figure 9.8).

The most common spatial dispersion pattern is *clumped*, in which individuals occur closer together than expected by chance. Because the population in Figure 9.6 is clumped, density varies widely among grid cells. Clumping often reflects positive interactions among individuals. Some species form social groups, such as schools of fish or flocks of birds. Plants that reproduce asexually often grow in clumps, as ramets extend outward from the parent (see Figure 9.1). Humans are clumped as a result of social behaviour, economics, and geography, reinforced by urbanization. Patchy habitat or resources are abiotic factors that may also cause clumping.

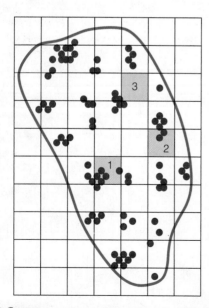

Figure 9.6 Difference between abundance and density for the population in Figure 9.3. Abundance = total number of individuals. Density = number of individuals per unit area. Assuming each grid cell = 1 m^2, density = 5/m^2 for cell 1, 2/m^2 for cell 2, and 0 for cell 3.

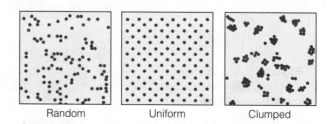

Random | Uniform | Clumped

Figure 9.7 Patterns of spatial dispersion within a population.

Figure 9.8 Shrubs in the Kara Kum Desert of central Asia are uniformly dispersed. Intense competition for water causes roots to extend laterally up to eight times the diameter of their canopy.

As with population distribution, we can describe spatial dispersion at multiple spatial scales. *Euclea divinorum*, a shrub occupying African savannahs, is highly clumped (Figure 9.9a). However, because *Euclea* grows under *Acacia tortilis* trees, *Euclea* clumps are uniformly spaced through the uniform dispersion of *Acacia* (Figure 9.9b), which reflects competition for water. In contrast, the moss in Figure 9.5 is clumped at two spatial scales. Populations are clumped along streams. Within these areas, individuals grow in clumps on conifer stumps.

9.5 SAMPLING POPULATIONS: Ecologists Use Various Methods to Estimate Abundance

As we have seen, population size is a function of density and area occupied: abundance = density × area. But how is density determined? When both the distribution and the abundance are

small, as with some rare or endangered species, a complete count (**census**) may be possible. Similarly, in habitats that are unusually open, such as antelopes on a plain, a direct count may be feasible. In most cases, however, density is estimated by sampling the population.

A method used for plants and sessile animals involves sampling units called **quadrats** (often square but sometimes rectangular or circular). Researchers divide the study area into a grid and count the organisms of concern, usually in a randomly selected subset (*sample*) of the quadrats (as in Figure 9.6). Quadrat shape and size vary with the organism type; ecologists use 10 m × 10 m quadrats for sampling trees, whereas a 1 m × 1 m quadrat is more appropriate for herbaceous (non-woody) species, and even smaller quadrats for mosses and lichens. Researchers determine the mean density of the species in the units sampled and multiply by the total area to estimate population abundance.

The accuracy of density estimates is affected by spatial dispersion. One might assume that clumped populations have a high density (or low, if sampling misses the clumps). In Figure 9.6, the population is clumped and density varies widely among grid cells. Yet overall, a clumped population may or may not be denser than random or uniform populations. Given the greater variation among quadrats in clumped populations, more quadrats must be sampled to get an accurate estimate. Estimates are most accurate for uniform populations.

Quadrat size also affects density estimates. Compare the variation in estimates for the same population depending on the dispersion pattern and the size of the sampling unit used (Figure 9.10, p. 188). The solution is not to use the largest quadrat possible. Rather, ecologists increase the number of quadrats and report a measure of variation (either standard error or confidence interval) for the density estimate.

For mobile animals, ecologists use other sampling methods. **Mark–recapture**, which involves capturing, marking, and recapturing individuals, is common. The method has many

(a)

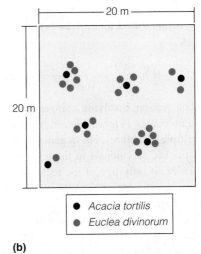

- ● *Acacia tortilis*
- ● *Euclea divinorum*

(b)

Figure 9.9 Spatial dispersion of *Euclea divinorum*. Shrubs are clumped under *Acacia* trees (a) as seen in (b) sample plots. The clumps are uniformly spaced as a result of uniform spacing of *Acacia*.

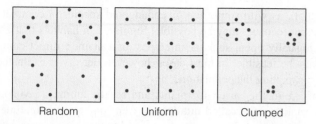

Figure 9.10 Effect of spatial dispersion on estimating density. The population size in each area is 16. We divide each area into four sampling units and choose one at random. Estimates differ with the unit selected. For the random population, estimates are 20, 20, 16, and 8. For the uniform population, any sampling unit gives a correct estimate (16). For the clumped population, estimates are the most variable: 32, 20, 0, and 12.

variations, and entire books are devoted to its application and analysis. Yet the basic concept is simple. A known number of animals (M) are trapped, marked, and released back into the population (N). After allowing sufficient time for the marked individuals to reintegrate into the population, more individuals are captured (n). Some of the individuals in this second capture will be marked (recaptured, R), and others unmarked. If we assume that the ratio of marked to total individuals in the second sample (R/n) represents the ratio for the population (M/N), we can estimate population size using the following relationship:

$$\frac{R}{n} = \frac{M}{N}$$

The unknown in this formula is N. We solve for N by rearranging the equation:

$$N = \frac{nM}{R}$$

Suppose that in sampling a rabbit population, we capture and tag 39 rabbits. After their release, the ratio of tagged or marked rabbits (M) to the number of rabbits in the entire population (N) is M/N. In the second sample, we capture 15 marked (R) and 19 unmarked rabbits—a total of 34 (n). Population size, N, is estimated as

$$N = \frac{nM}{R} = \frac{(34 \times 39)}{15} = 88$$

This simplest version, involving a single mark-and-recapture, is called the *Lincoln* (or *Petersen*) *index*. More advanced versions involve multiple recaptures, but in general, the smaller the proportion of marked individuals in the recapture, the greater the population size. If only five of the recaptured rabbits had carried marks, the estimate of the rabbit population would have been much larger.

Mark–recapture methods make many assumptions, some of which are problematic: (1) no individuals are born or die, or enter or leave, between captures; (2) no marks fall off, or are missed by the researchers; (3) marking neither harms the individual nor alters the probability that a marked individual

will be caught in the second or subsequent captures. We might assume that capturing a rabbit and releasing it might make its recapture less likely because the rabbit will try to avoid the trap, but marked individuals may seek out recapture for the food bribe.

Some methods of estimating animal abundance use indirect observations relating to the presence of organisms. Examples include counts of vocalizations, such as the number of drumming ruffed grouse heard along a trail; animal scat (feces) seen along a path; or tracks crossing a road. Assuming these observations have some reliable relationship to population size, the data are converted to the number of individuals seen per kilometre or heard per hour. Such counts, called **abundance indices**, cannot estimate absolute abundance, but a series of values collected from the same area over many years reveals abundance trends. For example, Christmas bird counts are held worldwide at the same time each year. Counts obtained from different areas during the same year also allow comparison of abundance among habitats. Most population estimates of birds and mammals derive from abundance indices. As valuable as they are, they are subject to bias, reflecting the impact of size, colour, and behaviour on the probability of detection.

9.6 AGE STRUCTURE: The Proportion of Individuals of Different Ages Affects Population Growth

Abundance describes the number of individuals in a population but does not indicate how individuals differ in age. Unless each generation reproduces and dies in a single season, not overlapping with the next generation (as do annual plants and insects), the population will have an **age structure (age distribution)**: the number or proportion of individuals in different age classes. Because reproduction is restricted to certain age classes and mortality is more prominent in others, age structure influences how quickly or slowly populations grow.

Populations have three ecologically important age classes: *pre-reproductive*, *reproductive*, and *post-reproductive*. How long individuals remain in each stage depends largely on the organism's size and life history (see Chapter 8). Among annuals, the length of the pre-reproductive stage has little impact on population growth, but for organisms with variable generation times, the length of the pre-reproductive stage is critical (see Section 8.10). Short-lived organisms with a short age to maturity have the potential to increase rapidly, with a short span between generations, whereas long-lived organisms with a long pre-reproductive period, such as elephants and whales, increase slowly, with a long span between generations.

Determining age structure requires obtaining the ages of individuals—easy for humans but problematic for many wild populations. Biologists age organisms in several ways, depending on the species (Figure 9.11). The most accurate method is to mark a cohort of individuals at birth and follow their survival

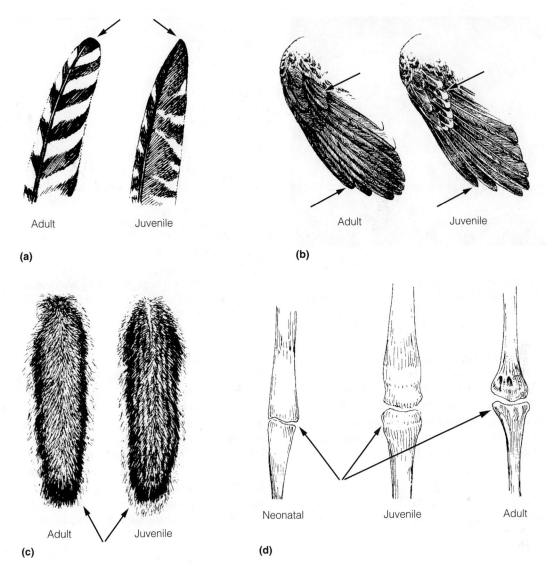

Figure 9.11 Methods for determining the age of birds and mammals. **(a)** The leading *primary feather* of wild turkeys is rounded for adults and pointed for juveniles. **(b)** Besides a pointed primary feather, juvenile bobwhite quail have buff-coloured *primary coverts* (arrows). **(c)** Juvenile grey squirrels have more distinctive bands of white and black along the tail edge than adults. **(d)** The bone structure of bat wings differs with age.

through time (see Table 10.1). Biologists often use other, less accurate methods, including examining carcasses to determine age at death and growth rings in the teeth of carnivores and *ungulates* (hooved mammals) or in the horns of mountain sheep. With birds, changes in plumage in living and dead individuals often separate juveniles from adults. Fish are aged by counting annual rings deposited on scales, *otoliths* (ear bones), and spines.

Studying the age structure of plant populations is difficult, in part as a result of their modularity. Foresters use trunk diameter to indicate tree age. However, although it may be valid to assume that diameter increases with age for dominant species in a forest, smaller trees and saplings often add little yearly to trunk diameter because their growth is suppressed by a lack of light or other resources. A more accurate method is to count annual growth rings (Figure 9.12a, p. 190), either with a sample of felled trees or by extracting cores from living trees with a *dendrometer*. As well as revealing the population's age structure, changes in ring width indicate growth response to abiotic factors over time (Figure 9.12b; see Research in Ecology: What Can Tree Rings Reveal about Forest Dynamics?, pp. 192–193). Aging herbaceous plants is harder. The best method is to follow marked seedlings over their lifetime.

Age pyramids (Figure 9.13, p. 190) are snapshots of population age structure at some point in time, illustrating the relative numbers in different age groups. Age structure reflects age-specific patterns of mortality and reproduction (see Section 10.5). In plants, age class distribution is often skewed in one of two ways. (1) Establishing populations have many saplings but few larger individuals, as in a balsam fir (*Abies balsamea*) population on the shores of Lake

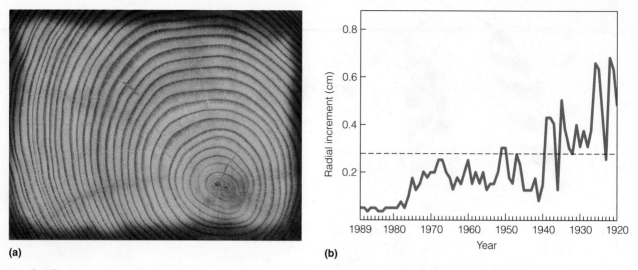

(a)

(b)

Figure 9.12 Tree growth ring analysis. **(a)** Cross section of a trunk showing annual rings of xylem. Ring width indicates the radial growth pattern through time. **(b)** Time series of radial increments for a beech tree (*Fagus grandifolia*). The dashed line is the overall mean.

Superior in Ontario (Figure 9.14a; in this case, age is inferred from size, which is less accurate than measuring age directly) (Hett and Loucks 1976). (2) Mature populations are dominated by mature trees that inhibit seedling establishment and sapling growth, as in an oak (*Quercus* spp.) forest in the Ozark Highlands of Missouri (Figure 9.14b) (Loewenstein et al. 2000). These individuals dominate until their death or removal, allowing individuals in younger age classes access to light, water, and nutrients.

Two populations can be identical in size but have very different futures depending on their age structure. If one pop-ulation has a disproportionately high number of mature individuals, it is likely heading for a decline compared with another population of the same (or smaller) size that has a high proportion of juveniles (Figure 9.15). Just as important is how the age structure changes over time. If the proportion of individuals in each class stays the same, it is a **stable age structure**. In actuality, stable age structures are relatively rare, occurring only in large mammals in non-fluctuating environments (Caughley and Sinclair 1994). Populations with stable age structures may be growing, declining, or staying the same. A **stationary age structure** is the special case

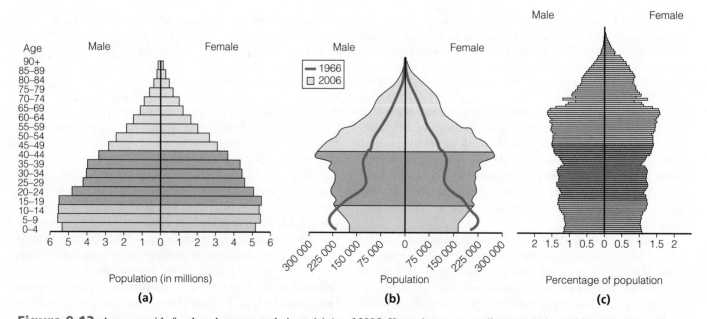

Population (in millions)

(a)

Population

(b)

Percentage of population

(c)

Figure 9.13 Age pyramids for three human populations. **(a)** As of 2005, Kenya has an expanding population, with a broad base of young entering the reproductive age classes. **(b)** As of 2006, Canada has a less tapered pyramid. The youngest age classes are no longer the largest as in the 1966 population (blue line). **(c)** As of 2006, the age pyramid for Finland is typical of a population that is approaching zero growth.

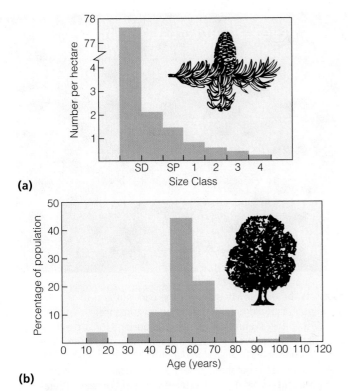

(a)

(b)

Figure 9.14 Size and age distributions in plants. **(a)** A balsam fir population near Lake Superior is dominated by seedlings (SD) and saplings (SP; < 10 cm diameter). Note the broken *y*-axis. Trees are grouped into 8-cm size classes. **(b)** An oak forest in Missouri is dominated by trees in the 40- to 80-year age classes. There has been little recruitment of younger age classes for the past 30 years.

(Adapted from (a) Hett and Loucks 1976 and (b) Loewenstein et al. 2000.)

of a stable age structure of a population that is constant in size because the absolute as well as the proportionate number of individuals do not change. What represents a stationary age structure for a given species depends on its survivorship in different age classes. Mortality of young age classes is lower in mammals than in insects, so an age structure that is stationary for black bears would foretell rapid decline for mosquitoes.

9.7 SEX RATIO: The Proportion of Males to Females Reflects Extrinsic and Intrinsic Factors

Age pyramids are often combined with a population's **sex ratio**—the proportion of males to females (see Figure 9.13). In theory, populations of sexually reproducing organisms tend towards a 1:1 *primary sex ratio* (ratio at conception). In most mammals, including humans, the *secondary sex ratio* (ratio at birth) is weighted towards males, but the population shifts towards females in older age classes. Males typically have a shorter life expectancy than females, for both physiological and behavioural reasons. Rivalries among males for

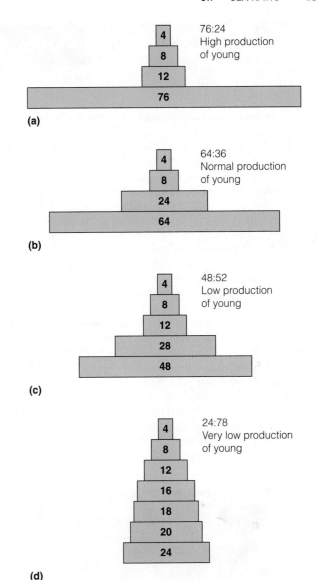

(a)

(b)

(c)

(d)

Figure 9.15 Age structures for mammals and birds. Values vary with species and habitats. **(a)** A growing population has a high proportion of young. **(b)** A stationary population has a more equitable distribution of size classes. The shape of its pyramid depends on mortality; the higher the juvenile mortality, the higher the proportion of young needed to sustain the population. **(c)** and **(d)** Declining populations have a top-heavy age structure, with a higher proportion of mature individuals.

dominant positions or for acquiring mates may not increase mortality directly, but they take an indirect toll. Bearing and raising young can increase mortality in female mammals. Male birds may outnumber females in reproductive age groups, as females are more susceptible to predation and attack while nesting.

Pollutants and other environmental hazards also affect sex ratios. Exposure to estrogen-like compounds causes feminization in fish (Lawrence and Hemingway 2003) and frogs (Pettersson et al. 2006). There are similar concerns about the effects of exposure of humans and other species to plastic residues.

RESEARCH IN ECOLOGY   What Can Tree Rings Reveal about Forest Dynamics?

Lori Daniels, University of British Columbia.

Tree ring analysis provides much more than a snapshot of the age structure of a population. By examining tree rings, scientists have learned much about climate change, both in the recent and in the distant past. Tree ring analysis is also a valuable tool for elucidating the biotic as well as the climatic history of a site. The University of British Columbia's Lori Daniels is challenging traditional models of *succession* (change in community structure over time; see Chapter 17) with studies using multiple lines of evidence, including stand reconstructions based on tree ring data. Her goal is to develop management strategies that sustain old-growth forests—their species composition, physical structure, and long-term resilience.

To sustain old-growth forests, we must first understand their structure and function. As part of a collaborative team, Daniels undertook a case study of Big Timber Park in Whistler, British Columbia (Figure 1). In preparation for the 2010 Olympics, the province established a Protected Areas Network to protect sensitive ecosystems. **Old-growth** forests were defined as forests that (1) are older than 250 years; (2) have characteristic structural features including large amounts of living and dead biomass (particularly dead standing trees or **snags**) as well as trees of a range of species, size, and age; (3) have developed between stand-replacing disturbances (in this case, fire); and (4) are disturbance-free (Daniels et al. 2008). The purpose was not only

to assess the old-growth status of the park but also to test the assumption that old-growth forests lack disturbance.

Why is this theoretical question important for a practical issue like ecosystem conservation? Traditionally, ecologists have regarded disturbance as an infrequent event that severely damages or removes vegetation, initiating a predictable successional sequence that culminates in re-establishment of "climax" vegetation in which little or no disturbance occurs—until the next major stand-replacing disturbance sets the process in motion again. In this view, old growth is in equilibrium and, if protected from human-initiated disturbances such as logging, can be preserved in a relatively unchanging state. However, if old-growth forest is instead prone to many and ongoing fine-scale disturbances of varying severity, then "preserving" such forests requires recognizing the importance of frequent disturbance. Fallen trees generate gaps in the canopy that may be colonized by various species, initiating a kind of mini-succession. The result is an ever-changing and not entirely predictable mosaic, as opposed to a static entity. So, in order to "save" these forests, we must recognize—and celebrate—their dynamic nature, not treat them as living museums. For a stimulating discussion of these opposing views, see Kimmins (2003).

Daniels's team sampled trees in a 30 m × 30 m plot in the centre of the Big Timber Park forest, extracting cores from all living trees with a diameter greater than 10 cm. For trees with decaying heartwood (and hence missing early rings), a conservative "minimum age" was estimated. Age structures were based on the proportion of individuals in 40-year age classes for each of the four tree species present (Figure 2). Living mature trees ranged from 121 to 305 years old, with most established from 1800 to 1880. For trees with decaying hardwood, minimum age ranged from 85 to 299 years, with establishment as far back as 1705. Populations of all four species (Douglas-fir, western hemlock, western redcedar, and Pacific silver fir) included trees that were over 246 years old, with a range of sizes as well as ages. Many younger trees were also present, most of which were the shade-tolerant Pacific silver fir (*Abies amabilis*).

Understanding the ecology of the different populations is critical. Despite its longevity, Douglas-fir (*Pseudotsuga menziesii*) is a shade-intolerant pioneer that regenerates after severe fires.

Figure 1 Big Timber Park forest in Whistler, British Columbia.

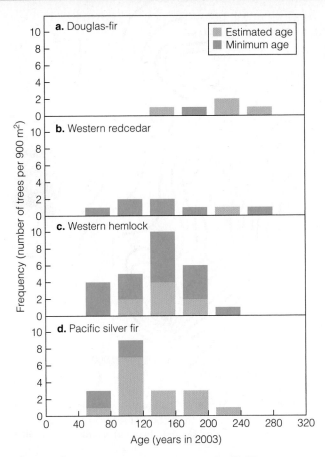

Figure 2 Age structure of **(a)** Douglas-fir, **(b)** Western hemlock, **(c)** Western redcedar, and **(d)** Pacific silver fir. "Minimum age" data represent trees with missing rings as a result of decayed heartwood.

(Adapted from Daniels et al. 2008.)

Its age structure is thus skewed towards older age classes (Figure 2a). The other species are more shade tolerant and regenerate after disturbances of many types and frequencies. Western hemlock (*Tsuga heterophylla*) has the greatest age class range, with more younger individuals (Figure 2b) than Douglas-fir. Although snags were present, they were not large trees but small individuals of understory species that likely died from suppression in deep shade. This fate is particularly likely for Pacific silver fir. Its age structure is skewed to younger age classes (Figure 2d), and its high sapling density makes it liable to high mortality. Nor were large logs present, another typical old growth indicator.

Based on these findings, Daniels concluded that, despite its considerable age, the Big Timber Park forest satisfies many but not all of the traditional criteria for old growth. Should this forest not be considered old growth, despite its age, or should we re-examine the criteria, particularly relating to the absence of disturbance? Whether it qualifies as old growth or not, what does the future hold for Big Timber Park? The age structure of its tree populations indicates that the answer will depend on the ongoing, interacting effects of tree aging, competition, and fine-scale disturbances from abiotic agents such as wind and biotic agents such as insects. Because biotic disturbances are host specific, small to medium-sized gaps arising from the deaths of one to several trees are more likely than widespread canopy loss. These gaps create opportunities for recruiting new individuals in younger age classes, rejuvenating existing populations.

The death of large trees will also allow some trees now in the understory to assume canopy status. Increased light in the gap will rescue some individuals from suppression and death as a result of self-thinning (see Section 11.4), allowing them to attain higher age classes. Daniels argues that the regime of frequent, fine-scale disturbance responsible for the diverse age structure of Big Timber Park is the norm for old-growth forests, not the exception. Rather than static, unchanging collections of old trees, old-growth forests are continually in flux.

As well as promoting an appreciation of the dynamic nature of old-growth forests as part of a conservation strategy, Daniels's findings have repercussions for management. Based on tree ring data, Daniels and her colleagues suggest that the interval between major stand-replacing canopy fires in British Columbia's coastal forests may be much longer (> 750 years) than once thought (~ 250 years) (Daniels and Gray 2004). In contrast to boreal forests where frequent fires maintain a mosaic of successional stages on the landscape, succession in west coast forests is largely a response to canopy gaps caused by the death of individual trees. Daniels advocates that ecologically sound management should encourage unevenly aged, structurally diverse forests (such as that of Big Timber Park) rather than monoculture plantings. Her work illustrates how basic population properties—age structure and density—not only enhance our understanding of forest succession but also apply to urgent issues such as forest conservation and management.

Bibliography

Daniels, L. D., and R. W. Gray. 2004. Disturbance regimes in coastal British Columbia. *BC Journal of Ecosystems and Management* 7:44–56.

Daniels, L. D., M. Yanagawa, K. L. Werner, L. Vasak, J. A. Panjer, R. A. Klady, J. Lade, and B. Brett. 2008. Preservation of old-growth forests: A case study of Big Timber Park, Whistler, BC. *The Canadian Geographer* 52:367–379.

Kimmins, J. P. 2003. Old growth forest: An ancient and stable sylvan equilibrium or a relatively transitory ecosystem condition that offers people a visual and emotional feast? Answer—it depends. *Forestry Chronicle* 79:429–440.

1. Age structures varied for the tree species growing in the park. Describe these differences and discuss what factors contribute to them.
2. What factors (abiotic and biotic) might cause tree ring size to vary among individuals of the same species, even in the same growing season?

Invertebrate sex ratios often deviate from 1:1. Many species harbour maternally inherited symbionts that bias sex ratios by feminization of males, induction of parthenogenesis, or killing of male offspring. Although the microbes responsible are most often *Spiroplasma* and *Wolbachia* bacteria, other microbial agents may be involved (Morimoto et al. 2001). However it is induced, parthenogenesis (see Section 8.1) typically produces female-biased sex ratios. Recall that in aphids, most of the population consists of wingless females produced asexually, with males produced only for sporadic sexual events.

For social insects such as bees, ants, and wasps, female-biased sex ratios play an integral role in group structure. Although sex ratios in social insects are used as a model system for exploring how selection shapes group behaviour, there may not be a single optimal strategy. A study of 47 species of the order Hymenoptera, including solitary and social species, indicates that sex ratio decisions are increasingly variable for species with complex social structure (Kümmerli and Keller 2011). Social insects alter ratios in response to various factors, including food supply, temperature, and number of queens, and decisions cannot always be assumed to be optimal.

9.8 DISPERSAL: Immigration and Emigration Involve Movements of Individuals

At some stage of their life cycle, most organisms are to some degree mobile, and their movements directly affect population density. Whereas *dispersion* describes the pattern of individuals in space, **dispersal** is the movement of individuals through space (and hence the pattern in time). The term *dispersal* may be used specifically for the movement of individuals away from each other, often from the place of birth to the place where they reproduce. When an individual disperses from a local subpopulation, it is called *emigration*, whereas movement into a subpopulation is *immigration*. Such dispersal is vital to metapopulation dynamics (see Chapter 12) and in maintaining gene flow between local subpopulations (see Section 5.6).

Many organisms, especially plants, depend on passive means of dispersal using either abiotic or biotic vectors, such as gravity, wind, water, and animals. The distance travelled depends on the properties of the organisms as well as the agents of dispersal. Seeds of most plants fall near the parent, with density falling off sharply with distance (Figure 9.16). Heavy seeds, such as oak acorns (*Quercus* spp.), disperse over much shorter distances than do light, wind-carried seeds of maples (*Acer* spp.), birch (*Betula* spp.), and dandelions (*Taraxacum officinale*). Some plants, such as wild plums (*Prunus* spp.), have heavy seeds that are dispersed over great distances because they use biotic vectors such as birds and mammals. These animals eat the fruits and deposit the seeds in their feces after passage through the digestive tract.

Dispersal mechanisms influence spatial dispersion. The clumped dispersion of *Euclea* in Figure 9.9 results from the use of *Acacia* as perches. As they perch atop *Acacia* trees, birds feed on *Euclea* fruits and deposit the seeds in their feces. Some seeds

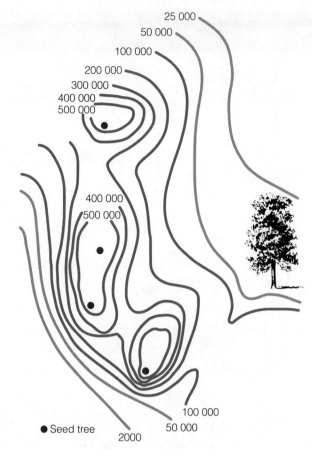

Figure 9.16 Annual seedfall of yellow poplar (*Liriodendron tulipifera*). Lines define areas of equal seed density. With this wind-dispersed species, seedfall drops off rapidly away from the parent trees.

(Adapted from Engle 1960.)

(or the fruits in which they develop) are armed with spines and hooks that attach to fur, feathers, or clothing. Although such dispersal is still passive because no energy is used for the movement itself, the plant uses energy to produce its specialized dispersal structures. Dwarf mistletoe (*Arceuthobium* spp.) is a rare example of active dispersal in plants. These parasites, which damage conifer forests across North America (see Section 15.1), disperse from one host tree to another by an explosive process that propels its seeds up to 10 m (Figure 9.17).

For mobile animals, dispersal is active, but many animals depend on passive, abiotic transport vectors, such as wind or water. Wind carries the young of some spiders, larval gypsy moths, and cysts of brine shrimp. In streams, larval forms of many invertebrates disperse downstream in the current. In the oceans, dispersal is often linked to currents and tides. In both land and aquatic habitats, some animals disperse by "piggy-backing" on other animals.

Among animals, there is no strict rule about what disperses. Dispersal may involve young and adults, males and females. Among birds and rodents such as deer mice (*Peromyscus maniculatus*), juveniles are most likely to disperse. Crowding, temperature change, food quality and abundance, and photoperiod have all been implicated in stimulating dispersal. Dispersing individuals are often seeking vacant habitat. In such cases, dis-

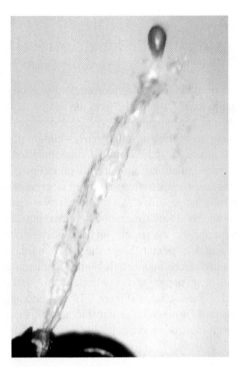

Figure 9.17 Explosive dehiscence (opening) of the fruits of the dwarf mistletoe.

tance travelled depends on the density of surrounding populations as well as the availability of suitable unoccupied habitat.

Dispersal entails ecological trade-offs. Dispersing individuals may benefit from reaching new habitats where resources are more available, especially if they are escaping a crowded site. But the new location may be less favourable than the previous, and they may encounter many risks en route. Individuals that remain in or return to the same location (**philopatric**) may possess locally adapted genes selected for over generations. They are more familiar with the habitat, which may help avoid predation, or secure prey if they are predators. However, overcrowding can cause intense competition, and excessive interbreeding can limit genetic variability.

9.9 MIGRATION: Some Species Exhibit Repeated Daily or Seasonal Movements

Unlike the one-way movement of dispersal, **migration** is an intentional, directional movement between two locations or habitats. It typically involves a round trip, either daily or seasonally. Zooplankton move to lower ocean depths by day and up to the surface by night, apparently in response to light intensity. Bats leave their daytime roosting places in caves and trees, travel to their feeding grounds, and return by daybreak.

Other migrations are seasonal, either short or long range, and normally involve individuals returning (often en masse) to their birthplace, where they reproduce. Earthworms make an annual vertical migration deep into the soil to spend the winter below frost level, and return to the upper soil in spring. Caribou (*Rangifer tarandus*) move from their summer calving range in the Arctic tundra to the boreal forests for the winter, where lichens are their major food source. Grey whales (*Eschrichtius robustus*) move

down from food-rich Arctic waters in summer to their warm wintering waters off the California coast, where they give birth (Figure 9.18). By far the most familiar migrations are those of waterfowl, shorebirds, and neotropical songbirds in spring to their nesting grounds and in fall to their wintering grounds.

New technologies are increasing our knowledge of long-distance fish migrations. The Pacific spiny dogfish (*Squalus acanthias*) is a shark that occurs off North America's west coast from California to Alaska and along the Aleutians to Europe and Japan. Its dorsal spines have been found in archaeological sites on the west coast of British Columbia, indicating that it has long been a food source for aboriginals. Many things about its biology are impressive; long-lived (80 years and over) and slow growing, it is a classic *K*-strategist (Section 8.10), with the longest gestation (22 months) of any animal. Equally impressive is its travel itinerary. In the most extensive program ever undertaken for a shark, 71 000 fish were tagged off the west coast of British Columbia from 1978 to 1988. Of the 4.1 percent recaptured by 2000, most were captured close to their release site, but some travelled up to 7000 km, including 30 that were captured near Japan (MacFarlane and King 2003).

Some migrations involve only one return trip, such as the Pacific salmon that spawn in freshwater streams (see introduction to Part Three). The least familiar type of migration is the one-way trip of some butterfly species. Adults travel one leg of the journey, reproducing after arriving. Although this sounds like dispersal, it is considered migration because subsequent generations complete the later legs of the journey, ultimately returning to the original location. Monarch butterflies (*Danaus plexippus*) are an example, taking several generations to travel between Canada and the United States and its breeding grounds in Mexico.

Figure 9.18 Migratory pathways of two vertebrates. **(a)** Ring-necked ducks (*Aythya collaris*) breeding in eastern Canada and the United States migrate along a coastal corridor to wintering grounds in South Carolina and Florida. **(b)** The grey whale summers in the Arctic and Bering seas and winters in the Gulf of California and the waters off Baja California (Mexico).

Migration is a complex behavioural adaptation, involving coordinating movement (often as a group) using a variety of signals, both visual and (in the case of birds) magnetic. It is an effective strategy for avoiding unfavourable seasons, but it entails trade-offs. The energy expended is substantial, and the risks en route, including predation and limited food supply, are high. Perhaps most seriously in a human-dominated world, migratory species are highly vulnerable to habitat loss or degradation. Even if a songbird species occupies relatively undisturbed summer habitat in the northern boreal forests of Canada, it must overwinter in the southern United States or Mexico, where pollution and/or habitat loss may pose substantial risks.

The risks associated with migration are substantive and hard to quantify. The black-throated blue warbler (*Dendroica caerulescens*) breeds in the forests of eastern North America—its range extends into southern Ontario, Québec, and the Maritimes—and overwinters in Jamaica and Cuba. More than 85 percent of annual mortality losses occur during the migratory period (Sillett and Holmes 2002). There are also survival differences among years in Jamaica, reflecting climatic variations associated with El Niño events (Sillett et al. 2000).

9.10 RANGE EXPANSION: Movements of Individuals Affect Population Distribution and Density

Movements of individuals have profound impacts on the properties of populations. Dispersal affects the spatial distribution of individuals and consequently affects local density. Emigration causes density to decline in some areas, while immigration into other areas increases the density of subpopulations or establishes new subpopulations in previously unoccupied sites.

In some cases, dispersal causes a shift or expansion of a species' geographic range. This effect is particularly evident in species that have been introduced (intentionally or not) into areas outside their native range. As the population establishes, individuals disperse into suitable habitat. Since the gypsy moth (*Lymantria dispar*) was introduced into eastern North America in 1869, it has expanded its range greatly and is projected to expand farther into Canada with climate change (Figure 9.19; see Ecological Issues: Human-Assisted Dispersal). The threat to native species posed by alien species is treated in more detail in Section 27.7.

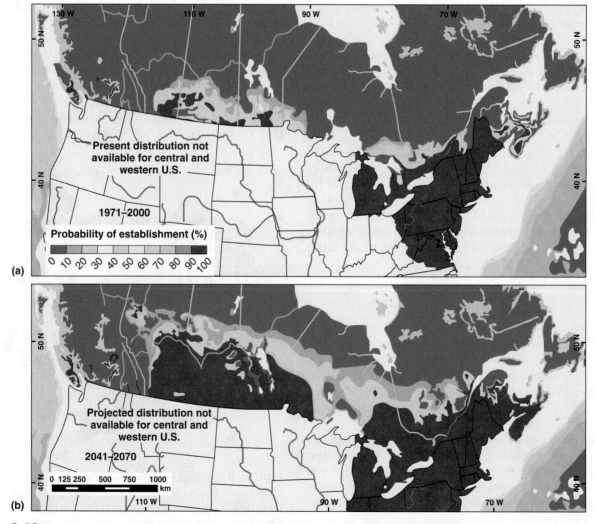

Figure 9.19 Range expansion of the gypsy moth in Canada by 2000 (**a**) and as projected with climate change by 2070 (**b**). Figure 9.19a has been modified to show the gypsy moth's distribution in the eastern U.S. as of 1994 (Krebs 2001); its spread into the central and western U.S. is not depicted.

(Source: Natural Resources Canada.)

Dispersal is a key feature of the life histories of all species, and many mechanisms have evolved to facilitate movement. In plants and fungi, seeds and spores are dispersed passively by wind or water, or actively by animals. In animals, dispersal of fertilized eggs, particularly in aquatic habitats, can transport offspring great distances. But animal dispersal more typically involves movement—either active or passive—of individuals, both juvenile and adult. In recent centuries, however, a new vector has redistributed species on a global scale: humans.

Increasingly, humans are on the move. As we move, we accidentally or intentionally introduce organisms into places where they have never occurred before. Sometimes these introductions are harmless. Many do not succeed, but of those that do, the introduced organisms may negatively affect native species and ecosystems. Many plants have been accidentally introduced along with agricultural products. Weed seeds are included in seed shipments, or on the bodies of domestic animals. Seed-carrying soil is loaded onto ships as ballast and then dumped elsewhere.

Humans have also introduced non-native plants intentionally. For example, purple loosestrife (*Lythrum salicaria*; see Figure 27.21), dubbed the "beautiful killer," has escaped from gardens to become a serious pest in Canadian wetlands. Most introduced plants do not become established, but some form extensive colonies. These invasives compete with native species for light, water, nutrients, and pollinators. They may also alter ecosystem function by changing patterns of water use or the frequency of disturbances such as fire—all of which can severely affect native species.

The gypsy moth (*Lymantria dispar*), native to temperate regions of central and southern Europe and Asia, northern Africa, and Japan, is a major insect pest of eastern North American hardwood forests. It was introduced into Massachusetts in 1869 in an effort to start a silk industry. Several escaped and became established. Some 20 years later, the first outbreak occurred, and despite all control efforts, it has not only persisted but extended its range north to eastern Canada, west to Wisconsin, and south to North Carolina (see Figure 9.19). Gypsy moth caterpillars defoliate millions of hectares of forests annually (Figure 1). According to the USDA Forest Service, losses to eastern forests exceeded $30 million annually from 1980 to 1996. The Asian strain that has invaded British Columbia has necessitated a $20 million control campaign that is unlikely to succeed.

Invasive plants can cause similar damage. We think of weeds as unwanted intruders into our front lawns (themselves dominated by non-native species), but the damage they cause to natural ecosystems is of far greater concern. There are countless examples, but leafy spurge (*Euphorbia esula*) (Figure 2) is among the most serious. Native to Europe and Asia, leafy spurge likely arrived in North America as a ballast contaminant in ships landing in New England

Figure 2 Leafy spurge is a common invasive weed across North America.

in the 1800s, and later as a seed contaminant to the prairies. Its geographic range in North America is broad. In Canada it occurs in all provinces except Newfoundland and Labrador, but is concentrated in southwestern Ontario and the Prairie provinces, extending as far west as British Columbia (Best et al. 1980).

Leafy spurge grows in open habitats, from native prairies and parklands to open woods, rangelands, and ditches. A herbaceous perennial, it establishes a deep root (up to 9 m) and spreads rapidly by clonal growth. Sexual reproduction is prolific, generating up to 8000 seeds per m^2 (Best et al. 1980). Its seed dispersal is explosive, catapulting seeds up to 5 m, and is aided by the presence of an elaiosome (see Section 8.9) that facilitates its dispersal by ants (Pemberton 1988). Given its combination of rapid vegetative growth in a diversity of habitats and its multifaceted seed dispersal, it is no wonder that leafy spurge has become such a problem.

Why should we care? In western mixed-grass prairie, leafy spurge reduces the diversity and abundance of native prairie species (Belcher and Wilson 1989). Particularly affected are rare species such as the already threatened western spiderwort (*Tradescantia occidentalis*). Nor are its effects restricted to natural ecosystems. It poses a serious hazard in rangeland because it produces a poisonous latex that can kill livestock (Best et al. 1980). Control measures have proved ineffective. Ecologists are exploring a range of biological controls using its predators and pathogens in its native habitat. While promising, these approaches are not without risk for the rest of the ecosystem, if the introduced predators diversify and affect native species.

Bibliography

Belcher, J. W., and S. D. Wilson. 1989. Leafy spurge and the species composition of a mixed-grass prairie. *Journal of Range Management* 42:172–175.

Best, K. F., G. G. Bowes, A. G. Thomas, and M. G. Maw. 1980. The biology of Canadian weeds. 39. *Euphorbia esula* L. *Canadian Journal of Plant Science* 60:651–663.

Pemberton, R. W. 1988. Myrmecochory in the introduced range weed leafy spurge (*Euphorbia esula* L.), *The American Midland Naturalist* 119:431–435.

1. On your campus or in your local community, can you identify examples of plant or animal species that owe their presence to active dispersal by humans?
2. How is agriculture an example of human-assisted dispersal?

(a) **(b)**

Figure 1 An oak forest **(a)** in summer has been completely defoliated by gypsy moths **(b)**.

In other cases, range expansion reflects temporal changes in environmental conditions, which shift the locations of suitable habitats. Ranges of North American tree species have shifted dramatically with climate change over the past 20 000 years (see Section 17.11). Jack pine (*Pinus banksiana*) and black spruce (*Picea mariana*), for example, have advanced northward since the end of the last ice age, albeit at different rates. Along the Grande Rivière in northern Québec, black spruce was the first to invade, followed a few millennia later by jack pine (Desponts and Payette 1993). Further changes in the ranges of plant, animal, and fungal species are likely to occur in response to human-induced changes in Earth's climate in the future.

Yet despite the impact of movements of individuals on population distribution and density over time, the primary factors driving population dynamics in the short term are simple—birth and death. These fundamental demographic processes are the subject of Chapter 10.

EcologyPlace

Visit EcologyPlace at www.pearsoncanada.ca/ecologyplace to access online resources that complement your textbook, and help you to apply and to review the information in this chapter. EcologyPlace includes

- an eText version of the book
- self-grading quizzes
- glossary flashcards
- and more!

Go to www.pearsoncanada.ca/ecologyplace and follow the registration instructions on the Student Access Code Card included with this text. If your book does not have a Student Access Code Card, you can purchase access to it at www.pearsoncanada.ca/ecologyplace.

SUMMARY

Types of Individuals 9.1

A population is a group of individuals of the same species living in a defined area. Most animal populations are composed of unitary organisms with a determinate growth form and longevity. In most plant and a few animal species, organisms are modular and consist of a series of repeated units. Modular populations consist of sexually produced genets and asexually produced ramets.

Distribution and Range 9.2

The distribution of a population describes the area over which it occurs and is influenced by the presence of suitable environmental conditions. The geographic range of a species encompasses the distributions of all its populations. Individuals are not typically distributed equally throughout the geographic range and can be described at different spatial scales.

Population Size 9.3

The number of individuals in a population defines its abundance, which is a function of (1) its population density (number of individuals per unit area) and (2) the area over which the population occurs. Because landscapes are not homogeneous, not all of the area is suitable habitat. The ecological density is the number of organisms per unit of suitable habitat.

Spatial Dispersion 9.4

Individuals within a population may be distributed differently in space. If the spacing of each individual is independent of others, then the dispersion pattern is random; if they are evenly distributed, dispersion is uniform. In most cases, individuals are clumped. These spatial dispersion patterns reflect a combination of abiotic and biotic factors.

Sampling Populations 9.5

For some species, a count or census is possible, but for most species, determining density and dispersion requires sampling. For sessile organisms, researchers often use sample plots (quadrats). For mobile organisms, researchers use mark–recapture techniques to estimate population size or determine relative abundance using indirect indicators of presence.

Age Structure 9.6

The age structure of a population is the proportion of individuals in each age class. There are three ecologically significant age classes: pre-reproductive, reproductive, and post-reproductive. In a stable age distribution, the proportion of individuals in each age class remains the same. A stationary age distribution is a special type of stable age distribution in which absolute as well as proportionate numbers remain constant. Trends in age structure through time indicate trends in population growth.

Sex Ratio 9.7

The sex ratio of sexually reproducing populations approaches 1:1 at conception and birth but often shifts as a function of sex-related differences in mortality in different age classes. Sex ratios in invertebrates are often female biased, as a result of bacterial endosymbionts and/or parthenogenesis. Social insects alter sex ratios in response to a variety of factors.

Dispersal 9.8

At some stage of their life cycle, most individuals are mobile. For plants, dispersal is usually passive and dependent on abiotic or biotic vectors. For mobile organisms, dispersal is typically active and occurs for various reasons, including the search for mates or for unoccupied habitat. Dispersal that moves individuals out of or into local populations is called emigration or immigration, respectively.

Migration 9.9

Whereas dispersal involves movement of individuals from the location where they were born to the location where they reproduce, migration is a systematic movement of individuals (often en masse) between different parts of their annual range. Migration can involve multiple return trips, one return trip, or a one-way trip. Migration entails ecological trade-offs. It has high energy and mortality costs, and makes species vulnerable to habitat loss in two locations.

Range Expansion 9.10

Dispersal alters the spatial distribution of individuals, affecting local population density and dispersion. Regionally, movements of individuals can cause range expansion, both naturally and through human-assisted dispersal.

KEY TERMS

abundance	distribution	meristem	population density	stable age structure
abundance indices	ecological density	migration	quadrat	stationary age
age structure (age	genet	modular organism	ramet	structure
distribution)	geographic range	old growth	sex ratio	unitary organism
census	local subpopulation	philopatric	snag	
dispersal	mark–recapture	population	spatial dispersion	

STUDY QUESTIONS

1. How does asexual reproduction make it difficult to define what constitutes an individual?
2. Many animals, such as fish and reptiles, are indeterminate in that they have the potential to continue growing throughout their lives. Yet these species are unitary, not modular. Explain.
3. Distinguish between the distribution of a population and the geographic range of a species.
4. What abiotic and biotic factors might contribute to a clumped dispersion in one species and a uniform dispersion in another?
5. You wish to estimate the density of two plant populations. Based on their life histories, you expect that one is uniform and the other clumped. How might your approach to estimating their density differ, based on their spatial dispersion?
6. You employ mark–recapture to estimate the abundance of red-backed voles. You bait the live traps with peanut butter, which is highly attractive to this species. Which of the method's assumptions is most likely to be violated? What other assumptions does the method make, and under what circumstances would they likely be violated?
7. Modern humans are highly mobile. Think of three locations in your local community that might be used as areas for estimating human population density. How might daily movement patterns change the density estimate at these sites?
8. Age structure can provide insight into whether the population is growing or declining. A proportionately large number of individuals in young age classes often indicates a growing population, whereas a large proportion of older individuals suggests a population in decline. What factors might invalidate this interpretation? When might a large proportion of individuals in young age classes not indicate a growing population?
9. How might sexual selection affect a population's sex ratio? How might the sex ratio affect sexual selection?
10. What are the ecological advantages and disadvantages to an individual that disperses versus an individual in the same population that does not? How might dispersal affect fitness?
11. Why is a migrating species more vulnerable to habitat loss than a non-migrating species?

FURTHER READINGS

Brown, J. H., D. W. Mehlman, and G. C. Stevens. 1995. Spatial variation in abundance. *Ecology* 76:2028–2043.
Excellent overview of the factors influencing geographic patterns of population abundance.

Cook, R. E. 1983. Clonal plant populations. *American Scientist* 71:244–253.
Introduction to modular growth in plants and its implications for studying plant populations.

Gaston, K. J. 1991. How large is a species' geographic range? *Oikos* 61:434–438.
Explores the methods used to define a species' geographic range and how range is influenced by life history.

Gompper, M. E. 2002. Top carnivores in the suburbs? Ecological and conservation issues raised by colonization of Northeastern North America by coyotes. *Bioscience* 52:185–190.
Fascinating story of the coyote's dispersal and enormous range expansion.

Krebs, C. J. 1999. *Ecological methodology.* 2nd ed. San Francisco: Benjamin Cummings.
Essential for those interested in sampling of natural populations, with many excellent examples.

Laliberte, A. S., and W. J. Ripple, 2004. Range contractions of North American carnivores and ungulates. *Bioscience* 54:123–138.
In contrast to the Gompper paper, which presents a case of range expansion, this paper explores range contraction of many large mammals once widely distributed across North America.

Mack, R., and W. M. Lonsdale. 2001. Humans as global plant dispersers: Getting more than we bargained for. *Bioscience* 51:95–102.
Discusses humans as agents of species dispersal, and illustrates the unexpected consequences.

Elephant (*Loxodonta africana*) herd in Amboseli National Park, Kenya. This small group contains adult females as well as juveniles of various ages. Because African elephants are largely restricted to national parks and conservation areas, these local populations are closed, with no immigration or emigration.

Population size is rarely static. **Population growth** is the increase or decrease in the number of individuals over time, as determined by the rate at which new individuals are added to a population through birth and immigration and subtracted through death and emigration. Populations in which immigration and/or emigration occur are **open**; those in which such movements do not occur (or are minimal) are **closed**. In this chapter, we explore growth in a closed population, the dynamics of which are a function only of the demographic processes of birth and death. (This treatment also applies to open populations in which immigration and emigration are equal.) In Chapter 12, we relax this assumption and examine how movements among subpopulations influence the dynamics of metapopulations.

10.1 BIRTH AND DEATH RATES: Growth of Closed Populations Reflects the Difference between Births and Deaths

Consider a population of an organism with a simple life history, such as freshwater hydra in an aquarium (see Figure 8.1). We define the population size at any time t as $N(t)$. Assume that the initial population is small, so that the available food supply is much more than needed to support the current population. How will the population size change through time?

No emigration or immigration can occur, so the population is closed. The number will increase with new "births" (recall that hydra reproduce by budding). The number will decrease as some hydra die. Because birth and death in this population are continuous, with no discrete period of synchronized birth or death, we can define the **per capita birthrate** (b) as the proportion of hydra producing a new individual per unit time, and the **per capita death rate** (d) as the proportion dying per unit time.

Assuming we have $N(t)$ hydra at time t, we calculate the total number of hydra reproducing in a given time period, Δt (the symbol Δ refers to the change in an associated variable, in this case, time), by multiplying the proportion reproducing per unit time by the total number of hydra and the length of the time period: $bN(t)\Delta t$. As each hydra adds only one individual at a time to the population, the number of births at time t is $B(t) = bN(t)\Delta t$—the same as the number reproducing. (This is not true for most organisms but it simplifies our calculations.)

Note that both B (number of births) and N (population size) change with time, but the per capita birthrate, b, is presumed constant. For this reason, we write $B(t)$ and $N(t)$, but simply b. The number of deaths at time t, $D(t)$, is calculated in a similar way: $D(t) = dN(t)\Delta t$.

The population size at the next time period ($t + \Delta t$) is then

$$N(t + \Delta t) = N(t) + B(t) - D(t)$$

or

$$N(t + \Delta t) = N(t) + bN(t)\Delta t - dN(t)\Delta t$$

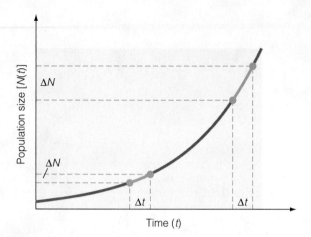

Figure 10.1 Exponential growth of a hypothetical hydra population. The change in size, ΔN, for a given interval, Δt, differs as a function of time (t), as indicated by the slope of the line segments (in orange). See **QUANTIFY it!** at **www.pearsoncanada.ca/ecologyplace** to review functions.

Repeating these calculations over a series of time units ($t + 2\Delta t$, $t + 3\Delta t$ and so on, assuming b and d remain constant), yields a pattern of steadily increasing population size called *exponential growth* (Figure 10.1). Exponential growth is typical of closed populations at low densities in favourable environments not subject to resource limitations.

10.2 EXPONENTIAL GROWTH: The Intrinsic Rate of Increase Determines Exponential Population Growth

Let's re-examine exponential growth, as depicted in Figure 10.1, from a mathematical perspective. We define the change in population size (ΔN) over the time interval (Δt) by moving $N(t)$ to the left-hand side of the equation and dividing both sides by Δt:

$$\frac{N(t + \Delta t) - N(t)}{\Delta t} = bN(t) - dN(t)$$
$$= (b - d)N(t)$$

If we substitute ΔN for $[N(t + \Delta t) - N(t)]$, we can rewrite the equation as

$$\frac{\Delta N}{\Delta t} = (b - d)N(t)$$

The term $\Delta N/\Delta t$ defines the unit change in population size per unit change in time, or the slope of the relationship between $N(t)$ and t—the "rise" over the "run" in Figure 10.1. The relationship is non-linear (a curve), so the slope changes continuously, with the rate of change depending on the time interval. The rate of change is best described by a derivative function (see Quantifying Ecology 10.1: Using Differential Equations to Describe Population Growth), written as

$$\frac{dN}{dt} = (b - d)N$$

QUANTIFYING ECOLOGY 10.1 Using Differential Equations to Describe Population Growth

Suppose we wish to measure the rate of change in a population through time. Assume that population size, $N(t)$, is a linear function of time, t. The graph will be a straight line (Figure 1). Consider two points on the graph at $t = t_1$ and $t = t_2$. We call $\Delta t = t_2 - t_1$ the "run" and $\Delta N = N(t_2) - N(t_1)$ the "rise." The rate of population change, ΔN, over the time interval, Δt, is given by the slope of the line (s), defined as the rise per unit of run:

$$s = \frac{\Delta N}{\Delta t}$$

If we choose another pair of points, the slope (rate of population change) will stay the same, because the slope of a straight line does not vary. The slope of a non-linear function (curve), however, does depend on the value of t (Figure 2).

Suppose we hold t_1 constant and move t_2 closer to t_1 (as in Figures 2a and 2b). As the points get closer, the slope (orange dashed line) will vary by smaller and smaller amounts as it approaches a constant "limiting value." When this happens, as in Figure 2c, we call the limiting value the slope of the tangent to the curve at point t_1 (note that the tangent intersects the function at the point $[t_1, N(t_1)]$ only). Mathematically, we can express this value as

$$\text{slope at } t_1 = \lim_{\Delta t \to 0} \frac{\Delta N}{\Delta t} = \lim_{\Delta t \to 0} \frac{N(t_1 + \Delta t) - N(t_1)}{\Delta t}$$

The slope of the function $N(t)$ at t_1 is the derivative of $N(t)$, written as $dN(t)/dt$ and defined as

$$\frac{dN(t)}{dt} = \lim_{\Delta t \to 0} \frac{N(t_1 + \Delta t) - N(t_1)}{\Delta t}$$

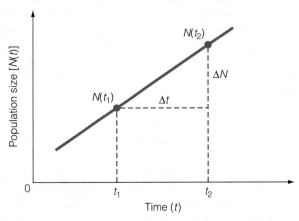

Figure 1 Change in population size over time, assuming a linear function.

Remember that when the derivative of $N(t)$ is evaluated at point t, t is held constant, whereas Δt is varied (approaches zero). For those who have not studied calculus, this process of taking limits will be confusing. Think of making the time interval Δt so small that decreasing it further will not affect the slope of the tangent—our estimate of the continuous rate of population change. An equation where the derivative appears on the left-hand side is called a *differential equation*, as with the exponential model of population growth: $dN/dt = rN$.

1. Why does the size of the time interval (Δt) matter when estimating the slope of a non-linear function (curve)?
2. What errors would arise in estimating the rate of population growth if the time interval (Δt) were large, as in Figure 2a?

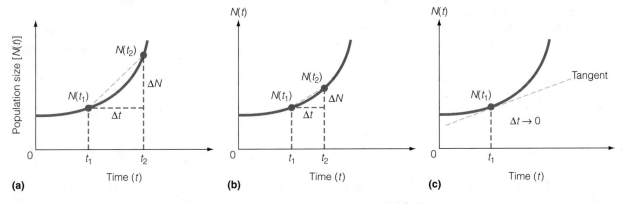

Figure 2 Change in population size over time as a non-linear function. Go to **QUANTIFYit!** at **www.pearsoncanada.ca/ecologyplace** to practise differential equations.

The term $\Delta N/\Delta t$ is replaced by dN/dt to indicate that Δt (the time interval) approaches zero, and that the rate of change is instantaneous. (Do not confuse the use of "d" [non-italicized] to denote a differential function with "*d*" [italicized] to denote per capita death rate, as defined in Section 10.1.) Values of b

and d represent the instantaneous per capita rates of birth and death; because we assume they are constants, we can define $r = b - d$ and rewrite the equation for continuous growth as

$$\frac{dN}{dt} = rN$$

This key term, r, is the **intrinsic rate of increase** (instantaneous per capita growth rate), and the resulting equation defines **exponential growth**. Positive feedback is fundamental to exponential growth. Individuals born into a population increase its reproductive capacity, which in turn leads to an ever larger population—much as compound interest generates an ever-increasing bank balance if interest earned is reinvested into the principal.

To use the exponential growth model to predict population change over time, we must integrate (another procedure involving calculus) the differential equation. The resulting equation is

$$N(t) = N(0)e^{rt}$$

where $N(0)$ is the initial population size at $t = 0$, and e is the base of the natural logarithm (~ 2.72).

Exponential growth for differing values of r is shown in Figure 10.2. Contrary to what many assume, exponential growth is not always associated with rapid population increase and may even lead to decline, depending on the value of r. When $r = 0$, there is no change in population size, but for values of $r < 0$, the population declines exponentially. Only when $r > 0$ does the population increase exponentially, with the steepness of the curve varying with the value of r.

Note that r is not equal to the slope of the growth curves in Figure 10.2. The slope equals $r \times N$, and is thus always increasing, whatever the direction. The key point is that exponential growth, like compound interest, causes a continuously accelerating rate of increase (or decelerating rate of decrease) as a function of population size. However, if the rate of growth (dN/dt) of an exponentially growing population is expressed as a function of N, then the slope *is* equal to r (Figure 10.3).

Positive exponential growth is typical of populations occupying favourable environments at low densities when resources are not limiting, such as during colonization of new habitats, or re-establishment after disturbance. For example, a small (4 males and 22 females) herd of reindeer (*Rangifer tarandus*)

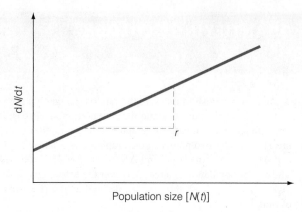

Figure 10.3 Population growth rate (dN/dt) expressed as a function of population size [$N(t)$] for exponential growth: $dN/dt = rN$. The growth rate increases continuously with $N(t)$. The slope of the line, defined as $(dN/dt)/N(t)$, is the intrinsic rate of increase, r.

was introduced in 1911 to the island of St. Paul in Alaska. By the 1940s, its numbers had grown to over 2000 (Figure 10.4; but see Figure 10.17 for its subsequent crash).

Another example is the trumpeter swan (*Cygnus buccinator*) (Figure 10.5a), the largest bird native to North America. In Canada, they breed from James Bay to the Yukon Territory, and although they once wintered as far south as Texas and California, hunting had decimated its southern populations by the early 20th century. The species persisted in a Pacific population in northern British Columbia, Alberta, and Alaska, and an interior population in eastern North America. Introductions re-established the species in the southern portions of its historic range, including the Rocky Mountains (Figure 10.5b). Based on survey data, the North American total has risen exponentially from 3722 in 1968 to a high of 23 647 by 2000 (USFWS 2001) (Figure 10.5c).

Ecologists may employ a **geometric model** instead of an exponential model for population growth. The difference is simple. A geometric model describes growth over discrete time intervals (often a year), whereas an exponential model uses a continuous time axis. As birth is not continuous for

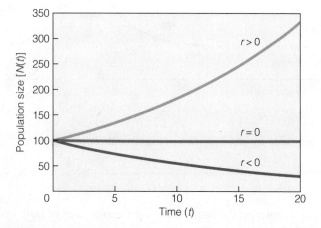

Figure 10.2 Exponential growth for differing values of r. When $r > 0$ ($b > d$), population size increases exponentially; for values of $r < 0$ ($b < d$), there is an exponential decline. When $r = 0$ ($b = d$), there is no change in population size over time.

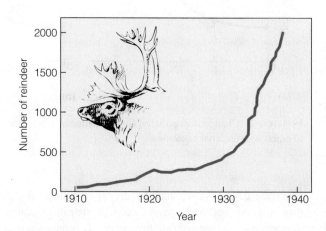

Figure 10.4 Exponential growth of the St. Paul reindeer herd.

(Adapted from Scheffer 1951.)

(a)

Figure 10.5 Exponential recovery of the trumpeter swan (a). (b) North American range as of 2000. (c) Population growth from 1968 to 2000.

(Adapted from U.S. Fish and Wildlife Service 2001).

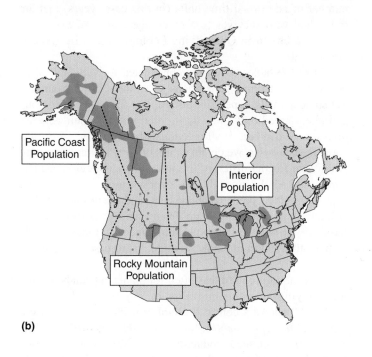

(b)

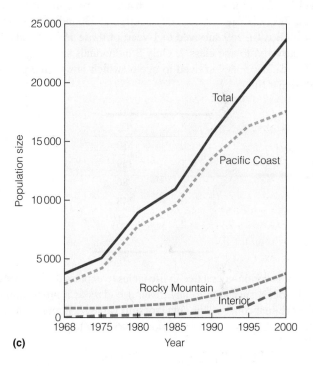

(c)

most species, the geometric model may be more appropriate. The net result is similar to exponential growth, but the equation differs: $N(t) = N(0)\lambda_t$. Compare this with the integrated equation of exponential growth: $N(t) = N(0)e^{rt}$. These two equations (finite and continuous) show the relationship between λ and r:

$$\lambda = e^r \quad \text{or} \quad r = \ln\lambda$$

10.3 LIFE TABLES: Life Tables Quantify Age-Specific Mortality and Survival

As we saw in Section 10.2, change in population size is a function of birth and death rates, as reflected in the instantaneous per capita growth rate, r. For the hydra population, where all individuals can be considered identical, and each "gives birth" to one individual at a time, simply counting the proportion of

individuals reproducing or dying per unit time estimates birth and death rates. But when birth and death rates vary with age, as they do for most populations, ecologists must take a different approach. They construct a **life table**, which quantifies mortality and survival in different age classes of a population. Although life tables were first developed by life insurance companies as the basis for evaluating age-specific mortality in humans, ecologists now use life tables to examine patterns of mortality and survivorship in natural populations.

Construction of the most common type, a *dynamic (cohort) life table*, begins with a **cohort**—a group of individuals born at the same time. Consider a species familiar to anyone living in most parts of North America—the grey squirrel (*Sciurus carolinensis*; the black squirrels that are so abundant in Ontario and Québec are colour variants of the same species). Ecologists followed a cohort of 530 female squirrels in a population in northern West Virginia (Table 10.1) until all its members had died some six years later.

Table 10.1 Grey Squirrel Cohort Life Table

x	n_x	l_x	d_x	q_x
0	530	1.0	371	0.7
1	159	0.3	79	0.5
2	80	0.15	32	0.4
3	48	0.09	27	0.55
4	21	0.04	16	0.75
5	5	0.01	5	1.0

The first column, labelled x, represents the age class in years. The second column, n_x, represents the number of individuals from the original cohort that are alive at the beginning of a specified age class (x). Of the original 530 individuals born (age class 0), 159 survived to 1 year; of those 159 individuals, 80 survived to age class 2. Only 5 individuals survived to age class 5, and none survived to age 6 (which is why there is no age class 6).

x	n_x
0	530
1	159
2	80
3	48
4	21
5	5

It is common practice in life tables to express the number of individuals surviving to any given age class as a proportion of the original cohort. This **survivorship** value (l_x), represents the probability at birth of an individual surviving to any given age.

x	n_x	l_x	
0	530	1.00	$n_0/n_0 = 530/530$
1	159	0.30	$n_1/n_0 = 159/530$
2	80	0.15	$n_2/n_0 = 80/530$
3	48	0.09	
4	21	0.04	
5	5	0.01	

The difference between the number of individuals alive in any age class (n_x) and the next older age class (n_{x+1}) is d_x: the number dying during that time interval.

x	n_x	d_x	
0	530	371	$= n_0 - n_1 = 530 - 159$
1	159	79	$= n_1 - n_2 = 159 - 80$
2	80	32	
3	48	27	
4	21	16	
5	5	5	

The **age-specific mortality rate**, q_x, is the number of individuals that died during any given time interval (d_x) divided by

the number alive at the beginning of that interval (n_x). Like l_x, it is a proportion and hence a more meaningful measure of age-specific mortality than the absolute number of deaths in any particular time interval (d_x):

x	n_x	d_x	q_x	
0	530	371	0.70	$= d_0/n_0 = 371/530$
1	159	79	0.50	$= d_1/n_1 = 79/159$
2	80	32	0.40	
3	48	27	0.55	
4	21	16	0.75	
5	5	5	1.00	

The complete life table for the cohort of female squirrels, including all of these calculations, is shown in Table 10.1. The calculation of age-specific life expectancy, e_x, the average number of additional time units (in this case, years) that an individual who survives to a given age class can expect to live, is described in Quantifying Ecology 10.2: Life Expectancy, p. 207.

Most life tables have been constructed for vertebrate species with overlapping generations, as in the squirrel example. Yet some animals, especially insects, live for only one season, and many breed only once during that season. Their generations do not overlap, so all individuals belong to the same age class. In such cases, we obtain values of n_x by estimating population size for the different life history stages (eggs, instars, pupae, and adults) several times over a season. If records are kept of weather, predator abundance, and occurrence of disease, we can estimate death from various causes. Go to QUANTIFYit! at **www.pearsoncanada.ca/ecologyplace** to work with another cohort life table example, in this case for an insect species.

Life tables are used less often for plants but can be useful for studying seedling mortality or population dynamics. Table 10.2 shows a cohort life table for the annual plant elf orpine (*Diamorpha smallii*; formerly *Sedum smallii*). The life cycle begins with seed production. Again, the l_x column shows the proportion of plants alive at the beginning of each developmental stage, but the d_x column shows the proportion rather than the actual number of individuals dying, as it did in the squirrel table. Because d_x is already expressed as a proportion, q_x (the age-specific mortality rate) is calculated simply as d_x/l_x.

Table 10.2 Life Table for a Natural Population of Elf Orpine

x	l_x	d_x	q_x
Seed produced	1.000	0.160	0.160
Available	0.840	0.630	0.750
Germinated	0.210	0.177	0.843
Established	0.033	0.009	0.273
Rosettes	0.024	0.010	0.417
Mature plants	0.014	0.014	1.000

(Source: Data from Sharitz and McCormick 1973.)

QUANTIFYING ECOLOGY 10.2   Life Expectancy

Not all of us are familiar with life tables, but we have all heard statements like: "The average life expectancy for a male born in Canada in 2005 is 78 years." What does this mean? **Life expectancy** (*e*) refers to the average number of years an individual is expected to live from the time of its birth. Life tables allow us to calculate the *age-specific life expectancy* (e_x) or average number of additional years that an individual that has attained a certain age is expected to live. Let's use data from the squirrel cohort in Table 10.1 to calculate age-specific life expectancies.

The first step in estimating e_x is to calculate L_x using the n_x column. L_x is the average number of individuals alive during the age interval from *x* to *x* + 1, calculated as the average of n_x and n_{x+1}. This estimate assumes that mortality within any age class is distributed evenly over the year.

x	n_x	L_x	
0	530	344.5	$= (n_0 + n_1)/2 = (530 + 159)/2$
1	159	119.5	$= (n_2 + n_3)/2 = (80 + 48)/2$
2	80	64.0	
3	48	34.5	
4	21	13.0	$= (n_5 + n_6)/2 = (5 + 0)/2$
5	5	2.5	

We then use these L_x values to calculate T_x, the total years lived into the future by individuals of age class *x*. T_x is calculated by summing the values of L_x cumulatively from the bottom of the column to age *x*. In the squirrel example, the value of T_0 is 578. In other words, the 530 individuals in the cohort lived a total of 578 years (some only 1 year, others up to 5 years).

x	L_x	T_x	
			$= L_0 + L_1 + L_2 + L_3 + L_4 + L_5$
			$= 344.5 + 119.5 + 64 + 34.5 + 13 + 2.5$
0	344.5	578.0	
1	119.5	233.5	
2	64.0	114.0	
3	34.5	50.0	$= L_4 + L_5 = 13 + 2.5$
4	13.0	15.5	
5	2.5	2.5	$= L_5$

To calculate the life expectancy for each age class (e_x), we divide the value of T_x (total number of years lived into the future by individuals of age *x*) by the corresponding value of n_x (total number of individuals in that age group).

x	n_x	T_x	e_x	
0	530	578.0	1.09	$= T_0/n_0 = 578/530$
1	159	233.5	1.47	
2	80	114.0	1.43	$= T_2/n_2 = 114/80$
3	48	50.0	1.06	
4	21	15.5	0.75	
5	5	2.5	0.50	

Note that life expectancy changes with age. On average, grey squirrels live only 1.09 years. However, for those that survive to one year, life expectancy increases to 1.47 more years, that is, they will be on average 2.56 years old at death. Additional life expectancy remains high for age class 2 and then declines for the remaining age classes.

1. Why does life expectancy increase for individuals that survive to age 1 (1.47 additional years as compared with 1.09 for newborn individuals)?
2. Which would have a greater influence on life expectancy of a newborn (age 0)—a 20 percent decrease in mortality rate of age class 0 (*x* = 0), or a 20 percent decrease in mortality rate of age class 4 (*x* = 4)? Why?

Tables 10.1 and 10.2 are both cohort life tables, in which a group of individuals born at a given time (e.g., in 2010) is followed from birth to death. A modification is a *dynamic composite life table*, based on individuals born over several periods—for example, from 2009 to 2011.

In contrast, a *time-specific life table* is based on sampling a population to obtain a distribution of age classes present during a single time period. An ecologist might sample the human population of New Brunswick and estimate survivorship, mortality, and life expectancy based on the age structure present. Because individuals are not tracked until the last one dies, this type is easy to construct, but it makes two assumptions: (1) each age class was sampled in proportion to its numbers in the population, and (2) age-specific birth and mortality rates remain constant. If these assumptions fail, the results are less reliable, but a time-specific table may be the only feasible type.

10.4 MORTALITY AND SURVIVORSHIP CURVES: Species Exhibit Differing Patterns of Mortality and Survivorship

We can graph any of the life table columns, but the two most common graphs are (1) a **mortality curve** based on the q_x column and (2) a **survivorship curve** based on the l_x column. Mortality curves plot age-specific mortality rates (q_x) against age class or developmental stage. For the grey squirrel cohort (Figure 10.6a, p. 208), the curve is J-shaped and has two parts: a juvenile phase, in which mortality is high, and a post-juvenile phase, in which mortality decreases with age until it reaches some minimum, after which it increases.

Plant mortality curves assume various patterns, depending partly on whether the plant is annual or perennial. For the *Diamorpha* population (Figure 10.6b), mortality is high initially,

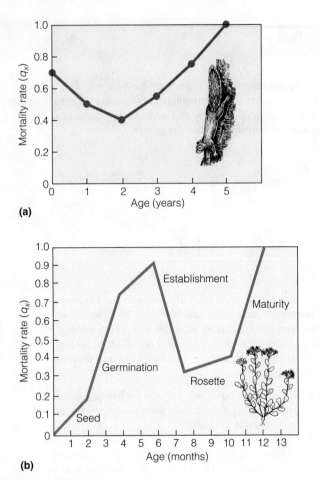

(a)

(b)

Figure 10.6 Mortality curves for **(a)** grey squirrels, based on Table 10.1, and **(b)** elf orpine, based on Table 10.2.

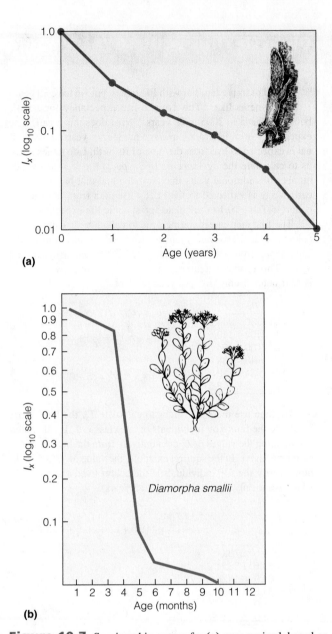

(a)

(b)

Figure 10.7 Survivorship curves for **(a)** grey squirrel, based on Table 10.1, and **(b)** elf orpine, based on Table 10.2.

declines after seedling establishment, and increases again with sexual maturity.

Survivorship curves plot l_x on the vertical axis against time or age class (x) on the horizontal axis. Unlike mortality curves, survivorship (l_x) is normally plotted on a log scale. For grey squirrels, survivorship declines linearly (Figure 10.7a), whereas for elf orpine, survivorship is variable at all life stages (Figure 10.7b).

Life tables (and the mortality and survivorship curves derived from them) reflect data from one population at a particular time and under particular environmental conditions. As snapshots of the population, they are most useful for comparing populations at one time or in one area, or for comparing one sex with another. For example, the kudu (*Tragelaphus strepsiceros*) is a woodland antelope native to South Africa. Males and females have similar survivorship patterns until three years of age, at which point male survivorship declines sharply (Figure 10.8), with females having much greater life expectancy (Owen-Smith 1993). Once a pattern is detected, ecologists explore its underlying reasons. In the kudu, expected factors such as mate competition were not responsible. There is a marked sexual size dimorphism (see Section 8.4) in this species, and the larger males are more vulnerable to both predation and malnutrition.

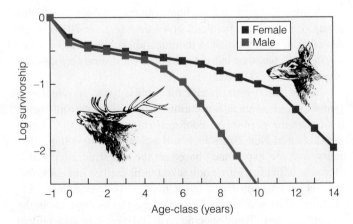

Figure 10.8 Survivorship curves for male and female kudu.

(Adapted from Owen-Smith 1993.)

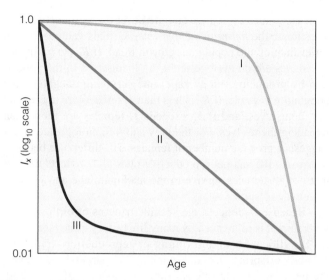

Figure 10.9 Three basic types of survivorship curves.

Survivorship curves fall into three basic types (Figure 10.9). When most individuals live out their lifespan, survivorship is high over most of their lives, with heavy mortality at the end. This type of curve, called *type I*, is strongly convex. A type I curve is typical of mammals, including humans. If survivorship stays constant with age, the relationship is linear or *type II*. A type II curve occurs in adult birds, rodents, some reptiles, and some herbaceous perennials. If mortality is high in early life—as in fish, invertebrates, and many plants, including most trees—the curve is concave, or *type III*. Type III is the most common in nature; few individuals survive past the first age class, but those that do have low mortality until the end of their lifespan.

These curves are idealized models to which survivorship of real populations can be compared. The human survivorship curve has become progressively more type I over time (Figure 10.10a). Many survivorship curves resemble different types at different life history stages (Figure 10.10b). Note that the curves reveal nothing about population size, as they are based on proportion. A population can have type III survivorship, as typical for insects, fish, and many plants (Figure 10.10c) and yet have much greater numbers than a population with a type I curve.

10.5 FECUNDITY TABLES: Age-Specific Birthrates and Survivorship Determine the Net Reproductive Rate

In Section 10.1, we discussed the per capita birthrate (*b*) in the context of exponential growth. For life table purposes, we need to refine our treatment. A standard convention in **demography** (the study of population growth) is to express birthrate as the number of births in a time interval (often a year) for every 1000 individuals in the population—the *crude birthrate*. A type of per capita rate, the crude birthrate does not take two factors into account: (1) in a sexually reproducing, non-hermaphroditic population, only females give birth, and (2) the birthrate of females varies with age. As only females

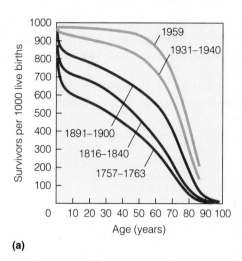

(a)

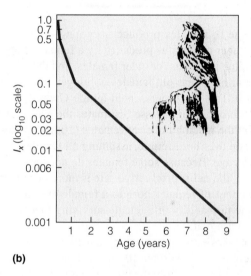

(b)

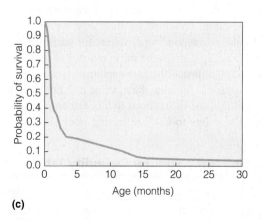

(c)

Figure 10.10 Examples of survivorship curves. **(a)** Type I curves for humans in Sweden over centuries. **(b)** The song sparrow's survivorship curve is typical of birds. After an initial type III phase, the relationship becomes linear (type II). **(c)** Sugar maple (*Acer saccharum*) has a pronounced type III curve, with high juvenile mortality.

(Adapted from (a) Yashin et al. 2002, (b) Tompa 1964, and (c) Cleavitt et al. 2011).

contribute to population growth, the most meaningful way to express birthrate is the **age- and sex-specific birthrate** (b_x, the mean number of females born to a female in a particular age group).

Here are the age- and sex-specific birthrates for the grey squirrel population in Table 10.1:

x	b_x
0	0
1	2
2	3
3	3
4	2
5	0
Σ	10

At age 0, females produce no young, so $b_x = 0$. The mean number of females produced by a female of age 1 is 2. The b_x value increases for older females, and then declines to 2 at age 4. Five-year-old females no longer reproduce, so b_x is again 0. The sum (represented by the Greek letter sigma, Σ) of b_x across all age classes estimates the **gross reproductive rate**: the mean number of females (10 in this case) born to a female over her lifetime, assuming all females survive to maximum age. Because some females do not survive to maximum age, the **net reproductive rate** is more meaningful: the mean number of females born to a female over her lifetime, taking the probability of a female surviving to a specific age class into account.

Ecologists estimate the net reproductive rate by constructing a **fecundity table** (Table 10.3), which incorporates the survivorship column, l_x, from the life table along with the age- and sex-specific birthrates (b_x). Although b_x may initially increase with age, as it does for here squirrels, survivorship (l_x) declines in each age class. To adjust for mortality, we multiply the b_x values by the corresponding l_x values for each age class. The resulting value, $l_x b_x$, gives the mean number of females born in each age class, adjusted for survivorship.

For 1-year-old females, the b_x value is 2, but when adjusted for survival (l_x), the value drops to 0.6. For age 2, b_x increases to 3, but $l_x b_x$ drops to 0.45, reflecting poor survival of adult

Table 10.3 Grey Squirrel Fecundity Table

x	l_x	b_x	$l_x b_x$
0	1.0	0.0	0.00
1	0.3	2.0	0.60
2	0.15	3.0	0.45
3	0.09	3.0	0.27
4	0.04	2.0	0.08
5	0.01	0.0	0.00
Σ		10.0	1.40

females. Values of $l_x b_x$ are summed over all reproductive ages to estimate the *net reproductive rate, R_0*. This value is critical for evaluating the population growth trend. If R_0 approximates 1, females are on average replacing themselves in the population by producing (on average) one surviving daughter—the population is stable. If R_0 is less than 1, females are not replacing themselves, and if R_0 exceeds 1, females are more than replacing themselves. For the grey squirrel, an R_0 value of 1.4 suggests a growing number of females. The difference between the gross (10) and net reproductive rates (1.4) reflects the fact that not all females survive to the maximum age and produce 10 female offspring.

Because R_0 reflects age-specific patterns of birth and survivorship, it is influenced by many life history traits (see Chapter 8): allocation to and timing of reproduction, trade-offs between offspring size and number, and parental care. Thus, the net reproductive rate (R_0) is a means of evaluating both individual fitness and the population effects of specific life history traits.

Although net reproductive rate is based on females born to females, it is a reliable indicator of a population's overall growth trend. Recall the concept of a population's *age structure*: the proportion of individuals in different age classes (see Section 9.6). Life and fecundity table data from several generations allow ecologists to construct an age pyramid. If the proportions remain constant over time, the population has a stable age structure. However, a stable age structure can occur in populations that are increasing, decreasing, or staying the same—it depends on what those proportions are, and what the mortality losses are between classes. Only if the stable age structure is also a *stationary age structure*, in which both absolute numbers and proportions in each class stay the same, is the population constant. Simply put, a population with a stationary age structure has a net reproductive rate (R_0) equal to 1.

10.6 GENERATION TIME: The Time Period between Generations Affects Population Growth

How do we progress from determining R_0 to calculating r, the intrinsic rate of increase, which is critical for predicting population growth? The net reproductive rate tells us if a population is increasing or decreasing, but because it quantifies the rate of increase per generation, it cannot predict growth over time. We must also factor in **generation time**: the mean time between when a female is born and when she reproduces. Even if females give birth to the same mean number of females in two populations, one population will grow faster than the other if females on average reproduce sooner. Reprising the compound interest analogy, the growth of an investment depends not only on the rate but on how often the investment is compounded.

Generation time (T_c) is approximated as the mean age of reproducing individuals, calculated by summing the lengths of time to reproduction for the entire cohort, divided by the total offspring:

$$T_c = \frac{\Sigma x l_x b_x}{\Sigma l_x b_x}$$

For the grey squirrel, substituting values for l_x and b_x from Table 10.3 into this equation, $T_c = [(1 \times 0.6) + (2 \times 0.45) + (3 \times 0.27) + (4 \times 0.08)]/1.4 = 2.63/1.4 = 1.88$ years. Now that we know the generation time for the grey squirrel population, we can estimate r. Recall that the exponential model predicts N at any given time:

$$N(t) = N(0)e^{rt}$$

Expressed for a single generation, i.e. where $t = T_c$, this equation becomes

$$N_G = N_0 e^{rT_c}$$

Rearranging gives

$$\frac{N_G}{N_0} = e^{rT_c}$$

N_G/N_0 is an approximation of R_0, because it represents the number of individuals after a single generation in proportion to the abundance at time 0. Thus the formula becomes

$$R_0 \sim e^{rT_c}$$

Taking the natural logarithm of both sides and rearranging, we can express r as

$$r \sim \ln (R_0)/T_c$$

For our grey squirrel cohort, $r \sim \ln (1.4)/1.88 = 0.18$.

Note that for an annual species, where $T_c = 1$, $r = \ln (R_0)$. So if we know the net reproductive rate, we can calculate the intrinsic rate of growth immediately. But to calculate r for a non-annual species, we must determine the generation time, as just outlined.

10.7 LOGISTIC GROWTH: Resource Limitations May Generate Logistic Population Growth

No population continues to grow indefinitely. Exponentially growing populations eventually confront the limits of their environment. As population density increases, interactions occur among its members—interactions that may regulate growth, converting it to **logistic growth**: an S-shaped model in which birth and death rates vary in a density-dependent manner. Interactions act as negative feedback loops that, unlike the positive feedback that drives exponential growth, cause population size to stabilize. In Chapter 11, we discuss these intraspecific interactions in detail, but now we consider the logistic growth pattern to which they contribute.

The exponential growth model makes two assumptions: (1) essential resources (space, food, etc.) are unlimited, and (2) the environment is constant, with no seasonal or annual variations that might influence birth and/or death. Under these assumptions, birth and death rates are constant, and the intrinsic rate of increase (r) is fully realized. But in the real world, neither of these assumptions holds up; resources *are* often limited, and the environment is *not* constant. As population density increases, the demand for resources increases. If the rate of consumption exceeds the rate at which resources are resupplied, then the resource base shrinks. Shrinking resources—and the potential for their unequal distribution—will increase mortality (death) rates, decrease fecundity (birth) rates, or both. These effects represent a departure from the constant birth and death rates assumed by the exponential model.

The simplest possible change in birth and death rates with increasing population size is a straight-line (linear) function (Figure 10.11). In this scenario, per capita birthrate (b) decreases and per capita death rate (d) increases with increasing population size (N). (In Section 11.1, we consider other possibilities.) We can describe the line expressing birthrate as a function of population size:

$$b = b_0 - aN$$

In this equation, b_0 is the intercept (value of b when $N = 0$), and a is the slope of the line ($\Delta b/\Delta N$; see Figure 10.11). The intercept, b_0, represents the birthrate under ideal conditions (no crowding or resource limitation), whereas b is the actual birthrate, which declines as a function of crowding. Note that the ideal birthrate, b_0, is the value used in the exponential growth model (see Section 10.2).

Similarly, we can describe the death rate as a function of population size:

$$d = d_0 + cN$$

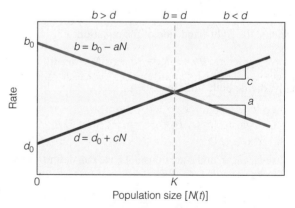

Figure 10.11 Rates of birth (b) and death (d), represented as linear functions of population size, $N(t)$. The values b_0 and d_0 represent "ideal" birth and death rates under conditions where population size is near 0 and resources are not limiting. The values a and c represent the slopes of the lines describing changes in birth and death rates as a function of N. The population density where $b = d$ and population growth is 0 is defined as K, the carrying capacity. For values of N above K, $b < d$ and growth rate is negative. For values of $N < K$, $b > d$, and growth rate is positive. (Go to **QUANTIFYit!** at **www.pearsoncanada.ca/ecologyplace** to review linear functions.)

Again, the constant d_0 is the death rate when the population size is close to 0 (no crowding or resource limitation), and the constant c represents the increase in death rate with increasing population size (the slope of the line in Figure 10.11).

Substituting in these expressions for b and d, we can now rewrite the exponential model developed in Section 10.2, ($dN/dt = (b - d)N$) to incorporate these variations in birth and death rates with population size:

$$\frac{dN}{dt} = \left[(b_0 - aN) - (d_0 + cN) \right]N$$

It is critical to note how logistic growth differs from exponential growth. As N increases, birthrate ($b_0 - aN$) declines, death rate ($d_0 + cN$) increases, and growth slows. If d exceeds b, growth is negative, and population size declines. When b equals d, the rate of population change (dN/dt) is 0. The population size at which $b = d$ is the **carrying capacity**: the maximum sustainable population size under the prevailing conditions. We solve for this value by setting the equation for population growth equal to 0 and solving for N:

$$\frac{dN}{dt} = \left[(b_0 - aN) - (d_0 + cN) \right]N = 0$$

$$(b_0 - aN)N - (d_0 + cN)N = 0$$

Move the term for death rate to the right side of the equation:

$$(b_0 - aN)N = (d_0 + cN)N$$

Then divide both sides by N:

$$b_0 - aN = d_0 + cN$$

Move d_0 to the left side of the equation and aN to the right side:

$$b_0 - d_0 = aN + cN$$

Rearrange the right-hand side of the equation:

$$b_0 - d_0 = (a + c)N$$

Finally, divide both sides by ($a + c$):

$$N = \frac{b_0 - d_0}{a - c}$$

Because b_0, d_0, a, and c are constants, we can define a new constant, K:

$$K = \frac{b_0 - d_0}{a + c}$$

We can now rewrite the equation for growth that includes rates of birth and death that vary with population size, substituting the value of K defined above:

$$\frac{dN}{dt} = rN\left(1 - \frac{N}{K} \right)$$

This equation describes the logistic model of population growth.

Let's recap. With exponential growth, change in population size is quantified by

$$N(t) = N(0)e^{rt}$$

However, if the population is following a logistic model, in which the rates of birth (b) and death (d) vary with population size, the formula for the change in population size is

$$\frac{dN}{dt} = rN\left(1 - \frac{N}{K} \right)$$

In both cases, population growth is affected by r (the intrinsic rate of increase) $\times$ N (the population size). In exponential growth, the growth potential represented by r is fully realized. But in logistic growth, there is a correction factor, $1 - N/K$, that reflects how close the population size is to K, the carrying capacity (see Ecological Issues: What Is the Carrying Capacity of Earth for the Human Species?, p. 214). Note that although K is assumed to be constant in a non-fluctuating environment, it will vary with resource supply (food, water, space, etc.).

So the logistic model has two components: the original exponential term (rN) and a second term ($1 - N/K$) that reduces population growth as population size approaches the carrying capacity. When population size (N) is low relative to carrying capacity (K), the term $1 - N/K$ approaches 1, and its growth approximates the exponential model (rN). However, as the population grows and N approaches K, the term $1 - N/K$ approaches 0, slowing growth more and more until the population stops growing entirely. Should population density exceed K, growth becomes negative and density declines back towards K.

In the logistic model, r is defined as $b_0 - d_0$, the birth and death rates under "ideal" conditions (low density, unlimited resources) and is no different than in the exponential model. In converting from exponential to logistic growth, the intrinsic rate of increase does not alter; rather, further population increase is prevented by density-dependent processes that become more and more intense as N approaches K. In essence, internal negative feedback loops involving intraspecific interactions "apply the brakes" to the population's inherent tendency to increase.

Figure 10.12a shows changing population size over time for the logistic model. When N is small, the population increases rapidly, at a rate only slightly lower than that predicted by the exponential model. The population growth rate (dN/dt) peaks when $N = K/2$ (the curve's inflection point) and then decreases as the population approaches K (Figure 10.12b). With the exponential model, in contrast, growth rate increases linearly with population size (see Figure 10.2).

To illustrate logistic growth, consider our grey squirrel example. Assume $K = 200$. Given the value of $r = 0.18$ calculated in Section 10.6 and an initial population size of 30, population growth differs depending on whether it follows an exponential or a logistic model (Figure 10.13). The concept of carrying capacity implies that negative feedback regulates population growth as density rises and per capita resource availability declines. We explore how density dependence is mediated by intraspecific interactions in Chapter 11.

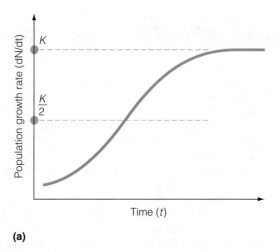

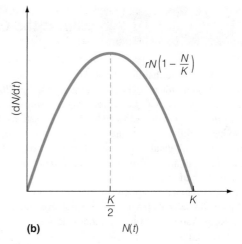

(a)   **(b)**

Figure 10.12 **(a)** Change in population size (N) as predicted by the logistic model ($dN/dt = rN(1 - N/K)$). Initially, the population grows exponentially, but as N increases, the growth rate decreases, eventually reaching 0 as N approaches K. **(b)** The population growth rate, dN/dt, peaks when the population size reaches one-half the carrying capacity ($N = K/2$).

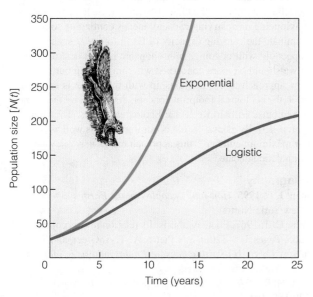

Figure 10.13 Predictions of exponential and logistic models for growth of the grey squirrel population in Tables 10.1 and 10.3. $r = 0.18$, $K = 200$, and $N(0) = 30$.

10.8 STOCHASTICITY: Intrinsic and Extrinsic Random Processes Affect Population Dynamics

The exponential and logistic models provide a theoretical framework for understanding the demographic processes governing population dynamics. But nature is rarely constant; stochastic (random) processes, both internal (demographic) and external (environmental), affect population dynamics. Populations may approximate exponential or logistic growth at different times or in different environments, but most deviate from these deterministic models.

How do we factor the indeterministic nature of the real world into our understanding of population growth? Thus far,

we have considered population growth as deterministic. Because the exponential model assumes birth and death rates are constant, it predicts only one outcome (N value) after a given number of years for given values of r and $N(0)$. The logistic model incorporates changes in birth and death rates with increasing population size, but in a similarly deterministic fashion. Neither model incorporates the impact of chance events.

Yet chance is always at work. Recall that age-specific rates of birth and survival in life and fecundity tables (see Tables 10.1 and 10.3) represent means derived from the cohort under study. Values of b_x are the mean number of females produced by a female of that age group. For one-year-old females, the average value is 2.0, but some females in this class may have four female offspring, whereas others may have none. The same holds true for *age-specific survival rates* (s_x, or $1 - q_x$): the probability of a female of that age surviving to the next age class.

Although fecundity and survival are expressed as probabilities, for any individual they are discrete events. An individual either survives to the next year or not, just as the outcome of a coin toss will be either heads or tails. With 10 coin tosses, we expect to get on average an outcome of 5 heads and 5 tails. But each outcome in the 10 tosses is independent, and it is possible to get 4 heads and 6 tails (probability $p = 0.2051$), or even 0 heads and 10 tails ($p = 9.77 \times 10^{-4}$). The same is true for the probability of survival when applied to individuals in an age class. The realization that population dynamics represents the combined outcome of many individual probabilities has led to the development of probabilistic, or *stochastic*, models of population growth. These models allow actual rates of birth and survival to vary about the mean estimate represented by b_x and s_x.

Random year-to-year variation in birth and death rates due to factors intrinsic to the population is called **demographic stochasticity** and causes populations to deviate from outcomes predicted by deterministic models. **Environmental stochasticity**—random variation in extrinsic factors such as temperature,

ECOLOGICAL ISSUES   What Is the Carrying Capacity of Earth for the Human Species?

Demographers estimate that the current human population of just over 7 billion will most likely almost double by 2200. *Can Earth support such an increase? What is the carrying capacity of Earth for the human species?* To address these questions we must add two key qualifiers: (1) at what standard of living (nutritional, cultural, technological, etc.) and (2) at what cost to the environment? In *How Many People Can the Earth Support?* (1995), Joel E. Cohen reviews estimates of human carrying capacity that use different assumptions relating to these two questions. The first estimate dates from 1891, the rest from the latter 20th century. Remarkably, they vary by a thousand-fold.

The Dutch agroecologist C. T. DeWit (1967) framed the question this way: *How many people can live on Earth if photosynthesis is the limiting process?* After examining geographic variation in the constraints on photosynthesis imposed by light, temperature, and growing season, DeWit estimated Earth's potential agricultural productivity in the absence of constraints of water and nutrients (i.e., assuming sufficient irrigation and fertilization). Based on estimates of primary production and mean annual per capita caloric requirements, DeWit estimated that 1 trillion (10^{12}) people could live from Earth—if photosynthesis were the limiting factor.

However, this estimate is for the population that could live *from* Earth, not *on* Earth, because it assumes that the entire land area is dedicated to crops. To account for the land area needed to support activities other than agriculture, DeWit assumed that each person would need at least 750 m^2 of land for urban use in addition to about 830 m^2 for crop production. This requirement of 1580 m^2 per person reduces the carrying capacity to 146 billion. Many problems are associated with the assumptions embedded in DeWit's calculations, and it is unlikely that environmental effects would allow humans to survive at such densities.

At the other extreme, H. R. Hulett (1970) set out to calculate the "optimum world population," using the standard of living and resource consumption for the average American as his definition of optimal lifestyle. Using the ratio of world production to average per capita American consumption, Hulett calculated an upper limit of 1 billion, supportable at the then-current agricultural and industrial production levels.

So there we have it. Given this range of estimates—1 billion to 146 billion—the answer is no doubt somewhere in between. And these studies are but two of many. A recent statistical analysis of 69 studies of the limits to the human population generated a best estimate of 7.7 billion, with 0.65 billion and 98 billion as the lower and upper bounds (Van den Bergh and Rietveld 2004). Most predictions for the human population by 2050 exceed these meta-estimates. Granted, technology may let us increase agricultural productivity further, and necessity will force us to shift from fossil fuels to alternative, renewable energy sources. But is the magnitude of the human carrying capacity even the right question to be asking? In a review of Cohen's book, F. Landis MacKellar (1996) argues that carrying capacity is a deterministic concept that is outmoded in the "new" ecology, which puts less emphasis on equilibrium states.

Perhaps the real issue concerning Earth's carrying capacity lies not in our ability to feed, clothe, and shelter as many people as possible but rather in the kind of Earth that we hope to inhabit. If maximizing *our* carrying capacity means continuing to reduce or eliminate the carrying capacity of Earth for *other* species—as is happening with ongoing, anthropogenic habitat destruction—then we should be more concerned with minimizing our impact. A new approach to our relationship with the planet is encapsulated in the ecological footprint concept, developed by the Canadian ecologist William Rees (see Ecological Issues: The Human Factor, p. 13). By asking "What is our impact?" as well as "What is our maximum number?" this approach brings new relevance to the population debate.

Bibliography

Cohen, J. E. 1995. *How many people can the Earth support?* New York: Norton.

DeWit, C. T. 1967. Photosynthesis: Its relation to overpopulation. Pages 315–320 in San Pietro, A., F. A. Greer, and T. J. Army (eds.). *Harvesting the sun: Photosynthesis in plant life.* New York: Academic Press.

Hulett, H. R. 1970. Optimum world population. *BioScience* 20:160–161.

MacKellar, F. L. 1996. On human carrying capacity: A review essay on Joel Cohen's *How many people can the Earth Support? Population and Development Review* 22:145–156.

Van den Bergh, J. C., and P. Rietveld. 2004. Reconsidering the limits to world population: Meta-analysis and meta-prediction. *BioScience* 54:195–204.

1. What are some factors (environmental, economic, etc.) that influence the carrying capacity of your region? Do you think the standards of North American consumption should be applied to the rest of the world as global population increases?

2. Extreme environmental events—hurricanes in Florida, heavy frosts in California, drought in the Canadian prairies—dramatically affect agricultural production. What does this imply about the carrying capacity of any particular region or country? How might a population respond to unpredictable variations in essential resources through time?

precipitation, and disturbances such as fire and drought—also influences birth and death. Both types of stochasticity contribute to fluctuations in population size.

10.9 FLUCTUATIONS AND CYCLES: Population Size Can Fluctuate Randomly or Oscillate in a Predictable Pattern

Under the influence of stochastic forces, natural populations tend to fluctuate. Some fluctuate more than others, but even populations that are established and approximating their carrying capacity rarely stay constant. Such fluctuations may be local phenomena. A species may be quite stable in numbers over its range or over many years, but highly variable locally or between years (Figure 10.14a). Populations may fluctuate with changes in carrying capacity resulting from extrinsic influences, especially weather. These fluctuations can be quite predictable, reflecting changes in food supply. For example, the population size of red-breasted nuthatches (*Sitta canadensis*) in Alberta (Figure 10.14b) (Toms et al. 2002) tracks the crop of conifer cones, as also occurs with cone-eating bird species in Québec (Sirois 2000). But even if the carrying capacity remains relatively constant, time lags in density-dependent processes (see Section 11.1), particularly birth and death rates as affected by predation and competition, can cause a population to over- or undershoot its carrying capacity.

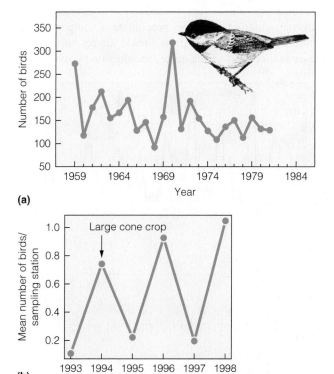

(a)

(b)

Figure 10.14 Fluctuation in population size. **(a)** Despite fluctuations, the numbers of black-capped chickadees (*Parus atricapillus*) in northwestern Connecticut remain relatively constant. **(b)** Fluctuation in numbers of red-breasted nuthatches in Alberta.

((a) Data from Loery and Nichols 1985 and (b) adapted from Toms et al. 2002.)

If estimates are based on census data collected at different times, fluctuations can reflect seasonal swings. The carrying capacity in the harshest season often determines population size in the breeding season. Fall numbers of many species, reflecting the short-term input of reproduction, may be larger or more variable than spring numbers, which reflect winter mortality.

Fluctuations reflect a population's **resilience**: the rate at which it returns to an equilibrium level after a disturbance. Resilience is a measure of how fast a population declines from above its equilibrium, or how quickly it increases from below it. (The same principle applies to homeostatic mechanisms in the body of an individual; see Section 7.5.) Resilience is one aspect of stability. The other is **resistance**: the tendency to remain unchanged (i.e., to maintain equilibrium) when faced by a disturbance. Both types of stability are also critical to ecosystems (see Section 19.4).

The resilience of a population is strongly influenced by reproductive rate. Body size is a valuable clue. Populations of small-bodied animals such as mice fluctuate more than populations of large-bodied animals such as deer. Because small animals typically have shorter lives, they have the potential to decrease in number dramatically from year to year. However, they also have a shorter pre-reproductive period than larger animals and tend to produce more offspring, allowing them to regain their numbers quickly. Such species, which are typical of *r*-strategists (see Section 8.10), have high resilience.

Large-bodied animals have more resistance; their numbers remain closer to equilibrium because they live longer and are less subject to environmental vagaries. However, their resilience is low; long-lived animals take longer to reach reproductive maturity and are much slower to return to equilibrium numbers. The measure of resilience is **return time**: the time required for the population to re-establish equilibrium size. The longer the return time, the lower the resilience. Return time is influenced by interactions with other species. No population occupies a habitat alone. If one species is disturbed, others are affected. If species A depends on species B for food, or if species A is in a mutually beneficial interaction with species B, both species A and species B will experience disturbance. Species A cannot return to equilibrium unless and until species B has done so (Pimm 1991). This point will resurface in our discussion of interactions in Part Four.

Fluctuations that are more regular than we would expect by chance are called **population cycles** (**oscillations**). Cycles vary in predictability. In a *stable limit cycle* (Figure 10.15a, p. 216), population size fluctuates around *K* in a regular manner between upper and lower limits. In other cases, population size shows *damped oscillations*, where fluctuations decrease over time (Figure 10.15b). Or, if density-dependence mechanisms are weak, population size may rise well above *K*, followed by a *crash* (Figure 10.15c), as is typical after rapid exponential growth.

In natural populations with stable limit cycles, the two most common intervals between peaks are 3 to 4 years, typified by sockeye salmon and voles (Figure 10.16a, p. 216), and 8 to 10 years, typified by snowshoe hares (Figure 10.16b). Shorter cycles occur in shorter-lived species, reflecting their shorter generation times. Peak densities can change over time, even when a cyclical pattern persists (Figure 10.16c). Cycles are most notable in ecosystems with fewer species, such as the boreal

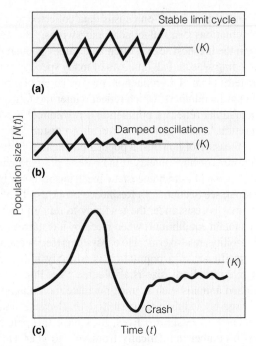

Figure 10.15 Types of fluctuations around *K*. **(a)** Stable limit cycle. **(b)** Damped oscillations. **(c)** Crash.

forest and tundra. In some cases only local or regional populations are affected, but broader synchronicities across the geographic range of a species can occur, as in lemmings (Krebs 2011) and among coexisting small rodents (Kormpimäki et al. 2005).

Based on observational and experimental studies, many hypotheses have been advanced to explain population cycles of small mammals. As summarized by Krebs (2011), these include (1) the *bottom-up model*, wherein food availability is the major driver; (2) the *top-down model*, in which predation and disease (including parasites) are primarily responsible; and (3) the *social behaviour model*, in which oscillations are attributed to interactions such as aggressive territorial behaviour and infanticide. Additional factors include alterations to the endocrine system induced by stress (see Section 11.5) and changes in the frequencies of genes that reduce resistance to aggressive behaviour (Krebs 1985). Dispersal may also be involved (Stenseth 1983), but not in species that peak synchronously throughout their range. As populations rarely oscillate in isolation, we examine cycles involving multiple interacting species in Section 14.11.

10.10 EXTIRPATION: Small Populations Are Highly Susceptible to Extinction

When deaths exceed births, populations decline: R_0 becomes less than 1.0 and r becomes negative. Unless the population can reverse the trend (as happens on the upswing of a population cycle) it may decline towards **extinction**. **Extirpation** is the local extinction or loss of a population as opposed to the extinction (total loss) of a species overall.

Many factors can cause extirpation. Environmental stochasticity often plays a role; extreme events, such as droughts, floods, or abnormal temperatures, can increase mortality and reduce population size. Should environmental factors exceed the tolerance limits of a species, the event could cause extirpation. A severe shortage of resources, caused by environmental variation or overexploitation, can also precipitate a sharp population decline and possible extirpation, should the resource base not recover in time to allow adequate reproduction by survivors.

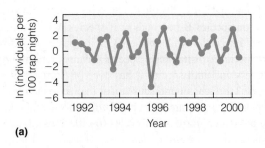

(a)

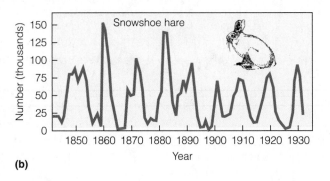

(b)

Figure 10.16 Population cycles in small mammals. **(a)** Voles (*Microtus* spp.) in northern Finland. Values are numbers caught per 100 trap nights (log transformed). **(b)** Snowshoe hares in northern Canada. **(c)** Decline in peak density of snowshoe hares at Kluane Lake, Yukon Territory, from 1980 to 2010.

(Adapted from (a) Korpimäki et al. 2005, (b) MacLulich 1937, and (c) Krebs 2011.)

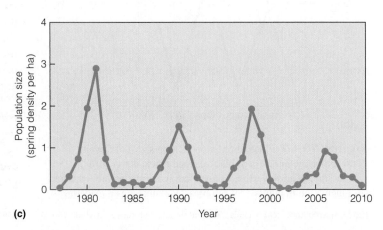

(c)

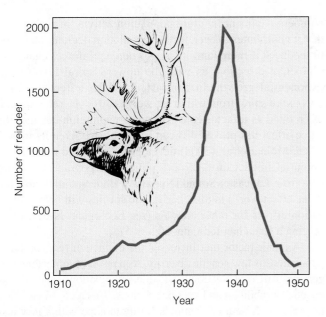

Figure 10.17 Decline of the St. Paul herd after overexploitation degraded the habitat. The initial exponential phase is depicted in Figure 10.4.

(Adapted from Scheffer 1951.)

In the example of exponential growth of reindeer on St. Paul Island (see Figure 10.4), the reindeer overgrazed their range so severely that the herd plummeted from a high of over 2000 in 1938 to 8 in 1950 (Figure 10.17; its numbers have only recovered slightly since). Its decline typifies a population that has exceeded its resource base, although recent comparative studies indicate that density-independent abiotic factors played a major role (Gunn et al. 2003). From a low point, the population may recover to undergo another exponential growth phase, which did not happen in this case, or it may remain low or disappear. Such a crash—whether followed by a recovery or not—contrasts with the fluctuation of population size around the carrying capacity typical of a population exhibiting logistic growth.

When a species enters a habitat, either through natural dispersal or introduced by humans, interactions with species already present can be detrimental. An introduced predator, competitor, or disease pathogen can have a devastating effect. In contrast, biotic interactions among naturally co-occurring species that have evolved together are more likely to regulate populations in a density-dependent manner. The result is logistic growth, which may reduce a population's extirpation risk.

The leading cause of current population extinctions is loss of habitat as a result of human activities, particularly clear-cutting of forests and clearing of land for agriculture and development. Not all species are equally at risk of extinction. In Section 27.3, we discuss various factors influencing the vulnerability of species to extinction. However, regardless of the species, small populations—in part as a result of their greater vulnerability to demographic and environmental stochasticity—are more susceptible than larger populations. If only a few individuals make up the population, the fate of each individual can be crucial to population survival.

The decline in either reproduction or survival under conditions of low population density is called the **Allee effect**. Many factors are involved. Declining population size may directly influence birthrates as a result of life history traits related to mating and reproduction. Among species that are widely dispersed, such as large cats, finding a mate becomes increasingly difficult once the population density falls below a certain point. Many insects use chemical odours or pheromones to communicate with and attract mates. As density falls, it is less likely that a chemical message will reach a potential mate, and reproductive rates may decrease.

Similarly, as a plant population declines and individuals are more scattered, the distance between plants increases and pollination becomes less likely. American ginseng (*Panax quinquefolius*), a perennial herb native to deciduous forests in eastern North America, has long been harvested for its medicinal value. At some locations, its numbers have been reduced to a few dozen plants. Experimental manipulation of population density demonstrated that fruit production per plant declines with decreasing population size (Figure 10.18) as a result of

(a)

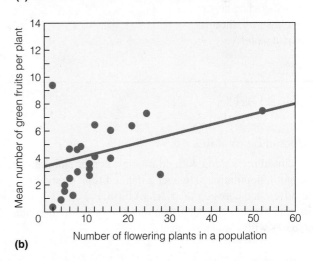

(b)

Figure 10.18 Allee effect, showing reduced fecundity at low population density in the herbaceous perennial American ginseng (a). (b) Fruit production per plant declines at small population sizes in 26 experimental populations. The line represents the general trend as defined by the linear regression $y = 3.36 + 0.077x$, where $x =$ number of flowering plants and $y =$ mean number of fruits per plant However, there is considerable variability around the line.

(Adapted from Hackney and McGraw 2001.)

fewer visits by pollinators (Hackney and McGraw 2001). Smaller populations of flowering plants are less obvious to potential pollinators and have a lower frequency of visitation.

What is puzzling is that the Allee effect does not pose more of a problem for small populations. Many species cope by reverting to self-pollination, allowing them to produce enough seeds to maintain their low numbers, as in the moth-pollinated South African orchid *Satyrium longicauda* (Johnson et al. 2009). Genetic variability may be compromised, but self-pollination confers resistance to the Allee effect.

In some animal species, small population size may contribute to the breakdown of social structures in species that practise cooperative behaviours relating to mating, foraging, or defence. Bird species that aggregate into groups on communal courtship grounds called *leks* (see Section 8.5) are particularly susceptible to disruption of mating behaviour with declining numbers. Similarly, many gregarious species live in herds or packs that enable individuals to defend themselves from predators or find food. Once the population is too small to sustain an effective group size, the population may decline from increased mortality as a result of predation or starvation.

All of these examples involve a *positive feedback loop* whereby a change in the population—in this case declining numbers—influences future population activity in such a way that it contributes to even further population decline. This type of feedback is more familiar to us when it reinforces change in a positive direction, as with the upward spiral involved in exponential growth. In contrast, the Allee effect entails a downward spiral from which recovery is difficult. Yet, although often cited as a factor contributing to population decline, the Allee effect has proved difficult to establish. Given its importance, Molnár et al. (2008) developed a model to quantify its potential impact on declining polar bear populations using data from Lancaster Sound (Nunavut). Understanding how the Allee effect works in this charismatic species will help conservation efforts for other animals (see Ecological Issues: Conserving a Canadian Icon, pp. 604–605).

Another factor that increases the extinction risk for small populations is low genetic diversity. As a result of inbreeding and genetic drift (see Section 5.6), small populations contain less genetic variability than larger populations. This reduced variation may compromise a population's ability to cope with a new disease or predator, or with climate change. Whatever the contributing factors, once population size falls below a critical level, recovery is difficult even if the factors that initiated the decline (such as habitat loss) have been alleviated by conservation efforts.

EcologyPlace

Visit EcologyPlace at www.pearsoncanada.ca/ecologyplace to access online resources that complement your textbook, and help you to apply and to review the information in this chapter. EcologyPlace includes

- an eText version of the book
- self-grading quizzes
- glossary flashcards
- and more!

Go to www.pearsoncanada.ca/ecologyplace and follow the registration instructions on the Student Access Code Card included with this text. If your book does not have a Student Access Code Card, you can purchase access to it at www.pearsoncanada.ca/ecologyplace.

SUMMARY

Birth and Death Rates 10.1

In a closed population with no immigration or emigration, the change in population size over a defined time interval reflects the difference between the rates of birth and death. When the birthrate exceeds the death rate, the rate of population change increases with population size.

Exponential Growth 10.2

As the time unit over which population size is measured approaches zero, the change in size of a growing population approximates a continuous exponential function. The difference between the instantaneous per capita rates of birth and death is defined as r, the intrinsic rate of increase. Exponential growth is typical of closed populations at low densities in favourable environments with no resource limitations.

Life Tables 10.3

Ecologists study mortality and survivorship by constructing a cohort life table. By following the fate of a group of individuals born at the same time until all have died, they estimate age-specific mortality and survival. A time-specific life table is based on sampling the population to obtain a distribution of age classes during a single time period.

Mortality and Survivorship Curves 10.4

Mortality and survivorship curves allow comparisons of demographic trends within and among populations of different species and in different environments. Survivorship curves can be type I, in which survival of juveniles is high; type II, in which mortality is constant for all ages; and type III, in which juvenile survivorship is low. Survivorship curves can differ not only among species and in different environmental conditions, but also between sexes.

Fecundity Tables 10.5

Birthrate greatly affects population growth and, like mortality, is age and sex specific. Some age classes contribute more offspring than others. A fecundity table summarizes gross reproduction, b_x, and survivorship, l_x, in each age class. The sum of $b_x \times l_x$ yields the net reproductive rate—the mean number of females produced over a lifetime by a female, taking survivorship into account.

Generation Time 10.6

Generation time is the mean length of time between when a female is born and when she reproduces. Two populations that have similar reproductive output will grow at different rates if their generation time differs. Using mean generation time, the intrinsic rate of increase (r) can be calculated from the net reproductive rate (R_0).

Logistic Growth 10.7

As populations increase, they encounter resource limitations, causing growth to shift from an exponential to a logistic pattern. In logistic growth, birth and death rates vary in a density-dependent manner. As population size approaches the carrying capacity (K), the birthrate decreases and the death rate increases until they equalize, preventing further population increase.

Stochasticity 10.8

Age-specific rates of mortality and birth represent probabilities. Random variations in birth and death rates due to intrinsic factors are called demographic stochasticity, whereas variations due to extrinsic factors are termed environmental stochasticity.

Fluctuations and Cycles 10.9

Few populations behave in a manner predicted by mathematical models. Most exhibit fluctuations. Some populations follow cycles, with varying predictability. In a stable limit cycle, numbers fluctuate around K in a regular manner between limits. In damped oscillations, fluctuations decrease over time. If density dependence is weak, population size may rise well above K, followed by a crash.

Extirpation 10.10

Many factors contribute to the extinction of local populations, including environmental stochasticity, introduction of new species, and habitat loss. Small populations are particularly susceptible to demographic stochasticity related to the Allee effect, including disruption of social structures that influence mating, feeding, and defence, and loss of genetic diversity.

KEY TERMS

age- and sex-specific birthrate	demographic stochasticity	fecundity table	logistic growth	population growth
age-specific mortality rate	demography	generation time	mortality curve	resilience
Allee effect	environmental stochasticity	geometric model	net reproductive rate	resistance
carrying capacity	exponential growth	gross reproductive rate	open	return time
closed	extinction	intrinsic rate of increase	per capita birthrate	survivorship
cohort	extirpation	life expectancy	per capita death rate	survivorship curve
		life table	population cycles (oscillations)	

STUDY QUESTIONS

1. What is the significance for population growth of whether a population is open or closed?
2. What is the difference between a discrete ($\Delta N / \Delta t$) and a continuous (dN/dt) model of population growth?
3. What two factors determine the rate of change in population size for an exponentially growing population?
4. What is a cohort life table, and what information is needed to construct one? Why are life tables often constructed only for females?
5. Why are survivorship curves based on l_x rather than n_x? What type of survivorship curve is the most common, and why?
6. The e_x value for squirrels reaching the age class of 4 years is 0.75 years. On average, how old will these squirrels be when they die? In contrast, what is the life expectancy of a squirrel born into the cohort (age class 0)? Why do these values differ so much?

7. How do the gross and net reproductive rates differ? What does the value of the net reproductive rate (R_0) indicate about the future trend in population size?

8. Explain why, for an annual species, knowing the net reproductive rate is sufficient to estimate the intrinsic rate of increase (r). What other information is needed to estimate r for a species with overlapping generations?

9. What is the difference between the exponential and logistic models of population growth, in terms of the factors influencing change in population size?

10. Define *carrying capacity* (K). What is its significance for density-dependent regulation?

11. What environmental factors might result in random yearly variations in the rates of survival and birth in a population?

12. What is the difference between a stable limit cycle and a damped oscillation? What ecological factors might contribute to these different patterns?

13. Identify three factors that could cause a population to decline towards extirpation (local extinction). Why are small populations more vulnerable to extirpation?

FURTHER READINGS

Begon, M., and M. Mortimer. 1996. *Population ecology: A unified study of animals and plants.* 3rd ed. Oxford: Blackwell Scientific Publications.
This well-written introductory text is an excellent resource for those wishing to read further about the structure and dynamics of natural populations.

Carey, J. R. 2001. Insect biodemography. *Annual Review of Entomology* 46:79–110.
Review of life tables in insect populations.

Gotelli, N. J. 2001. *A primer of ecology.* 3rd ed. Sunderland, MA: Sinauer Associates.
Explains in detail the most common mathematical models in population and community ecology.

Krebs, C. J. 2011. Of lemmings and snowshoe hares: The ecology of northern Canada. *Proceedings of the Royal Society of London* B 278:481–489.
Summarizes current understanding of population oscillations of lemmings and snowshoe hares.

INTRASPECIFIC POPULATION REGULATION

Pygmy sweep (*Parapriacanthus ransonetti*) form dense schools in openings in coral knolls in the South Pacific. They locate plankton at night using bioluminescent organs on their pectoral fins.

We saw in Section 10.7 that logistic growth of a population slows as it nears its carrying capacity. In response to limits imposed by the environment, negative feedback curtails population increase. Per capita resource availability drops to a level that prevents further growth. This regulation is mediated by **intraspecific** (within-species) interactions among individuals, including competition and social behaviour. Intraspecific interactions exhibit **density dependence**: their effect varies with population size, intensifying as a population nears its carrying capacity. This density dependence allows intraspecific interactions to regulate populations. **Interspecific** (between-species) interactions such as competition, predation, and mutualism (see Part Four) can also be density dependent and may also play a role in regulating populations.

Other factors can influence population growth independently of density. If adverse weather affects a population whatever its population size, or if the proportion of individuals affected is the same at any density, then the influence exhibits **density independence** (see Section 11.11).

11.1 DENSITY DEPENDENCE: Within-Species Regulation Is Linked to Population Density

Density dependence slows population growth by increasing the death rate (*density-dependent mortality*; Figure 11.1a), decreasing the birthrate (*density-dependent fecundity*; Figure 11.1b), or both (*full density dependence*; Figure 11.1c). Logistic growth typically involves full density dependence; birthrate (*b*) decreases and death rate (*d*) increases until they equal each other at the carrying capacity *K*, preventing further growth (see Figure 10.11).

Density-dependent processes can also influence birth and death rates at low densities (Allee effect; see Section 10.10). At low density, it may be hard to find mates, or social structures in species that use facilitative or cooperative behaviours for mating, foraging, or defence may break down. The result is a density dependence that decelerates growth. Mortality increases (Figure 11.2a), birthrate declines (Figure 11.2b), or both are affected (Figure 11.2c). Below some minimum density *A*, population growth is negative. Positive feedback causes an already declining population to decline further rather than regaining an equilibrium. Although not a means of population regulation, this type of density dependence is often an important consideration in species conservation (see Section 27.8).

11.2 INTRASPECIFIC COMPETITION: Resource Limitations May Induce Competition in Populations

As a population nears its carrying capacity, per capita resource availability declines and competition often ensues. In ecology, **competition** is an interaction in which two or more individuals seek out a common resource that is in limited supply relative to the number seeking it. Competition among individuals of the same species is **intraspecific competition**; we discuss *interspecific competition* (between individuals of different species) in Chapter 13.

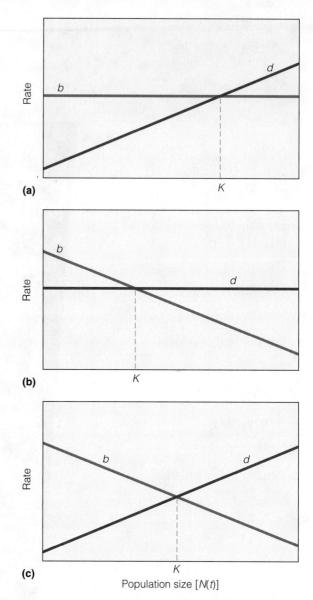

Figure 11.1 Types of density-dependent population regulation. **(a)** Birthrate (*b*) is independent of density, as indicated by the horizontal line. Death rate (*d*) increases with density. At *K*, equilibrium is maintained by increasing mortality. **(b)** Mortality is density-independent, but birthrate declines with density. At *K*, decreasing birthrate maintains equilibrium. **(c)** Full density dependence. Changes in both birth and death rates hold the population at or near *K*.

The fact that individuals share a resource does not prove competition is occurring. There may be enough resources to meet the needs of all individuals without adverse effects. In this case, population size is kept low by other (usually abiotic, density-independent) factors. But if resources are insufficient to meet the needs of all individuals, competition is likely, intensifying as the population increases or as individuals grow larger and increase their resource use.

Populations may experience two types of intraspecific competition, based on how resources are allocated. In **scramble competition**, limited resources are shared by individuals, all of which experience depressed growth as competition intensifies.

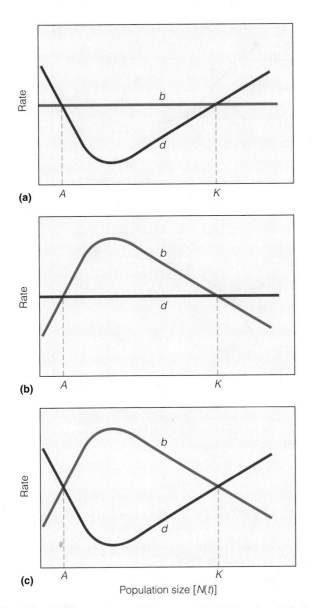

(a)

(b)

(c)

Population size [*N*(*t*)]

Figure 11.2 Impact of Allee effect on population growth. Below some minimum density *A*, death rate increases **(a)**, birthrate decreases **(b)**, or both **(c)**. The result is negative population growth. Unlike the negative feedback that occurs when *N* exceeds *K*, when *N* falls below *A*, positive feedback causes further decline in *N*, increasing the risk of extirpation.

In **contest competition**, dominant individuals claim enough resources for themselves while denying others a share. Typically, a species experiences either scramble or contest competition but may practise both at different life stages. Insect larvae usually undergo scramble competition, while adults often face contest competition.

The outcomes of scramble and contest competition vary. All individuals in intense scramble competition may receive insufficient resources to survive. Imagine a garden overplanted with carrot seeds. Many germinate, but all seedlings suffer resource limitations as they grow until, if the density is high enough, all die. In contest competition, successful individuals secure enough resources to meet their needs. Only a fraction of the population suffers. A relatively constant number of individuals survive, independent of initial density.

Competing individuals do not always interact with each other directly. They may respond to the decline in resources resulting from consumption. Grazing herbivores such as bison compete indirectly by reducing available forage. Similarly, as a tree absorbs water, it decreases the water left for the roots of other trees. This process, called **exploitation**, typifies scramble competition. In other cases, individuals interact directly, preventing some from occupying a site or acquiring its resources. Many birds defend their nest area in the breeding season, denying other individuals access to the site and its resources (*territoriality*; see Section 11.9). This mechanism, known as **interference**, typifies contest competition.

11.3 GROWTH AND DEVELOPMENT EFFECTS: Competition May Slow Individual Growth and Development

At first, intraspecific competition has minimal effects on growth. But as density increases and resource levels fall, some (contest type) or all (scramble type) individuals reduce their intake of resources below an optimal level. Growth and development slow, resulting in *density-dependent growth*: an inverse relationship between population density and individual growth that has been observed in many populations, experimentally and in the field. Consider white clover (*Trifolium repens*) grown at varying densities (Figure 11.3). Mean plant weight declines with increasing density. At low densities, all plants acquire enough resources, but at higher densities, demand exceeds supply. Growth rate and size decline (Chatworthy 1960).

A similar experimental response was observed in the salt marsh plant *Atriplex prostrata* (Wang et al. 2005). After four weeks, mean weight (Figure 11.4a, p. 224) and leaf area (Figure 11.4b) declined by 80 and 72 percent, respectively, from the highest- to lowest-density plants. Net photosynthesis fell by

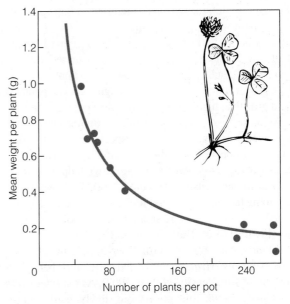

Figure 11.3 Relationship between density and growth of white clover.

(Adapted from Chatworthy 1960.)

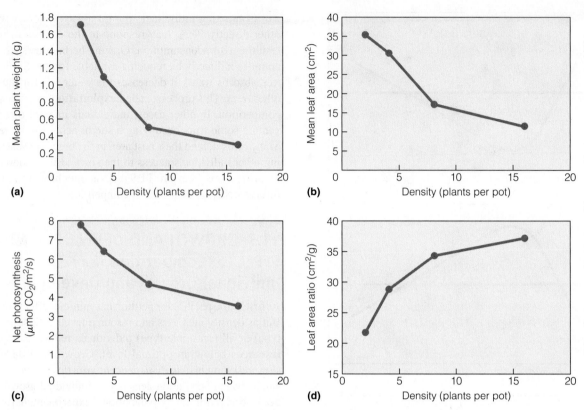

Figure 11.4 Effect of population density on mean **(a)** dry weight, **(b)** leaf area, **(c)** net photosynthesis, and (d) leaf area ratio of *Atriplex prostrata*. Whereas most growth parameters declined with density, the leaf area ratio increased.

(Adapted from Wang et al. 2005.)

INTERPRETING ECOLOGICAL DATA

Q1. What is the difference in weight between plants grown at densities of 2 and 8 per pot? How does this difference compare with that between plants grown at 8 and 16 plants per pot?

Q2. Why is the decline in mean weight with increasing density non-linear? Hint: Calculate the total biomass in each pot by multiplying density (2, 4, 8, and 16 plants per pot) by mean plant weight (1.7, 1.1, 0.5, and 0.3 g) for each treatment.

half (Figure 11.4c), suggesting that reduced carbon uptake is critical. But there was a revealing shift in allocation. The leaf area ratio (leaf area/total plant weight; see Quantifying Ecology 6.1: Calculating Plant Growth Rate, p. 111) increased with density (Figure 11.4d), indicating more investment in photosynthetic surface. As this response is typical in low light, reduced photosynthesis and growth at high density likely reflect competition for light.

As modular organisms, plants are highly plastic, making density-dependent growth effects especially dramatic. But similar responses occur in ectothermic vertebrates. In a study of the Indian bullfrog (*Hoplobatrachus tigrinus*), tadpoles grew more slowly at higher densities (Figure 11.5a). They were also less likely to metamorphosize, and took longer to do so. Those that reached threshold size were smaller than tadpoles at lower densities (Figure 11.5b) (Dash and Hota 1980).

A study of a common North American amphibian, the wood frog (*Lithobates sylvaticus*), revealed shifts in its behaviour and morphology with density (Relyea 2002). As expected, larvae in high-density populations grew more slowly, but they also had longer bodies (Figure 11.6a), shorter tails, and wider mouths, and were more active (Figure 11.6b)—an example of competi-

tion-induced phenotypic plasticity. Increased activity allowed crowded individuals to acquire more resources, as did the morphological shifts, suggesting that both responses are adaptive.

11.4 MORTALITY AND REPRODUCTION EFFECTS: As Competition Intensifies, Mortality Increases and Fecundity Declines

As intraspecific competition intensifies, it goes beyond affecting growth to impact survival and reproduction. For those that do survive, there is a payoff: the death of their competitors increases available resources, allowing increased growth. In a famous study, horseweed (*Erigeron canadensis*) seeds were planted at 100 000 seeds per square metre. Competition for limited resources ensued (Figure 11.7a), with survivor density declining to about 1000 per square metre. These deaths increased per capita resource availability, and mean survivor size increased as density declined (Figure 11.7b) (Yoda et al. 1963). The decline in density and increase in survivor biomass caused by density-

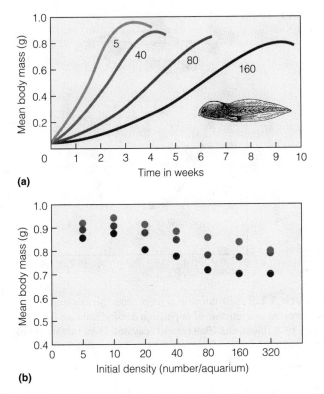

Figure 11.5 Effect of population density on **(a)** growth rate and **(b)** mean body mass at metamorphosis of Indian bullfrog tadpoles.

(Adapted from Dash and Hota 1980.)

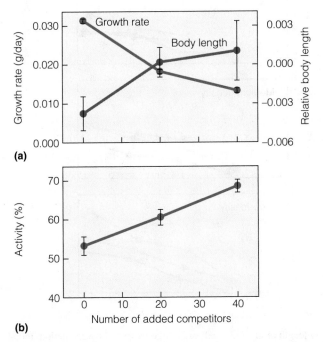

Figure 11.6 Effect of density on mean **(a)** growth rate and relative body length (±1 standard error) and **(b)** activity of larval wood frogs. To correct for size-related differences in length, relative body length was measured as the residual of the regression of length as a function of mass. Activity was measured as the percentage of individuals in motion.

(Adapted from Relyea 2002.)

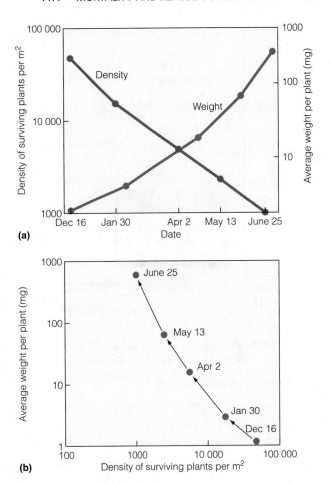

Figure 11.7 Self-thinning in plants. **(a)** Changes in survivor number and mean plant weight over time for horseweed. **(b)** Data re-plotted to show the relationship between density and mean plant weight. Competition increases mortality, which increases per capita resource availability, allowing increased growth of survivors.

(Adapted from Yoda et al. 1963.)

dependent mortality is called **self-thinning**. Note that, although survivors become larger as their competitors die, they are much smaller than when grown at lower initial density.

Often observed in plants, self-thinning is also common in sessile animals such as barnacles and mussels (Figure 11.8, p. 226). It even occurs in mobile animals, especially those with indeterminate growth. Mean size of brook trout (*Salmo trutta*) in two streams in the Sierra Nevadas was inversely related to survivor density (Figure 11.9, p. 226) (Jenkins et al. 1999). Decreased survival reflects intraspecific competition for food as body size increases. Similarly, in a study of steelhead trout (*Oncorhynchus mykiss*) in artificial streams, mortality increased (Figure 11.10a, p. 226) and growth decreased (Figure 11.10b) in response to both (1) increased density for a given food supply and (2) decreased food supply for a given density (Keeley 2003).

Competition can affect fecundity as well as survival and growth. In a study of harp seals (*Phoca groenlandica*), individuals became sexually mature at 87 percent of their mature mean weight of 120 kg. Reduced growth at high density delayed reproduction (increased the age at which females became reproduc-

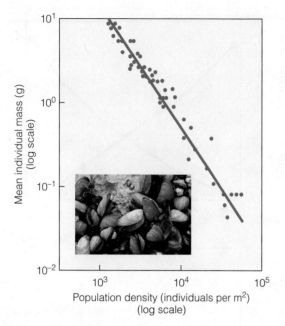

Figure 11.8 Relationship between mean individual mass and population density of the black mussel, *Choromytilus meridionalis*.

(Adapted from Hughes and Griffiths 1988.)

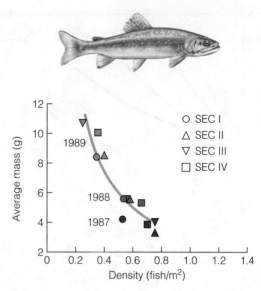

Figure 11.9 Self-thinning in brook trout. Mean survivor weight is expressed as a function of population density. Data are from 1987 (red), 1988 (blue), and 1989 (green) censuses. Points are means of 118 to 699 fish in specific stream subpopulations (Sections I-IV).

(Adapted from Jenkins et al. 1999.)

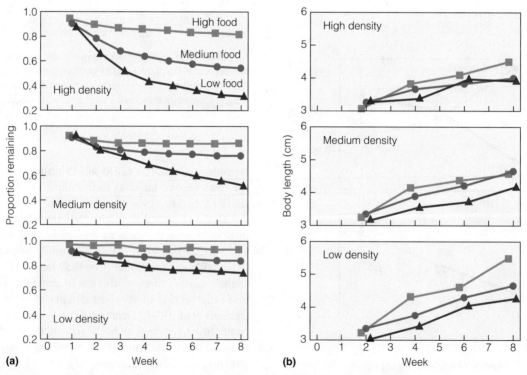

Figure 11.10 Impact of density on (a) survivorship and (b) growth in body length of juvenile steelhead trout over an eight-week period. Initial densities: 32/m² (low), 64/m² (medium), or 128/m² (high). Food abundance: 0.3 g/m²/day (low), 0.6 g/m²/day (medium), or 1.2 g/m²/day (high).

(Adapted from Keeley 2003).

INTERPRETING ECOLOGICAL DATA

Q1. How do (a) mortality and (b) growth differ for high-density populations with low, medium, and high food abundance? How do these differences change at medium and low densities?

Q2. Why do the greatest differences in mortality among the food treatments occur at high density, while the greatest differences in growth occur at low density?

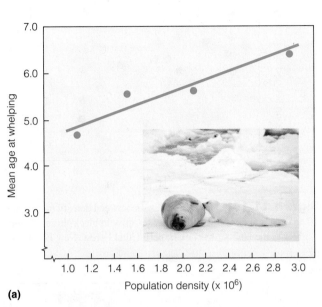

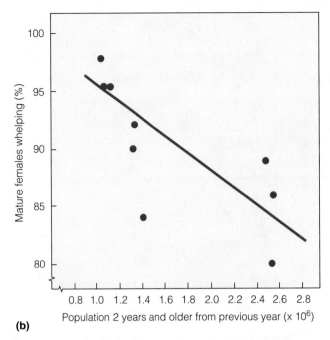

(a)

(b)

Figure 11.11 Density-dependent fertility in female harp seals. **(a)** Mean age (in years) at sexual maturity is related to weight. Seals attain sexual maturity at a younger age when population density is low. **(b)** As a result, fertility is density dependent; as the seal population increases, the percentage of reproducing females decreases. The large variation around the line suggests other factors influence this relationship.

(Adapted from Lett et al. 1981.)

tive; Figure 11.11a). Fertility (number of females giving birth) was density dependent, inversely related to population density in the previous year (Figure 11.11b) (Lett et al. 1981).

The reproductive impacts of intraspecific competition often interact with life history traits (see Section 8.10). When full siblings of the invasive South American freshwater apple snail (*Pomacea canaliculata*) were reared on a gradient of food supply, there were marked differences in size and age at sexual maturity (Tamburi and Martin 2009; the researchers studied competition by growing individuals at different resource levels rather than densities). This study highlights not only the role of phenotypic plasticity in intraspecific competition but also the relevance of life history traits. For females, a minimum size is needed for sexual maturity, so there is more variation in age at maturity than in size (Figure 11.12a, p. 228). The story differs for males. In this species, size is less relevant to male fitness, which is maximized by ~~...~~ ~~...~~ of size (Figure 11.12b). For both

~~...ty-~~
~~...~~ black-throated
~~...ens~~), a songbird that breeds in North American eastern deciduous forests (including southern Canada), is strongly density dependent. The inverse relationship between its density and fecundity (Figure 11.13, p. 228) reflects two interacting mechanisms: (1) a local crowding effect, involving negative interactions with other warblers or with natural enemies, and (2) a regional site effect involving subordinate individuals occupying poorer territories (Rodenhouse et al.

2003). These mechanisms interact with density-independent factors, specifically ENSO oscillations (see Section 2.10) to determine blue warbler population dynamics (Sillett et al. 2000).

In mammals, density-dependent effects on fecundity vary greatly. In large, long-lived species with low reproductive output, the lag time can be considerable. Density-dependent mechanisms may only become apparent as the population nears carrying capacity. The response may be non-linear, as in bison (*Bison bison*) (Figure 11.14a, p. 228), or linear, as in grizzly bears (*Ursus arctos*), whose fecundity declines at a constant rate with density (Figure 11.14b).

For species with indeterminate growth, density dependence can be a particularly powerful means of regulation, because fecundity is strongly related to body size. The density of fish populations affects individual growth, age at maturity, and fecundity. Density-dependent effects on reproduction are also critical for population control in plants. The number of seeds produced per plant in *Salicornia europaea* declines with ~~...~~ (~~...~~1.15) (Watkinson and Davy 1985).

11.5 DENSITY AND STRESS: Crowding-Induced Stress May Regulate Animal Populations

As density increases, crowding ensues and aggressive contacts may increase. Christian (1950) hypothesized that crowding triggers hormonal changes that suppress growth, delay maturity, and curtail reproduction in mice. Crowding also depresses the immune system, especially of subordinates (Salak-Johnson and McGlone 2007). In mammals, stress in pregnant females may

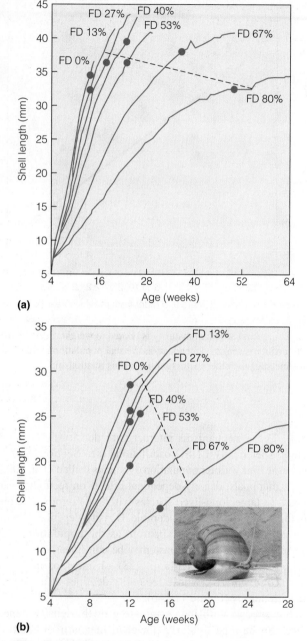

(a)

(b)

Figure 11.12 Mean growth curves for **(a)** females and **(b)** males of the freshwater apple snail, grown under a gradient of food deprivation (FD; 0 to 80 percent). Dots indicate age and size at sexual maturity. The dashed line represents the norm of reaction of phenotypic plasticity in response to the environment.

(Adapted from Tamburi and Martin 2009.)

increase fetus mortality and reduce lactation, stunting offspring survival, growth, and development. Stress thus acts in multiple ways to decrease births and increase juvenile mortality.

The physiological basis of crowding is complex. As well as hormones, **pheromones** (chemicals that facilitate communication) influence behaviour and physiology. In social insects, pheromones released by the queen control development and reproduction of colony members. Pheromones in the urine of stressed adult rodents delay female puberty, slowing population growth (Massey and Vandenbergh 1980). The effects may span

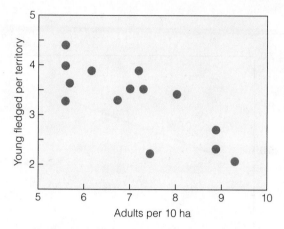

Figure 11.13 Inverse relationship between density (adults per 10 ha) and number of fledglings per territory in black-throated blue warbler in the Hubbard Brook Experimental Forest, New Hampshire.

(Adapted from Rodenhouse et al. 2003.)

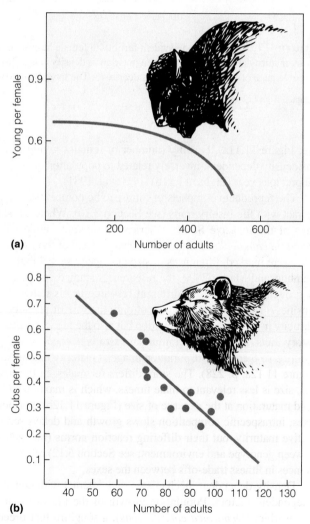

(a)

(b)

Figure 11.14 Non-linear and linear density-dependent effects on reproduction in **(a)** bison and **(b)** grizzly bears.

(Adapted from (a) Fowler 1981 and (b) McCullough 1981.)

generations. Adult pheromones trigger physiological responses in their young, including production and release of hormones that either accelerate or delay puberty (Feldhamer 2007).

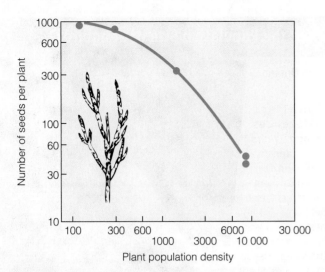

Figure 11.15 Seed production of *Salicornia europaea* at different densities.

(Adapted from Watkinson and Davy 1985.)

Although the population-regulating effects of crowding first reported in laboratory mice have been observed in enclosed wild populations of woodchucks (*Marmota monax*) (Bronson 1964), stress effects are harder to document in free-ranging animals. Factors other than crowding, particularly predation (see Research in Ecology: What Factors Affect the Population Size of Song-birds?, pp. 230–231), can invoke stress responses. So whereas ecologists acknowledge that a stress syndrome can occur in crowded populations, stress is not necessarily a major factor in natural population regulation. In an experimental study of snow-shoe hares, Boonstra and Singleton (1993) concluded that the pituitary–adrenocortical feedback system does not cause cyclical population decline, but may prolong its low phase. Stress is just one of a complex mix of possible contributing factors.

At extreme densities, crowding may cause social breakdown, as indicated by much-publicized experiments with rats and mice (Calhoun 1962). (This research inspired a popular book and the 1982 animated film *The Secret of NIMH*.) Yet it is unwise to extrapolate from the responses of animals kept in captivity at abnormally high densities to populations in the wild, where dispersal often occurs before numbers reach such levels. Moreover, posi-

tive social interactions can buffer the effects of stress. In laboratory studies of prairie voles (*Microtus ochrogaster*), a monogamous species often used to model human physiological response, individuals raised with social bonding (female siblings) suffered less respiratory arrhythmia and other effects from crowding than did individuals raised in isolation (Figure 11.16) (Grippo et al. 2010).

11.6 DENSITY AND DISPERSAL: Density-Dependent Dispersal May Play a Regulatory Role

Rather than cope with stress resulting from crowding, many animals seek vacant or less-crowded habitats. However, although dispersal is most apparent at high densities, it also occurs at other times. Some individuals leave the population whether it is crowded or not. Nor is there a hard-and-fast rule about who disperses (see Section 9.8). When a resource shortage forces some out, those that leave are often older juveniles driven out by adult aggression. Most perish, but some may establish in a suitable area. Because dispersal under high density is a response to over-population, it does not typically regulate populations, but it may have a "rescue effect" in metapopulations in which patches have been vacated by a local extinction (see Section 12.6).

More important to population regulation is dispersal when density is low or increasing, but before the population is in danger of overexploiting resources (i.e., when it is below its carrying capacity). In a study of a red squirrel (*Tamiasciurus hudsonicus*) population in the Yukon Territory, a varying proportion of reproductive females left their feeding territories to offspring, while others kept or shared them (Figure 11.17) (Berteaux and Boutin 2000). Squirrels practised this *breeding dispersal* (i.e., of reproductive individuals) more often in summers when food was plentiful. Whereas breeding dispersal had no effect on the survival of dispersing females, it increased the survival of their non-dispersing offspring over the winter. There was a clear survival advantage to the offspring of staying in the home range in this strongly philopatric species. Older females are more often involved,

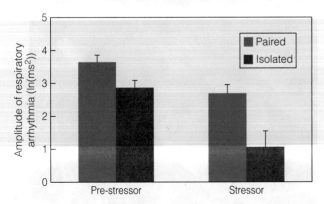

Figure 11.16 Amplitude of respiratory arrhythmia in paired and isolated prairie voles prior to and during crowding.

(Adapted from Grippo et al. 2010.)

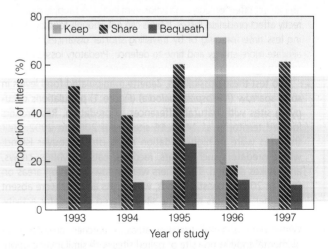

Figure 11.17 Proportion of litters in which female red squirrels kept, shared, or bequeathed (left) their territories to offspring from 1993 to 1997.

(Adapted from Berteaux and Boutin 2000.)

RESEARCH IN ECOLOGY   What Factors Affect the Population Size of Songbirds?

Liane Zanette, Department of Biology, University of Western Ontario.

The decline of songbirds is one of the most publicized conservation issues in North America. It was first highlighted by Rachel Carson, whose ground-breaking *Silent Spring* (1962) raised general awareness about the effects of pesticides as well as galvanizing support for banning DDT. But although pesticides were critical to songbird decline, habitat degradation and loss continue to take their toll. The goal of the University of Western Ontario's Liane Zanette is to identify the specific factors behind the ongoing decline of songbirds. Zanette and her colleagues combine field experiments with behavioural and physiological studies. Incorporating data from both individuals and populations, her approach exemplifies the "new ecology."

Two habitat-related factors are critical for reproductive success: (1) food supply and (2) losses to predation. The combined effects of supplemental food and reduced predation on snowshoe hare (*Lepus americanus*) reproduction were significantly greater than expected if the factors were acting independently (Krebs et al. 1995). Zanette asked if a similar synergistic relationship exists in birds, and if so, what are the proximate (first-hand) mechanisms governing it?

In birds, reproductive success reflects (1) the number of eggs laid over a breeding season, which is a function of *clutch number* × *clutch size* (number of eggs per clutch), and (2) the number of eggs that hatch into successful fledglings. Researchers have assumed that food supply is likely to affect the first factor through reproductive effort (see Section 8.6) and predation the second, through mortality losses (see Section 14.2).

As well as these direct effects, each factor may have subtle indirect effects that contribute to a synergistic response. Predation could affect not only the number of eggs lost but also the number laid (clutch size). In the mild climate of coastal British Columbia, many bird species have multiple clutches. If an entire clutch is lost, a pair will likely have more clutches over the season than if losses were lower. (This effect is less likely if nest losses are partial. A "successful nest" is one in which at least one young fledges, so if losses are partial, clutch number will likely be unchanged.) Alternatively, predation could affect clutch size. Either or both responses would constitute an indirect effect on egg production. Similarly, food supply may indi-

significant differences among years from density-independent environmental effects (see Section 11.11), nest survival was higher not only in the low-predation sites (as expected) but also in sites where the birds were fed (Figure 2b). The number of fledglings per nest did not differ in either treatment compared with the controls, but increased significantly in the combined treatment (Figure 2c). Along with the indirect effect of food on nest survival, these results indicate a synergistic response. Parents in the combined treatment had 1.8 times as many fledglings as expected if food and predation were operating independently.

This study suggests that both food and predation affect reproductive success, not as independent mechanisms but acting synergistically. Zanette then pursued follow-up questions: (1) *How do these factors affect total egg production?* (2) *What is the mechanism underlying the indirect effects of predation?*

With respect to the first query, food addition but not predation increased total egg production (Zanette et al. 2006b). As expected, predation increased clutch number by inducing re-nesting, but so did food addition, by shortening the time between nestings. Both factors affected clutch size, which increased with food addition but declined with predation. Although predation thus affected both factors determining egg production, it had no significant effect on egg production overall, given the negative correlation between clutch number and size (Figure 3) resulting from intraspecific competition. In essence, the two effects offset each other. Food supply seems to establish an upper limit to total egg production, whereas interactions between food and preda-

2a). Both feeding and predation affected nest survival. Despite

Figure 1 The song sparrow is a common songbird.

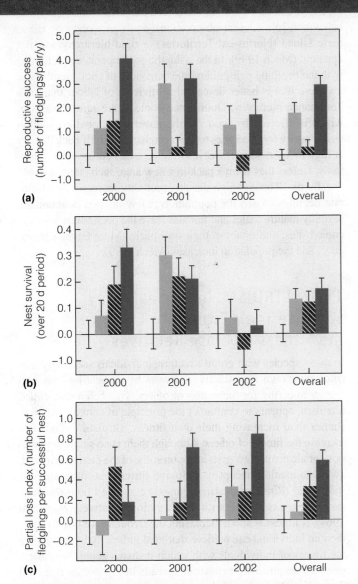

(a)

(b)

(c)

Figure 2 Effect of feeding and predation on **(a)** reproductive success, **(b)** nest survival, and **(c)** number of fledglings per successful nest in song sparrows. Treatments are: green (added food), red (low predation), and blue (combined). All results are expressed relative to controls.

(Adapted from Zanette et al. 2006a.)

With respect to the second query, it is easy to see how predation increases clutch number, if nest loss is complete. Assuming they have enough food, parents with unsuccessful nests will re-nest. But how does predation reduce clutch size? In collaboration with physiological and behavioural ecologists, Zanette studied the mechanisms underlying predation's indirect effect on egg production. They manipulated predation by trapping predators in the low-predation treatments and simulating predation in the high-predation treatments, while feeding all birds to eliminate food as a limiting factor.

Predatory pressure did not affect the size of the first clutch, but females in the high-predation treatments were in poorer condition and laid smaller subsequent clutches (Travers et al. 2010). Even

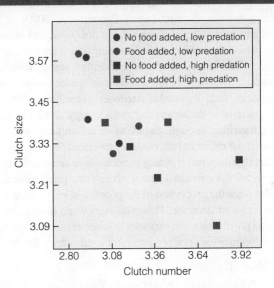

Figure 3 Relationship between clutch number and size.

(Adapted from Zanette et al. 2006b.)

though food was abundant, predation increased oxidative stress, depressed immune function, and reduced fat storage. We might expect a higher birthrate to compensate for predation losses if density dependence were operating. But predation seems to have an indirect, stress-related effect on birthrate. Response to predation is multi-faceted, involving many physiological and behavioural parameters that indirectly affect demography.

Many questions remain; but by combining innovative field studies with sophisticated physiological and behavioural analyses, Zanette is furthering songbird conservation efforts by expanding our understanding of their complex demography.

Bibliography

Carson, R. 1962. *Silent spring.* Boston: Houghton Mifflin.
Krebs, C. J., S. Boutin, R. Boonstra, A. R. Sinclair, J. N. Smith, M. R. Dale, K. Martin, and R. Turkington. 1995. Impact of food and predation

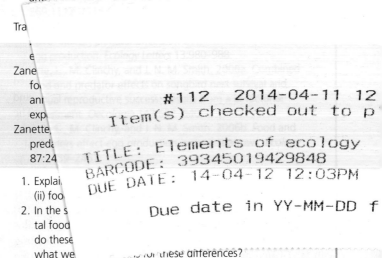

because experienced females are more likely to acquire multiple middens (food caches) in the fall. They then relinquish these middens to offspring the next spring, suggesting a parental investment that anticipates their offspring's needs (Boutin et al. 2000).

In other species, *natal dispersal* (i.e., of juveniles) is more common, wherein juveniles maximize their fitness by leaving the birthplace. When intraspecific competition is intense, dispersers relocate where resources are more accessible and/or breeding sites more available. Natal dispersal also reduces inbreeding. As always, there are trade-offs: dispersers use more energy and incur more risks from travelling through, and living in, unfamiliar terrain.

Given these different patterns, does dispersal regulate populations? Although dispersal typically increases with density, it is often not truly density dependent—that is, there is no predictable relationship between the proportion of the population leaving and population increase or decrease. However, even when dispersal does not regulate populations, it contributes to range expansion (see Section 9.10) and may enhance persistence of local subpopulations.

11.7 SOCIAL BEHAVIOUR: Species with Group Structure Limit Numbers by Behaviour

Intraspecific competition in animals can express itself in social behaviour, particularly the degree to which *conspecifics* (individuals of the same species) tolerate one another. Such behaviour may limit the number occupying a site, accessing shared resources, or engaging in reproduction.

Many animals live in groups with some form of social organization. Group structure may help acquire resources (as with predators that hunt in packs) and/or aid in defence. Social organization may be based on aggressiveness, intolerance, and dominance hierarchies. Two opposing forces are at work. One is mutual attraction of individuals, the other a negative reaction to crowding. Each individual occupies a position based on dominance and submissiveness. Individuals settle social rank by fighting, bluffing, and threatening. They maintain it by subordination of lower-ranked individuals, reinforced by threats and occasional punishment. Such organization stabilizes intraspecific competitive relationships and resolves disputes with minimal energy.

Social dominance is the dominance of one individual over another in a group, often maintained by aggressive behaviour within a dominance hierarchy. It can regulate a population when it affects reproduction or survival in a density-dependent way. For example, wolves (*Canis lupus*) live in a **pack**: an extended kin group consisting of a mated pair, one or more juveniles, and several related non-breeding adults (6–12 wolves in total). The pack is controlled by two hierarchies, one headed by an alpha female and the other by an alpha male, each of which dominates others of its sex. Below the alpha male is the beta male—closely related, often a full brother, who defends his position against any subordinate (omega) males below him in the hierarchy. Mating is rigidly controlled. The alpha male (or occasionally the beta male) mates with the alpha female, who prevents lower-ranking females from mating. The alpha male inhibits other males from mating with her. Thus, each pack has one reproducing pair and one annual litter, which is reared cooperatively.

However, this prevailing view may not apply to wolves in the wild. Most research on pack behaviour has involved wolves in captivity. Unrelated wolves are forced together, eliciting atypical behaviour. In a 13-year observational study of wolves on Ellesmere Island (Northwest Territories), a rigid hierarchy was not apparent (Mech 1999). In the wild, the pack operates as a family, with the breeding pair guiding the behaviour of pack members in a system that is better described as division of labour than as a dominance hierarchy. Dominance contests are rare. Pack size, which is affected by food supply, governs regional population size. Priority for food goes to the reproducing pair. If pack density is high or food scarce, wolves may be expelled or leave voluntarily. Unless they form a pack in a new area, such "lone wolves" usually die. Thus, at high density, mortality increases and birthrate declines. When the population is low or food is abundant, sexually mature males and females leave the pack, settle in unoccupied sites, and establish their own packs. More females reproduce, and the population increases (Mech 1999).

11.8 ALTRUISM: Individuals May Sacrifice Their Fitness to Enhance the Fitness of Close Relatives

In many species with group structure, individuals seemingly sacrifice their own reproductive interests by dedicating time and energy to caring for the young of others. This behaviour, called **altruism**, appears to contradict the principle of natural selection. Rather than increasing their own fitness, altruistic individuals increase the fitness of others. Although there is no single explanation for altruism, ecologists agree on some of the factors involved.

One useful concept is **inclusive fitness**: the fitness of the individual (*direct* or *classical fitness*) combined with the *indirect fitness* associated with helping close relatives (Hamilton 1964). If fitness is about increasing the frequency of one's genes, then an individual can achieve that goal indirectly by increasing the fitness of individuals with which it shares genes. Whether a gene for an altruistic behaviour spreads in a population should depend on the trade-offs between the fitness costs to the helper versus the fitness benefits to the recipient. The closer the relative, the more likely that they share genes, so helping behaviour should correlate not only with increased success of the offspring receiving help but also with their relatedness. Many observations support this view (see Research in Ecology: Sounding the Alarm in Richardson's Ground Squirrels, pp. 296–297).

But correlation does not establish cause. Confounding factors can be involved. Experimental removal of helpers did significantly reduce survival of offspring in the cooperatively breeding cichlid fish *Neolamprologus pulcher* (Figure 11.18) (Brouwer et al. 2005). Compared with that of younger, smaller helpers, the behaviour of older, larger helpers was not entirely altruistic: they were only allowed to stay in the group by "paying" with helping behaviour.

Inclusive fitness is related to **kin selection**: selection that favours the reproductive success of one's close relatives, increasing indirect fitness. Kin selection is most developed in social insects such as bees, ants, and termites. Their altruistic behaviour is not in any sense a choice but is enforced by **castes**: distinct groups of organisms within a population that perform

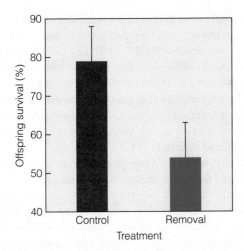

Figure 11.18 Mean percent survival (± 1 standard error) of cichlid fish offspring after seven days in groups from which helpers were removed compared with controls.

(Adapted from Brouwer et al. 2005.)

specific tasks. Species that combine caste structure with altruistic behaviour are termed **eusocial** (Batra 1966). The drone belongs to a reproductive caste, the worker bee to a helper caste. Eusociality also occurs in some aphids and crustaceans.

In social insects, eusociality entails a complex juggling act between inter- and within-group competition. Both are intraspecific, but competition with other hives selects for cooperative rather than competitive behaviour within a hive, enhancing its success (Reeve and Hölldobler 2007). Intergroup competition promotes development of a "superorganism" in which all individuals are subordinated to the whole. Reeve and Hölldobler argue that understanding the ecology of cooperation in eusocial insects has implications for human societies.

Eusociality is not restricted to invertebrates. The long-lived naked mole-rat (*Heterocephalus glaber*) of East Africa is a rare example of a eusocial mammal (Figure 11.19). Its social structure includes a queen, one to three males, and a reproductively suppressed worker caste. Social structure before and after queen removal is determined by a dominance hierarchy, with the next-highest ranking female assuming the queen status (Clarke and Faulkes 1997).

Figure 11.19 The naked mole-rat occupies desert burrows in East Africa. Almost blind, it is virtually hairless and regulates its temperature largely by behaviour.

Are these and other examples of altruism and kin selection means of intraspecific regulation? By refraining from reproducing while at the same time increasing the reproductive success of close relatives, altruistic individuals lower the population birthrate while increasing their inclusive fitness.

11.9 TERRITORIALITY: Individuals of Some Species Defend Part or All of Their Home Range

We have discussed intraspecific competition and social behaviour as processes that can regulate populations, but social organization itself can be a form of intraspecific competition—**territoriality**. Territorial individuals defend some portion of their home range for the exclusive use of the resources it contains. **Home range** is the area that an animal normally uses (usually over a year). The size of the home range varies with available resources, mode of food gathering, body size, and metabolic needs. In mammals, home range size increases with body size (Figure 11.20), reflecting the link between body size and energy needs. Carnivores require larger home ranges than herbivores and omnivores of similar size. Within a species, adult males often have larger home ranges than females and juveniles.

In any species, aggressive interactions may affect movements of individuals in another's home range. However, in territorial species, the individual actively defends part or all of its home range (Figure 11.21, p. 234). The defended area may be a **core area**, where the individual spends most of its time, or a feeding or nesting area. If the animal defends its entire home range, its home range and territory are identical. Individuals of all animal species have a home range, but only those that actively defend some portion of it are territorial. By defending a territory, the individual secures exclusive access to the resources it contains. Defence can involve many behaviours: song and call; intimidation displays, such as spreading wings in birds or baring

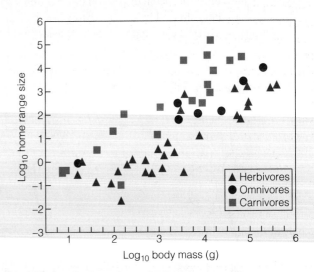

Figure 11.20 Relationship between home range and body weight of North American mammals. For a given body size, carnivores have larger home ranges than herbivores, because the range must be large enough to support their prey population(s).

(Adapted from Harestad and Bunnell 1979.)

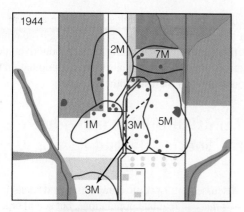

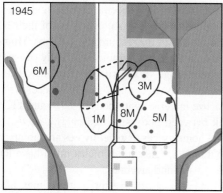

Figure 11.21 Territories of the grasshopper sparrow (*Ammodramus savannarum*) as determined by observations of male banded birds (1M, 2M, etc.). Dots indicate song perches, which are located near territory boundaries. Shaded areas indicate crop fields. Dashed lines indicate territory shifts before re-nesting during the same season. The tendency of individuals to return to the same territory is called *philopatry*.

(Adapted from Smith 1963.)

fangs in mammals; attack and chase; and marking with scents that repel rivals. As a result of these behaviours, territorial species typically have uniform spatial dispersion (see Section 9.4).

Total area available divided by mean territory size indicates the population size a habitat can support. When the available area is filled, owners evict excess individuals, denying them access to resources. Although their survival chances are low, these individuals are not necessarily doomed; they constitute a **floating reserve** of potential breeders that may increase in fitness if circumstances change and territory owners succumb to predation or disease. In a white-crowned sparrow (*Zonotrichia leucophrys*) population in California, 25 percent of territorial individuals had been floaters for two to five years before acquiring a territory (Petrinovich and Patterson 1982).

Similarly, when breeding pairs of great tits (*Parus major*) were removed from their territories in an English woodland, they were quickly replaced, largely by first-year individuals that moved into vacated territories from adjacent suboptimal hedgerow habitat (Figure 11.22) (Krebs 1971). In black-capped chickadees (*Parus atricapillus*), a related species that is one of the most common overwintering birds in Canada (Figure 11.23), floaters replaced only high-ranked territorial individuals, ignoring territories vacated by lower-ranked individuals (Smith 1987).

As a type of contest competition, territoriality limits access to the defended area. But does it regulate populations? If all pairs that settle in an area acquire a territory, territoriality influences spatial dispersion without regulating size. But if territory size has a lower limit, the number of pairs that can inhabit an area is limited, and individuals that fail to do so must leave. In that case, territoriality can regulate the population, assuming an excess of reproductive adults. Reproduction is then limited by territoriality, and density-dependent regulation occurs.

11.10 SPACE CAPTURE: Pre-emption of Space by Plants Is Analogous to Territoriality

Plants are not territorial in the same sense that animals can be, but they do "capture" and hold onto space. Some plants (especially trees) occupy space for a long time, preventing or inhibiting invasion by plants of the same or other species. Even at a microscale, the spreading leaf rosette of a dandelion eliminates other plants from the immediate vicinity. Plants that capture

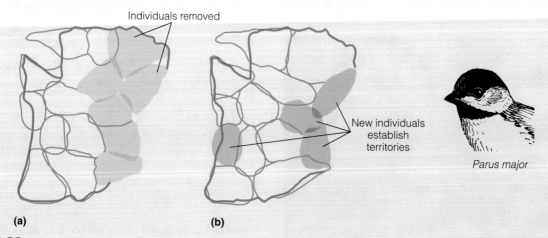

Parus major

Figure 11.22 Resettlement of vacant territories of great tits in an oak woodland. Lines show territory boundaries of breeding pairs. **(a)** Six pairs were removed from March 19–24, 1969 (shaded areas). **(b)** Within three days, some resident pairs had shifted (and sometimes expanded) their territories, and four new pairs had occupied new territories. The territories again formed a complete mosaic.

(Adapted from Krebs 1971.)

Figure 11.23 The black-capped chickadee occurs from coast to coast, from the northern United States to James Bay. It overwinters in much of Canada, but populations from the Northwest Territories often move south.

space may increase their fitness at the expense of others. This phenomenon of **pre-emption** resembles territoriality. Indeed, some ecologists define territoriality as the spacing out of individuals more than expected from random occupancy of suitable habitat. A uniform dispersion pattern is interpreted as an indicator that competition for space is occurring in a plant population.

Besides the space they actually occupy, plants create **resource depletion zones**—areas of reduced resources in the vicinity of their leaf canopies and root systems. Tall individuals intercept light, shading the ground below and inhibiting establishment by shade-intolerant species. Likewise, uptake of water and nutrients from soil reduces their availability to nearby plants with overlapping rooting zones. The root mass of neighbour plants has a major impact on belowground competition. In an experimental study, exclusion tubes were placed vertically in soil to separate roots of a target plant inside the tube from the roots of neighbours. Differing amounts of neighbour roots were allowed access by placing differing numbers of holes in the tube. Growth of target plants was significantly affected by the amount of overlap with neighbours (Figure 11.24) (Cahill and Caspar 2000).

Despite the wealth of evidence demonstrating intraspecific competition, most plant studies examine only short-term effects in herbaceous species. Yet over 150 woody species are known to form *root grafts* (unions) with neighbours of the same species. This behaviour raises the possibility of sharing minerals and carbohydrates among neighbours. Is this behaviour intraspecific parasitism, or a cooperative interaction that belies the view that competition is the major interaction affecting high-density populations? Evidence for such intraspecific facilitation is growing (Fajardo and McIntire 2011). Jack pine (*Pinus banksiana*) formed more graft unions in natural stands in Québec than in plantations. After an initial period of reduced growth assumed to reflect the costs of graft formation, trees with graft unions had higher growth rates than those without (Tarroux and DesRochers 2011). Possible benefits go beyond resource sharing to improving wind resistance and allowing a stand to persist by subsidizing smaller individuals. From an evolutionary viewpoint, root grafts (and a related phenomenon, *common mycorrhizal networks*; see Section 15.13) may be a kind of group behav-

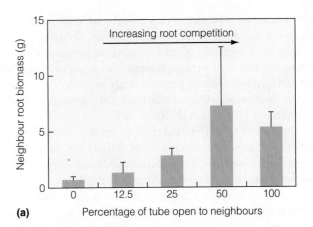

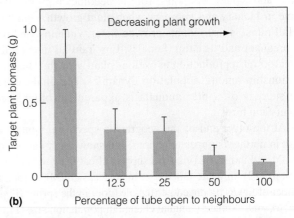

Figure 11.24 Root competition in *Amaranthus retroflexus*. **(a)** Root biomass from neighbour plants that grew into the tubes increased with the number of openings (percentage of tube open to neighbours). **(b)** Increase in belowground competition caused a decline in mean biomass of target plants.

(Adapted from Cahill and Caspar 2000.)

iour of plants that increases inclusive rather than direct fitness (see Section 11.8). In the northern Canadian forests where jack pine is common, such behaviour may be especially advantageous.

11.11 DENSITY INDEPENDENCE: Factors Unrelated to Density Also Affect Population Growth

As we have seen, population growth is subject to density-dependent responses. But there are other, sometimes overriding influences that are *density independent*. Factors such as temperature, precipitation, and natural disturbances often affect birth and death rates. Although critical, such factors do not regulate population growth, because regulation implies feedback.

If environmental conditions exceed an organism's tolerance limits, their impact can be far-reaching, affecting survival, growth, maturation, reproduction, and dispersal. The end effect can be extirpation of local populations. Even if the population persists, it can be greatly reduced. Populations of many insects are subject to density-independent effects. Regional outbreaks of spruce budworm (*Choristoneura fumiferana*) are usually preceded by an extended drought, ending when wet weather returns. Such density-independent factors rarely act in isolation. Long-term studies in New Brunswick reveal that intrinsic, density-dependent factors involving parasitoids (see Section 14.1) and disease affect mortality of spruce budworm larvae, establishing a basic oscillation in local population size (Royama 1984). Density-independent factors, predominantly weather, cause a secondary fluctuation, with the result that local populations oscillate in synchrony (Royama et al. 2005).

In deserts, precipitation directly affects the population growth of many rodents and birds. Merriam's kangaroo rat (*Dipodomys merriami*) inhabits the Mojave Desert in the southwestern United States. Although its water-conserving traits allow it to survive long dry periods, it needs enough seasonal moisture to stimulate growth of herbaceous desert plants in fall and winter. The kangaroo rat becomes reproductively active in January and February when plant growth, stimulated by fall rains, is lush. Plants provide water, vitamins, and food for pregnant and lactating females. Low rainfall thus inhibits kangaroo rat reproduction as well as plant growth. This close relationship among population dynamics, seasonal rainfall, and success of winter annuals is apparent in other desert rodents and birds.

At the other end of the geographic spectrum, winters are harsh in northern temperate regions, and snow accumulation can directly and indirectly (via food supply) affect many species (see Section 7.12). In white-tailed deer in northeastern Minnesota, the number of fawns produced per female (doe) in the spring (Figure 11.25a), and ultimately annual change in population size (Figure 11.25b), is inversely related to snow accumulation (Mech et al. 1978). Deep snow impairs winter foraging, which reduces spring fecundity. Changing depth and duration of snow cover in northern Canada with global warming is likely to alter population oscillations in lemmings and snowshoe hares by affecting reproduction (particularly of lemmings, which continue breeding in winter given adequate snow cover), food supply, and interactions with predators (Krebs 2011).

Populations that are subject to strong density-independent factors often exhibit relatively weak density-dependent regulation. Because their numbers are kept below carrying capacity by density-independent forces, the impact of natural selection on intrinsic regulatory control is less intense. This factor may contribute to the propensity of species occupying harsh, disturbance-prone habitats (including much of Canada) to undergo large swings in population size.

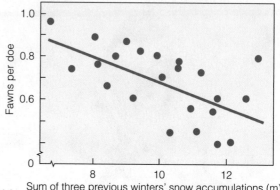

(a)

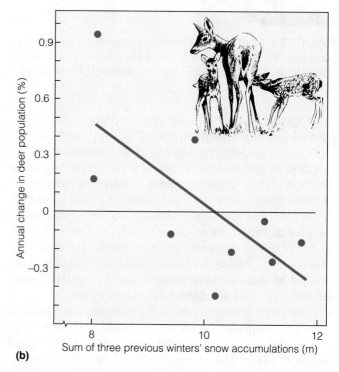

(b)

Figure 11.25 Impact of the previous three winters' snow accumulations in northeastern Minnesota on **(a)** fecundity (fawns per doe) and **(b)** annual change (%) in population size of white-tailed deer.

(Adapted from Mech et al. 1978.)

EcologyPlace

SUMMARY

Density Dependence 11.1

Populations do not increase indefinitely. As resources become limiting, birthrate decreases and/or mortality increases, slowing growth. If the population declines, negative feedback causes mortality to decrease and births to increase, allowing the population to recover. At very low density, positive feedback associated with the Allee effect can drive a population to local extinction.

Intraspecific Competition 11.2

Intraspecific competition can occur if resources are in short supply and/or population density increases. In scramble competition, growth and reproduction of all individuals are depressed as competition intensifies. In contest competition, dominant individuals claim resources while denying resources to others, which produce no offspring or perish. Competition can be direct, via interference, or indirect, via resource exploitation.

Growth and Development Effects 11.3

Density-dependent growth, whereby competition for scarce resources decreases growth and retards development, is common in many species. Highly plastic organisms such as plants respond to competition by modifying both form and size.

Mortality and Reproduction Effects 11.4

As competition intensifies, self-thinning commonly occurs. Higher mortality increases resource availability for the survivors, allowing their growth to increase. High density also delays reproduction and reduces fecundity.

Density and Stress 11.5

Crowding stress may delay reproduction, cause abnormal behaviour, and increase susceptibility to diseases and parasites. Crowding effects are mediated by hormones and/or pheromones. The extreme negative effects observed in crowded laboratory conditions are less likely in the wild.

Density and Dispersal 11.6

Dispersal is a frequent event. If dispersal occurs in response to declining resources or to overcrowding, it does not regulate populations. It can regulate populations if dispersal rate increases with density when a population is below its carrying capacity.

Social Behaviour 11.7

Social behaviour may regulate populations. The degree of tolerance for other individuals of the same species can limit population size in an area and affect access to resources. Social hierarchies in species with group behaviour are based on dominance. High-ranked individuals secure most of the resources, with subdominant individuals experiencing shortages. Rigid dominance hierarchies observed in captive populations may not be typical in the wild.

Altruism 11.8

Species that exhibit group behaviour often engage in altruistic behaviours that reduce their own fitness while increasing the fitness of close relatives. Kin selection is most developed in eusocial species, which combine altruism with a caste structure.

Territoriality 11.9

Members of territorial species defend part or all of their home range. The size of an animal's home range is influenced by body size and trophic level and is larger for carnivores than for herbivores. In territorial species, an individual (or a group) actively defends part or all of its home range for its exclusive use. Territoriality is a form of contest competition in which part of the population is excluded from reproduction. If non-reproducing individuals act as a floating reserve, territoriality can regulate populations.

Space Capture 11.10

Plants are not strictly territorial, but they do capture space, excluding other individuals of the same or smaller size. Plants pre-empt space by intercepting light, moisture, and nutrients, thereby creating resource depletion zones that reduce resource supply to their neighbours. Woody species may share resources with neighbouring individuals through root grafts.

Density Independence 11.11

Density-independent factors such as weather affect but do not regulate populations. Regulation implies feedback, and although density-independent forces often reduce and even eliminate populations, their effects do not vary with density. Density-dependent and density-independent processes interact to control population dynamics in many species.

KEY TERMS

altruism	density independence	interference	pack	self-thinning
caste	eusocial	interspecific	pheromones	social dominance
competition	exploitation	intraspecific	pre-emption	territoriality
contest competition	floating reserve	intraspecific competition	resource depletion zone	
core area	home range		scramble competition	
density dependence	inclusive fitness	kin selection		

STUDY QUESTIONS

1. In terms of its effects on birthrate and/or mortality, what are the three distinct ways in which density dependence can regulate population growth?

2. Explain how logistic growth involves negative feedback, whereas the Allee effect involves positive feedback.

3. Competition can function as a mechanism of density-dependent population regulation. How might scramble and contest competition differ in their regulatory effect?

4. Intraspecific competition can generate an inverse relationship between population density and growth (see Figures 11.3 and 11.8) and reproduction (see Figures 11.11 and 11.14). What condition must hold to establish that competition is responsible for these inverse relationships?

5. What is the role of shading in density-dependent growth of plants?

6. Age at sexual maturity increases with density for many animals. How does this play a role in density-dependent regulation? How is it related to (i) the importance of generation time for population growth (see Section 10.6) and (ii) the fast–slow continuum of life history traits (see Section 8.10)?

7. Why is dispersal more likely to play a regulatory role as density increases but is still below the carrying capacity rather than when density is at a maximum?

8. How might social dominance contribute to density-dependent population regulation?

9. How do ecologists explain the phenomenon of altruistic behaviour?

10. Should the wolf be considered a eusocial species?

11. Distinguish between *home range* and *territory*. What are the ecological trade-offs of territorial defence? What condition must hold for territoriality to exert a density-dependent influence on population growth?

12. Below-average rainfall in Kruger National Park (South Africa) causes a decline in the growth of grasses. Mortality rate rise in the African buffalo, causing their numbers to fall. Can you conclude that annual variation in rainfall is a density-dependent mechanism regulating the buffalo population? What additional information might change your answer?

13. How can root grafting of plants be interpreted as group behaviour? How does it compare with clonal stands of trees where new individuals have arisen from asexual offshoots? Can either be seen as involving altruism?

FURTHER READINGS

Dawkins, Richard. 1976. *The selfish gene*. Oxford: Oxford University Press.
   Provocative and influential book that changes the focus of evolution from the individual and the population to the gene. Discusses the evolution of altruism.

Fajardo, A., and E. J. MacIntire. 2011. Under strong niche overlap conspecifics do not compete but help each other to survive: Facilitation at the intraspecific level. *Journal of Ecology* 99:642–650.
   This important article challenges the traditional view that competition is the major intraspecific interaction affecting populations.

Krebs, C. J. 2011. Of lemmings and snowshoe hares: The ecology of northern Canada. *Proceedings of the Royal Society B* 278:481–489.
   Valuable paper that contrasts the regulation of population size in these two iconic species.

Sinclair, A. R. E. 1977. *The African buffalo: A study of resource limitations of populations*. Chicago: University of Chicago Press.
   Classic study of intraspecific population regulation. Provides an example of the interactions among various factors that function together to regulate the population of this large herbivore.

METAPOPULATIONS

In northern Canada and Alaska, caribou (*Rangifer tarandus*) belong to herds based on the location of their calving areas. The herds mix on their winter range, allowing interactions.

So far, we have considered populations that form a single spatial unit, in which individuals occupy a continuous local habitat. However, within the geographic range of many species, environmental factors are not uniformly favourable. Suitable habitat may consist of a network of **patches**—discrete spatial units of various shapes and sizes within a larger landscape of unsuitable habitat. Where these patches are large enough to support local breeding populations, the population of a species consists of a group of spatially discrete subpopulations. Richard Levins (1970) coined the term *metapopulation* to describe a population made up of interacting local subpopulations. In heterogeneous landscapes, metapopulations are a natural phenomenon, but their prevalence has increased tremendously in recent decades as a result of human activities. We fragment landscapes, creating metapopulations where none existed before. Understanding the ecology of metapopulations has thus become urgent to conservation strategies.

12.1 METAPOPULATION CONCEPT: Metapopulations Consist of Semi-isolated Subpopulations

Just as we defined a population as a group of individuals of a species occupying an area, so a metapopulation is a collection of *local subpopulations* (hereafter, *local populations*) interacting within a region (Figure 12.1). How do the dynamics of the collective differ from that of a single continuous population? It depends on the degree to which local populations are connected. Many models assume that populations are closed (i.e., with no emigration or immigration), and that growth is solely a function of birth and death. The local populations are independent, and the dynamics of the collective are the sum of the dynamics of the local populations.

However, if a significant number of individuals move between local populations, a broader framework is needed. For example, the checkerspot butterfly (*Euphydryas* spp.) (Figure 12.2a) occupies *serpentine soils* (alkaline soils high in magnesium and heavy metals) near San Francisco Bay. There are several populations in the region (Figure 12.2b), the best studied of which is the *E. editha* population in Jasper Ridge Biological Preserve. It consists of three local populations (C, G, and H) that are sufficiently isolated to have independent dynamics (Figure 12.2c) (Ehrlich and Murphy 1987). The G population has disappeared and re-established several times. Re-establishment after *local extinction* (*extinction* is used here in the sense of *extirpation*) followed emigration from area C or H. Had the populations been closed, area G would have stayed empty. Instead, the populations were open, linked by dispersal—a metapopulation with its own collective dynamics (Figure 12.2d). To understanding the regional dynamics of this species, we must consider processes at both the local population and the metapopulation levels.

12.2 METAPOPULATION CRITERIA: Four Criteria Define a Metapopulation

Many populations inhabit habitat patches, but not all act as metapopulations. Hanski (1999) suggested four criteria for the term *metapopulation* to apply to a collective of local populations. (1) Suitable habitat occurs in discrete patches, each of which may be occupied by a local breeding population. (2) Even the largest populations have a substantial risk of extirpation. (3) Habitat patches must not be so isolated that recolonization is impossible. (4) The dynamics of the local populations are not synchronized (i.e., do not peak or decline at the same time).

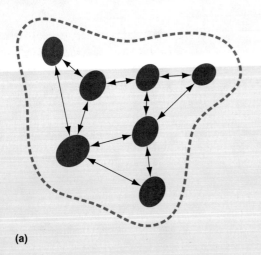

(a)

(b)

Figure 12.1 Metapopulation concept. **(a)** The distribution of a species (dashed line) is composed of local populations (red circles) linked by dispersal (arrows). **(b)** Ponds dotting the Arctic tundra are a network of habitat patches existing within a larger terrestrial landscape that is unsuitable for aquatic life. Inhabiting these ponds are discrete local populations that make up the larger regional metapopulation—assuming movement is possible between patches.

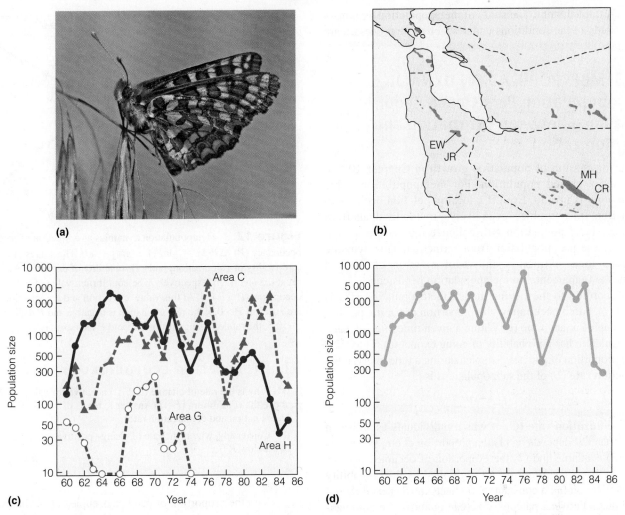

Figure 12.2 Local population dynamics of **(a)** the bay checkerspot butterfly. **(b)** Serpentine grassland habitats in San Francisco Bay area. JR = Jasper Ridge, EW = Edgewood Park, CR = Coyote Reservoir, and MH = Morgan Hill. **(c)** Local population dynamics in three areas (C, G, and H) of JR. Note the extinction and recolonization events in G. **(d)** Collective dynamics for the JR population (log scale).

(Adapted from Ehrlich and Murphy 1987.)

Not all uses of *metapopulation* meet these criteria. In many cases, a metapopulation consists of a larger **core population** that acts as the main source of emigrants to smaller **satellite populations**. The probability of extirpation of the core population may be very low, violating Hanski's second criterion. We explore this type of metapopulation structure in Section 12.6.

Metapopulation dynamics differ from those of closed populations in that they are governed by processes operating at two distinct spatial scales. (1) At a *local* (*within-patch*) *scale*, individuals move and interact during routine feeding and breeding activities. At this local scale, population growth and regulation are governed by the demographic processes of birth and death discussed in Chapter 10. (2) At a *metapopulation* (*regional*) *scale*, dynamics reflect the interactions of local populations via dispersal. Given that all local populations have a risk of extirpation, long-term metapopulation persistence depends on recolonization, which in turn involves dispersal from occupied to unoccupied patches to form new local populations. Typically, individuals moving between patches traverse habitats that are unsuitable for their feeding and breeding activities. They face substantial risk of failing to locate another suitable patch in which to establish.

Movement between local populations is key to metapopulation dynamics. If no individuals disperse between patches, local populations act independently. Alternatively, if movement between patches is frequent, then local populations act as a single population, not a metapopulation. The dynamics of the local populations would then be synchronized, violating Hanski's fourth criterion and making such populations equally susceptible to extirpation. Only at intermediate levels of dispersal are local extinction and recolonization in balance. The metapopulation persists as a shifting mosaic of occupied and unoccupied habitat patches. The metapopulation concept is thus linked with population turnover—extirpation and re-establishment. The study of metapopulation dynamics is the study of the conditions under which these processes are in balance.

12.3 METAPOPULATION DYNAMICS: Metapopulation Persistence Requires a Balance between Extirpation and Recolonization

In our discussion of population growth in Chapter 10, we assumed a closed population. For metapopulations, this assumption is invalid. Even if a local population declines to extinction, it is possible—even likely—that individuals from a nearby local population will colonize the vacated patch, "saving" the metapopulation from extinction. This dynamic between local extinction and recolonization of vacated patches is fundamental to metapopulation persistence.

According to the Levins model, metapopulation size is defined as **patch occupancy**: the proportion of habitat patches (P) occupied at any time (t). Within a given time interval, each subpopulation has a probability of going extinct (e). Thus, if P is the proportion of patches occupied during a time interval, the **extinction rate** (E) of the subpopulations is

$$E = eP$$

The **colonization rate** (C) at which individuals establish in empty patches depends on (1) the proportion of empty patches ($1 - P$) available and (2) the proportion of occupied patches that are providing colonists (P), multiplied by the probability of colonization (m; a constant that reflects the dispersal rate of individuals between patches). We can quantify the colonization rate as

$$C = mP(1 - P)$$

Metapopulation growth is analogous to population growth, where the change in population size over time is the difference between its rates of birth and death ($\Delta N/\Delta t = b - d$). Likewise, the change in metapopulation size, defined as the change in the proportion of patches occupied over time ($\Delta P/\Delta t$), is the difference between its rates of colonization (C) and extinction (E):

$$\Delta P/\Delta t = C - E$$

or

$$\Delta P/\Delta t = [mP(1 - P)] - eP$$

This metapopulation model is analogous to the logistic model in one key respect: growth is *density dependent*. This similarity is apparent from examining the equation graphically. For any given values of e and m, we can plot rates of extinction (E) and colonization (C) as a function of the proportion of habitat patches occupied (P; Figure 12.3). The extinction rate increases linearly with P, whereas the colonization rate forms a convex curve, initially rising with the proportion of occupied patches and then declining as P approaches 1.0 (all patches occupied). The value of P where the lines cross is the equilibrium value, P^*. At P^*, the extinction and colonization rates are equal ($E = C$) and the metapopulation growth rate is zero ($\Delta P/\Delta t = 0$).

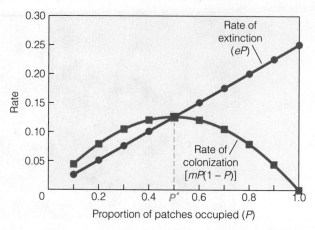

Figure 12.3 Metapopulation dynamics as a function of patch occupancy (P): $\Delta P/\Delta t = [mP(1 - P)] - eP$. The values of m (probability of colonization) and e (probability of extinction) are set at 0.5 and 0.25, respectively. The equilibrium value of patch occupancy (P^*) is 0.5. At this value, extinction and colonization rates are equal. If $P > 0.5$, the rate of change is negative and P declines. If $P < 0.5$, the rate of change is positive and P increases.

INTERPRETING ECOLOGICAL DATA

Q1. Why is the rate of change ($\Delta P/\Delta t$) negative for values of P above the equilibrium (P^*)? (Answer in terms of the influence on extinction and colonization rates.)

Q2. Conversely, why is the rate of change positive for values of P below P^*?

When the proportion of patches occupied (P) is below P^*, colonization exceeds extinction and the number of occupied patches increases. Conversely, if P exceeds P^*, extinction exceeds colonization and metapopulation size (proportion of occupied patches) declines. Just as with logistic growth, in which population size (N) tends to the equilibrium represented by carrying capacity (K), so too metapopulation size (P), tends to the equilibrium size represented by P^*.

The equilibrium value P^* is a function of the probabilities of extinction (e) and colonization (m):

$$P^* = 1 - \frac{e}{m}$$

We calculate the equilibrium value P^* by setting the metapopulation growth rate equal to zero: $\Delta P/\Delta t = 0$ (see Quantifying Ecology 12.1: Equilibrium Proportion of Occupied Patches, p. 243).

The Levins model makes three assumptions: (1) all patches are equal in size and habitat quality; (2) each patch contributes equally to the pool of emigrants, which have an equal probability of colonizing unoccupied patches; and (3) the dynamics of local populations are not synchronized (i.e., the probability of extinction of any local population is independent of that of other local populations).

Each of these assumptions may be unrealistic. Habitat patches differ in size and quality, and some contribute more than others to the emigrant pool. Coarse-scale features of the

When the rate of local population extinction (E) equals the rate at which unoccupied patches are colonized (C), the metapopulation size (proportion of patches occupied) reaches equilibrium. By setting $\Delta P/\Delta t$ equal to 0, we can derive the following equation:

$$\frac{\Delta P}{\Delta t} = \left[mP^*(1 - P^*) \right] - eP^* = 0$$

Using simple algebraic substitution, we can then solve for the equilibrium value P^*:

$$mP^*(1 - P^*) = eP^*$$

by dividing both sides of the equation by P^*:

$$m(1 - P^*) = e$$

then dividing both sides of the equation by m:

$$1 - P^* = \frac{e}{m}$$

and subtracting 1 from both sides:

$$-P^* = \frac{e}{m} - 1$$

then multiplying both sides of the equation by -1:

$$P^* = 1 - \frac{e}{m}$$

For a metapopulation to persist, P^* must be greater than zero, so the probability of extinction (e) cannot exceed the probability of colonization (m).

1. In what way is the P^* concept similar to that of carrying capacity (K) in Section 10.7?

regional environment often synchronize local population dynamics, and local populations often differ in their susceptibility to extinction. In the next sections we explore the consequences to metapopulation dynamics of relaxing each of these assumptions.

12.4 PATCH CHARACTERISTICS: Patch Size, Quality, and Isolation Affect Metapopulation Dynamics

Levins's model assumes all patches are similar in size and quality and contribute equally to the colonizer pool, but patches often vary greatly. Moreover, the ability of individuals to disperse between patches is related to patch arrangement on the landscape, particularly their isolation. For example, the bush cricket (*Metrioptera bicolor*) (Figure 12.4a, p. 244) is a flightless insect occupying grassland and heath patches of varying size and isolation (Figure 12.4b) in a pine-dominated landscape. Its metapopulation dynamics reveal the importance of patch size and isolation.

During three surveys from 1986 to 1990 in Sweden's Lake Vombsjön area, the proportion of patches occupied varied from 0.72 to 0.79. Patterns of occupancy correlated with patch traits. The probability of recolonization decreased with isolation (Figure 12.4c) (Kindvall and Ahlen 1992). Patch size, in contrast, affected the probability of extinction but not of recolonization. Of the 18 observed extirpations, habitat destruction or alteration from grazing, construction, or pesticides were responsible for 6. The other 12 occurred on patches that were significantly smaller than those with persisting populations (Figure 12.4d). Patch isolation did not affect extinction risk.

The effect of patch size on local population persistence was indirect, through the positive correlation between patch size and population size (Figure 12.5, p. 245). However, although the risk of local extinction increased for small patches (under 0.5 ha; see Figure 12.4d), the proportion of patches occupied (P) was higher than expected from their size and isolation. Indeed, P was fairly constant, although different patches were occupied each year (Kindvall and Ahlen 1992).

In the bush cricket study, extinction was affected only by patch size, and colonization only by isolation, but in some metapopulations, the impacts of patch area and isolation interact. For the skipper butterfly (*Hesperia comma*) (Figure 12.6a, p. 245), native to English grasslands, the probability of local extinction declined with increasing area, as expected, but increased with isolation. Conversely, the probability of colonization declined with isolation but increased with patch area. Thus, the probability of patch occupancy increased with declining isolation and increasing patch area (Figure 12.6b) (Thomas and Jones 1993). In this case, larger patch size compensated for isolation likely by providing a larger "target" for colonizers.

To explore how patch size and isolation interact to influence patch occupancy (P), we must modify the graphical model of metapopulation growth in Figure 12.3. No longer will we assume that the probabilities of extinction and colonization are constant. Instead, increasing isolation decreases the probability of colonization (m), thereby decreasing the colonization rate (C). Increasing patch size decreases the probability of extinction (e), thereby decreasing the extinction rate (E). Shifts in rates of colonization and extinction alter the equilibrium value (P^*) (Figure 12.7, p. 245). Increasing isolation reduces P^*, whereas increasing patch size increases P^*. Thus, increased patch size can offset increased isolation between patches. Conversely, reduced isolation can compensate for a decline in patch size. These same factors will recur in our discussion in Section 18.4 of island biogeography theory.

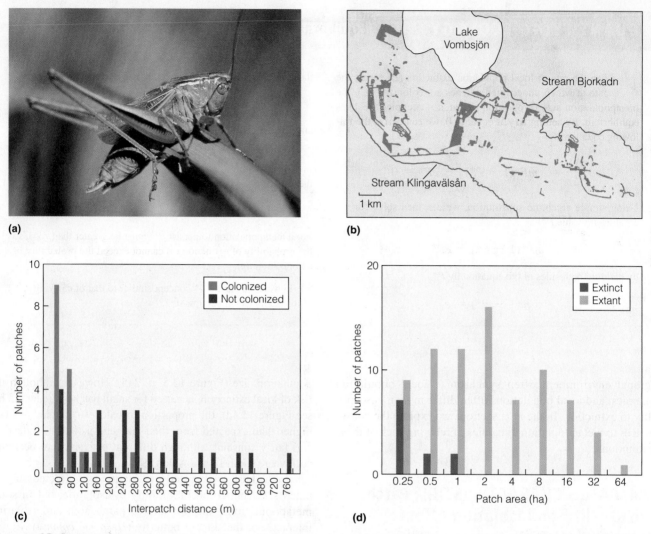

Figure 12.4 The Swedish metapopulation of bush cricket **(a)** is located south of Lake Vombsjön and surrounded by two streams **(b)**. Grassland patches with persistent local populations are shown in brown. Other patches were occupied either occasionally or not at all during the survey period. **(c)** Frequency distribution of interpatch distance between colonized and uncolonized patches. **(d)** Frequency distribution of patches with existing (extant) or extirpated (extinct) populations.

(Adapted from Kindvall and Ahlen 1992.)

INTERPRETING ECOLOGICAL DATA

Q1. Interpatch distance (Figure 12.4c) is a measure of spatial isolation. What is the maximum distance beyond which dispersal and colonization of bush crickets does not occur?

Q2. In Figure 12.4d, the total number of patches in each size category that were occupied during the survey period can be calculated by adding the values of currently occupied (extant) and vacated (extinct) patches. Which patch size category has the highest occupancy over the entire survey period? Which size category has the largest extant value for occupancy? Which patch size seems to be a threshold for long-term persistence of local populations?

12.5 HABITAT HETEROGENEITY:
Habitat Variation Enhances Population Persistence

Besides supporting larger populations by increasing carrying capacity, large patches may promote persistence of local populations by increasing environmental **heterogeneity** (variation). Larger patches are usually more diverse and more likely to include habitat types lacking in smaller patches. A critical consequence of environmental heterogeneity is that when conditions change, essential resources for a species might disappear from small patches, if only temporarily. In larger patches, favourable (or at least tolerable) conditions are more likely to remain somewhere in the patch. Increasing patch size should reduce the risk of extinction by supporting more habitat heterogeneity, thus reducing the impact of environmental stochasticity.

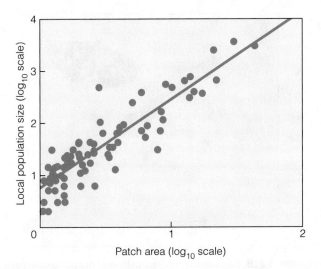

Figure 12.5 Relationship between local population size and patch size (both on log scale) in the bush cricket metapopulation. The line represents the general trend as defined by linear regression: $y = 1.70x + 0.74$ ($n = 83$, $r^2 = 0.81$). Go to **QUANTIFYit!** at **www.pearsoncanada.ca/ecologyplace** to review linear regression.

(Adapted from Kindvall and Ahlen 1992.)

How can habitat heterogeneity reduce fluctuations in local populations? For the bush cricket, habitat requirements change with weather. Wet conditions depress survival and fecundity, so in rainy years it is better for crickets to occupy

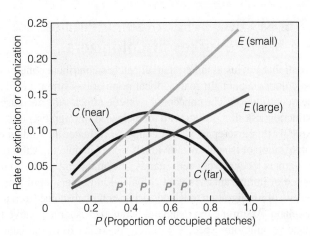

Figure 12.7 Changes in the equilibrium proportion of patches occupied (P^*; shown here in orange) for patches of differing size and isolation. Increasing patch size decreases the rate of extinction (E large), and increasing distance from neighbouring patches decreases the rate of colonization (C far).

sparsely vegetated sites on sandy soils that dry quickly. In contrast, during drought, dense, tall grassland enhances juvenile survival. Habitat patches that contained more types of vegetation supported populations that were less variable in size and subsequently less prone to extinction than populations inhabiting more homogeneous patches (Figure 12.8) (Kindvall 1996).

(a)

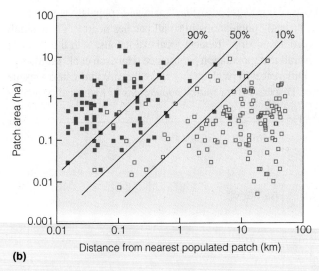

(b)

Figure 12.6 Distribution of occupied (solid) and vacant (open) habitat patches of the skipper butterfly (a) in southern England (1991) in relation to patch area and isolation from nearest populated patch (b).

(Adapted from Thomas and Jones 1993.)

INTERPRETING ECOLOGICAL DATA

The lines in Figure 12.6b define the combined values of patch isolation and patch size associated with patch occupancies of 90 percent, 50 percent, and 10 percent during the study period. Use these lines to answer the following questions.

Q1. What is the approximate probability that a 10-ha patch 1 km from the nearest populated patch will be occupied by skipper butterflies?

Q2. What is the area necessary to provide a 50 percent chance that a habitat patch 1 km from any neighbouring populated patches will be colonized?

12.6 RESCUE EFFECT: Some Patches Are Major Sources of Colonizers

Recall that Levins assumed that all patches contribute equally to the colonizer pool. In reality, local populations often differ in their contribution. The concept of a **rescue effect**—the decreased extinction risk that occurs with an increased immigration rate—explains this discrepancy. The concept originated in studies of colonization of islands by individuals from a mainland. Although movement occurs between islands, the mainland is the main source of immigrants. As with colonization in metapopulations, the rate of immigration to islands declines with distance from the mainland. Assuming the mainland population is large relative to island populations, species will not go extinct from the island network (i.e., will be "rescued") as long as there is continuing dispersal from the mainland (Brown and Kodric-Brown 1977).

This **mainland–island metapopulation structure** differs from other metapopulations in that a single (or several) habitat patch (the "mainland") is the major source of emigrants to other patches (the "islands") in the network. (This resembles the *core–satellite* distinction described in Section 12.2, although ecologists also apply this terminology to different species in a community.) Many metapopulations are structured similarly, reflecting variation in the size of habitat patches or populations. However, for mainland–island dynamics, there need not be a single large "mainland" or core population. Variation in patch or population size can have a similar effect. For example, the Morgan Hill (MH) population of the bay checkerspot butterfly (*Euphydryas editha bayensis*) in central California (Figure 12.9; see also Figure 12.2) consists of one large, persistent "mainland" population, with transient "island" populations in small patches nearby. These small populations undergo frequent local extinctions, with little impact on overall metapopulation persistence (Harrison et al. 1988).

Differences in patch quality as well as size may cause a rescue effect. Dispersal from *source populations* in high-quality habitat

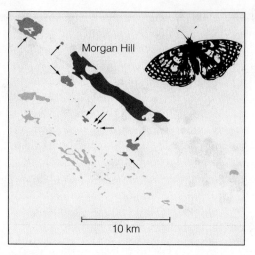

Figure 12.9 Serpentine grassland habitat of the bay checkerspot butterfly is fragmented by natural and human factors. Shaded areas are suitable habitat patches. Only patches closest to the Morgan Hill population are usually occupied, suggesting that the species is a poor disperser. Extinction and recolonization of small patches is common.

(Adapted from Harrison et al. 1988.)

may allow *sink populations* to persist in inferior habitat (Pulliam 1988). **Source populations** are defined by their ability to maintain positive growth ($r > 0$), whereas $r < 0$ for **sink populations**, which depend upon source populations to "rescue" them from extinction. The source–sink hypothesis (the terms can be applied to patches or populations) reinforces the fact that the presence and size of a population in a patch can give a misleading idea of both patch conditions and overall metapopulation status. Sink populations may persist solely through high rates of immigration from source populations. Although source populations often act as "mainlands" in some metapopulations, they are defined strictly by growth rate.

Understanding source–sink dynamics is vital in many conservation cases. A study of cougar (*Puma concolor*) demography in Washington State demonstrated that hunting was ineffective as a means of managing local cougar numbers in human-use areas as a result of increased immigration from nearby populations in the upper northwestern United States and southern British Columbia (Robinson et al. 2008). Young cougars were moving in from source populations to replace hunted animals in sink populations. Not only was hunting ineffective at reducing cougar–human conflicts, but it also masked the declining status of the population in the wider region.

12.7 SYNCHRONIZATION: Synchronized Local Dynamics Increase Metapopulation Extinction Risk

Asynchronous dynamics of local populations are key to metapopulation persistence. When the risk of extinction is independent in local populations, the probability that a metapopulation will go extinct falls rapidly with an increasing number of local populations. But if populations exhibit **synchronization** (similar timing of peaks and declines, as a result of correlated extinction

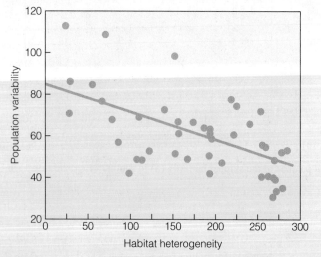

Figure 12.8 Relationship between temporal variability in population size of bush cricket populations (as indicated by the coefficient of variation in population size) and habitat heterogeneity (as indicated by an index measuring the diversity of vegetation cover types).

(Adapted from Kindvall 1996.)

probabilities), even metapopulations with many local populations are vulnerable to extinction.

Many factors promote synchronization among local populations. *Environmental stochasticity* often operates regionally (see Section 10.8). Weather can cause insect numbers to fluctuate in synchrony over broad geographic areas, with a single drought or freeze eliminating many local populations. A severe drought in central California from 1975 to 1977 caused a dramatic decline in the annual plant *Plantago erecta*, which in turn precipitated synchronized local extinctions of the bay checkerspot butterfly, whose larvae eat *Plantago* (Ehrlich et al. 1980). Similar synchronized local extinctions of ringlet butterflies (*Aphantopus hyperantus*) occurred in England in a severe drought (Sutcliffe et al. 1997). Such regional stochasticity reduces metapopulation persistence and increases the importance of mechanisms that promote persistence, such as refuge habitats (often provided by habitat heterogeneity) and dormant life stages.

On a longer timeline, landscape changes may be synchronized over large areas through changes in habitat availability. In the early 20th century, the skipper butterfly (*Hesperia comma*) inhabited most chalk (calcareous) grasslands of southern England (Figure 12.10). Despite its extensive range, the skipper butterfly has specific habitat needs, only occupying heavily grazed grasslands with short, sparse turf. Conversion to cropland and fewer domestic animals caused its numbers to drop,

with a southward retraction of its range. Grazing by introduced rabbits assured that the remaining grasslands persisted, and the species survived in the southern portion of its range.

However, when the myxomatosis virus was introduced in 1954 to control rabbits, chalk grasslands were overgrown. The skipper butterfly declined rapidly and by 1982 inhabited only refuges where domestic grazing continued (Thomas et al. 1996). Such long-term declines can be reversed. A 2000 survey revealed a fourfold increase in the number of populations and a tenfold increase in the habitat occupied by the skipper butterfly in Britain (Davies et al. 2005). This recovery was attributed to habitat management programs, recovering rabbit populations, and warming, which made more habitat available. Remnant metapopulations expanded, using dispersal to colonize a greater habitat range.

Changes in land use often occur over large areas in a relatively short time, causing widespread changes in the fortunes of associated species (see Ecological Issues: What Can the Metapopulation Concept Contribute to Conservation Ecology?, pp. 248–249). In Chapter 27, we discuss habitat fragmentation and species extinction in more detail.

12.8 SPECIES TRAITS: Species Vary in Their Colonization and Extinction Rates

In Section 12.7, we explored the influence of environmental heterogeneity on rates of extinction and colonization of habitat patches within a metapopulation. But species differ in their susceptibility to extinction and their ability to colonize habitats. The following traits are of particular importance: *dispersal ability* and *reproductive output*, which primarily affect colonization, and *body size* and *mode of thermoregulation*, which primarily affect extinction.

Dispersal rate greatly influences colonization ability. Life history classifications, such as the *r* versus *K*-strategists and the ruderal–competitor–stress tolerator classification (see Section 8.10), recognize dispersal ability as a key trait of species adapted to *ephemeral* (temporary) and/or disturbance-prone habitats. Dispersal is vital for insects that occupy temporary habitats or where variation in local carrying capacity is large. In contrast, less emigration occurs in stable or isolated habitats. The evolution of flightlessness in insects, which greatly reduces dispersal ability, is more likely in homogeneous, persistent habitats (Roff 1990).

Knowing the dispersal ability of a species is critical to understanding its ability to function as a metapopulation. The mountain caribou (*Rangifer tarandus caribou*), an endangered ecotype of the woodland caribou, occupies interior rain forests of British Columbia. Its distribution has become increasingly fragmented, and ecologists were unsure whether these remnant populations functioned as a metapopulation, linked by dispersal. Tracking studies indicated that dispersal rates were insufficient (1.4 percent) to rescue declining subpopulations (van Oort et al. 2010). Its progressive fragmentation is likely a nonequilibrium state signalling its impending extinction.

High reproductive output also facilitates colonization. Typically, species with enhanced dispersal traits produce many young. Flowering plants with wind-dispersed seeds

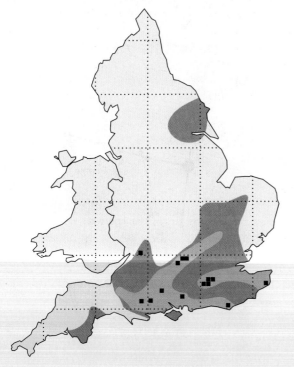

Figure 12.10 Long-term decline of the skipper butterfly in England. Tan shading represents the pre-1920 range. Green shading represents the known range between 1920 and 1961. By 1982, the distribution (black squares) consisted of several refuges. The decline was due to agricultural conversion and reduced grazing by rabbits after introduction of the myxomatosis virus. By 2000, the species had recovered significantly.

(Adapted from Thomas et al. 1996.)

ECOLOGICAL ISSUES | **What Can the Metapopulation Concept Contribute to Conservation Ecology?**

Human activity is having an ever-increasing impact on land use. Widespread conversion of forests and grasslands into fields and forest plantations causes habitat loss and fragmentation. Formerly continuous populations become networks of local populations with varying degrees of isolation and risk of extinction. The Iberian lynx (*Lynx pardinus*) provides one example.

The Iberian lynx (Figure 1a), the world's most endangered cat, is restricted to Europe's Iberian Peninsula. As reported by Rodriquez and Delibes (1992), this medium-sized cat (9–15 kg), which then numbered under 1000, occurred in 9 isolated metapopulations (Figure 1b), each consisting of local populations connected by dispersal. The best studied metapopulation is in Donana National Park in southwest Spain (Figure 1c).

The lynx metapopulation is distributed patchily over some 1500 km^2 and has been isolated from other metapopulations for over 50 years. Prior to 1950, its distribution was continuous, but human activities have fragmented its habitat. Agriculture and tourism now dominate the area, restricting movement and causing road mortality. A decline in its main prey, the European rabbit, from disease and overhunting is certainly important, but understanding the lynx metapopulation structure is crucial to conservation efforts. Its persistence hinges on maintaining exchange among local populations. Unfortunately, efforts to create natural habitat corridors to facilitate movement between patches have proved too little, too late. As of 2008, the International Union for the Conservation of Nature (IUCN) Red List classifies the Iberian lynx as critically endangered, with a maximum of 143 adults in only two isolated breeding populations.

We are hardly surprised to learn of habitat fragmentation in heavily populated Europe, but similar patterns are unfolding closer to home, and with similar consequences. The woodland caribou (*Rangifer tarandus caribou*) was once common in Canada's boreal forests and occurred as far south as Wisconsin in 1900. A subspecies of the more populous tundra-dwelling caribou, the woodland caribou has experienced widespread habitat fragmentation and range recession. From 1880 to 1990, half of its range was lost, at a rate of 34 800 km^2

(a)

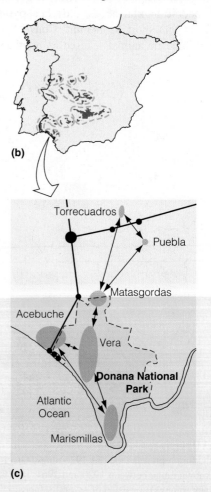

(b)

(c)

Figure 1 Distribution of the Iberian lynx (a) in the 1990s consisted of nine regional metapopulations (b). (c) Spatial distribution of local populations (in green) making up the Donana National Park metapopulation. Arrows indicate movement between subpopulations. The dashed line marks park boundaries. Black circles indicate human settlements, and thick lines roads.

(Adapted from Rodriguez and Delibes 1992 and Gaona et al. 1998.)

per decade, with its limits retreating north at 34 km per decade (Figure 2) (Schaefer 2003). Extrapolating from its present decline, Schaefer estimates a time to extirpation in Ontario of 91 years. Its decline may be accelerated by deer-borne parasites, accelerated human development, and vegetation changes related to global climate change.

The woodland caribou exists as a group of metapopulations, each of which constitutes a winter herd. As metapopulation density is about 1 per 1900 km^2, 18 winter herds have been lost per decade (Schaefer 2003). Although increased predation by wolves, black bears, and humans has contributed to the decline of this threatened species, there is a strong coincidence between its current southern limits and the northern extent of logging, suggesting that human activities are largely responsible.

The crisis facing woodland caribou is not yet as dire as for the Iberian lynx, but as the number of metapopulations diminishes, its decline could accelerate. As a precursor to an effective conservation strategy, Natural Resources Canada, in collaboration with Trent University, is studying woodland caribou metapopulation structure in Ontario. Using DNA profiling, researchers will evaluate population fragmentation, effective population size, movement between populations, and degree of isolation. Given the pace of habitat fragmentation, the metapopulation concept is key to species conservation worldwide (see Chapter 27).

BIBLIOGRAPHY

Cumming, H. G., and D. B. Beange. 1993. Survival of woodland caribou in commerical forests of northern Ontario. *Forestry Chronicle* 69:579-588.

Gaona, P., P. Ferraras, and M. Delibes. 1998. Dynamics and viability of a metapopulation of the endangered Iberian lynx *(Lynx pardinus). Ecological Monographs* 68:349–370.

Perera, A. H., and D. J. Baldwin. 2000. Spatial patterns in the managed forest landscape of Ontario. In Perera et al., eds., Ecology of a managed terrestrial ladscape. Vancouver, B.C.: University of British Columbia Press.

Rodríquez, A., and M. Delibes. 1992. Current range and status of the Iberian lynx *Felis pardina* Temminck 1824, in Spain. *Biological Conservation* 57:159–169.

Schaefer, J. A. 2003. Long-term range recession and the persistence of caribou in the taiga. *Conservation Biology* 17:1435–1439.

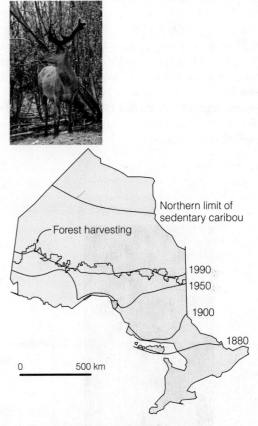

Figure 2 Range recession of woodland caribou in Ontario from 1890 to 1990.

(Adapted from Schaefer 2003 as based on Cumming and Beange 1993 and Perera and Baldwin 2000.)

1. Research a species that is native to your local region and whose population has been fragmented. What land-use changes have caused this fragmentation? What is the capacity of the species to disperse between patches?

produce more seeds than do those lacking specialized dispersal traits (Peat and Fitter 1994). For sessile organisms such as plants and fungi, dispersal is key to metapopulation persistence. Wind dispersal facilitates long-distance dispersal, whereas dispersal distances tend to be small (often mere centimetres) for species with unspecialized traits. Mode of reproduction also influences dispersal and colonization. Plants that reproduce asexually tend to have relatively rapid growth and a low risk of local extinction. However, reliance on clonal offshoots may limit widespread dispersal and compromise the ability to colonize new habitats.

Body size, and in particular its relationship with home range (see Section 11.9), affects extinction risk. Smaller organisms often support higher densities in the same habitat area than do larger-bodied species. Because a frequent cause of local extinc-

tions is the fact that small populations are vulnerable to demographic stochasticity, small body size may reduce extinction risk by allowing a larger population size. This relationship also affects the minimum patch size needed to support a viable local population, with larger-bodied species having a larger minimum.

Yet small body size can also increase extinction risk. Recall from Section 7.9 that mass-specific metabolic rate increases exponentially with decreasing body mass. As a result, smaller animals tend to starve more quickly when resources are scarce, making them more susceptible to environmental stochasticity. In a study of three shrews (*Sorex araneus*, *S. caecutiens*, and *S. minutus*) inhabiting 108 islands in 3 lakes in Finland, the probability of extinction varied significantly among species on islands under 8 ha and was negatively correlated with body size. The smallest species, *S. minutus*, had the highest extinction rate because it was more sensitive to environmental stochasticity and variable food supply (Peltonen and Hanski 1991).

An organism's mode of thermoregulation may also affect extinction risk. Ectotherms tend to be more affected by climatic anomalies (drought, extreme temperatures, etc.) than endotherms, causing greater flux in population size. In a review of studies of 91 terrestrial vertebrate and 99 terrestrial arthropod species, vertebrates had less variable populations and greater density-dependent regulation than invertebrates, making them less prone to local extinction (Hanski 1999).

In general, virtually all species traits can influence demographic processes, either directly or indirectly, and hence influence metapopulation dynamics.

12.9 POPULATION HIERARCHY: The Population Concept Is a Hierarchical Framework

The definition of a population as a group of organisms of the same species occupying a given area at a given time is so general that it applies to everything from a local population occupying a single patch to all members of a species within its geographic range. An alternative framework for the population concept recognizes a hierarchy of spatial units (Garton 2002). There are four levels of aggregation, each of which is associated with one of four key processes in population ecology: *demography*, *movement*, *geographic distribution*, and *evolution* (Figure 12.11).

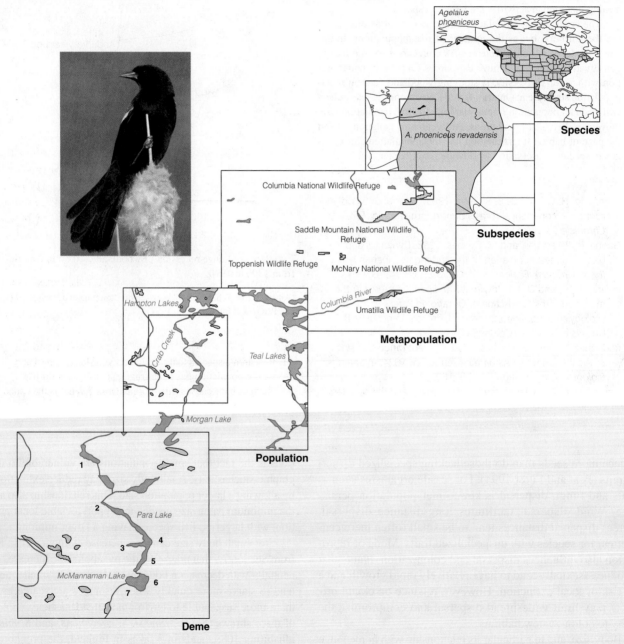

Figure 12.11 Hierarchy of spatial population units (shown in green), as applied to the red-winged blackbird at Columbia National Wildlife Refuge. (Adapted from Garton 2002.)

1. The *local population* (subpopulation) is the spatial unit for which ecologists can estimate rates of demographic processes such as birth and death. Ideally, its individuals are distributed within a single, continuous habitat area. In Garton's scheme, the local population in turn consists of smaller breeding units called *demes*.

2. The *metapopulation* is the collective of local populations with sufficient proximity that dispersing individuals can colonize empty habitat arising from local extinction, but with sufficient isolation that they function as asynchronized groups. Intermediate rates of dispersal maintain gene flow among local populations (see Section 5.6) and thus some degree of genetic similarity.

3. The *subspecies* is the collective of metapopulations (or populations, if metapopulation structure is absent) in a geographic region. Different metapopulations occupy habitats that may be separated by unsuitable habitat, resulting in demographic independence. This isolation can cause a divergence in selection pressures (especially if there are climatic differences between the areas), which contribute to the evolution of subspecies that are distinguishable by one or more genetically determined traits. Dispersal, although rare, maintains some gene flow among these potentially interbreeding populations.

4. The *species* is the collective of subspecies occupying the total geographic range of the species. Although natural selection and other evolutionary mechanisms are at work within other levels of the hierarchy, the species level is the culmination of such processes. It may encompass substantial variation in both phenotypes and genotypes.

Although differentiating among these spatial scales is critical to understanding population dynamics (both demographic and genetic), drawing a clear line between them is often difficult. For example, the Algonquin Park population of eastern timber wolves (*Canis lupus lycaon*) inhabits deciduous forests

Figure 12.12 Hybridization between both subspecies and species complicates wolf metapopulation dynamics in Ontario. This specimen is a wolf–coyote hybrid.

in the upper Great Lakes region. It numbers less than 200, but genetic evidence indicates that it belongs to a much larger metapopulation of eastern timber wolves that inhabit boreal forests from northwestern Ontario to Quebec (Wilson et al. 2009). In more northerly areas, timber wolves are hybridizing with the grey wolf (*C. lupus*), producing larger wolves, and in the south with coyote (*C. latrans*), producing smaller animals (Figure 12.12).

The ability of eastern timber wolves to hybridize with both grey wolves and coyotes promotes gene flow between the taxa, leading to a complex situation that the authors describe as an "interspecific metapopulation." But where does one species end and another begin? This demographic and evolutionary continuum has interesting repercussions for species conservation. Should we be trying to preserve "true" species, or should we accept this dynamic exchange as inevitable? It also sets the stage for the next rung on the ecological hierarchy—interactions between species.

EcologyPlace

Visit EcologyPlace at www.pearsoncanada.ca/ecologyplace to access online resources that complement your textbook, and help you to apply and to review the information in this chapter. EcologyPlace includes

- an eText version of the book
- self-grading quizzes
- glossary flashcards
- and more!

Go to www.pearsoncanada.ca/ecologyplace and follow the registration instructions on the Student Access Code Card included with this text. If your book does not have a Student Access Code Card, you can purchase access to it at www.pearsoncanada.ca/ecologyplace.

SUMMARY

Metapopulation Concept 12.1

When a population consists of spatially discrete local populations linked by dispersal, the collective of subpopulations is termed a *metapopulation*. Metapopulations become more common as landscapes are fragmented into non-continuous patches.

Metapopulation Criteria 12.2

Four conditions are necessary for the term *metapopulation* to apply. The local populations must (1) occupy discrete habitat patches; (2) have a substantial risk of extinction; (3) be linked by dispersal, which allows recolonization after local extinction; and (4) have asynchronous dynamics. Metapopulation dynamics operate at both the regional and the local scale.

Metapopulation Dynamics 12.3

Metapopulation persistence depends on a balance between extirpation and recolonization of vacated habitat patches. For extirpation (local extinction) to occur, populations must be sufficiently isolated that most recruitment comes from within the patch (birth) rather than from immigration. Recolonization involves movement of individuals from occupied patches to form new local populations. If recolonization exceeds extirpation, metapopulation size (proportion of occupied patches) increases. If recolonization is insufficient to offset extirpation, the proportion declines and the metapopulation may not persist. The model developed by Levins to quantify metapopulation dynamics makes several assumptions, which may not apply in nature.

Patch Characteristics 12.4

The ability to disperse between patches is affected by patch traits. Colonization declines with increasing isolation, and extirpation increases with decreasing patch size. A major cause of the inverse relationship between patch size and extinction risk is the fact that small patches support smaller populations, which have higher extinction risk as a result of demographic and/or environmental stochasticity. Patch area and isolation can interact, with reduced isolation compensating for patch size through an increased probability of colonization. Conversely, larger patch size and the associated lower risk of extinction can compensate for patch isolation and the associated lower probability of colonization.

Habitat Heterogeneity 12.5

Increasing patch size may influence the persistence of local populations by increasing habitat heterogeneity. When environmental conditions change, essential resources may disappear from small patches, whereas in larger areas, favourable conditions are more likely to remain somewhere within the patch and provide a refuge.

Rescue Effect 12.6

Immigration helps maintain local populations that would otherwise go extinct via the rescue effect, especially in metapopulations with a mainland–island structure. Assuming the mainland population is large relative to the island populations, species will not go extinct from the island network as long as there is ongoing emigration from the mainland. Mainland–island structure reflects variation in the sizes of habitat patches or populations. Differences in habitat quality may exert a similar rescue effect. Source populations maintain a positive growth rate ($r > 0$), whereas $r < 0$ for sink populations.

Synchronization 12.7

Asynchronous dynamics of local populations is essential for metapopulation persistence. When extinction risk is independent in each local population, the probability of metapopulation extinction declines with an increasing number of local populations. If extinction probabilities are synchronized, even metapopulations with many local populations may be prone to extinction. Demographic and environmental factors, including weather and changes in land use, can synchronize the dynamics of local populations.

Species Traits 12.8

Species differ in their susceptibility to local extinction and their ability to colonize habitats. Life history traits such as mode of reproduction, fecundity, dispersal type, and longevity can influence rates of colonization. Body size and mode of thermal regulation may influence the probability of local extinction, especially in response to environmental stochasticity.

Population Hierarchy 12.9

Four levels of spatial aggregation provide a framework for understanding key processes in population dynamics: (1) local population (demography), (2) metapopulation (dispersal), (3) subspecies (geographic distribution), and (4) species (evolution).

KEY TERMS

colonization rate	mainland–island	patch	satellite population	source population
core population	metapopulation	patch occupancy	sink population	synchronization
extinction rate	structure	rescue effect		
heterogeneity				

STUDY QUESTIONS

1. Define *metapopulation*. How does this concept relate to dispersion of a population over a geographic region? What four criteria define a metapopulation?

2. Discuss the role of local extinction and colonization in metapopulation dynamics. How does immigration rate influence the degree to which local population dynamics are synchronized? What effect does synchronicity have on metapopulation persistence?

3. How do the size and isolation of habitat patches influence metapopulation dynamics? Discuss in terms of the probabilities of extinction and colonization.

4. How does habitat heterogeneity influence the dynamics of local populations? What does this influence imply about the importance of heterogeneity among habitat patches?

5. Define *rescue effect*. Under what conditions might it be important.

6. Contrast *source* and *sink populations* with *mainland–island metapopulation structure*. How does the source–sink population concept relate to habitat quality?

7. What environmental factors synchronize local population dynamics over a region?

8. What influence do (i) body size and (ii) mode of thermal regulation have on extinction?

9. What traits are likely to affect metapopulation dynamics in plants?

10. Describe the four levels in Garton's population concept and the major processes associated with each.

FURTHER READINGS

den Boer, P. 1981. On the survival of populations in a heterogeneous and variable environment. *Oecologia* 50:39–53.
Early paper examining the importance of the metapopulation concept in understanding population persistence in variable habitats.

Hanski, I. 1999. *Metapopulation ecology*. New York: Oxford University Press.
Definitive reference written by a leading metapopulation ecologist. Provides a wealth of illustrated examples.

Holyoak, M., and C. Ray. 1999. A roadmap for metapopulation research. *Ecology Letters* 2:273–275.
Excellent discussion of the application of metapopulation research to conservation ecology.

Pulliam, R. 1988. Sources, sinks, and population regulation. *American Naturalist* 132:652–661.
Seminal paper that introduced the idea of source and sink populations.

PART FOUR SPECIES INTERACTIONS

Orchids are considered the most beautiful of flowers, but in the *Eucalyptus* woodlands of eastern Australia lives a small, nondescript orchid. Standing only 8 to 12 cm tall and bearing a single greenish flower some 20 mm wide, the broad-lipped bird-orchid (*Chiloglottis trapeziformis*) is a marked contrast to its showy relatives. Yet, despite its dull appearance, this orchid has a sex life that is a fascinating story of deception and exploitation (Gullan and Cranston 2010).

In most flowering plants, successful reproduction requires that pollen (containing the male gametes or sperm) from one individual be transported to the female organ (stigma) of another individual. Orchids, like many flowering plants, rely on animals to facilitate pollination. (Some plants self-pollinate, while others use wind to carry their pollen.) Most animal-pollinated plants secure the services of their animal vectors by offering rewards such as nectar (see Section 15.11). Not the broad-lipped bird-orchid. It secures the services of pollinators on false pretences.

This orchid practises its deception on male thynnine wasps (*Neozeleboria cryptoides*). In the breeding season, the smaller wingless female wasp climbs a plant stem, secures a perch, and releases a pheromone to attract males. Meanwhile, males seek out the odour of a receptive female. The orchid flower produces a chemical that mimics the pheromone. To enhance the deception, its enlarged lower petal (labellum) bears a structure resembling the female wasp. Males are attracted to the flower by its odour and attempt to mate with the structure. In this way, pollen is deposited on the male's back, and pollen from previously visited orchids lands on the stigma. The male continues his search, transferring pollen to other orchids as he goes.

After pollination, the orchid produces seeds, and another chapter unfolds. Orchids have the smallest seeds of any flowering plant. They lack endosperm, a nutritive tissue that typically supplies the carbon and minerals needed in germination until the seedling becomes photosynthetic. So germinating orchid seeds require the assistance of yet another unwitting partner. The seeds release a chemical that attracts certain fungi, which penetrate the seed coat, allowing organic carbon and minerals to pass from the fungus to the seed, thereby supplying the seed with all its energy needs. The seedling also benefits from the capacity of the fungus to absorb water and minerals. However, this benefit comes at a potential cost. Once the plant is photosynthetic, the fungus could become parasitic, extracting carbohydrates from the plant. But the orchid has evolved a mechanism for blocking this flow. In this exploitative interaction, the fungus apparently gains nothing.

This story illustrates two key ecological concepts: (1) Individuals of different species that coexist in a community interact in various ways. The orchid's relationship with both the wasp and the fungus is exploitative, benefiting only the orchid. These interactions provide services or resources necessary for its germination and reproduction. Its population dynamics can be understood only in the context of these interactions. (2) The traits of the orchid that enable these interactions are adaptations, phenotypic traits resulting from natural selection. The agents of selection here are not abiotic factors but interactions with other species—the biotic environment.

Thus far, we have focused on *intraspecific* (within-species) interactions. Chapter 8 considered intraspecific interactions relating to reproduction, such as mate selection and care of offspring. Chapter 11 discussed how intraspecific competition limits population growth. But no species exists in isolation. Species differ in their needs, but all photoautotrophs need light, CO_2, water, and minerals. *Interspecific* (among-species) competition can be intense, with resource acquisition by one species reducing resource availability to others.

Among heterotrophs, the possible interactions are many. Heterotrophs derive energy and nutrients from consuming organic matter, so the very act of feeding involves an interspecific interaction between predator and prey. If different predators consume the same prey, competition may ensue. Some invertebrates and

The broad-lipped bird-orchid (*Chiloglottis trapeziformis*).

A male thynnine wasp at the base of a broad-lipped bird-orchid flower. The black structure on the bottom petal resembles the smaller, wingless female wasp. The anther and stigma are located just above the black structure, under the top petal.

microorganisms are parasites that live in or on other organisms, securing habitat, energy, and nutrients from their hosts. Not all interactions are negative. Mutually beneficial interactions affect nutrition, defence, and reproduction. Some mutualisms are symbiotic, in which two organisms live together, while others involve free-living organisms. There are six possible interspecific interactions (Table 1), defined by their reciprocal impact on individuals: positive (+), negative (−), or neutral (0).

Table 1 Possible Interactions between Individuals of Two Species (A and B)

	Response	
Type of Interaction	Species A	Species B
Neutralism	0	0
Mutualism	+	+
Commensalism	+	0
Competition	−	−
Amensalism	−	0
Predation	+	−
Parasitism	+	−
Parasatoidism	+	−

The effects of interactions are not limited to individuals. By affecting birth and death, interactions alter population dynamics. An interaction that harms an individual can benefit its population. A prey population may benefit if a predator population keeps its numbers below its carrying capacity. Nonetheless, the predator–prey interaction is defined as +/− because of its impact on individuals. By differentially influencing survival and reproduction, interspecific interactions are powerful agents of natural selection. Some phenotypes are better at competing or avoiding competition. Some are better at capturing prey or avoiding capture. Such phenotypes will likely have higher fitness. As agents of selection, interactions play a major role in evolution.

In Part Four, we examine species interactions—their diversity, impact on demography, and role in natural selection. Chapter 13 discusses interspecific competition, building on the models of intraspecific competition developed in Chapter 11. Chapter 14 examines predation, including herbivory as well as carnivory. Finally, in Chapter 15, we explore the relationship between parasites and their hosts and consider how parasitic relationships can evolve into mutualisms. Our discussion of interactions lays the foundation for our treatment of communities in Part Five.

A yellow-necked mouse (*Apodemus flavicollis*) feeds on an acorn on a forest floor in Lower Saxony, Germany. This mouse is one of many species that depend on acorns as a food resource, setting the stage for interspecific competition.

n *The Origin of Species*, Charles Darwin wrote, "As more individuals are produced than can possibly survive, there must . . . be a struggle for existence, either one individual with another of the same species, or with the individuals of distinct species, or with the physical conditions of life." Interspecific competition is a cornerstone of evolutionary ecology. Indeed, Darwin based his idea of natural selection on competition, the "struggle for existence" among organisms. Because it is beneficial for organisms to avoid this struggle or minimize its effects, many consider competition the key force driving species divergence and specialization. In contrast, Alfred H. Wallace, who independently devised the theory of natural selection, stressed the role of what Darwin called "the physical conditions of life." Although both abiotic and biotic factors can be major drivers of evolutionary change, we focus here on the mechanisms and effects of interspecific competition.

13.1 TYPES AND MECHANISMS: Competition among Species Takes Many Forms

Interspecific competition is an interaction wherein individuals of two or more species seek a common resource that is in short supply and are both adversely affected. In competition among animals, the "winner" is the species with the greater relative fitness, but plant ecologists often compare growth or mortality effects, with the assumption that these differences affect fitness. Intra- and interspecific competition often occur simultaneously. Grey squirrels compete for acorns in a year when oaks produce fewer acorns than usual, or when squirrels are particularly abundant. Meanwhile, white-footed mice, white-tailed deer, and blue jays vie for the same scarce resource. To minimize interspecific competition, one or more of these species may alter their foraging by seeking alternative (and often less preferred) food items that are less in demand.

Like intraspecific competition, interspecific competition involves either (1) *exploitation*, in which individuals interact indirectly, as a result of their resource use, or (2) *interference*, in which individuals interact directly to gain access to a resource (see Section 11.2). Within this broad framework, Schoener (1983) proposed six possible mechanisms:

1. *Consumption:* This mechanism is a classic exploitation type, in which individuals consume a shared resource (e.g., animals eating the same food, or plants sharing water).

2. *Pre-emption:* This mechanism is a type of exploitation in which the scramble is for space. Established individuals prevent others from establishing, as in space capture by plants (see Section 11.10) or barnacles on a rock.

3. *Overgrowth:* In this mechanism, one organism grows over another, inhibiting access to a resource, such as a tall plant shading those below. It can involve both interference (if the larger organism directly inhibits access) and exploitation (if consumption enables overgrowth).

4. *Chemical:* This mechanism is an interference type in which growth inhibitors or toxins released by one individual kill or inhibit members of other species, as in *allelopathy* (see Section 13.7).

5. *Territorial:* In this classic interference type, individuals prevent other species from occupying a defended space, as in intraspecific territoriality (see Section 11.9).

6. *Encounter:* This mechanism involves non-territorial contact that harms the participants, such as scavengers fighting over a carcass. It involves interference, but is often accompanied by consumption.

There are other competition categories. In **asymmetric (one-sided) competition**, the per capita effects (expressed per individual rather than for the population) are greatly dissimilar, so that one individual is much more negatively affected than the other. If the superior competitor is hardly affected at all, the interaction approaches *amensalism* (see Table 1, p. 255). In **diffuse competition**, individuals of many species compete, and the impact on an individual of any one species reflects the combined effect of its interactions with individuals of many species. **Episodic competition** occurs intermittently, depending on resource supply. Clearly, interspecific competition is multifaceted, both reflecting and affecting the diverse species assemblages of natural communities.

13.2 MODELLING COMPETITION: The Lotka–Volterra Model Predicts Four Possible Outcomes

In the early 20th century, Alfred Lotka and Vittora Volterra independently developed a model for interspecific competition based on the logistic growth equation developed in Section 10.7: $dN/dt = rN(1 - N/K)$. They used the expanded version of this equation:

$$\frac{dN}{dt} = rN\frac{(K - N)}{K}$$

They modified this logistic equation by adding a term that quantifies the competitive effect of one species on the population growth of the other. For species 1, this term is αN_2, where N_2 is the population size of species 2 and α is the competition coefficient that quantifies the per capita effect of species 2 on species 1. Similarly, for species 2, the term is βN_1, where β is the competition coefficient that quantifies the per capita effect of species 1 on species 2. These coefficients effectively convert an individual of one species into the equivalent number of individuals of a competing species, based on their shared use of the resources that determine their carrying capacities. In terms of its resource use, a species 1 individual is equivalent to β individuals of species 2, and a species 2 individual is equivalent to α individuals of species 1.

Now we have two equations that quantify the effects of both intra- and interspecific competition:

$$\text{Species 1: } \frac{dN_1}{dt} = r_1N_1\left(\frac{K_1 - N_1 - \alpha N_2}{K_1}\right) \qquad (1)$$

Species 2: $\dfrac{dN_2}{dt} = r_2 N_2 \left(\dfrac{K_2 - N_2 - \beta N_1}{K_2} \right)$ (2)

With no interspecific competition—either $\alpha = 0$ or $N_2 = 0$ in Equation (1) and $\beta = 0$ or $N_1 = 0$ in Equation (2)—each population grows logistically to equilibrium at K, its carrying capacity, affected only by intraspecific competition. But with interspecific competition, the situation changes. The carrying capacity for species 1 is K_1, and as N_1 approaches K_1, its population growth (dN_1/dt) approaches 0. However, we must also consider the impact of species 2, which is vying for the same limited resource that determines K_1. Because α quantifies the per capita effect of species 2 on species 1, the effect of species 2 on species 1 is αN_2.

We now consider the effects of both species on growth. As the combined effect $(N_1 + \alpha N_2)$ approaches K_1, the growth rate of species 1 approaches 0. The greater the population of the competing species (N_2), the greater the reduction in its own growth. In essence, the per capita growth rate of either species is a linear decreasing function of both its own density and the density of its competitor. Depending on the values of K_1, K_2, α, and β, the **Lotka–Volterra competition model** predicts four outcomes, based on the **zero growth isoclines** (diagonal lines along which net growth is zero; Figure 13.1a, b). For an explanation of these graphs, see Quantifying Ecology 13.1: The Lotka–Volterra Competition Model. Here we focus on the outcomes.

QUANTIFYING ECOLOGY 13.1 The Lotka–Volterra Competition Model

Let's examine Figure 13.1 more closely. The x- and y-axes represent the population sizes of species 1 and 2 (N_1 and N_2). The diagonal line for species 1 in Figure 13.1a is its zero isocline. It represents the combined populations of species 1 and 2 that are equivalent to K_1 and thus where $dN_1/dt = 0$. Similarly, the zero isocline for species 2 in Figure 13.1b represents the combined populations of species 1 and 2 that are equivalent to K_2. For any point on the species 1 line, $N_1 + \alpha N_2 = K_1$. So, when $N_1 = K_1$, N_2 must be 0. Because α is the per capita effect of species 2 on species 1, the population of species 2 that is equivalent to the carrying capacity of species 1 $(\alpha N_2 = K_1)$ will be $N_2 = K_1/\alpha$. Thus, when $N_2 = K_1/\alpha$, N_1 must be 0. Similarly, the values where the species 2 line crosses the axes will be K_2 and $N_1 = K_2/\beta$, the population of species 1 that is equivalent to the carrying capacity of species 2.

In the region below the species 1 isocline in Figure 13.1a, combinations of N_1 and N_2 are below carrying capacity ($N_1 + \alpha N_2 < K_1$), and the population is increasing ($dN_1/dt > 0$), as represented by the arrows that are parallel to the x-axis and point to increasing values of N_1. Above the isocline, combinations of N_1 and N_2 exceed the carrying capacity, and the species 1 population is declining ($dN_1/dt < 0$). The arrows point towards decreasing population size. Figure 13.1b depicts the analogous situation for species 2, so the arrows now parallel the y-axis (N_2).

To predict the dynamics for any combination of α, β, K_1, and K_2, we must draw zero isoclines for both species on the same graph. For any combined values of species 1 and 2 (N_1, N_2), two arrows represent the direction of change (Figure 1). At each of four values of (N_1, N_2) (black dots), green arrows show the change in species 1, orange the change in species 2. Predicted values of N_1 and N_2 lie in the direction of the arrows. Thus, the next point representing combined values of N_1 and N_2 lies between the green and orange arrows (in the area defined by the dashed line—see inset), as indicated by the black arrow. In Figures 13.1c–13.1f, only the black arrows are shown.

Now we can interpret the possible outcomes. In Figure 13.1c, the species 1 isocline is parallel to and outside that of species 2. Even when species 2 is at its carrying capacity (K_2), its density cannot stop species 1 from increasing ($K_2 < K_1/\alpha$). As species 1 increases, species 2 is eventually excluded. In Figure 13.1d, the situation reverses—species 2 wins, excluding species 1.

In Figures 13.1e and 13.1f, the isoclines cross, and the outcomes differ. Note that in Figure 13.1e, the value of K_2/β along

the x-axis is less than K_1. Recall that the value $N_1 = K_2/\beta$ is the species 1 density that is equivalent to the species 2 carrying capacity (K_2). Because $K_2/\beta < K_1$, species 1 can achieve densities exceeding that required to drive species 2 to extinction (> K_2). Likewise, as $K_1/\alpha < K_2$ on the y-axis, species 2 can achieve densities high enough to drive species 1 to extinction (> K_1). Which species "wins" thus depends on their initial densities.

In Figure 13.1f, the isoclines also cross, but in this case the species coexist. Because $K_1 < K_2/\beta$, species 1 can never reach a density sufficient to exclude species 2. For this to occur, species 1 would have to reach $N_1 = K_2/\alpha$. Likewise, $K_2 < K_1/\alpha$, so species 2 can never achieve a density high enough to exclude species 1. As both species increase, intraspecific competition inhibits growth of each more than interspecific competition inhibits growth of the other.

1. Of the four possibilities outlined, what outcome is predicted by Figure 1?
2. Now assume that the species 2 isocline lies parallel to and outside that of species 1, which is present at its carrying capacity (K_1). A few species 2 individuals establish. According to the model, what will happen over time as a result of competition, and why?

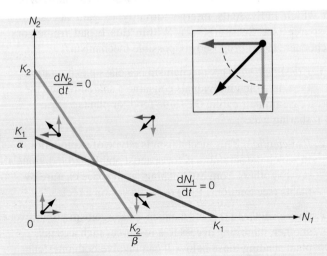

Figure 1 Zero isoclines for two competing species, showing changes in population size.

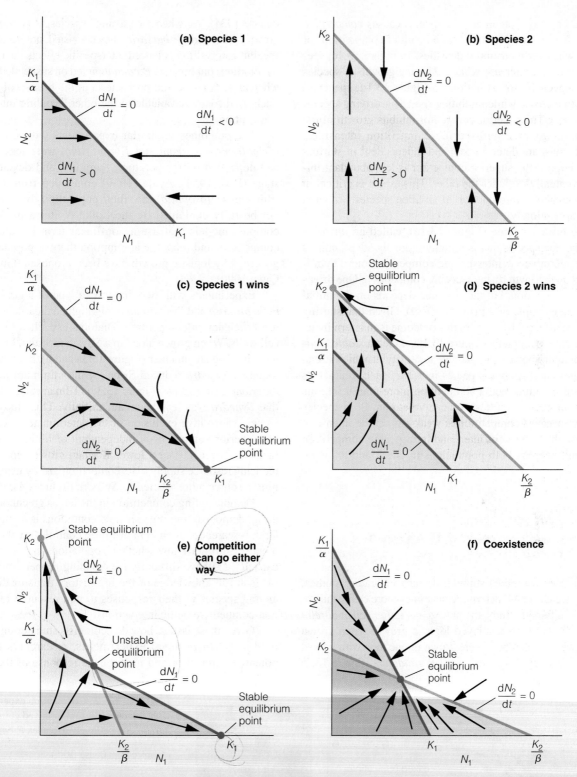

Figure 13.1 The Lotka–Volterra competition model. **(a, b)** Zero isoclines are the combinations of (N_1, N_2) for which $dN/dt = 0$ (zero growth). Here they are shown for the species independently; in (c)–(f), they are combined. For combined (N_1, N_2) values below the line (shaded area), the population rises; for combined (N_1, N_2) values above the line, the population drops. Vector arrows indicate the direction of growth. In (a), the arrows are horizontal because only species 1 (x-axis) changes; in (b), the arrows are vertical because only species 2 (y-axis) changes. **(c)** The species 1 isocline falls outside that of species 2. Species 1 wins, excluding species 2. **(d)** The reverse of (c) occurs—species 2 wins, excluding species 1. **(e)** The isoclines cross. Each species inhibits growth of the other more than it inhibits its own growth. The species that is initially more abundant wins. **(f)** The species coexist—each species inhibits its own growth more than that of the other species.

In the first two outcomes, one species causes *competitive exclusion* (loss of one population as a result of competition) of the other, whatever their initial densities. In Figure 13.1c, species 1 continues to increase while inhibiting growth of species 2, driving species 2 to extinction. In Figure 13.1d, species 2 continues to increase while inhibiting species 1, driving species 1 to extinction. The superior competitor inhibits growth of the other even as its own intraspecific competition intensifies. These outcomes are deterministic and independent of starting densities. Even if the superior competitor is less abundant initially, it eventually excludes the other. This aspect is critical in situations involving introduction of an alien species, which is usually scarce initially.

In the third outcome (Figure 13.1e), called an *unstable equilibrium*, the species that is initially more abundant inhibits growth of the other by interspecific competition more than it inhibits its own growth by intraspecific competition. One drives the other to extinction, but the outcome depends on the initial densities (an example of a **priority effect**). Given that starting densities are affected by chance, this outcome is indeterminate.

In the fourth outcome (Figure 13.1f), called a *stable equilibrium*, neither species can achieve a density at which it can exclude the other. As both species increase, each inhibits its own growth by intraspecific competition more than it inhibits growth of the other by interspecific competition. Both species persist, not because competition is weak—it can be intense—but because the impacts of inter- and intraspecific competition balance out, keeping both populations in check and allowing coexistence.

13.3 EXPERIMENTAL TESTS: Ecologists Have Tested the Lotka–Volterra Model in Many Settings

The Lotka–Volterra model stimulated many laboratory studies, where researchers could determine outcomes more easily than in the field. In a famous study, the protozoans *Paramecium aurelia* and *P. caudatum* each displayed logistic growth when grown alone. However, when both were grown in one tube with a fixed amount of food, *P. aurelia* excluded *P. caudatum* (Figure 13.2)

(Gause 1934). Yet when the "losing" species, *P. caudatum*, was grown with a third, *P. bursaria*, they coexisted, not because intraspecific competition balanced interspecific effects (as in a stable equilibrium) but because *P. caudatum* fed on suspended bacteria, whereas *P. bursaria* fed on bacteria at the bottom of the tube. Each used food unavailable to the other, avoiding interspecific competition.

The outcomes in similar experiments with flour beetles (*Tribolium castaneum* and *T. confusum*) were less clear cut and depended on temperature, humidity, and developmental stage (Park 1954). Ayala (1969) concluded from tests with fruit flies (*Drosophila*) that their population dynamics could not be fully explained by the Lotka–Volterra model. More complex models that use a non-linear term for quantifying competition and relax the assumption that the populations are governed by logistic growth have been proposed (Guowei and Qiwu 1990).

Experiments with two diatoms (a type of algae), *Asterionella formosa* and *Synedra ulna*, highlighted the role of resource use efficiency in competition. Diatoms use silica (Si) in their cell walls. When grown alone in a Si-rich medium, both species grew in a logistic manner (Figure 13.3a and b), but when grown together, *Synedra* reduced Si below the minimum needed for *Asterionella* to survive (Figure 13.3c) (Tilman et al. 1981). Note that *Synedra* was less abundant initially. This study demonstrates competitive exclusion of the determinate type, wherein the superior competitor is independent of initial densities. In fact, the "winner" had a lower K when grown separately. By reducing resource supply to its competitor, and by using the limiting resource more efficiently, *Synedra* excludes *Asterionella*.

Demonstrating competition in the lab or greenhouse is one thing; demonstrating it under field conditions is another. In the field, researchers often (1) have little control over the environment, (2) do not know whether populations are near carrying capacity, (3) have difficulty establishing proper controls, and (4) lack full knowledge of the life history traits or differences among species in their responses to non-resource factors and non-competitive biotic interactions.

Given these issues, how do ecologists study competition in field tests? In **removal experiments**, researchers remove a potential competitor and monitor the response of the "target"

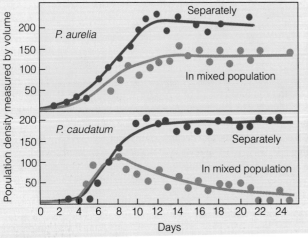

Figure 13.2 Competition experiments with *Paramecium* spp. In mixed culture, *P. aurelia* competitively excludes *P. caudatum*.

(Adapted from Gause 1934.)

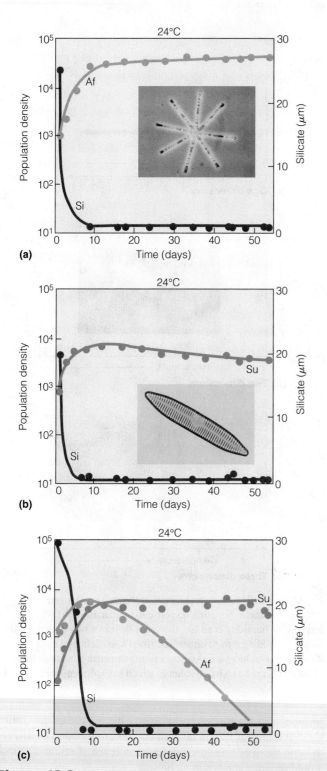

Figure 13.3 Competition between two diatoms: *Asterionella formosa* (Af) **(a)** and *Synedra ulna* (Su) **(b)**. When grown alone, both reach a carrying capacity that reduces Si to a minimum (shown in brown). *Synedra* draws Si lower. **(c)** If grown together, *Synedra* reduces Si to a level at which *Asterionella* dies out.

(Adapted from Tilman et al. 1981.)

species. Such tests seem straightforward, but removals may have unintended impacts that affect the response of the target species. Removing neighbours may increase soil aeration. What appears to be a response to removal of a competitor may

be partly a response to altered abiotic conditions. Such hidden treatment effects can hinder experimental interpretation.

In **transplant experiments**, potential competitors are removed from some plots and added to others, leading to varying relative densities of two or more species. If the total densities of both species remain constant, the study is called a *replacement series*. This approach also has problems. Transplanted individuals may die from the transplanting itself, not from competition with individuals in the mixture into which it was transplanted. Moreover, the densities may far exceed normal levels. If competitive effects have been forced, they may not be indicative of what happens in nature, even when coexisting species are at their carrying capacities.

Ideally, researchers monitor the relative impact of intraspecific and interspecific competition. Both may be occurring, and their relative strength affects the likelihood of a stable equilibrium. Many studies fail to separate their effects. Well-designed controls reduce this problem, but Connell (1983) concludes that less than half of studies that claim to show competitive effects actually do so. Also, the tendency for journals to reject "negative" results means that studies with no evidence of competition are less likely to be published.

Increasingly important is the use of *null models* (see Section 1.7) to detect patterns that are generated by random forces, in the absence of the ecological mechanism under investigation (Gotelli and Graves 1996). Null models help ecologists detect patterns that actually result from competition by acting as a kind of theoretical control (see the discussion of *limited similarity* in Section 13.12).

13.4 COMPETITIVE EXCLUSION: Although Theory Asserts That Complete Competitors Cannot Coexist, Exclusion Is Rare

In three of the four outcomes predicted by the Lotka–Volterra model, one species excludes the other. This model led Gause to propose a principle later formalized by Hardin (1960) as the **competitive exclusion (Gause's) principle**: complete competitors cannot coexist. (*Complete competitors* are non-interbreeding populations that live in the same place and have identical requirements for a limiting resource.) If species A increases its own numbers while inhibiting growth of species B, then A will outcompete B, eventually causing its local extinction. Yet, despite its theoretical predictability and common occurrence in laboratory tests (though not all; see Ayala 1969), competitive exclusion rarely occurs in nature. Most documented instances involve introduction of a non-native species (see Section 27.7). So, why doesn't competitive exclusion happen more often?

Competitive exclusion requires more than competition for a limited resource. Gause's principle makes assumptions about the species and their environment: (1) competitors have identical, unchanging requirements for a limiting resource; (2) abiotic factors are constant; and (3) no factor other than the shared resource prevents either population from reaching its carrying capacity. If one or more of these assumptions is violated, the predicted outcome may not occur and may even be reversed. One way to interpret the competitive exclusion principle (as with the Hardy–Weinberg principle regarding gene frequencies in evolutionary genetics) is as a baseline against which ecologists can assess all the reasons why competitive exclusion is unlikely.

Whatever its assumptions, the competitive exclusion principle raises critical questions about the role of competition. *How similar can two species be and still coexist? What conditions allow coexistence of species that consume similar resources?* Many factors affect competitive outcomes, including (1) competition for multiple (and often shifting) resources, (2) conditions such as temperature or pH that influence competitive ability but are not resources for which organisms can compete, (3) spatial and/or temporal variations in resources, and (4) resource partitioning. We consider the impact of each factor in allowing species coexistence as opposed to exclusion. But first, we introduce a concept that is fundamental to both the competitive exclusion principle and the broader topic of species interactions and community structure: *niche*.

13.5 ECOLOGICAL NICHE: The Niche Describes Species Response to and Impact on the Environment

The niche concept has a long and complex history (Chase and Leibold 2003). Some define the **niche** of a species as its response to the totality of all factors (abiotic and biotic) in its environment. Hutchinson (1957) famously described the niche as an *n*-**dimensional hypervolume**, in which the niche is a functional "space" occupying an indefinite (*n*) number of dimensions. Each axis quantifies the response of the species to an environmental factor that affects its survival, growth, and/or reproduction. Niche dimensions include resources (food, habitat, light, etc.) and conditions that affect performance (temperature, humidity, pH, etc.).

We can begin to imagine a multi-dimensional niche by creating a three-dimensional one. Consider three variables that affect a hypothetical organism: temperature, humidity, and food size (Figure 13.4).

Although ecologists talk about niche "space" or "volume," the niche is not a physical space. The niche includes habitat dimensions, such as the type of site that an animal uses for nesting, or the type of soil in which a plant grows, but the niche also includes non-habitat dimensions. The same habitat supports many niches because of the many ways in which species function within it.

The niche is often depicted as one-dimensional (see Figure 13.4a), but this depiction is misleading. Rarely do two species have exactly the same combination of responses to multiple dimensions. They may overlap on one niche axis (e.g., size of prey consumed) but not on another (e.g., foraging height). The potential for competition may thus be less than suggested by the overlap for any one dimension (Figure 13.5). The more dimensions we consider, the more likely it is that species that seem to be "complete competitors" are separated in niche space. In actuality, the number of dimensions needed to describe the

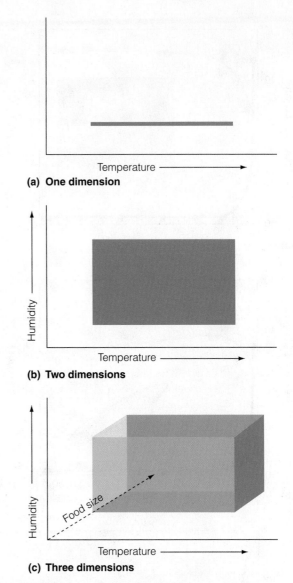

Figure 13.4 Conceptual model of the niche. **(a)** A one-dimensional niche depicting response to temperature. **(b)** When response to humidity is added, the space defines a two-dimensional niche. **(c)** Adding a third dimension (food size) defines a three-dimensional niche space. Adding a fourth dimension, such as time of foraging, creates a hypervolume, which has *n* dimensions.

niche of a given species is limitless. This infinite quality limits the testability of the niche concept in nature. One can always assume there is one more dimension (even if unknown) that would separate species in niche space.

An alternative view of the niche is that of Charles Elton, who stressed the impact of a species *on* its environment as much as or more than its response *to* the environment. Drawing on Darwin's idea that each species has a "line of life," Elton (1927) conceived of the niche as the functional role or occupation of a species in its habitat. Chase and Leibold (2003) propose a niche concept that integrates the Hutchinsonian and Eltonian perspectives. We revisit this topic in Section 16.2 in our discussion of the structuring of natural communities. For now, we focus on the fact that niche dimensions, particularly

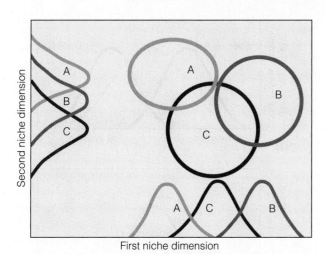

Figure 13.5 Niche relationships based on two dimensions. Species A, B, and C have considerable overlap on the second axis and less on the first. When their niche dimensions are combined (circles), total niche overlap is reduced.

(Adapted from Pianka 1978.)

for resources, determine the potential for interspecific competition. If two species have the same resource needs, then they may compete for that resource *if* its supply is limited relative to their population sizes.

To understand the role of niche in competition, we must distinguish two kinds of niche. An organism not subject to interactions with other species is free to use its **fundamental niche**: the full range of environmental factors within which it can survive, grow, and reproduce. Interspecific interactions (including competition) restrict a species to its **realized niche**: that portion of its fundamental niche that a species occupies in the presence of interacting species.

Consider two cattails that inhabit freshwater marshes and ponds across North America. The wide-leaved cattail (*Typha latifolia*) dominates in shallow water, the narrow-leaved cattail (*T. angustifolia*) in deeper water. Comparing this natural distribution with the depth response of these species if grown alone reveals how competition influences their realized niches (Figure 13.6). Both can survive in shallow water, which is part of the fundamental niche of each, but only *T. angustifolia* can grow in depths over 80 cm. When they grow together, their realized niches change. Although *T. angustifolia* can grow in shallow water and even on dry land (negative values indicate depth to the water table), *T. latifolia* outcompetes *T. angustifolia* in shallow water, limiting it to deepwater sites that are outside the tolerance limits of *T. latifolia* (Grace and Wetzel 1981).

Although each species has exclusive use of a subset of the habitats along the gradient, they coexist at intermediate depths. If two or more species use a portion of the same

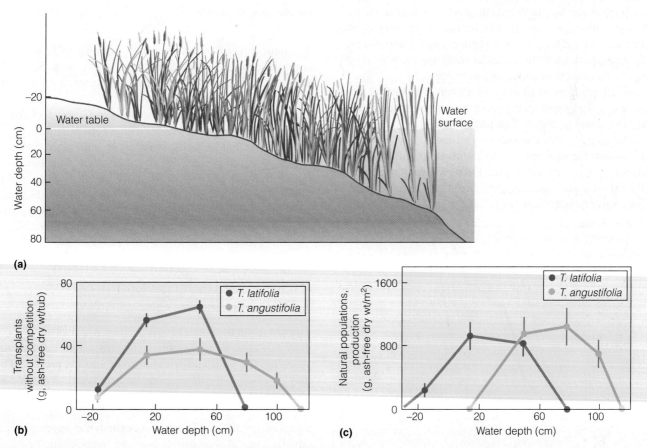

Figure 13.6 Distribution of two cattail species (**a**) along a water depth gradient; (**b**) grown separately in an experiment, indicating their fundamental niches; (**c**) grown together in natural populations, indicating their realized niches.

(Adapted from Grace and Wetzel 1981.)

resource simultaneously—be it habitat as in this case, or food, or light—the situation is called **niche overlap**. Some ecologists infer the likelihood of competition from the amount of overlap. But niche overlap may not always indicate intense competition. In fact, if resources are plentiful, extensive niche overlap may indicate just the opposite—that minimal competition is occurring because resources are non-limiting, allowing the species to tolerate the overlap. This contradictory interpretation is one of many contentious issues in competition theory.

Part of the problem is that ecologists often study only one niche dimension—in this case, water depth. As critical as depth is, there are likely other factors for which the species have differing responses and hence less overall overlap. Moreover, water depth is a habitat factor, so its impact involves species shifting their location. Niche dimensions that are not directly habitat related, such as the time of day when an animal forages, are less obvious. For example, in southern France, four weevil species (*Curculio* spp.) share a limiting and fluctuating resource—acorns, in which they lay their eggs. The potential for competition is great, but the species differ in the timing of their use of acorns (Figure 13.7) (Venner et al. 2011). Niche studies in the field are often restricted to habitat factors and may not incorporate temporal niche differences, as does the weevil study. This situation complicates both the assessment and interpretation of niche overlap.

Plastic shifts in where and how a species grows are typical of plants and are easy to detect. In animals, a shift in the realized niche often involves behaviour. An animal may change where and when it forages if competitors or predators are present (see Section 14.8). Behavioural shifts are more reversible than growth responses, and harder to detect.

Whether a niche shift involves growth or behaviour, competition may force species to restrict their use of space, food, or resource-related activities. The realized niche may contract to the optimal portion of the fundamental niche, in which conditions support the most growth and/or fitness (Figure 13.8a), or into suboptimal portions (Figure 13.8b) to avoid competition. In the cattail study, *Typha angustifolia* undergoes a greater shift in its realized niche than does *T. latifolia*, with its apparent

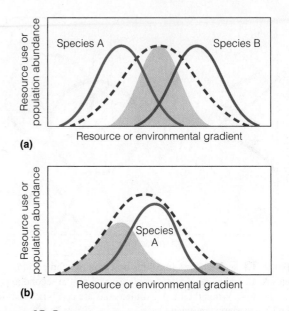

(a)

(b)

Figure 13.8 Two possible niche adjustments. The dashed red line indicates the fundamental niche of a hypothetical species, the orange shaded area its realized niche. **(a)** Species A and B compete with the focal species, causing its realized niche to compress to its optimum zone. The response curves are symmetric and bell shaped. **(b)** A dominant species A (blue line) restricts the focal species into suboptimal portions of its fundamental niche, creating a bimodal, asymmetric realized niche.

(Adapted from Austin 1999.)

optimum shifting to deeper water (see Figure 13.6). But its optimum depth, as revealed when grown alone, is 50 cm. It is thus misleading to infer a species preference from its performance when competitors are present.

For example, *Stipa neomexicana*, a C_3 grass of Arizona grasslands, grows only on dry ridge tops where total grass cover is low, not in moist, lower areas with more cover. One might assume it prefers dry sites, but when neighbours were removed, *Stipa* grew faster, produced more flowers, and had higher seedling survival in mid- and low-slope sites than on ridge tops (Figure 13.9) (Gurevitch 1986). Competition restricts *Stipa* to a suboptimal portion of its fundamental niche.

13.6 NICHE DYNAMICS: Realized Niches Can Undergo Compression or Release

The niche is a theoretical concept. How is it measured? Ecologists may choose one or more niche dimensions that they measure directly, with and without other species present. There may be a bias in the dimensions studied, based on assumptions about which are important. Response to food items and habitat is often measured for animals, and response to light, water, or minerals for plants. These choices may be appropriate, but equally critical dimensions may be overlooked. Feasibility often plays a role; some dimensions are easier to measure. In Section 13.12, we consider how ecologists infer niche dimensions indirectly by examining morphological traits.

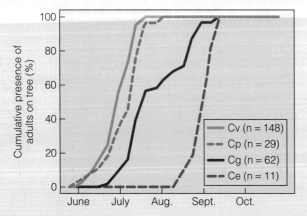

Figure 13.7 Temporal niche separation in exploitation of acorn resources of a single oak tree by four weevil species. Sample size (n) differed among species.

(Adapted from Venner et al. 2011.)

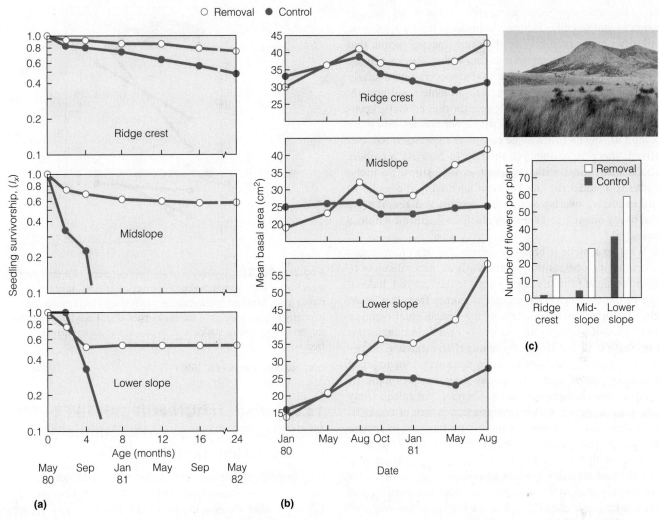

Figure 13.9 Impact of competition on *Stipa*. **(a)** Seedling survival. **(b)** Growth rate. **(c)** Flowers per plant of treatment (neighbours removed) and control plants.

(Adapted from Gurevitch 1986.)

INTERPRETING ECOLOGICAL DATA

Q1. How does competition affect seedling survival in ridge-crest and lower-slope sites?

Q2. What parts of *Stipa*'s life cycle are most influenced by competition, and how might this affect its distribution?

Based on its niche dimensions (measured or inferred), the **niche breadth** of a species (its tolerance range for one or more dimensions) is described as *broad* or *narrow*. If a species has a broad niche for several key factors, it is a **generalist**, or a **specialist** if its niche is narrow. Assessing niche breadth requires caution. Jack pine (*Pinus banksiana*) is a generalist for substrate, growing on bare rock, sandy soils, and boggy sites, but its shade intolerance makes it a specialist for high light, and its reliance on fire to release the seeds from its cones (*serotiny*; see Section 18.8) makes it a specialist for a disturbance regime involving frequent fires.

Comparing niche breadth on a regional basis is difficult. High-latitude species are more likely to be generalists. In variable, disturbance-prone habitats, specialization is risky. Harsh habitats have fewer species, so any one species may occupy a greater proportion of the possible niche space. Stable, favourable habitats usually contain more specialists, but such generalizations can obscure critical differences among species. In conifer forests in Eastern Canada, there is a significant latitudinal trend in the foraging plasticity of birds, with more feeding generalists in northern forests (Simon et al. 2003). The generalist feeding habit of three common species (the dark-eyed Junco, *Junco hyemalis*; the yellow-rumped warbler, *Dendroica coronata*; and the ruby-crowned kinglet, *Regulus calendula*) allows them to switch diet items seasonally and in response to disturbance. Yet their nesting and habitat needs are specialized, as are those of specialist feeders like the hermit thrush (*Catharus guttatus*).

What keeps a generalist niche from constricting in the absence of competing species? By keeping its niche broad, a species reduces intraspecific competition—but recall our discussion of intraspecific facilitation in Section 11.10. Genetic similarity among members of the same species means that their niche overlap is great, so reducing within-species overlap benefits populations in habitats with few species—particularly if conditions are favourable enough to promote rapid growth, which intensifies competition. In species-rich habitats, keeping niches narrow limits overlap with other species. But there is a limit to how narrow a niche can be. If a species is too specialized, there is so much similarity in how its members respond that intraspecific competition intensifies. So niche breadth is a balancing act between narrowing the niche to avoid excessive overlap with other species and keeping the niche broad enough to avoid too much competition within a species.

Whether a niche is broad or narrow, the short-term effect of interspecific competition (particularly of generalists) is to reduce the niche from the fundamental to the realized. Indeed, much of the evidence for competition comes from studies of **niche compression**: contraction of the fundamental niche in the presence of a competing species, as with *Typha angustifolia* in Section 13.5. Conversely, **competitive release** occurs when a species expands its niche in the absence or reduced presence of competitors. The species now occupies more (if not all) of its fundamental niche. Competitive release may occur when a species invades an area that is free of competitors or predators, allowing it to occupy habitats or utilize resources it never occupies or utilizes elsewhere. In such cases—as with the introduction of non-native species into Australia and Hawaii—species abundance can increase far more quickly than in native habitats, where competitors and predators keep it in check. Much of the impact of invasive species (see Section 27.7), even in non-island habitats, involves competitive release of the invader and niche compression of the native species.

Niche release can also occur when a competitor is removed, allowing species to move into microhabitats they could not previously occupy, or to utilize resources they could not previously access. In a study of the feeding niche of the three-spine stickleback (*Gasterosteus aculeatus*), researchers manipulated the presence or absence of two competitors: juvenile cut throat trout (*Oncorhynchus clarkii*), which feed at the surface and in the water column, and prickly sculpin (*Cottus asper*), a benthic (bottom) feeder. Examination of stomach contents revealed overlap of prey items with the three-spined stickleback, which feeds in both habitats. In experimental enclosures in Blackwater Lake on Vancouver Island, a 15-day release from competition with sculpin had no effect on the feeding niche of sticklebacks, as indicated by the diversity of prey consumed. However, when released from competition with juvenile cut-throat trout, sticklebacks experienced a significant expansion of their feeding niche (Figure 13.10) (Bolnick et al. 2010).

Niche compression and release are fluid processes, continually subject to flux. With this basic understanding of niche dynamics, we can now consider the many factors that complicate competitive outcomes in nature.

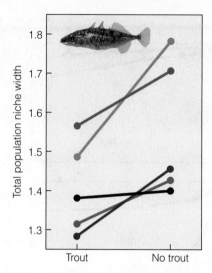

Figure 13.10 Expansion of the feeding niche of three-spine stickleback with the removal of a competing species, juvenile cut-throat trout. Total population niche width (*y*-axis) represents the diversity of prey species in the stickleback diet. Each line represents the mean response of stickleback in one of five experimental treatment blocks.

(Adapted from Bolnick et al. 2010.)

13.7 COMPETITION FOR MULTIPLE RESOURCES: Species May Compete for More Than One Resource

Interspecific competition may involve a single obvious resource, such as Tilman's experiment in which diatoms competed for silica. In other studies, the limiting resource is less apparent. The two *Typha* species vie for position along the depth gradient, but for what resource? In such cases, competition may involve multiple resources, and competition for one resource may influence an organism's ability to access other resources.

As an instance of competition for multiple resources, interspecific territoriality is relatively common in birds. Although it involves competition for space, it also affects access to food and nesting sites. Most often, species defend territories against related species, such as the grey (*Empidonax wrightii*) and dusky (*E. oberholseri*) flycatchers of western North America (Johnson 1966). Given that related species are likely to overlap in niche dimensions, this finding is not surprising. Territorial interactions among coral fish were thought to have little effect on population dynamics, but a removal study revealed that the damselfish *Stegastes planifrons* exerts strong, asymmetric competitive effects on two related species in Caribbean coral reefs. *S. partitus* and *S. variabilis* numbers doubled after removal of the larger, more aggressive *S. planifrons*. *S. partitus* used different microhabitats, suggesting niche release (Robertson 1996).

In contrast, relatively few species defend their territories against unrelated species. The acorn woodpecker (*Melanerpes formicivorus*) defends its "acorn trees" (Figure 13.11) against jays and squirrels as well as other woodpeckers. These interactions are often one-sided, and some argue that interspecific territoriality is non-adaptive (Murray 1971). Territorial aggression of the tundra

Figure 13.11 The acorn woodpecker defends trees in which it stores acorns against several unrelated species.

swan (*Cygnus columbianus*) towards the non-territorial snow goose (*Chen caerulescens*) in Alaska was attributed to misdirected intraspecific behaviour promoted by morphological similarity (Burgess and Stickney 1994). Yet interspecific territoriality is more common in North American songbirds than once thought (Greenberg et al. 1994), usually involving competition for localized food patches, such as flowering plants or insect-rich trees.

Some plants exhibit a type of chemical territoriality called **allelopathy**, in which individuals release compounds that deter the growth of neighbours. Allelopathy is hard to document, but the most famous example involves black walnuts and butternuts (*Juglans* spp.). Some invasive weeds share this strategy. Pollen of the invasive grass timothy (*Phleum pratense*) has allelopathic effects on co-occurring species (Murphy and Aarssen 1989).

Plants also provide non-territorial examples of how competition for a resource can influence the ability to exploit other resources. Ecologists have conducted many tests in which species are grown in monocultures and mixtures. In a landmark variant of these tests, Groves and Williams (1975) isolated the

effects of competition for aboveground (light) and belowground (water and nutrients) resources in subterranean clover (*Trifolium subterraneum*) and skeletonweed (*Chondrilla juncea*). In the monocultures, plants were grown with both canopies and roots intermingled. The mixtures contained (1) canopies and roots intermingled, (2) separated canopies and intermingled roots, and (3) intermingled canopies and separated roots (Figure 13.12). Competitive effects were strongly asymmetric. Clover was not significantly affected, but skeletonweed biomass was reduced by 35 and 53 percent, respectively, when either roots or canopies intermingled. This suggests **amensalism**, an interaction in which only one individual is harmed (see Table 1, p. 255). Yet when both shoots and roots intermingled, skeletonweed biomass was reduced by 69 percent, indicating an interaction in competition for above- and belowground resources (Groves and Williams 1975). Clover was the superior competitor for both resource types, enhancing the interactive effect.

Similar interactions occur in many situations. Fast-growing species grow taller, reducing light for slower growers. Faster root growth of the superior species in turn (via positive feedback) increases their access to soil resources, promoting more growth at the weaker competitor's expense. Eventually, competitive exclusion may occur, especially if conditions are unchanged. This pattern typifies plants in open habitats such as prairies, but not all species respond similarly to competition for multiple resources. In shady habitats, fast-growing species cannot meet their light needs and may be outcompeted by slow-growing species that use light more efficiently.

Some competition models incorporate the impact of multiple resources on competitive outcomes. As communities undergo *succession* (change over time), resource availability changes in both absolute and relative amounts. David Tilman (1985) proposed the *resource-ratio hypothesis*, which states that the changing relative amounts of light and nutrients (especially nitrogen) affect the relative competitive ability of species and are a major force driving successional change. We consider this hypothesis in more detail in Section 17.5.

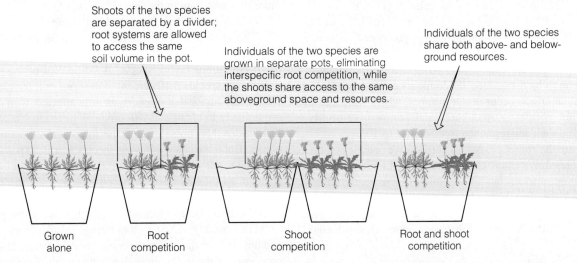

Figure 13.12 Experimental design for testing above- and belowground competition between subterranean clover and skeletonweed.

(Adapted from Groves and Williams 1975.)

13.8 IMPACT OF CONDITIONS: Varying Conditions Can Alter Competitive Outcomes

In interspecific competition, individuals of different species vie for the same limited resource, but factors other than the resource in question can affect the abilities of species to compete for it. As we have just seen, species with high rates of photosynthesis and shoot allocation often overgrow other species in open sites, pre-empting space and securing access to light. It might seem that adaptations that promote fast growth would always determine the superior competitor. Typically, this assumption holds true in open habitats; shade-intolerant, fast-growing species dominate, at least in the early stages of community development (see Section 17.6).

Does the "faster grower = stronger competitor" principle always prevail? The Lotka–Volterra model indicates that if a species can continue growing while inhibiting the growth of another, it will increase in relative abundance, even if circumstances stall exclusion. Most experiments use fast-growing species, partly for practical reasons, but also because of the interest in competition between (1) crop species and weeds and (2) forest trees and shrubs. Both scenarios unfold in open sites where consumption and/or overgrowth are common competitive mechanisms. Competition in less open sites and/or via other mechanisms has received less attention. In such cases, species that might be superior competitors in open sites cannot sustain rapid growth. Their relative growth—the measure of ecological success—is determined more by efficiency than by rapid growth potential. Recall the tortoise and the hare: it is not always the fastest-growing individual that wins an interspecific competitive "race."

Abiotic factors other than light influence establishment and growth, and hence competitive outcomes. Temperature, pH, and salinity directly affect growth and reproduction, even though they are not resources for which species compete. For example, in the grass community of the savannahs of southwest Zimbabwe during the 1970s, dominance shifted between two species (Figure 13.13a) in accordance with rainfall (Figure 13.13b). From 1971 to 1973, rainfall was far below normal. *Urochloa mosambicensis* grows faster than *Heteropogon contortus* in dry conditions, making it a better competitor during drought. When rainfall increased, the competitive advantage shifted back to *Heteropogon*, which regained its dominance (Dye and Spear 1982). Similar effects occur in salt marshes, where variations in salinity alter species growth rates and competitive ability, depending on their salt tolerance (see Section 16.10).

What are the consequences for competitive outcomes if different species thrive in different conditions? Low temperature may favour one species, but if temperatures increase, the advantage shifts. Here, then, is one reason why exclusion occurs less often than expected. The Lotka–Volterra model assumes that conditions remain constant and that populations approach equilibrium. But *if* conditions change, and *if* that change affects competitive ability, then a species that is a superior competitor in some conditions may lose its advantage. Competitive exclusion is avoided, at least for the time being. Indeed, the "winning"

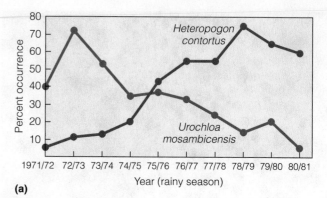

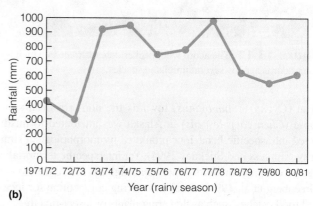

Figure 13.13 Shift in species dominance in a savannah in southwest Zimbabwe from 1971 to 1981 **(a)** in response to changing rainfall **(b)**. *Urochloa* competed more successfully under dry conditions. With increased rainfall, *Heteropogon* regained dominance.

(Adapted from Dye and Spear 1982.)

species may then be at risk of exclusion itself—unless conditions change again, in ways that favour its relative competitive ability. Under changing conditions, equilibrium outcomes rarely occur—assuming that the changing conditions alter long-term growth and mortality patterns. Environmental variation thus promotes coexistence of species that might be excluded under constant conditions. Clearly, competition in nature is much more dynamic than in the laboratory (see Quantifying Ecology 13.2: Competition in a Changing Environment).

The impact of environmental variation depends on the amount of change, and how these changes affect relative mortality. If conditions favour species A 90 percent of the time, and only favour species B 10 percent of the time, or if the changes are sufficiently minor that mortality is little affected, then outcomes are unlikely to change. Environmental variation would then be little more than noise around the long-term competitive outcomes. But if the changes favour species A or B 50 percent of the time, the result would be a saw-off. How likely is that to happen? Even if species A were favoured exactly half the time, the advantage may not switch to species B, which might be similarly disadvantaged, but to species C, increasing its relative growth.

How quickly abiotic factors change relative to the longevity and time to reproduction of the competitors is critical. Hutchinson, proponent of the multi-dimensional niche, posed the **paradox of the plankton** (1961): *How can so many phytoplankton species coexist, despite strongly overlapping niches and limiting*

QUANTIFYING ECOLOGY 13.2 Competition in a Changing Environment

In a given set of conditions, the outcome of competition reflects the relative abilities of species to acquire resources needed for survival, growth, and reproduction. In the Lotka–Volterra model, two factors interact to influence outcomes: the competition coefficients (α and β) and the carrying capacities (K_1 and K_2) of the species. As we saw in Section 13.2, competition coefficients quantify the per capita effect of a member of one species on the other and reflect two species traits: (1) overlap in resource use and (2) rate of resource uptake. In contrast, carrying capacities reflect the resource base for species in the prevailing environment. Changes in conditions that affect resource availability influence carrying capacities, thereby affecting competitive outcomes.

Consider two species that consume the same limiting resource: seeds (Figure 1a). If their seed intake per unit time is identical, we assume their competition coefficients are identical. Assume a value of 0.5 for both α and β. (This assumption is useful but will not always hold.) Now assume that seed size distribution and abundance vary with abiotic conditions. As seed size distribution changes from site A to site B to site C (Figure 1b), so will carrying capacity, as in this table:

Environment			
	A	B	C
Species 1 (K)	225	150	75
Species 2 (K)	75	150	225

Let's explore the changing nature of competition between species 1 and 2 at these sites, using the Lotka–Volterra equations and analysis of the zero isoclines, as in Section 13.2. Construction of the zero isoclines in each environment requires the following values:

Environment	K_1	K_1/α	K_2	K_2/β
A	225	450	75	150
B	150	300	150	300
C	75	250	225	450

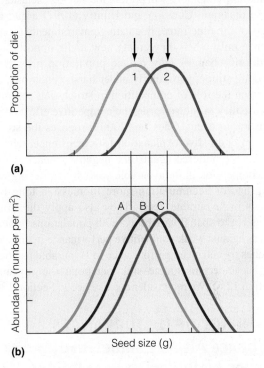

(a)

(b)

Figure 1 Proportion of different seed sizes (**a**) in diets of species 1 and 2. (**b**) Relative abundance of seed sizes in three environments.

Figure 2 depicts the predicted outcomes. In environment A, species 1 wins; in B, species 1 and 2 coexist; and in C, species 2 wins. As the relative abundance of seed sizes changes between sites, the resulting changes in carrying capacities alter the competitive outcomes.

1. Suppose that the competition coefficients were $\alpha = 0.5$ and $\beta = 0.25$. How would this influence the competitive outcome in environment A?
2. What factors might cause competition coefficients to change between sites?

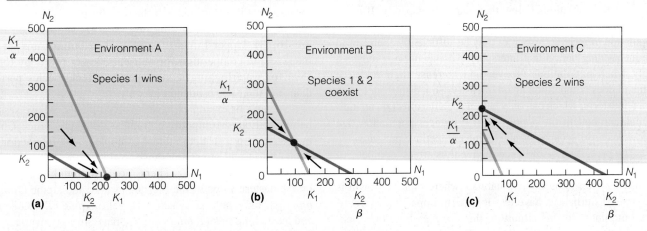

Figure 2 Competitive outcomes for the two species in different environments.

resources? Hutchinson acknowledged the impact of biotic factors, including mutualism and predation, but proposed that if abiotic factors change quickly relative to the short lifespan of plankton, non-equilibrium conditions will prevail, forestalling exclusion.

Some disagree. Chesson and Huntly (1997) argue that, regardless of the timeframe, fluctuating environments promote coexistence only when they create new niche opportunities. While abiotic changes may reduce population growth and weaken competition, particularly under harsh conditions, they do not lessen the role of competition in structuring communities because they may also lessen the competitive effect needed for exclusion to occur. Chesson (2000) proposes the **storage effect** as an alternative explanation for coexistence, whereby populations are buffered against the negative impact of interspecific competition in unfavourable years by producing more offspring and/or accumulating more biomass in long-lived individuals in favourable years. They also apply the storage effect concept to spatial variability, with populations buffering themselves against weak competitive performance in unfavourable patches by enhanced performance in favourable patches—an idea that underpins source–sink metapopulation dynamics (see Section 12.6). We reconsider coexistence in Section 13.13.

13.9 IMPACT OF DISTURBANCES: Disturbance Alters Competitive Outcomes by Reducing Populations or by Altering Conditions

If they are sufficiently severe, fluctuating conditions act as a **disturbance**: a discrete event that damages or eliminates a population or community. Disturbances (whether abiotic or biotic) occur at many spatial and temporal scales and typically affect mortality in a density-independent way (see Section 11.11). Even relatively stable habitats experience microscale disturbances that may affect competitive interactions. (In Section 18.7, we consider disturbance from a broader view, where it has mortality effects not just on one or a few populations but on the entire community).

Even if only part of the community is affected, extreme climatic variation can affect interspecific competition by causing density-independent mortality. Drought or extreme heat or cold may lower populations below their carrying capacity. If these events happen often enough relative to the time needed for the population to recover, resources may be sufficiently abundant in the intervening period that interspecific competition is reduced. Only when populations recover and increase their resource demand does competition again intensify. So a disturbance such as a windstorm, which can only have a negative effect on an individual, can benefit the population if it reduces population size below its carrying capacity, alleviating competitive pressures.

The prevailing disturbance regime thus affects competitive interactions by disrupting equilibrium conditions in the short term. However, in productive habitats where rapid regrowth is possible, disturbances have less long-term impact on competitive outcomes. In this situation, the *storage effect* (see Section 13.8) promotes coexistence by allowing temporarily less-favoured species a buffer. Alternatively, disturbances may affect competitive outcomes by causing long-term changes in abiotic factors that favour some species over others. These changes can be significant. In a grassland community, the pre-disturbance competitive hierarchy was more altered by changes in abiotic site factors caused by a simulated biotic disturbance event (burrowing of small mammals) than by changes in competitor density (Suding and Goldberg 2001).

13.10 ENVIRONMENTAL GRADIENTS: Relative Competitive Abilities Change along Abiotic Gradients

Thus far, we have discussed situations in which factors change over time, altering competitive ability and hence competitive outcomes. But abiotic factors also change in space, and as they change, so will relative competitive ability, reflecting changes either in carrying capacity with a changing resource base or in the ability to extract resources. In an **environmental gradient**, abiotic factors change along a continuum rather than in abrupt shifts.

Many studies have investigated plant competition along experimental gradients of resource availability. The relative competitive ability (measured by the growth response of a species in monoculture versus in mixture) of six annuals along a moisture gradient shifted with moisture level (Figure 13.14). No species were excluded, but some performed worse or better with more or less moisture (Pickett and Bazzaz 1978). Similar shifts occur along nutrient gradients. When five perennial C_4 grasses native to South Africa were grown in monoculture and in mixture at different soil fertility levels, the identity of the superior competitor switched (Fynn et al. 2005). Moreover, inversely related trends in competitive ability occurred only in species that contrasted in traits that are likely to affect competition, such as leaf width or plant height, suggesting that trade-offs are as important in interspecific interactions as they are in the adaptations of individual organisms (see Section 5.8).

Field studies along gradients reveal that multiple factors often interact to influence species response across a landscape, affecting overall community composition. In coastal areas, interspecific competition for nutrients is a critical factor determining salt-marsh zonation. Yet the relative competitive ability for nutrients is influenced by tolerance for physiological stress relating to salinity, waterlogging, and O_2 supply (see Figure 16.23). Upper limits of species distributions are set by interspecific competition for nutrients, lower limits by tolerance for waterlogging. Other species affect species performance at their upper boundaries only.

So far, our examples have involved plants, whose sessile habit facilitates competitive studies along gradients. Chipmunks furnish a striking animal example of the interaction of competition and stress tolerance in determining distribution along an environmental gradient. Four chipmunk species live on the eastern slope of the Sierra Nevadas: the alpine (*Tamias alpinus*), lodgepole (*T. speciosus*), yellow-pine (*T. amoenus*), and least chipmunk (*T. minimus*). All have strongly overlapping food requirements, and each occupies a different altitudinal zone (Figure 13.15).

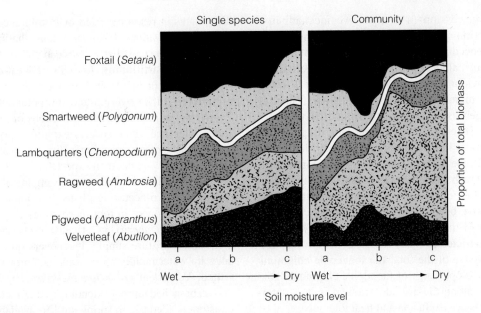

Figure 13.14 Responses of annuals grown in monoculture and mixtures along an experimental moisture gradient. Shaded areas represent the proportion of total biomass accounted for by each species.

(Data from Pickett and Bazzaz 1978.)

INTERPRETING ECOLOGICAL DATA

Q1. Which species accounts for the greatest proportion of biomass at moisture levels a, b, and c when grown in monoculture? If interspecific competition is the focus, why is it essential to grow the species in monoculture as well as in mixtures?

Q2. Which species shows the largest increase in relative biomass between monoculture and mixture treatments in dry soil (level c)? Which shows the largest decrease in relative biomass between monoculture and mixture treatments in wet soil (level a)?

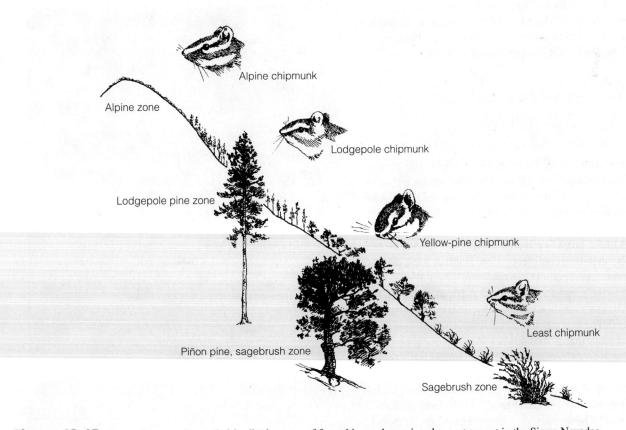

Figure 13.15 Vegetation zonation and altitudinal ranges of four chipmunk species along a transect in the Sierra Nevadas.

(Adapted from Heller and Gates 1971.)

Interspecific aggression (a competitive mechanism—*encounter*; see Section 13.1) plays a key role. Although the least chipmunk can occupy all habitats on the gradient (reflecting its broad fundamental niche), it is restricted to sagebrush by the aggressive behaviour of the yellow-pine chipmunk. Physiologically the least chipmunk is the most heat tolerant, enabling it to inhabit hot sagebrush sites that are outside the tolerance limits of the other species. If the yellow-pine chipmunk is removed, the least chipmunk undergoes niche release and moves into vacated pine woods. But if the least chipmunk is removed from sagebrush sites, the yellow-pine chipmunk does not invade, confirming its more restricted fundamental niche. In turn, the aggressive behaviour of the lodgepole chipmunk sets the upper limit of the yellow-pine chipmunk. The lodgepole chipmunk is restricted to shady forests by its low heat tolerance. Most aggressive of the four, the lodgepole chipmunk also limits the down-slope range of the alpine chipmunk. Distribution of species along the altitude gradient is thus determined by both aggressive exclusion and heat tolerance.

A long-term study of ecologically similar rodents in the Mojave Desert in the southwestern United States reveals similar patterns along an environmental gradient, against a dynamic backdrop of fluctuating population levels. Surveys of two kangaroo rats (*Dipodomys merriami* and *D. panamintinus*) conducted over almost two decades indicate that in years when population densities are high, *D. merriami* is more abundant at low elevation and *D. panamintinus* at high elevation. When drought reduces densities, *D. merriami* extends up the elevational gradient, suggesting that it is normally outcompeted by the larger *D. panamintinus* (Price et al. 2000). As with the chipmunk example, shifting distributions along gradients reflect a combination of physiological tolerance and competitive interactions.

13.11 RESOURCE PARTITIONING: Coexistence Often Involves Partitioning of Available Resources

Many species share resource needs. All terrestrial plants require light, water, and minerals. Given that these resources are often limited, competition among co-occurring species should be common. Although animals differ more in their resource needs (particularly diet), competition is also expected to be intense for insectivorous birds, large herbivores in African grasslands, and predatory fish inhabiting coral reefs. How is it that these diverse arrays of potential competitors coexist, even when abiotic conditions are relatively constant, and even if populations are not always kept below carrying capacity by density-independent abiotic factors? The competitive exclusion principle (see Section 13.4) suggests that if two species have identical resource needs, one will eventually displace the other, unless intraspecific competition offsets interspecific competition (a stable equilibrium). But how different do species have to be to avoid exclusion? Or, conversely, how similar can species be in resource use and still coexist? These questions are central to niche theory, species interactions, and ultimately community structure.

We have seen from examples presented thus far that coexistence can result from **niche differentiation**—evolution of

differences in resources used or in tolerance for non-resource factors. Observations of similar species sharing the same habitat suggest that one way that species can differentiate their niches is by **resource partitioning**. Animals may eat different kinds and sizes of food, or forage at different times or in different areas, minimizing niche overlap and the potential for competition. Plants may require different proportions of minerals or differ in shade tolerance. Each species exploits a portion of resources unavailable to others, resulting in differences among co-occurring species that are greater than expected from chance alone.

Field studies provide many examples of apparent resource partitioning. Co-occurring plants partition soil resources vertically by differences in root morphology (Figure 13.16). Bristly foxtail (*Setaria faberii*) has a shallow root system that copes with a variable moisture supply. It recovers quickly from drought, takes up water rapidly after a rain, and grows even when partly wilted. Velvetleaf (*Abutilon theophrasti*), in contrast, has a sparse, branched taproot extending to intermediate depths, where moisture is adequate in spring but less available later in the season. Smartweed (*Polygonum pensylvanicum*) has a taproot that is moderately branched in the upper soil but penetrates below the rooting zone of other species to access a continuous moisture supply. By accessing different soil regions, these species coexist side by side while avoiding the negative effects of intense belowground competition (Pickett and Bazzaz 1978).

Resource partitioning is also common among animals that share the same habitat and resources. Among coexisting wild

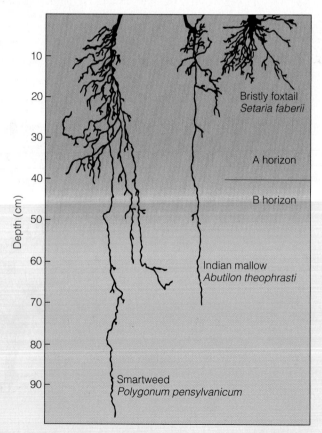

Figure 13.16 Vertical partitioning of soil resources by three annual plants, a year after disturbance.

(Adapted from Pickett and Bazzaz 1978.)

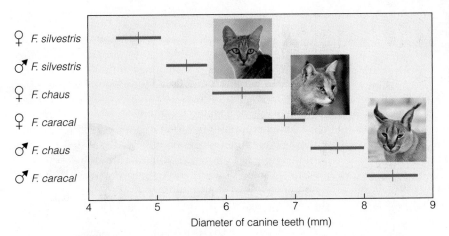

Figure 13.17 Size (diameter) of canine teeth of co-occurring cat species (females and males) in Israel. Teeth size is correlated with size of prey.

(Adapted from Dayan et al. 1990.)

cat species in the Middle East, there are systematic differences in canine teeth size, both between males and females of a species (*sexual dimorphism*; see Section 8.4) and among species (Figure 13.17). Canine teeth are critical for capturing prey. Intra- and interspecific competition has selected for feeding-related differences in teeth size, reducing overlap in prey type and size (Dayan et al. 1990).

Understanding niche differentiation via resource partitioning of coexisting species is complicated by the multi-dimensionality of the niche. We often represent the niche as one-dimensional in both theoretical depictions and experimental studies, where a single resource may be the focus. But the niche includes responses to many resources apart from the resource under study, as well as to non-resource factors. So even if species do share one resource, or react similarly to one condition, rarely will they have the same combination of responses to, and impacts on, the totality of factors affecting them. Species may overlap on one dimension (e.g., size of insect prey) but not on another (e.g., foraging height). Resource partitioning can be subtle, and competition may be less intense than suggested by the overlap for any one niche dimension.

Does this mean we can assume that co-occurring species that appear to be ecologically similar must be partitioning resources and hence that they must differ in niche? As many have pointed out, the open-endedness of the niche concept can be a weakness instead of a strength. If two species coexist even at high densities, it is too easy to assume that they *must* be differentiated in some dimension(s), even if we do not know which. And even if niche differentiation does exist, it may reflect factors other than competition, particularly predation. As we discuss in Section 13.13, there are many reasons why species with highly similar niches may coexist indefinitely.

13.12 EVOLUTIONARY EFFECTS: Competition May Select for Character Displacement

So far we have focused on competition in the present, but what about the impact of past competitive interactions? Whatever other factors may be at work to reduce competitive pressures and promote coexistence, resource partitioning does

reduce competition among co-occurring species. In turn, resource partitioning reflects specific physiological, morphological, and/or behavioural traits that facilitate access to resources. Although such traits reduce interspecific competition in the present, they may be an outcome of interspecific competition in the past. Recall that competition underpins Darwin's theory of natural selection: traits that enable an organism to reduce the negative impacts of competition will increase fitness. Competition thus acts as an agent of natural selection. **Character displacement** is the process whereby differences in genetically controlled traits emerge from the selective force of interspecific competition (and other interactions). Differences in root types in plants and canine teeth size in cats are examples (see Figures 13.16 and 13.17). The "ghost of competition past" (Connell 1980) expresses this idea that character displacement may reflect the selective impact of past competitive interactions.

Consider two seed-eating birds, species A and B. Their seed consumption is depicted as a bell-shaped curve, with seed size on the x-axis and proportion of the diet on the y-axis (Figure 13.18, p. 274). When two species overlap in the size of seeds they consume, there is the potential for competition, if seeds are limiting. If we assume that interspecific competition reduces the fitness of individuals of both species (even if the effects are asymmetric), then individuals choosing seeds from the tails of the distributions (where overlap is minimal) should encounter less competition and have greater fitness. Recall from Section 5.4 that the size of seeds eaten by seed-eating birds is constrained by beak size. If competition favours the choice of smaller and larger seeds by species A and B, respectively, then selection will favour (1) A individuals with small beaks and (2) B individuals with large beaks. Ultimately, this disruptive selection (see Section 5.5) will cause the species to diverge in beak size and the size of seeds eaten. By facilitating resource partitioning via character displacement, interspecific competition reduces niche overlap and promotes coexistence.

Although this scenario is consistent with resource partitioning observed in nature (such as the canine teeth example), it is difficult to prove as a direct result of competition. Differences among species have evolved over millennia and may reflect adaptations for the ability to tolerate abiotic factors independently of competition. More convincing support for the

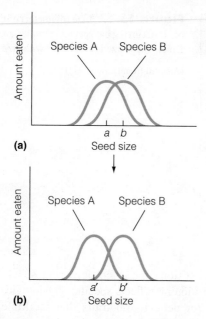

Figure 13.18 Possible impact of competition between two bird species. **(a)** In the absence of competition, the species feed on an overlapping range of seed sizes. Mean seed sizes consumed by species A and B are labelled *a* and *b*. **(b)** Competition for intermediate-sized seeds causes their diets to shift, reducing overlap in the size of consumed seeds. New means are labelled *a'* and *b'*.

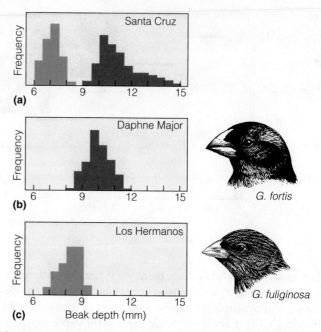

Figure 13.19 Apparent character displacement in beak size in the medium and small ground finch. **(a)** On Santa Cruz, where they coexist, their beak sizes do not overlap and mean beak size is significantly larger for the medium ground finch. On the smaller islands of **(b)** Daphne Major and **(c)** Los Hermanos, where the species are allopatric, their beak size is intermediate and overlapping.

(Adapted from Grant 1999, based on data from Schluter et al. 1985.)

selective impact of competition comes from studies examining differences in traits of populations with known competitive histories. Darwin's finches on the Galápagos furnish a well-documented example.

Both the medium and the small ground finch (*Geospiza fortis* and *G. fuliginosa*) feed on seeds with an overlapping size range. On the larger island of Santa Cruz, where the two species coexist (i.e., are *sympatric*), their beak size distributions do not overlap, and mean beak size is significantly larger for *G. fortis* (Figure 13.19a). On the nearby smaller islands of Los Hermanos and Daphne Major, where the species do not coexist (i.e., are *allopatric*), their beak size distributions overlap and are intermediate (Figure 13.19b and 13.19c) (Grant 1999). Where they coexist, competition favours medium ground finches with large beaks that exploit large seeds, and small ground finches with smaller beaks that exploit small seeds. The result is a shift in feeding niches that reduces overlap. Because the structural traits involved are heritable, this situation is an instance of character displacement.

As this example shows, interspecific competition can select for character displacement—divergence in phenotypic traits relating to exploitation of a shared, limiting resource. Character displacement indicates resource partitioning that is based not on a plastic shift in the realized niche but on heritable adaptive traits that reflect an altered fundamental niche. The use of morphological and physiological traits as indirect indicators of niche dimensions is called **ecomorphology**. Although it has advantages compared with direct niche observations (including being possible for fossils), ecomorphological analysis is prone to misinterpretation. Differences in traits may reflect biotic

interactions other than competition (e.g., predation and mutualism), abiotic factors, and/or chance. Moreover, niches of many species exhibit **niche conservatism**: retention of niche-related ecological traits over time (Wiens et al. 2010). This retention may seem paradoxical, given the rapid evolution of some traits, but Wiens argues it is a critical emerging principle.

A pioneer of the niche concept, G. E. Hutchinson (1959), proposed a corollary of the competitive exclusion principle called **limited similarity**: ecologically similar species can coexist only if they have evolved sufficient differences in their morphological traits to allow niche separation. Using aquatic insects called water boatmen (*Corixa* spp.) as an example, Hutchinson claimed that if they are too similar in size, they will be too similar in resource use, and their niche overlap will risk competitive exclusion. Further, Hutchinson argued for a specific quantitative limit to how similar species can be and still coexist: the "1.3:1 rule." The largest of the group, *Corixa punctata*, coexists with either *C. affinis* or *C. macrocephala*, he argued, because it differs in relative size from them by 1.16:1 to 1.46:1, respectively (on average by 1.3:1). The two smaller species, in contrast, never coexist.

As influential as this study was, ecologists have discredited it on several grounds: (1) The two species that do *not* coexist differ in size by 1.26, a ratio that should—by Hutchinson's own rule—allow coexistence (Morin 2011); (2) Studies show that even when *Corixa* spp. differ in size by over 1.3:1, they may still compete (Pajunen 1982); and (3) Size differences among coexisting species are no greater than in randomly assembled groups of non-competing species. The latter reason involves the use of a *null model* (see Section 13.3). Hutchinson's water boatman study provides a useful caution against overinterpreting pattern in the search for ecological "laws."

The fact that morphological similarity (or dissimilarity through character displacement) is a generally poor predictor of interspecific competition (Morin 2011) does not mean that character displacement never occurs in response to competition. Another ground finch study provides direct evidence of character displacement. Before 1982, *Geospiza fortis* was the only ground finch inhabiting Daphne Major. In 1982, the large ground finch, *G. magnirostris*, emigrated from the adjacent islands. *G. magnirostris* can potentially compete with *G. fortis* because both feed on seeds of the Jamaican feverplant (*Tribulus cistoides*). Initially, the *G. magnirostris* population was too small to have a substantive effect on *G. fortis*, but by 2003, its numbers had increased substantially. Low rainfall from 2003 to 2004 then caused both populations to decline as a result of food shortages. *G. magnirostris* depleted large seeds, forcing *G. fortis* to rely on smaller seeds.

This compression in its feeding niche caused strong directional selection for smaller beak size in *G. fortis* (Grant and Grant 2006)—the exact opposite of the trend after the earlier drought when the co-occurring species was the small, not the large, ground finch (see Figure 5.8). Thus, while most reported cases of character displacement have been inferred (sometimes mistakenly) as reflecting past competition, the decreased beak size of *G. fortis* on Daphne Major in 2003–2004 documents character displacement as a direct result of ongoing interspecific competition.

As noted earlier, character displacement involves a shift in the fundamental niche mediated by altered genotypic frequencies for specific heritable traits, in contrast with a plastic shift in the realized niche, as in the *Typha* example (see Figure 13.6). Yet the two phenomena are related: plastic shifts in realized niche occur first, followed by selection favouring individuals who do better in the altered niche. The mean beak size of medium ground finches did not instantaneously shift in response to competition with the large ground finch. Rather, medium ground finches shifted their feeding behaviour to include smaller seeds. Then, individuals with traits (smaller beaks) that increased their success in the altered niche had increased survival and presumably greater fitness, causing a genetically based alteration in the fundamental niche.

The circumstances of this important study highlight why competition-driven character displacement may occur infrequently. Competition involved only two species, allowing a kind of coevolution more similar to what occurs in parasitism and mutualism (see Chapter 15). In contrast, interspecific competition in most communities is diffuse, making coevolutionary pressure less focused on the traits of any two interacting species. Indeed, in species-rich communities (as opposed to species-poor island or high-latitude habitats), coevolution among multiple competing species may generate clusters of ecologically similar species (Scheffer and van Nes 2006). There may be two alternative strategies for species to live together—being sufficiently different (through character displacement associated with niche differentiation) or being sufficiently similar.

We reprise the community-level implications of competition in Chapter 16.

13.13 COEXISTENCE MECHANISMS: Competition Is Compatible with Species Coexistence

What can we conclusively say about competition? Despite the difficulties in proving it experimentally, competition is an important interaction affecting not only individuals but also the structure and dynamics of the populations and communities to which they belong. However, interspecific competition seldom involves a simple interaction between individuals of two species for a single limiting resource. It often involves many species seeking out multiple resources, and is influenced directly or indirectly by many factors that vary in time and space. Thus, the outcome of competition between two species for a specific resource under one set of conditions may differ markedly from the outcome under other conditions.

Many studies show that competition is an important structuring force in ecosystems, especially in the short term, through the alterations it causes in realized niches. Its long-term evolutionary impact on character displacement (and hence on fundamental niches) is less clear cut. The diffuse nature of interspecific competition in multi-species communities weakens its selective impact on any one species. What is clear is that, rather than inexorably leading to exclusion, competition is compatible with the long-term coexistence of many species. There is no one reason why, but rather an array. We conclude this chapter by summarizing some of them.

1. Species have evolved enough niche differentiation (via character displacement and/or behavioural differences) such that they are no longer "complete competitors." This factor is compatible with niche theory, because it invokes the selective force of competition in minimizing niche overlap. However, factors other than competition, particularly predation, may have contributed to the niche differentiation.

2. Both species are more affected by intra- than interspecific competition as their numbers increase, resulting in a stable equilibrium as predicted by outcome 4 of the Lotka–Volterra model. This reason is often ignored in studies where researchers fail to distinguish between intra- and interspecific competition. It typically follows a narrowing of the niche, because specialization intensifies intraspecific competition.

3. Populations of superior competitors are kept below carrying capacity by selective predation (which may be density dependent; see Sections 14.4 and 16.4), other interactions, or density-independent disturbances. This reason also applies if resources are non-limiting.

4. Changing abiotic factors shift relative competitive abilities. As Chesson (2000) points out, these shifts may not necessarily prevent exclusion, but if species can "store up" their growth advantage in good times they can be buffered against bad times. This non-linear *storage effect* can even out the impact of a shifting environment.

5. A population that has been excluded (or is about to be) is "saved" by immigration of individuals from a local population within the metapopulation, particularly a source (see Section 12.6). This reason highlights the importance of metapopulation dynamics for species interactions and local community structure, but it is less applicable at the regional level (Chesson 2000).

6. Indirect interactions resulting from diffuse competition influence outcomes. **Competitive mutualism** is an example. If species A and C do not compete with each other, they may both benefit from each other's presence if each competes with the other's competitor, species B. These and other indirect interactions (see Section 16.7) may reflect coevolutionary forces that generate clusters of ecologically similar species (Scheffer and van Nes 2006), rather than the assemblages of ecologically dissimilar species predicted by traditional niche theory.

More than one of these reasons may apply, and their strength will vary among species and communities. Coexistence mechanisms are *stabilizing* if they increase negative intraspecific relative to interspecific interactions (Chesson 2000), thereby maintaining diversity despite competition. Mechanisms are *equalizing* if they decrease fitness differences among species (Chesson 2000). In the absence of stabilizing mechanisms, equalizing mechanisms (such as disturbance) can only slow competitive exclusion, not prevent it. The more stabilizing mechanisms at work, the less variability in population trends and the more species that can coexist.

So, as pervasive and powerful as interspecific competition is, it is not only compatible with, but may also promote, species coexistence (and hence community diversity), in concert with other abiotic and biotic influences. Above all, competition is only one of many interactions that affect population dynamics, community structure, and ecosystem function.

EcologyPlace

Visit EcologyPlace at www.pearsoncanada.ca/ecologyplace to access online resources that complement your textbook, and help you to apply and to review the information in this chapter. EcologyPlace includes

- an eText version of the book
- self-grading quizzes
- glossary flashcards
- and more!

Go to www.pearsoncanada.ca/ecologyplace and follow the registration instructions on the Student Access Code Card included with this text. If your book does not have a Student Access Code Card, you can purchase access to it at www.pearsoncanada.ca/ecologyplace.

SUMMARY

Types and Mechanisms 13.1

In interspecific competition, individuals of two or more species share a limited resource, with adverse effects experienced by both. It can involve either exploitation or interference. Competitive mechanisms include consumption, pre-emption, overgrowth, chemical interaction, territoriality, and encounter. Competition can be symmetric or asymmetric, diffuse, or episodic.

Modelling Competition 13.2

The Lotka–Volterra model predicts four competitive outcomes. Species 1 excludes species 2 (or species 2 excludes species 1) if either species continues to increase while inhibiting growth of the other species. In an unstable equilibrium, the species that is initially most abundant excludes the other. In a stable equilibrium, the species coexist because each inhibits its own increase by intraspecific competition more than it inhibits growth of the other species by interspecific competition.

Experimental Tests 13.3

Laboratory studies of competition between two species for a single limiting resource provide some support for the Lotka–

Volterra model. However, the model makes assumptions that may not apply in nature. Field experiments often involve either removals or transplants.

Competitive Exclusion 13.4

The competitive exclusion principle states that two species with identical niche dimensions cannot coexist if their shared resource is limiting. This principle has stimulated many field studies of competition, especially regarding partitioning of resources. Despite its theoretical importance, competitive exclusion is relatively rare in nature.

Ecological Niche 13.5

Hutchinson defined the niche as a multi-dimensional hypervolume describing the response of a species to all the factors in its environment. Elton defined the niche as the functional role of an organism and stressed the impact of a species on its environment as much as its response. In the absence of interactions, an organism occupies its fundamental niche. In the presence of interactions, the fundamental niche is reduced to a realized niche. When two organisms use the same resource in the same way, their niches overlap, which may or may not indicate competition.

Niche Dynamics 13.6

Species are classed as generalists or specialists based on their niche breadth for critical niche dimensions. A species compresses or shifts its realized niche when competition forces it to alter its pattern of resource use. The realized niche may not represent the optimal niche dimensions for a species. In the absence of competition, the species undergoes competitive release. Intraspecific competition tends to broaden the niche, whereas interspecific competition tends to narrow it.

Competition for Multiple Resources 13.7

Competition may involve multiple resources. Competition for one resource may influence an organism's ability to compete for other resources. The impacts of multiple resources on competitive outcomes may interact.

Impact of Conditions 13.8

Non-resource environmental factors such as temperature, pH, and salinity may directly influence growth and reproduction but are not consumable resources for which species compete. By affecting relative competitive ability, environmental variability may give a species a temporary advantage, allowing it to avoid exclusion.

Impact of Disturbances 13.9

Density-independent environmental disturbances can influence competition indirectly by keeping population levels below carrying capacity or by altering abiotic factors. Their impact on competitive outcomes depends on whether species respond differentially to the disturbance.

Environmental Gradients 13.10

As environmental conditions change in space, so may relative competitive ability, either through changes in carrying capacities related to a changing resource base or changes in the abiotic environment that interact with resource availability. Many factors, both resource and non-resource, vary along environmental gradients. Species distributions on gradients often reflect a combination of interspecific competition and stress tolerance.

Resource Partitioning 13.11

Species sharing the same habitat may coexist by partitioning available resources, thereby reducing interspecific competition. Resource partitioning may be associated with character displacement: heritable differences in morphology or physiology.

Evolutionary Effects 13.12

Interspecific competition may lower fitness. If certain traits minimize competitive effects, individuals with those traits may have greater fitness. The result is a shift in the distribution of phenotypes within one or more populations. Character displacement has often been inferred from observations but may be misinterpreted as a result of other factors that influence trait selection.

Coexistence Mechanisms 13.13

Competition is more complex in nature than in simplified experimental studies. It seldom involves a simple interaction between two species for a single limiting resource and is influenced by many factors that vary in both time and space. Although competition is a pervasive and important interaction, there are many reasons why it rarely leads to exclusion. Coexistence mechanisms may be either stabilizing or equalizing. Interspecific competition is not only compatible with, but may also promote, species coexistence.

KEY TERMS

allelopathy	competitive release	interspecific	niche breadth	realized niche
amensalism	diffuse competition	competition	niche compression	removal experiment
asymmetric (one-sided)	disturbance	limited similarity	niche conservatism	resource partitioning
competition	ecomorphology	Lotka–Volterra	niche differentiation	specialist
character displacement	environmental gradient	competition model	niche overlap	storage effect
competitive exclusion	episodic competition	*n*-dimensional	paradox of the	transplant experiment
(Gause's) principle	fundamental niche	hypervolume	plankton	zero growth isoclines
competitive mutualism	generalist	niche	priority effect	

STUDY QUESTIONS

1. What conditions must be established before a researcher can conclude that two species are competing? Is establishing that two species overlap in their use of a resource enough to determine that interspecific competition is occurring?

2. In analyzing the Lotka–Volterra model of interspecific competition in Section 13.2, four outcomes between two species were identified. In three of the four, one species excludes the other. In the fourth, the two species coexist. What circumstance is necessary for this outcome?

3. How would a researcher determine that changes in species composition in a community with changing abiotic factors such as moisture are due to altered competitive outcomes rather than simply to different ecological tolerances?

4. As in Gause's experiments with *Paramecium*, researchers often use taxonomically related or ecologically similar species. Why?

5. If you observe a species in the wild, are you observing its fundamental or its realized niche? What is the difference? Does the realized niche always represent the optimum?

6. Explain Hutchinson's concept of the niche. Based on this concept, what is the relationship between the niche of a species and its habitat?

7. Explain how inter- and intraspecific competition interact to influence niche breadth.

8. Distinguish between *niche compression* and *niche expansion*. How might both be involved in an island habitat where a species is introduced from a mainland?

9. Abiotic conditions are not consumable resources, but species that occupy the same habitat often differ in their response to these factors. How might factors such as temperature and salinity influence the outcome of competition between two species occupying the same habitat?

10. Figure 13.18 presents a hypothetical case in which two species overlap in their use of food resources (seeds). Assume that as a result of interspecific competition, the two species cannot co-occur in the same habitat (say species A excludes B). Now assume that the distribution of seed sizes changes from year to year with rainfall, and that mean seed size varies from *a* to *b*. How might this temporal variation affect the outcome of competition between the species?

11. Resource partitioning that reflects natural selection associated with competition is often considered the cause of trait differences of related species that occupy the same area, such as the cats in Figure 13.17, or the ground finches in Figure 13.19. What non-competitive factors might account for the observed differences?

12. Describe the relationship among *niche shift*, *resource partitioning*, and *character displacement*. Which of these concepts apply to the Galápagos finch example in Figure 13.19?

13. How is the coexistence of many species in the same habitat compatible with the existence of interspecific competition? Consider as many factors as possible.

FURTHER READINGS

Bazzaz, F. A. 1996. *Plants in changing environments: Linking physiological, population, and community ecology.* New York: Cambridge University Press.
Excellent overview of plant competition, linking species traits and competitive interactions. Stresses the shifting nature of competitive interactions as conditions change in time and space.

Chase, J. M., and M. A. Leibold. *Ecological niches: Linking classical and contemporary approaches.* 2003. Chicago: University of Chicago Press.
Argues for the importance of a new synthesis of the historically disparate strands of niche theory.

Chesson, P. 2000. Mechanisms of maintenance of species diversity. *Annual Review of Ecology and Systematics* 31:343–366.
Thorough and thought-provoking article on coexistence mechanisms.

Connell, J. H. 1983. On the prevalence and relative importance of interspecific competition: Evidence from field experiments. *American Naturalist* 122:661–696.
Critical review of field experiments of competition.

Grace, J. B., and D. Tilman. 1990. *Perspectives on plant competition.* San Diego: Academic Press.
Excellent overview of competition in plant communities. Provides many examples from laboratory, greenhouse, and field experiments.

Hutchinson, G. E. 1959. Homage to Santa Rosalia, or why are there so many kinds of animals? *American Naturalist* 93:134–159.
Hutchinson introduced his niche concept in 1957, but this paper is the most influential of his writings on niche. Includes the paradox of the plankton and the limited similarity hypothesis.

Rohde, K. 2005. *Nonequilibrium ecology.* Cambridge: Cambridge University Press.
Contrasts the equilibrium and nonequilibrium perspectives on competition.

Schoener, T. W. 1983. Field experiments on interspecific competition. *American Naturalist* 122:240–285.
Reviews field experiments of competition, with different conclusions than Connell (1983).

Lioness attacking a female kudu in Etosha National Park, Namibia, Africa.

When the poet Tennyson described nature as "red in tooth and claw," he was evoking savage images of predation. The word *predation* brings to mind lions on the African savannahs, or great white sharks cruising ocean waters. However, *predation* is defined more generally as an interaction in which an organism eats another living organism in whole or in part. All heterotrophs consume organic matter, but predators are **biophages**, consuming living organisms, whereas decomposers and scavengers are **saprophages**, consuming dead tissues. The distinction is vital: as biophages, predators act as agents of mortality with the potential to regulate the population numbers of their prey. Similarly, as a food resource, prey have the potential to influence the population growth of their predators and to affect their competitive interactions. Predation thus affects community structure and acts as an agent of selection, influencing the evolution of both predator and prey species.

14.1 PREDATION TYPES: Predators Are Classified in Many Ways

Predation encompasses many subtypes. The simplest classification uses food type (see Section 7.2): *carnivory* (consumption of animal tissue), *herbivory* (consumption of plant or algal tissue), and *omnivory* (consumption of both plant and animal tissue). Carnivores are also called *true predators*, which kill and eat their prey upon capture. *Cannibalism* is intraspecific predation, in which a species preys on itself (Figure 14.1). Its impact interacts with that of intraspecific competition.

Predation involves more than just energy transfer. It is a complex interaction of two or more species—the eater and the eaten. As agents of mortality, predators may reduce, and possibly regulate, the growth of prey populations. In turn, prey availability may regulate predator population size. True predators typically consume multiple items of one or more prey species. In contrast, most herbivores (*grazers* and *browsers*) consume only part of their plant prey. Herbivory may harm a plant but rarely kills it directly. Seed predators

(*granivores*) and animals that eat plankton (**planktivores**) are exceptions; these herbivores (or omnivores, if the planktivore consumes both zooplankton and phytoplankton) act as true predators.

Parasites, as intestinal worms, feed on their prey (**host**) while it is alive, but unlike true predators, their effects are usually non-lethal in the short term. The parasite–host interaction (see Chapter 15) is more intimate than that of other predators, as many parasites live on or in their hosts for part of their life cycle. Finally, **parasitoids** are insects (usually flies or wasps) that attack their prey indirectly by laying eggs in or on the body of their host (another insect species). When the parasitoid eggs hatch, the larvae consume the host tissues. As with most internal parasites, parasitoids are intimately associated with a single host organism, but unlike most parasites they always cause its death.

This chapter focuses on true predators (called *predators* hereafter) and herbivores. We begin with a model that explores the link between the population dynamics of the hunter and the hunted.

14.2 MODELLING PREDATION: The Lotka–Volterra Model Predicts Mutual Regulation of Predator and Prey

In the 1920s, Lotka and Volterra independently proposed mathematical models to express the reciprocal impacts of predator and prey (see Quantifying Ecology 14.1: The Lotka–Volterra Predation Model). In essence, the **Lotka–Volterra predation model** uses two equations to quantify how each population functions as a density-dependent regulator of the other. Predators regulate the growth of their prey by affecting density-dependent mortality, and prey regulate the growth of their predators by affecting density-dependent fecundity. The growth of both populations is linked by a single term relating to prey consumption: $cN_{prey}N_{pred}$. For the prey, this term regulates population growth through mortality. For the predator, it regulates population growth through reproduction.

Predator growth reflects two distinct predator responses to changing prey density. (1) **A functional response** involves a change in the per capita rate of prey consumption with prey density. Predator growth depends on the rate at which prey are captured ($cN_{prey}N_{pred}$)—the more prey available, the more the predator eats. (2) **A numerical response** involves a change in predator reproduction ($b[cN_{prey}N_{pred}]$) with prey density. These responses (discussed in Sections 14.3 and 14.5) are connected; higher prey consumption (functional response) usually boosts predator reproduction (numerical response), assuming predators sustain their prey consumption over time.

The Lotka–Volterra model may overemphasize mutual regulation. Even if predator and prey do affect each other, other interactions, including intra- and interspecific competition and mutualism, may simultaneously exert density-dependent effects. The model also assumes a constant environment, which

Figure 14.1 Cannibalism occurs in some insects and fish, such as pike (*Esox lucius*), a common species in northern lakes. Pike is shown here preying on rudd, *Scardinius erythrophthalmus*.

Section 13.2 explored how two populations may affect each other's growth through interspecific competition. Now we consider how two populations may affect each other's growth through predation. We start with how prey numbers affect prey consumption and predator reproduction.

Let's assume that the per capita rate at which predators consume prey increases linearly with prey number (Figure 1a). It is represented as cN_{prey}, where c represents the predation efficiency as defined by the slope of the line. Total predation rate (prey captured per unit time) is the product of this per capita predation rate (cN_{prey}) and the number of predators (N_{pred}): $cN_{prey}N_{pred}$. This value is a source of mortality for the prey and is subtracted from the rate of prey population growth given by the exponential model. The rate of change in prey numbers (dN_{prey}/dt) becomes

$$\frac{dN_{prey}}{dt} = rN_{prey} - cN_{prey}N_{pred}$$

Now assume that the predator birthrate increases linearly with the rate of prey consumption: $cN_{prey}N_{pred}$ (Figure 1b). Birthrate is the product of b, the efficiency with which the predator converts prey consumed into reproduction (as defined by the slope of the line in Figure 1b) and predation rate ($cN_{prey}N_{pred}$): $b(cN_{prey}N_{pred})$. Predator mortality rate is assumed to be a constant proportion of predator numbers and is represented as dN_{pred}, where d is the probability of mortality. The rate of change in predator numbers then becomes

$$\frac{dN_{pred}}{dt} = b(cN_{prey}N_{pred}) - dN_{pred}$$

Taken together, and based on the same graphical approach used for competitive outcomes in Section 13.2, these predator–prey equations describe how each population exerts a density-dependent effect on the other. When predators are absent or scarce, the prey grows exponentially ($dN_{prey}/dt = rN_{prey}$). As predators increase, the prey mortality rate from predation ($cN_{prey}N_{pred}$) increases until it equals the inherent growth rate of the prey (rN_{prey}). The prey population growth (dN_{prey}/dt) is then 0. We can solve for the predator population at which this occurs:

$$cN_{prey}N_{pred} = rN_{prey}$$
$$cN_{pred} = r$$
$$N_{pred} = \frac{r}{c}$$

Growth of the prey population stops when the number of predators equals the per capita growth rate of the prey divided by the efficiency of predation. If predator numbers exceed this value, the prey's growth rate becomes negative—its numbers drop. Similarly, growth of the predator population ceases when the rate of predator increase resulting from prey consumption equals the predator mortality rate:

$$b(cN_{prey}N_{pred}) = dN_{pred}$$
$$bcN_{prey} = d$$
$$N_{prey} = \frac{d}{bc}$$

Now we plot both equations on the same graph (Figure 2a). Knowing the prey level at which growth of the predator is 0 ($N_{prey} = d/bc$) and the predator level at which growth of the prey is 0 ($N_{pred} = r/c$), we can draw the zero isoclines. If prey numbers are to the right of the predator isocline, predators increase; if to the left, predators decrease. If predator numbers are below the prey isocline, prey increase; if above, prey decrease. To predict the growth of predator and prey for any combination of population sizes, we combine the isoclines (Figure 2b).

These equations predict that the two populations will rise and fall in an oscillating pattern (Figure 2c). As predators increase, they consume more prey (i.e., prey mortality increases with rising predator birthrate), until the prey begin to decline. As prey numbers drop, they can no longer support a large predator population. Predators face a food shortage, and many starve or fail to reproduce (i.e., predator mortality increases and birthrate declines). Predator numbers then decline to a point where prey reproduction more than balances prey mortality from predation. Prey numbers

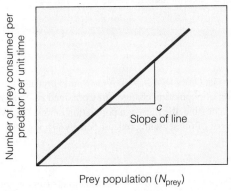

(a)

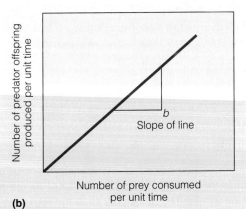

(b)

Figure 1 Impact of prey on predators. **(a)** Relationship between prey numbers (x-axis) and per capita consumption rate of predators (y-axis). The slope (c) represents predation efficiency. **(b)** Relationship between prey consumed (x-axis) and predator offspring (y-axis). The slope (b) represents the efficiency with which prey consumed is converted into predator reproduction.

continued on page 282

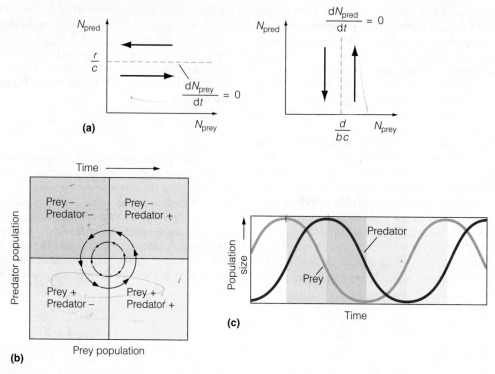

Figure 2 Patterns predicted by the Lotka–Volterra predation model. **(a)** Zero isoclines ($dN/dt = 0$) for prey and predator are defined for a fixed number of predators and prey, respectively. Arrows show the direction of population change. **(b)** Combined zero isoclines predict the trajectories of predator and prey. The minus sign indicates population decline, the plus sign population increase. **(c)** When the predicted changes are plotted over time, the populations continuously cycle out of phase, with predator density lagging behind prey density. Shaded regions relate to corresponding regions in (b).

increase, followed (after a lag) by an increase in predators. This pattern, which is the mathematical basis of the stable limit cycle, may continue indefinitely, driven by a negative feedback loop between predator and prey. Both populations experience highs and lows, but the prey is never quite destroyed and the predator never quite dies out.

1. In Figure 2b, which of the four quadrants (regions of the graph) correspond to the following conditions?

$$N_{prey} < d/bc \text{ and } N_{pred} > r/c$$

2. In which of the four shaded regions of Figure 2c does predator mortality rate exceed their birthrate as a result of low prey density?

is rarely the case. Climatic flux may exert density-independent effects (see Section 11.11) that are no less important than predation. As with the Lotka–Volterra competition model, ecologists have proposed alternatives, including *consumer-resource models*, that treat model parameters as random rather than fixed variables (Stow et al. 1995). But the continuing appeal of the Lotka–Volterra model is its ability to elucidate the *stable limit cycles* (see Section 10.9) that often occur between predator and prey. Modifications of the model that incorporate spatial diffusion and time delays (Cai et al. 2010) enhance its realism.

Studies of predator–prey interactions in nature reveal many factors that affect prey regulation: (1) availability of cover, (2) difficulty of locating prey when scarce, (3) selection among multiple prey, and (4) coevolution of predator and prey. We examine each of these factors in later sections.

14.3 FUNCTIONAL RESPONSE: Prey Consumption Increases with Prey Density

Predation is often treated as a *consumer–resource interaction*, in which the density-dependent response of the predator to changing availability of its prey (a biotic resource) provides a powerful mechanistic basis for understanding the impact of predation on consumer–resource dynamics (Murdoch et al. 2003). Solomon (1949) introduced the concept of a functional response as the relationship between per capita consumption rate and prey density. There are three general types (Holling 1959). Differences in these functional response types relate in part to time allocation (see Quantifying Ecology 14.2: Predator Time Budgets).

QUANTIFYING ECOLOGY 14.2   Predator Time Budgets

Different functional responses may reflect differences in a predator's time budget. What activities are involved, and how does the time spent on them affect the functional response?

The per capita predation rate (N_e) is the number of prey eaten by a predator during a period of **search time** (T_s) and is expressed as

Per capita rate of predation: the number of prey consumed during a given period of search time, T_s

Efficiency of predation

$$N_e = (cN_{prey})T_s$$

Prey population size

Period of search time

For a given search time (T_s), the relationship between the per capita predation rate (N_e) and prey density (N_{prey}) for the above equation is linear—a *type I functional response*. The prey mortality rate (proportion of prey eaten per unit time) is constant, a function of the efficiency of predation (c) (see Figure 14.2a, p. 284). In a type I response, virtually all the time allocated to feeding is spent searching. But in other functional responses, search time is typically less than total foraging time, because the predator needs **handling time** for chasing, killing, eating, and digesting its prey. If T_h is the time required to handle a prey item, then the time spent handling N_e prey will be the product N_eT_h. Total foraging time (T) spent in both searching and handling is then

$$T = T_s + (N_eT_h)$$

By rearranging the preceding equation, we can define search time as

$$T_s = T - N_eT_h$$

For a given total foraging time (T), search time decreases with increased handling time.

We now expand the original equation describing a type I functional response [$N_e = (cN_{prey})T_s$] by substituting the T_s equation just developed. This includes the extra handling time for N_e prey:

$$N_e = c(T - N_eT_h)N_{prey}$$

Note that N_e (number of prey consumed during time period T) appears on both sides of the equation, so to solve for N_e we rearrange the equation:

$$N_e = c(N_{prey}T - N_{prey}N_eT_h)$$

We move c inside the brackets:

$$N_e = cN_{prey}T - N_ecN_{prey}T_h$$

We add $N_ecN_{prey}T_h$ to both sides of the equation:

$$N_e + N_ecN_{prey}T_h = cN_{prey}T$$

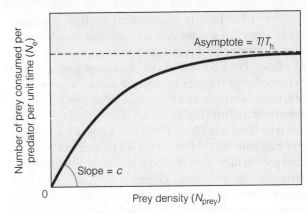

Figure 1 Relationship between prey density (*x*-axis) and per capita rate of prey consumed (*y*-axis) for a functional response that includes both search (T_s) and handling (T_h) time, where total time $T = T_s + N_eT_h$. At low prey density, the amount of prey consumed is low, as is handling time. As prey density rises, more prey are consumed and more of total foraging time is spent handling prey, reducing the time available for searching. As handling time approaches total foraging time, the per capita rate of prey consumed approaches an asymptote. The result is a type II response. Handling time also contributes to a type III response.

We rearrange the left-hand side of the equation:

$$N_e(1 + cN_{prey}T_h) = cN_{prey}T$$

We divide both sides of the equation by ($1 + cN_{prey}T_h$):

$$N_e = \frac{cN_{prey}T}{1 + cN_{prey}T_h}$$

Now we can plot the relationship between N_e and N_{prey} for a given set of values of c, T, and T_h (Figure 1; recall that c, T, and T_h are all assumed constant). In this relationship, which describes a *type II functional response*, the per capita predation rate (N_e) increases in a decelerating fashion up to a maximum at some high prey density (see Figure 14.2b). N_e approaches an asymptote as a result of the more complex time budget of a type II predator. Recall that total foraging time (T) includes both searching and handling ($T = T_s + N_eT_h$). As the number of prey captured during a time period (T) increases, the total handling time ($N_{prey}T_h$) increases, decreasing the time available for searching (T_s). The result is a declining prey mortality rate with increasing prey density.

1. Consider a lynx and two prey: a snowshoe hare and a ruffed grouse (a large, almost flightless bird). What factors might affect (i) searching and (ii) handling time for these prey items?
2. Given the relationship in Figure 1a of Quantifying Ecology 14.1, which functional response type is assumed by the Lotka–Volterra predation model? What assumption does this entail about total foraging time?

In a *type I functional response*, the prey consumed per predator increases linearly with increasing prey density. When expressed as a proportion of prey density, predation rate is constant and independent of prey density (Figure 14.2a). Type I characterizes **filter feeders** that extract prey from a constant volume of water that washes over their filtering mouthparts—a common behaviour of aquatic feeders from zooplankton to baleen whales. For a given flow rate, the rate of prey capture is a direct function of the prey density per unit of water volume, as with the marine copepod *Calanus pacificus* feeding on *Coscinodiscus angstii* (Figure 14.3a) (Frost 1972). Most spiders are type II feeders (Wise 1995), but a type I response can occur in web-building spiders, which are passive predators that intercept prey in their web. In field tests, the web-building spider *Grammanota trivittata*, common in North American marshes, exhibited a classic type I response when consuming planthoppers, *Prokelisia* spp. (Denno et al. 2004).

A type I response may occur in active predators if prey levels are low. The European kestrel (*Falco tinnunculus*) consumes voles (*Microtus* spp.) in linear proportion to their availability over the range of densities encountered (Korpimäki and Norrdahl 1991).

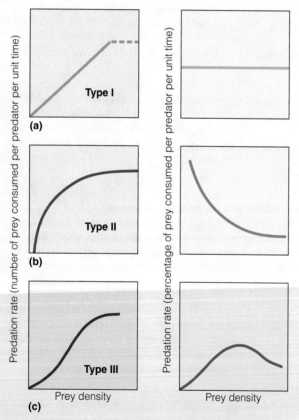

Figure 14.2 Three types of functional responses. **(a)** Type I: The number of prey taken per predator increases linearly with prey density. Predation rate is a constant proportion of prey density. **(b)** Type II: Predation increases at a decreasing rate to a maximum. Predation rate declines as a proportion of prey density with increasing prey density. **(c)** Type III: Predation rate increases in a non-linear, sigmoidal fashion, approaching an asymptote. Predation rate as a proportion of prey density is low at low prey density, and rises to a maximum before declining.

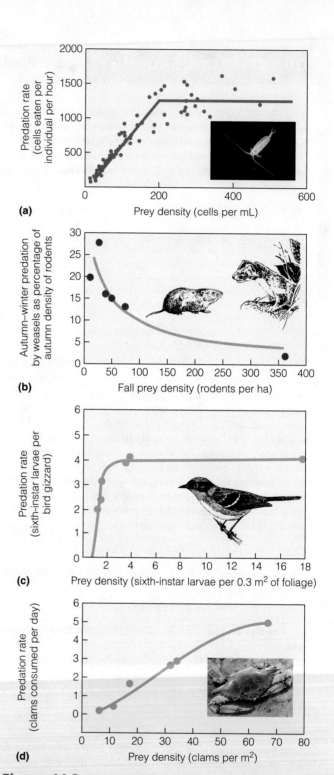

Figure 14.3 Examples of functional responses. **(a)** Response of the marine copepod *Calanus pacificus* (a zooplankton filter feeder) feeding on *Coscinodiscus angstii*. **(b)** Response of weasels to rodent density in deciduous forests in Poland. The *y*-axis depicts mortality rate as a proportion of prey density. **(c)** Response of bay-breasted warblers feeding on spruce budworm. **(d)** Blue crabs feeding on *Mya* clams.

(Adapted from (a) Frost 1972, (b) Jedrzejewski et al. 1995, (c) Mook 1963, and (d) Seitz et al. 2001.)

However, in a review of 814 functional responses from 235 studies, Jeschke et al. (2004) argue that a type I response is restricted to filter feeders because they are the only consumers that have minimal handling time and can perform other activities (such as migration) while foraging.

More common in terrestrial predators is a *type II functional response*, wherein prey consumed per predator increase at a decreasing rate to a maximum. Expressed as a proportion of prey density, predation rate declines continuously with increasing prey density (Figure 14.2b), as in the response of weasels to rodents in European forests (Figure 14.3b) (Korpimäki and Norrdahl 1991; note that mortality is expressed as a proportion of prey density). Goss-Custard et al. (2006) conclude that type II is the major functional response of shorebirds eating macroinvertebrates. It is also common in invertebrate predators. Three species of lady beetles (coccinellids) showed type II responses to densities of aphids, their major prey (Pervez and Omkar 2005).

Why does a type II curve level off? One reason relates to the predator's time budget, which (unlike a type I predator's budget) is divided between searching and handling. As more prey are captured, the total handling time (for chasing, killing, eating, and digesting) increases, decreasing the time available for searching. The result is a declining prey mortality rate with increasing prey density. At high prey densities, the predator cannot process prey any faster. If prey density increases further, the predator continues consuming at this maximum rate but cannot surpass it.

The *type III functional response* is the most complex and is common in generalists that consume many species. At high prey density, a type III response resembles a type II response, and for the same reason: increased handling time at high prey consumption rates prevents a further increase in rate. However, predation rate is much lower than a type II response at low prey densities, and increases in a sigmoidal (S-shaped) fashion. Expressed as a proportion of prey density, predation rate rises to a maximum and then declines to an asymptote (Figure 14.2c). A classic type III response involves bay-breasted warblers (*Dendroica castanea*) consuming spruce budworm (*Choristoneura fumiferana*) in New Brunswick (Figure 14.3c) (Mook 1963). The blue crab (*Callinectes sapidus*) consuming the clam *Mya arenaria* is an example of an invertebrate type III predator (Figure 14.3d) (Seitz et al. 2001).

Ecologists have studied the type III response extensively, as it has the most potential to regulate prey numbers. Why? Predation rate (and hence prey mortality) increases with prey density, instead of staying constant or decreasing, as in types I and II, respectively. Thus, when prey are scarce, mortality is negligible, allowing them to recover, but as prey numbers increase, prey mortality increases in a density-dependent fashion. This regulating effect is limited to the range of prey densities over which mortality increases. If prey density exceeds this range (i.e., if prey numbers increase faster than the predator can increase its consumption of them), the prey mortality rate declines. The prey then escape the regulatory effect of their predators. This situation is common; insect prey often increase faster than their vertebrate predators can keep up with them.

Holling's classification of functional responses of predators has stimulated ecologists to explore how the behaviour of predator and prey influences predation rate and hence predator–prey population dynamics. Because the model explicitly incorporates a predator's time budget, its framework has been expanded to examine questions about foraging efficiency (see Section 14.6).

14.4 SWITCHING: Choosing Alternative Prey Is One of Several Factors Underlying a Type III Response

Because the type III functional response is vital to density-dependent population regulation, we now examine its possible contributing factors: (1) availability of cover for prey, (2) formation of a search image, and (3) ability of the predator to switch between prey.

Availability of cover as a refuge for prey is critical. If the habitat provides few hiding places, it may protect most of the prey population when its numbers are low, but the risk of predation will increase as prey become more abundant. Without cover, prey become more obvious, causing the rate of prey consumption to increase in a non-linear fashion. Laboratory tests that do not simulate realistic habitat underestimate this factor.

Formation of a search image (Tinbergen 1960) also contributes to a type III response. When a new prey species enters an area, or when an existing prey species is scarce, its risk of being preyed upon is minimal. The predator does not encounter it often enough to form or reinforce a **search image**—a way to recognize a species (using visual, auditory, olfactory, or other cues) as suitable prey. Once the predator captures an individual, it may identify the species as desirable. The predator then has an easier time locating others of the same kind. The more adept the predator becomes at securing a preferred prey, the more it concentrates on it. In time, the number of prey becomes so small or its population so dispersed that encounters between it and the predator lessen. The search image fades, allowing the prey to recover.

A third factor underlying a type III response is **switching** or *frequency-dependent predation*, wherein a generalist predator turns to an alternative, more abundant prey. Although a predator may strongly prefer certain prey, it may switch to more abundant species that provide more profitable hunting. So, if rodents are more abundant than rabbits, foxes and hawks will concentrate on rodents. The predator pays little attention to the scarcer species, unless and until its relative abundance increases. Food preferences influence how abundant a prey must be before a predator switches. A predator may hunt longer and harder for a preferred species before turning to a more abundant but less preferred alternative. Conversely, the predator may switch from the less desirable species at a higher abundance than it would from a preferred species.

How is switching detected? The proportion of a prey item in a predator's diet is compared with its proportional availability in the habitat (Figure 14.4a, p. 286). If no switching occurs, the proportions are equal, as indicated by the straight line. Consider a type I filter feeder. If a plankton species makes up 15 percent of the total plankton community, then that species will make up 15 percent of the filter feeder's diet. If the

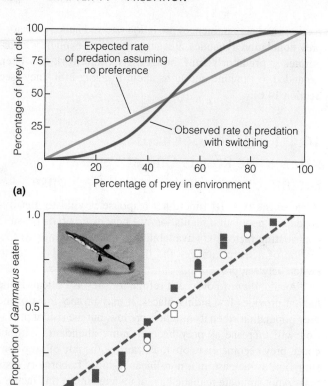

(a)

(b)

Figure 14.4 Prey switching. **(a)** The straight line represents the absence of switching. Prey are eaten in proportion to their availability. The curved line represents switching. At low densities, the proportion of the prey in the predator's diet is less than expected based on availability. Over this range of prey abundance, the predator selects alternative prey. At high prey density, the predator takes more of the prey than expected. Switching occurs where the lines cross. **(b)** Prey switching in the 15-spine stickleback. The proportion of *Gammarus* in the diet is plotted as a function of the proportion available. The dotted line represents the no-switching (frequency-independent) line. Boxes and circles represent different trials. Closed and open symbols denote trials with increasing and decreasing availability, respectively, of *Gammarus*.

(Adapted from Hughes and Croy 1993.)

species increases to 65 percent, it will make up 65 percent of the diet. But for a predator with a type III response, these proportions deviate. At low prey density, the proportion of that prey in the diet is lower than expected based on its availability, falling below the line. But as the prey species becomes more abundant and the predator switches to it, it becomes a higher proportion of the diet than expected and is above the line.

An experimental study demonstrated a switching response in predation by the 15-spine stickleback (*Spinachia spinachia*) on two prey: the amphipod (*Gammarus locusta*) and brine shrimp (*Aremia* sp.) (Figure 14.4b) (Hughes and Croy 1993). A combination of an altered attack efficiency and formation of search images caused the observed switching pattern.

Are these factors affecting switching independent, or do they interact? Cover helps a preferred prey avoid detection,

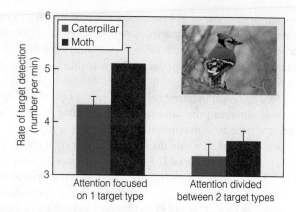

Figure 14.5 Impact of divided attention on prey detection by blue jays. When searching for a single prey, jays are more effective than when searching for two prey simultaneously. Blue bars represent an image of a caterpillar, red bars a moth.

(Adapted from Dukas and Kamil 2001.)

delaying switching until prey numbers rise to the point that they are more obvious. Switching is conditional on diet preference and availability, but also requires formation of an effective search image. There are biological limitations on this ability. In an unusual study, blue jays (*Cyanocitta cristata*) were exposed to digital images of caterpillars and moths with cryptic colouration (see Section 14.9). As corvids, blue jays are very intelligent, but when their attention is divided between two items, they are less effective at detecting prey than when focusing on a single prey (Figure 14.5) (Dukas and Kamil 2001). Limited attention constrains the predator's ability to form a search image, as new information interferes with information already stored in memory. Thus, predators searching for cryptic prey tend to focus on a single prey rather than switching to equally cryptic prey, even when the latter are abundant and of high quality.

As well as exerting a density-dependent influence on prey, switching can have community effects on species composition if the predator is a keystone species (see Section 16.4).

14.5 NUMERICAL RESPONSE: Predator Numbers May Increase with Prey Density

A functional response quantifies the change in a predator's per capita consumption, whereas a numerical response involves a change in predator numbers. As prey density increases, predator numbers often increase, after a time lag. The delay varies, but it reflects the fact that a numerical response involves one or both of two processes that cannot occur instantaneously: (1) predator reproduction (as quantified by the conversion factor b in the Lotka–Volterra equation) and (2) predator dispersal into areas of high prey density—an **aggregative response** (Figure 14.6). Aggregative responses greatly affect a predator's ability to regulate prey. Predators often reproduce more slowly than their prey, so an aggregative response shortens the time lag of a numerical response compared with an increase in predators that results from reproduction alone.

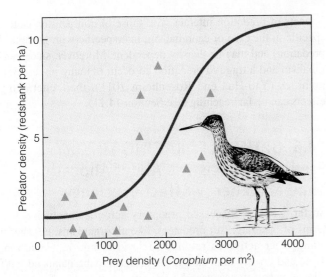

Figure 14.6 Aggregative response. Redshank (*Tringa totanus*) density increases with that of its arthropod prey (*Corophium* spp.).

(Adapted from Hassell and May 1974.)

Like most vertebrate predators, bay-breasted warblers (*Dendroica castanea*) increase slowly compared with their invertebrate prey. Mook (1963) quantified the response of warblers to a spruce budworm (*Choristoneura fumiferana*) outbreak in New Brunswick (see Ecological Issues: Science and Politics Interact in Spruce Budworm Control in Maritime Canada, pp. 400–401). The warblers exhibit a type III functional response (see Figure 14.2c) but also respond numerically (Figure 14.7a). At first, the response is aggregative. More nesting pairs enter the area, increasing the density of breeding territories. After a time lag, reproduction increases. The result is an increase in the total prey eaten per unit time by the warbler population (as opposed to per capita consumption) (Figure 14.7b): Total prey consumed = number of predators (numerical response) × prey consumed per predator (functional response). These combined responses regulate spruce budworm numbers—up to a point. As budworms increase further, mortality from predation drops and the predator's regulatory ability is outstripped by the prey's reproductive output.

Reproduction plays an even larger role in the numerical response of predators that have a similar reproductive time lag as their prey. Two rodents, the yellow-necked mouse (*Apodemus flavicollis*) and the bank vole (*Clethrionomys glareolus*), exploded in numbers in Poland's Biatowieza National Park in the 1990s. A heavy seed crop stimulated them to breed through the winter, raising their densities from 28 to 74 to nearly 300 per hectare, after which they dropped sharply (Figure 14.8, p. 288) (Jedrzejewski et al. 1995). Weasel (*Mustela nivalis*) density closely tracked that of their prey. Because all the species are small mammals, their time lags for gestation are similarly short. Weasels breed in the spring, and when food is abundant they have one large or two smaller litters, which in turn breed in their first year. In response to the rodent explosion, weasel density grew from 0.045 to 0.102 per hectare, dropping to 0.008 per hectare after the rodents crashed (Jedrzejewski et al. 1995). In

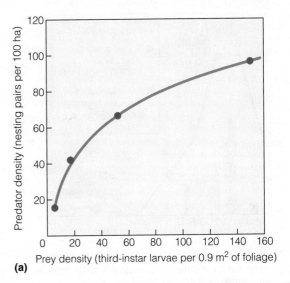

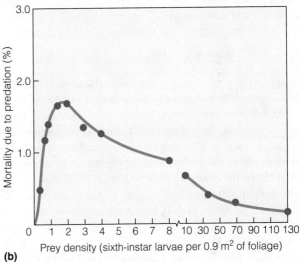

Figure 14.7 Response of bay-breasted warblers to spruce budworm in New Brunswick. **(a)** In a numerical response, predator density (*y*-axis) increases with prey density (*x*-axis). **(b)** Prey mortality rate (*y*-axis) expressed as a function of prey density (*x*-axis). Mortality rate was estimated by combining the numerical response in (a) with the functional response in Figure 14.3c.

(Adapted from Mook 1963.)

contrast, the reproductive time lag for the bay-breasted warbler is much longer than for its insect prey, making the aggregative component of its numerical response more critical.

Predator regulation of prey is far from simple. Moose (*Alces alces*) density is highly variable, both spatially and temporally. Wolves (*Canis lupus*) are their main predator. Messier (1994) analyzed 27 studies (mostly in the Yukon Territory and Alaska) where moose were the wolves' main prey. At low moose densities, predation is strongly density dependent (Figure 14.9, p. 288). Why is wolf predation so effective at reducing moose population growth at low densities, when we might expect predation risk to be lower? The answer lies in moose behaviour. As solitary, non-migratory herbivores, moose are well dispersed in their habitat year-round. Their size makes them more conspicuous than most prey at low density. At high moose

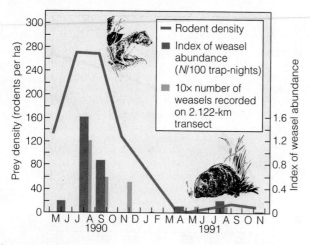

Figure 14.8 Numerical response of weasels to rodents. The *x*-axis represents the month, the *y*-axes rodent density (left) and predator density (right). Brown bars represent data from live trapping, orange bars data from observations and radio tracking. The weasel reproductive rate is high enough to track prey increases closely.

(Adapted from Jedrzejewski et al. 1995.)

density, wolf predation has an inversely density-dependent effect, suggesting that other factors regulate moose density.

As Messier notes, many factors complicate the moose–wolf interaction, including bear predation on moose calves. Habitat quality is also critical. In unproductive peat bogs, moose numbers will be low, making regulation by wolves more likely. Given the changing distributions of ungulates that result from human activities and climate change, the response of wolves to alternative prey is key. As deer become more abundant at high latitudes, they may reduce the functional response of wolves to moose, weakening its regulatory impact. Or, alternative prey may increase total predation by enhancing the wolf's numerical response, as has occurred in Wood Buffalo National Park in response to bison (Joly and Messier 2000).

The role of predation is complicated still further in situations involving cannibalism and/or **intraguild predation** (i.e., among members of a **guild**, a group of species that have a similar ecological function; see Section 16.6). In both cases, the impacts of predation interact with those of competition (intraspecific in the case of cannibalism, interspecific with intraguild predation) and may be density dependent. Moreover, since cannibalism and intraguild predation co-occur in many insect communities (Yao-Hua and Rosenheim 2011), their interacting effects can be far-reaching (see Section 14.11).

14.6 OPTIMAL FORAGING THEORY: Predators' Decisions About Allocating Time and Energy Affect Efficiency

So far, we have discussed predatory activities exclusively in terms of foraging. But predators, like all organisms, engage in many other activities relating to survival and reproduction. Time and energy allocated to foraging must be balanced with allocation to defence, searching for mates, or caring for young. These trade-offs between conflicting demands have spawned **optimal foraging theory**: the theory that natural selection favours efficient foragers, that is, individuals that maximize their energy or nutrient gain per unit of foraging effort.

Predators must make many decisions regarding foraging—what to eat, and where, how, and how long to search for it. Optimal foraging theory evaluates these decisions in terms of trade-offs. Costs are estimated in terms of time and energy expended, and benefits (ideally) in terms of fitness. As it is often difficult to quantify the consequences of a decision on fitness directly, benefits are often inferred from energy gain. Much foraging research focuses on what a predator eats from the choices available. Central to this decision is the predator's time budget. Recall that total foraging time (T) is divided into two activities: searching (T_s) and handling (T_h). Quantifying Ecology 14.2: Predator Time Budgets (p. 283), examined how foraging time affected functional response; now we consider its implications for foraging efficiency.

Consider a predator in a habitat containing two prey items: P_1 and P_2. Assume that the prey yield E_1 and E_2 units of net energy gain (benefits) and require T_{h_1} and T_{h_2} seconds to handle (costs). The profitability of the two prey is defined as the net energy gained per unit handling time: E_1/T_{h_1} and E_2/T_{h_2}. Now suppose that P_1 is more profitable than P_2: $E_1/T_{h_1} > E_2/T_{h_2}$. Optimal foraging theory predicts that the predator should prefer P_1. For example, the pied wagtail (*Motacilla alba*) feeds on beetles and dung flies in pastures in England, but prefers medium-sized prey (Figure 14.10a). Prey selected reflects the optimal-sized prey the birds can handle profitably: E/T_h (Figure 14.10b) (Davies 1977). Wagtails ignore smaller prey that, although easy to handle (low T_h), return minimal energy, and larger prey that require too much time and effort to handle relative to the energy gain. Such *size-selective predation* is common.

Optimal foraging theory implies that the predator should always choose the most profitable item. Are there ever situations where it is in the predator's interests to choose less profitable prey? Suppose that while searching for an individual of its most profitable prey, P_1, the predator encounters an individual of a less profitable prey, P_2. Should it eat it or continue searching for another P_1 item? The answer depends on the search time (T_s). The details of this calculation and an example

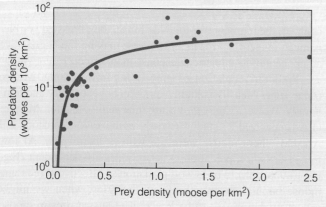

Figure 14.9 Relationship between moose and wolf density across 27 studies in North America.

(Adapted from Messier 1994.)

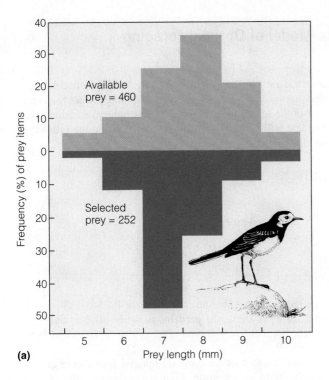

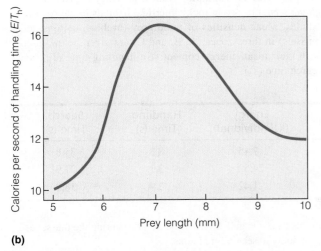

Figure 14.10 Optimal foraging in pied wagtails. **(a)** Frequency in the diet (*y*-axis) of prey of differing length (*x*-axis). Wagtails prefer mid-sized prey, which they take disproportionately. **(b)** Prey size chosen maximizes energy per unit handling time (E/T_h).

(Adapted from Davies 1977.)

are given in Quantifying Ecology 14.3: A Simple Model of Optimal Foraging, p. 290. Simply stated, the advisability of consuming the P_2 item considers not only the relative energy gain of the two items but also the extra time cost of searching for another P_1 item.

Numerous studies examine optimal foraging in many species and habitats, and patterns of prey selection generally follow the rules of efficient foraging. A landmark study on moose developed an optimization model for the foraging behaviour of this generalist herbivore. Diet selection by moose of various sexes and developmental stages was a better fit with an energy-maximization than a time-minimization strategy. Moreover, foraging

efficiency correlated with key reproductive parameters, such as size at weaning and at reproductive maturity (Belovsky 1978).

Activities other than foraging influence a predator's time budget, and factors other than energy content influence prey selection. One reason a predator consumes a varied diet is that it may not be able to meet its nutritional needs from a single prey species. Another reason—at least for predators that are not at the top of their food chain—is that pursuing a more energetically profitable prey may put the predator at greater risk of being itself preyed upon (see Section 14.8).

14.7 MARGINAL VALUE THEOREM: Predators Seek Out More Profitable Patches

Most predators occupy patchy environments. Some patches support high prey densities, others low densities or no prey at all. How should a predator allocate its time in patchy habitats? Again, optimal foraging theory suggests that the decision is based on maximizing profitability, where profitability is now defined as the net energy gain per unit time spent foraging in a given patch. The advantage of spending more time in profitable patches is obvious, but the question of when a predator should abandon one patch for another is more subtle. This question (which involves switching between habitat patches, not prey items) has generated a set of decision rules.

The **marginal value theorem** predicts that the length of time a predator (or herbivore) stays in a patch is related to the (1) richness of the food patch (prey density), (2) travel time needed to get there, and (3) time required to extract the prey resource (Charnov 1976). An initial time cost is needed to reach the patch (*t*). Once foraging begins, the energy gain per unit time in the patch is high. But over time, prey is depleted (N_{prey} declines) and the rate of gain drops, with cumulative gain (*G*) reaching an asymptote (Figure 14.11a, p. 291). At any point, the rate of energy return is defined as the cumulative value of *G* (*y*-axis) divided by the combined travel time (*t*) and foraging time (*T*; *x*-axis) (Figure 14.11b). So, how long should the predator forage in the patch?

Optimal foraging theory predicts that the predator should abandon the patch when the rate of energy gain is maximal, as represented by the tangent to the curve of cumulative gain (Figure 14.11c). Before this point, the rate of return is increasing; after it, the rate of return declines. The marginal value theorem also predicts that predators remain in a rich patch (i.e., with higher prey density) longer than in a poorer patch (Figure 14.12a, p. 291), and that for patches of the same quality, the time spent foraging (*T*) should increase with the time spent travelling to the patch (*t*; Figure 14.12b).

Do predators behave according to these predictions? As one of many examples, of which some but not all support the marginal value theorem, the parasitoid wasp *Trichogramma brassicae* exploits patches with varying densities of their host, the European corn borer (*Ostrinia nubilalis*). Females spend more time in higher-quality patches and reduce all patches to the same profitability level before leaving (Wajnberg et al. 2000).

QUANTIFYING ECOLOGY 14.3 | A Simple Model of Optimal Foraging

Faced with a choice of prey, predators must decide what items to eat, and where and how long to search for them. How are these decisions made? Do predators function opportunistically, pursuing prey as encountered, or do they pass up potential prey of lesser energy content to continue searching for more preferred prey? If the objective is to maximize energy gain per unit time, a predator should forage in a way that maximizes benefits (energy gained) relative to costs (energy expended). Maximizing net energy gain is the basis of optimal foraging theory.

Any food item has a benefit (energy content) and a cost (time and energy spent searching and handling). The benefit–cost relationship determines the item's profitability. Recall from Section 14.6 that the profitability of an item is the ratio of its energy content (E) to the time required to handle it (T_h) or E/T_h. Assume that a predator has two possible prey, P_1 and P_2, with energy contents of E_1 and E_2 kilojoules (kJ), handling times of T_{h_1} and T_{h_2} seconds, and search times of T_{s_1} and T_{s_2} seconds. Assume P_1 is more profitable (greater value of E/T_h).

As the predator searches for P_1, it encounters a P_2 individual. Should it capture P_2 or continue to search for another P_1 item? Which decision would maximize its energy gain? According to optimal foraging theory, the solution depends on the search time for P_1. The profitability of capturing and eating P_2 is E_2/T_{h_2}; and the profitability of continuing to search for, cap-

ture, and eat another P_1 item is $E_1/(T_{h_1} + T_{s_1})$. Notice that the decision to ignore P_2 and continue searching carries the extra cost of the search time (T_{s_1}) for P_1. The optimal decision—the one that will yield the most profit—is based on the following conditions:

If

$$\frac{E_2}{T_{h_2}} > \frac{E_1}{T_{h_1} + T_{s_1}}$$

then capture and eat P_2.

If

$$\frac{E_2}{T_{h_2}} < \frac{E_1}{T_{h_1} + T_{s_1}}$$

then ignore P_2 and continue to search for P_1. So, if the search time for P_1 is short, the predator is better off continuing to search, but if the search time is long, it is more profitable to consume P_2.

The benefit–cost trade-off for optimal prey choice is best grasped with an example. Glaucous-winged gulls (*Larus glaucescens*) forage in rocky intertidal habitats in the Aleutian Islands. Mean densities of three prey (urchins, chitons, and mussels) in three zones (A, B, and C) are given below, along with their mean energy content (E), handling time (T_h), and search time (T_s).

Prey Type	Density Zone A	Density Zone B	Density Zone C	Energy (kJ/individual)	Handling Time (s)	Search Time (s)
Urchins	0.0	3.9	23.0	7.45	8.3	35.8
Chitons	0.1	10.3	5.6	24.52	3.1	37.9
Mussels	852.3	1.7	0.6	1.42	2.9	18.9

In experiments where search and handling time were not a factor, chitons were the preferred prey and the optimal choice for maximizing energy gain. However, urchins are on average more abundant than chitons across the zones. If a gull happens on an urchin while hunting for chitons, should it capture the urchin or continue searching for chitons? The profit from capturing and consuming the urchin is $E/T_h = (7.45 \text{ kJ}/8.3 \text{ s})$ or 0.898 kJ/s, whereas the profit from ignoring it and searching, capturing, and consuming another chiton is $E/(T_h + T_s) = [24.52 \text{ kJ}/(3.1 \text{ s} + 37.9 \text{ s})]$, or 0.598 kJ/s. Because the profit from consuming the urchin is greater than that from searching for chitons, optimal foraging theory suggests that the gull should eat the urchin (Iron et al. 1986).

What if a gull foraging in zone A encounters a mussel? The profit from capturing and eating it is (1.42/2.9), or 0.490, while the gain from continuing to search for chitons remains at 0.598.

In this case, the gull would be better off ignoring the mussel and continuing to search for chitons.

We know what the gulls should do, but do they forage optimally, as predicted? If gulls were purely opportunistic, their selection would be in strict proportion to the relative abundance of prey. However, although their preferences for urchins and chitons reflected their profitability (E/T_h), gulls selected mussels less often than predicted by their relative E value (Irons et al. 1986).

1. How would reducing the energy content of chitons to 12.26 kJ influence the decision whether the gull should capture and eat the mussel or continue searching for a chiton?

2. Given that gulls cannot "calculate" energy returns when deciding whether to eat a prey item, how might natural selection lead to the evolution of optimal foraging?

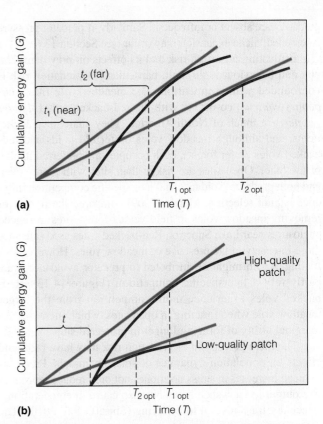

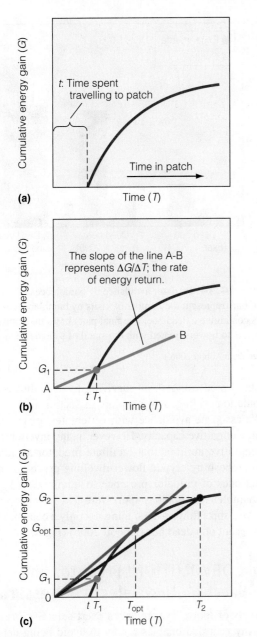

Figure 14.11 Marginal value theorem. **(a)** Total foraging time (T), including travel time to the patch (t), is on the x-axis; total energy gain (G) is on the y-axis. **(b)** For a given foraging time (T_1) the predator acquires an amount of energy (G_1). The slope of the line A-B from the origin to the point on the curve corresponding to $T_1 G_1$ represents the rate of energy gain for a predator that spends T_1 time in the patch. **(c)** The optimal time (T_{opt}) spent in a patch is the value of T at which gain is maximal, as depicted by the tangent to the curve (blue line). For values of T below or above T_{opt} the rate of gain increases or decreases, respectively, with increasing time in the patch.

Figure 14.12 According to the marginal value theorem, both time spent travelling (t) to a patch **(a)** and patch quality **(b)** influence foraging time (T_{opt}).

be preferable to remain in a less profitable but more secure area rather than visit a more profitable but predator-prone area.

Many studies demonstrate that the presence of predators influences foraging. Mixed-species flocks of willow tits (*Parus montanus*) and crested tits (*P. cristatus*) forage in spruce, pine, and birch in Finnish forests. Their main predator is the pygmy owl (*Glaucidium passerinum*), an ambush hunter. Its major food is voles, and when voles are plentiful the threat to tits declines. But when voles are scarce, tits become the owl's main prey—an instance of switching. During these high-risk times, tits forsake their preferred foraging sites on the outer branches. Adult willow tits and female crested tits forage in the inner protected parts of spruce trees. Male crested tits forage on the tops of the more open pine and leafless birches, where they command a good view of approaching predators, thereby protecting their young (Suhonen 1993).

As we have seen, where and how and for what a predator forages are vital niche dimensions. But negative niche dimensions—those that involve *not* engaging in a behaviour in order to minimize predation—represent a subtle aspect of niche space called **escape space** (Blest 1963) or **enemy-free space** (Holt and Lawton 1993). As well as hiding behaviour, the term *escape space* applies to any morphological or behavioural trait that reduces predation risk. Such decisions may restrict a prey species to suboptimal habitats. Predator–prey interactions are thus as likely as competition to cause niche compression. Indeed, they often interact, particularly if prey species compete for habitats that are suboptimal for other factors but are desirable because

14.8 PREDATION RISK: Foraging May Increase a Predator's Risk of Predation

Except for top-level predators, most predators are prey to other species and risk being preyed upon while foraging. Habitats vary in their foraging profitability and predation risk, and foragers must balance energy gain against the risk of being eaten. It may

predators are absent or infrequent. Similarly, if predator pressure is lessened, niche expansion may occur (see Section 13.6).

Evaluating predation risk and its effects on prey demography and behaviour is difficult, particularly if predation risk is confounded with competition. The meadow vole (*Microtus pennsylvanicus*) co-occurs with the red-backed vole (*Myodes gapperi*) in much of North America. They share several diet items, and although meadow voles prefer old fields and red-backed voles prefer forests, each occupies the preferred habitat of the other. Open sites are riskier than sites with more cover, and both predator avoidance and interspecific competition influence habitat selection. Morris (2009) compared the impact of removing meadow voles in field versus forest sites in fenced enclosures near Lake Superior. Red-backed voles used safe sites more intensely in the presence of meadow voles. However, this finding, which might be attributed to predator avoidance, may partly reflect interspecific competition (Figure 14.13). Red-backed voles experience more competition from the larger meadow vole when foraging in open sites, which increases the marginal utility of foraging longer in protected sites.

The mere perception of predation risk may have profound effects on population dynamics of small mammals. Predator-induced increases in stress hormones not only reduce reproductive output in snowshoe hares but also cause intergenerational effects by transmission to offspring (Sheriff et al. 2010). The

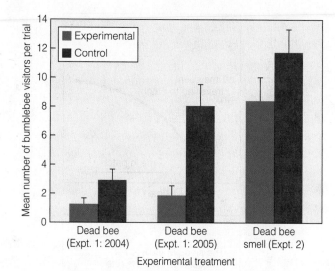

Figure 14.14 Predator avoidance by bumblebees. The first two pairs of bars represent the number of visits by bumblebees to rose flowers containing a dead bee. The final pair shows the number of floral visits to flowers treated with the smell of a dead bee.

(Adapted from Abbott 2006.)

impacts of predation risk on surviving females thus contribute to population cycling, as we explore in Section 14.13.

We associate avoidance with vertebrates, given their considerable cognitive capacity. However, many invertebrates also have cognitive abilities that facilitate predator avoidance. At risk of ambush by cryptic flower-dwelling predators, bees use indirect cues of predator presence to minimize risk. In field tests, bumblebees (*Bombus* spp.) were less likely to visit wild rose (*Rosa* spp.) flowers containing not only a dead bee but also just the scent of a dead bee (Abbott 2006) (Figure 14.14).

14.9 COEVOLUTION: Predator and Prey Exert Mutual Selective Pressures

As agents of mortality, predators exert selective pressure on prey. Any trait that enables a prey to avoid being detected or captured may increase its fitness. Thus, natural selection should produce "smarter," more evasive, prey. However, failure to capture prey increases predator mortality and decreases predator reproduction. So natural selection should also produce "smarter," more skilled, predators. Just as prey evolve traits to avoid being caught, predators evolve more effective ways of capturing prey. Predators will still catch prey individuals, but to survive as a species, the prey must present a moving (i.e., ever-evolving) target. This joint evolution of two or more non-interbreeding species with a close ecological relationship and involving reciprocal selective pressures is called **coevolution**.

Coevolution between predator and prey inspired Leigh Van Valen (1973) to propose the *Red Queen hypothesis*. In Lewis Carroll's *Through the Looking Glass*, there is a scene in which everything is always moving. No matter how fast Alice moves, the world about her stays still—to which the Red Queen remarks, "Now, here, you see, it takes all the running you can do, to keep in the same place." So it is with predator and prey. To avoid extinction by

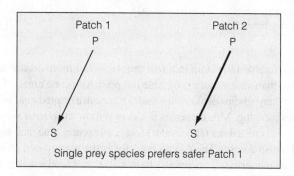

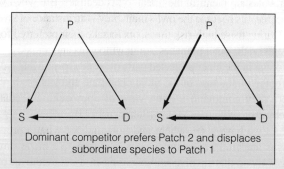

Figure 14.13 Apparent predation risk between two competing prey (S and D) with a common predator (P). Arrow thickness represents the magnitude of negative effects. Most studies of predation risk ignore competition and conclude that preferential use of safe patches (patch 1 in this case) indicates predation risk (upper panel). But if a dominant competitor (D) is present and prefers patch 2 (lower panel), then increased occupation of the safe patch by the subordinate (S) partly reflects competition with D as well as minimizing predation risk.

(© William Halliday and Douglas Morris.)

(a)

(b)

Figure 14.15 Chemical defences. **(a)** From a pair of glands on its back, the stinkbug discharges a volatile deterrent. **(b)** When confronted by a predator, the crested rat displays specialized hairs, which contain deadly cardiac glycosides acquired from the poison arrow tree.

predators, who are evolving ever more effective predatory strategies, prey must evolve ever more effective ways to avoid capture; they must keep "moving" just to stay where they are. This inevitable, never-ending, reciprocal process is a *coevolutionary arms race*—a concept explored in Section 8.4 with respect to intraspecific sexual selection (see Research in Ecology: Sexual Conflict and the Coevolutionary Arms Race, pp. 165–166). It can apply to any interspecific interaction, not only predation but also competition and mutualism, and can involve two species or many (*diffuse coevolution*).

Anti-Predator Defences: Prey Species Have Evolved Many Adaptations to Predators

As a result of this coevolutionary arms race, prey species have evolved **anti-predator defences**: traits that help prey avoid being detected, selected, and/or captured by predators. These traits are part of the escape space, and include one or more of the following:

1. *Chemical defences:* Some fish release *alarm pheromones* (chemical signals) that induce flight reactions in members of the same and related species. Arthropods, amphibians, and snakes employ odorous secretions to repel predators. The stinkbug (*Cosmopepla bimaculata*) discharges a volatile secretion from glands on its back (Figure 14.15a). The defensive role

of chemical emissions is often inferred rather than tested, but in controlled experiments, secretion by stinkbugs deterred feeding by both avian and reptile predators (Krall et al. 1999).

Although some arthropods as well as venomous reptiles and amphibians synthesize their own poisons, many acquire toxic substances by consuming plants and storing their toxins in their own bodies. For example, the monarch butterfly (*Danaus plexippus*) acquires a toxin by ingesting milkweed (*Asclepias syriaca*), which makes it unpalatable to predators. The only mammal known to acquire a toxin in this way is the African crested rat (*Lophiomys imhausi*) (Figure 14.15b). It acquires a deadly cardiac glycoside from chewing the bark of the poison arrow tree (*Acokanthera schimperi*; the compound is itself an anti-herbivore defence for the plant) and spreads the toxin-rich saliva onto its body, where it is drawn into specialized hairs that resemble those of a porcupine when displayed as a warning to predators (Kingdon et al. 2012).

2. *Colouration:* Many prey species use body colouring as a defence. **Cryptic colouration** involves colours and patterns that allow prey to blend into the background (Figure 14.16a). This trait is common in fish, reptiles, and ground-nesting birds. *Object resemblance*, a variant, is common in insects. Walking sticks (Phasmatidae) resemble twigs (Figure 14.16b), and katydids (Pseudophyllinae) resemble leaves. Some animals

(a)

(b)

Figure 14.16 Cryptic colouration. **(a)** Flounder (*Paralichthys* spp.) in shallow waters off eastern North America avoid detection by both predators and prey by changing colour and pattern to match sediments. **(b)** Walking sticks (Phasmatidae), which eat leaves, resemble twigs.

Figure 14.17 Flashing colouration serves as both an alarm and a distraction in white-tailed deer.

(a)

(b)

Figure 14.18 Warning colouration. The stripes of a skunk **(a)** and colouration of the strawberry poison dart frog **(b)** serve notice to predators.

have eyespot markings that distract or intimidate predators, or delude them into attacking a less vulnerable part of the body. **Flashing colouration** involves the display of highly visible colour patches when disturbed or put to flight, as in many butterflies, birds, and ungulates. It may distract and disorient predators, as with zebras, or, as with white-tailed deer, be a signal to promote group cohesion when confronted by a predator (Figure 14.17). When the animal is at rest, the bright or white colours vanish, and the animal's otherwise cryptic colouration allows it to blend into its surroundings.

Warning colouration uses bold colours or patterns that warn of a chemical defence, such as the stripes of a skunk (*Mephitis mephitis*) (Figure 14.18a) or the appearance of the strawberry poison dart frog (*Dendrobates pumilio*) (Figure 14.18b). (This trait, as in the bright orange of monarch butterflies and yellow and black markings of bees and wasps, is the basis of many mimicries.) The prey is protected by a negative search image (see Section 14.4) formed when the predator associates the prey's appearance with unpalatability or pain.

3. *Mimicry:* Some species have evolved to resemble other species. **Batesian mimicry** is evolution by an edible species of a resemblance to an inedible or toxic species. The edible *mimic* resembles the inedible *model*. After the predator learns to avoid the model, it avoids the mimic, too. Natural selection reinforces those traits of the mimic that resemble those of the model. To be an effective defence, the population size of the mimic must remain much lower than that of the model. Otherwise, the predator would be more likely to encounter the mimic and form a positive search image, causing it to seek out rather than avoid the mimic.

Batesian mimicry is most common in butterflies, notably the Monarch (*Danaus plexippus*) (Figure 14.19a), and its mimic, the Viceroy (*Limenitis archippus*), but it also occurs in snakes. In eastern North America, the non-venomous scarlet king snake (*Lampropeltis triangulum*) (Figure 14.19b) mimics the venomous eastern coral snake (*Micrurus fulvius*) (Figure 14.19c), and in the southwest, the mountain king snake (*Lampropeltis pyromelana*) mimics the western coral snake (*Micruroides euryxanthus*). Nor is mimicry limited to colour. Some snakes are acoustic mimics of rattlesnakes. The fox snake (*Elaphe vulpine*), pine snake, bull snake, and gopher snake—all

North American subspecies of *Pituophis melanoleucus*—vibrate their tails in leaf litter to produce a rattle-like sound.

A variant is *Müllerian mimicry*, wherein several toxic or unpalatable species evolve to resemble each other. Unlike Batesian mimicry, a predator forms a negative search image after feeding on any of the species. The black-and-yellow-striped bodies of social wasps, solitary digger wasps, and caterpillars of cinnabar moths warn predators that the organism is inedible (Figure 14.20). All are unrelated species with similar colouration.

4. *Structural defences:* Clams, armadillos, turtles, and many beetles withdraw into *protective armour* (coats or shells) when danger approaches. Porcupines, echidnas, and hedgehogs possess quills (modified hairs) that discourage predators.

5. *Behavioural defences:* Prey have evolved many behaviours (individual or group) that avoid detection, facilitate escape, or warn others. Animals may forage differently when predators are present (see Section 14.8). When a predator is sighted, some species emit high-pitched *alarm calls*. Such calls may communicate more specific information than once thought (see Research in Ecology: Sounding the Alarm in Richardson's Ground Squirrels, pp. 296–297). Alarm calls may induce flight, or mobilize large numbers of potential prey that mob the predator. An alarm sound can be non-vocal, such as the slapping of a beaver's tail. Some species use *distraction displays*. The kill-

Figure 14.19 Batesian mimicry. The colour pattern of the poisonous monarch butterfly **(a)** is mimicked by the Viceroy butterfly which is almost identical in appearance. Similarly, the non-venomous scarlet king snake **(b)** mimics the poisonous coral snake **(c)**.

deer (*Charadrius vociferus*) pretends to have a broken wing, directing attention away from the nest.

Group living is another behavioural defence. By maintaining a tight, cohesive group, prey make it hard for a predator to catch a victim (Figure 14.21, p. 298). Alternatively, sudden, explosive flight can confuse a predator, which may be unable to decide which animal to follow. In any group, those on the periphery are more at risk than those closer to the centre. For this reason, some group animals surround their young, but when a group is on the move, as in an ungulate herd, individuals who cannot keep up because they are either young or weak are at greatest risk.

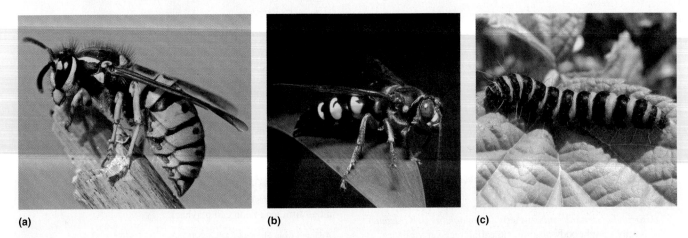

Figure 14.20 Müllerian mimicry. Black and yellow stripes of **(a)** social wasps (Vespidae), **(b)** solitary digger wasps (Sphecidae), and **(c)** cinnabar moth caterpillars (*Callimorpha jacobaeae*) warn predators that these organisms are inedible.

RESEARCH IN ECOLOGY   Sounding the Alarm in Richardson's Ground Squirrels

Jim Hare, University of Manitoba.

Nature documentaries often present anecdotal accounts of animal communication, in particular regarding the ability of some species to signal danger to members of their own species. Alarm signals take various forms, including volatile pheromones released by social insects. Among acoustic signals, some are non-vocal, such as the slapping of the beaver's tail. But many mammals emit alarm vocalizations; and in some groups these alarms are not only effective but complex. In collaboration with student co-researchers, Jim Hare has investigated animal communication in the context of mammalian social organization by studying signalling in Richardson's ground squirrels (Figure 1).

Ground squirrels are ideal subjects for field research in communication. As diurnal social animals, they inhabit populous colonies that are easy to monitor. They are easily handled and marked, and habituate well to humans. Care must be taken to minimize observer effects, but the colonies function in a relatively normal way under human surveillance. Early studies were strictly observational. Sound emissions were recorded in natural habitats in response to actual predator encounters. Results, though valuable, could only reveal correlation (see Section 1.6). To overcome this limitation, researchers introduce "model" predators that elicit similar responses as real predators. Recorded calls are then played back along with control calls, which—depending on the hypothesis tested—may include background noise or unmanipulated calls. The response measured is *receiver vigilance*, scored on a 4-point ordinal scale describing body position from prostrate to fully upright (see Figure 1).

Richardson's ground squirrels emit two major sounds: short "chirps" and long "whistles," often in response to avian and terrestrial predators, respectively. Hare and Jennifer Sloan described a third acoustic element—"chucks"—brief, low-amplitude trailing sounds that are correlated with the proximity of the caller to the perceived threat (Sloan et al. 2005). All three sounds are emitted as discrete elements separated by other sounds or (in repeated sequence) by silences appended to whistles. If these sounds are considered syllables, questions arise.

Does variation in the acoustic structure of alarm calls (which syllables are present) convey information? Chirp primary syllables elicit greater vigilance than whistles, but adding chucks to either chirps or whistles promotes greater and more lasting vigilance (Sloan et al. 2005).

Acoustic content can indicate the extent of a threat, but what about the role of syntax in repeated calls? (*Syntax* is the way in which acoustic elements are strung together to convey meaning.) To test for the effects of the order of acoustic elements, Hare and David Swan compared the response to playbacks of repeated calls with unaltered (controls) and randomized syllable order (treatment). Primary syllables acted as a general alert, priming the receiver for information to follow. This function persisted even when primary syllables were embedded in randomized sequences (Swan and Hare 2008). Vigilance varied with differing properties of the primary syllables, such as duration and frequency. Although these results do not establish syntax, they do indicate subtleties in ground squirrel communication.

Hare's work transcends the single caller–single receiver dynamic to explore the social dimension of ground squirrel communication. Having established that individuals not only (1) discriminate among callers (Hare 1998) but also (2) assess caller reliability (Figure 2) (Hare and Atkins 2001), Hare and Amy Thompson

Figure 1 Adult female Richardson's ground squirrel (*Urocitellus richardsonii*; formerly *Spermaphilus richardsonii*) in a fully upright body posture, indicating receiver vigilance.

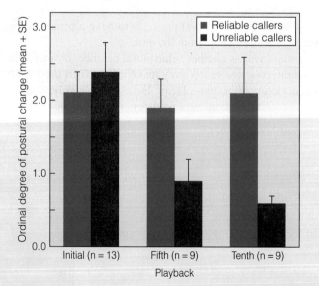

Figure 2 Degree of postural change (based on ordinal value of body position) of Richardson's ground squirrels to reliable callers (blue bars) versus unreliable callers (red bars) at the initial, fifth, and tenth call playback.

(Adapted from Hare and Adkins 2001.)

investigated receptiveness to multiple callers. In encounters with live predators, squirrels were more likely to track airborne than terrestrial predators with multiple caller bouts. To test whether recipients could perceive this information, they played back multiple caller bouts of either chirps or whistles from callers that were either familiar or unfamiliar to the recipients, in series that moved progressively towards or away from the receiver.

Overall, squirrels were more responsive to calls from familiar callers, but in response to chirps (generally associated with bird predators), receivers showed more vigilance when call bouts progressively increased in proximity (Thompson and Hare 2010). Ground squirrels thus not only respond to increased *quantity* of callers with increased vigilance (Sloan and Hare 2008), they also integrate *quality* of information from multiple callers to track avian predators. These findings suggest that alarm communication involves a coordinated social dimension.

Ground squirrels also produce ultrasonic signals (~ 50 kHz, "whisper calls") (Wilson and Hare 2004). Audible calls elicit more vigilance than ultrasonic calls, suggesting that ultrasonic calls are less important as warning signals. However, because ultrasonic calls do not travel as far and are more directional, they may give callers the selective ability to warn *kin* (more closely related individuals), which tend to be closer by, especially as juveniles. Fewer ultrasonic alarms were emitted, but the ratio of ultrasonic to audible alarms increased with distance from a threat stimulus (Figure 3) (Wilson and Hare 2006). As predators are less likely to detect ultrasonic signals over greater distances, whisper calls allow callers to warn nearby squirrels

without increasing their own predation risk. Given that kin selection is likely the main benefit to the caller, the ultrasonic signal may be of great value during the juvenile stage, when squirrels are not only vulnerable but also clumped near related adults (Wilson and Hare 2006).

Beyond its insights into animal communication, Hare's research exemplifies the multi-faceted and open-ended nature of ecological research. Observations prompt questions; questions prompt experiments; results provide partial answers, while opening up fascinating new avenues for exploration.

Bibliography

Hare, J. F. 1998. Juvenile Richardson's ground squirrels (*Spermophilus richardsonii*) discriminate among individual alarm callers. *Animal Behaviour* 55:451–460.

Hare, J. F., and B. A. Atkins. 2001. The squirrel that cried wolf: Reliability detection by juvenile Richardson's ground squirrels (*Spermophilus richardsonii*). *Behavioral Ecology & Sociobiology* 51:108–112.

Sloan, J. L., and J. F. Hare. 2008. The more the scarier: Adult Richardson's ground squirrels (*Spermophilus richardsonii*) assess response urgency via the number of alarm signalers. *Ethology* 114:436–443.

Sloan, J. L., D. R. Wilson, and J. F. Hare. 2005. Functional morphology of Richardson's ground squirrel (*Spermophilus richardsonii*) alarm calls: The meaning of chirps, whistles and chucks. *Animal Behaviour* 70:937–944.

Swan, D., and J. F. Hare. 2008. The first cut is the deepest: Primary syllables of Richardson's ground squirrel, *Spermophilus richardsonii*, repeated calls alert receivers. *Animal Behaviour* 76:47–54.

Thompson, A., and J. F. Hare. 2010. Neighbourhood watch: Multiple alarm callers communicate directional predator movement in Richardson's ground squirrels, *Spermophilus richardsonii*. *Animal Behaviour* 80:269–275.

Wilson, D. R., and J. F. Hare. 2004. Ground squirrel uses ultrasonic alarms. *Nature* 430:523.

Wilson, D. R., and J. F. Hare. 2006. The adaptive utility of Richardson's ground squirrel (*Spermophilus richardsonii*) short-range ultrasonic alarm signals. *Canadian Journal of Zoology* 84:1322–1331.

1. Why do the controls vary among experiments? For example, to test for the effect of syntax, the control consisted of calls in which syllable order was unaltered. What would the control be for testing the meaning of chucks?

2. Assume that, by emitting an alarm, a ground squirrel increases its predatory risk (by indicating its location) with no possible increase in fitness for itself or its offspring. From an evolutionary perspective, what explains such altruistic behaviour (see Section 11.8)?

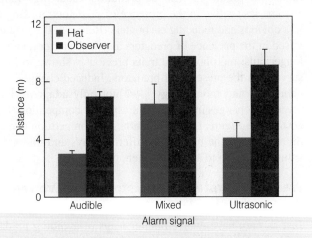

Figure 3 Mean distance (m) ± 1 standard error from a threat stimulus (blue bars depict response to a non-living stimulus, a hat; red bars depict response to a human observer) at which 103 Richardson's ground squirrels produced audible (n = 87), mixed (n = 6), and ultrasonic (n = 10) alarm signals.

(Adapted from Wilson and Hare 2006.)

Figure 14.21 Musk ox (*Ovibos moschatus*) form an outward-facing circle when threatened.

Figure 14.23 A dragonfly nymph feeding on a grey tree frog tadpole.

6. *Reproductive traits:* It may seem counterintuitive, but a prey may defend itself by providing its predator with an excess of food. **Predator satiation** involves timing of reproduction such that most offspring are born in a short period. Prey are then so abundant that predators can take only a fraction, allowing most offspring to escape and grow to a less vulnerable size. Periodic cicadas (*Magicicada* spp.) live most of their lives as nymphs underground, emerging as adults every 13 and 17 years in southern and northern portions, respectively, of their North American range. A local population emerges yearly somewhere in their range. Upon emergence, local densities can reach millions per hectare. The first cicadas to emerge are eaten by birds, but avian predators quickly became satiated. Birds consume 15 to 40 percent of the population at low cicada densities, but eat a much smaller proportion as densities increase (Figure 14.22) (Williams et al. 1993).

Anti-predator defences are either (1) **constitutive** (fixed traits such as object resemblance or warning colouration) or (2) **induced** (brought about by the presence or activity of predators, such as behavioural defences). Some chemical defences, such as alarm pheromones, are also induced, although the ability to synthesize them is heritable, whereas chemical defences that are always present are constitutive. Induced defences also include shifts in physiology or morphology.

It is tempting to view these categories as mutually exclusive, with constitutive defences representing the impact of selection and induced defences representing phenotypic plasticity. But many prey species have evolved more than one constitutive and/or induced defence, and the impacts of induced defences may interact with constitutive defences as well as with interactions other than predation. For example, dragonfly nymphs (*Anax longipes*) prey on grey tree frog tadpoles (*Hyla versicolor*) (Figure 14.23). Because tadpoles detect the presence of dragonfly nymphs by water-borne chemicals, researchers can monitor the impact of predators without predation occurring. Tadpoles become less active in the presence of predators, making them less obvious and reducing encounters (Relyea 2002a). Tadpoles raised in the presence of predators also had shorter bodies with longer tails, morphological traits previously shown to improve survival in the presence of predators. Induced changes in both behaviour and morphology were not only adaptive, but the reverse of changes induced by intraspecific competition (Relyea 2002b; see Figure 11.6). Predation and competition can cause the realized niche to shift in different ways and interact with natural selection to alter prey phenotypes.

Predatory Strategies: Predators Have Evolved Different Strategies for Hunting Prey

Just as prey have evolved better ways of avoiding predators, predators have evolved better ways of hunting. Predators use three general strategies:

1. *Ambush* involves lying in wait and is typical of frogs, alligators, crocodiles, lizards, and some insects. Although ambush has a low frequency of success, it requires minimal energy and is advantageous for ectotherms than cannot sustain extended periods of aerobic activity.

2. *Stalking*, typical of herons and some cats, is a deliberate form of hunting with a quick attack. The predator's search time is great, but pursuit time is minimal.

3. *Pursuit*, typical of hawks, lions, wolves, and insectivorous bats, involves minimal search time because the predator

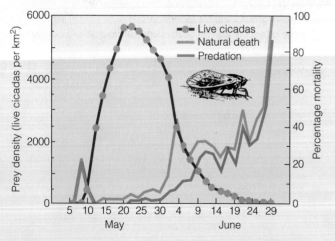

Figure 14.22 Predator satiation in cicadas. Estimated daily cicada density is on the left *y*-axis, estimated daily mortality from bird predation and natural causes on the right *y*-axis. At the height of predation, most cicadas escape predation.

(Adapted from Williams et al. 1993.)

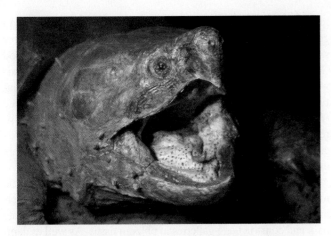

Figure 14.24 The alligator snapping turtle uses both cryptic colouration and aggressive mimicry to avoid detection and attract prey. Lying motionless on the bottom of a water body with its mouth open, it wiggles its worm-shaped tongue to attract and ambush prey.

usually knows the location of its prey, which are typically highly visible, but pursuit time can be considerable. Stalkers spend more time and energy searching for prey, whereas pursuers spend more time capturing and handling prey.

As well as hunting strategies, predators have evolved a range of adaptations relating to predation, including *cryptic colouration* to blend into the background or obscure their outlines (Figure 14.24). Some use **aggressive mimicry**; by evolving to resemble their prey, predators save energy by inducing their prey to approach them. For example, robber flies (*Laphria* spp.) mimic their bumblebee prey (Figure 14.25). Female fireflies (Lampyridae) imitate the mating flashes of other firefly species to attract males of those species, which she then kills and eats. Predators may also employ chemical poisons, as do venomous snakes, scorpions, and spiders. Other predators, such as wolves, use group behaviour to hunt prey. A highly specialized adaptation involves **echolocation**: the use of sound waves to determine prey location, as with insectivorous bats.

14.10 HERBIVORY: Herbivory Has Lethal and Non-Lethal Effects on Plants

Although the term *predator* typically refers to animals that feed on other animals, *herbivory* is a form of predation in which animals prey on autotrophs (plants and algae). Unlike most predators, herbivores typically do not kill their prey. Given that the ultimate source of energy for virtually all heterotrophs is carbon fixed by autotrophs, herbivory is critical to all communities.

On land, the amount of biomass consumed by herbivores is often small—only 6 to 10 percent of forest biomass, or 30 to 50 percent in grasslands (see Section 20.10). During outbreaks of herbivorous insects or when large herbivores are abundant, consumption is often much higher (Figure 14.26). But consumption is not the only measure of the impact of herbivory. Removal of tissue—leaves, bark, stems, roots, or sap—may affect a plant's survival, even if the plant is not killed outright. Loss of foliage or roots reduces vigour as well as biomass, potentially putting

Figure 14.25 The robber fly employs aggressive mimicry by resembling its bumblebee prey.

the plant at a competitive disadvantage. Ultimately, herbivory may lower plant reproductive effort and hence fitness. Herbivory has the most impact in juvenile stages, when the plant is typically most vulnerable and least competitive.

Given their modular growth, plants may respond to herbivory by **compensatory growth**, production of biomass that may equal or even exceed the lost tissue (McNaughton 1983). Ecologists have tested compensatory growth in many studies, using real or simulated grazing at differing intensities and frequencies on both herbaceous and woody species. Although it occurs in some species, particularly grasses, and seems clearly

(a)

(b)

Figure 14.26 **(a)** Impact of intense herbivory on oaks by gypsy moths in eastern North American forests. **(b)** Contrast between heavily grazed grassland in southeast Africa and an adjacent area where large herbivores have been excluded.

adaptive, the regrowth is vulnerable to ongoing herbivory. In fact, Crawley (1983) argues that because it mainly benefits herbivores, compensatory growth should only be selected for in species that experience herbivory seasonally or intermittently (as with migration of large grazers), allowing recovery. Its adaptive value has a broader community dimension, by allowing some species to coexist with superior competitors (Järemo and Palmqvist 2001). Herbivory would then interact with competition among plant prey—much as predator avoidance can interact with competition among animal prey.

Modelling studies suggest that many functional processes interact to influence the short-term impact of grazing on *primary productivity* (energy processed by primary producers; see Section 20.2). Biomass removal has both positive and negative, and direct and indirect, impacts on light absorption, available soil water, plant nitrogen content, and root/shoot allocation (Leriche et al. 2001). These factors vary in importance, with nitrogen content and root/shoot allocation having more impact than extrinsic abiotic conditions such as light and moisture. Grazing history is a critical intrinsic factor affecting compensatory growth. In African savannahs, where grazing has exerted ongoing selective pressure, biomass yield was maintained under moderate clipping frequency, suggesting that compensation (not overcompensation) is occurring (Leriche et al. 2003; ecologists use clipping to simulate grazing). In contrast, simulated grazing at all frequencies reduced aboveground primary production in a previously ungrazed wet tundra community in the High Arctic, even though nitrogen availability was enhanced at moderate clipping frequency (Elliott and Henry 2011; see Section 28.6).

Herbivory may damage plants by the loss of nutrients, depending on the age and amount of tissue removed. These effects may be critical in infertile soils. As dependent structures, young leaves are importers of minerals from reserves in roots and stems. Grazers, both vertebrate and invertebrate, often prefer young leaves because they are lower in cellulose and lignin. By selectively feeding on new growth, grazers reduce the plant's nutrient stores. Modularity allows many defoliated plants to respond with a flush of new growth. This response facilitates survival, but only by draining reserves that would otherwise be allocated to growth and reproduction. For example, herbivory by longhorn beetles (*Tetraopes* spp.) reduced milkweed (*Asclepias* spp.) biomass and fruit production by 20 to 30 percent (Agrawal 2004).

Effects on trees can be dramatic. If defoliation is complete, as often happens in outbreaks of gypsy moths (*Lymantria dispar*) or cankerworms (*Alsophila pometaria*), the new leaves are smaller, and the leaf canopy is much reduced. Regrown tissues may be immature at the onset of winter, compromising tree survival. Weakened trees are also more vulnerable to insects and disease. So even if a forest appears fully recovered from defoliation, it can suffer delayed damage and mortality from other stresses.

Severe defoliation may kill conifers, which are less able than hardwoods to produce new growth on old growth. Susceptibility varies with the species growth pattern. Height growth of northern pines such as white and red pine (*Pinus strobus* and *P. resinosa*) comes from elongation of buds formed the previous season. They respond poorly to defoliation, especially from large browsers that may remove more than the current year's

growth. In contrast, balsam fir (*Abies balsamea*), a common conifer in late successional boreal forests and a major food for spruce budworm, can produce late-season buds on old wood (Batzer 1973), facilitating its survival and recovery after budworm defoliation (Piene and Eveleigh 1996). Artificial defoliation treatments of another major boreal species, white spruce (*Picea glauca*), indicate that young (12-year-old) trees can also respond with compensatory growth. However, biomass recovery after 3 years was poorest when foliage from previous years as well as the current year was removed (Piene 2003).

Unlike grazers, browsers such as deer and mice consume the nutrient-rich growing tips of woody plants, killing them or altering their form into a more branched habit. Bark beetles penetrate bark and construct egg galleries in phloem tissue. As well as damage from larval and adult feeding, bark beetles may introduce lethal pathogens. Elm bark beetles carry the fungal agent of Dutch elm disease (*Ophiostoma* spp), responsible for the demise of the American elm (*Ulmus americana*). Similarly, the mountain pine beetle (*Dendroctonus ponderosae*) carries the blue-stain fungus (*Grosmannia clavigera*), a serious pathogen of pines in western North America (see Section 15.5). The fungus colonizes sapwood and disrupts water flow to the crown, hastening death. Genomic studies show that it circumvents the tree's antifungal compounds (resins), using them as a carbon source (DiGuistini et al. 2011).

Some herbivores, such as aphids and sapsuckers (Figure 14.27), do not consume plant cells but instead tap the sugar-rich fluid in phloem tissue, especially in new growth and young leaves. Sap-sucking insects can greatly decrease the growth and biomass of woody plants.

The impact of herbivory varies with growth form. Because the growing points of grasses are close to the ground, grazers first eat older tissue, leaving younger tissue with its higher nutrient content intact. This basal growth habit was once thought to reflect the selec-

Figure 14.27 The yellow-bellied sapsucker (*Sphyrapicus varius*), a type of woodpecker, is common in deciduous forests in northern North America.

tive impact of grazing, but grasses evolved under the combined (and interacting) influences of drought and fire as well as grazing (Anderson 2006). Whatever the origin of their basal habit, most grasses tolerate grazing, and may even be stimulated by it. Photosynthetic rate tends to decline with leaf age, and grazing can stimulate production by removing these older, less functional leaves and increasing light to the underlying younger leaves. In fact, some grasses remain vigorous only with moderate grazing, even though defoliation may reduce sexual reproduction and hence fitness. Intolerant grasses, in contrast, can be eradicated by heavy grazing.

Some effects of herbivory involve the entire community. Just as predators affect the animal community by their dietary preferences, so too can herbivores alter the species composition of the plant community. Herbivory gives a selective advantage to species with defence mechanisms. Given the basic trade-off between allocation to growth versus defence, herbivory may indirectly affect competitive outcomes and promote coexistence. Plants that invest in structural and chemical defences are often slow growing. Especially in open sites, they risk being competitively excluded by faster-growing plants whose tissue has less lignin and cellulose. Thus, herbivory, as a type of selective biotic disturbance, can help maintain community diversity.

Non-feeding effects of herbivores, particularly large grazers, can be substantial. Feces and other wastes are a local nitrogen source, although "burning" from concentrated urine can cause local die-off. There is often a ring of dead tissue at the point of contact with animal wastes, surrounded by an area of lush growth reflecting a fertilizer effect (Figure 14.28). Trampling and burrowing are influential, and not always negatively, as compaction by hooves can improve seed germination. Nor is trampling restricted to native grazers; humans, including researchers, have trampling effects on native vegetation. Experimental trampling did not affect germination of the dominant grass in North American tall-grass prairies, big bluestem (*Andropogon gerardii*), but it did depress both its above- and belowground growth (McNearney et al. 2002).

Ecotourism, considered a way of promoting ecosystem conservation, has unforeseen trampling effects. In tropical rain forests in Australia, compaction reduced the organic mat, increased soil density, and decreased soil permeability (Talbot et al. 2003). These alterations in soil properties affect not only hydrology but also germination and regrowth. Unlike grasses, tropical vegetation has not evolved under the selective pressure of ungulate herds, making it more vulnerable to trampling. In Costa Rican rain forests, diversity increased with distance from the centre of trails (Boucher et al. 1991). Trampling effects from recreational vehicles are substantial in temperate and tropical sites alike (Davenport and Switalski 2006). Depending on the frequency and intensity of use, boreal and tundra vegetation is damaged by off-road vehicles, snowmobiles, and tundra buggies, even if snow cover is abundant.

14.11 PLANT DEFENCES: Plants Have Evolved Adaptations to Deter Herbivores

Given their autotrophic lifestyle, plants do not need to move to obtain food. They remain rooted in soil, which is critical for obtaining water and minerals. Their sessile habit saves much energy, but it also exposes plants to herbivory. Modularity helps by allowing them to lose body parts without compromising survival, but reducing herbivory damage requires more specific traits. Not surprisingly, plants have evolved a wide array of such traits. As with animal defences, some plant defences are *constitutive* (built-in) whereas others are *induced* (stimulated by herbivore attack). They include the following:

1. *Structural defences*, such as hairy leaves, thorns, and spines, discourage feeding (Figure 14.29). As well as such specialized structures, cell walls toughened with lignin defend plants. For most herbivores, food quality is the main limitation. Given the difficulty of converting plant tissue into animal flesh, high-quality N-rich forage is preferred. Unless the herbivore harbours appropriate microbes, low-quality, fibrous foods are indigestible. Herbivores that rely on plants but lack digestive symbionts may suffer high mortality or reproductive failure. High-quality foods are young, soft, and green, or storage organs such as roots, tubers, and seeds.

Figure 14.28 Localized dieback of grasses caused by urine. Although this image shows the impact of dog urine, similar effects result from native grassland mammals.

Figure 14.29 The thorns of this *Acacia* tree deter or reduce herbivory. These trees also engage in a mutualistic relationship with ants for further protection (see Section 15.10).

2. *Chemical defences* are the main line of plant defence, using *secondary metabolites*—chemicals not involved in basic plant metabolism. Chemical defences are an alternative to an immune system, which plants lack. They either reduce digestibility or deter herbivory by their toxic or unpalatable properties. Plants produce a huge number, grouped into three classes: (i) N-based compounds, primarily *alkaloids* such as morphine, atropine, and nicotine; (ii) *terpenoids*, including essential oils, latex, and resins; and (iii) *phenolics*, aromatic compounds (i.e., with a benzene ring), including tannins and lignins.

Secondary compounds produced in large amounts are called *quantitative inhibitors* and may constitute up to 60 percent of leaf dry mass. For example, tannins stored in the vacuoles of oak leaf cells bind with proteins and inhibit digestion. In many species, tannins are constitutive defences, but tannin levels increase dramatically in some species upon herbivore attack (Heldt 1997)—hence an induced defence. Up to 35 percent of the carbon in the leaves of terrestrial plants (and much more in wood) is in the form of lignin. These complex phenolics not only support against gravity but also reduce food quality by their indigestibility, making the minerals bound in them unavailable. Lignin levels also increase with herbivory.

Other anti-herbivory compounds, called *qualitative inhibitors*, are present in minute quantities. Rather than merely reducing digestibility, these compounds are toxic, causing most herbivores to avoid eating them. This category includes many alkaloids, such as nicotine, that interfere with specific metabolic pathways. (We use some compounds, such as pyrethrin, as pesticides.) Although qualitative inhibitors protect against most herbivores, some specialized herbivores have evolved ways of breaching these defences—another example of coevolution. Some beetles and caterpillars sever leaf veins before feeding, stopping the flow of chemical defences. Some insects detoxify the compounds and may even store them either for their own defence, as do monarch butterfly larvae, or for production of insect pheromones.

3. *Herbivore satiation* (comparable to predator satiation) is common in species that employ **mast reproduction**: large and variable reproductive output that is not primarily due to abiotic factors. In mast years, oaks (*Quercus* spp.) produce far more acorns than their predators can consume, ensuring that some escape predation. To be effective, this high output cannot be sustained every year. If it were, herbivore levels would rise and seed predation would increase.

4. *Signalling* occurs when some plants "call for help." Parasitic and predatory insects often protect plants by killing herbivores. Some plants, when injured by herbivores, emit chemical signals that attract these herbivore predators. For example, corn seedlings eaten by caterpillars release volatile terpenoids that attract parasitoid wasps that then attack the caterpillars (Turlings et al. 1995). Induced emission is not limited to the damage site. The signalling role of such induced volatiles likely evolved secondarily from production of anti-herbivory compounds and has been reported in 20 species from 13 plant families (Schoonhoven et al. 2005).

Another type of signalling involves emitting chemical warnings. The active ingredient in Aspirin, salicylic acid, is a volatile phenolic that many plants release upon attack by pathogens. When detected by other parts of the plant, it triggers production of chemical defences. Jasmonic acid plays a similar role. This process, called **systemic acquired resistance (SAR)**, allows parts of the same plant to acquire resistance without direct contact. Its physiology is fascinating, but SAR also has many ecological implications, such as the trade-off between its defence benefits and its allocation costs for growth and fitness (Heil 2000). An intriguing issue concerns its effect on neighbour plants, which can "eavesdrop" on the volatile signal (Heil and Silva Bueno 2007).

Although salicylic acid stimulates protection against pathogenic fungi and bacteria, warnings can also induce defences against herbivores. Tannin production of *Acacia* leaves increases after grazing by kudu antelopes on South African savannahs. In this case, the gaseous warning signal is *ethylene*, a hormone that is detected by nearby plants downwind (Hughes 1990). This neighbourhood effect can greatly benefit other community members, allowing them to prepare for herbivore attack by producing induced defences.

What generalizations are possible about the type or combination of defences a plant species evolves? Ecologists have developed many hypotheses regarding anti-herbivory strategies (Stamp 2003). A major consideration is the trade-off between the allocation costs of defence to growth and fitness and the benefits of avoiding or minimizing losses. Constitutive defences such as structural defences and quantitative inhibitors provide built-in protection but are expensive to produce. They are selected for primarily in habitats where herbivory is intense and ongoing. Induced defences divert resources only when needed, but at the cost of not having protection in place. Qualitative compounds are often small, quickly produced, and inexpensive, but entail a time lag during which the plant is vulnerable. Yet the benefits may offset the costs if an induced defence provides protection against a range of stresses (Cipollini and Heil 2010).

Some hypotheses predict a negative correlation between induced and constitutive defences, with species that use one type being less likely to use the other. However, some species have a combined strategy. When attacked by bark beetles, conifers release large amounts of resin (a constitutive, quantitative defence) at the attack site to entomb the beetles. The tree may also mobilize induced, qualitative defences against the fungus at the wound site, as do Norway spruce (*Picea abies*) with induced production of terpenes (Zhao et al. 2011).

Resource availability may influence defence strategy, with plants in N-rich sites more likely to produce N-containing compounds such as alkaloids than C-rich compounds such as lignin. Finally, species with inherently fast growth may opt for replacing lost tissue and growing out of the reach of herbivores rather than investing in defence compounds. This strategy may be especially adaptive in favourable sites and/or sites with intermittent herbivory. But fast-growing species such as birch also benefit from investing in defence chemicals for their vulnerable juvenile stages. In response to herbivory, birch accumulate resins on their young twigs, increasing their resistance to browsing (Bryant et al. 1983) and in turn affecting the snowshoe hare cycle in boreal forests, as discussed in Section 14.12.

14.12 TROPHIC INTERACTIONS: Interactions among Trophic Levels Influence Population Cycles

Thus far, we have considered herbivory and carnivory as separate topics, linked by the theme of predation. However, because herbivores are themselves consumed by carnivores, we cannot understand a herbivore–carnivore system without understanding the plant–herbivore interaction. Nor can we understand plant–herbivore relations without understanding herbivore–carnivore relations. All three trophic levels are interrelated.

A well-studied three-level interaction involves plants, snowshoe hares (*Lepus americanus*), and Canada lynx (*Lynx canadensis*) (Figure 14.30) in boreal forests. In winter, snowshoe hares feed on woody browse, including twigs and nutrient-rich tips of conifers as well as aspen, birch, and willow. The hare–browse interaction becomes critical when the amount of browse is insufficient to support the hares over winter. Intense browsing reduces future woody growth, causing a food shortage as hare numbers increase. In turn, the shortage and poor quality of browse promote malnutrition and parasite infections. These conditions, as well as low temperatures and stress from heavy predation, weaken the hares, reducing their reproduction and making them more vulnerable to predation, which causes their numbers to drop rapidly. Now facing their own food shortage, predators reproduce less, and their numbers decline as

well. Meanwhile, released from browsing pressure, plant growth rebounds. Over time, with the growing abundance of winter food and the drop in predatory pressure, hare numbers recover, and another cycle begins.

In this interpretation, based on studies in Alberta, the major factor driving the cycle is *bottom-up* regulation by food supply, with predation delivering the final blow in the decline phase (Keith 1981). The cycle reflects a tripartite relationship of hares with both their predators and their plant prey. Yet, extensive experimentation and field studies in the Yukon Territory indicate that *top-down* regulation by predators (see Section 16.8) exerts the major influence (Krebs 2011). Food supply is important but does not affect hare mortality directly. Reduced food quality and quantity depress the vigour of snowshoe hares and make them more vulnerable to predation, but mortality reflects predation not only by lynx but also by great horned owls (*Bubo virginianus*) and coyote (*Canis latrans*). Reduced hare reproduction is also not directly due to food shortages, but rather to chronic stress induced by predation risk and inherited by the offspring of stressed females (Sheriff et al. 2009; see also Section 11.5). This subtle impact of perceived predation risk on population dynamics was dubbed "the ghosts of predators past" (Sheriff et al. 2010).

In contrast, a 13-year study in the Yukon Territory indicated that bottom-up regulation does cause fluctuations of small rodents, for whom overwinter survival is critical. Some 78 to 98 percent of the variation in the May and August numbers of red-backed voles (*Myodes rutilus*), deer mice (*Peromyscus maniculatus*), and field voles (*Microtus oeconomus* and *M. pennsylvanicus* combined) was explained by variation in the crops of berries and mushrooms produced the previous summer (Krebs et al. 2010). Clearly, different regulatory influences pertain to different species and even potentially to the same species in different habitats.

Three-way interactions involving plants and invertebrate herbivores and carnivores can be even more complex than those involving vertebrates, given the high incidence of both cannibalism and intraguild predation. For example, the tarnished plant bug (*Lygus hesperus*) feeds on cotton and alfalfa. Big-eyed bugs (*Geocoris pallens*) are density-dependent cannibals that feed on *Lygus* and are used as a biological control. The assassin bug, *Zelus renardii*, is an intraguild predator of *Geocoris* that also feeds on *Lygus*. In field experiments over two summers, *Zelus* reduced cannibalism in *Geocoris*, resulting in greater predatory suppression of *Lygus* and reduced losses to cotton plants (Yao-Hua and Rosenheim 2011).

14.13 NON-LETHAL EFFECTS: Predators Can Influence Prey Dynamics through Non-Lethal Effects

So far, we have stressed the impact of predators on prey mortality. But predators also alter prey traits by inducing non-lethal responses in prey morphology, physiology, and/or behaviour. These responses help defend prey, but at a cost.

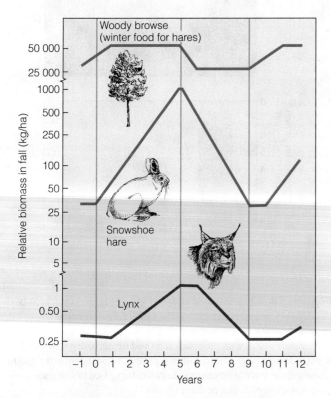

Figure 14.30 Three-way interaction of woody browse, snowshoe hare, and lynx. Note the time lag between cycles of the populations.

(Adapted from Keith 1974.)

Prey lose feeding opportunities by avoiding preferred but risk-prone habitats (see Section 14.8). Reduced activity in the presence of predators reduces prey foraging time and food intake, depressing prey growth and development.

Anti-predator behaviour can exact high fitness costs. Mayflies (Ephemeroptera) do not feed as adults, so their fitness depends solely on reserves acquired as larvae. Reduced feeding by larvae in the presence of predators slows their growth, resulting in smaller adults that produce fewer eggs (Figure 14.31) (Scrimgeour and Culp 1994).

Given their fitness costs, induced defences can affect prey population dynamics. We saw an example earlier, in the impact of chronic stress induced by predation on snowshoe hare fecundity (Sheriff et al. 2009). Predator-induced changes in behaviour also reduce population growth of pea aphids (*Acyrthosiphon pisum*), which feed on alfalfa phloem and produce 4 to 10 parthenogenetic offspring (see Section 8.1) per day. Damsel bugs (*Nabis* spp.), which pierce aphids with a long proboscis and ingest their body contents, influence prey not only by eating aphids but also by disturbing their feeding. When damsel bugs are present, aphids walk away or drop off the plant. Nelson et al. (2004) distinguished between lethal and non-lethal effects by removing the mouthparts of some damsel bugs, making them unable to kill aphids. Exposing aphids to these disarmed bugs let the researchers test the predators' ability to suppress aphid growth by non-lethal effects only. As expected, normal predators that both consumed and disturbed aphids reduced aphid growth the most, but growth was also reduced by non-consuming predators (Figure 14.32) (Nelson et al. 2004). Thus, predators reduce prey numbers by altering prey behaviour as well as by direct effects on mortality.

As with competition, it is hard to generalize about the impacts of predation. The predator–prey interaction is mediated by many morphological, behavioural, and physiological adaptations, some of which also relate to other interactions, including competition. Thus, any single trait of predator or prey reflects selection for many biotic and abiotic forces operating simultaneously. Nonetheless, many studies indicate that predators significantly alter prey abundance, behaviour, and traits—and vice versa. As we examine in Chapter 16, the direct impacts of predation on prey can indirectly influence interactions among prey, ultimately affecting community structure and function.

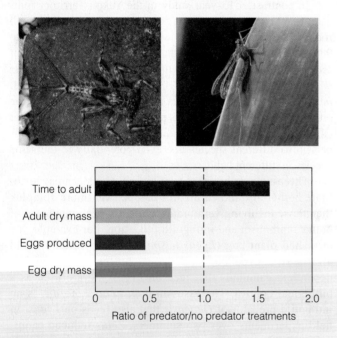

Figure 14.31 Impact of reduced activity on the mayfly *Baetis tricaudatus*. The ratio of the mean in the presence versus the absence of predators is given for time to maturity, adult dry mass, eggs produced, and egg dry mass.

(Adapted from Scrimgeour and Culp 1994, as in Lima 1998.)

INTERPRETING ECOLOGICAL DATA

Q1. How does the reduced activity of larval mayflies in the presence of predators influence the length of the juvenile period?

Q2. How and why does the presence of predators influence the fitness of adult mayflies?

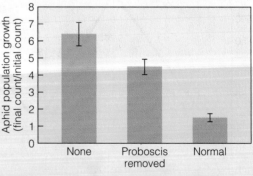

Figure 14.32 Non-lethal effects of damsel bugs on pea aphids. Per capita population growth (*y*-axis; ratio of final to initial population size) of pea aphids is shown for field cages with no damsel bugs, damsel bugs with proboscis removed (disturbance only), and normal damsel bugs. Vertical lines represent ± 1 standard error. Go to QUANTIFY it! at **www.pearsoncanada.ca/ecologyplace** to calculate confidence intervals and perform t-tests.

(Adapted from Nelson et al. 2004.)

EcologyPlace

Visit EcologyPlace at www.pearsoncanada.ca/ecologyplace to access online resources that complement your textbook, and help you to apply and to review the information in this chapter. EcologyPlace includes

- an eText version of the book
- self-grading quizzes
- glossary flashcards
- and more!

Go to www.pearsoncanada.ca/ecologyplace and follow the registration instructions on the Student Access Code Card included with this text. If your book does not have a Student Access Code Card, you can purchase access to it at www.pearsoncanada.ca/ecologyplace.

SUMMARY

Predation Types 14.1

Predation is the consumption of all or part of one living organism by another. Types include carnivory, parasitoidism, cannibalism, and herbivory. True predators kill their prey upon capture, whereas most parasites and herbivores do not kill their prey.

Modelling Predation 14.2

The Lotka–Volterra model quantifies the predator–prey interaction by linking the populations through their effects on birth and death rates. Predation is a source of mortality for prey, while predator reproduction is linked to prey consumption. If prey mortality and predator reproduction are both density dependent, the interaction can regulate both populations. The model predicts cycles, with the predator population lagging behind the prey.

Functional Response 14.3

In a functional response, a predator's per capita prey consumption changes with prey density. In a type I response, prey consumption increases linearly. In a type II response, consumption increases at a decreasing rate up to a maximum. In a type III response, consumption increases sigmoidally. All types involve search time, but only II and III are affected by handling time. Type III is the most important for regulating prey populations.

Switching 14.4

Generalist predators with a type III functional response often engage in switching, wherein the predator ignores a prey at low density but consumes it at a proportionately greater rate when it is more abundant. The level at which a predator switches to or away from a prey item depends on diet preferences, availability of cover for the prey, and formation of a search image.

Numerical Response 14.5

In a numerical response, predator number increases with prey number. It may involve an aggregative response, in which predators move into a food-rich area, or increased reproduction.

The more similar the lag times of the predator and its prey, the more important the reproductive component, and the more closely predator numbers track those of its prey.

Optimal Foraging Theory 14.6

Natural selection favours efficient foragers: individuals that maximize their net energy or nutrient gain per unit of effort. Decisions are based on the relative profitability of alternative prey. An optimal diet includes the most efficient size of prey for both handling and net energy return. When a predator encounters a less profitable prey, the decision whether to capture it or seek out a preferred item is affected by search time as well as relative profitability.

Marginal Value Theorem 14.7

Optimal foraging concentrates activity in the most profitable patches. After depleting the food to the mean profitability of the area as a whole, the predator should abandon the patch. The time a predator spends in a patch is influenced by patch quality and the time required to reach the patch.

Predation Risk 14.8

Many predators face predation risk while foraging. It may be to a predator's advantage not to visit a more profitable but predator-prone area and to remain in a less profitable but more secure habitat. Species can compete for enemy-free space, which involves any niche dimension by which a prey reduces its predation risk.

Coevolution 14.9

Prey species evolve traits to avoid being caught by predators, and predators evolve strategies for overcoming prey defences. In coevolution of predator and prey, each functions as an agent of natural selection for the other. Anti-predator defences are either constitutive or induced, and include distasteful or toxic compounds; cryptic colouration; structural defences; and behaviours including alarms, distraction, and group behaviour.

Warning colouration works in conjunction with a chemical defence. In Batesian mimicry, a palatable species mimics an unpalatable species. In Müllerian mimicry, unpalatable species mimic each other. In predator satiation, prey produce an excess of young. Predatory strategies include ambush, stalking, and pursuit. Some predators use cryptic colouration or aggressive mimicry. Others use group behaviour to hunt larger prey.

Herbivory 14.10

The amount of plant or algal biomass consumed by herbivores varies among communities. Plants are rarely killed by herbivory and often respond with a flush of regrowth, which lowers nutrient reserves. Such drawdown can weaken plants, especially woody species, making them more vulnerable to insects and disease. Moderate grazing may stimulate growth in grasses by removing older leaves. Herbivory has many indirect effects including trampling and wastes.

Plant Defences 14.11

Plants deter herbivory by structural defences or production of unpalatable or indigestible tissue or toxins that interfere with animal growth and reproduction. Some herbivores breach these defences by detoxifying compounds, blocking their flow, or sequestering them as anti-predator defences. Some plants release volatiles that attract predators of herbivores, or warn other plants. Plant defences are either constitutive or induced.

Trophic Interactions 14.12

Plant–herbivore–carnivore relationships are closely linked. In the 9- to 10-year cycle of snowshoe hare, declining food supply interacts with increased predator pressure. The major direct effect triggering the decline is predation, but hare reproduction declines as a result of the indirect effect of predation-induced stress. Three-way interactions involving invertebrates are particularly complex, given the high frequency of cannibalism and intraguild predation.

Non-lethal Effects 14.13

Besides influencing prey numbers directly through mortality, predators alter prey traits by inducing defensive responses in morphology, physiology, or behaviour. Reduced prey activity resulting from predator presence can reduce foraging and food intake, delaying growth and development.

KEY TERMS

aggregative response	cryptic colouration	handling time	mast reproduction	predator satiation
aggressive mimicry	echolocation	host	numerical response	saprophage
anti-predator defence	escape (enemy-free)	induced	optimal foraging	search image
Batesian mimicry	space	intraguild predation	theory	search time
biophage	filter feeder	Lotka–Volterra	parasite	switching
coevolution	flashing colouration	predation model	parasitoid	systemic acquired
compensatory growth	functional response	marginal value	planktivore	resistance
constitutive	guild	theorem	predation	warning colouration

STUDY QUESTIONS

1. Distinguish between predator types. Why are granivores considered "true" predators?
2. How might cannibalism interact with the effects of intraspecific competition?
3. The Lotka–Volterra model predicts mutual regulation of predator and prey, with the two populations oscillating through time. Why does the predator population lag behind the prey?
4. What is a functional response? Distinguish among type I, type II, and type III functional responses. Which term in the Lotka–Volterra model relates to a functional response? Which type of functional response is involved in the foraging behaviour of wagtails (Section 14.6)?
5. What is switching? How is it detected, and how could it regulate prey populations? What type of functional response does it involve? Is it more likely in a generalist or a specialist predator?
6. What is a numerical response? Which term does it relate to in the Lotka–Volterra model?
7. Compare the numerical responses in Figures 14.7a and 14.8. Which component (aggregative or reproductive) is more prominent in each, and why?
8. Optimal foraging theory suggests that a predator selects among prey based on their relative profitability. As predators do not calculate the profitability of prey before selecting or rejecting them, how might an optimal foraging strategy evolve?
9. What factors determine how long a predator should forage in a patch before leaving? While in the patch, it encounters a less-preferred prey. Should it eat it, or seek a preferred item?

10. Distinguish between *handling* and *search time*. Why is only handling time incorporated into the profitability calculation? How is search time incorporated into foraging theory?

11. How do predators and prey function as agents of natural selection for each other?

12. How can mimicry and cryptic colouration be either prey defences or predatory tactics?

13. Why is warning colouration always combined with another type of anti-predator defence?

14. Why is Batesian but not Müllerian mimicry affected by the relative population sizes of the mimic and model?

15. Is the interaction between spruce budworm and the bay-breasted warbler discussed in Section 14.5 an example of predator satiation? Defend your answer.

16. Compare *constitutive* versus *induced* defences. Provide an example of each in both plants and animals. What are their advantages and disadvantages? In what situations might an induced defence be more effective? A constitutive defence?

17. How do the three predatory tactics (ambush, stalk, and pursuit) affect handling and search time in a predator's time budget? Which tactic is more typical of ectotherms, and why?

18. Consider the curves in Figure 14.30. Why has browse such a prolonged plateau, compared with the curves of the animals? Consider (i) the difference between herbivory and true predation and (ii) the impact of modular plant growth. What eventually causes browse to decline?

19. Explain specific non-lethal ways in which predation can affect prey populations.

FURTHER READINGS

Clark, T. W., A. P. Curlee, S. C. Minta, and P. M. Kareiva (Eds.) 1999. *Carnivores in ecosystems: the Yellowstone experience*. New Haven: Yale University Press.
Covers interactions among canids and smaller predators and the three species systems of vegetation, herbivores, and predators.

Danell, K., R. Bergstrom, P. Duncan, and J. Pastor (eds.) 2006. *Large herbivore ecology, ecosystem dynamics, and conservation*. New York: Cambridge University Press.
Excellent synthesis of the impact of large herbivores on plant community structure and diversity.

Hay, M. E. 1991. Marine-terrestrial contrasts in the ecology of plant chemical defences against herbivores. *Trends in Ecology and Evolution* 6:362–365.
This review article contrasts the differences in chemical defences that have evolved in terrestrial and aquatic plants, and considers how adaptations to the abiotic environment constrain the evolution of plant defences.

Krebs, C. J., R. Boonstra, S. Boutin, and A. R. E. Sinclair. 2001. What drives the 10-year cycle of snowshoe hares? *Bioscience* 51:25–35.
Excellent overview of factors involved in the predator–prey cycle discussed in Section 14.12.

Stamp, N. 2003. Out of the quagmire of plant defense hypotheses. *Quarterly Review of Biology* 78: 23–55.
Comprehensive treatment of the many hypotheses concerning anti-herbivory defences.

Tollrian, R., and C. D. Harvell. 1999. *The ecology and evolution of inducible defenses*. Princeton, NJ: Princeton University Press.
An extensive, excellent review of the defenses exhibited by prey species (both plant and animal) that are induced by the presence and activity of predators. The text provides a multitude of il-lustrated examples from recent research in the growing area of coevolution between predator and prey.

Corn (*Zea mays*) infected with corn smut fungus (*Ustilago maydis*).

In Chapter 14, we examined coevolution in response to predatory interactions. Prey (animal or plant) evolve ways of defending against predators, and predators evolve ways to breach prey defences in an evolutionary game of adaptation and counteradaptation. Coevolution is even more evident in the interactions between parasites and their hosts. In most types, the parasite lives on or in the host for some part of its life in a relation called **symbiosis**: two or more organisms of different species living together in close and prolonged association.

Not all symbioses involve parasites; some are mutualistic, benefiting both species. Many people mistakenly equate *symbiosis* with *mutualism*, but the effects of symbiosis may be positive or negative. In parasitism, one species benefits at the expense of the other. The host, which provides both habitat and nutrients, suffers from the association. For the parasite, the relationship is obligatory: the parasite must secure a host for its survival and/or reproduction. As with predation, of which parasitism is a subtype, host species have evolved defences, which the parasite has evolved to counteract. Some parasitic interactions have evolved to the point that the host benefits from the association, which has then changed from parasitism to mutualism.

Ecologists have long recognized the role of parasites, but in contrast with competition and "true" predation, it was not until the late 1960s that ecologists began to appreciate their impact on population dynamics and community structure. Charles Krebs (2006) calls for more research in disease ecology, arguing that parasites are ideal model systems for studying the complex effects of species interactions. Development of mathematical models of infectious disease transmission that incorporate stochastic forces is particularly advanced (Grassly and Fraser 2008), and may serve a broader ecological context.

15.1 TYPES OF PARASITES: Parasites Differ in Size and Location on Hosts

Parasitism is an interaction (often but not always symbiotic) in which an individual of one species (the *parasite*) benefits from a prolonged association with one or more individuals of another species (the *host*), which is harmed by the interaction. Parasites increase their fitness by exploiting the host not only for food but also for habitat and often for dispersal. Although they derive nourishment from the host, parasites typically do not kill them, given the benefits of a living host. However, the host may die from secondary infection or suffer reduced fitness as a result of stunted growth, modified behaviour, or sterility. Whatever their type or effects, parasites share certain traits. (1) Parasites are much smaller than their hosts. (2) They reproduce more quickly and in greater numbers. (3) They are highly specialized for their way of life. A heavy load of parasites is called an **infection**, and the outcome of an infection is a **disease**.

This definition seems unambiguous. But as with predation, *parasitism* is often used to describe interactions that may satisfy some but not all of the above criteria. It may be hard to show harm to the host, specialization may not be apparent, and the interaction may be short-lived. For example, as a result of their episodic, non-symbiotic interactions, mosquitoes and hematophagic (blood-feeding) bats are sometimes not classified as true parasites, but rather as **intermittent parasites** or even *micropredators*.

Parasites include organisms from many taxonomic groups—viruses, bacteria, protists, fungi, plants, and invertebrates. They are often grouped by size. **Microparasites** or **pathogens** (viruses, bacteria, and protozoans) are small, with short generation times. Typically associated with the term *disease*, they multiply rapidly within the host. The infection lasts a short time relative to the host's lifespan, although many parasites remain long after the infection stage has past. The chicken pox virus, for example, can reappear in later life as shingles. Transmission is usually direct (see Section 15.2), although other species may serve as vectors. **Macroparasites**, which include flatworms, roundworms, flukes, lice, ticks, rusts, and smuts, are relatively large. They have comparatively long generation times and typically do not complete their life cycle in a single host. They spread by indirect or direct transmission, using intermediate hosts and/or vectors.

Although the term *parasite* is associated with animals and fungi, over 4000 plant species derive some or all of their sustenance from other plants. Parasitic plants have a *haustorium* (modified root) that penetrates the host, connecting to its vascular tissues. They may be **hemiparasites**, which are photosynthetic and obtain water and minerals by connecting to their host's *xylem* (water-conducting tissue). Dwarf mistletoe (*Arceuthobium americanum*), which infects many conifer species, is an example (Figure 15.1a, p. 310). Others are non-photosynthetic **holoparasites**, such as cancer root (*Conopholis americana*) (Figure 15.1b). Although they have photosynthetic ancestors, holoparasites are heterotrophs that rely on a plant host for all their carbon. Many hemiparasites also derive some organic compounds from their host.

Parasites are also classified by their use of their host as habitat. An array of parasites exploit every conceivable habitat on or within their hosts. **Ectoparasites** such as lice and ticks live on the host's skin, within the cover of feathers and hair. They are non-symbiotic because most (lice are an exception) do not live with their host for extended periods. **Endoparasites** live within the host, taking up residence in the blood, heart, brain, digestive tract, liver, stomach lining, spinal cord, nasal tract, lungs, gonads, bladder, pancreas, eyes, gills of fish, muscles, or other sites, depending on the species. Some parasites of insects live on the legs, body surfaces, and mouthparts. Fungal parasites of plants also divide up the host habitat. Some live on root and stem surfaces; others penetrate bark to live in the woody tissue beneath. Some live at the root collar or *crown*, where the shoot emerges from the soil. Others live on or in leaves, or on flowers, pollen, or fruits.

As with all things ecological, there are trade-offs. Compared with ectoparasites, endoparasites benefit from a protected internal environment in which food is easily accessed, but they have more difficulty dispersing to a new host and must deal with the host's immune system. Endoparasites share the twin problems of gaining access to a new host and escaping from a dead host. They enter and exit animal hosts through the mouth, nasal passages, skin, rectum, or urogenital system, and travel to

(a)

(b)

Figure 15.1 Plant parasites. **(a)** Dwarf mistletoe on jack pine (*Pinus banksiana*). As hemiparasites, mistletoes are photosynthetic but extract water, minerals, and some organics from their host. **(b)** Cancer root is a non-photosynthetic holoparasite on oak (*Quercus*) roots.

Figure 15.2 In this example of brood parasitism, a reed warbler (*Acrocephalus scirpaceus*) is feeding a much larger common cuckoo (*Cuculus canorus*) chick, which hatched from an egg deposited in the warbler nest by a female cuckoo.

the infection site through the pulmonary, circulatory, or digestive system. Parasites of plants enter through young roots or leaf stomata, bypassing the waxy cuticle and travelling through the xylem or phloem to reach the infection site.

Finally, parasitism can also describe a behaviour termed **kleptoparasitism**, in which an animal steals resources gathered by another. For example, a gull forces a puffin to drop a fish, which it then eats. A behaviour that is sometimes considered kleptoparasitism (particularly in insects; see Crowe et al. 2009) but that exacts a higher fitness cost, is **brood parasitism**. The parasitic female cases out the nest of a host during its construction and then lays an egg in it when the host is temporarily absent. The parasite typically ejects (and sometimes eats) one or more host eggs, but leaves at least one host egg to reduce the chance that the nest will be abandoned. Once hatched, the parasite young evicts

or outcompetes the host young, which are typically smaller. The parasite not only uses food gathered by the host but also usurps the host's time and effort in raising its own young (Figure 15.2). This behaviour occurs within some bird species, including robins, but some birds parasitize the nests of other species.

Cuckoos have evolved subpopulations whose eggs resemble those of their host, but the brown-headed cowbird (*Molothrus ater*) is a generalist, placing eggs in the nests of some 200 North American bird species (see Section 18.3). Given that there is little resemblance between cowbird eggs and those of their hosts, why don't host species learn to recognize and reject cowbird eggs?

Studies indicate that both acceptance and rejection strategies persist in yellow warblers (*Dendroica petechia*) (Sealy 1995). Both strategies have costs, and selection for acceptance or rejection oscillates as trade-offs shift with changing conditions. The cost to the host of accepting a parasitic egg is obvious—loss of its own nestlings as well as the time and energy spent raising the parasite's young. But what are the costs of rejection? It depends on the rejection strategy. Some species simply eject the egg, but the yellow warbler, whose beak is not large enough to grasp the egg, abandons the nest entirely, covering it with nest material and sacrificing its own eggs.

But even ejection has costs, if retaliation follows. In experimental tests, cowbirds attacked 56 percent of "ejector" nests of prothonotary warblers (*Protonotaria citrea*) and only 6 percent of "accepter" nests. Nor did the cowbirds forgive and forget; 85 percent of rebuilt nests were destroyed (Hoover and Robinson 2007). Although it is difficult to establish experimentally, such retaliatory behaviour of brood parasites could select against rejection.

15.2 PARASITE TRANSMISSION: Movement between Hosts Is Direct or Indirect

Hosts are habitat islands that eventually disappear upon the death of the host. Parasites must then escape from that host and locate another. Endomacroparasites typically move between hosts in their infective stage, as larvae. Transmission is direct or indirect and involves adaptations to many aspects of the feeding, social, and mating behaviours of the host species.

In **direct transmission**, a parasite needs no other organism to move between hosts. Transmission occurs by contact or by dispersal through air or water. Microparasites, including influenza viruses and bacteria that cause sexually transmitted diseases such as gonorrhea, often use direct transmission, as do many macroparasites, such as roundworms (*Ascaris* spp.). Females lay thousands of eggs inside the host. The eggs are then expelled in feces and dispersed in water, soil, or vegetation. If ingested by a susceptible host, the eggs hatch in the intestines. Larvae burrow into blood vessels, coming to rest in the lungs. They travel to the mouth by causing the host to cough, and are swallowed to reach the stomach, where they mature and enter the intestines.

Many ectoparasites also spread by direct contact, including lice, ticks, fleas, and mites that cause mange. Many lay eggs directly on the host. Fleas lay their eggs in the host's nest, from which the larvae hatch and feed on dead skin. The adults that emerge from the larvae then leap onto nearby hosts. Some plant holoparasites, such as cancer root, which parasitizes oak and beech roots (see Figure 15.1b), are also transmitted directly. Its seeds are dispersed locally. Upon germination, their roots attach to those of their host. Many fungal parasites spread through root grafts. For example, *Heterobasidion annosum*, a fungal parasite of white pine (*Pinus strobus*), spreads rapidly through pine stands when roots of one tree graft onto roots of a neighbour. Root grafts and mycorrhizal networks are potentially of great mutualistic benefit to plants (see Section 15.13), but the trade-off is that they provide easy routes for parasite transmission.

In **indirect transmission**, a parasite moves between hosts with the help of another organism, either as a biotic **vector** (means of transferring an entity) or as an *intermediate host*. The black-legged tick (*Ixodes scapularis*) is a biotic vector for transmitting Lyme disease, which is caused by a bacterium, *Borrelia burgdorferi*. The bacterium occupies the bloodstream of vertebrates, from birds and mice to deer and humans, but needs the tick for transmission. Similarly, the protozoan species (*Plasmodium*) that cause malaria are transmitted by the bite of a female mosquito (*Anopheles* spp.) (Figure 15.3). Mosquitoes transmit over half of the 102 **arboviruses** (arthropod-borne viruses) that produce human diseases, including dengue and yellow fever. In colder climates, mosquitoes pose little disease hazard for humans, apart from occasional outbreaks of encephalitis and West Nile virus. An associated hazard is the use of insecticides to control them.

Insect vectors also transmit parasites between plants. European and native elm bark beetles carry spores of the fungus *Ophiostoma ulmi*, responsible for Dutch elm disease. Birds transmit mistletoe by feeding on fruits and depositing the seeds (which pass through the digestive tract unharmed) on trees where the birds perch and defecate. The sticky seeds attach to limbs and send out roots that penetrate the host. Mistletoe also employs explosive dehiscence, catapulting its seeds in search of new hosts (see Figure 9.17). Using both direct and indirect transmission, dwarf mistletoe (*Arceuthobium* spp.) outbreaks cause serious damage to jack and lodgepole pine (*Pinus banksiana* and *P. contorta*) and white spruce (*Picea glauca*) forests. As hemiparasites, mistletoe does not

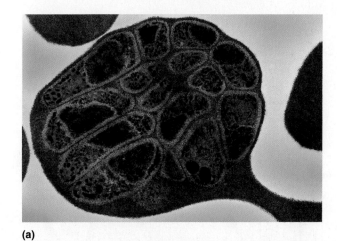

(a)

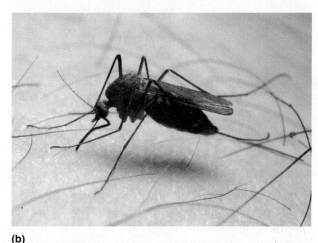

(b)

Figure 15.3 Malaria is a recurring infection caused by a protozoan parasite, shown here occupying an infected red blood cell. (a) The parasite is transmitted by female mosquitoes (b). Over 40 percent of humans are at risk, and more than 1 million die yearly.

rely entirely on its host for carbon, but does alter the tree's hormonal balance. Abnormal growths called "witches' brooms" (see Figure 15.1a) enhance mistletoe's ability to drain minerals from the host. Dwarf mistletoe threatens pine stands across the Prairies (Brandt et al. 2005). Because extreme cold impairs mistletoe seed survival, global warming will likely worsen infestations.

Some indirect transmissions involve multiple hosts, not merely a biotic vector. All organisms have a life cycle: phases associated with their development, divided into juvenile, reproductive, and post-reproductive (see Chapter 8). Some parasites cannot complete their life cycle in a single host. The species in which a parasite reaches maturity is the **definitive host**. All others are **intermediate hosts**, which harbour some developmental phase. For example, the meningeal worm (*Parelaphostrongylus tenuis*) parasitizes white-tailed deer, elk, and moose. Snails are its intermediate host. The deer ingests infected snails while grazing. Larvae puncture the stomach wall and travel via the spinal cord to the brain, where the worms mate and produce eggs. Eggs and larvae pass through the blood to the lungs, where the larvae penetrate air sacs and are coughed up, swallowed, and excreted in feces.

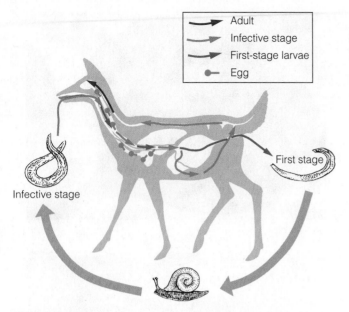

Figure legend:
→ Adult
→ Infective stage
→ First-stage larvae
•— Egg

Infective stage

First stage

Figure 15.4 Life cycle of the meningeal worm. Transmission is indirect, with snails or slugs as intermediate hosts.

(Adapted from Anderson 1963.)

Snails acquire the larvae through contact with deer feces, completing the cycle (Figure 15.4).

White pine blister rust (*Cronartium ribicola*) (Figure 15.5) is an introduced fungal parasite of five-needle pines. Asian pines are resistant, but mortality losses to North American pines, which have not coevolved with the rust, are high. As both saplings and mature trees are killed, regeneration is almost impossible. After decades of epidemic conditions, there is little seed stock left for regenerating eastern white pine (*Pinus strobus*) at its northern limits in Ontario and the Maritimes (Kinloch 2003). Elsewhere, patchy infections have created a metapopulation structure (see Chapter 12) in white pine. The rust uses currants (*Ribes* spp.) as an intermediate host. A campaign to eradicate currants was of limited success, as currants regrow quickly from roots. Damage to western white pine (*P. monticola*) is also severe, with over 90 percent losses reported in some areas.

Figure 15.5 White pine blister rust, shown here on a currant, its intermediate host, causes heavy mortality losses to five-needle pines in North America.

Some parasites require two or even three intermediate hosts. Parasite population dynamics are thus linked to the population dynamics, movements, and interactions of its host(s).

15.3 IMMEDIATE HOST RESPONSE: Hosts Avoid Infection or Minimize Its Effects

As with coevolution of predators and prey, host species have evolved adaptations that minimize the probability of infection, or combat infection once it has occurred. Some defences are behavioural. For example, birds and mammals rid themselves of ectoparasites by grooming. In birds, preening of plumage with the bill and feet removes lice adults and nymphs.

Once infection occurs, the first line of defence is an **inflammatory response**. Death or injury of host cells stimulates secretion of *histamines* (chemical alarm signals), which increase blood flow and cause local inflammation. This reaction mobilizes white blood cells to attack the infection. Scabs form, reducing entry. Internal reactions produce hardened cysts in muscle or skin that enclose and isolate the parasite. For example, cysts encase roundworms (*Trichinella spiralis*) in muscles of pigs and bears, causing trichinosis in humans from eating undercooked pork.

Plants also respond to bacterial and fungal invasion by forming cysts or scabs, reducing contact with healthy tissue. Plants react to gall wasps, bees, and flies by forming *galls*, abnormal growths unique to the gall insect (Figure 15.6). As

(a) Spruce cone gall

(b) Oak succulent gall

(c) Oak bullet gall

(d) Goldenrod ball gall

Figure 15.6 Gall formation in plants results from induced changes in genetic expression host cells. **(a)** Spruce cone gall caused by the aphid *Adelges abietis*. **(b)** Oak succulent gall induced by the gall wasp *Dryocosmus quercuspalustris*. **(c)** Oak bullet gall caused by the gall wasp *Disholcaspis globulus*. **(d)** Goldenrod ball gall induced by the gallfly *Eurosta solidaginis*.

an indirect effect, gall formation may expose parasite larvae to predation. For example, swollen knobs of the goldenrod ball gall (Figure 15.6d) attract downy woodpeckers (*Picoides pubescens*), which excavate the larvae (Confer and Paicos 1985).

In animals, the second line of defence is the **immune response**. When a virus or bacterium enters the blood, white cells called *lymphocytes* produce protein *antibodies* that target the *antigens* (the term *antigen* can also refer to the parasite itself) on the parasite's surface or released into the host, countering their effects. Antibodies are costly to produce and may damage the host's tissues. To be effective, the immune response need only reduce the parasite's feeding and reproduction to tolerable levels. The immune response is highly specific. It has a remarkable "memory" for antigens, allowing faster and more vigorous reactions on subsequent exposures.

The immune response can be breached. Some parasites vary their antigens continuously, keeping one jump ahead of the host. A chronic infection results. If the host suffers from protein deficiency, antibody production is impaired. Depleted energy reserves weaken the immune system, allowing viruses or other parasites to become pathogenic. The ultimate breakdown occurs in humans infected with the human immunodeficiency virus (HIV), the causal agent of AIDS. The virus itself is not lethal, but by attacking the immune system, it exposes the host to a range of potentially fatal infections. AIDS is a prime example of a parasite that kills indirectly by increasing the host's vulnerability to other diseases.

15.4 EFFECTS ON HOST INDIVIDUALS: Parasites Can Impair Host Reproduction and Survival

The many defences enabling hosts to prevent, reduce, or combat infection share one feature: they all require resources that the host might otherwise allocate to other purposes. Organisms have limited energy, so it is not surprising that parasitic infections often reduce host growth and reproduction. For example, clutch size (number of eggs) of the western fence lizard (*Sceloporus occidentalis*) in California is 20 percent less in malaria-infected females (Schall 1983). They store less fat in summer, reducing energy available for egg production the next spring.

Parasitic infection can also lower male fitness. Females often choose mates based on secondary sex traits, such as bright plumage (see Section 8.4). Infection may limit full expression of such traits, reducing the male's ability to attract a mate. For example, the bright red colour of the male zebra finch's beak derives from *carotenoids*, plant pigments associated with most of the colouration of autumn foliage. Birds obtain carotenoids in their diet. As well as providing colour, carotenoids stimulate antibody production and act as antioxidants, absorbing damaging free radicals generated during the immune response. Only male zebra finches with the fewest parasites can devote sufficient carotenoids to produce bright

red beaks. Parasitic infection thus lowers male fitness indirectly (Blount et al. 2003).

Parasites can also reduce survival. Whereas parasitoids always kill their host, most parasites do not kill their hosts directly. However, mortality can increase from indirect effects. Parasites may alter host behaviour, increasing their susceptibility to predation. Rabbits infected with the bacterial tularemia (*Francisella tularensis*) are sluggish and more vulnerable. Similarly, California killifish (*Fundulus parvipinnis*; Figure 15.7a) parasitized by trematodes (flukes) display abnormal behaviour

(a)

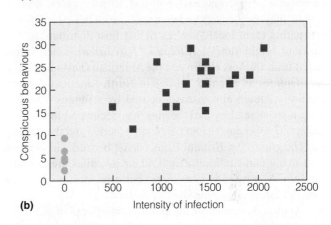

(b)

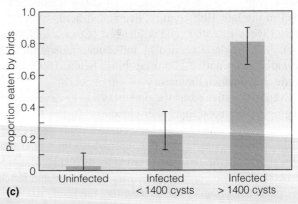

(c)

Figure 15.7 Indirect effect of parasites on survival. Infection of killifish **(a)** by trematodes causes abnormal behaviour that increases predatory risk. **(b)** In parasitized populations (■),frequency of conspicuous behaviours in a 30-min period (*y*-axis) increased with number of cysts per fish brain. Uninfected fish (●) had fewer abnormal behaviours. **(c)** Proportion of fish eaten after 20 days (*y*-axis). Heavily parasitized fish were more heavily preyed upon. Vertical lines represent 95 percent confidence intervals.

(Adapted from Lafferty and Morris 1996.)

such as surfacing and jerking. The frequency of these conspicuous behaviours correlated with the degree of parasitic infection (Lafferty and Morris 1996) (Figure 15.7b). These behaviours attract predatory birds, which prey more often on parasitized fish (Figure 15.7c). The interaction is even more complex. The fish-eating birds are the trematodes' definitive host, so that by making its intermediate host (killifish) more susceptible to predation, the trematode increases the chances of completing its life cycle.

15.5 POPULATION EFFECTS: Effects on Host Populations Can Be Density Independent or Density Dependent

Parasites may decrease host reproduction and increase mortality, but few studies have quantified the effect of a parasite on the population dynamics of a host with which it has coevolved. In contrast, introduced parasites often decimate host populations that have not evolved effective defences (see Ecological Issues: Plagues upon Us). In such cases, disease spread may be density independent, reducing populations, extirpating them locally, or restricting host distribution. The chestnut blight fungus (*Endothia parasitica*), when introduced from Europe, eliminated the American chestnut (*Castanea dentata*) as a major species in North American forests. Similarly, Dutch elm disease, caused by a fungus (*Ophiostoma ulmi*) spread by bark beetles, has decimated the American elm (*Ulmus americana*) in North America and the English elm (*U. glabra*) in Britain. Elms persist because (in contrast with many parasite interactions) their juveniles are not susceptible. Many seeds germinate, but elm saplings become prone to infection at reproductive maturity.

Animal examples of the population effects of parasites abound. Rinderpest, a viral disease of cattle introduced to Africa in the late 19th century, decimated herds of African buffalo (*Syncerus caffer*) and wildebeest (*Connochaetes taurinus*). Avian malaria carried by introduced mosquitoes has eliminated most native Hawaiian birds below 1000 m, the altitude above which the mosquito cannot persist.

Even with native parasites, outbreaks can be devastating if changing abiotic conditions alter outcomes. British Columbia is currently experiencing huge losses to lodgepole and ponderosa pine (*Pinus contorta* and *P. ponderosa*) from a blue-stain fungus (*Grosmannia clavigera*) carried by the native mountain pine beetle (*Dendroctonus ponderosae*). The beetle and fungus have coevolved a mutualistic relationship: the beetle has a specialized mouthpart that harbours the fungus, and the fungus prevents the tree from employing its normal defence of overwhelming the beetle with resin. The overwintering larvae can withstand midwinter temperatures as low as −40°C. Milder temperatures will also kill it in late fall before the larvae have acquired enough glycerol to be fully hardy, or in late winter when glycerol is metabolized (Thomson 2009). By reducing exposure to the extreme cold that formerly kept mountain pine beetles in check, climate change is encouraging spread of the fungus into northern Alberta.

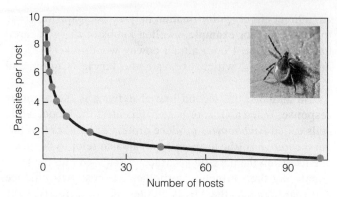

Figure 15.8 Clumped dispersion of the tick *Ixodes trianguliceps* on the European field mouse, *Apodemus sylvaticus*. A few mice carry most of the parasites.

(Adapted from Randolph 1975.)

Parasites may also exert a density-dependent influence on their host. Such cases typically occur with directly transmitted parasites that are maintained in the population by a small reservoir of infected carriers. Outbreaks occur only when host density is high. They reduce host populations sharply, causing cycles of host and parasite similar to those observed for predator and prey. For example, outbreaks of canine distemper, such as occurred in Toronto in 2010, kill raccoons (*Procyon lotor*) in urban areas where they have reached unnaturally high densities.

In other cases, parasites act as selective agents of mortality, affecting a subset of individuals. Spatial dispersion of macroparasites, especially those transmitted indirectly, is highly clumped. Individuals carrying more parasites through chance or genetic susceptibility are most likely to succumb to parasite-induced mortality or suffer reduced reproduction (Figure 15.8).

In contrast, some parasites are present throughout a population. Bighorn sheep (*Ovis canadensis*) may be infected with up to seven lungworm species. Infection is highest in spring, causing pneumonia, a secondary infection that may kill lambs. These infections may regulate sheep numbers by reducing their reproductive success (Woodard et al. 1974). However, in a study in Alberta, all bighorn sheep had *Protostrongylus* spp. larvae in their feces. Larval counts varied but were not reliable predictors of pneumonia outbreaks or herd health (Festa-Blanchet 1991).

15.6 EVOLUTIONARY EFFECTS: Coevolution of Parasite and Host May Take Many Paths

For parasite and host to coexist in a relation that is hardly benign even if not fatal, the host must either reduce the number of parasites or minimize their effects. Selection has usually resulted in a level of immune response where allocation of resources by the host minimizes the costs of parasitism without unduly impairing host growth and reproduction. Conversely, the parasite gains no advantage by killing its host. A dead host means dead parasites—unless they find a new host. The conventional wisdom is that virulence is selected against, with the result that parasites become less harmful over time. But does selection always work this way in host–parasite systems?

Natural selection does not always favour peaceful coexistence of host and parasite. To maximize fitness, a parasite must balance the trade-offs between virulence and ecological considerations, particularly transmissibility. As a result, natural selection may yield deadly (high-virulence) or benign (low-virulence) strains, depending on the parasite's requirements for reproduction and transmission. In **vertical transmission**, parasites are transmitted directly from mother to offspring during the period immediately before or after birth. Because the host must survive to maturity to transmit the parasite, species that use vertical transmission are typically less virulent than those transmitted by direct **horizontal transmission**, which occurs between host adults.

The situation is more complex in parasites with mixed transmission strategies. For example, the human papillomavirus and hepatitis, as well as many parasites of plants and insects, contain strains that use either vertical or horizontal transmission. Evolutionary pressures may select for maintenance of both. Whereas vertical transmission alone might select for progressively lower virulence, horizontally transmitted strains may allow vertically transmitted strains of moderate virulence to persist in the parasite population (Lipsitch et al. 1996).

Ultimately, the host's condition matters to a parasite mainly as it affects parasite reproduction and transmission. If

ECOLOGICAL ISSUES | Plagues upon Us

Humans have always been bedevilled by parasites, but more so in recent history. During our first 2 million years as hunter–gatherers, the most bothersome parasites were directly transmitted macroparasites such as roundworms. Only microparasites with high transmission rates that produced no immunity could persist in such small host groups. However, once humans became sedentary agriculturalists aggregated in villages, populations became large enough to sustain bacterial and viral parasites (Dobson and Carper 1996), many of which evolved from diseases of domesticated animals. Measles, for instance, evolved from canine distemper. Initially, human populations were too small to support ongoing disease without reinfection from a neighbouring settlement, but once a settlement grew sufficiently big, the population was dense enough to maintain a reservoir of infection. As commerce developed, people and goods moved long distances, introducing diseases to regions where populations lacked immunity. Periodic epidemics swept through cities.

A classic instance of the importation, transmission, and spread of a disease is bubonic plague, also called the Black Death because of the blackened extremities of its victims. Although some argue that other organisms were involved, bubonic plague is caused by a rod-shaped bacillus, *Yersinia pestis*. It is transmitted between hosts both indirectly, by the bite of a vector (fleas from infected rodents), and directly, by coughing. Infected individuals become ill in a few hours or days, developing high fever and swollen lymph nodes. Death quickly follows, often in days.

The bacillus is associated with burrowing rodents, especially the black rat (*Rattus rattus*) of India. An agile climber, the black rat boarded ships that carried it to port cities in Asia. Hidden in caravans, rats spread across the Asian steppes. In 1331, a plague epidemic swept through China. Mongol armies carried it across Asia to the Mediterranean. At the siege of Caffa in 1346 in the Crimea, the Mongol army was devastated by the plague and withdrew after catapulting infected corpses into town. Trade resumed, and ships carried infected rats to southern Europe.

Conditions were right for disease spread. Europe was undergoing rapid growth, the climate was worsening, and crops were failing. By 1347, the plague had spread to Italy and France; a year later, to Germany and England; and by December 1350, to Scandinavia. Between 1348 and 1350, one-third of the European

population, including entire villages, succumbed—upsetting the social, political, and economic stability of Europe. Later outbreaks occurred in Milan (1630), London (1665), and Marseilles (1721) (McNeill 1976). Sporadic outbreaks occurred until 1944, after which antibiotics cured the disease quickly, if diagnosed early. Harboured by rodents, the plague bacillus still thrives worldwide. The last North American outbreak was in Los Angeles in 1924, but isolated cases still occur in the American southwest, and the bacterium is present in British Columbia.

Other diseases introduced into populations lacking immunity have similar stories. Smallpox, measles, and typhus, carried to the Americas by explorers and settlers, decimated indigenous peoples, who lacked immunity. Diseases devastated the Aztecs in Mexico and nearly eliminated native Americans in New England, allowing uncontested settlement. More recently, a massive flu epidemic (the "Spanish Flu") killed 21 million people worldwide in 1918. Influenza remains a real threat, as evident from concerns about the H1N1 virus in 2009. Thanks to their high mutation rate and rapid reproduction, strains evolve faster than genetic resistance develops.

As many of the old plagues are contained by vaccinations and other measures, we often become complacent about disease. Even so, new diseases—Ebola in Zaire for example—warn us that plagues are still with us, as are new forms of old diseases. Tuberculosis, once considered eliminated, now kills 2 million yearly worldwide, according to Health Canada. In Canada it is of particular concern in aboriginal communities. Population increases, changing climate, and rapid movements of people and goods set the stage for future plagues. Add to that the complicating factor of antibiotic resistance, and the message is clear: humans are far from immune to the ecological impact of parasites.

Bibliography
Dobson, A. P., and E. R. Carper. 1996. Infectious diseases and human population history. *BioScience* 46:115–126.

McNeill, W. H. 1976. *Plagues and peoples*. New York: Random House.

1. Research the Irish potato famine as an ecological event involving parasites that has altered Canadian history.

the host species did not evolve, the parasite might well be able to achieve some optimal balance of host exploitation. But just as with predator and prey, host species also evolve, giving rise to a coevolutionary "arms race" between parasite and host (see also Research in Ecology: Sexual Conflict and the Coevolutionary Arms Race, p. 165). Yet, as we explore in Section 15.7, their intimate relationship can ultimately transform into symbiotic mutualism.

15.7 ORIGIN OF MUTUALISM: Parasitism Can Evolve into a Mutually Beneficial Relationship

Most parasites and their hosts live together in a symbiosis in which the parasite benefits at the host's expense. If the host evolves defences that counter the negative impacts entirely, the relationship becomes **commensalism**, in which one organism benefits without significantly affecting the other, as with epiphytes growing on rain forest trees (Figure 15.9). However, at some stage in host–parasite coevolution, the relationship may become beneficial to both. A host species may go beyond merely tolerating infection to exploiting the relationship, at which point it becomes **mutualism**, in which both organisms benefit. The most likely instances involve parasites with vertical transmission. Among vertically transmitted parasites, there will be selection to increase host survival and reproduction, because maximizing host reproductive success benefits both parasite and host.

There are some relationships that are still classified as parasitic in which there is a possible benefit to the host.

Figure 15.9 A tree trunk provides the substrate for epiphytic orchids and bromeliads, an example of commensalism in which the epiphyte benefits while the tree is largely unaffected.

Although the finding has been challenged by Bordenstein and Werren (2000), infection of the wasp *Nasonia vitripennis* by *Wolbachia* bacteria (which infect some 20 percent of insect species) may increase the number of offspring (Stolk and Stouthamer 1996). Similarly, rats infected with intermediate stages of the tapeworm *Spirometra* grow larger than uninfected rats because the larva produces an analogue of vertebrate growth hormone (Phares 1996). However, even though such effects are beneficial, the overall balance sheet can still be negative. Over evolutionary time, this balance can shift to a net positive outcome.

Some vertically transmitted parasites benefit their host indirectly by preventing or reducing infection by more virulent, horizontally transmitted parasites. This interaction is a kind of indirect mutualism deriving from competition between two parasites with different transmission modes (Lively et al. 2005). The host benefits through its reduced mortality. The ability of competition between parasites to select for host protection is a growing area of research (Jones et al. 2011).

But what about interactions that are clearly no longer parasitic? The most significant example is so integral to the biosphere that we take it for granted. According to the **endosymbiotic hypothesis**, first proposed in the 1920s by Ivan Wallin and later developed by Lynn Margulis (1982), chloroplasts, the organelles that perform photosynthesis, originated as bacteria that were ingested by heterotrophs. The relationship evolved first from true predation into parasitism, and ultimately into a symbiotic mutualism in which the eukaryotic cell gains energy from the organelle, while the organelle gains a favourable interior habitat. Chloroplasts resemble modern-day cyanobacteria, but they cannot live independently; and without them, their plant partners would be as heterotrophic as animals. Evolution has progressed so far that the two entities are no longer distinct. A similar origin is postulated for mitochondria.

Not all mutualisms may be this crucial to life, but this example illustrates a key point. Although it benefits both partners, mutualism may involve reciprocal exploitation rather than cooperative effort, reflecting its origins as a host–parasite interaction. The benefits for one or both participants are often environment dependent and may revert from being +/+ to being +/−. For example, most tree roots are associated with mycorrhizae (see Section 15.9). The fungus obtains carbon compounds, and in infertile soil the tree benefits from increased uptake of minerals (particularly phosphate) by the fungus. But in fertile soils, the fungus can exact a net cost. Depending on the environment, the association alternates between mutualism and parasitism.

15.8 MUTUALISM CATEGORIES: There Are Many Types of Mutualisms

Mutualism involves many diverse interactions, depending on the *type of benefit*, *degree of dependency*, *specificity*, and *duration*. Benefits include (1) provision of energy or minerals; (2) protection from predators, parasites, or herbivores; (3) reduced competition with other species; and (4) dispersal of gametes or offspring. Degree of dependency varies from **obligate** mutualists that cannot survive or reproduce without the partner, to **facultative** mutualists that can, although they do better in the presence of the partner. Specificity ranges from species-specific associations (*specialists*) to association with many partners (*generalists*). Duration varies from long-lasting and symbiotic, where the partners live together for some or all of their life cycle, to free living and non-symbiotic, with more fleeting associations. Symbiotic mutualisms are usually obligate, at least for one partner and often for both.

Some obligate mutualisms are so intimate that the distinction between partners is blurred. Chloroplasts and mitochondria are examples. Reef-forming corals are another. Corals secrete an external skeleton composed of calcium carbonate. Individual animals (*polyps*) occupy little cups (*corallites*) (Figure 15.10) in which they harbour single-celled algae (*zooxanthellae*). Although polyps are carnivores that feed on zooplankton suspended in water, they acquire about 90 percent of their energy from carbon compounds produced by the algae, without which they could not survive

Figure 15.11 A lichen consists of a fungus and an alga combined in a body or *thallus*.

in their nutrient-poor environment. In turn, the coral provides its algal partner with shelter and minerals, particularly N from bodily wastes.

Lichens are another symbiotic mutualism in which it is virtually impossible to distinguish the partners. Lichens (Figure 15.11) consist of a fungus and an alga or cyanobacterium coexisting within a body (*thallus*). The alga supplies carbon energy, and the fungus protects the alga from high light, produces a substance that accelerates photosynthesis, and absorbs and retains water and minerals for both organisms. There are about 25 000 known lichens of various forms, each composed of a unique combination of fungus and alga.

As we saw with parasites, not all symbiotic interactions are mutualistic. Conversely, not all mutualisms are symbiotic. In non-symbiotic mutualisms, the organisms benefit each other without living together. Non-symbiotic mutualisms may be obligatory, but most are facultative and involve mutual facilitation. Pollination and seed dispersal in flowering plants are examples. With some exceptions, these interactions are not confined to two species but involve a diffuse array of plants, pollinators, and seed dispersers. In the following sections we explore the diversity of mutualisms, symbiotic and non-symbiotic, focusing on the benefits received.

(a)

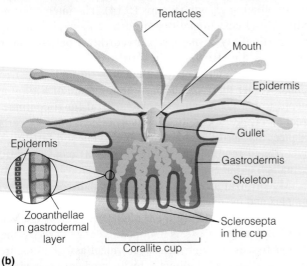

Tentacles

Mouth

Epidermis

Gullet

Epidermis

Gastrodermis

Skeleton

Zooanthellae in gastrodermal layer

Sclerosepta in the cup

Corallite cup

(b)

Figure 15.10 Coral structure. **(a)** Individual polyps of the Great Star (*Montastrea cavernosa*) coral. **(b)** Anatomy of a polyp, showing the location of the symbiotic zooanthellae.

15.9 NUTRIENT TRANSFER MUTUALISMS: Some Mutualisms Transfer Energy and Minerals

Ruminants are the most familiar and specialized mutualism in animal nutrition. The chambers of the ruminant stomach contain large numbers of anaerobic bacteria and protozoans that perform fermentation (see Section 7.2). Yet virtually all vertebrate herbivores and omnivores harbour an array of microbial mutualists that aid digestion of plant tissue (Stevens and Hume 1998).

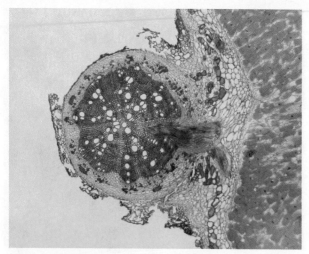

(a)

(b)

Figure 15.12 N-fixing *Rhizobium* bacteria infect **(a)** legume roots, forming nodules **(b)**. Bacteria provide N to the plant, which supplies energy in the form of carbohydrates.

Mutualistic interactions also facilitate plant nutrient uptake. Nitrogen is essential for protein and nucleic acids. Although it is the most abundant element in the atmosphere, gaseous N is unavailable to most life forms. Bacteria of the *Rhizobium* genus (collectively **rhizobia**) occur widely in soil. In a free-living state, they can grow and multiply, but they can only fix N_2 in association with the roots of legumes, a family (Fabaceae) that includes clover, beans, and peas. Using root exudates, legumes attract bacteria, which then enter their root hairs. "Infected" root hairs (the descriptor reflects the relationship's parasitic origin) form nodules (Figure 15.12) in which rhizobia reduce N_2 to ammonia (NH_3). The bacteria receive carbon from the plant and contribute nitrogen, allowing the plant to be independent of soil supplies (see Section 22.5 for more on N transformations by microbes).

Biological N-fixation is energy demanding. To fix 1 g of N, N-fixing bacteria require about 10 g of glucose, which is supplied by the plant partner. Legumes and other N-fixing plants rarely thrive in shade as a result of this high energy cost. They are often relegated to early succession, when light is abundant. As with mycorrhizae, the N-fixing interaction can fluctuate between mutualism and parasitism, given its energy demands. For example, alders (*Alnus* spp.) are common shrubs that form mutualistic associations with *Frankia*, filamentous N-fixing actinomycete bacteria. Alders develop resistance to infection by *Frankia* strains that confer fewer benefits than other strains relative to their energy costs (Markham 2008).

Another symbiotic mutualism involved in plant nutrition is mycorrhizae. The fungus increases mineral and water uptake, while the plant provides carbon energy. **Endomycorrhizae** have a broad range of hosts and associate with over 70 percent of plant species. Mycelia composed of fungal hyphae infect tree roots, penetrating the cells to form a branched network called an *arbuscle* (Figure 15.13a). The mycelia act as extended roots, absorbing N and P beyond the reach of the root hairs. **Ectomycorrhizae** produce shortened, thickened roots that resemble coral (Figure 15.13b). Fungal hyphae penetrate between, not into, root cells. Outside the root, they develop a network called a *Hartig net* that acts as an extended root system.

Ectomycorrhizae are restricted in their hosts, many of which are conifers. They contribute to the dominance of conifers in boreal forests, but the success of dipterocarps, a group of tropical tree species in the rain forests of southeast Asia, is also due in part to their ability to harbour ectomycorrhizae in infertile tropical soils (Ashton 1988).

Mycorrhizae are associated with the roots of most land plants and are especially critical in poor soil. Indeed, ecologists credit mycorrhizae with making it possible for ancestral plants to colonize land. The ability of a species to colonize and thrive often depends on the presence of a suitable mycorrhizal partner. In infertile soil, this relationship is extremely beneficial to the plant, increasing nutrient uptake and stimulating growth (Figure 15.14). The interaction may indirectly affect competitive outcomes between plant species, particularly if one has a mycorrhizal partner and the other does not. However, in fertile soil, plants are more likely to meet their nutrient demands through direct root uptake. Under these conditions, the fungi are of little if any benefit to the plant, whereas mycorrhizal fungi continue to exact an energetic cost on the plant, reducing its net carbon balance. So, across a landscape of varying soil fertility, the mycorrhizal interaction (like N_2-fixation) can revert from mutualism to parasitism.

Mycorrhizal associations can have indirect effects with complex and far-reaching outcomes. When we think of plants "preying" on animals, we think of species such as the Venus fly-trap that capture insects in N-poor habitats to access the N in their tissues. But through mycorrhizal associations, trees may access animal N indirectly, by the predatory behaviour of their fungal partners. Many soil arthropods eat mycorrhizal fungi, but less than 5 percent of the springtail *Folsomia*

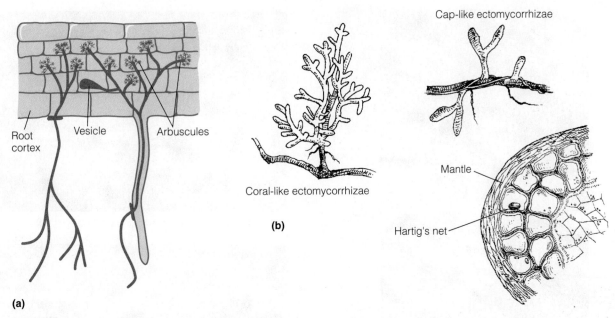

Figure 15.13 Two types of mycorrhizae. **(a)** Fungal hyphae of endomycorrhizae penetrate root cells. **(b)** Ectomycorrhizae form a fungal mantle around root tips and grow between root cells.

candida survived exposure to the fungus *Laccaria bicolor*, which apparently releases a toxin that paralyzes its arthropod prey. Yet the fungus is not the main beneficiary of this fungal predatory act. Radioactive tracer studies show that white pine (*Pinus strobus*) infected with *Laccaria* receives significant amounts of animal-origin N from its fungal partner, supporting faster growth relative to the non-infected controls (Klironomos and Hart 2001). As the authors state, if this ability of plants to swap carbohydrates for animal N proves widespread, it will greatly alter our understanding of nutrient cycling in forests.

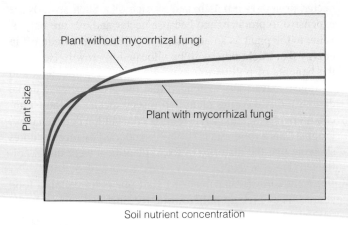

Figure 15.14 Growth of plants with and without mycorrhizal partners across a gradient of soil fertility. When soil nutrients are scarce, the plant with the fungal partner grows faster. As soil nutrient levels increase, the presence of the fungi switches from a net benefit to a net carbon cost. Now the plant that is lacking a fungal association has a higher growth rate.

15.10 DEFENCE MUTUALISMS: Some Mutualisms Protect One or Both Partners

Some mutualistic partners provide protection. Many nutrient transfer mutualisms also serve a defence function. For example, the algal partner in lichens cannot survive in a free-living state, but harboured within fungal tissues, it enjoys a sheltered habitat.

Other defence mutualisms are more targeted. The toxic effects of grasses such as perennial ryegrass and tall fescue pose a problem for livestock producers. These grasses harbour symbiotic **endophytic** ("inside the plant") **fungi** (Figure 15.15, p. 320) that produce bitter alkaloids (Clay 1988). These alkaloids are toxic to mammals (particularly domestic animals that have not evolved detoxifying defences) and to many insects. In mammals, the alkaloids constrict blood vessels in the brain, causing convulsions, tremors, stupor, and death. These fungi may also stimulate plant growth and seed production. In return, the plant supplies the fungus with energy.

A number of Central American ant species (*Pseudomyrmex* spp.) live in the swollen thorns of acacia (*Acacia* spp.) trees (see Figure 14.29). Besides shelter, the plants supply a virtually complete diet for all stages of ant development in *extrafloral* (outside the flower) nectaries and fat-rich bodies. In return, the ants protect the plants by swarming herbivores, emitting repulsive odours, and attacking intruders. They even trim back plant competitors (Janzen 1966). We examine other complex mutualisms that confer protection from negative interactions in Section 16.7.

A well-known defence mutualism is the non-symbiotic association between **cleaner organisms** and host fish in coral

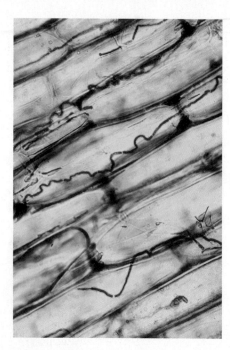

(a)

(b)

Figure 15.15 Endophytic fungi inside the cells of a blade of festuca (*Festuca* spp.) grass.

Figure 15.16 Cleaning mutualisms. **(a)** Bluehead wrasse eat ectoparasites on moray eel (Muraenidae). **(b)** The red-billed oxpecker feeds on parasites on the skin of an impala.

reefs. Cleaner shrimp and fish such as bluehead wrasse (*Thalassoma bifasciatum*) consume ectoparasites and diseased or dead tissue on host fish (Figure 15.16a), which presumably benefit from parasite removal. Cleaning mutualisms also occur in land habitats. The red-billed oxpecker (*Buphagus erythrorhynchus*) (Figure 15.16b) eats ticks and other parasites on the skin of large African mammals such as impala (*Aepyceros melampus*). Ecologists once assumed that the birds reduce the host's parasite load. However, experimental exclusion of the birds did not affect the tick load of cattle. If no parasites are available, oxpeckers peck a vulnerable area (often an ear) and draw blood. The relationship may thus be parasitic, not mutualistic (Weeks 2000).

15.11 REPRODUCTION MUTUALISMS: Many Plants Use Animals for Pollination

As sessile organisms, plants face difficulties in aspects of reproduction involving dispersal, not only of gametes but also of offspring. Whether a flower has both male and female parts or not, the goal of cross-pollination is to transfer pollen from the anther of one plant to the stigma of another plant of the same species. Some plants simply release their pollen—a cheap and efficient strategy for plants growing in open stands, such as grasses and pines. Wind dispersal is less reliable when individuals are scattered or in isolated patches. In such cases, plants are either self-pollinated or rely on insects, birds, or bats for pollen transfer in a non-symbiotic mutualism.

Plants entice pollinators by colour or odour, dusting them with pollen and rewarding them with rich foods: sugar-rich nectar, protein-rich pollen, and fat-rich oils. Such rewards are expensive to produce, and because nectar and oils are only of value to the plant as an attractant, they represent a trade-off in energy that might otherwise be allocated to growth. However, the benefits of pollination mutualisms are considerable, and biologists have long contended that the great increase in the diversity of flowering plant and insect species in the Cretaceous era (see Section 27.1) reflects coevolution between these groups. This hypothesis is strongly supported by recent studies showing that the majority of fossil plant taxa were insect pollinated (Hu et al. 2008).

Nectivores (nectar-feeders) visit plants to exploit a food source. They pick up pollen while feeding, and carry it to the next plant. The animal is thus a herbivore whose activity the plant capitalizes on for its own benefit. Most nectivores are generalists. Because each species flowers briefly, nectivores depend on a progression of flowers over the season, selecting against specialization. However, nectivores differ in body shape or ability to detect visual cues, and some plants have evolved specialized traits in response. Thus, bee-pollinated flowers are typically blue or yellow (bees cannot see red), whereas bird-pollinated flowers are typically red. Flowers that open at night (e.g., tobacco; *Nicotiana* spp.) are often luminously white and fragrant, to attract nocturnal pollinators such as moths. Nor do all flowers emit pleasant odours. Some, such as carrion flower (*Smilax* spp.), smell like rotting flesh, attracting flies and beetles that feed on carcasses. Some flowers have structural features such as landing platforms that facilitate insect visitors.

Many plants, such as blackberries, cherries, and goldenrods, are generalists with respect to pollinators. They flower profusely, providing an excess of nectar that attracts many insects, including bees, flies, and beetles. Others are more selective, screening their visitors to ensure efficient pollen transfer. These plants have traits such as long *corollas* (floral tubes), allowing access only to insects and hummingbirds with long tongues or bills and excluding small insects that eat nectar but do not transfer pollen (Figure 15.17). Some flowers have closed petals that only large bees can pry open. Orchids, which are often scattered, have evolved highly specialized mechanisms for pollen transfer and reception so that little pollen is lost when an insect visits flowers of other species. Some even involve aggressive mimicry, wherein the flower has evolved to mimic the female of its pollinating insect, as we described in the introduction to Part Four, p. 254.

Figure 15.18 Yuccas have a mutualistic interaction with adult yucca moths, whereas the moth larvae are parasitic on yucca seeds.

The risk of such extreme specialization is that the interaction can change from facultative (as are most pollination mutualisms) to obligate, with dependence on one pollinating partner. For example, the roughly 600 species of figs (*Ficus* spp.) have highly specialized wasp pollinators. Another plant group, yucca (*Yucca* spp.), depends exclusively on adult yucca moths (*Tegeticula* spp.) for pollination in a non-symbiotic, obligate mutualism that is combined with parasitism. Female moths lay eggs in the ovaries of yucca flowers, and the larvae are obligate, symbiotic parasites on developing yucca seeds (Figure 15.18). Although the yucca incurs seed losses, the pollination benefits outweigh these costs (Addicott 1986)—another case of ecological trade-offs.

Just as energy transfer mutualisms can evolve from parasitism, so too can pollination mutualisms evolve into other interactions. Some plants "cheat" their pollinators. They have odours or colours that attract pollinators and structural traits that facilitate pollen transfer, but offer no rewards. The mutualism has evolved into commensalism in which only the plant benefits. Some one-third of orchids (Orchidaceae, a highly evolved plant family) are cheaters (Galizia et al. 2005). The plant benefits from the energy saved, but if the flower is less likely to be pollinated, cheating might be a long-term disadvantage for the species. However, if the deceptive species evolves to resemble species that *do* offer nectar rewards—a variant of *Batesian mimicry* (see Section 14.9)—then pollinator visits are still likely. Some animals also "cheat" by robbing nectar without performing a pollinator function.

Figure 15.17 The extremely long bill of the sword-billed hummingbird (*Ensifera ensifera*) is uniquely adapted for extracting nectar from the long floral tubes of the passion flower (*Passiflora mixta*) in the Andes Mountains of South America.

15.12 DISPERSAL MUTUALISMS: Some Mutualisms Disperse Seeds or Fruits

Just as plants often rely on animals for pollen transfer, many plants with seeds too heavy to disperse by wind or water depend on animals to carry them away from the parent and deposit them in favourable sites. Some seed-dispersers are *granivores* (seed predators), eating the seeds for their own nutrition—another example of plants evolving to benefit from herbivory. Plants that depend on seed predators for dispersal produce a huge number of seeds. Recall the concept of *predator satiation* from Section 14.9; even if most seeds are eaten, the sheer number increases the chance that a few will find their way to suitable sites. Seed predator behaviour can facilitate the mutualism. Squirrels stash acorns for the winter by burying them—planting them, from the plant's perspective. Many will be forgotten and germinate the following season.

A close mutualistic interaction exists between wingless-seeded pines of western North America, such as whitebark pine (*Pinus albicaulis*), and several jay species. The geographic ranges of these species correspond, and is especially close between whitebark pine and Clark's nutcracker (*Nucifraga columbiana*). Only Clark's nutcracker has the morphology and behaviour to facilitate seed dispersal of whitebark pine. The bird carries up to 50 seeds in its cheek pouches, transports them as far as 20 km, and buries them deep enough to reduce predation by rodents. Only three to seven seeds are stored per cache, reducing competition for moisture and space after germination (Tomback 1982). As a result of the decline of whitebark pine from blister rust (see Section 15.2), this bird–pine dispersal mutualism is at risk. In sites where pine fall below a threshold density, Clark's nutcrackers leave the area, further threatening pine regeneration (McKinney et al. 2009).

Seed dispersal by ants is common among herbaceous species in deciduous forests, deserts, and shrublands. Such plants, called **myrmecochores**, have an oil-body on the seed coat called an *elaiosome* (see Figure 8.19b). Usually orange-yellow and shiny, elaiosomes contain energy-rich fats. Ants carry the seeds to their nests, where they sever the elaiosome and either eat it or feed it to their larvae. They discard the intact seed in abandoned nest galleries, which are richer in nitrogen and phosphorus than the surrounding soil and provide a good seedbed. By dispersing seeds, the ants also reduce plant losses to seed-eating rodents living near the parent plant.

Often it is not the seed but the fruit that is dispersed. All flowering plants enclose their seeds in fruits, but some have evolved fruit that is attractive to *frugivores* (fruit-eating animals) (Figure 15.19). Frugivores are not seed predators; they eat the tissue around the seed, normally without damaging the seed. Most frugivores consume more than fruits, which are deficient in protein and only seasonally available. Thus, to employ frugivores as dispersal agents, plants must attract them at the right time. *Cryptic colouration*, such as green unripened fruit amid green leaves, and chemical and structural defences, such as unpalatable texture, repellent taste, and hard surfaces, discourage consumption of unripe fruit. Later, when their seeds mature, fruits develop attractive odours, soft texture, increased sugar or oil content, and bright colours to attract frugivores.

Figure 15.19 Cedar waxwing (*Bombycilla cedrorum*) feeds on many tree fruits, including cherries (*Prunus* spp.).

As with seeds, fruit-dispersal mutualisms are non-symbiotic and usually facultative. Most fruits can be exploited by an array of animals. Such plants rely on quantity, producing many fruits to increase the odds that some seeds will find suitable sites. Such a strategy is typical in temperate regions, where frugivores are mostly generalists. Fruits dispersed by animals typically contain small seeds with hard seed coats that are resistant to digestive enzymes, allowing seeds to pass through the animal's gut unharmed. Indeed, exposure to digestive enzymes is sometimes needed to break seed-coat dormancy. Large numbers of small seeds are dispersed in feces, even if few land on suitable sites. In tropical forests, most tree species produce fleshy fruits dispersed by animals. Although frugivores are rarely obligates, exceptions include tropical fruit-eating bats (Figure 15.20), some of which rely on a single fig (*Ficus*) species. However, some bats eat more foods than once thought. The giant fruit-eating

Figure 15.20 The greater Indian fruit bat, *Pteropus giganteus*, is one of several forest-dwelling tropical bat species that specialize in eating fruits rather than insects.

bat, *Artibeus amplus*, also eats leaves year-round. Unlike grazers, they extract liquid from the leaves and discard the fibrous material (Ruiz-Ramoni 2011).

15.13 MULTI-LEVEL EFFECTS: Mutualisms Can Affect Communities and Ecosystems

Mutualism is easy to appreciate at the individual level. Consider an oak growing in a park. Even if we cannot see the partner, we can grasp the interaction between the oak and a mycorrhizal fungus. We can see squirrels and jays dispersing its acorns, and we can measure the cost of dispersal in terms of seeds consumed. Mutualism clearly benefits the fungus, the oak, and the seed predators. But what are the impacts on their populations, and on the community and ecosystem to which they belong? These higher-level effects are hard to quantify and even harder to predict.

At the population level, the effect is obvious for symbiotic, obligate mutualists. Remove species A, and species B perishes. Discerning the population effects of facultative, non-symbiotic mutualisms is more difficult. By definition, a mutualism exists only if the growth rate of species A increases with increasing density of species B, and vice versa (see Quantifying Ecology 15.1: Modelling Mutualism, p. 324). Such mutualisms are common with plants, which often depend on animals for pollination or seed dispersal. Although some pollination mutualisms are sufficiently specialized that loss of one population could cause extirpation of the other, the effects are usually subtler, requiring detailed demographic study to uncover. And when the mutualisms are diffuse, involving many species—as with pollinators and frugivores—the influence of a particular species–species interaction is especially difficult to quantify.

Some mutualisms involve **indirect interactions**, in which a relationship between two species is mediated by a third. Indeed, perhaps the greatest difficulty in evaluating the impact of mutualisms on community structure, is that many mutualisms involve these indirect interactions in which the species never come into direct contact. Consider the relations among conifers, mycorrhizae, and voles in the forests of the Pacific Northwest (Figure 15.21). As well as the typical mycorrhizal benefits, the fungus benefits from voles that feed on belowground fungal fruiting bodies (truffles) and disperse their spores, which then infect the roots of other conifers (Maser et al. 1978). This mutualism is not only *diffuse*, but *indirect*; the conifers receive no direct benefit from the voles. Also, because the vole–fungal interaction is *non-symbiotic* and *facultative*, its presence may go undetected.

We will examine others in Section 16.7, but such indirect interactions abound. Since the individuals never come into direct contact, their impact on each other is easily overlooked (see Research in Ecology: Do Fungal Associations Affect Plant Pollinators?, p. 325).

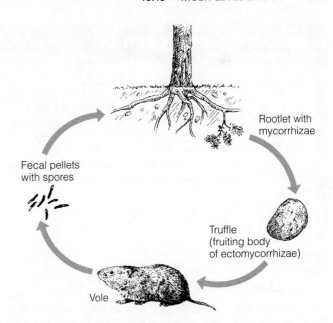

Figure 15.21 Indirect mutualism involving three species and both symbiotic and non-symbiotic interactions. Voles eat truffles, the belowground fruiting bodies of mycorrhizae. Through their feces, voles disperse the spores to locations where they infect new host plants.

Mutualisms (direct and/or indirect) may be as influential as either competition or predation on community structure and ecosystem function. These are new research frontiers. As reviewed by Simard and Durall (2004), **common mycorrhizal networks** are groups of plants linked by mycorrhizal connections. In laboratory studies, fungi colonize different species, providing a network for transfer of carbon, water, and minerals not just within one plant or species but among many. The direction and amount of transfer exhibits *source–sink dynamics*. A plant that is deficient acts as a "sink" and is a net importer from a "source" plant, much as we saw for source-sink dynamics for individuals in metapopulations (see Section 12.6). Transfer of organics has been demonstrated not only between Douglas-fir trees (*Pseudotsuga menziesii*) (Beiler et al. 2010) but also between Douglas-fir and a deciduous species, paper birch (*Betula papyrifera*) (Philip et al. 2011).

Such findings raise the possibility that an entire forest may be capable of cooperative exchange through mycorrhizal linkages, supplemented by plant-to-plant root grafts. This possibility lends new significance to mutualism, in virtue of its possible effects on seedling establishment, interspecific competition, species diversity, and community succession. Although extensive cooperative networks—humourously dubbed the *wood-wide web*—are possible, to what extent these networks function in the field is not yet established. In time, this research may support Lynn Margulis's claim that cooperation and mutual dependence, expressed through symbioses, are potent evolutionary forces: "Life did not take over the globe by combat, but by networking" (1998).

QUANTIFYING ECOLOGY 15.1   Modelling Mutualism

The simplest model for a two-species mutualistic interaction resembles the Lotka–Volterra model described in Chapter 13 for competing species. The crucial difference is that the two species influence each other's growth rate positively, not negatively. The negative competition coefficients α and β are replaced by positive coefficients, reflecting the beneficial per capita effect of an individual of species 1 on species 2 (α_{12}), and of an individual of species 2 on species 1 (α_{21}):

$$\text{Species 1:} \quad \frac{dN_1}{dt} = r_1 N_1 \left(\frac{K_1 - N_1 + \alpha_{21} N_2}{K_1} \right)$$

$$\text{Species 2:} \quad \frac{dN_2}{dt} = r_2 N_2 \left(\frac{K_2 - N_2 + \alpha_{12} N_1}{K_2} \right)$$

All the terms are analogous to those in the competition equations, except that $\alpha_{21} N_2$ and $\alpha_{12} N_1$ are added to, not subtracted from, the respective population densities (N_1 and N_2). This model only applies to facultative mutualisms in that the carrying capacities of the species are positive values (> 0)—that is, each population can grow in the absence of the other. In effect, the presence of one species increases the carrying capacity of the other.

To illustrate this model, we define values for the parameters r_1, r_2, K_1, K_2, α_{12}, and α_{21}:

$$r_1 = 3.22; K_1 = 1000; \alpha_{12} = 0.5;$$
$$r_2 = 3.22; K_2 = 1000; \alpha_{21} = 0.6$$

As with interspecific competition, we first calculate zero isoclines (Figure 1) by defining for species 1 the values of N_1 and N_2 where dN_1/dt equals zero. Given that the equation is a linear function, we can define the isocline by solving for two points. Likewise, we can solve for the species 2 zero isocline.

Unlike the outcomes of the competition equations (see Figure 13.1), the zero isoclines extend *beyond* the carrying capacities of the two species (K_1 and K_2), because the carrying capacity of each is increased by the presence of the other. If we use the equations to project population sizes over time (Figure 2), each attains a higher density when the other is present than on its own.

As with predation, more recent models treat mutualism as a consumer–resource interaction that is subject to stochastic forces (Holland and DeAngelis 2010).

1. On the graph displaying the zero isoclines (Figure 1), plot the four points listed below and indicate the direction of change for the two populations: $(N_1, N_2) = 500, 500$; $(N_1, N_2) = 3500, 3000$; $(N_1, N_2) = 3000, 1000$; and $(N_1, N_2) = 1000, 3000$. What outcome do the isoclines indicate for the interaction between these two species?

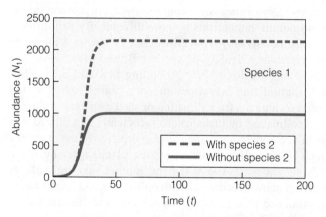

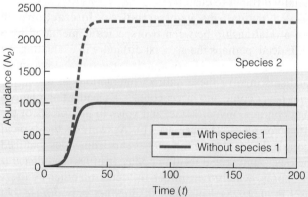

Figure 2 Population trajectories for a pair of facultative mutualists with and without their partners. Equation parameters are presented in the text.

(Adapted from Morin 1999.)

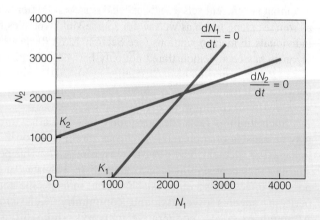

Figure 1 Zero isoclines for species 1 (N_1) and species 2 (N_2) based on the modified Lotka–Volterra model for two mutualistic species.

RESEARCH IN ECOLOGY Do Fungal Associations Affect Plant Pollinators?

Elizabeth Elle, Simon Fraser University.

Plants, animals, and fungi interact with each other in many and diverse ways. Some interactions are obvious—a fungus feeding on a decomposing log or an insect pollinating a plant. Although such direct interactions can be complex, they are the easiest to study, because the two species can be isolated in a field or laboratory setting. However, ecologists acknowledge that indirect interactions are not only ubiquitous in natural communities but also critical to population dynamics and ecosystem function. Early studies of multi-species interactions focused on the indirect effects of predation on populations at other trophic levels (see keystone predation and trophic cascades in Sections 16.4 and 16.8). Indirect interactions involving mutualisms can have similarly cascading effects. Two Canadian researchers recently combined forces to investigate one such interaction: the influence of mycorrhizae on insect pollinators.

Elizabeth Elle is a pollination biologist at Simon Fraser University. Her research centres on (1) selection pressures exerted by insects on the evolution of plant mating systems and (2) the impact of among-site variation in pollinator diversity on fragmented landscapes. She employs many methods to test the evolution of traits, including controlled experiments, observations of natural populations, and genetic markers. James Cahill, a plant ecologist at the University of Alberta, also uses a range of methods, from field and lab experiments to molecular markers, but his foci are (1) plant competition and its importance for community attributes and (2) plant foraging behaviour. Elle's and Cahill's research areas rarely overlap, but at a symposium on interactions, their paths intersected. Cahill was reporting on the impact of mycorrhizae on plant community structure, and Elle suggested that he investigate how changes in the community affect pollinators. The collaborative project that resulted was featured on the July 2008 cover of *Ecology*.

How can a belowground, symbiotic mutualism between fungi and roots affect an aboveground, non-symbiotic mutualism between flowers and insects? It is well known that mycorrhizae not only facilitate plant growth but also influence floral traits and species composition—two factors with the potential to affect plant–pollinator interactions. Moreover, a recent greenhouse study showed that mycorrhizal associations increase visitation rates of pollinators by increasing the number of flowers and/or pollinator rewards (Gange and Smith 2005). But the impact in a natural community is potentially more complex, so Elle and Cahill designed a field experiment to test whether these multi-species mutualisms were interdependent in co-occurring species.

The research team, including Cahill's graduate student, Bryon Shore, set up 20 blocks in a multi-species fescue grassland in Alberta in 2003 (Figure 1). Each block contained two 2 m × 2 m plots, separated by a 0.5-m buffer. One plot per block was treated biweekly with a fungicide (benomyl) to suppress mycorrhizal growth; the other was an untreated control. The researchers determined that benomyl did not attract or repel insects. In 2005, these variables were monitored biweekly for six plant species during the flowering season: (1) fungal colonization of roots; (2) floral display (number and size of flowers); and (3) number of insect visitors (grouped by body size and taxonomy).

Fungal colonization differed among species and was significantly reduced by benomyl. There were fewer visits by bumblebees to benomyl-treated plants overall, but visits by small-bodied bees

and flies increased (Figure 2). For all species combined, benomyl did not affect total visits per plot, but reduced visits per flowering stem by 67 percent (Figure 3). Total visits decreased for *Aster* and *Solidago* (sunflower) with benomyl treatment but increased for *Cerastium* (chickweed). Visits declined sharply per stem for *Aster*. Visits by larger pollinators such as bumblebees, which prefer *Aster* and *Solidago*, fell sharply. Suppressing mycorrhizae affected only floral morphology of *Cerastium*, which had larger flowers and more flowers per stem. Species number was unaffected, but *Cerastium* became a greater proportion of the community.

James Cahill, University of Alberta.

Clearly, disrupting belowground associations between roots and mycorrhizal fungi has aboveground consequences. Relative abundance of the pollinator community was altered, as was the number of visits per stem, which is likely to influence pollination success. These effects were not due to a direct impact on floral morphology. Mycorrhizal suppression did not make the flowers less attractive to pollinators, as it had in a previous study (Gange and Smith 2005).

What is the likely mechanism? Disruption of mycorrhizae could change competitive interactions among plants, particularly if some are more reliant on mycorrhizae than others. In control plots, *Cerastium* had among the lowest fungal colonization rates (5.6 percent), suggesting it has minimal reliance on mycorrhizae. When species that are more dependent on fungal symbionts (such as *Solidago*, 31.3 percent) are treated with fungicide, their growth will be compromised more, allowing *Cerastium* to increase in relative abundance. Thus, suppressing mycorrhizae may have indirect effects by influencing relative competitive ability. Whatever the cause, these changes in plant relative abundance affect community-level floral display, which influences composition of the pollinator community. Increased floral display of *Cerastium* may interfere with visual cues used by bumblebees to locate their preferred species, *Aster* and *Solidago*.

Many questions arise from this study, which examines the immediate impact of suppressing fungi. How do the differences in the pollinator community and the decline in per stalk visits affect seed-set, population stability, and long-term plant community composition? If some insects are more effective pollinators for different species, these short-term changes in pollen services could affect plant fitness. Are the relative competitive abilities of these plants, which rely heavily on clonal growth, actually affected by fungal suppression? Since mycor-

continued on page 326

RESEARCH IN ECOLOGY continued

Figure 1 Native fescue grassland in Kinsella, Alberta, with *Solidago* in foreground.

rhizae affect mineral uptake, does mycorrhizal suppression diminish root foraging efficiency? How would these possible impacts on plant fitness and/or competitive ability be affected by more complete and longer-term fungal suppression? Finally, are the results due in part to confounding factors, such as effects of benomyl on bacteria, non-mycorrhizal fungi, nematodes, and soil properties?

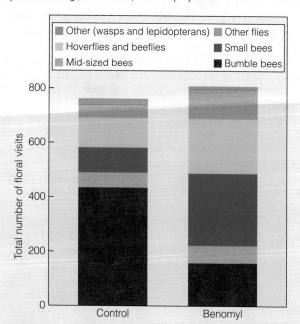

Legend:
- Other (wasps and lepidopterans)
- Hoverflies and beeflies
- Mid-sized bees
- Other flies
- Small bees
- Bumble bees

Figure 2 Impact of mycorrhizal suppression on the proportion of different groups of insect pollinators visiting six species in a fescue grassland.

(Adapted from Cahill et al. 2008.)

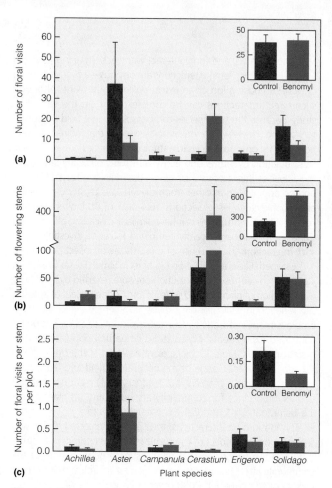

Figure 3 Impact of mycorrhizal suppression on number of floral stalks, total pollinator visits, and number of visits per stalk on six plant species (inset boxes show results for all species combined)

(Adapted from Cahill et al. 2008.)

Research into the effects of multi-species interactions is in its infancy, but Cahill and Elle's research explores the possible ecological and evolutionary consequences of altered pollination services caused by disrupting mycorrhizal associations. It points the way to a new direction in ecology, one attuned to the downstream effects of interactions once studied in isolation.

Bibliography

Cahill, Jr., J. F., E. Elle, G. R. Smith, and B. H. Shore. 2008. Disruption of a belowground mutualism alters interactions between plants and their floral visitors. *Ecology* 89:1791–1801.

Gange, A. C., and A. K. Smith. 2005. Arbuscular mycorrhizal fungi influence visitation rates of pollinating insects. *Ecological Entomology* 30:600–606.

1. This is a short-term study. How would the researchers test for the longer-term impact on the reproductive fitness of the six species?

2. *Cerastium* has among the lowest fungal colonization in the controls, and yet it is the species that (i) is affected most in floral morphology and (ii) benefits most (in terms of higher relative biomass) from mycorrhizal suppression. Explain why.

EcologyPlace

Visit EcologyPlace at www.pearsoncanada.ca/ecologyplace to access online resources that complement your textbook, and help you to apply and to review the information in this chapter. EcologyPlace includes

- an eText version of the book
- self-grading quizzes
- glossary flashcards
- and more!

Go to www.pearsoncanada.ca/ecologyplace and follow the registration instructions on the Student Access Code Card included with this text. If your book does not have a Student Access Code Card, you can purchase access to it at www.pearsoncanada.ca/ecologyplace.

SUMMARY

Types of Parasites 15.1

Parasitism is a symbiotic relationship between individuals of two species in which one benefits and the other is harmed. Microparasites, which include viruses, bacteria, and other disease agents, are small, multiply rapidly in the host, and spread by direct transmission. Macroparasites are relatively large and include parasitic worms, lice, ticks, fleas, smuts, and rusts. They have a long generation time, rarely multiply directly in the host, and spread by either direct or indirect transmission. Parasitic plants are classed as hemiparasites or holoparasites depending on whether they are photosynthetic or not, respectively. Ectoparasites live on external surfaces, while endoparasites live within their host. Their life cycles revolve around the twin problems of gaining entrance to and escaping from their host. In brood parasitism, a parasitic bird or insect deposits its eggs in the nest of a host species.

Parasite Transmission 15.2

Parasite transmission may occur directly between host organisms, either by physical contact or through air, water, or another substrate. Some parasites are transmitted between hosts via other organisms, which function as biotic vectors. Other parasites require multiple hosts. Indirect transmission takes them from definitive to intermediate to definitive host.

Immediate Host Response 15.3

Hosts respond to parasitic infections through behavioural changes, inflammatory responses at the infection site, and activation of the immune response. Chronic infection increases susceptibility to other diseases.

Effects on Host Individuals 15.4

A heavy parasite load can decrease host reproduction. Apart from parasitoids, most parasites do not kill their hosts directly, but mortality can result indirectly from increased vulnerability to predation and other factors.

Population Effects 15.5

Parasites can have density-independent effects. When introduced to a population that has not developed resistance, parasites can spread quickly, causing high mortality and even extinction. In some cases, parasites regulate host populations in a density-dependent manner.

Evolutionary Effects 15.6

The host–parasite relationship selects for traits that avoid infection or minimize its effects. Coevolution occurs between parasite and host. Parasites that require their host to survive to reproductive age (vertical transmission) may become less virulent over time.

Origin of Mutualism 15.7

Mutualism is a positive reciprocal relationship between two species that may evolve from predatory or parasitic relationships. According to the endosymbiotic hypothesis, chloroplasts and mitochondria evolved from an interaction between bacteria and eukaryotic cells. Where adaptations merely counter negative impacts, the relationship is commensalism. An interaction may fluctuate between parasitism and mutualism depending on abiotic factors.

Mutualism Categories 15.8

Mutualisms are classified according to types of benefits (nutrient transfer, protection or defence, gamete transfer, or dispersal); degree of dependency (obligate or facultative); specificity (generalist or specialist); and duration (life-long or during a particular developmental stage).

Nutrient Transfer Mutualisms 15.9

Many symbiotic mutualisms involve transfer of energy and/or minerals. Ruminants harbour microorganisms that aid digestion. Some plants associate with bacteria that fix N_2 in root nodules. Fungi form mycorrhizal associations with roots, facilitating uptake of minerals and water.

Defence Mutualisms 15.10

In many symbiotic energy transfer mutualisms, the host provides a protected, interior habitat. Cleaner mutualisms are a non-symbiotic mutualism that may protect the host from parasites. Some indirect mutualisms ameliorate the negative effects of interspecific competition.

Reproduction Mutualisms 15.11

Non-symbiotic mutualisms are involved in pollination of many plants. While extracting nectar, the pollinator transfers pollen to other plants. Some species have evolved traits that attract specialized pollinators or permit only certain animals to access nectar. Others cheat their pollinators by not offering a reward.

Dispersal Mutualisms 15.12

Non-symbiotic mutualisms facilitate seed dispersal. Plants that rely on seed predators for dispersal produce many seeds to increase the odds that some are carried to suitable sites.

Seeds dispersed by ants have oil-rich bodies called elaiosomes. Some plants develop fruit that attracts frugivores, which eat the tissue surrounding the seeds and pass them in their feces.

Multi-level Effects 15.13

Ecologists are investigating the impact of mutualisms—direct and indirect—on population dynamics, community structure, and ecosystem function. Complex networks involving mycorrhizae and root grafts may allow forests to share resources among individuals.

KEY TERMS

arbovirus	disease	facultative	infection	mutualism
brood parasitism	ectomycorrhizae	hematophagic	inflammatory response	myrmecochore
cleaner organism	ectoparasite	hemiparasite	intermediate host	obligate
commensalism	endomycorrhizae	holoparasite	intermittent parasite	rhizobia
common mycorrhizal	endoparasite	horizontal transmission	kleptoparasitism	symbiosis
network	endophytic fungi	immune response	macroparasite	vector
definitive host	endosymbiotic	indirect interaction	microparasite	vertical
direct transmission	hypothesis	indirect transmission	(pathogen)	transmission

STUDY QUESTIONS

1. In both true predation and parasitism, an individual of one species derives its energy and nutrients from consuming an individual of another species. How do these two interactions differ?

2. Distinguish among *symbiosis*, *parasitism*, and *mutualism*. How are they related from an evolutionary perspective?

3. Some parasites have complex indirect transmission involving multiple hosts (see Figure 15.4). This situation would seem to make the parasite more vulnerable, as it has to form associations with different species over its lifetime. How would such a life history evolve?

4. How might a clumped dispersion of hosts affect parasite transmission? What pattern of hosts (random, uniform, or clumped) would present the most difficulty for transmission?

5. As described in Section 15.1, yellow warblers reject cowbird eggs by burying them along with their own eggs. Other parasitized birds eject cowbird eggs by physically removing them with their beaks, sometimes dropping them some distance from their nest. The beaks of yellow warblers are not large enough to make ejection of the large cowbird eggs feasible. Relate this situation to two concepts discussed earlier: (i) "escape space" as part of a species niche (see Section 14.8) and (ii) ecomorphology, whereby morphological traits are used to infer niche (often feeding niche; see Section 13.12).

6. Distinguish between an *inflammatory* and an *immune response*. Why do some hosts have both?

7. Distinguish between *holoparasites* and *hemiparasites*. Why would a plant species evolve to lose its photosynthetic function?

8. Apart from parasitoids, parasites rarely kill their host directly. How can they regulate the population of their host?

9. Why are parasites that spread by vertical transmission less virulent than those that spread by horizontal transmission?

10. Distinguish between *obligate* and *facultative* mutualism. Which is more likely to involve symbiosis? Can you think of an example of an obligate, non-symbiotic mutualism?

11. Is mutualism reciprocal exploitation, or cooperation for mutual benefit? How would you distinguish between these two possibilities in a particular relationship?

12. Explain the four general categories of benefits in mutualisms. Can you think of examples in which two or more different kinds of benefits are combined?

13. How can orchids that "cheat" their pollinators be seen as analogous to Batesian mimicry?

14. Is reliance on frugivory a risky way of distributing seeds? Why?

15. Many interactions occur simultaneously. What social function could be facilitated by removal of ectoparasites by grooming?

16. The existence of root grafts and mycorrhizal networks does not prove that trees are sharing minerals and/or organics. How would you establish such transfer experimentally?

FURTHER READINGS

Barth, F. G. 1991. *Insects and flowers: The biology of a partnership.* Princeton, NJ: Princeton University Press.
Excellent overview of the ecology of plant–pollinator interactions.

Dobson, A. P., and E. R. Carper. 1996. Infectious diseases and human population history. *BioScience* 46:115–125.
Discusses the role of infectious disease in human history.

Handel, S. N., and A. J. Beattie. 1990. Seed dispersal by ants. *Scientific American* 263:76–83.
Fascinating discussion of the mutualistic relationship between ants and plants.

Hatcher, M. J. and A. M. Dunn. 2011. Parasites in ecological communities: from interactions to ecosystems, New York, Cambridge University Press.
Investigates the role parasites play in influencing interactions both detrimental and beneficial among competitors, predators, and their prey.

Klironomos, J. N., and M. M. Hart. 2001. Animal nitrogen swap for plant carbon. *Nature* 410:651–652.
Proposes that plants indirectly prey on animals through fungal associations.

Muscatine, L., and J. W. Porter. 1977. Reef corals: Mutualistic symbioses adapted to nutrient-poor environments. *BioScience* 27:454–460.
Discusses the symbiosis involved in coral reefs, the most diverse aquatic ecosystem on Earth.

Stachowicz, J. 2001. Mutualism, facilitation, and the structure of ecological communities. *BioScience* 51:235–246.
Reviews current research into the role of positive interactions in population dynamics and community structure.

When settlers first explored western North America, they encountered a landscape on a scale unlike any in Europe. The eastern forests gave way to a vast expanse of grass and wildflowers that seemed to extend endlessly westward. The prairies once covered a large portion of North America, from Illinois and Manitoba in the east to the foothills of the Rockies, extending south to Texas. Less than 1 percent of native prairie remains today, mostly in isolated patches, the result of widespread transformation to agriculture.

Canada's largest remaining tallgrass preserve is near Tolstoi, Manitoba. Covering only 20 km² of the biome's original extent of 6000 km² in the province, the Tolstoi preserve supports 200 plant and 90 bird species. Several species, including the western prairie fringe orchid (*Platanthera praeclara*) and the Poweshiek skipperling butterfly (*Oarisma poweshiek*) occur nowhere else in Canada. The world's largest tallgrass prairie preserve is in Oklahoma, covering 180 km² and supporting 750 plant and 300 bird species. A herd of 2500 bison graze the site, in recognition of the role of herbivory. There are larger preserves in North America for short- and mixed-grass prairies, largely because their dry conditions are less favourable for cropping. As of 2011, Saskatchewan's Grasslands National Park supports a bison herd of 253. The park has played a key role in bringing back several species from near extinction, such as the swift fox (see Section 27.8).

So little tallgrass prairie remains that efforts began in the 1960s to replace abandoned fields and pastures with native prairie—the plants, animals, fungi, and microbes that once occupied the area. But how is a community rebuilt? Can it be constructed just by assembling a set of species?

Early restoration efforts often met with failure. Whatever native species were available as seeds were planted, typically on small plots surrounded by croplands. The species established, but their populations declined over time. These first attempts underestimated the role of disturbances in maintaining prairies. Fire is an important selective force in prairies, and many of the species are adapted to periodic burning. Without fire, many native species were displaced by non-native weeds from adjacent pastures. Lacking a full complement of plant species, these restored sites failed to attract and support native prairie fauna. Prairies contain a diverse array of plants that differ in the timing of their germination, growth, and reproduction over the season. This shifting dynamic provides a shifting resource base for consumers, both biophages and saprophages.

The scale of these early restoration projects also contributed to their failure. Small, isolated prairie fragments support species in low populations that are prone to local extinction (see Section 10.10). Isolated patches often lack the necessary pollinators and are too distant from other patches for reliable dispersal of plants and animals. Yet much was learned, and many restoration efforts have since succeeded. After 40 years, restored sites at Fermi National Accelerator Laboratory in Illinois cover roughly 4 km²—the world's largest restored prairie habitat.

The restoration project on the land surrounding the Fermi National Accelerator Laboratory near Chicago, Illinois is the largest restored prairie habitat in the world.

The western prairie fringed orchid is an endangered species in Canada, occurring only in the Manitoba Tallgrass Prairie Preserve near Tolstoi. Because it does not normally reproduce asexually, its reproductive success is often limited by pollinator visits.

Attempting to preserve or restore prairie raises fundamental questions about community structure and dynamics. The community is the set of all species (producers and consumers, biophages and saprophages) that occupy an area. Some are abundant, while many are rare.

In Part Five, we explore the following questions, building on our treatment of populations in Part Three and interactions in Part Four: *What controls the relative abundance of different species in a community? Are all species equally important to community function and persistence? How do interactions among species restrict or enhance the presence of other species? How do communities change through time? How does the area occupied by a community influence the number of species it supports? Why does the same type of community (such as the remnant prairies in Manitoba and Oklahoma) support such different numbers and types of species? How do different communities on a landscape interact?*

In Chapter 16, we examine the biological and physical structure of communities and the factors that affect it. We consider how the concepts of the tolerance range and the fundamental niche of species influence their distribution and abundance, and how interspecific interactions modify species distribution and abundance, thereby affecting community structure. In Chapter 17, we discuss succession: how and why communities change through time. Finally, in Chapter 18, we examine how abiotic factors (topography, soils, climate, and geology); biotic forces; and disturbance interact to create a landscape mosaic. Above all, know this: the community is not just a collection of species with overlapping geographic ranges. It is a complex web of interacting entities whose structure and function change in time and space.

Coral reefs inhabiting shallow coastal waters of subtropical and tropical oceans are among the most biologically diverse communities on our planet.

Walking through a forest or canoeing across a lake, we encounter many organisms that share a habitat. In the broadest sense, **community** is the totality of all populations of all species occupying a given area, interacting directly or indirectly. But ecologists rarely study the entire community, so the term often refers to a subset, such as the plant, fungal, bird, small mammal, or fish community. This more restricted use suggests similarity among community members in their taxonomy or ecological function. Yet, no matter how narrowly or broadly the term is applied, the community is a spatial concept—the collective of co-occurring species within a defined boundary. Individuals of species living in such close association may interact. Some may compete for food or light. Some may prey on others. Some may provide mutual aid, and some may not affect each other at all.

Like any entity in the ecological hierarchy, a community has **emergent properties**—attributes that apply only to the collective. These attributes include aspects of its biological structure, such as the number and relative abundance of its constituent species, and its physical structure, as influenced by the growth forms of its members. In this chapter, we examine community structure. But science goes beyond description to explore the processes that shape patterns by posing questions and testing hypotheses. *Why are some communities more diverse than others? What is the relative importance of biotic interactions and abiotic factors for community structure? How do communities respond to the loss or introduction of a species?* Earlier, we examined topics that pertain to these issues, including adaptations to abiotic factors, evolution of life history traits and their impact on population growth, and regulation of populations by intra- and interspecific interactions. Now we integrate topics arising at lower levels of the hierarchy to comprehend the processes that control community structure.

16.1 SPECIES DIVERSITY: Species Number and Abundance Define Biological Community Structure

The species mix of a community—their number and relative abundance—defines its biological structure. Perhaps the most basic measure is **species richness**—the number of species present (*S*). Although species richness is conceptually simple, it can be difficult to determine (see Quantifying Ecology 16.1: Estimating Species Richness, pp. 335–336) and to interpret. And yet, the latitudinal gradient of increasing richness from the poles to the tropics is one of the most consistent patterns in global ecology. We discuss global richness trends in Section 27.2 and the implications of richness for ecosystem stability in Sections 19.6 and 19.7.

Whatever the richness of a community, not all of its species are equally abundant. To estimate the **relative abundance** of a species, ecologists count all the individuals of each species in a number of samples and determine what percentage each species contributes to the total number of individuals of all species. Consider the tree community in two deciduous forests in eastern North America. A total of 24 species are present in the sample from the first forest (Table 16.1). The two most

Table 16.1 Structure of a Mature Deciduous Forest in West Virginia

Species	Number of Individuals	Relative Abundance (Percentage of Total Individuals)
Yellow poplar (*Liriodendron tulipifera*)	76	29.7
White oak (*Quercus alba*)	36	14.1
Black oak (*Quercus velutina*)	17	6.6
Sugar maple (*Acer saccharum*)	14	5.4
Red maple (*Acer rubrum*)	14	5.4
American beech (*Fagus grandifolia*)	13	5.1
Sassafras (*Sassafras albidum*)	12	4.7
Red oak (*Quercus rubra*)	12	4.7
Mockernut hickory (*Carya tomentosa*)	11	4.3
Black cherry (*Prunus serotina*)	11	4.3
Slippery elm (*Ulmus rubra*)	10	3.9
Shagbark hickory (*Carya ovata*)	7	2.7
Bitternut hickory (*Carya cordiformis*)	5	2.0
Pignut hickory (*Carya glabra*)	3	1.2
Flowering dogwood (*Cornus florida*)	3	1.2
White ash (*Fraxinus americana*)	2	0.8
Hornbeam (*Carpinus carolinia*)	2	0.8
Cucumber magnolia (*Magnolia acuminata*)	2	0.8
American elm (*Ulmus americana*)	1	0.39
Black walnut (*Juglans nigra*)	1	0.39
Black maple (*Acer nigra*)	1	0.39
Black locust (*Robinia pseudoacacia*)	1	0.39
Sourwood (*Oxydendrum arboreum*)	1	0.39
Tree of heaven (*Ailanthus altissima*)	1	0.39
	256	100.00

Table 16.2 Structure of a Second Deciduous Forest in West Virginia

Species	Number of Individuals	Relative Abundance (Percentage of Total Individuals)
Yellow poplar (*Liriodendron tulipifera*)	122	44.5
Sassafras (*Sassafras albidum*)	107	39.0
Black cherry (*Prunus serotina*)	12	4.4
Cucumber magnolia (*Magnolia acuminata*)	11	4.0
Red maple (*Acer rubrum*)	10	3.6
Red oak (*Quercus rubra*)	8	2.9
Butternut (*Juglans cinerea*)	1	0.4
Shagbark hickory (*Carya ovata*)	1	0.4
American beech (*Fagus grandifolia*)	1	0.4
Sugar maple (*Acer saccharum*)	1	0.4
	274	100.0

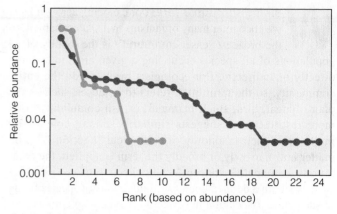

Figure 16.1 Rank–abundance curves for the forests in Tables 16.1 and 16.2. Relative abundance (*y*-axis; $\log_{10}$) of each species is graphed from most to least abundant. The forest in Table 16.1 (brown line) has greater richness and evenness than that in Table 16.2 (orange line).

INTERPRETING ECOLOGICAL DATA

Q1. How and why does the slope of the curve vary with species evenness?

Q2. What would the rank–abundance curve look like for a forest containing 10 species, where all species are equally abundant? Why is this situation ecologically unlikely?

abundant species, yellow poplar and white oak, make up 30 and 14 percent, respectively, of the total number of individuals. The four next most abundant species each make up about 5 percent, with the remaining 18 species contributing smaller proportions. The second forest is of the same general type, but differs in biological structure. It contains only 10 species, of which two (yellow poplar and sassafras) together make up 83.5 percent of the total individuals (Table 16.2).

Both forests illustrate the typical pattern of a few common species associated with many rare species, but they differ in species richness and relative abundance. One way to compare community patterns is with a **rank–abundance curve (Whittaker plot)**, which plots the relative abundance of each species against its rank. The most abundant species is plotted as the first point on the *x*-axis, with its relative abundance on the *y*-axis (log scale). The process is repeated for all species (Figure 16.1). The first forest has greater richness (hence a longer curve) and a more equitable distribution of individuals (**species evenness**), as indicated by the curve's more gradual slope.

Another way to compare communities is to calculate their **species diversity**, a measure that incorporates both richness and evenness, as quantified by **diversity indices** into a single value. (Section 27.2 explores how the diversity concept can also be applied at the regional and landscape scale.) Such indices are especially useful when richness and evenness exhibit different trends; that is, when one community has more species than another, but is less even in the relative abundance of its members.

An example is *Simpson's diversity index*, based on the probability that two individuals randomly chosen from a sample belong to the same species (Simpson 1949). There are several formulations of the index, but the most useful is the reciprocal (multiplicative inverse) version:

$$D = 1/\sum p_i^2$$

where p_i is the relative abundance of each species, expressed as a proportion:

$$p_i = n_i/N$$

In this formula, n_i is the number of individuals of each species *i*, and *N* is the total number of individuals of all species. (Although p_i is based on numbers for trees or animals, it can be based on cover for shrubs, low plants, fungi, or any species for which a count is infeasible.) The sum (Σ) of the p_i values of all the species present is then divided into 1.

The lowest possible value of *D* is 1, for a one-species community (monoculture). As richness and/or evenness increases, the index value increases, with its maximum value D_{max} = species richness (*S*). For example, there are 10 species in the forest in Table 16.2, so the maximum possible *D* for this community is 10. This maximum would only be reached in the unlikely event that all species were present in equal numbers. Species evenness (*E*) of the community is calculated as D/D_{max}. Evenness ranges from 0 to 1, where *E* = 1 is maximum evenness.

Another diversity index, the *Shannon index* or *Shannon–Wiener index* (*H*) (1963) also uses the relative abundance of each species (p_i), but is computed as

$$H = -\sum (p_i)(\ln p_i)$$

QUANTIFYING ECOLOGY 16.1   Estimating Species Richness

Species richness is the simplest measure of diversity. Unlike diversity indices, it does not incorporate evenness, but some consider richness the single most important community trait. Yet it is an elusive number. Robert May (1988) famously commented that if we were visited by aliens, they would likely ask how many distinct forms of life there are on Earth—a question that, he added, might embarrass us by our inability to answer. Measuring the richness of a single community is less daunting. But how do ecologists measure richness? They cannot identify each and every organism present, so they estimate richness by sampling.

Suppose our goal is to estimate the species richness of trees in a forest. First, we delineate the area of interest, in which we survey a number of sample plots (*quadrats*; see Section 9.5). Unless the community is a uniform plantation, we encounter more species as we survey more samples. How many samples do we need to get a reasonable estimate of richness? We compile a **species accumulation curve** (Figure 1), which plots the cumulative number of species encountered as more samples are added to the sample pool. Initially, the curve rises rapidly as we encounter the most common (abundant) species, and then more slowly as rarer species appear in later plots. Theoretically, the curve approaches an asymptote (not yet reached in Figure 1), with few if any additional species to be added. (In actuality, there will always be rare species that were missed in earlier plots.) The slope of the curve quantifies the rate at which new species are discovered as we increase our sampling effort, and can be used to extrapolate both species richness and the optimal (saturating) sample size, beyond which more sampling yields no more information about richness.

Suppose we wish to compare our findings with other studies. If an identical procedure was used, we can compare the results directly. But studies often differ in the number and/or size of samples. As illustrated by the accumulation curve in Figure 1, the richness estimate varies with sample number. We must subsample the larger survey to obtain a richness estimate that is comparable to the smaller survey, by using the same number of samples from each survey.

One method is to construct a **species rarefaction curve** by repeatedly re-sampling the total pool of N samples at random,

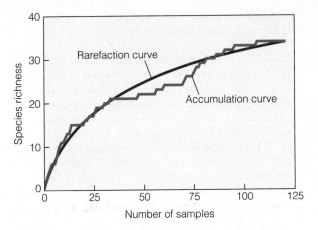

Figure 1 Species accumulation curve (blue) and rarefaction curve (purple) constructed from a survey of 125 samples. Sample size is on the *x*-axis; species richness on the *y*-axis.

plotting the number of species represented by 1, 2, . . ., N samples (see Figure 1). This rarefaction procedure generates the expected number of species in a subset of n samples drawn at random from the larger pool of N samples. As each sample contributing to the curve is equally likely to be included at any level of re-sampling, the rarefaction curve is related to the corresponding accumulation curve, but with its variation smoothed out. In essence, a rarefaction curve is an accumulation curve generated by random reorderings of the original samples. By constructing rarefaction curves for two surveys of differing sample size, ecologists can compare richness for any sample number up to the value of N samples that was employed for the smaller survey.

Let's apply this idea to real data. Table 1 lists the trees occurring in each of twenty 20 m × 20 m plots collected for a forest survey of Community A. (Codes are used for species names, but these data were the basis for Table 16.1.) Samples are presented from left to right, in order of collection. (Quadrat data are typically used in constructing rarefaction curves for plants, but animal data would derive from other methods, such as insect traps.

Table 1 Tree Species Surveyed on Twenty 20 m × 20 m Sample Plots (see Table 16.1)

1	2	3	4	5	6	7	8	9	10	11	12	13	14	15	16	17	18	19	20
dw	bg	mh	bg	bc	rc	bg	be	dw	br	ah	bc	be	Bh	bg	co	co	ah	bg	bg
rm	dw	rc	dw	mh	rm	mh	dw	hh	rm	hh	bg	br	Dw	bh	dw	rm	bw	dw	rm
ro	rc	rm	rc	rc	vp	rc	rb	rm	sg	rm	dw	dw	Po	dw	rm	sf	dw	mh	ro
tp	rm	tp	rm	so	wo	rm	rm	sg	sy	sg	wo	tp	Rm	ro	sf	sl	sg	rb	tp
wo	vp	wo	wo	vp		sl	ro	tp	wo	sy			Wo	so	vp	vp	tp	tp	wo
	wo					so	tp							vp			wo	wo	
						tp													
						vp													

continued on page 336

QUANTIFYING ECOLOGY 16.1   continued

If sampling involves individual captures, *number of individuals* rather than number of samples is on the *x*-axis.)

1. Using Table 1, first construct an accumulation curve. For the first data point, plot "1" for the number of samples (*x*-axis) and "5" for the number of species (*y*-axis). For two samples, combine the lists for the first two samples and count the number of unique species (8, since the samples have 3 species in common). Continue until all (*N* = 20) samples are included. With each new plot, you only increase the cumulative number of species if a new species is encountered.

2. How would you use these data to calculate a rarefaction curve? (This procedure is tedious by hand and is typically done by computer.) How would the rarefaction curve compare to the accumulation curve? Which curve

would provide a better basis for estimating the richness of Community A?

3. Instead of constructing a rarefaction curve, we will show how it works by calculating an estimate based on a subsample (*n*) of 5. Using 5 random numbers from 1 to 20 (available in any statistics textbook or online), re-sample 5 plots—e.g., plots 4, 7, 11, 16, and 19—and determine the cumulative number of species. Repeat twice more, using different random numbers. Take the mean of these 3 values as the estimate of richness based on a sample size of 5. How does your estimate compare with the estimate from the Community A accumulation curve at 5 and 20 sample points? What does this suggest about your sampling effort?

4. Our survey uses 20 m × 20 m plots. How would it affect our comparison if another survey used 10 m × 10 m plots?

When only one species is present, $H = 0$. Its maximum, when all species are present in equal numbers, is $H_{max} = \ln S$, where ln is the natural logarithm. The Shannon index gives more weight to rare species, so some favour it in situations where rare species are of particular concern. Species evenness (*J*) is calculated as H/H_{max} and, as with Simpson's *E* value, ranges between 0 and 1.

16.2 NICHE PACKING: Species Diversity Reflects Niche Number and Breadth

What factors determine the number, identity, and relative abundance of species in a community? The niche concept provides a useful starting point for exploring the factors responsible for biological community structure. Recall that the fundamental niche of a species (see Section 13.5) reflects in part its tolerance range for the totality of abiotic factors at work in a habitat (see Section 5.12). All organisms have a range of resources and conditions in which they can survive, grow, and reproduce. This range differs among organisms, both of the same and especially of different species, and reflects their structural, physiological, and behavioural traits.

Adaptations that allow an organism to thrive in some conditions limit its ability to do equally well in other conditions. For example, the traits that adapt plants to high light preclude them from succeeding in low light (see Section 6.9). Similarly, relying on ectothermy saves poikilotherms huge amounts of energy, but limits their daily and seasonal activity and their geographic range (see Section 7.9). Each suite of adaptations represents a solution (albeit imperfect) to the problems posed by one set of abiotic factors, but constrains the ability to adapt to another set. Considered in totality, such adaptations define the fundamental niche.

Environmental factors vary in time and space, and species differ in their tolerance ranges for these factors. This simple

fact goes a long way towards explaining biological community structure. For simplicity's sake, we represent the fundamental niches of species as bell-shaped curves along a gradient, such as availability of water or nutrients (Figure 16.2; note that whereas the tolerance range of the individual is defined in terms of its individual performance, the species response is depicted in terms of its abundance). Their fundamental niches overlap to varying degrees, and some niches will be narrow or broad, reflecting specialization for the factor in question, but each species has limits beyond which it cannot survive. The pattern of the number and relationships of niches in a community is called **niche packing**.

Distribution of fundamental niches along environmental gradients imposes a primary constraint on community structure. Given the factors prevailing in any site, only a subset of species can occur. As the factors change among locations, the possible species present and their relative abundances change too. Not all habitats have a single dominant environmental gradient, but the same principle applies to combinations of factors that change with location.

Consider the geographic ranges of three of the tree species occupying the forests in Tables 16.1 and 16.2 (Figure 16.3). Recall that geographic ranges reflect the occurrence of abiotic

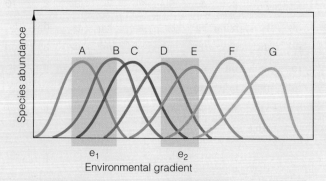

Figure 16.2 Fundamental niches of hypothetical species along an environmental gradient in the absence of interactions. As conditions change from e_1 to e_2, the set of species occurring changes.

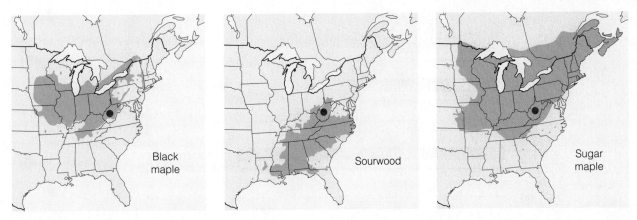

Figure 16.3 Geographic ranges of three tree species native to deciduous forests in eastern North America. The West Virginia site (red circle) falls within a limited region where their ranges overlap.

factors that fall within the tolerance limits of a species. These species have distinct ranges, and the West Virginia sites are in a small region where they overlap. Elsewhere, the set of species with overlapping ranges changes, and so too will the species composition of the forest. In deciduous forests in eastern Canada, sugar maple (*Acer saccharum*) is the only one of these three species that is common. Sourwood (*Oxydendrum arboreum*) is absent, and black maple (*A. nigra*), at its geographic northern limit, occurs only occasionally. Overall, richness is less in Canadian forests because winter lows fall below the tolerance limits of some species, but some species that are absent in southern forests, such as yellow birch (*Betula alleghaniensis*), join their ranks.

This view, that community composition is determined by fundamental niches, provides a starting point for our discussion. It assumes that species presence and abundance reflect the independent responses of species to prevailing abiotic factors, as well as to chance events. Interactions occur, but are assumed to have minimal effects on community structure. Given our discussions in Part Four, this assumption seems odd, but it is a useful default for comparing the actual patterns observed in a community. Indeed, this view functions as a *null model* (see Section 13.3) for experimental studies in which species interactions (competition, predation, parasitism, or mutualism) are tested by removing one species and examining the response(s) of the other(s). If the response does not differ after removal, we conclude that the interspecific interaction being tested does not affect the presence or abundance of the remaining species, which we assume to be governed by their response to abiotic factors and/or chance.

Yet much evidence indicates that interactions *do* influence species presence and abundance. Recall from Section 13.5 that interactions modify the fundamental niches of the species present into realized niches. Predation or competition may reduce abundance or even exclude a species, while commensalism and mutualism may enhance abundance or even expand the fundamental niche of a species. Because studies typically focus on only two (or at best a small subset of) species, they may underestimate the combined impact of these interactions on the community. As we explore in Section 16.7, indirect as well as direct interactions impact realized niches.

Many interactions are *diffuse*, involving multiple species. In a field study, selective removal of single species from test plots in

an old field (a crop field allowed to re-vegetate) had little community effect, suggesting that competitive interactions were weak (Fowler 1981). However, the response to removing groups of species was much stronger. Interactions between any two species may be weak, but collectively, diffuse competition may affect community composition. Similarly, in *diffuse mutualisms*, a species benefits from interacting with many species, as with plants that use many animals for pollination or dispersal. Many predators, including lynx, coyote, and horned owl, affect snowshoe hare cycles and hence plant community composition (see Section 14.12). Food webs represent only a subset of community interactions, but they exemplify their diffuse nature (see Quantifying Ecology 16.2: Quantifying Food Web Structure, p. 344).

As affected by the totality of such interactions—direct and indirect, one-on-one and diffuse—the realized niche represents not just the response of species to prevailing abiotic factors (what Chase and Leibold (2003) call the *requirements component*) but also its *impacts component*, reflecting the effect of a species on its environment. If the realized niche is compressed from the fundamental as a result of these requirements and/or impacts, then the pattern in Figure 16.2 is likely to alter, with more niches present and less overlap between them (Figure 16.4b; Figure 16.4a, p. 338 resembles Figure 16.2, the default state).

How does niche packing relate to community structure, especially richness? According to niche theory, narrower niches with less overlap and more specialization, such as in Figure 16.4b, typify communities with greater richness, such as tropical rain forests. Another way that communities can support more species is by having a greater diversity of resources available, allowing more niches even without closer packing (Figure 16.4a). In contrast, communities at high latitudes with harsher conditions are likely to have looser niche packing with more generalists, indicating more overlap (Figure 16.4c), and/or more vacant niches resulting from disturbance (Figure 16.4d). Species still interact, but abiotic factors impose more constraints than biotic interactions. Harsh conditions keep populations below their carrying capacity, minimizing the impact of overlapping niches while selecting for tolerance of extreme conditions.

In each portion of Figure 16.4, all niches are depicted as changing in the same way, whereas in real communities, there will be some generalists, some specialists, some with much

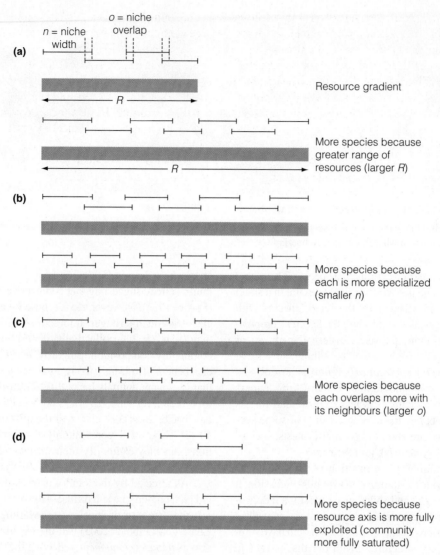

Figure 16.4 Niche packing and community structure. **(a)** Default state in which species richness reflects the number of fundamental niches a community can support, assuming interactions have minimal effect. **(b)** Greater specialization of fundamental niches and/or compression of realized niches resulting from interactions allow more niches, with less overlap. **(c)** Greater diversity of resources provides more niches, with or without altered overlap. **(d)** Harsh sites where abiotic factors fall outside tolerance limits of many species will have looser niche packing, with more generalists. In harsh sites, greater niche overlap is tolerated because populations are kept below carrying capacity by harsh and/or unpredictable abiotic factors. Similarly, disturbances can remove species from a habitat, lowering species richness and creating niche vacancies.

(Adapted from Begon et al. 1996.)

overlap, some with minimal, and so on. Nevertheless, niche packing provides a useful model for understanding community structure. We contrast the niche model with Stephen Hubbell's *neutral theory* (2001) in Section 16.13.

16.3 DOMINANCE: Ecologists Use Many Criteria to Evaluate Dominance

Although the species richness of the two forests in Tables 16.1 and 16.2 differs by more than twofold, they both contain a few common species and many species that are scarce. This feature is typical of most communities. If a single (or a few) species predominates, greatly influencing the identity and abundance of other species present, that species is called a **dominant** (the associated impact on other species is called

dominance). The less diverse a community, the greater its dominance by a few species. Recall that the forest in Table 16.2 is less diverse than that in Table 16.1, with fewer species, lower evenness, and more individuals belonging to the two most common species. Dominants are usually defined for different taxonomic or functional groups. Yellow poplar is a dominant in both forests, but we could also identify the dominant herbaceous (non-woody) plant, which might differ between the forests, or the dominant bird or small mammal.

How do ecologists identify dominants? We often assume dominance means the most numerous, as indicated by *relative abundance* (or *relative density*). This criterion may work well for animals, but for modular species that vary greatly in individual size (see Section 9.1), abundance alone is insufficient. Small trees may be more numerous, but a few large trees may account for most of the biomass and have the most impact on abiotic factors affecting the community. In this case, we define dominance based on criteria that include both number and individual size.

For trees, ecologists may calculate *relative basal area*, determined by calculating the total trunk basal area for each species as a proportion of the total basal area of all individuals of all species. For herbaceous species where density is hard to estimate and of limited meaning, *relative cover* (total leaf area of each species as a proportion of total leaf area of all herbaceous species) is more appropriate, or, if harvesting is feasible, *relative biomass*.

Two species may have similar density and individual size, but if one is scattered and the other clumped, they may affect the community differently. *Relative frequency*, or how often a species occurs relative to other species, may then be a dominance criterion. Like any criterion, frequency estimates are based on sampling, but unlike density or cover, frequency is highly sensitive to sampling procedure. If an ecologist uses quadrats, cover estimates are quite robust with different quadrat sizes, assuming enough quadrats are sampled in a statistically valid manner. However, frequency varies greatly with quadrat size; if an ecologist uses 10 m × 10 m quadrats, the frequency of a species will be higher than with 5 m × 5 m quadrats.

Where no single criterion is adequate, or when different criteria yield different answers to the question of which species is dominant, ecologists may calculate an **importance value**, a composite value based on the sum of the relative value of two or more of the following variables: density, cover, frequency, biomass, and basal area. Alternatively, for communities with many species, ecologists use more sophisticated methods involving multivariate statistics.

In harsh sites, stress tolerance may determine dominance, but in many communities, the dominant(s) may be the superior competitors under prevailing abiotic factors. Yet dominance hierarchies can change quickly. The American chestnut (*Castanea dentata*) was a dominant in oak–chestnut forests in eastern North America until the early 20th century, producing large nut crops that supported feeding niches for insects, birds, and mammals. Even in Canada, where it was restricted to the Carolinian forests of southwestern Ontario (just north of Lake Erie), it was widespread and often dominant (Ambrose 2004). Within a few short decades, the chestnut was decimated by an introduced fungus, chestnut blight, which spread rapidly with mail-order shipments of an ornamental, the Japanese chestnut (Tallamy and Darke 2009). Since its rapid demise, yellow poplar, oaks, and hickories have taken over the chestnut's niche. Such changes may seem inconsequential, as the appearance of the forest is unchanged, yet production of nuts and acorns is much reduced and less predictable than prior to the blight, with severe negative impacts on the niches of many animals (Diamond et al. 2000).

Not all communities have dominants. The more species present, the less likely is a strong dominant, even if some species are more common than others. For example, tropical rain forests have high species richness, but they rarely have dominants (see Section 23.2). Nor is it true that once a dominant, always a dominant. Communities are dynamic entities. A species that dominates at one stage of community development may be less common and even disappear at another stage as abiotic or biotic factors change during succession (see Chapter 17).

Dominants are usually plants, but in aquatic habitats other organisms can assume a dominant role. The coral *Oculina arbuscula* occurs along the east coast of North America as far north as North Carolina. It is the only coral in this temperate region with a complex morphology that provides shelter for a species-rich epifauna (organisms that live on and among the coral) (see Figure 16.8). Over 300 invertebrate species live among its branches, and many more complete their life cycle within the coral (Stachowicz and Hay 1996). In Section 16.4, we will see that its dominant status reflects an interaction with one of its invertebrate residents.

16.4 KEYSTONE SPECIES: Some Species Influence Community Structure Disproportionately

Relative abundance, whether based on numbers or on cover, is just one indicator of the importance of a species. Less abundant species may affect communities in crucial ways. A species whose effect on its community is disproportionate to its abundance is a **keystone species (functional dominant)**. Removal of a keystone triggers changes in the community that may lower diversity or impair function. The term was originally used for predators but can apply at any trophic level.

Keystone species have various roles. We expect dominants such as black spruce (*Picea mariana*) to modify their bog habitat—just as the coral *Oculina arbuscula* creates its reef habitat—but some keystones, called **bioengineers**, modify their habitat much more than expected given their numbers. Dam-building by beavers (*Castor canadensis*) creates new habitat, converting flowing water systems into ponds (see Figure 18.20) and modifying adjacent land habitats by their feeding. By increasing habitat heterogeneity, beavers increase the species richness of the herbaceous plant community in upstate New York by over 33 percent (Wright et al. 2002). Historical flux in beaver numbers associated with demand for their pelts has greatly altered the boreal landscape. We explore the multi-faceted impacts of beaver on stream habitat in Section 18.8, but their effects on ecosystem productivity and carbon cycling are substantive (Naiman et al. 1986).

In other cases, keystone species affect community structure not by modifying habitat but by influencing species interactions. Keystone herbivores modify the community by their feeding activity. African elephants feed on woody browse in the African savannahs, uprooting and destroying the shrubs and trees they feed on. In turn, a reduction in woody growth favours grasses, a change in species composition that is actually to the elephant's detriment, but benefits grazing herbivores (Huggett 2004). Similarly, prior to the decimation of its herds, the American bison (*Bison bison*) was a keystone herbivore on North American prairies (Figure 16.5, p. 340). Unlike the elephant, bison inhibit woody species invasion by trampling, not feeding, but their preference for grasses maintains the diversity of forbs (non-grass herbaceous species) (Anderson 2006).

Predators often function as keystone species. Sea otters (*Enhydra lutris*) inhabit kelp beds in coastal waters of the Pacific Northwest. Sea otters eat urchins, which feed on kelp, which in

Figure 16.5 Bison are keystone herbivores on the North American prairies. By trampling woody vegetation and feeding on grasses, they inhibit tree invasion and maintain forb diversity.

turn is a dominant, providing habitat for many species (see Figure 17.7). When otters declined in Alaska in the mid-1990s, possibly through increased predation by killer whales, *Orcinus orca* (Estes et al. 1998), urchins increased sharply, overgrazing kelp and reducing habitat for other species (Duggins 1980). In turn, fewer sea otter pups precipitated a diet shift in bald eagles (*Haliaeetus leucocephalus*) (Anthony et al. 2008).

Sea otters are restricted to the Pacific, but off Canada's east coast, massive die-offs of sea urchins from an amoebic disease reflect the increased frequency of "killer storms" (high-intensity tropical cyclones that pass close to the coast) with global climate change (Scheibling and Lauzon-Guay 2010). If they occur when water temperatures are warm enough to allow disease propagation, these events promote the spread of non-native pathogens by turbulent mixing. These die-offs shift the community from an urchin-dominated barren (Figure 16.6) to a kelp bed ecosystem—a shift that reflects a density-independent disturbance, not a keystone.

A keystone species may affect interactions between communities on a landscape. Predation of salmon by grizzlies

Figure 16.6 Aggregation of sea urchins (*Strongylocentrotus droebacientris*) in an experimental plot in St. Margaret's Bay off the coast of Nova Scotia.

(*Ursus arctos*) and black bears (*U. americanus*) transfers large amounts of energy and nutrients between aquatic and terrestrial systems in the Pacific Northwest. Stable isotope studies (which allow tracking of nutrients) conducted in 20 watersheds along the British Columbia coast indicate that up to 40 percent of the nitrogen used by riparian plants is derived from salmon, with the level varying with salmon density, bear abundance, plant species, and distance from the stream (Reimchen 2001). On average, bears eat only half of the carcass, leaving the residues on land, where they support an array of scavengers and decomposers and have a myriad of indirect food web effects. Nitrogen released by decomposition supplies 120 kg of N per hectare—a huge boost to the productivity of these coastal forests (Reimchen 2001). Bears act as bioengineers by promoting forest productivity rather than by creating habitat. Polar bears play a similar nutrient transfer role at the interface between polar seas and tundra (see Ecological Issues: Conserving a Canadian Icon, pp. 604–605).

Early accounts of keystone species were purely descriptive. However, where feasible, researchers remove a species presumed to be a keystone and then monitor the effects. A famous example involves rocky intertidal pools off the Pacific Northwest coast, which are home to many invertebrate herbivores, including mussels, barnacles, limpets, and chitons, all of which are eaten by the starfish *Pisaster* (Figure 16.7). Researchers removed starfish from some plots while leaving others as controls. After starfish removal, prey species richness dropped from 15 to 8 (Paine 1966).

Why is removing *Pisaster* so damaging to the prey community? Without predatory pressure, several mussel and barnacle species that are superior competitors exclude others. Community function is affected too; plants and animals rarely compete for space, but mussels, being sessile, can compete with algae, lowering productivity. The activity of the keystone predator, with its diet preference for superior competitors, promotes coexistence of inferior competitors and maintains the community as a whole. In effect, such **keystone predation** acts as a biotic disturbance, averting the equilibrium fate of competitive exclusion by keeping prey populations below their carrying capacities.

A similar keystone effect maintains temperate coral reefs. In tropical reefs, herbivorous fish suppress seaweed in well-lit, shallow waters. If unchecked, seaweed smothers the coral, depriving it of light. In temperate waters, herbivorous fish are less abundant, and seaweed biomass increases. The dependence of corals on keystone herbivory may explain why corals—even those that tolerate cooler waters—are rare in temperate latitudes. Temperate reefs dominated by *Oculina* owe their persistence to a crab, *Mithrax forceps* (Figure 16.8). In experimental plots, the presence of crabs was correlated with greater growth of corals and less seaweed (Stachowicz 2001), whereas coral survival was impaired without crabs. The crab, as a keystone herbivore, sustains the coral community, including its invertebrate epifauna. The crabs benefit too; crabs associated with coral were less vulnerable to predation (Stachowicz and Hay 1999), suggesting that the interaction is a facultative mutualism (see Section 15.8).

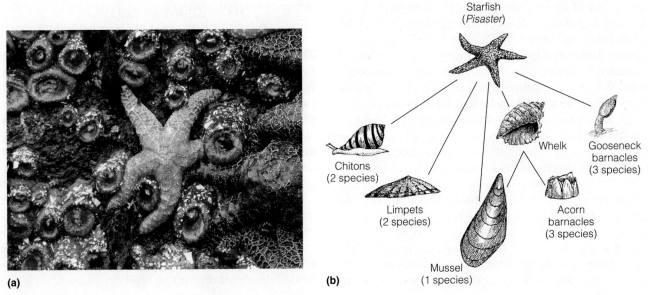

Figure 16.7 The starfish *Pisaster* in rocky pools in the Pacific Northwest (**a**). *Pisaster* feeds on many invertebrates (**b**). Starfish removal reduced prey richness.

(Adapted from Paine 1969.)

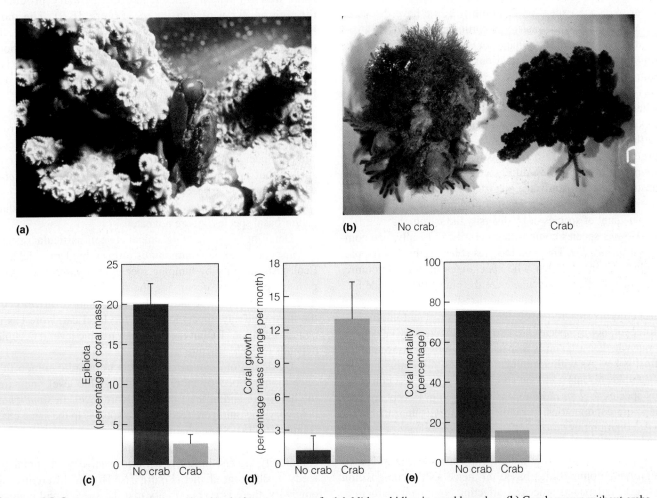

Figure 16.8 Interaction between coral and crabs in temperate reefs. (**a**) *Mithrax* hiding in coral branches. (**b**) Corals grown without crabs are smothered by seaweed. By reducing overgrowth (**c**), crabs increase coral growth (**d**) and survival (**e**).

(Adapted from Stachowicz 2001.)

As a type of indirect interaction (see Sections 15.13 and 16.7), such *keystone predation* (whether carnivory or herbivory) can amplify the impact of species loss or invasion. Loss of a keystone often has a greater community impact than loss of more numerous species. Conversely, if introducing a non-native disrupts the role of a native keystone, or if the non-native assumes a keystone role, the impact may be considerable.

Not all communities have a keystone. Some communities lack a competitive hierarchy upon which a keystone can act. Or, if a consumer has a diet preference for an inferior competitor, its presence can actually *reduce* diversity by accelerating competitive exclusion. Its prey suffers from a "one-two punch" of competition plus herbivory, as observed for sheep grazing on Welsh grasslands (Milton 1940). For a keystone predator to enhance diversity, it must be a generalist with a diet preference for the superior competitor(s). Nor is this sufficient; if the keystone is kept in check by abiotic factors or pathogens, it may be too scarce to exert its community effect.

16.5 SPECIES COMPOSITION: Species Identity Is as Important as Diversity

So far, with the exception of dominants, we have discussed overall richness and diversity, not particular species. But a species that is neither a dominant nor a keystone can affect the community through its interactions with other species. This reality highlights the importance of **species composition**—the identity of species present. Two communities can have identical richness and diversity, and even the same dominants or keystones, but differ in composition. In the southern boreal forest, the floor of several forest types that share dominance by black spruce (*Picea mariana*) is blanketed with either lichens or *Sphagnum* mosses. Such differences may reflect variations in substrate and/or disturbance history (Payette et al. 2000; compare Figures 23.35a and 25.17a) that are not revealed by a simple count of species or by diversity index value.

How is species composition described? Ideally, one compiles a *species list*. This task requires collaborative efforts, particularly if an ecologist who specializes in one community (e.g., birds) wishes to characterize the plants that provide bird habitat. Similarly, a plant ecologist studying the impact of insect herbivores relies on the expertise of entomologists. Moreover, not all species are known, and there may be debates about whether a taxon is a species or a subspecies. Sampling effort is key. As with accumulation curves, a minimal number of samples reveals common species, but rare or less conspicuous species require more extensive sampling. Except for species-poor communities, a complete inventory is unavailable for most communities.

Once a working species list is compiled, ecologists can assess how similar two or more communities are. Distinguishing among communities based on their species composition helps us understand processes affecting community structure. In Quantifying Ecology 16.3: Community Similarity, p. 355, we describe several similarity measures. Multivariate statistical methods allow more sophisticated comparisons (Pielou 1984).

16.6 FOOD WEBS: Species Are Classified into Functional Groups

The species present in a community are often classified into **functional groups**—species that perform similar community functions. The interactions discussed so far (predation, parasitism, competition, and mutualism) all contribute to the fundamental ecological process of acquiring energy and nutrients. Hence, the most vital functional grouping is into **trophic (feeding) levels**. Ecologists study how species in different trophic levels interact while acquiring food.

Food web classification describes groups of species that acquire energy in a similar way. A **food chain** is an abstract representation of community feeding relations—a diagram of "who eats whom," with arrows representing the flow of energy from prey (consumed) to predator (consumer). Grasshoppers eat grass; clay-coloured sparrows eat grasshoppers; marsh hawks (also called northern harriers) eat sparrows:

$$\text{grass} \rightarrow \text{grasshopper} \rightarrow \text{sparrow} \rightarrow \text{hawk}$$

Feeding relations in nature are rarely such simple, straight-line chains. Many chains are interwoven into a **food web**, in which links lead from primary producers through an array of consumers (Figure 16.9). The more links between a consumer and its prey (animal or plant), the more generalist the consumer and the greater the likelihood of diffuse interactions. For example, the prairie vole is a prey item for four predators in the prairie web depicted in Figure 16.9.

A simple model illustrates food web terminology (Figure 16.10). Each circle is a *node*, representing a species (or group of species). Arrows show the *links* from consumers to consumed. Species are (1) *basal species*, which do not eat other species but are food themselves; (2) *intermediate species*, which both eat other species and are themselves prey; or (3) *top species* (typically *top-level predators*), which eat intermediate and/or basal species but are not eaten.

Detailing direct feeding links between particular species becomes complex in communities of even moderate richness. Ecologists simplify by lumping species into *autotrophs* (A) or primary producers, occupying the first (basal) trophic level; *herbivores* (H), the second trophic level; and *carnivores* (C), the third (and any upper) levels. A top-level carnivore has its own designation (P), but the particular trophic level it occupies depends on web structure.

By definition, *omnivores* (which are more common than once thought) occupy more than one trophic level, including the herbivore and at least one carnivore level. In *same-chain omnivory*, the omnivore feeds on two species in the same chain (Figure 16.11a). Same-chain omnivores may interact with their prey as both a predator and a competitor, complicating their community effects. In *different-chain omnivory*, the omnivore feeds in different chains (Figure 16.11b). In different-chain omnivory, there is no possibility that the species will compete with its prey.

A *cannibal* is any species that feeds on itself. Cannibalism is more widespread than once thought, not just in insects such

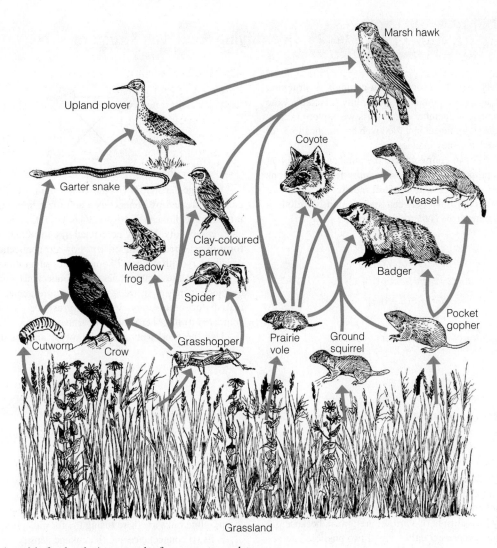

Figure 16.9 A prairie food web. Arrows point from prey to predator.

as backswimmers (*Notonecta hoffmanni*) but also in vertebrates such as northern pike (see Figure 14.1), and some birds and mammals. In cannibalism, the arrow loops back to the same box.

Another functional grouping subdivides each trophic level into *guilds*: groups of species that exploit a common resource in a similar way, such as seed-eating birds or shade-tolerant shrubs. In niche terms, guild members occur in clusters along a resource gradient, with considerable overlap. Although such clusters may seem to violate the idea of niche differentiation, they are common (Scheffer and van Nes 2006). As guild members use a shared resource, competition has the potential to be strong, at least if its member populations approach their carrying capacities.

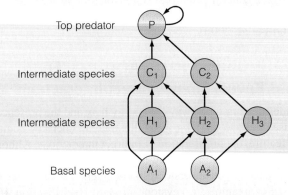

Figure 16.10 A hypothetical food web. A_1 and A_2 are basal species (autotrophs). H_1, H_2, and H_3 are herbivores. C_2 is a carnivore. C_1 is an omnivore, as it feeds on more than one trophic level, one of which is basal. H and C are intermediate because they are both predator and prey. P is a top predator, and in this case also a cannibal.

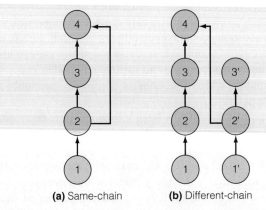

Figure 16.11 Two types of omnivory. **(a)** Same-chain omnivory. **(b)** Different-chain omnivory.

QUANTIFYING ECOLOGY 16.2 Quantifying Food Web Structure

ood webs raise many questions about community structure. Differences among communities in feeding links affect the population dynamics of member species. As we explore in Section 16.7, how species connect in a food web has implications far beyond direct predator–prey interactions.

Food webs differ in **connectance**: the number of links that are present, as a proportion of all possible *links* (feeding relations between species). In a web with S species, there are $S(S-1)/2$ possible unidirectional links between species pairs (excluding cannibalism). Connectance is then calculated as

Number of links
in the food web

$$C = L/[S(S-1)/2]$$

Index of
connectance

Number of species
in the food web

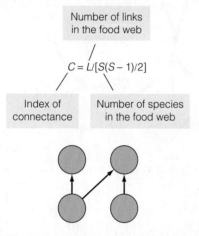

Consider the simple food web just presented. There are 4 species (circles) and 3 links (arrows). The number of possible links is $4(4-1)/2$ or 6, and the connectance is 3/6 or 0.5. All possible links are shown at right. No arrows are drawn, because the possible links could represent either one of two possible interactions (depending on the direction of the arrow):

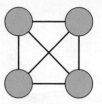

Linkage density, L/S, is the average number of links per species, as affected by connectance and the number of species. In this example, linkage density is 3/4 or 0.75. Linkage density indicates complexity. Highly connected systems contain many links for a given species number. But how do connectance and link density vary with richness? Some studies suggest that the number of links increases with richness, but connectance decreases. Connectance increases in webs that have more generalist than specialist consumers, and with the presence of omnivores as compared to strict herbivores or carnivores.

Use the food web shown below to answer the following questions.

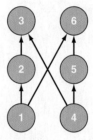

1. How many links are present in this web? How many possible links are there (using the formula provided)? What is its (i) connectance and (ii) linkage density?
2. Are any of the species omnivores?

Guild members may or may not be taxonomically related. For example, hummingbirds and other nectar-feeding birds and insects form a guild of unrelated species that exploit the same resource (nectar) in a similar way. Grouping species into guilds simplifies community study, allowing researchers to focus on manageable subsets. It also provides a framework for exploring questions about community organization. Just as we can explore interactions among guild members, we can also consider interactions among the various guilds that make up the community, which functions as a complex assembly of interacting guilds.

The guild concept has been expanded into *functional types*—groups of species based on life history, environmental response, or another attribute. Plants may be grouped by photosynthetic pathway (C_3, C_4, or CAM) (see Section 6.10) or shade tolerance. Grouping species as iteroparous or semelparous (see Section 8.7) reflects the timing of their reproductive effort. Functional types may also be based on morphological traits such as body mass. By reducing the

emphasis on one-to-one interactions between species that vary among communities, functional types allow ecologists to extract general principles about factors structuring complex communities from a wealth of location-specific data (McGill et al. 2006).

16.7 INDIRECT INTERACTIONS: Food Webs Reveal Many Indirect Interactions

Besides depicting pathways of energy flow and the presence of diffuse interactions, food webs illustrate another key community feature: the proliferation of indirect interactions. Although any two species are linked by only a single arrow, community function cannot be understood solely in terms of direct feeding links. For example, a keystone predator (see Section 16.4) reduces competition among its prey by keeping their populations below their carrying capacities. Understanding the processes controlling community structure must include consideration of such indirect effects.

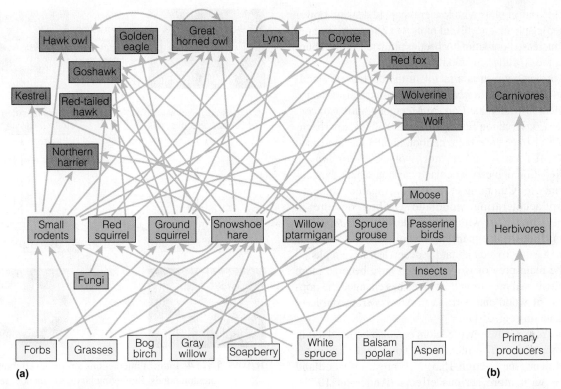

(a)   **(b)**

Figure 16.12 Aggregation of species in a boreal forest food web **(a)** in the Kluane region of the Yukon into generalized feeding groups occupying different trophic levels **(b)**.

(Adapted from Krebs et al. 2001).

As we saw in Section 15.13, *indirect interactions* are those in which a species does not interact with a second species directly, but affects a third species that directly interacts with the second. In the boreal food web (Figure 16.12), lynx and other predators do not interact directly with white spruce but, by preying on herbivores that feed on spruce, enhance survival of spruce seedlings and saplings. On Kent Island in New Brunswick, where snowshoe hares were introduced in 1959 in the absence of predators, spruce regeneration plummeted (Peterson et al. 2005), despite the fact that young spruce are not a favoured food item for hares because of their camphor content. Indirect interactions such as these can arise throughout an entire web, in potentially staggering numbers.

Keystone predation and *competitive mutualism* (see Sections 16.4 and 13.13) are indirect interactions. Another type, **apparent competition**, can occur when a predator feeds on two non-competing prey (Figure 16.13). Without the predator, each prey population is regulated by intraspecific competition. The predator's abundance depends on the combined abundance of both prey (a *numerical response*; see Section 14.5). With more predators, prey consumption increases, so both prey have lower densities when present together than separate. Lower density in the presence of the other prey species has the appearance of competition, but their lower numbers are caused by the greater abundance of the predator when supported by both prey (Holt 1977).

Apparent competition is an intriguing possibility, but does it occur in nature? Many observational studies have identified patterns consistent with it, and there is also experimental evidence in intertidal, freshwater, and terrestrial systems. For

example, nettle aphids (*Microlophium carnosum*) feed only on nettles (*Urtica* spp.), whereas grass aphids (*Rhopalosiphum padi*) feed on grasses. These aphids do not compete, but both are prey for lady beetles (Coccinellidae). Researchers placed potted nettle plants containing nettle aphid colonies in grassland plots with natural populations of grass aphids. On some plots, they applied fertilizer that stimulated grass growth and increased the grass aphid population. Nettle aphid colonies adjacent to the fertilized plots declined in density compared with colonies adjacent

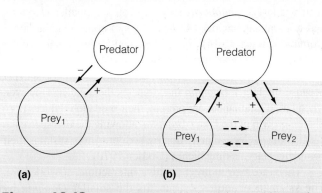

(a)   **(b)**

Figure 16.13 Apparent competition involving a predator and two non-competing prey. Solid and dashed arrows represent direct and indirect interactions, respectively. Circle size reflects population size. The interaction between a prey (P_1) and its predator **(a)** is modified by the presence of a second, non-competing prey (P_2) **(b)**. Combined prey populations increase predator numbers above those supported by either prey alone. Higher predation reduces both prey, giving the appearance of interspecific competition.

to unfertilized control plots with low grass aphid density. Lower density of nettle aphids in fertilized plots was due not to competition but to increased predation by beetles attracted by the abundant grass aphids (Müller and Godfray 1997).

Apparent competition is a factor in the decline of woodland caribou (*Rangifer tarandus caribou*) in northern Ontario (see Ecological Issues: What Can the Metapopulation Concept Contribute to Conservation Ecology?, pp. 248–249). Encroachment of moose (*Alces alces*) into caribou habitat has increased the numbers of a common predator, wolves (*Canis lupus*). Increased predation depletes woodland caribou where they co-occur with moose (Cumming et al. 1996). Apparent competition is also accelerating woodland caribou decline in northeastern Alberta, but with white-tailed deer (*Odocoileus virginianus*). Industrial expansion associated with oilsands development has improved habitat for deer. Deer have replaced moose as the main prey of wolves, which have become more numerous. Increased predation by wolves has changed the population status of woodland caribou in Alberta from stable to declining (Latham et al. 2011).

Parasites rather than true predators can also be involved in apparent competition. The meningeal worm that parasitizes white-tailed deer (see Section 15.3) also preys on woodland caribou, but with more serious effects (Anderson 1972). Increased contact between deer and caribou gives the appearance of competition, when the effects are largely due to two common predators—wolves and the meningeal worm. A similar parasite-mediated interaction occurs between deer and moose (Schmitz and Nudds 1994). Many such subtle apparent competition interactions likely exist in communities, but easily escape notice. These interactions contribute to the vulnerability of species to global climate change and human-induced landscape changes, which cause species that were formerly spatially separated to come into contact (Vors and Boyce 2009).

Indirect interactions can be beneficial, as with keystone predation and **indirect commensalism**, in which two predators each consume different but competing prey. Ponds in the Colorado mountains contain two competing herbivorous *Daphnia* species. A midge larva (*Chaoborus* sp.) preys on the smaller *Daphnia*, and a larval salamander (*Ambystoma* sp.) consumes the larger *Daphnia* (Figure 16.14). Ponds occupied by larval salamander contain few large and many small *Daphnia*. Ponds lacking larval salamander lack small *Daphnia*, and midges cannot survive because the larger *Daphnia* outcompetes the smaller. However, when larval salamander are present, their predation reduces population growth of the larger *Daphnia*, allowing the two species to coexist. Size-selective predation by both predators sustains their complementary feeding niches (Dodson 1970).

There are two indirect positive interactions at work here. The salamander indirectly benefits the smaller *Daphnia* by reducing the numbers of its larger competitor—keystone predation. But the midge also benefits from the presence of the salamander, because without it the smaller *Daphnia* would not be available. As the salamander is unaffected, the midge–salamander interaction is commensalistic, but if an indirect interaction benefits both species (as described for trees and voles in Figure 15.21), it is **indirect mutualism**.

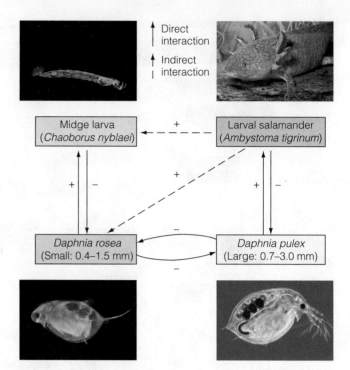

Figure 16.14 Indirect interactions among invertebrates in Colorado mountain ponds. Removing larval salamander resulted in competitive exclusion of *Daphnia rosea* by *D. pulex* and local extinction of the midge that preys on it.

(Adapted from Dodson 1970.)

We discuss another indirect interaction called a *trophic cascade* in Section 16.8, but overall there is a growing appreciation for the role of indirect interactions in communities. Experimentally, they require controlled manipulations of the populations involved, allowing a sufficient lag time for the indirect effects to express. Until recently, such experiments were rare, so the impact of indirect interactions, though likely great, was underestimated. Yet experiments suggest that indirect interactions—both beneficial and detrimental—do influence community structure. Understanding these interactions is more than an academic exercise: it has vital implications for ecosystem conservation and management.

Given the myriad of direct and indirect interactions of a species, its loss can have unforeseen effects. Local extinction of grizzly bears (*Ursus arctos*) and wolves (*Canis lupus*) from the Yellowstone region after decades of active predator control unexpectedly triggered a decline of bird species that use the vegetation along rivers. Elimination of large predators caused an increase in moose, which selectively consume willow (*Salix* spp.) and other species that flourish along riverbanks. Increased herbivory by moose precipitated the local extinction of some birds (Berger et al. 2001). Similar regions in western Canada that retain large predators continue to support these bird species, and following the politically controversial reintroduction of wolves into Yellowstone in 1995–1996 (using wolves from Alberta's Mackenzie Valley), recovery of riparian vegetation has been substantial (Beschta and Ripple 2010). Thus, large predators can exert top-down regulatory effects on terrestrial vegetation, just as sea otters do in kelp forests.

16.8 FOOD WEB CONTROLS: Trophic Interactions May Regulate Community Structure

Much evidence supports the role of direct and indirect interactions in community structure. Rejecting the default model presented in Section 16.2 seems justified. However, given the inherent complexity of food webs, how can we determine which of these possible interactions are most critical? Are all interactions important, or do some exert a major effect, with most having little impact beyond the species directly involved? The notion that all interactions are vital for maintaining structure suggests that the community is like a house of cards, and that removing any one species may touch off a community collapse. The notion of a smaller subset controlling community structure suggests a more loosely connected species assemblage.

These questions are critical to conservation ecology, given the sharp decline in diversity associated with human activity. We know that keystone species affect community structure, but the impact of many species is a mystery. Again, using functional groups can help. By aggregating species into guilds, researchers can explore the processes controlling community structure in more general terms. In Figure 16.12, which depicts a Canadian boreal forest food web, some species are listed separately, but most are grouped with functionally similar species: forbs, grasses, small rodents, insects, and passerine birds. When webs are summarized in this way, the questions are still complex, but more manageable. Results of studies are more generally comparable, because even though two sites differ in species composition, they may contain the same functional groups.

However, such lumping may obscure within-guild variation that can affect food web (and hence ecosystem) function. A bird may belong to a sap-feeders guild, but have niche dimensions and/or interactions that differ from other guild members in relevant ways. For example, lumping *forbs* (non-grass herbaceous species) masks ecologically critical differences, such as between a nitrogen-fixing legume and an invasive weed that is a heavy nitrogen user. We cannot assume that **functional redundancy** (Lawton and Brown 1993)—the idea that when one guild member is lost, others assume its role, reducing the impact of the loss—always applies. Functional redundancy is a useful null model for diversity studies (Loreau 2004), but we must be cautious in grouping species, given their possibly unique effects.

Within-species variation can also affect community structure and function. The planktivorous alewife (*Alosa pseudoharengus*) is a common fish along the northeast coast. (It used the Welland Canal to bypass Niagara Falls and invade the Great Lakes—another range expansion facilitated by human interference.) Alewives have both *anadromous* (marine but returning to freshwater to spawn) and landlocked forms that differ in morphology and foraging, making them ideal for studying the impact of within-species variation. Anadromous alewives, which select larger prey, reduced both zooplankton biomass and species richness in lake mesocosms in under two months, compared with landlocked alewives (Figure 16.15) (Palkovacs and Post 2009). This study provides direct evidence of the impact of within-species variation on community structure.

Grouping species into trophic levels raises questions about feeding-related control of community structure. Some ecologists assume **bottom-up control**: the productivity and abundance of populations at a given trophic level are controlled by the productivity and abundance of populations at the level below. Plant populations control herbivore abundance, which in turn controls carnivore abundance. Yet, as we saw in our discussion of predation, **top-down control** can also occur when predators control prey abundance. Both types can apply to different subsets of a food web. In a boreal food web in the Yukon, small rodents such as voles experience bottom-up control (Krebs et al. 2010), snowshoe hares top-down (Krebs 2011).

Some forms of top-down control involve indirect effects at trophic levels removed from that of the immediate prey. In streams in the U.S. Midwest, largemouth bass (*Micropterus*

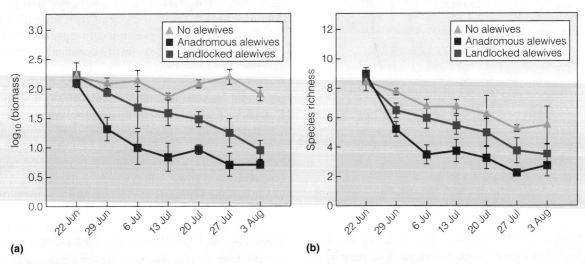

(a)   **(b)**

Figure 16.15 Impact of within-species variation in alewives on the biomass **(a)** and species richness **(b)** of the zooplankton community in lake mesocosms with anadromous alewives ($n = 3$ lakes), landlocked alewives ($n = 8$), and no alewives ($n = 8$).

(Adapted from Palkovacs and Post 2009.)

salmoides) has strong indirect effects called a **trophic cascade**, in which predation has a ripple effect through the web, ultimately affecting the abundance of primary producers (in this case, benthic algae). Herbivorous minnows (primarily *Campostoma anomalum*) graze on algae, and largemouth bass (along with two sunfish species) feed on the minnows. During low-flow periods, isolated pools form. The experimenters removed bass from some pools, which then developed high minnow numbers and low algal biomass. Control pools that retained bass had few minnows and abundant algae (Power et al. 1985). Top predators thus maintained primary producer abundance indirectly, by feeding on herbivores.

Top-down control via a trophic cascade underpins the *green world hypothesis* (Hairston et al. 1960), which posits that terrestrial systems accumulate plant biomass because predators keep herbivores in check. Although experimental support for trophic cascades is growing, most examples are in aquatic systems. On land, there are many reasons why "green" accumulates. Not all plant tissue is equally digestible or accessible, so herbivores may be limited by food supply, not predation. A survey of over 100 field studies revealed that trophic cascades are weaker in terrestrial systems and vary substantially in aquatic systems (Shurin et al. 2002). Interestingly, trophic cascades act differently with an even number of trophic levels; if there were a fourth level in the system described in the Power study, bass numbers would be kept in check, allowing herbivores to

rebound and reduce algal mass. Nor need a system always be subject to one influence; in some conditions, control may be bottom up, but may revert to top down if conditions change.

Salt marsh systems were once thought to be subject to bottom-up control, reflecting the impact of abiotic factors on productivity. The dominant autotroph in much of the marsh is cordgrass (*Spartina alterniflora*), which is heavily grazed by periwinkle snails (*Littoraria irrorata*). Snails in turn are prey for several crab species. After crabs were excluded for one year, snail densities were 100 times those in the controls (Figure 16.16) (Silliman and Bertness 2002). By controlling snail densities, predators indirectly benefit cordgrass—another example of top-down control involving a trophic cascade. We revisit food web regulation in Section 20.9 when discussing factors affecting primary productivity and the number of trophic levels.

16.9 PHYSICAL STRUCTURE: Growth Forms Determine Physical Community Structure

So far we have focused on a community's biological structure—the mix of species and the interactions among them. But a community also has a physical structure, reflecting both abiotic factors, such as water depth and flow in aquatic habitats, and biotic factors, particularly the growth forms and spatial arrangement of its largest residents. For example, tree size, height, density, and dispersion define the physical attributes of a forest, which in turn affect the environment of all others species present, even those with no direct interactions with trees.

The physical structure of most terrestrial communities is defined by vegetation. Plants may be tall or short, evergreen or deciduous, herbaceous or woody. These traits define groups based on **growth form**: trees, shrubs, and herbs. These categories are subdivided further, into needle-leaf evergreens, broad-leaf evergreens, broadleaf deciduous trees, thorn trees and shrubs, dwarf shrubs, ferns, grasses, forbs, mosses, and lichens. Ecologists often classify (and name) communities for their dominant growth forms because of their impact on physical structure: shrublands or grasslands. In aquatic habitats, dominants may also be used to classify communities: kelp forests, seagrass meadows, or coral reefs. However, because aquatic autotrophs are often single-celled and short-lived algae (phytoplankton), physical structure is more often defined by abiotic features such as water depth, flow rate, or salinity.

Based on its dominant vegetation, every community has an associated **vertical stratification**. On land, growth form (particularly height and type of branching) determines the number and type of vertical strata, which in turn influence—and are influenced by—the vertical light gradient. Vertical stratification provides a physical framework in which different animals are adapted to live. Most forests have multiple layers (Figure 16.17): canopy, understory, shrub layer, herb layer, and forest floor. (*Understory* can also refer to the vegetation in all strata under the canopy.)

The upper layer or **canopy** is the primary site of photosynthesis and exerts a major influence on the rest of the forest. If the canopy is fairly open, considerable light reaches the lower layers.

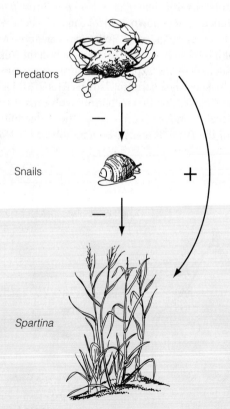

Predators

Snails

Spartina

Figure 16.16 A trophic cascade in a salt marsh. Overgrazing by snails damages cordgrass. By consuming snails, crabs indirectly benefit cordgrass.

(Adapted from Silliman and Bertness 2002.)

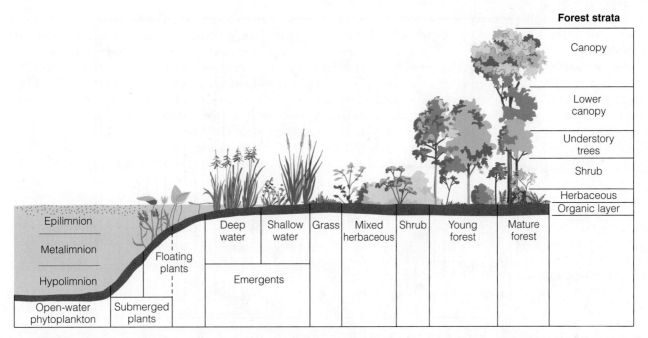

Figure 16.17 Vertical stratification of communities. The major zone of decomposition and regeneration is the bottom stratum, and the zone of energy fixation is the upper stratum. From left to right, stratification and complexity increase. Stratification in aquatic communities is determined by gradients of O_2, temperature, and light. On land, stratification is determined by the dominant growth forms of the vegetation, which affect microclimatic conditions of temperature, moisture, and light. Forests have four or five strata and typically support more diversity than grasslands. Floating and emergent aquatic systems support greater diversity than open water.

If ample water and nutrients are available, well-developed understory and shrub strata form. If the canopy is dense and closed, light levels are lower, and the understory and shrub layers are absent or poorly developed. In deciduous forests of eastern Canada, such as those dominated by sugar maple (*Acer saccharum*) and beech (*Fagus grandifolia*), the **understory** contains tall shrubs such as hobblebush (*Viburnum alnifolium*), small-statured trees such as ironwood (*Carpinus caroliniana*), and saplings of shade-tolerant canopy species. The **shrub layer**, which contains tall shrubs such as hazel (*Corylus* spp.), is prominent in deciduous forests but often sparse in coniferous forests. The growth forms and species composition of the **herb layer** (forbs, grasses, low shrubs, mosses, ferns, and lichens) depend not only on light penetration resulting from canopy openness, but also on soil traits (texture, moisture, pH, and nutrient content), slope position, and exposure. The final layer, the **forest floor**, is where litter accumulates and decomposition by fungi and microbes occurs, releasing minerals for reuse (see Chapter 21).

Although it is less obvious, grasslands also have vertical structure. The canopy of a tallgrass prairie refers to the taller grasses that exceed 1 m in height. The understory contains grasses and forbs of lower stature, and forbs with basal rosettes hug the soil surface in the ground stratum.

The strata of aquatic systems is determined largely by light penetration, and to a lesser extent by temperature and O_2 profiles. In summer, many lakes are stratified into the *epilimnion*, *metalimnion*, and *hypolimnion* (see Figure 16.17 and Section 3.3), plus the bottom sediments (**benthic zone**). Water bodies have an upper **photic** (or *euphotic*) **zone**, where light is sufficient to support photosynthesis, mostly by phytoplankton, and, in deeper waters, the **aphotic zone**, where light is insufficient. The depth of the photic zone varies with turbidity and particulate matter. In clear lakes, it may include part of the metalimnion, but in turbid lakes it may not even include all of the epilimnion.

Characteristic organisms inhabit each stratum. As well as the vegetation patterns just described, various consumers and decomposers occupy the strata, although decomposers are typically more abundant on the forest floor and in soil or sediment layers. Much interchange takes place, but even highly mobile animals may confine themselves to a few layers (Figure 16.18, p. 350). A ground-dwelling bird such as the Carolina wren has a very different activity pattern than a canopy bird such as the blue-grey gnatcatcher, which eats canopy insects. Species may move between layers over the day or season, reflecting variations in humidity, temperature, light, and O_2; shifts in the abundance of food resources; or differing needs for the completion of their life cycles. For example, zooplankton migrate vertically in the water column during the day in response to light.

16.10 ZONATION: Horizontal Patterns Reflect Abiotic Tolerance and Biotic Interactions

As we move across a landscape, the physical and biological structure of the community changes. Often the changes are small—subtle shifts in species composition or the height of vegetation. But as we travel farther, or as the substrate varies,

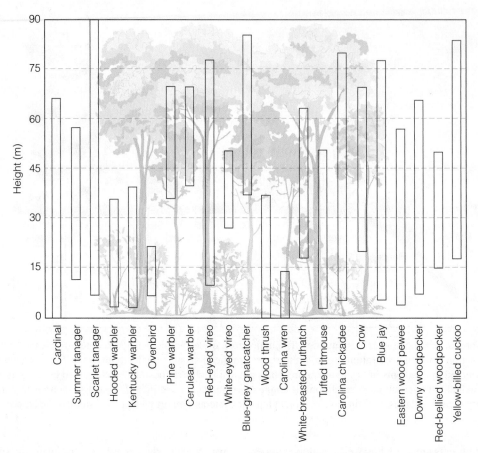

Figure 16.18 Vertical distribution of birds in a forest in Tennessee. Bars show height range, based on observations over a breeding season. (Adapted from Anderson and Shugart 1974.)

changes become more pronounced. For example, central Virginia just east of the Blue Ridge Mountains is a landscape of rolling hills, supporting a mosaic of forest and field. If we walk through this forest, its physical structure—its vertical stratification and dominant growth forms—appears much the same. Yet its biological structure—its mix of species and their relative abundance—changes dramatically.

As we move from a hilltop to the bottomland along a stream (Figure 16.19), species composition changes from dominance by oaks (*Quercus* spp.) to species associated with wetter habitats, such as ironwood (*Carpinus caroliniana*) and sweet gum (*Liquidambar styraciflua*). (Generalists such as red maple and yellow poplar are common throughout.) The animal community—insects, birds, and small mammals—changes too. **Zonation**—change in community structure across a landscape (see Section 18.2)—may be abrupt, but is often associated with an *environmental gradient*: progressive change in a single overriding abiotic factor such as elevation or moisture. Zonation is common in many habitats, both aquatic and terrestrial, and may form latitudinal, altitudinal, or horizontal belts within and among ecosystems.

Zonation in intertidal areas of a sandy beach relates to the changing tides, which alter animal distributions by affecting the gradient from dry to inundated (Figure 16.20,

p. 352). The zones are the *supratidal* (above high-tide line; dry); *intertidal* (between high- and low-tide lines); and *subtidal* (below low-tide line; continuously inundated). Each is home to a unique group of organisms. Ghost crabs (*Ocypode quadrata*) and beach amphipods (sand fleas, *Orchestia* and *Talorchestia* spp.) occupy the supratidal zone. True marine life begins in the intertidal zone, where species must adapt to alternating inundation and exposure. Many, such as mole crabs (*Emerita talpoida*), lugworms (*Arenicola cristata*), and hard-shelled clams (*Mercenaria mercenaria*), burrow into sand to avoid the extreme temperature and moisture fluctuations of this zone. The subtidal zone supports vertebrates and invertebrates that migrate into and out of the intertidal zone with the tides.

What causes zonation? We started our discussion of the community by assuming that its structure is constrained by the environmental tolerances of its species—their fundamental niches. These tolerances are then modified by direct and indirect interactions with other species, which define their realized niches. Competitors and predators restrict a species or lower its abundance, whereas mutualists facilitate its presence and abundance. Across a landscape, variations in abiotic factors affecting resource availability modify these constraints on distribution and abundance.

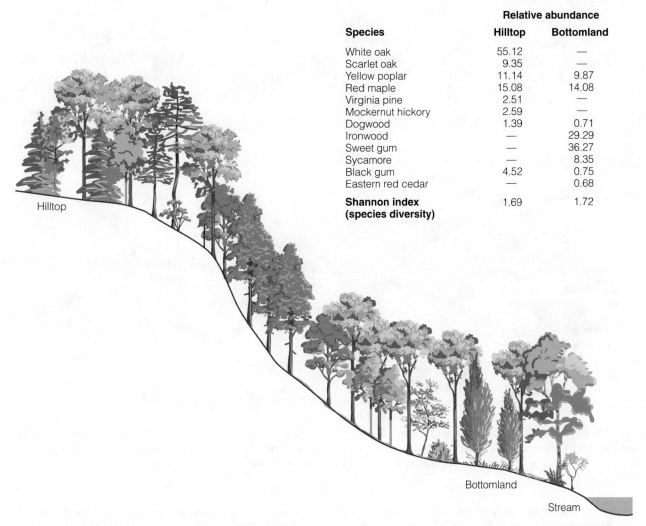

Species	Relative abundance	
	Hilltop	Bottomland
White oak	55.12	—
Scarlet oak	9.35	—
Yellow poplar	11.14	9.87
Red maple	15.08	14.08
Virginia pine	2.51	—
Mockernut hickory	2.59	—
Dogwood	1.39	0.71
Ironwood	—	29.29
Sweet gum	—	36.27
Sycamore	—	8.35
Black gum	4.52	0.75
Eastern red cedar	—	0.68
Shannon index (species diversity)	1.69	1.72

Figure 16.19 Zonation of forest stands along a topographic gradient in central Virginia. The table summarizes stand composition at two points: hilltop and bottomland. Transitional changes are continuous, but the end point communities are distinct.

Species differ in tolerance, and their interactions change in a changing environment.

Our discussion of plant adaptations helps us understand why competitive interactions vary along resource gradients. As discussed in Chapter 6, plant adaptations to light, nutrients, and water reflect trade-offs between the ability to survive when resources are scarce and the ability to grow rapidly when resources are abundant (Figure 16.21a, p. 353). Plant competitive success is often linked to growth and resource acquisition. Differences in traits that affect these processes translate into a competitive advantage for a species in that portion of a gradient where it has faster growth than other species (see Section 13.10). In part, zonation (Figure 16.21b) reflects these changing relative competitive abilities.

The lower boundary of each species is defined by its ability to *tolerate* resource limitations by efficient resource use; the upper boundary by its ability to *compete* by acquiring the greatest share of resources when they are abundant. Such trade-offs affect species distributions in semiarid grasslands (see Figure 13.13) and zonation along a moisture gradient (Figure 16.22, p. 353).

Plant species rarely compete for a single resource. Recall that in competition between clover and skeletonweed (see Figure 13.12), differences in adaptations for acquiring aboveground (light) and belowground (water and minerals) resources changed relative competitive ability along gradients where these two resources co-varied, largely as a result of trade-offs in allocation. Thus, as availability of belowground resources increases along a supply gradient (see Figure 16.22), the competitive advantage shifts from species adapted to low water (high root production) to those that allocate carbon to leaf production and height growth but require more water to survive.

This framework of trade-offs between tolerance and competitive ability, as affected by traits that favour performance at low versus high resource levels, is a powerful tool for understanding changes in communities along resource gradients. However, gradients may involve both resource and nonresource factors, as in salt marshes. In the marsh, the dominant growth forms are grasses and rushes, which give way to shrubs and trees on dry land. There are also differences within

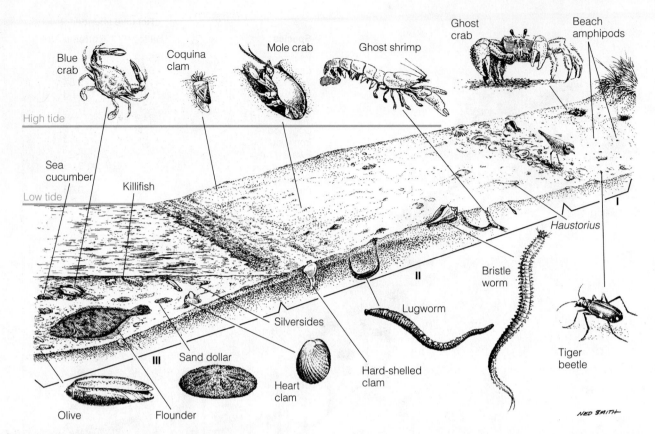

Figure 16.20 Zonation on a sandy beach on the mid-Atlantic is determined by faunal changes. Species distributions change with degree and duration of tidal inundation. Blue lines show tide levels. **I**—supratidal: ghost crabs, beach amphipods (sand fleas). **II**—intertidal: ghost shrimp, clams, lugworms, mole crabs. **III**—subtidal: flounder, blue crabs, sea cucumber.

each zone; the dominants shift as we move back from tidal areas, reflecting changes in salinity, waterlogging, and O_2. Because species differ in their fundamental niches, and because different interspecific interactions occur along the gradient and alter realized niches, species composition shifts. A dominant in one zone becomes less abundant in the next, changing structural features such as vegetation height, density, and dispersion.

Although the low marsh floods daily, the upper marsh is inundated only in high-tide cycles. Differences in inundation frequency and duration establish a gradient of salinity, waterlogging, and O_2. To investigate the interplay between resource and non-resource factors in salt marsh zonation, ecologists conducted tests involving fertilization, removal of potential competitors, and reciprocal transplants (Emery et al. 2001). Dominants in the middle and upper marsh (*Spartina patens*, a perennial turf grass, and *Juncus gerardii* [black needlegrass]) grew poorly and had higher mortality when transplanted to the lower marsh (Figure 16.23, p. 354). Their lower limits are determined by their tolerance of the stresses imposed by inundation. In contrast, *Spartina alterniflora*, a large perennial grass with extensive rhizomes that dominates in the lower marsh, flourished when it was transplanted into higher positions from which neighbours were removed (Emery et al. 2001). *S. alterniflora* is normally excluded from the higher marsh when other species

are present, suggesting its upper limit is determined by competition. Salt marsh zonation thus reflects an interaction between relative competitive ability and tolerance of physical stress.

At first, this example seems like just another instance of the trade-off between adaptations for tolerance of scarce resources and competitive ability, as in Figure 16.21. Yet such was not the case. Addition of nutrients changed competitive outcomes, but not in the predicted manner. Fertilization reversed the species' competitive abilities, allowing *S. alterniflora* and *S. patens* to shift their distributions to the higher marsh (see the bar graph insets in Figure 16.23). Why?

Juncus gerardii, the dominant under ambient (low) nutrient levels, allocates more carbon to roots than either *Spartina* species. This strategy allows *Juncus* to be the superior competitor when nutrients are scarce but restricts its ability to tolerate the high water of the low marsh. In contrast, *S. alterniflora* allocates more carbon to stems and leaves, an advantage in the low marsh by increasing access to atmospheric O_2. This trade-off in below- versus aboveground allocation creates the competitive hierarchy (and hence the zonation) normally observed. But when fertilization relieves nutrient limitation, competition for light dictates outcomes. Then, the greater height allocation of *Spartina* increases its competitive ability on the upper marsh by enhancing its ability to capture light. Thus, a trade-off

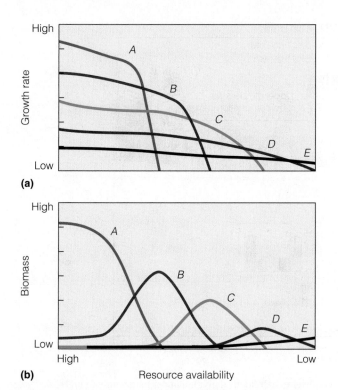

(a)

(b)

Figure 16.21 Trade-offs between tolerance of scarce resources and growth when resources are plentiful generate an inverse relationship between maximum growth rate and minimum resource requirements for hypothetical plant species *A* to *E*. **(a)** If the superior competitor at any point on the gradient is the species with the fastest growth, relative competitive ability will change with resource level. **(b)** In the resulting zonation, lower boundaries reflect tolerance for low resources, and upper boundaries reflect competition.

(Adapted from Smith and Huston 1989.)

> **INTERPRETING ECOLOGICAL DATA**
>
> **Q1.** In Figure 16.21a, under what resource levels is the growth of each species optimal?
>
> **Q2.** If species B were removed, how would the distribution of species A and C likely change in Figure 16.21b? Why?

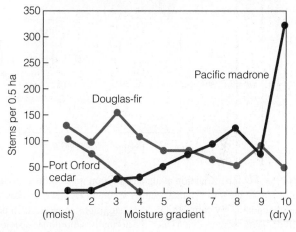

Figure 16.22 Distribution of tree species along a soil-moisture gradient in the central Siskiyou Mountains reflects competitive abilities at the moist end and tolerance of dry conditions at the dry end. Data are from 50 stands at elevations of 610 to 915 m.

(Adapted from Whittaker 1960.)

between belowground competitive ability and tolerance of the stresses of the low marsh drives salt marsh zonation. In this habitat, the stress gradient does not correspond to the resource gradient as in Figure 16.21, allowing traits that normally determine stress tolerance to enhance competitive ability when resources are abundant.

So far, we have discussed abiotic factors as distinct from biotic interactions. In **neighbourhood habitat amelioration**, organisms alter abiotic factors in ways that ameliorate their negative effects. The salt marsh is again a useful system for exploring these local effects. Removal of *Juncus gerardii*, a rush occupying the upper marsh (see Figure 16.23), more than doubled salinity and significantly decreased substrate O_2 (Bertness and Hacker 1994). In response, a coexisting shrub, *Iva frutescens*, grew more slowly and eventually disappeared (Figure 16.24, p. 354). Rather than competing, *Juncus*, with its greater salinity tolerance, improves local

habitat conditions for *Iva*, allowing it to grow in lower reaches of the marsh than it could otherwise tolerate. Nor were the facilitative effects confined to the plants. Initially, aphids were more numerous on *Iva* plants in the *Juncus*-removal plots (Hacker and Bertness 1996), suggesting that *Juncus* makes *Iva* less obvious to aphids. Yet, despite colonizing *Iva* better in the absence of *Juncus*, the aphids declined in number on *Iva* as a result of predation by lady beetles. Thus, the *Iva–Juncus* interaction helps maintain aphid populations in the marsh.

These examples are at a microhabitat scale, but spatial variation in abiotic factors also generates regional zonation patterns. Temperature and moisture gradients resulting from climatic variations are the major determinants of regional vegetation patterns and form the basis of most ecosystem classifications (see Chapter 23). Locally, climate interacts with topography and soils to influence temperature and soil moisture. The underlying geology of an area interacts with climate to influence soil traits such as texture, which in turn affects soil moisture-holding capacity and cation exchange (see Section 4.8). In aquatic habitats, water depth, flow rate, and salinity are the major gradients that determine the distribution and dynamics of communities at regional and local scales (see Chapter 24).

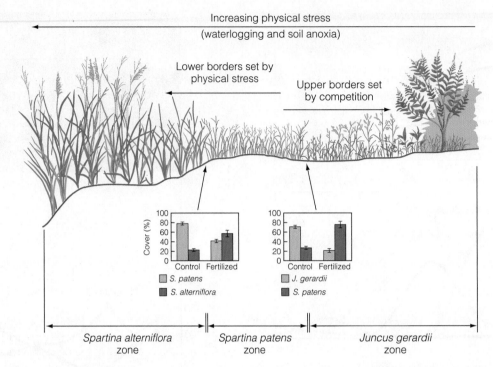

Figure 16.23 Zonation along an elevation gradient in a salt marsh. The lower marsh floods daily; the upper marsh only in high-tide cycles. High water in the lower marsh reduces sediment O_2 and increases salinity. Lower boundaries are determined by tolerance to physical stress; upper boundaries by competition. Inset bar graphs show shifts in percentage cover by the two adjoining species in controls and after fertilizer addition. With increased nutrients, the normally subordinate species became competitively superior.

(Adapted from Emery et al. 2001.)

INTERPRETING ECOLOGICAL DATA

Q1. In the transition zone between areas dominated by *Spartina alterniflora* and *S. patens*, which species was the superior competitor in the controls? Was the competitive outcome altered in the fertilized plots? Why? Repeat these questions for the transition zone between areas dominated by *S. patens* and *Juncus gerardii*.

Q2. What do these results suggest about the role of nutrients in limiting species distribution along the gradient from the low to the high marsh?

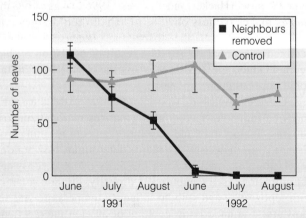

Figure 16.24 Neighbourhood habitat amelioration in a salt marsh. Mean total leaf number (± SE) on adult *Iva* plants, with and without *Juncus*.

(Adapted from Bertness and Hacker 1994.)

16.11 COMMUNITY CLASSIFICATION: Ecologists Classify Communities at Many Spatial Scales

As noted, the community has a spatial dimension, encompassing all the species in a given area. Ecologists distinguish between adjacent communities based on differences in structure, particularly the species assemblages typical of different physical habitats. But how different must assemblages be before we call them separate communities? This question has no simple answer. Consider the forest in Figure 16.19. Given the shift in species composition from hilltop to bottomland, most ecologists would call these distinct communities. Yet as we walk between them, the distinction seems less clear cut. If the transition between two communities is abrupt—as with a jack pine forest on a rocky outcrop with a sheer drop to a black spruce bog below—defining boundaries is straightforward. But, if species composition and abundance shift gradually and

QUANTIFYING ECOLOGY 16.3   Comparing Community Similarity

When we say that community structure changes as we move across the landscape, we imply that the set of species present differs from one place to another. But how do we quantify this change? Distinguishing between communities based on species composition is key to understanding the processes that control community structure. Ecologists use **community similarity measures** to compare two communities based on species composition.

Sorensen's coefficient of community (CC) is based on species presence or absence. Using species lists compiled for two sites, the index is calculated as

Number of species common to both communities

$$CC = 2c/(s_1 + s_2)$$

Number of species in community 1

Number of species in community 2

Consider the two forests presented in Tables 16.1 and 16.2:

$$s_1 = 24 \text{ species;}$$
$$s_2 = 10 \text{ species;}$$
$$c = 9 \text{ species}$$
$$CC = \frac{(2 \times 9)}{(24 + 10)} = \frac{18}{34} = 0.53$$

CC ranges from 0 (no species in common) to 1.0 (identical species composition). As *CC* does not consider relative abundance, it is useful when the focus is presence or absence. An index that does consider relative abundance is *percent similarity (PS)*. To calculate *PS*, we tabulate relative species abundance in each community as a percentage (as done for diversity indices; see Section 16.1). *PS* is the sum of the lowest percentage for each species found in both communities. In this example, 16 species occur in only one forest or the other. The lowest percentage for these species is 0, so they are excluded. For the remaining 9 species, the index is calculated as follows:

$$PS = 29.7\% + 4.7\% + 4.3\% + 0.8\% + 3.6\% + 2.9\%$$
$$+ 0.4\% + 0.4\% + 0.4\% = 47.2\%$$

This index ranges from 0 (no species in common) to 100 percent (identical in both species composition and relative abundance). When comparing more than two communities, a *similarity matrix* compiles all pairwise comparisons.

1. Calculate Sorensen's and percent similarity indices using the data in Figure 16.19 for the hilltop and bottomland forests.
2. Are these two forests more or less similar than those in Tables 16.1 and 16.2?

with much overlap, boundaries are less apparent. In Chapter 18, we consider community boundaries in the context of landscape ecology; here we focus on community classification.

Ecologists use various sampling and statistical techniques to classify communities. All employ some measure of community similarity or difference, be that a similarity index (see Quantifying Ecology 16.3: Comparing Community Similarity, above) or a multivariate measure. But how do ecologists determine where one community ends and another begins? If they have 70 percent of their species in common, but differ in their dominants, are they different communities? What if they share only 30 percent of species but have the same dominant? A forester might answer these questions differently than a conservationist.

The spatial scale at which vegetation is described is also relevant. The zonation in Figure 16.19 occurs over a few hundred metres. Over larger areas, differences in community structure increase. Consider zonation in Great Smoky Mountains National Park, straddling the border between Tennessee and North Carolina (Figure 16.25, p. 356), perhaps the most heterogeneous forest region in North America. The pattern is complex, reflecting an elevation gradient interacting with slope position and aspect. Yet community labels use the names of only a few species. Labels such as *hemlock forest* do not imply a lack of diversity; they are a

shorthand way of naming communities for the dominant tree. Only if dominants are absent is another type of label used; *cove forests* are the most diverse community in the area, but because they lack dominants, the name describes the habitat—protected coves and canyons where conditions are favourable for many species.

Within this landscape, each community could be described by a species list, along with their relative abundances, as in Table 16.1. If we expand the area of interest to include the entire eastern deciduous forest biome, the nomenclature for classifying communities becomes broader yet. In a landmark classification of forests in the eastern United States (Braun 1950), all of Great Smoky Mountains Park falls within a single forest type: *mixed mesophytic* (Figure 16.26, p. 356). Such simplification is inevitable in broad-scale classification.

The same principle applies when comparing the distribution of boreal forest, a major biome across Canada (see Figure 23.34), with the forest types of northwestern Ontario (Sims et al. 1989). A biome listed as *boreal forest* encompasses forest types as diverse as a dry jack pine forest, a black spruce bog, and a mixedwood forest dominated by white spruce and trembling aspen. These examples reinforce the principle that the community, like the population, is a spatial unit. Distinguishing among communities differs with the scale of a study and the classification criteria.

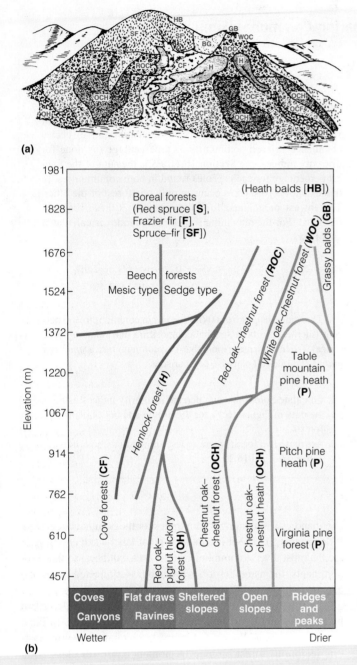

(a)

(b)

Figure 16.25 Forest communities in Great Smoky Mountains National Park. **(a)** Topographic distribution of communities on an idealized west-facing mountain and valley. **(b)** Idealized arrangement of communities according to elevation and aspect.

(Adapted from Whittaker 1956.)

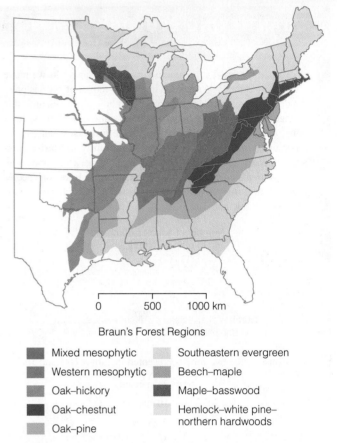

Braun's Forest Regions

- Mixed mesophytic
- Western mesophytic
- Oak–hickory
- Oak–chestnut
- Oak–pine
- Southeastern evergreen
- Beech–maple
- Maple–basswood
- Hemlock–white pine–northern hardwoods

Figure 16.26 Forest classification of the eastern United States as described by Braun (1950).

(Adapted from Dyer 2006.)

16.12 ENVIRONMENTAL HETEROGENEITY: Abiotic Variation Increases Diversity

As we have seen, community structure reflects both the direct responses of species to abiotic factors and their biotic interactions, direct and indirect. In turn, as abiotic factors change with location, so will the species that can inhabit the area, and the ways in which they interact. This framework helps us under- stand why community structure changes as we move from hill- top to valley, or from the shore into the open waters of a lake or ocean. However, **environmental heterogeneity** (variation) is typical *within* as well as *between* communities and is often patchy rather than along a gradient. For example, soil nitrate and moisture varied significantly (by more than an order of magnitude) across an agricultural field that was abandoned in the 1920s and reverted to an old-field community occupied by a variety of herbaceous and woody species (Figure 16.27) (Robertson et al. 1988). Similar studies show comparable fine- scale variation in forest, intertidal, and benthic communities.

Does this within-community environmental heterogeneity enhance diversity? Heterogeneity in light levels on the forest floor caused by death of canopy trees (a microdisturbance that causes a **gap** (opening) in the canopy) increases tree species diversity by facilitating the survival and growth of shade-intoler- ant species that otherwise might perish. Likewise, heterogeneity in prairie soils caused by burrowing of small mammals creates fine-scale variations in plant species composition, increasing overall species richness (Suding and Goldberg 2001).

Environmental heterogeneity also explains the link between vegetation structure and bird diversity. Structural features of veg- etation that influence habitat suitability for a given bird species reflect its needs for food, cover, and nesting. Because these needs vary among species, the heterogeneity of vegetation structure has

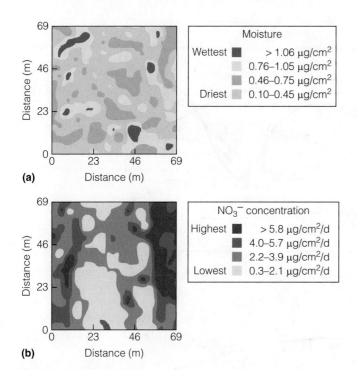

(a)

(b)

Figure 16.27 Variations in **(a)** soil moisture and **(b)** nitrate (NO_3) in an old-field community in Michigan.

(Adapted from Robertson et al. 1988.)

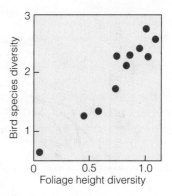

Figure 16.28 Relationship between bird species diversity and foliage height diversity in eastern North America.

(Adapted from MacArthur and MacArthur 1961.)

a pronounced impact on avian diversity. Increased vertical layering means more resources and a greater variety of animal habitats (see Figure 16.18). Grasslands, with fewer strata, support fewer bird species, all of which nest on the ground, whereas eastern North American deciduous forests support many species occupying different strata. For example, the wood peewee (*Contopus virens*) occupies the canopy, the hooded warbler (*Wilsonia citrina*) inhabits the shrub layer, and the ovenbird (*Seiurus aurocapillus*) forages and nests on the forest floor.

In a study of the impact of vegetation structure in 13 grassland and forest communities, bird diversity was measured with a modified Simpson's diversity index (see Section 16.1), and structural heterogeneity with an index of foliage height diversity. Bird diversity increased with the number of vertical layers and the relative amount of biomass within the layers. There was a strong positive correlation between bird diversity and foliage height diversity (Figure 16.28) (MacArthur and MacArthur 1961). Similar relationships between habitat heterogeneity and animal diversity have been reported in both terrestrial and aquatic habitats. Variation in physical structure opens up more niche opportunities, increasing niche packing (see Section 16.2).

16.13 CONTRASTING VIEWS OF COMMUNITY: Ecologists Interpret Communities as Discrete or Continuous

At the outset, we defined the community as the group of populations that occupy a given area, interacting either directly or indirectly. Whereas these interactions exert both positive and negative influences on populations, their impact on communities is more controversial. In the early 20th century, this question led to a debate that still influences community ecology.

Recall our discussion in Section 16.11 of the problem of assigning discrete community labels to the forests we encounter as we walk from hilltop to valley. But what if we continue our walk over the next hilltop and into the adjacent valley? We will likely notice that although the hilltop and valley communities are distinct, the two hilltop and two valley communities are similar. Botanists use the term **association** to describe this phenomenon: a community with (1) a relatively consistent species composition, (2) a uniform **physiognomy** (general appearance), and (3) a distribution that is typical of a particular habitat or set of abiotic factors. Wherever this habitat or set of factors repeats itself in a given region, the same association typically occurs.

Some argue that the existence of associations implies structuring processes. If clusters of species repeatedly co-occur, that constitutes indirect evidence for positive interactions among them, as well as negative interactions among species that do not co-occur. Such logic favours a view of communities as integrated units. In this view, called the **organismal (holistic) concept**, the association is compared to an organism, with each species representing an interacting, integrated component of the whole (Clements 1916). Change in the community over time (*succession*; see Chapter 17) is the development of that organism.

The organismal view posits that species in an association have similar limits along an environmental gradient, with many reaching their maximum abundance at the same point (Figure 16.29a). Associations are discrete and transitions between them are narrow, with few species in common. Some **ubiquitous species** (species with broad distributions) do occupy more than one association. For example, jack pine occurs on many substrates in the boreal forest, but the species growing with it differ substantively, from lichens on rock outcrops to feathermosses and low evergreen shrubs on sandy soils. A shared evolutionary history has selected similar responses and tolerances for co-occurring species. Interactions are critical; even though it might seem that similar responses would entail niche overlap, setting the stage for competition, mutualism and coevolution to play a facilitating role. The community has evolved as an integrated unit, with species interactions as the "glue" holding it together. Given that the association is a distinct entity with a predictable species mix, random events have minimal impact.

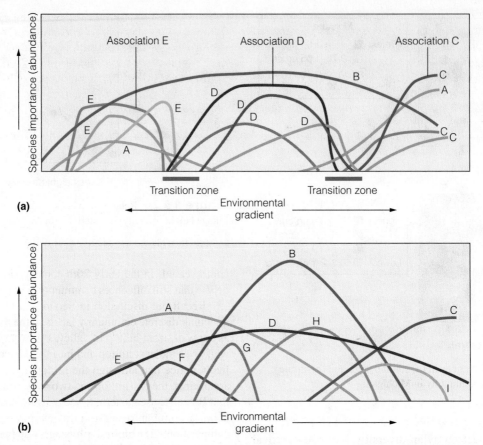

Figure 16.29 Two views of the community. **(a)** Organismal (holistic) view. Clusters of species show similar distribution limits and abundance peaks along the gradient. Each cluster defines an association. A few species (e.g., A) have sufficiently broad tolerance ranges that they occur in adjacent associations but in low numbers. Others (e.g., B) are ubiquitous. **(b)** Individualistic (continuum) view. Clusters of species do not exist. Peaks of abundance of dominants, such as A, B, and C, are arbitrary segments on a continuum.

In contrast, the **individualistic (continuum) concept** posits that species respond independently to differences in abiotic factors (Gleason 1939). The similar species composition of communities in similar habitats reflects their similar needs and tolerances, rather than biotic interactions. Changes in composition and abundance along gradients occur so gradually that it is arbitrary to divide vegetation into associations. Species distributions do not form discrete clusters on gradients, and transitions are gradual (Figure 16.29b). What we call a community is merely the group of species that coexist under any particular set of abiotic factors, with much overlap between communities.

The basic difference between these views is the importance of (1) interactions and (2) chance. Gleason acknowledged that interactions occur but argued that they are not critical in structuring communities. Also, because the continuum view posits that communities are loose assemblages not tightly regulated by biotic interactions, chance has a tremendous influence on community structure.

It is tempting to choose between these views, but current thinking incorporates elements of both. The organismal view dominated community ecology's early history. Gleason's view, which in essence describes the default model in Section 16.2, rose to favour in the 1960s when studies revealed that many communities that appeared discrete were in fact variable and subject to random events. The rise of niche theory in the 1970s could be seen as supporting both, depending on how

niche is interpreted. The Hutchinsonian idea of niche as the multiplicity of responses of a species to its environment supports the individualistic view, whereas Chase and Leibold's concept of niche as also incorporating reciprocal impacts of organisms both on each other and on their environment supports a more sophisticated version of the organismal view. The recent reprise of systems ecology with its emphasis on interspecific interactions (particularly indirect) affecting ecosystem stability and function reflects an organismal perspective, but one that acknowledges the impact of random events. On a global scale, where the entire biosphere rather than a single community is considered an integrated entity, the organismal view becomes the Gaia hypothesis (see Section 19.10).

Yet the individualistic view is far from dead. Describing communities with multivariate methods (Pielou 1984) is a better fit with its continuum perspective. Stephen Hubbell's **unified neutral theory** (2001), which posits that diversity patterns can be predicted independently of any assumptions of niche differences based on ecological trade-offs, suggests that population processes (in particular dispersal) are critical. Neutral theory differs from a Gleasonian view in that it assumes all species are ecologically equivalent. It proposes that *ecological drift* (random forces affecting population size), not niche packing, is the driver behind community diversity.

However at odds these perspectives may seem, understanding community structure requires remaining receptive to both.

David Tilman's **stochastic niche theory** (2004) attempts to bridge the gap between neutral and niche theories by stressing the impact of stochastic forces on the ability of species to establish and compete for resources with existing species. This "truce" is already bearing fruit; in a study of invertebrate communities in 10 streams in New Zealand that vary in abiotic factors and proximity, both niche *and* neutral processes structured the local community, with their respective strength varying with trophic level (Thompson and Townsend 2006). Tilman's stochastic niche is part of a conceptual synthesis (see Section 17.12) that is moving community ecology forward in a new and integrated direction.

EcologyPlace

Visit EcologyPlace at www.pearsoncanada.ca/ecologyplace to access online resources that complement your textbook, and help you to apply and to review the information in this chapter. EcologyPlace includes

- an eText version of the book
- self-grading quizzes
- glossary flashcards
- and more!

Go to www.pearsoncanada.ca/ecologyplace and follow the registration instructions on the Student Access Code Card included with this text. If your book does not have a Student Access Code Card, you can purchase access to it at www.pearsoncanada.ca/ecologyplace.

SUMMARY

Species Diversity 16.1

A community is the group of populations occupying a given area and interacting directly or indirectly. Species diversity, whether compared using rank–abundance curves or calculated by diversity indices, combines species richness (species number) and evenness (relative abundance).

Niche Packing 16.2

The range of abiotic factors tolerated by a species defines its fundamental niche. Constraints on the ability of species to survive and grow limit their distribution and abundance. As abiotic factors change, so does the biological structure of communities. This framework provides a null model against which community patterns can be compared. Species interactions, by altering fundamental niches into realized niches, alter community structure. Experiments that examine only pairwise interactions underestimate the impact of diffuse and indirect interactions.

Dominance 16.3

Dominants are species that predominate in a community, affecting conditions for associated species. Dominants are often the most numerous, but other criteria that consider both size and number may be more appropriate, especially for modular organisms. Dominants may or may not be superior competitors, and their identity may change over time. Not all communities have a dominant, and the identity of the dominant may change over time.

Keystone Species 16.4

Keystone species have an effect on community structure or function that is disproportionate to their numbers. Their removal initiates changes in community structure and often lowers diversity. Keystone species create or modify habitats (as bioengineers), or influence interactions among other species. Keystone species are most often generalist consumers with a diet preference for a weaker competitor. Not all communities have a keystone species.

Species Composition 16.5

Species composition is as important as species richness. Communities can have the same richness and dominants yet differ in the species present. Compiling a species list can be difficult, depending on taxonomic and sampling-related issues.

Food Webs 16.6

Feeding interactions are depicted by a food chain: a series of arrows linking species that are consumed to species that consume. Food chains interact in webs of varying complexity with links leading from primary producers (basal species) through intermediate to top-level consumers. Omnivores feed at more than one trophic level, in either the same or different chains. Guilds contain species that use shared resources in a similar way. Functional types group species based on their response to the environment, life history traits, or other attributes.

Indirect Interactions 16.7

Food webs illustrate the potential for indirect interactions, which occur when one species influences another through the action of a third species. Examples include keystone predation, apparent competition, indirect commensalism, indirect mutualism, and trophic cascades.

Food Web Controls 16.8

To understand the impact of interactions on community structure, food webs are simplified by placing species into functional groups based on similarity in resource use. Classification of species into trophic levels suggests the possibility of either bottom-up or top-down control.

Physical Structure 16.9

The physical structure of most terrestrial communities is defined by the dominant growth forms of the vegetation. Of terrestrial communities, forests have the most strata. In aquatic habitats, physical structure is determined by abiotic features such as light, temperature, and O_2. The upper strata of all communities perform photosynthesis, whereas decomposition occurs primarily in the bottom strata. Vertical layering provides the physical structure in which animals live.

Zonation 16.10

Zonation (changes in community structure across a landscape) has complex causes. Community structure is constrained by species tolerances (fundamental niche), modified by interactions with other species (realized niche). The trade-off between the ability of a species to tolerate scarce resources and its competitive ability when resources are abundant can cause zonation in sites with resource gradients. If a gradient also involves non-resource factors, the lower boundaries may be determined by stress tolerance, and the upper boundaries by competition. Some species alleviate the abiotic stresses of a habitat, making it more favourable for less tolerant species.

Community Classification 16.11

Transitions between communities are often gradual, making defining boundaries difficult. Communities are often named for their dominants, but species composition can vary greatly. Classification varies with spatial scale.

Environmental Heterogeneity 16.12

Spatial variation in environmental factors within communities increases diversity by supporting more species. Mini-disturbances create gaps and patchy habitats within communities. Increased vertical structure means more resources and a greater diversity of habitats for animals.

Contrasting Views of Community 16.13

In the organismal view, the community is a discrete association in which each species is an integrated part of the whole. In the individualistic view, communities exist on a continuum. Species co-occur as a result of similar tolerance ranges for abiotic factors. These views differ in the importance they assign to interactions and to chance. Ecologists are unifying classical niche theory with the stochastic forces affecting ecological drift, as stressed by the neutral theory.

KEY TERMS

aphotic zone	dominant	indirect commensalism	organismal (holistic)	species evenness
apparent competition	emergent properties	indirect mutualism	concept	species rarefaction
association	environmental	individualistic	photic (euphotic)	curve
benthic zone	heterogeneity	(continuum)	zone	species richness
bioengineer	food chain	concept	physiognomy	stochastic niche theory
bottom-up control	food web	keystone predation	rank–abundance curve	top-down control
canopy	forest floor	keystone species	(Whittaker plot)	trophic cascade
community	functional group	(functional	relative abundance	trophic (feeding) level
community similarity	functional redundancy	dominant)	shrub layer	ubiquitous species
measure	gap	linkage density	species accumulation	understory
connectance	growth form	neighbourhood habitat	curve	unified neutral theory
diversity index	herb layer	amelioration	species composition	vertical stratification
dominance	importance value	niche packing	species diversity	zonation

STUDY QUESTIONS

1. If two communities have the same number of species, is their species diversity always the same? Answer with reference to the concepts of species richness and species evenness.
2. Calculate richness, diversity (using Simpson's and/or Shannon's index), and evenness for the communities in Tables 16.1 and 16.2. Given that they occur in the same geographic area and have a similar climate, what factors might explain the differences in their biological structure?
3. How do the fundamental niches of species constrain community structure? In what sense does this idea function as a null model?
4. Distinguish between a *dominant* and a *keystone species*, and between a *keystone species* and a *bioengineer*. Is the coral *Oculina* described in Sections 16.3 and 16.4 a dominant? A keystone? A bioengineer? Why? Do all communities have a keystone species? If not, why not?

5. There can be 2500 species in 1 ha of rain forest whereas 1 ha of dry woodland may only contain 25 species. In which will diffuse competition likely be more prevalent? Why?

6. Are all carnivores top-level predators? What distinguishes a top predator? Can an omnivore be a top predator? If a fungus feeds on a living tree, what is its trophic level?

7. Define *indirect interaction* in the context of food webs. Give an example of how predation can be involved in an indirect interaction that has positive effects.

8. What type of indirect interaction is involved in the example of the moose and woodland caribou in Section 16.7?

9. Distinguish among *trophic cascade*, *keystone predation*, and *indirect mutualism*.

10. In Mary Power's research on largemouth bass (see Section 16.8), top predators control autotroph productivity by controlling herbivore abundance. Suppose we reduce primary productivity by applying a chemical that has no direct effect on consumers. If this reduced productivity reduces herbivore populations, in turn causing a decline in the top predator, what type of food web control (top-down or bottom-up) is suggested? How might the findings of these experiments be reconciled?

11. Contrast the vertical stratification of a lake and a forest. How does vertical stratification influence animal diversity?

What are the implications of this influence for animal species conservation?

12. Distinguish between *zonation* and *environmental gradient*. Consider cattail zonation along a depth gradient in Figure 13.6. Are the lower limits of the species set by their ability to tolerate low resource levels (as with moisture in Figure 16.19) or by their ability to tolerate abiotic stresses linked to a non-resource factor (as with soil anoxia in Figure 16.23)? How could you clarify these possibilities experimentally?

13. Based on the material presented in Chapter 6, what traits might enable a plant species A to tolerate low soil nutrients? How might these traits limit its growth rate in fertile soil? Conversely, what traits might enable a plant species B to maintain rapid growth in fertile soil? How might these traits limit its ability to tolerate low soil nutrients? Predict the outcome of competition between species A and B in two sites, one fertile and the other infertile.

14. In Figure 16.19, does diversity (as indicated by the index) change much from hilltop to bottomland? Does this mean there is little change in biological structure along the gradient?

15. Which view of the community does the classification in Figure 16.25 embody? Why? How would you determine which view was most appropriate for this particular environment?

FURTHER READINGS

Brown, J., T. Whitham, S. Ernest, and C. Gehring. 2001. Complex species interactions and the dynamics of ecological systems: Long-term experiments. *Science* 93:643–650.
Review of long-term experiments investigating complex community interactions.

Estes, J., M. Tinker, T. Williams, and D. Doak. 1998. Killer whale predation on sea otters linking oceanic and nearshore ecosystems. *Science* 282:473–476.
Explores the role of keystone species in coastal marine communities of Alaska.

McPeek, M. 1998. The consequences of changing the top predator in a food web: A comparative experimental approach. *Ecological Monographs* 68:1–23.
Excellent example of the application of experimental manipulations to understanding interactions.

Pimm, S. L. 1991. *The balance of nature.* Chicago: University of Chicago Press.

Applies theoretical studies of community structure to current issues in conservation ecology.

Power, M. E. 1992. Top-down and bottom-up forces in food webs: Do plants have primacy? *Ecology* 73:733–746.
Excellent discussion of the role of predation in structuring communities, including a contrast between bottom-up and top-down controls.

Power, M. E., D. Tilman, J. Estes, B. Menge, W. Bond, L. Mills, G. Daily, J. Castilla, J. Lubchenco, and R. Paine. 1996. Challenges in the quest for keystones. *Bioscience* 46:609–620.
Reviews concept of keystone species as presented by current leaders in community ecology.

Reimchen, T. 2001. Salmon nutrients, nitrogen isotopes, and coastal forests. *Ecoforestry* (Fall) 13–16.
Documents the links between bear predation on salmon and productivity of coastal forests.

CHAPTER

17 COMMUNITY DYNAMICS

A diverse grassland community occupies a former agricultural field. Studying the processes involved in colonizing abandoned croplands has yielded important insights into succession.

In Chapter 16, we examined community structure and the factors affecting it. Zonation reflects the shifting distributions of populations in response to changing abiotic factors, modified by biotic interactions and random forces. But a community changes over time as well as space, reflecting the population dynamics of its species. Its structure changes as plants establish, mature, and die. Birth and death rates change, altering species composition, dominance, and diversity. Changing community patterns over time is the subject of this chapter. We start by considering changes in the short term—changes any of us could observe in our lifetime. Then, at the end of the chapter, we consider how changes over a longer timescale have changed the face of the global landscape, particularly in countries like Canada where major glaciation events—ice ages—have promoted vegetation changes of massive proportions.

17.1 SUCCESSION DEFINED: Community Structure Changes over Time

Rather than moving across a landscape, suppose we stand in one spot and observe a community over time. Abandoned croplands are common in once-forested regions of eastern North America (see Ecological Issues: Climax versus Old-Growth: Implications for Conservation, p. 370). No longer cultivated, the land sprouts grasses, forbs such as goldenrod (*Solidago* spp.), and many annual and biennial "weeds" (aggressive species, typically introduced). In several years, shrubs such as blackberries (*Rubus* spp.), sumac (*Rhus* spp.), and hawthorn (*Crataegus* spp.) invade, followed by pin cherry (*Prunus pensylvanica*), pine (*Pinus* spp.), and aspen (*Populus* spp.). Decades later, the site supports a maple (*Acer* spp.) or oak (*Quercus* spp.) forest.

Succession is defined as change in community structure over time. In most sites, it is typically gradual and unidirectional. Clements (1916) defined a **sere** as a successional sequence of communities, for example, from grassland to shrubland to forest. Although each *seral stage* is a point in a continuum, it may be fairly distinct, with a characteristic structure and species mix. A seral stage may last a few years, or decades. Some may be cut short or missed entirely; if trees colonize a field (Figure 17.1), the shrub stage is often bypassed. Structurally, its role is assumed by tree saplings.

Succession occurs in aquatic habitats, too. In rocky intertidal pools, wave action overturns rocks, and algae colonize the cleared surfaces. To mimic this disturbance, experimenters placed concrete blocks in experimental pools. A predictable pattern of colonization and extinction ensued, with other species displacing those that first occupied the blocks (Figure 17.2, p. 364) (Sousa 1979).

Comparable (but more gradual) patterns occur on land. Consider succession after clear-cutting at New Hampshire's Hubbard Brook Experimental Forest (Figure 17.3, p. 364). Before logging, sugar maple (*Acer saccharum*) and beech (*Fagus grandifolia*) dominated both the canopy and the understory (as seedlings and saplings), indicating that they were replacing themselves. After logging, beech and maple

(a)

(b)

(c)

Figure 17.1 Successional changes over 30 years in an old field. **(a)** In 1942, it was still grazed. **(b)** The same area 21 years later. **(c)** By 1972, trembling aspen and maple had established.

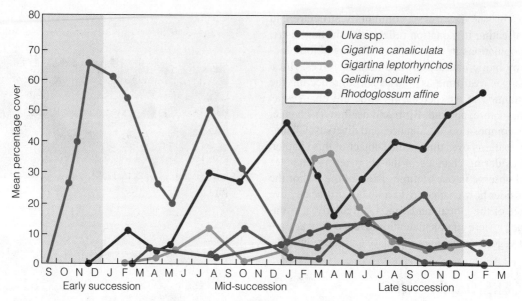

Figure 17.2 Mean cover of algal species colonizing concrete blocks in intertidal pools.

(Adapted from Sousa 1979.)

INTERPRETING ECOLOGICAL DATA

Q1. At what month and year does *Gigartina caniculata* first appear? During what period does it dominate?

Q2. Which species dominates in the first year? In the last year? Which species never dominate? Is species richness highest during early, mid-, or late succession?

seedlings declined, replaced by raspberry and shade-intolerant, fast-growing trees such as pin cherry and yellow birch (*Betula alleghaniensis*). Eventually, these species were replaced by beech and maple. With minor differences, this trajectory recurs in any beech–maple forest in Ontario, Québec, or Nova Scotia.

These examples illustrate the similarities of succession in different habitats. **Early-successional species (pioneers)** are typically small, with high rates of dispersal and population growth (recall *r*-strategists from Section 8.10). **Late-successional species** are larger and longer lived, with lower rates of dispersal, colonization, and growth (*K*-strategists). As *early* and *late* imply, species replacement is non-random.

These examples also represent different succession types. **Primary succession** occurs on a previously unvegetated site—a rocky plateau in Newfoundland, or (on a much finer scale) newly exposed surfaces such as cement blocks in a tidal pool as in Figure 17.2. **Secondary succession** occurs on previously vegetated sites after a disturbance has removed part or all of an existing community, as in the Hubbard Brook study. Disturbance does not always remove all individuals. Density, biomass, and composition of the surviving community affect successional dynamics.

17.2 PRIMARY SUCCESSION: Harsh Conditions Restrict Pioneers in Previously Unvegetated Sites

Because primary succession occurs on previously unvegetated sites, the site has not been influenced by life. No soil is present, and conditions are typically harsh. Consider succession on a sand dune. Sand is deposited by wind and water. Where deposits are extensive, as along shores of water bodies and on inland sites such as Manitoba's Carberry Sandhills (a remnant delta of the Assiniboine River where it flowed into glacial Lake Agassiz), sand piles up in windward slopes and forms dunes (Figure 17.4). With wind or water, the dunes shift, burying vegetation and making colonization difficult.

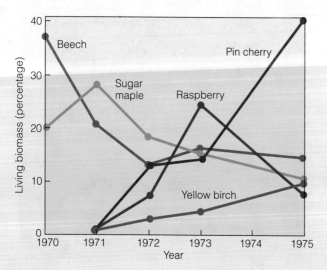

Figure 17.3 Changes in relative abundance (percentage biomass) of woody species in the Hubbard Brook Experimental Forest after clear-cutting.

(Adapted from Bormann and Likens 1979.)

Figure 17.4 Beach grass is a pioneer in primary succession of sand dunes.

Cowles (1899) first described succession on Lake Michigan sand dunes. Pioneer grasses, notably beach grass (*Ammophila breviligulata*), stabilize the dunes with their extensive rhizomes. Mat-forming shrubs then invade. After many decades, dominance shifts to trees—first pine, then oak. Given

sand's poor moisture retention, oak is rarely replaced by more moisture-requiring trees. Only on leeward slopes and in depressions where moisture accumulates does succession proceed to mesophytic species such as sugar maple. Trees shade the soil and accumulate litter, improving nutrient and moisture conditions on the site. This example emphasizes a typical feature of primary succession: pioneers ameliorate the environment, paving the way for later species.

Primary succession also occurs on newly deposited sediments on floodplains. Over the past 200 years, the glacier that once covered Glacier Bay National Park, Alaska, has been retreating (Figure 17.5a, b). As the glacier melts—at an accelerating rate with global warming—the exposed surfaces are colonized by mosses and lichens, followed a few decades later by shade-intolerant woody species such as alder (*Alnus* spp.) and cottonwood (*Populus* spp.). A century later, they are replaced by spruce (*Picea* spp.) and hemlock (*Tsuga* spp.) (Figure 17.5c). Ultimately, the forest resembles those in the surrounding landscape.

There are variations with substrate and climate, but site conditions are often harsh in primary successions. Only a few species can act as pioneers (usually *stress-tolerators*; see Section 8.10), so chance plays a minimal role, especially in the

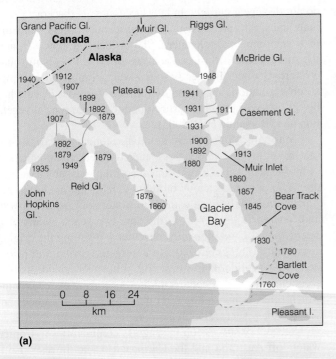

(a)

(b)

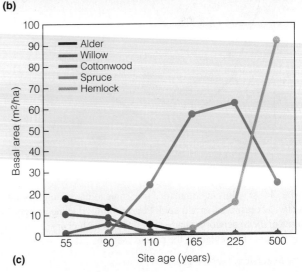

(c)

Figure 17.5 Primary succession on glacial sediments. **(a)** The Glacier Bay fjord complex in Alaska, showing ice retreat since 1760. As the ice retreats, it leaves moraines on which primary succession occurs. **(b)** Primary succession along riverine environments of Glacier Bay National Park. **(c)** Changes in community composition (basal area of woody species) after exposure of sediments. Non-woody pioneer stages are not shown.

(Adapted from Hobbie 1994.)

early stages. Primary successions tend to be slow, particularly at first, given the absence of soil. At 200 to 250 years, the Glacier Bay succession is relatively rapid; although the climate is cool, glacial sediments are a more favourable substrate than a rock outcrop. The site is alkaline initially, but exudates from plant roots lower the pH to more favourable levels. In time, the accumulating organic matter lightens the soil and improves its water- and mineral-holding capacity.

17.3 SECONDARY SUCCESSION: More Favourable Conditions Promote Revegetation after Disturbance

A classic secondary succession study documented changes in an abandoned crop field in North Carolina (Billings 1938). Initially, the field is claimed by annuals such as crabgrass (*Digitaria sanguinalis*), whose dormant seeds germinate in response to light and moisture. But its hold on the site is short-lived. By late summer, seeds of horseweed (*Lactuca canadensis*), a winter annual, ripen. Dispersed by wind—a trait typical of pioneers in secondary successions—the seeds settle, germinate, and by late fall produce vegetative rosettes. The next spring, horseweed, off to a head start, assumes dominance. In the second summer, herbaceous perennials such as white aster (*Aster ericoides*) and ragweed (*Ambrosia artemisiifolia*) invade, followed by broomsedge (*Andropogon virginicus*), a perennial bunchgrass, in year 3. Its ability to exploit soil moisture efficiently facilitates its increasing dominance. Short-leaf pine seedlings (*Pinus echinata*) establish amid the broomsedge clumps. In a decade, the pines are tall enough to shade broomsedge and other shade-intolerant herbs. Oaks (*Quercus* spp.) and hickories (*Carya* spp.) grow up through the pines, assuming dominance when the pines die (Figure 17.6). A deciduous forest continues to develop as shade-tolerants such as dogwood and redbud fill out the understory.

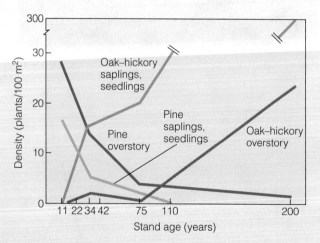

Figure 17.6 Shift in dominance from pine to hardwoods during secondary succession of an old field. Pine seedlings decline as light decreases in the understory. Shade-tolerant oak and hickory seedlings establish, eventually replacing the pine as they die.

(Adapted from Billings 1938.)

Figure 17.7 A kelp forest off the Aleutian Islands. Kelp beds have a complex physical structure, with many coexisting species.

Similar successions occur in any crop field allowed to "return to nature." The players differ—for example, in parts of southwestern Ontario and southern Manitoba, soils are too alkaline to support a pine stage—but the same shifts in dominants occur: from annuals to biennials to herbaceous perennials to shrubs to trees of varying shade tolerance. These growth forms do not disappear entirely. In late succession, herbaceous perennials and shrubs are present, but their identity differs, reflecting their environmental tolerances. Site history can, however, alter the successional path. In an analysis of 36 old fields in southwestern Québec, former pastures were dominated by spiny shrub vegetation, suggesting the selective impact of past herbivory, whereas crop fields succeeded to a more typical vegetation mix (Benjamin et al. 2005).

Marine ecologists have studied secondary succession in salt marsh, mangrove, seagrass, and coral reef communities. A year after removal of a kelp "forest" off the Alaskan coast (Figure 17.7), a mixed canopy of kelp (*Nereocystis luetkeana* and *Alaria fistulosa*) formed, with an understory of *Costaria costata* and *Laminaria dentigera*. In the next two years, stands of *L. setchellii* and *L. groenlandica* developed, restoring the original community (Duggins 1980). In contrast, *Agarum fimbriatum* dominates subtidal kelp forests off Vancouver Island. Its high secondary metabolite content minimizes grazing by sea urchins, which (unless checked by sea otters) overgraze kelp beds and set secondary succession in motion.

Secondary succession in seagrass communities in tropical waters off Australia and the Caribbean occurs in response to abiotic and biotic disturbances. Wave action or heavy grazing by sea turtles and urchins creates gaps in seagrass cover, exposing the sediments. Erosion on the down-current side of these gaps causes localized *blowouts* (Figure 17.8a). Colonizers are macroalgae such as *Halimeda* and *Penicillus* with root-like growths (Figure 17.8b). These algae have limited ability to bind sediments, but when they decompose they contribute organic matter. Shoal grass (*Halodule wrightii*) then colonizes from seed or clonal branching, stabilizing the sediments. Eventually, *Thalassia testudinum* invades. Its large leaves and extensive rhizome and root system trap sediments and greatly increase

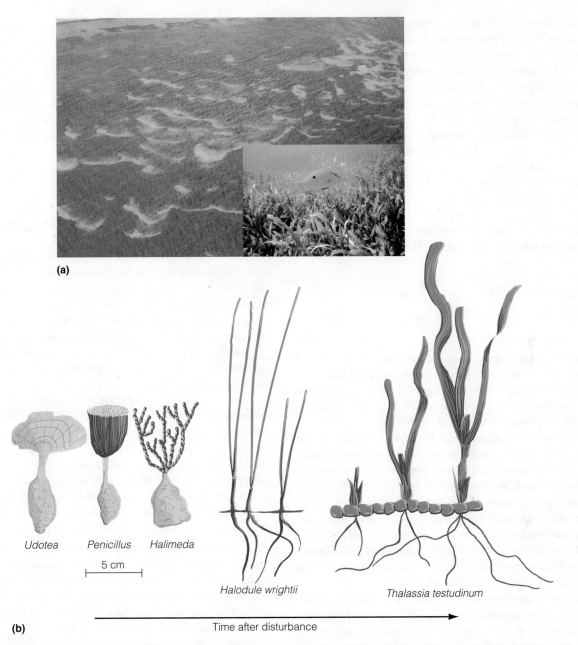

Figure 17.8 Secondary succession in seagrass communities. **(a)** Disturbed areas called blowouts (light-coloured) form in seagrass communities. Insert shows a mature community dominated by *Thalassia testudinum*. **(b)** Blowouts are colonized by rhizophytic macroalgae, which are displaced by the pioneer seagrass *Halodule wrightii* followed by *Thalassia*.

organic matter. Eventually, the disturbed area resembles the surrounding seagrass community.

In comparison with primary successions, secondary successions tend to be more rapid, unless the site has suffered severe damage—as happens if a severe ground fire damages soil structure (see Section 18.8). Conditions vary, but typically they are less harsh because of the presence of soil. Pioneers tend to be *ruderals* (see Section 8.10) such as annuals, not stress-tolerators. Conditions fall within the tolerance range of more species, so chance affects which of the possible pioneer species arrives first. In turn, differences in the identity of the pioneers may alter the successional path. Later stages of secondary successions may converge, given similar site and climate conditions.

17.4 DOCUMENTING SUCCESSION: Ecologists Study Succession by Substituting Space for Time

How do ecologists document succession? If succession is rapid, as in Figure 17.2, ecologists can study it in real time. For most successions, this is not feasible. At permanent research stations, generations of ecologists can monitor succession in the same location. But in most cases, ecologists use a "space-for-time" approach. There are two types. A **toposequence** is a series of sites that are adjacent to one another, as in Cowles's sand dune study, where areas back from the dunes were colonized earlier. Alternatively, ecologists may infer succession from a

chronosequence: a series of sites that are at different successional stages but are not physically adjacent. An ecologist studying post-fire succession might visit sites that burned 1, 25, 50, 100, and 150 years ago. For example, Boucher et al. (2006) used a chronosequence to investigate the impact of stand age on the structure of coniferous stands in northeastern Québec.

A chronosequence makes it feasible to study succession in a short timeframe, but finding comparable sites is essential to minimize confounding factors. If two sites burned at different times but their soil and climate differ, then their vegetation will differ for reasons other than succession. Despite these limitations, Walker et al. (2010) conclude that a chronosequence approach is valid for studying succession in communities with low diversity and rapid turnover. Changes in vegetation structure and richness may be more predictable than changes in species composition.

17.5 SUCCESSION MECHANISMS: Multiple Processes Drive Successional Change

Succession sounds simple enough—predictable, directional change in species composition over time. Species rise and fall in abundance, reflecting colonization, establishment, and (local) extinction, with the decline of one species followed by the rise of, and eventual replacement by, others. Early studies explored succession on a purely descriptive basis. Ecologists have since generated many hypotheses regarding the processes driving succession.

According to the **monoclimax hypothesis** (Clements 1936), succession is the progressive development of a highly integrated superorganism, the community (see Section 16.13), to its **climax stage**, after which there are no further significant changes. Each set of climatic factors generates a single outcome: the monoclimax. Tansley (1935) modified this idea into the **polyclimax hypothesis**: similar climates support different climax types, depending on substrate.

Both these hypotheses are deterministic. In contrast, the **initial floristic composition hypothesis** (IFC) (Egler 1954) posits that succession is at least partly indeterminate, and affected by which species arrive first. Replacement is neither orderly nor predictable. Some species suppress or exclude others from colonizing, so differences in the original species mix may send succession along different paths. Competition occurs, but no species has clear-cut competitive superiority. The pioneers inhibit later-arriving species from establishing. (Arguably, this effect involves *interference*, a mechanism of competition, or perhaps *amensalism*, if the pioneer species are unaffected.) The initial phase can be long or short, depending on pioneer longevity, but only when the original colonizers die is the site accessible to other species.

IFC is distinctly Gleasonian in flavour—succession is individualistic, dependent on the identity of colonizing species and the order in which they arrive. It is much affected by chance interacting with dispersal traits. Different outcomes may ensue, even with a similar substrate and climate. More broadly, **pattern climax theory** (Whittaker 1953) posits that local environmental variation generates regional differences in species composition, as opposed to Clements's homogeneous monoclimax. Indeed, some ecologists have discarded the term *climax* for the more

neutral concept of *old growth* (see Ecological Issues: Climax versus Old Growth: Implications for Conservation?, p. 370).

But what are the mechanisms that take a community to the end point of succession, be it a monoclimax, polyclimax, or pattern climax? Connell and Slatyer (1977) proposed a general framework that considers a range of interactions and responses. They suggested three mechanisms, all of which can occur in a single succession, albeit to different degrees and/or in different successional stages (Figure 17.9).

1. In **facilitation**, pioneers modify the environment in ways that make it more suitable for later species to invade and grow to maturity. Early species pave the way for later species, facilitating their success—a process called *relay floristics*. Pioneers also make the site less favourable to themselves, hastening their own demise. Facilitation, which is Clementsian in spirit, is crucial in primary successions, where pioneers ameliorate harsh conditions and initiate soil formation. For example, when lichens colonize bare rock, their exudates initiate chemical weathering of the rock, and their organic matter improves the water-holding capacity of the site for later arrivals.

Facilitation also occurs in secondary succession. In an old field, annuals germinate best on open sites with exposed mineral soil. Once annuals have completed their life cycles and their dead bodies accumulate, the seedbed is no longer exposed soil and no longer favours their germination. The seedbed is, however, optimal for later species that benefit from the moisture-holding capacity of organic litter. Similarly, late-successional trees can rarely establish in the open conditions of early succession. Once pioneers such as pine or aspen establish, conditions are more favourable for these later species to establish. In contrast, pine cannot establish in the shade of its own canopy (see Figure 17.6). Pioneers are thus bioengineers of their own demise.

2. In **inhibition**, which underpins the IFC hypothesis, strong competitive interactions of the interference type drive succession. No one species is competitively superior, but whichever arrives first holds the site against all invaders. Unlike facilitation, pioneers make the site less suitable for later species as well as themselves. As long as conditions stay within their tolerance limits, the pioneers remain dominant and relinquish the site only on death. Species composition gradually shifts as short-lived species give way to long-lived ones. Later species may be present at the onset, but are suppressed until the colonizers die. An example of a *priority effect*, this mechanism is subject to chance events that affect order of arrival. Inhibition can occur in both primary and secondary successions, but it is more common in secondary successions where many species are capable of colonizing because soil conditions are less harsh.

3. In **tolerance**, later species are neither suppressed (as in inhibition) nor aided (as in facilitation) by earlier species. Later species invade, establish, and grow to maturity independently of species that precede or follow them. They do so because they tolerate a lower level of the key resources that change in availability over time. The species present at any seral stage are those that are most efficient at exploiting limiting resources as conditions change over time. For example, Pacific silver fir (*Abies amabilis*) is a highly shade-tolerant species that can invade, persist, and grow under a canopy of Douglas-fir and Sitka spruce (both long-lived pioneers)

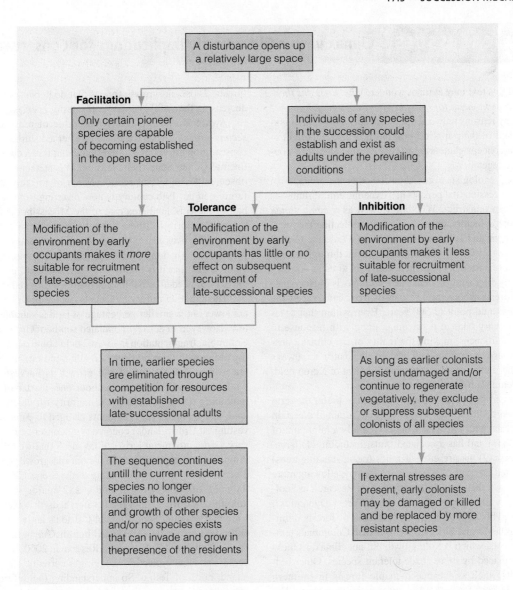

Figure 17.9 Comparison of the facilitation, inhibition, and tolerance mechanisms of succession. Secondary succession can be initiated at any stage.

(Adapted from Connell and Slatyer 1977.)

in British Columbia's temperate rain forests because silver fir uses light more efficiently (Agee 1995). As long as fires are infrequent, this late-successional specialist prevails, but does not require the death of the pioneers. Meanwhile, less shade-tolerant trees occupying the same site must await the death of canopy species, suggesting that inhibition and tolerance are working simultaneously.

In the tolerance mechanism, which combines Clementsian and Gleasonian elements, some species *are* competitively superior to others, as a result not of faster growth but more efficient resource use. Changing conditions with time alter relative competitive ability—much as along a spatial gradient in Section 16.10. Like inhibition, tolerance is more typical of secondary succession because initial site conditions allow both early and late species to establish early on. In contrast with facilitation, there is considerable overlap in species composition of seral stages.

Tilman (1985) supplemented the tolerance mechanism with the **resource-ratio hypothesis**, which posits that two major resources shift over succession: light decreases, and nutrients (par-

ticularly nitrogen) increase. Species come and go—establish and are replaced—because their relative competitive ability for these key resources changes with shifts in the ratio of their supply (Figure 17.10, p. 371). Each species reaches an equilibrium with the supply rates of limiting resources and, in doing so, lowers available resources to the point where other species cannot succeed.

The resource-ratio hypothesis assumes that nitrogen increases over time, as it often does in primary successions in which the site is often infertile initially (see Figure 17.12). But in secondary succession nitrogen may become less rather than more available, locked in living and dead biomass. However, moisture, another critical belowground resource, often does become more available, so the general principle of the impact of an altered ratio of two critical resources still applies. Note, too, that Tilman's competition model is based on the ability to use resources efficiently rather than to capture them (Grace 1991). The species that are strong competitors in late succession (i.e., those that make efficient use of limited light) differ from the rapid-growing,

ECOLOGICAL ISSUES   Climax versus Old Growth: Implications for Conservation?

A reader of this text may at times wonder: *What does this theoretical concept mean for everyday issues of concern to me?* There are many reasons why scientific concepts are worth pursuing independent of their applications, but it is also important to realize why a concept matters in practical terms. One such concept is climax vegetation.

Theoretical ecologists argue about whether the *climax*—an end-state of succession that persists in a dynamic equilibrium—is real or a human construct. Whether viewed as a *monoclimax* reflecting climatic factors, or as a *polyclimax* reflecting the interaction of climate and substrate (see Section 17.4), the climax concept assumes a view of nature that is filtered through human eyes—the end of vegetation change as obvious to us in our lifetimes. Yet study after study reveals that the climax is illusory, particularly in forests where the lifespan of trees can exceed the arbitrary climax end point of 500 years. Even within that timeframe, evolutionary change is ongoing, along with response to microscale disturbances, making the reality of the climax questionable. Species composition and relative abundance are always in flux, whether in the century after abandonment of a crop field, or in the millennium following a glaciation event.

Given this arbitrary quality, many ecologists prefer the term *old growth*: vegetation (often forest) that has attained a certain age (which varies with forest type but is typically a minimum of 175 to 250 years) and has associated traits, including (1) large, old, living trees; (2) the presence of *snags* (dead, standing trees) and accumulated litter; and (3) gaps in the canopy, allowing more understory growth than less mature stands (see Research in Ecology: What Can Tree Rings Reveal about Forest Dynamics?, pp. 192–193). This definition allows inclusion of forests dominated by Douglas-fir and Sitka spruce in British Columbia's temperate rain forests, which are old growth but not climax, as these species are replaced by more shade-tolerant species. Other old-growth forests, such as the beech–maple forests in southern Ontario and Québec, can regenerate under their own canopy. The term *old growth* encompasses both forest types. Some ecologists reserve the term for forests (often called *primeval*) that have not been disturbed by humans. However, site history may be known in many North American sites, but less so in Europe or Asia.

What is the relevance for habitat conservation? Old-growth forests are home to a high proportion of rare and endangered species. In part this reflects their structural complexity, which supports more animal biodiversity as a result of the many and varied niche opportunities. But it also reflects the fact that old-growth forests make up an ever-dwindling portion of the world's forests. Those old-growth forests that do persist are fragmented, decreasing their ability to support interior species.

Awareness of the ecological value of old-growth versus *second-growth* forests (regenerated after a disturbance) is vital to conservation efforts. It is often stated that forest cover in the continental United States is increasing. This statement is true, but it obscures an important reality. Much of the forest cleared for farming in the 19th century is now reverting to forest. Over one-half of the U.S. land area east of the Mississippi is forested, but the vast majority of these forests are less than 100 years old. Since 1972, over 110 000 km^2 of U.S. farmland has reverted to forest. This situation sounds encouraging for conservation—until one realizes that it is highly fragmented second growth, not the old growth needed as habitat for many endangered species.

The situation differs in Canada, where population pressures are lower and a smaller percentage of land is suitable for agriculture. However, in densely populated southern Ontario and British Columbia, fragmentation is severe, and in both countries, remaining old growth is being exploited at a rapid rate. In North America, some 10 000 km^2 of old growth is harvested annually. Although Canada is home to about one-quarter of the world's remaining old growth, the vast majority of which is on public land, low *stumpage fees* (amounts charged to companies for harvesting on Crown lands) coupled with conservation policies that vary across provincial jurisdictions mean that we fall far short of providing sufficient protection for our old-growth forests and the diversity they support. According to the Sierra Club, as of 2012, only 21 of Vancouver Island's 282 major watersheds are unlogged. Of the 7 unlogged watersheds that lack protection, 5 are in Clayoquot Sound, site of Canada's largest civil disobedience demonstration (1993). Although Clayoquot Sound was declared a UNESCO Biosphere Reserve in 2000, this designation confers no legal protection. Logging is currently slated for Flores Island, north of Tofino. So understanding the difference in ecological traits between second-growth and old-growth forest is critical to focusing conservation efforts in Canada and the world.

1. Visit the Global Forest Watch website, which discusses the state of old-growth forests in Canada. What are the major issues (ecological, social, and economic) and the possible solutions? How do stumpage fees contribute to the problem?
2. Foresters sometimes refer to old growth as "decadent" and argue that it should be logged. What does this term mean, and what bias does it entail?

---

resource-grabbing species that are good competitors of the exploitation type (see Section 13.1).

Since Connell and Slatyer's synthesis, the search for a general succession model has continued. Given the diversity of environments in which succession occurs, no single mechanism drives all successions. A pressing concern in succession theory today is the concept of *alternative stable states*: outcomes of succession, often in response to human disturbance, that involve significantly different configurations of a community than formerly present. For example, kelp beds off the Pacific coast of North America succeed to either kelp forests (see Figure 17.7) or algal crust barrens. Predation by sea urchins is the immediate factor influencing succession, but urchins in turn are influenced by storms and predation by sea otters, which are a keystone species in this system. Estes et al. (1998) posit that these alternative states are mediated by predation of otters by killer whales. What constitutes an alternative stable state is controversial (Petraitis and Dudgeon 2004), but the issue is central to long-term ecosystem diversity and stability (see Section 19.9).

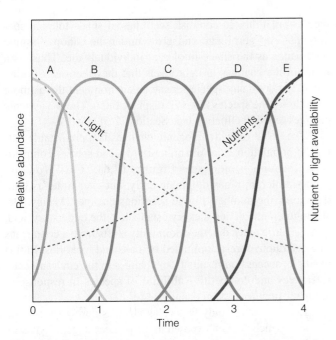

Figure 17.10 The resource-ratio hypothesis. Light becomes less available and nutrients become more available during succession. Altered ratios of these critical resources affect species performance and relative competitive ability.

(Adapted from Tilman 1988.)

17.6 CHANGES IN TRAITS: Properties of Ecosystems and Species Change during Succession

Rather than focus on mechanisms, some ecologists emphasize changes in the attributes of the ecosystem and its member species (Odum 1969). Community structure typically becomes more complex during succession. Biomass and total organic matter increase in most terrestrial successions, with progressively more nutrients stored in organic form. Food chains tend to grow in length and complexity, with more involvement of decomposers as a result of the increase in organic matter. Richness, which often increases from early to mid-succession, may decline in late succession.

Recently, the focus has shifted to how species traits (adaptations and life history) influence biotic interactions and ultimately community structure over time. In secondary successions, *r*-selected species (see Section 8.10) such as annuals and fast-growing herbaceous perennials are typical pioneers. They are effective colonizers, but over time are replaced by longer-lived and slower-growing *K*-strategists. These categories are also useful within a growth form. Aspen and birch are hardly *r*-strategists relative to annuals, but compared with long-lived oaks, they are *r*-selected. This change in species attributes highlights a key trait of late-successional species: their ability to replace themselves on site and resist replacement by other species. Shade tolerance is critical; a *K*-selected species such as beech produces large seeds with enough reserves to survive in shade until its shade-tolerant saplings are tall enough to support positive net carbon gain.

The dichotomy of *r*- versus *K*-selected species works less well in primary successions, where initial conditions are often

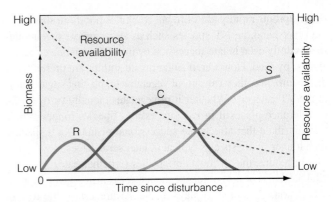

FIGURE 17.11 Shifts in dominance of Grime's three plant strategies during succession. R = ruderals; C = competitors; S = stress-tolerators.

(Adapted from Grime 1979).

too harsh for most *r*-strategists. Light is abundant, but nutrients may not be, and the rapid growth typical of *r*-strategists imposes high nutrient demands. In this case, Grime's model (see Section 8.10) is more useful. The pioneers in most primary successions are *stress-tolerators*, with their identity varying with the type of stress prevailing. Over time, depending on site conditions, they may be replaced by faster-growing *competitors* (Figure 17.11). In particularly harsh sites, mid-successional species may not be competitors but rather other stress-tolerators adapted to the altered but still stressful conditions. In late succession, competitors are replaced by species capable of tolerating deep shade. Grime's third category, *ruderals*, are typically *r*-strategists characteristic of the pioneer stage of secondary succession.

In Grime's view, in contrast to Tilman's, both light and nutrients decline with time. Note also that Grime considers species attributes other than competitive ability, particularly colonization ability and stress tolerance. His model assumes that a trade-off exists between attributes that enhance a species' capacity for acquiring resources (which he sees as fundamental to competitive ability) and its ability to tolerate stress. Tilman, who employs a different concept of competition (Grace 1991) in which the superior competitor is the one that makes most efficient use of limited resources, sees relative competitive ability under changing conditions as the major species trait that determines successional position (see Section 17.5).

Another approach categorizes species according to their **vital attributes**: traits that affect their role in succession (Noble and Slatyer 1980). The first trait involves recovery mode after disturbance: (1) V: vegetative spread (via clonal growth from roots or stems); (2) S: germination from an on-site seed bank; (3) D: germination from seeds that have dispersed to the site from surrounding communities; and (4) N: no specialized mechanism. The second trait reflects ability to reproduce in a competitive situation: T (tolerant) and I (intolerant). Species are described according to both traits—VI or NT, for example.

By assigning species to these categories, ecologists predict successional sequence. Grassland pioneers are typically SI or DI—annuals that quickly establish from an existing seed bank or by dispersal. In northern forests, in contrast, many post-fire pioneers are VI; they respond to disturbance by vegetative spread but compete poorly with larger individuals. An example is raspberry, which sprouts rapidly after fire but competes poorly in shade. On the other hand, an NT species such as sugar maple regenerates successfully even in late succession by invading existing canopies slowly, by seed. However, if sugar maple stumps are present after the disturbance, they can sprout vegetatively, placing sugar maple in the VT category. Whether it re-vegetates clonally or by seeds, its tolerance places it in late succession. Species longevity is a third attribute that affects the successional position of a species, with longer-living species typical of later seral stages.

In nature, the attributes of species (whether categorized by vital attributes or not) influence their success in any given stage of succession. Chance also plays a role. Awareness of the role of chance in succession led Horn (1975) to develop a Markov model that assigns probabilities to species replacements. The community is viewed as a checkerboard, with each space occupied by a mature tree (for a landscape version, see Figure 18.23). The model predicts the probability that a tree will be replaced by a tree of the same or a different species, based on the relative abundance of tree seedlings found beneath it. Like all models, it makes simplifying assumptions, but it illustrates the trend to make succession theory more quantitative and predictive.

17.7 AUTOGENIC CHANGES: Organisms Alter Abiotic Factors during Succession

Changes in environmental conditions that initiate shifts in community structure through time (and space) are of two general types. (1) **Autogenic** changes result directly from the presence and activities of organisms. For example, wind speed declines and humidity rises from early to late stages of a forest succession because of the presence of trees. (2) **Allogenic** changes result from extrinsic abiotic forces such as fire and altered precipitation, and are treated in Section 17.10.

Autogenic abiotic changes occur in all successions. The ability of any species to establish in any given seral stage reflects the tolerance limits of its fundamental niche. Once established, a species has reciprocal impacts on its environment, both directly and through its impact on other species. The inevitable consequence is an altered abiotic environment, which may or may not be favourable to that species. Consider alterations in light. In the pioneer stage, few plants are present, either because the site has never been occupied (primary succession) or because the plants have been removed by a disturbance (secondary succession). Either way, light is plentiful at ground level, and shade-intolerant species establish, assuming their other needs are met. As they grow, their leaves intercept light, reducing its availability to shorter plants and later arrivals. This reduction lowers photosynthesis, slowing growth of shaded individuals. Not all species photosynthesize and grow at the same rate, so those that grow tall the fastest have more access to light, reducing the supply to slower growers and outcompeting them in the short term.

However, as light declines over time, seedlings of these shade-intolerant dominants cannot survive, given the trade-off between adaptations that allow rapid growth in high light and the ability to continue growing in low light. So, although mature individuals still dominate, no new individuals are recruited into

their populations. In contrast, seedlings of shade-tolerant species that *can* germinate and grow under the canopy assume dominance as the shade-intolerant individuals die. Thus, as a result of the changes in light levels that they themselves bring about, the dominant species create an environment that is more suitable for the species that will displace them. These autogenic changes reflect facilitation (see Section 17.5) at work.

The result of such autogenic changes is a predictable pattern of population recruitment, mortality, and species composition through time (see Figure 17.6). Fast-growing, shade-intolerant pines dominate early, but over time, as light decreases, the number of pine seedlings declines. Meanwhile, shade-tolerant oak and hickory establish in the understory, and, as the mature pines die, the community shifts from a coniferous to a deciduous forest dominated by oaks and hickories. In this example, succession results from changes in the environmental tolerances and competitive abilities of species in response to autogenic environmental change.

Light is not the only factor that changes during succession. The soil, which is often exposed in the pioneer stage, is typically covered by an organic mat in later succession, affecting the ability of some seeds to germinate. Recall that in succession of seagrass communities, pioneers stabilize the sediments, adding organic matter that facilitates colonization of later species. Soil nutrients are also subject to autogenic influence. Consider primary succession on glacial till (see Figure 17.5). Given the absence of soil, little nitrogen is present initially, restricting establishment of most species. However, plants with mutualistic associations with N-fixing symbionts can dominate because they have access to atmospheric N_2 fixed by their partner. Alder (*Alnus* spp.), a pioneer shrub in virtually all boreal forests, has a mutualistic interaction with *Frankia*, an N-fixing actinomycete. As alders shed their leaves or die, the N they contain enters the soil after decomposition (Figure 17.12). Now other species can colonize and, in part because they lack the added energy cost of N fixation, dominate.

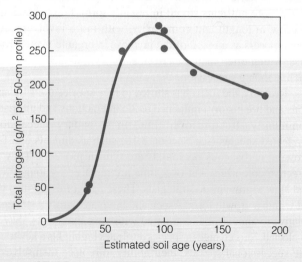

Figure 17.12 Changes in soil N during primary succession in Glacier Bay National Park. Initially, low N limits colonization to alder, which has a mutualistic association with N-fixing *Frankia* spp. As alder litter decomposes, N is released. With the buildup of organic matter and nitrogen, other species displace alder (see Figure 17.5c).

(Adapted from Oosting 1942.)

Succession is driven by changes in species tolerance and/or competitive ability in response to such autogenic changes.

Succession varies among communities and involves many mechanisms, including facilitation, competition, inhibition, and differences in tolerances. But whatever the mechanisms, the role of autogenic abiotic changes and differential responses of species to those changes is crucial.

17.8 DIVERSITY AND SUCCESSION: Colonization and Replacement Rates Affect Diversity

In addition to shifts in species composition and dominance, species diversity typically changes during succession. (*Diversity* is equated here with the number of species [*species richness*], as that is how it is treated in most succession studies. Different trends might emerge if evenness were also considered.) Using a chronosequence approach, ecologists have shown that species diversity in old-field successions typically increases with time since abandonment. However, a different pattern occurs in many deciduous forests (Figure 17.13). Species diversity increases into the late herbaceous stages and then declines in the shrub stage (Whittaker 1975). Diversity increases again in the young forest, only to decrease again later. Why?

Consider how colonization and replacement vary in time. Colonization increases the number of species, whereas replacement (whether resulting from competition or an inability to tolerate changing abiotic conditions) decreases it. The net balance of these two processes determines overall diversity. In early succession, diversity increases as new species colonize. Over time, some species are replaced. The diversity peak in mid-succession corresponds to a transition period after the arrival of shrubs but before the replacement of early-successional herbaceous perennials. The second, smaller peak corresponds to a similar transition between shrub- and tree-dominated stages, when both are still present. Later, diversity declines as shade-tolerant trees replace intolerant species.

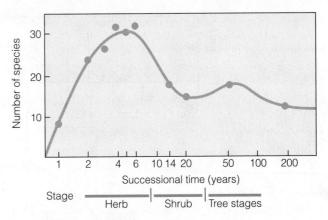

Figure 17.13 Changes in plant diversity during secondary succession of an oak–pine forest in Brookhaven, New York. Diversity is reported as species richness in 0.3-ha samples. Peaks in diversity correspond to transitions between stages, where species from both stages are present.

(Adapted from Whittaker 1975, Whittaker and Woodwell 1968, 1969.)

In tune with the focus on life history traits, growth rate and longevity affect the rate of replacement, which occurs more slowly for slow-growing, long-lived species than for fast-growing, short-lived species (hence the rapid algal succession in Figure 17.2). Over time, longer-lived species dominate, slowing replacement. Reflecting the strong correlation between growth rate and replacement, diversity trends during succession vary with abiotic factors (particularly resource supply) that influence growth (Huston 1994). By slowing growth of species that will eventually replace earlier species, resource limitations may prolong coexistence, keeping diversity high (Figure 17.14). Varying abiotic factors may also promote diversity by altering competitive advantages and preventing competitive exclusion.

Disturbance may have an effect similar to resource limitations by promoting species coexistence. By reducing or eliminating populations, disturbance resets the successional clock,

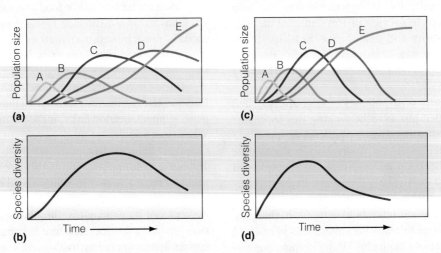

Figure 17.14 Model of **(a)** succession involving plant species (A–E) and **(b)** associated species diversity pattern. Diversity increases initially as species colonize, but then declines as autogenic changes and/or competition cause displacement. **(c)** When growth rates are doubled, succession speeds up. **(d)** Species diversity reflects earlier onset of competition and faster replacement, reducing the period of maximal diversity.

(Adapted from Huston 1994.)

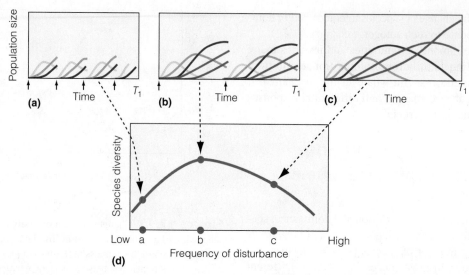

Figure 17.15 Succession patterns for species A to E in Figure 17.14 in different disturbance regimes. Time of disturbance is shown as an arrow on the *x*-axis. Note differences in species composition and diversity at time T_1. **(a)** With frequent disturbance, later species never colonize, reducing diversity. **(b)** At intermediate disturbance frequency, all species coexist, and diversity is maximal. **(c)** Without disturbance, later species displace earlier species, reducing diversity. **(d)** Maximum diversity occurs at intermediate disturbance frequency and magnitude. Examples (a) to (c) are labelled on the *x*-axis.

(Adapted from Huston 1994.)

allowing the site to once again be colonized by pioneer species. The twinned processes of colonization and replacement begin again. If disturbances are frequent, later species never have the opportunity to colonize, and diversity stays low. Without disturbance, later species displace earlier species, and diversity declines (Figure 17.15). Only at an intermediate frequency of disturbance can colonization still occur, while keeping competitive displacement to a minimum.

This relationship, called the **intermediate disturbance hypothesis** (Connell 1978), is under increasing attack. Chesson and Huntly (1997) contend that disturbance in and of itself cannot prevent exclusion or even make it less likely, and is not a mechanism for maintaining diversity through coexistence (see also Section 13.9). Empirically, the hypothesis does not help explain patterns in nature. In a review of 169 studies of the relationship between diversity and disturbance, non-significant results were the most common. Higher diversity at intermediate disturbance occurred in less than 20 percent of studies (Mackey and Currie 2001).

Whatever the validity of hypotheses linking diversity and succession, it is clear that just as niche packing was an important concept for understanding diversity patterns in space (see Chapter 16), it is equally relevant to diversity patterns over time. Many of the succession mechanisms can be explained using the niche concept, particularly Chase and Leibold's integrated version (2003). Niche opportunities change over time, and if fundamental niches are relevant as a result of changing abiotic conditions, even more critical are alterations in realized niches as a result of species interactions. If, by keeping populations below carrying capacity, disturbances prevent species from excluding others whose niches overlap, and if beneficial as well as detrimental interactions play a role, then more spe-

cies can coexist at any given seral stage. Hubbell's neutral theory (2001) is also central, as both ecological drift and dispersal affect species composition and turnover during succession.

17.9 HETEROTROPHIC SUCCESSION: Changes Occur in the Heterotrophic Community over Time

So far, we have focused on the plant community, but changes also occur in consumers. As succession advances, changes in vegetation structure and species composition initiate **heterotrophic succession**: changes in the animal and fungal species that depend on vegetation for habitat and/or food resources (Figure 17.16). Each stage has its own distinctive fauna. As animals are often influenced as much by structural traits as by species identity (see Section 16.9), successional stages in fauna may not correspond to stages identified by plant ecologists.

During *autotrophic* (plant) succession, animals can quickly lose habitat. North American grasslands and abandoned croplands support meadowlarks, meadow mice, and grasshoppers. When trees and shrubs invade, a new structural element appears. Grassland animals disappear, and shrubland animals take over. Towhees, goldfinches, and catbirds claim the thickets, and meadow mice give way to white-footed mice. As woody growth proceeds and the canopy closes, shrubland animals decline and are replaced by canopy-dwelling birds and insects. As succession proceeds, vertical structure becomes more complex. New species appear, including tree squirrels and woodpeckers as well as understory birds such as hooded warblers and ovenbirds.

As a result of human activities, fewer and fewer of the world's habitats are in late succession. This situation increases

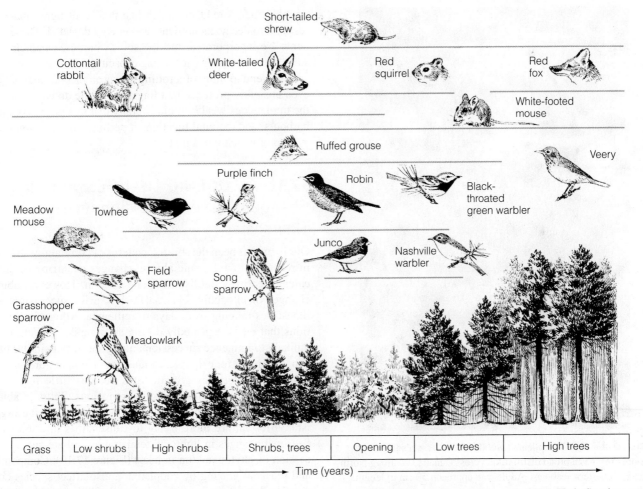

Figure 17.16 Changes in an animal community during succession from an old field to a conifer forest in central New York. Species appear or disappear as vegetation type, density, and height change. Brown lines represent vegetation stages inhabited by associated species.

the extinction risk for animals that are late-successional special-ists. The northern spotted owl (*Strix occidentalis caurina*), which nests in old-growth forests on the west coast, was the sub-ject of huge public controversy in the United States when its endangered status was used to halt logging on public land in the 1990s. The Canadian subspecies is even more at peril; according to British Columbia's Ministry of the Environment, it is down to under 100 breeding pairs. In contrast, the pileated woodpecker (*Dryocopus pileatus*; Figure 17.17, p. 376), which also uses old growth, has proved more flexible, expanding its realized niche to include second-growth forests and parks.

These changes occur in response to vegetation change, but in the heterotrophic succession involved in decomposition, ani-mals are the initiators. Dead trees, organic litter, carcasses, and fecal droppings (*scat*) are substrates on which decomposer communities thrive. Groups of fungi and invertebrates succeed each other, again via the twinned processes of colonization and replacement. Energy and nutrients are most abundant in the early stages and decline steadily as succession proceeds.

Consider the succession of organisms involved in decom-posing a rotting log (Figure 17.18, p. 376). When a storm uproots or breaks a tree, the newly fallen tree, its bark and wood intact, is a ready source of shelter and nutrients. Among the first

to exploit it are bark- and wood-boring *Ambrosia* beetles that drill into the bark and feed on the inner bark and cambium, reducing it to *frass* (insect excrement) and fragments, and cre-ating galleries in which to lay their eggs. Sugar fungi, which feed on simple carbohydrates, proliferate. The tunnels provide passageways, and the frass and softened wood provide sub-strates for other fungi and bacteria. Loosened bark provides cover for predatory invertebrates: centipedes, mites, pseudo-scorpions, and beetles. As decay proceeds, the rotting wood holds more moisture but accessible nutrients are depleted, leav-ing behind decay-resistant compounds. The arthropod pioneers leave for newer logs, while fungi able to digest cellulose and lignin remain. Mosses and lichens find the softened wood ideal habitat, and plant seedlings establish. Their roots penetrate the heartwood, providing a path for fungal growth.

Eventually, the bark and sapwood disappear, leaving only soft wood chips. At this advanced stage of decay, the log pro-vides the greatest array of microhabitats and supports the highest diversity. Invertebrates of many kinds shelter in its openings and passages, and salamanders and mice dig tunnels into the rotten wood. Fungi and microorganisms abound, and many species of mites feed on the wood fragments and fungi. We consider fungal succession in more detail in Section 21.3 but, ultimately, the log

Figure 17.17 The pileated woodpecker, Canada's largest woodpecker, has a broad distribution across Canada and along North America's eastern seaboard. Adults can approach 50 cm in length, with a wingspan of 75 cm.

Figure 17.18 A succession of fungi, animals, and plants participate in the decomposition of a log.

crumbles into a reddish, mulch-like mound of lignin materials resistant to decay, its nutrients and energy depleted. These remnants are incorporated into the soil as humus, enhancing the soil's water- and nutrient-holding capacity.

We tend to think of a rotting log as an inconsequential and even unsightly presence in a forest. Too little do we appreciate the tremendous wealth of biodiversity that this mini-ecosystem harbours, and the vital functions it performs for the larger ecosystem of which it is a part.

17.10 ALLOGENIC CHANGES: Abiotic Changes Cause Succession on Many Timescales

Our focus has been the shifting patterns of community structure in response to autogenic abiotic changes occurring at timescales affecting establishment and growth. However, abiotic changes of allogenic (external) origin drive succession over timescales ranging from days to millennia. Allogenic fluctuations that occur repeatedly during an organism's lifetime are unlikely to influence succession among species of similar lifespan. In forests, for example, annual flux in temperature and precipitation influences growth rates but has little impact on succession apart from slowing its transitions. Shifts in abiotic conditions on timescales as long as or longer than the organisms' lifespan are more likely to cause shifts in dominance. Seasonal changes in light and temperature initiate annual succession of phytoplankton in lakes (Figure 17.19). Competition and seasonal grazing by zooplankton also affect species composition, but the major shifts relate to the differing abiotic tolerances of plankton species (Crumpton and Wetzel 1982).

On a timescale of decades to centuries, sediment deposition influences succession of coastal estuaries. The marshlands of the River Fal in Cornwall, England, have expanded seaward 800 m over the past century with silt deposition. On the landward side, woody species invade the marshlands,

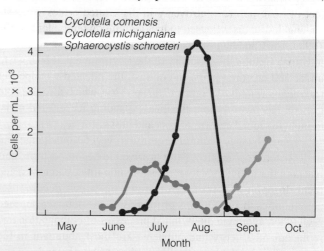

Figure 17.19 Seasonal changes in abundance of phytoplankton in Lawrence Lake, Michigan (1979). Generation times range from 1 to 10 days.

(Adapted from Crumpton and Wetzel 1982.)

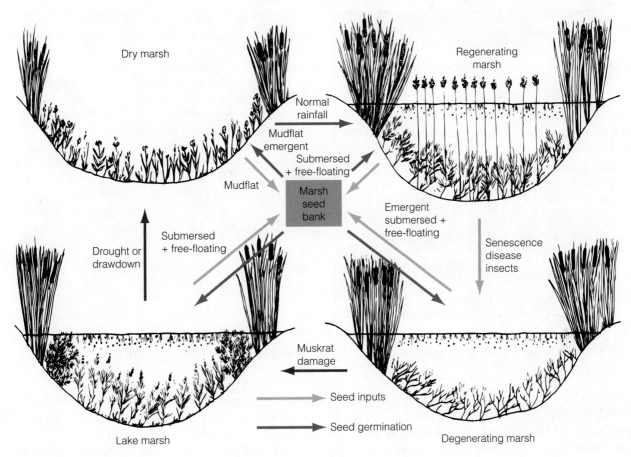

Figure 17.20 Cyclic succession in a freshwater marsh is driven by rainfall cycles interacting with autogenic forces (seed bank and muskrats). (Adapted from van der Valk and Davis 1978.)

leading to a successional sequence from marsh to woodland. A similar long-term transition occurs in freshwater habitats, again initiated by sediment deposition. Ponds and small lakes act as settling basins for inputs from the surrounding watershed. These sediments provide a substrate for rooted aquatics such as *Chara*, a branching green algae, and pondweeds, which bind the loose matrix and add to the accumulating organic matter. This accumulation reduces water depth and facilitates colonization by submerged and emergent plants, which add to the organic matter buildup and expand the area for colonization by larger, rooted plants. Eventually, the substrate, now supporting sedges and cattails, becomes a marsh. As drainage improves and the land builds higher, emergents are replaced by grasses. Depending on the region, the site may succeed to grassland, swamp, or peat bog. The allogenic process of sediment deposition thus interacts with autogenic changes initiated by the biota.

In succession of freshwater marshes and potholes across central North America, allogenic and autogenic changes interact to cause a cyclic pattern with no true beginning or end point (Figure 17.20) (van der Valk and Davis 1978). During droughts (every 5 to 20 years), shallow marshes enter the *dry marsh* phase. Accumulated organic debris decomposes rapidly upon exposure to air, releasing nutrients. Seeds in the **seed bank** (reservoir of seeds, usually in soil or sediments) germinate on

the exposed mud. As rainfall increases, the marsh enters the *regenerating marsh* phase. Emergents such as *Typha* thrive, and submerged and free-floating plants such as duckweed (*Lemna* spp.) establish. After a prolonged high-water period, the marsh enters the *degenerating marsh* phase, following the senescence (aging) of the emergents and buildup of disease. Eventually, spurred by heavy feeding from a rising muskrat population, the marsh becomes a *lake marsh*, with emergents confined to the edge. Although emergents such as *Typha* grow in standing water, they prefer to germinate on exposed mud, so only after the next drawdown can the marsh cycle restart. This succession is driven by allogenic forces (rainfall), but the key autogenic factor for marsh re-establishment is the seed bank. Humans, by "stabilizing" water levels, destabilize the normal pattern of change in freshwater marshes (van der Valk and Davis 1978).

17.11 LONG-TERM COMMUNITY CHANGE: Community Structure Changes over Geologic Time

By convention, succession is limited to significant changes occurring within a 500-year time span. Over a longer timescale, changes in regional climate influence community dynamics. The shifting distribution of tree species and forest communities

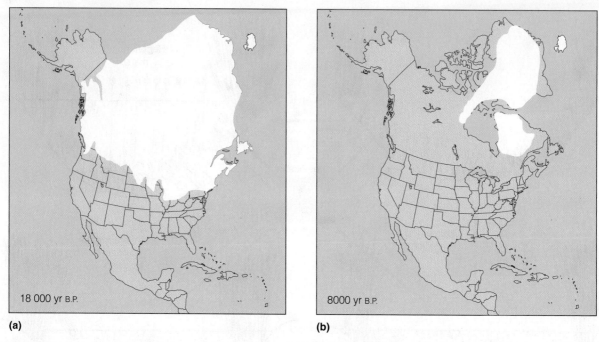

(a)

(b)

Figure 17.21 Glaciation of the North American continent at 18 000 **(a)** and 8000 **(b)** years before present (B.P.).

in the 18 000 years since the last glacial maximum in North America illustrates how dramatically long-term allogenic changes influence patterns of both succession and zonation at local, regional, and global scales.

Since its beginnings over 4.6 billion years ago, Earth has changed profoundly. Landmasses broke into continents; mountains formed and eroded; seas rose and fell; ice sheets advanced over large expanses of both hemispheres, and retreated. These changes affected climatic and substrate conditions worldwide. Plant, animal, fungal, protistan, and bacterial species evolved and were replaced. As conditions changed, so did species distribution and abundance. Records of past communities lie buried as fossils: bones; arthropod exoskeletons; and wood, pollen, and seeds.

Paleoecology (the study of the distribution and abundance of ancient species and their relationship to the environment) is of more than mere academic interest. Clues to the present-day distributions of species—as well as clues to their future fate—often emerge from paleoecological studies. Such studies allow reconstruction of tree species distributions in eastern North America after the last glacial maximum during the Pleistocene, an epoch of great climatic flux. At least four times in North America and three times in Europe, ice sheets advanced and retreated. Biota retreated and advanced as well, always with a slightly different species mix. In each interglacial period, the climate oscillated between cold and temperate phases. During the cold phase, tundra-like vegetation dominated, followed by advancement of shade-intolerant trees such as birch (*Betula* spp.) and pine (*Pinus* spp.). As soils improved and the climate warmed in the temperate phase, they were replaced by shade-tolerant species such as oak and ash (*Fraxinus* spp.). As the next glacial period began, species such as firs (*Abies* spp.) and

spruces (*Picea* spp.) assumed dominance. The climate cooled, and tundra-like vegetation again dominated.

The last great ice sheet, the Laurentian, reached its maximum about 20 000 to 18 000 years before present, during the Wisconsin glaciation (Figure 17.21). Canada was completely under ice. A narrow belt of tundra about 60 to 100 km wide bordered the sheet, extending south into the Appalachians. Boreal forest, dominated by conifers, covered most of the eastern and central United States as far south as Kansas. As the climate warmed and the ice sheet retreated north, vegetation re-invaded formerly glaciated areas. The advances of four representative tree genera after retreat of the Laurentian sheet have been reconstructed from pollen profiles in lake sediment cores dated with radiocarbon (Figure 17.22) (Davis 1981). Genera (and likely species) expanded their distribution at different rates, reflecting their differences in temperature tolerance, dispersal traits, and biotic interactions. Species distribution and abundance—and hence the structure of North American forests—have changed dramatically (Figure 17.23, p. 380).

17.12 COMMUNITY CONCEPT REVISITED: New Approaches Reconcile Contrasting Perspectives

Our discussion in Section 16.13 contrasted two views of the community. The organismal concept views the community as a quasi-organism in which interdependent species make up the community "body." The individualistic concept posits that each species responds independently to prevailing abiotic factors, making the concept of a discrete community arbitrary at best. In Chapter 16, we considered these views in the context of

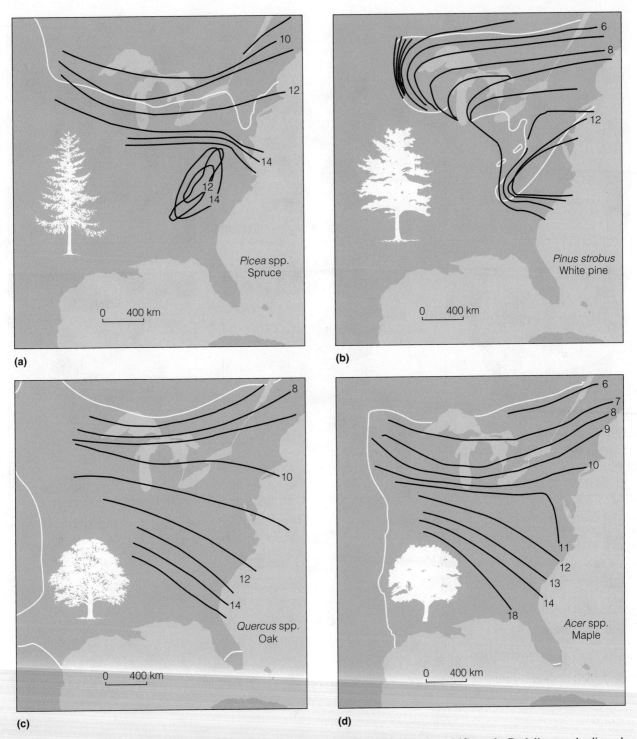

Figure 17.22 Post-glacial migration of four tree genera: **(a)** spruce, **(b)** white pine, **(c)** oak, and **(d)** maple. Dark lines are leading edges of northward-expanding ranges. White lines are present-day range limits. Numbers are 1000s of years before present (B.P.).

(Adapted from Davis 1981.)

changes in communities through space (*zonation*). In this chapter we have seen that the same contrasting perspectives also apply to changes in communities through time.

Clements saw seral stages as distinct, culminating in the final mature stage. Chance has little role to play, and for any given climate, only one possible climax (or several, depending on substrate) results. In contrast, Gleason saw seral stages as overlapping sets of species, reflecting their independent responses to changing abiotic factors over time. Interactions occur and are important insofar as they influence abiotic factors through autogenic effects. Chance plays a major role, particularly by affecting which species arrive first. Depending on the identity and order of arrival of the colonizers, different outcomes may occur for a given substrate and climate.

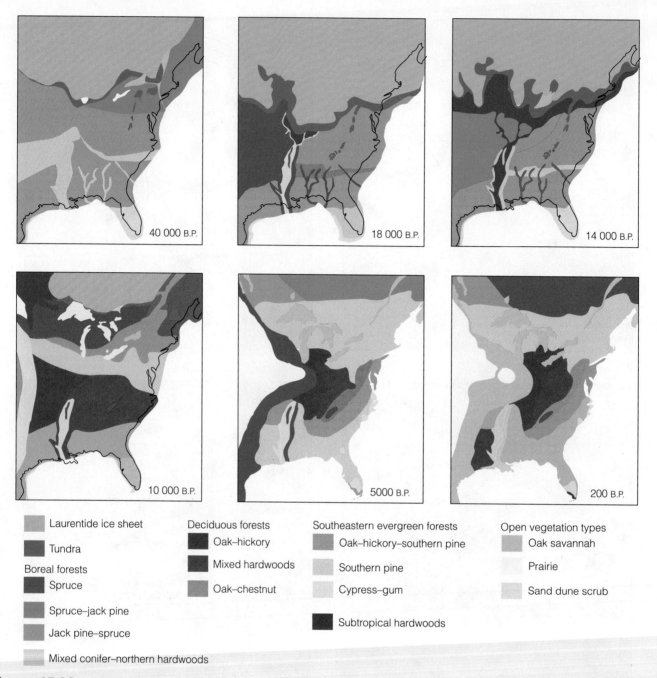

Figure 17.23 Changes in distribution of communities during and after retreat of the Laurentian ice sheet, based on pollen analysis. (Adapted from Delcourt and Delcourt 1981.)

Whether these contrasting views are applied spatially or temporally, research reveals that, as with most polarized debates, the reality is somewhere in between. The organismal view stresses biotic interactions, while the continuum view focuses on population dynamics in response to a changing abiotic environment. It is a matter of emphasis, and two ecologists examining the same data might find support for either view. Let's reconsider Great Smoky Mountains National Park (Figure 17.24a; see also Figure 16.25). Different combinations of elevation and slope (correlated with moisture) support different forest types, named for their dominants (Figure 17.24b).

When presented this way, community patterns appear to support an organismal view. Yet if we plot species abundance along a moisture gradient (Figure 17.24c), they seem to be distributed independently, supporting a continuum view. This pattern would also occur if time rather than moisture were on the x-axis.

In actuality, both perspectives help us understand communities in time and space. Remember, too, that many studies focus on one environmental factor (elevation in this case). Yet in any community, species respond to many factors that vary spatially and temporally, and interactions among organisms

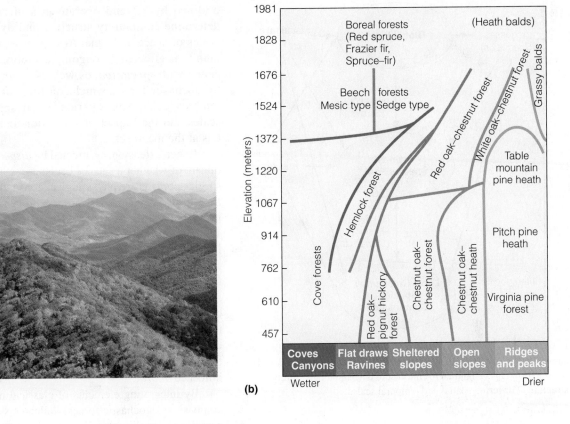

(a)

(b)

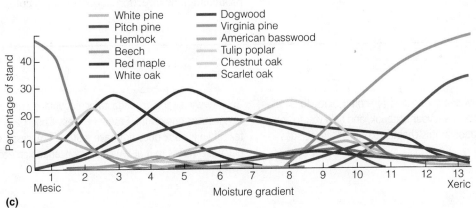

(c)

Figure 17.24 Two views of plant communities in **(a)** Great Smoky Mountains National Park. **(b)** Distribution of communities in relation to elevation (*y*-axis) and slope position and aspect (*x*-axis). Communities are classified based on dominant tree species. **(c)** Distribution and abundance of major tree species that make up vegetation communities in the park, plotted along the gradient of moisture availability.

((b) and (c) Adapted from Whittaker 1956.)

influence these responses. The result is a dynamic mosaic of communities on the landscape—the subject of Chapter 18.

With its myriad of theories and models, community ecology can be overwhelming. Mark Vellend (2010) proposes a unifying framework that promises not only to clarify the field but also to establish parallels with other areas of biology. He groups the many mechanisms and hypotheses that try to explain community patterns under four headings: (1) *selection*: deterministic, short-term outcomes of local interactions of species with their abiotic and biotic environment; (2) *speciation*: emergence and extinction of taxa on broader spatial

and temporal scales; (3) *drift*: impact of stochastic (chance) events on community composition and dynamics; and (4) *dispersal*: movement of individuals among communities. These four pillars may sound familiar, as they parallel the basic tenets of population genetics (see Section 5.6). Vellend's conceptual synthesis, which relates changes in populations (on timescales from the immediate to millennia) to changes in the community to which those populations belong, holds great promise.

Using a slight modification of Peter Morin's model (1999), Figure 17.25, p. 382 illustrates how the processes

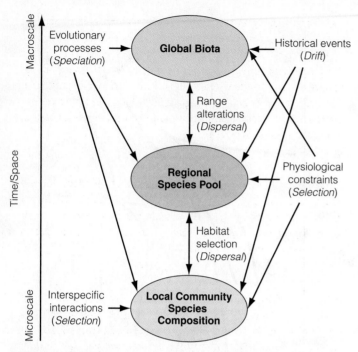

Figure 17.25 Morin's general model of community, modified to show how the processes of speciation, drift, selection, and dispersal (as suggested by Velland 2010) interact to determine community structure at differing spatial and temporal scales.

(Adapted from Morin 1999.)

outlined by Velend operate in a hierarchical manner to determine community structure and dynamics at different scales of space and time. At a macroscale, the global community is affected by ongoing evolutionary processes associated with *speciation*, as well as historical events affected by stochastic forces, which constitute *drift*. (All four of Vellend's mechanisms operate at all spatial and temporal scales, but the impact of speciation and drift is most obvious at the macroscale.)

Range extension, as affected by *dispersal*, is a major determinant of whether a species becomes a member of the regional species pool, as are physiological constraints that reflect *selection* (itself an evolutionary process influencing the fundamental niche, but not necessarily causing speciation in the short term). Which members of the regional species pool become part of the local community is a function of *dispersal*. Habitat selection and dispersal traits influence which species arrive at a site or in turn leave it at some later time, as dispersal is inherently bidirectional. Once established, the long-term persistence of a species is affected by *selection*, regarding both its physiological responses to abiotic factors and its interactions with other species. Meanwhile, historical events, including disturbance, continue altering the local community through the pervasive influence of *drift*.

By integrating elements of classical niche theory with an emphasis on stochastic forces, Vellend's synthesis represents a promising new direction in community ecology.

EcologyPlace

Visit EcologyPlace at www.pearsoncanada.ca/ecologyplace to access online resources that complement your textbook, and help you to apply and to review the information in this chapter. EcologyPlace includes

- an eText version of the book
- self-grading quizzes
- glossary flashcards
- and more!

Go to www.pearsoncanada.ca/ecologyplace and follow the registration instructions on the Student Access Code Card included with this text. If your book does not have a Student Access Code Card, you can purchase access to it at www.pearsoncanada.ca/ecologyplace.

SUMMARY

Succession Defined 17.1

Succession is the change in community structure over time. Opportunistic, early-successional species yield to late-successional species. The similarity of succession patterns in different environments suggests common processes are involved.

Primary Succession 17.2

Primary succession occurs on sites previously unoccupied by vegetation, such as sand dunes and glacial sediments. Primary succession tends to be slow, particularly in its early stages, because site conditions are typically harsh and soil is absent.

Secondary Succession 17.3

Secondary succession occurs on sites from which a community has been removed by disturbance, such as abandoned croplands or forests removed by logging or fire. In aquatic habitats, secondary succession is initiated by storms or herbivory. More favourable conditions mean that more species are capable of acting as pioneers.

Documenting Succession 17.4

Most successions cannot be followed in real time, so ecologists use space-for-time substitutions. They study either adjacent sites that are in progressive stages of a succession (toposequence) or sites that are at different stages of succession after a similar disturbance (chronosequence).

Succession Mechanisms 17.5

Ecologists have suggested several succession mechanisms: (1) facilitation: early-successional species make the environment more favourable for later species but less favourable for themselves; (2) inhibition: early-successional species inhibit growth of later species, which assume dominance after early species have died; and (3) tolerance: species replace each other over time as a result of their tolerance ranges for changing abiotic factors. According to the resource-ratio hypothesis, the ratio of changing abiotic factors affects relative competitive ability. Some regard the final stage as a monoclimax, polyclimax, or pattern climax; others use the term *old growth*.

Changes in Traits 17.6

Terrestrial communities become more structurally complex over time and accumulate organic matter. Early stages of secondary successions are often dominated by *r*-strategists; later stages by *K*-strategists. Pioneers of primary successions are typically stress-tolerators. Vital species attributes relating to mode of recovery, shade tolerance, and longevity, are critical.

Autogenic Changes 17.7

The presence and activities of organisms cause changes in abiotic factors, which influence succession. An example is the changing light levels and resulting shift in dominance from fast-growing, shade-intolerant plants to slow-growing, shade-tolerant plants. Autogenic changes in nutrient availability, organic matter, and sediment stabilization also affect succession.

Diversity and Succession 17.8

Changes in diversity during succession reflect the balance between colonization and replacement. Species number increases in early succession but may decline as later arrivals displace colonizers. Diversity often peaks during transitions, after the arrival of later species but before the replacement of earlier species. Diversity may be higher in sites with scarce resources, because the period in which species coexist is prolonged. Although the intermediate disturbance hypothesis suggests that diversity peaks with intermediate disturbance, empirical support is limited.

Heterotrophic Succession 17.9

Changes in vegetation composition and structure affect the heterotrophic community. Different species thrive at different seral stages, based on their resource and habitat needs and their tolerance ranges for abiotic factors. Decomposition involves a succession of fungi and animals.

Allogenic Changes 17.10

Allogenic changes in abiotic factors occur independently of organisms. Fluctuations that recur during an organism's lifetime are unlikely to influence succession among species of similar lifespan. Changes occurring over timescales greater than the longevity of the dominants can cause succession ranging from decades to millennia. A combination of allogenic and autogenic forces drive cyclic marsh succession.

Long-Term Community Change 17.11

The current pattern of vegetation distribution reflects past glacial events. Plants retreated and advanced with movements of the Pleistocene ice sheets. Rates and distances of advances affect present-day distributions of species and communities.

Community Concept Revisited 17.12

Contrasting community concepts also apply to succession. In the organismal view, seral stages are discrete communities, little affected by chance. In the individualistic view, seral stages are overlapping sets of species that change in response to abiotic factors. Outcomes depend on the impact of chance on order of arrival. A new synthesis summarizes community processes under four headings: selection, speciation, drift, and dispersal.

KEY TERMS

allogenic
autogenic
chronosequence
climax stage
early-successional (pioneer) species
facilitation model
heterotrophic succession
inhibition model
initial floristic composition hypothesis
intermediate disturbance hypothesis
late-successional species
monoclimax hypothesis
paleoecology
pattern climax theory
polyclimax hypothesis
primary succession
resource-ratio hypothesis
secondary succession
seed bank
sere
succession
tolerance model
toposequence
vital attributes

STUDY QUESTIONS

1. Distinguish between *zonation* and *succession*.
2. Distinguish between *primary* and *secondary succession*. Which type is usually slower, and why? In which type does chance play a greater role, and why?
3. The Sousa experiment in Figure 17.2 is described as a primary succession. How might its scale make it differ from primary successions that occur over a larger area?
4. Defoliation of oaks by gypsy moth causes death of extensive stands. Recovery involves growth of existing individuals that have escaped defoliation, as well as colonization by other tree species. Is this primary or secondary succession? Why?
5. Using the example of secondary succession in an abandoned crop field (Section 17.3), differentiate among *facilitation*, *inhibition*, and *tolerance*.
6. Consider the three plant life history strategies proposed by Grime (see Section 8.10). Explain which strategy is most likely to be typical of species colonizing (i) a newly disturbed site, (ii) a rocky outcrop, and (iii) the late stage of a secondary succession.
7. Why is the ability to tolerate low resource levels typical of species in late succession, according to Grime? How does this view differ from Tilman's resource-ratio hypothesis?
8. Explain how autogenic and allogenic changes interact during marsh succession.
9. Threats to the northern spotted owl include not only habitat loss but also barred owls (*Strix varia*), whose range expansion from eastern North America has been facilitated by tree planting on the prairies. Barred owls outcompete spotted owls for territories and sometimes interbreed with them. In its 2010 Recovery Plan for the spotted owl, the U.S. Fish and Wildlife Service recommended managing barred owls by trapping or shooting them in the spotted owl's range. Should Canada adopt this strategy? Consider both ethical and ecological issues.
10. Why might factors that slow growth increase diversity during succession?
11. How does the changing vertical structure of vegetation over time affect animal diversity?
12. Why do tree species differ in their rate of post-glacial advance? What impact does this have on community species composition? What influence will global climate change have on this long-term pattern?

FURTHER READINGS

Bazzaz, F. A. 1979. The physiological ecology of plant succession. *Annual Review of Ecology and Systematics* 10:351–371.
  Considers succession in terms of the varying adaptations of plants to changing abiotic factors.

Huston, M., and T. M. Smith. 1987. Plant succession: Life history and competition. *American Naturalist* 130:168–198.
  Discusses trade-offs between life history traits and the shifting nature of competition over time.

Tilman, D. 1988. *Plant strategies and the dynamics and structure of plant communities.* Princeton, NJ: Princeton University Press.
  Develops a theoretical framework for understanding pattern and process in plant communities.

Vellend, Mark. 2010. Conceptual synthesis in community ecology. *The Quarterly Review of Biology* 85:183–206.
  Bold proposal to unify the disparate theoretical strands of community ecology, written by one of Canada's top theoretical ecologists.

The boreal landscape, such as this one in the Yukon Territory, is a mosaic of patches of different forest types, interspersed with ponds and lakes.

ommunities have boundaries, but they also have a spatial context within the landscape. The landscape in Figure 18.1 contains forests, fields, hedgerows, plantation, road, pond, ocean, and human habitations (rural and urban). This patchwork of cover types is called a **mosaic**, using the analogy of a mosaic in which an artist combines many small pieces of variously coloured materials to create a larger pattern. As with mosaic art, the landscape pattern that emerges reflects the boundaries defined by changes in the physical and biological structure of the relatively distinct patches—communities— that are its elements. In mosaic art, the elements interact visually to create an image. In a landscape mosaic, patches and their boundaries interact in various ways depending on their size and spatial arrangement. Landscape ecology studies the patterns of habitat patches and boundaries, as well as the ecological consequences of these patterns.

We have used a human-altered landscape to illustrate the mosaic concept because it shows many different patch types in a small area, but a mosaic is also typical of most natural landscapes. The chapter cover page shows a boreal landscape, with patches of different forest types interspersed with ponds of various sizes. The same principles apply to pristine and human-altered landscapes alike, with the critical difference that humans have often fragmented natural landscapes into smaller and less connected patches than would normally be present.

18.1 HABITAT PATCHES: Environmental Processes Create a Variety of Patches on the Landscape

Habitat patches, the relatively homogeneous communities that make up a landscape, differ from their surroundings in physical structure and species composition. They vary in size, shape, and type, and are embedded within a mosaic of other patches from which they are separated by boundaries. The community surrounding a patch is its **matrix**—for example, a forest patch in a grassland. But how does a landscape mosaic develop?

Naturally occurring patches often reflect regional variations in geology, topography, soils, and climate. Upon this natural stage, human activity makes its mark. Ongoing **fragmentation** generates a mosaic of smaller, isolated patches of forest and grassland (Figure 18.2). In western Canada, landscape patterns reflect the impact of the Dominion Land Survey (begun in 1871) that divided land into sections. These surveys used straight lines rather than following natural topography, with straight corners where woods, fields, and other elements met. This checkerboard has left a lasting imprint. In contrast, in Québec, parts of the Maritimes, and eastern Manitoba, the French seigneurial system created long, narrow lots running back from major rivers (Harris and Warkentin 1991).

Although human activity is an obvious force affecting patch size and shape, natural variations in geology and soil and disturbances such as fire and herbivory also affect patch formation.

Figure 18.1 A landscape in Nova Scotia's Annapolis Valley, with a mosaic of patches of different cover types: crop fields, forest, plantation, road, rural and urban development, pond, and ocean.

influences and impacts in later sections, but first we consider the zones where patches meet.

18.2 TRANSITION ZONES: Borders, Edges, and Ecotones Offer Diverse Conditions and Habitats

Among a landscape's most conspicuous and pattern-forming features are the borders that mark patch perimeters. Some edges indicate an abrupt change in topography, substrate, soil type, or microclimate between adjacent communities. Where an **edge** (place where different communities meet) reflects long-term natural elements (called **enduring features**), it is often stable and permanent, creating an *inherent edge*. An example is a rock outcrop. Other edges are transitory, resulting from natural disturbances such as fires, storms, and flood, or from human activities such as logging, livestock, and urbanization. Such edges, which undergo succession, are *induced*.

The **border** is the region where the edges of two patches meet. It is an area of contact, separation, and transition. Some borders are narrow and abrupt with a sharp contrast, as between a forest and a crop field (Figure 18.3a). Others are wide, in terms of the distance between the border and the point where abiotic factors and vegetation do not differ significantly from the patch interior. Wide borders form an **ecotone**—a gradual

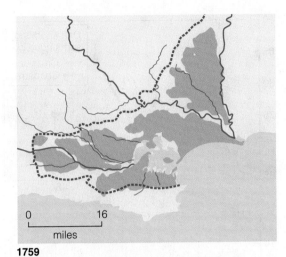

1759

1811

1978

Figure 18.2 Fragmentation of Poole Basin, Dorset, England. From 1759 to 1978, the area lost 86 percent of its heathland (from 40 000 to 6000 ha), changing from 10 large areas to 1084 patches, of which nearly half are under 1 ha, and only 14 are over 100 ha.

(Adapted from Webb and Haskins 1980.)

Patches may be square, round, elongate, or convoluted, covering many hectares or only a few square metres. Patch traits affect their suitability as habitat and influence processes such as wind flow, seed dispersal, and animal movements. We examine these

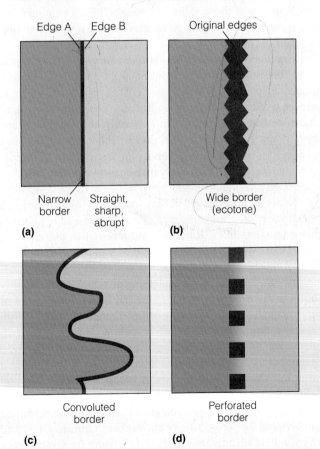

Figure 18.3 Types of borders: **(a)** a narrow, sharp border where the edges of two patches meet, **(b)** a wide border creating an ecotone, **(c)** a convoluted border, and **(d)** a perforated border.

transition zone between patches (Figure 18.3b). Borders may be tight or perforated, straight or convoluted, and of varying length (Figures 18.3c and d). Borders also have vertical structure, the height of which influences the steepness of the gradients in abiotic conditions between patches.

Functionally, borders connect patches by affecting flows of material, energy, and organisms. Borders may either restrict or facilitate seed dispersal and animal movements. Their height, width, and porosity influence gradients of wind, moisture, temperature, and light, particularly in landscapes where the same patch type (oak woodlands in this example) occurs in a different matrix (grassland or chaparral shrubland) (Sisk et al. 1997) (Figure 18.4).

Borders are also habitats in their own right. Because they blend elements from adjacent patches, borders often differ in structure and composition from either patch. Thus, borders offer unique habitats with easy access to nearby communities. Their abiotic conditions facilitate colonization by species with certain traits. Border plants tend to be shade-intolerants that can withstand the dry conditions associated with higher temperature and evapotranspiration. Border animals often require two or more habitats. Ruffed grouse (*Bonasa umbellus*) require forest gaps with an abundance of herbaceous plants and low shrubs for feeding their young, dense stands of saplings for nesting, and mature forests for winter food and cover. All three habitats must be contained within their home range, which is from 4 to 8 ha. Other species, such as the indigo bunting (*Passerina cyanea*), are true **edge species** that are restricted to border habitats (Figure 18.5).

The diverse conditions in borders allow them to support many (but not all) species from adjacent patches, as well as species adapted to the border itself. Thus, borders (especially ecotones) often support high biodiversity. This phenomenon, an example of an **edge effect**, is influenced by the border area and the amount of contrast. The greater the contrast between patches, the greater the diversity of the border. A border between forest and prairie typically supports more species than one between young and mature forest.

Although the edge effect may increase species diversity, it can also have negative population effects. Narrow, abrupt borders may attract predators. Raccoons and foxes often use edges as travel lanes. Avian predators such as crows and jays rob the nests of birds inhabiting edges. Borders alter interactions among species by either restricting or facilitating dispersal. Dense, thorny shrubs in the borders between forest and field block passage of some animals. Weedy species often proliferate in borders. With extensive fragmentation, there is much more edge relative to interior habitat, so interior species often suffer from habitat loss.

Borders are dynamic, often changing in space and time at a faster rate than adjacent patches. A border between forest and field typically has an abrupt change in vertical structure (Figure 18.6). As border vegetation grows taller, the abrupt vertical gap between the forest interior and border diminishes, forming a continuous vertical profile. Barring disturbance, the border expands horizontally as border vegetation invades the patch. Species composition changes as both abiotic factors and biotic interactions alter through time.

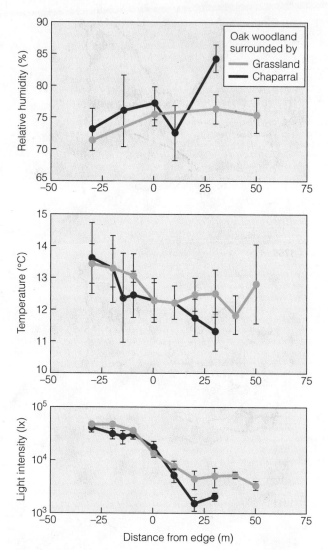

Figure 18.4 Habitat variation across edges between oak woodland patches in two matrix types (chaparral and grassland) in California's Santa Cruz Mountains. On the *x*-axis, positive values (means ± standard error) indicate distance from the edge into the woodland, negative values distance into the matrix. Light intensity: (lux) · 0.0185 = PAR (μmol photons/m²/s).

(Adapted from Sisk et al. 1997.)

INTERPRETING ECOLOGICAL DATA

Q1. How does air temperature change from the oak woodland into the chaparral habitat?

Q2. How do humidity and light intensity change from either the chaparral or grassland habitat into woodland patches?

18.3 PATCH TRAITS: Patch Size and Shape Affect Landscape Diversity

Patch size has a major impact on community species composition and diversity. Typically, large patches contain both more individuals (greater population size) and more species (greater richness) than small patches. Larger populations in larger patches reflect increased carrying capacities. The larger the patch, the more home

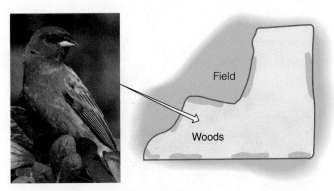

Figure 18.5 Territories of an edge species, the indigo bunting, which inhabits woodland edges, hedgerows, and roadside thickets. The male requires tall, open song perches and the female requires a dense thicket in which to build a nest.

(Adapted from Whitcomb et al. 1981.)

ranges (or territories, if defended) can be supported. Recall that home range increases with body size (see Section 11.9). Also, for a given body size, carnivores have larger home ranges than herbivores. So large predators such as grizzly bears and mountain lions are limited to large, contiguous habitat patches.

The relationship between patch size and richness is more complex (see also Section 18.4). Large patches are more likely to contain variations in topography and soil that support more plant diversity (taxonomic and structural), which in turn supports more animal diversity. A key aspect is the ratio of edge to interior. Only a patch large enough to be deeper than its border can develop interior conditions (Figure 18.7a, p. 390). At one extreme, a small patch is all edge, but as patch size grows, the edge-to-interior ratio decreases (Figure 18.7b), as a result of the same scaling principle responsible for the lower surface area–volume ratio of large animals (see Section 7.1). Altering patch shape can also alter its edge-to-interior ratio. Two patches may have similar area, but if one is elongated such that its depth is less than the width of its border, it will be all edge compared with a rounder patch (Figure 18.7c). Woodland strips, no matter how extensive, lack interior habitat.

Interior species require the conditions of interior habitats. The relative amounts of edge to interior in forest patches affect the probability of occurrence of many widespread North American birds (Robbins et al. 1989). *Edge species* such as the grey catbird (*Dumetella carolinensis*) and American robin (*Turdus migratorius*) prefer small forest patches dominated by edge. As patch size increases, the probability of edge species occurring declines (Figure 18.8, p. 390). In contrast, the ovenbird (*Seiurus aurocapilla*) and worm-eating warbler (*Helmitheros vermivorus*), both ground-dwelling species adapted to interiors of old forests, are unlikely to inhabit small patches. **Area-insensitive species**, such as the Carolina chickadee (*Parus carolinensis*) and wood pewee (*Contopus virens*), can use both edge and interior, so their

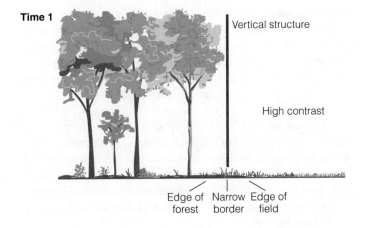

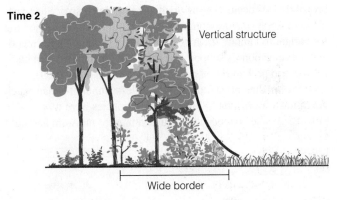

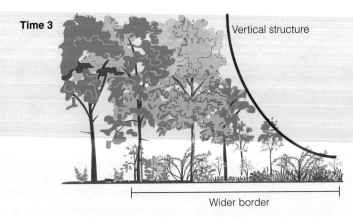

Figure 18.6 Changes in vertical and horizontal structure of a border over time. Border width increases and vertical structure changes from a high-contrast edge to a continuum.

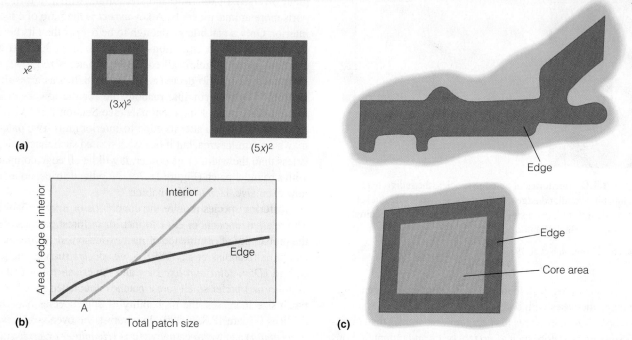

Figure 18.7 Influence of patch size and shape. **(a)** Assuming edge depth remains constant, the edge-to-interior ratio decreases as patch size increases. When the patch is large enough to maintain shaded conditions, an interior develops. **(b)** Relationship between patch size and edge. Below point A, the patch is all edge. As size increases, interior increases, and the edge-to-interior ratio declines. **(c)** This relationship holds for square or circular patches, but elongated patches whose widths do not exceed their edge depth are all edge, whatever their area.

probability of occurrence is unaffected by patch size. Such species can still be uncommon, depending on other factors. Allowing for variability among the members of each of these three categories, these responses support a general relationship between habitat area and probability of occurrence (Robbins et al. 1989).

For smaller plants, patch size is less important than patch conditions. Many shade-tolerant forest species have a *minimum patch size*—the area needed to sustain interior moisture and light

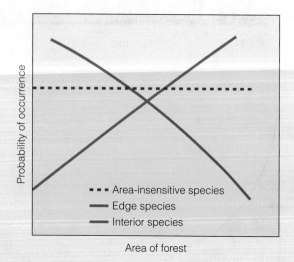

Figure 18.8 General model of response to forest area. As patch size increases, the probability of finding a species declines for edge species and increases for interior species. The probability of encountering area-insensitive species is independent of patch size. Many other factors affect the probability of occurrence for any particular species.

conditions, which depends in turn on the ratio of edge to interior and the type of border. If the patch is too small, light and wind penetrate, eliminating species that require humid, cool conditions and moist soil. Forest fragmentation causes moisture-requiring (*mesic*) species such as maples to decline, while encouraging growth of *xeric* (drought-tolerant) species such as oaks.

As plants are sessile, the impact of fragmentation on vegetation is often delayed. Imagine that a forest is fragmented for a housing development, but that forest patches are kept as landscaping. If an ecologist studies the area in the first year, interior animal species will likely already have vacated, but interior plant species will still be present in the residual patches. But in 10 or 20 years, interior plant species will have declined and even disappeared, as individuals die and cannot regenerate. This delayed loss of species over time is the **relaxation effect (extinction debt)**. It reflects time lags in reproductive and/or mortality effects as well as a failure to re-colonize. Nor is the relaxation effect restricted to plants. Barro Colorado Island was a hilltop vegetated by tropical forest that became an island in 1911–1914 during the making of the Panama Canal. Since that time, over 65 bird species have been extirpated—not immediately, but over many decades, and with interior species most affected (Robinson 1999).

In forest and grassland patches in agricultural regions of North America, bird species richness increases with patch size (Figure 18.9) (Whitcomb et al. 1981). Intermediate-sized patches may support more species by providing both edge and interior habitat. Two or more small patches may support more species than one large patch of similar area, depending on how small; very small patches cannot support interior species. Even if interior species are present, their numbers may be too low for the

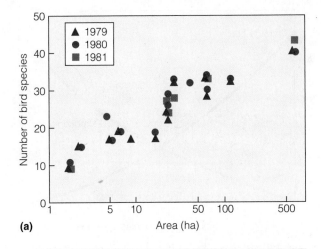

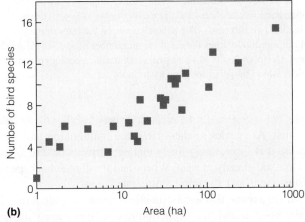

Figure 18.9 Species richness of a bird community as a function of patch area (log scale) in **(a)** woodland and **(b)** grassland. Different symbols in (a) refer to survey results from different years.

(Adapted from Whitcomb et al. 1981.)

species to persist. Thus, estimates of species richness alone do not give a complete picture of how fragmentation affects diversity. Large forest patches with much structural variation are needed to support the many species typical of both edge and interior habitats.

In observational studies, the effect of fragmentation may be confounded with that of forest cover. Trzcinski et al. (1999) separated these factors in an observational study of ninety-four 100-km² sites in southern Ontario and Québec that ranged in forest cover from 2.5 to 55.8 percent and varied independently in their degree of fragmentation. Responses of forest-breeding birds to cover were strong and positive, whereas responses to fragmentation were weak and variable. The impact of fragmentation did not increase with declining cover, suggesting that the primary focus of conservation efforts should be on retaining forest cover.

Manipulative experiments further enhance our understanding of the impacts of landscape changes on forest birds. In a total of 201 independent homing trials in Québec on the black-throated blue warbler (*Dendroica caerulescens*), ovenbird (*Seiurus aurocapilla*), and black-capped chickadee (*Poecile atricapillus*), birds took longer and were less likely to return to their territories as forest cover decreased. Landscape composition (cover type) was

more influential than landscape pattern, and neither the distance between patches nor the amount of edge per area affected homing time or success (Bélisle et al. 2001). Forest species that are less mobile than birds are likely to be more affected.

Fragmentation can have unexpected indirect effects. The brown-headed cowbird (*Molothrus ater*) is a brood parasite (see Section 15.1) of North American prairies. A generalist, it parasitizes over 200 other bird species. Prior to human habitation, it followed bison herds (its former common name was buffalo bird), feeding on insects and seeds stirred up by their hooves, but now it follows livestock herds. It places its eggs in open-cup nests of birds that nest in edge habitats. Fragmentation has created more edge, bringing the cowbird into contact with species that have not been selected for the ability to recognize foreign eggs. Cowbirds have become more numerous, increasing the impact of their parasitism on migratory birds (Brittingham and Temple 1983). In this case, the cause of fragmentation—human habitation—may be more critical than fragmentation itself, as the percentage of yellow warbler (*Dendroica petechia*) nests parasitized increased sharply with habitation density (Figure 18.10) (Tewksbury et al. 2006). Cowbirds are frequent visitors to bird feeders, making increased losses of songbirds to parasitism a possible unintended effect of a seemingly benign human interference.

Edge effects influence *extirpation* (local population extinction) risk. As fragmentation accelerates, interior species are more vulnerable, no matter how successful they might be in suitable habitat. Conversely, edge species such as robins and white-tailed deer thrive, giving the misleading impression that wildlife is not adversely affected. Research on the relation of patch size to diversity focuses on forests, but similar trends apply to fragmented grasslands, shrublands, and marshes. Grasshopper sparrows (*Ammodramus savannarum*) and prairie grouse (*Tympanuchus* spp.) and shrubland species such as sage grouse (*Centrocercus urophasianus*)—all interior species native to Canada—are in decline as their habitats are fragmented and patch size decreases. In a study of 24 grassland fragments in Illinois, fragment area accounted for a high percentage of the variation in breeding bird

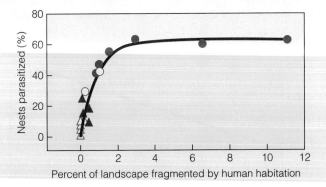

Figure 18.10 Percentage of parasitized yellow warbler nests in relation to percentage of landscape fragmented by human habitation in riverine areas in Idaho (triangles) and Montana (circles). Open symbols are sites buffered by forests; closed symbols are sites bordered by agriculture.

(Adapted from Tewksbury et al. 2006.)

species richness (Herkert 1994). The effect of fragmentation may again be confounded with habitat area, but all five area-sensitive species were affected, suggesting that fragmentation is contributing to the decline of grassland birds.

Teasing apart the impact of landscape-level processes requires careful attention to factors acting at multiple spatial and temporal scales. In a study of songbirds and ducks in 16 sites in southern Alberta, birds were more affected by local site factors, such as vegetation type and distance to edge, than by landscape-level factors such as fragmentation (Koper and Schmiegelow 2006).

18.4 ISLAND BIOGEOGRAPHY: Theories Pertaining to Island Diversity Also Apply to Habitat Patches

The influence of patch area on species richness did not escape the notice of early biogeographers. Johann Forster, a naturalist on Captain Cook's 1772–1775 voyage to the Southern Hemisphere, noted that large islands hold more species than small islands. This trend is supported quantitatively by a **species–area curve**: a type of species accumulation curve (see Quantifying Ecology 16.1: Estimating Species Richness, pp. 335–336) in which habitat area (or area sampled) is on the *x*-axis (Figure 18.11). Zoogeographer Philip Darlington (1957) generalized that species number doubles with a 10-fold increase in island area. By analogy, patches in a landscape mosaic resemble islands. They vary in size; some are near each other, or near larger patches (akin to a mainland); others are remote. A forest patch sits within a "sea" of grassland or cropland or housing, isolated from other patches. Patch size and the distance between patches influence the type and diversity of life that patches support.

The similarity between islands and isolated patches led ecologists to apply the **equilibrium theory of island biogeography** to terrestrial landscapes. The basic idea of the theory is simple: the number of species occupying an island represents a dynamic equilibrium between immigration of colonizing species and extinction of previously established species (Figure 18.12) (MacArthur and Wilson 1967). (In this context, *extinction* refers to *extirpation*—loss of a population.) Consider an uninhabited island. The species present on the mainland make up the **species pool**—the totality of available species in a region. Chance plays

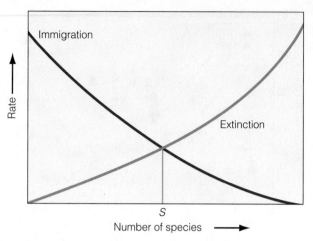

Figure 18.12 According to the island biogeography theory, immigration rate declines with increasing richness (*x*-axis), while extinction rate increases. The balance between the rates of extinction and immigration defines the equilibrium number of species (*S*) on the island. Go to **GRAPHit!** at **www.pearsoncanada.ca/ecologyplace** to explore island biogeography theory in more detail.

a role, but species with the greatest dispersal ability likely colonize first. As species richness rises, the immigration rate drops because it is increasingly likely that new arrivals belong to species that are already present. When (and if) all mainland species exist on the island, the immigration rate is zero.

If we assume that local extinctions occur at random (a null-model assumption), the extinction rate will increase with species number, based purely on chance. Other factors will amplify this effect. Later immigrants may be unable to establish if earlier arrivals have depleted or restricted access to habitats or resources. As richness increases, competition may increase, either because there is more niche overlap, or because more individuals use more resources. If competitive exclusion ensues, the extinction rate rises.

Equilibrium species richness is achieved when the immigration rate equals the extinction rate (see Figure 18.12). If the species number exceeds this value, extinction exceeds immigration, precipitating a drop in richness to the equilibrium level. If the species number is below this value, immigration exceeds extinction and species number increases. At equilibrium, the number of species is stable, but their identity may change. The rate at which one species is lost and a replacement species is gained in this equilibrium state is the **species turnover rate**.

MacArthur and Wilson's theory proposes that two major factors—distance from a mainland and island size—affect species richness. Immigration rate is determined by distance from the mainland; the greater the distance, the less likely that immigrants complete the journey, and the longer it takes them to arrive. The result is a lower immigration rate with increasing distance, and a corresponding decrease in equilibrium species number (Figure 18.13a). On the other hand, extinction rate is determined by island size. The larger the island, the lower the extinction rate, as larger islands generally contain more resources and habitats. Larger islands support larger populations as well as more species. Lower extinction rates on

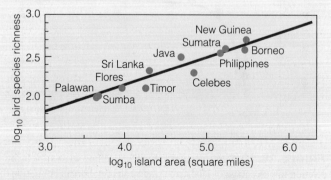

Figure 18.11 Species–area curve showing number of bird species on islands in the East Indies in relation to island area. Both axes are plotted on a $\log_{10}$ scale.

(Adapted from Preston 1962.)

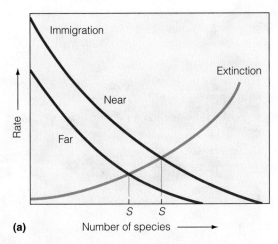

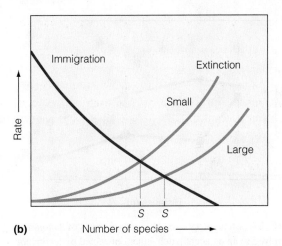

Figure 18.13 Effect of distance from mainland and island area on equilibrium species richness (S). **(a)** Immigration rate (and species richness) increases with proximity to a mainland. **(b)** Extinction rate decreases with increasing island area, leading to higher species richness.

large islands increase their equilibrium species richness compared with small islands (Figure 18.13b).

Although the theory posits that proximity alone determines immigration rate, island area may also have an influence. Large islands are bigger targets for dispersing organisms. A seed dispersed by water or an insect carried by wind is more likely to "hit" a larger than a smaller island. Similarly, although the theory posits that area alone determines extinction rate, distance from the mainland may have an effect. A species that might otherwise go extinct is more likely to be "rescued" from extinction by individuals of the same species arriving from a nearby source.

The equilibrium theory does not incorporate the effect of island shape. As we saw in Figure 18.7, shape affects the edge-to-interior ratio, and hence the number of species (especially interior species) that can establish and persist. Island orientation also affects immigration; an elongated island parallel to the mainland is an easier target for an immigrating species than if the narrow end were at right angles. Distance to species sources other than a mainland is also critical. A small island that is far from a mainland but close to a large island will have a higher immigration rate than a more isolated island of similar size.

Surprisingly (given that MacArthur was a strong proponent of niche theory), the theory disregards species identity, beyond acknowledging species turnover. (Like Hubbell's unified neutral theory, it treats species as ecologically interchangeable.) Yet, depending on the interactions that a species enters into upon arrival, changes in species composition may affect extinction. The model assumes that new arrivals affect the risk of competitive exclusion of existing species, but what if a new arrival is a facultative mutualist? By enhancing survival or growth of an existing species, the new arrival may lower, not increase, its extinction risk. Alternatively, the new arrival may be a keystone predator or engage in another indirect interaction that reduces the extinction risk of one or more existing species. Collectively these potential *rescue effects* (see Section 12.6) suggest that species composition may be as important as overall species richness.

The equilibrium theory is restricted to short-term diversity patterns, but in the longer term, natural selection—which studies indicate is particularly intense on islands—may increase richness

by *adaptive radiation* (see Section 5.9). Finally, the theory assumes equilibrium conditions apply, but many communities exist in a non-equilibrium state in which populations are below carrying capacity, making extinction less likely.

However, despite its limitations, island biogeography theory has tremendous intuitive appeal, with applications far beyond the original context of oceanic islands. Mountaintops, bogs, ponds, dunes, areas fragmented by humans, and even hosts of parasites can all, by analogy, be considered island habitats. In the words of one of the first ecologists to test the theory experimentally, "Any patch of habitat isolated from similar habitat by different, relatively inhospitable terrain traversed only with difficulty by organisms of the habitat patch may be considered an island" (Simberloff 1974, p. 162).

In applying the theory to habitat patches, we must remember that islands are land habitats surrounded by an aquatic barrier that varies among species in its difficulty of overcoming, depending on their size and dispersal mode (flying, swimming, or via wind). Habitat patches are surrounded by other land habitats, which may present fewer barriers to movement or dispersal.

18.5 CORRIDORS: Corridors Facilitate Movement among Patches

As we stressed at the outset, a mosaic pattern typifies many landscapes, particularly those like the boreal forest that contain varied substrates and are subject to patchy natural disturbances. But landscape mosaics become more pronounced when fragmentation of previously large contiguous habitats, whether forest, grassland, or water, creates habitat patches or "islands" between which movement is limited by an inhospitable intervening environment. The presence of **corridors** (strips of habitat connecting patches) enhances the ability of organisms to move among them. Corridors may contain similar vegetation as do the patches they connect, but differ from the matrix. Other corridors differ from the patches they connect, as in streams connecting lakes. A key aspect of any landscape is its **connectivity**, the extent to which patches are linked by corridors. Connectivity affects the ability of an individual or population to move between patches.

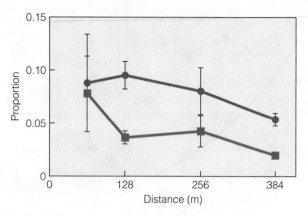

Figure 18.14 Mean proportion (± standard error) of *Junonia* males that moved to an adjacent patch either connected by a corridor (circles) or unconnected (squares).

(Adapted from Haddad 1999b.)

Corridors perform many functions. They are travel lanes for individuals moving within their home range in a fragmented landscape, but when corridors interconnect into networks, they offer dispersal routes for populations moving among patches. By facilitating movement beyond what might be possible in the adjacent matrix, corridors facilitate gene flow among subpopulations, and may re-establish species in patches from which they have been extirpated (see Chapter 12). Corridors also act as filters, allowing dispersal of some species but not others. Via the **filter effect**, different-sized gaps in corridors allow some organisms to cross while restricting others.

The role of corridors in species conservation has received much attention, but their benefits are often assumed rather than tested. For example, the common buckeye (*Junonia coenia*) and variegated fritillary (*Euptoieta claudia*) are common butterflies in early-successional habitat and rare in mature pine forest. Haddad (1999a, b) created five early-successional patches in each of eight 50-ha landscapes in mature forest in South Carolina by tree removal. In each group of five patches, a central patch was connected by a corridor to one of the patches, with the remainder unconnected. Both species were significantly more likely to move between patches via corridors than through mature forest (Haddad 1999a). When corridor length was altered, interpatch distance was inversely related to movement (Figure 18.14) (Haddad 1999b).

Corridors can also have negative effects. Like edges (which they often resemble), corridors allow predators to remain concealed while hunting in adjacent patches. They also create avenues for disease spread and facilitate invasion and spread of non-native species from the matrix to other patches. If they are too narrow, corridors can inhibit movement of social groups.

Corridors (like borders) are habitats in their own right. Vegetation along rivers provides vital habitat (Figure 18.15a). In suburban settings, corridors support edge species and provide stopover habitat for migrating birds. In North America, the Mississippi Flyway, which connects to the Mackenzie River system as well as other routes leading to central and eastern Canada, is a prime example.

(a)

(b)

Figure 18.15 Corridors include (a) riverbank vegetation (which are often residual from larger habitats that have been converted to human uses) and (b) hedgerows.

Incorporating many natural habitats along rivers, marshes, and residual forest reserves and urban parks, it facilitates migration of many species. Continuing human encroachment threatens its effectiveness. Besides breaks in connectivity, one of the biggest risks on this and other routes is exposure to high numbers of alien predators such as cats.

Many corridors are of human origin. They are often narrow, such as hedgerows and windbreaks (Figure 18.15b), bridges over rivers, highway medians, and ditches. In England, a long history of hedgerows (much wider than in North America and containing a variety of plant species) has encouraged development of distinctive hedgerow communities that support many native species. Wider bands of vegetation can contain both interior and edge habitat. Examples include forest strips along highways or between housing developments.

Roads—corridors designed as dispersal routes for humans—dissect the landscape and affect adjacent habitat. In an extensive review, Trombulak and Frissell (2000) concluded that roads have overwhelmingly negative effects on natural communities. Mortality effects are the most obvious, affecting animals from large mammals to tiny insects. Less obviously, roads partition populations and alter animal behaviour, either directly by affecting movements or reproduction, or indirectly, by altering home ranges or by inducing physiological stress. Roads damage nearby vegetation, whether from soil compaction and altered runoff; salt spread; particulate matter from tires and exhaust; or pollutants, including heavy metals. During storms and snowmelt, runoff carries pollutants and debris farther back from the road. Traffic noise discourages wildlife from occupying suitable habitat. Roads facilitate spread of invasive species. They allow access to remote areas, with widespread ecological effects, as exemplified by logging roads cut through boreal and tropical forests alike. Where roads invade, people, alien species, pollution, and fragmentation follow.

Hydro rights-of-way and pipelines are also corridors of human origin. Leakages from pipelines are common, and expansion of the oilsands in northern Alberta raises concerns about the impacts on forests, soils, and groundwater (see Ecological Issues: Oilsands: Collateral Damage of Our Fossil Fuel Dependence, pp. 626–627). To facilitate export of synthetic crude, the proposed Canadian Enbridge Northern Gateway pipeline will traverse British Columbia's coastal rain forests. Apart from the pollution risks, the corridor will fragment intact forest and may jeopardize salmon runs in the area. Nor is "green" power exempt from risks. The current controversy over whether to build a hydro transmission corridor on the west or east side of Lake Manitoba centres on its fragmenting impact. West-side proponents argue that the greater length and cost are justified by the protection afforded for First Nations' lands and wildlife habitat, which is part of a proposed UNESCO World Heritage Site. East-side supporters contend that the roads proposed for the west side would fragment boreal forest as much as a transmission corridor would.

Human-origin corridors can play a positive role in conservation of mammal habitat in multi-use areas. In Banff National Park, an extensive network of underpasses and vegetated overpasses (Figure 18.16) allows populations to persist in a landscape that, while containing much valuable habitat, has a high level of human use. According to park officials, there were over 200 000 wildlife crossings in 2011, including grizzlies, black bears, wolves, and ungulates, but the most noteworthy occurrence was the inaugural recorded use by a wolverine (*Gulo gulo*). This large weasel species is rare in the park, and its large home range makes its population particularly vulnerable to road mortality. Endangered in eastern Canada, the wolverine is listed by COSEWIC (2003) as of "special concern" in the west, and an indicator of overall ecosystem health. Taking steps to mitigate the negative effects of habitat fragmentation by facilitating animal movement across the landscape is of great importance.

18.6 METAPOPULATIONS AND METACOMMUNITIES: Population and Community Dynamics Affect Landscapes

Distribution of a species is limited to suitable habitat. In a landscape with a natural mosaic of patches, some species exist as *metapopulations*: distinct, partially isolated subpopulations, each with its own dynamics, and interconnected by movements of individuals. Other species that are normally present as large populations in continuous habitat become metapopulations following fragmentation. Whether natural or induced, metapopulations exhibit dynamics that are similar to those of the island biogeography model discussed in Section 18.4. In that model, the equilibrium species richness of an island (and, by analogy, of a habitat patch) is a balance between colonization and extinction. Likewise, in the metapopulation model, the equilibrium number of patches occupied by a species is the balance between colonization and extinction of local populations (see Section 12.3). Both models explicitly address how colonization and extinction are influenced by the size and isolation of habitat patches or islands in a landscape.

Indeed, the dynamics of species on islands can be seen as a consequence of the metapopulation dynamics of species occupying the larger landscape. This relationship led David Wilson (1992) to extend metapopulation models to multispecies communities, giving rise to the idea of a **metacommunity**: a community composed of many subcommunities located in patches over the landscape. As with metapopulations, each subcommunity has its own dynamics, but movement of populations between subcommunities along corridors allows communities to re-establish after they have gone "extinct" from one portion of the landscape. The metacommunity concept illuminates patterns in succession, species richness and composition, and food web structure (Holyoak et al. 2005). As Mark Vellend points out (2010), the metacommunity concept draws on the components of his general model (see Section 17.12)—most obviously dispersal, but also drift and selection. As in island biogeography theory, speciation is not incorporated, implying an ecological rather than a longer, evolutionary timeframe.

From an applied perspective, studies of how patch geometry (area and shape) affect the local patch environment (see

Figure 18.16 This 50-m-wide vegetated overpass is part of a network of corridors that facilitate animal movements in Banff National Park.

Section 18.3) facilitate predictions of the probability of colonization and extinction, based on the life history traits and habitat needs of species. They also illuminate how landscape patterns constrain dispersal, which is key to the dynamics of both metapopulations and metacommunities. But there are two vital ways in which metapopulations and metacommunities are not analogous: (1) When individuals establish in a patch, a subpopulation is automatically created. But how many populations are needed to establish a subcommunity? Do we base the existence of a metacommunity on the presence of its dominants, or must a certain proportion of its populations (autotrophs and heterotrophs) be present? (2) Dispersal involves movements of individuals, and although individuals can move en masse (as in migration), there is no analogous process for establishing subcommunities as there is for establishing subpopulations.

18.7 ROLE OF DISTURBANCE: Frequency, Intensity, and Scale Determine the Impact of Disturbances

Community patterns on the landscape, as well as of the populations inhabiting communities, are greatly influenced by disturbances, past and present. A *disturbance* is a relatively discrete event—fire, windstorm, flood, extremely cold temperatures, drought, or disease outbreak—that disrupts community structure and function. Disturbances both create and are influenced by landscape patterns. Hilltops are more susceptible to damage from wind, drought, and ice storms than are bottomland communities along rivers, which are more susceptible to flooding. The progression of fire over a landscape is influenced by the type and proximity of forest patches. In turn, these disturbances alter the mosaic of patches on the landscape.

Ecologists distinguish between a particular disturbance event—a single storm or fire—and the **disturbance regime**: the pattern of disturbance that characterizes a landscape over an ecological timeframe. A disturbance regime has spatial and temporal aspects, including (1) intensity, (2) severity, (3) scale, and (4) frequency. **Intensity** relates to the magnitude of the forces involved, such as wind strength or the energy released in a fire, and is often confused with **severity**, which is the proportion of total biomass or populations that is destroyed or eliminated. Severity is affected by intensity, but also by the susceptibility of the community. Two floods of similar intensity will differ in severity if the species present differ in flood tolerance. **Scale** refers to the spatial extent of the disturbance, and varies from a microhabitat to a global scale. **Frequency** is the mean number of disturbances that occur in a given time interval. Inversely related to frequency is the **return interval**: mean time between disturbances (Figure 18.17).

Disturbance frequency is often linked to intensity and scale. Natural disturbances that occur on a microscale, such

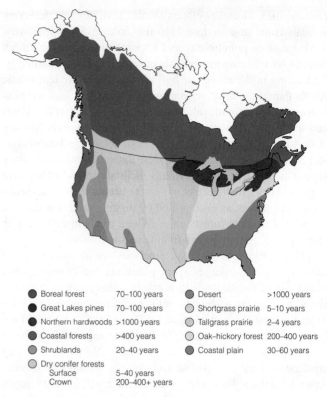

Boreal forest	70–100 years	Desert	>1000 years
Great Lakes pines	70–100 years	Shortgrass prairie	5–10 years
Northern hardwoods	>1000 years	Tallgrass prairie	2–4 years
Coastal forests	>400 years	Oak–hickory forest	200–400 years
Shrublands	20–40 years	Coastal plain	30–60 years
Dry conifer forests Surface Crown	5–40 years 200–400+ years		

Figure 18.17 Return interval for fire in different ecosystems across North America.

(Map from Aber and Melillo 1991. Data from Heinselman 1973, Heinselman and Casey 1981, Gray and Schleisinger 1981, Christensen 1981, and Bormann and Likens 1979.)

as the death of a single forest tree, occur frequently. By contrast, macroscale disturbances, such as fire, occur on average once every 50 to 200 years in a forest (much more frequently in a prairie), depending on the forest type and regional climate. In an intertidal zone, wave action tears away mussels and algae from rocks along the shore with high (and predictable) frequency. In grasslands, badgers and groundhogs dig burrows and expose small patches of soil. The outcome of such frequent, fine-scale disturbances is the creation of a *gap*—an opening that becomes a site of localized regeneration and regrowth.

Gap formation does more than provide access to physical space for colonizers. Within the gap, abiotic factors differ substantially from those of the surroundings. The microclimate in a gap created by the death of a canopy tree is brighter, warmer, and drier than the microclimate of the surrounding forest. These altered conditions encourage the germination and growth of shade-intolerant species and stimulate the growth of suppressed trees that are already present. In this mini-succession, the race is on as individuals grow in height. Their fate is determined by how rapidly the crowns of trees surrounding the gap expand to fill the opening, thereby reducing light availability.

Macroscale disturbances, such as fire, logging, or other forms of land clearing, initiate community responses that go beyond reorganizing the populations already present. Such disturbances reduce or eliminate local populations and significantly alter abiotic factors. Colonization by other species soon follows. Some may already be present but dormant, as seeds in a *seed bank*; as roots, rhizomes, or stump sprouts in a vegetative **bud bank**; or as surviving seedlings or saplings (called *advance regeneration* in forests). Other species disperse to the site by vectors such as wind or animals. Long-term recovery involves secondary succession, in which species typical of the original community replace the early colonizers (see Chapter 17).

18.8 NATURAL DISTURBANCES: Abiotic and Biotic Disturbances Alter Landscapes

Many disturbances arise from natural causes—abiotic or biotic—such as wind, ice storms, lightning fires, hurricanes, floods, grazing animals, and insect outbreaks. Such events not only cause the death of organisms and/or loss of biomass, but are also a powerful force for abiotic change. Storm floods cut away banks, alter the course of rivers, scour streambeds, deposit sediments, and carry away organisms. High winds and tides break down barrier dunes and allow seawater to invade shorelines, changing the geomorphology of barrier islands (Figure 18.18). By modifying an ecosystem in ways that favour survival of some species and elimination of others, disturbance can reduce or increase diversity in the short term.

Wind and water are two powerful disturbance agents. Large, inflexible forest trees, as well as trees growing in open fields and along forest edges, are vulnerable to *windthrow*. So are trees on poorly drained soils, as their shallow roots are not well anchored. Although caused by wind, toppling of one tree can in turn cause others to fall. Wind-related disturbances are

Figure 18.18 Erosion of coastal dunes by storm surges causes breaks in the front dunes, creating areas of inundation.

particularly severe when they interact with cold temperature and precipitation. The ice storms that often occur in eastern North America (notably hitting the Montréal area in 1998; see Figure 2.19) devastate entire forests through the combined effect of wind and the weight of ice.

On rocky intertidal and subtidal shores, waves overturn boulders and dislodge sessile organisms, clearing substrate for recolonization. Hurricanes are wind- and water-related disturbances that devastate ecosystems over extensive areas. In the Caribbean, hurricanes as intense as 1989's Hurricane Hugo, with wind speeds over 166 km/h, occur every 50 to 60 years. Higher latitudes experience them less often, making selection for wind resistance far weaker. In 2003, Hurricane Juan, though only a Category 2 hurricane, devastated Point Pleasant Park in Halifax.

As a major abiotic disturbance, fire greatly influences landscape patterns. Its most obvious short-term impacts are negative; losses of organisms are significant, reducing species diversity as well as biomass. Fire also consumes standing dead material, accumulated litter, and soil organic matter. The rapid release of nutrients locked in these substances by fire is called **pyromineralization**. But fire can have positive long-term effects. It prepares the seedbed for many species by exposing mineral soil. For fire-adapted species such as jack pine (*Pinus banksiana*) and black spruce (*Picea mariana*), fire has been a major selective force, promoting **serotiny**, whereby seeds are released from closed cones in response to the heat associated with fire. Some species in fire-prone grasslands employ similar strategies for opening their fruits. By removing some or all of the previous vegetation, fire increases the availability of light, water, and nutrients to both surviving and colonizing plants.

Fire is a factor worldwide, but the fire regime of any given region is influenced by many factors, including temperature, occurrence of droughts, accumulation and flammability of litter and living biomass, and human interference. Before European settlement, fires occurred about every three years in North American grasslands (see Figure 18.17) and prevented encroachment of woody species, particularly in tallgrass sites with sufficient moisture to support forests.

Among forests, fire frequency varies greatly with forest and fire type. Frequent, "cool" **surface fires** that burn only the litter layer have a return interval of 1 to 25 years, whereas **crown (canopy) fires** (Figure 18.19a, p. 398) have a return interval of 50 to 100 years, or longer, depending on the region. The most intense fire type is a **ground fire**, which penetrates the ground, burning organic matter down to the mineral substrate and damaging soil structure (Figure 18.19b).

Some natural disturbances are biotic. Beaver (*Castor canadensis*), discussed in Section 16.4 as *bioengineers*, greatly modify forested areas in North America. By damming streams, beaver not only alter the structure and dynamics of flowing water systems. They also convert forests into wetlands, when dammed streams flood lowland areas, creating pools that collect sediments. By feeding on aspen, willow, and

(a)

(b)

Figure 18.19 Fire effects in forests. **(a)** Crown fires create a mosaic of unburned and burned patches. **(b)** High-intensity ground fires can have damaging effects. After a spruce forest on the Alleghany Plateau in West Virginia was cut in the mid-1800s, ground fires, fuelled by logging debris, exposed bedrock and mineral soil. The forest has not yet recovered, and may be in an alternative stable state (see Section 19.9).

birch, they maintain stands of these species, which otherwise would be replaced by later-successional species. Beaver activity thus creates a diversity of patches—pools, meadows, thickets of willow and aspen—in the larger landscape (Figure 18.20). The landscape impact of beaver interacts with that of humans. In boreal mixed-wood forests in northeastern Alberta, beaver have increased from near-extirpation. Using a chronosequence spanning 50 years, Martell et al. (2006) examined their impact on riparian forests. Beaver increased the width and diversity of riparian zones along smaller streams, converting narrow, moving-water habitats into a mix of moving- and still-water habitats and reducing the width of buffer strips left along waterways after logging.

Grazing by large herbivores is a common biotic disturbance. In eastern North America, high numbers of white-tailed deer (*Odocoileus virginianus*) are decimating the forest understory (Figure 18.21), destroying wildlife habitat and disrupting succession by suppressing tree regeneration. Their activity threatens many herbaceous species as well, including American ginseng, native to eastern forests as far north as Québec (McGraw and Furedi 2005; in Section

10.10, we used ginseng as an example of the Allee effect. A classic edge species (see Section 18.2), deer have benefited greatly from fragmentation. Along with other large herbivores, they are important dispersal agents with the potential to affect landscape as well as population processes. Feces of white-tailed deer in central New York contained viable seeds from over 70 plant species, native and alien (Myers et al. 2004). The balance of their positive and negative impacts—recall from Section 16.7 their role in transmitting meningeal

Figure 18.21 Overabundance of white-tailed deer can eliminate understory plants and create a "browse line" at the highest point that deer can reach for food.

Figure 18.20 A pond created by a 2-m-high beaver dam along a stream in the Rocky Mountains.

worm to caribou and moose—is likely to be strongly density dependent.

Similarly, the African elephant (*Loxodonta africana*) is a keystone herbivore (see Section 16.4) that maintains the savannah ecosystem. When present at high numbers in areas with restricted movement, such as national parks, elephants can cause widespread destruction of woodlands. As with deer, humans have modified the impact of this natural disturbance agent on the landscape.

Birds may seem unlikely disturbance agents, but in the Hudson Bay lowlands, lesser snow goose (*Chen caerulescens caerulescens*) have affected brackish and freshwater marshes by grubbing for roots and rhizomes of grasses and sedges in spring, and grazing on their leaves in summer. A dramatic increase in geese has stripped large areas of vegetation, eroded peat, and exposed underlying glacial gravels (Kotanen and Jeffries 1997). Species colonizing these patches differ from the surrounding vegetation, creating a landscape mosaic.

Outbreaks of insects such as gypsy moths defoliate large areas of forest and cause death or reduced growth of affected trees. Gypsy moth (an introduced species, so indirectly a human disturbance) infestations of hardwood forests may cause mortality losses of 10 to 50 percent (see Ecological Issues: Human-Assisted Dispersal, p. 197). Infestation of spruce and fir stands by spruce budworm is even more devastating, and may kill entire stands (see Ecological Issues: Science and Politics Interact in Spruce Budworm Control in Maritime Canada, pp. 400–401). Insect outbreaks are most severe in large expanses of homogeneous forest occupied by older trees. In this case, disturbance types interact; human repression of fires increases vulnerability to budworm outbreaks.

18.9 ANTHROPOGENIC DISTURBANCES: Human Disturbances Have Lasting Landscape Effects

Some of the most lasting and damaging disturbances are **anthropogenic** (induced by humans). Because human activity is ongoing and involves continuous and severe effects (often under the name of "management"), it can affect ecosystems more profoundly than natural disturbances, which tend to be intermittent. One of the more radical changes involves replacement of natural communities for agricultural use. We think of agriculture as a relatively recent disturbance, but prehistoric humans cultivated crops and converted land to pasture as early as 5000 years B.P. (before present). Like eastern North America, much of Europe and Great Britain was clothed in forests prior to intensive human habitation, including areas that support little or no forest today. Human activities changed the landscape by extending or reducing ranges of woody and herbaceous plants, facilitating invasion and spread of opportunistic species, and altering the dominance structure in communities by selective removal.

We discuss agriculture in more detail in Chapter 26, but grazing by domestic animals is a common disturbance. In the American southwest, domestic cattle disperse seeds of mesquite (*Prosopis* spp.) and other shrubs in their feces, facilitating their invasion of already disturbed grasslands. Dung deposition increases community diversity and spatial heterogeneity, contributing to a microscale mosaic of patches in different successional stages (Brown and Archer 1987). However, although domestic animals facilitate weed transmission, they also play a valuable role as dispersal agents, particularly in tropical ecosystems where natural animal vectors have been reduced in number (Miceli-Mèndez et al. 2008). As well as affecting the plant community, domestic species can transmit disease to native herbivores.

Conversion of natural communities to urban use has even longer-lasting landscape effects than agricultural conversion. Vast amounts of land have been paved over as part of urban developments (particularly in low-density North American cities dedicated to a car culture). Such transformations may not be as permanent as they look—a visit to the ruins of Incan sites and Pompeii attests to the ability of vegetation to reassert itself, aided in the latter's case by the most catastrophic of all natural disturbances, a volcano—but for as long as they endure, urban conversions not only replace natural communities but also prevent or impede ecological functions. No photosynthesis can occur on pavement, and although the urban lawn monoculture does photosynthesize, releasing O_2 and transpiring water, the benefits of these basic processes must be weighed against the negative effects of fertilizer and pesticide application.

Logging is another macroscale human disturbance, treated in more detail in Sections 26.8 and 26.9. Its impact varies with the method used. In selective harvesting, foresters remove only mature single trees or groups scattered through the forest, producing canopy gaps that minimize the scale of disturbance. However, as benign as it may seem, selective harvesting changes species composition and diversity by favouring shade-tolerant over shade-intolerant species. It also necessitates more roads (themselves a disturbance; see Section 18.5) for a given amount of timber harvest. Clear-cutting, in contrast, removes entire blocks of trees (Figure 18.22),

Figure 18.22 Block clear-cutting in a western coniferous forest.

ECOLOGICAL ISSUES Science and Politics Interact in Spruce Budworm Control in the Maritimes

Spruce budworm is a major biotic disturbance agent in conifer forests across eastern Canada. It feeds on white spruce (*Picea glauca*), but its preferred prey is balsam fir (*Abies balsamea*), a shade-tolerant, late-successional species that is abundant in mature stands. A native species, its numbers fluctuate greatly, periodically reaching epidemic levels. During severe outbreaks, it causes widespread mortality across millions of hectares.

Forestry is an economic mainstay in both New Brunswick and Nova Scotia, but their budworm control strategies have differed greatly. York University's Anders Sandberg, and Peter Clancy of St. Francis Xavier University in Nova Scotia, explored the historical and scientific context of their differing strategies (2002). Their premise is that science (and its application to resource management) is not driven solely by experimentation and objective peer review, but also by (1) the "politics of science"—the combination of political, economic, and cultural perspectives that is the backdrop to all human endeavour, exerting influence through funding agencies, advocacy groups, and the institutional atmosphere in which scientists work—and (2) "science politics"—the contending scientific paradigms that underpin experimental design and interpretation. Conflicting perspectives are fundamental to science. Two controversies underpin the forest management debate. One is theoretical: *What is the role of disturbances in forest dynamics?* The second is applied: *What measures are safe and effective for controlling biotic disturbance agents?*

New Brunswick was an early player in the use of insecticides. Spraying for budworm began in the province in the early 1950s and peaked in the mid-1970s, when almost 4 million hectares were sprayed. DDT was the pesticide of choice until 1970, and New Brunswick became a poster child for the negative effects of DDT in Rachel Carson's *Silent Spring* (1962). When DDT was banned, fenitrothion took its place, until it too was banned in 1998 after being linked to high incidence of Reye's syndrome in children, sparking strong public protest.

Why was New Brunswick so committed to a chemical solution? After World War II, New Brunswick's pulp industry expanded rapidly, coincidentally during a major budworm outbreak. At that time, companies were vigorously marketing insecticides. In hindsight, it is hard to appreciate the prevailing climate of opinion. The environmental costs of pesticides had yet to surface; indeed, Paul Müller was awarded the Nobel Prize for medicine in 1948 for the advances against malaria and other insect-borne diseases that DDT, the "magic bullet," promised. It was in this cultural and scientific climate that an organization called Forest Protection Limited ("forest protection" was synonymous with insecticide use) was formed between pulp-and-paper companies and the New Brunswick government, with the goal of massive aerial spraying. The province's heavy reliance on resource extraction fuelled the desire to win the battle against a "pest" that seemingly threatened the province's economic well-being. Strategies that might ameliorate the budworm's impact were not considered viable alternatives. The overriding perception was that standing timber assets—"green storage"—must be protected from budworm attack.

Did the strategy work? No insecticide, no matter how toxic, can ever eliminate an insect. Indeed, by lowering their numbers in the short term, spraying can actually keep budworm levels artificially high by preventing the crash that would normally occur in its population cycle. Moreover, as with antibiotics, ongoing pesticide use exerts a strong selective pressure, increasing pesticide resistance. So New Brunswick seemed bent on pursuing a course of insect control that did little to eliminate the problem but much to increase the province's dependence on harmful chemicals—the *pesticide treadmill*. Collateral damage was noted, but often in euphemistic terms such as "knock-downs" (kills) that underplayed its ecological importance. There were protests, but opposition was ineffective until the impacts on human health became too serious to ignore.

How did the situation differ in Nova Scotia? After the war, the pulp industry was less important than in New Brunswick. Most pulp companies obtained wood from southwestern areas of the province where budworm was less of an issue because hardwoods such as aspen and birch were more prevalent. So the spraying lobby was weaker in the 1950s, when New Brunswick began its program, and only intensified in the 1970s, after the Nova Scotia pulp industry had grown considerably. Then, the Swedish company Stora, which began operations in Nova Scotia in the 1960s, applied great pressure, threatening to close down operations if spraying was banned.

Not only were pressures for spraying less intense in Nova Scotia early on, but public opposition was more effective, particularly by the 1970s when the damaging effects of insecticides were more evident than when New Brunswick had embraced DDT several decades before. So, while some New Brunswick scientists were arguing for continued chemical spraying on the grounds that its economic benefits outweighed its ecological and health costs, Nova Scotia resisted their use, and only began using a biological agent, Bt (*Bacillus thuringiensis*) in 1980.

Thus far we have focused on the "politics of science"—the ways in which the differing economic and social cultures of the provinces contributed to their management policies. But what about "science politics"? After the war, New Brunswick specialized in forestry research while Nova Scotia focused on agriculture. Indeed, New Brunswick was at the forefront of forest research in North America. Some scientists focused on spraying, but others made key contributions to modelling forest dynamics and plant–herbivore interactions. Research conducted by university and government scientists challenged the prevailing paradigm of stable forest systems, arguing that it was normal for spruce–fir forests to be subject to periodic crashes. Budworm outbreaks act in a similar way as fire, and the two often interact. Dense stands of dead fir are highly flammable, and as stands age and become more susceptible to budworm, their subsequent death (often initiated by drought interacting with age) increases fire intensity.

Yet, as much as this research enhanced our understanding of disturbance in boreal forests, it was often pressed into service as justification for continuing (and even intensifying) spray programs. The logic was simple: the forest is not stable; it undergoes frequent disturbance; humans act as just one more disturbance agent by using pesticides to keep the trees alive long enough to harvest them.

Meanwhile, ecologists in Nova Scotia, freed from ties to the pulp industry, were pursuing a different paradigm—arguably more conventional, and certainly less interventionist. Sandberg and Clancy contend that Nova Scotia's resistance to chemical spraying was based on a "balance of nature" paradigm, which supported the idea of "ecological pragmatism" (Worster 1994). If natural systems exist in a dynamic equilibrium, then the onus is on humans to interfere with that state as little as possible. Poisoning living organisms (even if the goal is to "protect" other organisms) violates that precept. This approach is based on acknowledging ignorance; we may not understand the regulatory impact of species as they interact in a food web, but to be on the safe side we should refrain from taking action that could upset that balance.

Is there an alternative to insecticides? According to Miller and Rusnock (1993), most forest ecologists in Nova Scotia supported the "silvicultural hypothesis." (*Silviculture* is the practice of planting and managing forest stands.) Conventional practices such as "high-grading" large trees, fire suppression, and clear-cutting increase outbreak severity by promoting large tracts of vulnerable species such as balsam fir. But although silvicultural practices can worsen the problem, they can also be part of the solution. By reducing the extent of uniform stands, establishing topographic or biological barriers to budworm spread, and restoring the natural forest mosaic, damage can be kept to manageable levels. The more variable age distribution that results lowers susceptibility. Outbreaks still occur, but they are patchy. However, although this approach seems more ecologically sensitive, it advocates more frequent harvests to prevent large tracts from succeeding to the mature stage in which fir dominates. As such, it may jeopardize old-growth habitat.

Where do the provinces stand now? Policies have converged. Insecticide use is down in New Brunswick and up in Nova Scotia. Both adhere to the Spruce Budworm Decision Support System, an integrated pest management approach that combines spraying with population monitoring and methods designed to lower outbreak severity, such as encouraging predators. Both provinces have reached a more ecological consensus, but the pressures of industrial forestry to maximize harvests remain, and have intensified in Nova Scotia. No longer polarized in their scientific perspective, forest ecologists from both provinces continue to make valuable research contributions. The paradigms have reconciled; the non-equilibrium view promoted in the New Brunswick context is now widely accepted, but the more cautious "balance of nature" view has spawned an appreciation of the vital role of disturbance.

Research advances continue in biocontrols, now the preferred option. Bt is not without side effects. It damages butterflies and moths (Scriber 2001), including many key pollinators. Other measures, such as introducing insect viruses (developed by the Great Lakes Research Centre in Sault Ste. Marie) may also have reverberating impacts on non-target species. A recent breakthrough from the collaborative efforts of the University of New Brunswick and federal researchers involves "escapers"—individual fir trees that are protected from budworm by anti-herbivore compounds produced by endophytic fungi (Calhoun et al. 1992).

Achieving a more sustainable use of forest resources within the economic and political constraints of modern forestry (especially amid growing concern for forest conservation) remains a challenge to academics in both provinces—and indeed in all of Canada.

Bibliography

Calhoun, L. A., J. A. Findlay, J. D. Miller, and N. J. Whitney. 1992. Metabolites toxic to spruce budworm from balsam fir endophytes. *Mycological Research* 96:281–286.

Carson, R. 1962. *Silent spring*. Boston: Houghton Mifflin.

Miller, A., and P. Rusnock. 1993. The rise and fall of the silvicultural hypothesis in spruce budworm (*Choristoneura fumiferana*) management in eastern Canada. *Forest Ecology Management* 61:171–189.

Sandberg, L. Anders, and P. Clancy. 2002. Politics, science and the spruce budworm in New Brunswick and Nova Scotia. *Journal of Canadian Studies* 37:164–191.

Scriber, M. J. 2001. Bt or not Bt: Is that the question? *Proceedings of the National Academy of Science* 98:12328–12330.

Worster, D. 1994. *Nature's economy: A history of ecological ideas*. Cambridge: Cambridge University Press.

favouring reproduction of shade-intolerant species. In accordance with patch dynamics, the effects of clear-cutting depend upon the size and shape of the cut areas. Small or elongated cuts reduce the damage and facilitate regeneration, whereas large cuts can trigger significant erosion, depending on slope and post-harvest treatment.

Disturbances affect other landscape elements than the area immediately affected. Logging in British Columbia forests has ripple effects on adjacent aquatic habitats by overloading streams with coarse woody debris. Streams adjacent to recently logged (less than five years) areas had nearly twice the macroinvertebrate and algal biomass (Fuchs et al. 2003) of streams in old-growth regions. Although the structural features of the stream habitat were unaffected, the impacts of this short-term pulse in productivity on stream ecosystems, particularly on higher-level consumers, are unclear.

Other anthropogenic disturbances with severe landscape effects are (1) mining related, including natural gas and oil as well as minerals; (2) transport related, including damage associated with gas and oil pipelines as well as roads and highways; (3) military related, including conventional and nuclear; and (4) industry related, including pollution of all sorts. The most dramatic human disturbances, which, given the severity and longevity of their effects, are better termed catastrophes, are nuclear events: bombs, as detonated by the United States during World War II; bomb testing in the South Pacific in the 1960s and 1970s; and nuclear accidents such as the 1980 Chernobyl incident (see also Section 22.12). When human and natural disturbances interact, such as the breakdown of nuclear plants in Japan after the 2011 earthquake and tsunami, the effects are especially devastating. Such events highlight the fact that we are the only species with the ability to cause disturbances that threaten not only particular ecosystems in the short and long term, but also the biosphere itself. We examine human impacts more specifically in Part Eight (including introduction of non-native species), but collectively we have much to answer for.

18.10 SHIFTING MOSAIC: Landscape Dynamics Generate a Shifting Mosaic of Patches

Unlike an artist's mosaic that can survive unchanged for centuries—witness the mosaic fragments from the ruins of Pompeii—the mosaic of communities on a landscape is ever changing. Disturbances—large and small, frequent and infrequent—and natural mortality alter the biological and physical structures of the community patches, initiating succession. The landscape is a **shifting-mosaic steady state** of patches, each in a different successional phase (Figure 18.23) (Bormann and Likens 1979).

The term *steady state* is a statistical description of the collection of patches and refers to the average state of the forest. The mosaic of patches in Figure 18.23 is far from

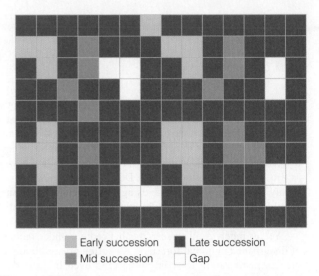

Early succession ■ Late succession
Mid succession □ Gap

Figure 18.23 A forested landscape as a mosaic of patches in various stages of succession. Each patch is always changing, but the average state of the forest may remain relatively constant.

static. Each patch is continuously changing. Disturbance causes patches in late succession to revert to early succession. Likewise, patches in early succession undergo shifts in species composition, eventually becoming late succession. So, although the overall mosaic is always changing, the average composition of the forest remains fairly constant—a dynamic steady state. This idea of a continuously changing mosaic that remains constant when viewed collectively (as a population of patches) resembles the concept of a stable age distribution presented in Section 9.6, wherein the proportion of individuals in each age class remains constant even though individuals are continuously entering and leaving the population through births and deaths.

Returning to the countryside in Figure 18.1, we now see the mosaic of patches—forest, fields, pond, hedgerows, and farms—as a dynamic entity, not a static image. Most of the forests were once fields or pasture, which had been cleared in past centuries. Farmers abandoned some of their fields, which reverted to forest. The current mosaic is maintained by active processes, many involving human disturbances. The communities occupying the patches are habitat islands, some bridged by corridors. Their populations are part of larger metapopulations linked by dispersal, and separated patches of similar communities belong to the larger metacommunity.

Over time, this mosaic will continue to change. Patch boundaries will shift; succession will alter community structure in existing patches; natural and human disturbances will interact to form new patches. Consistent with the growing realization of the importance of scale in ecological studies, Turner et al. (1993) argue that this concept of steady-state landscape equilibrium must be revised to account for the impact of contrasting temporal and spatial scales of recovery and disturbance. Given the scale of human disturbances relative to the time required for natural communities to recover, human-dominated landscapes such as that in Figure 18.1 are unlikely to attain the steady state that Bormann and Likens propose for natural landscapes.

Habitat transformation (loss and/or fragmentation) is a major focus of landscape ecology. Gonzalez et al. (2011) contend that to understand its impacts, ecologists must bridge the approaches of landscape and community ecology. While landscape ecologists traditionally focus on the spatial structure and dynamics of the habitat network—patches, corridors, and matrix—few consider the feedbacks between this physical network and the biological "interaction network" operating within it. The ability of habitat transformation to disassemble these ecological networks is of huge concern. Understanding it requires inputs not only from landscape and community ecology, but also from ecosystem ecology—the subject of Part Six.

EcologyPlace

Visit EcologyPlace at www.pearsoncanada.ca/ecologyplace to access online resources that complement your textbook, and help you to apply and to review the information in this chapter. EcologyPlace includes

- an eText version of the book
- self-grading quizzes
- glossary flashcards
- and more!

Go to www.pearsoncanada.ca/ecologyplace and follow the registration instructions on the Student Access Code Card included with this text. If your book does not have a Student Access Code Card, you can purchase access to it at www.pearsoncanada.ca/ecologyplace.

SUMMARY

Habitat Patches 18.1

Landscapes consist of mosaics of habitat patches interacting with each other within a matrix. Landscape ecology studies the causes and effects of these spatial patterns. Patches vary in origin, from remnants to introduced patches requiring human maintenance. Many factors contribute to landscape pattern: geology, topography, soils, climate, biotic processes and interactions, and disturbance.

Transition Zones 18.2

The place where the edges of two patches meet is a border. A border may be inherent (produced by a topographical feature or shift in soil type) or induced by a disturbance. Borders vary in width and contrast. Low-contrast, transitional borders form an ecotone. The vertical structure of borders influences abiotic gradients between patches. Borders connect patches through flows of material, energy, and organisms. Ecotones often have high species richness because they support some species from the adjoining communities as well as edge species.

Patch Traits 18.3

Diversity tends to increase with patch area. Many species require large, unbroken habitat areas. Larger areas encompass more microhabitats and support more species and larger populations. Some species are area sensitive. Interior species require conditions found in the interior of large patches. Fragmentation promotes edge species by producing smaller patches, which have a larger edge-to-interior ratio. Small patches may lack interior habitat. Some species are area insensitive.

Island Biogeography 18.4

The equilibrium theory of island biogeography proposes that the number of species occupying an island represents a balance between immigration and extinction. The greater the distance from a mainland, the lower the immigration rate. The smaller the island, the higher its extinction rate, because smaller islands support smaller populations and contain fewer habitats. The island biogeography theory can be applied to habitat patches.

Corridors 18.5

Corridors are strips of habitat that link patches. They differ from the surrounding matrix and may or may not differ from the habitat patches they connect. Corridors function as filters or barriers, facilitating or restricting dispersal among patches.

Metapopulations and Metacommunities 18.6

Habitat fragmentation and human exploitation have reduced many species to metapopulations—groups of semi-isolated subpopulations. Movement of individuals among patches is essential to sustain metapopulations. Metacommunities are groups of semi-isolated habitat patches occupied by similar communities.

Role of Disturbance 18.7

Disturbances are discrete events that remove or greatly alter a community. They initiate succession and affect diversity. Disturbances vary in frequency, intensity, severity, and scale. A regime of frequent, low-intensity, fine-scale disturbances creates gaps in the substrate or vegetation, which will be occupied by patches of different successional stages. Macroscale disturbances favour opportunistic species, and if severe, can replace the community altogether.

Natural Disturbances 18.8

Fire (ground, canopy, or surface) is an important natural disturbance, exerting both beneficial and adverse effects. Other abiotic natural disturbances include wind, floods, and storms. Animals can act as biotic disturbances by herbivory or as disease vectors.

Anthropogenic Disturbances 18.9

Human-induced disturbances, such as logging, mining, agriculture, urbanization, oil spills, and nuclear events, produce profound and often permanent changes in the landscape.

Shifting Mosaic 18.10

The landscape is dynamic, with patches in various stages of succession and disturbance. This process suggests a shifting mosaic whereby the landscape as a whole exists in a steady state, even though individual elements are continuously changing.

KEY TERMS

anthropogenic
area-insensitive
    species
border
bud bank
connectivity
corridor
crown (canopy) fire

disturbance regime
ecotone
edge
edge effect
edge species
enduring features
equilibrium theory of
    island biogeography

filter effect
fragmentation
frequency
ground fire
intensity
interior species
matrix
metacommunity

mosaic
pyromineralization
relaxation effect
    (extinction debt)
return interval
scale
serotiny

severity
shifting-mosaic steady
    state
species area curve
species pool
species turnover rate
surface fire

STUDY QUESTIONS

1. The study of spatial pattern is the defining characteristic of landscape ecology. Explain.
2. How do variations in the abiotic environment (geology, topography, soils, and climate) give rise to landscape patterns in your region? Contrast these patterns with those reflecting the influence of human activities.
3. Distinguish among *edge*, *border*, and *ecotone*. Why do ecotones often support more species than do (i) the adjoining communities and (ii) edge habitats?
4. How does the ratio of edge to interior change with patch size? What is the impact of this ratio on animal diversity? How is this trend affected by animal size and trophic position?
5. The island biogeography theory predicts the species richness of islands as a balance between colonization and local extinction. What two features of islands affect these rates? How might these features affect species turnover?
6. How is island biogeography theory applicable to habitat patches in terrestrial sites? How does the theory relate to metapopulation dynamics (see Section 12.3)?

7. How do corridors help maintain the diversity of habitat patches? What features of corridors affect how well they perform this function? Give examples.
8. You examine the plant species diversity of forest remnants the year that a housing project is completed, and conclude that it is unchanged. Criticize the validity of this conclusion, based on your knowledge of fragmentation and the relaxation effect.
9. Distinguish among these terms as they apply to disturbances: *scale*, *intensity*, *frequency*, and *severity*.
10. In 1954, Hurricane Hazel, which was only a Category 2, greatly damaged vegetation in eastern Canada. What does this fact suggest about hurricane disturbance and selection pressure?
11. How has the local landscape changed in your lifetime? What factors have been involved?
12. How can habitat patches be in flux, and yet the overall landscape remain unchanged?

FURTHER READINGS

Cadenasso, M. L., S. T. A. Pickett, K. E. Weathers, and C. D. Jones. 2003. A framework for a theory of ecological boundaries. *BioScience* 53:750–758.
Examines boundaries, focusing on the flows of organisms, materials, and energy in landscapes.

Forman, R. T. T. 1995. *Land mosaics: The ecology of landscapes and regions.* New York: Cambridge University Press.
Discusses the impacts of natural processes and human activities on the changing landscape mosaic.

Hilty, J. A., W. Z. Lidicker, Jr., and A. M. Merenlender. 2006. *Corridor ecology: The science and practice of linking landscapes for biodiversity conservation*. Washington, DC: Island Press.
Discusses the role of corridors in enhancing and maintaining connectivity between natural areas.

Holyoak, M., M. A. Leibold, and R. D. Holt. 2005. *Metacommunities: Spatial Dynamics and Ecological Communities*. Chicago: University of Chicago Press.
Develops the metacommunity concept as vital for both community and landscape ecology.

Lindenmayer, D. B., and J. Fischer. 2006. *Habitat fragmentation and landscape change: An ecological and conservation synthesis*. Washington, DC: Island Press.
Discusses the ways in which landscape change, fragmentation, and diversity are interrelated.

Worboys, G. L. W. L. Francis, and M. Lockwood (eds). 2010. *Connectivity conservation management: A global guide*. London: Earthscan.
Focuses on the establishment and management of connectivity conservation areas between areas of large scale parks and protected areas. Provides an excellent summary of all aspects of corridor ecology.

Wuerthner, G. (ed.). 2006. *Wildfire: A century of failed forest policy*. Washington, DC: Island Press.
Covers the topic of wildfire from ecological, economic, and sociopolitical perspectives.

In the summer of 2010, air traffic in the Northern Hemisphere was interrupted by clouds of ash emanating from Iceland's Eyjafjallajökull volcano. While the media dwelled on the disruption to air traffic, few considered the event from a broader, ecological perspective. Yet places like Iceland and Hawaii are among the most dynamic on Earth. Volcanic activity continuously forms new land, as lava flows to the ocean. Exposed to sun and wind, the hard surface of the cooled lava flows is a harsh site for succession. But where there are cracks, plants find more hospitable microsites.

In Hawaii Volcanoes National Park, stress-tolerating lichens, ferns, and the endemic evergreen tree *Metrosideros polymorpha* are the first to colonize. Temperature is moderated in the crevices, where moisture accumulates and the wind traps dust and organic matter. Nutrients are scarce, but both the lichen *Stereocaulon volcani* and the evergreen tree *Myrica faya* (one of many invasive species in Hawaii) fix N_2. As these plants grow, they modify the environment by shading the substrate and adding organic matter, vital for retaining moisture and nutrients. Their roots break down the substrate mechanically, by cracking the lava, and chemically, by excreting acids. Soon a sparse community develops, allowing other species to establish under its protective cover. This process continues as the colonizers ameliorate the site, facilitating the success of later arrivals until a forest eventually establishes. Succession after an earlier volcanic eruption (Surtsey) in Iceland from 1963 to 1967 followed a remarkably similar progression half a world away (Burrows 1990), despite a harsher climate and with different species as pioneers.

As we have discussed in previous chapters, the distribution and abundance of species and the biological structure of communities vary in response to abiotic environmental factors. In turn, organisms modify their abiotic environment, as in this example of succession on lava flows. It is this inseparable link between the biotic community and its abiotic environment that led Arthur Tansley (1935) to define the term *ecosystem* as the "basic unit of nature": "the whole system (in the sense of physics) including not only the organism-complex, but also the whole complex of physical factors forming what we call the environment."

By viewing biotic and abiotic components as a single interactive system, the ecosystem concept ushered in a new way of studying nature. The taxonomic perspective gave way to a functional approach. In discussing forest ecology with a population or community ecologist, one hears a story of species—the dynamics of populations: their interactions, food webs, and diversity. Discuss the same forest with an ecosystem ecologist, and a more abstract picture emerges: a story of energy and matter, in which the boundary between the abiotic and biotic components of the forest is often blurred. From an ecosystem perspective, the forest's biotic and abiotic components continually process and exchange energy and matter. Diversity is still important, but in terms of its consequences for ecosystem function rather than as a list of species and their abundances.

Recall the two types of biotic components in any ecosystem: (1) *autotrophs (primary producers),* predominantly green plants and algae, which use solar (or in some cases chemical) energy to transform inorganic substances into organic compounds, and (2) *heterotrophs (consumers)*, which consume the organic compounds produced by autotrophs as a source of food. In turn, there are two types of heterotrophs: (1) *biophages*, which consume living tissue, and (2) *saprophages (decomposers)*, which break down detritus into inorganic substances, to be used again by autotrophs. Solar energy drives the ecosystem. Once harnessed by photosynthesis, this energy flows from producers to consumers to decomposers, eventually dissipating as *entropy*. Abiotic components consist of air, water, soil, sediments, particulate matter, dissolved organic matter in water, and dead organic matter (*detritus*). This dead organic matter is derived from plant and consumer remains and is fuel for decomposers.

Like the community, the ecosystem is a spatial entity whose boundaries may be hard to define. At first glance, a pond is clearly separate from the surrounding land habitat. A closer look reveals a less distinct boundary. Emergent plants are rooted around the edge, tapping into the shallow water table. Amphibians move between the shore and the water. Nearby trees drop leaves into the

(a)

(b)

Figure 1 Plants characteristic of the pioneer stages of succession on lava flows in Hawaii. **(a)** *Metrosideros polymorpha.* **(b)** The lichen *Stereocaulon volcani.*

pond, adding to the detritus that feeds the decomposers on the pond bottom.

Allowing for the difficulty of assigning ecosystem boundaries, exchanges of substances and organisms into and out of an ecosystem are **inputs** and **outputs**, respectively. No ecosystem is wholly self-contained, but one with minimal exchange is a *closed system* while one with significant exchange is an *open system*. Inputs and outputs make feedback regulation possible, a topic central to our discussion of systems ecology in Chapter 19. Exchanges of energy and matter among components are discussed in Chapters 20 (energetics), 21 (nutrient cycling and decomposition), and 22 (biogeochemical cycles).

One might ask: *If the ecosystem concept is so central, why have we delayed a detailed treatment of it until now?* Systems ecology draws heavily on concepts that we have developed thus far. Indeed, it provides an overarching framework to integrate our understanding of adaptations, populations, communities, and the abiotic environment. Informed by these fundamental concepts, we can start to bring them together in our analysis of ecosystems.

Covering only 6 percent of Earth's land surface, tropical rain forests contain over half of all known terrestrial species.

We introduced the ecosystem concept in Chapter 1. Now we focus on it more directly, particularly the principles of system regulation. Pioneered by Paul Weiss in the 1920s, *systems ecology* rose to prominence in the 1950s and 1960s with the writings of Ludwig von Bertalanffy and Howard and Eugene Odum. Neglected in the 1980s, it has regained prominence as ecologists adopt an integrated approach to ecosystem function.

This chapter addresses the following key questions: *What is a system, and in what sense(s) does the ecosystem function as one? How do the components of natural systems relate to each other, and do these interrelations involve feedback regulation? What does it mean for an ecosystem to be stable, and do ecosystem properties (notably diversity and productivity) affect stability? Do ecosystems have alternative stable states, and what causes ecosystems to flip between them? Are natural ecosystems best studied with mathematical models, or through experimental model systems in the laboratory or field? Does the biosphere operate as a global system?* These are big questions, which some may call unanswerable, but they are questions that must be asked.

19.1 SYSTEM DEFINED: System Components Work Together as a Functioning Whole

We all have some sense of what a system is, but like many concepts, it has multiple meanings. Von Bertalanffy defined it as "a set of elements standing in interrelation among themselves and with [the] environment" (1969). For our purposes, we will define a **system** as a set of interrelated parts that work together to perform a function or functions within an environment. Systems ecology takes a holistic approach, focusing on the functioning entity as more than the sum of its parts. Indeed, the field emerged in part as a reaction to a mechanistic, reductionist view that studies a system's constituent parts—whether cells, organisms, populations, or communities—in isolation.

Not all systems are biological. A watch, car, or computer is a system whose components work together to perform particular functions. What distinguishes them from living systems is not just that they are abiotic, but that they are the product of design. An engineer designing a car decides what parts to use and how to organize them for optimal function. Some car designs work better than others, and the same car design works better or worse depending on its execution—how the design is carried out in terms of the quality of the parts and its manufacture. Finally, a car that is well designed and manufactured may work better or worse in different environments. A finely tuned Italian sports car is unlikely to work well in a northern Canadian climate.

With non-living systems—again because they are designed—all parts are there for a reason. (There may be some built-in redundancy to improve reliability or safety, so that if one part fails there is another to allow continued function.) If, after assembling a bicycle, there are parts left over, we can be fairly sure that, if the bike functions at all, it is unlikely to function as well as it would with all its parts assembled correctly.

Compare these human-designed systems with their living counterparts. A cell is a system; its parts relate to each other in ways that allow it to perform particular functions, depending on the type of cell and the tissue and organism in which it occurs. Chloroplasts and other subcellular organelles also have system properties. Both cells and organelles are systems nested within other systems. Indeed, we use the term *organ system* to describe the interdependent functioning of the circulatory or digestive system, for example. Farther up the hierarchy, the individual organism, whether animal, plant, or fungus, with its interrelated organ systems, tissues, and cells, is another living system.

In contrast to a car or computer, such biological systems—no matter where in the hierarchy from organelle to organism—result from evolution, not design. As vital as this difference is theoretically, what are its practical consequences? First, not all its components may be essential to its functioning. Parts of a cell or of a body may persist despite the fact that they no longer contribute—may never have contributed—to the functioning of the system of which they are a part. They may be accidents of evolutionary history, or residual from some prior function. Second, the biological system may not function optimally in any particular environment, even the one in which it has evolved. Evolutionary forces have increased, we presume, its adaptive properties, but there are many reasons why living systems that are not the product of design will be less than perfectly matched to their present (or future) environment (see Section 5.11).

Position in the biological hierarchy is relevant here. For systems at or below the organism level, two traits enhance our sense of it as a system: (1) an outer covering that provides a discrete boundary and (2) a genetic code that regulates its function. Neither applies to an ecosystem, which seems a weaker fit to the system concept. Yes, ecosystems have components (be they organisms, populations, or functional groups) that interact, both with other biotic components and with air, water, and soil. Yes, ecosystems perform functions, such as energy processing and matter cycling. But in contrast with either the designed system of the computer or the evolved system of the organism, an ecosystem is looser, with less predictability and consistency of its components. Not only are different species present at different abundances in different examples of an eastern deciduous forest; such forests also differ greatly in their overall ecosystem function, such as their productivity.

With such variability and looseness in composition and function, and lacking a discrete boundary and a genetic blueprint, is an ecosystem truly a system, subject to the principles of **cybernetics**, the science of system regulation? The question persists, especially regarding the importance of diversity to stability. For now, we will take a pragmatic view. Although we must relax our idea of a system, the increased understanding of ecosystem function that flows from stressing its system properties justifies a systems approach.

Von Bertalanffy stressed that, in contrast with closed mechanical systems, ecosystems are open, both to other ecosystems and to the biosphere. The components themselves, be they biotic (organisms) or abiotic (soil, water, or gases), move from one part of an ecosystem to another, or leave one ecosystem and enter another, as from a forest into a stream. Alternatively, a substance released by a component, as, for example, fecal matter or CO_2 or heat, becomes an input for some other component(s) of that or some nearby ecosystem. Overarching all these inputs and outputs within and among ecosystems is the pre-eminent respect in which all ecosystems are open—their reliance on continuing energy inputs from a source external to the biosphere, the Sun. Of course, organisms also rely on an external energy source, as indeed do cars or computers. But for an ecosystem, energy processing is a defining function rather than a means to another end.

Systems ecology depicts the ecosystem as a series of boxes inside a boundary (Figure 19.1a). The boxes represent biotic (populations or functional groups) and abiotic (air, water, soil) components, linked by arrows representing the relations (feeding and otherwise) by which they interact. Now consider Figure 19.1b.

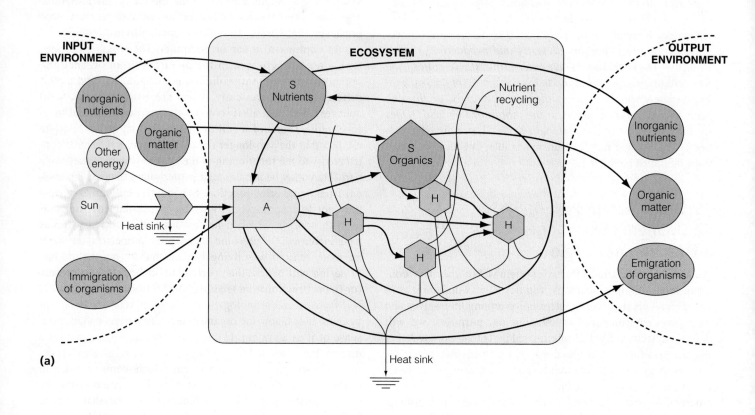

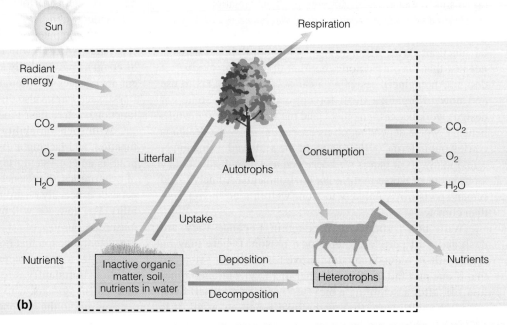

Figure 19.1 Ecosystem models. **(a)** General theoretical model. S = storage; A = autotroph; H = heterotroph. Component shapes inside the ecosystem box indicate roles in energy processing, as specified by Odum. Inputs and outputs link the ecosystem to its environment and to other ecosystems. **(b)** Simplified model of a terrestrial ecosystem.

(Adapted from (a) Odum and Barrett 2005, (b) O'Neill 1976.)

Representative organisms (trees, grass, deer, etc.) replace the boxes, but the concept is the same. Similarly, matter cycle diagrams (e.g., Figure 22.2) also assume the system concept, with images again replacing boxes. Such simplified models provide useful summaries, but they in no way represent the complexity of even the simplest ecosystem. They also depict fixed boundaries, whereas ecosystem boundaries are invariably "fuzzy" and fluid.

19.2 FEEDBACK: Outputs May Act as Inputs for the Same or Other System Components

The concept of feedback is integral to systems ecology, and in particular to ecosystem regulation. **Feedback** refers to a situation in which output from one system component becomes input for either the same or a different component—it "feeds back." So if a population (in response to some biotic or abiotic input) increases its reproductive output, this output becomes an input affecting the future activity of its own population and/or some other population or abiotic component with which it interacts directly or indirectly.

Theoretically at least, not all ecosystem outputs have feedback effects. If a plant population is stimulated to grow by some abiotic input—say, more rainfall—the change in its output (increased biomass) may have no impact on its own activity or that of other ecosystem components if the biomass is removed in a windstorm. (Presumably it will affect some component of the ecosystem to which it is carried as

detritus.) In this case, the output arrow does not feed back (Figure 19.2a).

In most situations, however, one of two types of feedback may occur. In **positive (reinforcing) feedback**, the output reinforces the original input, causing the system (or system component) to continue changing in the same direction. In our example of biomass production, the increased output (if allocated to leaves) adds to the photosynthetic capital of the plant population, allowing it to produce even more biomass. If the output is more offspring, its potential for a positive feedback effect on population size is even more obvious. We have discussed positive feedback as an upward spiral on several occasions, including allocation of carbon by plants in Section 6.7 and exponential growth in Section 10.2. In systems diagrams, positive feedback is shown by an arrow that circles back and links up with the original box (or alternatively some other box in the chain), transforming a linear system into a partially cyclical system (Figure 19.2b).

Positive feedback can also cause a downward spiral. This may seem counterintuitive (partly because we use the term *positive feedback* for a positive response to something), but in cybernetics the defining principle is that positive feedback reinforces the original stimulus, be it positive or negative. So if an input acts as a negative stimulus on a system component, positive feedback reinforces or enhances that negative impact. If a negative input (either abiotic, such as less moisture, or biotic, such as more competition or less prey) causes a population's output to decline, that output can reinforce the negative input by positive feedback, causing

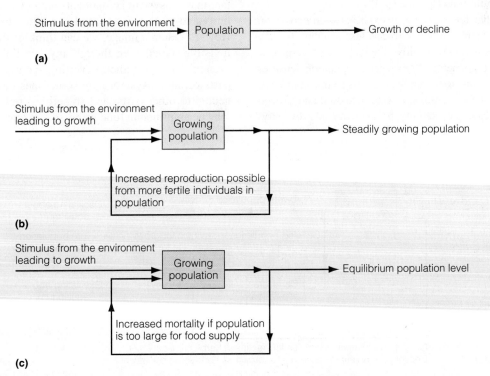

(a)

(b)

(c)

Figure 19.2 Feedback mechanisms in ecosystems. **(a)** No feedback. **(b)** Positive feedback. **(c)** Negative feedback.

(Adapted from Clapham 1983.)

further decline. Fewer offspring makes it harder for remaining individuals to find mates, exacerbating the original negative stimulus. Recall the Allee effect, in which positive feedback caused a small population to decelerate (i.e., decline at an accelerating rate) (see Section 10.10).

In **negative (correcting) feedback**, in contrast, the output acts counter to the input, damping its effect if not reversing it entirely. Again, the arrow "feeds back"—present output becomes future input for either the same or some other system component—but in a way that stabilizes the system rather than causing it to deviate further. If an input exerts a positive stimulus, causing output to increase, then negative feedback counters that change by altering activity in a way to reduce output (Figure 19.2c), bringing the system back in the direction of its original state, if not restoring it completely. For example, if the number of offspring increases in response to food supply, this increased output affects the population in the future, intensifying intraspecific competition and reducing future output.

In this instance, the same system component is affected, but negative feedback often involves several boxes in a system. Increased reproduction by one population may stimulate increased predatory activity, especially by generalists that switch to a prey when its numbers reach a threshold (see Section 14.4). These examples involve *density dependence* with respect to competition and predation, but other biotic interactions, including mutualism, may also be regulated by negative feedback. Whether the impacts are *direct* (as with a predator responding to prey density; see Section 14.3) or *indirect* (as with a trophic cascade; see Section 16.8), these interactions suggest the many negative feedback loops operating among populations (Figure 19.3).

Few doubt that feedback loops exist between ecosystem components, but their regulatory role is less evident. To what extent does negative and/or positive feedback regulate properties of (1) biotic components, for example, population densities; (2) abiotic components, for example, gaseous composition of air and pH, salinity, and other traits of water and soil; and (3) ecosystem traits, for example, stability, productivity, and diversity?

A non-living system again provides a useful contrast. A thermostat is a familiar example of the regulatory potential of negative feedback. A furnace is a mechanical system for generating heat when supplied with fuel. Without the regulatory action of a thermostat, it would generate more and more heat, until it broke down. (Positive feedback would accelerate its breakdown, much as one would experience if suffering from heat stroke or hypothermia if positive feedback were operating in a positive or negative direction, respectively.) But with a thermostat, room temperature is maintained around a set point. When sensors detect that the temperature has declined below, say, 20°C, they stimulate the furnace to increase its activity to counter the decline in heat output, increasing temperature to the set point. Similarly, when heat output raises temperature above the set point, sensors shut off the furnace to counter the increase. Similar negative feedback loops control thermoregulation in homeotherms (see Section 7.5).

Systems vary in how tightly their regulatory feedback loops operate. Some finely tuned devices (human-made or living) allow only minor deviations from a set point, whereas others have larger swings. The physiological systems of organisms (not just thermoregulation but control of many body states) are much more tightly regulated than are component populations in an ecosystem. Recall our discussion of cycling of plant and animal populations interacting in food webs; numbers undergo periodic oscillations that, in the case of boreal food webs, can be substantial (see Figure 10.16c).

Why are ecosystems so much more loosely regulated? Reproductive time lags are partly responsible, along with the fact that ecosystem components are not always in close proximity and often move around in relation to each other. Using Howard Odum's image, we can think of their interactions as invisible wires, even though any one link may be weak or even missing altogether. Reverting to our mechanical analogy, parts of an ecosystem may sometimes be functionally connected to other parts, and sometimes not. Feedback signals may be amplified in food webs, so that interactions that appear

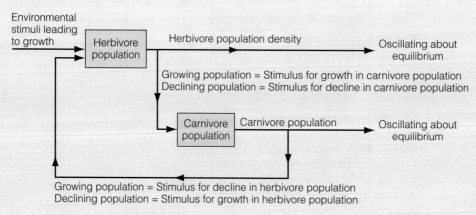

Figure 19.3 Negative feedback loops operating between a herbivore and a carnivore population and contributing to equilibrium population sizes. (Adapted from Clapham 1983.)

minor can have substantive impacts that are difficult to anticipate and to measure. In addition, ecosystems (like populations) experience erratic, density-independent abiotic influences (see Section 11.11). So, if negative feedback at the organism level is all about maintaining a controlled internal state within the discrete confines of a tightly managed system, no ecosystem—even in the most constant of environments—is ever so free of the effects of both internal variability and external abiotic forces.

Finally, it is vital to stress that feedback regulation in ecosystems is not restricted to interactions between biotic components. Increasingly, ecologists realize that feedback loops between organisms (often but not only involving microorganisms) and abiotic components, such as nitrogen, are critical to maintaining macroscale ecosystem function. Never is this linkage more vital than in the carbon sink activity of the biosphere (see Chapter 28 and Research in Ecology: Altered Matter Cycling in a Warmer Arctic, pp. 482–483).

19.3 HOMEORHESIS: Ecosystems May Exist in Pulsing States Rather Than Dynamic Equilibria

The presence of negative feedback loops seems to imply that a system is in a state of **dynamic equilibrium**: a state in which regulatory (negative feedback) devices are working to return a system to the set point, despite ongoing change in response to abiotic and biotic inputs. In fact, ecological theories can be differentiated into those that attempt to explain some aspect of the ecology of an individual, population, community, or ecosystem when it is in an equilibrium state (e.g., the Lotka–Volterra competition model in Section 13.2 or the island biogeography theory in Section 18.4) versus those that are concerned with non-equilibrium behaviour (such as the intermediate disturbance hypothesis in Section 17.8).

Yet, as central as the idea of dynamic equilibrium is to negative feedback regulation, the equilibrium state is likely the exception rather than the rule for most ecosystems. However, the fact that change is the norm does not nullify either the existence or the importance of negative feedback. It simply means that, unlike a thermostat-controlled furnace or a car with cruise control, the ecosystem spends most of its time away from the set point (if a set point exists). The focus then shifts from the set point itself to the behaviour of system components as they deviate away from or towards it.

Odum and Barrett (2005) proposed an alternative to dynamic equilibrium, suggesting that ecosystems exist in loosely controlled **pulsing states** rather than steady states, and that the concept of a set point does not apply above the level of the individual. As opposed to *homeostasis* (the behaviour of a system in a dynamic equilibrium), ecosystems may exhibit **homeorhesis** (Waddington 1975)—the tendency to return to a particular trajectory, even if deflected from that path and even if an end state is never reached. *Stasis* means *position*, and *rhesis* means *flow*, implying that the direction and nature of the path an ecosystem follows are more important than the destination.

The wide fluxes typical of biotic ecosystem components (populations) are often reduced at higher hierarchical levels. Some herbivore populations might increase while others decline, but overall herbivore abundance may stay relatively constant. It is for this reason, not just to simplify, that ecosystem diagrams often lump populations into functional groups. The system as a whole, taken as an assemblage of the pulsing states of its components, may be less complex than the sum of its parts. Rather than exemplifying a "balance of nature," ecosystems may be better described as in a state of robust imbalance.

19.4 RESISTANCE AND RESILIENCE: Ecosystems Differ in the Ability to Resist Change or to Re-Establish

A discussion of feedback regulation and states of ecosystems—whether dynamic steady state or pulsing states—inevitably brings us to the concept of stability. *Stability* is one of those words with a built-in value judgment. To be stable (loosely defined as to remain the same) seems to be a good thing, a property that we intuitively value. But what does stability mean in the context of the ecosystem, and how is stability related to other ecosystem properties, particularly productivity and diversity?

Ecologists employ two stability concepts, both derived from physics. **Resistance stability** is the ability of a system to remain the same when subjected to a force, whereas **resilience stability** is its ability to re-establish its former state after change has occurred (Figure 19.4). Consider a boulder at rest. We must apply considerable force to budge it—high resistance. However, once moved, it is hard to return to its original state—low resilience. An ecosystem is not a well-defined object like a boulder, but the analogy helps us distinguish between the ability to resist change and to re-establish after change.

As with a boulder, ecosystem resistance and resilience are often inversely related. A tropical rain forest is the classic example of a highly resistant ecosystem. Its many populations (system components) are linked in a complex web of feeding and other

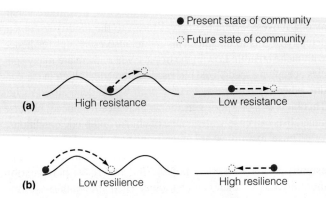

Figure 19.4 Two kinds of stability. **(a)** Resistance is the ability of a system to resist change. **(b)** Resilience is the ability of a system to return to its original state.

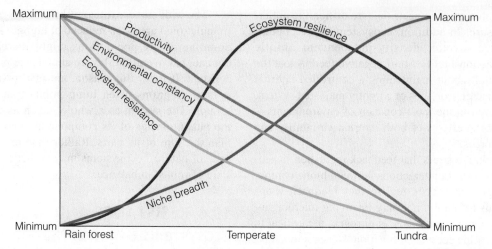

Figure 19.5 Variation in ecosystem parameters and niche breadth along an environmental gradient from tropical to polar. (Adapted from Clapham 1983.)

relations. Its high resistance derives from the combined effect of these many functional links, much as the strength of Velcro derives from its many points of attachment, even if any one link is weak. These between-species links are subject to regulatory negative feedback loops. However, if a critical number are severed, the system can no longer resist change. Its structure falls apart, or is substantively altered. Once altered, the community may have difficulty returning to its original state (low resilience). Many populations must re-establish, but these populations have not evolved with the selective pressure of disturbance, so their ability to bounce back (through positive feedback powering exponential growth) is weak.

In contrast, the component populations of a boreal forest ecosystem change rapidly in response to perturbation—low resistance. Fires, insect outbreaks, climate extremes, and other disturbances are frequent events. Because the boreal forest has fewer species, there are few linkages to re-establish, and their populations undergo large fluxes. Negative feedback operates between the component populations, but is less finely tuned. Instead, evolution has selected for the ability of boreal species to bounce back via positive feedback—high resilience.

This dichotomy between resilience and resistance can be depicted along a gradient from the tropics to the tundra (Figure 19.5). When the environment becomes even more harsh, as one moves beyond the boreal forest to the tundra, resilience actually begins to decline. This model also correlates resilience with niche breadth. Low in the tropics, where there are many species occupying specialized niches, niche breadth is high in the tundra, where there are few species, each with a generalized niche. This high proportion of generalists increases resilience, because generalists can cope with a range of abiotic and biotic conditions.

These examples are oversimplified. In a boreal forest, some biotic components are more resistant than others. Once established, black spruce populations can resist perturbation more effectively than can jack pine. Their individuals are longer lived, and their preference for boggy sites means they are less subject to fire, although infrequent severe fires will destroy them. By the same token, not all biotic components of a tropi-

cal rain forest are robust in the face of disturbance. As well as such within-system differences, many ecosystems are intermediate in stability, or more resistant or resilient to some stresses than others. Despite such simplifications, resilience and resistance stability are useful concepts, particularly in reference to *alternative stable states* (see Section 19.9).

So far, we have discussed stability only in reference to populations, which affect the stability of ecosystem structure. Both kinds of stability also apply to ecosystem functions, such as productivity or decomposition. In this case, deviation from the normal rate is a measure of the resistance of a function, whereas the time required for recovery of the function is a measure of its resilience (Figure 19.6). As the focus is function

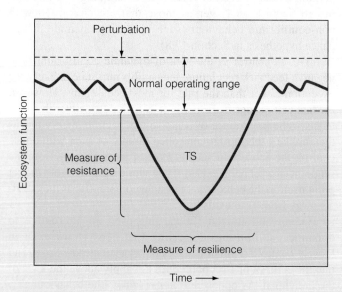

Figure 19.6 Resistance and resilience stability applied to ecosystem function. When a perturbation causes a function to deviate from its normal range, the amount of variation on the *y*-axis is a measure of its resistance, whereas the time to recover to the normal range on the *x*-axis is a measure of its resilience.

(Adapted from Odum and Barrett 2005.)

rather than structure, different populations would likely be present at different stages of recovery.

The engineering concept of resilience as the time needed for a system to re-establish a former state has limited applicability to systems that may occupy alternative stable states. Ecologists are developing a new concept of resilience that considers the trajectory of the system as well as its end state(s) (Walker et al. 2004), making it more relevant to a system in a state of homeorhesis rather than homeostasis (O'Neill et al. 1986).

19.5 DIVERSITY: Ecologists Interpret Diversity on Multiple Scales

The relationship between the stability of an ecosystem and its other properties is at the centre of a long-standing debate. This debate is not merely theoretical; it has profound implications for agroecology and sustainable forestry, both of which aim to manage landscapes in ways that are both productive and sustainable (see Sections 26.7 and 26.9). If, as many believe, diversity is at once the driver and the result of ecosystem stability, then it is key to the sustainability of natural and human-altered systems alike. Yet this debate is plagued by many issues: *What is the link between diversity and stability? Are some kinds of diversity more critical for stability than others? If diversity and stability are correlated in natural systems (positively or negatively), does diversity drive stability, or vice versa? Are diversity and complexity interchangeable? How are diversity and stability related to productivity? How can we test the relationship between these and other ecosystem properties in real and model systems?*

We first review the diversity concept. Like stability and productivity, diversity is an *emergent property*—a trait that is characteristic of the system as a whole. We consider global (macroscale) diversity patterns in Section 27.2, but quantifying diversity at any scale is hindered by the task of defining ecosystem boundaries, which shift over time.

Because the ecosystem as a whole exists at many scales, the modern concept of **biodiversity** (Harper and Hawksworth 1995) incorporates any or all of the following:

1. *Genetic diversity* (variation in genotypes within and among populations; see Section 5.4) is relevant to the stability debate because of its role as a buffer against abiotic change. The more genetically diverse a population, the greater its resistance stability is likely to be as an ecosystem component. Genetic diversity can be treated as a property of populations or of the ecosystem as a whole.

2. *Species diversity* is local variation in species composition, also known as **alpha diversity**. For example, two jack pine forests might differ in stability because they differ in alpha diversity. It may be quantified by diversity indices, which incorporate both species richness and species evenness (see Section 16.1), but richness alone is often used.

3. *Regional diversity* is the variation in diversity among ecosystems in a region, such as between a jack pine and a black spruce forest. Called **beta diversity**, it is correlated

with, but not the same as, *habitat diversity*. Different habitats may support different ecosystems, but the same habitat may support different ecosystems depending on temporal factors during succession, chance factors affecting isolation or order of arrival, and human manipulation. Natural habitats are often highly diverse, and the shifting mosaic steady state discussed in Section 18.10 may be critical to stability.

Ultimately, biodiversity represents the information base of natural ecosystems (Vogt et al. 2002), and should enhance both resistance and resilience by providing alternative pathways for energy flow in the event of a disturbance. Traditionally, as discussed in Section 19.6, species diversity has been the major focus of the stability–diversity debate, but increasingly other diversity levels are assuming more prominence. In particular, diversity among populations of the same species is emerging as being of critical importance to maintaining ecosystem function. Analysis of five decades of data from the salmon runs in Bristol Bay, Alaska, reveals that variability is substantially less than it would be if the species consisted of a single homogeneous population rather than the several hundred discrete populations of which it now consists (Schindler et al. 2010). Such a *portfolio effect* (Figge 2004), long recognized as important for different species in a community (Tilman 1996), is just as important among populations, and provides a strong argument for inclusion of within-species diversity in conservation programs.

19.6 STABILITY AND DIVERSITY: Stability May Be Correlated with Diversity

With this terminology in mind, let's start with the conventional wisdom, namely that diversity and stability are inextricably linked, with each promoting the other. Darwin was one of the first to weigh in, arguing that a more diverse grassland should be better able to maintain its productivity and biomass. Charles Elton (1958) was another early proponent of the **diversity–stability (insurance) hypothesis**, which argues that diversity (at all scales) is a buffer against environmental change. Even if the genetic diversity of any one population is insufficient to allow it to resist change or rebound, greater species diversity increases the probability that the ecosystem as a whole will resist change or rebound. Robert Macarthur (1955) stressed the role of productivity, arguing that the more possible pathways for sustaining energy flow, the less likely that the population density of any one species will change in response to altered population densities of other species.

The diversity–stability hypothesis was standard in textbooks until the 1970s, despite a lack of experimental or observational support. Why? An idea—especially if eloquently stated by an eminent scientist—quickly gains acceptance if it makes intuitive sense. Natural ecosystems are diverse. Natural ecosystems have been around for a while. Therefore, natural ecosystems are stable *because* they are diverse. However, although none would deny the diversity of most natural ecosystems, it is logically possible that ecosystems are stable *despite* being diverse, or that their diversity is not related to their stability. Others argue that natural ecosystems are much less stable than we once thought.

A more recent formulation is the **biodiversity–ecosystem functioning (BEF) hypothesis**, proponents of which include Shahid Naeem, David Tilman, and Michel Loreau. The underlying idea is similar to Darwin's, namely that ecosystems with more species are likely to be more stable in ecosystem functions such as productivity. However, BEF shifts the focus from the *causes* of taxonomic diversity to the *effects* of functional diversity (diversity of functional groups). As Hooper et al. (2005) explain, previous formulations of the diversity–stability link were rooted in a community-based notion of diversity as determined by biotic interactions:

S = f(B, A, ε); where S = species richness, B = matrix of biotic interactions, A = abiotic factors, and ε = error

In this view, diversity (again equated with richness) is the passive consequence of biotic interactions among an ecosystem's component populations, against a backdrop of abiotic factors and chance. This view explains the focus on negative feedback loops among populations, rather than between populations and abiotic components.

In contrast, BEF, as an ecosystem-based formulation, turns the equation on its head, making ecosystem function (F) a consequence of diversity (Hooper et al. 2005):

F = f(S, A, ε)

This turnabout represents a paradigm shift (Naeem 2002). No longer a passive effect, diversity becomes a driver not only of ecosystem functions such as productivity, but also of abiotic factors, particularly levels of atmospheric gases and soil nutrients. BEF also expands the scope of the diversity–stability link by considering the impact of diversity on the susceptibility of an ecosystem to invasion by non-native species. Moreover, functions other than productivity, such as nutrient and water cycling, are incorporated, as well as feedback loops involving abiotic components. Finally, instead of focusing just on plant diversity, BEF acknowledges the role of species at other trophic levels (including microbes), even when not incorporated experimentally.

Many ground-breaking studies have emerged from the BEF hypothesis (see Section 19.7), which derives its urgency from the rapid rate at which species are disappearing from natural systems. An instance of the growing tendency for cooperative international research networks, BioMERGE (Biotic Mechanisms of Ecosystem Regulation in the Global Environment) is dedicated to integrating taxonomic diversity studies with studies of ecosystem properties and processes. The presumption is that diversity *does* matter for ecosystems, even though the specific effects of changes in diversity will differ among ecosystems at differing spatial and temporal scales. Impacts of diversity losses will reverberate beyond ecosystems themselves to influence *ecosystem services* (see Section 27.12), which are vital not just for human well-being but for sustaining the biosphere.

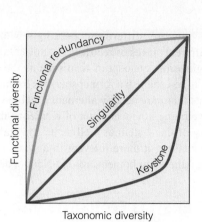

Figure 19.7 Three possible relationships between taxonomic diversity and ecosystem function: functional redundancy (green), singularity (blue), and keystone species (brown).

(Adapted from Naeem and Wright 2003.)

Taxonomic diversity and ecosystem function may be related in many ways (Figure 19.7) (Naeem and Wright 2003):

1. *Functional redundancy:* Ecosystem function increases with diversity but reaches a plateau as essential niches are filled.

2. *Singularity* (also called the *rivet model*): Each species has a unique role and contributes more or less equally to function, which declines after non-native species invade.

3. *Keystone:* Ecosystem function drops off sharply when diversity falls as a result of the loss of keystone species.

Other relations are possible, but in all cases taxonomic diversity is on the *x*-axis as the putative causal factor, rather than on the *y*-axis as the effect of some other factor(s).

Whatever the relationship between diversity and stability, it varies with successional stage. Early stages are inherently unstable, whatever their diversity, which varies with site harshness and abiotic conditions. Despite our tendency to value stability in and of itself, the instability typical of ecosystems in early succession is neither good nor bad—it just is. If pioneer communities were resistant to change, then succession would slow or cease. However, these early stages tend to have high resilience.

19.7 TESTING THE DIVERSITY—STABILITY HYPOTHESIS: Studies Reveal No Consistent Relationship between Diversity and Stability

In the 1970s, ecologists began testing the diversity–stability hypothesis. Robert May (1972) tested randomly assembled webs that varied in richness and the number and strength of their biotic interactions. (May was testing ecosystem complexity as well as richness; communities with the same richness may vary in *connectance*. See Quantifying Ecology 16.2: Quantifying Food Web Structure, p. 344.) In May's tests, increasing complexity and richness decreased rather than increased stability. However, these models included webs in which autotrophs prey upon herbivores, and herbivores upon carnivores. Re-running models after eliminating these webs indicates that logically possible communities are more stable than impossible ones (not surprising), but that stability again declines with complexity.

Results can reverse if one assumes consumers are determined by food supply in *bottom-up* rather than *top-down control*, where consumers are the regulatory components (see Section 16.8). (This ongoing debate about energy flow suffers from the "either-or" problem. In some habitats, or at some times of the year, either bottom-up or top-down control may be more or less likely, even in the same ecosystem.) So, depending on parameters and assumptions, early models supported, disputed, or were inconclusive about the diversity–stability link. Indeed, an early review (Goodman 1975) concluded that diversity is correlated with instability, not stability!

These early models rarely accounted for particular ecosystem components. Yet above a certain minimum diversity, it may not be diversity *per se* that is critical to stability, but rather particular kinds of producers, consumers, and/or interactions. Two ecosystems may have the same diversity, but differ in stability depending on the presence of free-living and/or symbiotic N-fixers, pollinating insects, mycorrhizal fungi, or generalist predators. Certain biotic components may have unique impacts on stability, particularly of ecosystem functions. Models thus require ecologically reasonable species weightings.

Ecologists have also performed field tests of the diversity–stability hypothesis. McNaughton (1977) tested the impact of nutrient addition and grazing on communities of differing diversity. Each factor reduced the diversity of the species-rich but not the species-poor community, suggesting that less rich communities are more resistant. However, diverse communities were more resistant to drought (Frank and McNaughton 1991), suggesting that the link between diversity and stability may vary with the type of stress, species composition, and stability criteria. McNaughton was employing the paradigm in which diversity was assumed to be an effect rather than a cause, and stability was evaluated by retention of diversity rather than maintenance of function.

The BEF hypothesis has sparked a plethora of more recent tests. David Tilman conducted a 7-year study of the functional response of artificial grassland communities of varying richness (1, 2, 4, 8, or 16 species). Species were chosen from a pool of 18 native species belonging to distinct functional groups: 4 C_4 (warm-season) grasses, 4 C_3 (cool-season) grasses, 4 legumes, 4 non-legume forbs, and 2 woody species. Both aboveground productivity and biomass increased with richness, with the trend strengthening and becoming more linear over time (Figure 19.8) (Tilman et al. 2001). These results support the idea of **niche complementarity**: with more species present of differing and complementary niche dimensions, ecosystems optimize overall resource use. In reply to the criticism (Huston 1994) that diverse test communities are more productive because they are more likely to contain one or more inherently more productive species—the **sampling effect**—Tilman et al. (2001) found that the high-diversity mixes consistently outperformed the monocultures, even of productive C_4 species.

How do Tilman's results relate to the models depicted in Figure 19.7? They most resemble the *singularity (rivet) model*,

wherein species make unique contributions to ecosystem function. However, since Tilman's maximal richness of 16 is far below that of native grasslands, whether ecosystem function plateaus at higher richness cannot be evaluated. The results are thus also compatible with the *functional redundancy model* (Walker 1992; see Section 16.8), wherein ecosystem function increases steeply at low richness, as more functional types are included, but saturates (plateaus) at higher richness. As more species are added, they have little or no incremental effect if they belong to a functional group that is already present. In this view, functional types are more important than species number. Diversity still represents information, but diverse communities are more likely to have many copies of the same information, with many species occupying similar niches. Additional species above the threshold can be lost without significant impact on function.

Although *redundant* has a negative connotation, Walker's hypothesis does not imply that "extra" species are unimportant. In fact, redundancy can be interpreted as supporting, rather than refuting, the value of diversity. The presence of a species that is redundant under one set of abiotic factors may become critical when abiotic factors change. Preserving overlap in functional types thus allows for different species to take up the slack under varying environments, better maintaining ecosystem function. Like Macarthur's idea of multiple energy pathways, functional redundancy has the potential to play an insurance or buffer role that enhances ecosystem resistance and/or resilience.

Studies of the impact of diversity (usually treated as richness) in many ecosystems and habitats support a growing consensus that diversity helps maintain ecosystem function (Naeem 2002). In particular, loss of biodiversity tends to depress productivity. The link to stability is more problematic. Pfisterer and Schmid (2002) tested the effect of drought on artificial grasslands of 1, 2, 4, 8, or 32 species (twice Tilman's maximal richness). As in Tilman's study, species-poor communities produced less biomass, but (in conflict with BEF) were more resistant to drought (Figure 19.9, p. 418). Pfisterer and Schmid (2002) argue

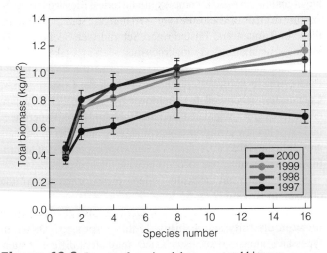

Figure 19.8 Impact of species richness on total biomass production of artificial grasslands. Values are means ± standard error.

(Adapted from Tilman et al. 2001.)

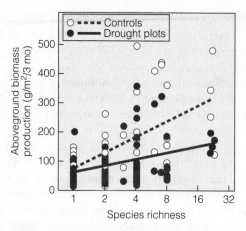

Figure 19.9 Impact of species richness (log scale) on aboveground productivity in control (o) and drought-stressed (•) plots (*n* = 120) in artificial grasslands. Drought reduced the slope of the diversity–production relationship evident in the control plots.

(Adapted from Pfisterer and Schmid 2002.)

that the very factor that makes diverse systems more productive—niche complementarity—lowers their resistance.

Not all experiments are in a field setting. Experimental set-ups of varying scale are often used to study ecosystem dynamics. A **microcosm** is a small, self-contained system, such as an aquarium or a bottle. Artificial microcosms are useful for illustrative purposes and for tests with small organisms in a laboratory setting. Exponential algal growth, for instance, can be easily demonstrated and manipulated, using differing inputs of nutrients and light. Natural microcosms, such as the pools inside pitcher plants or tropical bromeliads (see Research in Ecology: The Large and the Small of It: Ecosystem Experiments of Contrasting Scale) share the advantages of their artificial counterparts, but with the added virtue of realism.

A **mesocosm** is a larger artificial system, such as a field experimental enclosure. As well as enhancing realism by allowing factors such as light, temperature, and nutrients to fluctuate naturally, mesocosms allow experimentation with organisms that are larger and/or have more complex life histories. Mesocosms (with varying degrees of enclosure) are used in many studies, including those of Tilman, and Pfisterer and Schmid (see Section 19.7). Ecologists can also study **macrocosms**, such as a lake or forest. There are many famous examples, such as the Experimental Lakes Area in northwestern Ontario (which the Government of Canada has scheduled for shutdown by March 2014), where entire lakes are manipulated, and the Hubbard Brook Experimental Forest in New Hampshire, where logging and other treatments are applied to intact ecosystems. Even when experimental manipulation is infeasible, macrocosms provide vital baseline data.

No matter what the scale—micro, meso, or macro—or the ecosystem investigated, findings from models, observational studies, and experiments fail to yield a consistent relationship between diversity and stability. Is this surprising? Different types and intensities of disturbances may elicit different interactions between diversity and stability. Perhaps the most one can say from the current evidence regarding the link between diversity and stability is: *It depends!* Ecosystem variability and complexity make a single, one-size-fits-all answer unlikely.

19.8 PRODUCTIVITY AND DIVERSITY: Increasing Resources Has Contrasting Effects on Terrestrial Versus Aquatic Diversity

If the link between diversity and stability is controversial, so too is that between diversity and productivity. As we saw in Section 19.7, there is some consensus that diversity supports ecosystem function, including productivity. Yet there are noteworthy counterexamples: (1) low-diversity systems with high productivity, such as marshes, and (2) high-diversity systems with low productivity, such as the open ocean. Resource supply, not diversity, often limits productivity in both natural and agricultural ecosystems. Other factors can override the impact of resources on productivity, notably abiotic and biotic conditions and hazards, the impacts of which may be affected by diversity. Consider a hurricane or a tsunami. If a species-poor community does not survive whereas a diverse one does because of the greater chance that at least some of its member species can tolerate the hazard, then the diverse community will have higher resistance and be better able to support some productivity, however diminished.

Whereas increasing diversity as a treatment variable often increases productivity, as in both the Tilman and the Pfisterer and Schmid studies in Section 19.7, increasing nutrient availability (which typically increases productivity) often causes diversity to decline. This seems counterintuitive. After all, as discussed in Section 6.12, nutrients often limit photosynthesis and growth. Surely diversity should increase, not decline, when conditions become more favourable for more species. However, by reducing growth, low nutrient levels weaken competitive displacement, allowing more species to coexist. In contrast, adding nutrients intensifies competitive displacement. Diversity falls as faster-growing, taller species make light less available for slower, lower-growing species.

Field studies, both observational and experimental, show this inverse relation between nutrient levels and terrestrial plant diversity. Species richness and soil fertility were negatively correlated in 46 tropical sites in Costa Rica (Figure 19.10). Typically, tropical soils are highly weathered and infertile (see Section 4.9), and rain forests in low-nutrient sites contain more

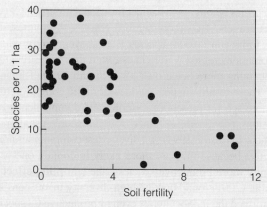

Figure 19.10 Relationship between tree species richness (per 0.1 ha) and soil fertility (index value) for 46 forests in Costa Rica.

(Adapted from Huston 1980.)

RESEARCH IN ECOLOGY   The Large and the Small of It: Ecosystem Experiments of Contrasting Scale

Diane Srivastava, University of British Columbia.

One of the themes of Chapter 1 was that ecologists tackle questions from many angles, using a variety of techniques. Nowhere is this more true than in systems ecology, where studies range from the miniaturized world of the microcosm to the Experimental Lakes Area, where entire lakes become laboratories. As disparate as these approaches are in test organisms, timeframe, and procedures, they approach similar questions from different angles and deepen our understanding of complex and controversial topics.

Consider the diversity–stability hypothesis. If two dissimilar studies generate similar results, the robustness of the hypothesis is enhanced. But if their results conflict, can we conclude that one is right and the other wrong? Or that different relationships apply among different ecosystems and/or in different conditions or scales? Conflicting results are not uncommon in any research area, even when using a similar approach. In the case of microcosms versus mesocosms, there are additional issues: *Can we scale up the results from microcosms to the real world, or are results from mesocosms inherently more realistic, with macrocosms the most realistic of all, despite their replication problems?* We consider this dilemma by highlighting the work of two prominent Canadian scientists.

Diane Srivastava uses the aquatic insect community inhabiting tropical bromeliads to study the relationship between diversity and ecosystem function. In contrast to an artificial microcosm, bromeliads are miniature natural ecosystems. As well as all the advantages of artificial microcosms—ease of replication, use of small organisms with quick generation times, well-defined boundaries that act as natural barriers for confounding factors such as dispersal—natural microcosms possess a realism that increases the applicability of their findings to less constrained systems. Srivastava manipulates treatment factors in artificial bromeliads, thereby expanding the scope of her research.

Bromeliads are non-parasitic *epiphytes* (plants that grow on other plants) that are common in tropical forests (Figure 1). A pool of water collects in the centre of their tightly interlocking leaf bases.

Insect larvae inhabit these pools, feeding on the plant detritus that accumulates there. This detritus is also a source of nutrients for the bromeliads. Species richness and trophic complexity of the insect community varies, depending on pool size and geometry. Using artificial bromeliads to which detritus was added, Srivastava tested (1) three levels of trophic diversity: detritivores (litter-feeding invertebrates) only, detritivores and predators, or no insects; (2) three levels of habitat complexity: one, three, or six leaves; and (3) two habitat sizes: small and 50 percent larger.

Larger bromeliads supported both more species and more trophic levels (Srivastava 2006). Pools in large bromeliads are less likely to dry up—a crucial factor for the top predator, a damselfly with a prolonged aquatic larval stage. However, larger and more complex bromeliad habitats decreased detrital processing by reducing foraging efficiency not only of the detritivores but also of the predators, presumably because more refuges were available for prey. Overall, more detritivores survived in large, complex bromeliads, but their per capita detrital processing declined. Habitat differences thus affected a vital ecosystem function—detrital processing—both directly and indirectly, via predation.

Catherine Potvin, McGill University.

Trophic diversity also affected another key system function—N cycling. Using detritus labelled with radioactive N in natural bromeliads, Ngai and Srivastava (2006) showed that N supply increased only when both detritivores and predators were present. Predators thus have an indirect facilitative impact on the bromeliads, which are N-limited. By consuming detritivores, predators convert some of the N in prey tissues into fecal pellets that are decomposed or leached into the pool. By facilitating nutrient supply, trophic diversity enhances ecosystem function. Yet the question remains whether these microscale processes—occurring in relative isolation from other, similar ecosystems and on short timelines—can be scaled up to ecosystems such as a forest or a lake.

Figure 1 Pool-forming bromeliad (*Guzmania donell-smithii*) in a tropical rain forest in Costa Rica.

continued on page 420

RESEARCH IN ECOLOGY   continued

Catherine Potvin also studies tropical biodiversity, but at different spatial and temporal scales. As part of an international group, she is investigating the impact of diversity in a tropical tree plantation in Sardinilla (Panama). As BEF proponents acknowledge, it is imperative to study systems other than grasslands, particularly in the tropics where loss of forest diversity is a major concern. However, experiments in tree plantations pose logistical problems (Potvin and Gotelli 2008). Large plots are needed to accommodate species mixtures. So whereas Tilman's plots were 9 m × 9 m (quite large for a grassland study), Potvin's were 45 m × 45 m, risking considerable within-plot environmental variability as well as making adequate replication difficult. (Compare Tilman's 29 to 39 replicates with Potvin and Gotelli's 6.) Even with this plot size, and some 230 trees per plot, the maximum species richness was 6, as opposed to 16 for Tilman and 32 for Pfisterer and Schmid. This low maximal richness reduces the study's realism.

After five years, the three-species mixtures yielded 30 to 58 percent more than the monocultures, with the six-species mixtures showing no further gains (Figure 2). Diversity enhanced the performance of individuals but did not affect survivorship; individual trees died, but mortality rates did not correlate with diversity (Potvin and Gotelli 2008). Environmental heterogeneity within and among plots explained more of the variation in productivity than did diversity, suggesting that abiotic factors can confound diversity effects in BEF experiments (Healy et al. 2008). Species identity explained much of the diversity effect, with one species, Mexican cedar (*Cedrela odorata*), associated with higher productivity. These findings highlight the need for more replication in heterogeneous sites and suggest that Huston's sampling effect is a real concern, requiring more species combinations.

The Sardinilla project's focus on productivity reflects an interest in the viability of mixed-species plantations as opposed to monocultures. But in natural ecosystems, productivity is not the only ecologically vital function. Researchers also reported a positive although inconsistent net effect of diversity on N- and P-use efficiency. Abiotic factors were again significant, as were species-specific effects (Zeugin et al. 2010). Soil CO_2 flux, in contrast,

varied with soil temperature but not with diversity, indicating that abiotic factors can outweigh diversity effects.

How should we compare the Srivastava and Potvin approaches? The bromeliad system, as an intact functioning ecosystem, has the advantage of realism, albeit on a much finer scale than most ecosystems. By incorporating trophic complexity, it considers the feedback loops so essential to system regulation. However, the bromeliad microcosm only uses plants as habitat (and indirectly as the source of detritus), rather than in relation to a key ecosystem function: primary productivity. In contrast, the Potvin approach studies an ecosystem that is of vital concern from the perspective of its contribution to biosphere function as well as its diversity. Yet, to make experimentation feasible, the richness tested is not even close to that of natural rain forests, in which 1 ha can contain over 200 tree species. When Ruiz-Jaen and Potvin (2011) compared plantation findings with trends in a nearby natural forest, stability (using carbon storage as the criterion) was not correlated with diversity. They concluded that natural forest functional stability could not be predicted from plantation data. Their results illustrate that sampling effects are real and that abiotic heterogeneity (whether as a determining factor of diversity or a consequence, or both) must be considered.

On balance, neither microcosms nor mesocosms are superior to the other. As with all things ecological, there are trade-offs in their theory and application. Ultimately, both approaches increase our understanding of the importance of diversity in ecosystems at multiple scales of space and time.

Bibliography

Healy, C., N. J. Gotelli, and C. Potvin. 2008. Partitioning the effects of biodiversity and environmental heterogeneity for productivity and mortality in a tropical tree plantation. *Journal of Ecology* 96:903–913.

Ngai, J. T., and D. S. Srivastava. 2006. Predators accelerate nutrient cycling in a bromeliad ecosystem. *Science* 314(10):916.

Potvin, C., and N. J. Gotelli. 2008. Biodiversity enhances individual performance but does not affect survivorship in tropical trees. *Ecology Letters* 11:217–223.

Ruiz-Jaen, M. C., and C. Potvin. 2011. Can we predict carbon stocks in tropical ecosystems from tree diversity? Comparing species and functional diversity in a plantation and in a natural forest. *New Phytologist* 189:978–997.

Srivastava, D. S. 2006. Habitat structure, trophic structure and ecosystem function: interactive effects in a bromeliad-insect community. *Oecologia* 149:493–504.

Zeugin, F., C. Potvin, J. Jansa, and M. Scherer-Lorenzen. 2010. Is tree diversity an important driver for phosphorus and nitrogen acquisition of a young tropical plantation? *Forest Ecology and Management* 260:1424–1433.

1. The timeframe of the bromeliad experiments was just over a month, whereas the Sardinilla project was just starting to show differences in trends at three years. Why? What general difference between microcosms and mesocosms does this illustrate?

2. If the Sardinilla plots are big enough to hold 230 trees, why is the maximal species richness only six? In replicates of the mixed-species treatments, would the same three or six species always be used? Or would you use replicates of different three- or six-species mixtures?

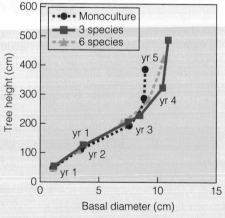

Figure 2 Impact of tree diversity on height (cm) and basal diameter (cm) of surviving trees over 5 years in monoculture (●), three-species (■), and six-species (▲) mixtures in the Sardinilla experiment.

(Adapted from Potvin and Gotelli 2008.)

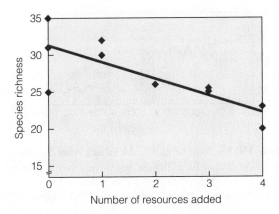

Figure 19.11 Impact of nutrient addition on grassland diversity in the Park Grass experiment (see Figure 1.6).

(Adapted from Silvertown 1980.)

tree species (Huston 1980). Infertile soils reduce growth and support less density and biomass. Species that might dominate in more fertile sites cannot realize their growth potential and so coexist with, rather than exclude, slower-growing species.

Many greenhouse studies support Huston's hypothesis. White mustard (*Sinapis alba*) and cress (*Lepidium sativum*) coexist on infertile soil, but *Sinapis* excludes *Lepidium* on fertile soils (Bazzaz and Harper 1976). Field support comes from the Park Grass experiment, begun at England's Rothamsted Station in 1859 to test the impact of fertilizers on hay production. This study, which is the longest-running ecological study worldwide, treated field plots containing a mix of grasses and forbs with various fertilizer types, amounts, and schedules. Changes in species composition began the second year, ultimately leading to a species-poor community. Unfertilized plots retained their high diversity. Fertilization reduced diversity, with the most fertilized plots occupied by the fewest species (Figure 19.11).

So greenhouse and field experiments indicate that increasing nutrients decreases diversity (both richness and evenness) of terrestrial ecosystems by promoting competitive exclusion. But what processes facilitate this exclusion? Field experiments indicate a shift in the importance of below- versus aboveground competition along a nutrient-supply gradient (Cahill and Caspar 2000). Cahill proposes that competition for belowground resources is *size symmetric* because nutrient uptake is proportional to the size of the root system. In size-symmetric competition, individuals compete in proportion to their size, with larger plants causing a greater decrease in the growth of smaller plants than smaller plants do in the growth of larger plants. In contrast, competition for aboveground resources is *size asymmetric*. Larger plants have a disproportionate advantage in competition for light by shading smaller ones, compounding the differences in their size over time. Thus, any factor that reduces aboveground growth initiates a positive feedback loop that further decreases the plant's chances of attaining a dominant position in the size hierarchy.

When nutrients are scarce, plant growth rate, size, and density decline. Competition occurs primarily belowground and is size symmetric. Competitive exclusion is unlikely, and diversity is maintained. As nutrients increase, so do growth rate, size, and density. Species that can maintain higher rates

of photosynthesis and growth increase disproportionately in size. As these species overtop others, creating a disparity in light availability, competition becomes size asymmetric. Fast-growing, tall species outcompete and eventually displace slower-growing, smaller-stature species, reducing diversity.

This relationship between nutrients and diversity seems to be reversed in aquatic systems. In a review of 44 studies, fertilization increases, not decreases, species richness of autotrophs in freshwater and marine systems (Hillebrand et al. 2000). Why the disparity? In contrast with terrestrial species in fertile soils, competitive exclusion is less common in open water. Rapid changes in nutrients in the water column make it unlikely that any one species will maintain competitive dominance for long. Recall Hutchinson's *paradox of the plankton*—the coexistence of many species with similar niches (see Section 13.8). If the environment changes quickly enough to reverse any competitive advantage (a *stabilizing coexistence mechanism*; see Section 13.13), more species will persist. Another reason relates to Cahill's theory: the single-celled plankton form means that size-asymmetric competition for light, which causes diversity to decline with more nutrients, is less likely in water. Hillebrand (2003) attributes the increased richness of *periphyton* (micro- and macroflora growing on submerged surfaces, including vegetation) in response to fertilization to a greater dominance of filamentous algae. This dominance shift increases the structural complexity of the community, creating more niches. Grazing has an opposing effect, reducing species richness while increasing species evenness.

However, there may be less disparity in the relationship between productivity and diversity in aquatic versus terrestrial systems than once assumed. Huston and Wolverton (2009) question the prevailing view that tropical rain forests are the most productive terrestrial ecosystems, arguing that less diverse ecosystems in temperate mid-latitudes have higher "ecologically relevant productivity." Evidence is growing from both observational and experimental studies that ecosystem size (area) is more critical than productivity in determining diversity, particularly in lakes (Post et al. 2000) The role of ecosystem size highlights the need for macrocosm studies.

19.9 ALTERNATIVE STABLE STATES: Tipping Points May Shift Ecosystems between States

Our discussion so far suggests that there may be no necessary and invariable relationships among the key ecosystem traits of stability, diversity, and productivity. Rather, these relationships are likely to vary with ecosystem type and size, abiotic factors, and seral stage. This conclusion does not mean these properties are unimportant, or that ecologists should stop investigating them via models, observational studies, or experiments. It simply means that we must expect and appreciate the complexity of natural systems and that different outcomes are not only possible, but likely.

Arising from stability theory is the concept of **alternative stable states** (Holling 1973): other, non-transitory states that arise when ecosystems are disturbed past a critical threshold and follow a path that does not restore them to their prior state. This idea may sound like succession, in which disturbance returns an ecosystem to a pioneer stage, from which it progresses to the same climax. Instead, an alternative stable state means that the ecosystem undergoes a phase shift towards a state other than its prior-disturbance state, which it may never regain.

Beisner et al. (2003) illustrate the theory with a simple diagram. Recall from Figure 19.4 that an ecosystem can be depicted as a boulder disturbed from its resting state by a perturbation. Depending on its resistance, it is either easy or hard to budge the ecosystem from this state and, depending on its resilience, easy or hard to re-establish that state. Now consider that the boulder may have two (or more) possible resting states (Figure 19.12). When a perturbation disturbs the system from one position, it might return to the original position—or to an alternative position, depending on chance, or changes in prevailing abiotic factors, or presence or abundance of species. These alternative positions are called "stable states" because they have some properties of persistence, but they are not steady states in the equilibrium sense; in keeping with the concept of *homeorhesis* (see Section 19.3), they could perhaps better be called *alternative pulsing states*.

Are there examples of natural systems occupying alternative stable states after catastrophic regime shifts? After a long period in which vegetative cover underwent a gradual decline, the Sahara region abruptly shifted to desert about 5000 to 6000 years BP (before present) (Demenocai et al. 2000) (Figure 19.13). A more recent example of a state shift involves Caribbean coral reefs, which rely on herbivory by sea urchins to keep algae in check. In 1983, urchin numbers crashed in the Caribbean, shifting reefs to an algae-dominated system that did not regain its former structure even after urchin numbers re-established (Mumby et al. 2007). In the salt marshes along Hudson Bay, overgrazing by lesser snow geese (*Chen caerulescens*) may have triggered a permanent change to an unvegetated state, with little recovery even after geese had been excluded for 15 years (Handa et al. 2002). In Canada's southern boreal forests, the combined

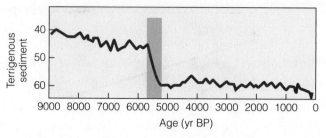

Figure 19.13 Regime shift in the Sahara. After fluctuating around a slow decline for millennia, vegetative cover underwent an abrupt collapse 5000 to 6000 years BP as indicated by deposition of dust in marine sediments at a site near the African coast.

(Adapted from Scheffer and Carpenter 2003, as modified from Demenocai et al. 2000.)

effects of spruce budworm and an increased fire frequency may have converted closed spruce–moss forests to spruce–lichen woodlands (Jasinski and Payette 2005; see Figure 23.35a).

Both the reef and salt marsh examples involve the impact of a single herbivore, but in some cases no one species may be responsible. Much of Great Britain was originally forested, but centuries of human occupation and intense agricultural use have altered conditions such that even when sites are allowed to revert to natural vegetation, the pre-existing ecosystems do not re-establish. Key species that regulate ecosystem function may no longer be present, but—just as likely—abiotic factors of soil or water may have changed substantively. Similarly, removal of large tracts of tropical rain forest may alter abiotic conditions (particularly declining rainfall as a result of the loss of vegetative cover, nutrient losses from biomass removal, and alterations in soil structure) such that rain forest will not re-establish (Wilson and Agnew 1992). Similar instances of alternative stable states exist in aquatic habitats in response to overexploitation of fisheries (see Section 26.10).

In many instances, these alternative states are degraded, or less ecologically valuable than the former states. Ecological value is difficult to assess given the problem of choosing and applying appropriate criteria, but these new states, even if little changed in productivity, are often lower in diversity, as in the case of tropical rain forest regrowth, and have a greater number of non-native, invasive species (see Section 27.7).

What causes an ecosystem to "flip" between alternative stable states? A disturbance or perturbation may trigger the change, but may be insufficient in and of itself. If nothing else differs, even a severe disturbance such as a crown fire does not cause a jack pine forest to switch to an alternative state. However, repeated disturbances that differ from the normal disturbance regime of the region, or severe and unnatural alterations to abiotic conditions, such as high levels of a pollutant, may be the trigger. Losses of, or damage to, vital yet unobtrusive microbial groups in soil or water may cause the shift. Ecologists call these game-changing circumstances **tipping points**—stress thresholds that, once surpassed, send the system on a different path. The tendency of an ecosystem to follow an alternative path once a threshold is passed reflects a parameter related to stability—**persistence**: the ability to remain viable despite repeated stress, oscillations in abiotic and biotic components, and occupation of a non-equilibrium state for prolonged periods (Harrison 1979). At present, there is much concern

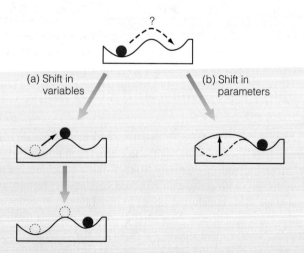

Figure 19.12 Alternative steady states. Ecosystems are thought to shift between stable states as a result of (a) shifts in variables such as population densities in a constant environment or (b) shifts in environmental parameters, as indicated by the transition from the dotted to the solid line.

(Adapted from Beisner et al. 2003.)

that human perturbation of global climate and soils puts many ecosystems at risk of flipping between alternative stable states, the consequences of which are difficult if not impossible to predict.

Can these catastrophic regime shifts be reliably predicted and their causes determined? In many cases, even close monitoring of system variables has not provided reliable cues forecasting regime shifts. Despite intensive ongoing monitoring of coral reefs in the Caribbean, for example, their sudden collapse was unforeseen (Scheffer and Carpenter 2003). After the fact, many causal factors surfaced. Nutrient loading from altered land use had stimulated algal growth, which was kept in check for some time by herbivorous fish. When fish numbers declined from increased fishing, sea urchins assumed their role. Only when a pathogen decimated urchin numbers did the sudden collapse occur. The short-term trigger—urchin decline—obscured the complex mix of causal factors that Scheffer and Carpenter suggest is typical of regime shifts.

Even a seemingly persistent ecosystem can become vulnerable to small incremental changes in conditions, once a threshold has been surpassed. Monitoring system variables may provide subtle early-warning signals of regime shifts, albeit rarely by the magnitude of a particular change. Rather, statistical measures, such as increasing variance and slowing return rates after perturbation, are more reliable signposts, as evidenced by a whole-ecosystem experiment in which top predators were gradually added to a lake (Carpenter et al. 2011). A combination of field observations (with appropriate statistical analysis), experimentation, and complex model simulations is needed to understand (and avoid) transitions between alternative stable states in natural ecosystems (Scheffer and Carpenter 2003).

As this discussion highlights, systems ecology can be approached from two perspectives. For some, the emphasis is on biotic system components. How do population densities and relative abundances change over time? With reference to Figure 19.12, the object itself is the focus; flipping between states is attributed to some change internal to the ecosystem, in its biotic components. The tipping point might be the loss or invasion of one or more species. For others, the emphasis is on abiotic system components. How do air, soil, and water change over time? The curve (physical environment) on which the object sits is then the focus. Has the environment changed sufficiently to flip the ecosystem from one state to another? The tipping point might be a pH change of soil or water, seasonal temperature, exposure to UV radiation, or the level of a pollutant.

Tipping points can also involve a combination of forcing functions involving both biotic and abiotic components. Whatever the tipping point(s) of natural systems, feedback loops and their regulatory effects are critical. The challenge will be, as ecologists and environmental scientists move forward in a cooperative spirit, to identify tipping points and feedback loops that link abiotic and biotic system components, particularly at the macroscale. Consider (1) the soil microbial loop involving bacteria, fungi, and microbivores (see Section 21.9) and (2) the carbon sink activity of the oceans (see Section 22.4). Both involve feedback between abiotic and biotic components, both are becoming more prominent as their importance to the biosphere becomes increasingly evident, and both represent major challenges for systems ecologists in the 21st century.

19.10 GAIA HYPOTHESIS: Microorganisms May Regulate Biosphere Conditions in Ways That Sustain Life

We end this chapter with the most extreme version of systems ecology. The **Gaia hypothesis** (Gaia was the ancient Greek goddess of Earth) is the idea that the entire biosphere is a superorganism. Attributed in the modern age to Lynn Margulis and James Lovelock, this idea has a long intellectual history. There are many formulations of Gaia, which is often considered a metaphor for connectedness rather than a testable hypothesis. Clearly the biosphere is not a organism in any literal sense. Rather, what the hypothesis signifies is that the feedback loops we have been discussing—the regulatory links that operate not only between populations in any given ecosystem but also between biotic and abiotic components of the world's ecosystems considered in their totality—are crucial to maintaining the biosphere. Perhaps the most useful formulation of the Gaia hypothesis maintains that organisms (particularly bacteria and fungi) have evolved in response to the abiotic environment in such a way as to create a complex, self-regulating global system that maintains (yet cannot guarantee) conditions that are favourable for life on Earth (Lovelock 1979).

Supporters of the Gaia hypothesis contend that it is impossible to account for the atmospheric conditions on Earth (notably its high O_2–low CO_2 content), and the moderate pH and temperature of water and soil, without reference to the activities of early forms of life, as well as the ongoing and coordinated activities of plants and microfauna that maintain these buffered conditions. This critical web of cybernetic controls that maintains not a steady state but a homeorhesis of pulsing states is located in the **brown belt**: the zone of soil and water dominated by microorganisms, encircling Earth. As a result of this control system (Gaia), Earth operates as single complex unified system.

The virtue of the Gaia philosophy is not that it is a testable scientific hypothesis—which it is not—but that it stresses the importance of regulatory biotic influences on abiotic conditions. More than a mere metaphor, it emphasizes that ecology isn't just about the big players—large trees, grazing mammals, and large carnivores. Yes, these conspicuous ecosystem components are important, but just as vital are the species (entire kingdoms in some cases) that are less obvious and often downplayed. A common ecological cliché is that "everything is connected to everything else." Although we cannot assume this level of interconnectedness, the fact remains that feedback loops among abiotic and biotic ecosystem components allow a degree of functional interdependence that ecological science is just beginning to appreciate.

The Gaia hypothesis has echoes of the North American aboriginal worldview, which reveres the biosphere as a web supporting all life, not only humans. Much aboriginal art incorporates symbols of biotic and abiotic components nested within others, evoking a systems concept (Figure 19.14, p. 424). Scale is often reversed, with larger elements such as the Sun depicted within organisms. A spirit of connectivity pervades throughout. In this spirit, the Clayoquot Sound UNESCO Biosphere Reserve has as its guiding principle the Nuu-chah-nulth saying "*Hishuk ish ts-awalk*," meaning "everything is one."

Figure 19.14 Much North American aboriginal art embodies a systems perspective by integrating individuals and ecosystems. This image, entitled "Honouring my Spirit Helpers—Baagitchigawag Manitou" is by Métis visual artist Christi Belcourt.

The Gaia hypothesis is an expression of an environmental philosophy that is focused on ecosystems (**ecocentrism**) rather than on organisms (**biocentrism**) or specifically humans (**anthropocentrism**). An early advocate of an ecocentric perspective was Aldo Leopold. Describing his "land ethic" in *A Sand County Almanac* (1949), Leopold argued that human actions are justifiable only insofar as they preserve the integrity and stability of natural ecosystems. Instead of focusing on the extrinsic value of ecosystems in terms of the ecological services they provide for humans, an ecocentric perspective assigns intrinsic value both to other life forms and to ecosystems.

Although such discussions may seem outside the realm of science, our values ultimately determine the research we as a society support, as well as the actions we as a society undertake. Philosophy and science are linked, both theoretically and in practice, and a more sustainable relationship between humans and nature may require a paradigm shift in worldview from an anthropocentric to an ecocentric perspective.

EcologyPlace

Visit EcologyPlace at www.pearsoncanada.ca/ecologyplace to access online resources that complement your textbook, and help you to apply and to review the information in this chapter. EcologyPlace includes

- an eText version of the book
- self-grading quizzes
- glossary flashcards
- and more!

Go to www.pearsoncanada.ca/ecologyplace and follow the registration instructions on the Student Access Code Card included with this text. If your book does not have a Student Access Code Card, you can purchase access to it at www.pearsoncanada.ca/ecologyplace.

SUMMARY

System Defined 19.1

A system is a set of interdependent parts that work together within an environment to perform a function. Biological systems share some traits of human-made systems, but are not the result of design. Because living systems at or below the individual are clearly bounded and under genetic control, they are more tightly integrated than an ecosystem, which is more variable and loosely regulated. An ecosystem is depicted as a series of boxes surrounded by a boundary. Inputs and outputs move between its components and its environment. All ecosystems are open in that they rely on energy inputs.

Feedback 19.2

Feedback occurs when an output from a system component becomes an input for the same or another component. In positive feedback, output reinforces the original input, causing the system to continue changing in the original direction. In negative feedback, output counters the effect of the original input and tends to have a stabilizing effect. Interspecific interactions indicate the potential for feedback regulation in ecosystems.

Homeorhesis 19.3

In human-made systems, negative feedback loops allow the system to operate within a narrow range around a set point. Many physiological processes maintain a dynamic equilibrium (homeostasis). Lacking a set point, ecosystems often exist in pulsing states known as homeorhesis. Nature is best described as a delicate imbalance.

Resistance and Resilience 19.4

Ecosystems differ in their ability to resist change (resistance) or re-establish their prior state (resilience). Resistance and resilience may be inversely related; an ecosystem that is highly resistant, such as a tropical rain forest, tends to have limited resilience. This relationship may reflect the amount and type of feedback. Populations in low-diversity ecosystems have been selected for resilience. Their negative feedback loops are less specialized and finely tuned than those of high-diversity, resistant ecosystems.

Diversity 19.5

Biodiversity incorporates genetic, species, and regional diversity. Quantifying diversity patterns is hindered by difficulties in defining ecosystem boundaries and changes in diversity during succession. However defined, biodiversity represents the information base of a population, community, or ecosystem.

Stability and Diversity 19.6

Ecologists initially assumed that more diverse ecosystems are more resistant because there are more pathways along which energy can travel, and more likelihood that some species will survive perturbation. A more recent formulation, the biodiversity–ecosystem function hypothesis, shifts the focus from taxonomic to functional diversity. Ecologists propose different relationships between diversity andecosystem function, including functional redundancy and singularity.

Testing the Diversity–Stability Hypothesis 19.7

Early mathematical models had varying results, but did not generally support the diversity–stability hypothesis. Experiments conducted with a range of organisms in different ecosystems support the general consensus that increasing diversity results in increased productivity. Niche complementarity may be responsible, although functional redundancy may occur. The impacts of diversity on stability are less clear cut. Experimental approaches include microcosms (artificial and natural), mesocosms, and macrocosms.

Productivity and Diversity 19.8

Experiments indicate that increasing diversity increases productivity, but stimulating productivity by increasing nutrients often reduces diversity. In terrestrial systems, increasing soil nutrients may alleviate size-symmetric belowground competition while intensifying size-asymmetric aboveground competition. Faster shoot growth leads to positive feedback that may competitively exclude slower-growing species. In aquatic habitats, competitive exclusion may be less likely, and increased productivity may increase richness by providing more biotic heterogeneity.

Alternative Stable States 19.9

Ecosystems may exist in more than one configuration, depending on the scale and frequency of perturbation and changes in prevailing abiotic factors. Tipping points may cause ecosystems to shift between alternative states. Ecosystems vary in persistence, but once such a shift has occurred, it may be difficult to regain the former state.

Gaia Hypothesis 19.10

The Gaia hypothesis proposes that Earth behaves as an integrated cybernetic system, analogous to a superorganism. Organisms (particularly microbes) have evolved in response to the abiotic environment to create a complex, self-regulating global system that maintains conditions favourable for life on Earth. An ecocentric perspective stresses the intrinsic value of ecosystems, independent of the services they provide to humans.

KEY TERMS

alpha diversity	brown belt	Gaia hypothesis	niche complementarity	resistance stability
alternative stable state	cybernetics	homeorhesis	output	sampling effect
anthropocentrism	diversity–stability	input	persistence	system
beta diversity	(insurance)	macrocosm	positive (reinforcing)	tipping point
biocentrism	hypothesis	mesocosm	feedback	
biodiversity	dynamic equilibrium	microcosm	pulsing states	
biodiversity–ecosystem	ecocentrism	negative (correcting)	resilience stability	
functioning hypothesis	feedback	feedback		

STUDY QUESTIONS

1. Compare a non-living system such as a computer with a living system such as an individual human.
2. Living systems differ in significant ways at levels above that of the organism. In terms of its system properties, how does an ecosystem differ from a cell?
3. Distinguish between *negative* and *positive feedback*. Explain the two types of positive feedback, giving ecological examples.
4. How does negative feedback regulate population levels within an ecosystem? How might negative feedback regulate atmospheric O_2 levels?
5. Distinguish between *homeostasis* and *homeorhesis*. How does this comparison relate to the idea of the "balance of nature"?
6. Which is more stable—a tropical rain forest or the tundra—and why? How do the different kinds of stability relate to the niche width of the component populations?
7. Why is genetic diversity important for ecosystem biodiversity?
8. Compare the BEF hypothesis with earlier formulations of the diversity–stability hypothesis.
9. Compare the relationship between diversity and ecosystem function proposed by the *functional redundancy* and *singularity hypotheses*.
10. Discuss the trade-offs (advantages and drawbacks) of experimental tests of the BEF hypothesis using natural microcosms, field mesocosms, and macrocosms.
11. According to Cahill, why is aboveground competition for light more likely than belowground competition for nutrients to result in competitive exclusion? How does this trend relate to the impact of increasing nutrient levels on grassland diversity?
12. Define *tipping point*, and explain its significance in systems ecology.
13. Explain the *Gaia hypothesis* and discuss its value from a scientific perspective. How could it contribute to a changing human attitude towards the biosphere?
14. Investigate the history of Biosphere 2. What lessons does it suggest about human ability to "manage" the biosphere?

FURTHER READINGS

Clapham Jr., W. B. 1983. *Natural ecosystems*. New York: Macmillan.
Excellent treatment of feedback.

Golley, F. B. 1994. *A history of the ecosystem concept in ecology*. New Haven, CT: Yale University Press.
Describes the development of the study of ecosystems.

Leopold, A. 1949. *A Sand County Almanac*. New York: Oxford University Press.
Credited with stimulating the birth of the environmental movement; along with Carson's *Silent Spring*, one of the most influential works in ecology.

Lovelock, J. 2000. (1979.) *Gaia: A new look at life on earth* (3rd ed.). Oxford: Oxford University Press.
Seminal statement of the Gaia hypothesis.

Naeem, S. 2002. Ecosystem consequences of biodiversity loss: the evolution of a paradigm. *Ecology* 83:1537–1552.
Insightful article concerning the paradigm shift in the study of diversity from a community to an ecosystem perspective, written by one of the leading BEF researchers.

Naess, A. 1973. The shallow and the deep, long-range ecology movement. *Inquiry* 16:95–100.
Written by a major proponent of the "deep ecology" movement, this paper provides the philosophical accompaniment to the Gaia hypothesis.

Schindler, D. W. 1998. Replication versus realism: The need for ecosystem scale experiments. *Ecosystems* 1:323–334.
Among the best-known of Canadian ecologists, Schindler considers the micro/meso/macrocosm issue, making a strong case for macrocosm studies.

ECOSYSTEM ENERGETICS

A Rocky Mountain alpine tundra ecosystem carpeted with cotton grass (*Eriophorum angustifolium*) in full bloom.

The sunlight that floods Earth is the ultimate source of energy for powering the biosphere. When photons of solar energy reach air, land, and water, some are transformed into heat energy that warms Earth and its atmosphere (see Chapter 2), drives the water cycle (see Chapter 3), and generates currents of air and water (see Chapters 2 and 3). A small portion of these photons are intercepted by plants and other photosynthetic organisms, which transform their radiant energy into chemical energy (see Chapter 6). With the exception of the small number of organisms that rely on chemosynthesis, solar energy stored in the covalent bonds of carbohydrates and other organic compounds is the source of energy for other organisms. The story of energy in ecosystems traces the path of carbon in living and dead tissues of plants, animals, fungi, and microorganisms.

All ecological processes involve energy transfer. In this chapter, we explore the pathways, efficiencies, and constraints that characterize energy flow through ecosystems. But first, we examine the physical laws that govern energy flow in living and non-living systems alike.

20.1 LAWS OF THERMODYNAMICS: Total Energy Remains Constant but Some Useful Energy Is Lost with Every Energy Transfer

Energy exists in two forms. **Potential energy** is stored energy or energy at rest; it is available for performing work but is not doing work at that particular time. **Kinetic energy** is energy in motion; it performs work at the expense of potential energy. In physics, "work" is force acting through a distance, but in biological terms, work can be thought of as making something happen, whether that be physical work, such as movement of an organism, a muscle, or a molecule inside a cell, or chemical work, such as reordering of matter in a chemical reaction.

Two laws govern energy expenditure and storage. The **first law of thermodynamics** states that energy (and matter, which Einstein established is a form of energy) cannot be created or destroyed. It may pass from one place to another, change form, or act upon matter, but whatever the transfer or transformation, no gain or loss in total energy occurs. When wood burns, the potential energy lost from the molecular bonds equals the kinetic energy released as heat.

This law also applies to chemical reactions in living organisms. If a reaction results in a loss of potential energy, it is spontaneous or *exergonic*. In contrast, *endergonic reactions* require an energy input to proceed. (Many but not all exergonic reactions release heat, i.e., are *exothermic*, and many but not all endergonic reactions absorb heat, i.e., are *endothermic*.) In photosynthesis, for example, the molecules of the sugar products store more energy than the reactants. The extra energy is acquired from solar radiation harnessed by chlorophyll (or other photosynthetic pigments in algae or bacteria), with no gain or loss in total energy. However, although the total energy remains constant, much of the potential energy in exergonic reactions such as burning wood or mitochondrial respiration degrades into a form of energy that is incapable of doing further work—**entropy**. Entropy is associated with transformation of matter to a disordered state.

Entropy is integral to the **second law of thermodynamics**, which states that with every energy transfer, some useful energy is lost—entropy increases. When coal is burned in a boiler, some energy creates steam, and some is dissipated as heat. The same thing happens to energy in ecosystems. As energy is transferred between organisms in food consumption, a portion is stored as energy in living tissue, but much is dissipated as heat—entropy increases. Thermodynamically, entropy is associated with waste heat, but we can also think of it as *disorder*.

At first glance, biological systems seem to defy the second law. Surely life produces order out of disorder, decreasing rather than increasing entropy. The explanation is simple: the second law applies to closed systems in which no energy or matter is exchanged with the surroundings. Over time, closed systems tend towards maximum entropy until eventually no energy is available to do work, and disorder triumphs. But living systems are open systems that rely on continuing inputs of solar energy to counteract entropy— at least in the short term. Upon death, when an organism stops processing energy, its body systems can no longer fend off the forces of entropy. Similarly, when the universe is considered as a closed system, even the biosphere is subject to the second law. So when the Sun dies, entropy—disorder—ultimately rules.

20.2 PRIMARY PRODUCTIVITY: Autotrophs Fix Energy in Photosynthesis

Energy flow through an ecosystem starts with the harnessing of sunlight by autotrophs. The rate at which solar energy is converted by photosynthesis (and/or chemosynthesis) into organic compounds is called *primary productivity* because it is the entry point of energy into food webs. **Gross primary productivity (GPP)** is the rate of photosynthetic energy assimilated by autotrophs per unit area per unit time. Like all organisms, autotrophs lose energy in *respiration*. **Net primary productivity (NPP)** is the rate of energy storage by autotrophs in organic matter after their respiratory losses (R) are subtracted:

$$\begin{array}{ccc} \text{Net primary} & \text{Gross primary} & \text{Respiration by} \\ \text{productivity} = & \text{productivity} - & \text{autotrophs} \\ \text{(NPP)} & \text{(GPP)} & \text{(R)} \end{array}$$

Productivity is a rate function, expressed in units of energy per unit area per unit time, for example, $kcal/m^2/yr$. Net primary productivity (from here on *productivity* or NPP) is more often expressed as dry matter, for example, $g/m^2/yr$. The terms *productivity* and *rate of production* are synonymous. Even when *production* is used, a time unit is often implied.

The amount of accumulated organic matter in an area at a given time is the **standing crop biomass** (*SCB* or *B*), expressed as grams per square metre or per hectare (g/m² or g/ha). Productivity generates biomass, but the concepts differ. Productivity is the rate at which organic matter is created, whereas biomass is a "snapshot" of the organic matter present at any given time. Consider a bank account: productivity is analogous to the interest paid on savings, whereas biomass is the amount of funds in the account at any one time. Two ecosystems can have similar productivity but differ greatly in biomass, if their dominant growth forms differ (see Table 20.1). Conversely, two ecosystems of similar biomass can differ greatly in productivity.

Ecologists estimate terrestrial productivity with infrared gas analyzers, which measure the CO_2 content of air. Vegetation is enclosed in a chamber that allows light entry. Conditions inside the chamber are controlled. The NPP estimate is based on the difference in CO_2 content of air entering and leaving the chamber. The greater the difference, the higher the NPP (see Figure 1.7 for a microscale example). This method is the most direct, because it quantifies a photosynthetic reactant. However, it is often infeasible, especially for ecosystem studies.

A simpler method is to estimate productivity using the change in standing crop biomass over a time interval $(t_2 - t_1)$, often one year: $\Delta SCB = SCB(t_2) - SCB(t_1)$, based on sampling in either natural ecosystems or experimental plots (Figure 20.1). There are two possible losses of biomass over the time period: (1) death of plants (*D*) and (2) consumption by heterotrophs (*C*). These losses, which are part of the current year's primary production, are missed by measuring the change in biomass. The corrected estimate of *NPP* is then $NPP = SCB + D + C$. Because it is difficult to harvest and distinguish root biomass from different years, ecologists often include leaves and stems only, in which case the estimate is of *aboveground net primary productivity (ANPP)*.

In plankton-dominated aquatic systems, the simplest method involves measuring the water's chlorophyll-*a* content, which is correlated with net primary productivity (Huston and Wolverton 2009). Another method is the *light/dark bottle method* (Figure 20.2). In a set of clear bottles, water samples incubate for a given time under prevailing conditions. If gross photosynthesis exceeds respiration, O_2 accumulates, yielding an estimate of NPP. Similar samples are incubated in dark bottles with no light entry. Because no photosynthesis can occur, the O_2 content of the dark bottles declines as a result of respiration. Gross primary productivity is estimated by adding the mean O_2 lost in the dark bottles (respiration) to the mean O_2 released in the light bottles (NPP).

The light/dark method has limitations. Some of the respiratory CO_2 losses may be of bacterial origin. Bacteria can be plentiful in aquatic systems, so this amount may be large. It also assumes that respiration in the dark is identical to that in the light, whereas light stimulates respiration. A more sophisticated way to estimate aquatic productivity involves measuring the rate of $^{14}C–CO_2$ uptake after the addition of radioactive carbon.

An important ecosystem parameter is **residence time**: the mean time that energy persists in a trophic level, calculated for primary producers as the energy stored in plant biomass (kJ/m²) divided by NPP (kJ/m²/yr). More commonly, the **biomass accumulation ratio (BAR)** is calculated, using biomass instead of energy units—in essence, standing crop divided by NPP. BAR can be as high as 25 years for forests and as low as 19 days for the open ocean (Whittaker and Likens 1973).

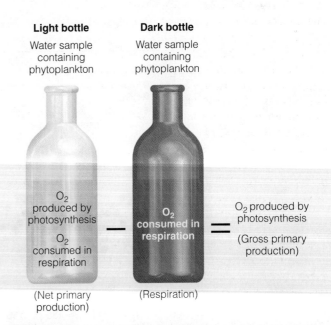

Figure 20.1 Field plots like these at the Cedar Creek Long Term Experimental Research (LTER) site in Minnesota are used to assess the impact of environmental factors on productivity as well as other ecosystem functions.

Figure 20.2 Light/dark bottle method for estimating aquatic productivity. A sample containing phytoplankton is incubated for a prescribed time. GPP is estimated by adding the mean O_2 lost in the dark bottles (respiration) to the mean O_2 released in the light bottles (NPP).

20.3 FACTORS AFFECTING PRIMARY PRODUCTIVITY: Light, Water, Nutrients, and Temperature Affect Productivity

Many abiotic factors affect primary productivity. Typically, terrestrial productivity increases with mean annual precipitation and temperature (Figure 20.3). Annual temperature increases with increasing annual solar radiation, which generates not only warmer temperatures but also a longer **growing season** (number of consecutive days that have favourable temperatures for photosynthesis). Warmer sites typically support faster photosynthesis as well as a longer time over which it occurs (Figure 20.4). So the effect of temperature on productivity is due in part to a seasonal effect as well as to temperature as a condition.

Correlation of productivity with precipitation (a resource) reflects the plant's need to open its stomata to absorb CO_2. When the stomata are open, water is lost in transpiration (see Section 6.3). For the plant to keep its stomata open, its roots must replace this lost water. Soil texture is also a factor (see Section 4.7), but the higher the rainfall, the more water that is typically available. Water availability thus limits both photosynthetic rate and the total amount of leaves (transpiring surface) that can be supported. Given these combined effects, precipitation exerts a strong influence on primary productivity.

Although Figure 20.3 depicts the independent effects of temperature and precipitation on NPP, in reality these two factors interact. Warm temperatures increase evapotranspiration and hence plant water demand. If temperatures are warm but little water is available, productivity decreases with stomatal closure. Conversely, at cool temperatures, productivity is low

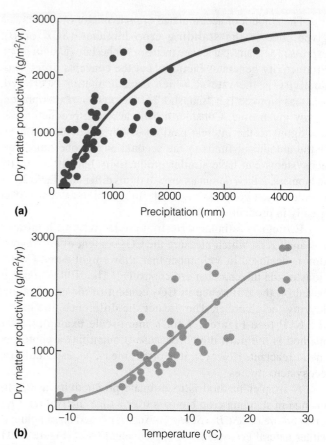

(a)

(b)

Figure 20.3 Net primary productivity of terrestrial ecosystems increases with both **(a)** mean annual precipitation and **(b)** mean annual temperature.

(Adapted from Lieth 1973.)

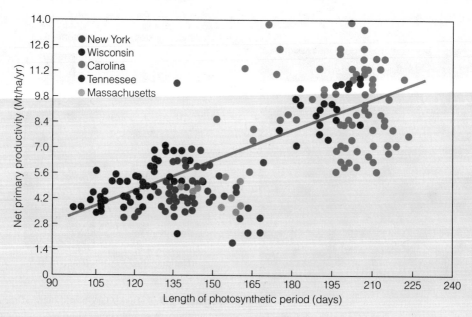

Figure 20.4 NPP increases with length of growing season for deciduous forests in North America. Each point represents a single site.

(Adapted from Lieth 1975.)

regardless of water supply. This interaction between water and temperature—an example of a shift in the *limiting factor* (see Section 5.12)—explains much of the variability in Figure 20.3b. In sites with a mean annual temperature of about 12°C, NPP ranges from 900 to over 2500 g/m^2/yr depending on annual precipitation. Similarly, variation in NPP for sites with similar precipitation reflects differences in temperature. Sites that are both warm and moist are the most productive, as reflected in the relationship between NPP and **actual evapotranspiration (AET)** (Figure 20.5). AET combines surface evaporation and transpiration, and reflects both demand for, and supply of, water. Demand varies with radiation and temperature, supply with precipitation.

These interacting influences generate strong global productivity trends (Figure 20.6 and Table 20.1, p. 432). Equatorial areas, with warm temperatures and abundant rain year-round, have the highest NPP, reflecting the long growing season and large leaf area of rain forests. However, Huston and Wolverton (2009) claim that the high productivity of tropical rain forests may be more apparent than real. In infertile tropical soils, productivity is nutrient limited, and much productivity is inaccessible or indigestible, compared with the more ecologically available productivity in mid-latitude regions. Moving north and south from the tropics, seasonality of rainfall increases as a result of the intertropical convergence zone, reducing the growing season and lowering NPP. Towards the poles, the growing season declines further and mean temperature falls. Temperature and rainfall also decline from coasts to continental interiors, reducing NPP.

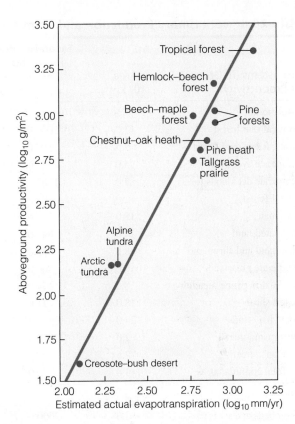

Figure 20.5 Terrestrial NPP increases with actual evapotranspiration, which varies with precipitation and temperature.

(Adapted from MacArthur and Connell 1966.)

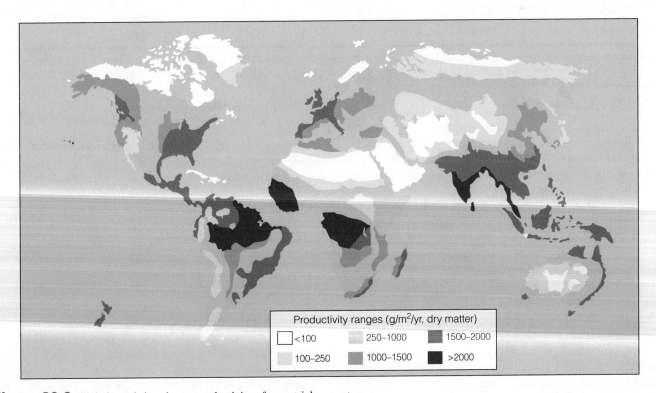

Figure 20.6 Global trends in primary productivity of terrestrial ecosystems.

(Adapted from Golley and Lieth 1972.)

Table 20.1 Net Primary Production and Plant Biomass of World Ecosystems

Ecosystems (in Order of Productivity)	Area (10^6 km²)	Mean Net Primary Production per Unit Area (g/m²/yr)	World Net Primary Production (10^9 Mt/yr)	Mean Biomass per Unit Area (kg/m²)	Relative Net Primary Productivity (g/g/yr)
Continental					
Tropical rain forest	17.0	2000.0/1251*	34.00	44.00	0.045
Tropical seasonal forest	7.5	1500.0	11.30	36.00	0.042
Temperate evergreen forest	5.0	1300.0/779**	6.40	36.00	0.036
Temperate deciduous forest	7.0	1200.0	8.40	30.00	0.040
Boreal forest	12.0	800.0	9.50	20.00	0.040
Savannah	15.0	700.0	10.40	4.00	0.175
Cultivated land	14.0	644.0	9.10	1.10	0.585
Woodland and shrubland	8.0	600.0	4.90	6.80	0.088
Temperate grassland	9.0	500.0	4.40	1.60	0.313
Tundra and alpine meadow	8.0	144.0	1.10	0.67	0.215
Desert shrub	18.0	71.0	1.30	0.67	0.106
Rock, ice, sand	24.0	3.3	0.09	0.02	0.170
Swamp and marsh	2.0	2500.0	4.90	15.00	0.167
Lake and stream	2.5	500.0	1.30	0.02	25.0
Total continental	149.0	720.0	107.09	12.30	0.058
Marine					
Algal beds and reefs	0.6	2000.0	1.10	2.00	1.0
Estuaries	1.4	1800.0	2.40	1.00	1.8
Upwelling zones	0.4	500.0	0.22	0.02	25.0
Continental shelf	26.6	360.0	9.60	0.01	36.0
Open ocean	332.0	127.0	42.00	0.003	42.3
Total marine	361.0	153.0	55.32	0.01	15.3
World total	**510.0**	**320.0**	**162.41**	**3.62**	

Relative net primary productivity (RNPP; g/g/yr; also called the **P:B ratio**) is the inverse of the biomass accumulation ratio (BAR). It is calculated by dividing NPP (column 3) by mean biomass (column 5; values converted to g/m²).*Tropical forest (wet and seasonal combined) NPP; **temperate forest (evergreen and deciduous combined) NPP.

(Adapted from Whittaker 1975 with additional data (*) from Saugier 2001. The 2001 data are not factored into totals.)

Minerals as well as climate may limit primary productivity. Higher nutrient levels increase plant growth (see Section 6.12). The trend of increasing NPP with increasing fertility often explains differences in NPP within and among ecosystems in similar climates. Soil N and aboveground productivity were strongly correlated in different forest types on Blackhawk Island, Wisconsin (Figure 20.7a), with deciduous species dominating high-N sites (Pastor et al. 1984). The same trend occurs in oak savannahs in transitional zones between eastern North American forests and western grasslands (Figure 20.7b) (Reich et al. 2001).

Light often limits photosynthesis of plants on the forest floor, but rarely limits overall community productivity. In contrast, productivity of aquatic systems is often light limited. The depth to which light penetrates (the *photic zone*) is crucial. Photosynthetically active radiation declines exponentially with depth (Figure 20.8). Gross productivity is highest at intermediate PAR, but respiration is far less depth sensitive, so for phytoplankton in deeper water, net photosynthesis

declines as light decreases. At the **compensation depth**, GPP equals respiration, and NPP is zero (see Figure 20.8). The compensation depth, which varies among aquatic systems and with the seasons, corresponds to the *light compensation point* discussed in Section 6.2.

Unless they are transported to the photic zone, nutrients often limit productivity. Based on a survey of 303 enrichment experiments in marine habitats, N stimulates phytoplankton growth the most, followed by iron (Fe) (Downing et al. 1999) (Figure 20.9a). These results reflect mean response, which varied significantly with habitat, particularly for P (Figure 20.9b). In polluted bays and harbours, added P often inhibits growth, likely because it is already abundant from sources such as sewage. In more pristine marine habitats, the positive response to P is nearly as great as for N.

In oceans, shallow coastal waters are the most productive (Figure 20.10, p. 434) because (1) light penetrates more of the water column; (2) nutrients are transported from sediments to surface waters, aided by wave and tidal action; and (3) nutrient inputs enter from terrestrial systems.

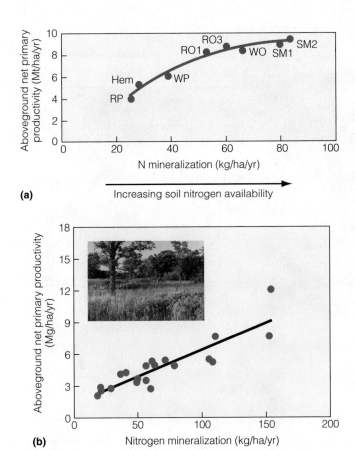

(a)

Increasing soil nitrogen availability

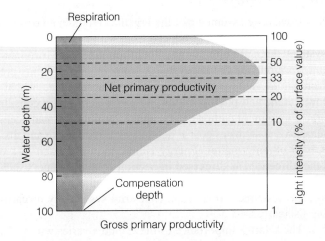

(b)

Figure 20.7 Aboveground NPP increases with N availability (mineralization rate) for **(a)** forests on Blackhawk Island, Wisconsin and **(b)** oak savannahs in Minnesota. ANPP is higher overall in savannahs. Figure 20.7a abbreviations refer to dominant trees: Hem, hemlock; RP, red pine; RO, red oak; WO, white oak; SM, sugar maple; WP, white pine. Go to QUANTIFYit! at **www.pearsoncanada.ca/ ecologyplace** to perform regression.

(Adapted from Pastor et al. 1984 (a) and Reich et al. 2001 (b).)

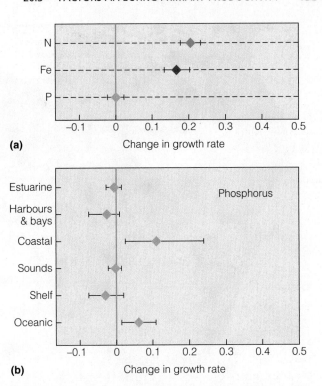

FIGURE 20.9 Effect of nutrient addition on marine phytoplankton. **(a)** Phytoplankton growth in 303 experiments (148 N, 114 P, and 35 Fe). Diamonds represent means; bars 95 percent confidence intervals. The solid orange line represents zero effect. **(b)** Response to P varied among marine habitats. Go to QUANTIFYit! at **www.pearsoncanada.ca/ ecologyplace** to perform confidence intervals and *t*-tests.

(Adapted from Downing et al. 1999.)

Light can also limit the primary productivity of lakes, depending on their depth and clarity. Nutrient effects on lake productivity are well established. Spring total P levels were strongly correlated with summer chlorophyll content in 19 lakes in southern Ontario, supporting a relationship that is characteristic of freshwater lakes worldwide (Figure 20.11, p. 434) (Dillon and Rigler 1974). Similar trends prevail in fertilization studies. Whole-system, long-term experiments conducted in Ontario's Experimental Lakes Area confirm that P determines lake productivity (Schindler 1974; Schindler et al. 2008). Nutrient loading is a serious concern, particularly when high P relative to N shifts dominance to N_2-fixing cyanobacteria (see Ecological Issues: Saving Lake Winnipeg, pp. 532–533).

Hazards also affect productivity. Disturbances such as fire can virtually eliminate primary production in the short term. However, NPP of regenerating post-fire vegetation, after an initial phase in which plants regain biomass, can surpass NPP of mature vegetation, which has high respiratory costs as a result of its high proportion of non-photosynthetic tissue. Less severe disturbances, such as insect outbreaks, often reduce productivity without eliminating it—unless, as with the invasive emerald ash borer (*Agrilus planipennis*; see Section 27.7), the species affected have little resistance.

In aquatic sites, pollutants compromise productivity either directly, by killing primary producers or decreasing their

Figure 20.8 Changes in light and productivity with water depth. GPP declines with depth, whereas respiration is relatively constant. At the compensation depth, GPP equals R (NPP = 0).

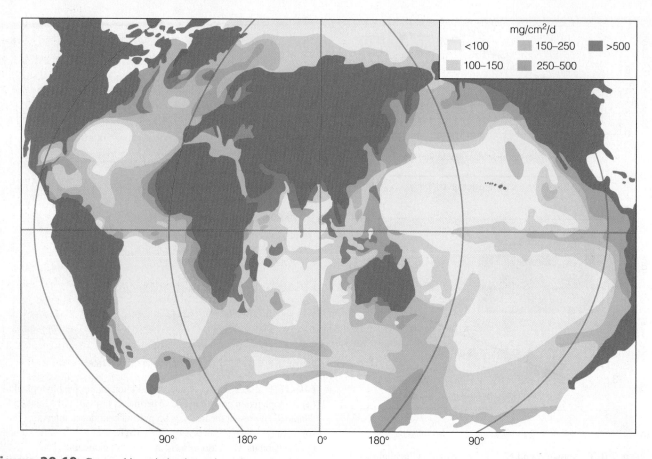

Figure 20.10 Geographic variation in marine primary productivity. The highest productivity is in coastal regions, whereas the lowest is in the open ocean.

growth, or indirectly, by altering food web relations. Recovering productivity after an event such as an oil spill depends on many factors, including residual toxicity. Although no results are yet available in scientific journals, productivity in the vicinity of the 2010 Deepwater Horizon spill seems to be recovering faster than after the 1989 *Exxon Valdez* spill in Alaska, likely reflecting the warmer water in the Gulf of Mexico. However, longer-term toxic effects may occur as a result of the greatly increased presence of polycyclic aromatic hydrocarbons.

20.4 EXTERNAL INPUTS: Aquatic Systems May Have Large Inputs of Organic Matter

So far we have assumed that the organic carbon in an ecosystem is **autochthonous** (produced from within), but in aquatic systems there are often large **allochthonous** inputs (originating from outside the ecosystem). Autochthonous inputs are from photosynthesis by aquatic plants, attached algae in shallow waters, and phytoplankton in open waters. But a considerable amount of carbon enters aquatic systems from adjacent terrestrial systems as coarse debris and as particulate and dissolved organic matter. Energy subsidies can also go in the other direction; as discussed in Section 16.4, consumers such as bears transfer large amounts of organic material from aquatic systems into adjacent terrestrial systems by discarding fish carcasses.

The relative importance of these inputs varies widely. In most marine systems, autochthonous inputs dominate, given large resident phytoplankton populations. In contrast, small streams flowing through forested areas derive most of their

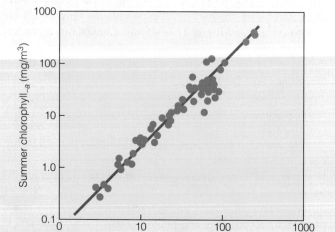

Figure 20.11 Relationship between summer mean chlorophyll content and spring total P concentration for temperate lakes worldwide.

(Adapted from Dillon and Rigler 1974; data from a number of studies.)

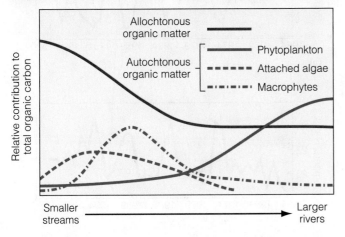

FIGURE 20.12 Relative contributions of allochthonous and autochthonous organic inputs in the transition from a stream to a river ecosystem.

(Adapted from Wetzel 1975.)

organic carbon from plant detritus entering from surrounding forests. Shading from overhanging trees limits available light, preventing any significant contributions by phytoplankton or attached algae. These external inputs support consumer levels beyond those that could be supported by internal sources. As the stream widens, shading becomes limited to the stream margins, and the increase in light supports more NPP by submerged plants, algae, and phytoplankton. The result is a continuum of the relative importance of external and internal carbon in the energy balance of flowing water systems from small headwater streams to large rivers (Figure 20.12).

In lakes, the relative importance of external carbon inputs varies with lake morphology (size and shape) and the surrounding catchment. In large lakes, autochthonous inputs dominate. Allochthonous inputs vary seasonally with the volume of water entering from streams and rivers. In smaller lakes, allochthonous inputs can be much greater.

20.5 ALLOCATION OF PRIMARY PRODUCTION: Abiotic Factors and Growth Form Affect Biomass Allocation

In Section 6.7, we discussed allocation of carbon by plants. Positive feedback is key: the more allocation to photosynthetic relative to non-photosynthetic tissues (leaves versus woody tissue and roots), the more net carbon gain and plant growth. That discussion centred on a single plant, but the pattern of decreas-

ing NPP with declining precipitation reflects changes in carbon allocation by the community as a whole. Reduced moisture increases allocation to roots, reducing leaf area and net carbon gain. Although species in any given ecosystem exhibit various adaptations to microclimate, overall allocation patterns among ecosystems mirror the response of individual plants to abiotic gradients such as moisture. The ratio of belowground to aboveground biomass (**root-to-shoot ratio, R:S**) is a meaningful indicator of the general trend of increased allocation to roots relative to leaves with decreasing precipitation. Estimates range from a low of 0.20 in tropical rain forests to 1.2 for arid shrublands and a high of 4.5 in deserts.

The decline in NPP from moist to dry environments in Figure 20.3a is paralleled by a reduction in standing crop biomass (SCB), the accumulation of NPP over time (Figure 20.13). More productive terrestrial systems tend to support more biomass, but the relationship between SCB and NPP differs among ecosystems depending on their dominant growth forms. Recall from Quantifying Ecology 6.1: Calculating Plant Growth Rate (p. 111) that larger plants generally have greater net carbon gain or absolute growth rate (grams per unit time) than smaller plants. But relative growth rate, or biomass gain per unit of plant mass (gram per gram plant mass per unit time) is often higher for smaller plants. The same trend holds true for collective growth of plants in an ecosystem. Relative net primary productivity (RNPP; see Table 20.1), the ratio of NPP to standing biomass is thus analogous to relative growth rate as expressed for the community, rather than the individual plant.

Comparing RNPP with mean standing biomass in terrestrial systems shows a pattern that is the inverse of that in Figure 20.13. Consider the data in Table 20.1. The NPP of a temperate forest is over twice that of a grassland (1200 vs. 500 g/m²/yr). But if we compare their productivity per unit biomass, grassland RNPP is an order of magnitude higher (0.31 vs. 0.04 g/g/yr). This difference reflects the higher relative growth rate of herbaceous species compared with trees. Trees accumulate large amounts of non-productive biomass (especially wood) over many years, so no matter how rapidly

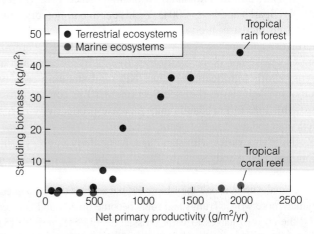

Figure 20.13 Relationship between NPP and standing crop biomass for the ecosystems in Table 20.1. Note the difference between this trend and trends in RNPP, as discussed in the text.

tree leaves photosynthesize, the forest will always have a lower RNPP than a grassland. And the older the forest, the higher its standing crop, and the lower its RNPP.

This trend of higher standing crop with higher NPP does not hold for marine systems. The inverse relation between standing crop and RNPP still applies, but the explanation differs. On land, longevity of the dominants is typically much greater than the period over which NPP is measured (a year). This trend is obviously true for trees, but even grasslands are dominated by perennials, except in pioneer stages of secondary succession when annuals briefly dominate. Such is not the case in most marine ecosystems. Phytoplankton, the dominant producers in open water, are short-lived. Turnover is high, with many generations in a single year. Given this high turnover, standing biomass at any one time is low compared to annual NPP, resulting in an extremely high RNPP value for the open ocean (42 g/g/yr) compared to terrestrial systems. Only marine ecosystems such as kelp forests and coral reefs accumulate enough biomass to have a substantially lower RNPP, though their NPP is much higher than that of the open ocean.

20.6 TEMPORAL VARIATION: Primary Productivity Varies Seasonally and Yearly

For any ecosystem, primary production varies over time and with community age. Both photosynthesis and growth are directly influenced by seasonal variations in abiotic factors. Regions with cold winters or dry seasons have a dormant period when primary productivity ceases entirely. Only in wet tropical regions, where conditions are favourable for plant growth year-round, is there minimal seasonal variation in primary productivity.

Year-to-year variation in primary productivity often reflects climatic variation. In the Park Grass experiment at the Rothamsted Experimental Station in England (see Section 19.8 and Figure 1.6), significant ups and downs have occurred since 1856, as measured with standard methods since 1965 (Figure 20.14) (Sparks and Potts 2004). Yields decline in hot, dry summers and increase in cool, moist years. Disturbances such as herbivory and fire also cause year-to-year variation. Overgrazing of grasslands or defoliation of forests by insects such as gypsy moth significantly reduce NPP and standing crop by removing leaf tissue (but see Section 14.10 re compensatory growth). Fire in grasslands may increase productivity in wet years but reduce it in dry years.

NPP also varies with stand age, particularly in ecosystems dominated by woody growth. Trees and shrubs can live a long time, which influences how they allocate energy. Early in life, leaves make up over half their biomass, but as trees age, they accumulate more wood. Trunks and roots become thicker and heavier, and the ratio of leaves to woody tissue declines (Figure 20.15). Eventually, leaves account for less than 5 percent of total tree mass. The production system (leaf mass) that supplies the tree's energy is much less than the non-photosynthetic biomass it supports. So, as

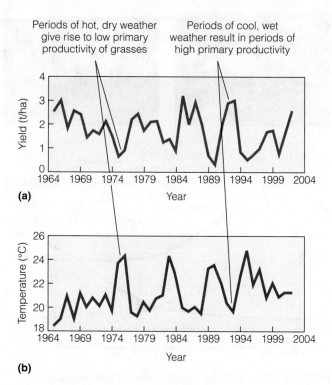

(a)

(b)

Figure 20.14 Seasonal variations in **(a)** productivity and **(b)** maximum summer temperatures in the Park Grass experiment in England's Rothamsted Experimental Station from 1965 to 2004.

(Adapted from Sparks and Potts 2004.)

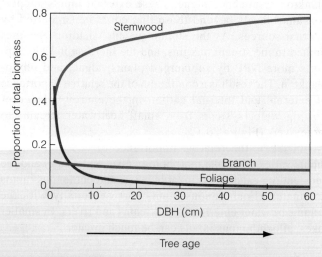

Figure 20.15 Change in proportion of biomass in foliage, branch, and stemwood (trunk) for white oak (*Quercus alba*) as a function of tree size (DBH—trunk diameter 1.5 m above ground).

INTERPRETING ECOLOGICAL DATA

Q1. Refer to Quantifying Ecology 6.1: Calculating Plant Growth Rate (p. 111). Assuming leaf area (cm²) increases linearly with leaf mass (g), how does the leaf area ratio of the tree change with age?

Q2. How would the relative growth rate change as the tree increases in size (DBH) and age?

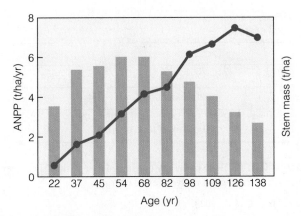

FIGURE 20.16 Changes in aboveground stem mass (blue line) and aboveground NPP (green bars) for stands of the boreal conifer *Picea abies* of differing ages.

(Adapted from Gower et al. 1996.)

the tree grows, more of its energy goes into support and maintenance (respiration), both of which increase with age, lowering photosynthetic gain.

This changing pattern of energy allocation affects forest NPP over time. As a stand ages, more and more biomass is in woody tissue, while leaf area remains relatively constant or declines. More of GPP goes towards maintaining woody tissue, leaving less for growth. NPP increases in the early stages, but declines as forest biomass increases (Figure 20.16) (Gower et al. 1996).

20.7 SECONDARY PRODUCTIVITY: Production of Consumers Is Limited by Primary Productivity

Net primary productivity (NPP) is the energy available to heterotrophs in an ecosystem—to the rest of the food web. Either herbivores or decomposers eventually consume all plant production, even if in a different location from where it was produced. (Fossil fuel deposits may seem an exception, but even they decompose extremely slowly.) Humans or natural agents such as wind or water may disperse NPP between ecosystems. For example, 45 percent of the NPP of a salt marsh is typically lost to adjacent estuaries.

Once organic matter is consumed, its energy has several fates. Some energy is not assimilated across the body wall, and exits as waste products in feces and urine. Of the energy that is assimilated into the gut, a portion is lost as respiratory heat. For homeotherms, most respiratory losses are due to thermoregulation, but for all consumers some respiratory losses reflect the energy demands of maintenance—capturing prey, performing muscular work, powering biochemical processes, and repairing wear and tear on the body.

Energy used for maintenance is lost to the surrounding environment as respiratory heat. Any energy left over (net energy gain) goes into *secondary production*, which includes growth of new tissues and production of offspring (allocation to

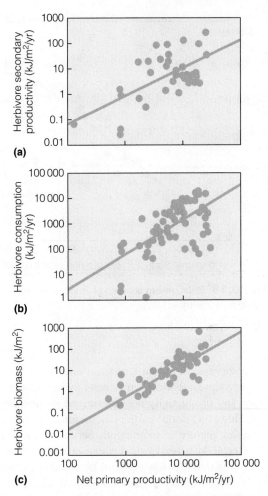

Figure 20.17 Relationship between aboveground NPP and **(a)** net secondary productivity of herbivores, **(b)** herbivore consumption, and **(c)** herbivore biomass. Each point represents a different terrestrial ecosystem.

(Adapted from McNaughton et al. 1989.)

reproduction). **Secondary productivity** is the amount of secondary production per unit time (grams per unit area per unit time), and is greatest when both the birthrate of the population and the growth rate of individuals are maximal.

Given that consumers rely on primary producers for energy, primary productivity constrains secondary productivity. A review of 69 studies from terrestrial systems ranging from Arctic tundra to tropical forests indicated that consumption of primary production by herbivores, herbivore biomass, and secondary productivity all increase with primary productivity (Figure 20.17) (McNaughton et al. 1989). Not surprisingly, given how much tree biomass is unpalatable to herbivores, forests have less consumption per unit primary productivity than grasslands. If only foliage consumption is compared, differences are less pronounced. Similarly, a survey of productivity trends in 43 lakes and 12 reservoirs from the tropics to the Arctic revealed a significant positive relationship between plankton productivity and productivity of both herbivorous and carnivorous zooplankton (Figure 20.18, p. 438) (Brylinsky and Mann 1973).

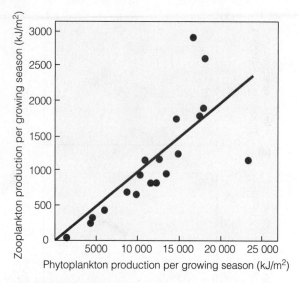

Figure 20.18 Relationship between phytoplankton and zooplankton productivity in lakes.

(Adapted from Brylinsky and Mann 1973.)

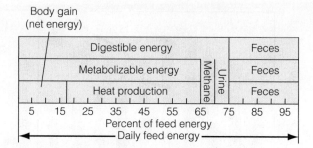

Figure 20.19 Fate of energy consumed by a white-tailed deer. Values are percentage of daily ingested energy.

(Adapted from Cowan 1962.)

These surveys, based on terrestrial and aquatic systems worldwide, suggest bottom-up control, wherein herbivore abundance and productivity are determined by primary productivity. However, in our earlier discussion of food webs, a more complex picture of interactions between trophic levels emerged. Recall the *green world hypothesis* (see Section 16.8), which suggests that top-down control by predators keeps herbivores in check, allowing plant biomass to accumulate. A growing body of experimental evidence suggests that NPP is influenced not only by abiotic factors but also by top-down regulation, especially in aquatic systems. But the fact remains that without primary productivity there would be nothing for consumers to regulate.

20.8 CONSUMER EFFICIENCY: Consumers Vary in How Efficiently They Convert Consumed Energy into New Tissue

Despite the general correlation between primary and secondary productivity across a range of ecosystems, within a given ecosystem consumers vary considerably in their efficiency at transforming consumed energy into secondary production. A simple model traces energy flow through a consumer, as shown for a white-tailed deer (Figure 20.19).

Of the food ingested by a consumer (I), a portion is assimilated across the gut wall (A), and the remainder expelled as waste products (W), including feces. Of this assimilated energy, some is used in respiration (R) for thermoregulation or maintenance, while the rest goes to secondary production (P), which includes new tissues and offspring. The overall process is summarized as

$$P = I - W - R; \text{ or, since } A = I - W, P = A - R$$

These values are used to calculate various efficiency ratios (usually reported as percentages), which vary with consumer type, species, and (for the same individual) age or developmental stage. The ratio of assimilation to ingestion (A/I), **assimilation efficiency** (AE), quantifies how efficiently a consumer extracts energy from food. In general, endotherms have higher assimilation efficiencies than ectotherms (Andrzejewska and Gyllenberg 1980). However, carnivores, even if ectothermic, have higher assimilation efficiencies (~ 80 percent) than herbivores (20–50 percent, depending on species and food item), because carnivores consume tissue that is more readily assimilated (see Section 7.2).

Net production efficiency (NPE) is the ratio of secondary production to assimilation (P/I). It is a measure of how efficiently a consumer incorporates assimilated energy into secondary production. Invertebrates tend to have high net production efficiencies (30–40 percent) (Table 20.2), losing little energy in respiratory heat and converting more energy into production. Among vertebrates, ectotherms such as fish have intermediate production efficiencies (~ 10 percent), but endotherms, given the expense of thermoregulation, convert only 1 to 3 percent of assimilated energy into production. Body

Table 20.2 Net Production Efficiency ($P/A \times 100$) of Various Animal Groups

Group	P/A (%)
Birds	1.29
Small mammals	1.51
Other mammals	3.14
Fish and social insects	9.77
Non-insect invertebrates	
Herbivores	20.8
Carnivores	27.6
Detritivores	36.2
Non-social insects	
Herbivores	38.8
Detritivores	47.0
Carnivores	55.6

Data from Humphreys 1979.

size also influences efficiency. Because mass-specific metabolic rate increases exponentially with decreasing body mass (see Figure 7.16), small endotherms have inherently low net production efficiency.

The overall efficiency of converting ingested energy into secondary production takes both A/I and P/A into account. This ratio, called the **gross production efficiency** (*GPE*, also *ecological growth efficiency*), is calculated as P/I, which is equivalent to $A/I \times P/A$. So, if an endothermic carnivore has a high *AE* (80 percent), but its *NPE* is low (2 percent) as a result of its high respiratory costs, its *GPE* is only 1.6 percent. It is easy to see why large endotherms often fail to meet their energy needs. In *Why Big Fierce Animals Are Rare*, Paul Colinvaux (1978) outlined many reasons why large top predators are at risk, but acquiring sufficient energy is certainly one of them.

20.9 FOOD CHAINS: Energy Flows through Interlinked Grazing and Detrital Food Chains

Energy fixed by autotrophs is the basis upon which life depends. As described in Section 16.6, energy passes through the ecosystem in a series of steps of eating and being eaten—a *food chain* or *food web*, which summarizes feeding relationships into trophic levels. **Primary producers** (*autotrophs*) occupy the first trophic level, and **secondary producers** (*heterotrophs* or *consumers*) the rest. Among secondary producers, **first-level consumers** (*herbivores*) occupy the second level, **second-level consumers** (*first-order carnivores*) occupy the third, and so on. All organisms that obtain their energy in the same number of steps removed from primary producers belong to the same trophic level, whatever their taxonomic identity. Some consumers occupy a single level, but omnivores and generalist predators occupy more than one.

Food chains represent abstract expressions of the food webs discussed in Chapter 16, with feeding groups based on organisms using a common energy source. Each group is linked to others in a manner representing energy flow. What is in reality a food web, as in Figures 16.9 and 16.10, is depicted as a single line-chain (Figure 20.20). Boxes represent autotrophs, herbivores, and carnivores; arrows linking the boxes indicate the direction of energy flow.

Every ecosystem has two major food chains: the grazing and the detrital, distinguished by the energy source of the first-level consumers. In the **grazing food chain**, the herbivores are *biophages* that consume living plant biomass, whereas in the **detrital (decomposer) food chain**, the herbivores are *saprophages* that consume dead organic matter (*detritus*). Cattle grazing on pasture, deer browsing on shrubs, insects feeding on leaves, and zooplankton feeding on phytoplankton are all first-level consumers in grazing food chains, whereas fungi and bacteria, as well as many invertebrates, including snails, beetles, millipedes, and earthworms, are first-level consumers in detrital food chains (see Chapter 21). (Some reserve the term *herbivore* for the grazing food chain, but the function is the same.) In turn, herbivores in each chain are the energy source for carnivores.

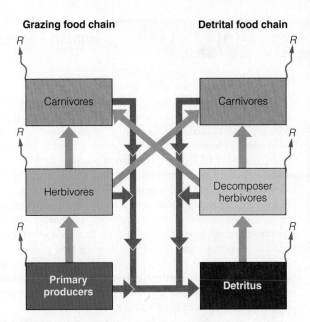

Figure 20.20 Grazing and detrital food chains. Orange arrows represent the flow of energy following ingestion. Blue arrows represent loss of energy through respiration. Brown arrows represent a combination of dead organic matter (unconsumed biomass) and waste products (feces and urine).

Figure 20.20 combines the grazing and detrital chains into a general model of energy flow. The two chains are functionally linked, because the source of energy for the detrital chain is dead organic matter and wastes from the grazing food chain. This linkage appears as brown input arrows from each trophic level in the grazing food chain, leading to the box designated as detritus. There is, however, a key difference between the chains. In a detrital chain, wastes and detritus from each consumer level return as inputs to the dead organic matter box at the base of the detrital food chain. This feature does not mean that energy is recycled in the detrital chain; thanks to the second law of thermodynamics, energy can only flow one way. It does mean that the detrital food chain can recycle its matter (wastes and dead tissues) until it is all used up, whereas the grazing food chain, which involves only living tissues, cannot.

The distinction between grazer and detrital food chains is blurred at higher trophic levels. Even highly selective carnivores rarely distinguish whether their prey are members of a grazing or a detrital food chain. The diet of an insectivorous bird might include beetles that feed on detritus, as well as caterpillars that feed on green leaves. This explains why the orange arrows crisscross between the two chains above the herbivore level.

20.10 QUANTIFYING ENERGY FLOW: Energy Flow through Grazing and Detrital Food Chains Varies among Ecosystems

To quantify energy flux through the ecosystem, we must take account of the processes involved in secondary production: ingestion, assimilation, respiration, and production. But as well

ECOLOGICAL ISSUES Human Appropriation of Net Primary Productivity

Although humans are only one of more than 1.9 million known species inhabiting Earth, we use a vastly disproportionate share of its resources. Vitousek et al. (1986) used three approaches to estimate the fraction of global terrestrial NPP appropriated by humans. A low estimate was derived by calculating the amount of NPP people use directly, as food, fuel, fibre, or timber. An intermediate estimate included the productivity of lands devoted entirely to human activities (such as the total NPP of croplands, including weeds and non-harvested plant parts, as opposed to the portion of crops consumed) and energy human activities consume, including land-clearing fires. A high estimate included productive capacity lost by converting land to cities and pastures, or associated with desertification, overgrazing, and erosion. These methods yielded estimates of 3, 27, and 39 percent, respectively—a remarkable level of use for a species that represents only 0.5 percent of Earth's total heterotrophic biomass.

Advances in satellite technology have enhanced our ability to monitor land use and productivity. More recent estimates of *human appropriation of net primary productivity* (HANPP), defined by Haberl et al. (2007) as the "combined effect of harvest and productivity changes induced by land use on the availability of NPP in ecosystems," are calculated as the difference between the NPP of the potential vegetation and that of the residual vegetation after harvest (Figure 1). Calculation details are available in their 2007 paper, but estimates were based on satellite GIS data integrated with (1) United Nations FAO (Food and Agriculture Association) data for 161 countries in 2000, and (2) Lund-Potsdam-Jena DGVM (a model of global vegetation) data for 1998 to 2002.

Using these data, Haberl and colleagues estimate that humans use 15.6 Pg of C per year, or 23.8 percent of global potential terrestrial NPP: 53 percent by direct harvest, 7 percent by clearing-related fires, and 40 percent by changes in productivity related to land use changes. If only aboveground NPP is considered, human impact is even greater: 28.8 percent. Compared with Vitousek's results, appropriation from the combined effects of land use and harvest are lower (23.8 percent vs. Vitousek's high estimate of 39 percent). In part this discrepancy reflects differences in the timeframe, but also in the nature of the data and Haberl's more conservative criteria.

Haberl's findings suggest substantial global appropriation by humans, particularly through lost NPP associated with land use, as well as tremendous regional variation in HANPP, expressed as a percentage of NPP (Figure 2). Some regions, including parts of Europe and southern Asia, consume more than half of regional NPP. In contrast, HANPP in other regions (typically the wet tropics) is under 15 percent. In the Canadian context, Daniel O'Neill and colleagues at Dalhousie University estimate (using Haberl's criteria) that Nova Scotia's HANPP is 25 percent overall. This value is remarkably close to the global average, but with substantial variation in different regions, with a high of 50 percent (O'Neill et al. 2007).

Population and per capita consumption interact to determine human impact at a regional scale. The role of population is clear from Figure 2, despite vast differences in consumption among nations. Southern Asia, with a substantial portion of the

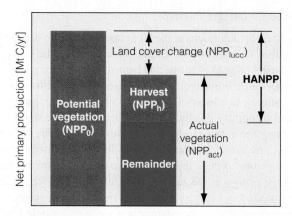

Figure 1 HANPP is defined as the difference between NPP of potential vegetation (NPP$_O$) and NPP of actual vegetation (NPP$_{act}$) remaining after harvest. Thus, HANPP equals NPP of harvested material (NPP$_h$) plus NPP lost through land use cover changes (NPP$_{lucc}$).

(Adapted from O'Neill et al. 2007; modified from Haberl et al. 2004).

as considering energy flow through a single organism consuming a single food item, we must examine flow through a single trophic compartment (Figure 20.21a, p. 442).

The energy available to a given trophic level (designated as level n) is the production generated by the next-lowest level ($n-1$). Net primary production (P_1) is the energy available for herbivores. Following energy flow through a single consumer, some portion of that energy is ingested (I), while the remainder becomes dead organic inputs for the detrital food chain. (Sooner or later this dead organic matter is ingested by a detrital herbivore and becomes I for the detrital chain.) Returning to the grazing chain, some portion of the energy ingested is assimilated (A) by the totality of organ-

isms at that trophic level, with the remainder lost as waste materials (W)—lost from the grazing chain, but available for ingestion by the detrital food chain. Of the energy that is assimilated, some is lost to respiration, as shown by the arrow labelled R exiting the upper left corner of the box. Finally, any remainder goes to production (P).

We can quantify this energy flow *within* a trophic level using the same ratios defined in Section 20.7: assimilation efficiency (A/I) and net production efficiency (P/A). But to track energy *between* one level and the next, another ratio is needed. **Consumption efficiency** is the ratio of ingestion to production (I_n/P_{n-1})—the proportion of energy available at one trophic level that is consumed by the next. Figure 20.21b

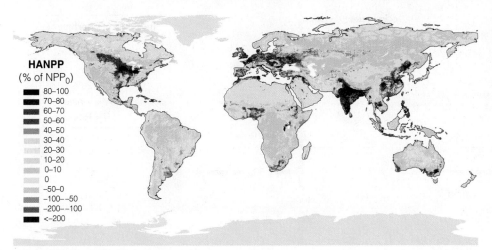

Figure 2 Total HANPP, expressed as percentage of potential NPP. Human-induced fires are excluded. See the source paper for details of how values were calculated.

(Adapted from Haberl et al. 2007.)

world's population, appropriates 63 percent of its regional NPP despite having low per capita consumption. Affluence also plays a role. Average annual per capita HANPP for North America and Europe far exceeds that of developing nations. If per capita HANPP of developing nations increases to match that of industrialized countries, global HANPP will climb far above Haberl's estimate of 23.8 percent. Remember that these estimates do not include human appropriation of marine productivity.

Nor are the impacts of HANPP restricted to diversion of energy from food webs for human use. Haberl documents many indirect effects, including impacts on biodiversity, water cycling, transfer of carbon between Earth and the atmosphere, and provision of ecosystem services. Land-use changes in particular have the potential not only to transform Earth's surface, but also to hamper its ability to provide the basic services so vital to human-dominated and natural ecosystems alike. Haberl warns that proposed solutions to our energy shortages that involve widespread conversion of land to harvesting biomass as a substitute for dwindling fossil fuel supplies have the potential to increase the human strain on global NPP. We might "solve" our energy problem only by creating a far more serious problem for the biosphere as a whole.

Bibliography

Haberl, H., K. H. Erb, F. Krausmann, V. Gaube, A. Bondeau, C. Plutzar, S. Gingrich, W. Lucht, and M. Fischer-Kowalski. 2007. Quantifying and mapping the human appropriation of net primary production in earth's terrestrial ecosystems. *Proceedings of the National Academy of Sciences* 104:12942–12947

Haberl, H., K. H. Erb, F. Krausmann, and W. Lucht. 2004. Defining the human appropriation of net primary productivity. *LUCC Newsletter* 10:16–17.

O'Neill, D. W., P. H. Tyedmers, and K. F. Beazley. 2007. Human appropriation of net primary production (HANPP) in Nova Scotia, Canada. *Regional Environmental Change* 7:1–14.

Vitousek, P. M., P. R. Ehrlich, A. H. Ehrlich, and P. A. Matson. 1986. Human appropriation of the products of photosynthesis. *BioScience* 36:363–373.

1. Compare Figure 20.6 and Figure 2 to explain why HANPP (expressed as a percentage) is low for tropical rain forests.
2. Since the mid-20th century, the efficiency of agricultural production (harvest produced per hectare of cultivated land) has increased dramatically. How would this increase have changed the HANPP values in Figure 2?

shows representative values for an invertebrate herbivore in a grazing chain. Using these efficiencies, we can track the fate of a given amount of energy (1000 kcal) available to herbivores as NPP as it passes through the second trophic level.

If, using similar procedures, we derive efficiency ratios for each level in the grazing and detrital food chains, we can calculate energy flow through the entire system. Production from each level provides input to the next level, and unconsumed production (dead individuals) and wastes become inputs into the detrital chain. (Recall that all energy entering the ecosystem as NPP is ultimately lost through respiration.) Practically, this task is daunting. We do not know the members, feeding links, and energy processed in most food webs.

And if our knowledge is incomplete for grazing webs, it is even more so for detrital webs, given the small size of many of its members and the innumerable species (many unidentified) involved.

Although the general model presented in Figure 20.21 applies to all ecosystems, the relative importance of the two food chains and the rate of energy flow through trophic levels varies widely. Recall that consumption efficiency defines the amount of energy produced by a trophic level (P_{n-1}) that is consumed by the next-highest level (I_n). Consumption efficiencies between trophic levels determine the path of energy flow, allowing ecosystem comparisons. Despite its conspicuousness, the grazing food chain is not the major chain in most terrestrial

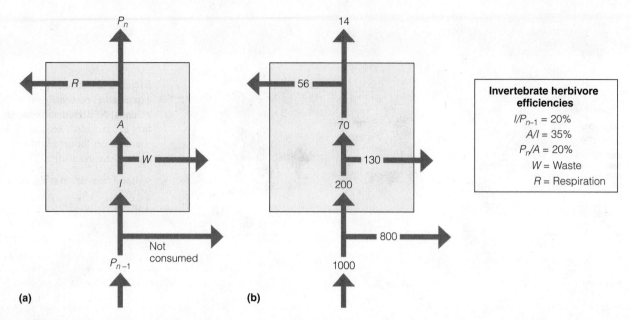

Figure 20.21 Energy flow within **(a)** a single trophic compartment. **(b)** A quantitative example of energy flow (kcal) for an invertebrate herbivore using efficiency ratios provided in the accompanying table.

and some aquatic systems. Only in open-water habitats does the grazing food chain predominate.

Surveys indicate that plankton-dominated aquatic systems have higher herbivory (79 percent) than those dominated by vascular plants (30 percent) (Figure 20.22). Only 17 percent of primary production is consumed by herbivores in terrestrial systems (Cyr and Pace 1993). In most terrestrial and shallow-water systems, with their high standing biomass and low grazing activity, the detrital food chain dominates. In open-water systems, with their low standing biomass, rapid turnover, and high harvest rate, the grazing food chain processes a greater proportion of primary productivity. These differences in grazing activity between aquatic and terrestrial systems (as well as in other aspects of food web structure and activity) persist across a range of productivities and reflect the nature of aquatic versus terrestrial primary producers (highly digestible plankton versus lignin-rich woody species) (Shurin et al. 2006).

Among terrestrial systems, forests and grasslands differ in consumption efficiency. Consumption efficiency averages 3.7 percent for herbivores in deciduous forests versus 9.3 percent in grasslands (Hairston and Hairston 1993; both values are lower than the median of 17 percent for terrestrial systems reported by Cyr and Pace). Variation is considerable even among similar ecosystems. Consumption efficiency for herbivores is lower in tallgrass than midgrass prairie (5.3 versus 16.5 percent) (Andrzejewska and Gyllenberg 1980). Smaller differences were reported by Hairston and Hairston (1993) in consumption efficiency of predators inhabiting forests versus grasslands (90 versus 77 percent).

Energy flow through flowing-water systems differs greatly from both terrestrial and standing-water systems. Streams and rivers have low NPP, and grazing activity is minimal. The detrital food chain dominates, typically fuelled by allochthonous inputs from adjacent terrestrial systems.

20.11 PYRAMID MODELS: Energy and Biomass Pyramids Differ among Ecosystems

Based on the preceding discussion, we can conclude that the energy flowing into a trophic level decreases with each successive level in a food chain. An **energy pyramid** constructed of the total energy processed at each trophic level is thus pyramidal in shape, with the slope reflecting the efficiency of energy transfer (Figure 20.23). The more efficient the transfer, the steeper the slope. This pattern occurs for two reasons: (1) not all energy is consumed between one level and the next, and (2) of the energy consumed, not all is transformed into production to be consumed by the next level. This reality is a simple consequence of the second law of thermodynamics.

As a rule of thumb, only 10 percent of the energy consumed as biomass in a given trophic level is converted to biomass at the next trophic level. If, for example, herbivores eat 1000 kcal of plant energy, about 100 kcal is converted into herbivore tissue, 10 kcal into first-level carnivore tissue, and so on. (Note that this *10 percent rule* assumes that all animal tissue is consumed, i.e., a consumption efficiency of 100 percent, which is unlikely.) But ecosystems are not governed by a simple principle that dictates a constant proportion of energy reaching successive levels. Rather, differences in consumption efficiency as well as in efficiencies of energy conversion (assimilation and production) among feeding groups influence energy transfer.

A useful overall measure for quantifying energy transfer between levels is **trophic efficiency** (*TE*): the ratio of productivity in a given trophic level (P_n) to the productivity of the level it feeds on (P_{n-1}): $TE = P_n/P_{n-1}$. Trophic efficiency incorporates all of the efficiency ratios discussed earlier, because it takes all possible losses into account:

$$TE = P_n/P_{n-1} = CE \times GPE$$

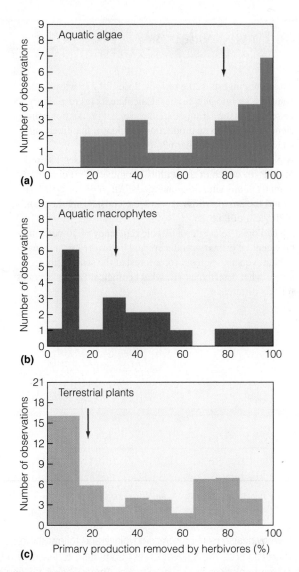

(a) Aquatic algae

(b) Aquatic macrophytes

(c) Terrestrial plants

Primary production removed by herbivores (%)

Figure 20.22 Consumption efficiencies in different ecosystems. Histograms (on *x*-axis) represent percentage NPP consumed by herbivores in systems dominated by **(a)** algae, **(b)** rooted aquatic plants, and **(c)** terrestrial plants. Number of observations (on *y*-axis) refers to the number of studies having a given level of consumption. Red arrows indicate median values.

(Adapted from Cyr and Pace 1993.)

A survey of 48 studies of aquatic systems strongly supported the 10 percent rule. Despite much variability, mean trophic efficiency was 10.1 percent (Pauly and Christensen 1995). Trophic efficiency is much lower for most terrestrial systems, as a result of their lower consumption efficiencies.

With decreasing energy transfer through a food web comes a corresponding decrease in standing biomass. A **biomass pyramid** sums the total biomass of living organisms (i.e., energy stored in standing crop) present at any one time. Because some energy is lost at each level, biomass at each level is limited by the rate at which energy is stored at the next-lower level. In general, the biomass of producers is greater than that of herbivores, and herbivore biomass is

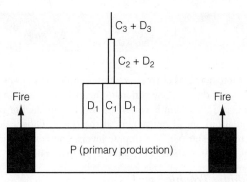

Figure 20.23 Generalized energy pyramid. P represents primary production; C_1, primary consumers (herbivores); C_2, and C_3 secondary consumers (carnivores); D_1, decomposers of plant tissue; D_2 and D_3, decomposers of animal tissue. Grazing and detrital chains have been collapsed. The open ocean, which has higher consumption efficiency and more efficient energy transfer between trophic levels, would have less dropoff in energy processed at each trophic level and hence a steeper slope. Some energy is lost from the system through fire.

greater than that of carnivores, resulting in a narrowing pyramid. The biomass pyramid for a Florida bog thus resembles a typical energy pyramid (Figure 20.24a).

This situation does not hold for all ecosystems. In lakes and open seas, short-lived phytoplankton cannot accumulate biomass and are heavily grazed by zooplankton that are larger and longer lived. No matter what their productivity, phytoplankton biomass is low compared to that of their herbivores (Figure 20.24b). An inverted pyramid results, with a lower standing crop of primary producers than herbivores. This situation might seem to violate the second law of thermodynamics, but recall that biomass is just a snapshot of the living tissue at any one time. An energy pyramid for this (as for any) ecosystem would indicate declining energy at each successive level.

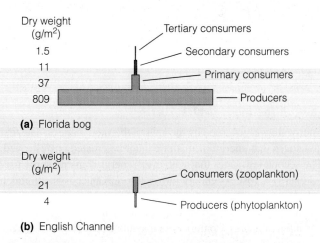

Figure 20.24 Biomass pyramids for the grazing food chain of **(a)** a Florida bog and **(b)** the marine ecosystem of the English Channel. The marine biomass pyramid is inverted given the fast turnover of phytoplankton and the high rate of consumption by zooplankton.

QUANTIFYING ECOLOGY 20.1 Summarizing Efficiency Ratios

Throughout this chapter, we have introduced a number of efficiency ratios. Some pertain to energy transfer *within* a trophic level, as a food item is consumed by an individual or by all the populations of all consumers at that level. Others pertain to energy transfer *between* trophic levels, and hence relate to the ecosystem as a whole. Let's summarize these ratios.

Within a Trophic Level

(1) Assimilation efficiency $(AE) = A/I$

where I = energy ingested; A = energy assimilated $(I - W)$; W = energy lost in wastes

(2) Net production efficiency $(NPE) = P/A$

where P = secondary production = $A - R$; R = energy lost in respiration

(3) Gross production efficiency $(GPE) = P/I = AE \times NPE$

Between Trophic Levels

(1) Consumption efficiency $(CE) = I_n/P_{n-1}$

where n = a given trophic level; $n-1$ is the next lower trophic level

(2) Trophic Efficiency $(TE) = P_n/P_{n-1} = CE \times GPE$

1. An animal ingests 120 g of food, loses 20 g in wastes, and converts 5 g into new tissues. Calculate its (a) respiratory losses, (b) assimilation efficiency, (c) net production efficiency, and (d) gross production efficiency. Is it likely an endotherm or an ectotherm?
2. A plant absorbs 80 units of light energy and uses 10 units in GPP. To which of the within-trophic level efficiencies for animals is this situation analogous: *AE*, *NPE*, or *GPE*? How would the answer differ if you were considering the plant's NPP yield of 8 units?
3. An ecologist calculates a trophic efficiency of 10 percent between the primary producers and herbivores in an ocean habitat, versus 1 percent for a forest at the same latitude. Given what determines *TE*, what ecological factors account for this difference?

EcologyPlace

Visit EcologyPlace at www.pearsoncanada.ca/ecologyplace to access online resources that complement your textbook, and help you to apply and to review the information in this chapter. EcologyPlace includes

- an eText version of the book
- self-grading quizzes
- glossary flashcards
- and more!

Go to www.pearsoncanada.ca/ecologyplace and follow the registration instructions on the Student Access Code Card included with this text. If your book does not have a Student Access Code Card, you can purchase access to it at www.pearsoncanada.ca/ecologyplace.

SUMMARY

Laws of Thermodynamics 20.1

Energy flow is governed by the laws of thermodynamics. The first law states that energy or matter cannot be created or destroyed. The second law states that with every energy transfer, some useful energy is lost. As energy moves through an ecosystem, much is lost as respiratory heat. Energy is degraded from a more organized to a less organized state (i.e., entropy increases). A continuous input of energy from the Sun maintains ecosystems.

Primary Productivity 20.2

The flow of energy through ecosystems starts with the harnessing of sunlight by autotrophs through photosynthesis. The total energy fixed by plants is gross primary production. The energy remaining after plants have met their respiratory needs is net primary production, which is stored as plant biomass. Net primary productivity is measured as dry mass/unit area/unit time.

Factors Affecting Primary Productivity 20.3

The primary productivity of terrestrial ecosystems is strongly affected by climate. The interacting influences of temperature and precipitation limit photosynthesis and the amount of leaves that can be supported. Warm, wet conditions make tropical rain forests the most productive terrestrial systems. Nutrients also influence terrestrial productivity. In aquatic systems, the depth to which light penetrates is crucial. Nutrient availability is a pervasive influence on marine productivity, which is highest in shallow coastal waters, coral reefs, and estuaries, where nutrients are more available. Nutrients, particularly phosphorus, limit the productivity of lakes. In flowing water, detrital inputs from adjacent ecosystems are the major energy source.

External Inputs 20.4

Organic compounds in most terrestrial ecosystems originate from internal sources (autochthonous), but many aquatic ecosystems rely on large external inputs of organic compounds (allochthonous).

Allocation of Primary Production 20.5

Energy fixed by plants is allocated to different plant parts and to reproduction, depending on life-form and abiotic factors. Allocation affects standing crop biomass and productivity. Investing more in roots under low water or nutrient levels enhances survival at the expense of carbon gain.

Temporal Variation 20.6

Primary production varies with seasonal and yearly fluxes in moisture and temperature. In forests, net primary production declines and standing biomass increases with stand age. As the ratio of woody biomass to foliage increases, more net production is allocated to maintenance.

Secondary Productivity 20.7

Net primary production is available to consumers directly as plant tissue or indirectly through consumer tissue. Of the energy consumed, some is lost to wastes; of the energy assimilated, some is diverted to maintenance, growth, and reproduction. Change in consumer biomass per unit time, including reproduction, is secondary productivity. Secondary productivity is limited by any abiotic constraint on primary productivity.

Consumer Efficiency 20.8

Consumers vary in their energy use efficiency. Homeotherms have high assimilation efficiency but low production efficiency because they expend so much energy in respiration. Poikilotherms have low assimilation efficiency but high production efficiency, allowing them to convert more energy into growth or offspring. Carnivores have higher assimilation efficiencies than herbivores.

Food Chains 20.9

All ecosystems process solar energy through the food chain: a series of energy transfers between producers and consumers. Members are grouped into trophic levels. Autotrophs occupy the first trophic level, herbivores the second, and carnivores the third and higher levels. Energy flows through either the grazing or detrital food chain, which are linked by inputs of dead organic matter and wastes, and by carnivores that consume members of both chains.

Quantifying Energy Flow 20.10

At each trophic level, the efficiency of energy exchange is affected by consumption efficiency, the proportion of available energy that is consumed; assimilation efficiency, the proportion of energy ingested that passes through the gut wall; and net production efficiency, the proportion of assimilated energy that is invested in growth of new tissues. The detrital food chain dominates in terrestrial ecosystems, whereas in lakes and oceans, more primary productivity enters the grazing food chain. Consumption efficiency of predators varies less among ecosystems than that of herbivores.

Pyramid Models 20.11

The amount of energy processed decreases with each trophic level because not all energy is consumed, and because not all consumed energy is converted into production. As a general rule, only 10 percent of the energy consumed at a given level is converted to biomass at the next trophic level. This rule applies best to aquatic systems dominated by phytoplankton; the percentage is lower in systems with lower consumption efficiencies. Biomass typically declines at each level, but aquatic systems may have an inverted biomass pyramid.

KEY TERMS

actual evapotranspiration	consumption efficiency	grazing food chain gross primary productivity	net production efficiency	second law of thermodynamics
allochthonous	detrital (decomposer) food chain	gross production efficiency	potential energy	second-level consumer
assimilation efficiency			primary producer	secondary producer
autochthonous	energy pyramid	growing season	relative net primary productivity	secondary productivity
biomass accumulation ratio	entropy first law of thermodynamics	kinetic energy net primary productivity	residence time root-to-shoot ratio	standing crop biomass trophic efficiency
biomass pyramid				
compensation depth	first-level consumer			

STUDY QUESTIONS

1. Define and relate the following terms: *gross primary productivity*, *autotrophic respiration*, *net primary productivity*.
2. Contrast *net primary productivity* and *standing crop biomass*. Explain how it is possible for two ecosystems—a prairie and a forest—to have identical NPP yet differ in biomass.
3. How do temperature and precipitation interact to influence NPP in terrestrial ecosystems?
4. How and why does NPP vary with water depth in lakes?
5. What environmental factors influence the light compensation depth of a lake? How does primary productivity constrain secondary productivity?
6. What does the top-down model of food chain structure imply about the role of secondary producers in controlling NPP and standing biomass within ecosystems?

7. How do assimilation efficiency and net production efficiency relate to the flow of energy through a trophic level?
8. How and why do energy allocation and net production efficiency differ between homeotherms and poikilotherms?
9. What are the two major food chains, and how are they related?
10. How and why does consumption efficiency differ between terrestrial and marine systems?
11. Why is the biomass pyramid of the open ocean often inverted? How would this differ from its energy pyramid?

FURTHER READINGS

Gates, D. M. 1985. *Energy and ecology*. Sunderland, MA: Sinauer Associates.
This classic text stresses the function of ecological systems in terms of energy flow.

Gosz, J. R., R. T. Holmes, G. E. Likens, and F. H. Bormann. 1978. The flow of energy in a forest ecosystem. *Scientific American* 238:92–102.
Detailed treatment of energy flow through a forest ecosystem.

Howarth, R. W. 1988. Nutrient limitation of net primary production in marine ecosystems. *Annual Review of Ecology and Systematics* 19:89–110.

This paper reviews research on the role of nutrients in limiting net primary productivity of the world's oceans.

Lieth, H., and R. H. Whittaker, eds. 1975. *Primary productivity in the biosphere*. Ecological Studies Vol. 14. New York: Springer-Verlag.
Classic work exploring global primary productivity in both terrestrial and aquatic ecosystems.

Colourful fungi such as honey mushrooms (*Armillaria mellea*) reside in forests across eastern North America. *Armillaria*, among the largest organisms on Earth because of their longevity and massive underground mycelia, function both as decomposers of decaying wood and as parasites of living deciduous trees, causing root rot.

As we saw in Chapter 20, one-way energy flow through an ecosystem is the story of carbon. Starting with CO_2 fixation in photosynthesis, carbon moves through the food web, ultimately returning to the atmosphere via respiration. But primary productivity depends on plant uptake not just of carbon but also of an array of other minerals. The atmosphere is the source of carbon, but what about the other essential elements? Each has its own story of origin and movement through ecosystems. We explore these pathways, called *biogeochemical cycles*, in Chapter 22, but their source is either the atmosphere or the weathering of rocks and minerals. Once in soil or water, minerals are absorbed by plants and move through the food web as part of living tissue.

What is the fate of these minerals once in the food chain? As living tissues senesce (age and die), nutrients return to the soil or sediments in dead organic matter (*detritus* or *litter*, a term often reserved for plant detritus), and eventually enter the detrital food chain. But unlike carbon, which is recycled by its release by aerobic respiration at all trophic levels, most mineral nutrients are recycled within local ecosystem compartments by specific decomposers. Microbial decomposers transform organic nutrients into mineral form, making them once again available for plants. This process, called **internal (nutrient) cycling**, is an essential function of all ecosystems.

This chapter examines nutrient cycling within ecosystems. Our central focus is on decomposition, nutrient mineralization, and the factors that affect the rate at which these processes proceed. We also explore how nutrient cycling varies between terrestrial and aquatic systems, setting the stage for discussing specific biogeochemical cycles in Chapter 22.

21.1 NUTRIENT CYCLING: Minerals Are Recycled within Ecosystems

In terrestrial systems, nutrients such as nitrogen cycle from soil to vegetation and back to soil (Figure 21.1). As with all nutrients, plants take up N in inorganic forms (although in N-poor boreal soils, some species absorb organic N; see Section 22.5). Nitrogen in the soil matrix (as ammonium, NH_4^+) or in the soil solution (as nitrate, NO_3^-) is taken up by roots and incorporated into structural proteins, enzymes, and other compounds. The N is now in organic form, as part of living tissues. The availability of N limits its uptake and can thereby affect primary productivity.

As plant tissues senesce, nutrients are returned to the soil surface in dead organic matter. Before shedding dead tissues, plants absorb and store some of their nutrients for future use. This nutrient recycling within the plant is called **retranslocation (reabsorption)**. Consider what happens in temperate deciduous forests as the days shorten in the fall. Synthesis of the green photosynthetic pigment chlorophyll stops. Yellow and orange accessory pigments already present in the leaf (carotenoids and xanthophylls) are revealed. Some senescing leaves also produce non-photosynthetic

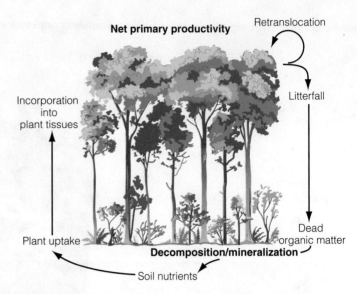

Figure 21.1 General model of nutrient cycling in a terrestrial ecosystem.

pigments (anthocyanins), which add red and purple hues. Leaves of some species, particularly oaks, contain high quantities of tannins that are brown in colour.

These chemical changes are visible to the eye. Less obviously, as senescence progresses, plants withdraw water and mobile nutrients from the dying leaves into their stem tissue, recovering from 50 to 70 percent of the N in leaves before they fall. Less mobile nutrients such as calcium (Ca) remain in the dying leaves, increasing the Ca content of the litter and ultimately reducing soil acidity compared with a coniferous forest. After leaf fall, decomposers break down the dead plant tissues, transforming the organic nutrients into inorganic form—**mineralization** (see Section 21.7). The cycle is complete, and the nutrients are again available for plant uptake.

21.2 DECOMPOSER GROUPS: Many Heterotroph Groups Participate in Decomposition

The key process in nutrient cycling is **decomposition**: breakdown of the chemical bonds of complex organic compounds. Whereas photosynthesis uses solar energy to convert inorganic compounds (CO_2 and H_2O) into organic compounds, decomposition converts these compounds back to inorganic form, releasing the energy captured by photosynthesis. Decomposition involves many processes, including leaching, fragmentation, alteration of physical and chemical structure, ingestion, and excretion. These processes are accomplished by decomposers.

All heterotrophs—even members of grazing food webs—function as decomposers to some degree. As they digest food, they break down organic matter, alter it structurally and chemically, and release it partially as wastes. Indeed, through respiration, every aerobic heterotroph mineralizes carbon by releasing CO_2. However, the term *decomposers* is typically reserved for **saprophages**: organisms that feed on dead organic matter, including (1) **decomposers** or **microflora** (a diverse group of bacteria and fungi), and (2) **detritivores**, animals (largely invertebrates) that fragment dead material. Other groups are *coprobivores*, which consume feces; **microbivores**, which feed on microflora; and **scavengers**. This term is difficult to define; often reserved for vertebrates such as gulls or vultures that consume animal remains, *scavenger* is sometimes used for any animal, vertebrate or invertebrate, that feeds on dead plant or animal material. In this usage, *scavenger* is functionally equivalent to *detritivore*.

Decomposer categories are based on size and function. Microbial decomposers are called "true" decomposers because they perform the crucial mineralization step. Based on their metabolism, microflora are either *obligate* or *facultative aerobes* or *anaerobes*. In aquatic sediments, bacteria and some fungi practice *anaerobic respiration*. The term is often considered synonymous with *fermentation* as practised by some yeasts and bacterial symbionts of ruminants, but anaerobic respiration involves an electron transport chain (with compounds such as nitrate or sulphate as the terminal acceptor), whereas fermentation does not. (Because fermentation also occurs in anaerobic conditions, it is included in the broader category of *anaerobic metabolism*.) Anaerobic respiration can generate half the energy of aerobic respiration, far more than fermentation.

Bacteria and fungi are the dominant decomposers of animal and plant tissue, respectively. (Bacteria and fungi that feed on living organisms are *parasites*; see Chapter 15.) Fungi range from single-celled yeasts to macroscopic fungi such as sugar fungi (Zygomycetes). Sugar fungi (e.g., bread mould, *Rhizopus* spp.) feed on simple compounds such as glucose, whereas other macroscopic fungi (e.g., Basidiomycetes and Ascomycetes) (Figure 21.2a, p. 450) utilize more complex compounds. These latter groups produce the conspicuous mushrooms and other spore-producing bodies common in forests. Given their size, it might seen odd that macrofungi are considered microbial, but the extracellular digestion practised by fungi involves individual microscopic hyphal filaments rather than the ingestion typical of multicellular consumers. A fungus is a fluid, branching colony of interacting hyphae that fuse or separate, rather than a single organism with an enclosing epidermis, as in plants or animals. We examine the many roles of fungi in Section 21.3, but first we consider the organisms that precede them in decomposition: *detritivores*.

An array of invertebrates aid decomposition by fragmenting detritus. Detritivores fall into four groups, based on body width: (1) *microfauna* (< 100 μm), including protozoans, rotifers, and smaller nematodes; (2) *mesofauna* (100 μm–2 mm), including mites (Figure 21.2b), pot worms, and springtails; (3) *macrofauna* (2–20 mm); and (4) *megafauna* (> 20 mm). The last two groups include millipedes, earthworms (Figure 21.2c), and snails in terrestrial habitats, and annelid worms, smaller crustaceans (especially amphipods and isopods), and molluscs and crabs in aquatic habitats (Figure 21.2d). Earthworms and snails dominate the megafauna.

Macro- and megafauna burrow into the substrate, affecting soil structure and acting as bioengineers (see Section 16.4). Although we generally regard earthworms as beneficial ecological agents, their impact can be negative in ecosystems that have evolved in their absence. The common earthworm, *Lumbricus terrestris*, is one of several alien invasive earthworms in North America. Their participation in detrital webs is initiating a cascade of effects on soil properties and the biological structure of forests (Figure 21.3, p. 451) (Frelich et al. 2006). Using tree rings to date earthworm invasion, Larson et al. (2010) established that stands of sugar maple (*Acer saccharum*) invaded by earthworms are more drought sensitive—a response of concern given that most climate-change models predict increasing drought in continental regions.

Why do earthworms increase plant susceptibility to drought? In greenhouse studies, the presence of three European earthworms reduced the thickness of the organic layer by accelerating decomposition (Hale et al. 2008). A thinner O horizon increases soil water loss, which will accelerate in a warmer climate. Faster decomposition also reduces carbon storage, further contributing to a warming world (see Chapter 28). An altered organic mat changes seedbed properties, likely affecting succession. The presence of earthworms increases nutrients in the short term, but this increase did not stimulate forest productivity (Hale et al. 2008). In the longer term, forests may become less fertile as a result of increased leaching. Other possible impacts include a shift in species composition of the forest understory, particularly towards species that are resistant to grazing by deer, which will likely increase in the more open forest stands; altered competitive interactions; and increased susceptibility to invasion by non-native plants (see Figure 21.3).

Do earthworms pose a threat to Canadian forests? Earthworm invasion of deciduous forests of eastern North America is essentially complete, and they are now making inroads into northern boreal forests. Coniferous-dominated forests are less vulnerable, as earthworms strongly prefer deciduous litter (Frelich et al. 2006), but mixed-wood and deciduous stands dominated by aspen and birch are highly susceptible. Although earthworms are slow to disperse, their spread is facilitated by humans. In northern Alberta, earthworm presence was significantly greater in forest stands near boat launches and roads (Cameron et al. 2007). Vehicle transport and bait abandonment are likely mechanisms of earthworm invasion.

Detritivores consume plant and animal remains, and although they do not mineralize tissue, they facilitate decomposition by increasing surface area for microbial decomposers. Meanwhile, energy and nutrients incorporated into bacterial and fungal biomass do not go unexploited. Feeding on bacteria and fungi are *microbivores*, including protozoans such as amoebas; invertebrates such as springtails (Collembola), nematodes, beetle larvae (Coleoptera), and mites (Acari); and other fungi. Smaller microbivores feed only on bacteria and fungal hyphae. Larger microbivores that consume both microflora and detritus overlap in function with detritivores.

Figure 21.2 Major decomposer groups. (a) Fungi and bacteria are microbial decomposers of plant and animal tissues. (b) Mites and springtails are abundant in the mesofauna. Common megafaunal members are (c) millipedes and earthworms in land habitats and (d) molluscs and crabs in water.

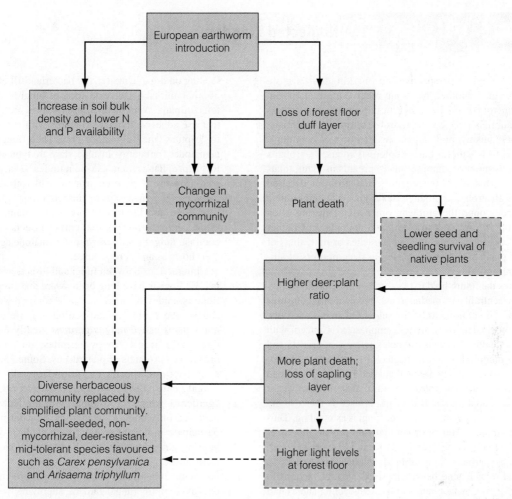

Figure 21.3 Proposed changes induced by invasion of European earthworms on soil properties and biological structure of a hardwood forest. Dashed lines and boxes indicate hypothesized effects with little data at this time.

(Adapted from Frelich et al. 2006.)

21.3 PLANT TISSUE DECOMPOSITION: A Sequence of Fungal Groups Breaks Down Plant Tissue

Now that we have met the major players, let's examine how they work together to decompose plant litter, which makes up the vast majority of detritus. Assume that detritivores (the identity of which varies with the ecosystem, climate, and soil) have fragmented the litter. What follows is a sequence of fungal decomposers that participate in overlapping activities based on the ecological "behaviour" of a particular fungal group and its ability to degrade specific organic substrates (Figure 21.4). Given the diversity of fungi (see Ecological Issues: The Neglected Kingdom, pp. 452–453), we use functional rather than taxonomic groupings (Deacon 2006).

   1. *Epiphytic fungi* are aerobic yeasts or yeast-like fungi that are adapted to tolerate the stressful conditions of the **phyllosphere** (environment of leaf surfaces), in which UV exposure is high, moisture and nutrients low, and temperature variable. This first group starts working while the organism is alive, feeding on dead cells and soluble nutrients leaking from the living host.

   2. *Endophytes*, *weak parasites*, and *pathogens*, which consume soluble substrates, colonize internal tissues. They also colonize living organisms, but use other ecological tactics. Endophytic fungi cause no obvious infection symptoms, and are facultative mutualists of many grasses. Occurring as hyphae between cells, they produce anti-herbivory compounds. Weak parasites and pathogens are less benign. Increasing in prevalence as plants senesce, they have an advantage while the host is alive over saprophytic fungi because they can tolerate the plant's resistance

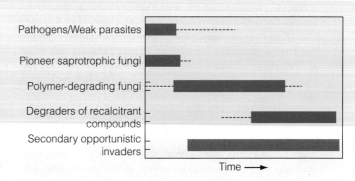

Figure 21.4 Generalized decomposition sequence, showing overlapping activities of fungal groups. Epiphytic fungi are not shown because they feed on living hosts.

(Adapted from Deacon 2006.)

ECOLOGICAL ISSUES · The Neglected Kingdom

From an ecological perspective, no one kingdom is more important than another, but some receive more attention. Most introductory ecology texts (including this one) stress plants and animals over bacteria, archaea, fungi, or protists, whether as organisms, populations, or members of communities. The reason is simple. Early ecologists focused on plants and animals as the most familiar macroorganisms. This focus persists, exacerbated by space limits that prohibit detailed treatment of all groups in a single textbook or undergraduate course. To compound the problem, other kingdoms are often considered only in relation to plants and animals. In Chapter 15, we discuss bacteria and fungi as parasites or mutualists of animals and plants, and in this chapter, as decomposers of animal and plant tissues.

What does the future hold for research in these "other" kingdoms? Their centrality to medicine and biotechnology ensures that bacteria will continue to be the subject of intense activity, even if their ecological roles are less emphasized. Concern with the declining health of oceans promotes interest in protists and archaea. But fungi risk scientific neglect. As parasites or pathogens of crops or trees, they are well studied by agricultural and forest scientists, but their roles in natural systems receive far fewer research dollars, even as their importance is increasingly apparent. Yet the ecological diversity of fungi is astounding. Thus far, we have stressed their most familiar roles, as decomposers and parasites of plants (see Sections 21.3 and 15.1, respectively), and as mutualist partners in mycorrhizae and lichens (see Section 15.8). Fungal roles extend beyond these species interactions to the ecosystem level, as facilitators of nutrient cycling and as part of *common mycorrhizal networks* that link organisms and populations in forests (see Section 15.13).

Even less familiar are their roles as parasites of other fungi and of animals, and as decomposers of keratin-containing animal tissue. Consider fungal parasites of animals. In humans and other mammals, fungal diseases are minor, at least in healthy individuals. Yeast infections and athlete's foot are well-known examples. But fungi have much greater parasitic effects on insects and nematodes. Many insects have fungal parasites, which penetrate the insect cuticle in much the same way as the fungal parasites called *rusts* infiltrate their plant host. Unlike many parasites, which are *biophages* (feed on living tissues), most fungal parasites of animals are **necrotrophic**; they typically kill their insect hosts with toxins, and then feed on the dead tissues. (They are thus *saprophages*, but unlike saprophagic fungi of plants, they kill their host.) Some produce an antibiotic after the insect dies, inhibiting bacterial invasion (Deacon 2006). Fungal parasites that attack agricultural insects have potential as biocontrol agents. One species even parasitizes mosquito larvae, raising the possibility of an alternative type of mosquito control.

Some fungi colonize another important invertebrate group—nematodes, which inhabit grassland soils in the millions per square metre. As well as being the most abundant belowground herbivores, consuming roots of both native and crop species, some nematodes belong to the detrital food web, feeding on organic matter and bacteria. Still others are parasites of other animals. A diverse array of fungi feed on these ubiquitous animals. *Nematode-trapping fungi* are "predatory" fungi that use adhesive hyphal structures with recognition capability to "capture" their nematode prey. These fungi do not consume nematodes for energy. Instead, they are typical cellulose-digesting saprophages, but use nematodes as a supplementary N source in the N-poor environment of rotting wood. Ecologically, they are analogous to "insectivorous" plants such as Newfoundland and Labrador's provincial plant, the pitcher plant (*Sarracenia purpurea*), that entrap insects as an N source. In contrast, *fungal endoparasites* of nematode eggs or cysts do use their hosts as an energy source.

Interactions between fungi and insects are equally fascinating. Ecologists have long been aware that fungal mutualisms in lichens and mycorrhizae have shaped—and continue to shape—the evolution and ecology of life on Earth. But the mutualisms called *fungus gardens* greatly facilitate access by three major insect groups (termites, ants, and beetles) to the breakdown products of plant decay. Some 330 of the over 2600 known termite species (Termitoidae) cultivate a specialized fungal genus, *Termitomyces*. Workers construct underground "gardens," which they supply with the feces of termites that have fed on wood and leaf litter. Non-farming termites also participate in decomposition, both as detritivores and by harbouring gut protozoans and bacteria that catalyze breakdown of cellulose.

Some 200 species of *attine* (fungal-growing) ants (Formicidae) also employ fungal gardens, supplying the fungus with plant litter rather than with feces. As with termite farmers, the interaction is obligate, with the ants rearing their young on a fungal diet. Leafcutter ants (*Atta* and *Acromyrmex* spp.) (Figure 1a) construct massive mounds, with radii over 75 m and containing millions of ants. Fungal-farming beetles (weevils or ambrosia beetles, Scolytidae) cultivate fungi in large galleries inside trees. The fungi consume the wood of dying or dead trees, and the beetles consume their hyphae. All three types of fungal farmers tend their gardens, protecting the fungi from contaminants. Fungus gardening most likely evolved from commensalistic use of insects as vectors for dispersing fungal spores. It evolved independently in the three groups—an example of *convergent evolution*, occurring some 40 million to 60 million years before humans "invented" agriculture (Mueller and Gerardo 2002).

Even in aquatic habitats where they are often considered insignificant, fungi perform many roles. The reliance of lignin-degrading fungi on oxidative breakdown means these groups play a minimal role in O_2-poor aquatic habitats. But some basidiomycetes are adapted to water (Figure 1b), and one early fungal group, chytrids (Chytridiomycota), is mainly aquatic. Once considered protists, most chytrids are single-celled or simply branched chains that consume organic material in wet soil or freshwater, degrading chitin and keratin. Some are internal parasites of plants, animals, or algae. They are the only true fungi with motile, flagellate zoospores, exemplifying

Figure 1 Examples of the diverse ecological roles of fungi. **(a)** Leaf-cutter ants transporting a leaf to their fungal garden. **(b)** This recently discovered basidiomycete, *Psathyrella aquatica*, is the first documented example of a mushroom fruiting under water. Note the bubbles coming from the gilled cap. **(c)** A cellular slime mould.

their aquatic adaptation. Spores of aquatic fungi that are not chytrids have appendages that aid dispersal in water, either by increasing buoyancy or by promoting adhesion. By latching on to plant debris that falls into fast-flowing streams, these fungi speed its decay.

Finally, fungus-like organisms occupy unique niches. Like fungi, *oomycetes* have hyphae with apical (tip) growth and enzymes that degrade plant cell walls. Causal agents of plant diseases such as potato blight (which precipitated the Irish potato famine of the 1840s, altering human history), sudden oak death, and downy mildew, oomycetes resemble plants, with cell walls containing glucose polymers; diploid nuclei (fungal nuclei are haploid, at least when non-reproductive); and similarities in subcellular structure, membranes, and storage compounds. Even more bizarre are *cellular slime moulds*—haploid, unicellular, amoeba-like organisms (Figure 1c) that engulf bacteria and organic particles. Only when nutrients are depleted do they aggregate into a large, multicellular fruiting body. Also ingesting bacteria and particles by phagocytosis (as opposed to the extracellular digestion of fungi) are *plasmodial slime moulds*

(Myxomycota), which consist of a protoplasm mass with no walls but containing many nuclei.

This brief summary of fungal ecological diversity shows that we cannot neglect research into any biological kingdom if we wish to grasp the complex interactions of nature. Fungi may seem of minor interest and peripheral importance, especially if we think of mushrooms as odd-looking structures that are either delicious or deadly. Instead, we must appreciate fungi (and members of all kingdoms) from an enlightened ecological perspective. The mycelial mass that dominates so many belowground and within-organism habitats performs an astonishing range of ecological roles, either independently or in collaboration with members of other kingdoms.

Bibliography

Deacon, J. 2006. *Fungal biology*. Oxford: Blackwell Scientific.

Mueller, U. G., and N. Gerardo. 2002. Fungus-farming insects: Multiple origins and diverse evolutionary histories. *Proceedings of the National Academy of Sciences* 99:15247–15249.

strategies, which weaken as the plant ages. They remain active after the plant dies, but are soon displaced. The line separating parasite from decomposer is blurred in fungi. A fungus such as *Armillaria* (see p. 447) feeds on the walls of dead cells in wood rather than attacking living plant cells as do fungal parasites, but continues consuming the same dead cells after the plant dies.

3. *Pioneer saprophytic fungi (sugar fungi)* colonize plant tissue after death, consuming sugars or storage compounds rather than structural polymers such as cellulose. Fast growing and with short life cycles, fungi such as bread moulds are typical *r*-strategists (see Section 8.10). Other examples are sapstain fungi that colonize fallen logs.

4. *Polymer-degrading fungi*, which degrade cellulose and *hemicellulose*, another structural polymer in plant cell walls, replace sugar fungi. Like most fungi, they employ **extracellular digestion**: secretion of enzymes that catalyze breakdown of organic compounds, followed by absorption of the breakdown products. However, not all fungal breakdown is enzymatic. Brown rot fungi (many of which produce the familiar shelf-like brackets on the trunks of dead trees) degrade cellulose via an oxidative process using hydrogen peroxide (H_2O_2) produced during enzymatic breakdown of hemicellulose. Because wood is N-poor (C:N ratio of ~ 500:1; see Section 21.7), this strategy conserves N by not relying on enzymes. Much longer lived than sugar fungi, polymer-degrading fungi are often highly specialized for different organic substrates or abiotic conditions. They exploit different detrital microhabitats. Some tolerate dry conditions, others low pH. Some are saline tolerant and can degrade plant tissues in estuaries and salt marshes.

5. *Lignin-degrading fungi*, which specialize in degrading *lignin*, a complex and highly resistant group of compounds, dominate later stages. Overlapping with polymer-degrading fungi, they rise to prominence as lignin becomes an increasing proportion of the remaining tissue. However, these fungi rarely use lignin as their major energy source. Too many enzymes are needed to degrade its many bonds. Instead, most lignin-degrading fungi rely on extracellular digestion of cellulose, but modify lignin with a few enzymes that release the cellulose (and other substrates) that are complexed with it. White-rot fungi, in contrast, use an oxidative process to degrade lignin. By generating powerful oxidants, they "burn" lignin.

Given its reliance on oxidative decay, lignin degradation is O_2-dependent and minimal in waterlogged conditions, allowing submerged wood to remain remarkably intact. This explains why dock supports are made of wood, and why the buildings of Venice rest on wooden piers. Also, as lignin-degrading fungi did not evolve until after the early seedless vascular plants had risen to dominance, large amounts of plant detritus accumulated as fossil fuel, and atmospheric O_2 rose to about 21 percent. So the past absence of lignin-degrading fungi as well as their present proliferation has profoundly affected biosphere structure and function.

Lignin-degrading fungi are long-lived and conservative in their mineral use. Indeed, release of lignin-degrading enzymes is stimulated by N shortage. Only after earlier-colonizing fungi have lowered available N are they stimulated to attack the residual material (Deacon 2006). Once established, basidiomycetes in particular maintain dominance by antagonistic behaviour towards hyphae of other fungi upon contact—an example of interspecific interference competition.

6. *Secondary opportunistic fungi*, which colonize dead fungal remains, insect exoskeletons, or living fungi, are the final group, overlapping with (and sometimes preceding) lignin-degraders. Not occupying a single niche, they function as scavengers, true decomposers, parasites, or commensalists. They must tolerate extremely low nutrient levels as well as the toxic by-products of other fungi.

This succession of six fungal groups decomposes plant organic matter until it is transformed into inorganic nutrients or highly resistant residual compounds in humus (see Section 21.8).

21.4 ANIMAL TISSUE DECOMPOSITION: Insect Detritivores and Bacteria Degrade Animal Tissue

What about animal tissue? As with plants, the surfaces of living animals are inhabited by millions of microbes, both bacterial and fungal. Normally they are harmless and may even suppress pathogens, but if the immune system is compromised by age or conditions such as AIDS, these microbes (including *fungal pathogens*, to which animals are normally resistant) can become life-threatening. Also as with plants, *internal parasites* and pathogens will be present, causing more damage as the animal nears death and its immune system weakens. For example, the chicken pox virus (*Varicella zoster*) resides in the body long after the childhood disease is past. Later in the host's life, the virus can rebound, causing the painful condition called shingles.

Scavengers such as vultures consume the more digestible parts of a carcass (muscles and organs). Given the similarity of animal tissues, consumption of recently dead tissue resembles carnivory in a grazing chain. *Insect detritivores*, notably maggots (larvae of flesh flies, Sarcophagidae, and blow flies, Calliphoridae) feed on the carcass once bloating has begun. Many also consume dung. Some plants have evolved to exploit this group. By mimicking the smell of rotting flesh, species such as carrion flower (*Smilax* spp.) attract blow flies as pollinating vectors—an example of *aggressive mimicry* (see Section 14.9).

Following fragmentation by scavengers, the initial phase of decomposition is **autolysis**, whereby enzymes released by the body's own lysosomes digest some cell contents. Oxygen levels fall, and anaerobic bacteria proliferate. This second stage, **putrefaction**, causes bloating from bacterial release of gases such as methane and hydrogen sulphide. Bacterial decay of proteins contributes other putrid-smelling compounds. Mass loss is greatest during putrefaction as a result of maggot feeding and liquefaction of body contents.

When only resistant materials such as bones, nails, and hair remain, decomposition slows. Fungi that degrade *keratin* (the major structural polymer of animal tissue) decompose hair and hooves. Some fungi also degrade *chitin*, the major structural polymer in insect exoskeletons and fungal walls. Finally, as with plant decay, minerals released by decomposition are returned to the soil.

Figure 21.5 In a litterbag experiment, a known quantity of detritus (senescent leaves in this case) is placed in mesh bags on the habitat floor. Bags are retrieved at various intervals, and the mass loss that results from consumption by decomposers is tracked through time.

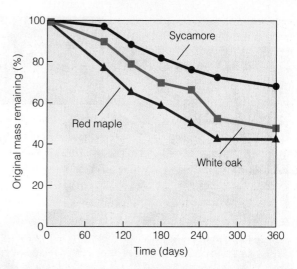

Figure 21.6 Decomposition of fallen leaves from red maple, white oak, and sycamore trees. The graph shows the percentage of the original mass remaining over the first year in a litterbag study.

INTERPRETING ECOLOGICAL DATA

Q1. What portion of the original mass was consumed for each of the three species by day 180?

Q2. Which of the three species has the lowest decomposition rate?

21.5 LITTERBAG STUDIES: Ecologists Measure Decomposition by Using Litterbags

As heterotrophs, decomposers derive their energy and most of their nutrients from organic compounds. Ecologists study decomposition by experiments that track this decay over time. A common approach uses **litterbags**—mesh bags made of decay-resistant material (Figure 21.5). Mesh holes must be large enough to allow decomposers to enter and feed on the litter but small enough to prevent decomposing material from leaving (typically 1–2 mm). Unfortunately, this mesh size alters decomposition by restricting the access of larger detritivores, reducing the ability of litterbag studies to generate realistic estimates of decomposition rate (Prescott 2005).

The researcher places a fixed amount of litter in each bag. In the experiment presented here, for each of 3 tree species, 30 bags were filled with 5 g each of leaf litter and buried in the forest litter layer. At 6 times over a year, 5 bags were collected per species, and their contents dried and weighed. The researchers then estimated the decomposition rate for each species (see Quantifying Ecology 21.1: Estimating Decomposition Rate, p. 457). The mass remaining decreases continuously over time (Figure 21.6) as carbon is released as CO_2.

Although the mass remaining in the bag includes not only residual plant material but also bacterial and fungal colonists of the litter, few studies quantify the changing contribution of primary (original plant litter) and secondary (microbial biomass) organic matter in the decaying litter. One such study estimated fungal growth during decomposition of sawdust by measuring the change in chitin content, which occurs in fungal but not plant cell walls. Sawdust mass decreased by 39 percent, but 58 percent of the remaining mass consisted of living and dead fungal biomass. The apparent decomposition rate (k) of the sawdust was 0.04/wk, but the rate more than doubled—to 0.09/wk—when recalculated to exclude fungal biomass

(Swift et al. 1979; for a discussion of k, see Quantifying Ecology 21.1: Estimating Decomposition Rate, p. 457). This shift in the proportion of remaining plant versus fungal biomass is crucial to understanding nutrient dynamics during decomposition (see Section 21.6), and supports the view that microbial mass becomes an increasingly important component of soil organic matter over time (Prescott 2010).

Aquatic ecologists use a similar method to quantify litter decomposition in streams. Leaf litter subsidies provide large energy and nutrient inputs into streams, which typically have low primary productivity. Leaf litter accumulates in deposition areas, forming **leaf packs** (Figure 21.7a, p. 456). To quantify decomposition, plant litter is placed in mesh bags (Figure 21.7b) (also called leaf packs), which are anchored in areas where deposition occurs. By varying mesh size, ecologists can determine the relative importance of microbes and detritivores. Fine-mesh screens allow only microbes to enter, while coarse screens also allow larger organisms to feed on the detritus (Figure 21.7c). As with litterbags, leaf packs are placed in the stream for several weeks to measure mass loss and chemical changes of the litter over time (Figure 21.8, p. 456). Decomposition in streams is discussed further in Sections 21.11 and 24.4. Leaf packs are also used in coastal wetlands, but the bags are tethered to stakes to stop them from being displaced by tidal currents.

21.6 FACTORS INFLUENCING DECOMPOSITION: Biotic and Abiotic Factors Affect Decomposition Rate

As illustrated by Figure 21.8, not all organic matter decomposes at the same rate. From many similar studies, some generalizations emerge. Rate of decay is related to (1) intrinsic biotic

(a)

(b)

(c)

Figure 21.7 Decomposition in streams. **(a)** Inputs of plant litter from surrounding terrestrial habitats form areas of deposition, called leaf packs, in streams. **(b)** Ecologists study decomposition using mesh bags to simulate natural leaf packs. **(c)** Smaller mesh bags allow access only to microbial decomposers, while larger mesh allows access to detritivores.

factors, particularly the quality of plant litter as a substrate for decomposers, and (2) extrinsic abiotic factors that influence decomposer populations, including soil texture, pH, and climate.

Intrinsic properties that influence the quality of plant litter relate to the types and amounts of organic compounds pres-

ent—their size, structure, and types of chemical bonds. Carbon is plentiful, making up 45 to 60 percent of the total plant litter dry mass, but (as is also true for herbivores feeding on living plant tissue) not all carbon compounds are of equal quality for decomposers. Glucose and other simple sugars are small. Breaking their chemical bonds yields much more energy than required to synthesize the enzymes needed to catalyze their breakdown. Cellulose and hemicellulose, the main constituents of plant cell walls, are more complex and difficult to decompose, making them of moderate quality as a substrate for microbial decay.

The much larger lignin molecules present in wood as a strengthening substance are among the most complex and variable organic compounds in nature. Lignin has no single chemical formula; rather, it is a class of large molecules folded into complex three-dimensional structures that shield their internal bonds from enzymatic attack. Thus, lignins are the slowest plant compounds to decompose and are of such low quality that they yield little net energy gain to microbial decomposers. Bacteria cannot degrade lignins, which are broken down only by basidiomycete (see Figure 21.2a) and ascomycete fungi, either independently or cultivated by insects in "fungal gardens" (see Ecological Issues: The Neglected Kingdom, pp. 452–453).

An experimental study of straw decomposition illustrates the variation in decomposition rates of organic compounds (Figure 21.9). Total C content, expressed as a percentage of the original mass, declined exponentially over the 80-day study. However, when C was partitioned into various classes, decomposition rates varied widely. Proteins, simple sugars, and other soluble compounds (15 percent of the original mass) decomposed quickly, disappearing in a few days. Cellulose

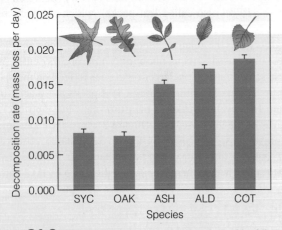

Figure 21.8 Decomposition (mass loss per day) of leaf litter from 5 tree species submerged in streams. Experiments used submerged leaf packs sampled at 5 intervals over 83 days. SYC: *Platanus wrightii*; OAK: *Quercus gambelii*; ASH: *Fraxinus velutina*; ALD: *Alnus oblongifolia*; COT: *Populus fremontii*.

(Adapted from LeRoy and Marks 2006.)

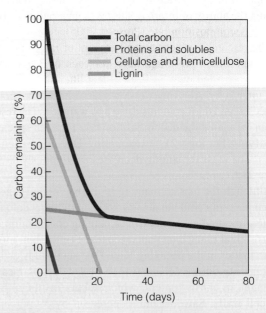

Figure 21.9 Variation in decay rates of different compounds during decomposition of straw. At any time, the sum of the three classes equals the total carbon.

(Adapted from Swift et al. 1979.)

QUANTIFYING ECOLOGY 21.1   Estimating Decomposition Rate

Litterbag studies are the major method of studying decomposition. After collecting multiple replicate bags at regular intervals, researchers plot the proportion of mass remaining over time (Figure 1). In this example, each data point represents the mean of five replicate bags collected on a given day. The x-axis plots time (weeks); the y-axis plots the percentage of the original mass that is remaining. Mass loss through time (decomposition rate) is expressed as a negative exponential function:

$$\text{Original mass remaining} = e^{-kt}$$

Here, e is the base of the natural logarithm, t is time, and k is the **decomposition coefficient** (the slope of the negative exponential curve). The value of k is estimated by fitting an exponential regression model: $y = e^{-kx}$, where y is the original mass remaining and x the corresponding time in weeks. When applied to the Figure 1 data, this formula estimates k as 0.0167 and 0.0097 for red maple and Virginia pine, respectively, in proportional mass loss per week. (Values are converted to percentages by multiplying by 100.) The higher the k value, the faster the decomposition.

Data for a third species, sycamore (*Platanus occidentalis*), are given in Table 1. Each value represents the remaining dry mass of the 7 g of leaf litter placed in the bag at day 0. Percentage of original mass remaining is calculated by dividing each value by 7 and multiplying by 100.

Recently, ecologists have questioned the validity of many aspects of litterbag studies, including their reliance on the early stages of decay and their use of regression analysis (Prescott 2005). Despite their limitations, litterbag studies remain invaluable, particularly for comparison purposes.

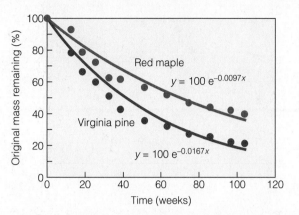

Figure 1 Decomposition of red maple (*Acer rubra*) and Virginia pine (*Pinus virginiana*) leaf litter over two years. Each point represents the mean percentage mass remaining in five replicate bags. Lines represent the predicted relationship based on estimating k using the exponential (nonlinear) model.

1. Using the data in Table 1, calculate the mean value for the five replicate samples at each time period, and convert the means to percentage original mass remaining (% OMR). Plot % OMR (y-axis) over time (x-axis), as in Figure 1. For comparison, convert days to weeks. How does the decomposition of sycamore compare to that of red maple and Virginia pine in Figure 1?
2. How does the lignin content of sycamore leaves likely compare to that of red maple or Virginia pine? Why?

Table 1 Mass of Decomposing Sycamore Leaf Litter in a Litterbag Study*

							Day					
Litterbag	0	90	131	179	228	269	360	445	525	603	680	730
Replicate 1	7	6.92	5.82	5.69	5.29	4.89	4.66	4.33	4.06	3.78	3.84	3.71
Replicate 2	7	6.84	6.08	5.59	5.28	4.99	4.87	4.24	3.99	4.09	3.71	3.83
Replicate 3	7	6.91	5.98	5.83	5.38	5.18	4.75	4.38	4.26	4.01	3.60	3.78
Replicate 4	7	6.75	5.74	5.88	5.44	5.13	4.92	4.50	4.21	3.88	3.92	3.58
Replicate 5	7	6.72	5.88	5.78	5.41	5.21	4.72	4.19	4.08	3.87	3.82	3.65

*All values in grams dry mass. Initial mass of litter in each bag was 7.0 g.

and hemicellulose (60 percent of the original mass) decomposed more slowly, but were gone in three weeks. Lignin (20 percent of the original mass) degraded slowly, with most still intact by day 80. As decomposition continued, the quality of the C resource declined, as the proportion of total C present as lignin rose. This decline in litter quality slows decomposition even further.

Given lignin's decay resistance, ecologists use the proportion of C present as lignin as an index of litter quality. The differing decomposition rates of the species in Figure 21.6

reflect their initial lignin content. Freshly fallen leaves of red maple (*Acer rubrum*) have a lignin content of 11.7 percent as compared to 17.7 percent for white oak (*Quercus alba*) and 36.4 percent for sycamore (*Platanus occidentalis*). This inverse relationship between lignin content and decomposition rate typifies many species in both terrestrial and aquatic habitats (Figure 21.10, p. 458).

Litter quality has a particularly important impact on decomposition in coastal marine sites. Phytoplankton contain no lignin and decompose quickly. However, vascular plants

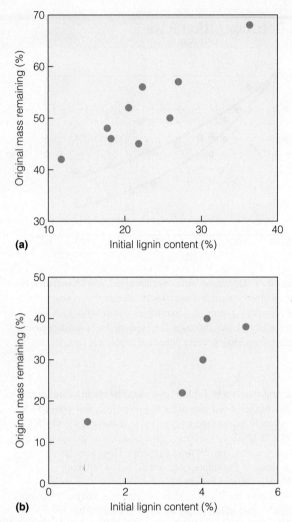

Figure 21.10 Initial lignin content and decomposition rate of plant litter from terrestrial (**a**) and aquatic (**b**) habitats. Each point represents a different plant species.

(Adapted from (a) Smith and Smith 2001 and (b) Klap et al. 1999.)

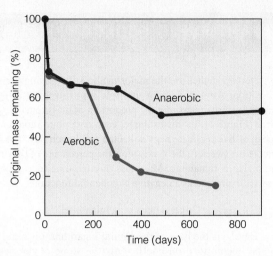

Figure 21.11 Decomposition of *Spartina alternifolia* litter exposed to aerobic (litterbags on marsh surface) and anaerobic (buried 5–10 cm below the marsh surface) conditions.

(Adapted from Valiela 1984.)

such as seagrasses and reeds that inhabit estuaries and marshes have lignin contents approaching that of land plants. Decomposition of their litter depends on the O_2 content of the water. In hypoxic or anoxic aquatic sediments, anaerobic bacteria perform most of the decomposition. Low abundance of fungi, most of which are aerobic, hinders breakdown of lignin, slowing decomposition (Figure 21.11).

Besides litter quality, abiotic factors directly affect decomposition. Low temperature inhibits microbial activity. Alternating wetting and drying and continuous dry spells also reduce microflora abundance and activity. Decomposition proceeds fastest in warm, wet climates with high humidity. Studies of decomposition at three sites in eastern North America (Figure 21.12) indicate that, although the lignin content of red maple (*Acer rubrum*) leaves does not differ significantly

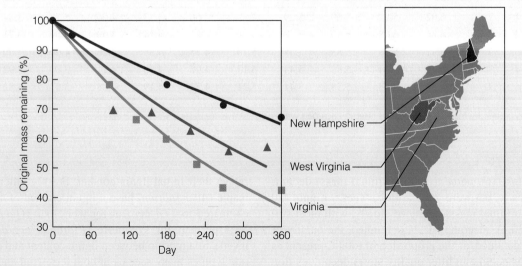

Figure 21.12 Decomposition of red maple leaf litter in New Hampshire (circles), West Virginia (triangles), and Virginia (squares). The declining decomposition rate from north to south reflects climatic differences.

(Adapted from Melillo et al. 1982, New Hampshire; Mudrick et al. 1994, West Virginia; Smith and Smith 2001, Virginia.)

Table 21.1 Mean Residence Time (years) of Organic Matter and Selected Nutrients in the Forest Floor of Different Forest Types

Region	Organic Matter	N	Ca	P
Boreal coniferous forests	350	230	150	324
Boreal deciduous forests	25	26	14	15
Subalpine coniferous forests	18	37	12	21
Temperate coniferous forests	17	18	6	1
Temperate deciduous forests	4	6	3	6
Mediterranean forests	3	4	4	1
Tropical rain forests	0.7	0.6	0.3	0.6

(Adapted from Andersson 2005; data from Cole and Rapp 1981.)

with latitude, decomposition occurs more rapidly in southern sites. These differences are directly attributable to climate. Mean daily temperature and annual potential evaporation ranged from 7.2°C to 14.4°C and from 621 to 806 mm from the most northerly to the most southerly site. The pervasive influence of temperature also causes a distinctive daily pattern of microbial respiratory activity, as indicated by CO_2 release from soil (Figure 21.13).

Decomposition rate affects litter **residence time**, calculated as litter accumulation (g/m^2) divided by the rate of litter production (g/m^2/yr). Litter residence time ranges from under a year in tropical rainforests to over 300 years in boreal coniferous forests (Table 21.1) (Andersson 2005; residence times for selected minerals are also listed). This long residence time explains why boreal forests (and Arctic tundra) are vital carbon sinks. If climate change accelerates decomposition—and particularly if deciduous species increase in dominance over coniferous—litter residence times will shorten, impairing carbon sink activity (see Section 28.7).

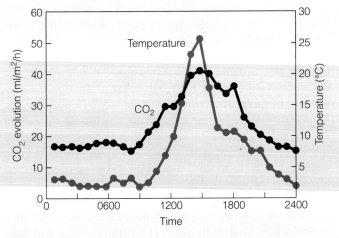

Figure 21.13 Diurnal changes in air temperature and decomposition in a temperate deciduous forest, as indicated by microbial release of CO_2 at different hours of a 24-hour period.

(Adapted from Whitkamp and Frank 1969.)

21.7 MINERALIZATION: Mineral Release Reflects the Balance between Immobilization and Mineralization

Detritus varies in nutrient as well as energy quality. Consider nitrogen. The higher the N content of plant litter, the higher the nutrient value, although not all organic N is equally accessible. As plant litter is consumed, bacteria and fungi transform organically bound N and other elements into inorganic form (*mineralization*). But the same decomposers that mineralize compounds also require N for their own needs. So, whenever mineralization occurs, **immobilization**—the uptake and assimilation of N (and other minerals) into the bodies of living organisms—runs counter to it (Figure 21.14, p. 460). Both mineralization and immobilization occur simultaneously as decomposers process litter, so the supply rate of minerals to the soil—the **net mineralization rate**—is the difference between the two.

Litterbag studies examine changes in nutrient content as well as in mass and carbon, again expressing it as a percentage of the nutrient content of the original litter. There are typically three stages (Figure 21.15, p. 460). Initially, N content declines as water-soluble compounds (including some proteins) are leached. This stage can be very short and, in terrestrial habitats, varies with rainfall. After this initial phase, microbial decomposers immobilize mineral N *external* to the litter, often causing the N content to rise above 100 percent, exceeding the litter's initial N content.

How is this possible? Recall that the organic matter in the bag includes living and dead microorganisms as well as residual plant litter. The N content of fungal and bacterial tissue is much higher than that of the plant litter they consume. The **C:N ratio** (grams of C per gram of N) of plant biomass typically ranges from 50:1 to over 500:1, whereas the C:N ratio of bacteria, animals, and fungi is some 10 to 15:1. As plant litter is consumed and N is immobilized in decomposer tissue, the C:N ratio of the remaining litter declines (Figure 21.16, p. 461), reflecting the changing proportion of plant and microbial biomass left in the bag, as well as the loss of plant C. As decomposition continues and litter quality declines (reflecting the higher proportion of lignin), there is a critical point at which mineralization exceeds immobilization, causing net release of N.

Figure 21.15 is an idealized model. Actual N dynamics depend on the initial nutrient content of the litter. If its N content is high, mineralization may exceed immobilization from the outset, and the N content of the litter will not rise above the initial level (Figure 21.17, p. 461). Similar patterns of immobilization and mineralization apply to other nutrients (Figure 21.18, p. 461).

21.8 SOIL ORGANIC MATTER: Residual Biomass Is Converted into Soil Organic Matter

Thus far, we have discussed decomposition as the process whereby decomposers consume plant litter, mineralizing nutrients and altering the composition of the residual organic matter.

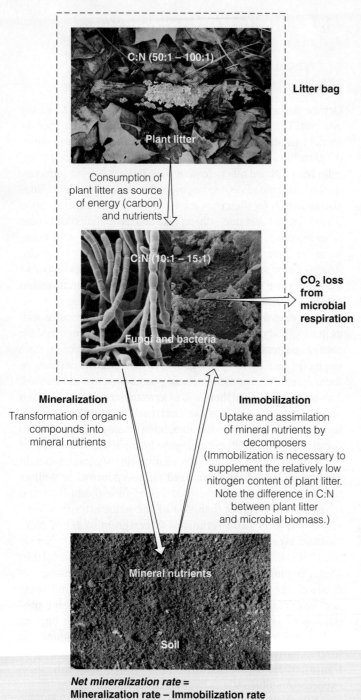

Figure 21.14 Exchanges between litterbag (dashed line) and soil during decomposition. As fungi and bacteria consume plant litter, much of the C is respired, and nutrients bound in organic matter are released by mineralization. At the same time, some minerals are immobilized into microbial tissue. Net mineralization rate is the difference between the rates of mineralization and immobilization. Over time, as the litter is converted into microbial biomass, the residual organic matter in the bag consists of a growing proportion of microbial biomass.

This process continues as the litter degrades into *humus*, dark brown or black organic matter residue that is highly resistant to further decay. As humus is embedded in the soil matrix, it becomes a major constituent of **soil organic matter**, along with compounds of microbial origin.

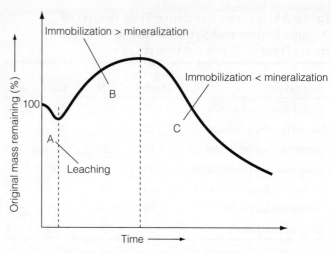

Figure 21.15 Idealized change in N content of plant litter during decomposition. In Phase A, soluble compounds are leached. In Phase B, N content increases above the initial level as N is immobilized. In Phase C, mineralization exceeds immobilization, causing net release of N.

Studying the development of soil organic matter requires a longer timeline than most litterbag studies, which often last only one or two years. Focusing on the early stages of decay, especially if detritivores are excluded by the mesh size, can generate misleading conclusions (Prescott 2005). In a long-term study of needle litter in a Scots pine (*Pinus sylvestris*) forest in Sweden, biomass continued to decline over five years (Figure 21.19a, p. 462). Meanwhile, the need for decomposers to convert plant litter with a C:N of 134:1 into microbial biomass with a C:N of 10:1 caused a prolonged N immobilization (Figure 21.19b). As a result, the N content of the residual material increased, and its C:N ratio fell (Figure 21.19c) (Berg et al. 1982). As decomposition proceeds, decomposition slows as *labile* (easily digested) compounds are consumed. Increasingly *recalcitrant* (resistant to break-down) compounds remain, with an increasing lignin fraction (Figure 21.19d).

Over time, compounds of plant origin make up an ever smaller proportion of soil organic matter. As fungi die, chitin and other recalcitrant components of fungal cell walls make up an increasing proportion. In fact, Prescott (2010) argues that the key to stable carbon stores in soil is not slowing the decomposition of plant litter but rather promoting the pro-duction of novel recalcitrant compounds by decomposers. The quality of soil organic matter as a substrate for further decomposition deteriorates over time as these compounds increase, stabilizing soil organic content. Its C:N ratio declines further as more C is respired and some of the miner-alized N is incorporated into humus, which forms complexes with mineral soil particles.

The low C:N ratio of humus is not an indicator of increased N availability, because the residual N is incorporated into com-pounds that resist further breakdown. By now, the original plant litter has been degraded to a point where decomposition proceeds very slowly. Soil organic matter typically has a resi-dence time of 20 to 50 years, although it ranges from 1 or 2

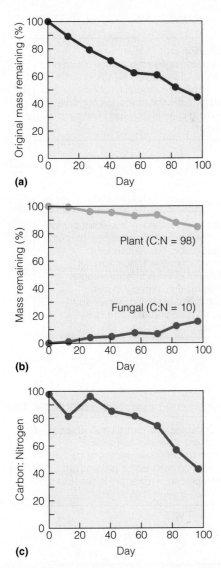

Figure 21.16 Changing composition of decomposing winter rye. **(a)** Mass loss over the 100-day field study. **(b)** Proportion of the remaining mass in plant and fungal biomass (living and dead). **(c)** Because the C:N ratio of fungal biomass is lower than that of the remaining litter, the C:N declines during decomposition.

(Adapted from Beare et al. 1992.)

INTERPRETING ECOLOGICAL DATA

Q1. Assume that fungal decomposers assimilate 100 units of C as plant litter, of which 40 units are converted into fungal biomass while the remaining 60 are respired as CO_2. Since the C:N ratio of fungi is 10:1, production of 40 units of biomass would require 4 units of N (40:4). Assuming the C:N ratio of the 100 units of plant litter is 98:1, and that the decomposers retain all the N in the litter they consume, how many units of N must be immobilized to produce 40 units of fungal biomass?

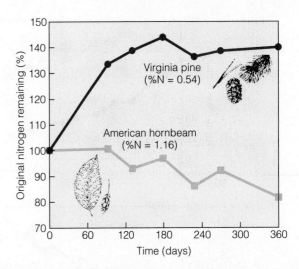

Figure 21.17 Change in N content of leaf litter from American hornbeam and Virginia pine, expressed as a percentage of original N remaining over year 1 of a litterbag study. Note the difference in initial N content and subsequent rates of immobilization.

(Adapted from Smith and Smith 2001.)

21.9 SOIL MICROBIAL LOOP: Plant Exudates Promote Decomposition of Soil Organic Matter

Over a century ago, Lorenz Hiltner used the term **rhizosphere** to describe the soil region that is an active zone of root growth and death, as well as the scene of intense bacterial and fungal activity. Decomposition is more rapid here than in bulk soil. The rhizosphere occupies virtually all of a grassland soil, where the mean distance between roots is 1 mm. Root density is far lower in forest soils (often 10 mm between roots). Roots alter the chemical environment of the rhizosphere by secreting carbohydrates and other compounds. These exudates can account for 40 percent of plant dry matter production. Such a large C expenditure must be of great importance for plants to justify the significant trade-off in allocation to other plant parts.

years in a cultivated field to millennia in habitats with slow decomposition (dry or cold or waterlogged or highly acidic, as in a peat bog). So humus decomposes slowly, but given its sheer abundance, it contains a significant portion of the carbon and nutrients released from soils.

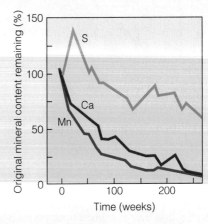

Figure 21.18 Nutrient dynamics of sulphur (S), calcium (Ca), and manganese (Mn) in decomposing Scots pine needles in a five-year litterbag study.

(Adapted from Staaf and Berg 1982.)

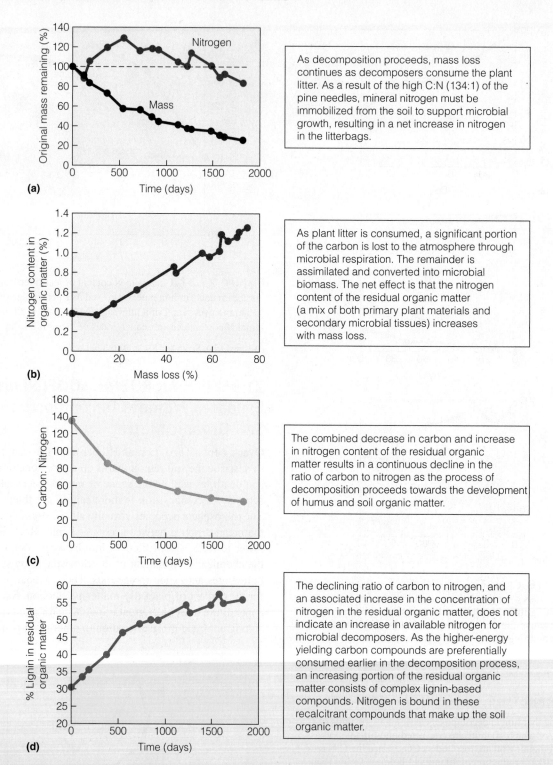

As decomposition proceeds, mass loss continues as decomposers consume the plant litter. As a result of the high C:N (134:1) of the pine needles, mineral nitrogen must be immobilized from the soil to support microbial growth, resulting in a net increase in nitrogen in the litterbags.

As plant litter is consumed, a significant portion of the carbon is lost to the atmosphere through microbial respiration. The remainder is assimilated and converted into microbial biomass. The net effect is that the nitrogen content of the residual organic matter (a mix of both primary plant materials and secondary microbial tissues) increases with mass loss.

The combined decrease in carbon and increase in nitrogen content of the residual organic matter results in a continuous decline in the ratio of carbon to nitrogen as the process of decomposition proceeds towards the development of humus and soil organic matter.

The declining ratio of carbon to nitrogen, and an associated increase in the concentration of nitrogen in the residual organic matter, does not indicate an increase in available nitrogen for microbial decomposers. As the higher-energy yielding carbon compounds are preferentially consumed earlier in the decomposition process, an increasing portion of the residual organic matter consists of complex lignin-based compounds. Nitrogen is bound in these recalcitrant compounds that make up the soil organic matter.

Figure 21.19 Decomposition of Scots pine leaf litter. **(a)** Mass and N loss. **(b)** Change in N content of residual litter. **(c)** C:N ratio. **(d)** Lignin content.

(Data from Berg et al. 1982.)

Bacterial growth in the rhizosphere is supported by these abundant, energy-rich exudates, which are non-cellular and hence easily digested. Bacterial growth is limited by nutrient supply because the exudates are low in N and other minerals. Bacteria must acquire these nutrients by breaking down soil organic matter. In essence, plant roots use C-rich exudates to facilitate decomposition of low-quality (C-poor) organic matter in the rhizosphere. Immobilization accompanies this growth, and nutrients would remain sequestered in bacterial biomass if predation by microbivorous protozoa and nematodes did not remobilize them for plant uptake.

Unlike the large discrepancy between microbial decomposers and plant litter, there is little difference in the C:N ratio between microbivores and their prey. Given their low feeding efficiency, protozoa and nematodes use only 10 to 40 percent and 50 to 70 percent, respectively, of their prey C for producing biomass, with the remainder lost as respiratory CO_2. They excrete excess N as NH_3, which is then available for root uptake. Interactions between microbivores and microbial decomposers thus affect the rate of nutrient cycling in the rhizosphere and thereby influence mineral availability to plants. This process whereby root exudates supply C to microbial decomposers in the rhizosphere, enhancing decomposition of soil organic matter and subsequently releasing minerals for plant uptake after consumption of microbes by microbivores, is called the **soil microbial loop** (Figure 21.20). Long downplayed in ecological studies, it is analogous to the microbial loop in aquatic ecosystems discussed in Section 21.11.

The amount of energy passing through the soil microbial loop is staggering. Populations of protozoa and nematodes

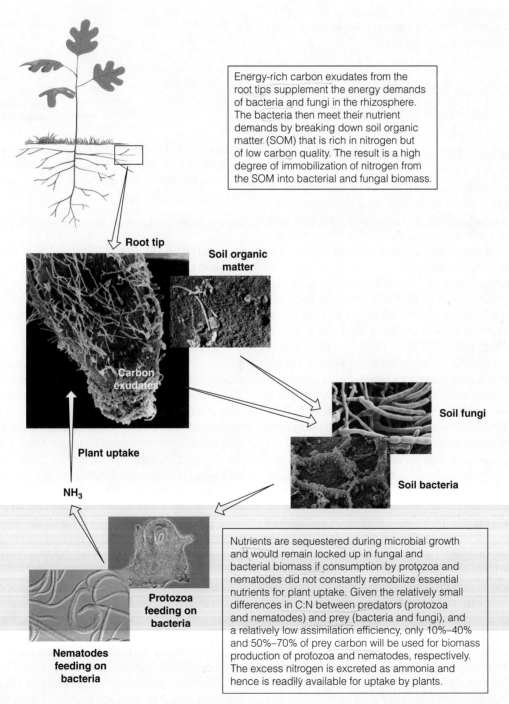

Energy-rich carbon exudates from the root tips supplement the energy demands of bacteria and fungi in the rhizosphere. The bacteria then meet their nutrient demands by breaking down soil organic matter (SOM) that is rich in nitrogen but of low carbon quality. The result is a high degree of immobilization of nitrogen from the SOM into bacterial and fungal biomass.

Root tip

Soil organic matter

Carbon exudates

Soil fungi

Soil bacteria

Plant uptake

NH_3

Protozoa feeding on bacteria

Nematodes feeding on bacteria

Nutrients are sequestered during microbial growth and would remain locked up in fungal and bacterial biomass if consumption by protozoa and nematodes did not constantly remobilize essential nutrients for plant uptake. Given the relatively small differences in C:N between predators (protozoa and nematodes) and prey (bacteria and fungi), and a relatively low assimilation efficiency, only 10%–40% and 50%–70% of prey carbon will be used for biomass production of protozoa and nematodes, respectively. The excess nitrogen is excreted as ammonia and hence is readily available for uptake by plants.

Figure 21.20 Soil microbial loop. Energy-rich exudates from roots enhance growth of microbial populations and breakdown of soil organic matter. Nutrients immobilized in microbial biomass are released through predation by microbivores, increasing mineral availability for plants.

fluctuate widely. As their numbers decline, their readily decomposed tissues enter the detrital food chain. Protozoan biomass can equal that of all other soil animal groups except earthworms. As much as 70 percent and 15 percent of soil respiration originates from protozoa and nematodes, respectively. Turnover of microbivores is rapid; their production rates can reach 10 to 12 times their standing biomass with a minimum generation time of 2 to 4 hours, indicating a significant effect on nutrient mineralization. On a global scale, rhizosphere processes utilize about 50 percent of the energy fixed by photosynthesis and contribute about 50 percent of the CO_2 emitted from terrestrial ecosystems. They mediate virtually all aspects of nutrient cycling. We underestimate their ecological significance at our peril.

21.10 FEEDBACK REGULATION: Nutrient Cycling Is Subject to Positive Feedback

It is clear from Figure 21.1 that internal cycling of nutrients through an ecosystem depends on the linked processes of primary production, which determines the rate of nutrient transfer from inorganic to organic form, and decomposition, which determines the rate of release of organic nutrients into inorganic form via net mineralization. But how do these two key processes interact to determine the rate of nutrient cycling? The answer lies in their interdependence.

Consider nitrogen cycling. The links among soil N supply, N uptake by roots, and N content of leaves were discussed in Section 6.12. Maximum photosynthetic rate is correlated with leaf N because compounds directly involved in photosynthesis, such as Rubisco and chlorophyll, contain much of leaf N. Thus, low N in soil (or in sediments or water in aquatic systems) reduces not only primary productivity but also the N content of the tissues produced. In turn, the amount and N content of litter inputs to the detrital food chain decline. The net effect is a lower return of N in litter. Reduction in the amount and nutrient content of organic matter as food for decomposers promotes nutrient immobilization from soil and water to meet decomposer needs. Immobilization reduces nutrient supply to plants by reducing mineralization, depressing productivity further.

We can now appreciate the positive feedback loop involved in internal cycling of nutrients (Figure 21.21). Low nutrient supply begets low nutrient content of plant tissues and low NPP, whereas high nutrient supply begets high nutrient content of plant tissue and high NPP. The resulting high quantity and quality of litter promote rapid net mineralization and release. Positive feedback is involved—in the one case, reducing nutrient levels, and in the other, increasing them.

Feedback among litter quality, nutrient cycling, and NPP is evident in a study of forests along a gradient of soil texture in Wisconsin. Deciduous species producing higher-quality litter (lower C:N ratio) dominated sites with finer soil texture (more silt and clay). This higher-quality litter increased net

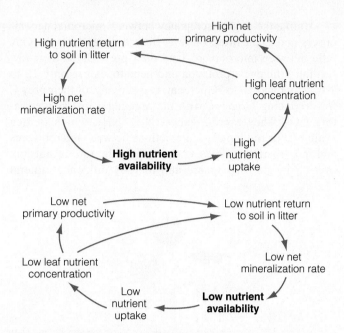

Figure 21.21 Positive feedback loop between nutrient availability, NPP, and nutrient release in decomposition in low- and high-nutrient sites.

(Adapted from Chapin 1990.)

mineralization (Figure 21.22a), which in turn increased NPP and nutrient return in litter (Figure 21.22b). N and P cycling increased (Pastor et al. 1984), with changes in species composition and litter quality along the gradient that were directly related to the influence of soil texture on available moisture (see Section 4.7).

Feedback loops involving nutrient cycling can lower productivity as well as increase it. The warm, wet conditions of tropical rain forests support the highest productivity and decomposition rates of any terrestrial system. However, factors affecting nutrient cycling can depress NPP in tropical sites with extremely high rainfall. In six montane sites in Hawaii, yearly rainfall ranged from 2200 mm to over 5000 mm. Other abiotic factors (altitude, soil, temperature, and topography) were similar. All sites were dominated by similar-aged stands of the evergreen *Metrosideros polymorpha*. The prevailing model suggests that NPP should increase with precipitation, levelling off with the high annual rainfall typical of the wet tropics. Instead, NPP decreased when annual precipitation exceeded 3000 mm/yr (Figure 21.23) (Schuur 2003).

How does increasing rainfall trigger a decrease in productivity? The chemical composition of leaves at the six sites offers some clues. Two traits varied along the moisture gradient: leaf N content decreased and lignin content increased with increasing rainfall. Given that both traits influence decomposition and mineralization (see Figures 21.16 and 21.17), these changes affect internal cycling and hence nutrient availability. In litterbag studies, leaf decomposition and N release decreased significantly with increasing precipitation

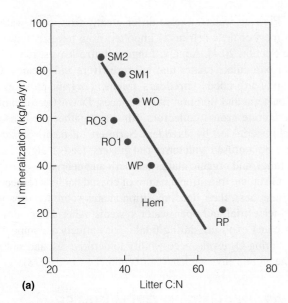

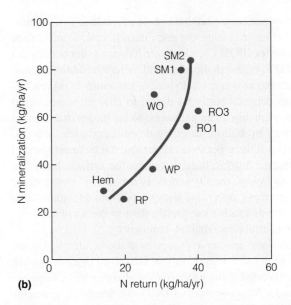

(a)

(b)

Figure 21.22 Relationship between **(a)** litter quality (C:N) and N mineralization rate, and **(b)** N mineralization rate and N returned in litter in forests on Blackhawk Island, Wisconsin. Hem, hemlock; RP, red pine; RO, red oak; WO, white oak; SM, sugar maple; WP, white pine.

(Adapted from Pastor et al. 1984.)

(Figure 21.24) (Schuur 2001). N release from decomposing litter declined with higher rainfall, as did O_2.

So lower N supply likely explains the decline in NPP with very high rainfall. The decline in decomposition and nutrient release may reflect lower soil O_2 and/or lower-quality litter at the wetter sites. These factors are interrelated. Low soil O_2 slows decomposition and N release. In turn, low soil N reduces leaf N content and litter quality, further depressing decomposition. Once again, positive feedback involving productivity, decomposition, and nutrient cycling—triggered by an abiotic stimulus, high rainfall—influences ecosystem function.

21.11 DECOMPOSITION IN WATER: Aquatic Habitats Differ in Organic Substrates and Decomposers

As in terrestrial habitats, decomposition in water involves leaching, fragmentation, colonization by bacteria and fungi, and consumption by microbivores. However, there are some major differences among habitats. First, we summarize the types of consumers and organic substrates in different aquatic habitats. The following sections then discuss nutrient-cycling patterns in each habitat type. Chapters 24 and 25 contain more detail on aquatic biota.

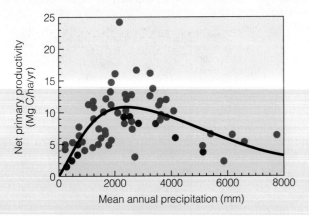

Figure 21.23 Relationship between net primary productivity and mean annual precipitation. Purple symbols represent data from the Maui sites; green symbols represent data from International Biological Programme and tropical surveys worldwide. Different colours represent different sites.

(Adapted from Schuur 2003.)

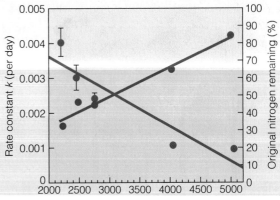

Figure 21.24 Rate of decomposition (as indicated by the decomposition coefficient, *k*; blue circles) and N loss (green circles) as a function of rainfall in six tropical rain forest sites in Hawaii.

(Adapted from Schuur 2001.)

In the open-water habitats of ponds, lakes, and oceans, dead organisms and other organic material, called **particulate organic matter (POM)**, gradually drift down to the bottom. En route, POM is ingested, digested, and mineralized until the residue settles out as humic compounds. How much POM reaches the bottom depends partly on depth. In shallow water, much arrives in relatively large packages, to be further fragmented and digested by bottom-dwelling detritivores such as crabs, snails, and molluscs. Bacteria decompose the bottom (*benthic*) organic matter. Aerobic bacteria inhabit the surface, but a few centimetres below, the O_2 supply is exhausted. Under these anoxic conditions, anaerobic bacteria perform decomposition, which proceeds much more slowly than in the aerobic conditions of the surface and shallow sediments.

Aerobic and anaerobic processes in the benthic layers are only part of aquatic decomposition. **Dissolved organic matter (DOM)** suspended in the water column also provides energy for decomposers. Major sources of DOM are free-floating macroalgae, phytoplankton, and zooplankton inhabiting open water. Upon death, plankton cells may break up and dissolve quickly, either by the action of enzymes released by the organisms themselves (*autolysis*) or through microbial action. Phytoplankton and other algae also excrete organic matter, particularly during rapid growth and reproduction. These C-rich exudates provide substrates for bacteria. In turn, ciliates and zooplankton eat bacteria and excrete nutrients in exudates and wastes. When food is abundant, zooplankton excrete half or more of their consumption as fecal pellets, which are abundant in the suspended material and provide a substrate for further bacterial activity.

This trophic pathway by which DOM is reintroduced into the food web by being incorporated into bacteria, which in turn are consumed by ciliates and zooplankton, is called the **microbial loop** (see Section 24.9). It resembles the soil microbial loop in function (see Figure 21.20). Lighter than POM, DOM remains longer in upper waters, so nutrients entering the microbial loop are more likely to remain in the photic zone, supporting further productivity.

Flowing-water systems differ greatly from open water in that most of their detritus is imported from terrestrial systems (see Section 20.4). Aquatic fungi colonize leaves, twigs, and other particulate matter that fall into rivers and streams. One group of arthropods, **shredders**, fragment organic particles and eat bacteria and fungi on litter surfaces. Downstream, another invertebrate group, **collectors**, filter and gather particles and fecal material left by shredders. **Scrapers** (also called *grazers*, not to be confused with terrestrial grazers) feed on algae, bacteria, fungi, and organic matter on rocks and debris.

Given the transitional nature of coastal habitats (especially salt marshes), their decomposition shares some features with both terrestrial and open-water systems. Alternating submergence and exposure during tidal cycles affects decomposition by altering O_2 levels, accessibility to detritivores, and stability of the abiotic environment (Odum and Haywood 1978).

21.12 CYCLING IN OPEN WATER: Thermocline Dynamics and Currents Affect Nutrient Mixing in Lakes and Open Seas

As the link between primary productivity and decomposition, nutrient cycling is a critical process in all ecosystems. How does this link vary between terrestrial and aquatic systems? In most ecosystems, there is a vertical separation between the zones of production (photosynthesis) and decomposition (Figure 21.25). On land, plants themselves bridge this separation by occupying both zones. Roots access nutrients released in the decomposition zone in the soil, while the vascular system transports them to production sites in the canopy.

In aquatic systems, plants do not always bridge these zones. In shallow shoreline sites, emergent vegetation such as cattails, cordgrasses, and sedges are rooted in the sediments. Here, as on land, plants directly link the zones of decomposition and production. Likewise, submerged vegetation is rooted

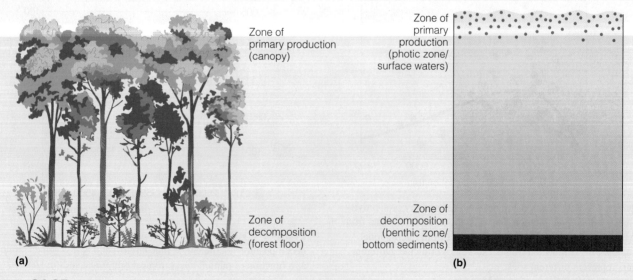

Figure 21.25 Comparison of zones of production and decomposition in **(a)** a terrestrial (forest) and **(b)** an open-water (lake) system. In terrestrial systems, these zones are linked by vegetation.

in the sediments, with the plants extending up into the photic zone, where light supports productivity. But in deeper waters, free-floating phytoplankton occupy the upper waters of the photic zone, seemingly far removed from the decomposition zone in the benthic layer. This physical separation between where nutrients become available by decomposition and where nutrients are needed for growth affects the productivity of open-water lakes and oceans.

The discrepancy is not as extreme as this description implies. Our increased understanding of the microbial loop, whereby significant decomposition and release take place in the water column as well as in the benthic layer, reveals that significant amounts of nutrients are available in the photic zone to support phytoplankton activity. Abiotic processes can also transport nutrients to the surface. Recall the vertical stratification of open-water systems (Figure 21.26; see Section 3.3). The shallow *epilimnion* is warmed by sunlight and has a relatively high O$_2$

content as a result of diffusion from air. The deeper *hypolimnion* is dense, cold, and O$_2$-poor. The transition zone between them is characterized by a steep temperature gradient or *thermocline*. This physical separation affects nutrient distribution and productivity. The colder, deeper waters where decomposition is more active are relatively nutrient rich, but too dark and cold to support productivity. The surface waters, where temperature and light are favourable, are nutrient poor.

Although winds mix the epilimnion, the thermocline prevents this mixing from extending to the hypolimnion. As fall approaches in temperate and polar zones, less radiation reaches the surface. The epilimnion cools until its temperature approaches that of the hypolimnion. The thermocline breaks down, and mixing occurs throughout the profile (Figure 21.27a). When surface waters become cooler than deeper waters, they sink, displacing deep waters to the surface (*turnover*). With thermocline breakdown and turnover, nutrients are brought to the surface. In spring, the increasing temperature and light of the epilimnion stimulate productivity, supported by the increased nutrients in surface waters (Figure 21.27b). As the season progresses, nutrients are depleted, and productivity declines. Thus, the annual productivity cycle in lakes (Figure 21.27c) directly reflects thermocline dynamics and its effect on the vertical distribution of nutrients.

In the open ocean, the thermocline is a permanent feature. Here, global current patterns influence water temperature, productivity, and nutrient cycling. The Coriolis effect

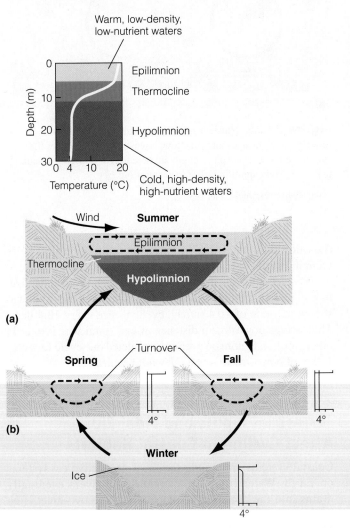

Figure 21.26 Seasonal dynamics of a temperate open-water system. Solid arrows track seasonal changes; dashed arrows circulation of water. In summer, winds mix epilimnion waters (a), but the thermocline isolates this mixing to surface waters. With seasonal thermocline breakdown, turnover allows mixing in the entire water column (b), moving nutrients up to surface waters.

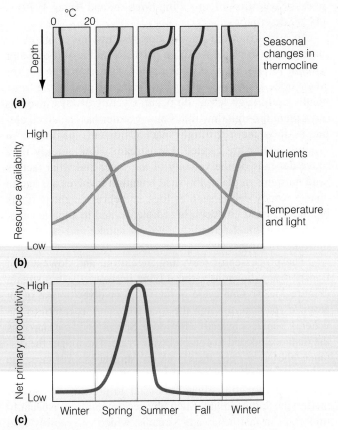

Figure 21.27 Seasonal changes in (a) thermocline, (b) availability of light and nutrients, and (c) net primary productivity of surface waters.

drives surface currents, but how deep does this lateral movement extend? Lateral flow is generally limited to the upper 100 m, but in some regions lateral movement causes vertical circulation (*upwelling*). Along western continental margins, surface currents flow along coastlines towards the equator (see Figure 2.14). At the same time, surface waters are pushed offshore by the Coriolis effect, bringing deeper, nutrient-rich waters to the surface (see Figure 3.16a). Surface currents generate similar upwellings in tropical waters. As the two equatorial currents flow west, they are deflected to the right and left north and south of the equator, respectively. Subsurface water is then transported vertically, bringing colder, nutrient-rich waters to the surface (see Figure 3.16b). These regions are highly productive and support some of the world's most important fisheries.

21.13 CYCLING IN FLOWING WATER: Nutrients Undergo Spiralling in Streams and Rivers

As we saw in Section 20.4, inputs in the form of dead leaves and woody debris from adjacent terrestrial systems, rainwater, and subsurface seepage bring nutrients into streams and rivers. Although nutrient cycling follows the same general path as on land and in open water, directional flow affects nutrient cycling in streams. Because nutrients are continuously transported downstream, **nutrient spiralling** (Webster and Patten 1979) is a better descriptor.

In terrestrial and open-water systems, nutrients cycle more or less in place. An atom passes from the soil or water column to plants and consumers and returns to the soil or water in dead organic matter. It is then recycled in the same location, although losses do occur. Cycling always involves movement through time, but flowing water has an added element—movement through space. Nutrients packaged in organic matter are carried downstream. How quickly they move depends on the velocity of the water and what factors hold nutrients in place. Physical retention involves storage in coarse woody debris such as logs, in debris caught in pools behind logs and boulders, in sediments, and in aquatic vegetation. Biological retention involves uptake and storage of nutrients in animal and plant tissue.

The processes of recycling, retention, and downstream transport are shown as a spiral lying horizontally in a stream (Figure 21.28). One cycle in the spiral involves uptake of an atom, its passage through the food web, and its return to the water, where it is available for reuse. Spiralling is measured as the distance needed to complete one cycle. The longer the distance, the more open the spiral; the shorter the distance, the tighter the spiral. If dead leaves and other debris are held in place long enough to allow organisms to process the organic matter, the spiral is tight. Such physical retention is especially important in fast headwater streams, which can rapidly lose organic matter downstream. Activities of organisms can either open or tighten the spiral. Organisms that shred and fragment organic matter open the spiral by facilitating transport of mate-

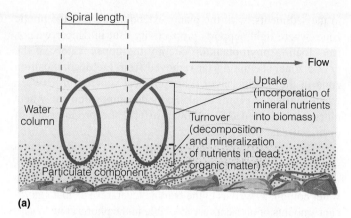

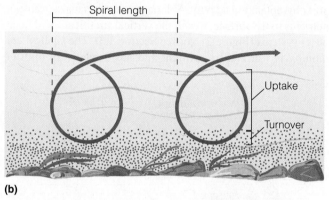

Figure 21.28 Nutrient spiralling in streams. Uptake and turnover take place as nutrients flow downstream. The tighter the spiral, the shorter the spiral length, and the longer nutrients remain in place. **(a)** Tight spiralling. **(b)** Open spiralling.

(Adapted from Newbold et al. 1982.)

rials downstream. Others tighten the spiral by physically retaining dead organic matter.

Distances travelled in nutrient spirals can be considerable. In a small forest brook in Tennessee, phosphorus moved downstream at a rate of 10.4 m/day, cycling once every 18.4 days. The average downstream distance of one spiral was 190 m. In other words, a P atom on average completed one cycle for every 190 m of downstream travel (Newbold et al. 1982).

21.14 CYCLING IN COASTAL WATERS: Terrestrial Inputs and Tidal Flux Affect Cycling in Coastal Habitats

Coastal systems are among the most productive ecosystems on Earth. Water from most streams and rivers eventually drains into the oceans. The place where the two meet is an **estuary**: a semi-enclosed part of the ocean where seawater is diluted and partially mixed with freshwater coming over land. As a river meets the ocean, its current velocity drops, and sediments are deposited within a short distance, creating a **sediment trap** (Figure 21.29). Sediment buildup produces an alluvial plain, with mudflats and salt marshes dominated by grasses and small shrubs (see Section 25.4). Nutrient

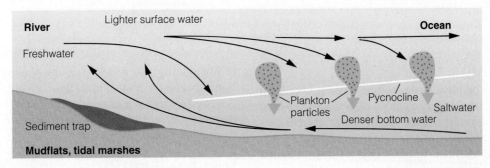

Figure 21.29 Circulation of freshwater and saltwater in an estuary creates a sediment trap. An incoming wedge of seawater on the bottom produces a surface flow of lighter freshwater and a counterflow of denser brackish water. The zone of maximum vertical difference in density, the pycnocline, functions like the thermocline in lakes. Living and dead particles settle through the pycnocline into the countercurrent and are carried up-estuary, conserving nutrients within the estuary rather than flushing them out to sea.

(Adapted from Correll 1978.)

cycling in estuaries combines features of terrestrial, open-water, and stream systems. As on land, the dominant plants are rooted in the sediments, linking the zones of decomposition and production. Submerged plants absorb nutrients from sediments as well as from water. As in streams and rivers, directional flow transports organic matter and nutrients into and out of the ecosystem.

Nutrients are carried into coastal marshes by precipitation, streams and rivers, and groundwater. The rise and fall of water depth with the tidal cycle flushes out salts and other toxins and brings in nutrients from coastal waters, a process called the **tidal subsidy**. The tidal cycle also replaces O_2-depleted waters in surface sediments with oxygenated water.

Like most terrestrial systems, the salt marsh is a detrital system with only a small portion of its primary production consumed by grazing herbivores. Almost 75 percent of salt marsh litter is broken down by bacteria and fungi. Nearly half of the NPP of a salt marsh is lost in microbial respiration. The low O_2 content of the sediments favours anaerobic bacteria that metabolize inorganic compounds such as sulphates. From 20 to 45 percent of the NPP of a salt marsh is exported to adjacent estuaries. The nature of this exchange depends on basin geomorphology (shape and type of opening to the sea) and the magnitude of tidal and freshwater fluxes.

Salt marshes differ in the way nutrients move through the food web as well as in the amount of exports. Some depend on tidal exchange, importing more than they export, whereas others are net exporters. Some material enters as mineral nutrients or as detritus; as bacteria; or as fish, crabs, and intertidal organisms. Although inflowing water from rivers and marshes carries nutrients into the estuary, primary productivity is regulated more by internal cycling than by external sources. Internal cycling involves release of nutrients within bottom sediments, and excretion of mineralized nutrients by herbivorous zooplankton.

As in tidal marshes, nutrients and O_2 are carried into the estuary by the tides. Typical estuaries maintain a "salt wedge" of intruding seawater on the bottom, producing a surface flow of freshwater and a counterflow of more brackish, heavier water (see Figure 21.29). These layers are separated by variations in water density arising from salinity and temperature differences. The zone of maximum vertical difference in density, the **pycnocline**, functions much like the thermocline in lakes. Living and dead particles settle through the pycnocline into the countercurrent and are carried up the estuary, thereby conserving nutrients within the estuary rather than flushing them out to sea. Regular movement of freshwater and saltwater into the estuary, coupled with its shallow and turbulent character, promotes vertical mixing. In deeper estuaries, a thermocline can form in summer, causing a seasonal pattern of vertical mixing and nutrient cycling similar to that described for open-water systems in Section 21.12.

EcologyPlace

Visit EcologyPlace at www.pearsoncanada.ca/ecologyplace to access online resources that complement your textbook, and help you to apply and to review the information in this chapter. EcologyPlace includes

- an eText version of the book
- self-grading quizzes
- glossary flashcards
- and more!

Go to www.pearsoncanada.ca/ecologyplace and follow the registration instructions on the Student Access Code Card included with this text. If your book does not have a Student Access Code Card, you can purchase access to it at www.pearsoncanada.ca/ecologyplace.

SUMMARY

Nutrient Cycling 21.1

Nutrients taken up by plants from soil or water are incorporated into living tissues as organic matter. As tissues senesce, dead organic matter returns to the soil or sediments. Decomposers transform organic nutrients into mineral form, which is again available for plant uptake. This process is called internal or nutrient cycling.

Decomposer Groups 21.2

Decomposition is the breakdown of chemical bonds in organic matter and involves leaching, fragmentation, digestion, and excretion. Ultimately, decomposers release nutrients in inorganic (mineral) form. Decomposers are classified based on their function and size. Microflora (bacteria and fungi) are "true" decomposers because they perform mineralization. Bacteria are either aerobic or anaerobic based on their requirement for O_2. Invertebrate detritivores are classified based on body size. Microbivores consume bacteria and fungi.

Plant Tissue Decomposition 21.3

A succession of organisms decompose plant litter. Before a plant dies, microbes inhabit its surfaces, absorbing simple compounds. Endophytes, parasites, and pathogens are also present. After death of the plant, detritivores fragment the material, increasing the surface area. Sugar fungi are the first fungal decomposers, consuming simple carbohydrates. Polymer-degrading fungi capable of digesting cellulose take over, producing enzymes for extracellular digestion. Fungi that can degrade lignin and secondary opportunistic fungi are the last groups to colonize the litter.

Animal Tissue Decomposition 21.4

After an animal dies, scavengers fragment the carcass. Autolysis is the first phase of breakdown, followed by putrefaction by bacterial decomposers. Blow fly maggots are important decomposers when the carcass begins to bloat.

Litterbag Studies 21.5

Decomposers derive their energy and most of their nutrients by consuming organic compounds. Ecologists study decomposition by using litterbags. A fixed amount of detritus is placed in mesh bags, and its rate of mass loss is followed through time. The proportion of residual compounds changes, with decay-resistant materials such as lignin increasing.

Factors Influencing Decomposition 21.6

Microbial decomposers use carbon compounds in dead organic matter for energy. Glucose and other sugars are easily broken down and are a high-quality energy source. Cellulose and hemicellulose, the main components of plant cell walls, are intermediate in quality, whereas lignins are of low quality and decompose the slowest. The higher the lignin content of plant litter, the slower its decomposition. Abiotic factors such as temperature and moisture influence decomposer activity. Decomposition occurs fastest in warm, wet climates.

Mineralization 21.7

Dead organic matter varies in nutrient quality. As microbes decompose detritus, they transform organically bound nutrients into an inorganic form. The organisms responsible for mineralization reuse some of the nutrients released, immobilizing them in their tissues. Release of nutrients is determined by the net mineralization rate, which is the difference between the rates of mineralization and immobilization. Net mineralization rate is affected by the nutrient content (particularly the C:N ratio) of decomposer tissue relative to that of the detritus.

Soil Organic Matter 21.8

Decomposition culminates in the production of humus, a black organic material that is embedded in the soil matrix as soil organic matter. Although high in nitrogen, humus decomposes slowly as a result of its high lignin content. Over time, fungal compounds become an increasing proportion of soil organic matter. Despite its low quality, humus represents a significant source of carbon and nutrients released from soils over time.

Soil Microbial Loop 21.9

The rhizosphere is an active zone of root growth and turnover, with intense microbial activity. Decomposition is more rapid than in bulk soil because plant roots release C-rich compounds to facilitate decomposition of low-quality organic matter. Nutrients immobilized in bacterial biomass are released to the soil by microbivores feeding on soil microflora.

Feedback Regulation 21.10

The rate at which nutrients cycle through an ecosystem is directly related to the rates of nutrient uptake for primary productivity and nutrient release via decomposition. Abiotic factors that influence these processes affect nutrient cycling. Positive feedback loops may increase or decrease productivity by either increasing or decreasing nutrient release.

Decomposition in Water 21.11

Aquatic systems vary in their decomposer groups and the substrates for decomposition. In open water, dead organisms and other particulate organic matter drift to the bottom. En route, POM is ingested, digested, and mineralized until much of the organic matter is in the form of humic compounds. Aerobic and anaerobic bacteria decompose organic matter on bottom sediments. Dissolved organic matter provides a source of carbon for decomposers, as part of a microbial loop. In flowing water, specialized detritivores, including shredders, collectors, and scrapers, fragment litter inputs from adjacent terrestrial systems.

Cycling in Open Water 21.12

In terrestrial and shallow-water systems, plants bridge the zones of primary production and decomposition. In open-water systems lacking rooted plants, this separation limits nutrient availability in surface waters, although the microbial loop leads to some nutrient release in upper waters. The thermocline limits movement of nutrients from deeper waters. Turnover allows vertical mixing, increasing lake productivity. In the open ocean, global patterns of surface currents bring deep, nutrient-rich waters to the surface in coastal areas. As surface currents move waters away from coastal margins, deep water moves to the surface, carrying nutrients.

Cycling in Flowing Water 21.13

Continuous, directional flow affects nutrient cycling in streams. This flow is best described as a spiral. One cycle in the spiral involves the uptake of an atom, its passage through the food chain, and its return to water, where it is available for reuse. Spiral tightness is related to flow rate and to physical and biological mechanisms of nutrient retention.

Cycling in Coastal Waters 21.14

Water from most streams and rivers drains into the seas, creating estuaries and coastal salt marshes. The tidal cycle flushes out salts and other toxins and brings in nutrients from coastal waters. The combined effect of inward movement of saltwater and outward flow of freshwater develops a countercurrent that carries living and dead particles and minerals back towards the coast, conserving nutrients within estuaries and salt marshes.

KEY TERMS

autolysis	dissolved organic matter	microbivore	phyllosphere	scavenger
C:N ratio	estuary	microflora	putrefaction	scraper
collector	extracellular digestion	mineralization	pycnocline	sediment trap
decomposer	immobilization	necrotrophic	residence time	shredder
decomposition	internal (nutrient) cycling	net mineralization rate	retranslocation	soil microbial loop
decomposition	leaf pack	nutrient spiralling	(reabsorption)	soil organic matter
coefficient	litterbag	particulate organic	rhizosphere	tidal subsidy
detritivore	microbial loop	matter	saprophage	

STUDY QUESTIONS

1. Define *decomposition*. Which organisms are the major decomposers of plant and animal material?

2. How do the types of carbon compounds present in dead organic matter influence its quality as an energy source? How and why does the lignin content of residual plant tissue change during decomposition?

3. Consider the sequence of processes taking place in a compost bin after plant material is added. What are the groups of organisms involved, and why do they change over time?

4. If detritivores do not actually decompose dead organic matter, why does their presence and abundance affect decomposition rate? How would you test this effect using litterbags?

5. Contrast *mineralization* and *immobilization*. What does an increase in the nitrogen content of decomposing plant tissues imply about the relative rates of these two processes? How does the initial ratio of carbon to nitrogen (C:N) of plant litter influence the relative rates of nitrogen mineralization and immobilization during decomposition?

6. How does the C:N ratio of plant litter differ from the C:N ratio of microbial decomposers? How does this difference affect nutrient dynamics during decomposition?

7. Tree species that inhabit boreal forests are characterized by low leaf nitrogen content. How might this trait influence nitrogen cycling? How is positive feedback involved?

8. Explain the *soil microbial loop*. How do plant exudates in the rhizosphere affect its activity?

9. Contrast nutrient cycling in terrestrial and open-water aquatic ecosystems, with particular attention to the zones of productivity and decomposition. How does the microbial loop make this contrast less extreme?

10. How do changes in the thermocline affect seasonal productivity patterns in lakes?

11. How does the continuous, directional flow of water influence nutrient cycling in streams? What factors affect whether a spiral is tight or open?

12. What mechanism conserves nutrients in estuaries?

13. Examine the global map of primary productivity of marine ecosystems in Figure 20.10. Identify areas of high productivity in the equatorial region that might reflect upwelling.

FURTHER READINGS

Aber, J. D., and J. M. Melillo. 1991. *Terrestrial ecosystems*. Philadelphia: Saunders College Publishing.

Valuable introduction to nutrient cycling in terrestrial ecosystems.

Deacon, Jim. 2006 (1980). *Fungal biology*. Oxford: Blackwell Publishing.

Comprehensive treatment of the ecological and structural diversity of fungi.

Likens, G. E., and F. H. Bormann. 1995. *Biogeochemistry of a forest ecosystem*. 2nd ed. New York: Springer-Verlag.

Long-term record of the structure and nutrient dynamics of the Hubbard Brook ecosystem, one of the most important long-term study sites in North America.

McNaughton, S. J., R. M. Ruess, and S. W. Seagle. 1988. Large mammals and process dynamics in African ecosystems. *BioScience* 38:794–800.

Discusses the role of consumers in cycling of nutrients in savannahs.

Newbold, J. D., J. W. Elwood, R. V. O'Neill, and A. L. Sheldon. 1983. Phosphorus dynamics in a woodland stream ecosystem: A study of nutrient spiraling. *Ecology* 64:1249–1265.

Influential early study of nutrient spiralling in flowing water.

Prescott, C. E. 2005. Do rates of litter decomposition tell us anything we really need to know? *Forest Ecology and Management* 220:66–74.

Provocative treatment of the limitations of litterbag studies, and some proposed new directions.

Wagener, S. M., M. W. Oswood, and J. P. Schimel. 1998. Rivers and soils: Parallels in carbon and nutrient processing. *BioScience* 48:104–108.

Useful comparison of decomposition and nutrient cycling in terrestrial and stream ecosystems.

BIOGEOCHEMICAL CYCLES

Impala (*Aepyceros melampus*) standing in the shade of acacia trees. Their urine and droppings make large mammals important contributors to the nitrogen cycle of savannahs.

n Chapter 21 we examined nutrient cycling within eco-systems, driven by the linked processes of productivity and decomposition. Internal cycling is a story of biological pro-cesses, but not every transformation of matter is mediated by organisms. Many reactions occur in abiotic ecosystem com-partments: air, water, soil, and rock. Weathering releases ele-ments, making them available for uptake. Lightning produces ammonia (NH_3) from N_2 and water in the atmosphere, provid-ing nitrogen inputs into aquatic and land systems. Other pro-cesses, such as sedimentation of calcium carbonate in oceans (see Section 3.6), remove elements from circulation.

Each element has its own story, but all elements flow from non-living to living and back to non-living ecosystem compo-nents in paths called **biogeochemical cycles**. To discuss these cycles, we must expand our treatment of nutrient flow beyond internal cycling to encompass a wider array of biotic and abiotic processes. We consider specific cycles from both a local and a global perspective. All obey the first law of thermodynamics: matter cannot be created or destroyed. Thus, matter cycles through ecosystems, in contrast with the one-way flow of energy.

22.1 TYPES OF CYCLES: Gaseous and Sedimentary Cycles Have Different Reservoirs

There are two general types of biogeochemical cycles, based on their major reservoirs (pools). In **gaseous cycles** (e.g., nitrogen and oxygen), the main pools are the oceans and the atmosphere, making their patterns strongly global. In **sedimentary cycles** (e.g., phos-phorus) the major pool is the lithosphere: soil, rocks, and minerals, with available forms adsorbed to soil particles or present as salts in soil water or water bodies. Sedimentary cycles vary, but all have a *rock phase* and a *salt solution* phase, and tend to be more localized. Mineral salts are derived from Earth's crust by weathering. Soluble salts enter the water cycle (see Section 3.7), which moves them through soil to streams and lakes and eventually to the oceans, where they remain indefinitely. Other salts return to the rock phase by sedimentation, becoming incorporated into salt beds, silt, and limestone. After weathering, they re-enter the cycle. Some cycles, such as sulphur, are a hybrid of gaseous and sedimentary, with major pools in Earth's crust as well as in the atmosphere.

Both gaseous and sedimentary cycles are driven by energy flow, and both are tied to varying degrees to the water cycle. Water is the medium for moving inorganic and organic matter through an ecosystem. Without water cycling, matter cycling would cease.

22.2 INPUTS AND OUTPUTS: Nutrient Cycles Have Inputs, Outputs, and Internal Cycling

Although the cycles of the minerals required by organisms differ in detail, they all share three aspects related to ecosystems: inputs, outputs, and internal cycling (Figure 22.1). Inputs depend on the cycle type. Nutrients with a gaseous cycle, such as oxygen (O) and nitrogen (N), enter the ecosystem from the atmosphere or hydro-sphere. Inputs from sedimentary cycles, such as calcium (Ca) and

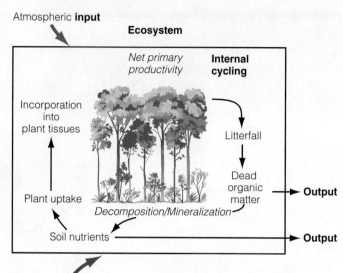

Figure 22.1 Generalized model of a biogeochemical cycle. Key processes in internal cycling (NPP and decomposition) are in italics.

phosphorus (P), are derived from weathering. Soil processes and properties greatly influence nutrient release and retention.

Supplementing these gaseous or sedimentary sources are nutrients carried by rain, snow, currents of air or water, and ani-mals. Some nutrients arrive in airborne particles and aerosols, collectively called *dryfall*. Precipitation also washes in large quantities, as *wetfall*. Some 70 to 90 percent of rain striking a forest canopy reaches the ground. As rain drips through the canopy (*throughfall*) and runs down stems (*stemflow*), it picks up and carries nutrients deposited as dust on leaves and stems together with soluble nutrients leached from them. Thus, rain reaching the forest floor is richer in Ca, sodium (Na), potassium (K), and other elements than rain falling in the open.

Major sources of nutrients for aquatic ecosystems are inputs from the surrounding land in drainage water, detritus, and sedi-ments, and from the atmosphere in precipitation. Streams and rivers depend on a steady input of detritus from their watersheds.

Nutrient outputs represent losses that must be offset by inputs to avert a net decline. Export occurs in many ways, depending on the cycle. C output as respiratory CO_2 by aerobic organisms is an obvious route, but many other biotic processes transform nutri-ents to gaseous forms that then exit the ecosystem into the atmo-sphere. Later sections provide examples for specific cycles.

Outputs also occur in organic matter. Woody debris from for-ested watersheds is carried by surface flow in streams and rivers (see Section 20.4). Moose feeding on aquatic plants transport nutri-ents to adjacent terrestrial systems in their feces. Conversely, the hippopotamus (*Hippopotamus amphibius*) consumes herbaceous vegetation near water bodies, transferring large amounts of nutri-ents as wastes into the water. Although such transport can be a sig-nificant output, organic matter plays a key role in recycling nutrients because it prevents rapid losses from the system. Nutrients bound tightly in organic matter are released slowly by decomposition.

Some nutrients are leached from the soil and exit an eco-system in underground water flow. These losses may be bal-anced by inputs from weathering of rocks and minerals. Large amounts of nutrients are withdrawn permanently by harvesting, in farming and logging. If these exports are not replaced, the ecosystem becomes impoverished, lowering productivity. As well as causing direct losses in biomass, logging can accelerate nutrient export from erosion (see Section 26.8).

Depending on its intensity, fire kills vegetation and converts biomass and soil humus to ash. Besides loss of nutrients by volatilization and in airborne particles, ash alters soil properties. Many nutrients become more available, and N in ash is rapidly mineralized via *pyromineralization*. If not absorbed by roots soon after the fire, nutrients may be lost to leaching and erosion. Particularly in regions with high rainfall, stream runoff increases after fire as a result of lower transpirational demand by plants.

Nutrient cycles are often studied locally, as internal cycling within ecosystems and as exchanges linking ecosystems. Outputs from one ecosystem become inputs to another, as in the export of organic matter from a forest into a nearby stream. Understanding nutrient exchange requires viewing cycles on a scale beyond that of a single ecosystem, especially for nutrients with gaseous cycles. As their reservoirs are the atmosphere or oceans, such nutrients circulate globally. In our treatment of several key elements, we first stress their movement through ecosystems at a local scale. We then expand our perspective to consider their global cycling.

22.3 LOCAL CARBON CYCLE: Local Carbon Cycling Is Linked to Energy Flow through Food Webs

Cycling of carbon, the basic constituent of all organic compounds, is so closely tied to energy flow that ecologists express ecosystem productivity as grams of C fixed per square metre per year. Earlier, we defined *net primary productivity (NPP)* as the difference between the rates of C uptake in photosynthesis and C release by autotrophic respiration. **Net ecosystem productivity (NEP)**, in contrast, is the difference between the rates of C uptake in photosynthesis and C loss in respiration of both autotrophs and heterotrophs (largely decomposers). The two often differ; in a study of sea ice off the coast of Greenland, NPP was positive but NEP was negative (Rysgaard and Glud 2004).

The ultimate source of organic carbon, in both living organisms and fossil deposits, is CO_2 in air or water. Photosynthesis incorporates CO_2 into the bodies of primary producers. From then on, C accompanies energy as it flows through the grazing food chain (Figure 22.2). Primary producers and consumers return it to the atmosphere as respiratory CO_2. Residual C is

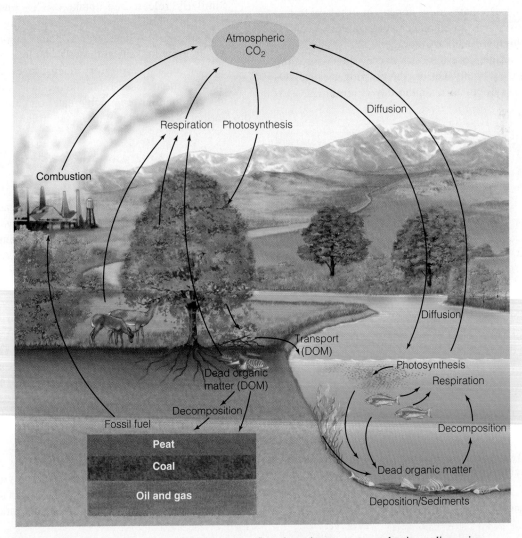

Figure 22.2 Local carbon cycle. Carbon's integral role in energy flow through ecosystems makes its cycling unique.

locked in living tissue. It eventually enters the biotic reservoir of dead organic matter, from which it is released, again as respiratory CO_2, by the detrital food chain. Thus, the rates of primary productivity and decomposition determine the rate of local C cycling. Both processes are influenced by abiotic factors, such as temperature and precipitation (see Sections 20.3 and 21.6). In warm, wet rain forests, productivity and decomposition rates are high, and C cycles rapidly. In cool, dry ecosystems, C cycles more slowly, and biomass accumulates. In swamps and marshes, detritus falls into the water and rarely decomposes completely. When stored as humus or peat, C circulates slowly. Over geologic time, this buildup of partially decomposed organic matter forms fossil fuels.

Similar cycling occurs in aquatic habitats. Phytoplankton use CO_2 that diffuses into upper waters (or is released from carbonates, as described in Section 3.6), incorporating it into their tissue. Carbon then passes through the aquatic food chain and is released as respiratory CO_2, which is reused, enters the bicarbonate reaction sequence, or diffuses into air. Significant amounts of C are bound as carbonates in the exoskeletons of molluscs and foraminifers. Some carbonates dissolve back into solution, but some are buried in bottom muds when the organisms die. Isolated from biotic activity, this C is removed from cycling. Upon incorporation into sediments, carbonates may appear in coral reefs and limestone deposits. For an element with a gaseous cycle, C has unusually high sedimentary storage.

As a result of its close ties with energy flow, C cycling varies daily and seasonally. During the growing season, CO_2 levels in a forest fluctuate over the day (Figure 22.3). At

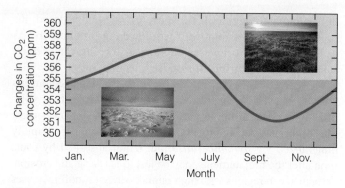

Figure 22.4 Annual CO_2 flux at Barrow, Alaska. Concentrations increase during the winter, followed by a decline with the onset of the growing season in May.

(Adapted from Pearman and Hyson 1981.)

dawn, CO_2 drops as plants withdraw it from air. By afternoon when the air warms and its relative humidity falls, partial stomatal closure causes photosynthesis to decline, and CO_2 rises. By evening, photosynthesis ceases entirely, respiration rises, and CO_2 increases sharply. A similar flux occurs in aquatic systems.

Similarly, release and uptake of CO_2 undergo seasonal fluxes reflecting both temperature and growing season (Figure 22.4). In spring, CO_2 decreases as plants begin photosynthesis. As the growing season draws to a close, photosynthesis declines or ceases, respiratory release exceeds uptake, and CO_2 rises. Although seasonal changes also occur in aquatic systems, fluctuations are much greater on land, particularly in the Northern Hemisphere with its larger land mass.

22.4 GLOBAL CARBON CYCLE: Carbon Follows Complex Global Patterns

Although the C cycle is usually considered gaseous, its large reservoirs in all spheres make it (like sulphur) a hybrid cycle. The global C cycle is closely linked to exchanges among air, land, and water, as affected by mass air movements (Figure 22.5). Earth contains about 10^{23} g, or 100 million Gt (gigatonne = 10^9 tonnes, or 10^{15} g), of C, all but a small fraction of which is buried in sedimentary rocks and does not circulate. Recoverable fossil fuels, created by the burial of partially decomposed organic matter, account for an estimated 10 000 Gt. The active C pool is an estimated 55 000 Gt, of which 70 percent is in the oceans, mostly as bicarbonate and carbonate. Of the remaining amount, detritus accounts for 1650 Gt, living matter, 3 Gt. Terrestrial systems combined contain an estimated 1500 Gt of C as dead organic matter and 560 Gt in living biomass. The atmosphere—the crucial link between vegetation and oceans—holds just 750 Gt.

Surface ocean waters are the major sites of C exchange with air. The ocean's ability to take up CO_2 is governed largely by bicarbonate formation (see Section 3.6). In surface waters, C circulates physically with the currents, and biologically

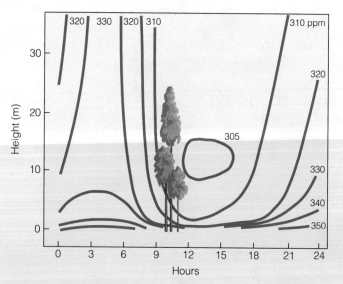

Figure 22.3 Daily flux of CO_2 in a forest. Isopleths (lines) define concentration gradients. Note the consistently high CO_2 level on the forest floor, the site of microbial respiration. Atmospheric CO_2 is lowest from mid-morning to late afternoon and highest at night, as photosynthesis ceases and respiration releases CO_2 into the air.

(Adapted from Baumgartner 1968.)

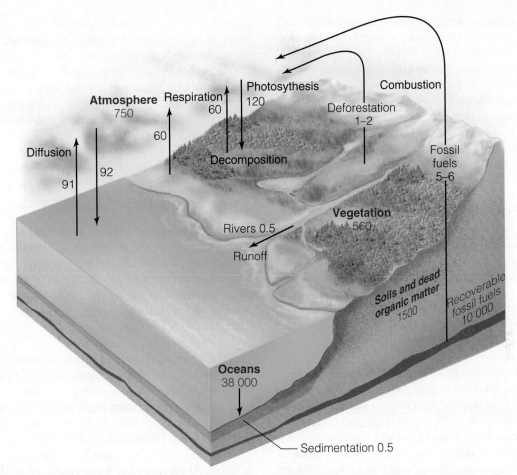

Figure 22.5 Global carbon cycle. Sizes of the major pools are labelled in red. Arrows indicate the major exchanges (fluxes) among them. All values are in Gt and exchanges are annual. The largest C pool (geologic) is excluded because of its slow rate of transfer with active pools.

(Adapted from Edmonds 1992.)

INTERPRETING ECOLOGICAL DATA

Q1. Which fluxes estimate global net primary productivity? Global net ecosystem productivity? What is the estimate for each?

Q2. Which major pool(s) estimate global standing crop biomass?

through the food web. Exchange of CO_2 between the oceans and the atmosphere by biotic and abiotic processes generates a net yearly uptake of approximately 2 Gt (Table 22.1). Some models predict oceanic uptake will reach 5 Gt annually by 2100 (Cox et al. 2000). Burial in sediments accounts for a net yearly loss of 0.5 Gt.

The ocean thus has the ability to act as a powerful C sink, ameliorating the impact of C inputs on global warming. However, humans compromise its sink potential by pollution and other disturbances that disrupt oceanic productivity and alter ocean chemistry. Even minor acidification overloads the carbonate equilibrium system, reducing calcification in coral reefs (Langdon 2002) and jeopardizing their future in an era of global change (Kleypas et al. 2001). Proposals to enhance the ocean's biotic sink activity by stimulating plankton productivity by fertilizing with iron (often a limiting factor) are controversial given concerns about its efficacy and possible side effects, including harmful algal blooms, destabilization of marine food webs, and increasing anoxic conditions on the ocean floor.

Table 22.1 Global CO$_2$ Sources and Sinks from 1980 to 2010

	1980–1990	1990–2000	2000–2010
Sources			
Fossil fuel emissions and cement	5.5 ± 0.3	6.4 ± 0.4	7.9 ± 0.5
[a]Land-use change	1.5 ± 0.7	1.6 ± 0.8	1.0 ± 0.7
Sinks			
Oceanic uptake	2.0 ± 0.4	2.2 ± 0.4	2.3 ± 0.5
Atmospheric increase	3.4 ± 0.1	3.1 ± 0.1	4.1 ± 0.1
[b]Residual terrestrial sink	1.6 ± 0.9	2.7 ± 0.9	2.5 ± 0.9

Values are in Gt C/yr.
[a]Land-use change is a net value including both release from deforestation and agricultural cultivation and uptake from regrowth in abandoned crop fields and after logging.
[b]Residual terrestrial sink is calculated as the difference between total sources and total sinks.

(Adapted from Le Quéré et al. 2009.)

Many marine ecologists argue that there is currently insufficient scientific basis for evaluating the benefits and risks of ocean fertilization (Buesseler et al. 2008).

Whereas abiotic processes are critical in aquatic systems, C uptake into terrestrial systems is governed by a biotic process: primary productivity. Losses reflect autotrophic and heterotrophic respiration, the latter dominated by decomposers. Biotic exchanges of CO_2 between Earth's landmasses and the atmosphere were once thought to be in equilibrium (as shown in Figure 22.5). However, recent research indicates that terrestrial landmasses act as a C sink, with a net annual uptake of some 2.5 Gt from the atmosphere from 2000 to 2010 (see Table 22.1) (Le Quéré et al. 2009; see also Section 28.2). In the short term, sink activity is promoted by warming and the fertilizer effect of CO_2 on young forests that are accumulating biomass after logging, or on abandoned cropland. But, as with the oceans, humans compromise terrestrial C sink activity by land-use changes (especially urbanization), that lower or eliminate productivity, and by degrading habitat. Enhancing C sequestration by facilitating natural processes is a critical imperative for the 21st century.

Of great importance is C storage in litter and soil organic matter (1500 Gt) and in vegetation (560 Gt). Mean C per soil volume increases from the tropics to the boreal forests and tundra. Low soil C in tropical soils reflects rapid decomposition, which counters the high productivity and litterfall of tropical rain forests. Frozen tundra soil and waterlogged muds of swamps and marshes have the most C accumulation in detritus because low temperatures, saturated soils, and anaerobic conditions inhibit microbial decay.

Release of massive amounts of methane (CH_4) and CO_2 with thawing of permafrost (Schuur et al. 2009) will amplify global warming by positive feedback. Both types of feedback are normally at work in the global C cycle, but human intrusions accelerate positive feedback loops (Figure 22.6) (Luo et al. 2001). So, although terrestrial C sink activity has increased in recent decades, positive feedback loops are poised to reduce C uptake and turn what is now a C sink into a C source. Such changes in the global C cycle are predicted to accelerate global warming (Matthews and Keith 2007). According to climate models that incorporate feedback loops and assume a "business as usual" scenario regarding C emissions, terrestrial systems will switch from being a C sink to a C source by approximately 2050 (Cox et al. 2000).

22.5 NITROGEN CYCLE: Bacteria and Fungi Transform Nitrogen Compounds

Nitrogen, an essential constituent of compounds such as proteins and nucleic acids, is available to plants in two major inorganic forms: ammonium (NH_4^+) and nitrate (NO_3^-). The atmosphere is almost 80 percent N_2, but this form is not accessible. N enters ecosystems via two routes, the relative importance of which varies (Figure 22.7). One route is atmospheric deposition, either in wetfall in rain, snow, cloud, and fog, or in dryfall as particulates or aerosols.

The second route is via **nitrogen fixation**, again in two ways. One is high-energy abiotic fixation. Cosmic radiation and lightning provide the energy to combine N_2 with H_2O, with the products reaching Earth in rain. Less than 0.4 kg N/ha reaches Earth annually from this source, with 2/3 deposited as NH_3 and 1/3 as nitric acid (HNO_3). The second method is biological, producing some 10 kg N/yr/ha of land surface, or about 90 percent of N fixed yearly. Biological fixation is performed by symbiotic bacteria living in mutualistic association with plants (see Section 15.9), as well as by some lichens, free-living bacteria, and cyanobacteria. Fixation splits N_2 into two atoms of free N, which then combine with H_2 to form NH_3. Fixation requires substantial energy, which in mutualistic interactions is supplied by the photosynthetic partner.

In agricultural systems, *Rhizobium* bacteria associated with some 200 legume species are the major N fixers. In natural systems, about 12 000 plant species fix N. Some are legumes, but others include alders (*Alnus* spp.), common boreal shrubs that associate with filamentous actinomycete bacteria (*Frankia* spp.) (Figure 22.8a, p. 480). *Frankia* associates with other plant groups as well, notably *Dryas* spp., small herbaceous members of the rose family that are common in Arctic and alpine tundra (Figure 22.8b). Some N-fixers are free-living soil bacteria in 15 known genera, including the aerobic *Azotobacter* and the anaerobic *Clostridium*. Cyanobacteria (blue-green algae) are important non-symbiotic N-fixers. Of some 40 species in soil and water, the most common are *Nostoc* and *Calothrix* spp.

Anthropogenic eutrophication of lakes favours blooms of toxin-producing blue-green algae, particularly if P is high relative to N. With their ability to fix N, blue-green algae gain a competitive advantage (see Ecological Issues: Saving Lake Winnipeg, pp. 532–533). Jelly lichens (*Collema* spp.) and dog lichens (*Peltigera* spp.; Figure 22.8c), which colonize rock outcrops in boreal sites, also fix N in symbiotic association with cyanobacteria.

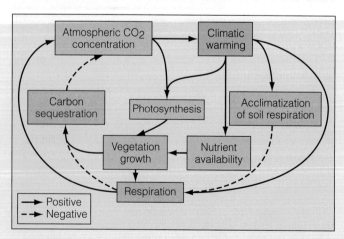

Figure 22.6 Involvement of feedback loops in the global C cycle, as affected by climatic warming.

(Adapted from Luo et al. 2001, as modified in Odum and Barrett 2005.)

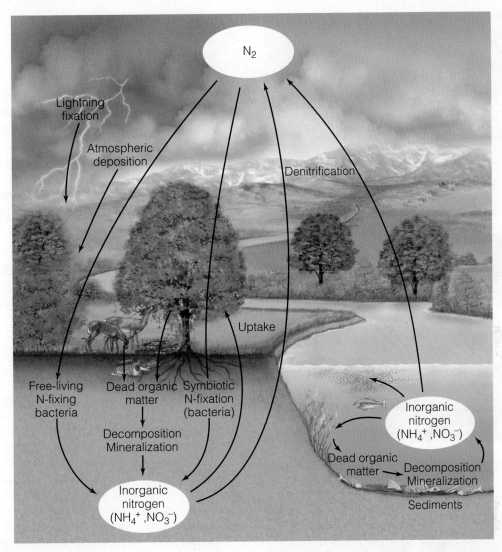

Figure 22.7 Local nitrogen cycle: Bacteria and fungi perform many of the transformations between nitrogen compounds.

Once formed, NH_3 is subject to complex transformations, many of which are mediated by bacteria (Figure 22.9, p. 480). Some NH_3 is released in microbial decomposition, via **ammonification**. Whatever its origin, NH_3 is converted to NH_4^+, particularly in acidic soils. In alkaline or neutral soils in which less H^+ is present, NH_3 returns to the atmosphere by **volatilization**. Volatilization is common in crop soils where both N fertilizers and lime (to lower acidity) are used extensively.

If it is not volatilized, NH_4^+ can be used directly by plants, although it can be toxic at high levels. Roots compete for NH_4^+ with aerobic bacteria, which use it in metabolism (see Figure 22.9). In the two-step process of **nitrification**, one bacterial group (*Nitrosomonas*) oxidizes NH_4^+ to NO_2^-, and another (*Nitrobacter*) oxidizes NO_2^- to nitrate (NO_3^-). NO_3^- has two major fates. (1) NO_3^- is the preferred uptake form for roots. (2) **Denitrification** by *Pseudomonas* bacteria reduces NO_3^- to two gases, N_2O and N_2, which return to the atmosphere. Although the low-O_2 conditions conducive to

denitrification are rare on land, they are common in wetlands and water bodies.

NO_3^- is the most common N compound exported from terrestrial systems in groundwater or streams. In undisturbed systems, loss is minimal as a result of high N demand. In such relatively closed systems, the amount of N recycled within an ecosystem far exceeds the amount entering or leaving. But if the ecosystem is "leaky," its outputs rise sharply. Clear-cutting in the Hubbard Brook Experimental Forest (a 3160-ha long-term study site in New Hampshire) increased N losses by from 3 to 20 times compared with uncut controls (Likens et al. 1977).

Although N inputs and outputs vary among ecosystems, internal cycling is similar. Following assimilation of NH_4^+ or NO_3^- by plants into organic compounds, decomposition returns N to the soil, sediments, and water.

Because many N-cycle processes are mediated by bacteria, the cycle is strongly influenced by abiotic factors. Temperature and moisture are important, but soil pH is especially

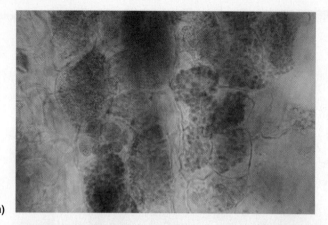

(a)

(b)

(c)

Figure 22.8 Non-legume N-fixers. **(a)** *Frankia* is a fungal-like actinomycete (shown here at the cellular level) that fixes N in association with alder bushes. It also associates with *Dryas octopetala* **(b)**, the official flower of the Northwest Territories and common in Arctic and alpine habitats throughout the Northern Hemisphere. An association with N-fixing cyanobacteria allows dog lichens (*Peltigera* spp.) **(c)** to colonize bare rock, facilitating soil formation.

critical given bacterial inhibition in acid soils (in contrast, fungi prefer low pH). Fixation is also affected by NO_3 levels; when NO_3^- is abundant, fixation declines, saving its high energy costs. In N-poor habitats such as boreal forests, most mosses as well as some flowering plants (including members of the Ericaceae such as blueberry) can utilize organic N

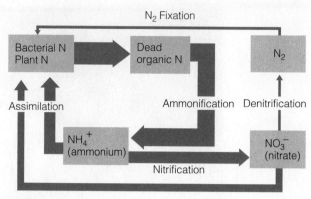

Figure 22.9 Bacterial processes involved in nitrogen cycling.

(amino acids). This ability—of great selective advantage in habitats with limited N and slow decomposition—is attributable to mycorrhizae associated with ericaceous species (Deacon 2006).

In the global N cycle (Figure 22.10) (Schlesinger 1997), the atmosphere is the largest pool (3.9×10^{21} g), with relatively small amounts in biomass (3.5×10^{15} g) and soils (95–140×10^{15} g). Global estimates of denitrification in terrestrial systems vary, but approach 200×10^{12} g/yr, of which more than half occurs in wetlands. Major oceanic N inputs are freshwater drainage from rivers (36×10^{12} g/yr), precipitation (30×10^{12} g/yr), and biological fixation (15×10^{12} g/yr). Denitrification accounts for an annual flux of 110×10^{12} g to the atmosphere (see Research in Ecology: Altered Matter Cycling in a Warmer Arctic, pp. 482–483). Small but steady losses occur to ocean sediments and sedimentary rocks, with small inputs arising from weathering and volcanic activity.

Human activity significantly influences global N cycling. Anthropogenic inputs from agriculture, industry, and automobiles are a serious concern. The first major intrusions likely began with agriculture, when people burnt forests and cleared land for crops and pasture. More recently, heavy applications of chemical fertilizers have disturbed the natural balance between N fixation and denitrification. Fertilizer nitrates are lost to groundwater and runoff, finding their way into aquatic systems. For more detail on the phenomenon of nitrogen saturation, see Section 22.10.

Vehicle exhaust and industrial high-temperature combustion add nitrous oxide (N_2O), nitric oxides (NOx), and nitrogen dioxide (NO_2) to the atmosphere, where they reside for up to 20 years, drifting slowly up to the stratosphere. There, UV light reduces N_2O to NO and atomic oxygen (O), which reacts with O_2 to form ozone (O_3) (see Section 22.8). NOx emissions are also an important source of acid rain, along with SOx (see Section 22.7).

22.6 PHOSPHORUS CYCLE: Phosphorus Has No Atmospheric Pool

As an element with a truly sedimentary cycle, phosphorus occurs in only minute amounts in the atmosphere. Thus, the P cycle follows the water cycle only partway, from land to sea (Figure 22.11). Because P lost from ecosystems via water is not returned, P is often in short supply in undisturbed habi-

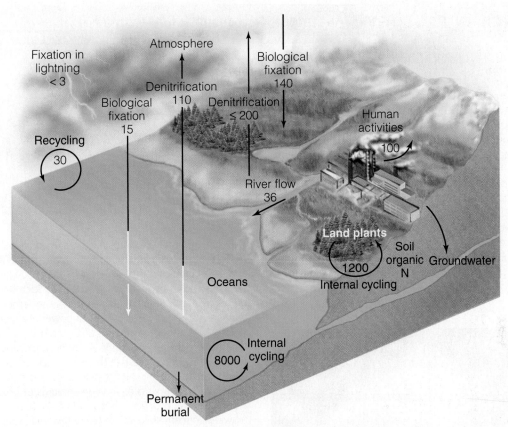

Figure 22.10 Global nitrogen cycle. Fluxes shown in units of 10^{12} g N/yr.

(Adapted from Schlesinger 1997.)

tats. The natural scarcity of P in water explains the explosive growth of algae in aquatic systems (especially freshwater lakes) that receive heavy discharges of P-rich wastes.

The main reservoirs of P are apatite rock and phosphate deposits, from which it is released by weathering, leaching, ero-sion, and mining. However, in most soils, only a small fraction of the P derived from weathering is available to plants. Thus the major process regulating P supply to plants is internal cycling from organic to inorganic forms. As the major uptake forms, phosphate (PO_4^{3-}) and HPO_4^{2-} are (like NO_3^-) negatively

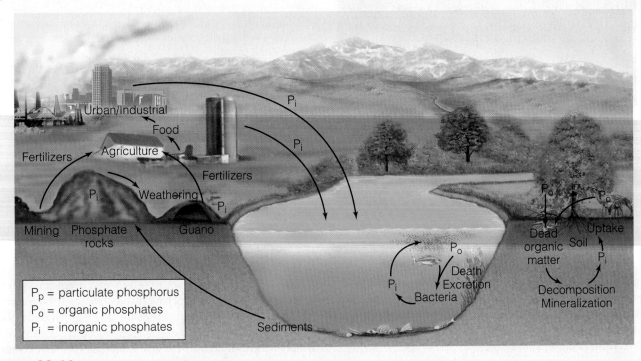

Figure 22.11 Local phosphorus cycle.

RESEARCH IN ECOLOGY | Altered Matter Cycling in a Warmer Arctic

Søren Rysgaard, Greenland Institute of Natural Resources
and the University of Manitoba.

Arctic sea ice has changed precipitously in recent decades (Figure 1). From 1978 to 1996, the decline was 3 percent per decade, but satellite data reveal it has accelerated to over 10 percent per decade from 1996 to 2006, with reduction in the multiyear portion of perennial ice exceeding 15 percent per decade (Comiso 2012). There was a modest recovery from 2007 to 2010 as part of a short-term cycle in multi-year ice cover (Comiso 2012), but northern polar ice is at its lowest extent since civilization began, raising concerns about its effect on Earth's energy balance.

Correlated with rising CO_2 emissions, these changes are accelerated by a positive feedback loop called *Arctic amplification*. As temperatures rise and the ice recedes, less radiation is reflected, causing further heating and further contraction of the ice cap. Moreover, as sea ice melts, it releases large amounts of additional heat. Mean Arctic temperatures are 3°C to 5°C warmer than 30 years ago, and if—as predicted—they become 6°C to 8°C warmer, significant thawing of permafrost would ensue. Such thawing would reinforce Arctic amplification by releasing large amounts of methane (CH_4), which is 25 times as powerful a greenhouse gas as CO_2. Given that the amount of C stored in terrestrial permafrost is more than twice that in the atmosphere and three times that of the world's forests (Schuur et al. 2009), such a release could be catastrophic. Nor is permafrost restricted to tundra; undersea permafrost overlies large amounts of frozen methane along the Arctic Ocean shelf. Additional C emanating from thawing permafrost is not yet being factored into most climate change models.

Abiotic ecosystem components are critical to sea ice decline. But these abiotic components interact with the organisms that inhabit sea ice, and understanding such biotic activity is essential. We often think of ice as devoid of life, yet sea ice supports an astounding variety of ecological niches that microorganisms have evolved to exploit. Many bacteria, fungi, algae, protozoa, and metazoa occupy its varied microhabitats. Narrow-diameter (nm to cm) brine channels contain diverse microbial communities, many of whose members are unidentified. Chemical reactions in these channels and their inhabitants "pump" CO_2 in and out of the ocean, with implications for C cycling and acidification in an altered Arctic climate.

Working with an international research team, Søren Rysgaard, of the Greenland Institute of Natural Resources and the University of Manitoba, is studying microbial activity in sea ice, including N cycling. Intermediates of the biological processes of the N cycle occur in sea ice, facilitated by a heterogeneous O_2 environment that promotes different microbial processes in adjacent microsites. Variability in O_2 levels of sea ice has complex sources, but partly reflects the patchiness of phytoplankton activity (Rysgaard et al. 2001).

Analyzing bubbles from ice cores, Rysgaard quantified bacterial denitrification in the lower 0.5 m of sea ice in a floe off

(a)

(b)

Figure 1 Cover of perennial ice in the Arctic in **(a)** 1980 and **(b)** 2012.

(Adapted from NASA.)

charged and repelled by negatively charged clay and humus particles, so they are readily leached from soils and exported to lakes and oceans where they contribute to eutrophication. Improving phosphate capture from sewage and other sources is key to rehabilitating aquatic systems.

The P cycle moves through three states in aquatic systems: (1) particulate organic P, (2) dissolved organically bound phosphates, and (3) inorganic phosphate (PO_4^{3-}). Organic phosphates are taken up by phytoplankton, which are consumed by zooplankton and detritus-feeders. Zooplankton may excrete as

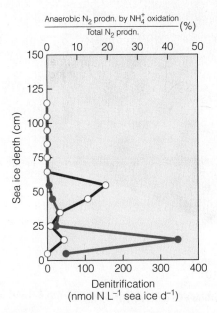

Figure 2 Anaerobic N_2 production by bacterial denitrification (closed symbols, blue line) and anaerobic NH_4 oxidation (open symbols, red line) in permanent ice off northeast Greenland.

(Adapted from Rysgaard and Glud 2004.)

northeast Greenland (Figure 2) (Rysgaard and Glud 2004). The combination of high levels of NO_3^- and dissolved organic C in anoxic ice promotes denitrification, previously considered the major process removing N from sea ice. But Rysgaard also documented considerable anaerobic bacterial oxidation of NH_4. Due to its lower temperature optimum, this process could be the favoured path for N removal in sea ice.

What are the implications of these N-cycle changes for global climate change? Rapid losses in perennial ice mean that both denitrification and anaerobic NH_4 oxidation will drop, especially as the latter process was only observed in perennial, not annual, ice. More N will remain in the oceans, potentially increasing the productivity of open-water systems and enhancing oceanic C uptake. Given an increased light penetration of up to 50 percent associated with reduced ice thickness and extent, productivity may triple by the end of the century (Rysgaard and Glud 2007). As positive as this sounds for C uptake, the overall picture is unclear. Other work by Rysgaard on C uptake by sea ice in Nordic waters indicates that sea ice acts as a C sink, removing

large amounts of atmospheric C by a sea-ice pump that transports inorganic C to deep and intermediate water masses (Rysgaard et al. 2007). With faster melting, the ability of sea ice to sequester C will decline, making it hard to anticipate the net effect of a warmer Arctic on global C pools.

Whatever the future changes in Arctic geomicrobiology, its repercussions will extend beyond the microbial community of sea ice to Arctic food webs, both aquatic and terrestrial. Rysgaard and Glud (2007) predict a shift from the current, copepod-dominated grazer community (see Section 24.5) to dominance by protozooplankton. Ultimately, large marine mammals such as polar bears and walruses will be affected. But the science of global climate change in the Arctic begins with the complex feedback loops that operate between its abiotic and biotic components. Scientists such as Rysgaard are revealing these linkages by cooperative research efforts that combine the expertise of biologists, biophysicists, and chemists to study sea ice as an integrated system.

Bibliography

Comiso, J. C. 2012. Large decadal declines of the Arctic multi-year ice cover. *Journal of Climate* 25:1176–1193.

Rysgaard, S., and R. N. Glud. 2004. Anaerobic N_2 production in Arctic sea ice. *Limnological Oceanography* 49:86–94.

Rysgaard, S., and R. N. Glud. 2007. Carbon cycling and climate change: Predictions for a High Arctic marine ecosystem (Young Sound, NE Greenland). *BioScience* 58:206–214.

Rysgaard, S., R. N. Glud, M. K. Seir, J. Bendtsen, and P. B. Christensen. 2007. Inorganic carbon transport during sea ice growth and decay: A carbon pump in polar seas. *Journal of Geophysical Research* 112:1–8.

Rysgaard, S., M. Kuhl, R. N. Glud, and J. W. Hansen. 2001. Biomass, production, and horizontal patchiness of sea ice algae in a high-Arctic fjord (Young Sound, NE Greenland). *Marine Ecology Progress Series* 223:15–26.

Schuur, E. A. G., J. G. Vogel, K. G. Crummer, H. Lee, J. O. Sickman, and T. E. Osterkamp. 2009. The effect of permafrost thaw on old carbon release and net carbon exchange from tundra. *Nature* 459:556–559.

1. We have focused on changes to the N and C cycles, but O_2 will play a key role in the biotic processes investigated by Rysgaard. Discuss the possible effects of O_2 on these processes.
2. Changes in Arctic productivity involve the concept of limiting factors. What factors might limit the increases in productivity predicted by Rysgaard and Glud?

much P as they store in biomass, returning it to the cycle. More than half of this excreted P is inorganic, which is reabsorbed by phytoplankton. The remaining P in aquatic systems is in organic compounds used by bacteria. In turn, these bacteria are consumed by microbivores, which excrete the PO_4^{3-} they ingest.

Some phosphates are deposited in shallow sediments and deep water. In ocean upwellings (see Sections 3.8 and 21.12), vertical transport brings phosphates to the surface, where they are absorbed by phytoplankton. Phosphorus locked in dead bodies sinks to the bottom and is deposited in sediments. Thus,

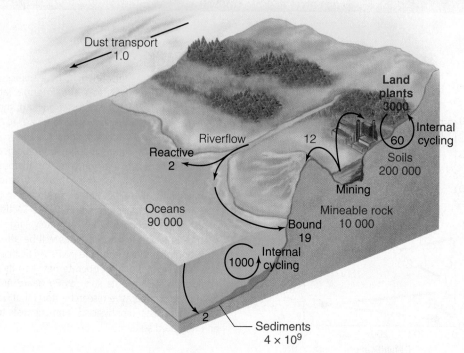

Figure 22.12 Global phosphorus cycle. Pools (in red) and fluxes (in black) are shown in units of 10^{12} g/yr.

(Adapted from Schlesinger 1997.)

surface waters may become P-depleted, while deep waters become saturated. Much of this P is locked up in the hypolimnion and bottom sediments, until some returns to surface waters by upwelling.

The global P cycle (Figure 22.12) (Schlesinger 1997) has no significant atmospheric pool, although airborne transport in dust and sea spray is 1×10^{12} g/yr. (A similar situation applies to other sedimentary cycles.) Rivers transport 21×10^{12} g/yr to oceans, but only about 10 percent is available for primary producers, with the rest deposited in sediments. The P content of the ocean is low, but the large volume constitutes a significant global pool. Turnover of organic P in surface waters occurs in days, and the vast majority of P absorbed by plankton is decomposed and mineralized in surface waters. However, 2×10^{12} g/yr is deposited in sediments or carried to deep waters, where organic P is decomposed into inorganic forms that are largely unavailable. On a geological timescale, uplifting and subsequent weathering return this P to active cycling.

22.7 SULPHUR CYCLE: The Hybrid Cycle of Sulphur Is Much Affected by Human Activity

The sulphur (S) cycle has both sedimentary and gaseous phases (Figure 22.13). In its long-term sedimentary phase, S is tied up in organic and inorganic deposits, released by

weathering and decomposition, and carried to terrestrial systems in salt solution. The gaseous phase permits circulation on a global scale.

Sulphur enters the atmosphere from both natural and anthropogenic sources: combustion of fossil fuels (particularly coal), volcanic eruptions, exchange at ocean surfaces, and release by decomposition. It enters the atmosphere as hydrogen sulphide (H_2S), which interacts with O_2 to form sulphur dioxide (SO_2). Atmospheric SO_2, soluble in water, returns to land and water in rainfall as weak sulphuric acid (H_2SO_4), the major component of **acid precipitation**. Whatever the source, S in a soluble form (sulphate, SO_4^{2-}) is absorbed by plants and incorporated into S-containing amino acids, which are transferred to consumers via the food web.

Earlier, we discussed the unique ecology of sulphur bacteria in hydrothermal vents (see Research in Ecology: The Extreme Environment of Hydrothermal Vents, pp. 46–47). Many bacterial transformations are also involved in local S cycling of all aquatic systems (Fossing et al. 1995). Excretion and death carry S from living tissue back to soil and the benthic (bottom) layer of water bodies, where anaerobic bacteria oxidize organic C by reducing SO_4^{2-} to H_2S. Another group, the colourless S bacteria (e.g., *Thioploca*), are chemoautotrophs, gaining energy by converting H_2S first to elemental S and then back to SO_4^{2-}, using NO_3^- stored in the vacuole as an oxidizing agent (Figure 22.14). In contrast, green and purple bacteria are photoautotrophs. These anaerobic organisms use H_2S as an electron source in photosynthesis and

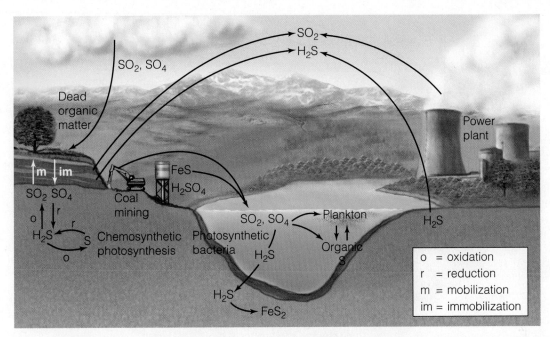

Figure 22.13 Local sulphur cycle. Note the sedimentary and gaseous phases. Major sources from human activity are burning of fossil fuels and acidic drainage from coal mines.

release elemental S, which is then oxidized to SO_4^{2-}. Best known are the purple bacteria of salt marshes, estuary mudflats, and sulphur hot springs.

As well as these and other microbial transformations, S precipitates as ferrous sulphide (FeS_2) in anaerobic conditions. FeS_2 is highly insoluble in neutral and acidic conditions, and is firmly held in mud and wet soil. Sedimentary rocks containing FeS_2 (pyritic rocks) often overlie coal deposits. When exposed to air by mining, FeS_2 reacts with O_2 and (in the presence of H_2O) yields ferrous sulphate ($FeSO_4$) and H_2SO_4. So, S in pyritic rocks, which would normally cycle very slowly if at all, is exposed to weathering by human activities, discharging H_2SO_4, $FeSO_4$, and other S compounds into aquatic systems, where they damage or destroy aquatic life.

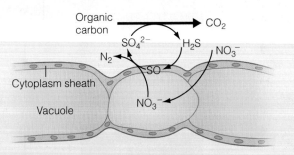

Figure 22.14 Sulphur transformations involving bacteria. In anoxic muds, sulphate-reducing bacteria oxidize organic compounds, releasing H_2S. Colourless S bacteria (e.g., *Thioplaca*) oxidize H_2S, generating elemental S that is embedded in their cytoplasm (blue-green dots) and ultimately oxidized to SO_4^{2-} using NO_3^- stored in vacuoles, releasing N_2.

(Adapted from Fossing et al. 1995.)

The indirect effects of acid precipitation are well known, but these less-publicized direct effects have converted hundreds of kilometres of streams in coal-mining areas to lifeless, highly acidic water.

Much research focuses on human impacts on the S cycle, but our understanding of the global S cycle (Figure 22.15, p. 486) (Schlesinger 1997) is limited. The gaseous phase allows global circulation, with annual flux in the atmosphere of 300×10^{12} g as SO_2, H_2S, and sulphate particles in dryfall. The gaseous forms combine with H_2O and are transported in wetfall. The oceans are a large source of sulphate-containing aerosols. Emissions of 16×10^{12} g/yr make dimethylsulphide ((CH_3)$_2$S) the largest source of S gases from natural biological processes.

There are various biological S emissions from terrestrial systems, but collectively they represent a minor flux. The dominant S gas emitted from freshwater wetlands and O_2-depleted soils is H_2S. Emissions from plants are poorly understood, but fires emit about 3×10^{12} g annually. Estimating biological turnover of SO_2 is difficult, but marine plants assimilate some 130×10^{12} g annually. Adding anaerobic oxidation of organic matter brings the total to an estimated 200×10^{12} g. Volcanic activity also contributes to global S cycling. Major events, such as the 1991 eruption of Mt. Pinatubo, release about $5-10 \times 10^{12}$ g. When volcanic activity is averaged over a long period, annual global flux is some 10×10^{12} g.

Human activity plays a dominant role in the global S cycle. Emissions of SOx (as well as NOx) from industrial sources and vehicles are the main sources of acid precipitation, which forms when these compounds combine with water to form sulphuric and nitric acids. We examine some of the

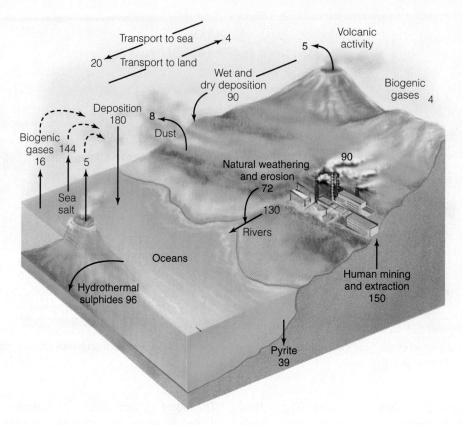

Figure 22.15 Global sulphur cycle. Fluxes are shown in units of 10^{12} g S/yr.

(Adapted from Schlesinger 1997.)

effects of acid precipitation (along with nutrient imbalances triggered by N saturation) in Section 22.10. Attempts to reduce emissions by the use of scrubbers and by reducing combustion temperatures have had some success, but overall it is still a growing problem, albeit one that the media pay much less attention to than previously.

22.8 OXYGEN CYCLE: Cycling of Oxygen Is Subject to Great Biological Influence

Oxygen has a gaseous cycle (Figure 22.16) (Schlesinger 1997), in which the atmosphere is the major pool. There are two major sources of free O_2. One is abiotic dissociation of H_2O vapour by sunlight. Most of the H_2 released escapes into space, while the rest recombines with O_2 to re-form H_2O. The other is biotic: photosynthesis, active only since life began. (The first O_2-producers were likely related to modern-day cyanobacteria.) Although O_2 is produced only by photoautotrophs, both aerobic autotrophs and heterotrophs use it in respiration.

Because photosynthesis and respiration alternately release and consume O_2, one would seem to balance the other, with no significant gain or loss. Yet at some time in

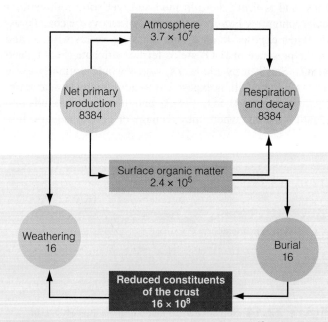

FIGURE 22.16 Global oxygen cycle. Units are 10^{12} moles O_2 per year or an equivalent amount of reduced compounds. An imbalance in the ratio of photosynthesis to respiration causes net storage of reduced organic materials in the crust and accumulation of O_2 in the atmosphere.

(Adapted from Schlesinger 1997.)

Earth's history, the O_2 released greatly exceeded the amount used in respiration (including decomposition) and in oxidation of sedimentary rocks. Present-day atmospheric O_2 levels reflect this past imbalance, as do the undecomposed organic matter present in fossil fuels and peat, and C in sedimentary rocks. The time gap between the evolution of seedless vascular plants and that of fungi capable of degrading lignin (see Section 21.3) is the key factor explaining this massive change in atmospheric O_2.

Other O reservoirs are H_2O and CO_2. These reservoirs are linked through photosynthesis, but O_2 is also biologically exchangeable from nitrates and sulphates, which organisms generate by oxidizing NH_3 and H_2S. Because O is so reactive, its cycling is particularly complex and almost entirely global. As part of CO_2, it circulates through ecosystems. In water, some CO_2 combines with Ca to form carbonates. O_2 also combines with N to form nitrates, with Fe to form ferric oxides, and with other minerals to form other oxides, temporarily removing it from circulation. In photosynthesis, freed O_2 is released from H_2O, only to re-form in respiration.

Ozone (O_3) is an ambivalent atmospheric gas. In the stratosphere, 10 to 40 km above Earth, it shields biota from harmful UV radiation. Yet in the troposphere, ground-level O_3 generated from the union of nitrogen and sulphur oxides in the presence of light is a damaging pollutant and a major component of smog. It irritates the eyes and the respiratory system and injures plants.

A reaction cycle requiring sunlight maintains O_3 in the stratosphere. Solar radiation breaks the O–O bond in O_2. Freed O atoms rapidly combine with O_2 to form O_3. At the same time, a reverse reaction consumes O_3 to form O and O_2. Under natural conditions, a balance exists between O_3 formation and destruction. Human activities have diminished stratospheric O_3 by disrupting this balance. Catalysts—some of which are human pollutants and some biologically derived—injected into the stratosphere reduce stratospheric O_3. Among them are chlorofluorocarbons (CFCs); methane (CH_4), both natural and arising from human activities; and nitric oxide (NO), from fertilizers, denitrification, and burning of fossil fuels. Of particular concern is chlorine monoxide (ClO), derived from CFCs and used in aerosol sprays (now banned in much of the world as a result of the 1987 Montréal protocol, but still used in some countries); refrigerants; solvents; and other sources.

The impact of humans on O_3 (increasing damaging ground-level O_3 and depleting protective stratospheric O_3) is perhaps the most dramatic of our effects on the O cycle. Another, more subtle long-term influences may occur. If humans lower O_2 production by reducing global primary productivity, while simultaneously increasing O_2 consumption by burning fossil fuels and exposing peat to oxidation by draining wetlands and thawing permafrost, the combined effects may reverse the positive O_2 flux into the atmosphere that life has achieved over millions of years.

22.9 STOICHIOMETRY: Relationships between Elements in Chemical Compounds Link Cycles

We have considered the major biogeochemical cycles independently, but they are linked in many ways. They may share membership in specific compounds that play important roles in their respective cycles. Examples are many, such as Ca and P in the mineral apatite, a phosphate of Ca; C and O in CO_2; and N and O in NO_3^-. More generally, nutrients are constituents of organic matter, travelling together in their odyssey through internal cycling within ecosystems.

Organisms require nutrients in different proportions for specific processes. Photosynthesis, for example, uses 6 moles of H_2O (net) and produces 6 moles of O_2 for every 6 moles of CO_2 reduced to 1 mole of sugar $(CH_2O)_6$. Likewise, a fixed amount of N is needed to produce Rubisco, the enzyme that catalyzes CO_2 fixation. The N content of Rubisco is the same in every plant, regardless of the species or environment. The same applies to amino acids, proteins, and other nitrogenous compounds. **Stoichiometry** is the branch of chemistry dealing with the quantitative relationships of elements in combination. Given these stoichiometric relationships, limitation of one nutrient can affect the cycling of others.

Consider the link between C and N. Although the N content of Rubisco is consistent, plants differ in the Rubisco content of their leaves and hence in their overall N content. Plants in low-N sites have less N available for the production of Rubisco. In turn, lower Rubisco results in lower photosynthetic rates and lower N content of leaf litter. Rates of immobilization, mineralization, and subsequent N release are also affected. In this way, N supply influences the rate at which C (and other nutrients) cycles through ecosystems. Conversely, other nutrients and abiotic factors that directly affect primary productivity (and hence N demand) influence the rate of N cycling.

22.10 ANTHROPOGENIC INFLUENCES: Forest Dieback May Reflect Nutrient Imbalances Caused by Humans

Thus far, we have mentioned human impacts on several cycles separately. Increasingly, ecologists are investigating the interactive effects of human-induced imbalances in a number of nutrients. Perhaps the most serious is the impact of human intervention in nutrient cycles on the vigour of forests worldwide. **Forest dieback** describes the general decline of forests in both Europe and North America, in which mortality is significantly above normal (Allen 2009). Damage is most severe in Germany, where most forests show symptoms such as crown thinning, leaf discoloration, premature leaf fall, and tip death. Coniferous forests are most susceptible, but deciduous trees are also affected. In

Figure 22.17 Sugar maple forests like this one in eastern Québec are prone to forest dieback.

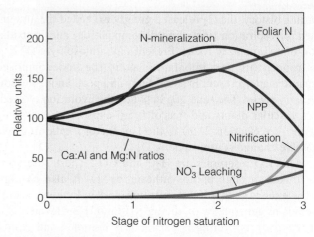

Figure 22.18 Hypothesized response of temperate forests to N addition. Stage 1: plant uptake and N-mineralization increase, stimulating NPP. Stage 2: NPP and mineralization decline as a result of decreasing Ca:Al and Mg:N ratios and soil acidification. Nitrification increases with excess NH_4^+. Stage 3: NO_3^- leaching increases sharply.

(Adapted from Aber et al. 1989.)

the Appalachians of the eastern United States, red spruce (*Picea rubra*) has been declining since the 1960s, with over 50 percent mortality in some areas. Similar losses are reported in the southern pine forests of Georgia. In eastern Canada, sugar maples (*Acer saccharum*) are showing crown dieback and bark peeling. In Québec, over half of sugar maples are affected (Figure 22.17). White pine (*Pinus strobus*) exhibits stunted growth and discoloured needles throughout its range in eastern Canada and the American northeast.

What are the causes of forest decline? Some attribute it to nutrient imbalances, triggered by stimulation of the N cycle. Net primary productivity in many forests (particularly in northern, N-poor sites) is often limited by availability of N. In recent decades, human activities have dramatically increased N deposition. Burning fossil fuels, industrial emissions, and high-intensity agriculture have increased atmospheric and soil inputs of nitrogen oxides (NO*x*) far above normal. These oxides undergo chemical transformations in the atmosphere and are deposited in the vicinity of the emissions. Deposition rates thus vary widely, and are highest in areas that are highly populated and/or in the path of industrial emissions.

Nitrogen oxides not only cause acid precipitation in both wetfall and dryfall, but also contribute to a state of excess N availability called **nitrogen saturation**. Soil N influences plant uptake and tissue N content, which in turn correlates with photosynthetic capacity (see Figure 6.25). Initially, N deposition has a fertilizer effect, increasing NPP. However, as water and other nutrients become more limiting relative to N, these ecosystems approach a saturated state, after which no further growth stimulation occurs. If N levels continue rising, complex changes to soil and plant processes may initiate soil acidification and forest decline (Figure 22.18) (Aber et al. 1989).

How does this sequence, also dubbed the *nitrogen cascade* because of its feedback loops (Galloway et al. 2003), work? Most N deposited is in the form of nitrate (NO_3^-) or

ammonium (NH_4^+), although NH_4^+ is dominant. In the early stages, plants take up much of this extra N, and forest productivity and growth are stimulated. (In Europe, additional NH_4^+ comes from application of treated human sewage, which is rich in NH_4^+, exacerbating the fertilizer effect.) However, as the limiting factor on plant growth shifts from N to other resources, more NH_4^+ becomes available in the soil, triggering release of other cations into the soil solution by overwhelming the exchange sites on soil particles (see Section 4.8). More of these cations then leach into groundwater, intensifying the nutrient limitations by creating an imbalance in supply.

Microbes also use some of the excess NH_4^+, stimulating decomposition and mineralization, adding to the fertilizer effect. Assuming conditions are favourable and nitrifying bacteria are present, much of the NH_4^+ is converted to NO_3^-, which is absorbed by plants or denitrified by microbes, reducing it to N_2 gas and completing the N cycle (see Figure 22.9). In contrast to NH_4^+, NO_3^- is highly mobile in soils because (as an anion) it is not adsorbed to soil particles. Hence, any NO_3^- not used by plants or microbes leaches into groundwater, causing eutrophication of water bodies. Further, when NO_3^- is leached, loss of base cations such as Ca^{2+} and soil acidification from release of H^+ from exchange sites may follow. Meanwhile, release of SO*x* and NO*x* from burning fossil fuels accelerates acidification. As soil pH drops from both causes, the buffering ability of soil cation exchange may be exhausted, triggering the release of aluminum (Al), which enters aquatic systems and has toxic effects. High Al in the soil solution also damages trees, causing dieback of root tips. Indeed, much of the damage of acid precipitation comes not from direct effects of lower soil pH but from these indirect effects of Al and other heavy metals.

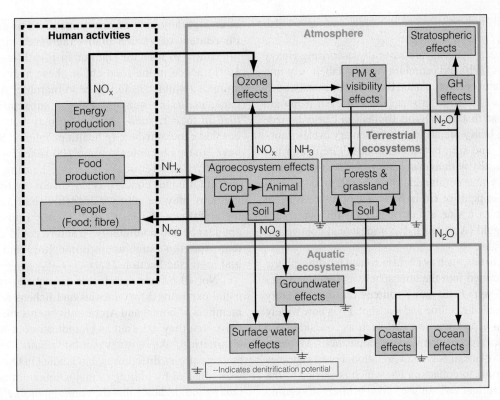

Figure 22.19 Nitrogen cascade, showing sequential effects of addition of reactive N. GH: greenhouse effect; PM: particulate matter.

(Adapted from Galloway et al. 2003.)

In the last stages of N saturation, productivity decreases and mortality increases. The causes are complex, but partly reflect nutrient imbalances stemming from excess availability of N. Plant health is affected by the relative concentrations of nutrients as much as by their absolute levels. As Al in the soil solution increases with acidification, Ca:Al and Mg:N ratios decline, lowering uptake of Ca and Mg. Calcium has many key roles in plants. It is a vital biochemical messenger and regulates membrane permeability. As a structural component of cell walls, Ca is used in producing sapwood, the outer xylem regions in which water moves. Less Ca restricts water flow, decreasing the leaf area that the tree can support. Ca shortages also inhibit cell division, stunting growth.

Similarly, Mg is part of many compounds, including chlorophyll. *Chlorosis* (yellowing) of leaves, a typical dieback symptom, is associated with Mg shortage. Reduced chlorophyll then limits photosynthesis. Faced with these nutrient imbalances, and stimulated to grow by high N, the plant tries to mobilize minerals from the leaves to the growing tips, exacerbating the deficiencies. The ultimate effect is reduced C uptake and growth.

Trees suffering from these nutrient imbalances are more prone to natural disturbances such as fungal parasites, herbivorous insects, and pollutants. In addition, ozone exposure in montane regions increases dieback damage (Dixon et al. 1998). In this complex sequence of cause and effect, we must distinguish between the immediate cause(s) of mortality and the factors that contribute to the weakened state. A particular pathogen or climate stressor might be the proximal factor, whereas the ultimate cause is a nutrient imbalance arising from human activities. In some regions, climatic factors, notably heat and drought, may be as important as nutrient imbalances (Allen 2009). We need long-term studies to tease apart the multiple interacting factors contributing to dieback in different regions, as well as possible remedies.

From a systems perspective, conversion by humans of non-reactive forms of nitrogen (N_2) into reactive forms, fuelled by the drive to increase productivity in agricultural ecosystems, has triggered many impacts on natural ecosystems, particularly forests (Figure 22.19). We have treated forest dieback in some detail, as a prime example of the interacting impacts of multiple factors—nutrient imbalances, acid precipitation, pollutants such as ozone, and biotic stressors—on ecosystem health. Some view it as a critical tipping point (see Section 19.9) for the ecosystem as a whole.

22.11 HEAVY METALS: Humans Increase Cycling of Toxic Heavy Metals

This chapter has focused on biogeochemical cycling of nutrients—substances that are needed by organisms. As well as those that we have discussed, all other nutrients have their own cycles, which resemble the cycles discussed to varying degrees. However, non-required and/or toxic substances also cycle, with many ecosystem effects.

Some of these substances are naturally occurring. Heavy metals such as lead (Pb) and cadmium (Cd) find their way into food webs when absorbed by plants. Some heavy metals, such as copper (Cu), are required in small amounts, but many are not, and even those that are micronutrients can cause damage at high levels. All heavy metals have sedimentary cycles, moving between rock and soil, but the amount in circulation has increased dramatically with human activities. The use of heavy metals in industry and commercial products (either deliberately as with Pb in paint or Cd in batteries, or as inadvertent contaminants) or as a side effect of mining for other substances, such as gold (Au), has led to widespread contamination of soil and water.

Some heavy metals, such as Pb, also spread by atmospheric contamination. Emitted into the atmosphere as small particles (smaller than 0.5 µm) from many sources but most notably from burning of leaded gasoline and leadshot, Pb is now widely distributed. Areas near point sources, such as roadsides and industrial emissions, receive the heaviest depositions. Pb particles settle on the surface of soil and vegetation. Forest canopies are particularly prone to collecting Pb, which is carried to the ground by rain and leaf fall. In the soil, Pb binds to organic matter in the litter layer and reacts with sulphate, phosphate, and carbonate anions. Although Pb additives have been banned in many areas, Pb persists in the upper soils for about 5000 years. As well, there are many other industrial sources, so Pb is an ongoing problem.

In aquatic systems, mercury (Hg) contamination is of great concern. Methylated mercury (CH_3-Hg) is released from industrial sources, or forms after release of Hg from burning of fossil fuels. Natural sources include volcanoes and weathering. In anaerobic aquatic habitats, bacteria convert Hg to the highly toxic methylated form. Minamata disease describes acute methyl-mercury poisoning, named for a 1956 incident in Minimata, Japan. A similar tragedy occurred in the Grassy Narrows area of northwestern Ontario in 1960s, from illegal release into the Wabigoon–English River system of industrial wastes associated with pulp bleaching. First Nations people were most affected, as a result of their high fish consumption.

Fortunately, amelioration is possible. Hrabik and Watras (2002) observed declines of 30 percent in Hg content in fish tissue in a Wisconsin lake between 1994 and 2000 as a result of decreased atmospheric release. Additional declines were attributed to reduced acid precipitation, as low pH stimulates methylation. Thus, modest changes in human activities can reduce Hg accumulation over the short term.

Organisms vary in their susceptibility to heavy metal damage. In Section 22.10, we discussed the role of Al toxicity in forest dieback, but many plants are relatively unaffected by other heavy metals. They sequester large amounts of Pb and Cd in their cell vacuoles, minimizing damage to their own tissues. Plants can thus appear vigorous, yet har-

bour high amounts of heavy metals. For example, Cd and Pb content of garden plants increased significantly with proximity to a metal smelter in Flin Flon, Manitoba (Pip 1991). Once in the food chain, these heavy metals damage animals, which are far more vulnerable. All living organisms, including humans, carry a substantial heavy metal load in their tissues—levels that may not cause symptoms but that can contribute to health problems. North Americans have tissue Pb contents some 100 times normal, contributing to mental retardation, palsy, hearing loss, and even death. Of equal concern is the discovery that soil microorganisms may be more vulnerable than animals to heavy metal effects (Giller et al. 2009b). Impacts on microbe abundance and community structure could alter key ecosystem processes, such as decomposition and the soil microbial loop (see Section 21.9).

Not all primary producers are equally tolerant of heavy metal exposure. Many mosses and lichens that are important members of boreal and Arctic ecosystems are sufficiently sensitive that they are used as bioindicators of heavy metal contamination. As with vascular plants, there are many among-species differences, but studies in Ontario and Québec sites affected by mining (a major source of heavy metal contamination) indicate that the common moss *Pleurozium schreberi* is highly sensitive, as are the reindeer lichens (*Cladonia* spp.) that are so prominent in spruce–lichen woodlands (Figure 22.20) (Tyler at al. 1989).

Many studies have documented heavy metal levels in various groups of organisms, but few have considered their impact from a systems perspective. In non-contaminated agricultural soil, heavy metal content was related to trophic position (Table 22.2) (Carter 1983). Within red clover plants, levels varied among both metals and plant parts (especially leaves), reflecting differences in soil content and plant strategies for exclusion or tolerance. Cd, the most toxic and the only one of the three metals measured that is not a micronutrient, was concen-

Figure 22.20 Spruce–lichen woodlands, such as this example in northeastern Québec, are highly sensitive to heavy metals.

Table 22.2 Levels of Heavy Metals in Tissues and Feces of Plants and Macrofauna in an Agricultural Field on Westham Island, British Columbia

	Mean ppm dry weight (standard deviation)		
	Zinc	Cadmium	Copper
Mineral soil	83 (8)	0.4 (0.1)	26 (6)
Living red clover			
Roots	38 (9)	0.5 (0.1)	39 (9)
Stems	36 (13)	0.1 (0.04)	16 (5)
Leaves	104 (114)	0.2 (0.1)	44 (34)
Red clover litter (mixed)	74 (24)	0.7 (0.2)	13 (17)
Earthworms			
Sexually mature *Lumbricus rubellus*			
Tissue	320 (129)	6.0 (2.0)	10 (3)
Feces	86 (10)	0.5 (0.1)	25 (4)
Slugs	214 (83)	2.0 (1.0)	100 (54)
Millipedes			
Tissue	321 (30)	0.2 (0.04)	221 (96)
Feces	340 (150)	1.0 (1.0)	28 (15)
Carabid beetles			
Adults	116 (28)	0.3 (0.2)	13 (4)
Larvae	218 (89)	3.5 (1.6)	21 (7)

Data from Carter 1983.

trated by herbivorous slugs, detrital-feeding earthworms, and predaceous carabid larvae. In contrast, millipedes (also litter feeders) had low Cd but concentrated Cu and Zn. These differences reflect not only trophic position but also structural and physiological differences affecting metal uptake and storage.

22.12 RADIONUCLIDES AND ORGANIC POLLUTANTS: Many Toxic Substances Bioaccumulate in Food Webs

Heavy metals are not the only hazardous substances that have increased in circulation with human activities. We associate nuclear pollution with Hiroshima and the post-war weapons testing era. Bikini Atoll (a group of islands in the South Pacific) was destroyed by a series of tests from 1946 to the late 1950s. But the impacts of radioactive fallout travelled much farther than the zone of direct annihilation. Global circulation carried it to remote areas, with the Arctic receiving a particularly high dose. Levels of strontium-90 (^{90}Sr) and cesium-137 (^{137}Cs) were so high that caribou meat was unsuitable for consumption. One reason for the post-war switch in North America from breast-feeding to formula was the dangerously high levels of radiation in human milk.

Humans were not the only species affected. ^{90}Sr and ^{137}Cs cycle similarly to Ca, following it through the food chain. Lichens, dominant primary producers in tundra, absorb most of the radioactive material to which they are exposed. Grazing herbivores such as caribou rely upon lichens for winter sustenance, and the radioactive load in their tissues accumulates over the winter. Once in the food web, radioactive substances pass to carnivores and humans. Even after the test era ended, the proliferation of nuclear arms means that the release of **radionuclides** (fission and non-fission products of nuclear reactions) is still a frightening possibility.

Of equal concern is the fact that increased demand for cheap energy, along with the desire to reduce dependence on fossil fuels, has increased our reliance on nuclear power. Ontario, for example, uses nuclear plants for about half of its power generation, and although many argue that nuclear power is part of the "green" solution to our energy problems, the as-of-yet unresolved nuclear aftermath of the 2011 earthquake in Japan (as well as the aging infrastructure of many nuclear plants worldwide, including in Canada) engenders sober second thoughts. After the accidents at Three Mile Island in Pennsylvania in 1979 and at Chernobyl in Russia in 1986 (the effects of which are still being felt in Eurasia), we became sanguine that improvements in design and safety technology meant that similar accidents were unlikely. The events in Japan in 2011 undermine that confidence. Extreme caution is paramount given the devastating direct effects of radionuclides as well as their cycling behaviour and long half-life. Some last for mere seconds; others remain active for millennia.

There is much controversy concerning the impact of radioactivity. In 2006, a report issued by the Chernobyl Forum, a group of scientists acting under the auspices of the United Nations, concluded that the Exclusion Zone set aside after the disaster had "paradoxically become a unique haven for biodiversity" (IAEA 2006). While acknowledging that individuals had suffered damage and increased mortality from radioactivity, the report argued that the cessation of human activities in the abandoned area had allowed populations and ecosystems to regain much needed habitat, albeit of suboptimal quality given the high radiation levels. However, such reports have yet to be supported by valid population studies for a range of species. Other studies show that populations in the 200 000-km^2 area affected have declined for many species, including birds of prey (raptors) (Møller and Mousseau 2008) and insects and spiders (Møller and Mousseau 2009).

Substances manufactured by humans constitute major pollutants in part as a result of their cycling behaviour. We associate the word *organic* with foods grown without the use of added chemicals, but it simply means carbon-bound. Humans manufacture many non-natural organic compounds for a variety of uses, many of them highly toxic. Chlorinated hydrocarbons used in pesticides (of which DDT is but one of many examples) and for other purposes (such as polychlorinated biphenyls—PCBs—used in electrical transformers, plastics, and solvents) cycle in complex ways among abiotic and biotic ecosystem components. Although many jurisdictions have banned their use, these compounds still persist in many ecosystems, as a result of their long half life (roughly 20 years). In addition, continued use in other parts of the world (as well as illegal use in countries where they have been banned) contributes new inputs of many toxic organic pollutants.

Complicating their cycling is the tendency of both radionuclides and many organic pollutants to concentrate at higher trophic levels. The amount of a substance may be negligible in soil or water and seem of little concern; but once in the food chain, it increases in concentration with every trophic

DDT levels

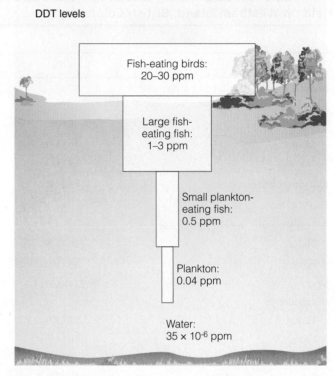

Figure 22.21 Biomagnification of DDT in a marine food web.

level via **biomagnification** (Figure 22.21) to orders of magnitude higher than background. (**Bioaccumulation**, in contrast, refers to the buildup in the tissues of a single organism, as with radiation in a caribou over its lifetime.) Biomagnification is particularly applicable to compounds such as DDT that are fat soluble, slow to degrade, and have a long half-life. It also occurs with some radionuclides and heavy metals, such as Hg, and is yet another reason why species at upper trophic levels are more vulnerable.

A host of other compounds used in pesticides and industrial purposes cycle in ecosystems. Their pathways and effects vary with their chemical properties and their interactions with biota, and in many cases are unknown or poorly understood. For our purposes, the general point is this: Any substance—of natural or of human origin, required or toxic—that cycles within and among ecosystems has the potential for both short- and long-term ecological effects.

EcologyPlace

SUMMARY

Types of Cycles 22.1

Nutrients cycle between living and non-living ecosystem components. Biogeochemical cycles are either gaseous, with major reservoirs in the atmosphere or oceans, or sedimentary, with major reservoirs in solid earth. Some elements have hybrid cycles. Sedimentary cycles have two phases: salt solution and rock. Minerals become available through weathering and enter the water cycle as salts. They take diverse paths through the ecosystem, ultimately returning to Earth's crust by sedimentation.

Inputs and Outputs 22.2

All cycles have inputs, internal cycling, and outputs. Inputs vary between sedimentary and gaseous cycles. Nutrient inputs in terrestrial systems depend on soil properties. Supplementing soil sources are inputs carried by precipitation, air currents, and animals. Major inputs for aquatic systems come from the surrounding land in drainage, detritus, and sediments. When outputs exceed inputs, net decline occurs. Major outputs are organic matter carried by streams and rivers and leaching from soils into surface and groundwater. Harvesting causes permanent loss. Fire increases nutrient exports. Biogeochemical cycles are linked, requiring a global perspective.

Local Carbon Cycle 22.3

The carbon cycle is tightly linked to energy flow. Assimilated as CO_2 by plants and other photoautotrophs in photosynthesis, C is consumed in organic tissue by heterotrophs, released through respiration, mineralized by decomposers, accumulated in biomass, and withdrawn into long-term reservoirs. The rate of C cycling varies with the rates of productivity and decomposition, which both increase in warm, wet ecosystems. In swamps and marshes, organic material stored as humus or peat circulates slowly. Similar cycling occurs in aquatic habitats. Cycling fluctuates daily and seasonally. CO_2 builds up at night and declines in the day as plants use it in photosynthesis. During the growing season, atmospheric levels drop.

Global Carbon Cycle 22.4

Earth's C budget is closely linked to the atmosphere, land, and oceans. In the oceans, surface water is the main site of C exchange with the air. The ability of surface waters to absorb CO_2 is governed by the abiotic carbonate reaction sequence and by its biotic uptake in photosynthesis. Uptake of CO_2 in terrestrial ecosystems is governed by gross productivity. Losses reflect autotrophic and heterotrophic respiration, particularly by decomposers. Earth and the oceans are currently net C sinks for excess CO_2 released by human activities.

Nitrogen Cycle 22.5

The nitrogen cycle begins with fixation of atmospheric N_2 by lightning or by free-living or symbiotic bacteria associated with the roots of some plants, and by cyanobacteria. Other processes include ammonification (breakdown of amino acids to produce NH_3), nitrification (oxidation of NH_3 to NO_3^-), and denitrification (reduction of NO_3^- to N_2). The atmosphere is the largest global pool, with small amounts in biomass and soil. Major sources of oceanic N are dissolved forms in drainage from rivers and precipitation.

Phosphorus Cycle 22.6

The main P reservoirs are rock and phosphate deposits. Phosphate is released by weathering and taken up by plants. In aquatic systems, the P cycle involves three states: particulate organic P, dissolved organic phosphates, and inorganic phosphates. Phytoplankton, zooplankton, bacteria, and microbial grazers participate in its cycling. The global P cycle has no significant atmospheric storage. Transfer of P from terrestrial to aquatic systems is normally low, but phosphate fertilizers and disposal of sewage and wastewater increase inputs to aquatic systems.

Sulphur Cycle 22.7

Sulphur has a hybrid cycle. S is released from weathering of rocks, runoff, and decomposition. Gaseous sources are evaporation, volcanoes, and burning fossil fuels. It enters the atmosphere mostly as H_2S, which quickly oxidizes to SO_2, which reacts with atmospheric water to form H_2SO_4, a major component of acid rain. Plants take up SO_4^{2-} and incorporate it into organic compounds. Consumption, excretion, and death return S to soil and aquatic sediments, where bacteria release it as H_2S or as SO_4^{2-}. The global S cycle has reservoirs in Earth's crust and in the atmosphere. In the sedimentary phase, S is bound in organic and inorganic deposits, released by weathering and decomposition, and carried to land and water in salt solution.

Oxygen Cycle 22.8

Oxygen, a by-product of photosynthesis, is chemically active, combining with many chemicals in Earth's crust, and reacting spontaneously with organic compounds and reduced substances. In respiration, it oxidizes carbohydrates to release energy and CO_2. Past biotic activity has increased atmospheric O_2. Ozone is an important part of the atmospheric O reservoir. Human activities have depleted stratospheric O_3 levels and increased damaging ground-level O_3.

Stoichiometry 22.9

The major biogeochemical cycles are linked. Elements from many cycles are combined in organic matter. Stoichiometric relations among the elements involved in biological processes related to C uptake and growth affect nutrient cycling.

Anthropogenic Influences 22.10

Forest dieback may reflect long-term impacts of human activities on nutrient cycles. Increasing N inputs have a short-term fertilizer effect. Artificially high NH_4^+ levels in the soil swamp cation exchange sites, increasing acidification and releasing toxic Al. Altered ratios of other nutrients, notably Ca and Mg, affect growth. The combined effects of stimulated growth, nutrient imbalances, and Al toxicity weaken trees, making them more vulnerable to biotic and abiotic hazards. High NH_4^+ increases NO_3^- release by nitrification, causing eutrophication of water bodies.

Heavy Metals 22.11

Human activities have increased circulation of heavy metals. Although most have sedimentary cycles, heavy metals such as Pb enter the atmosphere in particulate form and are deposited in soil and water. Heavy metals vary in toxicity, with Hg among the most lethal. Mosses and lichens are more affected than vascular plants. Trophic position and physiology affect exposure and accumulation in soil microfauna, which are highly vulnerable.

Radionuclides and Organic Pollutants 22.12

Cycling of radionuclides from military and non-military sources is of concern, given their high toxicity and long half-life. Toxic organic compounds of human manufacture in pesticides and industrial compounds not only cycle, but also often undergo biomagnification in food webs.

KEY TERMS

acid precipitation	biomagnification	net ecosystem	nitrogen saturation	stoichiometry
ammonification	denitrification	productivity	radionuclide	volatilization
bioaccumulation	forest dieback	nitrification	sedimentary cycle	
biogeochemical cycle	gaseous cycle	nitrogen fixation		

STUDY QUESTIONS

1. How are photosynthesis and decomposition linked in the carbon cycle?
2. In the temperate zone, is atmospheric CO_2 higher during the day or night? Why?
3. Distinguish the following processes in the nitrogen cycle in terms of their reactants, products, and biota responsible: *fixation, ammonification, nitrification,* and *denitrification.*
4. What biological and non-biological processes are responsible for N fixation? How is the biological mechanism affected by soil nitrate levels, and why?
5. What is the main reservoir of phosphorus in the phosphorus cycle? What is the major P input into aquatic ecosystems?
6. Saskatchewan has large phosphate deposits that are used to produce fertilizers. What are the long-term effects of phosphate fertilizers on the biosphere?
7. What are the main sources of sulphur in the sulphur cycle? In what respect does sulphur have a hybrid cycle?

Can the same be said of the carbon cycle, or is it a gaseous cycle?
8. What are the major impacts of human activity on the sulphur cycle?
9. What is the role of photosynthesis and decomposition in the oxygen cycle?
10. How has deposition of fossil fuels influenced the O_2 content of the atmosphere over geologic time? How does burning of fossil fuels influence this pattern?
11. With rising global temperatures, peat deposits in northern Canada are experiencing accelerated decomposition. What is the likely long-term impact on the oxygen cycle?
12. Increased levels of NH_4^+ stimulate microbial processes, including decomposition. How might faster decomposition affect the availability of other nutrients to plants?
13. Describe three distinct ways in which nitrogen saturation resulting from increased anthropogenic inputs contributes to forest dieback.

FURTHER READINGS

Aber, J. D. 1992. Nitrogen cycling and nitrogen saturation in temperate forest ecosystems. *Trends in Ecology and Evolution* 7:220–223.
Excellent introduction to the concept and consequences of nitrogen saturation in forests.

Houghton, R.A. 2002. Terrestrial carbon sinks—uncertain explanations. *Biologist* 49:155–159.
Discusses the changes in carbon sink activity in recent decades.

Sarmiento, J. L., and Gruber, N. 2006. *Ocean biogeochemical dynamics*. Princeton, NJ: Princeton University Press.
Overview of the circulation and interactions of major elements in the oceans.

Sprent, J. I. 1988. *The ecology of the nitrogen cycle.* New York: Cambridge University Press.
Valuable ecological perspective on the processes involved in the nitrogen cycle.

Vitousek, P. M., J. Aber, R. W. Howarth, G. E. Likens, P. A. Matson, D. W. Schindler, W. H. Schlesinger, and G. D. Tilman. 1997. Human alteration of the global nitrogen cycle: Sources and consequences. *Ecological Applications* 7:737–750.
Discusses the impact of human activities on the nitrogen cycle.

The 19th century witnessed the golden age of the naturalist–explorer. Young scientists such as Alfred Wallace, Henry Bates, Joseph Hooker, Alexander Von Humboldt, and Charles Darwin travelled the tropics—then the frontier of the natural sciences—cataloguing its amazing array of organisms. In 11 years in Brazil, Henry Bates collected an astounding 14 712 animal species, of which more than 8000 were previously unknown to science. To someone trained in a temperate zone, the biological variety of the tropics was awe-inspiring.

Besides being struck by the diversity of life, naturalists were impressed by the presence of **ecological equivalents**: organisms of similar form and function occupying geographically distinct regions. For example, the flying squirrel of eastern North America (*Glaucomys sabrinus*, a rodent) and the Australian sugar glider (*Petaurus breviceps*, a marsupial) are unrelated species inhabiting forests half a world apart, yet they closely resemble each other (Figure 1). Both have a flat, bushy tail and an extension of the skin between the foreleg and hindleg, enabling them to glide from one tree to another.

Links between form and function are not limited to animals. In hot deserts, unrelated plant species have converged on the same solution to life in dry, hot climates—the succulent habit. Cacti such as the prickly pear cactus (*Opuntia* spp.) occupy deserts in the American southwest. They combine a fleshy, water-storing photosynthetic stem with leaves modified into protective spines. Meanwhile, in the deserts of South Africa, members of another family, Euphorbiaceae, such as *Euphorbia horrida*, have evolved a remarkably similar morphology.

Convergent patterns are not confined to individual species. The naturalist–explorers also noted that geographically separate regions with similar climates support similar communities. Consider shrub deserts in southern Africa and Australia (Figure 2). Both regions have a hot, arid climate. Although the species occupying these ecosystems are distinct, the physical structures of their drought-tolerant shrub communities are remarkably similar in form and function. The early naturalists were thus faced with two conflicting observations in need of reconciliation: (1) the amazing diversity of species and (2) the convergence among unrelated species of similar

behaviours or morphologies when occupying similar niches (such as flying squirrels and sugar gliders) and/or when occupying similar but widely separated habitats (such as cacti and euphorbs).

Although the main goal of the theory of natural selection is to explain how diversity among species can arise from common ancestry, it can also explain the similarities of organisms in geographically distinct regions. **Convergent evolution** is the independent evolution of a similar trait in two species that do not share a recent common ancestor. Recall from Section 5.1 that natural selection is a two-step process: (1) generation of variation among individuals in a population in traits that (2) influence survival and reproduction. Genetic mutation and recombination generate variation, but the need to adapt to prevailing (and often variable) environments drives natural selection. (Ecological equivalence can also arise from **parallel evolution**, wherein two species with a recent common ancestor evolve along similar lines.)

Convergent evolution is controversial. Until recently, little quantitative data was available to tease apart the relative importance of adaptive convergence versus phylogenetic lineage. With the availability of molecular data, the debate is entering new territory. Lizards occupying deserts worldwide are being used as a model system in this long-standing debate (Melville et al. 2006).

The physical processes discussed in Chapters 2 to 4 generate the broad-scale patterns of climate and abiotic habitats on Earth. These processes transcend oceans, continents, mountains, and other geographic barriers that serve as isolating mechanisms in evolution. The similarity between shrub deserts in southern Africa and Australia reflects the similar form and function of the plant species that have adapted to these distant ecosystems—species that evolved independently, but under similar constraints imposed by similar abiotic factors.

Such correlations between vegetation and climate led 19th-century geographers such as Schimper and deCandolle to divide the world into **formations** (later called *biomes*)—similar vegetation types that arise in geographically distant yet similar habitats, as defined by their **physiognomy** (physical appearance) rather than

Figure 1 Convergent evolution of unrelated species showing similar relationships between form and function. A North American rodent, the flying squirrel, and an Australian marsupial, the sugar glider, both have a body form that enables them to glide between tree limbs.

their species composition. Examples are grasslands, deserts, and forests. Regions supporting a given formation have similar climates. Such observations were the beginnings of **biogeography**—the study of the geographical distribution of biota, past and present. Its goal is to describe and understand the distribution of taxa (species, genera, families, etc.). *Historical biogeography* reconstructs the origin, dispersal, and extinction of taxonomic groups, whereas *ecological biogeography* studies the distributions of present-day organisms.

Ecological biogeography is the focus of Part Seven. We begin by exploring the major terrestrial biomes in Chapter 23, followed by the major aquatic biomes in Chapter 24.

(a)

(b)

Figure 2 Shrub deserts. **(a)** Karoo desert in southern Africa. **(b)** New South Wales, Australia.

Chapter 25 considers coastal and wetland habitats that are transitional between land and water. Overall, Part Seven considers the ecology of place by examining how the processes and patterns discussed thus far are reflected in different parts of the world.

CHAPTER
23 TERRESTRIAL ECOSYSTEMS

This forest on the shores of George Lake, Ontario, is transitional between the temperate deciduous forests that dominate the eastern portion of North America and the northern coniferous (boreal) forests that dominate much of northern Canada.

n 1939, Frederick Clements and Victor Shelford introduced the **biome**, a classification unit that combined broad-scale distributions of plants and associated animals into a single entity. Labels vary, but there are at least nine major terrestrial biomes, based on their predominant vegetation: *tropical forests* (including *tropical rain forest* and *tropical seasonal forest*), *tropical savannahs, grasslands, deserts, shrublands (chaparral), temperate deciduous forests, temperate coniferous forests, northern coniferous forests*, and *tundra* (Figure 23.1; subdivisions are shown for some biomes). These categories reflect the relative dominance of three plant growth forms: trees, shrubs, and herbs (grasses and forbs). A closed tree canopy typifies forests, whereas trees and herbs co-dominate savannahs. Shrubs and herbs (particularly grasses) dominate shrublands and grasslands, respectively. Deserts have scarce plant cover of various types.

When plotted on gradients of mean annual temperature and annual precipitation, these biomes form a distinctive pattern (Whittaker 1970) (Figure 23.2, p. 500). Mean annual precipitation determines which *growth form* dominates (trees, shrubs, or herbs), while mean annual temperature determines *species composition* (tropical, temperate, or boreal species). In reality, biome boundaries are less distinct than this model implies, and non-climatic factors such as soil, topography, and disturbance influence which of several biomes occupies an area. But the model provides a useful overview of the climatic determinants of biomes.

Another general pattern emerges from Figure 23.2. The range of biomes defined by precipitation along the *y*-axis decreases with declining temperatures along the *x*-axis. In other words, there is less range in site conditions defined by moisture as one moves towards the poles. This pattern reflects the systematic latitudinal climatic patterns discussed in Chapter 2 that result from seasonal variation in the influx of solar radiation to Earth's surface. As one moves from the equator to the poles, both mean annual temperature and the length of the growing season decrease, with increased seasonal variation in temperature and day length. Nor is systematic variation in climate with latitude limited to temperature. Annual precipitation declines at higher latitudes (see Figure 2.16). Cooler air holds less moisture, reducing precipitation. Together, these systematic climatic patterns dictate biome distribution.

23.1 BIOME DISTRIBUTION: Terrestrial Biomes Reflect Their Dominant Plant Growth Forms

Given that the classification of biomes in Figure 23.1 reflects the dominance of trees, shrubs, and herbaceous species, the question then becomes: *What causes the consistent patterns in the distribution and abundance of plant growth forms in relation to climate?* The answer lies in the adaptations that these three growth forms possess, and in particular their advantages and trade-offs in different abiotic conditions.

Although the categories of herbs, shrubs, and trees each encompass a diversity of species and traits, members of each

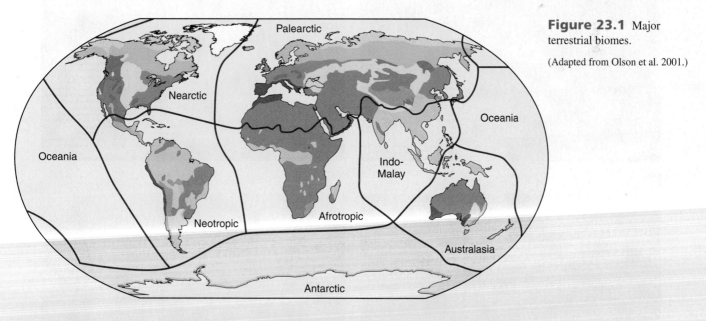

Figure 23.1 Major terrestrial biomes.

(Adapted from Olson et al. 2001.)

- ◯ Tropical and subtropical moist broadleaf forests
- ◯ Tropical and subtropical dry broadleaf forests
- ◯ Tropical and subtropical coniferous forests
- ◯ Temperate broadleaf and mixed forests
- ● Temperate coniferous forests
- ◯ Boreal forests/taiga
- ◯ Tropical and subtropical grasslands, savannahs, and shrublands
- ◯ Temperate grasslands, savannahs, and shrublands
- ◯ Flooded grasslands and savannahs
- ◯ Montane grasslands and shrublands
- ◯ Tundra
- ● Mediterranean forests, woodlands, and scrub
- ◯ Deserts and xeric shrublands
- ◯ Mangroves

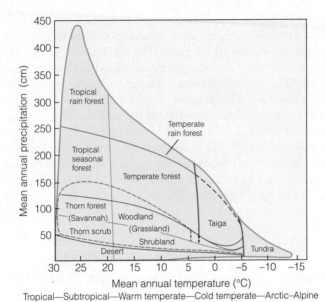

Figure 23.2 Pattern of terrestrial biomes in relation to temperature and moisture. Soil also affects biome type. The dashed line encloses environments in which either grassland or a biome dominated by woody plants may prevail.

(Adapted from Whittaker 1975.)

category share fundamental traits of morphology and carbon allocation. Grasses and forbs allocate less C to support tissues than do shrubs and trees, enabling them to invest more in photosynthetic tissues. Among woody plants, shrubs allocate less C to support than do trees. Although investing in woody stems confers the advantage of height and access to light, maintenance costs are high. If these costs are not offset by photosynthetic C gain, the plant cannot maintain a positive C balance. So, as abiotic conditions become increasingly adverse for photosynthesis (dry, cold, infertile, or short season), trees decline in size and density until they no longer persist.

To classify the several biomes in which trees dominate, biogeographers also use leaf form and longevity. Leaves that live for a single growing season are **deciduous**, whereas those that live beyond a year are **evergreen**. Deciduous species are typical of forests in habitats with a distinct growing season. Deciduous leaves are further divided into (1) *winter-deciduous* leaves, which are common in temperate regions where the dormant period occurs in winter (Figure 23.3a, b), and (2) *drought-deciduous* leaves, which are typical of tropical and subtropical habitats with seasonal rainfall, in which leaves are shed during the dry period (Figure 23.3c, d). Whatever the subtype, the deciduous habit confers a major benefit: the plant avoids the maintenance

Figure 23.3 Winter- and drought-deciduous trees. Temperate deciduous forest in Virginia during **(a)** summer and **(b)** winter. Semi-arid savannah in Zimbabwe in **(c)** rainy and **(d)** dry seasons.

Figure 23.4 Types of evergreen trees. **(a)** Broadleaf evergreen trees dominate the canopy of this tropical rain forest in Queensland, Australia. **(b)** Needle-leaf evergreen trees (foxtail pine) inhabit the high-altitude zone of the Sierra Nevada Mountains in western North America.

costs of supporting leaves during an unfavourable season. The trade-off is the yearly investment in energy and minerals required to produce new leaves.

Evergreen leaves also fall into two categories. (1) **Broadleaf evergreen** leaves (Figure 23.4a) dominate habitats in which growth continues year-round, such as tropical rain forests. They are also typical of many low-growing shrubs in boreal and shrub habitats. (2) **Needle-leaf** evergreen leaves (Figure 23.4b) of conifers are common in regions with a short season and/or very infertile soils, and in montane regions.

QUANTIFYING ECOLOGY 23.1 | Climate Diagrams

As Figure 23.2 illustrates, distribution of terrestrial biomes is closely tied to climate, especially mean annual temperature and precipitation. But biome distribution is also influenced by other aspects of climate, especially seasonality. In addition, topographic features such as mountains and valleys affect regional climate. To understand the relationship between climate and biome distribution, we present a map showing the global distribution of each biome discussed, along with diagrams describing the local climate at representative locations. Figure 1 is an example, annotated to explain the information presented. Take note of seasonal patterns.

1. In Figure 23.12, what is the distinctive feature of the climate diagrams for tropical savannahs? How do the patterns differ between sites in the Northern and Southern Hemispheres? What climate feature discussed in Chapter 2 causes these patterns?

2. In Figure 23.23, what climate feature is common to all shrublands?

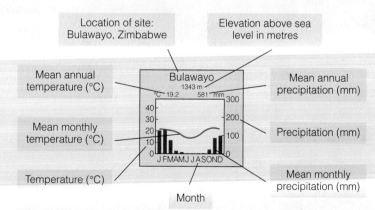

Figure 1 Climate diagram for Bulawayo, Zimbabwe. This city is in the Southern Hemisphere, where winter lasts from May to August. The dry season occurs in winter, with the rainy season beginning in October and lasting through the summer.

A simple economic model explains the adaptive value of needles. The time needed to pay back the cost of leaf production reflects the photosynthetic rate. Under unfavourable conditions, the payback time increases. If net photosynthesis is low, it may be impossible to pay back the cost in a single season. Then, the plant cannot "afford" a deciduous leaf form, which requires producing new leaves yearly. So the needle-leaf habit is an adaptation for survival in habitats with a short season and/or limited nutrients, where conditions limit the ability to produce enough C to subsidize the high cost of leaf production.

By combining this simple classification of plant life forms and leaf type with broad-scale climate patterns, we can better grasp the biome distribution pattern in Figure 23.2. Considering the *y*-axis (precipitation) first, regions with warm, wet **aseasonal** climates (no distinct seasons) are dominated by broadleaf evergreens in tropical rain forests. In regions with a distinct dry season, broadleaf evergreens give way to drought-deciduous species in tropical seasonal forests. As moisture declines still further, tree stature and density decline, giving rise to tropical dry forests, woodlands, and savannahs co-dominated by trees and grasses. With even less rain, trees cannot be supported, giving rise to arid shrublands and desert.

Now let's follow the *x*-axis (temperature), which represents the latitudinal gradient from the equator to the poles. Moving from the broadleaf evergreen forests of the wet tropics into the cooler, seasonal environments of the temperate zone, winter-deciduous species dominate in temperate deciduous forests. Temperate coastal regions support temperate rain forest, dominated by conifers. In temperate areas where precipitation is insufficient to support trees, grasses dominate, giving rise to temperate grasslands. Towards the pole, temperate deciduous and coniferous forests give way to northern coniferous boreal forests. As temperatures drop and the growing season shortens, the short-stature shrubs and herbs typical of the tundra dominate.

We now embark on an overview of the world's biomes. Relying on Archibold (1995), we relate each biome's global distribution to broad-scale constraints of regional climate and associated seasonal patterns of temperature and precipitation that determine the dominant life forms.

23.2 TROPICAL FORESTS: Favourable Climate Supports the High Biodiversity of Tropical Forests

Tropical rain forests are restricted primarily to the equatorial zone between latitudes 10°N and 10°S (Figure 23.5), where temperatures are warm year-round and it rains almost daily. The largest and most continuous rain forest region is South America's Amazon basin (Figure 23.6a). Other extensive regions are in Southeast Asia, particularly Malaysia (Figure 23.6b) and Borneo, and in West Africa around the Gulf of Guinea and in the Congo basin. Smaller rain forests occur on Australia's northeast coast, the windward side of the Hawaiian Islands, the South Pacific Islands, the east coast of Madagascar, and parts of Central America.

Climate varies geographically, but the mean temperature of all months exceeds 18°C (some use 20°C as a lower limit). Mean precipitation exceeds 2000 mm, with minimum monthly totals over 60 mm. In tropical lowland rain forests, mean tem-

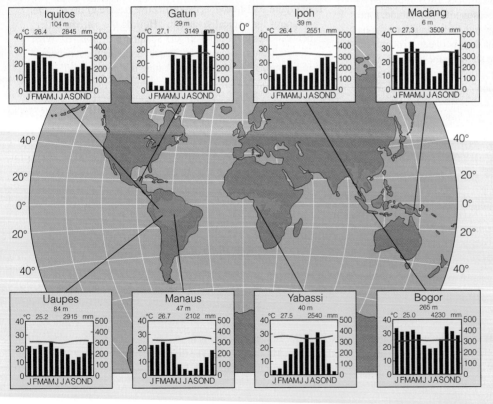

Figure 23.5 Geographic distribution and climate of tropical forests. Note the aseasonality in mean monthly temperatures, which exceed 20°C. Although rainfall in some tropical forests is seasonal, rain forests typically receive a minimum of 60 mm per month, and total annual precipitation above 2000 mm.

(Adapted from Archibold 1995.)

(a)

(b)

Figure 23.6 Tropical rain forests in **(a)** the Amazon Basin (South America) and **(b)** Malaysia (Southeast Asia). Despite being taxonomically distinct, these widely separated tropical rain forests are dominated by broadleaf evergreen trees and support vigorous growth year-round.

perature may exceed 25°C with an annual range of less than 5°C, and annual precipitation from 5000 to 9000 mm. These warm, moist conditions promote weathering and leaching. Typically, soils are highly weathered oxisols, with indistinct horizons (see Figure 4.12). Ultisols may develop in areas with seasonal rainfall. Some areas in tropical Asia are characterized by andisols formed from recent volcanic deposits.

The warm, wet conditions of tropical rain forests support high primary productivity and high rates of litter input to the forest floor (but see Section 21.10 regarding the decline in tropical productivity at extremely high precipitation, as reported by Schuur [2003]). Yet little litter accumulates, because decomposers consume dead organic matter almost as rapidly as it falls. Mean time for leaf litter decomposition is just over half a year (see Table 21.1). Most available nutrients derive from this rapidly decomposed organic matter, compensating for the low fertility of most tropical soils.

Tropical rain forests have a complex structure, with five vertical layers (Figure 23.7, p. 504): emergent trees, upper canopy, lower canopy, shrub and sapling understory, and a sparse ground layer of herbs and ferns. Particularly conspicuous are growth forms that traverse many strata: (1) *lianas* (woody vines) that hang from the canopy; (2) *herbaceous* (*non-woody*) *vines*; (3) *epiphytes*, such as bromeliads and orchids, which grow perched on trunks and in branches (an **epiphyte** is a plant that grows on another plant as a substrate); and (4) *stranglers* (figs, *Ficus* spp.), which grow down from the canopy and root in the ground, choking their host tree (Figure 23.8a, p. 504). Large trees often develop plank-like **buttresses** (Figure 23.8b) for support in shallow soil. The forest floor is occupied by a dense mat of roots with a sparse cover of herbs.

Tropical rain forests support an amazing biodiversity. Covering only 6 percent of Earth's land surface, they contain over half of all known plant and animal species, with more discovered each year. Tree species alone number in the thousands. A single 100-m² plot of lowland rain forest in Costa Rica contained 233 vascular species, of which 73 were trees (mostly seedlings) (Whitmore et al. 1986). A 10-km² area may contain 1500 flowering plant species, of which half are trees. The richest region is the Malaysian lowland forests, containing some 7900 species (Goudie 2001).

Animal diversity is impressive, too. Some 90 percent of primate species occupy tropical rain forests, including 64 species of New World primates, small mammals with prehensile tails. Indo-Malaysian forests are inhabited by many primates, often with limited distributions. The orangutan (*Pongo pygmaeus*; Figure 23.9a, p. 505) is an arboreal ape confined to Borneo. Peninsular Malaysia has seven primates—three gibbons, two langurs, and two macaques. The long-tailed macaque (*Macaca fascicularis*) occupies disturbed forests, and the pig-tailed macaque (*M. nemestrina*) is an omnivore adapted to human settlement. African rain forest is home to the mountain gorilla (a subspecies of *Gorilla gorilla*) and chimpanzees (*Pan troglodytes*; Figure 23.9b). The diminished Madagascar rain forest holds 39 lemur species. Invertebrate richness is unknown, but most new species identified yearly are insects, often parasitoids (see Section 15.1).

Figure 23.7 Vertical stratification of a tropical rain forest.

Moving from the equatorial zone to tropical regions with greater seasonality in rainfall, broadleaf evergreen rain forests are replaced by *tropical seasonal forests* (Figure 23.10, p. 505), with annual rainfall of 1000 to 2000 mm and a distinct dry season, and *tropical dry forests*. The farther from the equator, the longer the dry season—up to eight months. During the dry season, drought-deciduous trees such as teak (*Tectona grandis*) drop their leaves. New leaves emerge before the start of the rainy season, which may be wetter than the wettest time in rain forests. In monsoon regions of Southeast Asia, rainfall often exceeds 1500 mm, but virtually all of it falls in June.

The largest proportion of tropical seasonal forest is in Africa and South America, south of the rain forest zones. These regions are influenced by seasonal migration of the intertropical convergence zone (ITCZ; see Figure 2.17). Areas of Central America, northern Australia, India, and Southeast Asia are classified as tropical dry forest, although much of the original forest, especially in Central America and India, has been converted to agriculture.

(b)

Figure 23.8 Growth forms in tropical forests. **(a)** Strangler figs grow down from the canopy, eventually choking their host tree (in this case, a palm). **(b)** Buttresses provide support for tropical trees.

(a)

(a)

(b)

Figure 23.9 Primates that occupy tropical rain forests. **(a)** The orangutan is native to Borneo in Southeast Asia and **(b)** the chimpanzee to Central Africa.

23.3 TROPICAL SAVANNAHS: Trees and Grasses Co-dominate in Semi-arid Tropical Regions

The term **savannah** describes a range of vegetation types in the drier tropics and subtropics characterized by co-dominance of grasses with scattered trees or shrubs. Savannah includes a vegetation continuum of increasing cover of woody species, from open grassland to widely spaced shrubs or trees

to woodland (Figure 23.11, p. 506). South America has the densely wooded *cerrado* and the more open *campos* and *llano*. In Africa, *miombo*, *mopane*, and *Acacia* woodlands are distinguished from the more open *bushveld*. Scattered individuals of *Acacia* and *Eucalyptus* dominate the Australian *mulga* and *brigalow*.

Savannahs grow in warm continental climates, with distinct seasonality and large annual variation in rainfall (Figure 23.12, p. 506). Mean monthly temperature is rarely below 18°C, except

(a)

(b)

Figure 23.10 Tropical seasonal forest in Costa Rica during the **(a)** rainy and **(b)** dry season. Most tropical seasonal and dry forests in Central America have disappeared as a result of land clearing.

(a)

(b)

Figure 23.11 Savannah ecosystems such as the **(a)** cerrano of South America and **(b)** mulga woodlands of central Australia have a ground cover of grasses with scattered shrubs or trees.

in highland areas in the coldest months. Temperature regimes are seasonal, with the warmest temperatures at the end of the wet season. Vegetation is more influenced by precipitation than by temperature. Diversity in savannah physiognomy reflects the differing climates of this widely distributed biome. Moisture levels, affected by both rainfall (amount and seasonality) and soil properties (especially texture), determine the density of woody growth (Figure 23.13).

Despite their many regional differences, savannahs share certain features. They occur in low-relief regions, often plateaus interrupted by escarpments and dissected by rivers. Weathering has produced nutrient-poor oxisols, often low in phosphorus, although other soil types can occur. Given the recurrent fires, the dominant vegetation is fire-adapted. Savannahs have a simple, two-layer vertical structure consisting of an herbaceous ground cover and scattered shrubs or trees. Grasses and forbs are always present, and the woody component is short-lived—individuals seldom survive beyond a few decades. Productivity is controlled by seasonal rainfall and available soil moisture. Most leaf litter is decomposed during the wet season, and most woody debris is consumed by termites in the dry season. Microenvironments associated with tree canopies influence species distribution, productivity, and soil traits. Stem flow and litter accumulation increase nutrients and moisture under tree canopies, increasing productivity and facilitating the establishment of shade-adapted species.

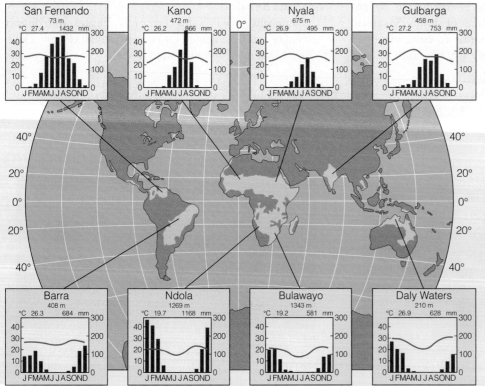

Figure 23.12

Geographic distribution and climate of tropical savannahs. Seasonal temperature patterns resemble those of tropical seasonal forests, but precipitation is lower and more seasonal. Note the shift in timing of the rainy season in Northern and Southern Hemisphere sites, reflecting seasonal shifts in the Intertropical Convergence Zone (ITCZ).

(Adapted from Archibold 1995.)

Figure 23.13 Interaction between precipitation and soil texture in the transition from woodland to savannah to grassland in southern Africa. Plants have less access to soil moisture on clay than on sandy soil, so more annual rainfall is needed to support woody species.

Savannahs support a varied herbivore community—invertebrate and vertebrate, grazing and browsing. The African savannah is visually dominated by a diverse ungulate fauna, some of which, such as wildebeests and zebras, migrate during the dry season. Living on the ungulate fauna is an array of carnivores including lions, leopards, cheetahs, hyena, and wild dogs. Scavengers, including vultures and jackals, subsist on leftover prey. Savannahs also support innumerable insects: flies, grasshoppers, locusts, crickets, carabid beetles, ants, and detritus-feeding dung beetles and termites. Mound-building termites excavate and move tonnes of soil, mixing mineral soil with organic matter and constructing extensive subterranean galleries in which they cultivate fungal gardens for decomposing lignin-rich plant tissues (see Ecological Issues: The Neglected Kingdom, pp. 452–453).

23.4 GRASSLANDS: Species Composition of Grasslands Varies with Climate and Geography

Grasslands occupy temperate regions where annual precipitation ranges from 250 to 1000 mm, but their existence is not solely climatically determined. Many persist through the intervention of fire and human activity. The pre-human extent of grasslands was highly variable, but the widespread conversion of forests to agriculture extended domestic grasslands into forest regions. Once covering about 42 percent of the land surface, natural grasslands have shrunk to under 12 percent through agricultural conversion.

The natural grasslands of the world occur in the mid-latitudes of mid-continental regions, where annual precipitation declines as air masses move inland from the coasts (Figure 23.14, p. 508). Temperate grassland experiences recurring drought, and the diversity of its vegetation reflects differences in the amount and reliability of precipitation. Grasslands do poorest where precipitation is lowest and temperatures highest, and are tallest and most productive where mean annual precipitation exceeds 800 mm and mean annual temperature exceeds 15°C.

North American grasslands are affected by declining precipitation from east to west and consist of three types based on the height of the dominants: tallgrass, mixed-grass, and shortgrass prairie (Figure 23.15, p. 508). A fourth type (fescue grassland) occurs in western Canada. *Tallgrass prairie* (Figure 23.16a, p. 509) is dominated by big bluestem (*Andropogon gerardii*), a C_4 grass growing over 1 m tall with flowering stalks to 3.5 m. Given its value as cropland, tallgrass prairie is the most endangered biome in Canada (see pp. 330–331). It occurs mainly in southern Manitoba, but was once extensive in southwestern Ontario, along with oak savannah (Bakowsky and Riley 1994). Only prairie remnants remain in Ontario, notably at Windsor and Walpole Island.

Mixed-grass prairie (Figure 23.16b), typical of the American Great Plains and in Canada from western Manitoba to Alberta, is dominated by needlegrass and grama grass (*Bouteloua* and *Stipa* spp). South and west of mixed-grass prairie and grading into desert is *shortgrass prairie*, dominated by sod-forming blue grama (*Bouteloua gracilis*) and buffalo grass (*Buchloe dactyloides*). As this region is typically too dry for crops, it is the most intact.

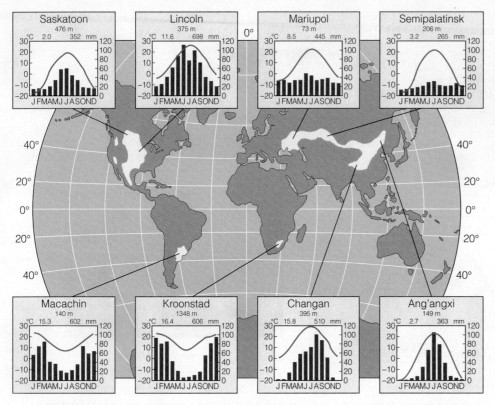

Figure 23.14 Geographic distribution and climate of temperate grasslands. Most grassland regions are mid-continental, with distinct seasonality in temperature. Mean annual precipitation is typically below 1000 mm—comparable to savannahs but less than most temperate deciduous forests.

(Adapted from Archibold 1995.)

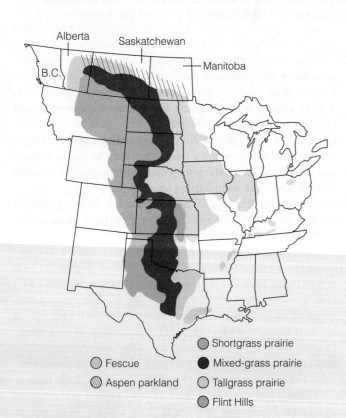

Figure 23.15 Original extent of prairie types in North America before the arrival of Europeans. Flint Hills prairie is similar to tallgrass, but grows on less fertile soils.

(Adapted from Reichman 1987; fescue grassland distribution in Canada based on Looman 1969.)

Another grassland type common in Alberta is *fescue grassland*, dominated by *Festuca scabrella*, a C_3 grass adapted to cool northern regions. Moister fescue grasslands, such as in Manitoba's Riding Mountain National Park, are dominated by plains rough fescue (*F. hallii*) and have higher productivity than do drier mixed-grass prairies. Finally, *palouse prairie*, dominated by mountain rough fescue (*F. campestris*) and bluebunch wheatgrass (*Agropyron spicatum*), occurs in dry valleys of the Rockies in British Columbia, extending south to Idaho and Washington.

From southeastern Texas to southern Arizona and south into Mexico lies *desert grassland*, similar in many respects to shortgrass, except that three-awn grass (*Aristida* spp.) replaces buffalo grass. *Annual grassland*, which is largely confined to California's Central Valley, has a Mediterranean climate, with rainy winters and hot, dry summers. Growth occurs in early spring, and most plants are dormant in summer, turning the hills a dry tan colour accented by the deep green foliage of scattered California oaks.

Grasslands once extended from eastern Europe to western Siberia south to Kazakhstan. These *steppes*, which are treeless except for forest patches, range from mesic meadows in the north to semi-arid grasslands in the south. Known as *pampas*, South American grasslands extend west in a large semi-circle from Buenos Aires to cover some 15 percent of Argentina. European forage grasses and alfalfa have been planted on the drier pampas, and the tallgrass pampas has been converted to cropland. In Patagonia, where annual rainfall averages only 25 cm, open steppe dominates.

In southern Africa, the *velds* (not to be confused with bushveld savannah) occupy the eastern part of a plateau 1500 to

(a)

(b)

(c)

Figure 23.16 North American grasslands. **(a)** Remnant tallgrass prairie in Tolstoi, Manitoba. **(b)** Mixed-grass prairie, with abundant wildflowers. **(c)** Shortgrass prairie in western Wyoming.

2000 m above sea level in the Transvaal and Orange Free State. Australia has four grasslands: *arid tussock grassland* in the north, where rainfall averages from 20 to 50 cm, mostly in summer; *arid hummock grasslands* in areas with less than 20 cm of rainfall; *coastal grasslands* in tropical regions; and *subhumid grasslands* in coastal areas with annual rainfall between 50 and 100 cm. Introduction of fertilizers, non-native species, and sheep has altered most of these grasslands. Smaller grasslands occur in drier portions of New Zealand.

The most visible feature of all grasslands is the tall herbaceous growth that develops in spring and dies back in fall. One of only two strata in a grassland, this herbaceous "canopy" arises from the crowns, nodes, and rosettes of plants hugging the soil. Grasses are strongly represented, with C_4 species dominating in warmer regions and C_3 in cooler sites, but many forbs are also present. In fact, mixed-grass prairies in North America are sometimes dubbed "daisyland" for the many species of the Asteraceae family, such as the familiar brown-eyed Susan (*Rudbeckia hirta*) (see Figure 23.16b). As well as a ground layer of shorter species, the well-developed root layer can contain over half the total plant biomass and typically extends deep into the fertile soil.

Depending on their history of fire and grazing or mowing, grasslands accumulate a mulch layer that retains moisture and, with continuous turnover of fine roots, adds organic matter to the soil. Most grassland soils are chernozems (see Figure 4.12), with a relatively thick, dark-brown to black surface horizon that is rich in organic matter. Soils typically become thinner and paler in drier regions because less organic material is incorporated into the topsoil. Grassland productivity is strongly related to annual rainfall (Figure 23.17), but temperature can complicate this relationship. Warmer temperatures stimulate photosynthesis (especially for C_4 grasses) but can reduce productivity by increasing demand for water.

Grasslands evolved under the combined selective pressures of fire and grazing (Anderson 2006). Indeed, moderate grazing can stimulate their productivity by compensatory growth. As well as high plant and fungal diversity, grasslands support many herbivores, both invertebrate and vertebrate. Large ungulates and burrowing mammals are the most conspicuous vertebrate

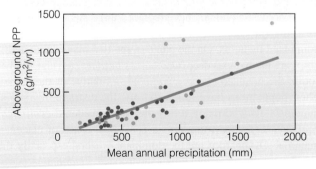

Figure 23.17 Relationship between aboveground NPP and mean annual precipitation for 52 grassland sites around the world. Each point represents a single site. North American grasslands are indicated by dark-green dots.

(Adapted from Lauenroth 1979.)

Figure 23.18 Burrowing rodents such as the prairie dog were once common in North American grasslands.

grazers. North American prairies were once dominated by migratory herds of bison (*Bison bison*) (see Figure 16.5) and the forb-eating pronghorn antelope (*Antilocapra americana*). Although the native tribes inhabiting the plains had a hunter–gatherer lifestyle based on coexistence, the presence of bison was deemed incompatible with farming by settlers, and bison were killed in huge numbers.

The most common burrowing rodent was the prairie dog (*Cynomys* spp.) (Figure 23.18), which along with gophers (*Thomomys* and *Geomys* spp.) and mound-building harvester ants (*Pogonomyrmex* spp.) were instrumental in developing and maintaining the ecological structure of mixed- and shortgrass prairie. In Canada's western provinces, the Richardson's ground squirrel (*Urocitellus richardsonii*) occupies a niche that is similar though not identical to that of the prairie dog (see Research in Ecology: Sounding the Alarm in Richardson's Ground Squirrels, pp. 296–297). Ground squirrel numbers remain strong in pastures and parklands, whereas the threatened black-tailed prairie dog (*Cynomys ludovicianus*) is confined to a protected population in Saskatchewan's Grasslands National Park.

The African veld also supported large herds of wildebeest (*Connochaetes taurinus*) and zebra (*Equus* spp.) along with their predators, the lion (*Panthera leo*), leopard (*Panthera pardus*), and hyena (*Crocuta crocuta*). Except in game reserves and parks, domestic animals have replaced ungulate herds. In contrast, the Eurasian steppes and Argentine pampas were never dominated by herds. On the pampas, the two major large herbivores are the pampas deer (*Ozotoceros bezoarticus*) and, farther south, the guanaco (*Lama guanicoe*), a relative of the camel. Their populations have fallen sharply. In Australia, marsupial mammals evolved many forms that are considered *ecological equivalents* (i.e., perform similar ecological roles) of placental mammals. Kangaroos are the dominant grazers, especially the red (*Macropus rufus*) and grey kangaroo (*M. giganteus*).

Although large herbivores are more conspicuous, the major grassland consumers are invertebrates. Insects abound and, as pollinators of many flowering species, may have subtle effects on the biological structure of prairies by affecting mycorrhizal interactions (see Section 15.13). The heaviest grazing occurs belowground, where the dominant herbivores are nematodes.

23.5 DESERTS: Deserts Occupy Arid Regions Worldwide

Arid regions occupy 25 to 35 percent of Earth's landmass (Figure 23.19). Most deserts occur from 15° to 30° latitude, where air carried aloft by the ITCZ subsides to form semipermanent high-pressure cells (see Figure 2.16). Cloudless skies and warming of air as it descends generate intense summer heat. Temperate deserts occupy the rain shadow of mountains or are inland, where moist maritime air rarely penetrates. Here, temperatures are high in summer but drop below freezing in winter. Lack of moisture, not heat, is thus the distinctive feature of deserts.

Infrequent rainfall coupled with rapid evaporation severely limits water supply, so primary productivity is low. Most desert soils are poorly developed, and sparse cover limits the ability of vegetation to modify the soil. Beneath established plants, "islands" of fertility can develop with higher litter input and enrichment by wastes from animals that seek shade under shrubs.

Most deserts are in the Northern Hemisphere. The Sahara, the world's largest desert, covers 9 million square kilometres of North Africa. It extends across to the Arabian deserts, eastward to Afghanistan and Pakistan, and terminates in the Thar Desert of northwest India. The temperate deserts of Central Asia lie to the north, of which the Kara Kum region of Turkmenistan is farthest west. Eastward lie the high-elevation deserts of China and the high plateau of the Gobi Desert.

A similar transition to desert occurs in western North America. The Sierra Nevadas block passage of moist air to the southwest interior. Mountain ranges run parallel to the Sierras throughout this region, with desert basins on their leeward east sides. Canada has relatively few deserts. The Osoyoos Desert at the southern tip of British Columbia's Okanagan Valley is an exception. Given its uniqueness, this small region (some of which is dry grassland) contains species found nowhere else in Canada. Desert species also grow in Alberta's Badlands region.

Apart from Patagonia in southern Argentina, the deserts of the Southern Hemisphere lie within the subtropical high-pressure belt. Cold ocean currents also contribute to arid coastal regions. Drought conditions are severe along a narrow strip of the coast that includes Chile and Peru. Drier parts of Argentina lie in the rain shadow of the Andes. There are three deserts in southern Africa. The Namib Desert occupies a narrow strip along the west coast from southern Angola to the border of South Africa's cape region. It continues south and east across South Africa as the Karoo, and merges with the Kalahari to the north, in Botswana. The most extensive arid region in the Southern Hemisphere is in Australia, of which more than 40 percent is desert.

Deserts are not the same everywhere. Differences in moisture, temperature, drainage, alkalinity, salinity, and topography create variations in dominant plants and associated species. There are hot deserts and cold deserts, extreme deserts and semi-deserts, with enough moisture to verge on grasslands or shrublands, and gradations between those extremes.

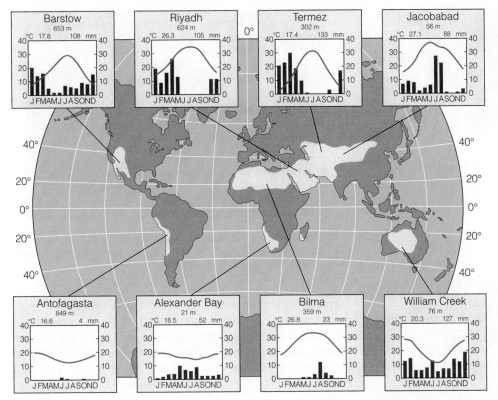

Figure 23.19 Geographic distribution and climate of deserts. Seasonal variations in temperature differ, but at all sites, annual precipitation is well below potential evaporative demand, resulting in low soil moisture.

(Adapted from Archibold 1995.)

Cool deserts—including North America's Great Basin, Asia's Gobi and Turkestan Deserts, and high elevations of hot deserts—are also called *shrub steppes* or *desert scrub*. Their climate is continental, with warm summers and cold winters. Vegetation falls into two main associations (Figure 23.20): (1) sagebrush, dominated by *Artemisia tridentata*, which often forms pure stands, and (2) saltbush or shadscale (*Atriplex confertifolia*), a C_4 shrub, and other chenopod shrubs that can tolerate saline soils (*halophytes*). Similar desert scrub exists in southwestern Australia, dominated by chenopods, particularly saltbushes (*Atriplex* and *Maireana* spp.)

Hot deserts range from those without any vegetation to ones with a combination of chenopods, dwarf shrubs, and succulents (Figure 23.21, p. 512). Creosote bush (*Larrea divaricata*) and bur sage (*Franseria* spp.) dominate deserts in southwestern North America—the Mojave, Sonoran, and Chihuahuan. Areas

(a)   **(b)**

Figure 23.20 Cool deserts. **(a)** Northern desert scrub in Wyoming is dominated by sagebrush (*Artemisia*). Considered a cool desert plant, sagebrush forms one of the most important arid shrub types in North America. **(b)** Saltbrush shrubland in Victoria, Australia, is dominated by *Atriplex* and is an ecological equivalent of the Great Basin shrublands in North America.

(a)

(b)

Figure 23.21 Hot deserts. **(a)** Chihuahuan Desert in Nuevo Leon, Mexico. The substrate is sand-sized gypsum particles. **(b)** Dunes in the Saudi Arabian desert near Riyadh.

Figure 23.22 A spadefoot toad. Named for the black, sharp-edged "spades" on its hind feet, this drought-evader emerges from its desert burrow to breed when the rains come.

ground area after occasional desert rains. Other perennials, such as cacti and other **succulents** (plants with specialized water-storage tissue), typically have wide-spreading shallow roots. After a rain, they soak up water from a wide area, storing it in their tissues and tolerating over a year without rain. Tall columnar cacti, in contrast, have deeper taproots for support. Many succulents also use CAM photosynthesis (see Section 6.10). By opening their stomata at night, when the air is cooler, they reduce their transpiration losses.

Despite their aridity, deserts support a surprising diversity of animal life, including beetles, ants, locusts, lizards, snakes, birds, and mammals. Desert mammals are mostly generalist herbivores that consume a wide range of species, plant types, and parts, depending on availability. Ants and desert rodents, such as kangaroo rats, feed largely on seeds. These granivores eat up to 90 percent of available seeds, affecting plant population dynamics and species composition. Desert carnivores, such as foxes and coyotes, eat some leaves and fruits. Even insectivorous birds and rodents eat some plant material. In deserts, omnivory seems to be the rule. An interesting behavioural pattern is that desert ectotherms (particularly lizards and snakes) are diurnal, while mammals such as desert rodents are nocturnal.

23.6 SHRUBLANDS: Xeric Woody Species Dominate Regions with a Mediterranean Climate

Shrublands—ecosystems in which shrubs are either dominant or co-dominant—are difficult biomes to categorize, largely because the term *shrub* is itself problematic. In general, a *shrub* is a plant with multiple woody stems but no central trunk, and a height up to 8 m. However, under severe conditions, many trees do not exceed that size. Moreover, some trees—particularly those that re-sprout from the stump after destruction of aboveground tissues by fire, browsing, or cutting—are multi-stemmed, as is paper birch (*Betula papyrifera*). Shrubs can also dominate tropical and temperate biomes, including tropical savannahs and desert scrub (see Sections 23.3 and 23.5, respectively). Despite these difficulties, five regions along the western margins of the continents between 30°

with more moisture support *Acacia* spp., saguaro cactus (*Cereus giganteus*), palo verde (*Cercidium* spp.), ocotillo (*Fouquieria* spp.), yucca (*Yucca* spp.), and ephemerals.

Both plants and animals cope with drought (and other abiotic stresses) by either (1) **avoidance** or (2) **tolerance**. (The combination of these strategies is called **resistance**). Drought-evading plants persist as seeds, ready to sprout, flower, and produce new seeds only when conditions are favourable. If no rains come, these *drought ephemerals* do not germinate. Drought-evading animals, like their plant counterparts, adopt an annual cycle of activities or enter *estivation* (a dormant stage) during the dry season. For example, the spadefoot toad (*Scaphiopus* spp.; Figure 23.22) remains underground in a burrow, emerging briefly to reproduce during occasional rains. If extreme drought occurs in the breeding season, many lizards and birds do not reproduce.

Drought-tolerating desert plants include deep-rooted shrubs, such as mesquite (*Prosopis* spp.) and *Tamarix*, whose taproots reach the water table, making them independent of rainfall. Others, such as *Larrea* and *Atriplex*, are deep-rooted perennials with superficial laterals that extend up to 30m from the stems. They rely on soaking up water from a large

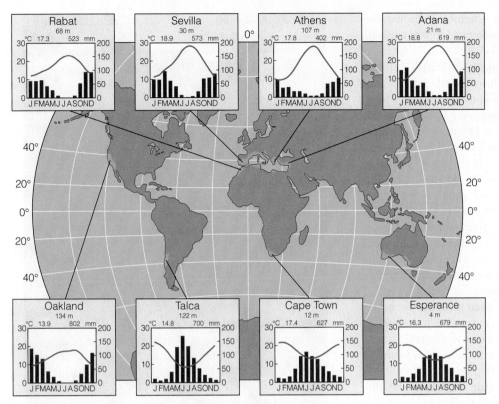

Figure 23.23
Geographic distribution and climate of shrublands. Despite variation among representative locations, the Mediterranean climate of shrublands is characterized by cool, wet winters and dry, hot summers.

(Adapted from Archibold 1995.)

and 40° latitude are classed as *Mediterranean shrublands*: semi-arid regions of western North America, the area bordering the Mediterranean Sea, central Chile, the cape region of South Africa, and southwest and southern Australia (Figure 23.23).

Shrublands are dominated by evergreen shrubs and sclerophyllous trees that are adapted to the distinctive Mediterranean climate of hot, dry summers, with at least one month of protracted drought, and cool, wet winters. About 65 percent of rainfall comes in the winter, when temperatures typically average 10°C to 12°C with a risk of frost. Summer conditions arise from seasonal changes in the semi-permanent high-pressure zones centred over the tropical deserts. Persistent flow of dry air in summer brings hot, dry weather and frequent fires. This climate supports communities of broadleaf evergreen shrubs and dwarf trees known as **sclerophyllous** ("hard-leaved") vegetation. Their small leaves, thick cuticles, glandular hairs, and sunken stomata are all **xeric** traits that reduce water loss during hot, dry periods (Figure 23.24).

The largest shrubland region forms a discontinuous belt around the Mediterranean Sea in southern Europe and North Africa. The benign climate and native biota (particularly the ancestors of wheat) played a vital role in the development of Western civilization. Greatly altered by human habitation, this region is currently or was once dominated by mixed evergreen woodland, supporting species such as holm oak (*Quercus ilex*) and cork oak (*Q. suber*; its outer bark is the source of cork) in mixed stands associated with strawberry tree (*Arbutus unedo*, unrelated to strawberries) and many shrubs. The region's easternmost limit is coastal Syria, Lebanon, and Israel, where it grades into the arid Middle East. Here, deciduous oaks are more abundant. Desert extends across North Africa as far as Tunisia, with Mediterranean shrub and woodland occupying the northern coastal areas of Algeria and Morocco.

There are many regional names for shrubland. In southern Africa, *fynbos* dominates montane regions and is composed primarily of broadleaf evergreen shrubs that grow to a height of

(a)   (b)   (c)

Figure 23.24 Sclerophyllous leaves of trees and shrubs inhabiting California chaparral: (**a**) chamise (*Adenostoma fasciculatum*), (**b**) scrub oak (*Quercus dumosa*), and (**c**) chinquapin (*Chrysolepis sempervirens*).

1.5 to 2.5 m (Figure 23.25a). In southwestern Australia, the *mallee* is dominated by low-growing *Eucalyptus*, 5 to 8 m tall, with broad sclerophyllous leaves. In North America, sclerophyllous shrublands are called **chaparral** (Figure 23.25b). California chaparral, dominated by scrub oak (*Quercus berberidifolia*) and chamise (*Adenostoma fasciculatum*), is evergreen and winter-active. Another type of chaparral occurs in the Rocky Mountain foothills, extending into the Okanagan. Dominated by Gambel oak (*Quercus gambelii*), it is winter-deciduous. In South America, *mattoral* shrublands grow in Chile, in the coastal lowlands, and on the west-facing slopes of the Andes. Most mattoral species are evergreen shrubs 1 to 3 m tall with small sclerophyllous leaves, but drought-deciduous shrubs also occur.

Despite regional differences in species composition, most shrubland species, including those in the sparse herbaceous layer, are adapted to fire and low nutrients. Many have seeds that require the heat and scarring action of fire to germinate. Without fire, chaparral species grow tall and dense, building up large fuel loads. In the dry season these shrubs, even though alive, nearly explode when ignited. After fire, the land returns to lush green sprouts arising from buried root crowns, or to grasses if a seed source is nearby. Eventually, chaparral once again becomes dense and the low canopy closes, setting the stage for another fire. As popular as this biome is worldwide for human habitation because of its benign climate, we occupy it at our own risk.

Diverse topography and geology give rise to a range of soil conditions, but most shrubland soils are infertile. Litter decomposition is limited by cool temperatures during winter and low soil moisture during summer. Constrained by abiotic factors, productivity is low but variable, depending on the severity of summer drought.

Marked similarity in the habitat structure of widely separated shrublands has generated strong patterns of similarity among bird and lizard species. Among mammals, different species occupy similar niches. In North America, shrublands support mule deer (*Odocoileus hemionus*), coyotes (*Canis latrans*), various rodents, jackrabbits (*Lepus* spp.), and sage grouse (*Centrocercus urophasianus*). The Australian mallee is rich in birds, including the endemic mallee fowl (*Leipoa ocellata*), which incubates its eggs in a large mound. Among the mammalian life are the grey kangaroo (*Macropus giganteus*) and various wallabies.

23.7 TEMPERATE DECIDUOUS FORESTS: Winter-Deciduous Forests Dominate Moist Temperate Regions

Climatic conditions in the humid mid-latitudes support forests dominated by broadleaf deciduous trees (Figure 23.26). Deciduous forest once covered large areas of Europe and China, parts of North and South America, and the Central American highlands. The deciduous forests of Europe and Asia have largely disappeared, cleared for agriculture and human habitation. A similar fate has befallen eastern North America, although the clearing was more recent and less extensive, and significant portions reverted to forest in the past century.

(a)

(b)

Figure 23.25 Shrubland vegetation. **(a)** Fynbos of the Western Cape region of South Africa. *Proteid* shrubs form a mat-like cluster of roots in leaf litter; unlike the ericaceous shrubs that dominate in the Northern Hemisphere, they do not form mycorrhizal associations. **(b)** Chaparral of southern California.

The North American temperate forest biome consists of several subtypes or *associations* (Figures 23.27 and 23.28, p. 516; distributions for the United States and Canada are shown separately). In Canada, the only temperate deciduous

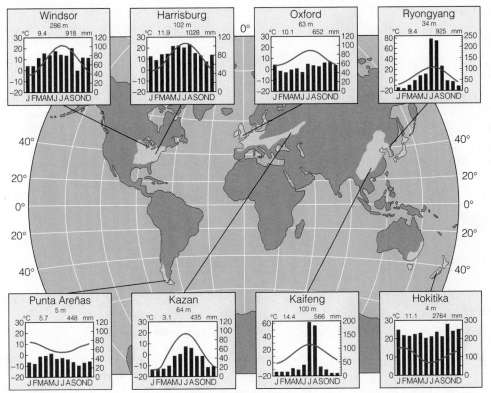

Figure 23.26 Geographic distribution and climate of temperate deciduous forests. Temperate deciduous forests have distinct seasonality in temperature with sufficient precipitation during the growing season to support a closed tree canopy.

(Adapted from Archibold 1995. Data from Windsor, Ontario, from Environment Canada.)

type is the Carolinian forest; Figure 23.28 shows the distributions of mixed-wood and evergreen forests as well. In the United States, as shown in Figure 23.27, the most southerly forests are also non-deciduous. The *subtropical evergreen forest* of the Gulf coast is dominated by broadleaf evergreens such as *Magnolia* spp. and "live" (evergreen) oaks. The *southern mixed forest*, dominated by southern hard pines and evergreen oaks, occupies poorly developed sandy or swampy soils.

The other subtypes are deciduous dominated, although some share dominance with conifers. Moving north, these are the *mesophytic forest* of the unglaciated Appalachian plateau, whose tree species richness is unsurpassed by any other temperate forest worldwide; the *oak–hickory forest* in drier western portions; the *oak–chestnut forest* (now called the *Appalachian oak forest* since die-off of the American chestnut), covering much of the Appalachians; the *beech–maple-basswood forest*, which extends into the Carolinian forests of southern Ontario as well as into southern Québec and the southern Maritimes; and the *alluvial plain forest* along the Mississippi, Red, and other rivers, its dominant species shifting with latitude.

Finally, the *northern hardwood forest* has shared dominance with northern conifers such as red and white pine (*Pinus resinosa and P. strobus*) in drier areas and eastern hemlock (*Tsuga canadensis*) in moister areas. Northern hardwood forests, which occur in northern Michigan and Wisconsin as well as in parts of Ontario, Québec, and the Maritimes as part of the Great Lakes–St. Lawrence and Acadian forest regions, are transitional to northern coniferous (boreal) forest. Deciduous forests are also common in the boreal forest, particularly in drier western areas where trembling aspen (*Populus tremuloides*) covers large portions of northern Alberta and Saskatchewan.

Temperate deciduous forests in eastern China, Japan, Taiwan, and Korea are often dominated by species of the same genera as in North America and Europe. Broadleaf evergreens become increasingly present in southern Asia and in the wet foothills of the Himalayas. Similarly, the deciduous-dominated forests of western Europe give way to broadleaf evergreens in the transition to Mediterranean shrublands.

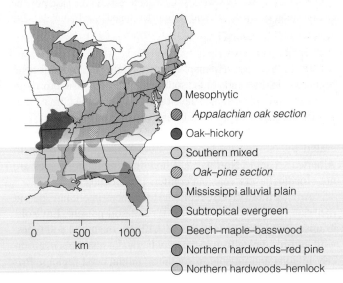

- ● Mesophytic
- ◐ *Appalachian oak section*
- ● Oak–hickory
- ○ Southern mixed
- ◐ *Oak–pine section*
- ● Mississippi alluvial plain
- ● Subtropical evergreen
- ● Beech–maple–basswood
- ● Northern hardwoods–red pine
- ○ Northern hardwoods–hemlock

Figure 23.27 Distribution of temperate forests (including evergreen as well as deciduous types) in the eastern United States, derived from contemporary data. Compare this map to Figure 16.26 (p. 356), which depicts the original forest distribution, prior to human habitation. See Fig. 23.28 (p. 516) for distribution of Canadian forests.

(Adapted from Dyer 2006.)

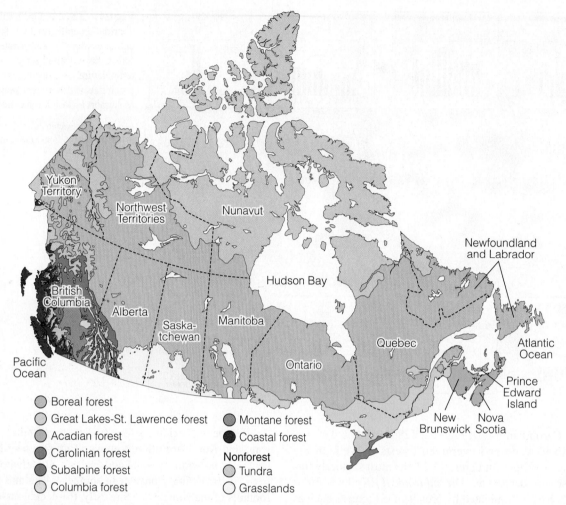

Figure 23.28 Distribution of forest regions in Canada. Among Canadian forests, only Carolinian forests are predominantly deciduous. Acadian and Great Lakes–St. Lawrence forests are largely coniferous or mixed deciduous–coniferous. However, purely deciduous stands occur in these forest regions, as well as in boreal forests. In the past, provinces have developed their own classification systems, but a national forest classification system is currently being developed. See Figure 23.32 (p. 519) for the sub-regions of the boreal forest.

(Source: Natural Resources Canada.)

In the Southern Hemisphere, temperate deciduous forests occur only in drier parts of the southern Andes. In Chile, broadleaf evergreen rain forests have developed in a maritime climate that is virtually frost free. Similar forests occur in New Zealand, Tasmania, and parts of southeast Australia where winters are moderated by the coastal location. Climate is similar to North America's Pacific Northwest, where the dominant species are not broadleaf evergreens but conifers (see Section 23.8).

Among these diverse regions, differences in climate, parent material, and drainage affect the type of soil present. Luvisols, brunisols, and ultisols (see Figure 4.12) are the major soil types, with luvisols typically associated with glacial materials in northerly regions and ultisols in southern, unglaciated regions. Primary productivity varies considerably, influenced by temperature and length of the growing season. Fall is marked by the characteristic changing colours of foliage (which vary among species) shortly before the trees enter their leafless winter phase (Figure 23.29). Leaf fall occurs over a short period, and nutrient supply reflects decomposition rates of the abundant leaf litter. In spring, trees resume growth in response to increasing temperatures and longer days. Some herbaceous species, such as Ontario's pro-

vincial flower, nodding wake-robin (*Trillium cernuum*; see Figure 8.19a), and other *spring ephemerals*, flower before the canopy leafs out, after which they become dormant.

Mature, unevenly aged deciduous forests usually have four or five vertical strata (see Figure 16.17): an *upper canopy* of dominant trees; a lower tree canopy or *understory*; a *shrub layer*; and *ground layer* of herbs, ferns, and mosses. The abundance and diversity of litter supports a diverse fungal community. Animal life is abundant too, in both species number and population size, in part reflecting the many niches provided by the complex vertical stratification (see Section 16.9). Some animals, particularly forest arthropods, spend most of their lives in a single stratum, while others range over several strata. The greatest concentration and variety of life occurs on and just below the ground. Many animals, particularly soil and litter invertebrates, remain belowground. Others, such as mice, shrews, ground squirrels, and salamanders, burrow into the soil or litter for shelter and food. Larger mammals feed on herbs, shrubs, and low trees. In older forests, hollow centres of living trees provide critical habitat for many animals. Birds move freely among strata but typically favour one layer over another (see Figure 16.18).

(a)   **(b)**

Figure 23.29 A temperate forest in the Appalachians. **(a)** Canopy during the fall and **(b)** interior of the forest during the spring. This forest is dominated by oaks (*Quercus* spp.) and yellow poplar (*Liriodendron tulipifera*), with an understory of redbud (*Cercis canadensis*) in bloom.

23.8 TEMPERATE CONIFEROUS FORESTS: Needle-Leaf Evergreens Dominate Cool Temperate and Montane Zones

Forests dominated by coniferous, needle-leaf evergreens grow in a broad circumpolar belt across the Northern Hemisphere and on mountain ranges, where low temperatures limit the growing season to a few months each year (Figure 23.30). The variable composition and structure of coniferous forests reflect the wide range of climatic conditions in which they grow.

In central Europe, extensive *montane forests* dominated by Norway spruce (*Picea abies*) cover the slopes up to the subalpine zone in the Carpathian Mountains and the Alps (Figure 23.31a, p. 518). In North America, coniferous forests of various subtypes blanket the western mountains. At high elevations in the Rockies grow subalpine forests dominated by Engelmann spruce (*Picea engelmannii*) and subalpine fir (*Abies lasiocarpa*). Middle elevations (Figure 23.31b) have stands of mountain Douglas-fir (*Pseudotsuga menziesii glauca*), while lower elevations are dominated by dense stands of lodgepole pine (*Pinus contorta*). Open stands of ponderosa pine (*P. ponderosa*) occupy the driest slopes. The giant sequoia (*Sequoiadendron giganteum*) grows in scattered groves on the western slopes of the California Sierra. Its range is a much-reduced relict of its pre-glacial extent.

As well as conifer dominance in mountainous areas, the mild, moist climate of the Pacific Northwest supports a highly productive *temperate coniferous forest* extending along the coastal strip of North America from southern Alaska, through British Columbia, and into northern California (Figure 23.31c). Dubbed "temperate rain forest" because of its abundant rainfall (from 2000 mm to over 4000 mm, much of it in winter and

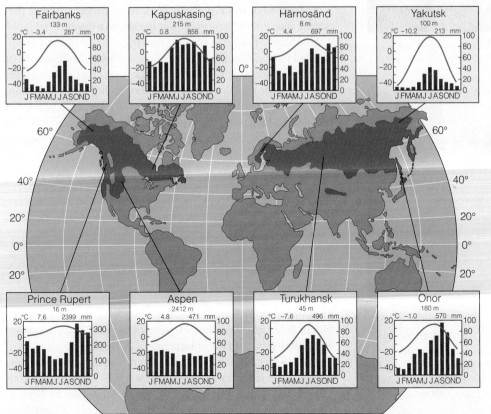

Figure 23.30 Geographic distribution and climate of coniferous forests. With the exception of coastal forests of the Pacific northwest, regions supporting conifer forests have lower mean annual temperature and shorter growing season than areas supporting temperate deciduous forest. Note that this biome also contains northern coniferous forest, which we treat in a separate section.

(Adapted from Archibold 1995.)

(a)

(b)

(c)

Figure 23.31 Three coniferous forest types. **(a)** A Norway spruce forest in the Carpathian Mountains of central Europe. **(b)** Montane coniferous forest in the Rocky Mountains. The dry lower slopes support ponderosa pine, while the upper slopes are cloaked in Douglas-fir. **(c)** Old growth temperate rain forests in Clayoquot Sound on Vancouver Island.

some in the form of fog), these forests are highly prized for their biodiversity. Another trait that resembles tropical rain forests is the strong presence of epiphytes and climbing forms. Although they differ structurally from tropical rain forests (achieving greater height) and are dominated by coniferous rather than broadleaf evergreens, their ground stratum is similarly sparse and occupied by shade-tolerant species, especially ferns.

Unlike temperate deciduous forests, temperate coniferous forests grow year-round, often attaining their highest photosynthetic rates during the mild winters, when moisture is most available. As in most conifer forests, soil fertility is low, but the rapid decomposition promoted by the mild, moist conditions recycles nutrients rapidly. In contrast to tropical rain forests, soils are not highly weathered but rather young brunisols (see Figure 4.12). Given the large size of the dominant trees (coastal Douglas-fir, *Pseudotsuga menziesii menziesii*, and Sitka spruce, *Picea sitchensis*), coarse woody debris is a key feature of its ecology. Fire is more common than in tropical rain forests but less frequent than in boreal forests. Many late-succession species prefer to germinate on the rotting carcasses of large "nurse" logs (see Figure 17.18).

Temperate rain forests abound in animal life, but (apart from their abundant bird life) most live on or near the ground, rather than in the canopy as is typical for many tropical rain forest animals. Small mammals and invertebrates are abundant, along with larger species such as black and brown bear. However, the animal that we most associate with this biome is actually one that lives not in the temperate rain forest itself, but in the streams and rivers found within it: the various species of Pacific salmon (*Oncorhynchus* spp.; see also introduction to Part Three, p. 180). Extensive clear-cutting in sloping coastal areas endangers not only the forests themselves but also adjacent stream habitat for salmon and other species.

Apart from isolated stands in national parks and other preserves, much of the temperate rain forests have disappeared in the United States. Extensive stands persist in mainland British Columbia and the Gulf Islands. The Clayoquot Sound region, a body of water containing many islands near the west coast of Vancouver Island (the name is also used for the surrounding watershed), contains large pristine forested areas that have been the subject of intense anti-logging protests since the 1980s. Although a UNESCO Biosphere Reserve since 2000, it is still not protected from logging.

23.9 NORTHERN CONIFEROUS FORESTS: Boreal Forests Are the Largest Biome on Earth

Despite the high profile of temperate rain forests, by far the largest expanse of conifer forest—and the largest biome on Earth—is the *boreal forest*, or *taiga* (this term is sometimes reserved for the northernmost portions, or for Eurasian boreal forests). This broad coniferous belt across the high latitudes of the Northern Hemisphere covers about 11 percent of Earth's land surface (estimates vary with the criteria used). In North America, boreal forest covers much of Alaska and Canada. It is Canada's largest biome, covering 35 percent of the country's landmass and constituting 77 percent of its forest cover, according to Natural Resources Canada. Boreal forest extends into northern New

England, with remnants persisting in parts of Wisconsin, Michigan, and New York. In Eurasia, the boreal forest begins in Scotland and Scandinavia and extends across the continent, covering much of Siberia and parts of northern Japan and Mongolia.

Occupying formerly glaciated sites, the boreal forest is dotted with cold lakes, rivers, and bogs. (We discuss its abundant wetlands in Section 25.6.) In North America, the boreal forest has three major zones (Figure 23.32): (1) *forest–tundra ecotone* with widely spaced, stunted black spruce (*Picea mariana*), lichens, and moss; (2) *open lichen woodland* with stands of lichens and black spruce (Figure 23.33a, p. 520); and (3) *southern boreal forest* with continuous stands of white spruce (*P. glauca*) and jack pine (*Pinus banksiana*) (Figure 23.33b) in upland locations, interspersed with aspen (*Populus tremuloides*) and paper birch (*Betula papyrifera*), particularly on deeper, more fertile soils or in disturbed sites.

Red pine (*Pinus resinosa*) was once abundant across the region, thriving on well-drained sites and tolerating windswept conditions from northwestern Ontario to the Maritimes. The old-growth stands of 300-year-old red pine in the Temagami area near Sudbury (Figure 23.33c) are protected from logging but are still at risk from mining. In the Maritimes and south along the eastern seaboard to the Appalachians, red spruce (*Picea rubens*) is a common dominant. Mixedwoods with shared dominance of deciduous and coniferous species, particularly white spruce and balsam fir (*Abies balsamea*), are common (Figure 23.33d). Poorly drained bog sites in the southern boreal are dominated by black spruce (see Figure 25.17). Similar zones exist in northern Europe and Asia, and are dominated by related species. At its southern edge, boreal forests grade into the northern hardwood forests of the temperate zone.

Vertical stratification varies with the canopy type. Mixedwood forests have a highly developed tall shrub layer of green alder (*Alnus viridis*), hazelnut (*Corylus cornuta*), and saskatoon (*Amelanchier* spp.), making them almost impenetrable. Herbaceous species such as bunchberry (*Cornus canadensis*), Canada mayflower (*Maianthemum canadense*), and clubmosses (*Lycopodium* spp.) occupy the forest floor (Figure 23.34,

p. 521). Forests dominated by jack pine have little tall shrub presence, but contain many low-growing, broadleaf evergreen shrubs, such as blueberry (*Vaccinium angustifolium*), bearberry (*Arctostaphylos uva-ursi*), other members of the Ericaceae, and coniferous shrubs such as juniper (*Juniperus communis*). Feathermosses abound, particularly *Pleurozium schreberii*. The understory of black spruce forests is dominated by lichens in drier northern forests and peat mosses (*Sphagnum* spp.) in bogs.

A cold continental climate with strong seasonal variation prevails. Summers are short, cool, and generally moist, with precipitation increasing from west to east. Winters are long, harsh, and dry, with a prolonged snow cover. The driest winters and greatest fluctuations are in interior Alaska and central Siberia, which experience a temperature range approaching 100°C. Boreal regions in Canada (particularly in mid-continental sites such as northern Ontario) experience only a slightly less extreme range.

Much of the northern boreal forest is under the influence of **permafrost**—a perennially frozen subsurface that may be hundreds of metres deep, impeding infiltration and maintaining high soil moisture. Permafrost develops where ground temperatures remain below 0°C for extended periods. Its upper layers may thaw in summer and refreeze in winter, but because it is impervious to water, permafrost forces all water to remain and move above it. Thus, the ground stays soggy even though precipitation is low, enabling plants (assuming they have sufficient cold hardiness) to exist in the driest regions, and inhibiting others with the resulting high water table.

Fires are recurring, ecosystem-shaping disturbances in the taiga. During drought, fires can sweep over hundreds of thousands of hectares. Virtually all boreal species, both deciduous and coniferous, are fire-adapted. Unless it is so severe that it destroys soil structure, fire provides a seedbed for regeneration of trees that prefer to germinate on exposed mineral soil. Light surface fires favour pioneer hardwoods such as aspen, whereas more severe fires eliminate hardwoods and favour black spruce and jack pine regeneration, both of which have serotinous cones (see Section 18.8). Woody and herbaceous understory species rapidly resprout after fire.

Figure 23.32 Major zones of North America's boreal forest.

(Adapted from Payette et al. 2001.)

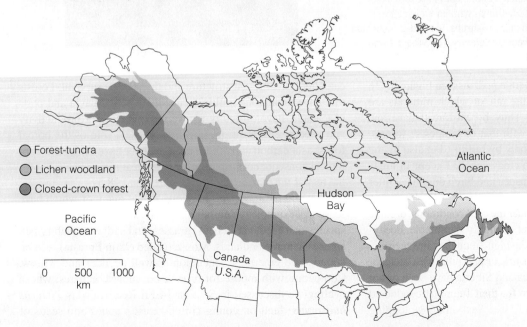

Forest-tundra
Lichen woodland
Closed-crown forest

Atlantic Ocean

Pacific Ocean

Hudson Bay

Canada
U.S.A.

0   500   1000
km

(a)

(b)

(c)

(d)

Figure 23.33 Representative boreal forest subtypes. **(a)** Open lichen woodland in Québec, with scattered clumps of black spruce **(b)** Jack pine forest in southeastern Manitoba, with an understory dominated by feathermosses and ericaceous shrubs. **(c)** Old-growth stand of red pine in the Wolf Lake Reserve near Sudbury. **(d)** Mixed-wood forest in northwestern Ontario with birch, aspen, and white spruce.

Soils are primarily acidic, infertile podzols (see Figure 4.12), and growth is often limited by the rate at which minerals are recycled through the ecosystem. Thus, boreal forests have low primary productivity compared with that of temperate forests. Their productivity is limited not only by fertility, but also by the cool temperatures and short growing season. Except in areas where deciduous species such as aspen and birch dominate, litter inputs are much lower than in temperate forests. However, because decomposition is slow under the cool conditions, organic matter accumulates in a thick mat of highly flammable, decay-resistant needle duff, facilitating the high-frequency fire regime.

Given the global demand for timber and pulp, vast areas of the boreal forest in North America and Siberia have been or are being clear-cut with little concern for their future. If conducted without considering forest ecology (see Section 26.8), logging can alter boreal forest structure and threaten its survival. Forests that have undergone multiple cuts, as in Scandinavia and parts of Canada, have difficulty recovering their former productivity and species composition. Given the critical role of the boreal forest as a carbon sink, these changes may have long-term ecological consequences, not just for the boreal region itself, but for the rest of the planet (see Sections 22.4 and 28.5).

The boreal forest has a unique fauna, many of which have served as examples in earlier chapters. Caribou, a wide-ranging species that feeds on lichens, grasses, and sedges, inhabit open spruce–lichen woodlands. Moose (called elk in Eurasia) feed on aquatic and emergent vegetation as well as alder and willow. Competing with moose for browse are snowshoe hares, whose cycling we discussed in Section 14.11. Red squirrels (*Sciurus hudsonius*) feed on young pollen-bearing cones and seeds of

Figure 23.35 Spruce trees near Churchill, Manitoba, exhibiting the Krummholz growth form. More extreme versions are bent over and hug the ground.

Figure 23.34 Boreal forests contain few tree species, but their herb layer is diverse. Shown here is a clubmoss, *Lycopodium obscurum*, with bunchberry (*Cornus canadensis*) in the foreground.

spruce and fir, and porcupines (*Erethizon dorsatum*) feed on the leaves, twigs, and inner bark of trees. Preying on these herbivores are various carnivores, including wolves, lynx, pine martens (*Martes americana*), and owls.

The boreal forest is the nesting ground of migratory neotropical birds and the habitat of northern seed-eating birds such as crossbills (*Loxia* spp.), grosbeaks (*Coccothraustes* spp.), and siskins (*Carduelis* spp.). Of great ecological and economic importance are herbivorous insects such as the spruce budworm. As major prey for insectivorous birds, these insects undergo periodic outbreaks during which they defoliate large expanses of forest. They have been the subject of much controversy among scientists and the general public (see Ecological Issues: Science and Politics Interact in Spruce Budworm Control in the Maritimes, pp. 400–401).

23.10 TUNDRA: Arctic and Alpine Tundra Experience Low Precipitation and Cold Temperatures

The boreal forest thins out towards the pole, until one reaches the **treeline**: the geographic limit beyond which forest does not grow. There are three ways of defining treelines: (1) a *timber line*, beyond which trees are not present at sufficient density and canopy closure to qualify as a forest; (2) a *treeline*, beyond which individuals do not achieve tree height; and (3) a *tree species line*, beyond which tree species are absent even in dwarf form. With climate change, the treeline (however it is defined) is slowly moving north (see Section 28.7).

Cold and moisture are not the only factors restricting trees. Species such as birch have enough cold hardiness to persist on the tundra, but the drying winds and physical stress of wind-driven ice prevent even the hardiest of species from attaining

tree height. Those tree species that do grow in the extreme north (such as black spruce) adopt a dwarf form called **Krummholz** (Figure 23.35), in which the trunk is bent and hugs the ground. This form allows protection by snow cover from winds and bombardment by ice particles. Aboveground growth is asymmetric, with stunted branches facing the wind. Even very old trees have a small girth as a result of the short season.

The treeline marks the transition to *Arctic tundra*—a circumpolar frozen plain, clothed in sedges, heaths, and willows; dotted with lakes; and crossed by streams (Figure 23.36, p. 522). Arctic tundra falls into two types: (1) *polar grassland*, with up to 100 percent plant cover and wet to moist soil (Figure 23.37a, p. 522), and (2) *polar desert*, with less than 5 percent plant cover and dry soil (Figure 23.37b). *Alpine tundra* is a similar biome found at high elevations worldwide.

Conditions on the Arctic tundra reflect three interacting forces: (1) a deep permafrost layer, (2) an overlying layer of organic matter and mineral soil that thaws each summer and then re-freezes, and (3) vegetative cover that reduces warming and retards thawing. Permafrost chills the soil, retarding plant growth, limiting soil microbe activity, and lowering soil aeration and fertility. Alternate freezing and thawing of the upper soil creates unique, symmetric landforms (Figure 23.38, p. 523). Frost action pushes stones and other material up and outward from the mass to form a patterned surface of frost hummocks, frost boils, earth stripes, and polygons. On sloping ground, soil creep, frost thrusting, and downward flow of supersaturated soil over the permafrost form **solifluction terraces**, or flowing soil. This gradual downward creep of soils and rocks rounds off ridges and other topographic irregularities. The resulting moulding of the landscape by frost action, called **cryoplanation**, is more important than erosion in wearing down the Arctic landscape.

Tundra vegetation is structurally simple. Species richness is low and growth is slow. Only stress-tolerators can withstand the extreme cold, constant soil disturbance, wind, and abrasion from soil particles and ice. Low ground is covered with a mix of cotton grasses, sedges, and *Sphagnum* mosses. Well-drained sites support ericaceous (broadleaf evergreen) shrubs, dwarf willows and birches, herbs, mosses, and lichens (Figure 23.39, p. 523). The driest, most exposed sites support scattered shrubs and crus-

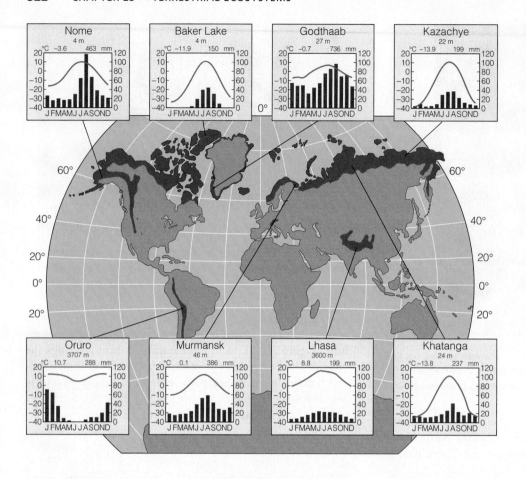

Figure 23.36
Geographic distribution and climate of tundra. Tundra is colder and drier than boreal forest, with a shorter growing season.

(Adapted from Archibold 1995.)

tose and foliose lichens growing on rock. Although Arctic plants rely on clonal (vegetative) reproduction, viable seeds hundreds of years old persist in the soil as a seed bank that is residual from periodic flowering events.

Tundra plants are active for only three months of the year. As snow cover disappears, plants become active. They make maximal use of the short season by photosynthesizing during the extended daylight period, even at midnight in midsummer when light is 1/10 that at noon. The nearly erect leaves of some species permit efficient interception of the low angle of the Arctic sun. Most carbon is invested in new shoot growth, but about one month before the season ends, plants withdraw nutrients from leaves and move them to belowground biomass, sequestering 10 times the amount stored by temperate grasslands. Tundra structure is thus unusual in that most of the biomass is underground. Root-to-shoot ratios of tundra vascular plants range from 3:1 to 10:1. Roots are concentrated in the upper soil that thaws during summer, and aboveground parts seldom grow above 30 cm. It is not surprising, then, that belowground biomass is often three times the magnitude of aboveground biomass.

This tendency of tundra vegetation to sequester large amounts of C belowground is critical to its role as a C sink.

(a)　　　　(b)

Figure 23.37 Types of tundra. **(a)** Plant cover typical of the Arctic tundra in the Northwest Territories is in stark contrast to **(b)** polar deserts, with dry soils and sparse cover.

(a)

(b)

Figure 23.38 Patterned landforms typical of tundra: **(a)** frost hummocks and **(b)** polygons.

Even after root tissue has died, the C remains organically bound as a result of the extremely slow decomposition. But most of the C sequestered in tundra is in the form of peat moss (*Sphagnum* spp.), both living and dead. Peat mosses build up over time in tundra (and in bogs in the boreal forest), where permafrost retards drainage. Global climate change risks converting the tundra (and the northern boreal forest) from a C sink into a C source, by releasing large amounts of methane and CO_2 as permafrost thaws (Schuur et al. 2009; see also Research in Ecology: Altered Matter Cycling in a Warmer Arctic, pp. 482–483).

Tundra supports a fascinating fauna. Its diversity is low, but its abundance is high, supported by the intense growth during the short season. Invertebrates concentrate near the surface, where there are abundant populations of segmented white-worms (Enchytraeidae); collembolas; and flies, chiefly crane flies. Summer brings hordes of blackflies (*Simulium* spp.), deer flies (*Chrysops* spp.), and mosquitoes (*Culex* spp.). Dominant vertebrates are herbivores, including lemmings, Arctic hare, caribou, and musk ox (*Ovibos moschatus*). Although caribou have the greatest herbivore biomass, lemmings, which breed year-round, reach densities as great as 125 to 250 per hectare and consume 3 to 6 times as much forage as caribou do (Krebs 2011). Arctic hares that feed on willows disperse over their range in winter and congregate in more restricted areas in sum-

Figure 23.39 Tundra vegetation. This well-drained tundra site near Churchill, Manitoba, contains a diversity of low-growing species, including lichens and members of the Ericaceae family.

mer. Caribou are extensive grazers, spreading out over the tundra in summer. Musk ox graze more intensively in localized areas, where they feed on sedges, grasses, and dwarf willow. Herbivorous birds are few, dominated by ptarmigan and migratory geese. Herbivory by snow geese is altering the structure of tundra in some regions, possibly leading to an alternative stable state (see Sections 17.5 and 19.9). Many migratory birds breed and nest in the tundra (and boreal forest), taking advantage of the abundant resources available in the short season.

The major Arctic carnivore is the wolf, which preys on musk ox, caribou, and lemmings (when abundant). Medium-sized to small predators include the Arctic fox (*Alopex lagopus*), which preys on Arctic hare, and several species of weasel, which prey on lemmings. Also feeding on lemmings are snowy owls (*Nyctea scandiaca*) and hawk-like jaegers (*Stercorarius* spp.). Sandpipers (*Tringa* spp.), plovers (*Pluvialis* spp.), longspurs (*Calcarius* spp.), and waterfowl, which nest on the wide expanse of ponds and boggy ground, feed heavily on insects.

At lower latitudes, *alpine tundra* occupies high-elevation mountain sites. Alpine tundra is a severe environment of rock-strewn slopes, bogs, meadows, and shrubby thickets (Figure 23.40). Like its Arctic counterpart, alpine tundra experiences

Figure 23.40 Alpine tundra in the Rocky Mountains.

strong winds, snow, cold, and fluctuating temperature. In summer, soil surface temperature can range from 40°C to 0°C. However, alpine tundra differs from Arctic tundra. The atmosphere is thin, so light intensity, especially ultraviolet radiation, is high on clear days. The long days of the Arctic tundra are not typical of alpine tundra. Although both experience temperature variation, alpine tundra has less seasonal but often greater daily flux. Permafrost only occurs at very high elevations. Lacking permafrost, alpine soils are drier, and ground surfaces are more sloped, enhancing drainage. Only in alpine wet meadows and bogs does soil moisture compare with Arctic tundra. Precipitation, especially snowfall and humidity, is higher in alpine regions, but steep topography induces rapid runoff. For these and other reasons, the species composition of alpine tundra is very different, although the growth forms are similar—ground-hugging low shrubs, herbaceous species, and mosses.

EcologyPlace

Visit EcologyPlace at www.pearsoncanada.ca/ecologyplace to access online resources that complement your textbook, and help you to apply and to review the information in this chapter. EcologyPlace includes

- an eText version of the book
- self-grading quizzes
- glossary flashcards
- and more!

Go to www.pearsoncanada.ca/ecologyplace and follow the registration instructions on the Student Access Code Card included with this text. If your book does not have a Student Access Code Card, you can purchase access to it at www.pearsoncanada.ca/ecologyplace.

SUMMARY

Biome Distribution 23.1

Terrestrial systems are grouped into biomes, based on the dominant vegetation. There are at least nine major terrestrial biomes with many subtypes: tropical forest, tropical savannah, grassland, shrubland, desert, temperate deciduous forest, temperate coniferous forest, northern boreal forest, and tundra. These categories reflect the relative presence of trees, shrubs, and herbs (forbs and grasses). Constraints on the adaptations of these plant life forms to abiotic factors determine their patterns of dominance along gradients of temperature and moisture.

Tropical Forests 23.2

Seasonality of rainfall determines the type of tropical forest. Tropical rain forests, associated with high aseasonal rainfall, are dominated by broadleaf evergreen trees and support many species. They have five strata: emergent trees, canopy, understory, shrubs, and a ground layer of herbs and ferns. Growth forms include herbaceous and woody vines, stranglers, and epiphytes. Large trees develop buttresses. Abundant rainfall and consistent warmth support high rates of productivity, decomposition, and nutrient cycling. Most primates live in tropical rain forests, and invertebrate diversity is high. Tropical seasonal forests undergo varying lengths of dry season, during which drought-deciduous trees and shrubs drop their leaves. Most tropical dry forests have been converted to agriculture.

Tropical Savannahs 23.3

Savannahs have shared dominance of grasses and woody plants and occupy regions with alternating wet and dry seasons. Depending on precipitation and soil texture, savannahs range from grasslands with occasional trees to communities where trees form an almost continuous canopy. Productivity and decomposition are closely tied to seasonality of precipitation. Savannahs support many invertebrate and vertebrate herbivores. The African savannah is dominated by a large, diverse ungulate fauna and associated carnivores.

Grasslands 23.4

Grasslands occupy regions with annual rainfall between 250 and 800 mm. They occupy a fraction of their original extent as a result of agricultural conversion. Grasslands vary with climate and geography. In North America, native grasslands, whose productivity is influenced by declining precipitation from east to west, consist of tallgrass, mixed-grass, and shortgrass prairie; desert grasslands; and fescue grasslands. Eurasia has steppes, South America the pampas, and southern Africa the veld. Grassland consists of a herbaceous layer arising from crowns, nodes, and rosettes; a ground layer; and a root layer. Depending on their fire and grazing history, grasslands accumulate a mulch layer. Grasslands, which once supported large ungulate herds, evolved under the selective pressure of fire and grazing. Although the most conspicuous grazers are large herbivores, the major consumers are invertebrates. Heaviest consumption is belowground, where the dominant herbivores are nematodes.

Deserts 23.5

Deserts occupy two global belts between 15° and 30° north and south latitude and result from dry descending air masses, rain shadows of coastal mountains, and remoteness from coastal regions. There are cool and hot deserts. Deserts are structurally simple, with scattered shrubs and ephemerals. Species have evolved strategies to avoid or tolerate drought. Desert animals are typically opportunistic herbivores and carnivores.

Shrublands 23.6

Shrubs have low stature and multiple-stemmed, woody structure. Shrublands are hard to classify, given the variety of climates in which shrubs are dominant or co-dominant. Mediterranean shrublands occur in five disjunct regions along the western margins of the continents between 30° and 40° latitude. Dominated by sclerophyllous evergreens, shrublands have adapted to a climate of summer drought and cool, moist winters. Shrubland species are fire-adapted.

Temperate Deciduous Forests 23.7

Deciduous forests occupy wetter portions of the warm temperate zone. They once covered large areas of Europe and China, but their distribution has been reduced by human activity. In North America, deciduous forests include subtypes such as beech–maple and oak–hickory forest. The most diverse is the meso-phytic forest of the unglaciated Appalachians. Well-developed deciduous forests have four strata: upper canopy, lower canopy, shrub, and ground layer. The Carolinian forest region is the major deciduous-dominated forest region in Canada. Temperate forests that are transitional to the southern boreal forest often have co-dominance between deciduous and coniferous species.

Temperate Coniferous Forests 23.8

Coniferous forests of temperate regions include montane forests and coastal temperate rain forests. Although temperate rain forests have high productivity and abundant rainfall, they differ from tropical rain forests in that moisture is more seasonal, and more abundant in winter. They grow on less-weathered soils, and fires do occur, although less frequently than in boreal forests. Coarse woody debris is an important aspect of their ecology.

Northern Coniferous Forests 23.9

The circumpolar boreal forest is Earth's largest biome. Characterized by a cold continental climate, the boreal forest consists of the forest–tundra ecotone, the open lichen woodland, and the southern boreal forest, which is bordered by the boreal–mixed forest ecotone. Permafrost affects northern boreal forests, as do recurring fires. Spruces and pines dominate, with birch and poplar common in some sites. Mosses dominate closed-canopy spruce forests, whereas lichens dominate open conifer stands. Low broadleaf evergreen shrubs and herbs are also present. Mammal herbivores include caribou, moose, and snowshoe hare. Predators include wolf, lynx, and marten.

Tundra 23.10

Arctic tundra extends beyond the treeline at high latitudes in the Northern Hemisphere. The climate is cold and dry, with a short season; permafrost; and a frost-moulded landscape dominated by low, slow-growing species. Dominant vegetation, such as cotton grass, sedges, and dwarf evergreen shrubs, exploits the short season by photosynthesizing during the long days. Most C is allocated belowground. In summer, insects abound, providing food for migratory birds. Dominant vertebrate herbivores are lemming, Arctic hare, caribou, and musk ox. Major carnivores are wolf, Arctic fox, and snowy owl. Alpine tundra occurs at high altitudes worldwide. It experiences fluctuating but warmer temperatures, high winds, snow, and high radiation. Growth forms resemble Arctic tundra, but the soils are well drained as a result of the sloped topography.

KEY TERMS

aseasonal	chaparral	evergreen	physiognomy	tolerance
avoidance	convergent evolution	formation	resistance	treeline
biogeography	cryoplanation	Krummholz	savannah	xeric
biome	deciduous	needle-leaf evergreen	sclerophyllous	
broadleaf evergeen	ecological equivalents	parallel evolution	solifluction terrace	
buttress	epiphyte	permafrost	succulent	

STUDY QUESTIONS

1. According to the Whittaker model in Figure 23.2, describe the general ways in which (i) precipitation and (ii) temperature determine the biome that occupies a given area.

2. What are the main climatic and vegetation differences between tropical rain forest and tropical seasonal forest?

3. Why are epiphytes so much more common in tropical rain forests than in temperate forests, even in temperate regions with abundant rainfall?

4. What types of leaves characterize tropical rain forest trees? How does this compare with the leaf type of trees

occupying (i) Mediterranean shrublands and (ii) temperate deciduous forests? What ecological factors explain these differences among biomes?

5. When tropical rain forest is removed, it is often difficult to re-establish, even after an extended time. What ecological factors contribute to this difficulty?

6. What distinguishes savannah from grassland? What climate conditions support savannah? How does seasonality influence net primary productivity and decomposition in savannahs?

7. What regional climatic factors lead to the formation of deserts?

8. What two distinct strategies allow organisms to inhabit deserts? Give examples of each strategy for both a plant and an animal species.

9. How does seasonality of temperature influence the structure and productivity of temperate deciduous forests? How does this compare with the influence of seasonality on the temperate coniferous forests of coastal British Columbia?

10. What are spring ephemerals, and why are they so much more common in temperate deciduous than in temperate coniferous forests?

11. Compare a tallgrass and a mixed-grass prairie in terms of climate, vegetation structure, and productivity.

12. Why are grassland soils typically more fertile than soils under forests, despite the much greater biomass of a forest?

13. How are temperate rain forests of coastal British Columbia similar to and different from tropical rain forests?

14. What type of trees characterizes boreal forest? To which of Grime's categories (ruderals, competitors, and stress-tolerators) do they belong? (Your instructor may ask you to apply Grime's categories to other biomes.)

15. What is permafrost, and how does it affect boreal forest structure and productivity?

16. Why does fire play such an important role in boreal forest? What are some of its effects?

17. What physical and biological features characterize Arctic tundra? How does Arctic tundra differ from alpine tundra?

18. What biome do you live in, and what are its dominant species? How much has it been altered or exploited by humans? Are there examples of it in a pristine state?

FURTHER READINGS

Archibold, O. W. 1995. *Ecology of world vegetation.* London: Chapman & Hall.
Outstanding reference for those interested in the geography and ecology of terrestrial biomes.

Bliss, L. C., O. H. Heal, and J. J. Moore (eds). 1981. *Tundra ecosystems: A comparative analysis.* New York: Cambridge University Press.
Major reference on the geography, structure, and function of tundra.

Bonan, G. B., and H. H. Shugart. 1989. Environmental factors and ecological processes in boreal forests. *Annual Review of Ecology and Systematics* 20:1–18.
Excellent review paper providing a good introduction to boreal forests.

Evenardi, M., I. Noy-Meir, and D. Goodall (eds). 1986. *Hot deserts and arid shrublands of the world.* Ecosystems of the World 12A and 12B. Amsterdam: Elsevier Scientific.
Major reference on the geography and ecology of deserts.

French, N. (ed). 1979. *Perspectives on grassland ecology.* New York: Springer-Verlag.
Good summary of grassland ecology.

Murphy, P. G., and A. E. Lugo. 1986. Ecology of tropical dry forests. *Annual Review of Ecology and Systematics* 17:67–88.
Overview of the distribution and ecology of tropical dry forests.

Quinn, R. D., and S. C. Keeley. 2006. *Introduction to California chaparral.* Berkeley: University of California Press.
Overview of chaparral with an excellent discussion of the role and impact of fire.

Reichle, D. E. (ed). 1981. *Dynamic properties of forest ecosystems.* Cambridge, UK: Cambridge University Press.
Major reference on forest ecology.

Reichman, O. J. 1987. *Konza prairie: A tallgrass natural history.* Lawrence: University Press of Kansas.
Excellent treatment of tallgrass prairies, stressing interactions of plants, animals, and landscape.

Richards, P. W. 1996. *The tropical rain forest: An ecological study.* 2nd ed. New York: Cambridge University Press.
Revised edition of a classic book on tropical rain forest ecology.

Sinclair, A. R. E., and P. Arcese (eds). 1995. *Serengeti II: Dynamics, management, and conservation of an ecosystem.* Chicago: University of Chicago Press.
Valuable reference on the ecology of tropical savannahs.

Thousands of freshwater lakes like this one occur throughout the boreal forest.

Whereas ecologists classify terrestrial ecosystems according to their dominant plant life forms, classification of aquatic ecosystems is based on features of the abiotic environment. Because salinity is such a major influence on aquatic life, ecosystems are classified as either *freshwater* or *marine* (saltwater). These two categories are further subdivided, based on substrate, depth and flow of water, and dominant organisms. Marine ecosystems are classified as either *coastal* or *open water*. Freshwater ecosystems are either flowing-water (**lotic**) systems of rivers and streams or non-flowing water (**lentic**) systems of ponds, lakes, and inland wetlands.

All aquatic systems, freshwater or marine, are linked directly or indirectly in the water cycle (Figure 24.1; see Section 3.7). Water evaporated from oceans and land falls as precipitation. Precipitation that does not infiltrate soil or evaporate follows a path determined by gravity and topography. Streams coalesce into rivers as they follow landscape contours, or collect in basins and floodplains to form ponds and lakes. Rivers eventually flow to the coast and form estuaries, which are transitional from freshwater to marine. In this chapter, we examine freshwater and marine ecosystems. Coastal and wetland ecosystems are discussed in Chapter 25.

24.1 LAKES AND PONDS: Lentic Systems Vary in Origin but Share Basic Features

Lakes and *ponds* are inland depressions containing standing water. They vary in maximum depth from 1 m to over 1600 m (Russia's Lake Baikal; Canada's deepest is Great Slave Lake at 614 m). They range in extent from small ponds of under 1 ha to large lakes covering thousands of square kilometres. Ponds can be so shallow that rooted plants grow over much of the bottom, and lakes can be so large that they mimic marine habitats. Most

ponds and lakes have outlet streams, and both may be temporary features on the landscape, on a geologic timescale.

Lakes and ponds have many origins. Many lakes arise from glacial erosion and deposition. Abrading slopes in high mountain valleys, glaciers carved basins that filled with water from rain and melting snow to form *tarns* (Figure 24.2a). Retreating valley glaciers left behind crescent-shaped ridges (*moraines*) of rock debris, which dammed up water behind them. Many shallow *kettle lakes* and *potholes* are residual from the glaciers that covered much of North America and Eurasia. Lakes also form when silt, driftwood, and other debris deposited in beds of slow-moving streams dam up water behind them, or when permafrost impedes drainage (Figure 24.2b). Loops of rivers that meander over flat valleys and floodplains may be cut off by sediments, forming crescent-shaped *oxbow lakes*, common along major rivers in the prairies (Figure 24.2c). Shifts in Earth's crust, uplifting mountains or displacing rock strata, may develop very deep water-filled depressions. Lakes that form in this way, as a result of subsidence associated with movements along fault lines, are called *rift lakes*, such as the Rift Valley Lakes of East Africa, and Lake Baikal. Craters of extinct volcanoes may also become lakes. Pingualuk Lake (maximum depth 252 m) in Québec and West Hawk Lake (115 m) in Manitoba occupy craters created by meteors. Landslides may block streams and valleys to form new lakes and ponds.

Some lentic systems arise from biological activity. Beavers act as bioengineers, damming streams to make shallow but often extensive ponds (see Sections 16.4 and 18.8). Humans create huge lakes by damming rivers and streams for power, irrigation, or water storage (Figure 24.2d; see Section 24.4), and construct smaller ponds and marshes for recreation, fishing, and wildlife. Quarries and surface mines can also leave behind ponds.

Whatever their origin, lentic systems share key features. Life in still water depends on light. Light penetration is influenced by natural attenuation, by silt and other material carried into the lake, and by growth of phytoplankton. As we saw in Section 3.3, temperature varies seasonally and with depth. O_2 can be limiting, especially in summer, as only a small portion of the water is in direct contact with air, and decomposer respiration consumes large amounts. Thus, variation in abiotic factors strongly influences the distribution and adaptations of life in lakes and ponds.

Lentic systems have horizontal and vertical strata based on light penetration and photosynthetic activity (Figure 24.3). Horizontal zones are obvious to the eye; vertical zones, influenced by light penetration, are not. Surrounding most lakes and engulfing some ponds completely is the **littoral (shallow-water) zone**, in which light reaches the bottom, stimulating growth of rooted plants. Beyond the littoral is the **limnetic (open-water) zone**, which extends to the depth of light penetration. Inhabiting this zone are microscopic autotrophic *phytoplankton* and **zooplankton** (floating or weakly swimming animal-like protists and small invertebrates) and **nekton** (free-swimming organisms such as fish). Next is the **profundal zone**, which begins at the *compensation depth*, where respiration balances gross photosynthesis (see Figure 20.8). The profundal zone depends for its energy on a rain of organic material from the limnetic zone. Underlying both the littoral and profundal zones is the **benthic (bottom)**

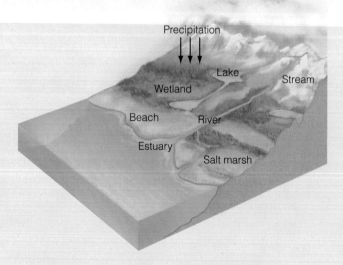

Figure 24.1 Linkages among various types of aquatic ecosystems via the water cycle. Go to **www.pearsoncanada.ca/ ecologyplace** to learn about global freshwater resources.

Figure 24.2 Lakes and ponds of different origins. **(a)** A glacial lake (tarn) in the Rockies. **(b)** The Siberian tundra is dotted with ponds and lakes. **(c)** An oxbow lake formed when a river bend was cut off from the main channel. **(d)** A millpond. Note the floating vegetation.

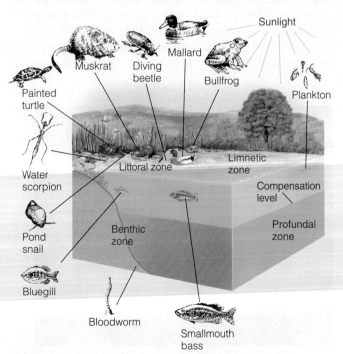

Figure 24.3 Major zones of a lake in midsummer: littoral, limnetic, profundal, and benthic, with typical organisms. At the light compensation depth, GPP equals respiration (NPP = 0).

zone, the major site of decomposition. All zones are closely interdependent in the dynamics of lakes.

24.2 LAKE BIOTA: Types of Organisms Vary in Vertical Lake Zones

There is a distinct zonation pattern to lake vegetation, reflecting response to water depth rather than succession. Aquatic life is most abundant in the shallow littoral waters at the edges of lakes and ponds, as well as in other places where accumulated sediments have decreased water depth (Figure 24.4, p. 530). Dominating such areas are *emergent plants* such as cattail (*Typha* spp.), reed grass (*Phragmites* spp.), and sedge (*Carex* spp.). An **emergent** is an aquatic plant whose roots are anchored in the substrate, but whose upper stems and leaves emerge above the water. In deeper water are *floating plants* such as duckweed (*Lemna* spp.) and *submerged plants* such as pondweed (*Potamogeton* spp.).

Associated with emergent and floating plants is a rich faunal community, including hydras, snails, protozoans, dragonfly nymphs, diving insects, amphibians, pickerel (*Esox* spp.), sunfish (*Lepomis* spp.), herons (Ardeidae), and blackbirds (*Agelaius* spp. and *Xanthocephalus xanthocephalus*). Pond fish have compressed bodies, letting them easily penetrate masses of aquatic

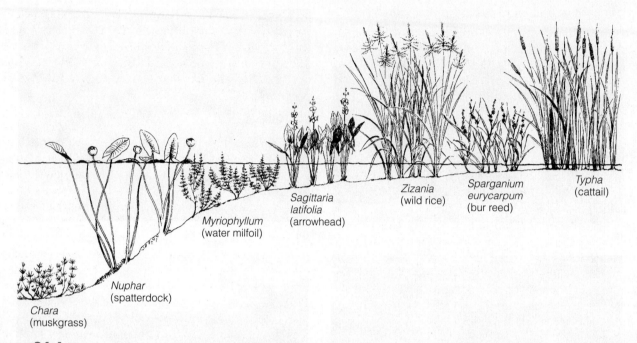

Typha
(cattail)

Sparganium eurycarpum
(bur reed)

Zizania
(wild rice)

Sagittaria latifolia
(arrowhead)

Myriophyllum
(water milfoil)

Nuphar
(spatterdock)

Chara
(muskgrass)

Figure 24.4 Zonation of emergent, floating, and submerged vegetation at the edge of a lake or pond reflects differing responses to water depth.

vegetation. Dominated by vascular species, the littoral zone contributes greatly to organic matter inputs. It also acts as a "scrubber," reducing nutrient and contaminant loads from rivers.

The limnetic zone is dominated by *phytoplankton* (Figure 24.5), including single-celled plant-like protists (algae) such as desmids and diatoms, but also filamentous forms. These organisms are the autotrophic base of open-water food webs. Also suspended in the column are *zooplankton*, animal-like protists and small invertebrates (mostly crustaceans) that consume phytoplankton. During seasonal turnover, nutrients released by decomposition are carried up to the surface. In spring when surface waters warm and stratification develops, phytoplankton have access to both light and nutrients

(see Section 20.3). A spring bloom occurs, followed by rapid depletion of nutrients and reduced plankton numbers.

Fish are the dominant *nekton* in the limnetic zone. Species distributions are influenced by food supply, O_2, and temperature. During summer, large predatory fish such as largemouth bass, pike, and muskellunge inhabit the warmer epilimnion, where food is abundant. In winter, these species retreat to deeper water. In contrast, lake trout require cold water and move to deeper water as summer advances. During seasonal turnovers, when O_2 and temperature are similar throughout, both warm- and cold-water species occupy all depths.

Life in the profundal zone depends not only on energy and nutrients from the limnetic zone, but also on temperature and O_2.

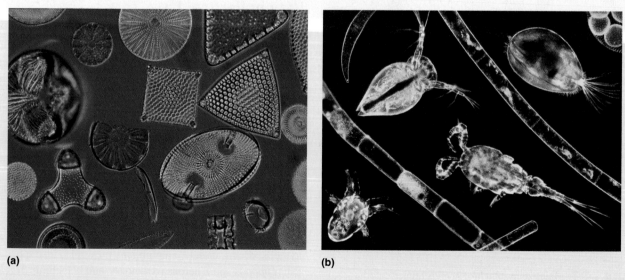

(a)   (b)

Figure 24.5 The limnetic zone is dominated by **(a)** phytoplankton and **(b)** zooplankton.

In nutrient-rich waters, decomposer O_2 demand limits aerobic life. Only during seasonal turnovers, when organisms from the upper layers descend, is life abundant in deep water. Easily decomposed substances are partly mineralized as they drift down through the profundal zone. The remaining debris—dead bodies of limnetic biota, and decomposing plant matter from littoral regions—settles on the bottom. Together with material washed in, these substances make up the bottom sediments. The benthic ooze is a site of great biotic activity—so great that O_2 curves drop sharply in profundal waters just above the bottom. Anaerobic bacteria dominate and decomposition is slow. When the organic matter at the bottom exceeds what can be used by bottom fauna, it forms a muck rich in hydrogen sulphide (H_2S) and methane (CH_4).

The **benthos** (biota occupying the benthic layer) changes in shallow littoral waters. The combined action of water, plant growth, and organic deposits modifies the typical bottom material of stones, gravel, sand, and clay. Increased O_2, light, and food support a greater diversity than in the benthos of profundal waters. Important members of the shallow-water benthic community are *periphyton*, algae that are attached to or move on a submerged substrate, including vegetation, without penetrating it. **Aufwuchs** (a broader term, including protozoans, animals, and bacteria as well as periphytic algae) colonize submerged plants, sticks, and rocks, making them slimy to the touch. On stones, wood, and debris, aufwuchs form a crust-like growth of cyanobacteria, diatoms, water moss, and sponges.

24.3 NUTRIENT INPUTS: Surrounding Communities Influence the Nutrient Status of Lakes

Reflecting the close link between land and water systems, the character of a lake is influenced by its surrounding landscape. Rain falling on land flows over the surface or moves through soil to enter springs, streams, and lakes, carrying with it silt and dissolved nutrients. Human activities including road construction, logging, mining, and farming add more silt and nutrients, especially P, N, and organic matter. These inputs cause nutrient enrichment (*eutrophication*) of lakes. Eutrophication can occur naturally in lakes, as a result of a slow accumulation of nutrients; when it results from human activities, it is called *cultural eutrophication*.

A typical **eutrophic** (high-nutrient) lake (Figure 24.6a) is shallow, with a high surface-to-volume ratio; that is, its surface area is large relative to its depth. Nutrient-rich farmland often surrounds it. An abundance of nutrients (particularly P) entering the lake stimulates growth of algae and aquatic plants. Increased primary productivity causes rapid recycling of nutrients and organic matter, stimulating even further growth via a positive feedback loop. Phytoplankton, including toxic cyanobacteria, concentrate in the warm upper waters.

Algae, inflowing organic debris and sediment, and remains of rooted plants drift to the bottom, where bacteria decompose them. Bacterial activities deplete the O_2 supply of the bottom sediments and deep water until these regions cannot support aerobic life. The number of benthic species drops, although the biomass and numbers of organisms remain high. In extreme cases, O_2 depletion causes massive die-offs of invertebrates and fish. So whereas the term *enrichment* sounds inherently positive, eutrophication sets in motion a string of negative consequences for lakes (see Ecological Issues: Saving Lake Winnipeg, pp. 532–533), Like forest dieback, it is another example of the ecological imbalances set in motion by human activities that cause too much of a good thing—nutrient inputs in both cases—upsetting the normal ecology of an ecosystem.

Oligotrophic lakes are nutrient poor. They have a low surface-to-volume ratio, so if they are large (such as Lake Superior), they are deep, with clear water that appears blue (Figure 24.6b) and in which P is limiting. Oligotrophic lakes are common in the Precambrian Shield, where exposed rock outcrops provide minimal nutrient inputs. Low nutrient levels reduce productivity, providing little input for decomposers and keeping O_2 levels in the hypolimnion high. The bottom sediments are largely inorganic.

(a)

(b)

Figure 24.6 Lake types. (a) Eutrophic lake. Note the floating mats. (b) Oligotrophic lake.

Although biomass in oligotrophic lakes and ponds is low, species diversity of fish and invertebrates is often high.

Oligotrophic lakes abound across the boreal forest, from Ontario's Muskoka Lakes to Saskatchewan's Lac la Ronge. Drawn to lakes for their beauty and recreational value, humans exert a heavy toll. Recreational impacts include nutrient loading from fertilizers and improperly maintained septic fields; pollutant inputs from boats and other motorized water vehicles, pesticides, and stains and paints; removal of natural vegetation along shorelines, increasing nutrient inputs; installation of "hard" shoreline protection devices such as walls and revetments that disturb longshore sand movement; introduction of non-native species such as zebra mussel (*Dreissena polymorpha*) and spiny crayfish (*Orconectes rusticus*); habitat loss or disturbance for many species including ducks, pelicans, eagles, and various fish (particularly spawning habitat affected by docks); alterations to behaviour of species such as loons and herons by water vehicles; alterations to fish community composition by fishing; and garbage dumping. The severity of these impacts varies with the lake and density of development.

Dystrophic lakes receive large amounts of organic matter in the form of humic materials that stain the water brown. These lakes occur on, or in contact with, peaty substrates in highly acidic bogs or heathlands (see Section 25.6). Dystrophic lakes typically have productive littoral vegetation but low open-water productivity, because little light penetrates the brown water.

24.4 STREAMS AND RIVERS: Lotic Systems Vary in Structure, Velocity, and Habitats

Even the largest rivers arise somewhere, either as outlets of ponds or lakes or as springs or seepage areas that become headwater streams. Some emerge fully formed from glaciers. As a stream drains from its source, it flows in a direction and manner dictated by topography and underlying rock formations. Joining the stream are other small streams, spring seeps, and surface water. Just below its source the stream may be small, straight, and swift, with waterfalls and rapids. Downstream, where the gradient is less steep, velocity decreases and the stream begins to meander, depositing sediment as silt, sand, or mud. At flood time, a stream drops its sediment load on surrounding level land, over which floodwaters spread to form floodplain deposits.

Where a stream flows into a lake or a river into the sea, its velocity is suddenly checked. The river then deposits its sediment load in a fan-shaped area about its mouth—a **delta** (Figure 24.7). Its course is carved into channels, which are blocked or opened with later deposits. The delta is a dynamic, ever-changing area of small lakes, swamps, and marshy islands. Material not deposited in the delta is carried out to open water and deposited on the bottom. By stabilizing a river's course, humans disrupt the natural dynamics of deltas. The Mississippi Delta is a prime example.

Streams become larger as they combine with others, allowing classification into orders (Figure 24.8). A small headwater stream with no tributaries is a *first-order stream*. Where two streams of the same order join, the stream increases in order.

Figure 24.7 River delta formed by sediment deposition.

Two first-order streams unite into a *second-order stream*, two second-order streams unite into a *third-order stream*, and so on. The order increases only when a stream of the same order joins it, not with entry of a lower-order stream. Typically, headwater streams are orders 1 to 3; medium-sized streams, 4 to 6; and rivers, over 6.

The velocity of a current moulds stream character and structure. The shape and steepness of the channel; the width, depth, and roughness of the bottom; rainfall intensity; and the rapidity of snowmelt all affect velocity. Fast streams have a velocity of over 50 cm/s. A current of this speed removes all particles under 5 mm in diameter, leaving a stony bottom. High water increases velocity; it moves stones and rubble, scours the bed, and cuts new banks and channels. As the gradient decreases and stream width, depth, and volume increase, silt and organic matter accumulate on the bottom. Velocity then slows (Figure 24.9), with an associated shift in species composition.

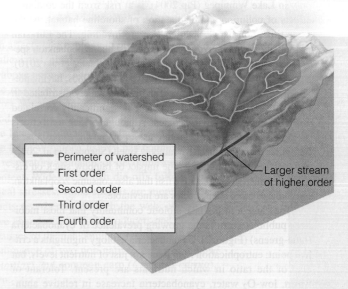

— Perimeter of watershed
— First order
— Second order
— Third order
— Fourth order

Larger stream of higher order

Figure 24.8 Stream orders within a watershed.

(a)

(b)

Figure 24.9 Stream velocity. **(a)** A fast mountain stream. The gradient is steep and the bottom is largely bedrock. **(b)** A slow stream is deeper, with a lower slope. Go to QUANTIFY it! at **www.pearsoncanada. ca/ecologyplace** to learn how ecologists quantify streamflow.

Current velocity or **streamflow**—the water discharge occurring within the natural streambed or channel—is the most ecologically important parameter of a stream or river. The rate at which a river or stream flows through its channel influences water temperature, O_2 content, nutrient spiralling (see Section 21.12), physical structure of the benthos, and types of inhabitants.

Lotic systems often alternate between two different but related habitats, distinguished by differences in velocity: turbulent *riffles (rapids)* and quiet *pools* (Figure 24.10). Processes occurring in riffles influence pools, and riffles are affected by events in upstream pools. Riffles are the major sites of primary production. Here the periphyton attached to or moving on submerged rocks and logs assume dominance. Periphyton, which are as important as phytoplankton in lakes and ponds, consist of diatoms, cyanobacteria, and water moss.

Above and below the riffles are the pools, which differ in chemistry, current, and depth. Just as riffles are the major sites of production, so pools are the major sites of decomposition.

Figure 24.10 Two stream habitats: a riffle (background) and a pool (foreground).

Current velocity slows enough for organic matter to settle. Pools, which release CO_2 via decomposer respiration, help maintain a supply of bicarbonate in solution. Without pools, photosynthesis in the riffles would deplete bicarbonates, resulting in smaller and smaller amounts of C downstream.

Humans disrupt the natural flow of water by constructing dams across rivers and streams (Figure 24.11), profoundly affecting river ecology. Dams damage lotic habitat. Under natural conditions, streams and rivers experience seasonal fluctuation in flow. Snowmelt and spring rains bring scouring high water; summer brings low water levels that expose some of the streambed and speed decomposition along the edges. Life has adapted to this seasonal flux. Dams interrupt both nutrient spiralling (see Section 21.12) and the river continuum. Downstream flow is greatly reduced as a pool of water fills behind the dam, developing features similar to those of a lake yet retaining some lotic traits, such as inflow. Fertilized by decaying material on the newly flooded land, the lake develops a heavy plankton bloom and, in tropical regions, dense growths of floating plants.

Figure 24.11 Dams greatly alter river ecology.

Species of fish, often lake-adapted exotics, replace fish that are adapted to flowing water.

The type of dam affects the type of pool that forms, as well as the resulting downstream conditions. During and just after floods, the river below a flood-control dam has a strong flow, scouring the riverbed. During normal periods, flow below the dam is stabilized, eventually returning to normal. If the dam is for water storage, drawdown is considerable in periods of water shortage. Large expanses of shoreline are exposed, stressing or killing littoral life. Downstream flow is minimal. With a series of such dams, the water volume moving downstream decreases with each dam, until eventually the river dries up. The Colorado River, the world's most regulated river, disappears entirely before reaching its mouth in the Gulf of California.

Hydroelectric and multipurpose dams hold varying volumes, depending on consumer needs. In peak periods, pulsed releases dislodge benthic life downstream. In Canada, dams associated with hydroelectric power have long been controversial. Large dams constructed on rivers such as the Manicougan

and La Grande in Québec and the Nelson and Churchill Rivers in Manitoba have not only altered the ecology of the flowing water system but also had damaging ecological effects on the lakes that are used as reservoirs (see Ecological Issues: Saving Lake Winnipeg, pp. 532–533).

24.5 FLOWING-WATER BIOTA: Many Species Have Adapted to Life in Flowing Water

Inhabitants of flowing water share the problem of remaining in place and not being swept downstream. They have evolved unique adaptations to solve this problem (Figure 24.12a). A streamlined form, which offers less resistance, typifies many animals in fast-water habitats. Larval forms of many insects have flattened bodies and broad, flat limbs that allow them to cling to the undersurfaces of stones, where the current is weaker. Caddisfly (Trichoptera) larvae construct protective

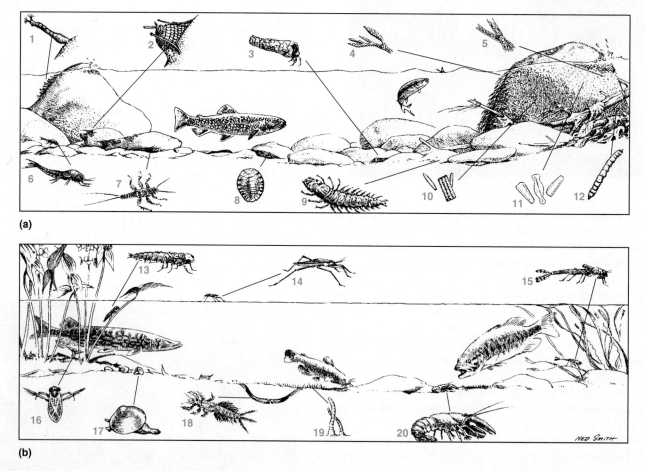

(a)

(b)

Figure 24.12 **(a)** Fast-stream biota: (1) blackfly larva (Simuliidae), (2) net-spinning caddisfly (*Hydropsyche*), (3) stone case of caddisfly, (4) water moss (*Fontinalis*), (5) algae (*Ulothrix*), (6) mayfly nymph (*Isonychia*), (7) stonefly nymph (*Perla* spp.), (8) water penny (*Psephenus*), (9) hellgrammite (Dobsonfly larva, *Corydalus cornutus*), (10) diatoms (*Diatoma*), (11) diatoms (*Gomphonema*), (12) crane fly larva (Tipulidae). **(b)** Slow-stream biota: (13) dragonfly nymph (Odonata, Anisoptera); (14) water strider (*Gerris*); (15) damselfly larva (Odonata, Zygoptera); (16) water boatman (Corixidae); (17) fingernail clam (*Sphaerium*); (18) burrowing mayfly nymph (*Hexagenia*); (19) bloodworm (Oligochaeta, *Tubifex*); (20) crayfish (*Cambarus*). The fish in the fast stream is a brook trout (*Salvelinus fontinalis*). The fish in the slow stream are, from left to right, northern pike (*Esox lucius*), bullhead (*Ameiurus melas*), and smallmouth bass (*Micropterus dolomieu*).

cases of sand or pebbles and cement them to stone bottoms. Sticky undersurfaces help snails and planarians cling tightly and move about on stones in the current. Among plants, water moss (*Fontinalis* spp.) and highly branched, filamentous algae cling to rocks with *holdfasts*. Other algae grow in cushion-like colonies or form slippery, gelatinous sheets that lie flat against the surfaces of stones and rocks.

Oxygen poses another problem. All animals living in fast-water streams require high, near-saturation levels of O_2 and moving water to keep their absorbing and respiratory surfaces in continuous contact with oxygenated water. Otherwise, a closely adhering film of O_2-depleted liquid forms a cloak about their bodies.

In slow-flowing streams, streamlined fish give way to species such as smallmouth bass, whose compressed bodies enable them to move through beds of aquatic vegetation. Pulmonate snails (Lymnaeacea) and burrowing mayflies (Ephemeroptera) replace rubble-dwelling insect larvae. Bottom-feeding fish, such as catfish (Akysidae), feed on life in the silty bottom, and backswimmers and water striders inhabit sluggish stretches and still backwaters (Figure 24.12b).

Invertebrate inhabitants are classed into four groups based on feeding habit (Figure 24.13): shredders, collectors, grazers, and gougers. *Shredders*, such as caddisflies (Trichoptera) and stoneflies (Plecoptera), make up a large group of insect larvae that fragment *coarse particulate organic matter* (CPOM; > 1 mm diameter), mostly leaves that fall into the stream. Shredders feed on CPOM less for the energy it contains than for the bacteria and fungi growing on it. They assimilate about 40 percent of the material they ingest, eliminating 60 percent as feces.

When broken up by shredders and partly decomposed by microbes, the leaves, along with invertebrate feces, become *fine particulate organic matter* (FPOM; between 0.45 μm and 1 mm diameter). Drifting downstream and settling on the bottom, FPOM is picked up by another group of insect larvae called *collectors*. *Filtering collectors* such as blackfly (Simuliidae) and net-spinning caddisfly larvae filter these particles from water flowing past them, whereas *gathering collectors*, such as

midge larvae, pick up particles from surfaces. Like shredders, collectors obtain much of their food from bacteria on detrital surfaces.

While shredders and collectors feed on detrital material, *grazers* feed on the periphyton coating stones and rubble. This group includes beetle larvae (water penny, *Psephenus* spp.) and caddisfly larvae. Much of the material they scrape loose enters the drift as FPOM. Another group, associated with woody debris, are **gougers**—invertebrates that burrow into water-logged limbs and trunks of fallen trees. Feeding on these detrital feeders and grazers are predaceous insect larvae and fish such as sculpin (*Cottus*) and trout. These predators rarely depend solely on aquatic insects; they also eat terrestrial invertebrates that wash into the stream.

Given the current, particulate organic matter and invertebrates are subject to **drift**: movement of material downstream, as a travelling benthos. Drift is a normal process in streams, even without high water or fast current, and may be subject to daily rhythms.

24.6 RIVER CONTINUUM: Flowing-Water Systems Constitute an Environmental Continuum

From its headwaters to its mouth, a flowing-water system constitutes a continuum of changing abiotic conditions (Figure 24.14, p. 538). Recall from Section 20.4 the transition from reliance on external (*allochthonous*) inputs to internal (*autochthonous*) energy production. Headwater streams are usually swift, cold, and found in shady forested areas. Primary productivity is low, and they depend on detrital input from streamside vegetation for almost all of their organic inputs. Even when headwater streams are exposed to sunlight and primary production exceeds inputs from adjacent ecosystems, most organic matter produced enters the detrital food chain. Dominant organisms are shredders, processing litter and feeding on CPOM, and collectors, feeding on FPOM. Grazer populations are minimal, given the low primary productivity. Predators are mostly small fish—sculpins, darters, and trout. Headwater streams, then, are accumulators, processors, and transporters of particulate organic matter of terrestrial origin. Their ratio of gross primary production to community respiration is less than 1.

In larger creeks, detrital inputs are less important. With more water surface exposed to sunlight, temperature increases, and as the slope declines, the current slows. These changes cause a shift from dependence on terrestrial inputs to dependence on inputs by algae and rooted plants. Gross primary production now exceeds community respiration. With less CPOM, shredders decline. Collectors, feeding on FPOM transported downstream, and grazers, feeding on autotrophs, are the major consumers. Predators increase little in biomass but shift from cold- to warm-water species, including bottom-feeders such as suckers (Catostomidae) and catfish.

As stream order increases from 6 through 10 and higher, river conditions develop. The channel is wider and deeper. Flow volume increases, current slows, and bottom sediments

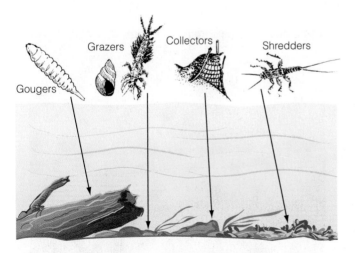

Figure 24.13 Major feeding groups in a stream: shredders, collectors, grazers, and gougers. Bacteria and fungi process leaves and other particulate organic matter.

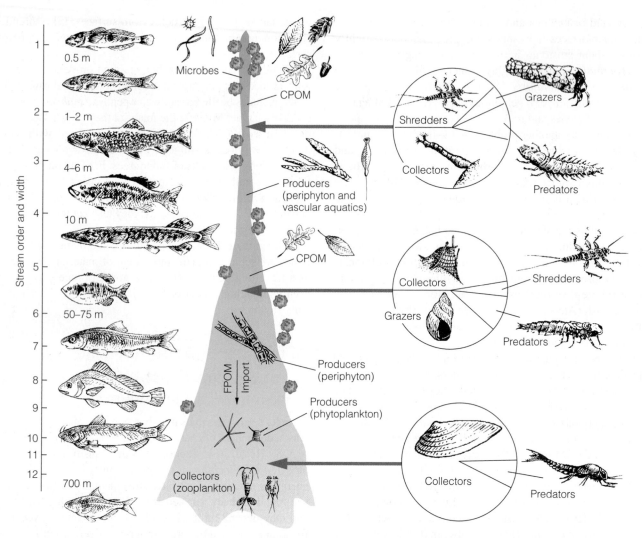

Figure 24.14 Changes in consumer groups along a river continuum. Stream order and width (m) are on the left axis. Headwater streams are heterotrophic, dependent on terrestrial organic inputs. Shredders and collectors are major consumers. As stream size increases, organic input shifts to primary production by algae and plants. Collectors and grazers dominate. In rivers, the system reverts to heterotrophy, although a small phytoplankton community may develop. Major consumers are gathering collectors. Fish shift as one moves downstream (from top to bottom: sculpin, darter, brook trout, smallmouth bass, pickerel, sunfish, sucker, freshwater drum, catfish, and shad).

accumulate. Both riverside and autotrophic production decrease. The main energy source is FPOM, used by the bottom-dwelling collectors that are the major consumers. Slower, deeper water and *dissolved organic matter* (DOM; < 0.45 μm) reduce phytoplankton and zooplankton populations. Along this downstream continuum, the community capitalizes on upstream feeding inefficiency. Downstream changes in production and abiotic factors are reflected in changes in consumers.

24.7 ESTUARIES: The Area Where Rivers Flow into Seas Is a Unique and Changeable Environment

Most streams and rivers eventually drain into the sea. The place where freshwater joins saltwater is an *estuary*: a semi-enclosed coastal area where seawater is diluted by and partially mixed with freshwater (Figure 24.15) (see Section 3.10). Here, the

Figure 24.15 An estuary on the east coast of Australia.

one-way flow of streams and rivers meets the alternating inflowing and outflowing tides. This confluence sets up a complex of currents that vary with the season, tidal oscillations, winds, and estuary size.

This mixing of waters of differing salinity and temperature creates a counterflow that acts as a sediment trap (see Figure 21.29). Inflowing river waters usually impoverish rather than fertilize the estuary, with the possible exception of P. Instead, nutrients and O_2 are carried into the estuary by the tides. If vertical mixing occurs, these nutrients are not swept back to sea but circulate vertically among organisms, water, and bottom sediments.

Organisms inhabiting estuaries share the problem with stream biota of maintaining their position, but face an additional issue—adjusting to changing salinity. Most estuarine life forms are benthic; they attach to the bottom, bury themselves in the mud, or occupy crevices. Mobile inhabitants are mostly crustaceans and fish, largely the young of species that spawn offshore in seawater. Plankton are at the mercy of the currents. Seaward movements and ebb tides transport them out to sea, with the rate of water movement determining the size of plankton populations.

Salinity dictates distribution of estuarine biota. Most estuary occupants are marine. Indeed, some cannot tolerate low salinity, declining along a salinity gradient from the ocean to the river's mouth. Sessile and slightly motile organisms have an optimum salinity range. When salinities vary outside this range, their populations decline.

Anadromous fish live most of their lives in saltwater, returning to freshwater to spawn. They are highly specialized to endure changes in salinity. Some fish species, such as the striped bass (*Morone saxatilis*), spawn near the interface of freshwater and low-salinity water, with the larvae and young fish moving downstream to more saline waters as they mature. Thus, for the striped bass, an estuary provides both a nursery and a feeding ground for the young. Anadromous species such as shad (*Alosa*) spawn in freshwater; the young fish spend their first summer in an estuary and then move out to open sea. The croaker (Sciaenidae) spawns at the mouth of the estuary, but the larvae are transported upstream to feed in plankton-rich, low-salinity areas.

The *oyster reef* is a unique estuarine community (Figure 24.16). Oysters may be attached to hard objects in the

Figure 24.17 Seagrass meadow in Chesapeake Bay dominated by eelgrass.

intertidal zone, or they may form reefs, where clusters grow cemented to the shells of past generations. Oyster reefs are usually at right angles to tidal currents, which bring in plankton, carry away wastes, and sweep the oysters clean of debris. Encrusting organisms such as sponges, barnacles, and bryozoans attach themselves to oyster shells. Oyster beds were among the communities most affected by the 2011 oil spill in the Gulf of Mexico.

In shallow estuarine waters, rooted aquatics such as widgeongrass (*Ruppia maritima*) and eelgrass (*Zostera marina*) dominate (Figure 24.17). These seagrasses support many epiphytic organisms. Seagrass communities are vital to vertebrate grazers such as geese, swans, and sea turtles, and provide nursery habitat for shrimp and bay scallops.

Estuaries face an added risk. Human settlement is often concentrated along coasts, and estuaries have long been favoured spots for habitation. Unfortunately, most coastal cities still dump many tons of untreated sewage into the ocean, causing cultural eutrophication of estuaries. Many countries are starting to address this issue, and some cities in Canada that used to dump raw sewage are now rectifying this situation. Victoria is a major coastal city that is continuing this ecologically untenable practice, although in July 2012 federal funds were pledged for a secondary treatment plant. Moreover, technical solutions can fail; Halifax's sewage treatment plant (opened only a few years before, in 2007) malfunctioned in 2009, causing raw sewage to be dumped into the Atlantic Ocean until it was fixed and reopened in 2010.

24.8 OPEN OCEAN: Oceanic Zones Differ in Depth and Physical Parameters

The marine environment differs from freshwater in several ways. It is large and contiguous, occupying 70 percent of Earth's surface, and deep, in places exceeding 10 km. The surface area lit by solar radiation is small in comparison with the total water volume. Its small volume of sunlit water and dilute nutrient solution limit marine primary production. Unlike freshwater systems, only some of which are interconnected, all of Earth's seas are interconnected by currents, influenced by wave action and tides, and characterized by salinity.

Figure 24.16 Oyster reef in an estuarine environment.

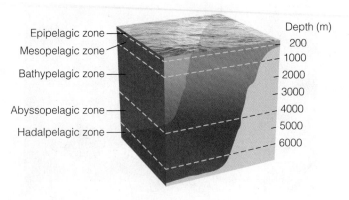

Figure 24.18 Major oceanic regions.

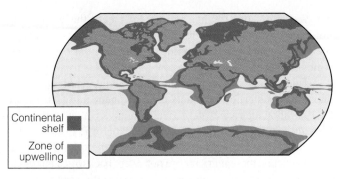

Figure 24.19 Location of continental shelf and upwellings.

(Adapted from Archibold 1995.)

Like most lakes, seas exhibit stratification and zonation. The ocean has two main divisions: the **pelagic zone**, or whole body of water, and the *benthic zone*, or bottom region. The pelagic zone contains two horizontal provinces: the *neritic province*, overlying the continental shelf, and the *oceanic province*. Because conditions change with depth, the pelagic zone is divided into vertical strata (Figure 24.18). From the surface to 200 m is the *epipelagic (photic) zone*, in which there are sharp gradients in illumination, temperature, and salinity. From 200 to 1000 m is the *mesopelagic zone*, where little light penetrates and temperature changes gradually, with little seasonal variation. This zone often contains the highest levels of nitrate and phosphate. Below the mesopelagic from 1000 to 4000 m is the *bathypelagic zone*, where darkness is virtually complete, except for bioluminescent organisms; temperature is low; and water pressure is great. The *abyssopelagic zone* extends from 4000 m to the seafloor. The deepest zone, the *hadalpelagic*, includes deep-sea trenches and canyons.

Differences in abiotic conditions in these zones affect productivity, which is limited to regions where there is enough light and nutrients for photosynthesis. Vertical light attenuation limits productivity to the epipelagic zone. However, the thermocline limits movement of nutrients to surface waters where light levels are adequate, especially in tropical seas, which have a permanent thermocline. The rate at which nutrients return to the surface (and hence increase productivity) is controlled by (1) seasonal breakdown of the thermocline and subsequent turnover and (2) nutrient-rich upwellings (see Section 21.12). Maximal productivity occurs in coastal waters (see Figure 20.10). There, the shallow waters of the continental shelf allow for turbulence and seasonal turnover (where it occurs) to increase vertical mixing, and coastal upwelling brings nutrient-rich water to the surface (Figure 24.19).

In open waters, productivity is low in tropical seas because the permanent thermocline slows upward diffusion of nutrients. In these regions, phytoplankton activity is controlled by nutrient cycling in the photic zone. Productivity remains relatively constant year-round (Figure 24.20a). The highest productivity in open tropical seas occurs where upwellings arise from divergence of surface currents (see Figure 3.16). As well as P, iron (Fe) often limits marine productivity.

Primary productivity in temperate oceans (Figure 24.20b) is strongly related to seasonal variation in nutrients, driven by changes in the thermocline.

Productivity is also low in the Arctic, in part because of low light. Much light is reflected as a result of the low solar angle, or absorbed by the snow-covered sea ice covering up to 60 percent of the Arctic Ocean in summer. There is very little thermocline, and a short biomass peak in summer (Figure 24.20b). Ongoing changes in the Arctic (especially reduced ice cover) will likely increase productivity, with unforeseeable effects on food webs (see Research in Ecology: Altered Matter Cycling in a Warmer Arctic, pp. 482–483). In contrast, Antarctic waters are more productive, given the continuous upwelling of nutrient-rich water around the continent. However, productivity is limited by the short summer.

24.9 OCEAN BIOTA: Life Varies in Different Oceanic Zones

The oceanic zones support different biota. When viewed from an airplane, the open sea appears uniform. Nowhere can one detect any strong pattern of life or well-defined communities, as is apparent on land. Why not? Pelagic systems lack a framework of large, conspicuous plant life. The dominant autotrophs are *phytoplankton*, and their major herbivores are *zooplankton*. Both terms are ecologically, not taxonomically, defined, and relate to small size and floating (or weakly swimming) habit.

Why are plankton so small? Surrounded by a medium that contains (in varying amounts) the nutrients necessary for life, they absorb nutrients directly from the water. The smaller the organism, the greater its surface-to-volume ratio and the more efficient its absorption of nutrients and/or light. In addition, seawater is so dense that there is little need for support structures. Because they require light, phytoplankton are restricted to the upper surface waters of the open ocean where light penetration varies from tens to hundreds of metres.

Each ocean or oceanic region has its own dominant phytoplankton. Littoral and neritic waters and upwellings are richer in plankton than is the mid-ocean. In upwelling regions, the dominant phytoplankton are *diatoms*. Enclosed in a silica case, diatoms are particularly abundant in cold Arctic waters. In downwelling regions, *dinoflagellates*—a large, diverse group characterized by two whip-like flagella—concentrate near the surface in areas of low turbulence, and are most abundant in warm waters. In summer, they concentrate in such numbers that they colour the surface water red or brown. These "red tides" are toxic to vertebrates.

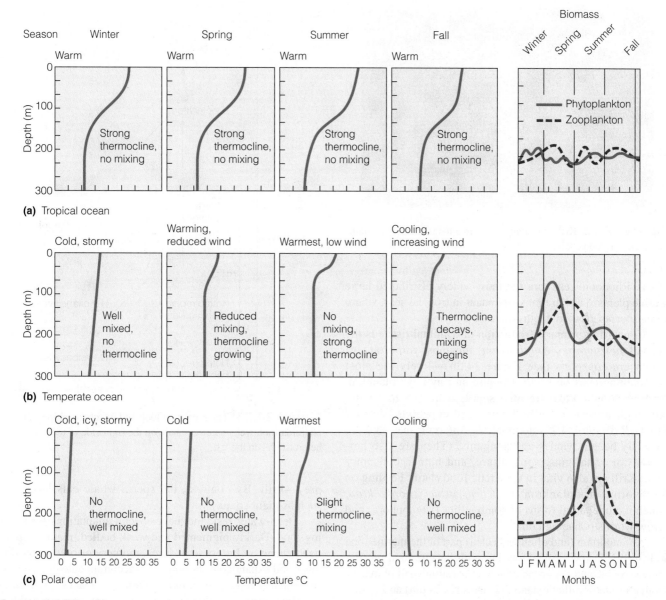

Figure 24.20 Thermal profiles, vertical mixing, and associated patterns of productivity in **(a)** tropical, **(b)** temperate, and **(c)** polar oceans during the four seasons.

(Adapted from Nybakken 1997.)

Nanoplankton, smaller than diatoms, dominate the epipelagic zone of open temperate and tropical seas. Most abundant are tiny *cyanobacteria*. Finally, *haptophytes*—a group of over 500 species of primarily unicellular algae—occupy all waters except polar seas. The most important members, *coccolithophores*, are major marine producers. They have an armoured appearance from the calcium carbonate platelets (*coccoliths*) covering their cell exterior (Figure 24.21).

In contrast to the dominance of phytoplankton in the open ocean, the dominant autotrophs in shallow coastal marine waters are attached *macroalgae*, restricted by light availability to a maximum depth of about 120 m. *Brown algae* (Phaeophyceae) are the most abundant, associated with rocky shorelines. Included in this group are large kelps such as *Macrocystis*, which grows to 50 m and forms dense subtidal forests in tropical and subtropical regions (see Figure 4.1a). *Red*

Figure 24.21 Coccolithophores have $CaCO_3$ platelets (coccoliths) covering their cells.

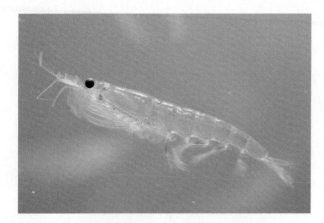

Figure 24.22 Krill are eaten by baleen whales and are essential to the marine food chain.

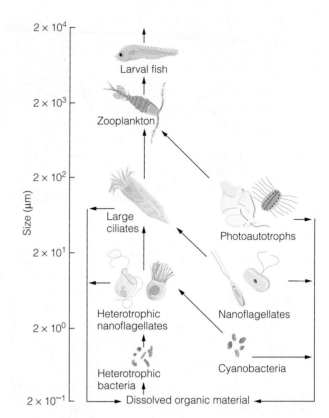

Figure 24.23 The microbial loop and its relationship to the plankton food web. Autotrophs are on the right side of the diagram, heterotrophs on the left.

algae (Rhodophyceae) are the most widely distributed large marine plants. They are most abundant in tropical seas, where some species grow to depths of 120 m.

Converting primary production into animal tissue is the task of *herbivorous zooplankton*, the most important of which are *copepods* (see Figure 24.5b)—likely the most abundant animals on Earth. Feeding on tiny phytoplankton, grazing zooplankton are also small, from 0.5 to 5 mm. Although most grazing herbivores are copepods, shrimp-like *krill* (Figure 24.22) dominate in Antarctic seas, and are eaten by baleen whales and penguins. (They are also harvested for aquariums, aquaculture, and human consumption.) Krill are also vital in the Arctic food chain. Feeding on herbivorous zooplankton are *carnivorous zooplankton*, which include larval forms of comb jellies (Ctenophora) and arrow worms (Chaetognatha).

An important (and often neglected) part of the marine food web begins not with plankton, but with bacteria and protists (heterotrophic and autotrophic), which constitute half of marine biomass. Photosynthetic nanoflagellates (2–20 μm) and cyanobacteria (1–2 μm), responsible for much marine photosynthesis, excrete a substantial amount of the dissolved organic material that heterotrophic bacteria consume. In turn, heterotrophic nanoflagellates consume heterotrophic bacteria. This **microbial loop** (Figure 24.23; see also Section 21.11), which adds several trophic levels to the oceanic food web, is functionally analogous to the *soil microbial loop* (see Section 21.9).

Like phytoplankton, most zooplankton move with the currents, but some have enough swimming power to exert some control. Some zooplankton migrate daily to preferred depths. At dusk, they rapidly rise to the surface to feed on phytoplankton, moving back down at dawn. Feeding on zooplankton are *nekton*, swimming organisms that can move at will. They range from small fish to large predatory sharks and whales, seals, and marine birds such as penguins. Some predatory fish, such as tuna, are largely restricted to the photic zone. Others occupy deeper zones or move between them, as does the sperm whale. Although the size ratio of predator to prey tends to fall within limits, some of the largest nekton—baleen whales—feed on disproportionately small prey—krill. By contrast, the sperm whale eats large prey such as giant squid.

Deep-water residents have special adaptations for securing food. Darkly pigmented and weak bodied, many deep-sea fish depend on luminescent lures, mimicry of prey, extensible jaws, and expandable abdomens (to accommodate large items). In the mesopelagic zone, *bioluminescence* reaches its greatest development—two-thirds of the resident species produce light. Some fish have rows of luminous organs along their sides and lighted lures that enable them to bait prey and recognize other individuals of the same species. Squid and euphausiid shrimp possess searchlight-like structures complete with lens and iris. Some discharge luminous clouds to escape predators. Nowhere is the remarkable adaptability of life more apparent than in deep-water oceanic habitats.

24.10 BENTHOS: The Ocean Floor Is a Unique Environment

The term *benthic* refers to the floor of the sea (or other water body), and *benthos* to organisms that live there. In a dark world, no photosynthesis occurs, so the benthos is heterotrophic, reliant on a rain of organic matter. Dead plankton and carcasses of marine mammals, invertebrates, birds, and fish provide an array of food items. Despite the darkness and depth, benthic habitats support a high species diversity. In shallow benthic regions, polychaete worms may exceed

Figure 24.24 A black smoker in full flow, some 2250 m down on the ocean floor west of Vancouver Island, on the Juan de Fuca Ridge.

250 species and pericarid crustaceans over 100. Deep-sea benthos has been reported to be even more diverse. A total of 798 species occurred in samples collected over a surface of 21 m² at depths of 1500 to 2100 m (Grassle and Maciolek 1992). However, the high diversity of the deep-sea benthos may be an area effect, and may not differ from that of coastal benthos (Gray et al. 1997).

Heterotrophic bacteria are vital members of the benthic food web. Common where organic matter accumulates, bacteria are most abundant in the topmost sediments. Bacteria synthesize protein from dissolved nutrients and are a source of protein, fat, and oils for other organisms.

Recent discoveries have revealed a fascinating oceanic community based on chemoautotrophic bacteria. In 1977, oceanographers first discovered high-temperature, deep-sea *hydrothermal vents* along volcanic ridges in the ocean floor of the Pacific near the Galápagos (Figure 24.24; see Research in Ecology: The Extreme Environment of Hydrothermal Vents, p. 46). These vents spew jets of superheated fluids that heat the surrounding water to 8°C to 16°C, considerably above the 2°C ambient water temperature. Since then, oceanographers have discovered similar vents on other volcanic ridges of the ocean floor in the mid-Atlantic and eastern Pacific. Vents form when cold seawater flows through cracks in the basaltic lava floor deep into the underlying crust. The waters react with the hot basalt, giving up some minerals but becoming enriched with others, such as copper, iron, sulphur, and zinc. The heated water re-emerges through mineralized chimneys rising up to 13 m above the sea floor. *White smokers* are rich in zinc sulphides and issue a milky fluid with a temperature of under 300°C. *Black smokers*, rich in copper sulphides, issue jets of clear water from 300°C to 450°C through chimneys blackened by precipitation of sulphur-mineral particles.

Associated with these vents is a rich and unique biota, confined to within a few metres of the vents. The primary producers are chemosynthetic bacteria that oxidize reduced S compounds such as H_2S to release energy used to form organic matter from CO_2. Consumers include giant clams, mussels, and polychaete worms that filter bacteria from water and graze on the bacterial film on rocks (see Figure 2, p. 47).

24.11 CORAL REEFS: Coral Reefs Are Colonial Ecosystems Inhabiting Localized Oceanic Habitats

In the warm, shallow waters around tropical islands and continental landmasses are *coral reefs*—colourful, rich oases within nutrient-poor seas (Figure 24.25). They are composed of dead skeletal material built up by carbonate-secreting organisms, mostly living coral (Cnidaria, Anthozoa), but also coralline red algae (Rhodophyta, Corallinaceae); green calcareous algae (*Halimeda*); foraminifera; and molluscs. Although various types of corals occur to depths of 6000 m, reef-building corals generally occur at depths less than 45 m because their symbiotic relationship with algae (zooxanthellae; see Figure 15.10) limits their distribution to depths where light is available for photosynthesis. Precipitation of Ca to form coral skeletons only occurs when water temperature and salinity are high and CO_2 content is low, limiting distribution of reef-building corals to shallow, warm tropical waters (20°C–28°C, but see Section 16.4 for a discussion of the role of herbivory in maintenance of temperate coral reefs).

There are three types of coral reefs: (1) *fringing reefs*, which grow seaward from rocky shores of islands and continents; (2) *barrier reefs*, which parallel shorelines, separated from land by shallow lagoons; and (3) *atolls*, rings of coral reefs and islands surrounding a lagoon. These lagoons, which form when a volcano subsides beneath the surface, are about 40 m deep. They usually connect to the open sea by breaks in the reef, and may have small islands of patch reefs.

As we saw in Section 16.3, the structural complexity of coral reefs provides heterogeneous habitat for a great diversity of associated organisms. This complexity derives from the corals themselves. Corals are modular animals—anemone-like polyps, with prey-capturing tentacles surrounding the opening or mouth. Most corals form sessile colonies on dead ancestors and cease growth when they reach the surface. In the tissues of the gastrodermal layer live zooxanthellae—symbiotic, photosynthetic, dinoflagellate algae. On the calcareous skeletons live other kinds

Figure 24.25 A rich diversity of coral species, algae, and colourful fish occupy this reef in Fiji in the South Pacific.

of algae (encrusting red and green coralline species and filamentous species, including turf algae) and many bacteria.

Coral reefs act as a sediment trap (see Section 21.14), making them highly productive oases (1500–5000 g $C/m^2/yr$) within the nutrient-poor, low-productivity sea (15–50 g $C/m^2/yr$). Their varied habitats support a great diversity of life, including thousands of invertebrates, such as sea urchins, that feed on coral and algae. Other coral invertebrates include molluscs, such as giant clams (*Tridacna, Hippopus*), and echinoderms, crustaceans, polychaetes, and sponges. Many herbivorous fish graze on algae, including the zooxanthellae within coral, and predatory fish consume invertebrate and vertebrate prey. Some, such as puffers (Tetraodontidae) and filefish (Monacanthidae), are corallivores that feed on polyps. Others lie in ambush for prey in coralline caverns. There are also many symbionts such as cleaning fish and crustaceans that pick parasites and detritus from larger fish and invertebrates.

Coral reefs are under intense pressure from human activities, including poisoning from chemicals used to capture reef fish, in high demand for aquariums; damage caused by divers and other forms of ecotourism; and acidification from increased CO_2 (see Section 22.4). Non-native lionfish (*Pterois* spp.) (Figure 24.26) have recently invaded the Caribbean and western North Atlantic (Schofield 2009). Equipped with poisonous spikes that deter other predators, lionfish may have severe long-term impacts on coral communities by increasing mortality of herbivorous species that keep algae in check (Albins and Hixon 2011). An additional stress

Figure 24.26 Invasion by lionfish is one of many stressors causing damage to coral reef communities in the Caribbean.

that is indirectly related to human activities is high temperature–induced bleaching, a phenomenon that is degrading coral reef communities worldwide (Baker et al. 2008). Given this litany of stressors, many of which interact, the very future of coral reefs may be in jeopardy (Kleypas et al. 2001). Yet there are some encouraging signs that coral reefs may be more resilient than once thought. Although ecologists had feared that the damage to coral reefs in the Caribbean was irreversible, Idjadi et al. (2006) reported significant recovery in coral reefs in Jamaica that had undergone a phase shift to a macroalgae-dominated system.

EcologyPlace

Visit EcologyPlace at www.pearsoncanada.ca/ecologyplace to access online resources that complement your textbook, and help you to apply and to review the information in this chapter. EcologyPlace includes

- an eText version of the book
- self-grading quizzes
- glossary flashcards
- and more!

Go to www.pearsoncanada.ca/ecologyplace and follow the registration instructions on the Student Access Code Card included with this text. If your book does not have a Student Access Code Card, you can purchase access to it at www.pearsoncanada.ca/ecologyplace.

SUMMARY

Lakes and Ponds 24.1

Lakes and ponds are bodies of water that fill depressions in the landscape. Many arise from glacial action, while others reflect geological processes or biological activities. On a geological timescale, lakes and ponds are temporary features. In time, most lakes fill, get smaller, and may be replaced by terrestrial systems. Lakes exhibit both vertical and horizontal gradients. Seasonal stratification in light, temperature, and dissolved gases influences the distribution of life.

Lake Biota 24.2

The littoral zone, in which light penetrates to the bottom, is occupied by floating and rooted plants, both emergent and submerged. The open-water (limnetic) zone is inhabited by phytoplankton, zooplankton, and fish. Below the depth of effective light penetration is the profundal region, where diversity varies with temperature and O_2. The bottom (benthic) zone is the major decomposition site. Anaerobic bacteria dominate the benthic zone beneath profundal water, whereas the benthos of the littoral is rich in invertebrate detritivores and decomposers.

Nutrient Inputs 24.3

Lakes are strongly influenced by the surrounding landscape. Lakes are eutrophic (nutrient rich), oligotrophic (nutrient poor), or dystrophic (acidic and rich in humus). Many lakes undergo cultural eutrophication, which is the rapid addition of nutrients (especially N and P) from sewage, industrial wastes, and agricultural runoff. Recreational use of lakes imposes multiple stresses.

Streams and Rivers 24.4

Flowing-water systems are characterized by the presence of current. They often depend on detrital inputs from surrounding terrestrial systems. Currents shape life in streams and rivers and carry nutrients and biota downstream. Lotic systems change in flow and size from headwater streams to rivers. They may be fast or slow, and may contain both riffles and pools.

Flowing-Water Biota 24.5

Organisms adapted to currents inhabit fast-water streams. They may be streamlined or flattened to conceal themselves in crevices and beneath rocks, or they may attach to substrates. Slow-flowing streams are inhabited by fish with compressed bodies that allow them to penetrate aquatic vegetation. Burrowing invertebrates inhabit the silty bottom. Four groups of stream invertebrates feed on detrital material: shredders, collectors, grazers, and gougers.

River Continuum 24.6

Life in flowing water reflects a continuum of changing conditions from headwater streams to the river mouth. Headwater streams depend heavily on detrital inputs. As stream size increases, algae and rooted plants become important producers, affecting the species composition of fish and invertebrates. Large rivers depend on particulate matter and dissolved organic matter for energy and nutrients. River life is dominated by filter feeders and bottom-feeding fish.

Estuaries 24.7

Rivers eventually reach the sea. The place where the inflowing freshwater meets the incoming and outgoing tides is an estuary. Intermingling of freshwater and saltwater creates a nutrient trap. Salinity determines estuarine life. Marine species decline as salinity drops from the estuary up into the river. Estuaries are nurseries for many marine species. The young are protected from predators and from competing species unable to tolerate lower salinity.

Open Ocean 24.8

Oceans are characterized by salinity, waves, tides, depth, and vastness. Like lakes, oceans exhibit stratification of temperature and light, which affect species distributions. The open sea has several vertical zones. The hadalpelagic zone includes deep-sea trenches. The abyssopelagic zone extends from 4000 m to the sea floor. Above is the bathypelagic zone, devoid of light and inhabited by darkly pigmented, bioluminescent animals. The next zone is the dimly lit mesopelagic, inhabited by species such as sharks and squid. Both the bathypelagic and mesopelagic zones depend on detrital inputs from the epipelagic zone, where light and nutrients allow photosynthesis. Coastal regions and upwellings are the most productive. The productivity of tropical oceans is low because the permanent thermocline slows upward diffusion of nutrients. The productivity of temperate oceans reflects seasonal variation in nutrients, driven by the thermocline. The Antarctic is more productive than the Arctic Ocean, as a result of nutrient upwellings.

Ocean Biota 24.9

Phytoplankton dominate surface waters. The littoral and neritic zones are richer in plankton than the open ocean. Tiny nanoplankton, which make up the largest biomass in temperate and tropical waters, are the major primary producers. Feeding on phytoplankton are herbivorous zooplankton, especially copepods. They are preyed upon by carnivorous zooplankton. The greatest diversity of zooplankton, including larval forms of fish, occurs over coastal shelves and upwellings. Larger life forms are free-swimming nekton, ranging from small fish to sharks and whales.

Benthos 24.10

Benthic organisms vary with depth and substrate and depend on organic matter drifting to the bottom. They include filter feeders, collectors, deposit feeders, and predators. Along volcanic ridges are hydrothermal vents inhabited by unique species of crabs, clams, and worms. Chemosynthetic bacteria that use sulphates as an energy source are the primary producers.

Coral Reefs 24.11

Coral reefs are nutrient-rich oases in nutrient-poor tropical waters. They are complex systems based on anthozoan coral and coralline algae, involving a symbiosis between coral polyps and photosynthetic zooxanthellae. New growth is built on the skeletons of dead organisms. These productive and varied habitats support a rich diversity of invertebrate and vertebrate life. Coral reefs are under great stress.

KEY TERMS

anadromous fish	drift	lentic	lotic	profundal zone
aufwuchs	dystrophic	limnetic (open-water)	microbial loop	streamflow
benthic (bottom) zone	emergent	zone	nekton	watershed
benthos	eutrophic	littoral (shallow-water)	oligotrophic	zooplankton
delta	gouger	zone	pelagic zone	

STUDY QUESTIONS

1. Distinguish the *littoral*, *limnetic*, and *profundal zones*.
2. What conditions distinguish the benthic zone from other lake zones? What is its major role?
3. Distinguish among *oligotrophy*, *eutrophy*, and *dystrophy*. Describe the ecological consequences of eutrophication of a formerly oligotrophic lake.
4. What characteristics are unique to flowing-water ecosystems? Contrast these conditions in a fast- and slow-flowing stream. How do organisms adapt to these differing conditions?
5. How do environmental conditions change along a river continuum?
6. What is the cause and ecological significance of changing salinity in an estuary?
7. What are the ecological advantages and disadvantages of reliance on hydroelectric power? (You will need to do additional research to answer this question fully.)
8. Characterize the major zones of the ocean, both vertical and horizontal.
9. How does temperature stratification differ in tropical and temperate seas? How do these differences affect primary productivity?
10. The open ocean is sometimes described as a marine desert. Should humans "fertilize" the sea to increase marine productivity, as a way of removing excess atmospheric CO_2?
11. What are hydrothermal vents, and what makes their biota unique?
12. Describe the formation of coral reefs. How are they affected by acidification of the oceans (see also Section 22.4)?

FURTHER READINGS

Allan, J. D. 1995. *Stream ecology: Structure and functioning of running waters*. Dordrecht: Kluwer Academic Press.
Essential reference on stream ecology.

Grassle, J. F. 1991. Deep-sea benthic diversity. *Bioscience* 41:464–469.
Excellent introduction to the strange and wonderful world of the deep ocean.

Gross, M. G., and E. Gross. 1995. *Oceanography: A view of Earth*. 7th ed. Englewood Cliffs, NJ: Prentice-Hall.
Valuable reference for the physical aspects of oceans.

Jackson, J. B. C. 1991. Adaptation and diversity of reef corals. *Bioscience* 41:475–482.
Important review relating species distribution in reefs to life history and disturbance.

Nybakken, J. W., and M. D. Bertness. 2005. *Marine biology: An ecological approach*. 6th ed. Menlo Park, CA: Benjamin Cummings.
Well-illustrated reference on marine life and ecosystems.

Wetzel, R. G. 2001. *Limnology: Lake and river ecosystems*. 3rd ed. San Diego: Academic Press.
Outstanding introduction to freshwater ecology.

COASTAL AND WETLAND ECOSYSTEMS

Eastern white cedar (*Thuja occidentalis*) and a diversity of floating and emergent aquatic plants and water-loving ferns dominate this protected cedar swamp in Kouchibouguac National Park, New Brunswick.

Wherever land and water meet, a transitional zone supports a variety of unique ecosystems. In coastal regions, this zone lies between terrestrial and marine habitats. These transitional areas are classified based on their geology and substrate—sediment type, size, and shape. Rocky shores, which result from marine erosion, have been altered the least. Sandy beaches, found in wave-dominated, depositional settings, are dynamic environments subject to continuous and often extreme change. Associated with estuaries or in protected regions of coastal dunes are tidal mudflats, salt marshes, and mangrove forests.

Between freshwater and land, transitional zones are occupied by **wetlands** dominated by specialized plants that occur where soils remain saturated for most or all of the year. These wetlands include marshes, swamps, bogs, fens, and riparian vegetation along rivers and lakes.

(a)

25.1 INTERTIDAL ZONE: Tidal Action Dominates the Transition between Land and Sea

Whether rocky, sandy, or muddy, and either protected or pounded by incoming waves, intertidal zones have one feature in common—they are alternately exposed and submerged by tides. Seashores are bounded by the limits of extreme high and low tide. Within these limits, conditions change hourly (see Section 3.9). At high tide, the seashore is a water world, at ebb tide, a land habitat. Uppermost areas experience temperature fluctuations, intense radiation, and desiccation for a prolonged time, whereas the lowest fringes are exposed only briefly before the tide submerges them again. Seashore dwellers are essentially marine organisms adapted to withstand some degree of exposure to air for varying periods. These conditions generate the most striking feature of coastal shorelines, whether rocky or sandy: zonation of life.

25.2 ROCKY SHORELINES: Rocky Coasts Exhibit Distinct Biotic Zones

Rocky shores have three zones (Figure 25.1), distinguished by their dominant organisms (Figure 25.2). The approach to a rocky shore involves a gradual transition from land plants and lichens to marine life dependent on tidal inundation. Moving from the terrestrial or **supralittoral (supratidal) zone**, the first major change occurs at the *supralittoral fringe*, where saltwater comes only once every two weeks on the spring tides. It is marked by the *black zone*, a thin black layer of cyanobacteria (e.g., *Calothrix* spp.) growing on the rock along with lichens (e.g., *Verrucaria* spp.) and green algae (Chlorophycophyta) above the high-tide line. Albeit living in conditions that few plants could survive, these organisms are semi-terrestrial. Grazing on wet algae on the rocks are peri-

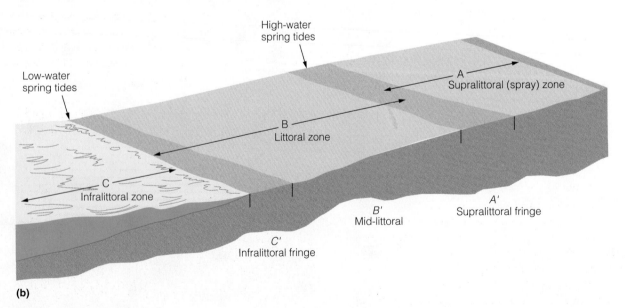

(b)

Figure 25.1 Rocky shorelines. **(a)** Rocky coast off Newfoundland. **(b)** Zonation of a rocky coast. See Figure 25.2 for associated biota.

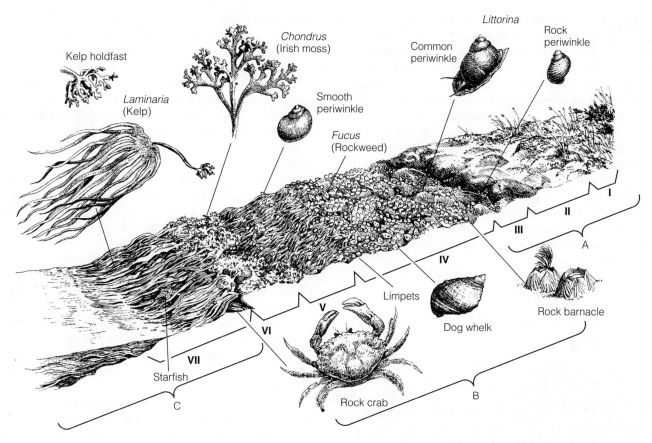

Figure 25.2 Zonation on a rocky shore on the North Atlantic. Compare with Figure 25.1b. I, land: lichens, herbs, grasses; II, bare rock; III, black algae and periwinkle snails (*Littorina*); IV, barnacle (*Balanus*) zone: barnacles, dog whelks, common periwinkles, mussels, limpets; V, fucoid zone: rockweed (*Fucus*) and smooth periwinkles; VI, Irish moss (*Chondrus*) zone; VII, kelp (*Laminaria*) zone.

winkle snails (*Littorina* spp.), from which the term *littoral* derives.

Below the black zone lies the **littoral (intertidal) zone**, which is covered and uncovered daily by the tides. In the upper reaches, barnacles (*Balanus* spp.) dominate. Oysters, blue mussels, and limpets appear in the mid- and lower littoral, along with periwinkle snails. Occupying the mid-littoral of colder climates, and in places overlying the barnacles, is brown algae or rockweed (*Fucus* spp.) and wrack (*Ascophyllum nodosum*). On hard-surfaced shores partly covered by sand and mud, blue mussels may replace brown algae. The lowest part of the littoral, uncovered only at the spring tides and not even then if wave action is strong, is the *infralittoral fringe*. This zone consists of kelp forests of large brown algae (*Laminaria* spp.) with a diverse undergrowth of smaller plants and animals among the kelp holdfasts. Below the infralittoral fringe is the **infralittoral (subtidal) zone**, which is fully marine.

Wave action greatly influences life on rocky shores. Waves bring in a steady supply of nutrients and carry away organic material. They move seaweed fronds in and out of sunlight, allowing even light distribution and more efficient photosynthesis. By dislodging plants and invertebrates from rocks, waves open up space for colonization and reduce interspecific competition by lowering population densities. Heavy wave

action can reduce the activity of predators such as starfish (sea stars) that feed on sessile intertidal invertebrates.

In addition to abiotic disturbance, biotic factors such as grazing, predation, and competition strongly influence rocky intertidal communities. In Section 16.4, we examined the role of keystone predation by starfish on the diversity of the invertebrate community in intertidal pools. These pools form when ebbing tides leave behind water in crevices and depressions (Figure 25.3), forming

Figure 25.3 Tidal pools fill depressions on rocky shores.

distinct habitats that differ not only from exposed rock and the open sea, but also among themselves. At low tide, the pools undergo sudden, wide fluxes in temperature and salinity, particularly if shallow. In summer, temperatures may exceed the tolerance limits of many species. As the water evaporates, salt crystallizes at the edges. When rain or drainage from adjacent land brings freshwater into the pools, salinity decreases. In deep pools, this freshwater influx forms a top layer, developing a salinity stratification in which the bottom layer and its inhabitants are little affected. If algal growth is considerable, O_2 will be high in daylight but low at night, a situation that rarely occurs at sea. Rising CO_2 at night lowers pH.

On the rising tide, most pools abruptly return to marine conditions, with large changes in salinity, temperature, and pH. Life in the tidal pools must be able to withstand these extreme fluctuations.

25.3 SANDY AND MUDDY SHORES: Sandy Beaches and Mudflats Are Harsh Habitats

In contrast to life-filled rocky shores, sandy and muddy shores often appear devoid of life at low tide (Figure 25.4). But sand and mud are not as barren as they seem. Beneath them lurks life, waiting for the next tide. Sandy shores are particularly harsh environments, given the relentless weathering of rock, both inland and along the shore. Rivers and waves carry the products of weathering, depositing sand along the margin. Sand-particle size influences the nature of the beach, water retention at low tide, and the ability of animals to burrow in it. On sheltered beaches, the slope may be very gradual. Outgoing tidal currents are slow, leaving behind organic residues and fine sediments that settle out. In these situations, mudflats develop.

Life on sand is almost impossible. Unlike rock, sand provides no surface for attachment of seaweeds and their

Figure 25.4 Blooming Point Beach, Prince Edward Island. Although sandy beaches appear barren, life is abundant beneath the sand.

associated fauna. The crabs, worms, and snails typical of rocky shores find no protection. So, apart from a minor presence of **epifauna** (organisms living on sediment surfaces), life is forced to live beneath the sand. Most **infauna** (organisms living in sediments) are able to burrow rapidly into the substrate, where they occupy permanent or semi-permanent tubes. Other infauna live between particles of sand and mud. Called **meiofauna**, these tiny (0.05 to 0.5 mm) organisms include copepods, ostracods, nematodes, and gastrotrichs.

Organisms living within sand and mud avoid the temperature flux typical of rocky shores. Although surface temperatures of sand may be 10°C or more higher than the returning seawater at midday, temperatures a few centimetres beneath are much less variable. Nor is there much salinity flux below 25 cm, even when freshwater runs over the surface.

Sandy beaches also exhibit tidal zonation (see Figure 16.20), revealed only by digging. Sand-coloured ghost crabs and sand fleas occupy the upper beach (*supralittoral*). The intertidal beach (*littoral*) is where true marine life appears. Although sandy shores lack the variety of rocky shores, populations of burrowing animals are often enormous. Invertebrates such as starfish and sand dollars live above the low-tide line in the littoral zone.

Just below the low-tide line live predatory gastropods, which prey on bivalves beneath the sand. Predatory portunid (swimming) crabs such as the blue crab (*Callinectes sapidus*) and green crab (*Carcinus maenas*; native to the Atlantic but an invasive species in other parts of the world) move back and forth with the tides, feeding on mole crabs, clams, and other organisms. Incoming tides also bring small predatory fish, including killifish and silversides. As the tide recedes, gulls and shorebirds scurry across the sand and mudflats to hunt for food.

The energy base for the sandy shore food web is accumulated organic matter. Detritus (seaweed, dead animals, and feces) is brought in by the tides and accumulates in sheltered areas. It is decomposed by bacteria, which are most active at low tide when more O_2 is available. Some detrital-feeders ingest organic matter largely to feed on the bacteria found on its surfaces. Examples are copepods; polychaete worms (*Nereis* spp.); molluscs; and lugworms (*Arenicola* spp.), which leave the coiled and cone-shaped casts that are commonly seen on beaches. Other fauna are filter-feeders, such as mole crabs (*Emerita* spp.) and coquina clams (*Donax* spp.). They obtain food by sorting organic particles from tidal water as it advances and retreats.

Erosion on sandy shorelines is an increasing problem in both marine and large lake systems such as the Great Lakes, exacerbated by rising sea levels with global climate change (see Section 28.8) and the use of large lakes as hydro reservoirs. Ironically, anthropogenic "protection" structures designed to prevent beach loss exacerbate the problem by altering the natural erosional and depositional equilibrium. These structures interfere with **longshore (littoral) drift**—transportation of sediments (particularly sand) along a coast at an angle to the shoreline.

25.4 SALT MARSHES: Tides and Salinity Dictate Salt Marsh Structure

Salt (tidal) marshes occur in temperate latitudes where coastlines are protected from wave action within estuaries and deltas, or by barrier islands and dunes (Figure 25.5). Salt marsh structure is dictated by tides and salinity, which create a sequence of distinct communities. From the sea edge to high land, vegetation zones develop, reflecting a microtopography that lifts plants to various heights within and above high tide (see Figure 16.23). On the seaward edge of salt marshes and along tidal creeks of North America's east coast is the *low marsh*, dominated by tall, dense growths of salt marsh cordgrass, *Spartina alterniflora*. Cordgrass forms a marginal strip between the open mud to the front and the high marsh behind. It is extremely tolerant of both salt and semi-submergence. To get air to its roots, which are buried in anaerobic mud, cordgrass (like most emergents) has air-filled *aerenchyma* (see Section 6.13).

Above the low marsh and closer to the upland is the *high marsh*. Tall salt marsh cordgrass abruptly gives way to a short, yellowish form of *S. alternifolia*—an example of phenotypic plasticity in response to abiotic conditions (see Section 5.12). The explanation is simple: water is shallower in the high marsh, so plants do not need to grow as tall to emerge above the water. The high marsh is also more saline and less nutrient rich than the low marsh, as a result of less tidal exchange. Here also grow translucent glassworts (*Salicornia* spp.; Figure 25.6); sea lavender (*Limonium carolinianum*); spearscale (*Atriplex patula*); and sea blite (*Suaeda maritima*).

At microelevations about 5 cm above high tide, short *Spartina* and its associates are replaced by salt meadow cordgrass (*S. patens*) and saltgrass (*Distichlis spicata*). At higher elevations, particularly with an influx of freshwater, *Spartina* and *Distichlis* may be replaced by black needlerush (*Juncus roemerianus* and *J. gerardii*). Beyond is shrubby growth of marsh elder (*Iva frutescens*) and groundsel (*Baccharis halimifolia*).

Figure 25.6 Glasswort dominates highly saline areas in salt marshes. The plant, which turns red in fall, is a major food source for overwintering geese.

On the upland fringe grow bayberry (*Myrica pensylvanica*) and sea holly (*Hibiscus palustris*).

Two notable features of salt marshes are salt pans and pools interspersed among meandering tidal creeks, which form an intricate system of channels carrying tidal waters back out to sea (Figure 25.7). Their exposed banks support dense growths

Figure 25.5 A coastal salt marsh.

Figure 25.7 A tidal creek at high tide. Tall *Spartina* grows along the banks.

of mud-dwelling algae, diatoms, and dinoflagellates. Salt pans are circular to elliptical depressions that are flooded at high tide. At low tide, they remain filled with saltwater. If they are shallow enough, the water evaporates, leaving a salt accumulation. These salt flats are invaded by glasswort and spikegrass.

Although highly productive (see Table 20.1), the salt marsh is not noted for its faunal diversity. Some animals are permanent residents of sand and mud, and others are seasonal visitors, but most are transients that feed on a tidal rhythm. Dominant invertebrates of the low marsh are ribbed mussels (*Modiolus demissus*), filter-feeders that lie half-buried in the mud, becoming active at high tide; fiddler crabs (*Uca* spp.), detrital-feeders that run across the marsh at low tide; and marsh periwinkle snails (*Littorina* spp.), which move up and down *Spartina* stems and onto the mud to feed on algae. Conspicuous vertebrates of the low marsh of eastern North America are the diamond-backed terrapin (*Malaclemys terrapin*) and birds such as the clapper rail (*Rallus longirostris*) and seaside sparrow (*Ammospiza maritima*).

In the high marsh, animal life changes as abruptly as the vegetation. The small, coffee-coloured pulmonate snail (*Melampus* spp.) is a dominant invertebrate, while the willet (*Catoptrophorus semipalmatus*) and seaside sharp-tailed sparrow (*Ammospiza caudacuta*) are common birds. With low tide, various predators enter the marsh to feed. Herons, egrets, gulls, terns, ibis, raccoons, and others spread over the exposed marsh bed and banks of tidal creeks. At high tide, the food web changes as tidal waters flood the marsh. Small predatory fish such as silversides (*Menidia menidia*), killifish (*Fundulus heteroclitus*), and four-spined stickleback (*Apeltes quadracus*), restricted to channel waters at low tide, enter the marsh, as does the blue crab.

25.5 MANGROVES: Mangrove Forests Replace Salt Marshes along Tropical Coasts

Replacing salt marshes in tropical regions are **mangrove forests** or **mangals** (Figure 25.8), which cover 60 to 75 percent of tropical coastal tidal flats. Mangals develop where wave action is minimal, sediments accumulate, and anoxic conditions prevail. They extend landward to the highest tidal range, where they may be only periodically flooded. The dominant plants are mangrove trees, including members of 12 genera in 8 families dominated by *Rhizophora*, *Avicennia*, *Bruguiera*, and *Sonneratia* spp., as well as other salt-tolerant species, mostly shrubs.

Mangroves range from short, prostrate forms to trees 30 m high. All mangroves have shallow, spreading roots. Many also have *prop roots* arising from trunks (Figure 25.9) and root extensions called *pneumatophores* (see Figure 6.30) that grow out of the water, absorbing O_2. The tangle of prop roots and pneumatophores slows tidal movements, allowing sediments to settle. Land moves seaward, followed by colonizing mangroves. Many mangrove species germinate on the parent tree, after which their seedlings fall into the water and take root.

Mangals support a unique mix of terrestrial and marine fauna. Nesting in the upper branches are birds, including herons

Figure 25.8 Mangrove forests replace salt marshes in tropical regions.

and egrets. As in salt marshes, *Littorina* snails are common, living on mangrove prop roots and trunks along with oysters and barnacles. Detritus-feeding snails occupy the bottom muds. Fiddler crabs burrow into the mud at low tide and live on prop roots and high ground during high tide. In the Indo-Malaysian mangrove forests live fish called mudskippers (*Periophthalmus* spp.), with modified eyes set high on the head (Figure 25.10). They occupy mud burrows and crawl about, acting more like amphibians than fish. The sheltered waters about the roots provide a nursery and refuge for the larvae and young of crabs, shrimp, and fish.

As well as their high productivity (see Table 20.1), mangroves perform another vital ecological function. They act as **bioshields**, buffering adjacent terrestrial systems from the damaging effects of *tsunami* (tidal waves). In recent times, humans have removed large stretches of mangroves in Central America and Southeast Asia, partly for recreational development and partly for shrimp farms. These losses have compromised this bioshield function, and the damage associated with tropical storms and tsunami has increased accordingly, as witnessed in

Figure 25.9 Interior of a red mangrove stand, showing prop roots.

Figure 25.10 Mudskippers are adapted to burrowing in the mud of mangrove forests.

(a)

(b)

Figure 25.11 The Okavango delta **(a)** as photographed from space. **(b)** The dry northwest corner of Botswana is the starting point for the annual summer floods of the Okavango River (also called the Kavanga), which spills down through a vast network of narrow waterways, lagoons, oxbow lakes, and floodplains covering some 22 000 km².

the 2004 event off the Indonesian coast. When we take a tropical holiday to escape the rigours of a cold winter, the resort we visit may have displaced a mangrove forest.

25.6 FRESHWATER WETLANDS: Marshes, Swamps, and Peatlands Are Important Freshwater Habitats

Wetlands occupy transition zones between freshwater and land, forming ecotones that share traits of both adjacent habitats. Freshwater wetlands cover 6 percent of Earth's surface and occur in every climate zone, but tend to be local in extent. Only a few are extensive, such as the Florida Everglades, the Pantanal in Brazil, the Okavango in Africa (Figure 25.11), and England's fens.

Wetlands range along a gradient from permanently flooded to periodically saturated soil (Figure 25.12). They support specialized **hydrophytes**—plants adapted to grow in water or in soil that is periodically anoxic from being saturated for most or all of the year. Hydrophytes include (1) *obligate wetland plants* that require saturated soils, such as submerged pondweeds, water lilies, cattails and other emergents, and trees such as bald cypress (*Taxodium distichum*); (2) *facultative wetland plants* that can grow in either saturated or upland soil but rarely grow in non-wetland sites, such as some sedges (Cyperaceae), alders (*Alnus* spp.), and cottonwood (*Populus deltoides*); and (3) *occasional wetland plants* that are usually found elsewhere but can tolerate wetlands. This third group is critical in indicating the upper limit of a wetland.

Freshwater wetlands typically occur in three topographic situations (Figure 25.13, p. 554), which create differences in the direction of water flow and hence of nutrients and sediments. (1) **Basin wetlands** develop in shallow basins, from upland depressions to filled-in lakes and ponds. Water flow is vertical as a result of precipitation and downward infiltration of water into the soil. (2) **Riverine wetlands** develop along shallow and periodically flooded banks of rivers and streams. Water flow is

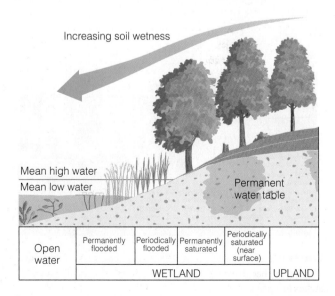

Figure 25.12 Location of wetlands along a soil-moisture gradient.

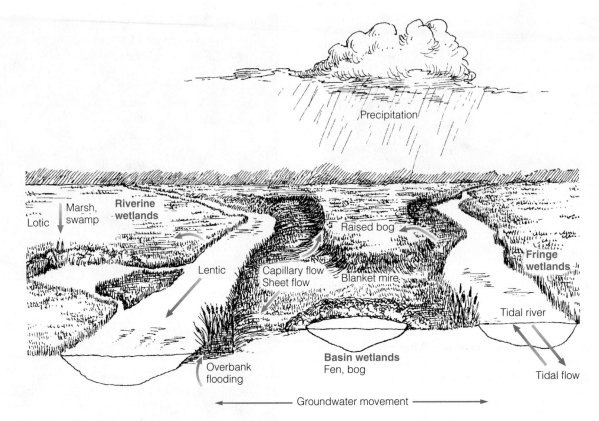

Figure 25.13 Water flow in various types of freshwater wetlands.

lateral and unidirectional. (3) **Fringe wetlands** occur along the coasts of large lakes. Water flows in two directions, as a result of rising lake levels and/or tidal action.

Whatever their topography, wetlands dominated by emergent vegetation, such as reeds, sedges, grasses, and cattails, are **marshes** (Figure 25.14). Forested wetlands are **swamps**. They may be deep-water swamps dominated by cedar (*Thuja occidentalis*; see p. 547), cypress (*Taxodium* spp.), or swamp oaks (*Quercus* spp.), or shrub swamps dominated by alder (*Alnus*

spp.) and willow (*Salix* spp.) (Figure 25.15). Along many large rivers are extensive **riparian woodlands** (also *gallery* or *bottomland forests*) (Figure 25.16). They are occasionally flooded but (unlike swamps) are dry for most of the growing season.

Peatlands (mires) are wetlands with large amounts of partly decayed organic matter. Organic matter accumulates because it is produced faster than it is decomposed. The water table is at or near the surface, generating anaerobic conditions that slow microbial activity. **Fens** are peatlands that are fed by groundwater

Figure 25.14 Northern freshwater marshes, such as this one in southern Québec, have well-developed emergent vegetation and patches of open water.

Figure 25.15 An alder (*Alnus*) shrub swamp with a herbaceous understory dominated by skunk cabbage (*Symplocarpus foetidus*).

Figure 25.16 A riparian forest along the Red River in southern Manitoba.

moving through mineral soil, from which they obtain most of their nutrients, and are dominated by sedges (*Carex* spp.). Compared with most peatlands, they are relatively nutrient rich and less acidic. Fens dominate large parts of southeast England.

Bogs are peatlands dependent largely on precipitation for water and nutrient inputs and dominated by peat mosses (*Sphagnum* spp.) (Figure 25.17). Supporting a canopy of black spruce, and hence often classed as forest, *Sphagnum* bogs are a dominant ecosystem across northern Canada. **Blanket mires (raised bogs)** (Figure 25.18) are peatlands that develop in uplands, where decomposed, compressed peat forms a barrier to downward water movement, causing a *perched water table* (saturated zone above an impermeable horizon). Also called *moors*, raised bogs are common in Great Britain and Scandinavia. Because bogs depend on precipitation for inputs, they are highly infertile and acidic.

Bogs also develop when a lake basin fills with sediments and organic matter from inflowing water. These sediments divert water around the basin, raising the surface of the bog above the influence of groundwater. Other bogs, called **quaking bogs**, form when a lake basin fills in from above rather than from below, creating a floating mat of peat over open water and supporting a typical sequence of vegetation from aquatic to terrestrial (Figure 25.19, p. 556).

In total, some 12 percent of Canada's land surface is covered by peatlands, which represent 90 percent of Canadian wetlands (Keys 1992). North America contains 40 percent of the world's peatlands. The distribution of Canadian peatlands parallels that of the northern boreal forest, which cuts a diagonal swath across the country (Figure 25.20, p. 556). Thus, peatlands dominate not only northern Ontario and Québec, making up 25 and 9 percent of land area, respectively, but also large portions of the prairies (Table 25.1, p. 557). Manitoba has the highest

(a)

(b)

Figure 25.17 Types of bogs. (**a**) A black spruce bog in Northwestern Ontario. (**b**) A moor (raised bog) in Yorkshire, England.

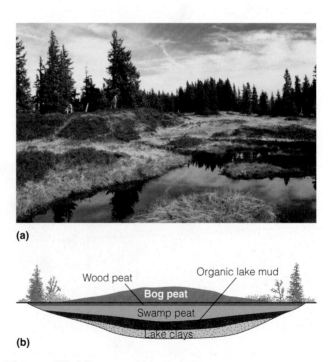

(a)

(b)

Figure 25.18 A raised bog (**a**) develops when an accumulation of peat rises above the surrounding landscape (**b**).

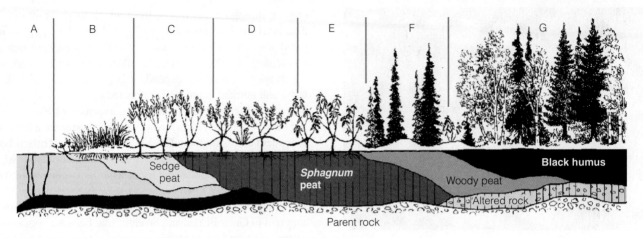

Figure 25.19 Transect through a quaking bog, showing vegetation zones and substrates. A, water lily in open water; B, buckbean (*Menyanthes trifoliata*) and sedge (*Carex*); C, sweetgale (*Myrica gale*); D, leatherleaf (*Chamaedaphne calyculata*); E, Labrador tea (*Rhododendron groenlandicum*); F, black spruce; G, birch–black spruce–balsam fir forest. Alternative vegetation sequences dominated by deciduous species can also occur.

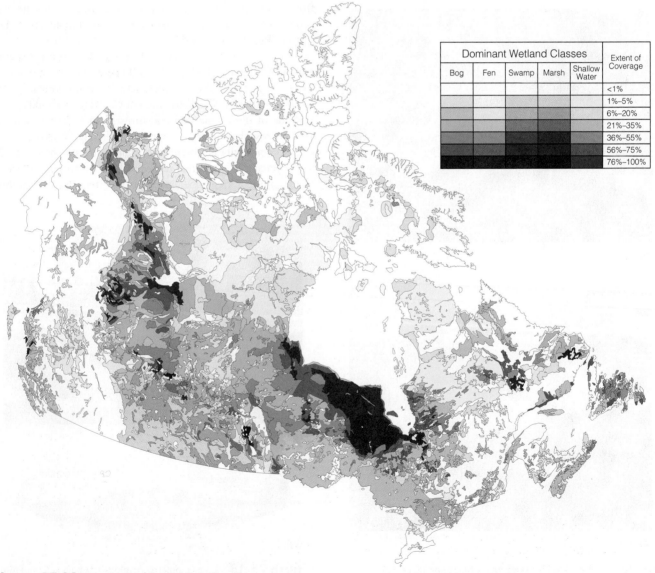

Figure 25.20 Distribution of wetlands in Canada.

(Adapted from Tarnocai and Lacelle 2001.)

Table 25.1 Extent of Peatlands and Other Wetlands Across Canada

Province/Territory	Peatland (ha × 1000)	% Land Area	Total Wetland (ha × 1000)	% Land Area
Alberta	12 673	20	13 704	21
British Columbia	1 289	1	3 120	3
Manitoba	20 664	38	22 470	41
New Brunswick	120	2	544	8
Newfoundland and Labrador	6 429	17	6 792	18
Northwest Territories	25 111	8	27 794	9
Nova Scotia	158	3	177	3
Ontario	22 555	25	29 241	33
Prince Edward Island	8	1	9	1
Québec	11 713	9	12 151	9
Saskatchewan	9 309	16	9 687	17
Yukon	1 298	3	1 510	3
Canada	111 327	12	127 199	14

Total wetlands includes peatlands.
(Source: National Wetland Working Group, 1988.)

proportional land area covered by peatlands (38 percent), while the Northwest Territories has the largest absolute area.

Peatlands typically form under oligotrophic or dystrophic conditions (see Section 24.3). Most abundant in boreal regions of the Northern Hemisphere, peatlands also develop in tropical and subtropical regions, particularly in mountainous and coastal areas where hydrological conditions encourage accumulation of organic matter. Examples include the Florida Everglades, which includes many wetland types, such as swamps, marshes, and peatlands, and the pocosins (evergreen shrub bogs) on the coastal plains of the southeastern United States.

25.7 WETLAND HYDROLOGY: Water Movement and Flooding Define Wetland Structure

Wetland structure is determined by **hydrology**: the properties, distribution, and effects of water on Earth's surface. Hydrology has two components, both of which vary among wetlands: (1) physical aspects, particularly with respect to water movement (precipitation, surface and subsurface flow, flow direction and kinetic energy, and water chemistry), and (2) **hydroperiod**—duration, frequency, depth, and season of flooding.

Basin wetlands typically have a long hydroperiod. They flood during wet periods and draw down during dry periods. Both phenomena are essential to the wetland's long-term persistence. To control flooding in nearby agricultural areas, humans often construct diversion structures that reduce flooding cycles in basin wetlands, threatening their viability (see Ecological Issues: The Continuing Decline of Wetlands, pp. 558–559). In contrast, riverine wetlands have shorter floods associated with peak stream flow, usually in spring following snowmelt. Floodways that protect urban areas from seasonal flooding (such as Winnipeg's Red River Floodway) reduce silt deposition on bottomlands, altering fertility and species composition.

The hydroperiod of fringe wetlands is influenced by wind and lake waves, and—if unaltered—is typically short and lacking the seasonal flux common in basin and riverine marshes. However, humans often maintain lake levels at artificially high levels to maximize associated river flow for hydroelectric power generation and/or navigation. Levels of Lake Ontario, which empties into the St. Lawrence River, have been regulated since 1960 to ensure adequate flow for power plants in both Canada (at Cornwall) and the U.S. (in upstate New York). Prolonged high water (along with other factors including invasive species) has had substantive negative effects on Lake Ontario's Cootes Paradise Marsh (Chow-Fraser 2005; see Section 25.8).

Hydroperiod influences plant composition by affecting germination, survival, and mortality at various life cycle stages. The impact is perhaps most pronounced in **potholes**—basin wetlands in prairies. In potholes large and deep enough to have standing water during drought, submergents dominate (Figure 25.21). If the pothole goes dry annually or during

Figure 25.21 Prairie potholes.

ECOLOGICAL ISSUES   The Continuing Decline of Wetlands

For centuries, we have considered wetlands as forbidding, mysterious places. Aesthetically unpleasing to many, they are often associated with pestilence-causing insects and sinister creatures that arise from swamps. When Sir Arthur Conan Doyle wrote *The Hound of the Baskervilles*, he was drawing on this persistent human fear, as does the movie *Cape Fear*. Too often they are dismissed as wastelands that should be drained for uses deemed more important: agriculture, dumps, housing, industrial developments, and roads. The Romans drained the great marshes around the Tiber to accommodate Rome. Similarly, the very name of the Great Dismal Swamp on the Virginia–North Carolina border reveals the negative attitudes of the early settlers.

Rationales for draining wetlands are many. The most pervasive involve agriculture. Draining wetlands opens large expanses of rich organic soil for crop production. In the prairies, the innumerable potholes are considered an impediment to efficient farming. Draining them tidies up fields and allows unhindered use of large machinery. Landowners and local governments view wetlands as a liability that produces no economic return and generates little tax revenue. Many regard the wildlife that wetlands support as threats to grain crops. Typically, wetlands are considered without value, at best filled in and used for development. Some wetlands have been in the way of dam projects. Peat bogs in the northern United States, Canada, Ireland, and northern Europe are excavated for fuel and peat. Such exploitation threatens to wipe out peatlands in many areas.

Residual wetlands in agricultural areas are contaminated by materials carried by surface and subsurface drainage and sediments from surrounding croplands. Although inputs of N and P increase marsh productivity, pesticides and heavy metals poison the water; destroy invertebrate life; and have debilitating effects on vertebrates, including deformities, reduced reproduction, and death. Waterfowl in scattered wetlands are more exposed to predation and breed less successfully without access to upland vegetation. The impact of declining wetlands on adjacent systems takes time to surface and is often ignored. The extent and duration of flooding of rivers and lakes, for example, greatly increase when marshes and potholes are drained and filled. With loss of marshes around lakes, the abil-

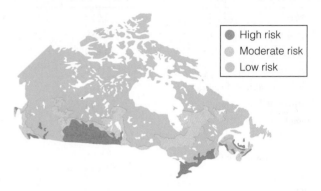

Figure 1 Wetlands lost or at risk in Canada.

(Source: Natural Resources Canada.)

ity of the marsh to filter contaminants and nutrients is lost, accelerating eutrophication (see Ecological Issues: Saving Lake Winnipeg, pp. 532–533). Water level regulation, urban encroachment, and invasion by alien species also degrade freshwater wetlands.

Population pressures pose a particular threat to coastal wetlands. By 2001, some 38 percent of Canadians lived within 20 km of a coast, in only 2.6 percent of Canada's land area (Manson 2005). The trend is continuing, and, worldwide, 70 percent of people live within 80 km of a coastline. With so much humanity clustered near coasts, it is obvious why coastal wetlands are disappearing so rapidly. Despite efforts through regulation and acquisition to slow the loss, coastal wetlands in the United States, for example, are disappearing at a rate of 8000 ha yearly. Coastal salt marshes are ditched, drained, and filled for real estate, industrial development, and agriculture. Reclamation for agriculture is most extensive in Europe, where the high marsh is enclosed within sea walls and drained. Most of Holland's marshes and tidelands have been reclaimed in this way. Coastal cities such as London; Boston; and Sackville, New Brunswick, occupy filled-in marshes. Of surviving salt marshes, those close to urban and industrial developments may become polluted with spillages of oil, which is easily trapped within the vegetation.

drought, tall or mid-height emergents such as cattails dominate. If the pothole is shallow and flooded only briefly in spring, grasses, sedges, and forbs form a wet-meadow community. If the basin is sufficiently deep and large towards its centre, vegetation zones may develop. Those areas with a long hydroperiod support submerged species and deep-water emergents; those with a short hydroperiod support shallow-water emergents and wet-meadow species. In turn, pothole hydrology affects their ability to support wildlife. Populations of migratory ducks and geese depend on these often overlooked wetlands. Along with marshes, potholes reduce flooding on riverine and fringe wetlands, thereby affecting their hydrology.

Alternating periods of drought and high water induce vegetation cycles in freshwater marshes (see Figure 17.20). Periods of above-normal rainfall drown emergents, creating a lake marsh dominated by submerged plants. During drought, the marsh bottom is exposed, stimulating germination of emergents and mudflat annuals. When water levels rise, mudflat species drown, while emergents spread vegetatively, joined by floating species such as duckweed. If the water remains high, the marsh eventually re-enters the lake marsh phase (van der Valk 1981).

Peatland hydrology differs from that of other freshwater wetlands. Acid-forming *Sphagnum* mosses accumulate new tissue on the remains of past growth. Their sponge-like properties increase water retention. As the peat blanket thickens, the

Loss of coastal wetlands damages associated estuaries, which are a nursery ground for commercial and recreational fisheries as well as for natural communities. There is a positive correlation between the expanse of coastal marsh and shrimp production in coastal waters. Decline of oysters and blue crabs also relates to loss of salt marshes. Coastal marshes are also major wintering grounds for waterfowl. One-half of the migratory waterfowl that follow the Mississippi Flyway to nesting sites in Canada depend on Gulf Coast wetlands, and the bulk of the Arctic snow goose population winters on coastal marshes from Chesapeake Bay to North Carolina. Through grazing or uprooting, these geese may remove nearly 60 percent of the belowground production of marshes. Ironically, the forced concentration of these migratory birds into shrinking habitats jeopardizes marsh vegetation and the future of remaining salt marshes.

Just how serious is the rate and extent of wetland loss and degradation across North America? Accurate figures are hard to obtain, particularly for Mexico, where survey data is less complete. However, despite the difficulties of inventorying remote areas and agreeing on criteria for classifying wetlands, scientists estimate that losses have been most severe in the United States, which has lost some 30 percent of its total wetland area, from 160 million ha in colonial times to 110 million ha by the 1980s (Davidson et al. 1999). These values include Alaska; losses in the rest of the country approached 50 percent by the 1980s (Davidson et al. 1999), and greatly surpassed that level by 2010. Many of the remnants are badly degraded, and losses vary greatly among regions. Less populated states such as New Hampshire have lost a mere 9 percent, versus 91 percent in California (Davidson et al. 1999).

Although the total wetland area is small (3.3 million ha), Mexican wetlands are vital for many migratory birds. Losses are estimated at 35 percent since the 1800s (Davidson et al. 1999). Canada has fared slightly better, having lost 14 percent of its total wetlands, from 148 million to 127 million ha (NWGG 1988). Davidson et al. (1999) peg the Canadian total at 150 million ha. Whatever the absolute area, wetland losses are disproportionate, with densely populated regions of Ontario and Québec and the Fraser River valley most affected (Figure 1). Conversion rates in Ontario exceed 70 percent. Wetland loss and degradation are especially severe along Lakes Erie and Ontario, with the Point Pelee marsh the largest marsh (~ 10 km^2) left in the Great Lakes system. Conversion of prairie potholes has also been extensive.

Elsewhere in the world, wetlands have met a worse fate. Coastal Europe has lost 65 percent of its original tidal marshes, and 75 percent of the remaining are intensively managed. Since the 1980s, 35 percent of tropical mangroves have been diked for aquaculture (pond-rearing of shrimp and fish) and cut for wood and charcoal. Loss of wetlands worldwide has reached a point where the ecosystem services they provide—waterfowl habitat, groundwater supply and quality, floodwater storage, and sediment trapping—are in serious jeopardy. Although some countries have made some progress towards preserving their remaining wetlands through legislative action and land purchase, the future of freshwater and coastal wetlands is far from secure. Apathy or hostility towards wetland preservation, political manoeuvring, court decisions, and disputes over what constitutes a wetland allow wetland destruction to continue at an alarming pace.

Bibliography

Davidson, I., R. Vanderkam, and M. Padilla. 1999. Review of wetland inventory information in North America. In Finlayson, C. M., and Spiers, A. G. (eds.). *Global Review of Wetland Resources and Priorities for Wetland Inventory*. Ottawa: Wetlands International.

Manson, G. K. 2005. On the coastal populations of Canada and the world. *Canadian Coastal Conference 2005*:1–11.

National Wetland Working Group (NWWG). 1988. *Wetlands of Canada*. Environment Canada and Polyscience Publications.

1. Using the Internet and other sources, investigate the major wetland areas that are under threat in your province or territory. What ecological services do these wetlands provide?

2. Conservation efforts for non-wetland ecosystems often elicit a more positive public response based on their aesthetic appeal. What strategies can you suggest to promote a greater understanding of the values of preserving wetlands?

---

saturated moss mat is raised above, and insulated from, the mineral soil. Thus, although precipitation is the main source of water and nutrients, bogs are less sensitive to hydroperiod and do not exhibit the cyclic succession typical of marshes.

25.8 WETLAND BIODIVERSITY: Freshwater Wetlands Support a Rich Biodiversity

Among the most productive of all the world's ecosystems (see Table 20.1), freshwater wetlands support a diverse community of benthic, limnetic, and littoral invertebrates, especially crustaceans and insects. These invertebrates, along with small fishes, provide a food base for waterfowl, herons, and gulls. Many passerine (perching) birds, such as the red-winged blackbird, also rely on wetland habitat. Wetlands supply the fat-rich nutrients ducks need for egg production and growth of their young. Frogs, toads, salamanders, snakes, and turtles inhabit the emergent growth, soft mud, and open water of marshes and swamps.

The dominant vertebrate herbivore in prairie marshes is the muskrat (*Ondatra zibethicus*). Its numbers rise during the regenerating marsh phase—so much so that its "eat-outs" contribute to the decline of emergent vegetation during the subsequent degenerating phase. Muskrats are the major prey of

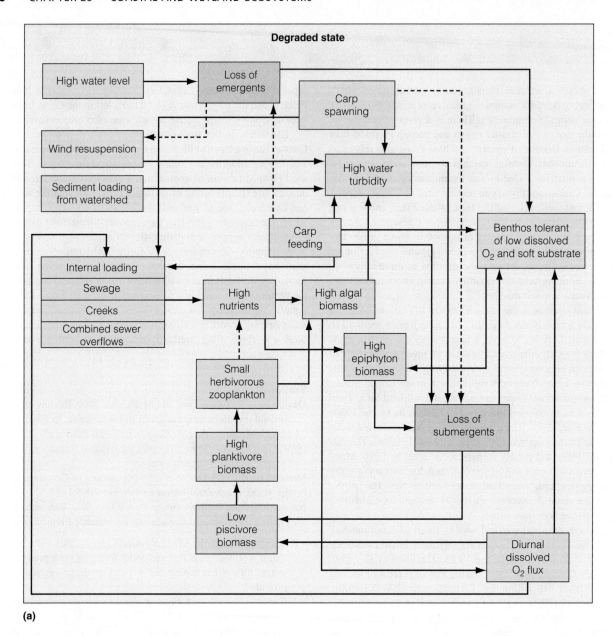

(a)

Figure 25.22 Model of interacting factors affecting functioning of Cootes Paradise Marsh, Lake Ontario as **(a)** a degraded and **(b)** a healthy ecosystem (see following page).

(Adapted from Chow-Fraser 1998.)

mink (*Mustela vison*), the dominant marsh carnivore. Other predators, including raccoon, fox, weasel, and skunk, can seriously reduce the reproductive success of waterfowl on small marshes. In boreal peatlands, moose are the most conspicuous vertebrate herbivores.

Freshwater marshes support a diverse biota, and, in turn, a healthy biotic community helps maintain the health of wetlands. As in all ecosystems, wetland stability and function depend on a dynamic interplay between abiotic and biotic factors. Chow-Fraser (1998) identifies 17 interacting components that have contributed to the degraded state of Cootes Paradise Marsh, a coastal marsh on Lake Ontario. Although the factor

triggering the degradation was abiotic—high water levels in the 1940s and 1950s that drowned emergent vegetation—subsequent factors include many biotic influences (Figure 25.22a) (Chow-Fraser 1998).

Without emergent growth to trap sediments, the marsh became turbulent and windswept, which caused a decline in submerged vegetation. Excessive nutrient loading from sewage and surface runoff stimulated algal growth, which further maintained the high turbidity, as did feeding and spawning activities of an invasive species, the common carp (*Cyprinus carpio*). In its present unvegetated state, the substrate no longer supports a diverse assemblage of aquatic insect larvae.

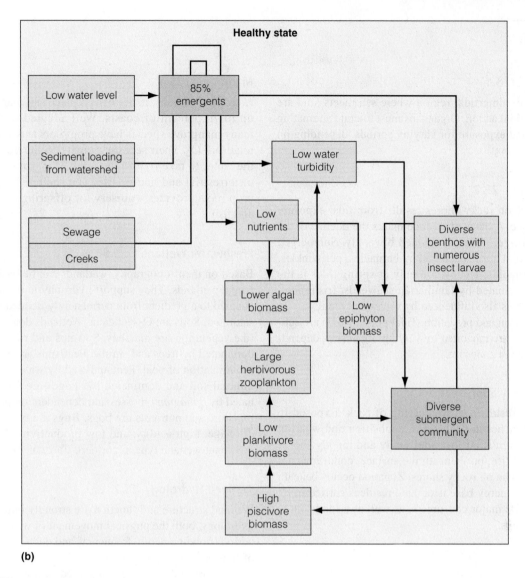

Healthy state

(b)

Figure 25.22 (*continued*)

Predation by carp has reduced benthic grazers, allowing epiphytic algae to flourish. Piscivores such as pike are inhibited by the fluctuating O_2 levels resulting from eutrophication. Planktivores dominate, virtually eliminating large herbivorous zooplankton such as *Daphnia*. Moreover, this degraded system appears to be an alternative stable state (see Section 19.9) and will only revert to a healthy state (Figure 25.22b) if restoration efforts now underway (including promotion of emergent and submerged plants, control of nutrient loading, and carp exclusion) are successful.

EcologyPlace

Visit EcologyPlace at www.pearsoncanada.ca/ecologyplace to access online resources that complement your textbook, and help you to apply and to review the information in this chapter. EcologyPlace includes

- an eText version of the book
- self-grading quizzes
- glossary flashcards
- and more!

Go to www.pearsoncanada.ca/ecologyplace and follow the registration instructions on the Student Access Code Card included with this text. If your book does not have a Student Access Code Card, you can purchase access to it at www.pearsoncanada.ca/ecologyplace.

SUMMARY

Intertidal Zone 25.1

Conditions in the intertidal region where sea meets land are determined by tidal action. Organisms must tolerate alternating submergence and exposure for varying periods, depending on the zone they occupy.

Rocky Shorelines 25.2

Zonation of life on rocky shores results from tidal exposure and submergence. A black algal zone marks the tideline of the supralittoral fringe, which is flooded biweekly. Submerged daily is the littoral zone, occupied by barnacles, periwinkles, mussels, and fucoids. Uncovered only at spring tides is the infralittoral, dominated by laminarian seaweeds, Irish moss, and starfish. Life is also influenced by waves and biotic factors such as competition and predation. Tidal pools, which are subject to daily flux, are inhabited by varying organisms, depending on the degree of exposure.

Sandy and Muddy Shores 25.3

Sandy beaches result from weathering of rock. Exposed to wave action, beaches are subject to deposition and wearing away of the substrate. At low tide, sandy and muddy shores appear barren of life, but beneath the surface, conditions are more amenable than on rocky shores. Zonation occurs beneath the surface. The energy base is organic residues carried in by tides. Bacteria are major consumers, as well as a food source for other organisms.

Salt Marshes 25.4

Interactions of salinity, tidal flow, and height produce the distinctive zonation of salt marshes. Salt marsh cordgrass dominates marshes flooded by daily tides. Higher elevations that are flooded only by spring tides and subject to higher salinity support salt marsh cordgrass and spikegrass. Salt marsh animals are adapted to tidal rhythms. Detrital feeders such as fiddler crabs and their predators are active at low tide, while filter-feeding mussels are active at high tide.

Mangroves 25.5

In tropical regions, mangroves replace salt marshes, covering up to 70 percent of coasts. Well adapted to tidal habitats, many mangrove species have prop roots and pneumatophores to absorb O_2. Their seeds germinate on the tree and drop into the water to take root in the mud. Mangroves support a mix of terrestrial and marine life. The sheltered water about the prop roots provides a nursery for offspring of crabs, shrimp, and fish.

Freshwater Wetlands 25.6

Based on their topography, wetlands can be basin, riverine, or fringe wetlands. They support hydrophytic plant communities adapted to a gradient from permanently flooded to periodically saturated. Soils are O_2-deficient. Wetlands dominated by grass-like vegetation are marshes. Swamps and riparian forests are dominated by trees and shrubs. Peatlands are characterized by accumulation of peat. Peatlands fed by water moving through mineral soil and dominated by sedges are fens; those dominated by *Sphagnum* mosses and dependent on precipitation for moisture and nutrients are bogs. Bogs are typified by blocked drainage, high acidity, and low productivity. Peatlands are a dominant wetland type in northern Canada.

Wetland Hydrology 25.7

Wetland structure and function are strongly influenced by their hydrology, both the physical movement of water and its hydroperiod (flooding depth, frequency, and duration). The influence of hydroperiod is most evident in basin wetlands that exhibit zonation from deep-water submerged vegetation, to emergents, to wet-meadow species.

Wetland Biodiversity 25.8

Freshwater wetlands support a diversity of wildlife. They provide habitat for frogs, toads, turtles, and many invertebrates, as well as nesting, migrant, and wintering waterfowl. Biotic and abiotic factors interact to influence wetland stability and function.

KEY TERMS

basin wetland	fen	infralittoral (subtidal)	marsh	riverine wetland
bioshield	fringe wetland	zone	meiofauna	salt (tidal) marsh
blanket mire	hydrology	littoral (intertidal) zone	peatland (mire)	supralittoral
(raised bog)	hydroperiod	longshore (littoral) drift	pothole	(supratidal) zone
bog	hydrophyte	mangrove forest	quaking bog	swamp
epifauna	infauna	(mangal)	riparian woodland	wetlands

STUDY QUESTIONS

1. Describe the three major zones of the rocky shore.
2. What are the major abiotic stresses for organisms in the rocky intertidal zone? What is an example of a biotic influence on this community?
3. Compare life on sandy shores with that on rocky coastlines.
4. What influences the major structural features (zonation) of a salt marsh?
5. Compare salt marshes with mangrove forests. Why do mangrove trees have prop roots whereas salt marsh species do not? How do prop roots affect habitat for animals?
6. What is the major ecological importance of mangroves for adjacent ecosystems?

7. What is a freshwater wetland? Describe the three major types of freshwater wetlands in terms of their (i) topography and (ii) dominant vegetation.
8. Among peatlands, how does the hydrology of a fen differ from that of a bog? What are the consequences for fertility?
9. Define *hydroperiod*, and explain how it influences wetland structure. How and why does the hydroperiod of a fringe wetland differ from that of a basin or riverine wetland?
10. How do biotic and abiotic factors interact to influence the health of a freshwater wetland?

FURTHER READINGS

Bertness, M. D. 1999. *The ecology of Atlantic shorelines.* Sunderland, MA: Sinauer Associates.
    Well-written and illustrated introduction to coastal ecology.

Keddy, P.A. 2010. *Wetland ecology: Principles and conservation.* Cambridge, UK: Cambridge University Press.
    A comprehensive treatment of the ecology of wetlands, written by one of Canada's most eminent ecologists.

Mathieson, A. C., and P. H. Nienhuis (eds.) 1991. *Intertidal and littoral ecosystems.* Ecosystems of the world 24. Amsterdam: Elsevier.
    Detailed survey of the intertidal and littoral zones of the world.

Moore, P. G., and R. Seed (eds.) 1986. *The ecology of rocky shores.* New York: Columbia University Press.
    Comprehensive, worldwide review of the rocky intertidal zone.

Niering, W. A., and B. Littlehales. 1991. *Wetlands of North America.* Charlottesville, VA: Thomasson-Grant.
    Exceptionally illustrated survey of North American wetlands, both freshwater and salt.

Teal, J., and M. Teal. 1969. *Life and death of the salt marsh.* Boston: Little, Brown.
    This beautifully written classic is an excellent introduction to salt marsh ecology.

Valiela, I., J. L. Bowen, and J. K. York. 2001. Mangrove forests: One of the world's threatened major tropical environments. *BioScience* 51:807–815.
    Overview of the impact of mariculture on mangroves.

Van der Valk, A. (ed.) 1989. *Northern prairie wetlands.* Ames, IA: Iowa State University Press.
    Detailed studies of this unique wetland ecosystem that is rapidly disappearing.

Williams, M. (ed.) 1990. *Wetlands: A threatened landscape.* Cambridge, MA: Blackwell Publishers.
    This comprehensive appraisal of global wetlands covers their occurrence and composition, physical and biological dynamics, human impact, management, and preservation.

All organisms modify their environment, but no species has altered Earth as much as humans. As our numbers have grown and the power of our technology has expanded, the nature and scope of our impacts have changed dramatically. The story of our species is one of continuously redefining our relationship with the environment—a relationship ultimately based on energy.

With the melting of the polar ice caps after the last ice age, humans spread over the globe, with the Americas and Australia the last continents to be inhabited, 25 000 to 10 000 years ago. At that time, the global population approached 5 million. Further growth was constrained by our direct dependence on other species for energy. Hunter–gatherer societies consisted of small bands of several hundred individuals who depended on the productivity and abundance of species inhabiting natural ecosystems. These bands of humans were nomadic, tracking resources in space and time, and vulnerable to environmental change.

Some 10 000 years ago, a change occurred that redefined forever the relationship between humans and their environment. The Neolithic period (8000–5000 BCE) saw the rise of agriculture—cultivation of plants and domestication of animals. Although this shift from a hunter–gatherer to an agricultural society did not alter the dependence of humans on primary productivity, it shifted the dependence from the productivity of natural ecosystems to that of managed agricultural systems. The result was a larger and more predictable food supply, accompanied by the rise of permanent villages, division of labour, and emergence of new social structures.

The transition to an agricultural society greatly eased the abiotic constraints imposed on human carrying capacity. In a burst of effort, the human body can muster some 100 W of power (1 joule second (J · s) = 1 watt (W)). The most any society could devote to a given task with humans or animals as the energy source was a few hundred thousand watts. Expanding territory could increase energy supply but not raise the total that could be applied to a single task. It is impossible to concentrate more than a few thousand bodies on a given project, be it construction or battle. But by the 18th century, the mechanical energy of animal and human labour was replaced by a much more concentrated form of energy: coal.

The Industrial Revolution, which began in the mid-1700s, saw the rise of the steam engine and with it a shift in labour from humans and animals to machines. Steam engines, which transform the heat energy of steam into mechanical energy, were inefficient at first, losing over 99 percent of their energy. But by 1800, their efficiency rose to about 5 percent, with a capacity of 20 kW in a single engine (the equivalent of 200 people). By 1900, engines could handle high-pressure steam and were 30 times as powerful (6000 people). Their portability was as vital as their power. Steam engines could be used in ships and trains, facilitating widespread transport of coal and establishing a positive feedback loop that fueled further industrialization.

With mechanization, manufacturing once conducted in homes and small workshops became centralized in factories. By the late 1800s, mechanization was transforming agriculture, too. Field size increased, and the amount of land dedicated to farm animals fell sharply. Machines more efficiently performed the work once done by human and animal labour. A single large 150- to 200-hp tractor (1 hp = 735 W) (Figure 1a) can do the work of 1000 workers or 200 draft animals. Mechanization shifted the labour force, reducing the need for farm labour and providing a workforce for industrialization. According to Statistics Canada (2006), 32 percent of Canadians lived on farms in 1931, the year of the first farm census. By 2006—less than a lifetime—that number had dropped to 2.2 percent.

So the 20th century witnessed a transition from a rural, agricultural economy to an urban, industrial economy. In turn, this change triggered a shift in population distribution. By 1900, just over 35 percent of people lived in urban areas. As we entered the 21st century, that number had risen to almost 80 percent in both Canada and the United States, and continues to grow. Urbanization (Figure 1b) demands an ever-increasing infrastructure of transportation and trade, and an increasing energy supply. By the 1990s, the average global citizen used 20 "human equivalents" in energy (equal to a human working 24 hours a day, 365 days a year), primarily as fossil fuels. This transformation allowed industrial labour efficiency to

(a)

(b)

Figure 1 Humans have removed natural ecosystems by (**a**) industrialized agriculture and (**b**) urbanization.

increase 200-fold between 1750 and 1990. Modern workers produce as much in a week as their 18th-century counterparts did in four years. In the 20th century alone, global industrial output grew 40-fold.

Fuelled by these dual transformations (the rise of agriculture and the Industrial Revolution), the human race has multiplied by a factor of 1000 over the past 10 000 years, while at the same time sharply increasing our per capita use of resources. Our number is now just over 7 billion, and our collective "ecological footprint" (see Ecological Issues: The Human Factor, p. 13) continues to grow. Nearly 40 percent of Earth's terrestrial net primary productivity is used directly, co-opted, or forgone as a result of human activity (see Ecological Issues: Human Appropriation of Net Primary Productivity, pp. 440–441). We use over 50 percent of all freshwater resources, of which 70 percent are used in agricultural production. In all, our activities have transformed 40 to 50 percent of the terrestrial surface to produce food, fuel, and fibre, and our alterations of natural systems have triggered the extinction of thousands of species.

Profound as they are, these and other changes humans have brought about in the environment are not the result of thoughtless or malicious behaviour. They are partly due to the needs of a growing population, but also reflect the consumption pattern that has accompanied industrial development. Little did we realize the extent and severity of the impacts of our collective actions. As recently as the 1970s, the mantra was "the solution to pollution is dilution."

Times have changed, and we are beginning to appreciate the consequences of our past and present activities. **Human ecology** is a new field, focusing on the interactions between humans and the environment. Like ecology itself, it is interdisciplinary, drawing upon many sciences as well as anthropology, sociology, and history. In Part Eight, we explore three topics that underlie current issues regarding human environmental impacts: population growth and sustainability (Chapter 26), threats to biodiversity and habitat conservation (Chapter 27), and global climate change (Chapter 28).

CHAPTER

26

POPULATION GROWTH AND SUSTAINABILITY

The human population entered the 20th century with 1.6 billion people and left it with 6.1 billion. Demographers estimate that this number could rise to 9 billion by 2043, according to the *2010 Revision of the World Population Prospects*, released by the United Nations in 2011.

n Canada, it is easy to think of overpopulation as someone else's problem. Covering a land area of some 9 million square kilometres, Canada is one of the world's most sparsely populated countries, next only to Russia. Granted, most Canadians inhabit a narrow swath within 150 km of the 49th parallel, but even southern Canada enjoys a luxury of space that is unimaginable in most countries. Indeed, our politicians are more concerned with increasing our population. Growth in numbers, and the economic activity associated with it, is a major index of prosperity. In the introduction to Part Eight, we discussed human population as a global issue, but why should we as Canadians be concerned, if this issue mainly affects China, India, and other regions remote from our boundaries?

As global citizens, we ignore the ecological consequences of continued human population growth at our peril. For much of our history, the human population increased very slowly, with periodic downturns associated with events such as the Black Plague in the Middle Ages. But with the onset of the industrial age, it took a sharp exponential rise, and from 1900 to 2011, the global population increased more than six-fold to

7 billion. Like any population, more humans must exploit more resources to meet their needs for food, water, and shelter. Resource consumption is driven by two factors: (1) total number of individuals (population size) and (2) per capita rate of consumption. Both factors have increased steadily over the past century (Figure 26.1), with little sign of abating.

Nor is per capita resource use uniform. Our neighbour to the south, whose economic structure and lifestyle we have emulated, uses far more resources per capita in comparison not only with developing countries but also with other industrialized countries in Europe. Coupled with their population size of 312 million, the United States has a massive ecological footprint (see Ecological Issues: The Human Factor, p. 13). Yet Canadians have no reason to be smug; our population of 34 million may be small, but our per capita resource use is similar to that of the United States and outstrips it in some categories, including energy consumption, according to the World Resources Institute (2001). Even if the industrialized world were to achieve zero growth, our consumption patterns would cause resource use to continue to rise. Because our resource-hungry lifestyle

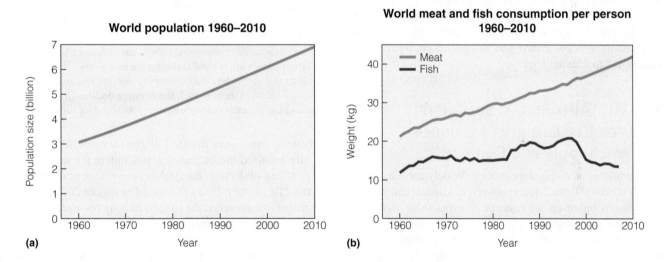
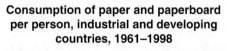
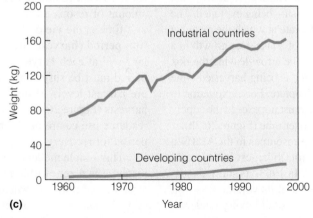

Figure 26.1 Trends in global population **(a)** and per capita consumption of meat and fish **(b)** and paper and paperboard **(c)**.

(Source: U. S. Census Bureau and FAO.)

contributes to the global problem, it behooves Canadians not only to understand the problem but also to contribute to its solution.

A growing human population coupled with a drive for economic growth has brought sustainability to the forefront of current economic, political, and environmental debates. Although the term *environmental sustainability* is broad, covering many topics, including population growth, energy and resource use, and economic development, most discussions focus on the use of **natural capital**: the natural resources provided by ecosystems—air, water, soil, forests, grasslands, etc.—that are essential to humans. **Environmental sustainability** is the ability to sustain exploitation of natural ecosystems (and their natural capital).

In this chapter, we examine resource management and extraction in the context of basic ecological processes and patterns. We then consider environmental sustainability as it relates to agriculture, forestry, and fisheries. All three activities exploit populations to provide the most essential of human resources: food and shelter. Clearly, these are not the only human activities with ecological repercussions. However, other resource exploitation activities, such as mining, while having harmful ecological effects, do not exploit other populations directly. They reflect less basic human "needs" in that they are partly dictated by lifestyle and influenced by social pressures exerted by advertising. These factors also come into play in agriculture, forestry, and fisheries, but to a lesser degree.

26.1 SUSTAINABILITY: Sustainable Resource Use Requires a Balance between Supply and Demand

Sustainability is an ambiguous concept. Widely used in a broad sense, it is rarely defined quantitatively. Environmental sustainability seems based on the concept of *sustainable yield*, first used in German forestry during the late 18th century—matching periodic harvests to the rate of biological growth. Theoretically, such harvesting should not undermine the long-term ability of forests to regenerate. However, as we will see, applying this concept has met with only limited success.

In its simplest form, sustainable resource use is constrained by supply and demand (Figure 26.2a). The box represents the amount of a resource—water, trees, or fish—being exploited. The arrow leading to the box represents the rate at which the resource is being supplied—the rate of recharge of a lake, tree growth in a forest, or growth of a fish population. The arrow leaving the box represents the rate at which the resource is being harvested—the rate of water use, tree harvest, or fish capture. For exploitation to be sustainable, the rate of resource use must not exceed the supply rate. Otherwise, the resource declines over time (Figure 26.2b).

Let's apply this principle to water resources in the Aral Sea (actually a freshwater lake, once the fourth largest in the world by area) in central Asia (Figure 26.3). In 1963, its surface measured 66 100 km². By 1987, 60 percent of its volume was lost, transforming 27 000 km² of former sea bottom to dry land, and its salt concentration had doubled. By 2007, it occupied just 10 percent of its original area, and is now divided into smaller water bodies (Aladin et al. 2008). What caused its demise?

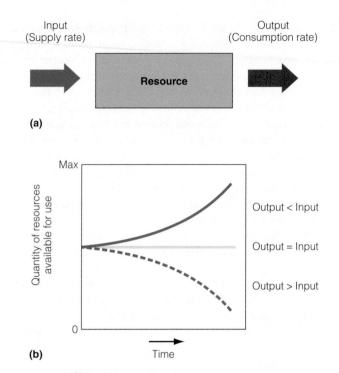

Figure 26.2 A simple model of resource use. **(a)** The amount of a resource available (green box) reflects the difference between rates of supply (green arrow) and consumption (red arrow). **(b)** If the consumption rate is less than the supply rate, the resource increases. If consumption exceeds supply, the resource declines. Sustainable resource use depends on consumption not exceeding supply.

Inflowing rivers were diverted to irrigate crops, a practice that in turn reflected the increased population in the region. At the current rate of decline, the Aral Sea may disappear completely in the 21st century. Using the model in Figure 26.2, the rate of water harvest exceeded the supply, causing resource decline—a case of unsustainable resource use.

In the Aral Sea example, water resources are continuously harvested. But other resources are harvested periodically because they require an extended time between harvests to regenerate. Consider a tree plantation. After saplings are planted or after a forest regenerates naturally, time is needed for the forest to attain a certain biomass (Figure 26.4a). The amount of resource (in this case, tree biomass) harvested per unit time is the **yield**. The time between harvests is the **rotation period (harvest interval)**. If the goal is to ensure a similar yield at each harvest (**sustained yield**), then the rotation period must be sufficiently long for the resource to recover to pre-harvest levels. If not, the yield diminishes in successive harvests (Figure 26.4b). Many of the conflicts over sustainable resource use centre on the need to allow an adequate rotation period for recovery.

This simple model of sustainable resource use assumes the resource is a **renewable resource**—one that can be regenerated. Use of **non-renewable resources** is by definition unsustainable, and the rate of resource decline is a function of the rate at which it is extracted. Minerals (e.g., zinc, aluminum, and copper) are non-renewable. Other resources are considered non-renewable even though they are being resupplied, if their

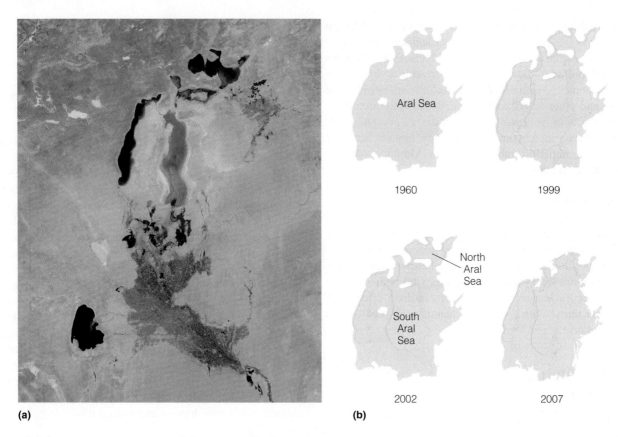

(a)

(b)

Figure 26.3 Unsustainable water use in the Aral Sea. **(a)** Aerial image taken in 2006. **(b)** Changes in area from 1960 to 2007. As a result of diversion of water for irrigation, the Aral Sea decreased in volume by 60 percent between 1963 and 1987. Its decline continues.

resupply rate is infinitesimally low compared to their consumption rate. Fossil fuels are an example. Coal, oil, and natural gas are non-renewable resources because their formation requires millions of years, making their resupply rate effectively zero on the timescale of human consumption. Unlike fossil fuels, many non-renewable resources (such as aluminum) can be recycled, reducing the harvest rate for the resource and extending its effective lifetime for human use.

But sustainability isn't just about rates of *direct* resource harvest and replenishment. The adverse consequences of resource use play a critical indirect role. This is particularly applicable to **ecosystem services**, whereby the environment produces natural capital resources such as clean air, water, timber, or fish. Although the rate at which ecosystems supply these renewable resources imposes a fundamental limit, sustainable use can also be constrained indirectly as a result of negative impacts arising from resource extraction and/or use. A prime example is waste. Dealing with domestic, industrial, and agricultural waste is a growing issue with implications both for ecosystem and human health. Wastes contaminate air, water, and soil with **pollutants**—harmful substances that limit the ability of ecosystems to provide resources and services.

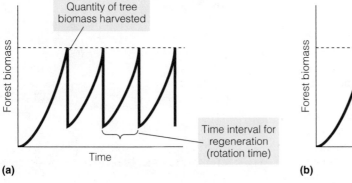

(a)

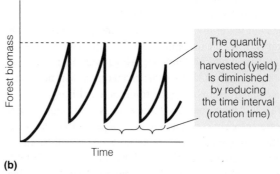

(b)

Figure 26.4 Concept of sustainable yield. **(a)** A sufficient rotation period is necessary for biomass to return to pre-harvest levels. The rotation period depends on species growth rate and site conditions that influence productivity. **(b)** If the rotation period is reduced, the stand will not recover fully. Harvest rate then exceeds regeneration rate, and the resource declines.

The Aral Sea case illustrates such indirect effects all too well. Daily, about 200 000 tonnes of salt and sand are carried by winds from the Aral Sea and dumped within a 300-km radius. Salt pollution destroys pasture, creating a shortage of forage for domestic animals. Fishing has ceased entirely, and shipping and other activities have declined. Water quality is deteriorating with increased salinity, bacterial contaminants, and the presence of pesticides and heavy metals. Rates of typhoid fever, hepatitis, tuberculosis, and throat cancer are much higher than in the general region.

Clearly, resource use has impacts beyond the direct effects of resource depletion. We must consider not only the ability to sustain future resource consumption, but also the indirect impacts arising from resource extraction and use. Such impacts compromise an ecosystem's ability to provide not only the resource, but also ecosystem services. Labelling such impacts *indirect* does not imply that they are less vital, either for the ecosystem or for human health. In fact, indirect effects are often longer term, more difficult to quantify and predict, and harder to reverse.

26.2 NATURAL ECOSYSTEMS: Sustainable Resource Use Is Characteristic of Natural Ecosystems

When we try to manage natural resources sustainably, we attempt to mimic natural ecosystems. Recall from Section 19.2 that natural ecosystems function as sustainable units. Consider the link between primary productivity and microbial decomposition. Plant uptake of nutrients such as nitrogen is constrained by the rate at which they become available. The rate of nutrient "harvest" by plants (and hence the rate of productivity) cannot exceed the rate at which nutrients bound in organic matter are released by decomposition. As nutrients are mineralized, they are quickly taken up by plants, minimizing their loss to ground and surface waters. To relate natural ecosystem function to the model in Figure 26.2, consider the rate of resource use of a mineral such as nitrogen. The rate at which the resource is used is balanced by the rate at which it is supplied—in this case, mineralized. The ecosystem is "managing" nitrogen use sustainably, even though the size of the box representing the pool of available resources is very small.

Sustainability does not imply static rates of resource use. When supply varies over time, the rate of resource use likewise varies. Consider water use by a prairie during drought. If insufficient water is available to replace water lost through transpiration, stomata close, and productivity and growth decline. If the drought is particularly severe, some plants may wilt and even die, depending on their drought tolerance. Although water stored in the soil may ameliorate the effects of drought, plant growth is often closely correlated to annual precipitation. Similarly, variations in the supply rate of resources needed to maintain productivity is a major constraint on sustainable yield. Working with these constraints without triggering negative impacts is the challenge for sustainable agriculture and forestry. Meeting this challenge will likely require a reduction in resource use, at least on a per capita basis.

26.3 AGRICULTURE AND ENERGY: Agricultural Systems Differ in Their Energy Inputs

The vast majority of human food resources derive from agriculture—crops and livestock. Even though botanists estimate that as many as 30 000 wild plant species have edible plant organs (seeds, roots, leaves, fruits, etc.), only 15 cultivated plant and 8 animal species produce 90 percent of our global food supply. Seeds of 3 annual grasses—wheat, rice, and corn (maize)—constitute over 80 percent of the cereal crops consumed by humans. Although initially derived from wild species, cereal varieties cultivated today are the products of intensive breeding and genetic modification. The same is true for domestic animals.

About 11 percent of Earth's ice-free land area is under cultivation. Another 25 percent is used as pasture for grazing livestock, primarily cattle and sheep. Whatever the crop or cultivation method, agriculture involves replacing diverse natural ecosystems—grassland, forest—with a community consisting of a single crop species (**monoculture**) or a mixture (**polyculture**). Although many diverse practices are used worldwide, agricultural methods fall into two broad categories: traditional and industrialized (Figure 26.5).

Traditional agriculture is dominated by *subsistence agriculture*, which uses human and animal labour to produce only

(a)

(b)

Figure 26.5 Examples of **(a)** traditional and **(b)** industrialized agriculture, in which mechanization replaces human and animal labour.

enough crops or livestock for a family to survive. Examples of this low-input approach are shifting cultivation and nomadic herding. **Industrialized agriculture** (*mechanized* or *high-input*) depends on large energy inputs in the form of fossil fuels for mechanization, chemical fertilizers, irrigation systems, and pesticides. Although energy demanding, industrialized agriculture produces large yields, albeit with significant environmental impacts. Industrialized agriculture is practised on about 25 percent of all cropland, mainly in developed countries. It has spread to developing regions in recent decades, with mixed results.

In reality, industrialized and traditional agriculture define two points on a continuum. We will focus on these two end points as background for a discussion of sustainable agriculture.

26.4 SWIDDEN AGRICULTURE: Shifting Cultivation Is the Main Form of Tropical Agriculture

Swidden (slash-and-burn) agriculture (*shifting cultivation*) is the main form of subsistence farming in the tropics. It involves a rotating cultivation in which trees are first felled and burned to clear land for planting (Figure 26.6). Burning trees and brush removes debris, clearing the land for planting and making the plot relatively weed free. It also has a fertilizing effect, as the resulting ash is high in minerals through *pyromineralization*. The plot is seeded, often without tillage, or only with draft animals. Weeding is by hand, and no irrigation or pesticides are used.

Productivity typically declines with each successive harvest (Figure 26.7). Why? With each harvest, nutrients are removed in plant biomass. Even organic fertilizers are rarely used, so soil nutrients (which are deficient already in most tropical soils) decline (Ewel et al. 1981). The site is abandoned and reverts to natural vegetation through secondary succession. If left undisturbed for sufficient time, the site's nutrient status recovers to pre-cultivation levels. Levels of some nutrients, notably phosphorus, may actually increase over several cultivation cycles, thanks to the nutrient-pumping action of the fine roots of colonizing trees (Figure 26.8; Lawrence and

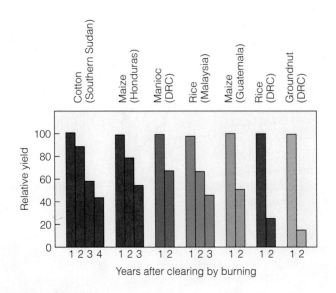

Figure 26.7 Declining productivity in swidden agriculture for various crops in tropical regions. (DRC = Democratic Republic of the Congo.)

Schlesinger 2001). After about 20 years, the site can be cleared again. Meanwhile, other areas have been cleared, burned, and planted. Swidden agriculture creates a heterogeneous patchwork of plots in various stages of cultivation and secondary growth (Figure 26.9, p. 572), similar to the shifting mosaic typical of many natural landscapes (see Section 18.10).

Swidden cultivation is a sustainable method when sufficient time is allowed for regrowth of natural vegetation and recovery of soil nutrients. But it also requires enough land area to permit the appropriate rotation time. The current problem in many tropical regions is that a growing human population is placing ever-increasing demands on crop production, and sufficient recovery periods are not always observed. Land then quickly degrades, and yields decline. Moreover, it is questionable whether regrowth of tropical forest ever achieves a similar species composition as primary forest. In Sri Lanka, for example, most of the forests are regrowth after swidden agriculture, and are badly degraded (Perera 2001).

Figure 26.6 Swidden agriculture. A plot of forest is cleared and burned to allow planting of crops. The ashes are an important source of nutrients. Fertilizers are typically not used.

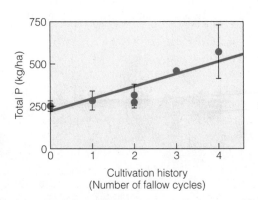

Figure 26.8 Changes in P levels (mean ± 1 SD for six sites) in the top 30 cm of soil during the first four cycles in swidden plots in Indonesia.

(Adapted from Lawrence and Schlesinger 2001).

Figure 26.9 Mosaic resulting from swidden agriculture in the Yucatán region of Mexico. The foreground has recently been cleared, the area beside it has been abandoned for one year, and the forest in the background has been regenerating for several years.

26.5 INDUSTRIALIZED AGRICULTURE: High-Input Methods Dominate Agriculture in Temperate Regions

Industrialized agriculture is widely practised in North America, Europe, Russia, and parts of South America, Australia, and Asia. Machines and fossil fuel energy replace human and animal labour. Mechanization requires large tracts of land for machines to operate effectively and economically. As different crops require specialized equipment for planting and harvest, farmers typically plant a monoculture or a few varieties season after season (*continuous rotation*).

With crops such as wheat and corn, little organic matter remains after harvest. In the Canadian prairies, wheat straw stubble is often burned to facilitate subsequent planting, igniting conflicts between urban and rural communities. Tilling

exposes the soil to wind and water erosion (Figure 26.10), resulting in annual soil losses of some 40 tonnes per hectare. Harvesting directly removes large quantities of nutrients (Table 26.1). Given these indirect and direct effects, mechanized agricultural systems lack functional internal cycling. To maintain productivity, farmers add large quantities of chemical fertilizers, typically in a mineral form that is readily available for uptake—and just as readily leached to surface and groundwater (Table 26.2).

Because the same crop varieties are planted over large, contiguous areas, pests and diseases spread readily. Pests are

Table 26.1 Yield and Nutrient Content of Various Crops (kg/ha)

Crop	Yield	N	P	K	Ca	Mg
Wheat						
Grain	2 690	56	12	15	1	7
Straw	3 360	22	3	33	7	3
Corn						
Grain	9 416	151	26	37	18	22
Stover	10 080	112	18	135	31	19
Rice						
Grain	5 380	56	10	9	3	4
Straw	5 610	34	6	65	10	6
Loblolly pine (22 yr)	84 000	135	11	64	85	23
Loblolly pine (60 yr)	234 000	344	31	231	513	80

"Stover" refers to leaves and stems of corn left in field after harvest. Data from Buol 1995.

Figure 26.10 A recently tilled field. Given their lack of ground cover, tilled fields have a high rate of water and wind erosion.

Table 26.2 Nutrient Inputs (fertilizer and precipitation) and Outputs (harvest and runoff) for a Corn Field in the Central United States (kg/ha/yr)

	In precipitation	Fertilizer	Harvest	Runoff to streams
Nitrogen	11.0	160.0	60.0	35.0
Calcium	3.2	190.0	1.0	47.0
Phosphorus	0.03	30.0	12.0	3.0
Potassium	0.2	75.0	13.0	15.0

Soil loss from erosion: 44 t/ha/yr.

typically controlled to limit yield losses, using insecticides, herbicides, and fungicides. These chemicals cause many environmental problems, both for humans and for wild species sharing the agricultural habitat. Some pests, particularly fungal diseases, are managed by ongoing breeding programs that select for crop resistance. In Canada, wheat rusts are a prominent example. However, resistance to one pest does not guarantee resistance to a new pest or new strain of an existing pest, and the decline in genetic variability associated with many breeding programs compromises the ability to cope with other environmental stresses. Besides traditional breeding, genetically modified crops play an increasingly large role in industrialized agriculture, giving rise to serious ecological concerns (see Ecological Issues: Genetically Modified Crops, p. 574).

26.6 AGRICULTURAL TRADE-OFFS: Farming Methods Entail a Trade-Off between Sustainability and Productivity

The two very different agricultural systems just discussed—traditional and industrialized—represent a trade-off between energy input in production and energy harvest in food resources. Table 26.3 compares energy inputs and yields for corn production in Mexico using swidden agriculture and in the United States using industrial techniques. Energy inputs in the traditional system are dominated by labour (approximately 92 percent of total inputs), with small inputs in tools and seed. Total yield is just over 1900 kg/ha. In contrast, labour is a negligible portion (0.05 percent) of total energy inputs in the industrialized system. Major inputs are in farm machinery (3.2 percent), fossil fuels (4 percent), irrigation (7.1 percent), chemical fertilizers (13.6 percent), and pesticides (3.5 percent). Crop yield is 7000 kg/ha, more than 3.5 times that of the traditional methods used in Mexico.

However, a different story emerges if we examine the ratio of kilocalories of energy input in production to kilocalories of energy in harvested food: 13.6 for the traditional system versus 2.8 for the industrialized. Suddenly our modern techniques look much less efficient; although industrialized agriculture produces 3.5 times the corn yield per unit land

Table 26.3 Comparison of Energy Inputs in Production and Yields for Corn Harvested Using a Traditional Agricultural System in Mexico and an Industrial Agricultural System in the United States

Item	Mexico (kcal/ha)	United States (kcal/ha)
Inputs		
Labour	589 160	5 250
Axe and hoe	16 570	
Machinery		1 018 000
Gasoline		400 000
Diesel		855 000
Irrigation		2 250 000
Electricity		100 000
Nitrogen		3 192 000
Phosphorus		730 000
Potassium		240 000
Lime		134 000
Seeds	36 608	520 000
Insecticides		300 000
Herbicides		800 000
Drying		660 000
Transportation		89 000
Total	642 338	11 036 650
Outputs		
Total corn yield	1944 kg	7000 kg
	8 748 000 kcal	31 500 000 kcal
kcal output/kcal input	13.62	2.85

area, it does so at the cost of over 17 times the energy input. In addition, these inputs are in the form of materials and services that entail heavy environmental costs. Loss of fertilizer compounds such as nitrates and phosphates from crop fields to adjacent streams, lakes, and coastal waters causes massive enrichment, which in turn damages ecosystems via *cultural eutrophication* (see Section 24.3).

Besides these impacts on nearby ecosystems, widespread fertilizer use damages human health. Groundwater provides drinking water for many households and communities. Nitrate is one of the most common contaminants in rural areas. U.S. Environmental Protection Agency surveys indicate that 1.2 percent of community water supplies and 2.4 percent of rural domestic wells contain nitrate levels exceeding public health standards. Similarly, Environment Canada reports that 1.5 percent of private wells in the British Columbia lower mainland have nitrate levels above national guidelines. Although this value may seem low, contamination is concentrated in agricultural areas, where the percentage of contaminated wells is much higher.

In humans, high nitrate levels have been implicated in birth defects and nervous system impairment. They may also indicate the presence of other contaminants, such as bacteria and pesticides. Nor are humans the only species at risk;

ECOLOGICAL ISSUES   Genetically Modified Crops

Over the past half century, crop yields have steadily increased. In part, this reflects the increasing use of chemical fertilizers, irrigation, and pesticides. However, increases in yield are also due to varieties developed by selective breeding. Plants with desired traits are interbred to introduce the desired traits of one variety into another. For example, a rust-resistant variety of wheat is crossed with a high-yielding but rust-susceptible variety, resulting in a plant that has rust resistance as well as high yield. Humans have practised such selective breeding for thousands of years. For example, some 8000 years ago, farmers in Central America crossed two mutant strains of the annual grass *Balsas teosinte* to produce the ancestor of modern maize.

The diversity of modern crops attests to the success of selective breeding, but biotechnological innovations allow scientists to select specific genes from one organism and introduce them directly into the genome of another to confer a desired trait. Compared with conventional breeding, this technology not only produces new varieties more quickly but also introduces traits not possible using traditional methods. Genes from organisms as dissimilar as bacteria and plants can be successfully inserted into each other. Combining genes from different organisms is called *recombinant DNA technology*, and the resulting organism is **genetically modified (transgenic)**.

The principal application of this technology has been the development of genetically modified (GM) crops engineered to tolerate herbicides and resist pests. Crops carrying herbicide-tolerant genes allow farmers to spray their fields to eliminate weeds without damaging the crop. Similarly, pest-resistant crops contain a gene for a toxic protein from the soil bacterium *Bacillus thuringiensis*. This protein, called Bt, is then produced by the plant, making it resistant to pests like the European corn borer. Other crops have been engineered for resistance to plant viruses.

In 1996, some 1.7 million hectares (ha) were planted with GM crops worldwide. By 2011, that number had grown to almost 170 million ha—a 100-fold increase in 15 years (James 2011). Soybeans and corn are the crops most affected (Figure 1a), and herbicide resistance the trait most commonly incorporated (Figure 1b). The United States is the largest user of GM crops (69 million ha) but although Canada's GM use is much less (10.4 million ha) (James 2011), Canada is no stranger to the phenomenon. According to Agriculture Canada, the most common GM crop in Canada is canola (over 70 percent), but GM sugar beets have also been introduced, first in Ontario in 2008 and later in Alberta and Prince Edward Island. Because beets are wind pollinated, there are concerns about infiltration of modified genes into related species. In fact, Prince Edward Island debated a ban on cultivation and importation of GM crops. This step was not taken, and Canada and the United States, unlike Europe, do not require labelling of products derived from GM crops. Other major producers are Argentina and Brazil, which has recently legalized GM soybeans; China, where acreage of GM cotton is increasing; and South Africa, where cotton is also the major GM crop.

Why are GM crops an issue? Some concerns involve intellectual property rights, whereas others are ecological. Planting crops in which herbicide resistance is built-in allows far more liberal application of pesticides, which may then enter the environment and affect other organisms, including humans. Moreover, given the ability of plants to share genes by crossing with wild relatives (such as canola with wild mustard), genes for herbicide resistance

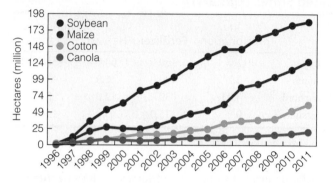

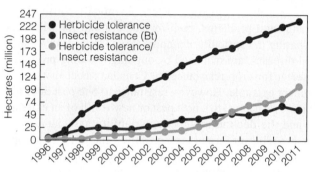

Figure 1 Growth in global area of genetically modified crops from 1996 to 2011. **(a)** Listed by crop. **(b)** Listed by trait.

(Adapted from James 2011.)

may escape from crop fields. If a weed species acquires herbicide resistance, then ever more noxious substances are needed to control it. This consequence has already surfaced in many regions, with some farmers advocating a return to formerly used, more toxic herbicides.

Some argue that GM crops actually reduce pesticide use, specifically for crops that have acquired the gene for Bt. Less insecticide should be needed, and less should escape into the environment. However, if this trait is transferred to wild species, then widespread disruption of natural systems, including damage to insect pollinators that play a key role in both agricultural and natural systems, may ensue. For these and other reasons, GM crops are considered by many ecologists to be a Pandora's box of potentially dangerous and unforeseen ecological consequences. Indeed, the public debate is heating up rather than abating. In July 2011, the European Parliament backed proposals for a European Union restriction that would allow member countries to ban cultivation of GM crops. Such a move has tremendous public support across Europe, where labelling of GM products is required. In North America, the future of GM-crops is an open question.

Bibliography

James, C. 2011. *Brief 43: Global status of commercialized biotech/GM crops: 2011.* ISAAA.

1. Does use of GM crops constitute a threat to biodiversity? Why or why not?
2. Do you think Canada should change its regulations regarding labelling of GM products? Why or why not?

Environment Canada reports that of over 8500 samples from the Great Lakes area on either side of the border, 19.8 percent had nitrate levels that cause developmental aberrations in amphibians, and 3.1 percent had levels that can be lethal to tadpoles. In freshwater animals (which are more susceptible than marine animals), the major toxic effect of nitrate is to convert O_2-carrying pigments to non-functional forms. Damage increases with concentration and exposure time (Camargo et al. 2005).

How can an input so basic to plant growth, whether in agricultural or in natural ecosystems, have become such a problem? Recall that in natural systems, there is a tight link between the uptake of nitrogen to support primary productivity and its release via mineralization by decomposers (see Section 21.10). In agricultural systems, this balance is disrupted. Plants, and the nutrients they contain, are harvested, and the organic matter locked in their biomass does not return to the soil to be decomposed. To supply adequate nutrients for continued production, fertilizers must be added.

Humans have long used organic fertilizers such as manures and ground-up bones, and still use them in traditional agriculture. But it was not until the modern scientific study of plants and soils began in the 19th century that an understanding of the specific elemental needs of plants emerged. With that understanding came the first production of fertilizers from inorganic sources. Originally, the three elements that plants need in large quantities (N, P, and K) were all derived from mineral deposits. Phosphate rocks and potash still furnish adequate supplies of P and K. Such is not the case with N. One mineral source, saltpeter ($NaNO_3$), in deposits in Chile, accounted for over 60 percent of the world's supply for most of the 19th century.

As the world population increased, the growing demand for food made it imperative to develop a way to tap the almost limitless source of N_2 gas in the atmosphere. The reaction between N_2 and H_2 to produce ammonia (NH_3) had been known for years, but yields were small and the reaction slow. In the early 20th century, the German chemists Fritz Haber and Carl Bosch developed a process that made NH_3 manufacture economically feasible. They received a Nobel Prize for their discovery, and since the 1920s, fertilizers derived from their process have revolutionized agriculture. Demand for N fertilizers has grown exponentially, and ammonia fertilizers were instrumental in the "Green Revolution" (see Section 26.7). Scientists estimate that the N content of over one-third of the protein that humans consume worldwide derives from the Haber–Bosch process.

Yet the bounty of food produced with synthetic fertilizers comes at a heavy cost. As we have seen, nitrate pollutes drinking water and, through farm runoff, infiltrates lakes, streams, and eventually the oceans. There, it disrupts the normal constraints on productivity that result from seasonal cycling of N, and supports explosive algal growth. High organic matter inputs stimulate decomposition and heterotrophic respiration, reducing O_2 levels in the water. Populations that cannot survive under anoxic conditions decline. A report released by the Pew Oceans Commission (2003) concluded that N fertilizer is the major oceanic pollutant. (In freshwater lakes, P is the greater issue.) So the ingenuity that developed the Haber–Bosch process that enables us to feed the world population is faced with an equally difficult task—reducing the environmental impacts of the production and application of this most critical element, N, for plant growth.

Along with these and other problems caused by widespread fertilizer and pesticide use, large inputs of fossil fuels for mechanization, irrigation, and fertilizer production add to the growing atmospheric inputs of CO_2 and other greenhouse gases (see Chapter 28). Yet, despite the serious problems of energy use and the ecological consequences of pesticide and fertilizer use, the way to make agriculture more sustainable is not to revert to traditional methods, which cannot produce enough to feed the human population. We can increase production to meet the growing demand in two ways: (1) increasing land area under cultivation and/or (2) increasing production per unit land area. Historically, land under cultivation worldwide has risen, keeping pace with the growing

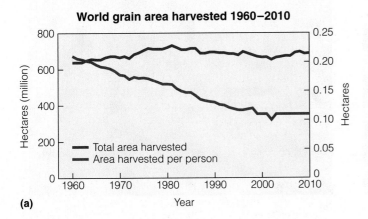

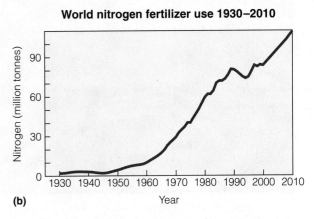

(a)

(b)

Figure 26.11 Total global land area in crop production (a) has begun to stabilize, while the area harvested per capita has declined. This decline reflects gains in productivity per hectare resulting largely from increased use of chemical fertilizers over the same period (b).

(Source: FAO.)

population. In the later 20th century, this trend slowed, and per capita land area under production declined (Figure 26.11a). This decline is partly due to more productive varieties and increased irrigation, which itself has ecological repercussions, including salinization and groundwater depletion. But by far the factor most responsible for the increased productivity is the growing use of chemical fertilizers, particularly nitrogen (Figure 26.11b)—which brings us back full circle to the problem with which we started.

The reality is that we depend on industrialized agriculture to feed the world's growing numbers. Expanding the amount of cultivated land area is no longer a viable option. To do so would only exacerbate the single greatest cause of Earth's declining biodiversity—habitat loss arising from human land transformations (see Section 27.6). The challenge is to develop methods of mechanized agriculture that minimize ecological impacts while sustaining long-term production.

26.7 SUSTAINABLE AGRICULTURE: Sustainable Agriculture Requires Methods Specific to Crop and Location

As with the broader concept of sustainability, **sustainable agriculture** refers to maintaining agricultural production while conserving resources and minimizing negative environmental impacts. There is no universal set of quantitative criteria. In general terms, the goals of sustainable farming techniques are to (1) minimize erosion; (2) reduce the use of chemical fertilizers, pesticides, and energy; and (3) conserve and protect water and soil resources. Agroecologists recommend different methods

for different crops in different locations, but examples of sustainable practices include the following.

1. *Soil conservation methods.* Contour and strip cropping (Figure 26.12a) and reduced tillage or "no-till" cultivation (Figure 26.12b) minimize soil loss from wind and water erosion. Planting shrubs and trees as fencerows provides wind barriers, further reducing soil erosion.

2. *Reduced pesticide use.* Crop rotations (e.g., planting wheat one season and canola the next) and strip cropping with multiple crops or varieties help reduce the spread of pests and disease. Timing planting to avoid pest outbreaks is encouraged. These practices are part of a broader approach called **integrated pest management**, which aims to minimize losses to pests not by eradication but by applying knowledge of their ecology. For example, hedgerows not only reduce erosion but also provide habitat for predators that help regulate insect pest populations.

3. *Alternative sources of soil nutrients.* Increased use of on-farm sources, such as manure and N-fixing legume crops that are ploughed under as a **green manure**, are alternatives to chemical fertilizers. Although expensive, slow-release mineral fertilizers reduce leaching.

4. *Water conservation and protection.* Many strategies have been developed to conserve and improve the quality of drinking and surface water, and to protect wetlands. As well as providing wildlife habitat, wetlands filter nutrients and pesticides.

Nowhere is sustainable agriculture a more vital goal than in developing countries. In their haste to modernize traditional

(a)

(b)

Figure 26.12 Methods that promote sustainable agriculture include **(a)** contour and **(b)** no-till cultivation. Go to QUANTIFYit! at www.pearsoncanada.ca/ecologyplace to graph food production efficiency.

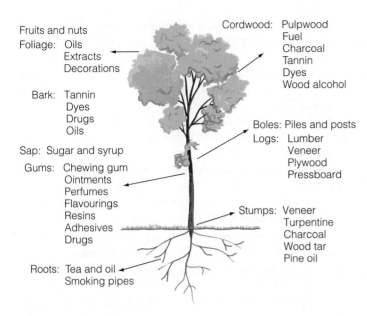

Fruits and nuts
Foliage: Oils
          Extracts
          Decorations

Bark: Tannin
       Dyes
       Drugs
       Oils

Sap: Sugar and syrup

Gums: Chewing gum
       Ointments
       Perfumes
       Flavourings
       Resins
       Adhesives
       Drugs

Roots: Tea and oil
        Smoking pipes

Cordwood: Pulpwood
           Fuel
           Charcoal
           Tannin
           Dyes
           Wood alcohol

Boles: Piles and posts
Logs: Lumber
        Veneer
        Plywood
        Pressboard

Stumps: Veneer
         Turpentine
         Charcoal
         Wood tar
         Pine oil

Figure 26.13 A variety of products derived from forests.

agriculture, Western agricultural scientists involved in international aid projects introduced high-input, industrialized practices in the so-called **Green Revolution** of the 1960s and 1970s. Farmers in developing countries were encouraged (and in some cases coerced) to replace local varieties with dwarf, high-yielding hybrid crops that required large energy inputs in fertilizers, water, seeds, and pesticides. Yields soared, but with negative long-term effects, both ecological and social. No longer relying on traditional, self-seeding varieties, farmers in developing countries became less self-sufficient and more dependent on buying seed and other inputs from large agrochemical companies.

As a result of these problems, agroecologists are now pursuing a more ecologically sensitive approach by working with local farmers in developing countries (Sanginga et al. 2009). Local farmers are knowledgeable about the prevailing soil and climate, as well as the advantages and limitations of crops and pasture methods traditionally used in an area. By respecting this knowledge, scientists can encourage methods that sustain high yields while minimizing negative impacts.

26.8 CONVENTIONAL FORESTRY: Logging Aims for Sustained Yield with Limited Success

Forests cover some 35 percent of Earth's surface and provide a wealth of resources, including fuel, building materials, and food (Figure 26.13). Although plantations provide a growing percentage (more so in the United States than in Canada), over 90 percent of global forest resources are still harvested from native forests. Sustainable forestry must thus be concerned not only with maintaining yields, but also with the integrity of forests as natural ecosystems. This is particularly true in Canada, where plantations are still relatively rare.

Globally, about half of the forest present under modern (post-Pleistocene) climatic conditions has disappeared, largely through human activities. The spread of agriculture, harvesting of forests for timber and fuel, and expansion of populated areas have all taken their toll. The causes and timing of forest loss differ among regions and forest types, as do current changes in forest cover (see Section 27.6). In the face of increasing demand for timber and declining forest cover, forestry must achieve a balance between net growth and harvest. To achieve this end, foresters use various methods that vary in the damage caused to the forest as a natural ecosystem.

1. Clear-cutting involves removing the forest and reverting the site to an early stage of succession (Figure 26.14a, p. 578). The area harvested can range from thousands of hectares to small patch cuts of a few hectares designed to create habitat for wildlife species that require openings within the forest (see Section 18.3). Post-harvest management varies widely. When natural forest stands are clear-cut, there is often no follow-up management. Stands are left to regenerate naturally from existing seed and sprouts on the site, along with input of seeds from nearby stands. This practice may sound ecologically irresponsible, but in some regions, including parts of Canada's boreal forest that are prone to fire and other disturbances, clear-cuts followed by natural regeneration can re-establish well, if they mimic post-fire site conditions. Natural regeneration may be preferable to planting, especially when cuts are small and elongated to facilitate establishment from nearby stands.

However, foresters are questioning whether clear-cutting is the best option for the boreal forest (Youngblood and Titus 1996), and clear-cutting on sloping sites in the temperate rain forests of the Pacific Northwest causes severe damage. Large clear-cuts erode badly, inhibiting recovery as well as damaging adjacent aquatic communities. In such cases, appropriate site treatment, including post-harvest planting, is critical.

Even if some types and applications of clear-cutting can be sustainable, the type of clear-cutting known as *whole-tree* is always devastating. The entire tree, including the root mass, is harvested, causing maximum disturbance to the forest floor and soil environments.

Clear-cutting is typically practised on tree plantations, followed by intensive site management. Materials that are not harvested (branches and leaves) are usually burned to clear the site for planting. After clearing, seedlings are planted and fertilizer applied. Herbicides are used to discourage weedy species that could compete with seedlings. These practices enhance tree growth, but generate similar ecological problems as do agricultural fertilizer and pesticide use.

2. The **seed-tree harvesting** system removes all trees from an area except for a small number of mature trees that are left as a seed source for natural regeneration (Figure 26.14b). Seed trees (5–30 per ha) are scattered uniformly or in small clumps, and they may or may not be harvested later. The seed-tree system resembles a clear-cut because not enough trees are left standing to affect the local microclimate.

(a)

(b)

Figure 26.14 Examples of **(a)** a clear-cut and **(b)** a seed-tree harvesting system. Go to QUANTIFYit! at **www.pearsoncanada.ca/ecologyplace** to graph deforestation rates.

Its advantage is that the seed source is not limited to adjacent stands. This feature can improve seedling frequency and density (stocking) and produce a more natural species mix. Certain criteria must be met for successful re-establishment. Seed trees must produce enough seed, and be strong enough to withstand winds. Seedbed conditions must be conducive to seedling establishment, which may require a preparatory treatment during or after harvest. Follow-up management may be needed to establish the regeneration fully.

3. The **shelterwood cutting** system uses a series of three cuts, with the goal of establishing an even-aged stand under shelter provided by mature trees. First, a preparatory cut removes some mature and dead and dying trees. After seed-lings have established, a second cut leaves enough mature trees to shelter the desired species, which is a shade-tolerant late-successional specialist. The final cut removes the desired species. The key difference from the seed-tree method is that enough mature trees are left to affect the microclimate of the regenerating trees.

Although shelterwood harvesting sounds inherently more ecological than other harvesting methods, it is not appropriate for forests dominated by shade-intolerant species such as pine, and it greatly alters the age and size distribution of most forests—in effect, converting them into plantations. Whereas seed tree and shelterwood methods are commonly used in the United States, they represent a minor component of Canadian logging operations.

4. In **selection (selective) cutting**, mature single trees or groups of trees scattered through the forest are removed. Selective cutting produces small openings or gaps in the forest canopy, which remains intact. Although this method minimizes the scale of disturbance caused by direct tree removal, the network of trails and roads necessary for access can be a major disturbance to both plants and soils. Much more area must be logged to obtain the same harvest. Selection cutting can also change species composition and diversity because only certain species are removed, altering biotic interactions and affecting habitat suitability for wildlife.

Whatever the method, some general principles emerge regarding sustained yield. Whether a forest is planted or allowed to regenerate naturally, establishment begins with a population of seedlings that may compete for light, water, or minerals. As biomass increases, tree density decreases through self-thinning, while mean tree size increases (see Section 11.4). For a stand to be economically viable for harvest, minimum thresholds must be satisfied for both harvestable timber volume per hectare and mean tree size (Figure 26.15). These thresholds vary with species.

After harvest, sufficient time must pass for the forest to regenerate. For sustained yield, the time between harvests must allow the forest to regain the biomass at the previous harvest (see Figure 26.4). Rotation period depends on many factors related to species, site conditions, management type, and intended use of the trees. Wood for paper products (pulp) is harvested from fast-growing species such as aspen, allowing a short rotation time (15–40 years). These species may be grown in plantations where trees are spaced to reduce competition and fertilized to maximize growth. Trees harvested for timber (saw logs) require a much longer rotation. Hardwood species used for furniture and cabinetry (such as oak) are slower growing and may have a rotation time of 80 to 120 years. Sustained forestry of these species works best in extensive areas where blocks of land can be maintained in different age classes.

As with crops, many nutrients are lost with tree harvest (see Table 26.1), compounded by losses from erosion and post-harvest management. In addition to nutrients removed directly in biomass, logging removes nutrients by altering internal cycling. After logging, more radiation reaches the soil. Soil temperatures rise, speeding decomposition of remaining soil organic matter and increasing net mineralization (Figure 26.16). This short-term burst in nutrient supply coincides with low nutrient demand because plants have been removed and net primary productivity is

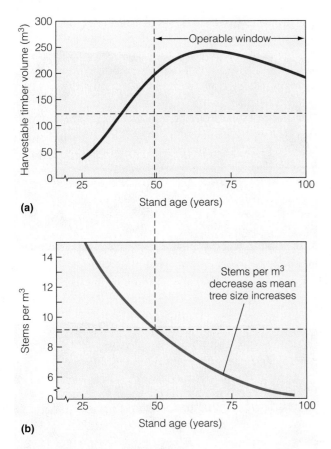

(a)

(b)

Figure 26.15 Two criteria determine the operable window, i.e., the time when a stand is suitable for harvest: (1) harvestable wood volume (m³/ha) and (2) mean tree size (stems per m³ wood volume). In this example, the dashed horizontal lines represent constraints of **(a)** minimum saleable volume of 125 m³/ha and **(b)** minimum tree size that is associated with ~ 9 stems/m³. Dashed vertical lines indicate the earliest stage at which both constraints are met. Go to QUANTIFY it! at **www.pearsoncanada.ca/ecologyplace** to explore the effects of initial stand density on the operable window.

INTERPRETING ECOLOGICAL DATA

Q1. What does graph (a) imply about the change in mean tree size and number with stand age?

Q2. Assume the stand is harvested as soon as it reaches the minimum harvestable wood volume (125 m³/ha). How many tree stems per m³ wood volume would be harvested? Would this decision be viable? Hint: Note from graph (b) that as tree size increases with age, the number of stems per wood volume decreases.

(Adapted from Orians et al. 1986.)

low. As a result, leaching into ground and surface waters increases (Figure 26.17). Higher nutrient outflow is a direct result of decoupling the linkage between nutrient release in decomposition and uptake in primary productivity that is so vital to ecosystem function. In the long term, smaller amounts of nutrients are available, which in turn reduces plant growth. A longer rotation period is required for subsequent harvests. If the rotation period is kept unchanged, yields fall. In conventional management systems, foresters often counter nutrient loss by applying chemical fertilizers, creating ecological problems for adjacent aquatic ecosystems.

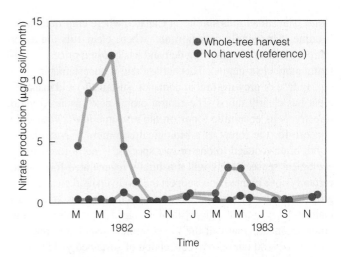

Figure 26.16 Nitrate release after whole-tree clear-cutting of a loblolly pine (*Pinus taeda*) plantation compared with an unharvested stand over a 2-year period (every other month is labelled; J = July, S = September, etc.).

(Adapted from Vitousek 1992.)

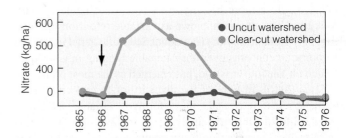

Figure 26.17 Changes in nitrate concentration of streamwater of two watersheds in Hubbard Brook, New Hampshire. The forest on one watershed was clear-cut, the other undisturbed. Increased nitrate after clear-cutting reflects increased decomposition and mineralization, followed by leaching into surface and groundwater.

(Adapted from Likens and Borman 1995.)

Sustained yield is practised to varying degrees, but all too often, industrial forestry sees trees as a crop rather than part of an integrated forest ecosystem. This approach resembles industrialized agriculture: they clear-cut, spray herbicides, plant or seed the site to one species, clear-cut, and plant again. Clear-cutting practices in parts of the Pacific Northwest (coastal regions of British Columbia, Oregon, and Washington) and Alaska hardly qualify as sustained yield when below-cost sales are mandated by the government to meet politically determined harvest quotas. As the timber supply dwindles in the Pacific Northwest, the industry that moved west after depleting the eastern hardwood and pine forests of the Great Lakes region is returning east to exploit the secondary regrowth of eastern hardwood forests, especially the diverse central hardwood forest. From Virginia and eastern Tennessee to Arkansas and Alabama, timber companies have built more than 140 chip mills that cut up trees of all sizes into chips for paper pulp and particle board, which is increasingly used for cabinets and furniture. Feeding these mills requires clear-cutting 500 000 ha annually in the United States alone. The situation is similar in Canada, where chip mills have become increasingly common and where clear-cuts are as or more extensive. The growing demand has boosted prices, stimulating more clear-cutting. This harvest rate is unsustainable.

In face of growing timber demands, sustained-yield management has clearly failed. The central problem of sustained-yield forestry is its economic focus on the resource itself, with little concern for the forest as a biological community. A managed stand, often reduced to one or two species, is not a forest in an ecological sense. Rarely will a naturally regenerated forest, and certainly not a planted one, support the biodiversity typical of old-growth forests. By the time the trees reach economic (financial) maturity, they are cut again, with little consideration for the many other organisms that call the forest home. Sustainable forestry must go beyond our current conception of sustained yield to take the forest's broader ecological significance into account.

26.9 SUSTAINABLE FORESTRY: Variable Retention Aims to Protect Attributes of Mature Forests

What would a more broadly conceived sustainable forestry look like? The approach known as **variable retention harvesting**, first promoted by the Clayoquot Scientific Panel in 1995 in response to concerns over unsustainable logging in Clayoquot Sound on Vancouver Island, has emerged as the most promising (D'Eon 2006). In light of the diversity of forest ecosystems, variable retention can take many forms—"variable" alludes to the need for flexibility in applying the method—but its goal is to retain ecologically significant components of forests during and after harvest. This strategy not only promotes tree regeneration but also conserves biodiversity.

The components to be retained, as well as their frequency and pattern on the landscape, depend on the structure of a particular forest (Figure 26.18). Obtaining this basic knowledge is a key first step. In coastal British Columbia, where variable retention was developed, critical components include (1) *fine and coarse woody debris*, particularly fallen logs that provide habitat for animals and fungi and are nurse logs for regenerating tree species that prefer to germinate on rotting organic matter; (2) *forest floor and understory species*, which maintain soil processes and site conditions; (3) *dead-standing trees (snags)*, which provide habitat and food for late-successional species such as woodpeckers; and (4) *living mature trees*, which maintain microclimate conditions for regenerating trees as well as acting as a seed source. Retaining living trees may superficially resemble either a seed-tree system or a shelterwood system (depending on their density and pattern), but the reasons for retaining them include not only their roles in tree regeneration but also their value for non-target species.

Ecological, not economic, objectives take precedence, with the goal of increasing structural complexity by retaining critical attributes of old-growth stands. Of course, harvesting still reduces stand complexity, but as regeneration continues, a diversified stand structure will (it is hoped) re-establish more rapidly. In addition, variable retention better maintains the visual aesthetics

(a)

(b)

Figure 26.18 Contrasting variable retention patterns. **(a)** 12 percent group (aggregated) retention. **(b)** 12 percent individual tree (dispersed) retention.

of forests compared with clear-cutting. Conventional forestry has been criticized for leaving "beauty strips" (sometimes called "idiot strips") of intact forest along highways to hide the impacts of logging from the public. Variable retention addresses this concern by employing logging methods that achieve economic ends while also meeting ecological objectives.

Since its development in British Columbia, scientists have studied and encouraged implementation of variable retention harvesting elsewhere in Canada. In Alberta, logging companies have begun using it, and stand retention is now required in Ontario and Nova Scotia. Other jurisdictions will likely follow suit, but the vital components of stand retention in these forests differ from those in west coast temperate rain forests. In particular, altering the retention strategy to emulate the widespread role of fire will prove critical to long-term success in boreal forests.

An essential element in evaluating the success of variable retention is its impact on wildlife. Animals respond variably to both the amount of stand retention and its pattern on the landscape. Some songbirds that rely on particular elements of old growth benefit from as little as 10 percent retention of components such as snags for cavity-nesting species, or live mature trees for foliage-feeders. More sensitive species require more retention. To minimize the impact of open areas on tree regeneration and wildlife, many recommend that cuts not exceed 1 ha.

In addition to monitoring wildlife response, evaluating nutrient cycling and other key ecosystem processes mediated by fungi and bacteria is essential if variable retention is to fulfill its promise as an ecosystem-based approach. But the question remains whether variable retention methods (and other sustainable approaches) can meet the pressures of increasing demand for forest products. However it is managed, forestry must still balance harvested biomass with regrowth. Rotation times must be sufficiently long to allow this balance to be reached. If pressures from a growing population coupled with consumer demands continue to rise, adopting the more ecologically based variable retention approach will not guarantee sustainable forest use.

Some think the answer is to leave natural forests alone and rely on tree plantations. But just as agricultural land conversion has made tallgrass prairie the most endangered ecosystem in Canada, so too would tree plantations inevitably replace more and more forests—especially if such plantations were relied upon not only for paper and wood, but also for biofuels. We might "solve" one environmental problem (dependence on fossil fuels and its associated impacts on pollution and global climate change) with another, equally serious, problem—loss of forest ecosystems and their associated biodiversity and ecosystem services. Which would you choose? The bottom line is this: We can and must adopt more sustainable methods of harvesting forests, but eventually we will have to consume less and recycle more if our forests are to survive the pressures of human exploitation.

26.10 CONVENTIONAL FISHERIES: Humans Have Exploited Fish Stocks in Unsustainable Ways

Although some question the reliability of global estimates, aquaculture now provides half of the annual fish and shellfish harvest for human consumption (Naylor et al. 2009). But this increase by no means indicates that human pressures on natural aquatic populations have diminished. Indeed, raising carnivorous fish requires large inputs of wild fish (Naylor et al. 2000), exacerbating rather than ameliorating our unsustainable use of aquatic habitats.

There are many historical accounts of overexploitation and declines, but until the late 1800s there was no attempt to manage fisheries to ensure their continuance. At that time, wide fluctuations in North Sea catches began to affect the commercial fishing industry. Debates ensued over whether commercial fishing was responsible for the decline. Only when Danish fish biologist C. D. J. Petersen developed the mark–recapture method for estimating population size (see Section 9.5) could biologists begin assessing fish populations. Along with data from egg surveys and aging of fish from commercial catches, these studies suggested that overharvesting was indeed the culprit. But the con-

troversy was not laid to rest until after World War I (1914–1918). During the war, fishing in the North Sea ceased. After the war, fishermen experienced sizable increases in their catches. Biologists suggested that renewed fishing would once again reduce populations, and that catches would stabilize and eventually decline. Their predictions were correct, and with time, attention turned to the issue of sustainable harvest.

The basis of sustainable fishing is the logistic model of population growth, first presented in Section 10.7. Under the logistic model, growth rate (number of offspring produced per year) is low not only when a population is small (Figure 26.19) but also when a population nears its carrying capacity (K) because of density-dependent processes, particularly competition for limited resources. Intermediate-sized populations have the greatest growth capacity and ability to produce the most harvestable fish. The key realization is that fisheries can optimize harvest of a species by keeping its population at an intermediate level and harvesting it at a rate equal to its annual growth rate—a strategy called **maximum sustainable yield**.

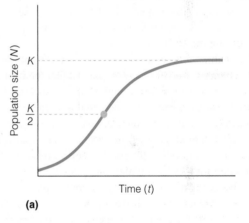

(a)

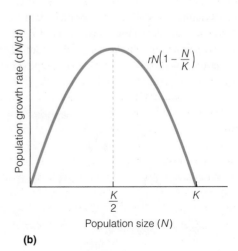

(b)

Figure 26.19 Graphical basis of maximum sustained yield. Assume growth follows the logistic model: $dN/dt = rN(1 - N/K)$. **(a)** In the absence of fishing, the population reaches carrying capacity, K. **(b)** The relationship between population growth rate, dN/dt, and population size, N, takes the form of a parabola, reaching a maximum when $N = K/2$.

Simply put, the concept of sustainable yield is to try to be a "smart predator" by maintaining the prey at a density where reproduction of new individuals just offsets mortality from harvest. The higher the rate of population increase, the higher the harvest rate at the maximum sustainable yield. Species with rapid population growth often lose much of their production to density-independent mortality caused by abiotic factors (see Section 11.11). The objective is to reduce this "waste" by harvesting individuals that would otherwise succumb to natural mortality.

As straightforward as this strategy sounds, fish populations have proved hard to manage. Stocks can deplete quickly if reproduction is disrupted by abiotic conditions. Consider the Pacific sardine (*Sardinops sagax*). Exploitation in the 1940s and 1950s shifted the age structure to younger age classes. Before exploitation, reproduction was distributed among the first five age classes (years). In the exploited population, the pattern shifted, with almost 80 percent of reproduction occurring in the first two age classes. Two consecutive years of environmentally induced reproductive failure (as a result of natural climate variations associated with El Niño; see Section 2.10) triggered a population collapse from which the species never recovered (Figure 26.20).

Sustainable yield requires a detailed understanding of population dynamics. Recall from Chapter 10 that the intrinsic rate of population growth, r, is a function of age-specific birth and mortality rates. Unfortunately, the usual approach to sustained yield fails to give adequate consideration to size and age classes and their differential rates of growth, survival, and reproduction; sex ratio; and environmental uncertainties—all difficult data to obtain for natural populations.

Compounding the problem is the common-property nature of the resource. Because the resource is perceived as belonging to no one, we act as if it belongs to everyone to use as each sees fit. Fisheries provide an example of what Garrett Hardin (1968) called the **tragedy of the commons**. An individual vessel (or total fishing by a country) gets all the economic benefit from an extra unit of harvest, whereas the long-term costs of depleting stocks are distributed among all participants. Without enforced limits, it "pays" the individual vessel or country to overexploit the resource.

Perhaps the greatest problem with sustainable harvest models (in forestry and agriculture as well as in fisheries) is their failure to incorporate economic factors (see Section 26.12). Once commercial exploitation begins, the pressure is on to increase it to maintain the underlying economic infrastructure. Attempts to reduce exploitation meet strong opposition. People argue that reduction will mean unemployment and bankruptcy—that, instead, the harvest effort should increase. This argument is short-sighted. An overused resource will inevitably fail, and the livelihoods it supports will collapse. In the long run the resource will be depleted. Abandoned fish-processing plants and rusting fleets in the Maritimes and Newfoundland and Labrador support this view. With more conservative exploitation, the resource could be maintained.

From an ecological perspective, a major stumbling block with the sustainable yield concept is that conventional management considers stocks of species as isolated biological entities rather than as components of an ecosystem. Each stock is managed to maximize economic return, ignoring the need to leave behind a certain portion to continue its ecological role(s) within the community, be it predator or prey. This attitude encourages a huge discard problem, euphemistically called **bycatch** (Figure 26.21). Employing large purse seines and gill nets (Figure 26.22) that cover many square kilometres of ocean, fishers haul in not only commercial species but also other marine life, including sea turtles, dolphins, and scores of fish. To harvest bottom-dwelling groundfish, such as halibut, haddock, and flounder, fishers use trawls that sweep over the ocean floor, scraping up all forms of benthic life. This

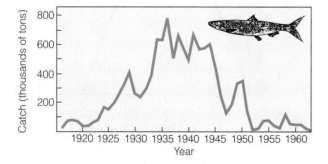

Figure 26.20 Annual catch of Pacific sardines along the Pacific coast of North America. Overfishing, environmental changes, and increased numbers of a competing fish, the anchovy, caused a collapse of the sardine population.

(Adapted from Murphy 1966.)

Figure 26.21 Bycatch can outweigh target species (in this case, shrimp) by 5- to 20-fold. This bycatch in the Gulf of California includes bottom-dwelling brittlestars, crabs, small fish, and skates.

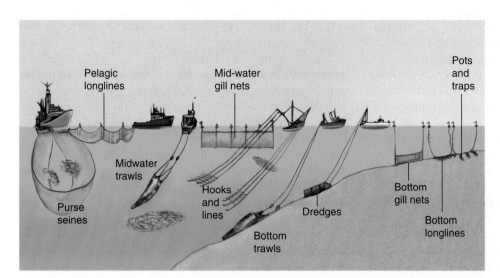

Figure 26.22 Commercial fishing gear. Purse seines, mid-water gill nets, and pelagic longlines cover many square kilometres of ocean. Dredges and bottom trawls scrape the seabed.

(Adapted from Chuenpadgee et al. 2003.)

practice damages critical bottom habitats and alters community structure at all trophic levels.

The North Atlantic cod (*Gadus morhua*) fishery is another an example of an unsustainable fishery (Kurlansky 1997). For 500 years, Atlantic waters from Newfoundland to Massachusetts supported one of the world's greatest fisheries. In 1497, the Italian explorer Giovanni Caboto (dubbed John Cabot by his English employers) marvelled at the abundance of cod off Newfoundland. He spoke of seas "swarming with fish that could be taken not only with nets but with baskets weighted down with stone." Some cod were 2 m long and weighed up to 80 kg. The news sparked a frenzy of exploitative fishery. The Portuguese, Spanish, English, and French sailed to Newfoundland to take between 100 000 and 200 000 tonnes a year. In the 1600s, England took control of Newfoundland and its waters, establishing numerous coastal posts where merchants salted and dried cod before shipping it to England. So abundant were cod that people thought it was an inexhaustible resource.

Catches remained rather stable until after World War II (1939–1945), when demand increased dramatically and fishing efforts intensified. Large factory trawlers that could harvest and process the catch at sea replaced smaller vessels. Equipped with sonar and satellite navigation, fishing fleets could locate spawning schools and engulf them with huge purse nets, sweeping the ocean floor clean of fish and associated marine life. In the 1950s the annual catch was 500 000 tonnes of cod but by the 1960s had almost tripled to 1 475 000 tonnes. In 15 years from the mid-1950s through the 1960s, 200 factory ships off Newfoundland took as many northern cod as were caught over the prior 250-year span since Caboto's arrival (Kurlansky 1997).

The fishery could not endure such intense exploitation. By 1978 the catch had declined to 404 000 tonnes. To protect their commercial interests, Canada and the United States excluded all foreign fisheries in a 200-mile (320-km) zone. But instead of capitalizing on this chance to allow cod stocks to recover, the Canadian government provided industry subsidies for huge factory trawlers. After a brief surge in catches during the 1980s,

the catch in Canadian waters fell to 13 000 tonnes by the mid 1990s. In 1997 the fishery collapsed entirely (Figure 26.23). The Gulf of Maine and Grand Banks fisheries controlled by the United States followed a similar trend. In 2005, combined total commercial cod landings for Canada and the United States were 6957 tonnes, down 17 percent from 2004. Cod is still being overfished, even under a management program designed for stock recovery. Despite closures, gear restrictions, minimum size limits, limits on days at sea, and a moratorium on permits, cheating on catch limits and a substantial illegal trade contribute to overfishing.

The collapse of the Atlantic cod fishery created an ecological, economic, and social disaster. In Newfoundland it left 40 000 fishers and fish processors out of work and devastated nearly half of the region's 1300 fishing communities. In New England, it led to fleet reductions, processing plant closures, and job losses. The industry blamed the collapse on

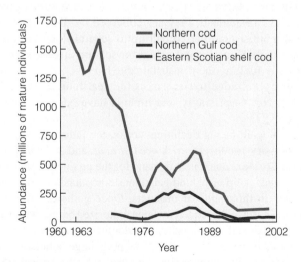

Figure 26.23 Historical trends in Canadian Atlantic cod fishery, based on three cod stocks. Note the spike in the 1980s, followed by eventual collapse.

(After Hutchings and Reynolds 2004.)

seal predation of the cod's preferred food items, Atlantic herring (*Clupea harengus*) and capelin (*Mallotus villosus*), among other factors. After a 10-year moratorium, cod has yet to recover. Many scientists who once denied that overfishing was the root cause have now concluded that centuries of overfishing cod and devastating other species through discard are indeed responsible.

As with forestry, emphasis on managing commercial fish using a single-species model, thereby viewing the dynamics of that species in isolation from its community, ignores the multitude of direct and indirect interactions that affect populations. As we saw in Section 16.8, an impact on one species can reverberate through the food web. The cod is a top predator (Figure 26.24a). After cod populations collapsed, their prey—herring, crabs, sea urchins, etc.—increased dramatically. Spiny dogfish (*Squalus acanthias*) replaced cod as a top predator. With the increase in spiny dogfish, the fishery shifted to it, fuelled by an increasing European demand. A slow-growing, slow-reproducing shark, the spiny dogfish (Figure 26.24b) became an important commercial species subject to the same unsustainable pattern of overfishing and decline.

Sadly, this history of overexploitation and an unwillingness of governments to regulate fisheries in the interests of marine ecosystems continues to repeat itself. Following the collapse of the cod and dogfish populations, the industry moved to the next-lower trophic level, concentrating on a species that was once the cod's main prey—herring. Intense fishing pressure caused a sharp decline in that species, too. So the industry went farther down the food chain to the dominant herbivores, (1) the kelp-grazing spiny green sea urchin (*Strongylocentrotus droebachiensis*; Figure 26.24c), prey for lobsters, crabs, and seabirds, and (2) green-algae-feeding periwinkles (*Littorina* spp.; Figure 26.24d), a critical food for many invertebrates, fish, and waterfowl. Periwinkle snails, a delicacy in Europe and Japan, were formerly harvested from rocky coasts at low tide. Now, given the growing demand, they are collected by diver-operated suction harvesters that collect up to several thousand kilograms daily. A growing demand for sea urchins for sushi is currently fuelling their unsustainable harvest, accomplished either by a dredge towed across the sea floor or by scuba divers. Not surprisingly, sea urchins have been in decline since 1993.

Yet again facing declining stocks, the industry turned to the primary producers, rockweed (*Fucus* and *Ascophyllum*; Figure 26.24e), which is in demand for the production of fertilizer, mulch, kelp meal, and gelatinous compounds used as stabilizers in thousands of products. Once gathered by hand, rockweed is now harvested by large motorized barges with vacuum devices and cutter blades. As a dominant primary producer of coastal ecosystems, rockweed supports large populations of invertebrates, including periwinkles, and provides critical nursery habitat for many species of fish. The long-term impact of rockweed harvesting is of great concern.

The Atlantic fishery illustrates all too clearly that managing for a single species has direct and indirect impacts on the rest of the food web. Fishing removes not just the target species, but many other species at all trophic levels as discard, while damaging or destroying benthic habitat. The problem grows more acute as the industry moves down the web to the primary producer level, as in the harvesting of rockweed. Fortunately, there is a growing realization of the long-term effects of current harvesting activities, coupled with a recognition that the long-prevailing model of maximum sustained yield for individual species must be replaced with a model of ecologically sustainable yield that considers the role of the harvested species in the broader ecosystem. This approach will entail stricter catch limits, as well as changes in the timing and methods of fishing.

Implementation of ecologically sustainable yield will require political, economic, and social changes that involve forgoing short-term profits for long-term sustainability. Whether it will be too little, too late remains to be seen. The ultimate outcome of human overexploitation of oceanic ecosystems—not just on the Atlantic coast of North America but around the world—is as yet unknown, but many believe that we may have already pushed the world's oceans beyond a tipping point (see Section 19.9).

26.11 SUSTAINABLE FISHERIES: New Approaches in Fisheries and Aquaculture Are Ecosystem Based

Setting aside the worst-case scenarios and assuming recovery is possible, what would a sustainable fishery and aquaculture system look like? An inherent difference between implementing sustained management in fisheries versus forestry is that population levels of both target and non-target species (competitors, prey, and predators) are harder to estimate, despite improved methods of sampling and analyzing aquatic populations. Other relevant information regarding seasonal movements and feeding and reproductive activities is also more difficult to obtain, despite increasingly sophisticated tracking devices.

Given this situation, many support aquaculture as a way of harvesting fish while preserving aquatic systems—just as some promote tree plantations as a way to preserve natural forests. Certainly reliance on aquaculture has increased greatly. About half of the fish and seafood we consume derives from aquaculture (Naylor et al. 2009). By "cultivating" aquatic species—the term reveals the agricultural model—ecologists can monitor and manipulate the growth of individuals and populations, allowing predictable harvests.

But, as with plantations, there are ecological consequences. The first aquaculture systems (as well as some existing systems) displaced natural ecosystems in sensitive transition habitats such as mangroves and coastal marshes. These systems not only reduced the numbers of non-target species, but also compromised or eliminated the services that these displaced ecosystems provide, such as the buffering effect of mangroves. Moreover, cultivating predatory species uses large amounts of fishmeal, so the drain on natural systems is still substantial

Figure 26.24 Major species involved in the North Atlantic harvest. (**a**) Atlantic cod, (**b**) spiny dogfish, (**c**) green sea urchin, (**d**) common periwinkle, and (**e**) rockweed, a brown seaweed.

(Naylor et al. 2000). Efforts are underway to reduce the use of fishmeal, but the problem remains.

Other effects are more subtle. As with agricultural species, cultivated aquatic species, such as farmed salmon, may escape and mate with wild species, introducing bred traits (and also genetically modified traits, where applicable) into wild populations, with unknown impacts on natural biodiversity. Interactions with disease organisms are affected; disease outbreaks are more likely to occur under crowded fish farm conditions, which in turn

expose wild populations not only to disease organisms but also to the chemicals used to treat them. Open-net pens generate large and concentrated amounts of wastes, much as livestock operations do on land. These wastes can affect both aquatic habitats and adjacent terrestrial sites, if the assimilative capacity of the system is exceeded. It is estimated that the Clayoquot Sound UNESCO Biosphere Reserve is exposed to the equivalent sewage of 150 000 people from salmon farms. As well as feces, these wastes include contaminated feed and toxic chemicals.

Recently, attempts to establish a more sustainable aqua-culture have gained ground. This movement incorporates eco-system-based principles in order to minimize the adverse effects of conventional aquaculture. One approach is *polyculture*: rearing more than one species in the same site. Akin to raising more than one crop in an agricultural field, polyculture has a long history. Over 1000 years ago, fish farmers in China stocked ponds with carp, supported by an algal food web "fed" with agricultural by-products. However, if polyculture involves cultivating several species of fish, all the organisms undergo similar environmental and physiological processes, and the issues arising in conventional aquaculture remain unsolved.

Modern versions are based on a deeper understanding of trophic structure. An example is **integrated multi-trophic aquaculture** (IMTA), which cultivates organisms at different trophic levels, with different and complementary functions in the ecosystem. In the example depicted in Figure 26.25, the farming of salmon is combined with that of mussels and sea-weed, which utilize some of the organic and inorganic wastes of the salmon. Wastes are recaptured, and all three species are harvested. Research is ongoing into the addition of other inver-tebrates to increase diversity and improve efficiency. In the interest of sustainability, yields for individual "crops" are lower than they would be in monoculture, but total yield is enhanced and adverse effects are minimized. Outputs of one species become inputs for another, mimicking natural ecosystem func-tion (Chopin et al. 2008). Economic as well as ecological sta-bility improves, as a result of product diversification, risk reduction, and job creation in coastal communities. Instead of overexploiting one trophic level after another, IMTA aims for a balance among all cultivated populations.

IMTA can be conducted in either open-water (freshwater or marine) or terrestrial (ponds or tanks) sites. Use of waste-water in pond-based aquaculture has the added benefit of reducing eutrophication. Another approach to sustainable aquaculture involves closed containers rather than the open-net technology of most monoculture fish farms. Designed to

eliminate release of fish, disease organisms, wastes, and con-taminants, closed-container systems are often land based. However, escapes can occur and can lead to invasive species introductions (see Section 26.7). Also, the issues of effluent treatment remain.

Instead of aquaculture, some ecologists advocate an approach more like variable retention in forestry—harvesting fish in their natural habitat, but retaining enough components of the natural system such that populations can be regenerated more successfully while maintaining biodiversity. This approach often draws on the traditional ecological knowledge of aboriginal peoples, many of whom have harvested fish popu-lations sustainably for centuries (see Research in Ecology: Using Traditional Ecological Knowledge to Manage the Pacific Lamprey). However, although more ecologically sustainable methods of harvesting wild species are a realizable goal, feed-ing the growing human population will require multiple approaches. When fisheries have reached their sustainable lim-its or have been overexploited, the gap will have to be filled by some form of aquaculture.

26.12 ENVIRONMENTAL ECONOMICS: Economics Play a Critical Role in Resource Management

Although a large portion of the global population produces its own food and forest products by subsistence farming, fishing, and wood gathering, these resources are part of the global mar-ketplace. Economic considerations are thus central to the pro-duction and management of natural resources.

A **cost–benefit analysis**, which involves measuring and comparing all the benefits and costs of a particular project or activity, is a critical economic tool used in making decisions regarding natural resource management. If it costs a farmer $100 per hectare to produce corn this year, and the expected value of that corn is $200 per hectare, the benefits ($200) out-weigh the costs ($100), and the decision will most likely be to plant the corn. If the corn was worth only $80 per hectare, the costs would outweigh the benefits, so the farmer would be unlikely to plant corn and incur a loss. In this example, the dol-lar value of costs and benefits can be compared directly because both expenses and revenue occur during the same time period—the year the crop is planted and harvested. When costs and ben-efits extend over a longer period, it is necessary to use a procedure called **discounting** to compare costs and benefits that occur at different times.

As central as discounting is to the economics of natural resource management, it often runs counter to the objectives of sustainability. The problem is simple to explain, but hard to solve. In farming, forestry, and fisheries, there are substantial initial costs in acquiring land, equipment, permits, and so forth. These costs must be weighed against expected earnings from the production and harvest of crops, trees, or fish. However, a dollar earned in the future does not have the same value as a dollar in hand today. Inflation depreciates the value of future

Figure 26.25 Commercial IMTA site (Cooke Aquaculture Inc.) in the Bay of Fundy. Salmon cages are on the left, a mussel raft in right foreground, and a seaweed raft in right background.

RESEARCH IN ECOLOGY | Using Traditional Ecological Knowledge to Manage the Pacific Lamprey

David Close,
University of British Columbia.

Attempts to manage fisheries, even when based on principles of sustained yield, have had a lamentable track record. In search of a better way, scientists are exploring alternative approaches that draw on the **traditional ecological knowledge (TEK)** of aboriginals, many of whom have relied on fish populations for millennia. Collaborative efforts among aboriginal groups, academics, and governments promise a more sustainable fishery than decades of management based on maximum sustained yield theory have ever delivered.

At the forefront of this effort is the University of British Columbia's David Close. A member of the Umatilla Nation, Close is in the vanguard of scientists who are seeking a new synthesis between TEK and conventional science. Focusing his research on the Pacific lamprey (*Entosphenus tridentatus*; formerly *Lampetra tridentata*) (Figure 1), Close asks these questions: *Can traditional knowledge of this species, important to the peoples of the Pacific Northwest both as food and for its spiritual significance, forge a new understanding of its population dynamics and role in the larger ecosystem? Can this understanding inform more sustainable harvesting of this species, allowing it to continue to play these roles in the future?*

The species itself may seem an unlikely candidate for inspiring public concern. Humans tend to focus on so-called charismatic species—species that, because they are large in size, physically attractive, or have some traits of interest in their appearance or behaviour, hold a special appeal. The panda bear in China or the polar bear in Canada are obvious examples (see Ecological Issues: Conserving a Canadian Icon, pp. 604–605). In general, fish hold less appeal, with the exception of the Pacific salmon. Yet the importance of a species should surely not be dictated by its charisma.

About half of the world's 34 lamprey species are vulnerable, endangered, or extinct (Close et al. 2009). Pressures exerted on lamprey include overexploitation, but go far beyond direct harvest. In Scandinavia and Europe, industrial pollution, lower pH, and hydroelectric dams and other flow diversions also contribute to their decline. Physical obstructions threaten the lamprey in the Columbia River basin, where large dams were constructed from the 1930s to the 1970s. Although, like salmon and other anadromous species, the lamprey returns to freshwater to spawn, it has relatively weak *philopatry* and does not always return to the stream of its birth. Less than half of adult lampreys pass above the dams using fishways and other bypass structures. On lesser tributaries, devices installed to divert water for irrigation have proved as limiting to upriver movement as the larger dams on major rivers. Installing screens on irrigation canals prevents adult fish from being marooned on nearby fields, but larvae are still killed or impeded.

So little were lamprey valued by non-aboriginals that Umatilla River populations were deliberately poisoned by rotenone applications in 1967 and 1974 (Close et al. 2004) to reduce problems for dam operators and farmers. Subsequently, there were restoration efforts for salmon and trout, but not lamprey. Considered a threat to commercial and recreational salmon fishing, the lamprey was left to re-colonize the Umatilla River naturally—not easy, given the barriers that limit its ability to migrate from the Columbia River, where its numbers were in decline.

To facilitate re-establishment of Pacific lamprey in the Umatilla River, a coalition between the Confederated Tribes of the Umatilla Indian Reservation and the Oregon Department of Fish and Wildlife initiated a reintroduction program. Between 1999 and 2007, Close and his colleagues released over 2600 adult lamprey. Monitoring indicated success in the initial stages. The fish were able to find suitable spawning habitat; construct nests; and deposit viable eggs, which developed into healthy larvae. However, although distribution of larvae increased downstream, larvae failed to establish in the lower reaches of the river, where flows are heavily regulated for irrigation (Close et al. 2009). These reintroduction attempts, even when not entirely successful, identify additional factors affecting the recovery of species whose life history is not fully known.

So far, this research conforms to a conventional scientific approach. Yet preliminary work done by Close used ethnographic methods to gather traditional knowledge by interviewing tribal members. This knowledge derives from generations of eelers that have relied on lamprey as a subsistence fishery. In the past, conventional science has dismissed such knowledge as anecdotal. Although observational and non-experimental, TEK is increasingly recognized as of value, in part because it transcends the population approach to encompass the larger ecosystem. Firkes Berkes, a forerunner in developing the concept, defines TEK as "a cumulative body of knowledge, practice, and belief, evolving by adaptive processes and handed down through generations by cultural transmission, about the relationship of living

Figure 1   The Pacific lamprey, showing the "breather hole" that aboriginals believe is used during muddy conditions.

continued on page 588

RESEARCH IN ECOLOGY continued

beings (including humans) with one another and with the environment" (Berkes 1999).

In the case of the Pacific lamprey, traditional knowledge provides insights into its habitat, behaviour, and life history; trends in population size in the short and long term; and local ecological roles. At the very least, this information can cast doubt on previously held biases, as well as generate alternative hypotheses that can then be tested using conventional methods. Some ecologists promote a more radical approach to TEK. Working with Pacific lamprey in the Klamath River, Robin Petersen Lewis (2009) suggests that TEK can do more than feed the mill of conventional science—it can provide a parallel path towards understanding that can be intertwined with, not superseded by, conventional science. Complicating the use of TEK is the fact that such knowledge is viewed as sacred, with specific rules regarding its application and transmission, especially to non-aboriginals. Because disclosure can violate its sacredness, TEK is subject to intellectual property rights, akin to those that complicate use of hybrid or genetically modified crops.

Petersen Lewis based her composite picture of TEK of the Pacific lamprey on semi-directed interviews, observation sessions, focus groups, and archival research. In essence, the "secret" to the aboriginals' sustainable fishery involves the ability to adjust to the different microhabitats determined by stream morphology and hydrology (river mouth, falls, channels, riffles, banks, etc.) by altering both the timing of fishing and the type of fishing device (basket, net, hook, platform, hand) (Figure 2) in accordance with detailed knowledge of lamprey distribution, abundance, and behaviour in changing conditions.

A strong spiritual commitment to allow the river to replenish the lamprey population has fostered a deep appreciation for the lamprey's roles in the larger ecosystem, as well as an awareness of factors that jeopardize its ability to perform those roles. From Petersen Lewis's study, the factors identified by TEK as contributing to the lamprey's decline go beyond those stressed

by conventional science—overexploitation, contaminants, and physical obstructions to movement caused by dams and irrigation structures. They include more subtle influences associated with human management of adjacent terrestrial systems, including fire suppression, logging, herbicide use, and drainage of wetlands. These practices have combined to increase the severity of episodic floods, which in turn damage spawning habitat. Such a broad-based perspective is invaluable for furnishing a direction for future reintroductions, in keeping with Close's call to cast a wider net for the causal factors contributing to lamprey decline.

From a broader ecosystem perspective, TEK identifies the declines in the number of both living and dead lamprey since the 1960s as ecologically significant. Dismissed as "trash fish" by non-aboriginals, lamprey are seen by aboriginals as an indicator of ecosystem quality. As living prey for piscivores or as dead matter for scavenging salmon fry, lamprey are a major food source within the river system, and as dead biomass, they contribute nutrients derived from marine sources. In the past, the role of dead lamprey biomass has been ignored or downplayed by conventional science, but pulses of nutrients from decomposing fish can stimulate freshwater productivity and the number of benthic invertebrates in nutrient-poor river systems.

Insights gleaned from TEK, if integrated with conventional science as an equal partner, can foster a more far-reaching, systems-based approach to managing both the Pacific lamprey and other species at risk from the potent combination of overexploitation and ecosystem mismanagement.

Bibliography

Berkes, F. 1999. *Sacred Ecology*. Philadelphia: Taylor and Francis.

Close, D. A., A. D. Jackson, B. P. Conner, and H. W. Li. 2004. Traditional ecological knowledge of Pacific lamprey (*Entosphenus tridentatus*) in northeastern Oregon and southeastern Washington from indigenous peoples of the Confederated Tribes of the Umatilla Indian Reservation. *Journal of Northwest Anthropology* 38:141–162.

Close, D.A., K. P. Currens, A. Jackson, A. J. Wildbill, J. Hansen, P. Bronson, and K. Aronsuu. 2009. Lessons from the reintroduction of a noncharismatic, migratory fish: Pacific lamprey in the Upper Umatilla River, Oregon. *American Fisheries Society Symposium* 72:233–253.

Petersen Lewis, R. S. 2009. Yurok and Karuk traditional ecological knowledge: Insights into Pacific lamprey populations in the Lower Klamath Basin. *American Fisheries Society Symposium* 72:1–32.

1. Acquiring TEK involves techniques such as interviews and group sessions that are more at home in the social than the natural sciences. How would evidence derived in this way be evaluated from a conventional scientific perspective?

2. The approach discussed here involves collaboration between scientists and non-scientists. If the findings of science and TEK correspond, then there is no conflict, but how should this approach proceed when the findings contradict each other?

Figure 2 Lamprey harvest by Karuk tribal members using bicycle rim eel baskets along the Klamath River in California.

earnings. In addition, a dollar invested today is worth more in the future as a result of compound interest. Thus, in comparing present-day costs with expected benefits (revenue), the benefits are discounted to reflect the reduced value of future dollars. When applied to forestry or fisheries, discounting may support the conclusion that it is better to overexploit the resource now and invest the profits rather than to harvest the resource in a sustainable way over a longer period.

The economist Colin W. Clark (1973) made this point persuasively in the case of the blue whale, *Balaenoptera musculus* (see Figure 7.1b). Over 30 m long and weighing upward of 150 tonnes at maturity, the blue whale is the largest animal on Earth. It is also among the easiest to hunt and kill. More than 300 000 blue whales were harvested during the 20th century, with a peak harvest of 29 000 in 1930–1931. By the early 1970s, the population had plummeted to several hundred. Although international talks were held to address the problem and discuss regulatory policies, some countries were eager to continue the hunt even at the risk of extinction. So Clark asked which practice would yield the whalers the most money: cease hunting, let the population recover, and then harvest them sustainably in the future, or kill the rest off as quickly as possible and invest the profits in the stock market. The troubling answer was that if the whalers could achieve an annual rate of return on their investments of 20 percent or more, it would be economically advantageous to harvest all of the blue whales.

The problem with this purely economic approach is that the value of a blue whale (or any natural resource) is based only on measures relevant to the existing market, that is, on the going price per unit weight of whale oil and meat. It ignores other services provided by the blue whale, such as ecotourism (whale-watching), or the value of blue whales to future genera-tions. Above all, it views blue whales, trees, and even whole ecosystems as without inherent value beyond those calculated in economic terms. None of the functions and impacts of the blue whale in its aquatic ecosystem are factored into the calculations.

Another economic concept essential to understanding sustainable resource use is that of **externalities**: the impacts of the actions of one individual or group on another's well-being, but for which the costs (or benefits) are not reflected in market prices. For example, clear-cutting a forest will likely have adverse effects on adjacent areas. Erosion transports silt to adjacent streams. Fertilizers and pesticides used in site preparation pollute groundwater. The reduced water quality will affect drinking water, recreational value, and aquatic biota. Less obvious externalities include the lost ecosystem services associated with the forest itself. The economic costs of these impacts are typically not borne by the logging company. As externalities, they are not reflected in market prices. The company reaps the profits of the harvested trees, but the environmental costs of the harvest are borne by the public (and by the ecosystems themselves). If these externalities were included in the cost analysis, the benefits (profits) might no longer outweigh the costs, and the clear-cut might not take place. Another option is to pass on these costs to the consumer to reflect the actual environmental costs of the goods and services.

Recent decades have seen the emergence of **environmental economics**: the study of environmental problems with the perspective and analytic tools of economics. Incorporating economic principles into environmental decision-making is critical. Until the true ecological value of natural resources and the costs of their extraction and use are incorporated into economic decisions, sustainable management of natural resources is an impossible goal.

EcologyPlace

Visit EcologyPlace at www.pearsoncanada.ca/ecologyplace to access online resources that complement your textbook, and help you to apply and to review the information in this chapter. EcologyPlace includes

- an eText version of the book
- self-grading quizzes
- glossary flashcards
- and more!

Go to www.pearsoncanada.ca/ecologyplace and follow the registration instructions on the Student Access Code Card included with this text. If your book does not have a Student Access Code Card, you can purchase access to it at www.pearsoncanada.ca/ecologyplace.

SUMMARY

Sustainability 26.1

For resource exploitation to be sustainable, the rate of resource use (consumption rate) must not exceed its supply (regeneration) rate. If the resource is non-renewable, its use is not sus-tainable, and the rate of resource decline is a function of the rate at which the resource is exploited. Sustainable resource use can also be constrained indirectly by the negative impacts of resource extraction and use. Wastes contaminate air, water, and soil with pollutants that impair the regeneration of natural resource capital.

Natural Ecosystems 26.2

When we try to manage natural resources sustainably, we mimic natural ecosystems, which function as sustainable units. The rate at which they use resources such as nitrogen is limited by resupply via mineralization. Sustainability does not imply static rates of resource use.

Agriculture and Energy 26.3

Whatever the crop or cultivation method, agriculture replaces diverse natural ecosystems with a community consisting of a single species or a simple mixture. Agricultural methods differ in the type and amount of energy inputs. Traditional agriculture involves subsistence agriculture in which energy from human and animal labour produces only enough for a family to survive. Industrialized agriculture depends on large energy inputs in fossil fuels, chemical fertilizers, irrigation systems, and pesticides.

Swidden Agriculture 26.4

Swidden agriculture, a traditional method in the tropics, involves shifting cultivation in which trees are felled and burned to clear land for planting. Production declines with successive harvests. The plot is then abandoned, and the forest allowed to regenerate. Eventually, the nutrient status of the site recovers, and it can again be cultivated.

Industrialized Agriculture 26.5

In industrialized agriculture, machines and fossil fuel energy replace human and animal labour. Mechanization requires large expanses of land for machines to operate effectively. Because different crops require specialized equipment, farmers typically plant one or a few crops season after season. Tilling and removal of organic matter reduce soil nutrients and promote erosion. Large quantities of chemical fertilizers and pesticides are used to maintain productivity.

Agricultural Trade-offs 26.6

Traditional and industrialized agriculture involve a trade-off between energy input in production and energy harvest in food resources. Industrialized agriculture produces high yields at the expense of large inputs in fertilizer, pesticides, and fossil fuels. Each input has ecological impacts. Traditional agriculture yields more energy per unit of input.

Sustainable Agriculture 26.7

Sustainable agriculture attempts to maintain production while minimizing negative impacts. Its methods conserve soil and water resources, reduce pesticide use, control pest populations by ecological means, and use alternative sources of nutrients.

Conventional Forestry 26.8

The goal of sustained yield in forestry is to balance net growth and harvest using various techniques, from clear-cutting to selective harvest. A harvestable stand must satisfy minimum thresholds for volume of timber per hectare and mean tree size. Time between harvests must be sufficient for the forest to regain its pre-harvest biomass. Because harvesting and post-harvest site treatments remove nutrients, it is often necessary to fertilize, which increases nutrient losses to adjacent aquatic systems.

Sustainable Forestry 26.9

An enlightened approach to sustainable forestry considers the viability of the forest ecosystem, not just populations of commercial species. The goal of variable retention harvesting is to retain components of mature forests, such as live mature trees, snags, and understory species. These components not only promote tree regeneration but also support forest diversity and function.

Conventional Fisheries 26.10

Central to sustainable harvest is the logistic model. Theoretically, optimal harvesting maintains the population at an intermediate density (where its growth rate is highest) and harvests it at a rate equal to its annual growth. In practice, maximum sustainable yield theory has not prevented overexploitation because it requires a greater understanding of population structure and dynamics than is often available, and does not factor in the impacts of density-independent factors and economics. Stocks are treated as single units and managed to maximize economic return. As target species are depleted, fisheries deplete stocks at lower trophic levels.

Sustainable Fisheries 26.11

Ecological sustained yield considers harvested species as part of an ecosystem. Aquaculture can have serious environmental impacts, but if based on ecosystem principles, multi-trophic aquaculture may prove sustainable. Others advocate harvesting species in a natural setting, while integrating conventional scientific methods with the traditional ecological knowledge of aboriginal peoples.

Environmental Economics 26.12

Cost–benefit analysis is a economic tool for making decisions about resource management. When costs and benefits extend over a long time, future benefits are discounted to reflect the reduced value of future dollars. Discounting often promotes decisions that run counter to sustainability. Externalities occur when the actions of one individual affect another's well-being, but the relevant costs or benefits are not reflected in market prices. Many production costs (pollution, habitat degradation, and impacts on human health) are not built into the price structure. These externalities must be considered if resource use is to be truly sustainable.

KEY TERMS

bycatch	externalities	integrated pest	renewable resource	swidden (slash-and-
clear-cutting	genetically modified	management	rotation period	burn) agriculture
cost–benefit	(transgenic)	maximum sustainable	(harvest interval)	traditional agriculture
analysis	green manure	yield	seed-tree harvesting	traditional ecological
discounting	Green Revolution	monoculture	selection (selective)	knowledge
ecosystem services	human ecology	natural capital	cutting	tragedy of the commons
environmental	industrialized	non-renewable	shelterwood cutting	variable retention
economics	agriculture	resource	sustainable	harvesting
environmental	integrated multi-trophic	pollutant	agriculture	yield
sustainability	aquaculture	polyculture	sustained yield	

STUDY QUESTIONS

1. What relationship between supply and consumption rates must exist for consumption of a resource to be sustainable?
2. Contrast *renewable* and *non-renewable resources* in the context of sustainable resource use.
3. How might sustainable use of resources be limited indirectly by adverse impacts from the management, extraction, and consumption of resources? Provide examples.
4. Contrast *traditional* and *industrialized agriculture*. What are the major energy inputs in each? How do their nutrient cycles differ?
5. Which agricultural method (industrialized or traditional) produces the greatest crop yield per unit area? Which produces the greatest yield per unit of energy input? Why?
6. Based on information in this chapter as well as your own research, what were some of the positive and negative outcomes of the Green Revolution?
7. Discuss practices that can increase the sustainability of industrialized agriculture. What obstacles do you foresee to their widespread implementation on Canadian farms?
8. Contrast *sustained yield* with *maximum sustainable yield*. Why has application of maximum sustainable yield theory not prevented the overexploitation of resources?
9. Discuss two sources of nutrient losses during forest harvest and post-harvest management. How could these losses be minimized?
10. Some argue that clear-cutting is an ecologically sound method in Canada's boreal forest. What is the theoretical basis of this position, and what features of clear-cutting would need to be employed to support it?
11. Compare *variable retention harvesting* with *shelterwood* and *selective cutting* in terms of their objectives and methods.
12. Discuss (based on this chapter as well as independent research) some of the ecological issues involved in conventional, open-container aquaculture. How does multi-trophic aquaculture address these issues? What issues remain unresolved?
13. Why is it important to take an ecosystem approach to fisheries management rather than approaching the management of each species as a separate unit? How would traditional ecological knowledge contribute to this new approach?
14. We have discussed TEK in relation to fisheries, but what contributions can it make to agriculture in the developing world?
15. Why do economists discount future benefits? What is the effect of discounting on sustainable resource management?
16. Why is air pollution caused by coal-fired power plants considered an externality? How could externalities best be incorporated into resource management decisions?

FURTHER READINGS

Berkes, F. 1999. *Sacred ecology: Traditional ecological knowledge and resource management*. Philadelphia: Taylor and Francis.
This seminal work, written by an eminent Canadian cultural ecologist, explores the importance of local and indigenous knowledge as a complement to conventional scientific ecology.

Clover, C. 2006. *End of the line: How overfishing is changing the world and what we eat*. New York: New Press.

Excellent treatment (not restricted to cod) of the global impacts of overfishing.

Daily, G. (ed.) 1997. *Nature's services: Societal dependence on natural ecosystems*. Washington, D.C.: Island Press.
Eminent scientists explain the critical role of natural ecosystems in meeting basic human needs.

Farley, J., J. Erickson, and H. E. Daly. 2003. *Ecological economics*. Washington, D.C.: Island Press.
Places traditional economic concepts in the context of economic growth, environmental degradation, and social inequity.

Gliessman, S. R. (ed.) 1990. *Agroecology: Researching the ecological basis for sustainable agriculture*. Ecological Studies Series no. 78. New York: Springer-Verlag.
Discusses agricultural systems employed in different regions, including tropical and temperate zones, and discusses the ecological issues associated with agricultural production.

Jaeger, W. K. 2005. *Environmental economics for tree huggers and other skeptics*. Washington, D.C.: Island Press.
Explains how to apply economics to environmental causes.

Jenkins, M. B. (ed.) 1998. *The business of sustainable forestry: Case studies*. Chicago: J. D. and K. T. MacArthur Foundation.
Integrates and analyzes 21 case studies to provide a composite snapshot of sustainable forestry.

Kurlansky, M. 1997. *Cod: The biography of the fish that changed the world*. New York: Walker and Company.
Documents the history of cod, from abundance through scarcity through determined governmental short-sightedness.

McNeeley, J. A., and S. J. Sheerr. 2002. *Ecoagriculture*. Washington, D.C.: Island Press.
Discusses the management of tropical landscapes for production of food and conservation of biodiversity.

Pesek, J., ed. 1989. *Alternative agriculture*. Washington, D.C.: National Academy Press.
Includes 11 case studies describing the practices and performance of alternative farming systems.

Roughgarden, J., and F. Smith. 1996. Why fisheries collapse and what to do about it. *Proceedings of the National Academy of Sciences USA* 93:5078–5083.
Applies maximum sustainable yield theory to the Newfoundland fishery collapse.

CONSERVATION ECOLOGY

Roadways, such as this one through Brazil's Mato Grotto region, function as both a barrier to dispersal and a source of mortality for many species.

Scientists believe that 65 million years ago—at the end of the Cretaceous era, in Mexico's Yucatán region—a massive meteorite struck Earth, leaving a crater 180 km across under the Caribbean. Deep-sea sediment cores provide a record of the impact. Debris, blasted high into the atmosphere, triggered a major temperature decline. The impact was largely responsible for the extinction of 70 percent of all species then inhabiting Earth, including non-avian dinosaurs. As we explore in Section 27.1, there have been other such *mass extinction events*. They have changed the course of evolution, inducing dramatic shifts in the types of organisms on Earth.

As strange as it may seem, Earth is in the midst of a mass extinction event on a par with previous events, with thousands of species lost yearly. The current event differs in critical ways from past events. Its cause is not extraterrestrial, such as a meteorite, or terrestrial, such as natural climate change. Instead, the extinctions are due to our activities. In North America's recent past, hunting caused the extinction of mammals such as Stellar's sea cow (about 1767), the New England sea mink (about 1880), and the Caribbean monk seal (about 1952). Globally, overkill has been the main cause in virtually all 46 modern extinctions of large land mammals. Among birds, overkill has been responsible for 15 percent of the extinctions of 88 species and 83 subspecies. Affected species include the great auk (1844) and passenger pigeon (1914). Sometimes, overkill reflected a mistaken belief that a species was a threat to crops or domestic animals. Examples include the Carolina parakeet, the only native North American parakeet (Figure 27.1). Yet, when compared to current extinction rates, the number of species exterminated by hunting and overexploitation is relatively small. By far, the greatest cur-

rent threat to Earth's biotic diversity is alteration and destruction of habitat by humans.

In this chapter, we explore global diversity patterns in space and time, considering the possible factors and processes that have contributed to those patterns. We then explore current threats to the world's biodiversity, as well as the strategies that are underway to conserve biodiversity and restore damaged ecosystems.

27.1 GLOBAL DIVERSITY PATTERNS: The Number of Species Varies over Time and Space

As well as its controversial relationship to stability and productivity (see Chapter 19), diversity is of great interest in and of itself, given the strong diversity patterns in time and space. Earth supports an amazing biodiversity. By the end of the 20th century, scientists had identified 1.4 million species (Wilson 1999) (Figure 27.2). By 2011, that number approached 1.9 million, but the task is far from complete. Quantifying global species richness is an ongoing exercise (Figure 27.3). Whatever that number is—Wilson (1999) contends it approaches 10 million, whereas Mora et al. (2011) peg it at 8.7 million eukaryotic species, of which 2.2 million are marine—diversity is not static. New species evolve while others disappear. Over 99 percent of species that have ever existed have gone extinct. Earth's diversity is thus a story of continual change. Nor is diversity similar everywhere. Geographic diversity patterns reflect abiotic factors that have affected habitat suitability in the short term and evolution in the long term.

In Section 17.8, we explored changes in diversity during succession (within 500 years). In succession, local diversity patterns reflect short-term changes in the distribution and abundance of **extant** (existing) species. Over a longer timeline, climatic changes have influenced regional diversity by shifting

Figure 27.1 The Carolina parakeet (*Conuropsis carolinensis*) went extinct in the early 20th century from hunting and extermination.

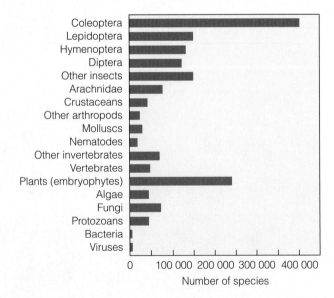

Figure 27.2 Number of currently known species, classified by major taxonomic group. Insects and plants dominate known diversity, but many microbial species are as yet unidentified.

(Adapted from Wilson 1999.)

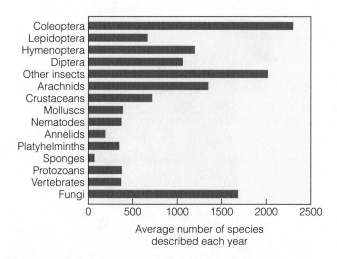

Figure 27.3 Many previously unknown species are discovered each year. Most belong to taxonomic groups in which the organisms are small in size. Archaea and bacteria (both highly diverse groups for which there are no accurate estimates) are not included.

(Adapted from Wilson 1999.)

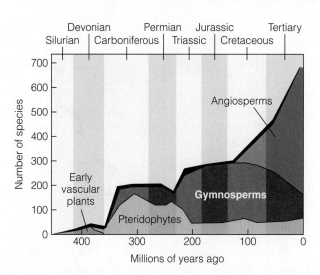

Figure 27.5 Expansion and contraction of major terrestrial plant groups over the past 400 million years.

(Adapted from Niklas et al. 1983.)

geographic ranges. In eastern North America, shifts in tree distributions after the last glacial maximum are an example (see Figure 17.22). Species ranges continue to shift, altering regional and local diversity. Changes in distributions of species (and ecosystems) in response to future global climate change is a topic of much concern (see Section 28.6).

On an even longer timeline, evolution rather than range shifts drives changes in diversity. Over the past 600 million years, among most groups for which fossil data exist, species number has increased almost continuously (albeit not steadily) since the group first appeared in the fossil record. For example, the number of invertebrate species has increased sharply, with only slight declines during the late Devonian and Permian eras (Figure 27.4).

Evolution of land plants follows an interesting pattern (Niklas et al. 1983) (Figure 27.5). Since their appearance over 400 million years ago, vascular species have greatly increased in number, with large shifts in the dominating groups. The earliest

vascular plants, the rootless and leafless psilopsids, went extinct by the end of the Devonian. They were replaced by pteridophytes (ferns), which flourished in the Carboniferous era. Ferns decreased by the early Triassic, coincident with diversification of the gymnosperms (which include cycads and conifers). Although conifers still dominate many ecosystems, gymnosperms declined in abundance and diversity over the most recent 100 million years as angiosperms (flowering plants) rose to prominence.

Although the history of Earth's biodiversity is generally a story of increasing richness, it has undergone periods of decline. Most extinctions are clustered in brief periods (Figure 27.6, p. 596) called **mass extinction events**, during which many species are lost. The most devastating occurred at the end of the Permian, 225 million years ago, when over half of all species and over 90 percent of shallow-water marine invertebrates disappeared. Another occurred at the end of the Cretaceous, 65 million to 125 million years ago, when the dinosaurs vanished. Most believe that the cause was an asteroid striking Earth, interrupting oceanic circulation, altering climate, and triggering volcanic and mountain-building activity.

One of the great mammalian extinctions occurred relatively recently, during the Pleistocene, when such species as the woolly mammoth, giant deer, mastodon, giant sloth, and sabre-toothed cat vanished. Climate changes associated with glaciation likely contributed, but overexploitation of large mammals, as humans swept through North and South America between 11 550 and 10 000 years ago, may have been a factor. Large grazing herbivores may have succumbed to the combined effects of abiotic changes and predatory pressure of humans.

A group of prominent evolutionary biologists contends that we are at this moment undergoing the world's sixth mass extinction event (Barnosky et al. 2011). Most present-day extinctions have occurred since 1600. Humans are responsible for the majority, through (1) habitat destruction and degradation; (2) introduction of predators, parasites, and competitors;

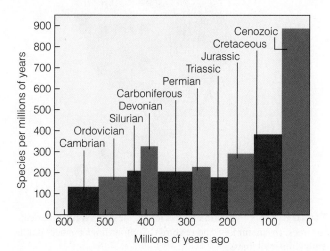

Figure 27.4 Estimated species richness of fossilized invertebrates over geologic time.

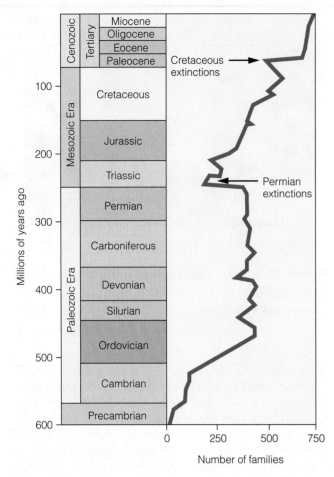

Figure 27.6 Mass extinction events, based on the fossil record. The most recent event was in the Cretaceous era and wiped out over half of all species, including non-avian dinosaurs.

and (3) exploitative harvesting (see Chapter 26). Unlike the possible role of humans in the Pleistocene extinctions, the current event threatens to affect more species and compromise the ability of the biosphere to recover from the damage we have caused to land and seascapes.

27.2 LATITUDINAL DIVERSITY TRENDS: Diversity Decreases at Higher Latitudes in Both Terrestrial and Marine Ecosystems

There are strong diversity patterns in space as well as in time. Using species richness as the measure of diversity, terrestrial diversity decreases from the equator to the poles, with by far the most species in the tropics. In North America, there are distinct geographic patterns of species richness in trees, mammals, and birds (Figure 27.7), with the number decreasing from south to north. This pattern is even more obvious when richness is plotted as a function of latitude (Figure 27.8). Note that regional estimates of bird diversity in eastern North America are seasonally dependent. Over half of the species that nest in this region migrate in fall, altering seasonal diversity patterns and confounding overall estimates.

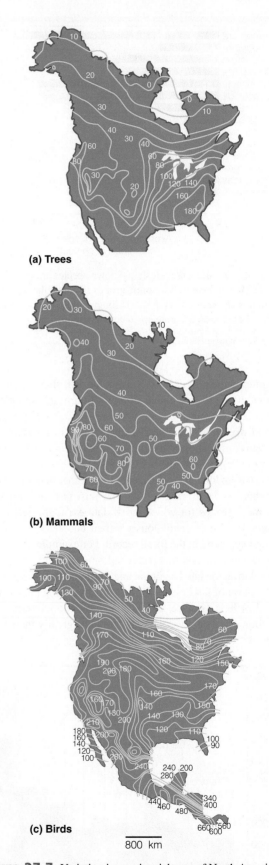

(a) Trees

(b) Mammals

(c) Birds

800 km

Figure 27.7 Variation in species richness of North American **(a)** trees, **(b)** mammals, and **(c)** birds. Contour lines connect points with similar richness.

(Adapted from (a) and (b) Currie 1991 and (c) Cook 1969.)

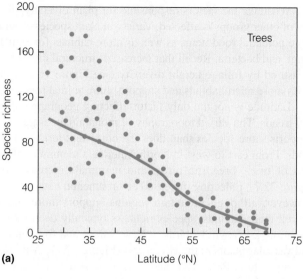

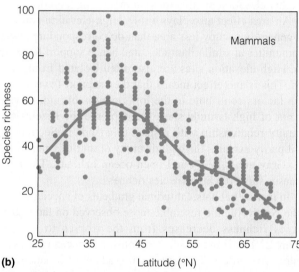

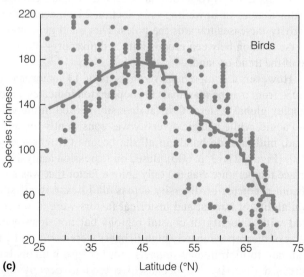

(a) **(b)** **(c)**

Figure 27.8 North American latitudinal gradients in species richness for **(a)** trees, **(b)** mammals, and **(c)** birds. Data are based on cells of 2.5″ × 2.5″ latitude/longitude, derived from range maps for individual species.

(Adapted from Currie 1991.)

These continental trends are consistent with a global pattern of declining diversity (both terrestrial and marine) from the equator towards the poles. Scientists have suggested many hypotheses to explain this pattern, including the following:

1. *Evolutionary time:* The tropics are older, with more time to evolve more species.

2. *Ecological time:* The tropics have been less interrupted by periodic disturbances (particularly glaciation) that drive some species to extinction or alter geographic ranges.

3. *Climatic stability:* Tropical climates are more favourable and stable, falling within the tolerance limits of more species.

4. *Heterogeneity:* The tropics have higher biotic heterogeneity, which increases the variety of habitats and niche space for more species.

5. *Productivity:* The tropics support more species because they are more productive.

These (and other) hypotheses are not mutually exclusive; diversity in any one location reflects many factors acting in concert, with the relative importance of abiotic and biotic factors shifting in space and time. Some factors are more relevant to some groups than others. Climatic stability may have the most influence on plant diversity, whereas biotic heterogeneity may have more impact on animals. In the tropics, the insect community, particularly parasitoids (see Section 15.1), is especially diverse. With more plant species supporting more insect consumers at various trophic levels, a huge range of hosts is available for parasites in tropical systems.

At a continental scale, the richness of North American trees correlates more strongly with actual evapotranspiration (AET) (Figure 27.9) than with other climate variables

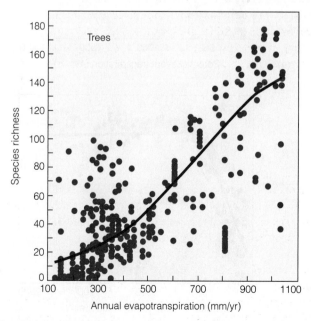

Figure 27.9 Relationship between actual evapotranspiration (AET, annual basis) and tree species richness in North America.

(Adapted from Currie and Paquin 1987.)

(Currie and Paquin 1987). AET is the flux of water from land to air through the combination of evaporation and transpiration (see Section 2.6). It reflects water demand, and in turn is correlated with net primary productivity (see Figure 20.5). Abiotic factors favourable to photosynthesis may increase plant species richness.

Currie (1991) also reports a positive correlation between *potential evapotranspiration* (PET), an index of energy availability, and species richness of mammals and birds (Figure 27.10). Although the correlation between PET and vertebrate richness suggests an explanatory factor other than latitude, PET is correlated with temperature, precipitation, radiation, and other factors that vary with latitude. Also, like AET, PET correlates with plant productivity. (In fact, PET equals AET when water supply is not limiting vegetation.) Thus, vertebrate and plant richness are positively correlated, as are primary and secondary productivity (see Figure 20.17). So animal diversity may reflect secondary productivity in much the same way that plant diversity reflects primary productivity. However, although these results seem to support the productivity hypothesis (also called *richness–energy*) and climatic stability hypotheses, there is much counterevidence of highly productive ecosystems with low richness, such as freshwater marshes.

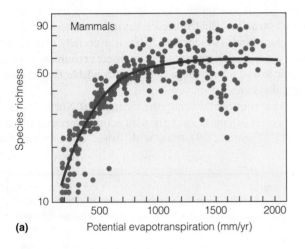

(a)

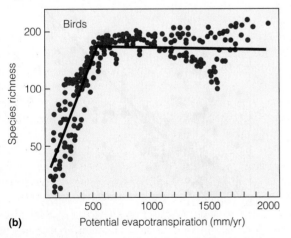

(b)

Figure 27.10 Relationship between potential evapotranspiration (PET, annual basis) and species richness of **(a)** mammals and **(b)** birds in North America.

(Adapted from Currie 1991.)

Whatever the factors responsible for plant diversity, diversity of other groups is affected. Variety in plant species provides more potential food items as well as more habitats for animals, fungi, and bacteria. Recall that increased structural diversity, as measured by foliage height diversity (see Figure 16.18), provides more microhabitats and supports more animal species.

Latitude is not the only factor affecting geographic diversity trends. The varied topography of mountain areas generally supports more species than does the consistent terrain of flatlands. From east to west in North America, the number of species of trees, breeding birds, and mammals increases (see Figure 27.7), reflecting increased environmental heterogeneity. However, although mountain regions support more species overall than flatlands, species richness typically declines with elevation, as shown for birds in New Guinea and for mammals and vascular plants in the Himalayas (Figure 27.11). Variations in temperature, PET, AET, and vegetation structure that occur at higher elevation parallel those at increasing latitude.

An area effect also plays a role. High-elevation communities generally occupy less area than do corresponding lowland communities at similar latitudes, and hence support fewer species. High-elevation sites are also often isolated from similar sites. This island effect means that they support fewer species. Such factors contribute to the low richness of high-elevation but not of high-latitude sites. We explored the species area–diversity relationship with respect to islands in Section 18.4, but it partly explains the high diversity of marine benthos habitats (Gray et al. 1997) and open ocean habitats. Their large expanses support a high species richness.

In marine habitats, latitudinal gradients of species richness in the North Atlantic resemble those observed on land; that is, species richness decreases from the tropics to the poles (Figure 27.12). However, the relationship between richness and productivity differs. The latitudinal gradient of productivity in the oceans is the reverse of that observed on land. Except in localized areas of upwelling (see Section 24.8), oceanic productivity increases towards the poles. This trend generates an inverse relation between productivity and diversity—the opposite of the trend on land.

However, a comprehensive analysis of 13 major species groups from marine mammals to zooplankton indicates a more complex global pattern. Coastal diversity peaked in the western Pacific, while pelagic diversity was consistently highest in broad, mid-latitude bands in all the oceans (Tittensor et al. 2010) (Figure 27.13, p. 600). Based on regression analysis, sea surface temperature was the only abiotic factor that was a significant predictor of diversity across all 13 taxa in all sites. Availability of habitat and historical factors were also correlated with diversity in coastal regions but not open-water. Areas of disproportionately high richness were correlated with moderate to high human impacts, not because humans have increased diversity but because we tend to occupy high-diversity coastal areas. The importance of temperature as a predictor of diversity suggests that global climate change, in concert with other more direct human impacts, is likely to significantly affect marine biodiversity patterns (Tittensor et al. 2010).

The role of temperature suggests that seasonality, not productivity, may be a critical factor affecting diversity of

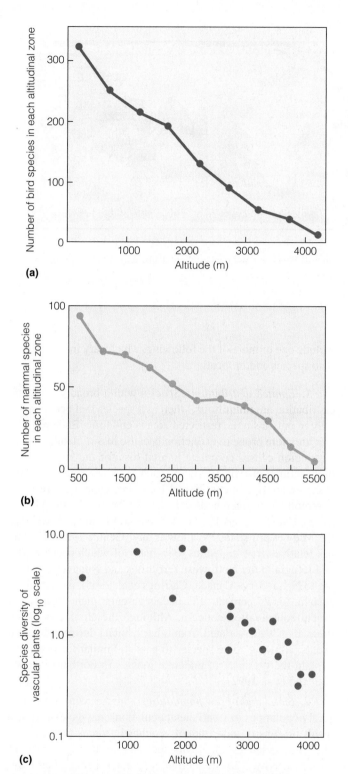

(a)

(b)

(c)

Figure 27.11 Relationship between altitude and species richness for **(a)** birds in New Guinea and **(b)** mammals and **(c)** vascular plants in the Himalayas.

(Adapted from (a) Kikkawa and Williams 1971, (b) Hunter and Yonzon 1992, and (c) Whittaker 1977.)

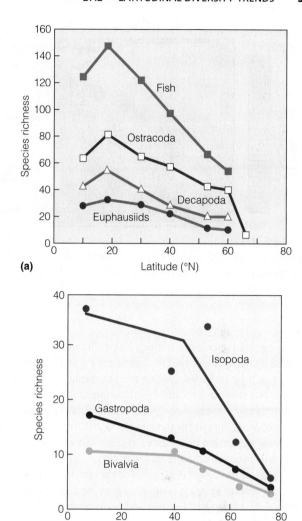

(a)

(b)

Figure 27.12 Latitudinal gradient of marine species richness in the North Atlantic. **(a)** Four groups of pelagic organisms caught at six stations along 20° W (longitude) over a two-week period from the top 2000 m of the water column. **(b)** Three groups of benthic organisms. Across all marine regions, maximum richness occurred at depths of 1000 m to 1500 m.

(Adapted from (a) Angel 1991 and (b) Rex et al. 1993.)

pelagic (open-water) and *benthic* (bottom) communities. The amount of seasonal temperature flux increases from the tropics to the poles (see Figure 24.20). Aquatic productivity reflects the seasonal dynamics of the thermocline and vertical transport of nutrients (see Figure 21.26). In northern latitudes,

seasonal turnover causes productivity of surface waters to range from high in spring and summer to low in winter (see Figure 24.20). In contrast, the permanent thermocline in tropical waters generates low productivity year-round. Seasonal temperature variation in surface waters makes temperate waters more productive, whereas the lack of variation supports more species in tropical waters.

Broad-scale marine diversity patterns can be confounded by local factors, as indicated by the pelagic diversity trends in Figure 27.12a (Angel 1991). At the time of sampling, the major boundary between the South and North Atlantic Central Waters was at 18° N. Because these two major regions differ in abiotic factors, each supports a different fauna, which converge in this zone. Thus, the peak in species diversity observed for the four groups of organisms sampled at 20° N may be partly due to an edge effect (see Section 18.2), although it is also consistent with the global pattern described by Tittensor et al. (2010; see Figure 27.13).

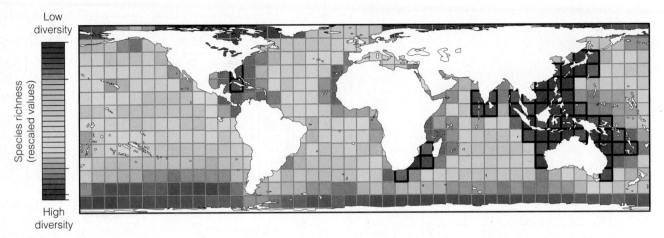

Figure 27.13 Global species richness for 13 major animal taxa in oceanic and coastal habitats combined. Cells with a bold outline are hotspots.

(Adapted from Tittensor et al. 2010.)

As on land, geologic history has influenced latitudinal trends in marine diversity. Glaciation may contribute to low diversity at high latitudes. During the Quaternary, sea ice covered the Norwegian Sea and North Atlantic. The decline in regional diversity towards the poles may reflect recovery from glaciation, as discussed for trees in Section 17.11. Slow northward range expansion of species suggests that recovery is still in progress, accelerated by global warming.

Whatever the causes of global trends in diversity of different taxonomic groups, ecologists may be arguing at cross purposes. According to Huston and Wolverton (2009), what appears to be a paradox wherein the most productive terrestrial systems contain the most species and the most productive marine systems in high latitudes contain the least, is resolved if (as they argue) tropical rain forests are actually less productive than mid-latitude terrestrial systems. The peak in marine diversity at mid-latitudes reported by Tittensor et al. (2010) (see Figure 27.13) also makes what was once a sharp dichotomy in marine versus terrestrial systems much less clear cut.

No matter the spatial or temporal scale at which we study diversity—global or local, geological or successional—and no matter our focus—terrestrial or aquatic, marine or freshwater, autotrophs or heterotrophs—diversity is critical. Many abiotic and biotic processes affect it, and although diversity in turn affects ecosystem structure and function, its full significance remains elusive.

27.3 EXTINCTION RISK: Many Factors Affect Susceptibility to Extinction

We noted earlier that it is the nature of evolution that all species will eventually go extinct, but of all the species that contribute to Earth's present-day biodiversity, not all are equally susceptible to extinction in the short term. Susceptibility is related to factors (including life history traits) that affect vulnerability to human activities as well as to natural hazards. Risk factors include one or more of the following, which vary in importance with species and/or location:

1. *Limited distribution:* Species with a broad geographic distribution are **ubiquitous**, whereas a species that occurs naturally in only a single, restricted area is **endemic**. Endemic species are more prone to extinction because loss of habitat in that one region causes complete habitat loss for the species. The high extinction risk of islands such as Hawaii reflects their many endemics. In continental locations, endemics are more common in warmer and/or isolated regions. In the United States, California has 1517 endemics and Florida 385, whereas the Midwestern states have few if any (Gentry 1986). Canada has relatively few endemics (68), many of which occur in glacial refugia in the Northwest Territories and Nunavut. According to NatureServe Canada (Carrings et al. 2005), these regions account for 25 percent of Canada's vascular plants that are of global conservation concern. Although the climate is harsh, these areas were isolated from other habitats during glaciation events, allowing some species to persist. Similar factors likely explain the presence of endemic mosses in northern Québec (Belland et al. 1992).

2. *Small and/or few populations:* Species with one or few local populations (a small metapopulation) may go extinct as a result of density-independent, stochastic forces (see Section 11.11) associated with events such as fire, flood, disease, or human activity. The Allee effect also plays a role when numbers get very small (see Section 10.10). Species with many local populations may be less vulnerable than a single small population (see Section 12.6).

3. *Migratory habit:* Species that migrate seasonally depend on two or more habitats in different regions. If either habitat disappears, the species cannot persist. The over 120 species of neotropical migrant birds that migrate each year between eastern North America and the tropics of Central and South America and the Caribbean need suitable habitat in both places (Figure 27.14), as well as stopover habitat.

Figure 27.14 Range map of the Nashville warbler (*Vermivora ruficapilla*). The spring and summer breeding range is coloured purple, and the winter habitat is blue.

Barriers to migration (such as dams blocking upriver movements of fish) can also prevent an individual from completing its life cycle.

4. *Specialized habitat:* Species (often endemics) with specialized needs are vulnerable to habitat change. Their preferred habitats may be scattered and rare across a region. For example, Long's braya (*Braya longii*) (Figure 27.15) is adapted to the limestone barrens of Newfoundland's Great Northern Peninsula. This diminutive plant (under 10 cm tall) colonizes gaps in the tundra-like vegetation of these cool, wet, windy coastal sites. Under natural conditions, Long's braya (and a related endemic, Fernald's braya, *B. fernaldii*) is maintained by fine-scale disturbances such as frost action. Human impacts have created a more homogeneous substrate that has encouraged a braya monoculture. In the short term, this change may seem favourable, but the altered growth pattern exposes Long's braya to other threats, such as an introduced agricultural insect pest, the diamondback moth (*Plutella xylostella*), as well as hybridization with Fernald's braya. Both brayas are currently the focus of a recovery strategy (Hermanutz et al. 2002).

5. *Large home range:* Species requiring a large home range (particularly large carnivores) are endangered by habitat fragmentation. Even if suitable habitat is abundant, fragmentation may restrict the availability of contiguous patches that are large enough to support breeding populations. The risk is higher if the home range must be in old growth, which is typically less available.

6. *Direct interactions with humans:* Species that are hunted or otherwise exploited, or that come into conflict with human activities, are especially vulnerable. Hunting for consumption or for body parts (such as ivory) has caused the endangerment or extinction of many species, while others have been eradicated because they threaten human lives or interfere with human activities. Large carnivores are prime examples. In parts of North America, the wolf, grizzly bear, and mountain lion were hunted to near extinction because they were considered a threat to livestock and humans. Programs to reintroduce these species into areas of their former range—as with wolves in Yellowstone National Park—have met with strong public resistance (Williams 2007). Herbivores, large and small, are also a target, of which the American bison and black-tailed prairie dog are examples.

27.4 RISK CATEGORIES: Biologists Assess Extinction Risk with Specific Criteria

Given these multiple risk factors, the International Union for the Conservation of Nature (IUCN) has developed a classification system for **threatened species** based on their probability of extinction: (1) *Critically endangered species* have a risk of extinction of 50 percent or more in the wild within 10 years or 3 generations, whichever is longer. (2) *Endangered species* have a risk of extinction of at least 20 percent within 20 years or 5 generations. (3) *Vulnerable species* have a 10 percent probability of extinction within 100 years. Non-threatened species include those that are (1) *conservation dependent* (likely threatened if an ongoing conservation program were to

Figure 27.15 Long's braya is restricted to Newfoundland's limestone barrens.

Table 27.1 Trends in Number of Threatened Species Worldwide in Various Taxonomic Groups

"Threatened" includes all three IUCN categories (critically endangered, endangered, and vulnerable).

	Number of described species, as of 2011	Number of species evaluated in 2006	Number of species evaluated in 2011	Number of threatened species in 2006	Number of threatened species in 2011
Vertebrates					
Mammals	5 494	4 856	5 494	1 093	1 134
Birds	10 027	9 934	10 027	1 206	1 240
Reptiles	9 362	664	3 004	341	664
Amphibians	6 711	5 918	6 312	1 811	1 910
Fishes	32 000	2 914	9 353	1 173	2 011
Subtotal	**63 654**	**24 284**	**34 189**	**5 622**	**6 959**
Invertebrates					
Insects	1 000 000	1 192	3 338	623	746
Molluscs	85 000	2 163	4 419	975	1 570
Crustaceans	47 000	537	2 399	459	596
Others	173 250	86	956	44	287
Subtotal	**1 305 2500**	**3 978**	**11 122**	**2 102**	**3 199**
Plants					
Mosses	16 236	93	101	80	800
Ferns and allies	12 000	212	293	139	158
Gymnosperms	1 052	908	963	306	374
Flowering plants	268 000	10 688	12 761	7 865	8 477
Green algae	4 242	n/a	13	n/a	0
Red algae	6 144	n/a	58	n/a	9
Subtotal	**307 674**	**11 901**	**14 189**	**8 390**	**9 098**
Fungi and protists					
Lichens	17 000	2	2	2	2
Mushrooms	31 496	1	1	1	1
Brown algae	3 127	n/a	15	n/a	6
Subtotal	**51 623**	**3**	**18**	**3**	**9**
TOTAL	**1 728 201**	**40 168**	**59 508**	**16 117**	**19 265**

(Source: IUCN 2011.)

cease); (2) *near threatened* (close to qualifying as vulnerable); or (3) of *least concern*. Species are *data deficient* if there are insufficient data to assess their status, or if they have not been evaluated.

Assigning a species to a category requires at least one of the following data types: (1) *population size* (number of breeding individuals); (2) *population trends*, including observed decline; projected decline, if habitat loss or other risk factors are unabated; and fluctuation in numbers; and (3) *geographic factors*, including area occupied; extent of occurrence in occupied areas; and fluctuation, fragmentation, and quality of habitat. Not all criteria need be met. If the number of breeding individuals is less than 250, the species would be considered critically endangered even if those individuals occurred over a wide area. Similarly, if habitat for a specialized species were declining precipitously, its risk category would be affected even if its numbers had been relatively stable in the past decade.

Despite its limitations—for example, critical population size varies greatly among species, and the criteria are hard to apply to modular organisms such as plants and fungi—the IUCN system provides a standardized, quantitative method for evaluating conservation decisions. Assigning categories based on habitat loss is particularly useful for species whose life history and population trends are little known. The IUCN Red List summarizes the conservation status and distribution information of species that have been evaluated using their criteria (Table 27.1) (IUCN 2011).

The percentage of known animal species that are threatened is especially high for vertebrates (13–41 percent, depending on the group). Unfortunately, relatively few invertebrates have been evaluated, so estimates of the percentage threatened are unreliable. Among plants, gymnosperms are the most threatened, with 40 percent of known species at risk. Scientists are evaluating more and more species in all groups, but mosses, lichens, fungi, and algae are still underrepresented.

In Canada, COSEWIC (Committee on the Status of Endangered Wildlife in Canada) is charged with assessing the conservation status of native species. Using data from government and academic scientists, aboriginal groups, and other community sources, COSEWIC makes decisions regarding which taxa

Table 27.2 Summary of COSEWIC Assessments of Global Conservation Status of Canadian Species at Risk as of May 2011

	Extinct	Extirpated	Endangered	Threatened	Special concern	Total	Not at risk	Data deficient
Vertebrates								
Mammals	2	3	20	16	29	70	45	8
Birds	3	2	29	26	20	80	36	2
Reptiles	0	4	17	11	9	41	5	2
Amphibians	0	2	9	5	6	22	15	0
Fishes*	7	3	48	37	49	144	45	26
Subtotal	12	14	123	95	113	357	146	38
Invertebrates								
Arthropods	0	3	29	6	6	44	0	1
Molluscs	1	2	19	3	6	31	2	4
Subtotal	1	5	48	9	12	75	2	5
Vascular plants	0	3	94	48	40	185	21	4
Mosses	1	1	8	3	4	17	1	1
Lichens	0	0	5	3	7	15	0	3
Grand total	14	23	278	158	176	644	170	50

"Total" includes species that are extirpated from Canada (not including those species that are also extinct worldwide) and all three COSEWIC categories. COSEWIC uses similar criteria as the IUCN, but its categories differ slightly. "Endangered" corresponds to the IUCN category "critically endangered," "threatened" to "endangered," and "special concern" to "vulnerable." Data were not available for subcategories of vascular plants.
*COSEWIC's 2011 summary does not list marine fish, but COSEWIC 2008 lists freshwater fish as making up 76 percent of total fish species in the combined risk categories (Hutchings and Festa-Bianchet 2009).

should be listed under Canada's Species at Risk Act and recommends actions. As of 2011, there are 635 species listed: 278 endangered, 158 threatened, 176 of special concern, and 23 extirpated (the total of 658 includes species that are already extirpated from Canada). Among taxonomic groups, Canadian trends mirror global trends, albeit with lower numbers (Table 27.2) (COSEWIC 2011). Freshwater fish are particularly affected (Figure 27.16). Some 27 percent of Canada's 205 freshwater and *diadromous fish* (fish that spend considerable time in freshwater, not just to breed) are at risk in all or part of their geographic range (Hutchings and Festa-Bianchet 2009).

DNA barcoding is another important recent advance in conservation biology. Pioneered by the Biodiversity Institute of Ontario at the University of Guelph, this system uses short, specific DNA segments as markers to aid in identification and to clarify taxonomic issues. For conservation purposes, barcoding provides an "identity card" and is particularly valuable for species (or subspecies) of similar appearance. As of 2011, the International Barcode of Life Project had over 1.3 million records, with 108 386 species formally described.

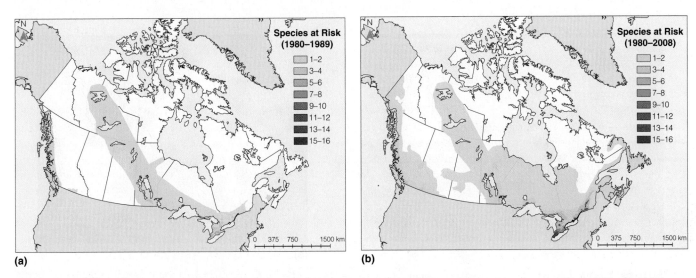

Figure 27.16 Distribution of freshwater and diadromous fish assessed by COSEWIC as at risk in Canada, **(a)** 1980 to 1989, **(b)** 1980 to 2008.

(Adapted from Hutchings and Festa-Bianchet 2009).

ECOLOGICAL ISSUES Conserving a Canadian Icon

No image is more quintessentially Canadian than that of the polar bear (*Ursa maritimus*) (Figure 1). Every year, thousands of tourists visit the shores of Hudson Bay to see this magnificent animal in its native habitat. As of 2002, the IUCN estimated that there were 21 500 to 25 000 polar bears worldwide. Yet the species is listed as "vulnerable" on their Red List ("special concern" by COSEWIC), in light of the projected drop in numbers with habitat decline. What factors threaten the polar bear? What steps can be taken to avert its extinction? Canadian scientists are at the forefront of this pressing conservation challenge.

Understanding the factors threatening the polar bear involves understanding the species itself—its demography, life history, trophic interactions, behaviour, etc.—as well as its environment. The polar bear exhibits many of the traits discussed in Sections 10.10 and 27.3 as increasing the risk of extinction. A classic *K*-strategist (see Section 8.10), the polar bear is large, matures late, and has a low reproductive output. As expected for such a large predator in a high-latitude environment, it requires a very large home range. Its habitat needs are also quite specialized. Polar bears require sea ice as a platform upon which to travel and hunt, making them vulnerable to changes in sea-ice distribution, properties, and timing of formation and breakup.

As described by Canadian Wildlife Service biologist Ian Stirling, polar bears favour *polynyas* (small areas of open water surrounded by ice, where seals surface to breathe) and *leads* (rifts in sea ice) (Stirling 1997). Southern populations also use land habitats for their maternal dens, retreating to land in regions where the sea ice melts completely during summer. These behaviours increase their contact with humans—contact that is more hazardous for bears than for humans. Thus, alterations to the land–sea interface from natural and anthropogenic climate change have ecological repercussions for feeding, reproduction, and movement.

Central to understanding the vulnerability of the polar bear is its status as a top-level predator. Although it can feed on various marine mammals—indeed, the plasticity of its diet and hunting strategies may prove critical to its fate—the polar bear is a specialized predator of phocid seals. The western Hudson Bay population feeds mainly on ringed seals (*Phoca hispida*), but they also eat bearded seals (*Erignathus barbatus*). Their preferred feeding grounds are near-shore annual ice overlying the continental shelf, the most productive region of the Arctic.

What are the consequences for the polar bear of spatial and temporal changes in ice with global warming? Seals also require sea ice, so polar bear prey will become less abundant. With longer periods of open water, bears will undergo longer summer fasts, either on land or (for northern populations) on residual multi-year ice, as they wait for annual ice to re-form. Ice floes will be smaller, more dispersed, and composed of thinner ice that will move more with winds and currents. As a result, bears will expend more energy walking or swimming to feeding grounds and dens. To compound the energy problem, as sea ice retreats farther, it will be positioned over deeper and less productive polar waters (Derocher et al. 2004), reducing the ability of the food web to support large numbers of top predators. Intraspecific competition may intensify, as remaining polar bears (even if fewer in total) become concentrated in smaller areas of suitable habitat.

Ultimately, these changes in both prey abundance and the energy expended to acquire prey will affect polar bear reproduction. Females rely on laying down large fat deposits to sustain themselves and their cubs during the winter fast. With a shorter period in late spring when sea-ice conditions are optimal for feeding, females will grow more slowly, deposit less fat, reproduce at a later age (or not reproduce at all if they do not reach a minimum mass), and have fewer and less vigorous cubs (Derocher et al. 2004). Cubs that do survive will risk falling prey to hungry males. As bear numbers drop, positive feedback will likely kick in, accelerating their further decline.

(a)

(b)

Figure 1 (a) A polar bear on thin ice near Churchill, Manitoba, and (b) on high cliffs on Coats Island, Hudson Bay. Given the ongoing changes in the North, the plasticity of the polar bear in its use of habitat and/or prey—here, the species of interest is a thick-billed murre, *Uria lomvia*—will likely prove critical to its long-term survival as a species.

Given the many factors increasing their risk of extinction under a warming scenario, what is the polar bear's current status? Intensive studies began in 1970, when Stirling began investigating the ecological relationships of bears in the eastern Beaufort Sea. At that time, the main concern was managing the polar bear hunt. Attention focused on the bear's interactions with their ringed seal prey. The assumption was that the polar sea environment was stable, and that by documenting the present population status in relation to its prey, it would be possible to establish a sustainable harvesting strategy. However, when the numbers of both polar bears and seals declined with unusually heavy ice conditions in the mid-1970s, stability—either of polar bear numbers or of their habitat—could no longer be assumed. These fluctuations highlighted the need to study the scale and frequency of natural fluctuations in the polar habitat, and how these fluctuations interact with polar bear–seal predation in the long term. Changes in sea ice from anthropogenic causes were not yet on the radar.

Since then, the scope of polar bear research has broadened greatly. Teams of researchers with differing expertise are involved, employing standardized methods that allow integration of short-term projects by a variety of agencies, in a variety of locations. Managing the polar bear hunt is still a concern, but as the rate of ice retreat accelerates, response of polar bears to climate-based habitat change has become the major focus. At present, there are 19 subpopulations designated for management purposes worldwide, of which 13 are under at least partial Canadian jurisdiction. Both IUCN and COSEWIC treat the species as a single unit for conservation purposes. However, given the species' circumpolar distribution, different subpopulations may be at greater risk than others, and may benefit from different conservation strategies.

Thiemann et al. (2008) used genetic, life history, and biogeographic data to propose five polar bear subpopulations or "designatable units" (Figure 2). Arguing that threats to polar bears are not spatially uniform, the authors contend that this reclassification provides a more biologically sound basis for conservation. The Hudson Bay subpopulation, for instance, is at greater risk from both longer ice-free periods and human contact than is the High Arctic subpopulation, for which the Allee effect (see Section 10.10) is likely more critical. But, even if these subpopulations are given separate legal status, the question remains: *Can we take the steps necessary to address the polar bear conservation challenge?*

For some species, the answer is scientifically simple, even if politically and socially difficult. Grizzly bears, for example, need large habitat areas set aside in protected wilderness. Even in existing protected areas, conservation efforts encounter obstacles, as with reintroduction of wolves into Yellowstone National Park. But with polar bears, establishing protected areas will not guarantee the needed habitat, if the factors triggering habitat loss—greenhouse gas emissions and the consequent climate change that is destroying sea-ice platforms—continue unabated. Given the linkages between the three spheres that make up the biosphere (air, water, and land), such causal factors are outside the control of any one country. So even though Canada is one of five nations (along with the United States, the former Soviet

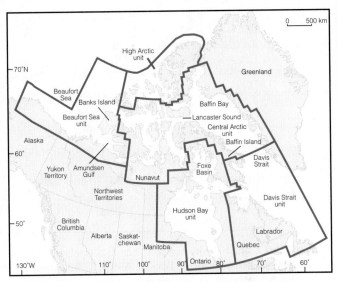

Figure 2 Proposed designatable units (DUs) for polar bear conservation based on genetic, ecological, and life history data.

(Adapted from Thiemann et al. 2008.)

Union, Norway, and Denmark) that signed the 1973 Agreement on the Conservation of Polar Bears, which committed to protect polar bear habitat, it remains to be seen whether we can make good on that promise.

Bibliography

Derocher, A. E., N. J. Lunn, and I. Stirling. 2004. Polar bears in a warming climate. *Integrative and Comparative Biology* 44:163–176.

Stirling, I. 1997. The importance of polynyas, ice edges, and leads to marine mammals and birds. *Journal of Marine Systems* 10:9–21.

Thiemann, G. W., A. E. Derocher, and I. Stirling. 2008. Polar bear *Ursus maritimus* conservation in Canada: An ecological basis for identifying designatable units. *Oryx* 42: 504–515.

1. We have only considered the tip of the iceberg regarding the problems facing polar bears. Given their top predator status, what other factors relating to pollutants will intensify as the Arctic warms? (See Sections 22.11 and 22.12 as well as your own research.)

2. We have included this segment under the banner of conservation ecology, but it could just as easily be placed in many other chapters. Make a list of the different topics involving all aspects of the ecological hierarchy—individuals, populations, communities, landscapes, ecosystems—as well as the abiotic environment, that come into play. Many are mentioned explicitly (e.g., home range, Allee effect, climate change), whereas some are implied (abiotic features of aquatic habitats), and others you will devise on your own.

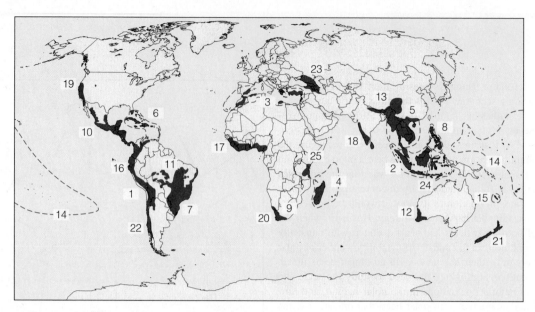

Figure 27.17 Biodiversity hotspots. Actual hotspots make up 3 to 30 percent of the red areas, with the larger regions shown to depict global richness patterns. 1, Tropical Andes; 2, Sundaland; 3, Mediterranean Basin; 4, Madagascar and Indian Ocean Islands; 5, Indo-Burma; 6, Caribbean; 7, Atlantic Forest Region; 8, Philippines; 9, Cape Floristic Province; 10, Mesoamerica; 11, Brazilian Cerrado; 12, Southwest Australia; 13, Mountains of south-central China; 14, Polynesia/Micronesia; 15, New Caledonia; 16, Chocó-Darién of Western Ecuador; 17, Guinean forests of West Africa; 18, Western Ghats and Sri Lanka; 19, California Floristic Province; 20, Succulent Karoo; 21, New Zealand; 22, Central Chile; 23, Caucasus; 24, Wallacea; 25, Eastern Arc Mountains and coastal forests of Tanzania and Kenya.

(Adapted from Myers et al. 2000.)

27.5 BIODIVERSITY HOTSPOTS: Conservation Efforts Focus on High-Diversity Areas

As we saw in Section 27.2, biodiversity is not evenly distributed. Richness increases towards the equator in both terrestrial and aquatic ecosystems. Although tropical rain forests cover only 6 percent of the land surface, they support over half of all known species. In addition, regions with more topographic variation support more species than flatter areas, as a result of their habitat variation. Complicating these global and regional patterns is the fact that many species are endemic, with restricted geographic ranges. Of Earth's 10 000 bird species, over 2500 are endemic, restricted to ranges under 50 000 km². Over half of the world's plants are endemic to a single country. Of the thousands of new species identified yearly, virtually all are endemic to tropical regions. Their restricted distributions make endemics highly vulnerable to human activities that degrade or destroy their habitat. Of the species red-listed by the IUCN, 91 percent are endemic.

Nor are endemic species distributed evenly within their narrow range. **Biodiversity hotspots** are regions with unusually high richness and endemism (Myers 1988). Hotspot designation is based on (1) overall diversity and (2) the impact of human activities on available habitat. Plant diversity is the basis; a hotspot must support 1500 or more endemic plant species (0.5 percent of the global total) and have lost over 70 percent of its original habitat. Plants are used not just because they are easier to monitor but because they foster diversity in other groups.

The 25 biodiversity hotspots designated by the IUCN (Figure 27.17) contain 44 percent of all plant species and 35 percent of all terrestrial vertebrate species in only 1.4 percent of Earth's land area. By far the largest proportion of the 121 000 threatened tropical species are endemic to these 25 locations, where high diversity and catastrophic habitat loss coincide. Several are island chains, such as the Caribbean and the Philippines, or relatively large islands, such as New Caledonia. Others are continental "islands"—isolated regions surrounded by deserts, mountain ranges, or seas, or landlocked regions in high mountains and mountain ridges. For ridge communities in the Andes (South America) and the Caucasus (Asia), the lowlands are insurmountable barriers. Considered in total, and acknowledging the difficulty of coordinating conservation efforts across different political jurisdictions, these areas are the most important for protecting Earth's biodiversity.

27.6 HABITAT LOSS: Destruction of Habitat Is the Leading Cause of Current Extinctions

Much of the concern of ecologists and the general public regarding diversity focuses on the factors that threaten it. The major cause of the current mass extinction event is habitat loss and degradation. Globally, half of the forest present in the

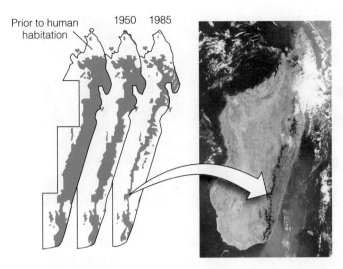

Figure 27.18 Decline of rain forest in eastern Madagascar from the arrival of humans to the present. Dark areas are existing rain forest, which once extended from coast to coast.

modern (post-Pleistocene) era is gone, largely through human impacts. Given their diversity and the pressures of growing populations, tropical regions have been the main focus. Indeed, loss of rain forest has become synonymous with declining biodiversity. In Amazonia, over 100 million hectares (15–20 percent) of its original extent has vanished. Madagascar has lost over 90 percent of its forest cover (Figure 27.18), and 95 percent of forest cover in Ecuador west of the Andes has been destroyed since 1960.

The main cause of land transformation has been agricultural expansion to feed a growing population. According to the Global Forest Resources Assessment (FRA) 2010 (FAO 2011), 130 million hectares of forest were converted to other uses or lost to natural causes from 2000 to 2010. Although this deforestation rate was down from the 1990s, it is still unsustainably high. Global statistics obscure changes in forest cover among regions and countries (Table 27.3). Net loss was highest in South America, West Africa, and Oceania. Asia, which had high rates of loss in the 1990s, reported a net gain as a result of tree planting in China. Deforestation remained high in Southeast Asia. Among countries, Nigeria had the highest annual loss rate (3.7 percent), but Brazil had the most absolute loss (2.6 million hectares, 0.5 percent). Fortunately, Brazil's loss rate abated sharply after 2004, with introduction of a protection program by the United Nations. Indonesia, which had annual losses of 1.8 percent in the 1990s, continued to lose 0.5 percent of its forest annually.

Quantity alone is an inadequate indicator of forest status. Many tropical and temperate forests that were once continuous are now highly fragmented (Figure 27.19, p. 608). Fragmentation has received less attention than deforestation as a conservation issue, yet it jeopardizes both populations and communities. Change in total cover also obscures changes in forests that are primary or naturally regenerated versus planted. According to the FRA 2010 report, primary forests represent 36 percent of forest area, a decrease of over 40 million hectares from 2000 to 2010. Even if total forest cover were constant, continued loss of natural forest would pose an extinction risk

Table 27.3 Forest Cover and Annual Change for the Period 2000 to 2010 by Region and Subregion

	Forest area (2010)		Annual change (2000–2010)	
	1000 ha	% total land area	1000 ha/yr	%
Region/subregion				
East and Southern Africa	267 517	27	−1 839	−0.66
North Africa	78 814	8	−41	−0.05
Western and Central Africa	328 088	32	−1 535	−0.46
Total Africa	**674 419**	**23**	**−3 414**	**−0.49**
East Asia	254 626	22	2 781	1.16
South and Southeast Asia	294 373	35	−677	−0.23
Western and Central Asia	43 513	4	131	0.31
Total Asia	**592 512**	**19**	**2 235**	**0.39**
Russia	809 090	49	−18	ns
Europe (except Russia)	195 911	34	694	0.36
Total Europe	**1 005 001**	**45**	**676**	**0.07**
Caribbean	6 933	30	50	0.75
Central America	19 499	38	−248	−1.19
North America	678 961	33	188	0.03
Total North and Central America	**705 393**	**33**	**−10**	**ns**
Total Oceania	**191 384**	**23**	**−700**	**−0.36**
Total South America	**864 351**	**49**	**−3 997**	**−0.45**
WORLD	**4 033 060**	**31**	**−5 211**	**−0.13**

(Source: FAO 2011.)

(a) **(b)**

Figure 27.19 Forest clearing in the Rondonia region of Brazil in the Amazon basin. **(a)** Broad-scale view shows light-coloured linear clearings associated with access roads, against a backdrop of forest cover (dark green). **(b)** Forest is then cleared for agriculture, primarily pasture.

for species, particularly those needing old-growth habitat. One quarter of planted forests use non-native species, which support less native biodiversity. Even native trees cannot regenerate a naturally functioning ecosystem, if planted in monoculture.

Some sources report more severe deforestation rates than indicated by FRA 2010. A study using globally consistent satellite data and employing more stringent criteria for forest cover found that loss from 2000 to 2005 was highest in the boreal forest. Among countries, Brazil had the largest loss (165 000 km^2), but Canada was a close second (160 000 km^2), through the effects of fire, insects, and logging (Hansen et al. 2010). As this study reported gross and not net loss, it does not factor in regenerated forest, but does indicate that North America is also losing mature forest, with its ability to support biodiversity.

Changes in land use that destroy habitat and endanger species are not limited to tropical rain forests. Tropical dry forest (see Section 23.2) has been almost eliminated from the Americas, India, and Africa. Current dry forest cover in the Pacific coastal area of Central America is less than 2 percent of its original extent.

Temperate grasslands are also much affected. Once covering 42 percent of Earth's land surface, grasslands have shrunk to less than 12 percent of their original extent. In North America, remaining prairie is highly fragmented and scattered, effectively becoming habitat islands (see Section 18.4) in a sea of agricultural land.

Aquatic systems have fared no better. Pollution of waterways, dredging and filling of coastal wetlands, and damage to coral reefs by pollution and siltation are having an effect similar to forest clearing on freshwater and marine habitats. Fertilizers, detergents, industrial wastes, and sewage add large nutrient inputs to aquatic systems, causing eutrophication (see Section 24.3). Although typically associated with freshwater, cultural eutrophication also affects coastal marine systems. The Caribbean and Mediterranean seas are both facing severe problems that are due in part to nutrient inputs from developed coastal areas that have encouraged a phase-shift to an algal-dominated system (Figure 27.20) (Hughes 1994). Many other stressors exacerbate the damage, including various diseases (both known and as yet unidentified) and "bleaching" of coral caused by expulsion of the algal

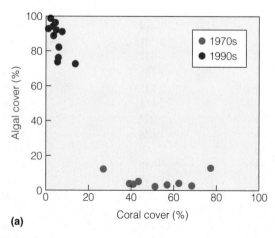

(a) **(b)**

Figure 27.20 Destruction of coral habitat. **(a)** In the 1970s, coral dominated Jamaica's reef ecosystems; 20 years later, algae had taken over, as a result of pollution and overharvesting of algae-eating fish. **(b)** Algae growing on soft corals.

(Adapted from Hughes 1994 (a); as in Primack 1998.)

partner in response to rising water temperatures. Recent decades have seen encouraging recoveries in some coastal systems, including coral reefs in Jamaica (Idjadi et al. 2006; see Section 24.11), but worldwide, coastal habitat is under siege.

27.7 INTRODUCED SPECIES: Exotic Species Threaten Many Native Species

Intentionally or not, humans have acted as dispersal agents for countless species, transporting them outside their natural geographic ranges (see Ecological Issues: Human-Assisted Dispersal, p. 197). Many species fail to survive in their new homes, but others flourish. Freed from the constraints of their native competitors, predators, and parasites, they establish and spread. These nonnatives are called **introduced (alien) species** (also *exotic* or *nonindigenous*). Introduced species are often *invasive*, but the meaning differs. **Invasive species** have the ability to spread aggressively and displace dominants, and although some alien species are invasive, so also are some native species, such as trembling aspen. Other introduced species, including most crops, are not invasive. Some species are both; invasive aliens pose the most serious threat.

Although the exact number is unknown, Environment Canada estimates that about 25 percent of Canada's vascular plant species are exotics. Among known exotic vertebrates in Canada, there are 24 birds, 26 mammals, 55 freshwater fish, 2 reptiles, and 4 amphibians. Among invertebrates, there are 181 exotic insects and various molluscs, although scientists acknowledge that these numbers are vast underestimates. There are many more exotics in the United States—about 50 000, according to Pimentel et al. (2005)—largely because the more benign climate supports more species, native and exotic. Of this number, some 4300 are considered invasives (Corn et al. 1999).

Animal invaders may cause extinction of vulnerable native species through predation, grazing, competition, and habitat alteration. Indeed, the IUCN lists exotics as the second-greatest threat to species at risk of extinction, after habitat loss. Island taxa suffer most. In Hawaii, where many native species are endemic and over half of animal species are exotics, 271 native species have disappeared in the past two centuries. Over 300 others are endangered and many more are in trouble, largely as a result of interactions with exotics. Among Hawaii's 111 native birds, 61 are extinct and 40 endangered. Similarly, on the island of Guam, the brown tree snake (*Boiga irregularis*), a native of New Guinea, accidentally arrived around 1950, likely aboard military equipment. The snake has eliminated 9 of 12 native birds, 6 of 12 native lizards, and 2 of 3 native fruit bats (Fritts and Rodda 1998).

Despite these dramatic examples from island habitats, Gurevitch and Padilla (2004) contend that evidence of a direct and primary effect of invasive exotics on the extinction of native species is as yet limited. While acknowledging the critical impact of invasives on community structure, they argue that most extinctions reflect multiple factors. Ricciardi (2004) suggests that population loss (extirpation) rather than species extinction is a more appropriate focus for the impact of invasive exotics. Displacement of a native population reduces biodiversity even if species extinction does not occur.

Plant invaders, many introduced as ornamentals, may outcompete native species and alter fire regimes, nutrient cycling, energy budgets, and hydrology. Some hybridize with related native species, contributing to their demise. As with animal species, the negative impacts are most severe on islands. Exotic species have caused 95 percent of plant species loss and endangerment in Hawaii. Among Hawaii's 1126 native flowering plants (of which 89 percent are endemic), 93 are extinct and 270 endangered.

In North America, the ornamental herb purple loosestrife (*Lythrum salicaria*; Figure 27.21a), brought from Europe in the mid-1800s, has reduced cover of native wetland plants, to the detriment of wildlife. Although once presumed to be infertile, garden cultivars contribute to the spread of this problem plant by producing viable seed or pollen (Ottenbreit and Staniforth 1994). In Florida, another introduced ornamental, Australian

(a)

(b)

Figure 27.21 Two exotic plant species that have damaged native communities in North America: **(a)** purple loosestrife and **(b)** Australian paperbark tree.

Figure 27.22 Cheatgrass grassland, converted from a sagebrush community by fire, in Oregon. Cheatgrass and smooth brome now cover thousands of hectares of rangeland in North America.

Figure 27.23 Zebra mussels, so called because of the striping on the shell, cover solid substrate, water intake pipes, and shells of native mussels.

paperbark tree (*Melaleuca quinquenervia*; Figure 27.21b), is displacing cypress and other species in the Everglades, drawing down water levels and fostering more frequent or intense fires.

An especially notorious plant invader is cheatgrass or downy brome (*Bromus tectorum*) (Figure 27.22), a winter annual accidentally introduced from Europe in the 1800s. It spread rapidly across rangeland in the western United States and by 1930 had replaced native species. Highly flammable, cheatgrass promotes fire throughout the region. It is also a major problem in western Canada, along with a related exotic, perennial smooth brome (*Bromus inermis*). Other invasive aliens in Canada include the misnamed Canada thistle (*Cirsium arvense*) and white sweet-clover (*Melolitus alba*). Like most weeds, they thrive on sites disturbed by humans.

The problem of invasive species is not restricted to land habitats. Over 139 exotics have invaded the Great Lakes by way of shipping. Most infamous is the zebra mussel (*Dreissena polymorpha*; Figure 27.23), introduced from the ballast of ships traversing the St. Lawrence Seaway. Since it first appeared in 1998, the zebra mussel has spread to many rivers and lakes and has accelerated local extinction of unionid bivalves by a factor of 10 (Ricciardi 2004). Similarly, the San Francisco Bay Area is occupied by 96 alien invertebrates, from sponges to crustaceans. According to conservation agencies, exotic fish, introduced on purpose or accidentally, have caused 68 percent of fish extinctions in North America over the past century as well as the decline of 70 percent of endangered fish.

A classic example of how an alien species can devastate community structure is the peacock bass (*Cichla pleiozona*), an Amazon native introduced into Panama's Gatun Lake. A popular sport and food fish, peacock bass feeds mostly on adult *Melaniris* spp. Other species that eat *Melaniris*, such as the Atlantic tarpon, black terns, and herons, have all declined. Six or eight common fish have been eliminated or greatly reduced by introduction of this top predator (Helfman 2007).

North America's forests are under siege from a succession of non-native insect and fungal pests. In Sections 15.5 and 16.3, we described the devastating impact of chestnut blight, *Chryphonectria parasitica*, a pathogenic fungus accidentally introduced from China around 1900. Within a few decades, chestnut blight exterminated 40 billion chestnuts (*Castanea dentata*) throughout its historic range, from Ontario to Mississippi. In the Appalachians, one in four hardwoods were chestnuts, some over 4 m in diameter. The species survives, but as short-lived root sprouts that may attain 6 m in height before dying. Another fungus, Dutch elm disease, *Ophiostoma novi-ulmi*, has decimated the American elm (*Ulmus americana*) in North America and the English elm (*U. procera*) in Europe (see Sections 14.10 and 15.5). Although spread locally by elm bark beetles, the Dutch elm fungus was first introduced by shipments of elm logs.

There are other, less publicized introduced pests. Two species of woolly adelgids (a type of wingless bug) feed on the phloem sap of their host plants, with devastating effects. The balsam woolly adelgid (*Adelges piceae*), an invasive European species, has destroyed 99 percent of Fraser firs (*Abies fraseri*) in the Great Smoky Mountains of North Carolina and Tennessee, giving rise to "ghost forests" of standing dead trees. Its spread in Canada is restricted by its vulnerability to winter lows below −20°C, but it is a serious mortality risk to balsam fir (*Abies balsamea*) in the Atlantic region (Adam and Ostoff 2006).

Another woolly adelgid, the Asian species *Adelges tsugae*, arrived in British Columbia in the 1920s and in the eastern United States by the 1960s. Like the balsam adelgid, it is not a serious pest in its native range, where its host is resistant through natural selection, but is damaging eastern and Carolina hemlocks (*Tsuga canadensis* and *T. caroliniana*). It too is less of a problem in Canada, but as winters become warmer, its impact will likely increase. A predatory beetle has been introduced as a biological control, but its success as a control measure is as yet unknown.

Most devastating of all introduced forest pests may be the emerald ash borer (*Agrilus planipennes*) (Figure 27.24). Another Asian native, it reached North America in wooden packing material. First appearing in Michigan as recently as 2002, it has already killed millions of trees from Ontario to West Virginia. Unlike the other pests discussed, which typically attack a single tree species, the emerald ash borer threatens the entire North American *Fraxinus* genus, including green ash (*F. pennsylvanica*), common from southern Manitoba to the Atlantic provinces; black ash (*F. nigra*), which inhabits wet boreal sites; and white ash (*F. americana*). Continent-wide, ash number some 75 billion. Their loss would have ecological and economic impacts no less than the loss of the chestnut or elm.

Ecological journals are full of well-documented examples of the ways in which exotics can not only increase the extinction risk for particular native species, but also alter habitat for countless others with which they have no direct interactions. The common carp (*Cyprinus carpio*) (Figure 27.25), native to Eurasia and the same species as the koi that is so popular in backyard ponds, is a problem species in lakes and marshes across North America. It not only competes with native fish such as ciscos (*Coregonus* spp.) but

Figure 27.25 The common carp affects habitat for many freshwater lake and marsh species.

also indirectly affects the breeding habitat of fish and ducks by uprooting vegetation and disturbing benthic habitat (Matthews 1998). Indeed, their main impact may be as a bioengineer (see Section 16.4). Other Asian carp species are poised to invade the Great Lakes, where they would likely have a range of detrimental impacts (Herborg et al. 2007). The fact that carp (and other invasive exotic fish such as snakeheads, Channidae) are facultative air-breathers facilitates their ability to invade water bodies. Electric barriers have been installed between the Great Lakes and the Mississippi system, but their efficacy is questionable.

27.8 POPULATION STRATEGIES: Protecting Populations Is Crucial to Conservation Efforts

Strategies aimed at preventing species extinctions are focused at different levels of the ecological hierarchy. Because endangered species consist of a few (or even a single) local populations, protecting populations is critical to protecting the species. Often, these populations are restricted to protected areas, such as nature reserves. An adequate conservation plan protects as many individuals as possible within the greatest possible area. But with limited land and resources, ecologists must ask: *What population size is needed to save the species?*

Mark Shaffer (1981) defined the **minimum viable population (MVP)** as the number of individuals needed to ensure that a species can persist in the long term, given the foreseeable impacts of stochasticity (demographic, environmental, and genetic). Models suggest that vertebrate populations with an **effective population size** (number of individuals actually contributing to future generations) of 100 or less and an actual population of less than 1000 are highly vulnerable. For species with variable population size (such as invertebrates and annual plants) and/or high rates of inbreeding, the MVP can be substantially higher. Traill et al. (2007) report a median MVP of 4169 from 30 years of published estimates.

The difference between the MVP of a population and its effective population size is influenced by its mating system. In

Figure 27.24 The emerald ash borer is killing ash trees across North America.

polygamous populations (see Section 8.3), a few dominant males are responsible for all mating. Their alleles contribute disproportionately to subsequent generations. From a genetic standpoint, non-breeding males might as well not exist. (Ecologically, however, they can have a significant impact on interactions such as competition.) MVP of a species also depends on longevity and the ability of individuals to disperse among habitat patches. Despite difficulties among species, the MVP concept is of great importance in conservation biology. A review of 21 long-term ecological studies indicates that population viability analysis, which uses computer simulations to predict MVP, has high predictive value for evaluating prediction risk and assessing alternative strategies for endangered species (Brook et al. 2000).

Once a minimum viable population size has been estimated, the **minimum dynamic area** (MDA; area of suitable habitat required to support the MVP) is determined. Estimating MDA begins with quantifying the home range of individuals, family groups, or colonies. Recall that an individual's home range increases with body size (see Figure 11.20). As well, for a given body size, the home range of a carnivore is larger than that of a herbivore. With knowledge of the area needed per individual and an estimate of species MVP, the total area needed to sustain a viable population can be calculated (Figure 27.26).

For large carnivores, the area required to sustain an MVP can be enormous. Biologists estimate that maintaining an MVP of 2000 grizzly bears in the northern U.S. Rockies requires 129 495 km² of secure, interconnected habitat (Metzgar and Bader 1992). The area required is higher in more northerly sites, given the larger home ranges. This factor explains why many large carnivores (such as African lions, Asian tigers, and North American grey wolves) are endangered, restricted to the largest and most isolated public lands and preserves, most of which are insufficient to guarantee their survival. In Alberta, for example, where grizzlies numbered some 6000 in past centuries, the Canadian Parks and Wilderness Society (CPAWS) estimates that under 500 remained by 2008 (Shier 2009), mostly restricted to mountains and foothills.

One of the best-documented cases of MVP size involves a study of 120 populations of bighorn sheep (*Ovis canadensis*) (Figure 27.27a) in deserts of the American southwest. All populations with 50 individuals or fewer went extinct within 50 years, whereas virtually all populations of 100 or more persisted over the same period (Figure 27.27b) (Berger 1990). No one cause was identified for the local extinctions; rather, a

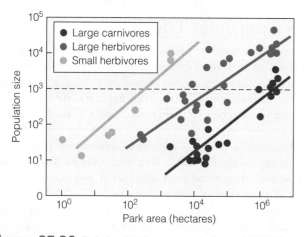

Figure 27.26 Relationship between park size and MVP. Larger parks support larger populations; only the largest parks may contain the MVP of large vertebrates. Each symbol represents an animal population. If the MVP of a species is 1000 (dashed line), parks of at least 100 ha are needed to protect small herbivores.

(Adapted from Schonewald-Cox 1983.)

INTERPRETING ECOLOGICAL DATA

Q1. What is the approximate park size needed to support large herbivores? Large carnivores?

Q2. Why is a larger area required to support a viable population of carnivores than a similar-sized population of herbivores?

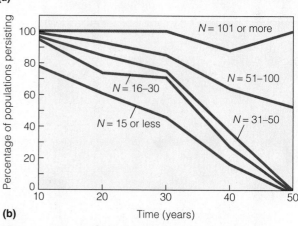

Figure 27.27 Relationship between the population size of **(a)** bighorn sheep (*N*) and percentage of populations persisting over time **(b)**. Most populations of 100 or more persist beyond 50 years, whereas those with fewer than 50 do not.

(Adapted from Berger 1990.)

range of factors was responsible. Extinction trajectories may vary in steepness (Berger 1999), but the general principle holds.

Species rarely occur as a single contiguous population. Given the fragmentation of most landscapes and the habitat needs of a given species, species often consist of semi-isolated subpopulations connected by dispersal (*metapopulations*; see Chapter 12). Rates of birth, death, immigration, and emigration of each subpopulation interact with patch size and spatial arrangement to determine metapopulation dynamics and persistence. In many cases, some habitat areas function as *source* populations (or patches), and others as *sinks* (see Section 12.6). In source patches, reproductive rate exceeds mortality, producing a surplus of individuals to colonize other patches. In sink patches, mortality exceeds reproduction, and the subpopulation is extirpated unless it regularly recolonizes. For such metapopulations, identifying source patches and the corridors that link them with sinks is fundamental to conservation. Destroying one population might trigger extirpation of many smaller subpopulations that depend on it as a source. The woodland caribou case (see Ecological Issues: What Can the Metapopulation Concept Contribute to Conservation Ecology?, pp. 248–249) illustrates the importance of metapopulation structure to species conservation.

If populations have been lost, saving the species from what seems an inevitable slide to extinction may require direct action—establishing new populations by moving individuals between locations (Figure 27.28). Efforts to reintroduce the two species of African rhinoceros, the white (*Ceratotherium simum*) and black (*Diceros bicornis*), have met with differing results.

The southern white rhino was at the brink of extinction, with fewer than 50 alive by 1900. Farmers and hunters slaughtered it, and poaching is still a problem in protected areas. The KwaZulu–Natal Board of South Africa launched "Operation Rhino" in 1961. This project transported surplus white rhinos from viable populations in Hluhluwe–Umfolozi Park in KwaZulu–Natal to other protected areas. By 2000, 2367 white rhinos had been redistributed, of which 1262 had been rehabilitated in protected areas. Thanks to these and other efforts, numbers

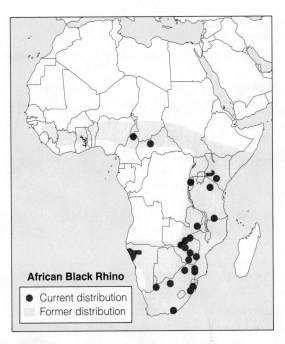

Figure 27.29 Former and current distribution of the African black rhino.

have recovered to an estimated 17 480 in 2008 according to the IUCN. (The northern subspecies, unfortunately, has not recovered, and now numbers fewer than 6.)

The black rhino story is less encouraging. From a total of 65 000 in 1970, fewer than 2500 remained by the mid-1990s (Figure 27.29). Black rhinos are scattered across central and southern Africa in populations too small to be self-sustaining, or in large populations that are vulnerable to poaching, disease, or overpopulation relative to the size of the reserve. In a massive capture and release program, black rhinos are being moved from parks in Namibia and South Africa, which have been their havens, and sent to new sites on private, public, and communal lands. It is hoped that both old and new populations will grow and interbreed as they did before poaching decimated the species. However, black rhinos are reluctant to use unfamiliar habitat. When individuals are introduced into new areas, aggressive interactions cause over half of the deaths after release. Overall, modest recoveries have been reported, but the black rhino is still classed by the IUCN as critically endangered, with some subspecies likely extinct.

Restoration programs for whooping crane (*Grus americana*), masked bobwhite (*Colinus virginianus*), Hawaiian goose (*Branta sandwicensis*), peregrine falcon (*Falco peregrinus*), and California condor (*Gymnogyps californianus*; Figure 27.30, p. 614)—and, among mammals, the wolf and European wisent (*Bison bonasus*)—have introduced organisms from captive-bred populations. Introducing such individuals to the wild requires pre- and post-release conditioning, including activities such as acquiring food, finding shelter, interacting with individuals of their own species, and learning to avoid humans. Despite these difficulties, reintroduction programs have succeeded in halting, and in some cases reversing, the downward spiral to extinction.

A successful Canadian reintroduction involves the swift fox (*Vulpes velox*) (Figure 27.31, p. 614). Although listed by

Figure 27.28 Black rhino captured at Umfolozi Reserve in Natal, South Africa, being released into a newly established reserve at Pilansburg, South Africa.

(a)

(b)

Figure 27.30 Conservation efforts for the California condor. **(a)** Staff at the San Diego Wild Animal Park Condor Breeding Facility release adult condors into a holding area. **(b)** A 12-hour-old, captive-hatched California condor chick that will eventually be released into the wild.

IUCN as of "least concern" globally given its stable numbers in the United States, it was extirpated from the northern portion of its range in the shortgrass prairies of Alberta and Saskatchewan by the 1970s by land-use transformation. In the 1980s, private groups and government agencies initiated a reintroduction project, using wild foxes from American populations as well as captive-bred individuals. Numbers are steadily increasing, and the program is considered one of the most successful canid reintroductions in the world. COSEWIC upgraded the swift fox's status from "extirpated" to "endangered" in 1998.

27.9 HABITAT CONSERVATION: Conserving Habitats Protects Entire Ecosystems

Despite the need to focus efforts on particular threatened species, the most effective way to protect biodiversity in the long run is to protect habitats and thereby entire ecosystems. Efforts to reintro-

Figure 27.31 Once extirpated in Canada, the swift fox has been successfully reintroduced to the western prairies.

duce the swift fox into Canada would not likely have succeeded without protection of extensive habitat, as in Saskatchewan's Grasslands National Park. Habitat protection is most likely the only way we can successfully conserve biodiversity, given our limited understanding of the natural history of most species and the complex interactions among species in a community. Unlike the population approach, which focuses on protecting the habitat of particular species, an ecosystem-based conservation approach requires understanding the relationship between overall patterns of biological diversity and landscape features. One key element in designing a program to protect a region's overall species diversity is the relationship between area and species richness.

As noted in Section 18.3, large areas generally contain more species. The reasons are many. (1) Larger areas are often more heterogeneous, encompassing more habitat variety and increasing the probability that a given species finds suitable habitat. (2) Some species require larger areas to meet their resource needs. Larger organisms have larger home ranges (see Figure 11.20) and need more area to maintain a minimum viable population. (3) Smaller areas have a greater edge-to-area ratio. Edge habitats have unique constraints involving microclimate and contact with predators and disease, and interior species require the conditions found only in larger, contiguous habitats. (4) Many species are locally rare and require a large area to be present even in small numbers.

Given this general relationship between richness and area, conserving diversity requires protecting as large an area as possible. However, an early debate questioned whether species richness is maximized by protecting one large area or several smaller patches of similar total area. Proponents of large reserves argue that such areas minimize edge effects, encompass the most habitat diversity, and are the only reserves that can support the MVPs of large, low-density species such as large carnivores. Yet, once an area reaches a certain size, the number of new species added with each successive increase in area declines (see Quantifying Ecology 16.1: Estimating Species Richness, pp. 335–336). In that case, establishing a second area some distance away may help protect more species than adding area to an existing reserve. In addition, a network of smaller areas positioned over a larger region may include more variety of habitats and more rare species. It may also be less prone to catastrophic events than a single, contiguous area.

The consensus among conservation ecologists now favours a mixed strategy. Larger areas are needed to maintain large species, but a network of connected reserves (large and small) may optimize long-term protection of all species. A major force behind this shift has been the development of metapopulation biology, and more recently the metacommunity concept.

In parts of the world where the environment has been altered by human activities, establishing historical baselines to aid conservation efforts is essential. The question often arises: *Are we conserving the "natural" habitat, or a habitat (and community structure) that is the product of human manipulation?* In some parts of the world where human habitation has been extended and intense, such as Great Britain and much of western Europe, this issue is, in a way, less problematic, simply because the landscape has been so obviously and extensively modified for human use, particularly agriculture. But in some regions, its impacts are more subtle. Some parts of the Brazilian rain forest, for example, have soils known as *anthrosols* that were greatly modified by the "slash-and-char" agriculture practised by the Incas (Thomas and Packham 2007). With incorporation of charcoal from charring as opposed to burning, these soils have a higher organic matter content and are more fertile than most tropical oxisols (see Figure 4.12).

Closer to home, alterations of the fire regime by aboriginal peoples in many parts of North America, followed by fire suppression by European settlers, have greatly influenced community development. On Vancouver and Saltspring Islands, records indicate that aboriginal peoples regularly burned forests to create open habitats (Bjorkman and Vellend 2010). With subsequent fire suppression, there have been shifts in species composition and tree age distribution, and a decline in open areas, which contain most of the region's threatened species. Establishing this baseline is not strictly of academic interest. Until now, ecologists have focused on conserving open areas associated with Garry oak (*Quercus garryana*), whereas the evidence indicates that before European settlement, open areas were fewer and more associated with invading Douglas-fir (Bjorkman and Vellend 2010). Is the threatened status of these open-habitat species itself an artefact of past human activities?

Or should we resume burning to simulate past human alterations to the habitat? The issue of which ecosystem is more "natural," and therefore which type of habitat we should conserve and by what means, is far from clear cut.

27.10 PROTECTED AREAS: Legal Protection Is Vital for Habitat Conservation

Whatever the configuration of the preserves, ever-growing pressures on habitat from human activities means that protecting biodiversity increasingly depends on establishing legally designated protected areas, through government action or by purchase of lands by private individuals or organizations such as the Nature Conservancy and the Audubon Society. The IUCN has established six categories (Table 27.4), ranging from strictly protected nature reserves and wilderness areas (Category I), where protection of biodiversity is the sole objective, to managed resource areas (Category VI), where protecting biodiversity is just one of several management objectives, including sustainable resource extraction. Although the latter category seems less desirable for conservation purposes, these areas are often much larger than more strictly protected sites and can play a key role in protecting biodiversity.

As of 2005, 10 942 strictly protected areas (Categories Ia, Ib, and II) have been designated worldwide, covering some 6.16 million square kilometres. An additional 55 774 partially protected areas (Categories III–V) cover a combined 5.67 million square kilometres (see Table 27.4). Although this amount may seem extensive, most protected areas are small, with half 100 km^2 or less. Total protected lands account for only 11.9 percent of Earth's land surface. Of this total area, only 65 percent has been classified using the IUCN system. Marine protection efforts have lagged behind conservation of terrestrial environments, with only 1 percent currently protected.

In Canada, the number and total area of protected sites have grown steadily. According to Environment Canada, 9.8 percent (977 621 km^2) of Canada's terrestrial area and 0.66 percent (46 658 km^2) of its marine territory were protected as of 2010 (Figure 27.32, p. 616). Although this represents a doubling since

Table 27.4 Global Number and Extent of IUCN Classified Protected Areas as of 2005

Category	Number of sites	Proportion of total number of protected areas (%)	Area covered (1000 km^2)	Proportion of total area protected (%)
Ia	5 549	4.9	1 048	5.4
Ib	1 371	1.2	639	3.3
II	4 022	3.5	4 475	23.1
III	19 813	17.3	271	1.4
IV	27 466	24.0	3 005	15.5
V	8 495	7.4	2 393	12.3
VI	4 276	3.7	4 284	22.1
No category	43 304	37.9	3 267	16.9
Total	**114 296**	**100**	**19 382**	**100**

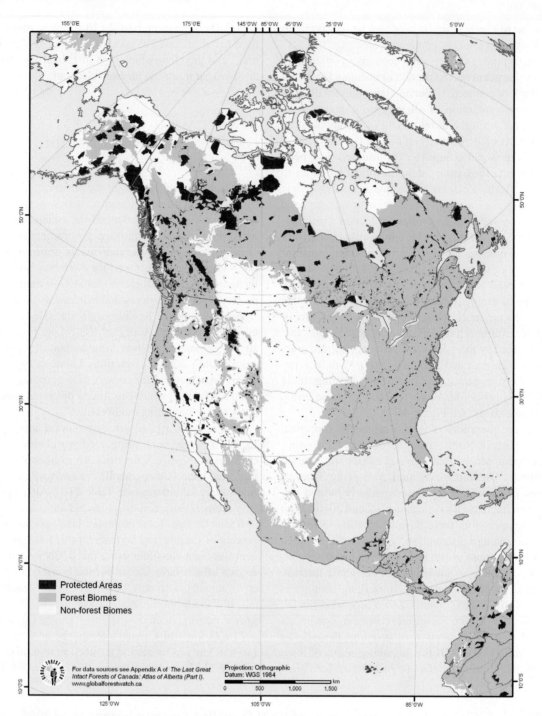

Protected Areas
Forest Biomes
Non-forest Biomes

For data sources see Appendix A of *The Last Great Intact Forests of Canada: Atlas of Alberta (Part I)*. www.globalforestwatch.ca

Projection: Orthographic
Datum: WGS 1984

km
0   500   1,000   1,500

Figure 27.32 Protected areas within North and Central America, as of 2010.

(Source: Environment Canada.)

1990, it is still less than the global average of 12 and 1 percent for terrestrial and marine habitat, respectively. Of this total, 89 percent is in Categories Ia, Ib, and II. Among provinces, protection varies from a low of 3 percent in Prince Edward Island and New Brunswick to highs of 12.4 and 14.4 percent in Alberta and British Columbia, respectively (Table 27.5).

In the world overall, much of the protected area is in the less-protected Category VI and in small rather than large parks (Figure 27.33). In the United States, for example, in addition to national parks and wildlife refuges, there are extensive national forests and other lands governed by the Department of Agriculture or the Bureau of Land Management. Covering 20 percent of the country's land area, these lands are managed for multipurpose use, including logging, grazing, and extraction of minerals and water. As critical as these areas are to maintaining biodiversity, they are far from fully protected.

What about the status of protected areas in the future? New reserves are being established worldwide, and lands under limited protection are being reclassified for greater protection. For example, in 2002 over 4000 ha of the George Washington

Table 27.5 **Total Land Protected by Canadian Provinces and Territories, as of 2010**

Province/territory	Protected area (km²)	Provincial/territorial area (km²)	Percentage of area protected
Alberta	82 361	661 848	12.4
British Columbia	135 992	944 735	14.4
Manitoba	57 840	647 797	8.9
New Brunswick	2 268	72 908	3.1
Newfoundland and Labrador	18 535	405 212	4.6
Northwest Territories	119 794	1 346 106	8.9
Nova Scotia	4 608	55 284	8.3
Nunavut	209 690	2 093 010	10.0
Ontario	106 463	1 076 395	9.9
Prince Edward Island	161	5 660	2.8
Québec	131 007	1 513 879	8.7
Saskatchewan	50 459	651 036	7.8
Yukon	57 459	482 443	11.9
Canada Total	976 473	9 984 670	9.8

(Source: Environment Canada.)

National Forest in Virginia was designated as wilderness. Pressure to designate more marine preserves is also growing. As well as new areas being designated, existing protected lands are being improved by adding buffer zones and corridors. Small reserves are being aggregated into larger blocks. Reserves are often embedded in a matrix of habitats managed for resource extraction, such as logging or grazing. If protecting diversity becomes a secondary priority in managing these areas, more habitats and species can be protected. Whenever possible, a protected area should include a contiguous block of land or water, such as a watershed, lake, or mountain range. This strategy allows managers to better control the spread of fire, pests, and destructive influences related to human activity.

A new approach to managing a system of nature reserves is to link isolated protected areas into one large system by establishing habitat corridors—areas of protected land running between the reserves (see Section 18.5). Such corridors facilitate dispersal of organisms and assist species that migrate seasonally to different habitats to forage or breed. In Costa Rica, two wildlife reserves, Braulio Carillo National Park and La Selva Biological Station, were linked by a 7700-ha forest corridor several kilometres wide. At least 35 bird species can now migrate between the reserves. The idea of corridors is intuitively appealing, but there are drawbacks. Corridors may also facilitate the movement of fire, pests, or disease (see Section 18.5).

In some cases, more extensive actions are being taken to link established protected areas. In southern Africa, the Great Limpopo Transfrontier Park (Figure 27.34) will link existing national

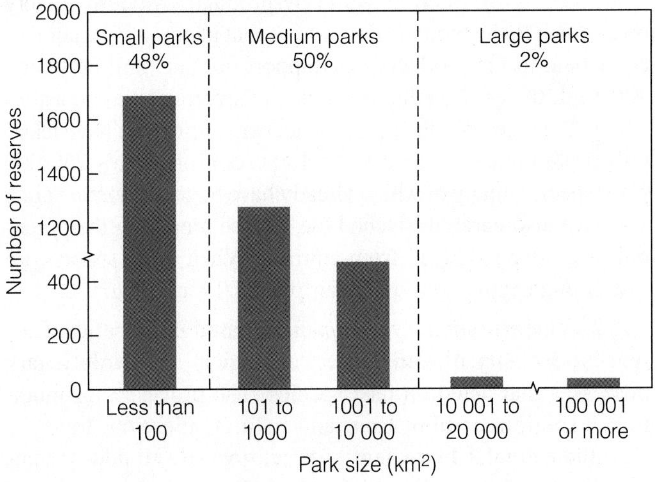

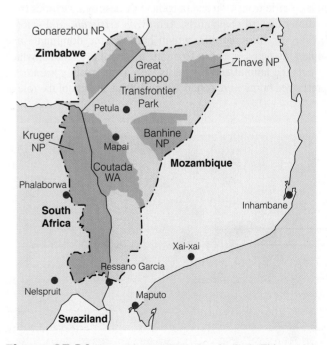

Figure 27.34 Great Limpopo Transfrontier Park. This international conservation effort links existing national parks in South Africa (Kruger National Park), Zimbabwe (Gonarezhou National Park), and Mozambique (Coutada Wildlife Area, Banhine and Zinave National Parks).

Figure 27.33 Size distribution of the world's nature reserves.

(Adapted from IUCN 2005).

parks in South Africa, Zimbabwe, and Mozambique. Establishing this international park will create a conservation area of 100 000 km², one of the world's largest contiguous nature reserves.

27.11 RESTORATION ECOLOGY: Restoring Habitat Is Essential to Conservation Efforts

As well as protecting existing habitat, considerable efforts are underway to restore natural ecosystems affected by human activities. This work has stimulated the field of *restoration ecology*. Its goal is to return an ecosystem to a close approximation of its state prior to disturbance by applying ecological principles. Restoration ecology involves various approaches ranging from reintroducing species and restoring habitats to re-establishing entire ecosystems.

Some restoration efforts involve rejuvenating existing communities by addressing specific (and often limited) issues, such as removal of exotic species, replanting of native species, and reintroduction of disturbances such as fires in grasslands and pine forests. Lake restoration involves reducing nutrient inputs (especially phosphorus) from surrounding lands, restoring aquatic plants, and reintroducing native fish. Wetland restoration involves re-establishing the natural hydrology so that the wetland is flooded on a natural cycle, and replanting aquatic species (Figure 27.35).

More intensive restoration involves re-creating the community from scratch. This type of restoration involves preparing the site, introducing native species, and employing appropriate management to maintain the community, especially against invasion by exotic species. In a project initiated in 2004 in Saskatchewan's Grasslands National Park, 700 ha was restored to native mixed-grass prairie from crop and rangeland by destroying exotics (particularly crested wheatgrass, *Agropyron cristatum*, and smooth brome, *Bromus inermis*, which had become dominant) and replanting native species. To mimic a natural fire regime, which affects the animal and fungal as well as the plant community, prescribed burns were conducted on 300 ha. In light of the role of

Figure 27.35 Volunteers prepare seagrass shoots for planting in the Florida Keys. The plantings enhance recovery of sites where die-off of seagrass communities has occurred.

large herbivores in maintaining prairie community structure, 44 000 ha have been fenced for grazing by bison. Presence of other herbivores and predators (invertebrate as well as vertebrate) has also been encouraged, including reintroduction of the swift fox. Although this ongoing restoration effort is led by Parks Canada, other public and private groups are active participants, including aboriginal stakeholders.

27.12 ENVIRONMENTAL ETHICS: Ethical Issues Underpin Conservation Ecology

At the United Nations Earth Summit in Rio de Janeiro in 1992, over 150 nations signed the Convention on Biodiversity. Canada had the distinction to be the first country to sign the convention, although it has since been criticized for lagging behind other countries in its conservation efforts. (As of 2012, Canada has officially announced its intention to repeal the 1999 Canadian Environmental Protection Act.) The Rio de Janeiro convention makes preservation of biodiversity an international priority. (Because *preservation* means maintaining biodiversity or habitat as it is, *conservation* is a more viable goal, as it recognizes the reality of ongoing ecological forces.) Arguments for the importance of maintaining biodiversity fall into three categories: economic, evolutionary, and ethical.

1. The *economic argument* is based on self-interest. Many of the products we use come from organisms with which we share the planet. Obviously, our foods are all derived from other organisms. Less obviously, every time we buy a pharmaceutical, there is a high probability that some of its essential constituents derive from a wild species. According to evolutionary biologist E.O. Wilson (2003), 40 percent of pharmaceuticals in the United States alone were originally derived from wild species, including nine of the 10 leading prescription drugs. The value of medicinal products from such sources now totals over $40 billion yearly. We derive rubber, solvents, and paper from trees. We use cotton, flax, leather, and a host of other natural materials for clothing. The houses we live in and the furniture in them are made largely of wood or wood by-products. Modern society owes much to the biological resources that in one way or another contribute to the products that support our standard of living. Although today's benefits from Nature's cornucopia are astonishing, they are only the tip of the iceberg. Scientists have taken only a preliminary look at some 10 percent of Earth's 250 000 plant species, many of which already have huge economic value. We have also barely scratched the surface regarding the potential of products derived from animals. When these species are lost through extinction, so is their potential for human use.

2. The *evolutionary argument* is based on genetics. Current biodiversity patterns reflect ecological and evolutionary processes that acted on past species. The processes of mutation, genetic recombination, and natural selection, together with the essential ingredient of time, give rise to new species (see Section 5.6). All species eventually go extinct, many of them leaving no traces other than fossilized remains. Others fade into extinction after having given rise to new species. For example, all modern birds trace their evolutionary history to the earliest known bird, *Archaeopteryx*, which lived during the Jurassic period (145 million years ago). If *Archaeopteryx* had been driven to extinction before acting as the evolutionary seed of modern birds, the variety of life in our backyards would be quite different today. Likewise, the ongoing mass extinction of modern species limits the potential evolution of species diversity in the future.

3. The *ethical argument* recognizes that humans are but one of millions of species inhabiting Earth. Although it is the nature of all organisms to both respond to and alter their environment, it is unlikely that any other species in Earth's history has so dramatically affected its environment in such a short time. The fundamental question facing humanity is thus an ethical one: *To what degree should we allow human activities to continue causing the extinction of the thousands of species with which we share this planet?*

Ultimately, debate on the value of biodiversity centres on this core ethical question. Economic arguments fall to the wayside as technology allows us to synthesize medicines and other products currently made from natural products. Evolutionary arguments, while convincing to biologists, may seem abstract to the public when considered against the demands of a growing world population. Science helps us understand biodiversity and how to conserve it, but the solution to the biodiversity problem involves social, economic, and ethical issues that influence all our lives. Unlike so many problems facing society, it is a problem that science can only illuminate, and that technology cannot resolve. It is up to society—to you and me—to find a solution.

EcologyPlace

Visit EcologyPlace at www.pearsoncanada.ca/ecologyplace to access online resources that complement your textbook, and help you to apply and to review the information in this chapter. EcologyPlace includes

- an eText version of the book
- self-grading quizzes
- glossary flashcards
- and more!

Go to www.pearsoncanada.ca/ecologyplace and follow the registration instructions on the Student Access Code Card included with this text. If your book does not have a Student Access Code Card, you can purchase access to it at www.pearsoncanada.ca/ecologyplace.

SUMMARY

Global Diversity Patterns 27.1

Although the fossil record indicates that diversity has increased over the past 600 million years, Earth's history is marked by several mass extinction events. We are currently in the midst of a mass extinction event caused by human activities. Regional diversity differs on shorter timescales, as affected by succession and changes in geographic ranges with climate.

Latitudinal Diversity Trends 27.2

Species are not distributed evenly. Richness declines from the equator to the poles in both land and aquatic habitats. Various hypotheses have been proposed to explain these global patterns, including evolutionary time, ecological time since disturbance, climatic stability, heterogeneity, and productivity. Regional plant species richness is correlated with actual evapotranspiration, suggesting a positive relationship with productivity. Species richness of terrestrial vertebrates is correlated with potential evapotranspiration, a measure of energy input. Species diversity of marine organisms is influenced by seasonality of productivity.

Extinction Risk 27.3

Many factors influence extinction risk. Endemic species are vulnerable because loss of habitat in their restricted range leads to a complete loss of habitat. Migrating species depend on two or more habitats in different regions. Other factors include small populations, specialized habitats, and species that are in direct conflict with human activities.

Risk Categories 27.4

Classifying threatened species is critical to conservation efforts. The IUCN classifies threatened species as critically endangered, endangered, or vulnerable, based on a variety of criteria. The number of threatened species is increasing in all taxonomic groups, with vertebrates most affected. In Canada, COSEWIC uses similar categories as the IUCN to evaluate extinction risk.

Biodiversity Hotspots 27.5

Biodiversity hotspots have a high species richness and a disproportionate number of endemic species. They are at high risk of habitat damage or loss from human activities. Identifying hotspots helps conservationists decide which areas are the most important for conservation efforts. The 25 designated hotspots contain 44 percent of all plant species and 35 percent of all terrestrial vertebrate species in only 1.4 percent of Earth's land area.

Habitat Loss 27.6

The main cause of species extinctions is habitat damage and loss, largely from expansion of agricultural lands to meet the needs of a growing population. Given their high diversity and the pressures of economic development, tropical regions have been the primary concern. Tropical forests are being lost at a rapid rate, but quantity alone is an inadequate indicator because many forests are highly fragmented and face continued pressure from humans. Mature boreal forest is also being lost at an unsustainable rate. Plantations do not support natural diversity levels.

Introduced Species 27.7

Non-native species, introduced either intentionally or accidentally, may cause extinction of native species through predation, grazing, competition, and habitat alteration. Invasive exotics are considered to be the second major cause of species extinctions after habitat loss.

Population Strategies 27.8

Protecting populations is key to species preservation. An adequate conservation plan protects as many individuals as possible within the greatest possible area. The minimum viable population (MVP) is the number of individuals necessary to ensure long-term survival. The area of suitable habitat needed to maintain the MVP is the minimum dynamic area (MDA). Species often exist as metapopulations, whose persistence depends on a complex dynamic. Saving species from extinction may require establishing new populations through reintroductions. Success of reintroductions varies with species and location.

Habitat Conservation 27.9

Despite the need to focus conservation efforts on endangered species, the best way to protect biodiversity in the long term is to protect habitats. An ecosystem-based conservation approach requires understanding the relationship between overall diversity patterns and landscape features. As large areas contain more species, it is generally better to protect as large an area as possible.

Protected Areas 27.10

Given the growing pressures exerted by humans, protecting biodiversity is increasingly dependent on establishing legally designated protected areas, most of which are public lands. Protected lands differ in their degree of protection. Many categories serve multiple purposes, including resource extraction. Regions within North America differ in their amount and distribution of protected areas. Current efforts focus on working with and expanding existing protected lands and providing buffer zones and corridors to enhance their habitat quality and conservation value.

Restoration Ecology 27.11

By applying ecosystem principles, restoration ecology attempts to restore natural communities affected by human activities to an approximation of their natural state. Approaches range from reintroducing species and rehabilitating existing communities to re-establishing ecosystems.

Environmental Ethics 27.12

The economic argument for conserving biodiversity is based on self-interest, focusing on the products and services (current and potential) that species provide. The evolutionary argument states that species extinction limits the evolution of species diversity in the future. The ethical argument asks whether it is justifiable for humans to be the main cause of species extinctions.

KEY TERMS

biodiversity hotspot	extant	mass extinction	minimum viable	threatened species
effective population	introduced (alien)	event	population	ubiquitous species
size	species	minimum dynamic		
endemic species	invasive species	area		

STUDY QUESTIONS

1. How does the current mass extinction event differ in cause and possible consequences from previous events?
2. Describe the different factors that have been proposed to explain the high terrestrial species richness of the tropics.
3. Diversity declines with higher latitudes in both terrestrial and marine ecosystems, but the relationship to productivity differs. Explain.

4. What is a *biodiversity hotspot*? What role can hotspots play in conservation efforts? No biodiversity hotspots have been identified in Canada. What does this imply for conservation efforts in Canada? Does it take us off the hook?
5. Discuss several ways in which introduced species may disrupt a community, leading to the decline of native species. Consider the role of species interactions in

We are all familiar with it, but the term *global climate change* is redundant. Change is inherent in Earth's climate. Although the tilt of Earth's axis relative to the Sun is 23.5°, Earth is actually wobbly. The tilt varies from 22.5° to 24° and affects the amount of sunlight striking different parts of the globe. This variation occurs over a 41 000-year cycle and is responsible for the ice ages. In turn, climatic variations greatly affect life on Earth. Paleoecology has revealed the responses of populations and ecosystems to climate changes during glacial expansions and retreats over past millennia (see Figures 17.22 and 17.23). On a geologic timescale, the fossil record depicts evolutionary changes reflecting the dynamics of Earth's climate (see Figures 27.4 and 27.5).

This text has described many examples of the influences of climate, from uptake of CO$_2$ in photosynthesis to the distribution and productivity of Earth's ecosystems. But now we have entered a new era, in which a single species—humans—has the ability to alter Earth's climate. In this chapter, we examine how human activities are changing atmospheric chemistry and how these changes are affecting Earth's climate. We explore how these changes may shift species distributions, alter biotic interactions, and ultimately influence ecosystems, and how these changes in climate and ecosystems may affect the health and well-being of the agent of change—humans.

28.1 GREENHOUSE GASES: Many Atmospheric Gases Influence Earth's Energy Balance

Many compounds naturally present in Earth's atmosphere—H$_2$O vapour, CO$_2$, methane (CH$_4$), and ozone (O$_3$)—absorb thermal (longwave) radiation emitted by Earth's surface, preventing it from being immediately re-radiated back to the thermal sink of outer space. The atmosphere is thereby warmed and, in turn, emits thermal radiation, which warms Earth's surface and atmosphere further (see Figure 2.3). Earth's mean surface air temperature is about 30°C higher than it would be without this atmospheric absorption and re-radiation. This phenomenon is called the **greenhouse effect**, and the gases responsible for it are **greenhouse gases**.

The greenhouse effect helps make Earth habitable for life. In the past, the atmosphere was much higher in CO$_2$ than at present, allowing Earth to warm. But what about the role of the greenhouse effect in the future? If the amount of energy reaching Earth's surface is about the same as the amount radiated back into space, the temperature of Earth's surface should remain roughly constant (see Section 2.2). However, since the industrial era began, atmospheric concentrations of greenhouse gases have increased greatly. Given their impact on Earth's energy balance, concerns have arisen over the impact of these rising concentrations on global climate.

28.2 ATMOSPHERIC CO$_2$ LEVELS: Evidence Indicates That Atmospheric CO$_2$ Is Rising

Although human activities have increased atmospheric levels of many greenhouse gases, most attention centres on CO$_2$, which has increased over 25 percent in the past century. The evidence comes from atmospheric observations started in 1958 at Mauna Loa, Hawaii (Figure 28.1). Evidence prior to 1958 comes from various sources, including air bubbles trapped in glacial ice in Greenland and Antarctica. In reconstructing atmospheric CO$_2$ trends over the past 300 years, values fluctuate between 280 and 290 ppm until the mid-1800s (Figure 28.2a, p. 624). With the onset of the Industrial Revolution, the value increases steadily, rising exponentially by the mid-19th century with adoption of fossil fuels as an energy source (Figure 28.2b).

In 2008, developed countries generated over 70 percent of the total C emissions from burning fossil fuels (Figure 28.2c) (Boden et al. 2011). China was the largest single source, having overtaken the United States for the first time. Canada is a smaller source, but only because of our smaller population. Per capita, our annual emissions in 2008 of 16.4 tonnes rival those of Americans (17.5 tonnes). In the next few decades, 90 percent of population growth will occur in developing countries, many of which are undergoing rapid economic growth. Per capita energy use in developing countries—once a small fraction of North America's—is rising. By 2008, China's per capita emissions had grown to 5.3 tonnes, compared with just 1.4 tonnes in India.

Fossil fuels are not the only cause of rising CO$_2$. Cement production contributes surprising amounts. First used by the

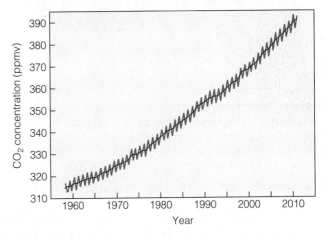

Figure 28.1 Atmospheric CO$_2$ as measured at Mauna Loa Observatory, Hawaii.

(Source: NOAA 2009.)

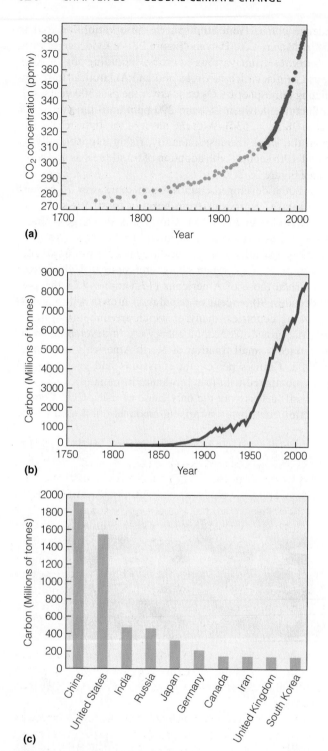

(a)

(b)

(c)

Figure 28.2 Relationship between atmospheric CO_2 and carbon emissions. **(a)** Atmospheric CO_2 over the past 300 years. Pre-1958 data is estimated (ppmv = parts per millilitre volume). **(b)** Annual CO_2 input from burning fossil fuels since 1750. **(c)** Carbon emissions from burning fossil fuels by the top 10 countries in 2008.

(Data from (a) IPCC 2007, (b) CDIAC 2009, and (c) Boden et al. 2011.)

Romans, cement became common in construction around 1900 and is currently responsible for 10 percent of CO_2 emissions worldwide. It is of particular concern as an emissions source in China, which produces half the world's total (Gregg et al. 2008). Another source is deforestation (Figure 28.3). When

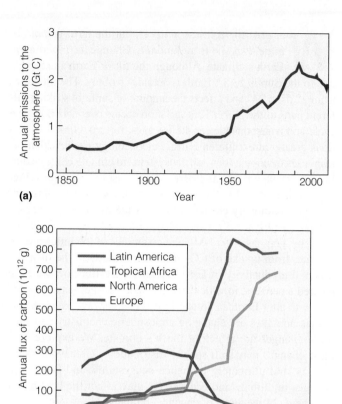

(a)

(b)

Figure 28.3 Annual input of CO_2 to the atmosphere from clearing and burning of forest **(a)** globally and **(b)** in selected regions: Latin America, tropical Africa, North America, and Europe.

(Adapted from Houghton 1997.)

forests are cleared for farming, the trees are used for timber or pulp but much of the biomass, litter, and soil organic matter is burned, releasing CO_2.

Estimating the contribution of land clearing to atmospheric CO_2 is complex. After logging on lands managed for forestry or swidden agriculture (see Section 26.4), vegetation and soil organic matter re-establish. The net atmospheric contribution is calculated as the difference between the CO_2 released during clearing and burning and the CO_2 absorbed by photosynthesis and growth during re-establishment. Originally, scientists used estimates of population growth and land use (forestry and agriculture), along with models of vegetation and soil succession, to estimate the contribution of changing land use to atmospheric carbon. Recent estimates use satellite images.

Scientists estimate that the mean annual amount of C released to the atmosphere during the 1990s was 8.5 gigatonnes (Gt = 10^9 tonnes), of which some 6.3 Gt was from combustion of fossil fuels and 2.2 Gt from forest clearing (Prentice et al. 2001). (To put this number into perspective, if humans weigh on average 70 kg, 1 Gt would be the weight of 14 billion people.) Yet direct measurements of atmospheric CO_2 over this period show an annual accumulation of only 3.2 Gt. The difference (5.3 Gt) must have exited the atmosphere into other pools in the global cycle (see Figure 22.5)—the oceans and terrestrial systems. Determining the fate of CO_2 released from fossil fuels requires input from many scientific disciplines, plus considerable detective work.

Although estimates vary with the source, diffusive uptake of CO_2 into oceans is more predictable than on land because it is determined primarily by physical processes. From 2000 to 2010, the oceans absorbed 2.4 Gt of C annually (Le Quéré et al. 2009). In contrast, quantifying terrestrial C exchange at a global scale is more difficult, even though the biotic processes controlling C exchange between terrestrial systems and the atmosphere are well understood. Therefore, scientists use a simple formula to estimate global terrestrial uptake. Carbon that has been emitted over a given period that cannot be accounted for by changes in atmospheric C or by estimates of ocean uptake is attributed to terrestrial systems. Below are estimated values for the 1990s (Prentice et al. 2001):

Emissions from fossil fuels	−	Atmospheric increase	−	Ocean uptake	=	Net uptake by terrestrial systems
6.3 Gt		3.2 Gt		2.4 Gt		0.7 Gt

Using this formula, terrestrial systems were a net C sink, with an annual net uptake of 0.7 Gt (Houghton 2003). However, deforestation was responsible for an annual release of 2.2 Gt during the 1990s, not a net uptake of 0.7 Gt. The discrepancy of 2.9 Gt per year (2.2 + 0.7) has been dubbed the "problem of the missing carbon" (Figure 28.4). Estimates of this *residual terrestrial sink* vary widely (Houghton 2003). According to the estimates of global CO_2 sources and sinks for the decades from 1980 to 2010 in Table 22.1, combined oceanic and terrestrial sink activity was more or less constant from the 1990s to the 2000s (4.9 to 4.8 Gt C annually). This levelling-off has caused more of the increased C emissions from fossil fuels, cement, and land-use change to accumulate in the atmosphere (3.1 to 4.1 Gt annually) (Le Quéré et al. 2009).

Net C uptake by terrestrial systems is largely due to reforestation in temperate regions of the Northern Hemisphere, following abandonment of lands cleared for agriculture during the late 19th and early 20th centuries. However, although reforestation plays a key role in balancing the global C cycle, tropical forests may represent a larger sink than once thought. Boreal forests and peat are also part of the puzzle. Current rates of C sequestration in organic soils in the sub-Arctic and low Arctic regions are reduced from long-term averages (Koven et al. 2011). Although these soils are still sequestering C, and may account for some of the missing amount, their ability to do so in the future will be compromised by thawing of permafrost. Rather than acting as a C sink, they will then be a C source, compounding the impact of C enrichment (see Section 22.4).

28.3 UPTAKE BY OCEANS: Oceans Respond to Elevated CO_2 by Increasing Uptake

As we saw in Section 3.6, CO_2 that diffuses into oceanic surface waters dissolves and undergoes chemical reactions, including transformation to carbonate and bicarbonate. Its diffusion rate is a function of the concentration gradient. As atmospheric CO_2 rises, so does oceanic uptake. Given their volume, the oceans have the potential to absorb much of the C released by fossil fuel burning and land clearing. This potential is not fully realized, because the oceans are not a homogeneous sponge that absorbs CO_2 equally into its entire volume. Recall that the oceans have two distinct layers. Solar radiation warms surface waters. Depending on the amount of radiation, this warm-water zone ranges from 75 to 200 m, with a mean temperature of 18°C. Deeper waters are much colder (mean temperature of 3°C). The transition between the zones (thermocline) is abrupt and permanent. The ocean is thus a thin layer of warm water floating on a much deeper, colder layer.

The temperature difference between the layers separates many processes. Winds mix surface waters, transferring absorbed CO_2 into the waters immediately below. Given the thermocline, this mixing does not extend into deeper waters. Mixing between surface and deep waters (Figure 28.5, p. 628) depends on ocean currents caused by the sinking of surface waters as they move towards the poles (see Sections 2.5 and 3.8). This process occurs over centuries, severely limiting uptake of CO_2 into deep waters. Thus, the amount of CO_2 absorbed by oceans in the short term is limited, despite their large volume. Some argue that we should facilitate the biotic C sink activity of oceans by promoting marine productivity. Uptake of CO_2 and bicarbonate into marine food webs pulls the reaction sequence to the right, allowing more uptake by increasing the diffusion gradient.

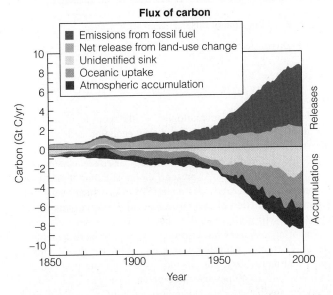

Figure 28.4 Carbon release and accumulation from 1850 to 2000. The missing carbon is represented here as the "unidentified sink," the activity of which is in decline. Go to **QUANTIFYit!** at **www.pearsoncanada. ca/ecologyplace** to see emission trends.

(Adapted from the Woods Hole Research Center, www.whrc.org.)

ECOLOGICAL ISSUES   Oilsands: The Collateral Damage of Our Fossil Fuel Dependence

In this chapter, we stress fossil fuel combustion as a source of atmospheric carbon. Although the climatic impact of these emissions is of great concern, less obvious is the collateral damage resulting from our dependence on oil. This damage includes much-publicized spills from tankers and off-shore drilling rigs that have polluted our oceans and coastlines. But there is a more insidious yet potentially devastating side effect: pollution from developing and transporting fuel from oilsands.

Canada is no stranger to fossil fuel exploration. The oil fields of Alberta have long been a source of economic wealth. But as conventional sources of oil and gas are depleted and prices rise, exploiting alternative sources of fossil fuels becomes more attractive. New technologies made it possible to exploit dilute and formerly economically infeasible sources. So, what are the oilsands, and what ecological hazards does their exploitation pose?

According to Environment Canada, the oilsands are one of the world's largest single oil accumulations. Underlying 140 000 km^2 of boreal forest, peatlands, and prairie in three regions of Alberta (Figure 1), they hold an estimated 1.7 trillion barrels of crude. The deposits consist of bitumen (a semi-solid) mixed with sand, clay, and water. Instead of gushing to the surface like conventional oil, the deposits require specialized technologies for extracting and processing. With current technology, about 10 percent is recoverable, and much more with emerging technology. Of the 170 billion barrels of recoverable reserves, about 20 percent (in the Athabasca deposit) are accessible by surface mining. Over time, more of the region's production will involve *in situ* processing (45 percent by 2008). Using new technologies, production is projected to grow from 1.3 million barrels a day in 2009 to 3.5 million by 2025. Canada would seem to be "sitting pretty," in terms of both our own fuel needs and exports to the U.S. market.

But there is a darker side. Open-pit mining uses 2 to 4 barrels of freshwater per barrel of oil recovered, imposing an additional stress on already strained river and groundwater resources (see Ecological Issues: Our Disappearing Groundwater Resources, pp. 51–52). It exposes large expanses of natural ecosystems (Figure 2), severely fragmenting habitat for large mammals, including the boreal subspecies of the woodland caribou. The tailings, which are collected in artificial ponds, contain toxic levels of heavy metals, polycyclic aromatic hydrocarbons (PACs), and solvent residues. Many birds and waterfowl that land in the ponds are poisoned. Despite ongoing efforts, reclamation of mine sites and tailing ponds is extremely slow. Leakage of toxins from the ponds and use of heavy transport equipment have caused significant damage to surrounding areas.

Aboriginal peoples in the region rely on fish as a food source, and contamination is a grave and controversial concern. Erin Kelly, a research biologist with the Northwest Territories government, and the University of Alberta's David Schindler are investigating the contribution of contaminants from oilsands development. PAC levels in tributaries of the Athabasca increased from 0.009 µg/L upstream of oilsands development to 0.202 µg/L downstream in summer. PAC-loading of snowmelt was also considerable. Although PACs are naturally present in the river sys-

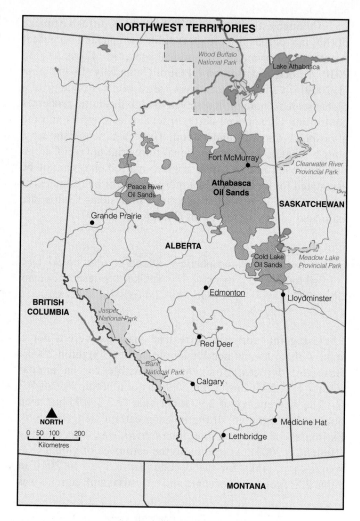

Figure 1 Location of oilsands deposits in Canada.

(Source: Environment Canada.)

tem, these elevated levels are likely to damage fish embryos (Kelly et al. 2009).

In situ extraction technologies would seem to incur less site damage than surface mining. Less vegetation is destroyed directly, large expanses of soil are not exposed, and no tailings ponds are required. However, the ecological effects depend on the specific technology. At present, the most common method is *steam-assisted gravity drainage* (SAG-D), which heats the bitumen by injection of pressurized steam. Projected increases in SAG-D will place huge demands not only on groundwater and aquifer resources but also on energy sources to generate steam. Currently, natural gas is used, adding to the C emissions. Estimates vary, but oilsands extraction and processing emit some 15 to 40 percent more greenhouse gases than conventional fossil fuels. Nuclear power, with all of its concomitant ecological concerns, is a possible alternative for steam generation.

Figure 2 A open-pit site in the Athabasca oilsands region, near Fort McMurray, Alberta.

Newer technologies may solve some of SAG-D's problems, but at the risk of creating new ones. A process using vapour extraction rather than steam injection improves energy efficiency, but employs hazardous hydrocarbon solvents. The most publicized emerging technology, *toe-to-heel air injection*, ignites the bitumen in order to burn the heavier components of the oil. Less water is needed and fewer greenhouse gases are released, but at the risk of setting wells on fire.

Even if more efficient and less polluting technologies are implemented, another major issue remains: transporting the upgraded crude (called *synthetic oil*). Currently, a debate is raging regarding construction of the Keystone XL pipeline from the Alberta oilsands to refineries in the United States, ultimately extending some 3500 km to the Gulf of Mexico. Concerns include the risk of leakages en route and a growing reliance on "dirty" oil. Canadians are not comfortable with promoting dirty technology. Yet our environmental track record is not strong—we are under fire for selling asbestos to parts of the world where its use is permitted despite its carcinogenic properties. So, do we want to develop a technology that has the potential not only to increase our contributions to global C emissions but also to pollute ground and surface waters and disturb or destroy extensive habitat?

The bottom line is that the impacts of fossil fuel use are not restricted to release of CO_2. Even if new technologies counter the impacts of greenhouses gases, there are other far-reaching impacts that we ignore at our peril. In a 2010 article in the influential journal *Nature*, David Schindler issued a call to slow the pace of oilsands development until sufficient scientific data are available to address these serious concerns.

Bibliography

Kelly, E. N., J. W. Short, D. W. Schindler, P. V. Hodson, M. Ma, A. K. Kwan, and B. L. Fortin. 2009. Oil sands development contributes polycyclic aromatic compounds to the Athabasca River and its tributaries. *Proceedings of the National Academy of Sciences* 106:22346–22351.

Schindler, D. W. 2010. Tar sands need solid science. *Nature* 468:499–501.

1. Oilsands development highlights a central issue: *What should we as a country do when our environmental interests seem to conflict with our economic interests?* Discuss in terms of the direct and indirect impacts of oilsands development on habitats, climate, and ecosystem services in the short and long term.

2. Based on research, do you think the United States's increasing reliance on coal technology is more or less "dirty" than oilsands technology?

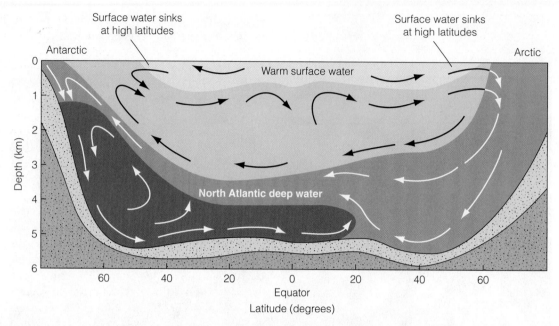

Figure 28.5 Circulation pattern in the Atlantic Ocean. Surface waters flow north from the tropics. They cool and sink when they reach sub-Arctic latitudes and become part of the huge, deep, southward countercurrent extending to the Antarctic.

However, fertilizing the ocean by adding limiting nutrients such as iron is ill advised, given the unpredictable effects that such a massive intervention might trigger.

28.4 PLANT RESPONSE: Elevated CO_2 Increases Photosynthesis in the Short Term, but Long-Term Impacts Are Unclear

To understand how elevated CO_2 affects productivity of terrestrial ecosystems, we must consider plant response to CO_2 enrichment. Elevated CO_2 has two direct, short-term effects on plants. First, photosynthetic rate increases. Recall that CO_2 diffuses into leaves through stomata on leaf surfaces. The higher the CO_2 level in the air, the greater the diffusion rate into the leaf. Faster diffusion increases the availability of CO_2 in the leaf mesophyll, thereby increasing photosynthetic rate. Enhanced photosynthesis with elevated CO_2 is called the **CO_2 fertilization effect**. Second, elevated CO_2 causes stomata to close partially, reducing transpirational H_2O loss. Thus, at higher CO_2, plants increase their *water-use efficiency* (C uptake per unit H_2O loss; see Section 6.10).

These short-term effects are straightforwardly beneficial. Long-term impacts are more complex. According to a review of over 600 experiments, C_3 species are most stimulated by elevated CO_2, with a mean increase in biomass of 47 percent (Figure 28.6) (Poorter and Pérez-Soba 2002). Within C_3 species, crops show the most enhancement (59 percent) and wild herbaceous plants the least (41 percent). Stimulation of woody seedlings averaged 49 percent. However, most experiments with woody species were conducted with seedlings, considering

only a small part of their life cycle. Data on CAM species were limited to six species, but the mean increase was 21 percent. C_4 species had the most modest increase (11 percent) (Poorter and Pérez-Soba 2002).

However, these enhancement effects are often short-lived (Figure 28.7). Over time, some plants produce less Rubisco at elevated CO_2, reducing photosynthesis to rates that are comparable to rates at ambient CO_2—an example of **downregulation**. This compensatory response conserves nitrogen, which often limits plant growth. In other studies, plants grown at elevated

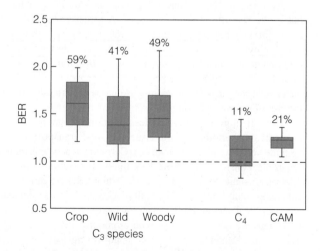

Figure 28.6 Biomass enhancement ratio (BER; ratio of biomass at elevated versus ambient CO_2) for plant functional types. Trends are based on 280 C_3 (crop, wild herbaceous, and woody); 30 C_4; and 6 CAM species. Boxes show the observation range. The line represents the median, the lower box the 25th percentile, and the upper box the 75th percentile. Error bars give the 10th and 90th percentiles.

(Adapted from Poorter and Pérez-Soba 2002.)

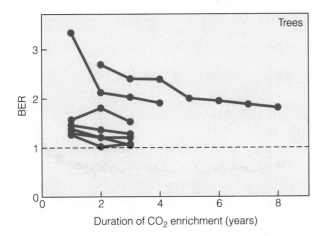

Figure 28.7 Time course of biomass enhancement ratio in elevated CO_2. Each line represents results of an experiment with a different tree species.

(Adapted from Poorter and Pérez-Soba 2002.)

CO_2 allocate less C to leaves and more to roots, and develop fewer stomata. Reduced leaf area and fewer stomata lower H_2O loss, and more root investment increases H_2O uptake, but these changes eventually reduce productivity.

How do trends observed for leaves or single plants scale up to the ecosystem level? Supply of H_2O or nutrients may limit enhancement of NPP at elevated CO_2. Ongoing since 1996, the Free Air CO_2 Experiment at Duke Forest in North Carolina (Figure 28.8a) is a macroscale study of CO_2 effects on intact ecosystems. Elevated CO_2 has consistently increased NPP over

ambient CO_2 controls (Figure 28.8b) (Finzi et al. 2006). Field studies in grasslands and agricultural systems indicate an increase in biomass of 14 percent at twice ambient CO_2 levels. However, estimates among sites ranged from an increase of 85 percent to a decline of 20 percent (Koch and Murray 1995). These results stress the importance of the interactions of elevated CO_2 with other factors, particularly temperature, moisture, and nutrients. In Arctic tundra, productivity initially increased at elevated CO_2, followed by a return to original levels after three years of continuous exposure (Oechel and Vourlitis 1996), indicating that downregulation can also occur at the ecosystem level.

The largest and most persistent responses to elevated CO_2 occur in seasonally dry sites, where productivity is enhanced only during years of average or below-average rainfall. In tallgrass prairie in Kansas, there was no significant increase in aboveground NPP during years with above-average rainfall for plots exposed to doubled CO_2. Yet aboveground NPP increased 40 and 80 percent during years with average and below-average rainfall, respectively (Owensby et al. 1996). (Even though these relative increases are large, they occur in years of low NPP, so the absolute changes were low.)

Why is the impact of CO_2 greater in dry years? Enhancement of NPP in dry habitats is due mainly to reduced transpiration with the partial stomatal closure made possible in elevated CO_2. Reduced transpiration increases soil moisture, particularly during prolonged drought. Increased soil moisture extends the growing season and increases microbial activity, decomposition, and mineralization.

(a)

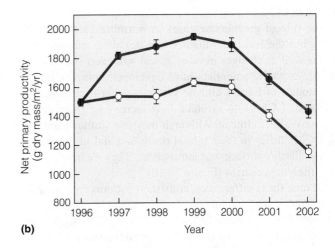

(b)

Figure 28.8 The Free Air CO2 Experiment at Duke Forest in North Carolina. (a) Towers release CO_2, allowing scientists to evaluate the long-term response of forests. (b) Net primary productivity (NPP) under ambient (open circles) and elevated (closed symbols) CO_2.

(Adapted from Finzi et al. 2006.)

INTERPRETING ECOLOGICAL DATA

Q1. How does NPP change over the observation period for ambient and doubled CO_2 levels?

Q2. How does the response of NPP to elevated CO_2 for Duke Forest relate to the biomass enhancement results shown in Figure 28.7? Are the results for individuals consistent with the patterns observed for NPP of forests?

28.5 CLIMATE CHANGE: Increases in Greenhouse Gases Are Altering Global Climate

So far, we have documented changes in atmospheric CO_2 and the direct responses of plants to that change. Of equal concern is the response not to CO_2 *per se* but to the climate change it may generate. According to the National Oceanic and Atmospheric Administration (NOAA), 2010 tied with 2005 as the warmest year on record since 1880, with global mean surface temperature approximately $0.62 \pm 0.07°C$ above the 20th-century average. 2010 was also the wettest year on record. Moreover, the decade from 2000 to 2010 ranks as the warmest decade on record (see Ecological Issues: Who Turned Up the Heat?, p. 632). The cause has been the focus of much debate, but a general consensus is emerging. According to the Intergovernmental Panel on Climate Change (IPCC 2007), an intergovernmental body set up by the World Meteorological Organization and the United Nations, most of the observed increase in mean global temperature since the mid-20th century is "very likely" due to changes in atmospheric levels of greenhouse gases, primarily CO_2.

As human activities continue to increase atmospheric CO_2, how will it alter global climate further? Scientists estimate that at current emission rates, the pre-industrial atmospheric level of 280 ppm CO_2 will double by 2020. Nor is CO_2 the only greenhouse gas that is increasing because of human activities. Others include methane (CH_4) (Figure 28.9a), nitrous oxide (N_2O) (Figure 28.9b), chlorofluorocarbons (CFCs), hydrogenated chlorofluorocarbons (HCFCs), ozone (O_3), and sulphur dioxide (SO_2). Although much lower in concentration, these gases contribute significantly to the greenhouse effect because most trap heat more effectively than CO_2.

The role of greenhouse gases in warming Earth is well known, but the specific influence that doubling atmospheric CO_2 (as well as changes in other gases) will exert is less certain. Atmospheric scientists have developed complex computer models of Earth's climate system—**general circulation models**, or GCMs—to estimate how increasing greenhouse gases may affect climate. Although they use similar parameters, GCMs differ in their spatial resolution and in how they describe Earth's surface and atmosphere. They therefore generate different scenarios (Figure 28.10).

Despite these differences, consistent patterns emerge. All GCMs predict a global increase in both temperature and precipitation. Findings published in 2007 by the IPCC suggest an increase in global mean surface temperature from 1.1°C to 6.4°C by 2100 (the actual range depends on the emissions scenario). These changes would not be evenly distributed. Warming is expected to be greatest during winter and in northern latitudes (Figure 28.11), while rainfall will likely decrease in some regions in summer. Although in popular speech *greenhouse effect* is synonymous with *global warming*, the models predict more than just hotter days. Many predict more variable rain and snowfall, as well as more hurricanes and other storms.

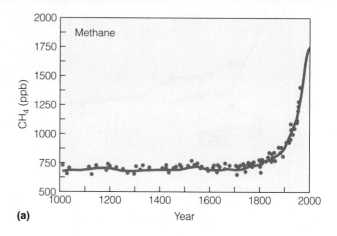

(a)

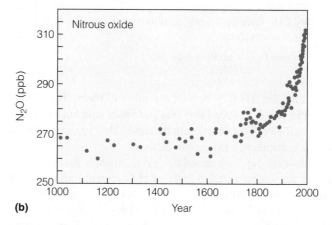

(b)

Figure 28.9 Historic trends in atmospheric levels of **(a)** CH_4 and **(b)** N_2O.

(Adapted from IPCC 2007.)

A recent development that has influenced climate models is the impact of aerosols on Earth's energy balance. **Aerosols** (small particles suspended in the atmosphere) absorb radiation and scatter it back to space, reducing the amount reaching Earth. Aerosols arise from many sources, including dust particles blown into the air over deserts or from sea spray over oceans, as well as from burning of forests and grasslands. Occasionally, large amounts enter the atmosphere from volcanic eruptions, as occurred with Mt. Pinatubo in 1991. Burning fossil fuels is a major anthropogenic source. Sulphate particles form from SO_2 released by coal-burning power stations. These particles remain in the atmosphere for only five days on average, so their distribution is concentrated near their source (Figure 28.12a, p. 633). In the Northern Hemisphere, their levels are significant and partly offset the effects of greenhouse gases, reducing estimates of warming (Figure 28.12b). The very fact that global temperatures are rising despite the cooling effect of aerosols suggests how much more intense warming would be in their absence (Andreae et al. 2005). Lowering aerosol emissions from pollution reduces their ameliorating effect, as indicated by an analysis of the intense 2005 drought in the Amazon (Cox et al. 2008).

As models improve, there will no doubt be further changes in the patterns and severity of changes they predict. However,

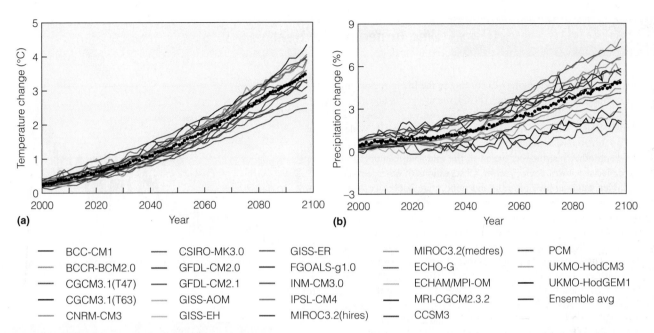

Figure 28.10 Changes in globally averaged **(a)** surface air temperature and **(b)** precipitation as predicted by various GCMs under a scenario (scenario A2) of increasing atmospheric concentrations of greenhouse gases, as developed by IPCC. Percentage changes are relative to means for the period from 1980 to 1999. Abbreviations refer to the research programs that have developed the various GCMs used in the analyses.

(Adapted from IPCC 2007.)

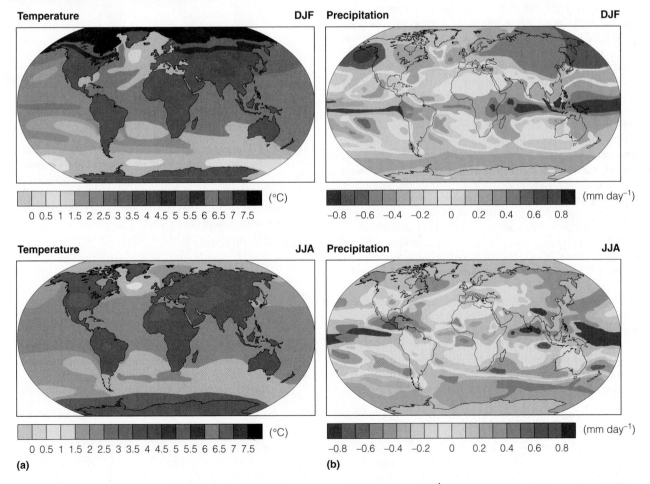

Figure 28.11 Changes in **(a)** surface air temperature (°C) and **(b)** precipitation (mm · day⁻¹) for Northern Hemisphere winter (DJF—December, January, and February: top) and summer (JJA—June, July, and August: bottom) as predicted for rising levels of greenhouse gases. Results are the mean of patterns predicted by the GCMs in Figure 28.10 for 2080 to 2099 relative to 1980 to 1999.

(Source: IPCC 2007.)

ECOLOGICAL ISSUES   Who Turned up the Heat?

Is Earth's climate changing? According to the Intergovernmental Panel on Climate Change (IPCC), the answer is unequivocally yes. This conclusion is supported by observations that track global climatic changes over the past century (Figure 1). Widespread direct measurements of surface temperature by thermometers and other instruments began in the mid-19th century, and are called the *instrumental record*. Observations of other surface variables, such as precipitation and wind, date back a century. Sea-surface temperatures have been measured from ships since the mid-19th century, supplemented since the 1970s by a network of instrumented buoys. Systematic measurements of the upper atmosphere began only in the late 1940s, but since the late 1970s, satellites have provided continuous records for many variables.

What do these records reveal? As of 2010, mean global surface temperature has increased by 0.74 ± 0.2°C since the early 20th century. According to NASA scientists, the six warmest years since 1850 occurred between 1998 and 2010. Analyses of daily land-surface temperatures from 1950 to 1993 show that the daily temperature range is decreasing. Minimum temperatures are increasing at about twice the rate of maximum temperatures (0.2°C versus 0.1°C per decade)—night-time lows are rising faster than daytime highs. Global ocean temperatures have also increased since the late 1950s. Over half of the increase has occurred in the upper 300 m, which has warmed by 0.04°C per decade. Adding to the concern is the fact that oceanic regions have been differently affected. Because of cool upwellings, the eastern Pacific has warmed less than the western Pacific, increasing the probability of strong El Niño events (Hansen et al. 2006).

Although scientists agree that the climate has changed significantly over the past century, debate continues over the more difficult question of why it is changing. The debate centres on two points. The first relates to the instrumental data measuring trends in land-surface temperatures. Most weather stations are located in cities, which are typically warmer than surrounding rural areas (see Ecological Issues: Urban Microclimates, pp. 34–35). Recent studies have removed this potential bias from the data. Current findings by the IPCC affirm that the warming trend of the past century is independent of the effects of urbanization.

The second point relates to the difficulty of determining long-term trends from an instrumental record covering less than two centuries. Climate varies on many timescales, and Earth has experienced many periods of warming and cooling in the past. In fact, the Northern Hemisphere is still recovering from the last glacial maximum, some 20 000 years before the present. However, climate reconstructions of the past 1000 years suggest that the warming trend of the past century is consistent with that expected from rising levels of greenhouse gases. Although a scientific consensus has been reached, the debate will likely continue, but the real question is: *Will the warming continue, and with what effects?*

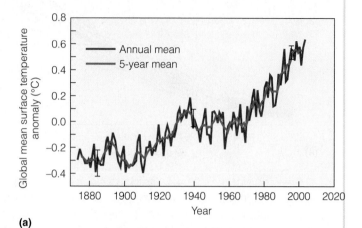

(a)

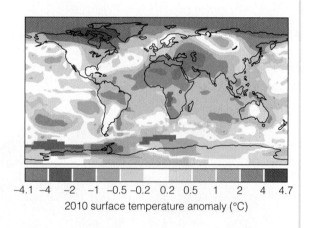

(b)

Figure 1 Climate change data. **(a)** Combined anomalies in annual land-surface air and sea-surface temperature (°C) from 1880 to 2010. Anomalies are the difference between annual temperature for any given year and mean annual temperature for the period 1951 to 1980. **(b)** Global pattern of surface temperature anomalies, as defined in (a), for 2010.

(Source: NASA GISS 2010.)

Bibliography

Hansen, J., Mki. Sato, R. Ruedy, K. Lo, D. W. Lea, and M. Medina-Elizade. 2006. Global temperature change. *Proceedings of the National Academy of Sciences* 103:14288–14293.

1. Canadians often assume that global warming will be a good thing. What are the possible advantages and disadvantages of climate change in your region of Canada?
2. Based on Figure 28.20, how might these changes affect the specific natural ecosystems now found in your area?

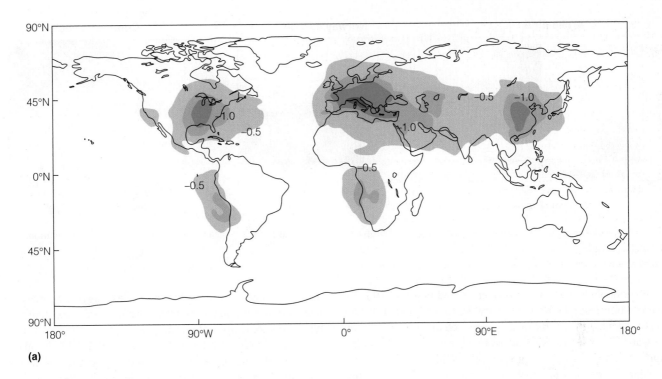

(a)

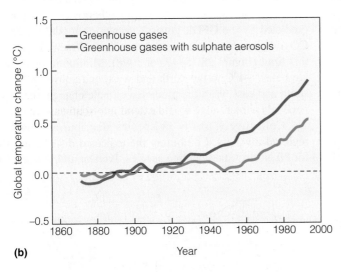

(b)

Figure 28.12 Impact of sulphate aerosols.
(a) Estimates of reduction in radiation (values on the graph are in units of W m^{-2}) resulting from anthropogenic sulphate aerosols. Reduction is greatest closest to the source. **(b)** Predicted changes in mean global temperature for United Kingdom Meteorological Office GCM, with and without inclusion of sulphate aerosols.

(Adapted from Mitchell et al. 1995.)

the physics of greenhouse gases and the consistent qualitative predictions of the GCMs strongly suggest that rising CO_2 will significantly affect global climate.

28.6 POPULATION AND COMMUNITY EFFECTS: Climate Change Will Alter Species Abundance, Distribution, and Diversity

As we have seen throughout this book, climate influences every aspect of ecology: physiology and behaviour of organisms (Chapters 5-8); birth, death, and growth rates of populations (Chapters 9-12); relative competitive abilities and other interactions of species (Chapters 13-15); community structure and succession (Chapters 16-18); ecosystem structure, productivity,

and nutrient cycling (Chapters 19-22); and biome distribution (Chapters 23-25). Current research on global climate change focuses on the response at all levels of the ecological hierarchy. Changes in temperature and precipitation directly affect species distribution and abundance. For example, mean annual temperature and rainfall affect the abundance of three widely distributed European tree species (Figure 28.13, p. 634) (Miko 1996). Given their differing responses, these species will likely change in distribution and abundance with a change in regional climatic patterns.

The potential impact of regional climate change on plant species distribution is tremendous. Statistical models developed to predict the distribution of 80 tree species native to eastern North America as a function of climate, soils, and topography allow investigators to predict shifts in species distribution and abundance based on climatic changes predicted by various

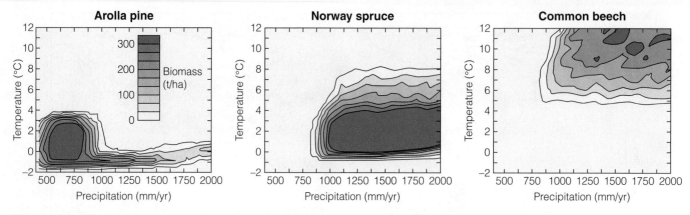

Figure 28.13 Abundance (biomass) of three common European tree species as related to mean annual temperature and precipitation. (Adapted from Miko 1996.)

GCMs at doubled levels of CO_2 (Iverson and Prasad 2001). To take just one example from this study, recall the broad distribution of red maple (*Acer rubrum*), as depicted in Figure 9.4. With the warmer temperatures predicted by climate models, populations of this moisture-loving species will likely become less abundant in the eastern deciduous forests of the United States (Figure 28.14) (Iverson et al. 1999). However, a similar study conducted in Canada (McKenney at al. 2007) indicates that the distribution of red maple (as well as several other species in the *Acer* genus) will shift substantially northward (from 2° to 7.6° latitude, depending on dispersal) into regions now typically occupied by boreal coniferous forest. However, although some tree species will increase in distribution across North America with climate change, others will decrease, notably American beech (*Fagus grandifolia*). A prominent feature of both the American and Canadian studies is the broad spectrum of

responses predicted for different species, suggesting that community composition will be altered rather than an existing forest type undergoing a simple shift.

Climate also affects animal distribution and abundance. For example, the northern limit of the winter range of the Eastern phoebe (*Sayornis phoebe*) correlates with a mean minimum January temperature of −4°C. Two isotherms (one current, one predicted by the GFDL general circulation model for doubled CO_2) define the portion of eastern North America that meets this limit (Figure 28.15) (Root 1998). Minimum temperatures are below −4°C to the north and west, and above −4°C to the south and east. With the predicted climate change, the Eastern phoebe's winter range would extend into southeastern Canada.

Collectively, shifts in species distributions will alter regional diversity. Combining the expected distribution shifts for 80 North American tree species, Iverson and Prasad (2001)

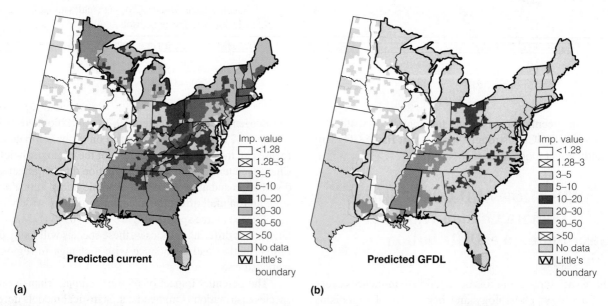

(a)

(b)

Figure 28.14 Distributions of red maple under current **(a)** and doubled CO_2 levels as predicted by the GFDL general circulation model **(b)**. Species abundance is expressed by importance value (sum of relative density, basal area, and frequency; see Section 16.3). Little's boundary refers to species distribution as reported by Little (1977).

(Adapted from Iverson et al. 1999.)

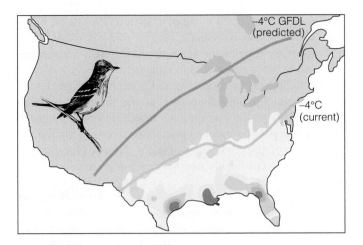

Figure 28.15 Existing range of Eastern phoebe along the mean minimum January isotherm, and with predicted isotherm under a changed climate, assuming doubled CO_2.

(Adapted from Root 1988.)

predict changes in local and regional species richness, most notably a decline in the southeastern United States (with the exception of Florida). The impact of global climate change on Canadian forests depends on the interplay among temperature, moisture, and fire. Warmer winter lows should allow species with more southerly ranges (such as red maple) to move north, but other factors (particularly moisture) may limit their success. Thompson et al. (1998) suggest that fire-adapted species such as jack pine and aspen will become more abundant in Ontario, and that white pine will decline. However, these predictions are controversial; according to McKenney et al. (2007), white pine may or may not increase its distribution in Canada, depending on dispersal patterns.

For most taxonomic groups, we lack sufficient information on the factors controlling species distribution to allow a detailed analysis. We must therefore depend on general relationships between abiotic factors and overall diversity patterns, which

may or may not apply in specific situations. Species richness of most terrestrial animal groups, including vertebrates, co-varies with abiotic factors related to the energy and water balance of organisms: temperature, evapotranspiration, and incident radiation (Currie 1991; see Section 27.2). In a follow-up study, Currie used the relationship among January and July temperatures, precipitation, and regional species richness to predict changes in bird and mammal diversity for the continental United States under climate change (Figure 28.16). His analyses predict a northward shift in the regions of highest diversity, with richness declining in the southern United States and increasing in New England, the Pacific Northwest, the Rocky Mountains, and the Sierra Nevadas (Currie 2001). Extrapolation suggests that richness will also increase in southern Canada, assuming adequate moisture. However, given that many models predict less moisture in interior regions, such increases may not occur.

For example, in the drier western regions of Canada, northward expansions of tree distributions and increases in forest productivity are much less likely than the increased temperatures brought about by global warming might suggest. A long-term study using permanent plots has revealed that drought-induced mortality is pervasive across the boreal forest, and that mortality rates have risen more sharply in western than in eastern portions (4.9 percent per yr versus 1.9 percent per yr, respectively, in the period from 1963 to 2008) (Peng et al. 2011). These results put in doubt an expanded role for Canadian forests as a carbon sink.

These regional analyses of diversity changes (whether of plants or animals) reflect responses to abiotic climatic factors. But species distribution and abundance also reflect biotic interactions (competition, mutualism, predation, etc.). Altered population growth and reproduction in response to climate change may influence these interactions, affecting community patterns of zonation and succession. Given the difficulty of manipulating field conditions, few ecologists have examined these effects experimentally. One example is the International Tundra Experiment (ITEX). Researchers from 13 countries alter season

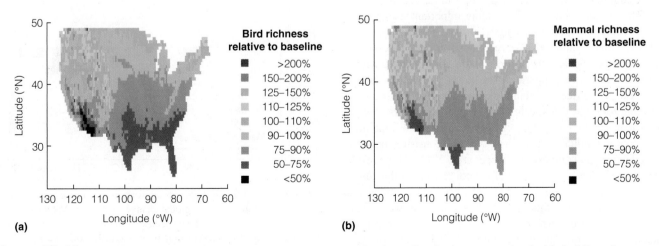

Figure 28.16 Changes in (a) bird and (b) mammal species richness resulting from climatic changes associated with doubling of atmospheric CO_2. Richness was predicted using five GCMs.

(Adapted from Currie 2001.)

Figure 28.17 The International Tundra Experiment (ITEX) uses small, clear, plastic, open-top chambers to warm the tundra and extend the growing season. The chambers raise the daily temperature of the plant canopy by 1.5°C to 1.7°C, within the range predicted by climate models. Depicted here is a long-term experiment at Alaska's Barrow Environmental Observatory.

length using standard techniques, such as passive warming and manipulation of snow depth (Figure 28.17). These studies examine species-, community-, and ecosystem-level responses to summer warming in the Arctic. As of 2012, there are seven ITEX sites in Canada. In each site, studies have been established in one to seven plant communities.

In addition, there are nine community-based sites near Nunavut, Nunavik, and Nunatsiavuk that are linked to ITEX, with the primary objectives of monitoring vegetation change and berry production. The field work at these sites is done by high school students and is scheduled to become part of the school curriculum. Given its integration of research scientists with local communities, this project exemplifies a new approach to tackling the many ecological issues that transcend the interests of the academic community. We are all stakeholders in Earth's future.

This research initiative is yielding important insights into ecosystem functioning in a changing climate. For example, at an ITEX site in the southern Yukon, ecologists monitored responses of four target species (three forbs and a shrub, Arctic willow [*Salix arcticus*]) to temperature variation. As well as natural variation over a decade, plants were exposed to experimental warming for eight years, using open-topped chambers. Despite significant annual variation in reproduction and vegetative growth of individual species, there were no consistent trends in plant community composition (Pieper et al. 2011). These results suggest considerable resilience of the community at this tundra site. However, the open-topped chambers were less effective at simulating warming at this site, which was alpine as well as sub-Arctic. They were warmer during the day but cooler at night than nearby tundra areas, with minimal net daily warming.

Impacts of warming on vegetation will influence trophic interactions. A warming Arctic will likely expose the plant community to altered herbivory pressure, particularly in regions

that have been ungrazed in the past. In an ITEX site on Ellesmere Island, scientists simulated grazing on a wet sedge tundra community by clipping the vegetation at different frequencies, followed by litter removal. Soil temperature was not affected by any treatment, but available soil nitrogen was greatest at intermediate clipping frequencies, as was the shoot nitrogen content of two of the four species studied. However, simulated grazing lowered aboveground primary productivity in all treatments, suggesting that compensatory growth (see Section 14.10) is not occurring in these previously ungrazed sites (Elliot and Henry 2011).

The impacts of a changing climate on populations and communities interact with other human impacts on the world's atmosphere and soils, particularly increases in nitrogen deposition from fertilizer use (see Section 26.5). Not only does N enrichment interact with CO_2 to affect nutrient balance, but N_2O, a by-product of human activities, is itself a greenhouse gas, thereby affecting climate directly. These combined changes in climate and atmospheric chemistry may alter diversity by changing resource levels and affecting species performance and interactions.

In a study investigating these effects on Mediterranean grasslands in California's Jasper Ridge Biological Preserve, plots were exposed to four treatments (singly and combined) associated with global change: (1) elevated CO_2 (ambient plus 300 ppm); (2) soil-surface warming (0.8°C–1.0°C); (3) elevated precipitation (50 percent increase, with a growing-season extension of 20 days); and (4) N-deposition (increased by 7 g/m^2/day). After three years, N-deposition and higher CO_2 reduced richness by 5 and 8 percent, respectively. Higher rainfall increased richness by 5 percent, and temperature had no significant effect (Figure 28.18) (Zavaleta et al. 2003a).

Different functional groups reacted differently. Although these grasslands are dominated in biomass by annual grasses, which were relatively unaffected, forbs (non-grass herbaceous

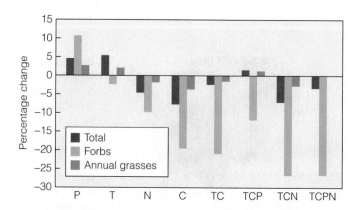

Figure 28.18 Changes in total, forb, and annual grass diversity under single and combined global change treatments. Values are percentage difference in mean species richness between controls and elevated levels. Treatments (used alone and in combination): P, increased precipitation; T, warming; N, nitrogen fertilization; and C, elevated CO_2.

(Adapted from Zavaleta et al. 2003a).

species) make up most of their native diversity. All three native N-fixing forbs disappeared with N-deposition. The combined treatments lowered forb diversity by over 10 percent. This functional group, which includes many of the rare species in many ecosystems, seems especially susceptible to global change. Yet the combined treatments had no significant effect on total plant diversity, because increases in perennial grass diversity partly offset forb losses. Whether greater effects will occur elsewhere and/or after more prolonged exposure is under investigation.

An interesting result emerged from the combined elevated CO_2 and warming treatment (TC). Although most models predict that global precipitation will increase, warming is expected to increase aridity in drought-prone ecosystems such as the California grasslands by accelerating evapotranspiration. Yet the reverse occurred. Simulated warming increased spring soil moisture by 5 to 10 percent under both ambient and elevated CO_2 (Zavaleta et al. 2003b). This effect was not due to lower leaf area or productivity but to earlier senescence, which lowered transpiration losses. This biotic link between warming and water balance may ameliorate the response of grasslands and savannahs to climate change.

28.7 ECOSYSTEM EFFECTS: Climate Change Will Affect Ecosystem Function and Distribution

Climate change may also affect vegetation indirectly, through impacts on ecosystem processes such as decomposition and nutrient cycling. In terrestrial systems, these processes are influenced by temperature and moisture, with decomposition promoted in warm, moist conditions. A long-term experiment at Harvard Forest in Massachusetts is examining the impact of elevating soil temperatures by 5°C (using buried heat cables) on CO_2 emission, decomposition, and nutrient cycling. In the first year of the study, CO_2 emission increased by almost 60 percent (Peterjohn et al. 1994). These results, which the authors document as consistent with patterns observed in forest soils worldwide, suggest that warming increases soil CO_2 emissions by promoting decomposition. This effect may generate a positive feedback loop, accelerating further warming.

Yet this stimulation of emissions may be short-lived. In the Harvard Forest and other ongoing field experiments, elevated soil respiratory rates drop to pre-warming levels within several years (Bradford et al. 2008). Ecologists have proposed two hypotheses to explain this phenomenon: (1) depletion of easily decomposed soil organic matter as a respiratory substrate and (2) thermal adaptation. Thermal adaptation may include selection of microbial genotypes with differing temperature responses and/or a phenotypic response involving physiological acclimation to warmer temperatures. After 15 years of observation in the Harvard Forest study, there is evidence supporting both hypotheses. As well as lower soil carbon and microbial biomass, the heated plots have lower mass-specific respiratory rates. Global warming may thus stimulate CO_2 release less in the long-term than first anticipated (Bradford et al. 2008).

Climate change will affect not only processes occurring within ecosystems, but also their location and geographic extent. Altered ecosystem distribution in response to climate is by no means unusual. By studying past climate flux, ecologists have learned much about the responses of ecosystems to changing climate. Pollen samples from sediment cores in lake beds allow paleoecologists to reconstruct vegetation over past millennia. The shifting distributions of tree species in North America since the last glacial maximum (see Section 17.11) are a good example. Tree genera migrated north at different rates after glacial retreat, depending on how a species' physiology, dispersal ability, and interactions with other species responded to climate change. In fact, the composition of existing forests is a relatively recent result of differential responses of tree species to a changing climate. As Earth's climate has changed in the past, so the distribution and abundance of organisms and the communities and ecosystems to which they belong have changed accordingly (see Figure 17.22). No doubt they will continue to change in the future.

Ecologists have long recognized the link between climate and plant distribution. According to L.R. Holdridge's simple biogeographical model (1947), tropical rain forests are limited to areas where mean annual temperatures are above 24°C and annual precipitation is above 2000 mm. How will rain forest distribution change with global warming? Under the climatic patterns predicted by the United Kingdom Meteorological Office (UKMO) GCM for doubled CO_2 (a dynamic model that factors in feedback effects), tropical rain forests will shrink significantly (Figure 28.19, p. 638) (Smith et al. 1992). The cause will be either (1) higher temperatures accompanied by lower precipitation or (2) higher precipitation that cannot meet the increased demand for water induced by warmer temperatures (Cox et al. 2000). Together with the pressures of agriculture and forestry, this scenario would devastate tropical rain forests along with the biodiversity they support. Tropical deforestation is currently the single major cause of species loss, with annual extinction rates of thousands of species (see Section 27.6). The loss of rain forest predicted by the UKMO model would accelerate this pace of extinction, as well as inhibit terrestrial carbon sink activity.

Ecosystems in temperate regions would also be affected. According to Environment Canada, doubling atmospheric CO_2 would cause major shifts in the ecoclimatic provinces and their associated vegetation by 2050 (Figure 28.20, p. 638). Arctic and sub-Arctic regions will decline by 20 percent and boreal regions by 14 percent, whereas temperate regions will increase by 16 percent, grasslands by 15 percent, and semi-desert by 2 percent (Rizzo 1990). Ecosystem zones are expected to shift approximately 100 km for every 1°C temperature increase.

Changes in global temperature patterns also alter the distribution of aquatic ecosystems. Coral reefs (see Section 24.11) are restricted to tropical waters where mean surface

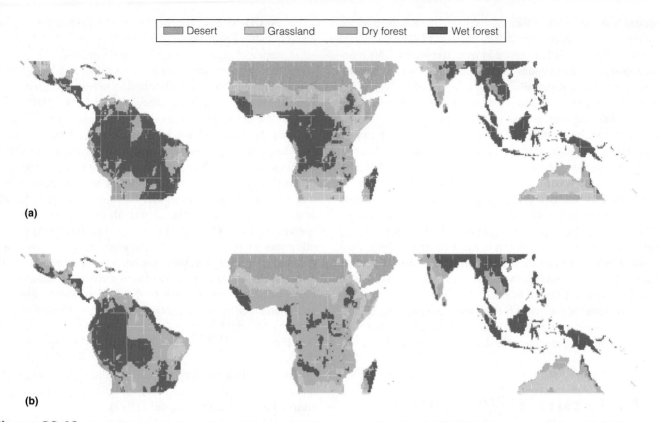

Figure 28.19 Areas of the tropical zone that could support rain forest, as predicted by the Holdridge biogeographical model. **(a)** Current ecosystem distribution. **(b)** Predicted ecosystem distribution under climate conditions predicted by the UKMO GCM for a doubled atmospheric CO_2 concentration.

(Adapted from Smith et al. 1992.)

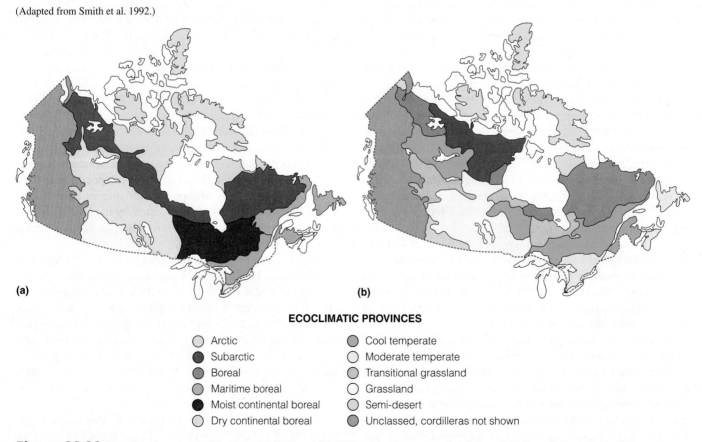

Figure 28.20 Ecoclimatic zones in Canada in **(a)** 1990 and **(b)** 2050, with changes predicted for a doubling of atmospheric CO_2.

(Adapted from Rizzo, Environment Canada, 1990.)

temperatures are at or above 20°C, and reef development is optimal at mean annual temperatures of 23°C to 25°C. Oceanic warming may allow reef formation farther up the eastern coast of North America. However, such shifts may not occur, as a result of human-induced changes in marine ecology, such as overexploitation of algae-feeding fish.

Besides altering reef distribution, global warming can cause immediate damage to coral reefs. Since the 1980s, bleaching episodes induced by high temperature have been an almost annual occurrence (Baker et al. 2008). The most severe events have been coupled with ENSO phenomena and have altered community structure in many cases, including instances of complete collapse. Although it has sometimes set the stage for further decline, bleaching has been followed by rapid recovery in some regions, particularly the Indian Ocean. In the western Atlantic, in contrast, further bleaching events and other secondary stresses have led to continued decline.

The complex interaction of temperature-induced bleaching with other factors of both natural and anthropogenic origin, coupled with regional variation, is yet another instance of the difficulty of predicting the ecosystem impacts of global warming. There is little question, however, that changes in temperature and precipitation patterns of the magnitude predicted by climate models will significantly influence the distribution and function of terrestrial and aquatic systems.

28.8 COASTAL EFFECTS: Global Warming Will Raise Sea Level and Affect Coastal Habitats

Coastal habitats are likely to be affected by an aspect of global change that is an indirect impact of warming—rising sea levels. During the last glacial maximum about 20 000 years ago, sea level was 100 m lower than today. Highly productive coastal waters, such as the continental shelf of eastern North America, were above sea level and covered by terrestrial systems (see Figure 17.21). As the climate warmed and the glaciers melted, sea levels rose—over the past century, by 1.8 mm yearly (Figure 28.21). This rise reflects the general pattern of global warming and the associated thermal expansion of oceans and melting of glaciers. The 2007 IPCC report estimates that global mean sea level will rise by 0.18 to 0.59 m between 1990 and 2100, with considerable regional variation. A rise of this magnitude will seriously affect coastal habitats from the perspective of both natural ecosystems and human populations.

Many of us inhabit coastal areas; 13 of the world's 20 largest cities occupy coasts. Particularly vulnerable are delta regions; small low-lying oceanic islands; and low-elevation countries such as Netherlands, Surinam, and Nigeria. For example, Bangladesh, with some 120 million people, occupies the delta of several large rivers (Figure 28.22). One-quarter of its people live in areas that are situated less than 3 m above sea level, and 7 percent of its habitable land and 6 million people reside in areas that are less than 1 m above sea

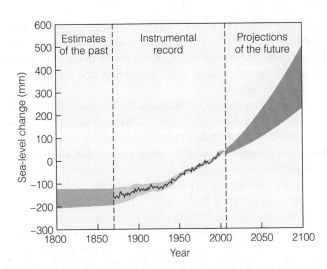

Figure 28.21 Change in global mean sea level (deviation from 1980–1999 mean) in the past and as projected for the future. Pre-1870, global measurements are not available. Grey shading shows uncertainty in estimated long-term sea-level change. The red line reconstructs global mean sea level from tide gauges, and red shading denotes range of variation. The green line (the short segment just prior to 2000) shows global mean sea level observed from satellite data. Blue shading represents the range of model projections for a scenario of rising levels of greenhouse gases developed by IPCC (scenario A1B).

(Source: IPCC 2007.)

level. Estimates of sea-level rise in this region from a combination of land subsidence (caused by groundwater removal) and global warming are 1 m by 2050 and 2 m by 2100. Despite the considerable uncertainty in these estimates, the effect would be devastating.

Other regions would be similarly affected. In Egypt, about 12 percent of the arable land, with a population over 7 million, would be submerged by a 1-m rise. In coastal areas of eastern China, a sea-level rise of just 0.5 m would inundate an area of some 40 000 km^2 inhabited by over 30 million people. Small islands are particularly vulnerable. Over 500 000 people live in archipelagos of small islands and coral atolls,

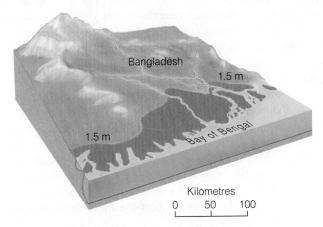

Figure 28.22 Land area in Bangladesh that would be submerged (dark green area) if sea level rose by 1 m.

(Adapted from Nicholls and Leatherman 1995.)

as in the Maldives in the Indian Ocean and the Marshall Islands of the Pacific. These island chains lie almost entirely below 3 m above sea level. A 0.5 m or more rise would greatly reduce their land area and have a drastic impact on groundwater supply.

Apart from its dire implications for humans, sea-level rise will also affect coastal ecosystems, including inundation of low-lying wetlands and dryland areas, shoreline erosion, increased salinity of estuaries and aquifers, rising water tables, and increased flooding and storm surges. Estuaries and mangroves would be highly susceptible. Salt marshes depend on the twice-daily tidal inundation of saltwater mixing with freshwater. Water depth, temperature, salinity, and turbidity are critical to maintaining salt marshes. Invasion of saltwater farther into the estuary could cause salinization of adjacent land. In turn, estuaries and mangroves are critical to coastal fisheries. Many of the fish caught for human use, as well as many birds and animals, depend on coastal marshes and mangroves for part of their life cycles.

28.9 AGRICULTURAL EFFECTS: Climate Change Will Affect Agricultural Production

Despite improvements in crop varieties and irrigation methods, climate and soils remain the key factors determining agricultural production. Changes in global climate patterns will exacerbate an already increasing problem of feeding the world population, which is predicted to double over the next half century. The major cereal crops that feed people—wheat, corn (maize), and rice—are domesticated species. Like native species, these crops have tolerance limits to temperature and moisture that control their survival, growth, and reproduction. Changes in regional climates will influence their suitability and productivity. These effects will be complex, with economic and social factors interacting to influence food production and distribution.

To predict these effects, ecologists must consider responses to increasing CO_2 levels as well as to climate change. Many studies suggest that most crops will benefit from a rise in CO_2, at least in the short term (see Figure 28.6). In five years of field experiments, cotton yield increased on average by 60 percent under elevated CO_2 (accompanied by irrigation) than when grown under identical conditions and ambient CO_2 (Kimball et al. 1993). Production of cool-season grains such as barley and spring wheat will be less stimulated or may even decrease, particularly if moisture stress ensues, but new opportunities (such as conversion from spring to winter wheat) may compensate (Smit et al. 1988). Note that cotton and wheat are C_3 species; C_4 crops such as corn and sugar cane are less stimulated, and may become even less productive if C_3 weeds with which they coexist are stimulated, thereby affecting competitive interactions. Even if crops are stimulated, the effect may be short-lived as a result of downregulation (see Section 28.4).

One way to evaluate the potential impacts of climate change on agriculture is to examine shifting geographic ranges

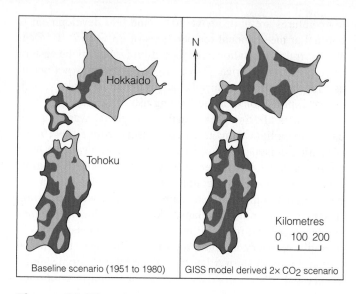

Figure 28.23 Regional shifts in areas suitable for irrigated rice production in Japan (dark green) under climate change as predicted by the Goddard Institute for Space Studies GCM.

(Adapted from Yoshino et al. 1988.)

of crop species. A mean daily increase of 1°C during the growing season would shift the North American corn belt about 175 km to the northeast (Newman 1980), extending it into southern Canada. A similar shift may occur in areas suitable for growing irrigated rice in Japan (Figure 28.23) (Yoshino et al. 1988). Such shifts involve major changes in land use, with associated economic and social impacts. Although such analyses help predict changing patterns of regional production, evaluating the actual effect on global production and markets requires a detailed interdisciplinary approach.

Extrapolating from such regional changes to a global scale is difficult. The Environmental Change Unit at England's Oxford University conducted a collaborative study of the regional and global impacts of climate change on production of wheat, corn, rice, and soybeans at 112 sites in 18 countries. Researchers made assumptions about farmers' ability to adapt to changing conditions by changing crop species, varieties, or practices such as irrigation. The analysis also assumed continuation of current economic growth, certain changes in trade restrictions, and projected population growth. The study concluded that the negative effects of global warming may to some extent be compensated for by increased productivity from elevated CO_2, but that climate change (as predicted by GCMs assuming doubled CO_2) will reduce global cereal production by up to 5 percent (Rosenzweig and Parry 1994). These reductions will not likely be evenly distributed. Production will rise in developed countries in temperate regions but drop in developing nations (Figure 28.24), increasing the portion of their populations that are at risk of hunger. In many regions, climatic variability and marginal conditions will intensify. Changes in production will be compounded by political and social factors, making developing countries more vulnerable to global warming (Rosenzweig and Parry 1994).

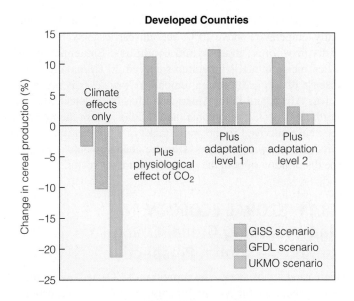

Developed Countries

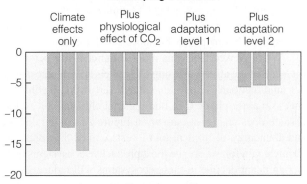

Developing Countries

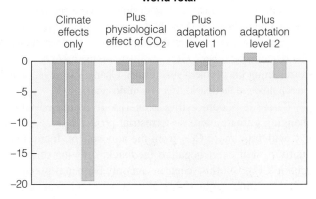

World Total

Figure 28.24 Predicted changes in cereal (wheat, corn, and rice) production in response to climate change predictions based on three GCMs. Changes are relative to baseline estimates of production for 2060. Four scenarios are evaluated, involving changes in production in response to (1) climate effects only, (2) climate changes together with expected increases in productivity as a result of the impact of elevated CO_2 on crop physiology, (3) scenario 2 plus level 1 adaptations, and (4) scenario 2 plus level 2 adaptations. Level 1 adaptations include changes in crop varieties only, changes in planting date of less than 1 month, and application of additional water to already irrigated lands. Level 2 adaptations also include changes in the type of crop planted, additional fertilizers, changes in planting dates by over 1 month, and use of irrigation in lands not currently under irrigation. Analyses were conducted at a country and regional level, but have been grouped for presentation. Note that most developed countries are in temperate regions, while most developing countries are in tropical and subtropical regions.

(Adapted from IPCC 2007.)

INTERPRETING ECOLOGICAL DATA

Q1. Why are the trends so different in developed versus developing countries? What factors might explain the large differences among the trends predicted by different GCMs?

Q2. Given that wheat and rice are C_3 grasses and corn is a C_4 grass, which crop species do you think would show the greatest responses to climate change, and why?

Q3. Across all the models, the level 2 adaptations mitigate the negative effects of climate change on cereal production to the greatest degree. Given our discussion of sustainable agriculture in Section 26.7, what ecological trade-offs do you predict might happen if they were implemented?

hurricanes, etc.); and changes in diet and nutrition in response to changed agricultural production.

Extreme heat causes more deaths than most other environmental hazards, such as storms, floods, and earthquakes, combined (Pengelly et al. 2005). Several studies have revealed the direct relationship between maximum summer temperatures and mortality. Climate change is expected to increase the frequency of very hot days. For example, if the average July temperature in Chicago were to rise by 3°C, the probability that the heat index would exceed 35°C would increase from 1 in 20 to 1 in 4. Warm and humid conditions cause the most mortality. The highest death toll in North America occurred during the summer of 1936, when 4700 excess deaths were recorded in the United States and 780 in Canada (225 in Toronto alone) as a result of heat-related causes. In Toronto, the frequency of heat alerts is highly variable, but has increased significantly over the past five decades (Pengelly et al. 2005; unlike many others, this study corrected for pollution effects).

Climate change scenarios predict a significant rise in heat-related mortality across North America over the next several decades, particularly in the United States. Although impacts will be less pronounced in Canada, projected increases in the number of hot days in Toronto (Figure 28.25, p. 642) will take their toll. Other Canadian cities will also be affected; see Ecological Issues: Urban Microclimates, pp. 34–35.

28.10 HUMAN HEALTH EFFECTS: Climate Change Will Affect Human Health Directly and Indirectly

Climate change will have many direct and indirect effects on human health. Direct effects include increased heat stress, asthma, and other cardiovascular and respiratory ailments. Indirect health effects are less predictable, but would likely include increased incidence of communicable diseases; increased mortality and injury from natural disasters (floods,

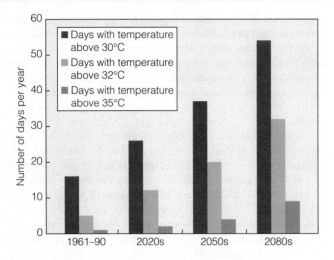

Figure 28.25 Predicted increases in the number of hot days in Toronto under climate change associated with doubling of atmospheric CO_2.

(Adapted from Environment Canada 2002.)

During heat waves, cardiovascular and respiratory illnesses are the major causes of death. Children, the elderly, and the poor are at greatest risk. Air conditioning alleviates the risk for some, but in prolonged heat waves, energy shortages can cause brownouts and blackouts, eliminating their use. Once again, the combined effects of energy use interact in their impacts on human well-being.

As well as direct heat-related mortality, distribution and transmission of diseases will be affected. Disease involves pathogens—viruses, bacteria, or protozoa—and hosts. However, many human diseases are transmitted by animal vectors (see Section 15.2). Insects are a primary vector of human disease. Of the roughly 102 insect-borne viruses (arboviruses) that produce disease in humans, about half have been isolated from mosquitoes. Other organisms that transmit arboviruses are blood flukes and ticks. Disease-transmitting insects are adapted to specific ecosystems, exhibiting specific tolerances to abiotic factors, especially temperature. Climate change will affect their distribution and abundance.

An example is malaria, a recurring infection produced in humans by protozoan parasites transmitted by the bite of an infected female mosquito of the genus *Anopheles* (see Section 15.2). The optimal breeding temperature for *Anopheles* is 20°C to 30°C, with relative humidity over 60 percent. Mosquitoes die at temperatures above 35°C and relative humidity under 25 percent. Forty percent of the world's population is now at risk, and over 2 million are killed annually by this disease. The current distribution of malaria will be greatly modified by climate change. Expansion of *Anopheles*'s geographic range is expected to increase the proportion of the world's population at risk of malaria to over 60 percent by the latter part of the 21st century. The severe North American outbreak of West Nile virus in 2012 (over 1000 reported cases in the U.S. and 58 in Canada, with 30 in Toronto alone) is yet another instance of a mosquito-borne disease spread by mosquitoes.

Dengue and yellow fever, two related viral diseases, are transmitted by the mosquito *Aedes aegypti*, which is adapted to urban habitats. This mosquito can colonize areas with a mean daily temperature above 10°C, but only carries the yellow fever virus in warmer, more humid conditions. Epidemics occur when mean annual temperatures exceed 20°C, making this a disease of tropical forested regions. Yellow fever is prevalent across Africa and Latin America but has been detected as far north as Bristol, Philadelphia, and Halifax, where the mosquitoes have survived in the water tanks of ships that have travelled from tropical regions. Climate change would influence the distribution of both the virus and its insect vector.

28.11 GLOBAL ECOLOGY: Understanding Global Change Requires a Global Perspective

Increasing atmospheric concentrations of CO_2 and other greenhouse gases, and the potential changes in global climate patterns that may result, represent a new class of ecological problems. To understand the effect of rising CO_2 emissions, we must examine carbon cycling on a global scale (see Figure 22.5), linking the atmosphere, hydrosphere, biosphere, and lithosphere.

When viewed from this more comprehensive perspective, it is clear that the effects of rising CO_2 levels and climate change are not unidirectional. Ecosystems may exert strong feedback influences on atmospheric CO_2 and regional climate. For example, if climate does change as shown in Figure 28.10, the distribution and abundance of tropical rain forests will decline dramatically. Tropical rain forests are the most productive ecosystems on Earth. A significant decline in these ecosystems will reduce global primary productivity, uptake of CO_2 from the atmosphere, and CO_2 storage in biomass. As tropical rain forests shrink, atmospheric CO_2 will increase further. The drying of these regions will kill trees, increase fires, and transfer C stored in living biomass to the atmosphere as CO_2 in much the same way as does forest clearing. The rise in atmospheric CO_2 will increase the greenhouse effect, exacerbating the problem still further. Changes in vegetation thus exert a positive feedback effect on atmospheric CO_2.

There is a contrasting scenario. If increased CO_2 and changing climate increase terrestrial productivity, ecosystems will withdraw more CO_2 from the atmosphere. Increased productivity will exert negative feedback, drawing down atmospheric CO_2. But this scenario can only happen if precipitation does not limit productivity. However, many climate models suggest that changes in the distribution of rain forests will directly alter regional precipitation. In some rain forests, much of the rainfall is water that has been transpired from the vegetation itself. In effect, vegetation drives the local hydrologic cycle (see Figure 3.12). Removing the forest (either through deforestation or by shifts in the distribution of ecosystems as a result of climate change) reduces transpiration and increases runoff to rivers that transport water away from the area.

As summarized by Costa and Foley (2000), many studies have simulated the potential impacts of deforestation in the Amazon Basin. Loss of forest cover would reduce rainfall by reducing internal water cycling. Using the GENESIS GCM, Costa and Foley (2000) estimate that deforestation alone would lower mean precipitation by 0.73 mm per day. Doubling CO_2, meanwhile, would increase precipitation by 0.28 mm per day, for a net overall decline of 0.42 mm per day and a temperature increase of 3.5°C. These changes would further alter the region's climate and make it harder for rain forest to re-establish, particularly in areas where rainfall is borderline.

Another telling example of the direct influence of terrestrial systems on regional climate is in northern latitudes, where the largest amount of warming is predicted (see Section 28.5). This warming would significantly reduce snow cover and shift boreal forests farther north. A major factor influencing absorption and reflection of solar radiation by Earth's surface is its **albedo** (reflectance). Snow has a high albedo whereas vegetation (with its darker colour) has a low albedo. Both reduced snow cover and northern movement of the boreal forest would reduce regional albedo and increase the solar radiation absorbed by Earth. This increase in absorption would further increase regional temperatures—another positive feedback loop (Bonan 2002).

As residents of a cold country, Canadians may seem less affected by climate change than other parts of the world. The boundaries of our natural ecosystems will likely shift (see Figure 28.20), the types of crops grown will likely change, and there will be more hot days (see Figure 28.25). However, in addition to alterations in albedo, changes in the remote northern regions of Canada are predicted to play an increasingly important role in the global carbon cycle, with unknown long-term impacts. Thawing of organic soils in the sub-Arctic and low Arctic, coupled with increased decomposition rates, may effectively halt terrestrial carbon sequestration, triggering positive feedback loops that will accelerate warming (Koven et al. 2011).

As well as releasing methane and reducing albedo, loss of Arctic sea ice may contribute yet another piece to the increasingly complex puzzle of global climate change: remobilization of organic pollutants. A joint research venture of Canada, China, and Norway is investigating release of *persistent organic pollutants* (POPs; organic compounds that are resistant to breakdown) from Arctic sea ice. A variety of industrial chemicals, including polychlorinated biphenyls (PCBs) and various pesticides, are being remobilized from melting Arctic sea ice (Ma et al. 2011). Transported long distances and deposited in Arctic sites, these toxins may pose yet another stress on ecosystems coping with rapid climate change—and illustrate yet again that the biosphere is one interconnected system.

These are not simple linkages. To understand the interactions among the atmosphere, oceans, and terrestrial systems, ecologists must study Earth as a single, integrated system. Much of the uncertainty that surrounds predictions of climate change reflects our uncertainty about the relative strengths of the positive and negative feedback loops at work in global systems. Only by pursuing a global perspective will ecologists, working with oceanographers and atmospheric scientists, understand the short- and long-term impacts of increasing CO_2 levels. It is a challenge that we all must face, as scientists and as global citizens.

EcologyPlace

Visit EcologyPlace at www.pearsoncanada.ca/ecologyplace to access online resources that complement your textbook, and help you to apply and to review the information in this chapter. EcologyPlace includes

- an eText version of the book
- self-grading quizzes
- glossary flashcards
- and more!

Go to www.pearsoncanada.ca/ecologyplace and follow the registration instructions on the Student Access Code Card included with this text. If your book does not have a Student Access Code Card, you can purchase access to it at www.pearsoncanada.ca/ecologyplace.

SUMMARY

Greenhouse Gases 28.1

Greenhouse gases in the atmosphere absorb thermal radiation emitted by Earth's surface and atmosphere. The atmosphere is warmed by this greenhouse effect. Concentrations of greenhouse gases have been rising, causing concern about its impacts on Earth's climate.

Atmospheric CO_2 Levels 28.2

Direct observations since 1958 reveal an exponential increase in atmospheric levels of CO_2, as a direct result of fossil fuel combustion and clearing of land for agriculture. Of the CO_2 released, about 60 percent remains in the atmosphere. The remainder is absorbed by the oceans and terrestrial ecosystems. Oceanic uptake is estimated based on CO_2 diffusion into surface waters. Carbon uptake by terrestrial systems is calculated as the difference between inputs to the atmosphere, atmospheric levels, and oceanic uptake.

Uptake by Oceans 28.3

CO_2 diffuses from the atmosphere into surface waters. A rise in atmospheric concentrations increases CO_2 uptake into surface waters. The thermocline limits vertical mixing and CO_2

transfer to deep waters, which make up over 85 percent of the ocean volume.

Plant Response 28.4

Plants respond to increased atmospheric CO_2 with higher rates of photosynthesis and partial stomatal closure. These responses increase water-use efficiency, and are greatest in C_3 species. Responses to long-term exposure vary, including increased allocation to roots at the expense of leaves, reduced stomatal density, downregulation, and changes in primary productivity.

Climate Change 28.5

Rising atmospheric levels of greenhouse gases could raise mean global temperature by 1.1°C to 6.4°C by 2100. The greatest warming is predicted for northern latitudes during winter. Increased climatic variability is predicted, including changes in precipitation and storm frequency. Sulphates and other aerosols from human sources reduce the input of solar radiation to Earth's surface, and may reduce warming.

Population and Community Effects 28.6

Species distribution and abundance will shift as temperature and precipitation change. Climate changes will influence the competitive ability of species and alter community zonation and succession. Climate change will also affect ecosystem processes such as decomposition and nutrient cycling, which are sensitive to temperature and moisture.

Ecosystem Effects 28.7

Although warming increases soil CO_2 emissions in the short term, the long-term effects are less predictable, given thermal adaptation of microbial decomposers. Throughout Earth's history, climate change has caused shifts in ecosystem distribution and abundance. The climate change projected for the next century will cause similar changes, notably a decline in tropical rain forests and a northward movement of the boreal forest. These

shifts will affect global patterns of biodiversity as well as ecosystem function.

Coastal Effects 28.8

Sea level is currently rising globally at an average rate of 1.8 mm per year. Global warming will likely raise sea level by 0.18 to 0.59 m by 2100, as the polar ice caps melt and warmer ocean waters expand. This sea-level rise will affect people inhabiting coastal areas, as well as coastal ecosystems such as beaches, estuaries, and mangroves.

Agricultural Effects 28.9

Climate change will affect global agriculture. Decreases in production from drier conditions will be partially offset by increased photosynthesis under elevated CO_2 levels. Current models project a 5 percent decline in global cereal crop production. This decline is not distributed evenly. Developed countries in the middle latitudes will realize a slight increase, whereas production in developing countries will likely decline. The likely result will be increased hunger.

Human Health Effects 28.10

Climate change will have direct and indirect effects on human health. Mortality rates will likely rise from heat-related deaths. Indirect effects include increased mortality and injury from climate-related natural disasters, as well as changes in diet and nutrition from changes in agriculture. Distribution and transmission of insect-borne diseases will also be affected.

Global Ecology 28.11

To understand the effect of rising levels of greenhouse gases and global climate change, we must study Earth as a single, complex system. Both positive and negative feedback loops may affect climate change. Thawing of Arctic permafrost will reduce its ability to sequester carbon. Loss of Arctic ice will reduce albedo and may release toxins.

KEY TERMS

aerosols	CO_2 fertilization effect	general circulation model	greenhouse gas
albedo	downregulation	greenhouse effect	

STUDY QUESTIONS

1. Why is CO_2 called a greenhouse gas? What are the major sources of greenhouse gases, especially CO_2?
2. Not all CO_2 released to the atmosphere remains there. What happens to the rest? What is the problem of the "missing carbon," and what might explain it?
3. How will present trends in human population growth in different parts of the world affect carbon emission patterns?

4. In oceans, what factors limit transfer of CO_2 from surface to deeper waters? How does this limitation affect the ability of oceans to act as an abiotic carbon sink? What process is responsible for the ocean's biotic sink activity?
5. How does elevated CO_2 influence photosynthesis and transpiration? Why are C_3 species likely to experience more enhancement than C_4 species? (Review Section 6.10.)

6. The impact of elevated CO_2 on photosynthesis is often short-lived. What might be the ecological advantage to a plant of downregulation? How does downregulation relate to carbon allocation, as discussed in Section 6.7?

7. What are aerosols, and what are their effects on global climate change?

8. How might changes in climate (temperature and precipitation) influence the distribution of animal species? Consider both direct and indirect effects.

9. Consider Figure 28.13. Both common beech and Norway spruce produce more biomass with more precipitation, but respond differently to higher temperatures. Describe this difference, and discuss how it will affect their response to expected global climate change.

10. In the combined treatments presented in Figure 28.18, how does increased precipitation influence the effects of elevated CO_2 and temperature on the diversity of forbs? Based on the discussion of plant response to elevated CO_2 in Section 28.4, how might changes in stomatal opening and transpiration affect soil moisture over a growing season?

11. How might climate change affect the distribution and abundance of terrestrial systems?

12. As depicted in Figure 28.19, the UKMO model predicts that the drying effect of higher temperatures will reduce the extent of tropical rain forest. How might the unexpected results of the Zavaleta experiment on soil moisture (see Figure 28.18) alter this prediction?

13. How and why will global warming influence sea level? How might rising sea levels influence human populations? Coastal environments?

14. How might (i) climate change and (ii) rising CO_2 levels influence crop production? Why are these trends not necessarily the same?

15. How might climate change influence human health, both directly and indirectly?

FURTHER READINGS

Bazzaz, F. A. 1996. *Plants in changing environments: Linking physiological, population, and community ecology.* New York: Cambridge University Press.
Provides several excellent examples of experimental approaches that examine the response of plant species to elevated CO_2 levels.

Bonan, G. B. 2002. *Ecological climatology.* Cambridge: Cambridge University Press.
Provides an ecological perspective on climate, including feedback loops.

Dessler, A. E., and E. A. Parson. 2006. *The science and politics of global climate change: A guide to the debate.* Cambridge: Cambridge University Press.
Written by a leading atmospheric scientist, this book is a highly readable treatment of the complexity of the global warming debate.

Graedel, T. E., and P. J. Crutzen. 1995. *Atmosphere, climate, and change.* Scientific American Library. New York: W. H. Freeman.
Outstanding introduction to the issue of global climate change.

Houghton, R. A. 2002. Terrestrial carbon sinks: uncertain explanations. *Biologist* 49:155–160.
Highly accessible account of the mechanisms involved in the terrestrial C sink, and the problems associated with estimating its activity, written by a senior scientist at Woods Hole.

Intergovernmental Panel on Climate Change (IPCC). 2007. *Fourth Assessment Report—Climate Change.* Available at www.ipcc.ch. Accessed 12 May, 2012.
The summary of the IPCC assessment report provides a comprehensive overview of the climate change issue. Detailed scientific reports are also available at the website.

Peters, R. L., and T. E. Lovejoy (eds.). 1992. *Global warming and biological diversity.* New Haven: Yale University Press.
Sobering examination of the potential effect of global warming on vegetation, soils, animals, diseases, and ecosystems.

References

Abbott, K. R. 2006. Bumblebees avoid flowers containing evidence of past predation events. *Canadian Journal of Zoology* 84:1240–1247.

Aber, J. D., and J. M. Melillo. 1991. *Terrestrial ecosystems*. Philadelphia: Saunders College Publishing.

Aber, J. D., K. N. Nadelhoffer, P. Steudler, and J. M. Melillo. 1989. Nitrogen saturation in northern forest ecosystems. *BioScience* 39:378–386.

Adam, C. I., and D. P. Ostoff. 2006. Balsam wooly adelgid. *Natural Resources Canada*. Fredericton, NB: Canadian Forest Service, Atlantic Forestry Centre, Pest Note 3.

Addicott, J. F. 1986. Variations in the costs and benefits of mutualism: The interaction between yucca and yucca moths. *Oecologia* 70:486–494.

Agee, James. 1995. *Fire ecology of Pacific northwest forests*. Victoria, BC: Island Press.

Agrawal, A. A. 2004. Resistance and susceptibility of milk-weed to herbivore attack: Consequences of competition, root herbivory, and plant genetic variation. *Ecology* 85:2118–2133.

Aladin, N. V., I. S. Plotnikov, P. Micklin, and T. Ballatore. 2008. The Aral Sea: Water level, salinity and long-term changes in biological communities of an endangered ecosystem—past, present and future. In Oren, A., Naftz, D. L., and Wurtsbaugh, W. A. (eds.). *Saline lakes around the world*. Logan, UT: Utah State University College of Natural Resources.

Albins, M. A., and M. A. Hixon. 2011. Worst case scenario: Potential long-term effects of invasive predatory lionfish (*Pterois volitans*) on Atlantic and Caribbean coral-reef communities. *Environmental Biology of Fishes*. doi:10.1007/s10641-011-9795.

Allen, C. D. 2009. Climate-induced forest dieback: An escalating global phenomenon? *Unasylva* 60:43–49.

Ambrose, J. D. 2004. *Update COSEWIC status report on American chestnut* (Castanea dentata) *in Canada*. Ottawa: COSEWIC.

Anderson, E. 1948. Hybridization of the habitat. *Evolution* 2:1–9.

Anderson, R. C. 1963. The incidence, development, and experimental transmission of *Pneumostrongylus tenuis* Dougherty (Metastrongyloidea: Protostrongylidae) of the meninges of the white-tailed deer (*Odocoilus virginianus borealis*) in Ontario. *Canadian Journal of Zoology* 41:775–792.

Anderson, R. C. 1972. The ecological relationships of meningeal worm and native cervids in North America. *Journal of Wildlife Diseases* 8:304–309.

Anderson, R. C. 2006. Evolution and origin of the Central Grassland of North America: Climate, fire, and mammalian grazers. *Journal of the Torrey Botanical Society* 133:626–647.

Anderson, S., and H. H. Shugart. 1974. Avian community analyses of Walker Branch Watershed. *ORNL Publication No. 623*. Oak Ridge, TN: Environmental Sciences Division, Oak Ridge National Laboratory.

Andersson, F. 2005. *Coniferous forests*. Amsterdam: Elsevier.

Andreae, M. O., C. J. Jones, and P. M. Cox. 2005. Strong present-day aerosol cooling implies a hot future. *Nature* 435:1187–1190.

Andrzejewska, L., and G. Gyllenberg. 1980. Small herbivore subsystem. Pages 201–267 in Breymeyer, A. I., and van Dyne, G. M. (eds.). *Grasslands, systems analysis and man*. Cambridge: Cambridge University Press.

Angel, M. V. 1991. Variations in time and space: Is biogeography relevant to studies of long-time scale change? *Journal of the Marine Biological Association of the United Kingdom* 71:191–206.

Anthony, R. G., J. A. Estes, M. A. Ricca, A. K. Miles, and E. D. Forsman. 2008. Bald eagles and sea otters in the Aleutian archipelago: Indirect effects of trophic cascades. *Ecology* 89:2725–2735.

Archibold, O. W. 1995. *Ecology of world vegetation*. London: Chapman and Hall.

Arnold, M. L. 2004. Transfer and origin of adaptations through natural hybridization: Were Anderson and Stebbins right? *The Plant Cell* 16:562–570.

Ashmole, N. P. 1963. The regulation of numbers of tropical oceanic birds. *Ibis* 103:458–473.

Ashton, P. S. 1988. Dipterocarp biology as a window to the understanding of tropical forest structure. *Annual Reviews of Ecology and Systematics* 19:347–370.

Austin, M. P. 1999. A silent clash of paradigms: Some inconsistencies in community ecology. *Oikos* 86:170–178.

Avery, R. A. 1975. Clutch size and reproductive effort in the lizard *Lacerta vivipara* Jacquin. *Oecologia* 19:165–170.

Awada, T. M., E. L. Perry, and W. H. Schact. 2003. Photosynthetic and growth responses of the C_3 *Bromus inermis* and the C_4 *Andropogon gerardii* to tree canopy cover. *Canadian Journal of Plant Science* 83:533–540.

Ayala, F. J. 1969. Experimental invalidation of the principle of competitive exclusion. *Nature* 224:1076–1079.

Ayala, F. J., M. E. Gilpin, and J. G. Ehrenfeld. 1973. Competition between species: Theoretical models and experimental tests. *Theoretical Population Biology* 4:331–356.

Ayotte, J. B., K. L. Parker, J. M. Arocena, and M. P. Gillingham. 2006. Chemical composition of lick soils: Functions of soil ingestion by four ungulate species. *Journal of Mammalogy* 87:878–888.

Baker, A. C., P. W. Glynn, and B. Riegl. 2008. Climate change and coral reef bleaching: An ecological assessment of long-term impacts, recovery trends and future outlook. *Estuarine Coastal and Shelf Science* 80:435–471.

Baker, M. C., T. K. Bjerke, H. Lampe, and Y. Espmark. 1986. Sexual responses of female great tits to variation in size of male song repertoires. *American Naturalist* 128:491–498.

Bakker, R. T. 1972. Anatomical and ecological evidence of endothermy in dinosaurs. *Nature* 238:81–85.

Bakowsky, W., and J. L. Riley. 1994. A survey of the prairies and savannas in southern Ontario. *Proceedings of the Thirteenth North American Prairie Conference* 7:8–13.

Barnes, B. M. 1989. Freeze avoidance in a mammal: Body temperatures below 0°C in an Arctic hibernator. *Science* 244:1593–1595.

Barnosky, A. D., N. Matzke, S. Tomiya, G. O. U. Wogan, B. Swartz, T. B. Quental, C. Marshall, J. L. McGuire, E. L. Lindsey, K. C. Macguire, B. Mersey, and E. A. Ferrer. 2011. Has the Earth's sixth mass extinction already arrived? *Nature* 471:51–57.

Barry, R. G., and R. J. Chorley. 1998. *Atmosphere, weather, and climate.* 7th ed. New York: Routledge.

Basolo, A. L. 1990. Female preference for male sword length in the green swordtail, *Xiphophorus helleri* (Pisces: Poeciliidae). *Animal Behaviour* 40:332–338.

Basolo, A. L., and G. Alcarez. 2003. The turn of the sword: Length increases male swimming costs in swordtails. *Proceedings of the Royal Society B* 270:1631–1636.

Batra, S. W. T. 1966. Nests and social behavior of halictine bees of India (Hymenoptera: Halictidae). *Indian Journal of Entomology* 28:375–393.

Batzer, H. O. 1973. Defoliation by the spruce budworm stimulates epicormic shoots on balsam fir. *Environmental Entomology* 2:727–728.

Baumgartner, A. 1968. Ecological signficiance of the vertical energy distribution in plant stands. Pages 367–374 in Eckhardt, F. E. (ed.). *Functioning of terrestrial ecosystems at the primary production level.* Natural Resources Research V. Paris: Unesco.

Bazzaz, F. A. 1996. *Plants in changing environments.* New York: Cambridge University Press.

Bazzaz, F. A., and J. L. Harper. 1976. Relationship between plant weight and numbers in mixed populations of *Sinapis arvensis* (L.) Rabenh. and *Lepidium sativum* L. *Journal of Applied Ecology* 13:211–216.

Beare, M. H., R. W. Parmelee, P. F. Hendrix, W. Cheng, D. C. Coleman, and D. A. Crossley. 1992. Microbial and fungal interactions and effects on litter nitrogen and decomposition in agroecosystems. *Ecological Monographs* 62:569–591.

Begon, M., J. L. Harper, and C. R. Townsend. 1996. *Ecology: Individuals, populations and communities.* New York: Blackwell Scientific.

Beiler, K. J., D. M. Durall, S. W. Simard, S. A. Maxwell, and A. M. Kretzer. 2010. Mapping the wood-wide web: Mycorrhizal networks link multiple Douglas-fir cohorts. *New Phytologist* 185:543–553.

Beisner, B. E., D. T. Haydon, and K. Cuddington. 2003. Alternative stable states in ecology. *Frontiers in Ecology and the Environment* 1:376–382.

Bélisle, M., A. Desrochers, and M.-J. Fortin. 2001. Influence of forest cover on the movements of forest birds: A homing experiment. *Ecology* 82:1893–1904.

Belland, R. J., W. B. Schofield, and T. R. Hedderson. 1992. Bryophytes of Mingan Archipelago National Park Reserve, Quebec: A boreal flora with arctic and alpine components. *Canadian Journal of Botany* 70:2207–2222.

Belovsky, G. E. 1978. Diet optimization in a generalist herbivore: The moose. *Theoretical Population Biology* 14:105–134.

Beneragama, C. K., and K. Goto. 2010. Chlorophyll *a:b* ratio increases under low light in "shade-tolerant" *Euglena gracilis. Tropical Agricultural Research* 22:12–25.

Benjamin, K., G. Domon, and A. Bouchard. 2005. Vegetation composition and succession of abandoned farmland: Ecological, historical and spatial factors. *Landscape Ecology* 20:627–647.

Berg, B., K. Hannus, T. Popoff, and O. Theander. 1982. Changes in organic-chemical components of needle litter during decomposition. *Canadian Journal of Botany* 60:1310–1319.

Berger, J. 1990. Persistence of different-sized populations: An empirical assessment of rapid extinctions in bighorn sheep. *Conservation Biology* 4:91–98.

Berger, J. 1999. Intervention and persistence in small populations of bighorn sheep. *Conservation Biology* 13:432–435.

Berger, J., P. B. Stacey, L. Bellis, and M. P. Johnson. 2001. A mammalian predator-prey imbalance: Grizzly bear and wolf extinction affect avian neotropical migrants. *Ecological Applications* 11:947–960.

Berteaux, D., and S. Boutin. 2000. Breeding dispersal in female North American red squirrels. *Ecology* 81:1311–1326.

Bertness, M. D., and S. D. Hacker. 1994. Physical stress and positive associations among marsh plants. *American Naturalist* 148:363–372.

Beschta, R. L., and W. J. Ripple. 2010. Recovering riparian plant communities with wolves in Northern Yellowstone, U.S.A. *Restoration Ecology* 18:380–389.

Billings, W. D. 1938. The structure and development of old-field shortleaf pine stands and certain associated physical properties of the soil. *Ecological Monographs* 8:437–499.

Björkman, A. D., and M. Velland. 2010. Defining historical baselines for conservation: Ecological changes since European settlement on Vancouver Island, Canada. *Conservation Biology* 24:1559–1568.

Björkman, O. 1981. Responses to different quantum flux densities. Pages 57–107 in Lange, O. L., Nobel, P. S., Osmond, C. B., and Zegler, H. (eds.). *Encyclopedia of plant physiology, Vol. 12.4. Physiological plant ecology. I, Responses to the physical environment.* Berlin: Springer-Verlag.

Blest, A. D. 1963. Relations between moths and predators. *Nature* 197:1046–1047.

Block, W., N. R. Webb, S. Coulson, I. D. Hodkinson, and M. R. Worland. 1994. Thermal adaptation in the arctic collembolan *Onychiurus arcticus* (Tullberg). *Journal of Insect Physiology* 40:715–722.

Blount, J. D., N. B. Metcalfe, T. R. Birkhead, and P. F. Surai. 2003. Carotenoid modulation of immune function and sexual attractiveness in zebra finches. *Science* 300:125–127.

Boag, P. T. 1983. The heritability of external morphology in Darwin's ground finches (*Geospiza*) on Isla Daphne Major, Galápagos. *Evolution* 37:877–894.

Boag, P. T., and P. R. Grant. 1981. Intense natural selection in a population of Darwin's finches (*Geospiza*) in the Galápagos. *Science* 214:82–85.

Boag, P. T., and P. R. Grant. 1984. The classical case of character release: Darwin's finches (*Geospiza*) on Isla Daphne Major, Galápagos. *Biological Journal of the Linnean Society* 22:243–287.

Boden, T. A., G. Marland, and R. J. Andres. 2011. *Global, regional, and national fossil-fuel CO_2 emissions.* Oak Ridge, TN: U. S. Department of Energy, Carbon Dioxide Information Analysis Center. doi:10.3334/CDIAC/00001_V2009.

Bolnick, D. I., T. Ingram, W. E. Stutz, L. K. Snowberg, O. L. Lau, and J. S. Paull. 2010. Ecological release from interspecific competition leads to decoupled changes in population and invidividual niche width. *Proceedings of the Royal Society B* 277:1789–1787.

Bonan, G. B. 2002. *Ecological climatology*. Cambridge: Cambridge University Press.

Bonnell, M. L., and R. K. Selander. 1974. Elephant seals: Genetic variation and near extinction. *Science* 184: 908–909.

Boonstra, R., and G. R. Singleton. 1993. Population declines in the snowshoe hare and the role of stress. *General and Comparative Endocrinology* 91:126–143.

Bordenstein, S. B., and J. H. Werren. 2000. Do *Wolbachia* influence fecundity in *Nasonia vitripennis*? *Heredity* 84:54–62.

Bormann, F. H., and G. E. Likens. 1979. *Pattern and process in a forested ecosystem*. New York: Springer-Verlag.

Boucher, D., S. Gauthier, and L. De Grandpré. 2006. Structural changes along a chronosequence and a productivity gradient in the northeastern boreal forest of Québec. *Écoscience* 13:172–180.

Boucher, D. H., J. Aviles, and R. Chepote. 1991. Recovery of trail side vegetation from trampling in a tropical rain forest. *Environmental Management* 15: 257–262.

Boutin, S., K. W. Larsen, and D. Berteaux. 2000. Anticipatory parental care: Acquiring resources for offspring prior to conception. *Proceedings of the Royal Society B* 267:2081–2085.

Bradford, M. A., C. A. Davies, S. D. Frey, T. R. Maddox, J. M. Melilo, J. E. Mohan, J. F. Reynolds, K. K. Treseder, and M. D. Wallenstein. 2008. Thermal adaptation of soil microbial respiration to elevated temperature. *Ecology Letters* 11:1316–1327.

Bradshaw, A. D., M. J. Chadwick, D. Jowett, and R. W. Snaydon. 1964. Experimental investigations into the mineral nutrition of several grass species. IV. Nitrogen level. *Journal of Ecology* 52:665–676.

Brandt, J. P., Y. Hiratsuka, and D. J. Pluth. 2005. Germination, penetration, and infection by *Arceuthobium americanum* on *Pinus banksiana*. *Canadian Journal of Forest Research* 35:1914–1930.

Braun, E. L. 1950. *Deciduous forests of eastern North America*. New York: McGraw-Hill.

Brittingham, M. C., and S. A. Temple. 1983. Have cowbirds caused forest songbirds to decline? *BioScience* 33:31–35.

Broders, H. G., C. S. Findlay, and L. Zheng. 2004. Effects of clutter on echolocation call structure of *Myotis septentrionalis* and *M. lucifugus*. *Journal of Mammalogy* 85:273–281.

Bronson, F. H. 1964. Agonistic behaviour in woodchucks. *Animal Behaviour* 12:470–478.

Brook, B. W., J. J. O'Grady, A. P. Chapman, M. A. Burgman, H. R. Alcakaya, and R. Frankham. 2000. Predictive accuracy of population viability analysis. *Nature* 404:385–387.

Brouwer, L., D. Heg, and M. Taborsky. 2005. Experimental evidence for helper effects in a cooperatively breeding cichlid. *Behavioral Ecology* 16:667–673.

Brown, J. H., and A. Kodric-Brown. 1977. Turnover rates in insular biogeography: Effect of immigration on extinction. *Ecology* 58:445–449.

Brown, J. R., and S. Archer. 1987. Woody plant seed dispersal and gap formation in a North American subtropical savanna woodland: The role of domestic herbivores. *Vegetatio* 73:73–80.

Bryant, J. P., F. S. Chapin, and D. R. Klein. 1983. Carbon/nutrient balance of boreal plants in relation to vertebrate herbivory. *Oikos* 40:357–368.

Brylinsky, M., and K. H. Mann. 1973. An analysis of factors governing productivity in lakes and reservoirs. *Limnology and Oceanography* 18:1–14.

Buchanan, K. L., and C. K. Catchpole. 1997. Female choice in the sedge warbler, *Acrocephalus schoenobaenus:* Multiple cues from song and territory quality. *Proceeedings of the Royal Society B* 264:521–526.

Buesseler, K. A., S. C. Doney, D. M. Karl, P. W. Boyd, K. Caldeira, F. Chai, K. H. Coale, H. J. W. de Baar, P. G. Falkowski, K. S. Johnson, R. S. Lampitt, A. F. Michaels, S. W. A. Naqvi, V. Smetacek, S. Takeda, and A. J. Watson. 2008. Ocean iron fertilization: Moving forward in a sea of uncertainty. *Science* 319:162.

Buol, S. W. 1995. Sustainability of soil use. *Annual Review of Ecology and Systematics* 26:25–44.

Burgess, R. M., and A. A. Stickney. 1994. Interspecific aggression by tundra swans toward snow geese on the Sagavanirktok River delta, Alaska. *Auk* 111:204–207.

Burns, R. M., and B. H. Honkala. 1990. *Silvics of North America. Volume 2: Hardwoods*. Washington, D.C.: Forest Service USDA.

Burrows, C. 1990. *Processes of vegetation change*. London: Routledge.

Cahill, J. F., and B. B. Caspar. 2000. Investigating the relationship between neighbor root biomass and belowground competition: Field evidence for symmetric competition belowground. *Oikos* 90:311–320.

Cai, L., X. Jia, J. Zhao, and Z. Cheng. 2010. Bifurcation and chaos of a three-species Lotka-Volterra food-chain model with spatial diffusion and time delays. *Scientific Research and Essays* 5:4068–4076.

Calhoun, J. B. 1962. Population density and social pathology. *Scientific American* 206:139–148.

Camargo, J. A., A. Alonso, and A. Salamanca. 2005. Nitrate toxicity to aquatic animals: A review with new data for freshwater invertebrates. *Chemosphere* 58:1255–1267.

Cameron, E. K., E. M. Bayne, and M. J. Clapperton. 2007. Human-facilitated invasion of exotic earthworms into northern boreal forests. *Écoscience* 14:482–490.

Campbell, N. A., and J. B. Reece. 2005. *Biology*. 7th ed. New York: Benjamin Cummings.

Careau, V., O. R. P. Bininda-Emonds, D. W. Thomas, D. Réale, and M. M. Humphries. 2009. Exploration strategies map along fast-slow metabolic and life history continua in muroid rodents. *Functional Ecology* 23:150–156.

Carfagno, G. L. F., and P. J. Weatherhead. 2006. Intraspecific and interspecific variation in use of forest-edge habitat by snakes. *Canadian Journal of Zoology* 84:1440–1452.

Carpenter, S. R., J. J. Cole, M. M. Pace, R. Batt, W. A. Brock, T. Cline, J. Coloso, J. R. Hodgson, J. F. Kitchell, D. A. Seekell, L. Smith, and B. Weidel. 2011. Early warnings of regime shifts: A whole-ecosystem experiment. *Science* 332:1079–1082.

Carrings, S., M. Anions, R. Reiner, and B. Stein. 2005. *Our home and native land: Canadian species of global conservation concern*. Ottawa: NatureServe Canada.

Carson, R. 1962. *Silent spring*. Boston: Houghton Mifflin.

Carter, A. 1983. Cadmium, copper, and zinc in soil animals and their food in a red clover system. *Canadian Journal of Zoology* 61:2751–2757.

Caughley, G., and A. R. E. Sinclair. 1994. *Wildlife ecology and management*. Cambridge, MA: Blackwell Scientific.

Chapin, F. S. 1990. The mineral nutrition of wild plants. *Annual Review of Ecology and Systematics* 11:233–260.

Charnov, E. L. 1976. Optimal foraging: The marginal value theorem. *Theoretical Population Biology* 9:129–136.

Chase, J. M., and M. A. Leibold. 2003. *Ecological niches: Linking classical and contemporary approaches.* Chicago: University of Chicago Press.

Chatworthy, J. N. 1960. *Studies on the nature of competition between closely related species.* D. Phil. dissertation, University of Oxford.

Chesson, P. 2000. Mechanisms of maintenance of species diversity. *Annual Review of Ecology and Systematics* 31:343–366.

Chesson, P., and N. Huntly. 1997. The roles of harsh and fluctuating conditions in the dynamics of ecological communities. *American Naturalist* 150:519–533.

Chopin, T., S. M. C. Robinson, M. Troell, A. Neori, A. H. Buschmann, and J. Fang. 2008. Multitrophic integration for sustainable marine aquaculture. Pages 2463–2475 in Jørgensen, S. E., and Fath, B. D. (eds.). *The Encyclopedia of Ecology, Ecological Engineering (Vol. 3).* Oxford: Elsevier.

Chow-Fraser, P. 1998. A conceptual ecological model to aid restoration of Cootes Paradise Marsh, a degraded coastal wetland of Lake Ontario, Canada. *Wetland Ecology and Management* 6:43–57.

Chow-Fraser, P. 2005. Ecosystem response to changes in water level in Great Lakes marshes: Lessons from the restoration of Cootes Paradise Marsh. *Hydrobiologia* 539:189–204.

Christensen, N. L.1981. Fire regimes in southeastern ecosystems. Pages 112–136 in Mooney, H. A., Bonnicksen, T. M., Christensen, N. L., Lotan, J. E., and Reiners, W. A. (eds.). *Proceedings of the conference fire regimes and ecosystem properties.* USDA Forest Service, General Technical Rep.

Christian, J. J. 1950. The adrenopituitary system and population cycles in mammals. *Journal of Mammalogy* 31:247–259.

Chuenpadgee, R., L. E. Morgan, S. M. Maxwell, E. A. Norse, and D. Pauly. 2003. Shifting gears: Assessing collateral impacts of fishing methods in U.S. waters. *Frontiers of Ecology and the Environment* 10:517–524.

Cipollini, D., and M. Heil. 2010. Costs and benefits of induced resistance to pathogens and herbivores in plants. *CAB Reviews: Perspectives in Agriculture, Veterinary Science, Nutrition and Natural Resources.* doi:10.1079/PAVSNNR20105005.

Clapham, W. B. 1983. *Natural ecosystems.* 2nd ed. London: Macmillan.

Clark, C. 1973. The economics of overexploitation. *Science* 181:630-634.

Clarke, F. M., and C. G. Faulkes. 1997. Dominance and queen succession in captive colonies of the eusocial naked mole-rat, *Heterocephalus glaber. Proceedings of the Royal Society B* 264:993–1000.

Clausen, J., D. D. Keck, and W. M. Hiesey. 1948. Experimental studies on the nature of species. III. Environmental responses of climatic races of *Achillea. Carnegie Institute of Washington Publication* 581.

Clay, K. 1988. Fungal endophytes of grasses: A defensive mutualism between plants and fungi. *Ecology* 60:10–16.

Cleavitt, N. L., T. J. Fahey, and J. J. Battles. 2011. Regeneration ecology of sugar maple (*Acer saccharum*): Seedling survival in relation to nutrition, site factors, and damage by insects and pathogens. *Canadian Journal of Forest Research* 41:235–244.

Clements, F. E. 1916. Plant succession: An analysis of the development of vegetation. *Carnegie Institute of Washington Publication* 242.

Clements, F. E. 1936. Nature and structure of the climax. *Journal of Ecology* 24:252–284.

Clements, F. E., and V. E. Shelford. 1939. *Bio-ecology.* New York: John Wiley and Sons.

Cody, M. L. 1966. A general theory of clutch size. *Evolution* 20:174–184.

Cole, D. W., and M. Rapp. 1981. Elemental cycling in forest ecosystems. Pages 341–419 in Reichle, D. E. (ed.). *Dynamic properties of forest ecosystems.* Cambridge: Cambridge University Press.

Colinvaux, P. A. 1978. *Why big fierce animals are rare.* Princeton, NJ: Princeton University Press.

Confer, J. L., and P. Paicos. 1985. Downy woodpecker predation at goldenrod galls. *Journal of Field Ornithology* 56:56–64.

Connell, J. H. 1978. Diversity in tropical rain forests and coral reefs. *Science* 199:1302–1310.

Connell, J. H. 1980. Diversity and the co-evolution of competitors, or the ghost of competition past. *Oikos* 35:131–138.

Connell, J. H. 1983. On the prevalence and relative importance of interspecific competition: Evidence from field experiments. *American Naturalist* 122:661–696.

Connell, J. H., and R. O. Slatyer. 1977. Mechanisms of succession in natural communities and their role in community stability and organization. *American Naturalist* 111:1119–1144.

Cook, R. E. 1969. Variation in species density of North American birds. *Systematic Zoology* 18:63–84.

Corn, L. C., E. H. Buck, J. Rawson, and E. Fischer. 1999. Harmful non-native species: Issues for Congress. *Congressional Research Service Issue Brief,* RL30123.

Correll, D. L. 1978. Estuarine productivity. *BioScience* 28: 646–650.

COSEWIC (Committee on the Status of Endangered Wildlife in Canada). 2003. *COSEWIC assessment results.* Ottawa: Environment Canada.

COSEWIC. 2011. *Canadian wildlife species at risk.* Ottawa: Environment Canada.

Costa, M. H., and J. A. Foley. 2000. Combined effects of deforestation and doubled atmospheric CO_2 concentrations on the climate of Amazonia. *Journal of Climate* 13:18–34.

Cowan, R. L. 1962. Physiology of nutrition of deer. Pages 1–8 in *Proceedings 1st National White-tailed Deer Disease Symposium.*

Cowles, H. C. 1899. The ecological relations of the vegetation on the sand dunes of Lake Michigan. *Botanical Gazette* 27:95–117.

Cox, P. M., R. A. Betts, C. D. Jones, S. A. Spall, and I. J. Totterdell. 2000. Acceleration of global warming due to carbon-cycle feedbacks in a coupled climate model. *Nature* 408:184–187.

Cox, P. M., P. P. Harris, C. Huntingford, R. A. Betts, M. Collins, C. D. Jones, T. E. Jupp, J. A. Marengo, and C. A. Nobre. 2008. Increasing risk of Amazonian drought due to decreasing aerosol pollution. *Nature* 453:212–215.

Crane, A. L., and B. D. Greene. 2008. The effect of reproductive condition on thermoregulation in female *Agkistrodon piscivorus* near the Northwestern range limit. *Herpetologica* 64:156–167.

Crawley, M. J. 1983. *Herbivory: The dynamics of animal-plant interactions.* Berkeley: University of California Press.

Crawley, M. J., A. E. Johnston, J. Silvertown, M. Dodd, C. de Mazancourt, M. S. Heard, D. F. Henman, and G. R. Edwards. 2005. Determinants of species richness in the Park Grass experiment. *American Naturalist* 165:179–192.

Crowe, M., M. Fitzgerald, D. L. Remington, and J. Rychtar. 2009. On deterministic and stochastic models of kleptoparasitism. *Journal of Interdisciplinary Mathematics* 12:161–180.

Crumpton, W. G., and R. G. Wetzel. 1982. Effects of differential growth and mortality in the seasonal succession of phytoplankton populations in Lawrence Lake, Michigan. *Ecology* 63:1729–1739.

Cumming, H. G., D. B. Beange, and G. Lavoie. 1996. Habitat partitioning between woodland caribou and moose in Ontario: The potential role of shared predation risk. *Rangifer* 9:81–94.

Currie, D. J. 1991. Energy and large-scale biogeographical patterns of animal and plant species richness. *American Naturalist* 137:27–49.

Currie, D. J. 2001. Projected effects of climate change on patterns of vertebrate and tree species richness in the conterminous United States. *Ecosystems* 4:216–225.

Currie, D. J., and V. Paquin. 1987. Large-scale biogeographical patterns of species richness in trees. *Nature* 329:326–327.

Cyr, H., and M. L. Pace. 1993. Magnitude and patterns of herbivory in aquatic and terrestrial ecosystems. *Nature* 361:148–150.

Danks, H. 2004. Seasonal adaptations in Arctic insects. *Integrative and Comparative Biology* 44:85–94.

Darlington, P. J. 1957. *Zoogeography: The geographical distribution of animals.* New York: John Wiley and Sons.

Darwin, C. 1859. *The origin of species.* Philadelphia: McKay. [Reprinted from 6th U.K. edition.]

Darwin, C. 1871. *The descent of man and selection in relation to sex.* London: Murray.

Dash, M. C., and A. K. Hota. 1980. Density effects on the survival, growth rate, and metamorphosis of *Rana tigrina* tadpoles. *Ecology* 61:1025–1028.

Davenport, J., and T. A. Switalski. 2006. Environmental impacts of transport, related to tourism and leisure activities. *Environmental Pollution* 10:333–360.

Davies, N. B. 1977. Prey selection and social behavior in wagtails (Aves: Motacillidae). *Journal of Animal Ecology* 46:37–57.

Davies, S. J. 1998. Photosynthesis of nine pioneer *Macaranga* species from Borneo in relation to life-history traits. *Ecology* 79:2292–2308.

Davies, Z. G., R. J. Wilson, T. M. Brereton, and C. D. Thomas. 2005. The re-expansion and improving status of the silver-spotted skipper butterfly (*Hesperia comma*) in Britain: A metapopulation success story. *Biological Conservation* 124:189–198.

Davis, M. B. 1981. Quaternary history and the stability of forest communites. In West, D. C. (ed.). *Forest succession: Concepts and applications.* New York: Springer-Verlag.

Dayan, T., D. Simberloff, E. Tchernov, and Y. Yom-Tov. 1990. Feline canines: Community-wide character displacement among the small cats of Israel. *American Naturalist* 136:39–60.

Deacon, J. 2006 (1980). *Fungal biology.* Oxford: Blackwell Scientific.

DeCoursey, P. J. 1960. Daily light sensitivity rhythm in a rodent. *Science* 131:33–35.

Delcourt, P. A., and H. R. Delcourt. 1981. Vegetation maps for eastern North America, 40,000 yr bp to present. Pages 123–166 in Romans, R. (ed.). *Geobotany.* New York: Plenum Press.

DeLucia, E. H., J. G. Hamilton, S. L. Naidu, R. B. Thomas, J. A. Andrews, A. Finzi, M. Lavine, R. Matamala, J. E. Mohan, G. R. Hendry, and W. H. Schlesinger. 1999. Net primary production of a forest ecosystem under experimental CO_2 enrichment. *Science* 284:1177–1179.

Demenocai, P., J. Ortiz, T. Guilderson, J. Adkins, M. Sarnthein, L. Baker, and M. Yarusinsky. 2000. Abrupt onset and termination of the African Humid Period: Rapid climate responses to gradual insolation forcing. *Quaternary Science Reviews* 19:347–361.

Denno, R. F., M. S. Mitter, G. A. Langellotto, C. Gratton, and D. L. Finke. 2004. Interactions between a hunting spider and a web-builder: Consequences of intraguild predation and cannibalism for prey suppression. *Ecological Entomology* 29:566–577.

D'Eon, R. 2006. Variable retention: Maintaining diversity through planning and operation practices. *Research Note Series 25.* Edmonton, AB: Sustainable Forest Management Network.

Desponts, M., and S. Payette. 1993. The Holocene dynamics of jack pine at its northern range limit in Québec. *Journal of Ecology* 81:719–727.

Diamond, S. J., R. H. Giles, Jr., and R. L. Kirkpatrick. 2000. Hard mast production before and after the chestnut blight. *Southern Journal of Applied Forestry* 24:196–201.

DiGuistini, S., Y. Wang, N. Y. Liao, G. Taylor, P. Tanguay, N. Feau, B. Henrissat Simon K. Chan, U. Hesse-Orce, S. M. Alamouti, C. K. M. Tsui, R. T. Docking, A. Levasseur, S. Haridas, G. Robertson, Inanc Birol, R. A. Holt, M. A. Marra, R. C. Hamelin, M. Hirst, S. J. M. Jones, J. Bohlmann, and C. Breuil. 2011. Genome and transcriptome analyses of the mountain pine beetle-fungal symbiont *Grosmannia clavigera*, a lodgepole pine pathogen. *Proceedings of the National Academy of Sciences* 108:2504–2509.

Dillon, P. J., and F. H. Rigler. 1974. The phosphorus–chlorophyll relationship in lakes. *Limnology and Oceanography* 19:767–773.

Dixon, M., D. Le Thiec, and J. P. Garrec. 1998. Reactions of Norway spruce and beech trees to 2 years of ozone exposure and episodic drought. *Environmental and Experimental Botany* 40:77–91.

Dodson, S. I. 1970. Complementary feeding niches sustained by size-selective predation. *Limnology and Oceanography* 15:131–137.

DOE-CDIAC (Department of Energy—Carbon Dioxide Information Analysis Center). 2009. Annual Report.

Downing, J. A., C. W. Osenberg, and O. Sarnelle. 1999. Meta-analysis of marine nutrient-enrichment experiments: Systematic variation in the magnitude of nutrient limitation. *Ecology* 80:1157–1167.

Duggins, D. O. 1980. Kelp beds and sea otters: An experimental approach. *Ecology* 61:447–453.

Dukas, R., and A. Kamil. 2001. Limited attention: The constraint underlying search image. *Behavioral Ecology* 12:192–199.

Duman, J. G. 2001. Antifreeze and ice nucleator proteins in terrestrial arthropods. *Annual Review of Physiology* 63:327–357.

Dumont, A., J.-P. Ouellet, M. Crête, and J. Huot. 2005. Winter foraging strategy of white-tailed deer at the northern limit of its range. *Écoscience* 12:476–484.

Dye, P. J., and P. T. Spear. 1982. The effects of bush clearing and rainfall variability on grass yield and composition in south-west Zimbabwe. *Zimbabwe Journal of Agricultural Research* 20:103–118.

Dyer, J. M. 2006. Revisiting the deciduous forests of Eastern North America. *BioScience* 56:341–352.

Edmonds, J. 1992. Why understanding the natural sinks and sources of CO_2 is important: A policy analysis perspective. *Water, Air and Soil Pollution* 64:11–21.

Egler, F. 1954. Vegetation science concepts. I. Initial floristic composition, a factor in old field vegetation development. *Vegetatio* 14:412–417.

Ehrlich, P., and D. Murphy. 1987. Conservation lessons from long-term studies of checkerspot butterflies. *Conservation Biology* 1:122–131.

Ehrlich, P., D. Murphy, M. Singer, C. Sherwood, R. White, and I. Brown. 1980. Extinction, reduction, stability and increase: The responses of checkerspot butterfly (*Euphydryas*) populations to the California drought. *Oecologia* 46:101–105.

Eis, S., E. H. Garman, and L. F. Ebel. 1965. Relation between cone production and diameter increment of Douglas fir (*Pseudotsuga*), grand fir (*Abies grandis*), and western white pine (*Pinus monticola*). *Canadian Journal of Botany* 43:1553–1559.

Elliott, T. L., and Henry, G. H. R. 2011. Effects of simulated grazing in ungrazed wet sedge tundra in the High Arctic. *Arctic, Antarctic, and Alpine Research* 43:198–206.

Elton, C. 1927. *Animal Ecology.* London: Sidgwick and Jackson.

Elton, C. S. 1958. *The ecology of invasions by animals and plants.* London: Methuen.

Emery, N. C., P. J. Ewanchuk, and M. D. Bertness. 2001. Competition and salt marsh plant zonation: Stress tolerators may be dominant competitors. *Ecology* 82:2471–2485.

Engle, L. G. 1960. Yellow-poplar seedfall pattern. *Central States Forest Experimental Station Note* 143.

Estes, J. A., M. T. Tinker, T. M. Williams, and D. F. Doak. 1998. Killer whale predation on sea otters linking oceanic and nearshore ecosystems. *Science* 282:473–476.

Ewel, J., C. Berish, B. Brown, N. Price, and J. Raich. 1981. Slash and burn impacts on a Costa Rican wet forest site. *Ecology* 62:816–829.

Fajardo, A., and E. J. MacIntire. 2011. Under strong niche overlap conspecifics do not compete but help each other to survive: Facilitation at the intraspecific level. *Journal of Ecology* 99:642–650.

FAO (Food and Agriculture Organization). 2002. *World agriculture: Toward 2015/2030. Summary report.* Available at ftp://ftp.fao.org/docrep/fao/004/y3557e/y3557e.pdf.

FAO. 2011. *Global Forest Resources Assessment 2010.* New York: United Nations.

Feduccia, A., T. Lingham-Sollar, and J. R. Hinchliffe. 2005. Do feathered dinosaurs exist? Testing the hypothesis on neontological and paleontological evidence. *Journal of Morphology* 266:125–266.

Feldhamer, G. A. 2007. *Mammalogy: Adaptation, diversity, ecology.* Baltimore, MA: Johns Hopkins University Press.

Festa-Blanchet, M. 1991. Numbers of lungworm larvae in feces of bighorn sheep: Yearly changes, influence of host sex, and effects on host survival. *Canadian Journal of Zoology* 69:547–554.

Field, C. B., and H. Mooney. 1986. The photosynthetic-nitrogen relationship in wild plants. Pages 25–55 in T. J. Givnish (ed.). *On the economy of plant form and function.* Cambridge: Cambridge University Press.

Figge, F. 2004. Bio-folio: Applying portfolio theory to biodiversity. *Biodiversity and Conservation* 13:827–849.

Finzi, A. C., D. J. Moore, E. H. Delucia, J. Lichter, K. S. Hofmockel, R. B. Jackson, H.-S. Kim, R. Matamala, H. R. McCarthy, R. Oren, J. S. Pippen, and W. H. Schlesinger. 2006. Progressive nitrogen limitation of ecosystem processes under elevated CO_2 in a warm-temperate forest. *Ecology* 87:15–25.

Fitzpatrick, L. C. 1973. Energy allocation in the Allegheny mountain salamander, *Desmognathus ochrophaeus. Ecological Monographs* 43:43–58.

Forman, R. T. T. 1964. Growth under controlled conditions to explain the hierarchical distribution of a moss, *Tetraphis pellucida. Ecological Monographs* 34:1–25.

Fossing, H., V. A. Gallardo, B. B. Jørgenson, M. Hüttel, L. P. Nielsen, H. Shulz, D. E. Canfield, S. Forster, R. N. Glud, J. K. Gundersen, J. Küver, N. B. Ramsing, A. Teske, B. Thamdrup, and O. Ulloa. 1995. Concentration and transportation of nitrate by the mat-forming sulphur bacterium *Thioploca. Nature* 374:713–715.

Fowler, C. W. 1981. Density dependence as related to life history strategy. *Ecology* 62:602–610.

Fowler, N. L. 1981. Competition and coexistence in a North Carolina grassland II: The effects of the experimental removal of species. *Journal of Ecology* 69:843–854.

Frank, D. A., and S. J. McNaughton. 1991. Stability increases with diversity in plant communities: Evidence from the Yellowstone drought. *Oikos* 62:360–362.

Frelich, L. E., C. M. Hale, S. Scheu, A. R. Holdsworth, L. Heneghan, P. J. Bohlen, and P. B. Reich. 2006. Earthworm invasion into previously earthworm-free temperate and boreal forests. *Biological Invasions* 8:1235–1245.

Fritts, T. H., and G. H. Rodda. 1998. The role of introduced species in the degradation of island habitats: A case history of Guam. *Annual Review of Ecology and Systematics* 29:113–140.

Frost, B. W. 1972. Effects of size and concentration of food particles on the feeding behavior of the marine planktonic copepod *Calanus pacificus. Limnology and Oceanography* 17:805–814.

Fry, F. E. J. 1947. Effects of the environment on animal activity. *Publications of the Ontario Fisheries Research Laboratory* 68:1–62.

Fuchs, S. A., S. G. Hinch, and E. Mellina. 2003. Effects of streamside logging on stream macroinvertebrate communities and habitat in the sub-boreal forests of British Columbia, Canada. *Canadian Journal of Forest Research* 33:1408–1413.

Fynn, R. W. S., C. D. Morris, and K. P. Kirkman. 2005. Plant strategies and trait trade-offs influence trends in competitive ability along gradients of soil fertility and disturbance. *Journal of Ecology* 93:384–394.

Galizia, C. G., J. Kunze, A. Gumbert, A.-K. Borg-Karlson, S. Sachse, C. Markl, and R. Menzel. 2005. Relationship of visual and olfactory signal parameters in a food-deceptive flower mimicry system. *Behavioral Ecology* 16:159–168.

Galloway, J. N., J. D. Aber, J. W. Erisman, S. P. Seitzinger, R. W. Howarth, E. B. Cowling, and B. J. Cosby. 2003. The nitrogen cascade. *BioScience* 53:341–356.

Gaona, P., P. Ferraras, and M. Delibes. 1998. Dynamics and viability of a metapopulation of the endangered Iberian lynx (*Lynx pardinus*). *Ecological Monographs* 68:349–370.

Garton, E. O. 2002. Mapping a chimera? Pages 663–666 in Scott, J. M., Heglund, P. J., Hauffler, J. B., Raphael, M. G., Wall, W. A., and Samson, F. B. (eds.). *Predicting species occurrence: Issues of accuracy and scale.* Covelo, CA: Island Press.

Gause, G. F. 1934. *The struggle for existence.* Baltimore, MD: Williams and Wilkins.

Geist, V. 1998. *Deer of the world: Their evolution, behavior, and ecology.* Mechanicsbur, PA: Stackpole Books.

Gentry, A. H. 1986. Endemism in tropical versus temperate communities. Pages 153–181 in Soule, M. E. (ed.). *Conservation biology: The science of scarcity and diversity.* Sunderland, MA: Sinauer.

Giller, K. E., E. Witter, and M. Corbeels. 2009a. Conservation agriculture and smallholder farming in Africa: The heretics' view. *Field Crops Research* 114:23–34.

Giller, K. E., E. Witter, and S. P. McGraph. 2009b. Heavy metals and soil microbes. *Soil Biology and Biochemistry* 41:2031–2037.

Gillette, J. R., R. G. Jaeger, and M. G. Peterson. 2000. Social monogamy in a territorial salamander. *Animal Behaviour* 59:1241–1250.

Givnish, T. J. 1988. Adaptation to sun and shade: A whole-plant perspective. *Australian Journal of Plant Physiology* 15:63–92.

Gleason, H. A. 1939. The individualistic concept of the plant association. *The American Midland Naturalist* 21:92–110.

Global Footprint Network. 2010. *Ecological Footprint Atlas.* www.footprintnetwork.org/en/index.php/GFN. Accessed 10 April 2012.

Golley, F. B., and H. Lieth. 1972. Basis of organic production in the tropics. Pages 1–26 in Golley, P. M., and Golley, F. H. (eds.). *Tropical ecology with an emphasis on organic production.* Athens, GA: University of Georgia Press.

Gonzalez, A., B. Rayfield, and Z. Lindo. 2011. The disentangled bank: How loss of habitat fragments and disassembles ecological networks. *American Journal of Botany* 98:503–516.

Goodman, D. 1975. The theory of diversity-stability relationships in ecology. *Quarterly Review of Biology* 50:237–266.

Goss-Custard, J. D., A. D. West, M. G. Yates, R. W. Caldow, R. A. Stillman, L. Bardsley, J. Castilla, M. Castro, V. Dierschke, S. E. A. Le V. dit Durell, G. Eichhorn, B. J. Ens, K.-M. Exo, P. U. Udayangani-Fernando, P. N. Ferns, P. A. R. Hockey, J. A. Gill, I. Johnstone, B. Kalejta-Summers, J. A. Masero, F. Moreira, R. V. Nagarajan, I. P. F. Owens, C. Pacheco, A. Perez-Hurtado, D. Rogers, G. Scheiffarth, H. Sitters, W. J. Sutherland, P. Triplet, D. H.

Worrall, Y. Zharikov, L. Zwarts, and R. A. Pettifor. 2006. Intake rates and the functional response in shorebirds (Charadriiformes) eating macroinvertebrates. *Biological Reviews* 81:501–529.

Gotelli, N. J., and G. Graves. 1996. *Null models in ecology.* Herndon, VA: Smithsonian Institute.

Goudie, A. 2001. *The nature of the environment.* Oxford: Blackwell Scientific.

Gower, S. T., R. E. McMurtie, and D. Murty. 1996. Above-ground net primary productivity declines with stand age: Potential causes. *Trends in Ecology and Evolution* 11:378–383.

Grace, J. B. 1991. A clarification of the debate between Grime and Tilman. *Functional Ecology* 5:583–587.

Grace, J. B., and R. G. Wetzel. 1981. Habitat partitioning and competitive displacement in cattails (*Typha*): Experimental field studies. *American Naturalist* 118:463–474.

Grant, P. 1999. *Ecology and evolution of Darwin's finches.* Princeton, NJ: Princeton University Press.

Grant, P. D., and R. R. Grant. 2006. Evolution of character displacement in Darwin's finches. *Science* 313:224–226.

Grassle, J. F., and N. J. Maciolek. 1992. Deep-sea species richness: Regional and local diversity estimates from quantitative bottom samples. *American Naturalist* 139:313–341.

Grassly, N. C., and C. Fraser. 2008. Mathematical models of infectious disease. *Nature Reviews Microbiology* 6:477–487.

Gray, J. S., G. C. Poore, K. I. Ugland, R. S. Wilson, F. Olsgard, and O. Johannessen. 1997. Coastal and deep-sea benthos compared. *Marine Ecology Progress Series* 159:97–103.

Gray, J. T., and W. H. Schleisinger. 1981. Nutrient cycling in Mediterranean-type ecosystems. In P. C. Miller (ed.). *Resource use by chaparral and matorral.* New York: Springer-Verlag.

Greenberg, R., J. S. Ortiz, and C. M. Caballero. 1994. Aggressive competition for critical resources among migratory birds in the Neotropics. *Bird Conservation International* 4:115–127.

Gregg, J. S., R. J. Andres, and G. Marland. 2008. China: Emissions pattern of the world leader in CO_2 emissions from fossil fuel consumption and cement production. *Geophysical Research Letters* 35:L08806. doi:10.1029/2007GL032887.

Griffith, S. C., I. P. F. Owens, and K. Thuman. 2002. Extra pair paternity in birds: A review of interspecific variation and adaptive function. *Molecular Ecology* 11:2195–2212.

Grime, J. P. 1973. Competitive exclusion in herbaceous vegetation. *Nature* 242:344–347.

Grime, J. P. 1977. Evidence for the existence of three primary strategies in plants and its relevance to ecological and evolutionary theory. *American Naturalist* 111:1169–1194.

Grime, J. P. 1979. *Plant strategies and vegetation processes.* London: John Wiley and Sons.

Grippo, A. J., A. Sgoifo, F. Mastorci, N. McNeal, and D. M. Trahanas. 2010. Cardiac dysfunction and hypothalamic activation during a social crowding stressor in prairie vole. *Autonomic Neuroscience* 156:44–50.

Gross, J. E., Z. Wang, and B. A. Wunder. 1985. Effects of food quality and energy needs: Changes in gut morphology and capacity of *Microtus ochrogaster*. *Journal of Mammalogy* 66:661–667.

Groves, R. H., and J. D. Williams. 1975. Growth of skeleton weed (*Chondrilla juncea* L.) as affected by growth of subterranean clover (*Trifolium subterraneum* L.) and infection by *Puccinea chondrillina* Bubak and Syd. *Australian Journal of Agricultural Research* 26:975–983.

Grünke, S., A. Lichtschlag, D. de Beer, M. Kuypers, T. Lösekan-Behrens, A. Ramette, and A. Boetius. 2010. Novel observations of *Thiobacterium*, a sulphur-storing gammaproteobacterium producing gelatinous mats. *ISME Journal* 4:1031–1043.

Gullan, P. J., and P. S. Cranston. 2010. *The insects: An outline of entomology*. 4th ed. New York: John Wiley and Sons.

Gunn, A., F. L. Miller, and S. J. Barry. 2003. Conservation of erupting ungulates on islands—a comment. *Rangifer* 24:3–12.

Guowei, S., and C. Qiwu 1990. A new mathematical model of interspecific competition: An expansion of Lotka-Volterra competition equations. *Chinese Journal of Applied Ecology* 1:31–9.

Gurevitch, J. L. 1986. Competition and the local distribution of the grass *Stipa neomexicana*. *Ecology* 67:46–57.

Gurevtich, J. L., and D. Padilla. 2004. Are invasive species a major cause of extinctions? *Trends in Ecology and Evolution* 29:470–474.

Hacker, S. D., and M. D. Bertness. 1996. Trophic consequences of a positive plant interaction. *American Naturalist* 148:559–575.

Hackney, E., and J. B. McGraw. 2001. Experimental demonstration of an Allee effect in American ginseng. *Conservation Biology* 15:129–136.

Haddad, N. 1999a. Corridor and distance effects on interpatch movements: A landscape experiment with butterflies. *Ecological Applications* 9:612–622.

Haddad, N. 1999b. An experimental test of corridor effects on butterfly densities. *Ecological Applications* 9:623–633.

Hairston, N. G., and C. H. Pope. 1948. Geographic variation and speciation in Appalachian salamanders (*Plethodon jordani* group). *Evolution* 2:266–278.

Hairston, N. G., F. E. Smith, and L. B. Slobodkin. 1960. Community structure, population control, and competition. *American Naturalist* 94:421–425.

Hairston, N. J., Jr., and N. G. Hairston, Sr. 1993. Cause-effect relationships in energy flow, trophic structure, and interspecific interactions. *American Naturalist* 143:379–411.

Hale, C. M., L. E. Frlelich, P. B. Reich, and J. Pastor. 2008. Exotic earthworm effects on hardwood forest floor, nutrient availability and native plants: A mesocosm study. *Oecologia* 155:509–518.

Halverson, H. G., and J. L. Smith. 1979. Solar radiation as a forest management tool. *U.S. Forestry Service General Technical Bulletin* PSW-33.

Hamilton, J. G., E. H. DeLucia, K. George, S. L. Naidu, A. C. Finzi, and W. H. Schlesinger. 2002. Forest carbon balance under elevated CO_2. *Oecologia* 131:250–260.

Hamilton, W. D. 1964. The genetical evolution of social behaviour. *Journal of Theoretical Biology* 7:1–52.

Hamilton, W. D., and M. Zuk. 1982. Heritable true fitness and bright birds: A role for parasites? *Science* 218:384–387.

Hamme, R. C., P. W. Webley, W. R. Crawford, F. A. Whitney, M. D. DeGrandpre, S. R. Emerson, C. C. Eriksen, K. E. Giesbrecht, J. F. R. Gower, M. T. Kavanaugh, M. A. Peña, C. L. Sabine, S. D. Batten, L. A. Coogan, D. S. Grundle, and D. Lockwood. 2010. Volcanic ash fuels anomalous plankton bloom in subarctic northeast Pacific. *Geophysical Research Letters* 37:L19604.

Handa, I. T., R. Harmsen, and R. L. Jefferies. 2002. Patterns of vegetation change and the recovery potential of degraded areas in a coastal marsh system of the Hudson Bay lowlands. *Journal of Ecology* 90:86–99.

Hansen, M. C., S. V. Stehman, and P. V. Potapov. 2010. Quantification of global gross forest cover loss. *Proceedings of the National Academy of Sciences* 107:8650–8655.

Hanski, I. 1990. Density dependence, regulation and variability in animal populations. *Philosophical Transactions of the Royal Society of London* 330:141–150.

Hanski, I. 1999. *Metapopulation ecology*. Oxford: Oxford University Press.

Hardin, G. 1960. The competitive exclusion principle. *Science* 131:1292–1297.

Hardin, G. 1968. The tragedy of the commons. *Science* 162:1243–1248.

Harestad, A. S., and F. L. Bunnell. 1979. Home range and body weight—a reevaluation. *Ecology* 60:389–402.

Harper, J. L., and D. L. Hawksworth. 1995. Preface in Hawksworth, D. L. (ed.). *Biodiversity—measurement and estimation*. London: Chapman and Hall.

Harpole, W. S., and D. Tilman. 2007. Grassland species loss resulting from reduced niche dimension. *Nature* 446:791–793.

Harris, R. C., and J. Warkentin (eds.). 1991. *Canada before confederation: A study of historical geography*. Ottawa: Carleton University Press.

Harrison, G. W. 1979. Stability under environmental stress: Resistance, resilience, persistence, and variability. *American Naturalist* 113:656–669.

Harrison, S., D. Murphy, and P. Ehrlich. 1988. Distribution of the Bay checkerspot butterfly, *Euphydryas editha bayensis:* Evidence for a metapopulation model. *American Naturalist* 132:360–382.

Hassell, M. P., and R. M. May. 1974. Aggregation of predators and insect parasites and its effect on stability. *Journal of Animal Ecology* 43:567–594.

Heard, D. C., and T. M. Williams. 1990. Ice and mineral licks used by caribou in winter. *Rangifer* 3:203–206.

Heil, M. 2000. Different strategies for studying ecological aspects of systemic acquired resistance (SAR). *Journal of Ecology* 88:707–708.

Heil, M., and J. C. Silva Bueno. 2007. Herbivore-induced volatiles as rapid signals in systemic plant responses. *Plant Signaling Behavior* 2:191–193.

Heimberger, M., D. Euler, and J. Barr. 1983. The impact of cottage development on common loon reproductive success in central Ontario. *Wilson Bulletin* 95:431–439.

Heinselman, M. L. 1973. Fire in the virgin forest of the Boundary Waters Canoe Area, Minnesota. *Quaternary Research* 3:329–382.

Heinselman, M. L., and T. M. Casey. 1981. Fire and succession in the confier forests of northern North America. Pages 374–405 in D. E. West, Shugart, H. H., and Botkin, D. B. (eds.). *Forest succession: Concepts and applications*. New York: Springer-Verlag.

Heldt, H.-W. 1997. *Plant biochemistry and molecular biology*. Oxford: Oxford University Press.

Helfman, D. 2007. *Fish conservation: A guide to understanding and restoring global aquatic biodiversity and fishery resources.* Washington, D.C.: Island Press.

Heller, H. C., and D. M. Gates. 1971. Altitudinal zonation of chipmunks (*Eutamias*): Energy budgets. *Ecology* 52: 424–443.

Hendry, A. P. 2009. Ecological speciation! Or the lack thereof? *Canadian Journal of Fisheries and Aquatic Sciences* 66:1383–1398.

Hendry, A. P., P. G. Grant, B. R. Grant, H. A. Ford, M. J. Brewer, and J. Podos. 2006. Possible human impacts on adaptive radiation: Beak size bimodality in Darwin's finches. *Proceedings of the Royal Society B* 273: 1887–1894.

Herborg, L.-M., N. E. Mandrak, B. C. Cudmore, and H. J. MacIsaac. 2007. Comparative distribution and invasion risk of snakehead (Channidae) and Asian carp (Cyprinidae) species in North America. *Canadian Journal of Fisheries and Aquatic Science* 64:1723–1735.

Herkert, J. R. 1994. The effect of habitat fragmentation on Midwestern grassland bird communities. *Ecological Applications* 4:461–471.

Hermanutz, L., H. Mann, M. F. E. Anions, D. Ballam, T. Bell, J. Brazil, N. Djan-Chékar, G. Gibbons, J. Maunder, S. J. Meades, N. Smith, and G. Yetman. 2002. National recovery plan for Long's braya (*Braya longii* Fernald) and Fernald's braya (*Braya fernaldii* Abbe). *National Recovery Plan No. 23.* Recovery of Nationally Endangered Wildlife. Ottawa.

Hett, H., and O. L. Loucks. 1976. Age structure models of balsam fir and eastern hemlock. *Journal of Ecology* 64:1029–1044.

Hill, R. W., and G. A. Wyse. 1989. *Animal physiology.* New York: Harper & Row.

Hillebrand, H. 2003. Opposing effects of grazing and nutrients on diversity. *Oikos* 100:592–600.

Hillebrand, H., B. Worm, and H. K. Lotz. 2000. Marine microbenthic community structure regulated by nitrogen loading and grazing pressure. *Marine Ecology Progress Series* 204:27–38.

Hobbie, E. A. 1994. *Nitrogen cycling during succession in Glacier Bay, Alaska.* Masters thesis, University of Virginia.

Hodkinson, I. D. 2003. Metabolic cold adaptation in arthropods: A smaller-scale perspective. *Journal of Functional Ecology* 17:562–567.

Holdridge, L. R. 1947. Determination of world plant formations from simple climatic data. *Science* 105:367–368.

Holland, G. M., and D. L. DeAngelis. 2010. A consumer-resource approach to the density-dependent population dynamics of mutualism. *Ecology* 91:1286–1295.

Holling, C. S. 1959. The components of predation as revealed by a study of small mammal predation of the European sawfly. *Canadian Entomologist* 91:293–320.

Holling, C. S. 1966. The functional response of invertebrate predators to prey density. *Memoirs of the Entomological Society of Canada* 48:1–86.

Holling, C. S. 1973. Resilience and stability of ecological systems. *Annual Review of Ecology and Systematics* 4:1–23.

Holt, R. D. 1977. Predation, apparent competition and the structure of prey communities. *Theoretical Population Biology* 12:197–229.

Holt, R. D., and J. H. Lawton. 1993. Apparent competition and enemy-free space in insect host-parasitoid communities. *American Naturalist* 142:623–645.

Holyoak, M., M. A. Leibold, and R. D. Holt (eds.). 2005. *Metacommunities: Spatial dynamics and ecological communities.* Chicago: University of Chicago Press.

Hooper, D. U., P. S. Chapin, J. J. Ewel, A. Hector, P. Inchausti, S. Lavorel, J. H. Lawton,. D. M. Lodge, M. Loreau, S. Naeem, B. Schmid, H. Setälä, A. J. Symstad, J. Vandermeer, and D. A. Wardle. 2005. Effects of biodiversity on ecosystem functioning: A consensus of current knowledge. *Ecological Monographs* 75:3–35.

Hoover, J. P., and S. K. Robinson. 2007. Retaliatory mafia behavior by a parasitic cowbird facilitates host acceptance of parasitic eggs. *Proceedings of the National Academy of Sciences* 104:1479–1483.

Hopkins, W. G., and N. P. A. Huner. 2009. *Introduction to Plant Physiology.* 4th ed. Toronto: John Wiley and Sons.

Horn, H. 1975. Markovian properties of forest succession. In Cody, M. L., and Diamond, J. R. (eds.), *Ecology and evolution of communities.* Cambridge, MA: Bellknap Press.

Houghton, J. T. 1997. *Global warming: The complete briefing.* Cambridge: Cambridge University Press.

Houghton, R. A. 2002. Terrestrial carbon sinks—uncertain explanations. *Biologist* 49:155–159.

Houghton, R. A. 2003. Why are estimates of the terrestrial carbon balance so different? *Global Climate Change* 9:500–509.

Hrabik, T. R., and C. J. Watras. 2002. Recent declines in mercury concentration in a freshwater fishery: Isolating the effects of de-acidification and decreased atmospheric mercury deposition in Little Rock Lake. *Science of the Total Environment* 297:229–237.

Hu, S., D. L. Dilcher, D. M. Jarzen, and D. W. Taylor. 2008. Early steps of angiosperm–pollinator coevolution. *Proceedings of the National Academy of Sciences* 105:240–245.

Hubbell, S. P. 2001. *The unified neutral theory of biogeography.* Princeton, NJ: Princeton University Press.

Huey, R. B., C. R. Peterson, S. J. Arnold, and W. P. Porter. 1989. Hot rocks and not-so-hot rocks: Retreat-site selection by garter snakes and its thermal consequences. *Ecology* 70:931–944.

Huggett, R. J. 2004. *Fundamentals of Biogeography.* Hove, U.K.: Psychology Press.

Hughes, R. N., and M. I. Croy. 1993. An experimental analysis of frequency-dependent predation (switching) in the 15-spined stickleback, *Spinachia spinachia. Journal of Animal Ecology* 62:341–352.

Hughes, R. N., and C. L. Griffiths. 1988. Self-thinning in barnacles and mussels: The geometry of packing. *American Naturalist* 132:484–491.

Hughes, S. 1990. Antelopes activate the acacia's alarm system. *New Scientist* 127:19.

Hughes, T. P. 1994. Catastrophes, phase shifts and large-scale degradation of a Caribbean coral reef. *Science* 265:1547–1551.

Humphreys, W. F. 1979. Production and respiration in animal populations. *Journal of Animal Ecology* 48:427–453.

Humphries M. M., D. W. Thomas, and D. L. Kramer. 2003. The role of energy availability in mammalian hibernation: A cost-benefit approach. *Physiological and Biochemical Zoology.* 76:165–179.

Hunter, M. L., and P. Yonzon. 1992. Altitudinal distributions of birds, mammals, people, forests and parks in Nepal. *Conservation Biology* 7:420–423.

Huston, M. A. 1980. Soil nutrients and tree species richness in Costa Rican forests. *Journal of Biogeography* 7:147–157.

Huston, M. A. 1994. *Biological diversity: The coexistence of species on changing landscapes.* New York: Cambridge University Press.

Huston, M. A., and S. Wolverton. 2009. The global distribution of net primary productivity: Resolving the paradox. *Ecological Monographs* 79:343–377.

Huston, M. A., and S. Wolverton. 2011. Regulation of animal size by eNPP, Bergmann's rule, and related phenomena. *Ecological Monographs* 81:349–405.

Hutchings, J. A., and M. Festa-Bianchet. 2009. Canadian species at risk (2006–2008), with particular emphasis on fishes. *Environmental Reviews* 17:53–65.

Hutchings, J. A., and J. D. Reynolds. 2004. Marine fish population collapses: Consequences for recovery and extinction risk. *BioScience* 54:297–309.

Hutchinson, B. A., and D. R. Matt. 1977. The distribution of solar radiation within a deciduous forest. *Ecological Monographs* 47:185–207.

Hutchinson, G. E. 1957. Concluding remarks. *Cold Spring Harbour Symposium on Quantitative Biology* 22:415–427.

Hutchinson, G. E. 1959. Homage to Santa Rosalia, or why are there so many kinds of animals? *American Naturalist* 93:134–159.

Hutchinson, G. E. 1961. The paradox of the plankton. *American Naturalist* 95:137–145.

IAEA (International Atomic Energy Agency). 2006. *The Chernobyl Forum: Major findings and recommendations.* Vienna: IAEA.

Idjadi, J. A., S. C. Lee, J. F. Bruno, W. F. Precht, L. Allen-Requa, and P. J. Edmunds. 2006. Rapid phase-shift reversal on a Jamaican coral reef. *Coral Reefs* 25:209–211.

IPCC (Intergovernmental Panel on Climate Change). 2007. *Climate Change 2007: Fourth Assessment (AR4).* Cambridge: Cambridge University Press.

Irons, D. B., R. G. Anthony, and J. A. Estes. 1986. Foraging strategies of glaucous-winged gulls in a rocky intertidal community. *Ecology* 67:1460–1474.

IUCN. 2011. *The IUCN Red List of Threatened Species. Version 2011.2.* www.iucnredlist.org. Downloaded on 8 October 2011.

Iverson, L. R., and A. M. Prasad. 2001. Potential changes in tree species richness and forest community type following climate change. *Ecosystems* 4:186–199.

Iverson, L. R., A. M. Prasad, B. J. Hale, and E. K. Sutherland. 1999. *Atlas of current and potential future distributions of common trees of the Eastern United States. General Technical Report NE-265.* Radnor, PA: USDA Forest Service, Northeastern Research Station.

James, F. 1971. Ordination of habitat relationships among birds. *Wilson Bulletin* 83:215–236.

Janzen, D. H. 1966. Coevolution of mutualism between ants and acacias in Central America. *Evolution* 20:249–275.

Järemo, J., and E. Palmqvist. 2001. Plant compensatory growth: A conquering strategy in plant-herbivore interactions? *Evolutionary Ecology* 15:91–102.

Jasinski, J. P., and S. Payette. 2005. The creation of alternative stable states in the southern boreal forest, Québec, Canada. *Ecological Monographs* 75:561–583.

Jedrzejewski, W., B. Jedrzejewski, and L. Szymura. 1995. Weasel population response, home range, and predation on rodents in a deciduous forest in Poland. *Ecology* 76:179–195.

Jenkins, T. M., S. Diehl, K. W. Kratz, and S. D. Cooper. 1999. Effects of population density on individual growth of brown trout in streams. *Ecology* 80:941–956.

Jenny, H. 1980. *The soil resource.* New York: Springer-Verlag.

Jeschke, J. M., M. Kopp, and R. Tollrain. 2004. Consumer-food systems: Why type I functional responses are restricted to filter feeders. *Biological Reviews* 79:337–349.

Johnson, N. K. 1966. Bill size and the question of competition in allopatric and sympatric populations of dusky and gray flycatchers. *Systematic Zoology* 15:70–87.

Johnson, S. D., E. Torninger, and J. Ågren. 2009. Relationships between population size and pollen fates in a moth-pollinated orchid. *Biology Letters* 5:282–285.

Johnston, R. F. 1956. Predation by short-eared owls in a *Salicornia* salt marsh. *Wilson Bulletin* 68:91–102.

Joly, D. O., and F. Messier. 2000. A numerical response of wolves to bison abundance in Wood Buffalo National Park, Canada. *Canadian Journal of Zoology* 78:1101–1104.

Joly, S., and A. Bruneau. 2004. Evolution of triploidy in *Apios americana* (Leguminosae) revealed by genealogical analysis of the histone HD-3 gene. *Evolution* 58:284–295.

Jonasson, K. A., and C. K. R. Willis. 2011. Changes in body condition of hibernating bats support the thrifty female hypothesis and predict consequences for populations with white-nose syndrome. *PLoS ONE* 6(6):e21061.

Jones, E. O., A. White, and M. Boots. 2011. The evolution of host protection by vertically transmitted parasites. *Proceedings of the Royal Society B* 278:863–870.

Jones, M. B. 1978. Aspects of the biology of the big-handed crab, *Heterozius rotundifrons* (Decapoda: Brachura), from Kaikoura, New Zealand. *New Zealand Journal of Zoology* 5:783–794.

Jorgenson, S. E., B. Fath, S. Bastiononi, J. C. Marques, F. Muller, S. N. Nielsen, B. C. Patten, E. Tiezzi, and R. E. Ulanowicz. 2007. *A new ecology: Systems perspective.* Amsterdam: Elsevier.

Karpinski, S., H. Reynolds, B. Karpinska, G. Wingsle, G. Creissen, and P. Mullineaux. 1999. Systemic signalling and acclimation in response to excess excitation energy in *Arabadopsis. Science* 284:654–657.

Keeley, E. R., 2003. An experimental analysis of self-thinning in juvenile steelhead trout. *Oikos* 102:543–550.

Keith, L. B. 1974. Some features of population dynamics in mammals. *Proceedings of the International Congress of Biology* 11:17–58.

Keith, L. B. 1981. Population dynamics of hares. Pages 395–440 in Myers, K., and MacInnes, C. D. (eds.). *Proceedings of the World Lagomorph Conference.* Guelph, ON: University of Guelph.

Keys, D. 1992. Canadian peat harvesting and the environment. *North American Wetlands Conservation (Canada) Sustaining Welands Issues* paper 1992–3.

Kikkawa, J., and W. T. Williams. 1971. Altitudinal distribution of land birds in New Guinea. *Search* 2:64–69.

Kimball, B. A., J. R. Mauney, F. S. Nakayama, and S. B. Idso. 1993. Effect of increasing atmospheric CO_2 on vegetation. *Vegetatio* 104:65–73.

Kindvall, O. 1996. Habitat heterogeneity and survival in a bush cricket metapopulation. *Ecology* 77:207–214.

Kindvall, O., and I. Ahlén. 1992. Geometrical factors and metapopulation dynamics of the bush cricket, *Metrioptera bicolor. Conservation Biology* 6:520–529.

Kingdon, J., B. Agwanda, M. Kinnaird, T. O'Brian, C. Holland, T. Gheysens, M. Boulet-Audet, and F. Vollrath. 2012. A poisonous surprise under the coat of the African crested rat. *Proceedings of the Royal Society B* 279:675–680.

Kinloch, B. B., Jr. 2003. White pine blister rust in North America: Past and prognosis. *Phytopathology* 93:1044–1047.

Klap, V., P. Louchouarn, J. J. Boon, M. A. Hemminga, and J. van Soelen. 1999. Decomposition dynamics of six salt marsh halophytes as determined by cupric oxide oxidation and direct temperature-resolved mass spectrometry. *Limnology and Oceanography* 44:1458–1476.

Kleypas, J. A., R. W. Buddenmeier, and J. P. Gattuso. 2001. The future of coral reefs in an era of global change. *International Journal of Earth Science* 90:426–437.

Klironomos, J. N., and M. M. Hart. 2001. Animal nitrogen swap for plant carbon. *Nature* 410:651–652.

Koch, G. W., and H. A. Mooney (eds.). 1995. *Carbon dioxide and terrestrial ecosystems.* New York: Academic Press.

Koper, N., and F. K. A. Schmiegelow. 2006. A multi-scaled analysis of avian response to habitat amount and fragmentation in the Canadian dry mixed-grass prairie. *Landscape Ecology* 21:1045–1059.

Korpimäki, E., and K. Norrdahl. 1991. Numerical and functional responses of kestrels, short-eared owls, and long-eared owls to vole densities. *Ecology* 72:814–826.

Korpimäki, E., K. Norrdahl, O. Huitu, and T. Klemola. 2005. Predator-induced synchrony in population oscillations of coexisting small mammal species. *Proceedings of the Royal Society B* 272:193–202.

Kotanen, P. M., and R. L. Jeffries. 1997. Long-term destruction of wetland vegetation by Lesser Snow Geese. *Écoscience* 4:1895–1898.

Koven, C. D., B. Ringeval, P. Friedlingstein, P. Ciais, P. Cadule, D. Khvorostyanov, G. Krinner, and C. Tarnocai. 2011. Permafrost carbon-climate feedbacks accelerate global warming. *Proceedings of the National Academy of Sciences* 108:14769–14774.

Kozlowski, T. T. 1982. Plant responses to flooding of soil. *BioScience* 34:162–167.

Krall, B. S., R. J. Bartelt, C. J. Lewis, and D. W. Whitman. 1999. Chemical defense in the stink bug *Cosmopepla bimaculata. Journal of Chemical Ecology* 25:2477–2494.

Krebs, C. J. 1985. Do changes in spacing behaviour drive population cycles in small mammals? Pages 295–312 in Sibly, R. M. and Smith, R. H. (eds.). *Behavioural ecology.* Oxford: Blackwell Scientific.

Krebs, C. J. 2006. Ecology after 100 years: Progress and pseudo-progress. *New Zealand Journal of Ecology* 30:3–11.

Krebs, C. J. 2011. Of lemmings and snowshoe hares: The ecology of northern Canada. *Proceedings of the Royal Society B* 278:481–489.

Krebs, C. J., S. Boutin, and R. Boonstra. 2001. *Ecosystem dynamics of the boreal forest, the Kluane Project.* New York: Oxford University Press.

Krebs, C. J., K. Cowcill, R. Boonstra, and A. J. Kenney. 2010. Do changes in berry crops drive population fluctuations in small rodents in the southwestern Yukon? *Journal of Mammalogy* 91:500–509.

Krebs, J. R. 1971. Territory and breeding density in the great tit *Parus major. Ecology* 52:2–22.

Kreulen, D. A. 1985. Lick use by large herbivores: A review of benefits and banes of soil consumption. *Mammal Review* 15:107–123.

Kubien, D. S., and R. F. Sage. 2003. C4 grasses in boreal fens: Their occurrence in relation to microsite characteristics. *Oecologia* 137:330–337.

Kümmerli, R., and L. Keller. 2011. Between-year variation in population sex ratio increases with complexity of the breeding system in Hymenoptera. *American Naturalist* 177:835–846.

Kurlansky, M. 1997. *Cod: The biography of the fish that changed the world.* New York: Walker and Company.

Lack, D. 1954. *The natural regulation of animal numbers.* Oxford: Oxford University Press.

Lafferty, K. D., and A. K. Morris. 1996. Altered behavior of parasitized killifish increases susceptibility to predation by bird final hosts. *Ecology* 77:1390–1397.

Lamb, H. H. 1995. *The little ice age: Climate, history and the modern world.* London: Routledge.

Landahl, J. T., and R. B. Root. 1969. Differences in the life tables of tropical and temperate milkweed bugs, genus *Oncopeltus* (Hemiptera). *Ecology* 50:734–737.

Langdon, C. 2002. Review of experimental evidence for effects of CO_2 on calcification of reef builders. *Proceedings of the 9th International Coral Reef Symposium* 2:1091–1098.

Larcher, W. 1996. *Physiological plant ecology.* 3rd ed. New York: Springer-Verlag.

Larson, E. R., K. F. Kipfmueller, C. M. Hale, L. E. Frelich, and P. B. Reich. 2010. Tree rings detect earthworm invasions and their effects in northern Hardwood forests. *Biological Invasions* 12:1053–1066.

Latham, A. D. M., M. C. Latham, N. A. McCutchen, and S. Boutin. 2011. Invading white-tailed deer change wolf-caribou dynamics in northeastern Alberta. *Journal of Wildlife Management* 75:204–212.

Lauenroth, W. K. 1979. Grassland primary production: North American grasslands in perspective. Pages 3–24 in French, N. R. (ed.). *Perspectives in grassland ecology.* New York: Springer-Verlag.

Lawrence, A. J., and K. Hemingway. 2003. *Effects of pollution on fish: Molecular effects and population responses.* New York: John Wiley and Sons.

Lawrence, D., and W. A. Schlesinger. 2001. Changes in the distribution of soil phosphorus during 200 years of shifting cultivation. *Ecology* 82:2769–2780.

Lawton, J. H., and V. K. Brown. 1993. Redundancy in ecosystems. Pages 255–270 in Schulze, E. D., and Mooney, H. A. (eds.). *Biodiversity and ecosystem function.* New York: Springer-Verlag.

Lee, T. M., and I. Zucker. 1990. Photoperiod synchronizes reproduction and growth in the southern flying squirrel, *Glaucomys volans. Canadian Journal of Zoology* 68:134–139.

Leishman, M. R. 2001. Does the seed size/number trade-off model determine plant community structure? An assessment of the model mechanisms and their generality. *Oikos* 93:294–302.

Leopold, A. 1949. *A Sand County Almanac*. New York: Oxford University Press.

Le Quéré, C., M. R. Raupach, J. G. Canadell, G. Marland, L. Bopp, P. Clais, T. J. Conway, S. C. Doney, R. A. Feely, P. Foster, P. Friedlingstein, K. Gurney, R. A. Houghton, J. I. House, C. Huntingford, P. E. Levy, M. R. Lomas, J. Majkut, N. Metzl, J. P. Ometto, G. P. Peters, I. C. Prentice, J. T. Randerson, S. W. Running, J. L. Sarmiento, U. Schuster, S. Sitch, T. Takahashi, N. Viovy, G. R. van der Werf, and F. I. Woodward. 2009. Trends in the sources and sinks of carbon dioxide. *Nature GeoScience* 2:831–836.

Leriche, H., X. LeRoux, F. Desnoyers, A. Tuzet, D. Benest, G. Simione, and L. Abbadie. 2003. Grass response to clipping in an African savanna: Testing the grazing optimization hypothesis. *Ecological Applications* 13:1346–1354.

Leriche, H., X. LeRoux, J. Gignoux, A. Tuzet, H. Fritz, L. Abbadie, and M. Loreau. 2001. What functional processes control the short-term effect of grazing on net primary production in grasslands? *Oecologia* 129:114–124.

LeRoy, C. J., and J. C. Marks. 2006. Litter quality, stream condition, and litter diversity influence decomposition rates and macroinvertebrate communities. *Freshwater Biology* 51:601–617.

Lett, P. F., R. K. Mohn, and D. F. Gray. 1981. Density-dependent processes and management strategy for the northwest Atlantic harp seal populations. Pages 135–158 in Fowler, C. W., and Smith, T. D. (eds.). *Dynamics of large mammal populations*. New York: John Wiley and Sons.

Levins, R. 1970. Extinction. Pages 77–107 in Gerstenhaber, M. (ed.). *Lectures on mathematics in the life sciences*. Vol 2. Providence, RI: American Mathematical Society.

Liebhold, A. M., J. A. Halverson, and G. A. Elmes. 1992. Gypsy moth invasion in North America: A quantitative analysis. *Journal of Biogeography* 19:513–520.

Lieth, H. 1973. Primary production: Terrestrial ecosystems. *Human Ecology* 1:303–332.

Lieth, H. 1975. Primary productivity in ecosystems: Comparative analysis of global patterns. Pages 67–88 in van Dobben, W. H., and Lowe-McConnell, R. H. (eds.). *Unifying concepts in ecology*. The Hague: W. Junk.

Likens, G. E. (ed.). 1985. *An ecosystem approach to aquatic ecology: Mirror Lake and environment*. New York: Springer-Verlag.

Likens, G. E., and F. H. Bormann. 1995. *Biogeochemistry of a forested ecosystem*. 2nd edition. New York: Springer-Verlag.

Likens, G. E., F. H. Bormann, R. S. Pierce, J. S. Eaton, and N. M. Johnson. 1977. *Biochemistry of a forested ecosystem*. New York: Springer-Verlag.

Lillywhite, H. B. 1970. Behavioral temperature regulation in the bullfrog, *Rana catesbeiana*. *Copeia* 1970:158–168.

Lima, S. L. 1998. Nonlethal effects in the ecology of predator-prey interactions. *BioScience* 48:25–34.

Lipsitch, M., S. Stiller, and M. A. Nowak. 1996. The evolution of virulence in pathogens with vertical and horizontal transmission. *Evolution* 50:1729–1741.

Little, E. L. 1977. *Atlas of United States trees*. Minor eastern hardwoods. Publicaton 1342. Washington, D.C.: U.S. Department of Agriculture, Forest Service.

Lively, C. M., K. Clay, M. J. Wade, and C. Fuqua. 2005. Competitive co-existence of vertically and horizontally transmitted parasites. *Evolutionary Ecology Research* 7:1183–1190.

Loery, G., and J. D. Nichols. 1985. Dynamics of a black-capped chickadee population, 1958–1983. *Ecology* 66:1195–1203.

Loewenstein, E. F., P. S. Johnson, and H. E. Garrett. 2000. Age and diameter structure of a managed uneven-aged oak forest. *Canadian Journal of Forest Research* 30:1060–1070.

Looman, J. 1969. The fescue grasslands of western Canada. *Vegetatio* 19:128–145.

Loreau, M. 2004. Does functional redundancy exist? *Oikos* 104:606–611.

Lovelock, J. 1979. *Gaia: A new look at life on Earth*. Oxford: Oxford University Press.

Luo, Y., E. Weng, X. Wu, C. Gao, X. Zhu, and L. Zhang. 2001. Parameter identifiability, constraint, and equifinality in data assimilation with ecosystems models. *Ecological Applications* 19:571–574.

Ma, J., H. Hung, C. Tian, and R. Kallenborn. 2011. Revolatilization of persistent organic pollutants in the Arctic induced by climate change. *Nature Climate Change* 1:255–260.

MacArthur, R. H. 1955. Fluctuations of animal populations and a measure of community stability. *Ecology* 33:533–536.

MacArthur, R. H., and J. H. Connell. 1966. *The biology of populations*. New York: John Wiley and Sons.

MacArthur, R. H., and J. W. MacArthur. 1961. On bird species diversity. *Ecology* 42:594–598.

MacArthur, R. H., and E. O. Wilson. 1967. *The theory of island biogeography*. Princeton, NJ: Princeton University Press.

MacCulich, D. A. 1937. Fluctuations in the numbers of varying hare (*Lepus americanus*). *University of Toronto Biological Series* 43.

MacFarlane, G. A., and J. R. King. 2003. Migration patterns of spiny dogfish (*Squalus acanthius*) in the North Pacific Ocean. *Fisheries Bulletin* 101:358–367.

MacKellar, F. L. 1996. On human carrying capacity: A review essay on Joel Cohen's *How Many People Can the Earth Support? Population and Development Review* 22:145–156.

Mackey, R. L., and D. J. Currie. 2001. The diversity-disturbance relationship: Is it generally strong and peaked? *Ecology* 82:3479–3492.

MacLean, S. F. 1981. Introduction: Invertebrates. In Bliss, L. C., Heal, O. W., and Moore, J. J. (eds.). *Tundra ecosystems: A comparative analysis*. Cambridge: Cambridge University Press.

Mallet, J. 2005. Hybridization as an invasion of the genome. *Trends in Ecology and Evolution* 20:229–237.

Margulis, L. 1982. *Early life*. Boston: Science Books International.

Margulis, L. 1998. *Symbiotic planet*. New York: Basic Books.

Markham, J. 2008. Variability of nitrogen-fixing *Frankia* on *Alnus* species. *Botany* 86:501–510.

Martell, K. A., A. L. Foote, and S. G. Cumming. 2006. Riparian disturbance due to beavers (*Castor canadensis*) in northeastern Alberta's boreal mixedwood forests: Implications for forest management. *Écoscience* 13:164–171.

Maser, C., J. M. Trappe, and R. A. Nussbaum. 1978. Fungal–small mammal interrelationship with emphasis on Oregon coniferous forests. *Ecology* 59:799–809.

Massey, A., and J. D. Vandenbergh. 1980. Puberty delay by a urinary cue from female house mice in feral populations. *Science* 209:821–822.

Matthews, H. D., and D. W. Keith. 2007. Carbon-cycle feedbacks increase the likelihood of a warmer future. *Geophysical Research Letters* 34:L09702.

Matthews, W. J. 1998. *Patterns in freshwater fish ecology.* New York: Springer-Verlag.

May, R. M. 1972. Will a large complex system be stable? *Nature* 238:413–414.

May, R. M. 1988. How many species are there on Earth? *Science* 241:1441–1449.

Mayr, E. 2000. Darwin's influence on modern thought. *Scientific American* 283:78–83.

McCorquodale, D. B., and R. W. Knapton. 2003. Changes in numbers of wintering American Black Ducks and Mallards in urban Cape Breton Island, Nova Scotia. *Northeastern Naturalist* 10:297–304.

McCullough, D. R. 1981. Population dynamics of the Yellowstone grizzly bear. Pages 173–196 in C. W. Fowler and Smith, T. D. (eds.). *Dynamics of large mammal populations.* New York: John Wiley and Sons.

McDowell, S. C. L., N. G. McDowell, J. D. Marshall, and K. Hultine. 2000. Carbon and nitrogen allocation to male and female reproduction in Rocky Mountain Douglas-fir (*Psuedotsuga menziesii* var. *glauca*, Pinaceae). *American Journal of Botany* 87:539–546.

McGill, B. J., B. J. Enquist, E. Weiher, and M. Westoby. 2006. Rebuilding community ecology from functional traits. *Trends in Ecology and Evolution* 21:178–185.

McGraw, J. B., and M. A. Furedi. 2005. Deer browsing and population viability of a forest understory plant. *Science* 307:920–922.

McKenney, D. W., J. H. Pedlar, K. Lawrence, K. Campbell, and M. F. Hutchinson. 2007. Potential impacts of climate change on the distribution of North American trees. *BioScience* 57:939–948.

McKibben, W. 1989. *The end of nature.* New York: Random House.

McKibben, W. 2010. *Eaarth: Making a life on a tough new planet.* Toronto: Alfred A. Knopf.

McKinney, S. T., C. E. Fiedler, and D. F. Tomback. 2009. Invasive pathogen threatens bird-pine mutualism: Implications for sustaining a high-elevation ecosystem. *Ecological Applications* 19:597–607.

McNaughton, S. J. 1975. r- and K-selection in *Typha*. *American Naturalist* 109:251–261.

McNaughton, S. J. 1977. Diversity and stability of ecological communities: A comment on the role of empiricism in ecology. *American Naturalist* 111:515–525.

McNaughton, S. J. 1983. Compensatory plant growth as a response to herbivory. *Oikos* 40:329–336.

McNaughton, S. J., M. Oesterheld, D. A. Frank, and K. J. Williams. 1989. Ecosystem-level patterns of primary productivity and herbivory in terrestrial habitats. *Nature* 341:142–144.

McNearney, P., J. Riley and A. Wennersten. 2002. Trampling increases soil compaction; soil compaction depresses vigor of *Andropogon gerardii*. *Tillers* 3:25–28.

Mech, L. D. 1999. Alpha status, dominance, and division of labor in wolf packs. *Canadian Journal of Zoology* 77:1196–1203.

Mech, L. D., R. E. McRoberts, R. O. Peterson, and R. E. Page. 1978. Relationship of deer and moose populations to previous winter's snow. *Journal of Animal Ecology* 56:615–627.

Melillo, J., J. Aber, and J. Muratore. 1982. Nitrogen and lignin control of hardwood leaf litter decomposition dynamics. *Ecology* 63:621–626.

Melville, J., L. J. Harmon, and J. B. Lobos. 2006. Intercontinental community convergence of ecology and morphology in desert lizards. *Proceedings of the Royal Society B* 273:523–530.

Messier, F. 1994. Ungulate population models with predation: A case study with the North American moose. *Ecology* 75:478–488.

Metzgar, L. H., and M. Bader. 1992. Large mammal predators in the Northern Rockies: Grizzly bears and their habitat. *Northwest Environmental Journal* 8:231–233.

Miceli-Mèndez, C. L., B. G. Ferguson, and N. Ramirez-Marcial. 2008. Seed dispersal by cattle: Natural history and applications to neotropical forest restoration and agroforestry. Chapter 7 in Myster, R. W. (ed.). *Post-agricultural succession in the neotropics.* New York: Springer-Verlag.

Michaels, H. J., and F. A. Bazzaz. 1989. Individual and population responses of sexual and apomictic plants to environmental gradients. *American Naturalist* 134:190–207.

Miko, U. F. 1996. Climate change impacts on forests. In Watson, R. T., Zinyowera, M. C., and Moss, R. H. (eds.). *Climate change 1995: Impacts, adaptations and mitigation of climate change.* New York: Cambridge University Press.

Millien, V., S. K. Lyons, L. Olson, F. A. Smith, A. B. Wilson, and Y. Yom-Tov. 2006. Ecotypic variation in the context of global climate change: Revisiting the rules. *Ecology Letters* 9:853–869.

Milton, W. E. J. 1940. The effect of manuring, grazing, and cutting on the yield, botanical and chemical composition of natural grasslands. *Journal of Ecology* 28:326–356.

Mitchell, J. F. B., T. J. Johns, J. M. Gregory, and S. B. F. Tett. 1995. Climate response to increasing levels of greenhouse gases and sulfate aerosols. *Nature* 376:501–504.

Møller, A. P., and T. A. Mousseau. 2008. Reduced abundance of raptors in radioactively contaminated areas near Chernobyl. *Journal of Ornithology* 150:239–246.

Møller, A. P., and T. A. Mousseau. 2009. Reduced abundance of insects and spiders linked to radiation at Chernobyl 20 years after the accident. *Biology Letters of the Royal Society* 5:356–359.

Molnár, P. K., A. E. Derocher, M. A. Lewis, and M. K. Taylor. 2008. Modelling the mating system of polar bears: A mechanistic approach to the Allee effect. *Proceedings of the Royal Society B* 275:217–226.

Monteil, S., A. Estrada, and P. Leon. 2011. Reproductive seasonality of fruit-eating bats in Northwestern Yucatan, Mexico. *Acta Chiropterologica* 13:139–145.

Mook, L. J. 1963. Birds and the spruce budworm. *Memoirs of the Entomological Society of Canada* 95:269–291.

Mooney, H. A., O. Bjorkman, J. Ehleringer, and J. Berry. 1976. Photosynthetic capacity of in situ Death Valley plants. *Carnegie Institute Yearbook* 75:310–413.

Mora, C., D. P. Tittensor, S. Adi, A. G. B. Simpson, and B. Worm. 2011. How many species are there on earth and in the oceans? *Plos Biology* 9:e1001127.

Morimoto, S., M. Nakai, A. Ono, and Y. Kunimi. 2001. Late male-killing phenomenon found in a Japanese population of the oriental tea tortix, *Homona magnanima* (Lepidoptera: Tortricidae). *Heredity* 87:435–440.

Morin, P. J. 1999. *Community ecology.* New York: Wiley-Blackwell.

Morris, D. 2009. Apparent predation risk: Tests of habitat selection reveal unexpected effects of competition. *Evolutionary Ecology Research* 11:209–225.

Morton, E. G. 1990. Habitat segregation by sex in the hooded warbler: Experiments on proximate causation and discussion of its evolution. *American Naturalist* 135:319–333.

Mudrick, D., M. Hoosein, R. Hicks, and E. Townsend. 1994. Decomposition of leaf litter in an Appalachian forest: Effects of leaf species, aspect, slope position and time. *Forest Ecology and Management* 68:231–250.

Mueller, U. G., and N. Gerardo. 2002. Fungus-farming insects: Multiple origins and diverse evolutionary histories. *Proceedings of the National Academy of Sciences* 99:15247–15249.

Müller, C. B., and H. C. J. Godfray. 1997. Apparent competition between two aphid species. *Journal of Animal Ecology* 66:57–64.

Mumby, P. J., A. Hastings, and H. J. Edwards. 2007. Thresholds and the resilience of Caribbean coral reefs. *Nature* 450:98–101.

Murchie, E. H., and P. Horton. 1997. Acclimation of photosynthesis to irradiance and spectral quality in British plant species: Chlorophyll content, photosynthetic capacity, and habitat preference. *Plant Cell and Environment* 20:438–448.

Murdoch, W. M., C. J. Briggs, and R. M. Nisbet. 2003. *Consumer-resource dynamics.* Princeton, NJ: Princeton University Press.

Murphy, G. I. 1966. Population biology of the Pacific sardine (*Sardinops caerulea*). *Proceedings of the California Academy of Science, 4th Series* 34:1–84.

Murphy, S. D., and L. W. Aarssen. 1989. Pollen allelopathy among sympatric grassland species: *In vitro* evidence in *Phleum pratense* L. *New Phytologist* 112:295–305.

Murray, B. G. 1971. The ecological consequences of interspecific territorial behavior in birds. *Ecology* 52:414–423.

Myers, J. A., M. Vellend, S. Gardescu, and P. L. Marks. 2004. Seed dispersal by white-tailed deer: Implications for long-distance dispersal, invasion, and migration of plants in eastern North America. *Oecologia* 139:35–44.

Myers, N. 1988. Threatened biotas: "Hotspots" in tropical forests. *Environmentalist* 8:187–208.

Myers, N. 1991. The biodiversity challenge: Expanded hotspots analysis. *Environmentalist* 10:243–256.

Myers, N., R. A. Mittermeir, C. G. Mittemeir, G. D. B. de Fonseca, and J. Kent. 2000. Biodiversity hotspots for conservation priorities. *Nature* 403:853–858.

Naeem, S. 2002. Ecosystem consequences of biodiversity losses: The evolution of a paradigm. *Ecology* 83:1537–1552.

Naeem, S., and J. P. Wright. 2003. Disentangling biodiversity effects on ecosystem functioning: Deriving solutions to a seemingly insurmountable problem. *Ecology Letters* 6:567–579.

Naiman, R. J., J. M. Melillo, and J. E. Hobbie. 1986. Ecosystem alteration of boreal forest streams by beaver (*Castor canadensis*). *Ecology* 67:1254–1269.

National Wetland Working Group (NWWG). 1988. *Wetlands of Canada.* Ottawa: Environment Canada and Polyscience Publications.

Naylor, R. L., R. J. Goldburg, J. H. Primavera, N. Kautsky, M. C. Beveridge, J. Clay, C. Folke, J. Lubchenco, H. Mooney, and M. Troell. 2000. Effect of aquaculture on world fish supplies. *Nature* 405:1017–1024.

Naylor, R. L., R. W. Hardy, D. P. Bureau, A. Chiu, M. Elliott, A. P. Farrell, I. Forster, D. M. Gatlin, R. J. Goldburg, K. Hua, and B. D. Nichols. 2009. Feeding aquaculture in an era of finite resources. *Proceedings of the National Academy of Sciences* 106:15103–15110.

Nelson, E. H., C. E. Matthews, and J. A. Roenheim. 2004. Predators reduce prey population growth by inducing change in behavior. *Ecology* 85:1853–1858.

Neori, A., O. Holm-Hansen, B. G. Mitchell, and D. A. Kiefer. 1984. Photoadaptation in marine phytoplankton. *Plant Physiology* 76:518–524.

Newbold, J. D., R. V. O'Neill, J. W. Elwood, and W. Van Winkle. 1982. Nutrient spiraling in streams: Implications for nutrient and invertebrate activity. *American Naturalist* 20:628–652.

Newman, J. E. 1980. Climate change impacts on the growing season of the North American corn belt. *Biometeorology* 7:128–142.

Nicholls, R. J., and S. P. Leatherman. 1995. Global sea-level rise. In Strzepek, K., and Smith, J. B. (eds.). *As climate changes: International impacts and implications.* Cambridge: Cambridge University Press.

Niklas, K J., B. H. Tiffney, and A. H. Knoll. 1983. Patterns in land plant diversification. *Nature* 303:293–299.

Noble, I. R., and R. O. Slatyer. 1980. The use of vital attributes to predict successional changes in plant communities subject to recurrent disturbances. *Vegetatio* 43:5–21.

Nybakken, J. W., and M. D. Bertness. 2005. *Marine biology: An Ecological Approach.* 6th ed. Menlo Park, CA: Benjamin Cummings.

Odum, E. P. 1969. The strategy of ecosystem development. *Science* 164:262–270.

Odum, E. P., and G. W. Barrett. 2005. *Fundamentals of Ecology.* 5th ed. Belmont, CA: Thomson Brooks/Cole.

Odum, W. E., and M. A. Haywood. 1978. Decomposition of intertidal freshwater marsh plants. Pages 89–97 in Good, R. E., Whigham, D. F., and Simpson, R. L. (eds.). *Freshwater wetlands.* New York: Academic Press.

Oechel, W. C., and G. L. Vourlitis. 1996. Direct effects of elevated CO_2 on arctic plant and ecosystem function. Pages 163–176 in Koch, G., and Mooney, H. A. (eds.). *Carbon dioxide and terrestrial ecosystems.* San Diego: Academic Press.

Oksanen, T. A., E. Koskela, and T. Mappes. 2002. Hormonal manipulation of offspring number: Maternal effort and reproductive costs. *Evolution* 56:1530–1537.

Oli, M. K. 2004. The fast-slow continuum and mammalian life-history patterns: An empirical evaluation. *Basic and Applied Ecology* 5:449–463.

Olson, D. M., E. Dinerstein, E. D. Wikramanayake, N. D. Burgess, G. V. N. Powell, E. C. Underwood, J. A. D'Amico, I. Itoua, H. E. Strand, J. C. Morrison, C. J. Loucks, T. F. Allnutt, T. H. Ricketts, Y. Kura, J. F. Lamoreux, W. W. Wettengel, P. Hedao, and K. R. Kassem. 2001. Terrestrial ecoregions of the world: A new map of life on Earth. *BioScience* 51:933–938.

O'Neill, R. V. 1976. Ecosystem persistence and heterotrophic regulation. *Ecology* 57:1244-1253.

O'Neill, R. V., D. L. DeAngelis, J. B. Waide, and T. F. Allen. 1986. *A hierarchical concept of ecosystems.* Princeton, NJ: Princeton University Press.

Oosting, H. 1942. An ecological analysis of the plant communities of Piedmont, North Carolina. *American Midland Naturalist* 28:1–26.

Orians, G. H., J. Buckley, W. Clark, M. Gilpin, J. Lehman, R. May, G. Robilliard, and D. Simberloff. 1986. *Ecological knowledge and environmental problem-solving.* Washington, D.C.: National Academy Press.

Ottenbreit, K. A., and R. J. Staniforth. 1994. Crossability of naturalized and cultivated *Lythrum* taxa. *Canadian Journal of Botany* 72:337–341.

Owensby, C. E., J. M. Ham, A. Knapp, C. W. Rice, P. I. Coyne, and L. M. Auen. 1996. Ecosystem-level responses of tallgrass prairie to elevated CO_2. Pages 147–162 in Koch, G., and Mooney, H. A. (eds.). *Carbon dioxide and terrestrial ecosystems.* San Diego: Academic Press.

Owen-Smith, N. 1993. Comparative mortality rates of male and female kudus: The costs of sexual size dimorphism. *Journal of Animal Ecology* 62:428–440.

Paine, R. T. 1966. Food web complexity and species diversity. *American Naturalist* 100:65–75.

Paine, R. T. 1969. The *Pisaster-Tegula* interaction: Prey patches, predator food preference, and intertidal community structure. *Ecology* 50:950–961.

Pajunen, V. 1982. Replacement analysis of non-equilibrium competition between rock pool corixids (Hemiptera, Corixidae). *Oecologia* 52:153–155.

Palkovacs, E. P., and D. M. Post. 2009. Experimental evidence that phenotypic divergence in predators drives community divergence in prey. *Ecology* 90:300–305.

Palmer, J. D. 1990. The rhythmic lives of crabs. *BioScience* 40:352–358.

Park, T. 1954. Experimental studies of interspecies competition: 2. Temperature, humidity and competition in two species of *Trilobium. Physiological Zoology* 27:177–238.

Pastor, J., J. D. Aber, C. A. McClaugherty, and J. M. Melillo. 1984. Aboveground production and N and P cycling along a nitrogen mineralization gradient on Blackhawk Island, Wisconsin. *Ecology* 65:256–268.

Patel, N. H. 2006. Evolutionary biology: How to build a longer beak. *Nature* 442:515–516.

Pauly, D., and V. Christensen. 1995. Primary production required to sustain global fisheries. *Nature* 374:255–257.

Payette, S., N. Bhiry, A. Delwaide, and M. Simard. 2000. Origin of the lichen woodland at its southern range limit in eastern Canada: The catastrophic effect of insect defoliators and fire on the spruce-moss forest. *Canadian Journal of Forest Research* 30:288–305.

Payette, S., M.-J. Fortin, and I. Gamache. 2001. The subarctic-forest tundra: The structure of a biome in a changing climate. *BioScience* 51:709–718.

Pearcy, R. W. 1977. Acclimation of photosynthetic and respiratory CO_2 to growth temperature in *Atriplex lentiformis. Plant Physiology* 61:484–486.

Pearman, G. I., and P. Hyson. 1981. The annual variation of atmospheric CO_2 concentration observed in the northern hemisphere. *Journal of Geophysical Research* 86:C10.

Peat, H. J., and A. H. Fitter. 1994. Comparative analyses of ecological characteristics of British angiosperms. *Biological Reviews* 69:95–115.

Peltonen, A., and I. Hanski. 1991. Patterns of island occupancy explained by colonization and extinction rates in shrews. *Ecology* 72:1698–1708.

Peng, C., Z. Ma, X. Lei, Q. Zhu, H. Chen, W. Wang, S. Liu, W. Li, X. Fang, and X. Zhuo. 2011. A drought-induced pervasive increase in tree mortality across Canada's boreal forests. *Nature Climate Change* 1:467–471.

Pengelly, D., C. Cheng, and M. Campbell. 2005. *Influence of weather and air pollution on mortality in Toronto.* Toronto: Toronto Public Health Department.

Perera, G. A. D. 2001. The secondary forest situation in Sri Lanka: A review. *Journal of Tropical Forest Science* 13:768–785.

Pervez, A., and Omkar. 2005. Functional responses of coccinellid predators: An illustration of a logistic approach. *Journal of Insect Science* 5:5.

Pessarakli, M. (ed.). 2002. *Handbook of plant and crop physiology.* New York: Marcel Dekker.

Peterjohn, W. J., J. M. Melillo, P. A. Steudler, K. M. Newkirk, F. P. Bowles, and J. D. Aber. 1994. Responses of trace gas fluxes and N availability to experimentally elevated soil temperatures. *Ecological Applications* 4:617–625.

Peterson, C. R., A. R. Gibson, and M. E. Dorcas. 1993. Snake thermal ecology: The causes and consequences of body-temperature variation. Pages 241–314 in Seigel, R. A., and Collins, J. T. (eds.). *Snakes: Ecology and behavior.* New York: McGraw-Hill.

Peterson, T. S., A. Uesugi, and J. Lichter. 2005. Tree recruitment limitation by introduced snowshoe hares, *Lepus americanus*, on Kent Island, New Brunswick. *Canadian Field-Naturalist* 119:569–572.

Petraitis, P. S., and S. R. Dudgeon. 2004. Detection of alternative steady states in marine communities. *Journal of Experimental Marine Biology and Ecology* 300:343–371.

Petrie, M. 1994. Improved growth and survival of offspring of peacocks with more elaborate trains. *Nature* 371:598–599.

Petrinovich, L., and T. L. Patterson 1982. The White-crowned sparrow: Stability, recruitment, and population structure in the Nuttall subspecies (1975–1980). *Auk* 99:1–14.

Pettersson, I., A. Aruke, K. Lundstedt-Enkel, A. S. Mortensen, and C. Berg. 2006. Persistent sex-reversal and oviducal agenesis in adult *Xenopus (Silurana) tropicalis* frogs following larval exposure to the environmental pollutant ethynylestradiol. *Aquatic Toxicology* 79:356–365.

Pew Oceans Commission. 2003. *America's living oceans: Charting a course for sea change.* Arlington, VA: Pew Charitable Trusts.

Pfisterer, A. B., and B. Schmid. 2002. Density-dependent production can decrease the stability of ecosystem functioning. *Nature* 416:84–86.

Pfitsch, W. A., and R. W. Pearcy. 1989. Daily carbon gain by *Adenocaulon bicolor* (Asteraceae), a redwood forest understory herb, in relation to its light environment. *Oecologia* 80:465–470.

Phares, K. 1996. An unusual host-parasitoid relationship: The growth hormone-like factor from plerocercoids of spirometrid tapeworms. *International Journal for Parasitology.* 26:575–588.

Philip, L. J., S. W. Simard, and M. D. Jones. 2011. Pathways for belowground carbon transfer between paper birch and Douglas-fir seedlings. *Plant Ecology and Diversity* 3:221–233.

Pianka, E. R. 1972. *r* and *K* selection or *b* and *d* selection? *American Naturalist* 100:65–75.

Pianka, E. R. 1978. *Evolutionary ecology.* 3rd ed. New York: HarperCollins.

Pickett, S. T. A., and F. A. Bazzaz. 1978. Organization of an assemblage of early successional species on a soil moisture gradient. *Ecology* 59:1248–1255.

Pielou, E. C. 1984. *The interpretation of ecological data: A primer on classification and ordination.* New York: John Wiley and Sons.

Piene, H. 2003. Growth recovery in young, plantation white spruce following artificial defoliation and pruning. *Canadian Journal of Forest Research* 33:1267–1275.

Piene, H., and E. S. Eveleigh. 1996. Spruce budworm defoliation in young balsam fir: The "green" tree phenomenon. *The Canadian Entomologist* 128:1101–1107.

Pieper, S. J., V. Loewen, M. Gill, and J. F. Johnstone. 2011. Plant responses to natural and experimental variations in temperature in alpine tundra, southern Yukon, Canada. *Arctic, Antarctic, and Alpine Research* 43:442–456.

Pierotti, R. 1982. Habitat selection and its effect on reproductive output in the herring gull in Newfoundland. *Ecology* 63:854–868.

Pietrini, F., M. A. Iannelli, and A. Massacci. 2002. Anthocyanin accumulation in the illuminated surface of maize leaves enhances protection from photo-inhibitory risks at low temperature, without further limitation to photosynthesis. *Plant, Cell and Environment* 25:1251–1259.

Pimentel, D., R. Zuniga, and D. Morrison. 2005. Update on the environmental and economic costs associated with alien-invasive species in the United States. *Ecological Economics* 52:273–288.

Pimm, S. L. 1982. *Food webs.* New York: Chapman & Hall.

Pimm, S. L. 1991. *The balance of nature? Ecological issues in the conservation of species and communities.* Chicago: University of Chicago Press.

Pip, E. 1991. Cadmium, copper, and lead in soils and garden produce near a metal smelter at Flin Flon, Manitoba. *Bulletin of Environmental Contamination and Toxicology* 46:790–796.

Poorter, H., and M. Pérez-Soba. 2002. Plant growth at elevated CO_2. Pages 489–496 in Mooney, H. A., and Canadell, J. G. (eds.). *Encyclopedia of global change.* Vol. 2: *The Earth system: Biological and ecological dimensions of global environmental change.* Chichester, U.K.: John Wiley and Sons.

Porceddu, A., E. Albertini, G. Barcaccia, E. Falistocco, and M. M. Falcinelli. 2002. Linkage mappng in apomictic and sexual Kentucky bluegrass (*Poa pratensis* L.) genotypes using a two way pseudo-testcross strategy based on AFLP and SAMPL markers. *Theoretical and Applied Genetics* 104:273–280.

Post, D. M., M. L. Pace, and N. G. Hairston, Jr. 2000. Ecosystem size determines food-chain length in lakes. *Nature* 405:1047–1049.

Power, M. E., M. J. Matthews, and A. J. Stewart. 1985. Grazing minnows, piscivorous bass, and stream algae: Dynamics of a strong interaction. *Ecology* 66:1448–1456.

Prentice, I. C., G. Farquhar, M. Fasham, M. Goulden, M. Heimann, V. Jaramillo, H. Kheshgi, C. L. Quéré, R. Sholes, and D. Wallace. 2001. The carbon cycle and atmospheric carbon dioxide. Pages 183–237 in Houghton, J. T., and Yihui, D. (eds.). *Climate change 2001: The scientific basis.* New York: Cambridge University Press.

Prescott, C. E. 2005. Do rates of litter decomposition tell us anything we really need to know? *Forest Ecology and Management* 220:66–74.

Prescott, C. E. 2010. Litter decomposition: What controls it and how can we alter it to sequester more carbon in forest soils? *Biogeochemistry* 101:133–149.

Preston, F. W. 1962. The canonical distribution of commonness and rarity. *Ecology* 43:185–215.

Pretzlaw, T., C. Trudeau, M. M. Humphries, J. M. Lamontagne, and S. Boutin. 2006. Red squirrels (*Tamiasciurus hudsonicus*) feeding on spruce bark beetles (*Dendroctonus rufipennis*): Energetic and ecological implications. *Journal of Mammalogy* 87:909–914.

Price, M. V., N. W. Waser, and S. A. McDonald. 2000. Elevational distributions of kangaroo rats (genus *Dipodomys*): Long-term trends in a Mojave desert site. *American Midland Naturalist* 144:352–361.

Primack, R. B. 1998. *Essentials of conservation biology.* 2nd ed. Sunderland, MA: Sinauer.

Pulliam, R. 1988. Sources, sinks, and population regulation. *American Naturalist* 132:652–661.

Rajakumar, R., D. San Mauro, M. B. Dijkstra, M. H. Huang, D. E. Wheeler, F. Hiou-Tim, A. Khila, M. Cournoyea, and E. Abouheif. 2012. Ancestral developmental potential facilitates parallel evolution in ants. *Science* 335:79–82.

Ramsey, J., and D. W. Schemske. 2002. Neopolyploidy in flowering plants. *Annual Review Ecology and Systematics* 33:589–639.

Randolph, S. E. 1975. Patterns of the distribution of the tick *Ixodes trianguliceps birula* on its host. *Journal of Animal Ecology* 44:451–474.

Raven, P. H., R. F. Eichorn, and S. E. Eichorn. 2005. *Biology of Plants.* 7th ed. New York: W. H. Freeman.

Reeburgh, W. S. 1997. Figures summarizing the global cycle of biogeochemically important elements. *Bulletin of the Ecological Society of America* 78:260–267.

Reekie, E. G., and Bazzaz, F. A. (eds.). 2005. *Reproductive allocation in plants.* New York: Academic Press.

Reeve, H. K., and B. Hölldobler. 2007. The emergence of a superorganism through intergroup competition. *Proceedings of the National Academy of Sciences* 104:9736–9740.

Reich, P. B., D. A. Peterson, K. Wrage, and D. Wedin. 2001. Fire and vegetation effects on productivity and nitrogen cycling across a forest-grassland continuum. *Ecology* 82:1703–1719.

Reich, P. B., M. G. Tjoelker, M. B. Walters, D. Vanderklein, and C. Buschena. 1998. Close association of RGR, leaf and root morphology, seed mass and shade tolerance in seedlings of nine boreal tree species grown in high and low light. *Functional Ecology* 12:327–338.

Reich, P. B., M. B. Walters, and D. S. Ellsworth. 1992. Leaf lifespan in relation to leaf, plant and stand characteristics among diverse ecosystems. *Ecological Monographs* 62:365–392.

Reichman, O. J. 1987. *Konza prairie: Tallgrass natural history*. Lawrence, KS: University Press of Kansas.

Reimchen, T. 2001. Salmon nutrients, nitrogen isotopes, and coastal forests. *Ecoforestry* Fall:13–16.

Relyea, R. A. 2002a. Local population differences in phenotypic plasticity: Predator induced changes in wood frog tadpoles. *Ecological Monographs* 72:77–93.

Relyea, R. A. 2002b. Competitor-induced plasticity in tadpoles: Consequences, cues, and connections to predator-induced plasticity. *Ecological Monographs* 72:523–540.

Rex, M. A., C. T. Stuart, R. R. Hessler, J. A. Allen, H. A. Sanders, and G. D. Wilson. 1993. Global-scale latitudinal patterns of species diversity in the deep-sea benthos. *Nature* 365:636–639.

Reznick, D., F. H. Shaw, F. H. Rodd, and R. G. Shaw. 1997. Evaluation of the rate of evolution in natural populations of guppies (*Poecilia reticulata*). *Science* 275:1934–1937.

Ricciardi, A. 2004. Assessing species invasions as a cause of extinctions. *Trends in Ecology and Systematics* 19:619.

Richards, J., and M. M. Caldwell. 1987. Hydraulic lift: Substantial nocturnal water transport between soil layers by *Artemesia tridentata* roots. *Oecologia* 73:486–489.

Rizzo, B. 1990. The ecosystems of Canada in 2050: A scenario of change. *State of the Environment Reporting. Newsletter No. 5*. Ottawa: Environment Canada.

Robbins, C. S., D. K. Dawson, and B. A. Dowell. 1989. Habitat area requirements of breeding forest birds of the Middle Atlantic States. *Wildlife Monographs* 103:3–34.

Robertson, D. R. 1996. Interspecific competition controls abundance and habitat use of territorial Caribbean damselfishes. *Ecology* 77:885–899.

Robertson, G. P., M. A. Huston, F. C. Evans, and J. M. Tiedje. 1988. Spatial variability in a successional plant community: Patterns of nitrogen availability. *Ecology* 69: 1517–1524.

Robinson, H. G., R. B. Wielgus, H. S. Cooley, and S. W. Cooley. 2008. Sink populations in carnivore management: Cougar demography and immigration in a hunted population. *Ecological Applications* 18:1028–1037.

Robinson, W. D. 1999. Long-term changes in the avifauna of Barro Colorado Island, Panama, a tropical forest isolate. *Conservation Biology* 13:85–97.

Rodenhouse, N. I., T. S. Sillett, P. J. Doran, and R. T. Holmes. 2003. Multiple density-dependence mechanisms regulate a migratory bird population during the breeding season. *Proceedings of the Royal Society B* 270:2105–2110.

Rodríguez, A., and M. Delibes. 1992. Current range and status of the Iberian lynx *Felis pardina* Temminck 1824, in Spain. *Biological Conservation* 61:189–196.

Roff, D. A. 1990. The evolution of flightlessness in insects. *Ecological Monographs* 60:389–421.

Root, T. 1988. Energy constraints on avian distributions and abundances. *Ecology* 69:330–339.

Rosenzweig, C., and M. L. Parry. 1994. Potential impact of climate change on world food supply. *Nature* 367:133–138.

Rowe, S. 2006. *Earth alive: Essays on ecology*. Edmonton: NeWest Press.

Royama, T. 1984. Population dynamics of the spruce budworm *Choristoneura fumiferana*. *Ecological Monographs* 54:429–462.

Royama, T., W. E. MacKinnon, E. G. Kettela, N. E. Carter, and L. K. Hartling. 2005. Analysis of spruce budworm outbreaks in New Brunswick, Canada, since 1952. *Ecology* 86:1212–1224.

Ruiz-Ramoni, D., M. Múnoz-Romo, P. Ramoni-Perazzi, Y. Aranguren, and G. Fermin. 2011. Folivory in the giant fruit-eating bat *Artibeus amplus* (Phyllostomidae): A non-seasonal phenomenon. *Acta Chiropterilogica* 13:195–199.

Rysgaard, S., and R. N. Glud. 2004. Anaerobic N_2 production in Arctic sea ice. *Limnology and Oceanography* 49:86–94.

Salak-Johnson, J. L., and J. J. McGlone. 2007. Making sense of apparently conflicting data: Stress and immunity in swine and cattle. *Journal of Animal Science* 85:E81–E88.

Sandilands, A. P. 2005. *Birds of Ontario: Habitat requirements, limiting factors, and status*. Vancouver: University of British Columbia Press.

Sanginga, P. C., A. Waters-Bayer, S. Kaaria, J. Njuki, and C. Wettasinha (eds.). 2009. *Innovation Africa: Enriching farmers' livelihoods*. London: Earthscan.

Saugier, B., J. Roy, and H. A. Mooney (eds.). 2001. *Terrestrial global productivity: Past, present and future*. London: Academic Press.

Schaefer, J. A. 2003. Long-term range recession and the persistence of caribou in the taiga. *Conservation Biology* 17:1435–1439.

Schafer, K. V. R., R. Oren, D. S. Ellsworth, C.-T. Lai, J. D. Herrick, A. C. Finzi, D. D. Richter, and G. G. Katul. 2003. Exposure to an enriched CO_2 atmosphere alters carbon assimilation and allocation in a pine forest ecosystem. *Global Change Biology* 9:1378–1400.

Schall, J. J. 1983. Lizard malaria: Cost to vertebrate host's reproductive success. *Parasitology* 87:1–6.

Scheffer, M., and S. R. Carpenter. 2003. Catastrophic regime shifts in ecosystems: Linking theory to observation. *Trends in Ecology and Evolution* 18:648–656.

Scheffer, M., and E. H. van Nes. 2006. Self-organized similarity, the evolutionary emergence of groups of similar species. *Proceedings of the National Academy of Sciences* 103:6230–6235.

Scheffer, V. C. 1951. Rise and fall of a reindeer herd. *Scientific Monthly* 73:356–362.

Scheibling, R. E., and J.-S. Lauzon-Guay. 2010. Killer storms: North Atlantic hurricanes and disease outbreaks in sea urchins. *Limnology and Oceanography* 55:2331–2338.

Schindler, D. E., R. Hilborn, B. Chasco, C. P. Boatright, T. P. Quinn, L. A. Rogers, and M. S. Webster. 2010. Population diversity and the portfolio effect in an exploited species. *Nature* 465:609–612.

Schindler, D. W. 1974. Eutrophication and recovery in experimental lakes: Implications for lake management. *Science* 184:897–899.

Schindler, D. W., X. Augerot, E. Fleishman, N. J. Mantua, B. Riddell, M. Ruckelshaus, J. Seeb, and M. Webster. 2008. Climate change, ecosystem impacts, and management for Pacific salmon. *Fisheries* 33:502–506.

Schlesinger, W. H. 1997. *Biogeochemistry: An analysis of global change*. 2nd ed. London: Academic Press.

Schluter, D. 1982. Seed and patch selection by Galápagos ground finches: Relation to foraging efficiency and food supply. *Ecology* 63:1106–1120.

Schluter, D. 2000. *The ecology of adaptive radiation*. Oxford: Oxford University Press.

Schluter, D., T. D. Price, and P. R. Grant. 1985. Ecological character displacement in Darwin's finches. *Science* 227:1056–1059.

Schmidt-Neilsen, K. 1997. *Animal physiology: Adaptation and environment.* 5th ed. New York: Cambridge University Press.

Schmitz, O. J., and T. D. Nudds. 1994. Parasite-mediated competition in deer and moose: How strong is the effect of meningeal worm on moose? *Ecological Applications* 4:91–103.

Schoener, T. W. 1983. Field experiments on interspecific competition. *American Naturalist* 122:240–285.

Schofield, P. 2009. Geographic extent and chronology of the invasion of non-native lionfish (*Pterois volitans* [Linnaeus 1758] and *P. miles* [Bennett 1828]) in the Western North Atlantic and Caribbean Sea. *Aquatic Invasions* 4:473–479.

Scholtz, G., A. Braband, L. Tolley, A. Reimann, B. MIttmann, C. Lukhaup, F. Steuerwald, and G. Vogt. 2003. Parthenogenesis in an outsider crayfish. *Nature* 421:806.

Schonewald-Cox, C. M. 1983. Conclusions: Guidelines to management: A beginning attempt. Pages 414–445 in Schonewald-Cox, C. M., Chambers, S. M., MacBryde, B., and Thomas, L. (eds.). *Genetics and conservation: A reference for managing wild animal and plant populations.* Menlo Park, CA: Benjamin Cummings.

Schoonhoven, L. M., J. J. A. van Loon, and M. Dick. 2005. *Insect-plant biology.* Oxford: Oxford University Press.

Schulze, E. D., M. M. Caldwell, J. Canadell, H. A. Mooney, R. B. Jackson, D. Parson, R. Scholes, O. E. Sala, and P. Trimborn. 1998. Downward flux of water through roots (i.e., inverse hydraulic lift) in dry Kalahari sands. *Oecologia* 115:460–462.

Schuur, E. A. G. 2001. The effect of water on decomposition dynamics in mesic to wet Hawaiian montane forests. *Ecosystems* 4:259–273.

Schuur, E. A. G. 2003. Productivity and global climate revisited: The sensitivity of tropical forest growth to precipitation. *Ecology* 84:1165–1170.

Schuur, E. A. G., J. G. Vogel, K. G. Crummer, H. Lee, J. O. Sickman, and T. E. Osterkamp. 2009. The effect of permafrost thaw on old carbon release and net carbon exchange from tundra. *Nature* 459:556–559.

Scrimgeour, G. J., and J. M. Culp. 1994. Feeding while evading predators by a lotic mayfly: Linking short-term foraging behaviours to long-term fitness consequences. *Oecologia* 100:128–134.

Sealy, S. G. 1995. Burial of cowbird eggs by parasitized yellow warblers: An empirical and experimental study. *Animal Behavior* 49:877–899.

Seitz, R. D., R. N. Lipcius, A. H. Hines, and D. B. Eggleston. 2001. Density-dependent predation, habitat variation, and the persistence of marine bivalve prey. *Ecology* 82:2435–2451.

Shaffer, M. L. 1981. Minimum population sizes for species conservation. *BioScience* 31:131–134.

Shannon, C. E., and W. Wiener. 1963. *The mathematical theory of communications.* Urbana, IL: University of Illinois Press.

Sheriff, M. J., C. J. Krebs, and R. Boonstra. 2009. The sensitive hare: Sublethal effects of predator stress on reproduction in snowshoe hares. *Journal of Animal Ecology* 78:1249–1258.

Sheriff, M. J., C. J. Krebs, and R. Boonstra. 2010. The ghosts of predators past: Population cycles and the role of maternal programming under fluctuating predation risk. *Ecology* 91:2983–2994.

Shier, C. 2009. Alberta's grizzly on the front lines. *Alberta Wild* Spring/Summer 2009:1.

Shurin, J. B., E. T. Borer, E. W. Seabloom, K. Anderson, C. A. Blanchette, B. Broitman, S. D. Cooper, and B. S. Halpern. 2002. A cross-ecosystem comparison of the strength of trophic cascades. *Ecology Letters* 5:785–791.

Shurin, J. B., D. S. Gruner, and H. Hillebrand. 2006. All wet or dried up? Real differences between aquatic and terrestrial food webs. *Proceedings of the Royal Society B* 273:1–9.

Sillett, T. S., and R. T. Holmes 2002. Variation in survivorship of a migratory songbird throughout its annual cycle. *Journal of Animal Ecology* 71:296–308.

Sillett, T. S., R. T. Holmes, and T. W. Sherry. 2000. Impacts of a global climate cycle on population dynamics of a migratory songbird. *Science* 288:2040–2042.

Silliman, B. R., and M. D. Bertness. 2002. A trophic cascade regulates salt marsh primary production. *Proceedings of the National Academy of Sciences* 99:10500–10505.

Silvertown, J. 1980. The dynamics of a grassland ecosystem: Botanical equilibrium in the Park Grass Experiment. *Journal of Applied Ecology* 17:491–504.

Simard, S. W., and D. M. Durall. 2004. Mycorrhizal networks: A review of their extent, function, and importance. *Canadian Journal of Botany* 82:1140–1165.

Simberloff, D. S. 1974. Equilibrium theory of island biogeography and ecology. *Annual Review of Ecology and Systematics* 5:161–182.

Simon, N. P., A. W. Diamond, and F. E. Schwab. 2003. Do northern forest bird communities show more ecological plasticity than southern forest bird communities in eastern Canada? *Écoscience* 10:289–296.

Simpson, E. H. 1949. Measurement of diversity. *Nature* 163:688.

Sims, R. A., W. D. Towill, K. A. Baldwin, and G. M. Wickware. 1989. *Forest ecosystem classification for northwestern Ontario.* Ottawa: Ministry of Natural Resources.

Sirois, L. 2000. Spatiotemporal variation in black spruce cone and seed crops along a boreal forest-tree line transect. *Canadian Journal of Forest Research* 30:900–909.

Sisk, T. D., N. M. Haddad, and P. R. Ehrlich. 1997. Bird assemblages in patchy woodlands: Modelling the effects of edge and matrix habitats. *Ecological Applications* 7:1170–1180.

Smit, B., L. Ludlow, and M. Brklacich. 1988. Implications of a global climatic warming for agriculture: A review and appraisal. *Journal of Environmental Quality* 17:519–527.

Smith, R. L. 1963. Some ecological notes on the grasshopper sparrow. *Wilson Bulletin* 75:159–165.

Smith, R. L., and T. M. Smith. 2001. *Ecology and field biology.* 6th ed. Boston: Benjamin Cummings.

Smith, S. M. 1987. Responses of floaters to removal experiments on wintering chickadees. *Behavioral Ecology and Sociobiology* 20:363–367.

Smith, T., D. A. Wharton, and C. J. Marshall. 2008. Cold tolerance of an Antarctic nematode that survives intracellular freezing: Comparisons with other nematode species. *Journal of Comparative Physiology* 178:93–100.

Smith, T. M., and M. Huston. 1989. A theory of the spatial and temporal dynamics of plant communities. *Vegetatio* 83:49–69.

Smith, T. M., R. Leemans, and H. H. Shugart. 1992. Sensitivity of terrestrial carbon storage to CO_2-induced climate change: Comparison of four scenarios based on general circulation models. *Climatic Change* 21:367–384.

Smith, T. M., and R. L. Smith. 2008. *Elements of ecology.* 7th ed. Boston: Benjamin Cummings.

Snow, A. A., S. E. Travis, R. Wildová, T. Fér, P. M. Sweeney, J. E. Marburger, S. Windells, B. Kubátová, D. E. Goldberg, and E. Mutegl. 2010. Species-specific SSR alleles for studies of hybrid cattails (*Typha latifolia* X *T. angustifolia*; Typhaceae) in North America. *American Journal of Botany* 97:2061–2067.

Solomon, M. E. 1949. The natural control of animal populations. *Journal of Animal Ecology* 19:1–35.

Sousa, W. P. 1979. Disturbance in marine intertidal boulder fields: The nonequilibrium maintenance of species diversity. *Ecology* 60:1225–1239.

Sparks, T. H., and J. M. Potts. 2004. *Late summer grass production. Agriculture and Forestry (Indicators of Climate Change in the U.K.).* www.ecn.ac.uk/iccuk/indicators/24.htm. Downloaded 5 December 2011.

Staaf, H., and B. Berg. 1982. Accumulation and release of plant nutrients in decomposing Scots pine needle litter. II. Long-term decomposition in a Scots pine forest. *Canadian Journal of Botany* 60:1516–1568.

Stachowicz, J. 2001. Mutualism, facilitation, and the structure of ecological communities. *BioScience* 51:235–246.

Stachowicz, J., and M. Hay. 1996. Facultative mutualism between an herbivorous crab and a coralline alga: Advantages of eating noxious seaweeds. *Oecologia* 105:377–387.

Stachowicz, J., and M. Hay. 1999. Mutualism and coral persistence: The role of herbivore resistance to algal chemical defense. *Ecology* 80:2085–2101.

Stamp, N. 2003. Out of the quagmire of plant defense hypotheses. *Quarterly Review of Biology* 78:23–55.

Statistics Canada. 2006. *Canada's farm population: Agriculture-population linkage data for the 2006 census.* Ottawa: Statistics Canada.

Stenseth, N. C. 1983. Causes and consequences of dispersal in small mammals. Pages 63–101 in Swingland, I. R., and Greenwood, P. J. (eds.). *The ecology of animal movement.* Oxford: Oxford University Press.

Stevens, C. D., and I. E. Hume. 1998. Contributions of microbes in vertebrate gastrointestinal tracts to production and conservation of nutrients. *Physiological Reviews* 78:393–427.

Stolk, C., and R. Stouthamer. 1996. Influence of a cytoplasmic incompatibility-inducing *Wolbachia* on the fitness of the parasitoid wasp *Nasonia vitripennis. Entomologia Experimentalis et Applicata* 7:33–37.

Stow, C. A., S. R. Carpenter, and K. L. Cottingham. 1995. Resource vs. ratio-dependent consumer-resource models: A Bayesian perspective. *Ecology* 76:1986–1990.

Strathdee, A. T., J. S. Bale, W. C. Block, N. R. Webb, I. D. Hodkinson, and S. J. Coulson. 1993. Extreme adaptive life-cycle in a high arctic aphid, *Acyrthosiphon svalbardicum. Ecological Entomology* 18:254–258.

Stutchbury, B. J. 1998. Extra-pair mating effort of male hooded warblers (*Wilsonia citrina*). *Animal Behaviour* 55:553–561.

Suding, K., and D. Goldberg. 2001. Do disturbances alter competitive hierarchies? Mechanisms of change following gap formation. *Ecology* 82:2133–2149.

Suhonen, J. 1993. Predation risk influences the use of foraging sites by tits. *Ecology* 74:1197–1203.

Sultan, S. E. 2010. Plant developmental responses to the environment: Eco-devo insights. *Current Opinion in Plant Biology* 13:96–101.

Sultan, S. E., and F. A. Bazzaz. 1993. Phenotypic plasticity in *Polygonum persicaria.* II. Norms of reaction to soil moisture and the maintenance of genetic diversity. *Evolution* 47:1032–1049.

Sutcliffe, O., C. Thomas, T. Yates, and J. Greatorex-Davies. 1997. Correlated extinctions, colonizations and population fluctuations in a highly connected ringlet butterfly metapopulation. *Oecologia* 109:235–241.

Swift, M. J., O. W. Heal, and J. M. Anderson. 1979. *Decomposition in terrestrial ecosystems.* Berkeley: University of California Press.

Taillon, J., and S. D. Côté. 2007. Social rank and winter forage quality affect aggressiveness in white-tailed deer fawns. *Animal Behaviour* 74:265–275.

Talbot, L. M., S. M. Turton, and A. W. Graham. 2003. Trampling resistance of tropical rainforest soils and vegetation in the wet tropics of north east Australia. *Journal of Environmental Management* 69:63–69.

Tallamy, D. W., and R. Darke. 2009. *Bringing nature home: How you can sustain wildlife with native plants.* Portland, OR: Timber Press.

Tamburi, N. E., and P. R. Martin. 2009. Reaction norms of size and age at maturity of *Pomacea canaliculata* (Gastropoda: Ampullariidae) under a gradient of food deprivation. *Journal of Molluscan Studies* 75:19–26.

Tansley, A. 1935. The use and abuse of vegetational terms and concepts. *Ecology* 16:284–307.

Tarnocai, C., and B. Lacelle. 2001. *Wetlands of Canada database.* Ottawa: Agriculture and Agri-Food Canada, Research Branch.

Tarroux, E., and A. DesRochers. 2011. Effect of natural root grafting on growth response of jack pine (*Pinus banksiana*; Pinaceae). *American Journal of Botany* 98:967–964.

Tattersall, G. J., P. C. Eterovick, and D. V. deAndrade. 2006. Skin colour and body temperature changes in basking *Bokermannohyla alvarengai* (Bokermann 1956). *Journal of Experimental Biology* 209:1185–1196.

Teeri, J. A., and L. G. Stowe. 1976. Climatic patterns and the distribution of C_4 grasses in North America. *Oecologia* 23:1–12.

Tewksbury, J. J., L. Garner, S. Garner, J. D. Lloyd, V. Saab, and T. E. Martin. 2006. Tests of landscape influence: Nest predation and brood parasitism in fragmented ecosystems. *Ecology* 87:759–768.

Thomas, C., and T. Jones. 1993. Partial recovery of a skipper butterfly (*Hesperia comma*) from population refuges: Lessons for conservation in a fragmented landscape. *Journal of Animal Ecology* 62:472–481.

Thomas, C., M. Singer, and D. Boughton. 1996. Catastrophic extinction of population sources in a butterfly metapopulation. *American Naturalist* 148:957–975.

Thomas, P. A., and J. Packham 2007. *Ecology of forests and woodlands: Description, dynamics and diversity.* Cambridge: Cambridge University Press.

Thompson, I. D., M. D. Flannigan, B. M. Wotton, and R. Suffling. 1998. The effects of climate change on landscape diversity: An example in Ontario forests. *Environmental Monitoring and Assessment* 49:213–233.

Thompson, R., and C. Townsend. 2006. A truce with neutral theory: Local deterministic factors, species traits and dispersal limitation together determine patterns of diversity in stream invertebrates. *Journal of Animal Ecology* 75:476–484.

Thomson, A. J. 2009. Climate indices and mountain pine beetle-killing temperatures. *Forestry Chronicle* 85:105–109.

Tilman, D. 1985. The resource-ratio hypothesis of plant succession. *American Naturalist* 125:827–852.

Tilman, D. 1988. *Plant strategies*. Princeton, NJ: Princeton University Press.

Tilman, D. 1996. Biodiversity: Population versus ecosystem stability. *Ecology* 77:350–363.

Tilman, D. 2004. Niche tradeoffs, neutrality, and community structure: A stochastic theory of resource competition, invasion, and community assembly. *Proceedings of the National Academy of Sciences* 101:10854–10861.

Tilman, D., M. Mattson, and S. Langer. 1981. Competition and nutrient kinetics along a temperature gradient: An experimental test of a mechanistic approach to niche theory. *Limnology and Oceanography* 26:1020–1033.

Tilman, D., P. B. Reich, J. Knops, D. Wedlin, T. Miekle, and C. Lehman. 2001. Diversity and productivity in a long-term grassland experiment. *Science* 294:843–845.

Tinbergen, L. 1960. The natural control of insects in pinewoods. I. Factors influencing the intensity of predation by songbirds. *Archives Néerlandaises de Zoologie* 13:265–336.

Tittensor, D. P., C. Mora, W. Jetz, H. K. Lotze, D. Ricard, E. Vanden Berghe, and B. Worm. 2010. Global patterns and predictors of marine biodiversity across taxa. *Nature* 466:1098–1101.

Tomback, D. F. 1982. Dispersal of white-barked pine seeds by Clark's nutcracker: A mutualism hypothesis. *Journal of Animal Ecology* 51:451–457.

Tompa, F. S. 1964. Factors determing the number of song sparrows, *Melospiza melodia* (Wilson) on Mandarte Island, B.C., Canada. *Acta Zoologica Fennica* 109:3–73.

Toms, J. D., S. J. Hannon, and F. K. Schmiegelow. 2002. Quantifying temporal variation in abundance of birds. Pages 87–94 in Chamberlain, D. E., and Wilson, A. M. (eds.). *Avian landscape ecology: Pure and applied issues in the large-scale ecology of birds*. U.K.: IALE.

Tracy, C. R. 1976. A model of the dynamic exchanges of water and energy between a terrestrial amphibian and its environment. *Ecological Monographs* 46:293–296.

Traill, L. W., J. A. Bradshaw, and B. W. Brook. 2007. Minimum viable population size: A meta-analysis of 30 years of published estimates. *Biological Conservation* 139:159–166

Trombulak, S. C., and C. A. Frissell. 2000. Review of ecological effects of roads on terrestrial and aquatic communities. *Conservation Biology* 14:18–30.

Trzcinski, M. K., L. Fahrig, and G. Merriam. 1999. Independent effects of forest cover and fragmentation on the distribution of forest breeding birds. *Ecological Applications* 9:586–593.

Turlings, T. C. J., J. H. Loughrin, P. J. McCall, U. S. R. Rose, W. J. Lewis, and J. H. Tomlinson. 1995. How caterpillar-damaged plants protect themselves by attracting parasitic wasps. *Proceedings of the National Academy of Sciences* 92:4169–4174.

Turner, M. G., W. H. Romme, R. H. Gardner, R. V. O'Neill, and T. K. Kratz. 1993. A revised concept of landscape equilibrium: Disturbance and stability in scaled landscapes. *Landscape Ecology* 8:213–227.

Tweddle, D., D. H. Eccles, C. B. Frith, G. Fryer, P. B. Jackson, D. S. Lewis, and R. H. Lowe-McConnell. 1998. Cichlid spawning structures—bowers or nests? *Environmental Biology of Fishes* 51:107–109.

Tyler, G., A.-M. Balsberg Pahlsson, G. Bentsson, E. Baath, and L. Tranvik. 1989. Heavy-metal ecology of terrestrial plants, microorganisms, and invertebrates. *Water, Air, and Soil Pollution* 47:189–216

United Nations. 2011. *World population prospects, the 2010 revision*. New York: United Nations.

Urban, M. C., L. De Meester, M. Vellend, R. Sotks, and J. Vanoverbeke. 2012. A crucial step toward realism: Responses to climate change from an evolving metacommunity perspective. *Evolutionary Applications*. doi:10.1111/j.1752-4571.2011.00208.x.

U.S. Fish and Wildlife Service. 2001. *Trumpeter swan: Population status, 2000*. Laurel, MA: USFWS.

Valiela, I. 1984. *Marine ecological processes*. New York: Springer-Verlag.

Valladares, F., and U. Niinemets. 2008. Shade tolerance, a key plant feature of complex nature and consequences. *Annual Review of Ecology, Evolution and Systematics* 39:237–257.

Van der Valk, A. G. 1981. Succession in wetlands: A Gleasonian approach. *Ecology* 62:688–696.

Van der Valk, A. G., and C. B. Davis. 1978. The role of seed banks in the vegetation dynamics of prairie marshes. *Ecology* 59:322–335.

van Oort, H., B. N. McLellan, and R. Serrouya. 2010. Fragmentation, dispersal and metapopulation function in remnant populations of endangered mountain caribou. *Animal Conservation* 14:215–224.

Van Schaik, C. P., and R. I. M. Dunbar. 1990. The evolution of monogamy in large primates: A new hypothesis. *Behaviour* 115:30–62.

Van Valen, L. 1973. A new evolutionary law. *Evolutionary Theory* 1:1–30.

Vargas, H., C. Lougheed, and H. Snell. 2005. Population size and trends of the Galápagos penguin, *Spheniscus mendiculus*. *Ibis* 147:367–374.

Vellend, M. 2010. Conceptual synthesis in community ecology. *Quarterly Review of Biology* 85:183–206.

Venner, S., P.-F. Pélisson, M.-C. Bel-Venner, F. Débias, E. Rajon, and F. Menu. 2011. Coexistence of insect species competing for a pulsed resource: Toward a unified theory of biodiversity in fluctuating environments. *PLoS ONE* 6(3):e18039.

Vitousek, P. M. 1992. Global environmental change: An introduction. *Annual Review of Ecology and Systematics* 23:1–14.

Vogt, K. A., M. Grove, H. Asbjornsen, K. B. Maxwell, D. J. Vogt, R. Sigurdardottir, B. C. Larson, L. Schibli, and M. Dove. 2002. Linking ecological and social scales for natural resource management. Pages 143–175 in Liu, J., and Taylor W. M. (eds.). *Integrating landscape ecology into natural resource management*. Cambridge: Cambridge University Press.

Von Bertalanffy, L. 1969. *General system theory: Foundations, development, applications.* New York: Braziller.

Vors, L. S., and M. S. Boyce. 2009. Global declines of caribou and reindeer. *Global Change* 15:2626–2633.

Wackernagel, M., and W. E. Rees. 1996. *Our ecological footprint: Reducing human impact on the Earth.* Gabriola Island, BC: New Society Publishers.

Waddington, C. H. 1975. *A catastrophe theory of evolution: The evolution of an evolutionist.* Ithaca, NY: Cornell University Press.

Wajnberg E., X. Fauvergue, and O. Pons. 2000. Patch leaving decision rules and the marginal value theorem: An experimental analysis and simulation model. *Behavioral Ecology* 11:577–586.

Wakowsky, W., and J. L. Riley. 1992. A survey of the prairies and savannas of southern Ontario. 13th North American Prairie Conference.

Walker, B. H. 1992. Biodiversity and ecological redundancy. *Conservation Biology* 6:18–23.

Walker, B., C. S. Holling, S. R. Carpenter, and A. Kinzig. 2004. Resilience, adaptability and transformability in social-ecological systems. *Ecology and Society* 9:5.

Walker, L. R., D. A. Wardle, R. D. Bardgett, and B. D. Clarkson. 2010. The use of chronosequences in studies of ecological succession and soil development. *Journal of Ecology* 98:725–736.

Walters, R. G. 2005. Towards an understanding of photosynthetic acclimation. *Journal of Experimental Botany* 56:435–447.

Wang, L.-W., A. M. Showalter, and I. A. Unger. 2005. Effects of intraspecific competition on growth and photosynthesis of *Atriplex prostrata*. *Aquatic Botany* 83:187–192.

Waters, I., and J. M. Shay. 1990. A field study of the morphometric response of *Typha glauca* shoots to a water depth gradient. *Canadian Journal of Botany* 68:2339–2343.

Watkinson, A. R., and A. Davy. 1985. Population biology of salt marsh and dune annuals. *Vegetatio* 62:487–497.

Watts, P. C., K. R. Buley, S. Sanderson, W. Boardman, C. Ciofi, and R. Gibson. 2006. Parthenogenesis in Komodo dragons. *Nature* 444:1021–1022.

Wauters, L., and A. A. Dohondt. 1989. Body weight, longevity, and reproductive success in red squirrels (*Sciurus vulgaris*). *Journal of Animal Ecology* 58:637–651.

Webb, N. R., and L. E. Haskins. 1980. An ecological survey of heathlands in the Poole Basin, Dorset, England in 1978. *Biological Conservation* 17:281–296.

Webster, J. R., and B. C. Patten. 1979. Effects of watershed perturbation on stream potassium and calcium dynamics. *Ecological Monographs* 49:51–72.

Weeks, P. 2000. Red-billed oxpeckers: Vampires or tickbirds? *Behavioral Ecology* 11:154–160.

Werner, P. A., and W. J. Platt. 1976. Ecological relationships of co-occurring goldenrods (*Solidago*: Compositae). *American Naturalist* 110:959–971.

Wetzel, R. G. 1975. *Limnology.* Philadelphia: Saunders.

Whitcomb, R. E., J. F. Lynch, M. K. Klimkiewicz, C. S. Robbins, B. L. Whitcomb, and D. Bystrak. 1981. Effects of forest fragmentation on avifauna of the eastern deciduous forest. Pages 125–205 in R. L. Burgess and Sharpe, D. M. (eds.). *Forest island dynamics in man-dominated landscapes.* New York: Springer-Verlag.

Whitham, T. G. 1980. The theory of habitat selection: Examined and extended using *Pemphigus* aphids. *American Naturalist* 115:449–466.

Whitkamp, M., and M. L. Frank. 1969. Evolution of carbon dioxide from litter, humus and subsoil of a pine stand. *Pedobiologia* 9:358–365.

Whitmore, T. C., R. Peralta, and K. Brown. 1986. Total species count in a Costa Rica tropical rain forest. *Journal of Tropical Ecology* 1:375–378.

Whitney, C. G., M. M. Farley, J. Hadler, I. H. Harrison, C. Lexau, and A. Reingold. 2000. Increasing prevalence of multidrug-resistant *Streptococcus pneumoniae* in the United States. *New England Journal of Medicine* 343:1917–1924.

Whittaker, R. H. 1953. A consideration of climax theory: The climax as a population and pattern. *Ecological Monographs* 23:41–78.

Whittaker, R. H. 1956. Vegetation of the Great Smoky Mountains. *Ecological Monographs* 26:1–80.

Whittaker, R. H. 1960. Vegetation of the Siskiyou Mountains, Oregon and California. *Ecological Monographs* 23:41–78.

Whittaker, R. H. 1975. *Communities and ecosystems.* New York: Macmillan.

Whittaker, R. H. 1977. Evolution of species diversity in land communities. *Evolutionary Biology* 10:1–67.

Whittaker, R. H., and G. E. Likens. 1973. Carbon in the biota. In G. M. Woodwell and Pecan, E. V. (eds.). *Carbon and the biosphere conference 72501.* Springfield, VA: National Technical Information Service.

Whittaker, R. H., and G. M. Woodwell. 1968. Dimension and production relations of trees and shrubs in the Brookhaven forest, New York. *Ecology* 56:1–25.

Whittaker, R. H., and G. M. Woodwell. 1969. Structure, production, and diversity of the oak-pine forest at Brookhaven, New York. *Journal of Ecology* 57:155–174.

Wiens, J. J., D. D. Ackerly, A. P. Allen, B. L. Anacker, L. B. Buckley, H. V. Cornell, E. I. Damschen, T. J. Davies, J.-A. Grytnes, S. P. Harrison, B. A. Hawkins, R. D. Holt, C. M. McCain and P. R. Stephens. 2010. Niche conservatism as an emerging principle in ecology and conservation biology. *Ecology Letters* 13:1310–1324.

Wigginton, J. D., and F. S. Dobson. 1999. Environmental influences on geographic variation in body size of western bobcats. *Canadian Journal of Zoology* 77:802–813.

Williams, K., K. G. Smith, and F. M. Stevens. 1993. Emergence of 13-year periodical cicada (Cicadidae: *Magicicada*): Phenology, mortality, and predator satiation. *Ecology* 74:1143–1152.

Willson, M. F., T. L. De Santo, and K. E. Sieving. 2003. Red squirrels and predation risk to bird nests in northern forests. *Canadian Journal of Zoology* 81:202–207.

Wilson, D. 1992. Complex interactions in metacommunities, with implications for biodiversity and higher levels of selection. *Ecology* 73:1984–2000.

Wilson, E. O. 1999. *The diversity of life.* New York: Norton.

Wilson, J. B., and A. D. Q. Agnew. 1992. Positive feedback switches in plant communities. *Advances in Ecological Research* 23:263–336.

Wilson, P. J., S. K. Grewal, F. F. Mallory, and B. N. White. 2009. Genetic characterization of hybrid wolves across Ontario. *Journal of Heredity* 100:S80–S89.

Winemiller, K. O., and K. A. Rose. 1992. Patterns of life history diversification in North American fishes: Implications for population regulation. *Canadian Journal of Fisheries and Aquatic Sciences* 49:2196–2218.

Wise, D. H. 1995. *Spiders in ecological webs*. Cambridge: Cambridge University Press.

Wolf, M., G. Sander van Doorn, O. Leimar, and F. J. Weissing. 2007. Life-history trade-offs favour the evolution of animal personalities. *Nature* 447:581–584.

Woodard, T. N., R. J. Gutierrez, and W. H. Rutherford. 1974. Bighorn lamb production, survival and mortality in south-central Colorado. *Journal of Wildlife Management* 28:381–391.

Woodward, I., and T. Smith. 1993. Predictions and measurements of the maximum photosynthetic rate, A_{max}, at a global scale. Pages 491–508 in Schultz, E. D., and Caldwell, M. M. (eds.). *Ecophysiology of photosynthesis*. Vol. 100. Berlin: Springer-Verlag.

World Resources Institute. 2001. *World resources 2000–2001: People and ecosystems: The fraying web of life*. New York: United Nations Environment Programme.

Wright, J. P., C. G. Jones, and A. S. Flecker. 2002. An ecosystem engineer, the beaver, increases species richness at the landscape scale. *Oecologia* 132:96–101.

Yamazaki, J., S. Takahisa, M. Emiko, and K. Yasumara. 2005. The stoichiometry and antenna size of the two photosystems in marine green algae, *Bryopsis maxima* and *Ulva pertusa*, in relation to the light environment of their natural habitat. *Journal of Experimental Botany* 56:1517–1523.

Yao-Hua, L., and J. A. Rosenheim. 2011. Effects of combining an intraguild predator with a cannibalistic intermediate predator on a species-level trophic cascade. *Ecology* 92:333–341.

Yashin, A. I., A. S. Begun, S. I. Boiko, S. V. Ukraintseva, and J. Oeppen. 2002. New age patterns of survival improvement in Sweden: Do they characterize changes in individual aging? *Mechanisms of Aging and Development* 123:637–647.

Yoda, K., T. Kira, H. Ogawa, and K. Hozumi. 1963. Self-thinning in overcrowded pure stands under cultivated and natural conditions. *Journal of Biology* 14:107–129

Yoshino, M., T. Horie, H. Seino, H. Tsujii, T. Uchijima, and Z. Uchijima. 1988. The effects of climate variations on agriculture in Japan. In Parry, M., Carter, T. R., and Konijn, N. T. (eds.). *The impacts of climate variation on agriculture*. Vol 1: *Assessments in cool temperate and cold regions*. Dordrecht, Netherlands: Kluwer.

Yotova, V., D. Labuda, E. Zietkiewicz, D. Gehl, L. Lovell, J. F. Lefebvre, S. Bourgeois, E. Lemieux-Blanchard, M. Labuda, H. Vézina, L. Houde, M. Tremblay, B. Toupance, E. Heyer, T. J. Hudson, and C. Laberge. 2005. Anatomy of a founder effect: Myotonic dystrophy in Northeastern Quebec. *Human Genetics* 117:177–187.

Youngblood, A., and B. Titus. 1996. Clearcutting—a regeneration method in the boreal forest. *Forestry Chronicle* 72:31–36.

Zavaleta, E. S., B. D. Thomas, N. R. Chiariello, G. P. Asner, and C. B. Field. 2003a. Plants reverse warming effect on ecosystem water balance. *Proceedings of the National Academy of Sciences USA* 100:9892–9893.

Zavaleta, E. S., M. R. Shaw, N. R. Chiariello, H. A. Mooney, and C. B. Field. 2003b. Additive effects of simulated climate changes, elevated CO_2, and nitrogen deposition on grassland diversity. *Proceedings of the National Academy of Sciences USA* 100:7650–7654.

Zhao, T., P. Krokene, J. Hu, E. Christiansen, N. Björklund, B. Långström, H. Solheim, and A.-K. Borg-Karlson. 2011. Induced terpene accumulation in Norway spruce inhibits bark beetle colonization in a dose-dependent manner. *PLoS ONE* 6(10):e26649.

Glossary

A horizon (topsoil). Surface stratum of mineral soil, characterized by maximum biological activity, accumulation of organic matter, and loss of iron and aluminum oxides and clays; compare *B horizon*.

Abiotic. Nonliving; abiotic components of ecosystems include soil, water, air, light; compare *biotic*.

Abundance. Total number of individuals in a population in a given area; compare *population density*.

Abundance index. Estimate of relative population size based on counts of animal signs, calls, or number observed; compare *census* and *mark–recapture*.

Acclimation. Reversible phenotypic plasticity in response to changing environmental conditions; also *acclimatization*; compare *developmental plasticity*.

Acid precipitation. Atmospheric fallout with pH < 5.6, resulting from water vapour combining with hydrogen sulphide and nitrous oxide vapours released by burning fossil fuels; called *wetfall* or *wet deposition* if associated with precipitation and *dryfall* or *dry deposition* if associated with airborne particles and aerosols.

Acidity. Concentration of H^+ ions in a solution; acid solutions have pH ($-\log [H^+]$) less than 7; alkaline solutions have pH greater than 7.

Actual evapotranspiration (AET). Total moisture lost from vegetation and land surfaces by transpiration and evaporation combined; *potential evapotranspiration* (PET) is total moisture that would be evaporated and/or transpired under optimal conditions of soil moisture and plant cover.

Adaptation. Heritable trait that improves an organism's ability to survive and reproduce in prevailing environmental conditions; results from natural selection.

Adaptive radiation. Evolution from a common ancestor of different species adapted to distinct features of the environment, such as food items or habitats.

Adiabatic lapse rate. Rate of cooling with elevation where cooling is due to expansion of rising air mass rather than to heat loss; rate of cooling that is due to heat loss is the *environmental lapse rate*.

Adipose fat. Highly vascularized brown fat used to generate heat in hibernating homeotherms.

Adventitious roots. Roots arising from non-root structures, often from stem base; facilitate access to O_2 in waterlogged soils.

Ae horizon. Lighter-coloured soil stratum at border between A horizon and B horizon, from which eluviation and leaching have removed particles and solutes.

Aerenchyma. Specialized air-storage tissue in submerged tissues of flood-tolerant plants; facilitates O_2 transport.

Aerobic (cellular) respiration. Oxidative breakdown of sugars in the presence of O_2 via mitochondrial electron transport; compare *anaerobic respiration* and *fermentation*.

Aerosols. Small particles suspended in atmosphere that absorb solar radiation and scatter it back to space, reducing the amount of radiation reaching Earth.

Age- and sex-specific birthrate. Mean number of females produced per female per unit time in a given age class; compare *per capita birthrate*.

Age-specific mortality rate. Mean number of deaths occurring in a time interval in a given age class, expressed as the proportion alive at the outset of the interval; can also be sex-specific; compare *per capita death rate*.

Age structure. Number or proportion of individuals in various age classes in a population; also *age distribution*; see *stationary age structure* and *stable age structure*.

Aggregative response. Movement of predators into areas of high prey density; one component of a numerical response.

Aggressive mimicry. Adaptation wherein a predator resembles its prey; compare *Batesian mimicry*.

Albedo. Reflectivity of an object or surface.

Allee effect. Decline in reproduction or survival at low population density.

Allele. One of two or more alternative forms of a gene that occupies the same relative position (locus) on homologous chromosomes.

Allele (gene) frequency. Proportion of a given allele in a population among all alleles present at that locus; compare *genotype frequency*.

Allelopathy. Detrimental effect of the metabolic products of plants on the growth and development of nearby plants.

Allochthonous. Of organic carbon, entering an ecosystem from another ecosystem, e.g., plant debris entering a stream; compare *autochthonous*.

Allogenic. Of successional vegetation change, brought about by a change in the abiotic environment; compare *autogenic*.

Allopatric speciation. Speciation that occurs as a result of geographic separation of a population into two or more isolated subpopulations; compare *sympatric speciation*.

Alpha (local) diversity. Species diversity within a local area of homogeneous habitat; compare *beta (regional) diversity*.

Alternative (alternate) stable states. In systems ecology, different configurations of an ecosystem than formerly present, often in response to human impacts; see *tipping point*.

Altitude. Distance above Earth's surface, usually in reference to a mountain; compare *elevation*.

Altricial. Condition among birds and mammals of being hatched or born too weak to survive without post-natal parental care; compare *precocial*.

Altruism. In species with group structure, tendency of individuals to sacrifice their fitness by dedicating time and energy to caring for the young of other group members; see *kin selection*.

Amensalism. Interspecific interaction in which one individual is harmed while the other is unaffected; compare *commensalism*.

Ammonification. Breakdown of proteins and amino acids, especially by fungi and bacteria, with ammonia as a by-product; compare *volatilization*.

Anadromous fish. Fish that live most of their lives in saltwater, returning to freshwater only to spawn; *diadromous fish* divide their time more equitably between both habitats.

Anaerobic metabolism. Metabolism that occurs in the absence of O_2; includes both *anaerobic respiration* and *fermentation*.

Anaerobic respiration. Respiration of organic compounds in the absence of O_2; uses compounds other than O_2 as the final electron acceptor; compare *fermentation* and *aerobic (cellular) respiration*.

Anoxic. Condition of a complete absence of O_2; compare *hypoxic*; only compatible with anaerobic metabolism.

Anthropocentrism. In environmental philosophy, approach that stresses impacts on the well-being of humans; compare *ecocentrism* and *biocentrism*.

Anthropogenic. Caused by human activities, often in reference to disturbance or climate change.

Anti-predator defence. Any morphological or behavioural trait that allows prey to avoid or reduce predation.

Aphotic zone. Deep portion of the water column of a lake or ocean in which light is insufficient to support positive net photosynthesis; see *compensation depth*; compare *photic zone*.

Apomixis. Type of asexual reproduction in plants in which seeds form without meiosis and sexual recombination; compare *parthenogenesis*.

Apparent competition. Indirect interaction in which a single predator species feeds on two non-competing prey species, supporting higher predator density; resembles competition because the density of both prey is reduced.

Aquifer. Porous, rock formation functioning as long-term water storage; compare *groundwater*.

Arbovirus. Disease-causing virus that uses arthropods (typically insects) as vectors.

Area-insensitive species. Species that use both edge and interior habitat and are relatively unaffected by patch size; compare *edge species* and *interior species*.

Artificial selection. In evolution, manipulation by humans of the gene frequency of domesticated species by selective breeding; compare *natural selection*.

Aseasonal. Of climate, lacking distinct seasons; typical of tropical areas.

Asexual reproduction. Any form of reproduction that does not involve fusion of gametes; see *budding* and *parthenogenesis*.

Assimilation. Incorporation of a substance by an organism; involves absorption and conversion of energy and nutrients into living tissues.

Assimilation efficiency (AE). Ratio of assimilation to ingestion; measure of the efficiency with which a consumer extracts energy from food; compare *gross production efficiency* and *net production efficiency*.

Association. Community with a consistent species composition, uniform physiognomy, and distribution typical of a habitat type or set of abiotic factors; vegetation subtype within a biome.

Assortative (non-random) mating. Process whereby individuals choose mates based on phenotypic traits; either *positive* (with similar traits) or *negative* (with dissimilar traits).

Asymmetric (one-sided) competition. Type of interspecific competition in which per capita effects are much greater for one species than the other.

Atmosphere. Layers of air surrounding Earth; the closest layers are the troposphere and the stratosphere.

Atmospheric (air) pressure. Downward force exerted by the mass of overlying atmosphere.

Aufwuchs. Community of organisms that colonize submerged surfaces; often used interchangeably with *periphyton*, which is sometimes restricted to algal components.

Autecology. Study of the response of individual organisms to their environment; includes structural and physiological ecology.

Autochthonous. Of organic carbon, produced within an ecosystem; compare *allochthonous*.

Autogamy (self-fertilization). Fusion of gametes from the same individual; see *inbreeding*.

Autogenic. Self-generated; of successional vegetation change, brought about by the organisms themselves; compare *allogenic*.

Autolysis. (1) Initial stage of decomposition of animal tissue, involving release of enzymes from the lysosomes of a dead organism's cells; compare *putrefaction*. (2) Breakdown of plankton by its own enzymes.

Autotroph. Organism that produces organic material from inorganic chemicals and an energy source; see *photoautotroph* and *chemoautotroph*; compare *heterotroph*.

Available water capacity (AWC). Supply of water available to plants in well-drained soil; equals the difference between *field capacity* and *wilting point*.

Avoidance. Ability of an organism to evade an abiotic stress; compare *tolerance*.

B horizon (subsoil). Soil stratum beneath Ae horizon, characterized by an accumulation of silica, clay, and iron and aluminum oxides and possessing blocky or prismatic structure.

Basin wetland. Freshwater wetland that develops in shallow basins, with vertical water flow; compare *riverine wetland* and *fringe wetland*.

Batesian mimicry. Anti-predator defence in which a palatable or harmless species (the mimic) evolves to resemble an unpalatable or poisonous species (the model).

Behavioural ecology. Study of the behaviour of a species, usually in its natural habitat.

Benthic zone. Bottom sediments in a water body; underlies the littoral and profundal zones of an ocean.

Benthos. Organisms inhabiting the bottom of a water body.

Bergmann's rule. Trend among related homeotherms for species inhabiting higher latitudes to have larger-bodied individuals.

Beta (regional) diversity. Species diversity of different habitats in a region; compare *alpha (local) diversity*.

Binary fission. Asexual reproduction in which a bacterium divides in half; compare *budding*.

Bioaccumulation. Buildup of substances (often pollutants) in an organism over its lifetime; compare *biomagnification*.

Biocentrism. In environmental philosophy, approach to environmental issues that stresses impacts on non-human organisms; compare *anthropocentrism* and *ecocentrism*.

Biodiversity. Composite measure of biological variation; incorporates genetic, species, and regional diversity.

Biodiversity–ecosystem functioning (BEF) hypothesis. View that biodiversity is a driver of ecosystem function rather than a passive consequence of biotic interactions; compare *diversity-stability hypothesis*.

Biodiversity hotspot. Region with high species richness and endemism, and in which species are vulnerable to extinction.

Bioengineer. Type of keystone species whose activities create or modify its habitat in disproportion to its number.

Biogeochemical cycle. Movement of matter through abiotic and biotic ecosystem components; see *gaseous cycle* and *sedimentary cycle*; compare *internal (nutrient) cycling*.

Biogeography. Study of the past and present geographical distribution of species and their ecological relationships.

Biological clock. Internal mechanism of an organism that controls periodicity; see *circa-annual rhythm* and *circadian rhythm*.

Biological species. Group of potentially interbreeding populations that are reproductively isolated from other populations.

Bioluminescence. Production of light by organisms via chemical reactions.

Biomagnification. Increase in the concentration of fat-soluble substances (often pollutants) at progressively higher trophic levels; compare *bioaccumulation*.

Biomass. Weight of living organisms, expressed as dry mass per unit area; includes dead tissue if part of living organisms; see *standing crop biomass*; compare *detritus*.

Biomass accumulation ratio. Measure of biomass accumulation in a community; see *standing crop biomass*; inverse of relative net primary productivity; compare *residence time*.

Biomass pyramid. Graph depicting the amount of biomass at each trophic level; compare *energy pyramid*.

Biome. Broadscale regions dominated by similar ecosystems; usually correspond to *formations*; compare *association*.

Biophage. Heterotroph that consumes living organisms; compare *saprophage*.

Bioshield. Protective effect of vegetation, e.g., the buffering role of mangroves against *tsunami* (tidal waves).

Biosphere. Narrow interface at Earth's surface that contains and supports living organisms.

Biotic. Living; biotic components of ecosystems include organisms of all types; compare *abiotic*.

Bisexual. Having both male and female organs on the same individual; compare *unisexual*.

Blanket mire (moor). Large upland peatland dominated by *Sphagnum* moss and dependent upon precipitation for its water supply.

Blubber. Layer of fat just below the skin of many marine mammals; provides insulation, energy storage, and buoyancy.

Bog. Wetland with accumulation of peat, highly acidic conditions, and dominance of *Sphagnum* moss; compare *fen*.

Border. Region where the edges of two patches meet; compare *edge* and *ecotone*.

Bottom-up control. Influence of primary producers on trophic levels above them in a food web; compare *top-down control*.

Boundary layer. Layer of still air or water close to or at a surface; affects heat exchange with the environment.

Broadleaf evergreen. Non-coniferous woody plants that retain leaves for several years; compare *needle-leaf evergreen*.

Brood parasite. Behavioural parasite that lays one or more eggs in the nest of its host, which raises the parasite's young.

Brown belt. Zone of earth and water dominated by microorganisms; plays a regulatory role in maintaining the biosphere; see *Gaia hypothesis*.

Browsers. Herbivores that feed on *browse* (leaf and twig growth of trees, shrubs, and woody vines); compare *grazers*.

Bud bank. Presence in the soil or substrate of plant structures such as roots or rhizomes that can reproduce asexually after disturbance; compare *seed bank*.

Budding. Type of asexual reproduction in which buds pinch off to form new individuals; compare *binary fission*.

Buffer. In reference to chemical reactions, a substance that reduces the change in pH by absorbing or releasing H^+ ions.

Bundle sheath cells. Cells surrounding veins in leaves or stems; involved in photosynthesis in C_4 plants.

Buoyancy. Ability of a liquid to exert upward force on a body placed in it.

Buttress. Plank-like projection at the trunk base of tropical trees; lends support in shallow soil.

Bycatch. Non-target fish and other aquatic life that are caught in fishing devices and discarded.

C horizon (regolith). Unconsolidated mineral debris overlying unweathered parent material that will give rise to soil; little affected by biological activity or soil-forming processes.

C_3 plant. Plant that produces the three-carbon compound PGA in the first step of the light-independent reactions of photosynthesis; compare C_4 *plant*.

C_4 plant. Plant that produces a four-carbon compound (malic acid or aspartic acid) in the first step of the light-independent reactions of photosynthesis; adapted to bright, hot, dry conditions; compare C_3 *plant*.

C:N ratio. Dry mass of carbon divided by the dry mass of nitrogen; higher for plant tissue than for animal or fungal tissue; indicator of litter quality and decomposition rate.

Calcification. Soil formation process characterized by an accumulation of calcium in lower soil horizons; common in temperate grasslands.

CAM plant. Plant (cactus or other succulent) adapted to dry habitats that separates CO_2 uptake and fixation by taking up CO_2 at night when stomata are open, and fixing it during the day when stomata are closed; similar steps as in C_4 plants but performed in the same cell.

Canopy. (1) Uppermost layer of vegetation formed by trees. (2) Top layer of any community where taller vegetation forms a distinct habitat.

Canopy fire. See *crown fire.*

Capillary water. Portion of water held between soil particles by capillary forces.

Carbon allocation. Allocation of net photosynthetic carbon by a plant to different plant parts, such as leaves, stems, roots, and reproductive organs.

Carbon balance model. Model that describes the balance between photosynthetic CO_2 uptake and respiratory CO_2 loss.

Carnivore. (1) Organism that feeds only on animal tissue; compare *herbivore* and *omnivore*; *carnivory* is act of killing and eating of animals by another animal. (2) Mammal of the order Carnivora.

Carrying capacity (*K*). Maximum population size that can be sustained in a given environment; typically limited by the resources available during the most unfavourable period.

Caste. Distinct group in a population that performs specific tasks; see *eusocial species.*

Categorical data. Qualitative data of distinct non-numerical types; includes *nominal data* and *ordinal data.*

Cation exchange capacity (CEC). Total number of negatively charged sites on soil particles per standard soil volume; indicator of soil fertility.

Cellular respiration. See *aerobic respiration.*

Census. Method of quantifying population size by counting all individuals; compare *abundance index* and *mark–recapture.*

Chaparral. Vegetation consisting of broadleaf evergreen shrubs, found in regions with a Mediterranean climate.

Character displacement. Principle that two species differ more when co-occurring than when separated geographically, as a result of the selective force of interspecific competition; see *limited similarity* and *niche differentiation.*

Chemical weathering. Action of chemical processes such as oxidation, hydrolysis, and reduction that break down and re-form rocks and minerals, affecting soil formation; compare *mechanical weathering.*

Chemoautotroph. Organism that converts CO_2 into organic matter by oxidizing inorganic molecules; see *chemosynthesis*; compare *photoautotroph.*

Chemosynthesis. Synthesis of organic compounds using energy derived from the oxidation of reduced compounds, as by bacteria inhabiting hydrothermal vents; compare *photosynthesis.*

Chlorophyll. Light-absorbing pigment in chloroplasts of green plants and most algae that powers the light-dependent reactions of photosynthesis; mainly chlorophyll *a* or chlorophyll *b.*

Chromosome. Thread-like structures on which genes are located; composed of DNA and histone proteins; found in nuclei of eukaryotic cells; see *homologous chromosomes.*

Chronosequence. Sites in the same geographic area that are in different successional stages because they experienced the same disturbance type at different times in past; compare *toposequence.*

Circa-annual rhythm. Seasonal responses to changing photoperiod; compare *circadian rhythm.*

Circadian rhythm. Internal rhythm of physiological or behavioural activity with a duration of approximately 24 hours; compare *circa-annual rhythm.*

Cleaner organism. Organism that consumes ectoparasites on the surfaces of other organisms; type of non-symbiotic mutualism, although may be commensalistic.

Clear-cutting. Forest-harvesting procedure in which all trees are cut and removed; compare *seed-tree harvesting* and *selection cutting.*

Climate. Long-term patterns of local, regional, or global weather; compare *weather.*

Climax. In succession, stable end-point community that is capable of self-perpetuation under prevailing conditions; see *monoclimax* and *polyclimax hypothesis*; compare *old growth.*

Cline. (1) Gradual change in the mean of a phenotypic trait over the geographic range of a species. (2) Gradient in *genotype frequency.*

Closed. (1) Of a population, with minimal emigration or immigration. (2) Of a system, with minimal exchange of inputs and outputs with nearby systems; compare *open.*

CO_2 fertilization effect. Increase in net primary productivity with elevated CO_2; strongest in C_3 species.

Coevolution. Joint evolution of two or more non-interbreeding species with a close ecological relationship; involves reciprocal selective pressures; called *coevolutionary arms race* if predation is the interaction involved.

Cohesion. Ability of a molecule to stick to other molecules of the same type, often by hydrogen bonding; restricts breakage of bonds by external forces; in contrast, *adhesion* is ability to stick to molecules of a different type.

Cohort. Group of individuals born into a population at the same time; basis of a cohort life table.

Cold hardiness. Ability of plants to resist freezing stress without injury; see *deep hardening.*

Collectors. Invertebrates that filter organic particles from flowing water (*filtering collectors*) or collect organic particles from a stream bottom (*gathering collectors*); compare *shredders*, *scrapers*, and *grazers.*

Colonization rate. In metapopulations, rate at which individuals establish in vacated patches; compare *extinction rate.*

Commensalism. Interspecific interaction in which one species benefits while the other is unaffected; compare *amensalism.*

Common mycorrhizal network. Community-wide network in which plants (often trees) are linked by fungal hyphae, allowing nutrient transfer and cooperative regulatory effects.

Community. Totality of all populations of all species inhabiting a given area at a given time.

Community ecology. Study of communities, particularly the description and analysis of patterns and processes operating within them.

Community similarity measure. Index of similarity between the species composition and/or abundance of two communities, e.g., Sorenson's coefficient and percent similarity.

Compensation depth. Water depth at which light is just sufficient for photosynthesis to balance respiration; see *photic zone*; compare *light compensation point*.

Compensatory growth. Plant growth in compensation for tissue lost to herbivory; either under-, exact, or overcompensation.

Competition. Interaction in which two or more individuals seek out shared limited resources and in which both individuals experience negative effects; see *intraspecific competition* and *interspecific competition*.

Competitive exclusion (Gause's) principle. Hypothesis that two or more species with closely overlapping niches cannot coexist if resources are limited; see *niche differentiation*.

Competitive mutualism. Indirect interaction in which two non-competing species benefit from each other's presence because each competes with the other's competitor, keeping it in check.

Competitive release. Expansion of realized niche of a species in absence or reduction of competitors or predators; compare *niche compression*; may occur with introduced species.

Competitors. (1) Individuals that are competing with each other. (2) In Grime's theory, species that inhabit predictable, stable habitats; typically long-lived with delayed and limited reproduction; cope well with competition in crowded habitats; similar to *K-strategists (species)*; compare *ruderals* and *stress-tolerators*.

Condition. Environmental factor that influences an organism's survival, growth and/or reproduction, but is not consumed; compare *resource* and *hazard*.

Conduction. Direct transfer of heat energy from one substance to another; compare *convection*; *conductivity* is ability of an object to exchange heat directly with its environment.

Confounding factor. Factor whose effects may be confused with the effects of another factor.

Connectance. Proportion of all possible feeding links that is present in a food web; compare *linkage density*.

Connectivity. In landscape ecology, extent to which habitat patches are connected by corridors; affects the ability of individuals to move between patches.

Conservation ecology. Application of the principles of ecology, biogeography, population genetics, economics, and sociology to the maintenance of biodiversity.

Constitutive. Of an anti-predator defence, a fixed trait that deters predators; compare *induced*.

Consumption efficiency. Ratio of consumption at one trophic level to production at the next-lowest level; compare *trophic efficiency*.

Contest competition. Type of competition in which dominant individuals deny other individuals access to a limited resource; compare *scramble competition*.

Continuous data. Type of numerical data in which values are limited only by measurement device; compare *discrete data*.

Convection. Transfer of heat by circulation of a liquid or gas; compare *conduction*.

Convergent evolution. Development of similar traits in different species occupying different areas under similar selective pressures; compare *parallel evolution*.

Coprophagy. Feeding on feces; *coprobivores* are organisms (e.g., lagomorphs) that practise coprophagy.

Core area. Portion of its home range where an animal spends most of its time.

Core population. Large local population that acts as the main source of emigrants to satellite populations.

Coriolis effect. Deflection of objects by Earth's rotation; causes moving objects to appear to veer to the right in the Northern Hemisphere and to the left in the Southern Hemisphere.

Corridor. Strip of vegetation that differs from the surrounding matrix; its vegetation may or may not be similar to patches it connects; see *connectivity*.

Cost–benefit analysis. Procedure that compares the benefits of an activity or adaptation with its costs or disadvantages.

Counteradaptation. Structural, physiological, or behavioural trait that evolves in one species to counter the effect of an adaptation of another species; see *coevolution*.

Countercurrent exchange. General principle whereby efficient transfer of energy or materials is maximized by having incoming flow run counter to outgoing flow, e.g., between outgoing warm arterial blood and cool venous blood returning to the body core; important in maintaining physiological homeostasis in many animals.

Critical day length. Period of daylight that triggers a response in organisms exhibiting photoperiodism; short-day species respond to days shorter than the critical length; long-day species to days longer than the critical length; see *photoperiod*.

Critical temperatures. Upper and lower limits of a homeotherm's thermoneutral zone, above and below which its metabolic rate increases rapidly.

Crown fire. Fire that burns a forest canopy; compare *ground fire* and *surface fire*.

Cryoplanation. Moulding of a landscape by frost action; important in Arctic regions.

Cryptic colouration. Anti-predator defence involving colouration that blends a prey organism into its surroundings; also used by some predators to avoid detection by prey; compare *warning colouration*.

Current. Water movements that result in horizontal transport of water masses; compare *gyre*.

Cybernetics. Science of system regulation, usually focusing on feedback mechanisms.

Deciduous. (1) Of leaves, shed during a certain season (winter in temperate regions, dry season in the semi-tropics). (2) Of trees, having deciduous parts; compare *evergreen*.

Decomposer. Saprophyte (bacterium or fungus) that obtains energy from the breakdown of dead organic matter; also *microflora*; compare *detritivore*.

Decomposition. Breakdown of complex organic substances into simpler compounds, as performed by decomposers; compare *mineralization*.

Decomposition coefficient. In litterbag studies, slope of negative exponential curve depicting decomposition over time.

Deep hardening. Multi-step process whereby plants adapted to extremely cold climates acquire the ability to tolerate intercellular ice formation; see *cold hardiness*.

Definitive host. In indirect transmission, host in which a parasite reaches sexual maturity; compare *intermediate host*.

Delta. Fan-shaped deposition of sediments at the mouth of a river where it enters a lake or sea; compare *estuary*.

Demographic stochasticity. Random variations in birth and death rates that occur in a population as a result of intrinsic factors; compare *environmental stochasticity*.

Demography. Statistical study of changes in population size and/or structure.

Denitrification. Reduction of nitrates and nitrites to nitrogen gas by microorganisms.

Density. See *population density* and *ecological density*.

Density dependence. Regulation of population growth whereby factors affect a population in proportion to its size.

Density independence. Lack of regulation of population growth by population size.

Dependent variable. Experimental variable (on *y*-axis) that responds to an independent (treatment) variable (on *x*-axis).

Detrital food chain. Food chain in which detritivores consume detritus (mostly plant litter), with subsequent energy transfer to other trophic levels; compare *grazing food chain*.

Detritivore (detritus feeder). Invertebrate that consumes dead organic matter, fragmenting but not mineralizing it; compare *decomposer*.

Detritus (necromass). Dead tissues in varying stages of decomposition; also *litter* (plant origin); *detritus feeders* consume detritus.

Developmental plasticity. Irreversible phenotypic plasticity that occurs during development; compare *acclimation*.

Dew point temperature. Temperature at which saturation vapour pressure is reached for an air parcel of a given water content; below this temperature, water condenses.

Diapause. Period of dormancy, usually seasonal, in an insect life cycle in which growth and development cease and metabolism is minimal; compare *estivation and hibernation*.

Diffuse competition. Interspecific competition in which a species competes with many other species that consume the same resources; *diffuse* also applies to other interactions.

Diffusion. Spontaneous movement of molecules from an area of higher to lower concentration; compare *osmosis*.

Dioecious. Having male and female reproductive organs on separate plants; compare *monoecious* and *bisexual*.

Diploid. Having two copies of each chromosome in homologous pairs; *haploid* refers to having only one copy.

Direct transmission. Movement of a parasite between hosts without involvement of other species; either by contact or dispersal through air or water; compare *indirect transmission*.

Directional selection. Selection favouring individuals at one extreme of the range of phenotypes in a population; compare *stabilizing selection* and *disruptive (divergent) selection*.

Discounting. In environmental economics, practice of computing costs and benefits occurring at different times; often runs counter to sustainable resource management.

Discrete data. Numerical data with a finite number of data points; compare *continuous data*.

Disease. (1) Any deviation from a normal state of health. (2) Outcome of infection by a disease organism, particularly endomicroparasites such as bacteria or viruses; *epidemic* refers to rapid disease spread in a human population, *epizooic* to rapid disease spread in an animal population.

Dispersal. Movement of individuals in a population, usually leaving their area of birth; compare *migration*; not to be confused with *spatial dispersion (distribution)*.

Disruptive (divergent) selection. Selection that favours phenotypes at either extreme in a population; compare *stabilizing selection* and *directional selection*.

Dissolved organic matter (DOM). Residual organic matter dissolved in water; derives from plankton autolysis and algal excretions; compare *particulate organic matter*.

Distribution. Spatial extent of a population; compare *geographic range* and *spatial dispersion (distribution)*.

Disturbance. Discrete event that damages or removes a population or community, often changing substrates and resource availability, and initiating *secondary succession*.

Disturbance regime. Pattern of disturbance that is typical of a landscape over an ecological timeframe; see *intensity*, *severity*, and *frequency*.

Diversity. Variability in natural systems; see *biodiversity*; *species diversity*; *alpha diversity*; and *beta diversity*.

Diversity index. In community ecology, a mathematical index value that incorporates both *species richness* and *species evenness*, e.g., Simpson's index.

Diversity–stability (insurance) hypothesis. View that the diversity of an ecosystem is correlated with its stability; compare *biodiversity–ecosystem functioning hypothesis*.

Dominance. (1) In a community, control by one or several species over abiotic conditions influencing other species. (2) In a population, control that gives higher-ranking individuals priority of access to resources. (3) In genetics, ability of an allele to mask expression of another allele when in a heterozygous condition.

Dominant. (1) Of a community, a species possessing ecological dominanc, thereby influencing the identity and abundance of other species present. (2) Of an allele, expressed in either the homozygous or heterozygous state.

Dormancy. Seasonal state of suspended biological activity, during which life is maintained; refers to plants (seeds and vegetative buds) but also used for animals, especially insects; compare *diapause* and *estivation*.

Downregulation. In photosynthesis, decline in Rubisco levels after prolonged exposure to elevated CO_2, reducing photosynthesis to rates comparable to rates at ambient CO_2.

Drift. Downstream movement of material; *drift rate* is an index of stream productivity.

Drought-deciduous. Habit of some woody species in tropical seasonal regions of dropping their leaves prior to the rainy season; compare *winter-deciduous*.

Drought tolerance. Ability of a plant to maintain physiological activity despite low water supply, or to survive drying of tissues; differs from *drought avoidance*, in which a plant escapes dry periods by dormancy; *drought resistance* is the sum of drought tolerance and drought avoidance.

Dynamic equilibrium. State of a biological entity (from an individual to an ecosystem) in which change is occurring but in which feedback mechanisms are working to return it to a set point; see *homeostasis*; compare *homeorhesis*.

Dystrophic. Of a water body, having high humic or organic matter content, often with high littoral and low planktonic productivity; compare *oligotrophic* and *eutrophic*.

Early-successional species. Plant species that occupy the early stages of succession; usually characterized by high dispersal, ability to colonize disturbed sites, short lifespan, and shade intolerance; see *pioneer species* and *r-strategist*; compare *late-successional species*.

Echolocation. Use of sonar for detecting prey, as in bats.

Ecocentrism. Approach to environmental issues that stresses impacts on ecosystems and the biosphere; compare *biocentrism* and *anthropocentrism*.

Ecological density. Number of individuals in a population per unit area of suitable habitat; compare *population density*.

Ecological equivalents. Two or more species that perform similar ecological roles in different environments; see *convergent evolution* and *parallel evolution*.

Ecological footprint. Demand imposed by humans on Earth's systems on a per capita or country basis.

Ecological speciation. Speciation (allopatric or sympatric) in which traits that prevent interbreeding result from natural selection for ecological adaptations.

Ecology. Study of the relationships between organisms and their biotic and abiotic environments; compare *environmental science*.

Ecomorphology. Use of morphological traits as surrogates of fundamental niche dimensions.

Ecosystem. Biotic community and its abiotic environment, functioning as a system.

Ecosystem ecology. Study of natural ecosystems, emphasizing energy flow and nutrient cycling.

Ecosystem services. Processes by which ecosystems provide resources that are essential for life, such as clean air and water.

Ecotone. Wide border that forms transitional zone between structurally different communities; compare *edge*.

Ecotype. Population that is genetically distinct and adapted to a particular set of environmental conditions, often on a continuum; also *ecological race*; compare *geographic isolates*.

Ectomycorrhizae. Symbiotic mutualistic association in which fungal hyphae form sheaths (Hartig net) around the outside of plant roots; compare *endomycorrhizae*.

Ectoparasite. Parasite that lives on the external surfaces of its host in a prolonged or intermittent association; compare *endoparasite*.

Ectothermy. Reliance on external heat sources; ectotherms typically have variable body temperature; see *poikilothermy*; compare *endothermy*.

Edge. Place where two or more communities meet; called an *induced edge* if due to disturbance or an *inherent edge* if due to presence of an enduring feature; compare *border* and *ecotone*.

Edge effect. Response of individuals or the community to environmental conditions present at an edge.

Edge species. Species restricted to edge habitats; compare *interior species* and *area-insensitive species*.

Effective population size. Number of individuals in a population that are actually contributing to future generations; compare *minimum population size*.

El Niño–Southern Oscillation (ENSO). Major climatic variations arising from broadscale interactions between the Pacific Ocean and the atmosphere; compare *La Niña*.

Elevation. Distance above sea level; only equal to altitude if a landform arises at sea level; compare *altitude*.

Eluviation. Movement of suspended materials in water through soil; see *illuviation*; compare *leaching*.

Emergent properties. Attributes of an entity that arise only with respect to the collective; usually refers to community properties, but applies to any level of biological hierarchy.

Emergents. (1) Aquatic plants rooted in the substrate that emerge above water in lakes, oceans, or wetlands (e.g., cattails), as opposing to floating or submerged species. (2) trees whose crowns rise above the general forest canopy.

Emigration. Movement of individuals out of a population; compare *immigration*.

Endemic species. Species with a highly localized geographic range; compare *ubiquitous species*.

Endomycorrhizae. Symbiotic mutualistic association in which fungal hyphae penetrate the cortex of plant roots, forming vesicles and arbuscles; compare *ectomycorrhizae*.

Endoparasite. Parasite that lives in the body of its host; compare *ectoparasite*.

Endophytic fungi. Defensive mutualism in which herbaceous species (mostly grasses) harbour fungi in their shoots that produce anti-herbivore compounds.

Endosymbiotic hypothesis. Hypothesis that chloroplasts originated as photosynthetic bacteria that were ingested by heterotrophs, eventually establishing a symbiotic mutualism; also applied to the origin of mitochondria.

Endothermy. Reliance on internal heat generation; most endotherms regulate their body temperature; see *homeothermy*; compare *ectothermy*.

Enduring feature. Long-term natural element (such as a rock outcrop) that has a determining influence on the type of ecosystem present.

Energy pyramid. Model depicting the amount of energy processed at each trophic level; compare *biomass pyramid*.

Entropy. Measure of randomness or disorder in a system resulting from transformations of matter and energy; form of energy that is incapable of doing work; see *second law of thermodynamics*.

Environment. All components and/or factors (abiotic and biotic) external to an organism that influence its survival, growth, and/or reproduction; compare *habitat*.

Environmental economics. Incorporation of economic principles into environmental decision-making.

Environmental gradient. Situation in which one or more abiotic factors changes in physical space along a continuum rather than in abrupt shifts.

Environmental heterogeneity. Amount of variation in environmental factors within and among communities.

Environmental science. Study of the impact of humans on the environment; compare *ecology*.

Environmental stochasticity. Random variation in the abiotic environment that directly affects birth and death rates; compare *demographic stochasticity*.

Environmental sustainability. Ability to sustain the exploitation of natural ecosystems or resources.

Ephemeral. (1) Species that grows and reproduces in a short period when conditions are favourable and then goes dormant; *spring ephemerals* are common in deciduous forests, and *drought ephemerals* in deserts when moisture is available. (2) Of a habitat or resource, of short duration.

Epifauna. Benthic animals that live on or move over surfaces; compare *infauna* and *meiofauna*; *epiflora* are benthic plants living on surfaces.

Epilimnion. Warm, O_2-rich upper layer above the thermocline in a lake or ocean, often seasonal; compare *hypolimnion*.

Epiphyte. Plant that lives wholly on the surface of other plants, deriving support but not nutrients (e.g., many tropical orchids); *epiphytic fungi* are fungi that occupy plant surfaces.

Episodic competition. Competition that occurs intermittently, often as a result of varying resource levels.

Equilibrium theory of island biogeography. Theory proposing that the species richness of islands reflects a dynamic equilibrium between immigration of colonizing species and extirpation of existing species.

Escape space (enemy-free space). Any niche dimension or adaptation that minimizes predation risk, including the avoidance of habitats or behaviours that entail risk.

Estivation. Dormancy in ectothermic animals during drought; compare *diapause*.

Estuary. Partially enclosed coastal area where freshwater in rivers and streams drains into seas, mixing with saltwater; location of a sediment trap; compare *delta*.

Eusocial. Describes species (such as social insects) that combine caste structure with altruism, allowing the group to function as a single cooperative entity.

Eutrophic. Of a water body, having high nutrient content and productivity; also *hypertrophic*; compare *oligotrophic* and *dystrophic*; *eutrophication* is the nutrient enrichment of water bodies; called *cultural eutrophication* if due to human activity.

Evaporation. Transformation of water from liquid to gas from soil, open water, or other surfaces, including organisms.

Evapotranspiration. Total water lost by combined evaporation and transpiration; see *actual evapotranspiration*.

Evergreen. Pertaining to trees and shrubs that do not lose their leaves seasonally; see *needle-leaf evergreen* and *broadleaf evergreen*; compare *deciduous*.

Evolution. Change in gene frequency and hence properties of a population over generations; results from various mechanisms, including but not only natural selection.

Evolutionary bottleneck. Situation arising when disturbance wipes out a large proportion of a population, leaving a few individuals from which the population re-establishes; reflects genetic drift, not natural selection, because survival is due to chance rather than adaptive traits.

Evolutionary ecology. (1) Study of evolution, genetics, natural selection, and adaptations in an ecological context. (2) Evolutionary interpretation of ecology at any level of the biological hierarchy.

Experiment. Test of an hypothesis conducted under controlled conditions in either the field or laboratory.

Exploitation. Indirect competition in which individuals respond to a decline in shared resource(s) arising from consumption by other individuals; also *scramble competition*; compare *interference*.

Exponential growth. Pattern of growth in which a population increases (or decreases) at a steadily increasing rate over time; compare *logistic growth*.

Extant. Of species, currently existing; not extinct.

Externalities. Impacts of an individual or group on another individual or group's well-being, but for which environmental costs or benefits are not reflected in market prices.

Extinction. (1) Loss of a species. (2) Local extinction (*extirpation*) of a subpopulation. (3) Decline in light level; see *light extinction coefficient*.

Extinction rate. Rate at which a local population is extirpated in a metapopulation; compare *colonization rate*.

Extirpation. Loss of a population; also *local extinction*.

Extracellular digestion. Process whereby organisms (notably fungi) release enzymes, which break down compounds outside the cell, followed by absorption of breakdown products.

Facilitation. (1) Process whereby one individual or population benefits from the presence or actions of another. (2) In succession theory, the *facilitation model* proposes that species at one stage facilitate the success of species in the next stage, while making the site less favourable for themselves; compare *inhibition model* and *tolerance model*.

Facultative. (1) In interspecific interactions, able to form an interaction, but not requiring it. (2) In metabolism, able to function in either aerobic or anaerobic conditions; compare *obligate*.

Fall turnover. Vertical circulation in water bodies (especially lakes) during fall, when cooling surface waters sink, displacing warmer water to the surface and causing mixing; see *thermocline*.

Fecundity table. Table showing the number of offspring produced per unit time; combines age- and sex-specific birth-rates with survivorship data.

Feedback. System regulation whereby output from one component affects the future functioning of that and/or another component; see *positive feedback* and *negative feedback*.

Fen. Slightly acidic wetland, dominated by sedges, in which peat accumulates; compare *bog*.

Fermentation. Breakdown of organic matter under anaerobic conditions; type of anaerobic metabolism that lacks an electron transport chain; compare *anaerobic respiration*.

Field capacity (FC). Amount of water held by soil against the force of gravity; expressed as the percentage of soil mass occupied by water when a soil is saturated compared to an oven-dried soil at standard temperature.

Filter effect. Differential impact of corridors in a landscape by providing dispersal routes for some species but restricting the movement of others; compare *connectivity*.

Filter feeder. Organism that extracts prey from a constant volume of water that washes over a filtering apparatus in its mouthparts.

First law of thermodynamics. Principle that energy is neither created nor destroyed; in any transfer or transformation, the total energy is constant; compare *second law of thermodynamics*.

First-level consumer. Organism (herbivore) that feeds on primary producers, either in a grazing or a detrital food web; *first-level carnivore* is an animal that feeds on herbivores.

Fitness. Relative genetic contribution of an individual to future generations; also *direct (classical) fitness*; compare *inclusive fitness*.

Flashing colouration. Anti-predator defence involving hidden markings on animals that, when quickly exposed, startle or divert potential predators; compare *cryptic colouration* and *warning colouration*.

Floating reserve. Individuals in a population of a territorial species that do not hold a territory and remain unmated, but may refill territories vacated by others.

Food chain. Linear sequence of feeding relations through trophic levels, as depicted by arrows depicting the flow of energy and nutrients from consumer to consumed; compare *food web*.

Food web. Totality of interconnecting food chains in an ecosystem.

Forest dieback. General decline in forests, attributed to various causes including acid precipitation, ozone pollution, and nutrient imbalances.

Forest floor. Ground layer of leaves and other detritus in a forest.

Formation. Broad category of vegetation occurring in different regions, defined on the basis of similar physiognomy, not species composition; see *biome*; compare *association*.

Founder effect. Impact of a population arising from a small number of colonists, which contain a small and often biased sample of the genetic variation of the parent population and which can generate a markedly different new population; see *genetic drift*.

Fragmentation. (1) In landscape ecology, reduction of large continuous habitat area into small, scattered remnant patches. (2) In decomposition, reduction of detritus into smaller pieces.

Frequency. (1) In disturbance theory, mean number of disturbances in a given time interval; see *return interval*. (2) In community ecology, proportion of sample plots in which a species occurs relative to the total number of plots; probability of finding a given species in any one plot.

Freshwater. Pertaining to non-marine aquatic habitats, as in lakes, ponds, and rivers; compare *marine*.

Freezing point reduction. Lowering of freezing point in ectotherms by accumulation of solutes, often glycerol in animals and sucrose or proline in plants; compare *supercooling*.

Fringe wetland. Freshwater wetland along the shores of large lakes; compare *basin wetland* and *riverine wetland*.

Frugivores. Herbivores that consume fruit; compare *granivores, nectivores,* and *grazers*.

Functional group. Species that perform similar functions in a community; typically have similar environmental responses, life histories, or other attributes; see *trophic level* and *guild*.

Functional redundancy. Theory that because communities contain several species performing similar ecological functions, loss of one species has little effect on stability.

Functional response. Change in the per capita rate of prey consumption by a predator with changing prey density; compare *numerical response*.

Fundamental niche. Range of environmental factors under which an organism (or species) can survive, grow, and reproduce in the absence of interspecific interactions; compare *realized niche*.

Gaia hypothesis. View that the biosphere is a superorganism that maintains itself through complex feedback loops between its biotic and abiotic components; see *brown belt*.

Gap. Opening in a forest canopy (or other community) created by a microdisturbance such as a windfall or death of a tree; typically initiates a local secondary succession.

Gaseous cycle. Biogeochemical cycle for which the main reservoir is the atmosphere and/or the ocean; compare *sedimentary cycle*.

Gene. Unit of inheritance; small length of DNA molecule on a chromosome that codes for a specific protein (or part of a protein) and gives rise to a particular trait.

Gene flow. Movement of genes between populations as a result of immigration or emigration; compare *genetic drift*.

Gene frequency. See *allele frequency*.

Gene pool. Sum of all genes of all individuals in a population; compare *genome*.

General circulation model (GCM). Complex computer model that predicts the impact of greenhouse gases and other factors on future climate.

Generalist. Species with a broad fundamental niche for one or more key environmental factors; compare *specialist*.

Generation time. Mean time between when a female is born and when she reproduces; estimated by the mean age of reproducing individuals.

Genet. In reference to organisms with clonal growth, genetically distinct, free-living individual that develops from a zygote, parthenogenetic gamete, or spore, and that produces ramets vegetatively during growth; compare *ramet*.

Genetic differentiation. Genetic variation among local populations of a species.

Genetic drift. Random fluctuation in allele frequency over time that is due to chance alone, not natural selection; important in small populations; see *founder effect*.

Genetically modified (GM). Containing genes transferred from other organisms by recombinant DNA technology; also *genetically engineered* or *transgenic*.

Genome. Total genetic complement of a cell.

Genotype. Genetic constitution of an organism for a given trait, as determined by the alleles present at a given locus on homologous chromosomes; compare *phenotype*.

Genotype frequency. Proportion of a given genotype among all genotypes for that trait in a population; compare *allele frequency*.

Geographic isolates. Populations that are isolated from one another by a geographic barrier; compare *subspecies* and *ecotype*.

Geographic range. Total area over which a species occurs; compare *distribution*.

Geometric model. Pattern in which a population increases or decreases steadily over time; similar to *exponential growth* but described for discrete time intervals.

Gill. Structural adaptation of fish and many aquatic invertebrates for uptake of O_2 in water; evagination that uses countercurrent exchange; compare *lung*.

Gleization. Process in waterlogged organic soils in which iron, as a result of inadequate O_2 supply, is reduced to ferrous compounds, giving a dull grey or bluish appearance.

Global ecology. Study of ecological systems on a global scale.

Gougers. Stream-dwelling insect larvae that burrow into waterlogged limbs and trunks of fallen trees.

Granivores. Herbivores that eat seeds; considered true predators.

Grazers. (1) Herbivores that eat leafy material, especially grasses; compare *browsers*. (2) Invertebrates that consume periphyton on rocks and other substrates in streams.

Grazing food chain. Food chain in which the primary producers are eaten by grazing herbivores; compare *detrital food chain*.

Green manure. Practice of ploughing under N-fixing legumes or grass-legume mixtures as an alternative to chemical fertilizer.

Green Revolution. Introduction of industrialized agriculture to developing countries, particularly dwarf hybrid varieties and heavy fertilizer use; compare *sustainable agriculture*.

Greenhouse effect. Selective absorption by greenhouse gases of longwave radiation emitted by Earth and atmosphere, followed by re-radiation back to Earth as heat.

Greenhouse gas. Gas that absorbs longwave radiation and contributes to the greenhouse effect when present in the atmosphere; includes H_2O vapour, CO_2, CH_4, nitrous oxides, and ozone.

Gross primary productivity (GPP). Energy fixed per unit area per unit time in photosynthesis (or chemosynthesis) by primary producers prior to respiration; compare *net primary productivity*.

Gross production efficiency (GPE). Measure of efficiency with which a consumer converts food ingested into production of new tissues; also *ecological growth efficiency*; compare *net production efficiency*.

Gross reproductive rate. Mean number of females born to a female over her lifetime, assuming all females survive to maximum age; compare *net reproductive rate*.

Ground fire. Fire that consumes organic matter down to mineral substrate or bare rock; compare *surface fire* and *canopy fire*.

Groundwater. Water occurring below Earth's surface in pore spaces within bedrock and soil and free to move under the influence of gravity; compare *aquifer*.

Growing season. Number of consecutive days during which conditions are favourable for photosynthesis and growth.

Growth form. Different forms of vegetation based on whether species are shrubs, trees, herbs, etc.

Guild. Functional group in which species utilize shared resources in similar ways; more specific than trophic level.

Gyre. Large circular pattern of water movement in ocean basins; compare *current*.

Habitability. Ability of an environment to support life.

Habitat. Place where an organism lives; often the microhabitat or fine-scale portion of a general habitat; compare *niche*.

Habitat selection. Processes whereby an individual actively chooses a location to live; in animals, involves behavioural responses to environmental cues.

Halophyte. Terrestrial plant adapted to saline soil.

Handling time. Total time that a predator spends chasing, killing, eating, and digesting prey; compare *search time*.

Hardy–Weinberg principle. Proposition that allele and genotype frequencies remain unchanged in successive generations, provided evolutionary mechanisms are not occurring.

Harvest interval. See *rotation period*.

Hazard. Environmental factor that is not required by an organism but that causes damage if present; compare *resource* and *condition*.

Heat island. Local area (often in urban microclimates) that is warmer than its surroundings as a result of factors that increase energy gains over losses.

Heat-stress proteins. Specialized proteins that protect living cells of hot-climate ectotherms (including plants and animals) from high-temperature damage.

Heat transfer coefficient. Measure of how easily heat travels through a gas or liquid.

Heliothermy. Type of ectothermy in which an organism gains most of its heat from basking in the sun.

Hematophagic. Feeding on blood; type of ectoparasitic behaviour.

Hemiparasite. Parasitic plant that has chlorophyll and performs photosynthesis, yet derives some nutrients from its host; compare *holoparasite*.

Herb layer. Grouping of lichens, moss, ferns, herbaceous plants, and small woody species on the forest floor.

Herbivore. Organism that feeds exclusively on plant tissue; *herbivory* is the act of consuming plant tissue; compare *carnivore*.

Hermaphroditic. Having both male and female reproductive organs on the same individual; see *bisexual*; in animals, either as *sequential hermaphrodites* (involving a sex change) or *simultaneous hermaphrodites* (at same time).

Heterogeneity. Variation in the physical environment; contributes to community diversity.

Heterotherm. Organism that practises both ectothermy and endothermy during its life history; undergoes rapid, drastic, repeated changes in body temperature; see *hibernation*.

Heterotroph (consumer). Organism that derives its energy and nutrients from consuming other organisms (living or dead); also *secondary producer*.

Heterotrophic succession. (1) Change in animal community as a result of vegetation changes. (2) Succession of consumers feeding on detritus.

Heterozygous. Containing two non-identical alleles at corresponding loci of homologous chromosomes; compare *homozygous*.

Hibernation. Prolonged seasonal torpor in heterotherms characterized by greatly reduced metabolism and near cessation of bodily activity; compare *diapause*.

Holoparasite. Parasitic plant that lacks chlorophyll and depends on its plant host for energy as well as water and nutrients; compare *hemiparasite*.

Home range. Area over which an animal ranges; compare *core area*.

Homeorhesis. Tendency of a system to return to a trajectory rather than a set point; see *pulsing state*; compare *homeostasis*.

Homeostasis. (1) Maintenance of a nearly constant internal environment in varying external environment. (2) Tendency of a system to maintain an equilibrium state; compare *homeorhesis*.

Homeostatic plateau. Set-point range of homeostasis; limited range of physiological tolerances within which an organism operates optimally.

Homeothermy. Maintenance of a relatively constant body temperature by endothermy coupled with homeostatic thermoregulation; *homeotherm* is an organism that practises homeothermy; compare *poikilothermy*.

Homologous chromosomes. In diploid organisms, corresponding chromosomes from male and female parents that pair during meiosis and that have alleles for the same genes.

Homozygous. Containing two identical alleles at corresponding loci of homologous chromosomes; compare *heterozygous*.

Horizontal transmission. Direct transmission of a parasite between host adults; compare *vertical transmission*.

Host. Organism that provides food or other benefits to an organism of a different species; usually refers to parasitic interactions, but is sometimes used for mutualisms.

Human ecology. Branch of ecology concerned with the interactions of humans with their environment.

Humus. Brown to black residual organic material that resists further decomposition as a result of its high lignin content; compare *soil organic matter*.

Hybridization. Matings between genetically divergent individuals of the same species; compare *interspecific hybridization*.

Hydraulic redistribution. Redistribution of water in soil resulting from water leaving roots in upper soil regions at night when no transpiration is occurring.

Hydric. Wet conditions, particularly of soil; compare *xeric* and *mesic*; see *hydrophyte*.

Hydrologic cycle. See *water cycle*.

Hydrology. In wetlands, all aspects of water movement or flooding; see *hydroperiod*.

Hydroperiod. In wetlands, combination of the duration, frequency, depth, and season of flooding.

Hydrophyte. Plant adapted to grow in water or soil that is saturated for most or all of the year.

Hydrosphere. Water at or near Earth's surface, including water bodies and the soil solution; compare *lithosphere*.

Hydrothermal vent. Site on the ocean floor where water issues from fissures; vent water contains sulphides oxidized by chemosynthetic bacteria, providing energy for a unique food web; see *hyperthermophilic*.

Hyperosmotic. Having a higher concentration of salts in body tissue than in the surrounding water; compare *hypoosmotic* and *isoosmotic*.

Hyperthermia. Rise in body temperature; in heat-adapted animals, reduces heat inflow and the need for evaporative cooling; compare *hypothermia*.

Hyperthermophilic. Pertaining to bacteria that can tolerate high temperatures of hydrothermal oceanic vents.

Hypolimnion. Region of cold, dense water in a lake or an ocean, below the thermocline; compare *epilimnion*.

Hypoosmotic. Having a lower concentration of salts in body tissue than in the surrounding water; compare *hyperosmotic* and *isoosmotic*.

Hypothermia. Reduction of body temperature; in heterotherms, controlled drop associated with relaxation of thermoregulation during hibernation; compare *hyperthermia*.

Hypothesis. Testable explanation for a phenomenon; compare *model* and *theory*.

Hypoxic. Condition of low O_2 but not its complete absence; depresses *aerobic respiration* and promotes *anaerobic respiration*; compare *anoxic*.

Illuviation. Accumulation of materials such as clays, iron oxides, and salts in the B horizon of a soil as a result of leaching and/or eluviation.

Immigration. Movement of individuals into a population; compare *emigration*.

Immobilization. Conversion of an element from an inorganic to organic form in living tissues (often of microbial decomposers), rendering it temporarily unavailable to other organisms; see *net mineralization rate*.

Immune response. In parasitic infections, second line of defence of a host involving production of antibodies by the immune system; compare *inflammatory response*.

Importance value. Measure of species dominance in a community based on the sum of two or more of relative density, basal area, cover, biomass, and frequency.

Inbreeding. Mating of individuals that are more closely related than expected by chance; see *autogamy*; compare *outcrossing*; *inbreeding depression* refers to the detrimental effects of inbreeding, often as a result of the increased frequency of homozygous recessives.

Inclusive fitness. Fitness that considers not only the direct fitness of the individual but also indirect fitness associated with helping close relatives (kin); see *kin selection*.

Independent variable. Experimental (test) variable that is manipulated; on *x*-axis; compare *dependent variable*.

Indeterminate growth. Pattern in which an organism continues to grow throughout its adult life, with a wide range of final adult size; see *modular organism*; compare *unitary organism*.

Indirect commensalism/mutualism. Indirect interaction that involves two predators feeding on two competing prey; interaction is commensalistic if only one predator benefits, and mutualistic if both benefit.

Indirect interaction. Interspecific interaction in which an individual of a species does not interact with an individual of a second species directly, but influences a third species that has a direct interaction with the second species.

Indirect transmission. Movement of a parasite between hosts with the help of another species, either as a vector or as an intermediate host; compare *direct transmission*.

Individualistic (continuum) concept. View that species respond independently to changing abiotic factors; e.g., that vegetation is a continuous variable in a continuously changing environment; similar communities arise only from similar requirements for and responses to abiotic factors; compare *organismal (holistic) concept*.

Induced. Of an anti-predator defence, brought about by the presence or action of a predator (e.g., alarm pheromones); compare *constitutive*.

Industrialized agriculture. Agricultural methods based on high inputs and mechanization; compare *traditional (subsistence) agriculture* and *sustainable agriculture*.

Infauna. Animals living within a substrate; compare *epifauna* and *meiofauna*.

Infection. (1) Diseased condition arising when pathogenic microorganisms enter a body, establish, and multiply. (2) Condition involving a heavy load of parasites.

Infiltration. Downward movement of water into soil; see *water cycle*; compare *percolation* and *surface runoff*.

Inflammatory response. First line of defence in a parasitic infection, involving the death or injury of host cells, causing local inflammation; compare *immune response*.

Infralittoral (subtidal) zone. Fully marine region below littoral zone, from which it is separated by the *infralittoral fringe*.

Infrared radiation. Radiation from 740 to 100 000 nm; includes near-infrared and thermal radiation.

Inhibition model. Model of succession proposing that dominants occupying a site inhibit colonization by plants of next successional stage; compare *facilitation* and *tolerance model*.

Initial floristic composition hypothesis. View that a successional sequence is partly dependent on which species arrive first; compare *monoclimax hypothesis* and *pattern climax theory*.

Inputs. Movement of materials, energy, or organisms into an ecosystem from surrounding environment; compare *outputs*.

Insolation. Radiation of any wavelength that reaches Earth's surface directly or indirectly and is not reflected; compare *photosynthetically active radiation*.

Integrated multi-trophic aquaculture (IMTA). Form of aquaculture that incorporates organisms at different trophic levels, with goal of imitating the function of natural marine systems.

Integrated pest management. Approach that considers the ecological, economic, and social aspects of pest control, with the aim of avoiding or minimizing significant economic damage.

Intensity. Measure of the force of a disturbance, e.g., wind speed or energy released in a fire; compare *severity*.

Interception. Capture of precipitation by structures (particularly vegetation), from which water evaporates or reaches the ground; see *water cycle*.

Interference. Mechanism of competition in which individuals interact directly by limiting access to a shared resource; typical of *contest competition*; compare *exploitation*.

Interior species. Organisms that require interior habitat; compare *edge species* and *area-insensitive species*.

Intermediate disturbance hypothesis. View that species richness is greatest in habitats with moderate disturbance, allowing coexistence of early- and late-successional species.

Intermediate host. Host that harbours a developmental phase of a parasite that is spread by indirect transmission; the infective stage can only develop when the parasite is independent of its definitive host; compare *definitive host*.

Intermittent parasite. Parasite (usually an ectoparasite) that has a non-symbiotic, episodic interaction with its host.

Internal (nutrient) cycling. Pathway of a nutrient through the biotic components of an ecosystem, from assimilation by organisms to release by decomposition; compare *biogeochemical cycle*.

Intersexual selection. Natural selection in which mate choice is affected by traits of the opposite sex; see *sexual dimorphism*; compare *intrasexual selection*.

Interspecific. Of interactions, occurring between individuals of different species; compare *intraspecific*.

Interspecific competition. Competitive interaction between two or more individuals of different species; compare *intraspecific competition*.

Interspecific hybridization. Mating between individuals of two species that produces viable offspring that can themselves reproduce; important in plant speciation.

Intertidal zone. See *littoral zone*.

Intertropical convergence zone (ITCZ). High-precipitation zone separating northeast trade winds of Northern Hemisphere from southeast trade winds of Southern Hemisphere.

Intraguild predation. Predation by one member of a guild on another; may interact with interspecific competition.

Intrasexual selection. Natural selection in which competition occurs among members of the same sex for mates; compare *intersexual selection*.

Intraspecific. Of interactions, occurring between individuals of the same species; compare *interspecific*.

Intraspecific competition. Competitive interaction between two or more individuals of the same species; compare *interspecific competition*.

Intrinsic rate of increase (*r*). Instantaneous per capita growth rate of a population, expressed as the proportional increase per unit time; characterizes the growth rate of an exponentially growing population with a stable age distribution in a non-limiting environment.

Introduced (alien) species. Species occupying locations other than where it is native, as a result of human activity; also *exotic species*; compare *invasive species*.

Invasive species. Species (often introduced) that colonizes new habitat, spreads, and outcompetes native species.

Ion. Atom or molecule that is electrically charged as a result of the loss or gain of electrons; either an *anion* (negatively charged) or *cation* (positively charged).

Isoosmotic. Of body tissues, having a salt concentration that is identical to that of the surrounding medium; compare *hyperosmotic* and *hypoosmotic*.

Iteroparity. Reproductive strategy in which the individual has repeated reproductive events over its lifetime; compare *semelparity*.

K-strategist (species). Species in which individuals are relatively large and long-lived, with slow growth and stable populations; typically produce few offspring, and provide parental care; compare *r-strategist (species)*.

Keystone predation. Predatory interaction that maintains community structure by reducing the abundance of superior competitors and preventing competitive exclusion of inferior competitors.

Keystone species. Species whose activities have an effect on community structure or function in disproportion to their number, often through keystone predation; also *functional dominant*; compare *bioengineer*.

Kin selection. Natural selection that increases fitness indirectly by favouring selection of close relatives; see *inclusive fitness*.

Kinetic energy. Energy associated with motion; performs work at the expense of *potential energy*.

Kleptoparasitism. Behavioural parasitism in which a parasite appropriates resources gathered by an individual of another species; usually non-symbiotic; compare *brood parasite*.

Krummholz. Stunted and/or asymmetric tree form characteristic of the transition zone between alpine tundra and subalpine coniferous forest; also occurs in Arctic tundra.

La Niña. Global climate phenomenon characterized by strong winds and cool ocean currents flowing westward from the coastal waters of South America to the tropical Pacific; compare *El Niño–Southern Oscillation*.

Landscape. Area of land composed of a number of adjoining ecosystems; *seascape* if marine; see *patch*.

Landscape ecology. Study of the spatial extent and arrangement of interacting ecosystems in a landscape.

Late-successional species. Long-lived, shade-tolerant plant species, typically with low rates of dispersal, colonization, and growth; compare *early-successional species*.

Latent heat. Energy released or absorbed during a phase change, such as evaporation.

Laterization. Soil-forming process in humid tropical climates, characterized by intense oxidation and leaching; results in highly weathered soils.

Leaching. Movement of solutes in aqueous solution through soil; compare *eluviation* and *illuviation*.

Leaf area index (LAI). Total leaf area exposed to incoming radiation relative to ground surface area; usually defined for the total plant community.

Leaf area ratio (LAR). Total leaf area as a proportion of total plant mass; compare *specific leaf area*.

Leaf pack. (1) Aggregation of leaf material in areas of deposition, usually in streams. (2) Type of *litterbag* used to quantify decomposition of litter in streams.

Lek. Communal courtship area in which males aggregate to attract and mate with females; contains temporary mating territories.

Lentic. Pertaining to standing water, such as lakes and ponds; compare *lotic*.

Life expectancy. Mean number of years to be lived; usually age-specific; i.e., mean number of additional years that an individual of a certain age is expected to live.

Life history. Organism's lifetime pattern of growth, development, and reproduction.

Life table. Tabulation of age-specific mortality of a population; a *dynamic (cohort) life table* is based on a group of individuals born into a population at the same time; a *dynamic–composite life table* is based on a group of individuals born over several time intervals; a *static life table* is based on the distribution of age classes in a population during a single time period.

Light compensation point (LCP). Light level (value of PAR) at which gross photosynthetic gain equals respiratory loss, i.e., net carbon gain is 0; compare *light saturation point*.

Light-dependent reactions. Sequence of photosynthetic reactions in which chlorophyll absorbs light energy and produces ATP and NADPH via electron transport in the chloroplast; compare *light-independent reactions*.

Light extinction coefficient. In Beer's law, amount of light attenuated per unit of leaf area index or with depth in a water column.

Light-independent reactions. Sequence of photosynthetic reactions that uses the products of the light-dependent reactions to reduce CO_2 to sugars in the chloroplast; also *Calvin–Benson cycle*.

Light saturation point. Light level (value of PAR) at which a plant achieves its maximum net photosynthetic rate; compare *light compensation point*.

Limited similarity. Theory that ecologically similar species can only coexist if they evolve sufficient differences in morphological traits to reduce niche overlap and prevent competitive exclusion; see *niche differentiation*.

Limiting factor. Environmental factor that limits growth of an individual or abundance and distribution of a population, by being present in either too low or excessive amounts.

Limnetic zone. Open-water zone of a lake or ocean; compare *littoral zone*.

Linkage density. Mean number of links per species in a food web; compare *connectance*.

Lithosphere. Solid earth and soil; compare *hydrosphere* and *atmosphere*.

Litterbag. Mesh bag used to quantify the rate of litter decomposition; compare *leaf pack*.

Littoral zone. (1) Shallow-water zone of a lake or ocean, in which light reaches the bottom, permitting growth of rooted plants; compare *limnetic zone*. (2) Intertidal zone between high and low water.

Local (sub)population. Subpopulation of a metapopulation occupying a local habitat patch.

Locus. Site on a chromosome occupied by the allele of a specific gene; see *homologous chromosomes*.

Logistic growth. Mathematical model of population growth describing the S-shaped growth curve in which the rate of increase decreases linearly as population size increases, reaching a plateau at carrying capacity; compare *exponential growth*.

Longshore (littoral) drift. Transportation of sediments (e.g., sand) along a coast at an angle to the shoreline.

Longwave (thermal) radiation. Infrared radiation of wavelengths greater than 4 μm; compare *shortwave radiation*.

Lotic. Pertaining to flowing water; compare *lentic*.

Lotka–Volterra competition model. Model that predicts the outcome of interspecific competition, based on the impact of each species on logistic growth of the other species.

Lotka–Volterra predation model. Model that predicts the regulatory impact of a predator–prey interaction on the logistic growth of both species.

Lung. Internal cavity lined with air sacs that increase the surface area for O_2 uptake; compare *gill* and *tracheae*.

Lyotropic series. Sequence of cations in order of their strength of bonding to soil exchange sites.

Macrocosm. Natural system that is used for ecosystem experimentation, such as a lake or forest; compare *microcosm* and *mesocosm*.

Macronutrients. Essential nutrients needed by an organism in large amounts for some portion of its life cycle; compare *micronutrients*.

Macroparasite. Large parasite (e.g., parasitic worms, lice, and fungi) that has a comparatively long generation time and is spread by direct or indirect transmission, often involving intermediate hosts or vectors; compare *microparasite*.

Mainland–island metapopulation structure. Metapopulation network in which one or several habitat patches (mainland) supply emigrants to other patches (islands).

Mangrove forest (mangal). Tropical inshore community dominated by mangrove trees and shrubs capable of growth and reproduction in areas inundated daily by seawater.

Marginal value theorem. Theory predicting the amount of time that a predator spends in a patch, based on the marginal rate of return on its energy expenditure; see *optimal foraging theory*.

Marine. Pertaining to saltwater habitats; compare *freshwater*.

Mark–recapture. Method of estimating population size of mobile animals by capturing, marking, releasing, and recapturing individuals; compare *census* and *quadrat*.

Marsh. Fresh or saltwater wetland dominated by emergent vegetation such as cattails and sedges; compare *swamp* and *peatland (mire)*.

Mass extinction event. Era in which the species extinction rate far exceeds the normal background rate, often as a result of cataclysmic environmental change.

Mast reproduction. Large and variable reproductive output by plants that is not due to abiotic factors; defends against herbivory by satiating herbivores.

Mating system. Pattern of mating in a population; see *monogamy* and *polygamy*.

Matric potential. Component of water potential associated with adhesion of water to *hydrophilic* (water-loving) substances such as cellulose in cell walls or clay in soil; compare *pressure potential* and *osmotic potential*.

Matrix. Background land-use type in a landscape mosaic, characterized by extensive cover and high connectivity; compare *patch*.

Maximum sustainable yield. Maximum rate at which individuals can be harvested from a population without reducing population size; achieved when recruitment balances harvesting.

Mechanical weathering. Breakdown of rocks and minerals by physical processes such as freezing; compare *chemical weathering*.

Meiofauna. Benthic organisms ranging in size from 0.1 to 1 mm; also *interstitial fauna*; compare *epifauna* and *infauna*.

Meristem. Region of active growth retained over the lifetime of a plant; see *modular organism*.

Mesic. Moderately moist, particularly of soil; compare *xeric* and *hydric*; a *mesophyte* is a plant adapted to mesic sites.

Mesocosm. Relatively large, artificial system used to conduct ecosystem experiments in a field setting; compare *microcosm* and *macrocosm*.

Mesophyll. Specialized photosynthetic tissue located in the middle region of a leaf; in C_3 species, differentiated into *palisade* and *spongy mesophyll*.

Metacommunity. Groups of similar local communities located in patches over a landscape.

Metapopulation. Groups of local populations linked by the dispersal of individuals.

Microbial loop. In marine systems, feeding loop in which bacteria take up dissolved organic matter produced by plankton, and are in turned consumed by nanoplankton; adds several trophic levels to a plankton food chain; compare *soil microbial loop*.

Microbivore. Organism that consumes bacteria and fungi in soil and litter.

Microclimate. Local climate, which may differ from the general climate of an area; influences organism presence and abundance; major defining feature of the microhabitat in which an organism lives.

Microcosm. Small, self-contained system designed to simulate an ecosystem for testing purposes; compare *mesocosm* and *macrocosm*.

Microflora. (1) Bacteria and fungi inhabiting soil, particularly those involved in microbial decomposition. (2) Microorganisms inhabiting the gut of an animal.

Micronutrients. Essential nutrients needed by an organism in small (trace) quantities for some portion of its life cycle; compare *macronutrients*.

Microparasite (pathogen). Small parasite (virus, bacterium, or protozoan), typically with short generation time, rapid increase, and direct transmission; causes disease; compare *macroparasite*.

Migration. (1) Movement of individuals in or out of a population. (2) Intentional, directional, often seasonal movement of animals (usually en masse) between two regions; involves individuals returning to their birth location to reproduce; see *philopatric*; compare *dispersal*.

Mineral lick. Site where animal licks or eats mineral-rich soil or other substrate.

Mineralization. Final stage of decomposition involving microbial breakdown of organic matter to inorganic (mineral) substances; see *net mineralization rate*; compare *immobilization*.

Minimum dynamic area (MDA). Area of suitable habitat necessary for maintaining a minimum viable population.

Minimum viable population (MVP). Population size that, for a given probability, ensures a population's existence for a given period of time; see *minimum dynamic area*.

Model. (1) Abstraction or simplification of a natural phenomenon, developed to predict and/or understand the phenomenon; see *null model*. (2) In mimicry, organism mimicked by another organism.

Modular organism. Organism that exhibits indeterminate growth involving repeated iteration of parts, such as plant branches or shoots, some of which may become independent; see *ramet* and *genet*; compare *unitary organism*.

Monoclimax hypothesis. View of succession culminating in a single, predictable climax community for a given climate, with minimal impact of other environmental factors or chance; compare *polyclimax hypothesis* and *pattern climax theory*.

Monoculture. Planting of a single plant species; compare *polyculture*.

Monoecious. Type of hermaphroditism in plants in which male and female organs are produced in different flowers on the same plant; compare *dioecious*.

Monogamy. Animal mating system involving maintenance of a pair bond with one member of the opposite sex at a time; compare *polygamy*.

Moor. See *blanket mire*.

Mortality curve. Graph of age-specific mortality rates at different ages; compare *survivorship curve*.

Mosaic. Pattern of patches, corridors, and matrices in a landscape.

Mutation. Heritable change in the structure of a gene (*micromutation*) or chromosome (*macromutation*).

Mutualism. Interspecific interaction in which individuals of two or more species benefit; compare *symbiosis*.

Mycorrhizae. See *ectomycorrhizae* and *endomycorrhizae*.

Myrmecochores. Plants that possess ant-attracting substances on seed coats; facilitates *myrmecochory* (seed dispersal by ants).

N-dimensional hypervolume. View of the niche as a hypothetical space with multiple dimensions, each describing species response to an environmental factor.

Natural capital. Natural resources provided by ecosystems; see *ecosystem services*.

Natural experiment. Non-experimental study that monitors response to a natural disturbance or event that has not been manipulated by the researcher; compare *experiment*.

Natural selection. Differential success (survival and reproduction) of individuals that reflects their interactions with the environment; generates adaptations and reduces maladaptive traits in populations; compare *artificial selection*.

Nectivores. Herbivores that feed on plant nectar; compare *granivores*.

Necrotrophic. An organism feeding on the dead tissue of another organism that it has killed; typical of parasitic fungi.

Needle-leaf evergreen. Conifer that retains its needles for several years; compare *broadleaf evergreen*.

Negative (correcting) feedback. Regulatory mechanism in which the output of a component counters the effect of a stimulus, causing the component to change in the opposite direction and returning a system to its set point; important in *density dependence*; see *homeostasis*; compare *positive (reinforcing) feedback*.

Neighbourhood habitat amelioration. Influence of organisms (usually plants) on abiotic factors in their immediate surroundings in ways that lessen the negative effects of these factors on their neighbours.

Nekton. Free-swimming aquatic animals; compare *zooplankton*.

Net assimilation rate (NAR). In plant growth, assimilation of new tissues per unit leaf area per unit time.

Net ecosystem productivity (NEP). Difference between net primary productivity and carbon lost in heterotrophic respiration; compare *net primary productivity*.

Net energy balance. Difference between the amount of radiation an organism or other object receives and the amount it reflects and/or emits.

Net mineralization rate. Difference between the rates of mineralization and immobilization during decomposition; determines the supply rate of minerals to soil.

Net photosynthesis. Difference between the rate of CO_2 uptake in gross photosynthesis and CO_2 loss in plant respiration; net carbon gain; compare *net primary productivity*.

Net primary productivity (NPP). Rate of net photosynthesis for the autotrophic community; measured as energy stored as organic matter by primary producers per unit area per unit time; compare *gross primary productivity*.

Net production efficiency (NPE). Measure of efficiency with which a consumer incorporates assimilated energy into secondary production; compare *gross production efficiency*.

Net reproductive rate. Mean number of females produced by a female over her lifetime, taking the probability of surviving to a specific age into account; compare *gross reproductive rate*.

Niche. Functional roles of a species in its community, including its activities and responses to and impacts on its environment; see *n-dimensional hypervolume*, *fundamental niche*, and *realized niche*; compare *habitat*.

Niche breadth. Range of niche dimensions occupied by a species; see *generalist* and *specialist*.

Niche complementarity. Tendency for co-occurring species in a community to have complementary niche dimensions, thereby optimizing the efficiency of resource use; compare *functional redundancy*.

Niche compression. Contraction of one of more realized niche dimensions in the presence of an interacting species; compare *competitive release*.

Niche conservatism. Tendency for a species to retain a similar fundamental niche over evolutionary time; compare *niche differentiation*.

Niche differentiation. Evolution of differences in fundamental niches as a result of natural selection; see *resource partitioning*; compare *niche conservatism*.

Niche overlap. Similarity among species in one or more niche dimensions; compare *niche differentiation*.

Niche packing. Pattern of the number and relationships (especially overlap) among niches of co-occurring species in a community; if all niches are occupied, the pattern is called *niche saturation*.

Nitrification. Breakdown of nitrogenous organic compounds into nitrites and nitrates; compare *ammonification*.

Nitrogen fixation. Abiotic or biotic conversion of atmospheric N_2 into forms that are usable by biota.

Nitrogen saturation. Excess of nitrogen leading to a cascade of effects related to nutrient imbalances; see *forest dieback*.

Nominal data. Data in which objects fall into unordered categories; compare *ordinal data*.

Non-random mating. See *assortative mating*.

Non-renewable resource. Resource that is not replenished by an ecosystem, or is replenished at a rate that is infinitesimally small relative to its consumption; compare *renewable resource*.

Null (neutral) model. Model that assumes observed patterns are generated by random processes, in the absence of a particular ecological mechanism.

Numerical data. Data in which quantitative traits of an entity are measured; either *discrete data* or *continuous data*.

Numerical response. Change in predator population size in response to a change in prey density; compare *functional response*.

Nutrient cycling. See *internal cycling*.

Nutrient spiralling. Mechanism of nutrient retention in flowing-water systems, involving the combined processes of recycling and downstream transport.

O horizon. Organic layer of soil, consisting of plant material at various stages of decomposition; subdivided into L, F, and H layers.

Obligate. (1) Having no alternative in response to a particular condition or way of life, e.g., an obligate anaerobe. (2) Pertaining to an interaction (often symbiotic) in which one species cannot survive and/or reproduce without the other species; compare *facultative*.

Old growth. Vegetation (usually forest) that has been undisturbed by humans for centuries; criteria include specific age (> 250 years) and the presence of structural features such as *snags* (dead standing trees); compare *climax*.

Oligotrophic. Of a water body, having a low nutrient content and productivity; compare *eutrophic* and *dystrophic*.

Omnivore. Organism that consumes both plant and animal tissue; sometimes restricted to biophages but may also be used for detritus feeders; *ominvory* is the act of feeding on plant and animal tissue; compare *carnivore* and *herbivore*.

Open. (1) Of a population, with significant emigration or immigration. (2) Of a system, with significant exchange of inputs and outputs with its environment; compare *closed*.

Operative temperature range. Range of body temperatures over which a poikilotherm is active; compare *thermoneutral zone*.

Optimal foraging theory. Theory that natural selection favours efficient foraging, i.e., behaviour that maximizes energy and nutrient gain for energy expended.

Optimum range. Portion of the tolerance range of an individual or population that is most favourable for its growth and development; compare *zone of physiological stress*.

Ordinal data. Type of data in which objects fall into ordered categories; compare *nominal data*.

Organismal (holistic) concept. View that species, especially plants, are integrated into an interdependent unit with a predictable developmental sequence; compare *individualistic concept*.

Ornamentation. Secondary sex traits involving bright colours and/or elaborate external structures that influence mate selection; see *intersexual selection*.

Oscillations. See *population cycles*.

Osmosis. Movement of a solvent (often water) across a differentially permeable membrane in response to a concentration or pressure gradient; compare *diffusion*.

Osmotic potential. Component of water potential associated with the solute concentration of the cell or soil solution; a higher concentration means a lower (larger negative number) osmotic potential and a higher rate of water uptake across a membrane; compare *pressure potential*.

Outcrossing. Fusion of gametes from different individuals; called *outbreeding* if individuals are distantly related; can enhance offspring vigour or cause outbreeding depression if it disrupts co-adapted gene complexes, making the offspring less adapted to local conditions; compare *inbreeding*.

Outputs. Export of materials, energy, or organisms from an ecosystem to the surrounding environment; compare *inputs*.

Oxygen minimum zone. Water depth at which O_2 reaches a minimum; 500 to 1000 m in oceans.

Pack. Extended kin group, often with a mated pair and a defined social hierarchy.

Paleoecology. Study of ecology of past communities based on the fossil record.

Paradox of the plankton. Theory regarding why so many phytoplankton species coexist despite resource shortages and strongly overlapping niche dimensions.

Parallel evolution. Evolution of related taxa that are exposed to similar conditions after separation; often generates *ecological equivalents*; compare *convergent evolution*.

Parasite. Organism that feeds on or in a living host in an intimate relationship (often symbiotic) that harms but usually does not kill the host directly; compare *parasitoid*.

Parasitoid. Parasite (typically a larval stage of an insect) that kills its host by consuming the host's soft tissues before pupating or metamorphosizing into an adult; some restrict the term to insects, but it can refer to other invertebrate parasites with a similar strategy.

Parent material. Mineral substance from which soil forms, either as underlying bedrock or deposited by wind, water, ice, or gravity; compare *A horizon* and *B horizon*.

Parietal eye. Photosensory organ or light-sensitive spot connected to the pineal gland in some animals; involved in the functioning of the biological clock, not vision.

Parthenogenesis. Asexual reproduction in which an individual develops from an unfertilized egg; compare *apomixis*.

Particulate organic matter (POM). Dead organisms and other organic matter suspended in the water column, gradually drifting to the bottom; compare *dissolved organic matter*.

Patch (habitat patch). Discrete area of habitat that differs from the surrounding landscape matrix and that supports a local population; see *metapopulation*.

Patch occupancy. Proportion of habitat patches that are occupied by local populations in a metapopulation.

Pattern climax theory. View that local environmental variation (both abiotic and biotic) generates differences in climax species composition during succession; compare *monoclimax hypothesis* and *polyclimax hypothesis*.

Peatland (mire). Ecosystem with large accumulation of *peat* (unconsolidated material consisting of undecomposed and slightly decomposed organic matter derived from *Sphagnum* mosses under very moist conditions); see *bog* and *fen*.

Pelagic zone. Open-water zone of a sea or ocean, divided into neritic and oceanic provinces and several vertical strata.

Per capita birthrate. Number of individuals born into a population per unit time as a proportion of total population size; called *crude birthrate* if expressed per 1000 individuals.

Per capita death rate. Number of individuals dying in a population per unit time as a proportion of total population size; called *crude death rate* if expressed per 1000 individuals.

Percolation. Movement of water downward through subsurface soil, often entering groundwater; compare *infiltration* and *surface runoff*.

Permafrost. Permanently frozen soil; found in both tundra and portions of the northern boreal forest.

Persistence. Ability of ecosystem to remain viable despite repeated stress, oscillations in its populations, and departure from equilibrium for long periods.

Phenotype. Expression of a trait, as determined by the genotype and influenced by the environment.

Phenotypic plasticity. Ability of an genotype to alter the phenotypic expression of one or more traits in different environments; see *reaction norm*.

Pheromone. Chemical released by an animal that influences the behaviour (often but not only reproductive) of other individuals of the same species.

Philopatric. Of an individual, tending to remain in the same location or in a migratory species, to return to the same location.

Photic (euphotic) zone. Portion of the water column with enough light to support positive net photosynthesis; see *compensation depth*; compare *aphotic zone*.

Photoautotroph. Organism (primary producer) that derives energy for converting CO_2 into organic molecules through photosynthesis; compare *chemoautotroph*.

Photoinhibition. Negative effect of high light on plant processes, especially photosynthesis.

Photoperiod. Daily period of light and dark; *photoperiodism* is the response of organisms to seasonal changes in photoperiod; see *critical day length*.

Photorespiration. Release of recently fixed CO_2 by plants in light, as a result of Rubisco catalyzing the oxygenation of RuBP; reduces the photosynthetic efficiency of C_3 plants.

Photosynthesis. Process whereby green plants, algae, and some bacteria use light energy to convert CO_2 and water into carbohydrates; see *net photosynthesis*.

Photosynthetically active radiation (PAR). Portion of the solar electromagnetic spectrum between wavelengths of approximately 400 and 700 nm that is used by plants in photosynthesis; compare *insolation*.

Phyllosphere. Environment of leaf surfaces, colonized by epiphytic fungi; compare *rhizosphere*.

Physiognomy. Outward appearance of vegetation, as determined by the dominant growth forms.

Phytochrome. Non-photosynthetic pigment involved in many plant physiological processes, including photoperiodic timing of flowering and germination.

Phytoplankton. Small, floating, autotrophic algae and bacteria in aquatic systems; compare *zooplankton*.

Pioneer species. See *early-successional species*.

Planktivore. Aquatic animal that consumes plankton.

Plankton. Small, floating or weakly swimming algae and animal-like organisms as well as some small larval invertebrates; see *phytoplankton* and *zooplankton*; compare *nekton*.

Pneumatophore. Modified root that protrudes above a water-logged soil to access O_2; typical of mangrove species.

Podzolization. Soil-forming process in which acid leaches the A horizon, and iron, aluminum, silica, and clays accumulate in lower horizons; typical of non-organic boreal forest soils.

Poikilothermy. Strategy of allowing body temperature to vary with external temperature; *poikilotherm* is an organism whose body temperature varies with external temperature and that lacks homeostatic temperature control; see *ectothermy*; compare *homeothermy*.

Pollutant. Harmful substance released directly or indirectly by human activities; includes wastes and by-products of industrial production, as well as pesticides.

Polyandry. Type of polygamy in which one female mates with several males at the same time; compare *polygyny*.

Polyclimax hypothesis. View that succession can culminate in different climax vegetation types in a similar climate, depending on the substrate; compare *monoclimax hypothesis*.

Polyculture. Planting or raising of several species; compare *monoculture*.

Polygamy. Mating system in which an individual acquires two or more mates at the same time, none of which is mated to others; see *polyandry* and *polygyny*; compare *monogamy*.

Polygyny. Type of polygamy in which one male mates with several females at the same time; compare *polyandry*.

Polymorphism. Existence of distinct forms in a population, usually as a result of divergent selection in the same location; may generate new species if reproductive isolation occurs.

Polyploidy. Multiplication of the normal chromosome number; promotes plant speciation if it follows interspecific hybridization; the *polyploid* condition involves three or more copies of a chromosome, as opposed to being *haploid* (one copy) or *diploid* (two copies).

Population. Group of individuals of the same species living in the same area at the same time.

Population cycle (oscillation). Recurrent fluctuation in population size that is more regular than expected from chance.

Population density. Number of individuals in a population per unit area or volume; also *crude density*; compare *ecological density*.

Population ecology. Study of how populations grow, spread, and interact with their own and other populations.

Population genetics. Study of changes in gene frequency and/or genotypes in populations.

Population growth. Change in population size (positive or negative) as a result of birth, death, immigration, and emigration.

Positive (reinforcing) feedback. Feedback that reinforces a process, causing a system (or subsystem component) to continue changing in the same direction as triggered by the original stimulus; compare *negative feedback*.

Potential energy. Stored energy that is available to do work; compare *kinetic energy*.

Pothole. Basin wetland in prairies; may retain water year-round depending on its size and depth.

Precipitation. All forms of water that reach Earth, including rain, snow, hail, sleet, fog, and mist.

Precocial. In birds and mammals, being born in a relatively advanced stage, with the ability to see and move independently; compare *altricial*.

Predation. Interaction in which one living organism consumes another, in whole or in part; *true predation* involves one animal consuming another (*carnivory*).

Predator satiation. Anti-predator defence involving the timing of reproduction by prey to maximize output within a short period to satiate predators and allow more offspring to survive; see *mast reproduction*.

Pre-emption. Mechanism of competition whereby the presence of an individual prevents or inhibits establishment by individuals of the same or different species.

Pressure potential. Component of water potential that typically involves positive pressure exerted by the plant cell wall on the cell contents; compare *osmotic potential*.

Primary producer. Green plant, alga, or bacterium that converts light or chemical energy into organic matter; see *photoautotroph* and *chemoautotroph*; compare *secondary producer*.

Primary succession. Sequence of vegetation changes on a previously unvegetated site; compare *secondary succession*.

Priority effect. Situation in which the outcome of an interaction is affected by the initial population densities.

Productivity. Rate of energy fixation or storage by an individual, population, or community per unit time; see *gross primary productivity*, *net primary productivity*, and *secondary productivity*.

Profitability. In foraging, energy per unit time spent searching for and handling a diet item.

Profundal zone. Deep-water zone in lakes or seas, below the *limnetic zone* and above the *benthic zone*.

Promiscuity. Mating system in which males and females mate with more than one member of the opposite sex, without forming a pair bond; compare *monogamy* and *polygamy*.

Pulsing state. Loosely regulated state of an ecosystem in which components tend to return to a trajectory rather than to a set point; see *homeorhesis*; compare *homeostasis*.

Putrefaction. Second stage in the decomposition of animal tissue, involving bloating and decay from the action of anaerobic bacteria; compare *autolysis*.

Pycnocline. Area in a water column in which the greatest change in density occurs for a given change in depth; compare *thermocline*.

Pyromineralization. Release by fire of minerals bound in organic compounds.

Quadrat. Square, rectangular, or round plots used for estimating population size or community composition for plants or sessile animals; compare *census* and *mark–recapture*.

Quaking bog. Bog with a floating mat of peat and vegetation over water in a basin; compare *blanket mire (moor)*.

Qualitative trait. Phenotypic trait that falls into a limited number of discrete, non-numerical categories; compare *quantitative trait*.

Quantitative trait. Phenotypic trait with a continuous, numerical distribution; compare *qualitative trait*.

r-strategist (species). Species that is short-lived, with small body size, rapid growth, large reproductive output at low densities, and minimal parental care; compare *K-strategist (species)*.

Radionuclides. Fission and non-fission products of nuclear reactions.

Rain shadow. Dry region on the leeward side of a mountain resulting from precipitation at upper altitudes on the windward side.

Raised bog. See *blanket mire (moor)*.

Ramet. In modular organisms, individual that arises as a clonal (asexual) offshoot of a genet, and is capable of independent existence; compare *genet*.

Rank–abundance curve (Whittaker plot). Plot of the relative abundance of each species in a community against its rank order from most to least abundant.

Reabsorption. See *retranslocation*.

Reaction norm. Set of phenotypes expressed by a single genotype across a range of environments; see *phenotypic plasticity*.

Realized niche. Portion of its fundamental niche occupied by an organism (or species) in presence of interacting species.

Recessive allele. Allele whose phenotypic effect is expressed in the homozygous state and masked in organisms that are heterozygous for that gene (assuming complete dominance).

Regolith. See *C horizon*.

Relative abundance. Number of individuals in a single species as a proportion of the total number of individuals of all species in a community; see *importance value*.

Relative growth rate (RGR). Mass gained during a given time interval, relative to the mass of the individual at the beginning of the time interval; compare *net assimilation rate*.

Relative humidity. Water vapour content of air at a given temperature, expressed as percentage of saturation water vapour.

Relative net primary productivity (RNPP). Net primary productivity of a community divided by its standing crop biomass; also *P:B ratio*; compare *biomass accumulation ratio* and *residence time*.

Relaxation effect. Delayed loss of a species as a result of fragmentation; reflects time lags in birth and/or death rates or a failure to recolonize; also called *extinction debt*.

Removal experiment. Experiment that tests the impact of an interspecific interaction (often competition) by monitoring the response of remaining species to removal of one or more potentially interacting species; compare *transplant experiment*.

Renewable resource. Resource that can be replenished by an ecosystem; compare *non-renewable resource*.

Reproductive effort. Proportion of its time and resources that an organism allocates to reproduction.

Reproductive isolation. Genetic separation of populations (by pre- or post-mating isolation mechanisms) as a result of the inability to produce viable offspring.

Rescue effect. In metapopulation dynamics, increase in population size and decrease in extirpation risk as a result of the immigration of individuals.

Residence time. (1) Length of time that energy persists in a trophic level, calculated for primary producers as the energy stored in standing crop biomass divided by net primary productivity; compare *biomass accumulation ratio*. (2) Length of time before litter is decomposed; calculated as the litter accumulation divided by the rate of litter production.

Resilience stability. (1) Ability of system component or function to re-establish after change. (2) Measure of how quickly a population returns to equilibrium after a disturbance; see *return time*; compare *resistance stability*.

Resistance. Ability of an organism to cope with an abiotic stress, by avoidance and/or tolerance.

Resistance stability. (1) Ability of a system component or function to resist change. (2) Measure of a population's ability to remain at equilibrium; compare *resilience stability*.

Resource. Environmental factor or component that is consumed by an organism and that is necessary for its survival, growth, and/or reproduction; compare *condition* and *hazard*.

Resource depletion zone. Region of reduced resource availability located in the immediate vicinity of plant roots and/or leaf canopies as a result of resource uptake.

Resource partitioning. Niche differentiation in which species divide up available resources by using either different resources or similar resources in different ways or at different times; promotes *coexistence* and lowers the risk of *competitive exclusion*.

Resource-ratio hypothesis. Theory that succession is due to altered relative competitive abilities of species as a result of shifts in available resources over time; see *tolerance model*.

Respiration. See *aerobic respiration* and *anaerobic respiration*.

Restoration ecology. Application of principles of ecosystem ecology to restoration of natural communities, often on disturbed sites.

Rete. Network or bundle of intermingling veins and arteries that acts as a heat exchanger in mammals and some fish; see *countercurrent exchange*.

Retranslocation (reabsorption). Recycling of nutrients from senescing (dying) tissues within a living organism.

Return interval. Mean time that elapses between disturbance events; see *frequency*.

Return time. Time for a population to regain equilibrium after a disturbance; see *resilience stability*.

Rhizobia. Collective term for *Rhizobium* bacteria that are capable of fixing atmospheric N_2 in mutualistic interaction with plants in the legume family.

Rhizosphere. Environment in the immediate vicinity of roots, occupied by microflora; see *soil microbial loop*; compare *phyllosphere*.

Riparian woodland. Woodland along the bank of a river or stream; riverbank forests are called *gallery* or *bottomland forests*; only flooded for part of season; compare *swamp*.

Riverine wetland. Non-wooded wetland along shallow or periodically flooded banks of rivers and streams, with lateral water flow; compare *basin wetland* and *fringe wetland*.

Root-to-shoot ratio. Ratio of the total dry mass of roots to shoots of a plant or plant community.

Rotation period. (1) Interval between disturbance events. (2) Interval between harvests of a crop, such as trees; also *harvest interval*.

Rubisco. Enzyme in the light-independent reactions of photosynthesis that catalyzes the initial step in the reduction of CO_2 to sugar.

Ruderal. In Grime's model, species inhabiting disturbed sites; typically small, short-lived, with high reproductive output and specialized for dispersal; compare *competitors* and *stress-tolerators*.

Ruminant. Ungulate (hooved mammal) with a three- or four-chamber stomach; bacteria ferment plant matter in the large first chamber (*rumen*).

Salinity. Salt concentration of water or soil; measured as *practical salinity units* (parts per thousand, ‰) and estimated as grams of Cl per kilogram of H_2O.

Salinization. Process whereby soluble salts accumulate in soil, usually by upward capillary movement from groundwater in deserts and irrigated lands.

Salt (tidal) marsh. Community of emergent vegetation rooted in a soil that is alternately flooded and drained by marine tides; consists of a low marsh and a high marsh.

Sample. That part of a statistical population that is observed or measured; *sampling effort* is the number and size of sampling units in a study.

Sampling effect. View that species-rich experimental communities are more productive because they are more likely by chance to contain one or more highly productive species.

Sap-feeders. Herbivores that tap sugar-rich phloem tissue; compare *nectivores*, *grazers* and *browsers*.

Saprophage. Heterotroph that eats dead tissue; see *detritivore* and *decomposer*; compare *biophage*; not to be confused with *saprophyte* (plant that draws its nourishment from dead plant tissue) or *saprophytic fungus* (fungus that draws its nourishment from dead plant tissue).

Satellite population. Small local population maintained by emigration from a larger core population.

Saturation vapour pressure. Maximum water vapour that a volume of air can hold at a given temperature; *saturated* pertains to air or soil that contains the maximum water (as a vapour or liquid) possible at a given temperature and pressure; see *relative humidity*.

Savannah. Vegetation in drier tropics and subtropics, characterized by co-dominance of grasses with scattered trees or shrubs.

Scale. (1) Level of resolution within dimensions of time or space. (2) Spatial proportion as a ratio of length on a map to actual length.

Scaling. Predictable changes in morphological or physiological features with body size; see *surface area/volume ratio*.

Scavenger. Animal that eats dead animals or animal products; often reserved for vertebrates; functionally equivalent to a *detritivore*; may include *coprobivores*.

Sclerophyllous. Of vegetation, dominated by drought-tolerant broadleaf evergreen shrubs and dwarf trees; see *chaparral*.

Scramble competition. Type of competition in which limited resources are shared by all individuals, all of which experience detrimental effects; also *exploitation*; compare *contest competition*.

Scrapers. Aquatic invertebrates that feed on the algal coating on stones and rubble in streams; also *grazers*; compare *shredders* and *collectors*.

Search image. Mental image formed in predators that enables them to find prey faster and focus on preferred prey.

Search time. Amount of time a predator spends seeking prey; compare *handling time*.

Second law of thermodynamics. Principle that in any energy transfer or transformation, some useful energy is lost as *entropy*; compare *first law of thermodynamics*.

Second-level consumers. Organisms that eat first-level consumers (*herbivores*); a second-level carnivore is also called a *first-level carnivore* and a *third-level consumer*.

Secondary producer. See *heterotroph*.

Secondary productivity. Rate of biomass production by consumers, including new tissues and offspring; measured as grams per unit area per unit time; compare *net primary productivity*.

Secondary succession. Development of vegetation on a previously vegetated site after a disturbance has removed part of all of the community; compare *primary succession*.

Sediment trap. In coastal systems, deposition of sediments within a short distance of the area where a river meets the sea; see *delta* and *estuary*.

Sedimentary cycle. Biogeochemical cycle in which the main reservoir is in solid rock or soil; compare *gaseous cycle*.

Seed bank. Reservoir of dormant seeds, usually in soil or sediments; compare *bud bank*.

Seed-tree harvesting. Method in which a small number of trees are left to provide a seed source for natural regeneration; compare *clear-cutting*, *shelterwood cutting*, and *selection cutting*.

Selection. Differential survival or reproduction of individuals as a result of phenotypic differences; *selection pressure* is any force acting on individuals that affects which individuals leave more descendants than others; see *fitness*, *natural selection*, and *artificial selection*.

Selection (selective) cutting. Method in which only single trees or groups of trees are harvested; compare *clear-cutting*, *shelterwood cutting*, and *seed-tree harvesting*.

Selective agent. Environmental cause of fitness differences among organisms with different phenotypes; see *natural selection*; compare *target of selection*.

Self-thinning. Progressive decline in density and increase in survivor biomass in a population as a result of density-dependent mortality caused by intraspecific competition.

Semelparity. Reproductive strategy involving a single reproductive effort in the lifetime of an organism, followed by death; compare *iteroparity*.

Sere. Sequence of successional stages on a site; *seral stage* is a stage in a vegetation continuum over time.

Serotiny. Condition in which heat (usually associated with fire) is needed to release seeds from the cones of some coniferous trees or desert species.

Severity. Proportion of the total biomass or numbers of a population that a disturbance eliminates; compare *intensity*.

Sex ratio. Proportion of males to females in a population; *primary sex ratio* is the ratio at conception; *secondary sex ratio* is the ratio at birth.

Sexual dimorphism. Presence of differences in secondary sex traits (traits not involved directly in reproduction) between males and females in a species; often resulting from *intra-* or *intersexual selection*.

Shade-intolerant. Growing and reproducing best under high light, with little tolerance for shade; also *sun-adapted*; compare *shade-tolerant*.

Shade-tolerant. Able to grow and reproduce under low light; compare *shade-intolerant*.

Shelterwood cutting. Method involving a series of three cuts that establishes an even-aged stand of a desired tree species; compare *clear-cutting*, *seed-tree harvesting*, and *selection cutting*.

Shifting-mosaic steady state. Changing pattern of patches on a landscape in which each patch passes through successional stages but the overall pattern remains constant.

Shivering. Involuntary muscular activity that increases heat production.

Shortwave radiation. Solar radiation of wavelengths from 0.1 to 4 μm (cutoffs vary); includes visible, near-ultraviolet, and near-infrared spectra; compare *longwave radiation*.

Shredders. Aquatic invertebrates that consume particulate organic matter and microbes on litter in streams; compare *collectors* and *scrapers*.

Shrub layer. Vertical stratum dominated by *shrubs* (woody species with multiple stems) within a forest understory; compare *canopy*.

Siblicide. Killing of an offspring by a sibling, often after asynchronous hatching.

Sink population. Local population that produces fewer individuals than needed for replacement and requires immigration from source populations to persist; *sink patches* are marginal habitats in which sink populations occur.

Snag. Dead (or partially dead) standing tree of at least 10-cm diameter at breast height and 1.8 m tall; typical of old-growth forests and important habitat for cavity-nesting birds and mammals; see *old growth*.

Social dominance. Dominance of one individual over another, usually maintained by aggressive behaviour and resulting in a dominance hierarchy.

Soil. Substrate (usually mineral) formed by the weathering of rocks and the action of living organisms.

Soil erosion. Movement of soil particles by wind, water, frost action, or gravity; often results from *surface run-off*.

Soil horizon. Horizontal layer in a soil with characteristic thickness, colour, texture, structure, acidity, and nutrient composition; see *soil profile*.

Soil microbial loop. Process whereby root exudates supply C to microbial decomposers in the rhizosphere, enhancing decomposition of soil organic matter and subsequently releasing minerals for plant uptake after consumption of microbes by microbivores; compare *microbial loop*.

Soil order. Basic unit of soil classification system; different types of soil as determined by parent material, soil-forming processes, climate, and biotic action.

Soil organic matter. Residual organic matter (largely humus) in a soil from microbial decomposition; affects water and cation-holding capacity.

Soil profile. Distinctive layering of horizons in a soil.

Soil solution. Aqueous solution held in the soil matrix.

Soil texture. Relative proportions of particle size classes (sand, silt, and clay) in a soil; affects soil water-holding and cation exchange capacity.

Solar constant. Rate at which solar energy is received at point just outside Earth's atmosphere.

Solar radiation. Electromagnetic energy emitted by the Sun; see *shortwave radiation* and *longwave radiation*.

Solifluction terrace. Landscape formation caused by downhill movement of a water-saturated soil; typically form over permafrost in Arctic tundra.

Solution. Liquid that is a homogeneous mixture of two or more substances (solutes); called an *aqueous solution* if water is the solvent.

Source population. Local population that produces more individuals than needed for replacement, thus contributing emigrants to sink populations; *source patches* are habitats (typically high-quality) in which source populations occur.

Spatial dispersion (distribution). Spatial pattern of organisms in a population; can be *random*, *clumped*, or *uniform (regular)*.

Specialist. Species with a narrow fundamental niche for one or more key environmental factors; compare *generalist*.

Speciation. Evolution of new species, often through separation of a population into two or more reproductively isolated subpopulations; see *allopatric* and *sympatric speciation*.

Species accumulation curve. Plot of the cumulative number of species encountered against the total number of samples; estimates *species richness*; compare *species rarefaction curve*.

Species area curve. Type of species accumulation curve in which the total area sampled rather than the total number of samples is used on the *x*-axis.

Species composition. Identity of all species present in a community; compare *species richness*.

Species diversity. (1) In community ecology, concept that combines species richness and evenness, as quantified by *diversity indices*. (2) In conservation ecology, total number of species present in a community (richness).

Species evenness. Measure of the distribution of individuals among all the species in a community; component of *species diversity*; compare *species richness*.

Species pool. Totality of species that are present in a region, a subset of which occurs in a given community.

Species rarefaction curve. Curve produced by repeatedly re-sampling the pool of samples at random and plotting the number of species in each re-sampling; allows comparing surveys with differing sample size; compare *species accumulation curve*.

Species richness. Number of species in a community; component of *species diversity*; compare *species evenness*.

Species turnover rate. In the equilibrium theory of island biogeography, rate at which species are lost to extinction and replacement species are gained by immigration when a community is at equilibrium; alters species composition but not richness.

Specific heat. Amount of energy that must be added or removed to raise or lower temperature of 1 gram of a substance by 1°C.

Specific leaf area (SLA). Ratio of the surface area to mass of a single leaf; compare *leaf area ratio*.

Spiracles. Openings in the insect body wall that connect to tracheal tubes.

Stabilizing selection. Selection that favours individuals with phenotypes near the mean at the expense of the extremes; compare *divergent selection* and *directional selection*.

Stable age structure. Age structure in which the proportion of individuals in different age classes remains constant; compare *stationary age structure*.

Standing crop biomass. Total biomass per unit area at a given time; compare *net primary productivity*.

Stationary age structure. Subtype of stable age structure in which the birthrate equals the death rate, and population size remains constant.

Statistical population. Set of entities about which inferences are drawn.

Stochastic niche theory. Theory stressing the impact of stochastic forces on the ability of species to establish and compete for resources with existing species.

Stochasticity. See *demographic stochasticity* and *environmental stochasticity*.

Stoichiometry. Branch of chemistry dealing with the quantitative relationships of elements in combination.

Stomata. Pores on a plant leaf or stem that allow gas exchange between internal tissues and the atmosphere.

Storage effect. Means by which populations are buffered against the negative effects of interspecific competition by producing more offspring and/or accumulating more biomass in long-lived individuals in favourable years or habitats; promotes species coexistence.

Streamflow. Current velocity of a river or stream within its natural channel.

Stress-tolerators. In Grime's model, species that inhabit sites where resources are limited as a result of harsh conditions; typically small, long-lived but slow-growing, with limited reproduction; allocate resources to maintenance over growth; compare *ruderals* and *competitors*.

Subspecies. Subdivisions of a species, distinguishable by morphological, behavioural, and/or physiological traits; may or may not be geographically isolated.

Succession. Change in community structure over time, involving replacement of one seral stage by another; see *primary succession* and *secondary succession*.

Succulence. Type of perennial plant growth form (e.g., cacti) with specialized water storage tissues; adapted to dry habitats and often combined with CAM photosynthesis.

Sunflecks. Light that penetrates canopy openings and reaches the ground without passing through leaves.

Supercooling. Cooling of body temperature of ectotherms below freezing without ice formation, by means of antifreeze proteins; compare *freezing point reduction*.

Supralittoral zone. Highest zone on an intertidal shore, bounded below by the upper limit of barnacles and above by the upper limit of *Littorina* snails.

Surface area to volume ratio (SA/V). Organism's surface area as a ratio of its volume; scaling relationship affecting the exchange of energy and materials of an organism with its environment.

Surface fire. Frequent, fast-moving fire that burns litter layer of forests or grasslands; compare *ground fire* and *canopy fire*.

Surface runoff. Movement of water across the ground as overland flow; eventually makes its way into groundwater or water bodies; causes *soil erosion*.

Surface tension. Tautness of the surface of a liquid caused by greater attractive forces between the liquid molecules than between the liquid and the air.

Survivorship. Probability that an individual in a cohort will survive to a given age.

Survivorship curve. Graph depicting the age-specific survivorship of a cohort from birth to the maximum age reached by cohort members; compare *mortality curve*.

Sustainable agriculture. Farming practices that maintain production while minimizing negative environmental impacts and maintaining the natural environment and its resources; compare *industrialized agriculture*.

Sustained yield. Maintaining a similar yield at each harvest of an exploited population.

Swamp. Wooded wetland in which water is near or above the ground level; compare *marsh* and *peatland*.

Swidden agriculture (slash-and-burn). Dominant form of traditional agriculture in the tropics; alternates periods of annual cropping with extended periods when land is abandoned.

Swim (gas) bladder. Structural adaptation of many fish to control buoyancy, allowing depth adjustment.

Switching. Change in a predator's diet from a less to a more abundant prey; often underlies a type III functional response.

Symbiosis. Interspecific interaction in which members of two dissimilar species live together in close and prolonged association; may be mutualistic or parasitic.

Sympatric speciation. Evolution of new species within an existing population living in the same area (*sympatric*), without geographic barriers; compare *allopatric speciation*.

Synchronization. In metapopulations, tendency for local populations to rise and fall in step with other local populations.

System. Set of interdependent parts or subsystems enclosed within a defined boundary and linked by fluxes of energy and nutrients, both within the system and with the environment; *systems ecology* is the study of ecosystems; see *cybernetics*.

Systemic acquired resistance. Induced plant chemical defence involving emission of a volatile warning chemical (often salicylic acid) that is detected by other parts of the plant or by other plants.

Target of selection. Phenotypic trait that selection acts upon; see *natural selection*; compare *selective agent*.

Territoriality. (1) Behaviour in which an individual actively defends part or all of its home range (*territory* is that part of a home range that is defended by an animal); may be temporary or permanent, and for purposes of mating, nesting, and/or feeding. (2) Any situation in which individuals of any species, including plants, are spaced more evenly than expected based on random occupancy of suitable habitat.

Theory. Integrated set of hypotheses that attempts to explain a broader set of observations than does a single hypothesis; compare *model*.

Thermal conductivity. Ability of an object to conduct or transmit heat; *thermal conductance* is rate at which heat travels through an object.

Thermal window. Structures (such as ears) that allow an animal to dissipate heat.

Thermocline. Transition layer in a thermally stratified water body in which temperature changes rapidly; separates epilimnion from hypolimnion; sometimes called *metalimnion*.

Thermoneutral zone. In homeotherms, range of external temperatures over which the metabolic rate is constant and relatively minimal; compare *operative temperature range*.

Threatened species. Species with significant extinction risk, usually from human activities; in the IUCN system, includes *endangered*, *critically endangered*, and *vulnerable species*.

Tidal overmixing. Mixing of freshwater and seawater when a tidal wedge moves upstream in a tidal river faster than freshwater moves seaward; seawater sinks as the lighter freshwater rises to the surface.

Tidal subsidy. Nutrients carried to coastal systems and wastes carried away by tidal cycles.

Tide. Water movements in response to the gravitational pulls of the Sun and Moon.

Tipping point. Changes (abiotic or biotic) that trigger an ecosystem to take a different trajectory, potentially leading to an *alternative stable state*.

Tolerance. Ability of an organism to survive an abiotic stress; compare *avoidance* and *resistance*.

Tolerance model. View that succession leads to a community composed of species that are the most efficient in exploiting resources at different stages; colonists neither facilitate nor inhibit the recruitment or growth of later species; compare *facilitation* and *inhibition model*.

Tolerance range. Range of an environmental factor in which an individual can survive, grow, and reproduce; see *zone of intolerance* and *zone of physiological stress*.

Top-down control. Influence of predators on the structure and productivity of lower trophic levels in a food web; compare *bottom-up control*.

Topography. Physical structure of a landscape, including landforms.

Toposequence. Sequence of sites at different successional stages that are adjacent to one another; used in succession studies; compare *chronosequence*.

Torpor. State in which the body temperature of a heterotherm drops to near ambient; entails a large reduction in respiration, with loss of motion and feeling; reduces energy costs in unfavourable conditions; either daily or seasonal (*hibernation*).

Tracheae. Adaptation in insects, consisting of a system of tubes that open to the air through spiracles on the body wall and that carry O_2 to body tissues; compare *lung* and *gill*.

Traditional ecological knowledge (TEK). Body of knowledge and practices handed down through cultural transmission about the relationships between organisms and their environment.

Traditional (subsistence) agriculture. Farming methods used in less developed countries with the goal of feeding the family unit; see *swidden agriculture*; compare *industrialized agriculture*.

Tragedy of the commons. Problem arising in exploitation of common-property resources, wherein an individual reaps all the benefits from each extra harvested unit, but the long-term costs of resource depletion are distributed among all individuals.

Transgenic. Of genetically modified organisms, combining genetically determined traits from different species as a result of genetic engineering.

Transmittance. Passage of radiation without absorption or reflection.

Transpiration. Loss of water vapour from plant tissues to the atmosphere through stomata; compare *evapotranspiration*.

Transpirational pull. Theory that water movement in plants is based on a water potential gradient established by transpiration and maintained by water cohesion in xylem (water-conducting tissue); compare *hydraulic redistribution*.

Transplant experiment. Experiment that tests the impact of an interspecific interaction (usually competition) by monitoring the response of a remaining species when one or more potentially interacting species are transplanted into treatment plots; compare *removal experiment*.

Treeline. Geographic limit above which forest does not grow; defined variously by the presence of a forest stand, individuals of tree height, or tree species individuals of any size.

Trophic cascade. Indirect interaction in which activities at one tropic level have a compounding effect on levels more than one removed; usually applied to top predators maintaining primary producers by feeding on herbivores; see *top-down control*.

Trophic efficiency. Ratio of productivity in a given trophic level with productivity of the trophic level on which it feeds; often follows the 10-percent rule in open-water habitats.

Trophic level. Functional classification of organisms in a food web according to their feeding relationships, ranging from autotrophs through herbivores and carnivores.

Turgor pressure. Pressure exerted by the protoplast of a plant cell on the cell wall due to osmotic intake of water; compare *pressure potential*.

Ubiquitous species. Species with a widespread geographic range; compare *endemic species*.

Ultraviolet radiation (UV). Radiation of 280 to 380 nm, immediately adjacent to *photosynthetically active radiation*.

Understory. (1) Layer of medium-height and small trees beneath a forest canopy. (2) All vegetation under a canopy, including low-growing species as well as tall shrubs.

Unified neutral theory. Theory proposing that community diversity patterns can be predicted independently of any assumptions of niche differences based on ecological trade-offs; suggests that population processes (in particular dispersal) affected by random forces determine community composition.

Unisexual. Having separate male and female individuals; compare *bisexual*.

Unitary organism. Organism that grows from birth to adult as a single integrated unit, often with determinate development; compare *modular organism*.

Upwelling. (1) Movement of surface waters that brings nutrient-rich colder water to the surface. (2) In open oceans, regions where surface currents diverge into deep waters, which rise to the surface.

Vapour pressure. Amount of pressure that water vapour exerts independent of dry air; *vapour pressure deficit* is the difference between saturation and actual vapour pressures at a given temperature.

Variable retention harvesting (VRT). Forest harvesting method that retains the attributes of mature forests.

Vector. In indirect transmission, organism of another species that transmits a pathogen between hosts; compare *intermediate host*.

Vertical stratification. Division of an aquatic or terrestrial community into distinct layers based on vegetative growth forms and/or abiotic factors; creates habitat zones for other organisms.

Vertical transmission. In parasitism, direct transmission from mother to offspring immediately before or after birth; compare *horizontal transmission*.

Viscosity. Ability of a fluid to resist a force that causes it to flow or to allow objects to pass through it.

Vital attributes. Traits that affect the role of a species in succession, including its recovery mode and reproductive ability in competitive situations.

Volatilization. Release of ammonia to the atmosphere in areas with alkaline or neutral soils; compare *ammonification*.

Warning colouration. Anti-predator defence involving conspicuous colour or markings; combined with chemical or structural defences; compare *cryptic colouration*.

Watershed. (1) Region drained by a waterway into a lake or reservoir. (2) Total area above a given point on a stream that contributes water to flow at that point. (3) Topographic dividing line from which surface streams flow in different directions.

Water balance. Balance of water inputs and outputs between an organism and its environment.

Water (hydrologic) cycle. Totality of processes whereby water moves between the atmosphere, hydrosphere, and Earth.

Water potential. Measure of the free energy of water as a result of the combination of pressure, osmotic, and matric potentials; predicts water movement in plants and between plants and the environment; water moves from areas of higher (less negative) to areas of lower (more negative) water potential.

Water-use efficiency (WUE). Ratio of net carbon fixed in photosynthesis per unit of water lost in transpiration; indicator of drought tolerance.

Weather. Combination of temperature, humidity, precipitation, wind, cloudiness, and other atmospheric conditions at a specific place and time; compare *climate*.

Wetfall. Any material deposited in moisture; see *acid precipitation*.

Wetland. General term encompassing seasonally or permanently waterlogged ecosystems; see *marsh, bog, fen, swamp*.

Wilting point (WP). Moisture content of soil at which plants can no longer extract water; at the *permanent wilting point*, the plant wilts permanently.

Windthrow. Local wind disturbance that topples trees (or other vegetation).

Winter-deciduous. Of woody species, dropping leaves prior to winter; compare *drought-deciduous*.

Xeric. Dry, especially of soil conditions; compare *mesic* and *hydric*; a *xerophyte* is a plant adapted to xeric sites.

Yield. Individuals or biomass harvested from a population per unit time; see *sustained yield* and *maximum sustainable yield*.

Zero growth isocline. In the Lotka–Volterra competition model, isocline along which net population growth is zero.

Zonation. Changes in physical and biotic structure of vegetation across a landscape; may be abrupt or associated with gradients that form latitudinal or altitudinal belts.

Zone of intolerance. Range of an environmental factor in which an individual cannot survive.

Zone of physiological stress. Portion of an individual's tolerance range in which it can survive, but is inhibited in its growth and/or reproduction; compare *optimal range*.

Zooplankton. Small floating or weakly swimming animal-like protists and small invertebrates that inhabit open-water areas of lakes and seas; compare *phytoplankton* and *nekton*.

Credits

Text/Illustration Credits

Chapter 2 2.7 Copyright 1992 from *Atmosphere, Weather and Climate*, 6th ed., by Barry and Chorley. Reproduced by permission of Routledge/Taylor & Francis Books.

Chapter 3 3.4 Robert G. Wetzel, Limnology, Third Edition: *Lake and River Ecosystems*, fig 5.7, pg. 56, Academic Press. Copyright Elsevier (2001).; 3.9 From Gene Likens, *An Ecosystem Approach To Aquatic Ecology: Mirror Lake And Its Environment* (New York: Springer-Verlag, 1985). Permission granted by G.E. Likens.

Chapter 4 4.5 Reprinted from William A. Pfitsch, "Daily Carbon Gain by Adenocaulon bicolor (Asteraceae) a Redwood Forest Understory Herb, in Relation to Its Light Environment", *Oecologia* 80(4): 465–470, (1989), fig. 1E, Springer-Verlag © 1989, with kind permission from Springer Science+Business Media B.V.; 4.6 *Ecological Monographs* by ECOLOGICAL SOCIETY OF AMERICA Copyright 1977 Reproduced with permission of ECOLOGICAL SOCIETY OF AMERICA in the format Textbook via Copyright Clearance Center.; p. 68 "Human Transformation of the Land", AAAS Atlas of Population and Environment, Source ESRI, pp. 72–73. © 2000 by American Association for the Advancement of Science.; 4.13 Soil Order Map of Canada, 2012, http://atlas.agr.gc.ca/agmaf/index_eng.html#context=soil-sol_en.xml Agriculture and Agri-Food Canada, 2012. Reproduced with the permission of the Minister of Public Works and Government Services, 2012.

Chapter 5 5.4 and 5.5 From *Ecology and Evolution Of Darwin's Finches* by Peter R. Grant. © 1986 Princeton University Press; 5.6a P.T. Boag and P.R. Grant, "Intense Natural Selection in a Population of Darwin's Finches in the Galapagos", *Science* 214: 82–85, 1981. Copyright © 1981. Reprinted by permission of AAAS.; 5.8 *Ecology and Evolution of Darwin's Finches*, Peter Grant, p. 205, fig 59b; 5.10 From O.H. Frankel and M.E. Soule, *Conservation and Evolution* (Cambridge: Cambridge University Press,1981), p. 32. Reprinted by permission.; 5.11 R.H. Baker, "Origin, Classification, and Distribution", *White-Tailed Deer: Ecology and Management*, ed. By L.K. Halls, © Wildlife Management Institute (WMI). Jane Kaminski, artist. Reprinted by permission of WMI.; 5.14 From *Ecology and Evolution Of Darwin's Finches* by Peter R. Grant. © 1986 Princeton University Press; p. 87 Courtesy of Laura Nagel. Used with permission.; p. 88 BERNER, D., ADAMS, D. C., GRANDCHAMP, A.-C. and HENDRY, A. P. (2008), "Natural selection drives patterns of lake–stream divergence in stickleback foraging morphology", *Journal of Evolutionary Biology*, 21: 1653–1665. With permission of John Wiley & Sons.; 5.15 Reprinted by permission from Macmillan Publishers Ltd: N. Patel, "Evolutionary biology: How to build a longer beak", *Nature* 442: 515–516 (Aug. 3, 2006). Copyright © 2006.

Chapter 6 6.9 Reprinted from Walter Larcher, *Physiological Plant Ecology*, 4e, fig. 2.28, p. 112, © Springer-Verlag Berlin Heidelberg 1975, 1980, 1991, 1995, 2003, with kind permission from Springer Science+Business Media B.V.; 6.10 *Ecology* by ECOLOGICAL SOCIETY OF AMERICA Copyright 1998 Reproduced with permission of ECOLOGICAL SOCIETY OF AMERICA in the format Textbook via Copyright Clearance Center.; 6.13 Reich, P. B., Walters, M. B., Tjoelker, M. G., Vanderklein, D. and Buschena, C. (1998), "Photosynthesis and respiration rates depend on leaf and root morphology and nitrogen concentration in nine boreal tree species differing in relative growth rate", *Functional Ecology*, 12: 395–405. With permission of John Wiley & Sons.; 6.19 Adapted from Mooney et al., "Photosynthetic capacity of Death Valley plants", *Carnegie Institute Yearbook* 75: 310–413, 1976. Used with permission of Carnegie Institution; 6.21 *Plant Physiology* by AMERICAN SOCIETY OF PLANT PHYSIOLOGISTS Copyright 1977 Reproduced with permission of AMERICAN SOCIETY OF PLANT BIOLOGISTS in the format Textbook via Copyright Clearance Center.; 6.25 Thomas Givnish, *On the Economy of Plant Form and Function: Proceedings of the Sixth Maria Moors Cabot Symposium*. Copyright © 1986. Reprinted by permission of Cambridge University Press.; 6.26 Reprinted from Woodward and Smith, "Global photosynthesis and stomatal conductance: Modeling the Controls by Soil and Climate", *Advances in Botanical Research*, vol. 20: 1–20 (1994), copyright 1994, with the permission from Elsevier; 6.28 *Ecological Monographs* by ECOLOGICAL SOCIETY OF AMERICA Copyright 1982 Reproduced with permission of ECOLOGICAL SOCIETY OF AMERICA in the format Textbook via Copyright Clearance Center.

Chapter 7 7.11 Hill and Wyse, *Animal Physiology*, 2/e, fig 6.7 p. 83. Copyright © 1989. Reprinted by permission of Pearson Education.; 7.12 Peterson, et al., "Snake Thermal Energy", *Snakes: Ecology and Behavior* edited by R.A. Seigel and J. T. Collins (Blackburn Press, 1993). Reprinted by permission of the publisher; 7.14 Peterson, et al., "Snake Thermal Energy", *Snakes: Ecology and Behavior* edited by R.A. Seigel and J. T. Collins (Blackburn Press, 1993). Reprinted by permission of the publisher; 7.15 K. Schmidt-Nielsen, *Animal Physiology* 5E; 7.17 Geist, V. 1998. *Deer of the world: their evolution, behavior, and ecology*. Stackpole Books, Mechanicsburg, Pennsylvania, USA. Fig. 1.9.; p. 144 Reprinted with permission from Macmillan Publishers Ltd: Makarieva, A.M., V.G. Gorshkov, and B.-L. Li. 2009. "Re-calibrating the snake palaeothermometer", *Nature* 460: E2-E3. © Copyright 2009.; 7.29 From P. J. Decoursey, *Cold Spring Harbor Symposia on Quantitative Biology XXV*, fig. 1, p. 50 (1961). Copyright 1961 Cold Spring Harbor Laboratory Press. Reprinted by permission.; 7.31 "Tidal rhythm of a fiddler crab in the laboratory in constant light at a constant temperature" John D. Palmer, "The Rhythmic Lives of Crabs", *BioScience* vol. 40, no. 5 (May 1990), pp. 352–358. © 1990 by the American Institute of

age at maturity of Pomacea canaliculata (Gastropoda: Ampullariidae) under a gradient of food deprivation". *J. Mollus. Stud.* (2009) 75(1): 19–26 first published online October 14, 2008. By permission of Oxford University Press.; 11.13 Nicholas L. Rodenhouse, T. Scott Sillett, Patrick J. Doran, and Richard T. Holmes, "Multiple density–dependence mechanisms regulate a migratory bird population during the breeding season", *Proc. R. Soc. Lond.* B October 22, 2003 270 1529 2105–2110. By permission of the Royal Society.; 11.14a *Ecology* by ECOLOGICAL SOCIETY OF AMERICA Copyright 1981 Reproduced with permission of ECOLOGICAL SOCIETY OF AMERICA in the format Textbook via Copyright Clearance Center., 11.14b "Population Dynamics of the Yellowstone Grizzly", *Dynamics of Large Mammal Populations*, C.W. Fowler and T.D. Smith, eds. (New York: John Wiley & Sons, 1981), p. 177; 11.16 Reprinted from *Autonomic Neuroscience* Vol. 156. Grippo, A.J., A. Sgoifo, F. Mastorci, N. McNeal, and D.M. Trahanas. "Cardiac dysfunction and hypothalamic activation during a social crowding stressor in prairie vole", pp. 44–50. © 2010, with permission from Elsevier.; 11.17 *Ecology* by ECOLOGICAL SOCIETY OF AMERICA Copyright 2000 Reproduced with permission of ECOLOGICAL SOCIETY OF AMERICA in the format Textbook via Copyright Clearance Center.; p. 231 (fig2) With kind permission from Springer Science+Business Media: Zanette et al., "Combined food and predator effects on songbird nest survival and annual reproductive success: results from a bi-factoral experiment", *Oecologia* (2006) 147: 632: 640. © 2006, p. 231 (fig3) *Ecology* by ECOLOGICAL SOCIETY OF AMERICA Copyright 2006 Reproduced with permission of ECOLOGICAL SOCIETY OF AMERICA in the format Textbook via Copyright Clearance Center.; 11.18 Lyanne Brouwer, Dik Heg, and Michael Taborsky, "Experimental evidence for helper effects in a cooperatively breeding cichlid", *Behavioral Ecology* (May/June 2005) 16(3): 667–673. By permission of Oxford University Press.; 11.21 Adapted from R.L. Smith, "Some Ecological Notes on the Grasshopper Sparrow", *Wilson Bulletin* 75, 1963.; 11.22 *Ecology* by ECOLOGICAL SOCIETY OF AMERICA Copyright 1971 Reproduced with permission of ECOLOGICAL SOCIETY OF AMERICA in the format Textbook via Copyright Clearance Center.; 11.24 Cahill, Jr., J. F. and Casper, B. B. (2000), "Investigating the relationship between neighbor root biomass and belowground competition: field evidence for symmetric competition belowground", *Oikos*, 90: 311–320. With permission from John Wiley & Sons.; 11.25 Mech et al., "Relationships of Deer and Moose populations to previous winters' snow", *Journal of Animal Ecology* 56: 615–627, 1987. Copyright © 1987. Reprinted by permission of Blackwell Publishing Ltd.

Chapter 12 12.2 b, c and d EHRLICH, P. R. and MURPHY, D. D. (1987), "Conservation Lessons from Long-Term Studies of Checkerspot Butterflies", *Conservation Biology,* 1: 122–131. With permission of John Wiley & Sons.; 12.4b, c and d Kindvall, O. and Ahlén, I. (1992), "Geometrical Factors and Metapopulation Dynamics of the Bush Cricket, Metrioptera bicolor Philippi (Orthoptera: Tettigoniidae)", *Conservation Biology*, 6: 520–529. With permission of John Wiley & Sons.; 12.6b and 12.10 C. Thomas and T. Jones, "Partial recovery of a Skipper Butterfly (Hesperia comma) from Population Refuges: Lessons for Conservation in a Fragmented Landscape", *Journal of Animal Ecology* 62: 472–481, (July 1993). Copyright © 1993. Reprinted by permission of Blackwell Publishing Ltd.; 12.8 *Ecology* by ECOLOGICAL SOCIETY OF AMERICA Copyright 1996 Reproduced with permission of ECOLOGICAL SOCIETY OF AMERICA in the format Textbook via Copyright Clearance Center.; 12.9 Susan Harrison et al., "Distribution of the Bay Checkerspot Butterfly, Euphydryas editha bayensis: Evidence for a metapopulation model", *The American Naturalist*, 132 (3): 360–382, (Sept 1988). Copyright © 1988 The University of Chicago Press.; p. 248 Adapted from A. Rodriguez and M. Delibes, "Current range and status of the Iberian lynx Felis paradina in Spain", *Biological Conservation* 61: 189–196. Used with permission.; p. 249 Schaefer, J. A. (2003), "Long-Term Range Recession and the Persistence of Caribou in the Taiga", *Conservation Biology*, 17: 1435–1439. With permission of John Wiley & Sons.; 12.11a Silvy, Nova J., ed.. *The Wildlife Techniques Manual: Volume 1: Research.* p. 9, Figure 1.3. © 2012 The Johns Hopkins University Press. Reprinted with permission of The Johns Hopkins University Press.

Chapter 13 13.3a D. Tilman et al., "Competition and Nutrient Kinetics Along a Temperature Gradient: An Experimental Test of a Mechanistic Approach to Niche Theory," *Limnology and Oceanography*, 26 (6): 1020–1033, (Nov. 1981). © American Society of Limnology and Oceanography (ASLO). Reprinted by permission fo ASLO; 13.7 Venner, S., P.-F. Pélisson, M.-C. Bel-Venner, F. Débias, E. Rajon, and F. Menu. 2011. Coexistence of insect species competing for a pulsed eesource: Toward a unified theory of biodiversity in fluctuating environments", *PLOS ONE* 6(3): e18039.; 13.9 *Ecology* by ECOLOGICAL SOCIETY OF AMERICA Copyright 1986 Reproduced with permission of ECOLOGICAL SOCIETY OF AMERICA in the format Textbook via Copyright Clearance Center.; 13.10 D.I. Bolnick, et al., "Ecological release from interspecific competition leads to decoupled changes in population and individual niche width", *Proceedings of The Royal Society* B277: 1789–1797 (17 Feb 2010). © The Royal Society.; 13.15 *Ecology* by ECOLOGICAL SOCIETY OF AMERICA Copyright 1971 Reproduced with permission of ECOLOGICAL SOCIETY OF AMERICA in the format Textbook via Copyright Clearance Center.; 13.16 *Ecology* by ECOLOGICAL SOCIETY OF AMERICA Copyright 1975 Reproduced with permission of ECOLOGICAL SOCIETY OF AMERICA in the format Textbook via Copyright Clearance Center.; 13.17a Tamar Dayan et al., "Feline canines: community-wide character displacement among the small cats of Israel", *The American Naturalist* 136, pp. 39–60, 1990. Copyright © 1990 The University of Chicago Press. Used with permission.; 13.19 Grant, Peter R., *Ecology and Evolution*

fig. 10.5, FAQ 5.1, fig. 1. Cambridge University Press. Used by permission of IPCC.; 28.25 Environment Canada (2002). *Science & Impacts of Climate Change.* In Toronto Smog Report Card 2005, Toronto Environmental Alliance, Toronto, Ontario. Environment Canada. Reproduced with the permission of the Minister of Public Works and Government Services, 2012.

Photo Credits

Part Openers Part One p. 18 Courtesy NASA/JPL-Caltech; p.19 (left) © Jptenor | Dreamstime.com; p. 19 (right) javarman/Shutterstock; Part Two p. 74 Wendy Dennis/Frank Lane Picture Agency Limited; p. 75 Thorsten Milse/ Photolibrary; Part Three p. 181 (left) Sunset/Frank Lane Picture Agency Limited, p. 181 (right) Jim Wark/Photolibrary Inc.; Part Four b and c Ron Peakall; Part Five p. 330 Paul Lawrence/Photolibrary; p. 331 © Terrance Klassen/Alamy; Part Six p. 407 (left) © Frans Lanting Studio/Alamy, p. 407 (right) © Bill Brooks/Alamy; Part Seven p. 497 Thomas M. Smith; Part Eight p. 565 (left) Glowimages/Getty Images, p. 565 (right) © wherrett.com/Alamy

Chapter 1 Chapter Opener Rapho Agence/Photo Researchers, Inc.; 1.1 NASA/Johnson Space Center; 1.2 Thomas M. Smith; 1.3 (in descending order) Dr. Ryan Brook, University of Saskatchewan, © Ocean/Corbis, Courtesy of Ashleigh Westphal, Dr. Ryan Brook, University of Saskatchewan, © Danita Delimont/Alamy, NASA/Goddard Space Flight Center; 1.6 © Copyright Rothamsted Research Ltd.; 1.7 Thomas M. Smith

Chapter 2 Chapter Opener Dr. Paul A. Zahl / Photo Researchers, Inc.; 2.9 Graeme Shannon/Shutterstock; 2.19 © Christopher Morris/Corbis; 2.21 Courtesy of Darwyn Coxson; pp. 34–35 Courtesy of Camilo Perez Arrau

Chapter 3 Chapter Opener © Beyond/SuperStock; 3.3 © Felix Büscher/Age Fotostock; 3.4 Franco Banf/ Photolibrary; 3.8 Garry Robinson (Winnipeg, Manitoba); p. 46 V.Tunniclifee, University of Victoria; p. 47 DR KEN MACDONALD/SCIENCE PHOTO LIBRARY; 3.14a Hydromet/Shutterstock, 3.14b Jens Stolt/Shutterstock; 3.15 © Kcphotos | Dreamstime.com

Chapter 4 Chapter Opener Radius Images/Jupiter Images; 4.1a Gregory Ochocki/Photo Researchers, 4.1b © Riekefoto | Dreamstime.com; 4.8 Joyce Photographics/Photo Researchers; 4.12 (Entisol, Mollisol, Alfisol, Aridisol, Inceptisol, Vertisol, Spodosol and Gelisol) Agriculture and Agri-Food Canada, (Andisol and Histosol) Loyal A. Quandt/USDA National Soil Survey Center, (Oxisol and Ultisol) U.S. Department of Agriculture USDA National Soil Survey Center; 4.14a Mark Edwards/Peter Arnold Inc., 4.14b Leslie D. McFadden

Chapter 5 Chapter Opener 5a © Kim Taylor/naturepl.com, Chapter Opener 5b © Adam Gasch/Age Fotostock, CO5c Frank Greenaway/DK Images; p. 87 (right) Beren Robinson, Kathryn Peiman, p. 87 (left) Photo by Andrew P. Hendry; 5.16 Gerald D. Carr; 5.18a Elliotte Rusty Harold/Shutterstock, 5.18b Jim Zipp / Photo Researchers, Inc.; 5.19 Courtesy of Jennifer Kling

Chapter 6 Chapter Opener Francois van der Merwe/ Shutterstock; 6.14 Eric V. Grave/Photo Researchers, Inc.; 6.16 © National Geographic Image Collection/Alamy; 6.23 Courtesy of Patrick Hill; 6.24 Allison Krause Danielsen; 6.29a Eric J. W. Visser, Radboud University Nijmegen, The Netherlands, 6.29b © Murray Clarke/Alamy; 6.30a © DCR-DNH/Gary P. Fleming, 6.30b Superstock/agefototstock

Chapter 7 Chapter Opener Hans Thomashoff/Photo Library; 7.1a Biosphoto/Leroy Christian/Peter Arnold, 7.1b David B. Fleetham/Photolibrary, 7.1c Lukas Hudec/Shutterstock, 7.1d Daniel Heuclin/agefotostock; 7.3 Comstock/Thinkstock; 7.4 Ian R. MacDonald; 7.7 NancyS/Shutterstock; 7.13 Courtesy of Glenn Tattersall; p. 143 Robert Caputo/Getty Images; 7.18a © Sazonoff | Dreamstime.com, 7.18b © Alltatts | Dreamstime. com; 7.19 Craig Willis; 7.20 Scott Camazine / Photo Researchers, Inc.; 7.21a Outdoorsman / Dreamstime.com/ GetStock.com, 7.21b © Musat | Dreamstime.com; 7.24a image 100/SuperStock; 7.25 Tom Brakefield/Getty Images; 7.26a Karl Ammann/Getty Images; 7.28 Melissa Kaplan

Chapter 8 Chapter Opener © jason crader/iStock; 8.1 Roland Birke/OKAPIA/Photo Researchers, Inc.; 8.4 George Grall/ Getty Images; 8.5 Federica Grassi/Getty Images; 8.6 Stockbyte/Getty Images; 8.7 Tony Heald/Nature Picture Library; 8.8 Eky Studio/Shutterstock; 8.9 Craig K. Lorenz / Photo Researchers, Inc.; 8.10a Alexandra Basolo; p. 165 (top) Courtesy of Locke Rowe, p. 165 (bottom) © B. Borrell Casals/ Frank Lane Picture Agency/Corbis; 8.12 Joel Sartore/National Geographic Stock; 8.14b (insert) © imagebroker/Alamy; 8.16 David Anderson; 8.19a Fungus Guy/Wikipedia, 8.19b Benoit Guenard; 8.20a © Rolf Nussbaumer Photography/Alamy, 8.20b Derrick Hamrick/Photolibrary

Chapter 9 Chapter Opener Mazzzur/Shutterstock; 9.2a Ralf Hettich/agefotosotock, 9.2b © Marty Snyderman/Corbis; 9.5 Michael Leuth; 9.8 Thomas M. Smith; 9.9b Thomas M. Smith; 9.12a Zastol`skiy Victor Leonidovich/Shutterstock; p. 192t Photo taken by Dave Andison, 192b Lori Daniels; 9.17 F.G. Hawksworth & D. Wiens; p. 197 (bottom left) Mark Robinson, USDA Forest Service, p. 197 (bottom right) D. Kucharski & K. Kucharska/Shutterstock, p. 197 (top) Ralf Neumann/Shutterstock

Chapter 10 Chapter Opener Exactostock / SuperStock; 10.5a © Rhigley | Dreamstime.com; p. 214 Mark Edwards/ Photolibrary; 10.18a © john t. fowler/Alamy

Chapter 11 Chapter Opener © Georgie Holland/Age Fotostock; 11.8 (insert) Joy Prescott/Shutterstock; 11.11 (insert) © Zanskar | Dreamstime.com; 11.12 (insert) Stijn Ghesquiere, www.applesnail.net; p. 230 (top) Michael Clinchy, p. 230b Marek C. Allen; 11.19 © Frans Lanting/ Corbis; 11.23 © Wayneduguay | Dreamstime.com

Chapter 12 Chapter Opener Michael Melford/National Geographic Stock; 12.1b Serg Zastavkin/Shutterstock; 12.2a Ron Wolf; 12.4a Petr Kocarek; 12.6a Matteo photos/ Shutterstock; p. 248 © imagebroker/Alamy; p. 249 © Cybernesco | Dreamstime.com; 12.11b © Crookid | Dreamstime.com; 12.12 Jonathan Way/Eastern Coyote Research

© Corbis/Photolibrary, 23.9b © Jeryl Tan/iStock; 23.10a and b Erin Bohman; 23.11a Luis C. Marigo/Photolibrary, 23.11b Robin Smith/Photolibrary; 23.13 (top left) Exactostock/Superstock, 23.13 (top right) Villiers Steyn/Shutterstock, 23.13 (bottom) Thomas M. Smith; 23.16a © Mike Grandmaison/Corbis, 23.16b © Clint Farlinger/Alamy, 23.16c © Bill Brooks/Alamy; 23.18 Galina Dreyzina/Shutterstock; 23.20a NASA Earth Observing System, 23.20b Australia Department of Lands; 23.21a Robert Leo Smith, 23.21b © Yasmin Gahtani/iStock; 23.22 Chris Mattison/agefotostock; 23.24a Lynne Watson, 23.24b Charles E. Jones, 23.24c Gary A. Monroe; 23.25a Martin Harvey/Getty Images, 23.25b Michael Almond/Shutterstock; 23.29a © Mark Karrass/Corbis, 23.29b Robert Leo Smith; 23.31a Danilo Donadoni/AGE Fotostock, 23.31b John S. Sfondilias/Shutterstock, 23.31c Darwyn Coxon; 23.33a Serge Payette, 23.33b and d Allison Krause Danielsen, 23.33c Sophia Laine Cosby; 23.34 and 23.39 Allison Krause Danielsen; 23.35 © Kevin Schafer/Alamy; 23.37a Paul Nicklen/National Geographic Stock, 23.37b Neil Nightingale/naturepl.com; 23.38a Kristin Piljay/Wanderlustphotos, 23.38b George Burba/Shutterstock; 23.40 © Tom Uhlman/Alamy

Chapter 24 Chapter Opener Allison Krause Danielsen; 24.2a Alan Majchrowicz/Photolibrary, 24.2b © imagebroker/Alamy, 24.2c James Steinbeg/Photo Researchers, Inc., 24.2d Robert Leo Smith; 24.5a Darlyne A. Murawski/Photolibrary, 24.5b Laguna Design/Photo Researchers, Inc.; 24.6a Michael P. Gadomski/Photo Researchers, Inc., 24.6b Robert Leo Smith; p. 533 NASA; 24.7 Geoff Higgins/Photolibrary; 24.9a Hydrome/Shutterstock, 24.9b vkimages/Shutterstock; 24.10 Darren K. Fisher/Shutterstock; 24.11 © David R. Frazier Photolibrary, Inc./Alamy; 24.15 Thomas M. Smith; 24.16 South Carolina Department of Natural Resources; 24.17 © Marevision/Age Fotostock; 24.21 Steve Gschmeissner/Photo Researchers, Inc.; 24.22 National Marine Fisheries Service; 24.24 Verena Tunnicliffe; 24.25 Darryl Leniuk/Getty Images; 24.26 Bjorn Stefanson/Shutterstock

Chapter 25 Chapter Opener © Parks Canada; 25.1a Nagel Photography/Shutterstock, 25.1b Thomas M. Smith; 25.3 Maine Department of Environmental Protection; 25.4 Yasehtor/Wikipedia; 25.5 Michael P. Gadomski/Photo Researchers, Inc.; 25.6 © Florida Images/Alamy; 25.7 Dr. Carleton Ray/Photo Researchers, Inc.; 25.8 and 25.9 © Mark Conlin/Alamy; 25.10 Lip Kee Yap/Wikipedia; 25.11a

NASA/Johnson Space Center, 25.11b © Robert Harding Picture Library Ltd/Alamy; 25.14 Caroline Savage, Environment Canada; 25.15 Michael P. Gadomski/Photo Researchers, Inc.; 25.16 © Steveheap | Dreamstime.com; 25.17a Allison Krause Danielsen, 25.17b Joe Cornish/Getty Images; 25.18a G. Schwabe/Photolibrary; 25.21 Larry Mayer/Jupiter Images

Chapter 26 Chapter Opener © Mira/Alamy; 26.3a NASA Earth Observing System; 26.5a Santokh Kochar/Getty Images, 26.5b Kletr/Shutterstock; 26.6 © Aflo/naturepl; 26.9 Thomas M. Smith; 26.10 Rostyslav.H/Shutterstock; 26.12a USDA/NRCS/Natural Resources Conservation Service, 26.12b © AgStock Images, Inc./Alamy; 26.14a NASA/Johnson Space Center, 26.14b Stockbyte/Getty Images; 26.18 Michael Meitner; 26.21 Paul Kay/Getty Images; 26.24a © blickwinkel/Alamy, 26.24b © Doug Perrine/naturepl.com, 26.24c © Peter Scoones/naturepl.com, 26.24d © Gary K. Smith/naturepl.com, 26.24e Andrew J. Martinez/Photo Researchers, Inc.; 26.25 Dr. Thierry Chopin; p. 587 (top) David Close, p. 587 (bottom) U.S. Fish and Wildlife Service; p. 588 Robin S Petersen Lewis

Chapter 27 Chapter Opener Georg Gerster/Photo Researchers, Inc.; 27.1 Julie Dermansky/Photo Researchers, Inc.; 27.14 (left) © Neil Julian/Alamy, 27.14 (right) © Bill Brooks/Alamy; 27.15 Michael Burzynski; p. 604 (left) © SuperStock/Alamy, p. 604 (right) Courtesy of Ashleigh Westphal; 27.18 NASA Earth Observing System; 27.19a NASA/Johnson Space Center, 27.19b © Stock Connection Blue/Alamy; 27.20b © KIKE CALVO VWPics / SuperStock; 27.21a Andrew J. Martinez/Photo Researchers, Inc., 27.21b Bill Alexander/Photolibrary; 27.22 National Park Service; 27.23 Ed Reschke/Photolibrary; 27.24 Photo Researchers, Inc.; 27.25 Picavet/Getty Images; 27.27a © National Geographic Image Collection/Alamy; 27.28 Thomas M. Smith; 27.30a KEN BOHN/AFP/GETTY IMAGES/Newscom, 27.30b AP Photo/San Diego Wild Animal Park; 27.31 © Wave Royalty Free/Alamy; 27.35 National Marine Fisheries Service

Chapter 28 Chapter Opener National Oceanic and Atmospheric Administration/NASA; 28.1 Simon Fraser/Mauna Loa Observatory/Photo Researchers, Inc.; p. 626 NormanEinstein/Wikipedia; 28.8a Chris Hildreth/Duke Photography; p. 632 NASA GISS 2010; 28.17 Greg Henry

Index

Note: Page numbers followed by f represent figures and page numbers followed by t represent tables.